MW00578906

Diagnostic Microbiology *of the* Immunocompromised Host

Diagnostic Microbiology *of the* Immunocompromised Host

Edited by

Randall T. Hayden
St. Jude Children's Research Hospital
Memphis, Tennessee

Karen C. Carroll
The Johns Hopkins Hospital
Baltimore, Maryland

Yi-Wei Tang
Vanderbilt University School of Medicine
Nashville, Tennessee

Donna M. Wolk
The University of Arizona
Tucson, Arizona

Washington, DC

ASM Press
American Society for Microbiology
1752 N St., N.W.
Washington, DC 20036-2904

Library of Congress Cataloging-in-Publication Data

Diagnostic microbiology of the immunocompromised host / edited by Randall T. Hayden
... [et al.].
p. ; cm.
Includes bibliographical references and index.
ISBN-13: 978-1-55581-397-0 (hardcover : alk. paper)
ISBN-10: 1-55581-397-6 (hardcover : alk. paper) 1. Diagnostic microbiology—Technique. 2. Immunological deficiency syndromes—Diagnosis. 3. Opportunistic infections. I. Hayden, Randall T.
[DNLM: 1. Infection. 2. Bacterial Infections. 3. Immunocompromised Host. 4. Microbiological Techniques. 5. Mycoses. 6. Virus Diseases. WC 195 D536 2009]

QR67.D55 2009
616.9′0475—dc22

2008022215

Printed in the United States of America

10 9 8 7 6 5 4 3 2 1

Address editorial correspondence to ASM Press, 1752 N St., N.W., Washington, DC 20036-2904, U.S.A.

Send orders to: ASM Press, P.O. Box 605, Herndon, VA 20172, U.S.A.
Phone: 800-546-2416; 703-661-1593
Fax: 703-661-1501
E-mail: books@asmusa.org
Online: http://estore.asm.org

Cover images (from top): *Nocardia* species (Color Plate 8) (see Chapter 13); *Sporothrix schenckii*, culture isolate (photograph courtesy of G. Roberts); biopsy specimen of patient with CMV colitis, showing "owl's-eye" intranuclear inclusion (Color Plate 2) (see Chapter 3); *Aspergillus* species in tissue section (photograph courtesy of G. Roberts); typical adenovirus cytopathic effect in hybrid cell line composed of A549 and mink lung (Mv1Lu) (photograph courtesy of M. Bankowski) (see Chapter 6).

CONTENTS

CONTRIBUTORS

Justin A. Bishop
Department of Pathology, The Johns Hopkins Medical Institutions, Baltimore, MD 21287

Karen C. Bloch
Department of Medicine, Vanderbilt University School of Medicine, Nashville, TN 37232

Karen C. Carroll
Division of Medical Microbiology, The Johns Hopkins Hospital, Meyer B1-193, 600 N. Wolfe St., Baltimore, MD 21287

James H. Clark III
Department of Pathology, Children's Medical Center Dallas, 1935 Motor St., Dallas, TX 75235

Phyllis Della-Latta
Clinical Microbiology Services, Columbia University Medical Center, New York-Presbyterian Hospital, 622 West 168th St., CHC 3-325, New York, NY 10032

Lesia K. Dropulic
Laboratory of Clinical Infectious Diseases, National Institutes of Health, Bethesda, MD 20892

James J. Dunn
Department of Pathology, Cook Children's Medical Center, 801 Seventh Ave., Fort Worth, TX 76104

Molly E. Eaton
Department of Medicine, Emory University School of Medicine, 341 Ponce de Leon Avenue, Atlanta, GA 30308

Lynne S. Garcia
LSG & Associates, 512 12th Street, Santa Monica, CA 90402-2908

Thomas E. Grys
Division of Clinical Microbiology, Department of Laboratory Medicine and Pathology, Mayo Clinic and Mayo Clinic College of Medicine, Rochester, MN 55905

Barbara L. Haller
San Francisco General Hospital, 1001 Potrero Ave., San Francisco, CA 94110-3518

Xiang Y. Han
Department of Laboratory Medicine, Unit 84, The University of Texas M. D. Anderson Cancer Center, 1515 Holcombe Boulevard, Houston, TX 77030

Hafsa Hassan
Department of Pathology, University of Utah School of Medicine, Salt Lake City, UT 84312

Randall T. Hayden
Department of Pathology, St. Jude Children's Research Hospital, Memphis, TN 38105

Heather R. Hensler
Department of Infectious Diseases and Microbiology, University of Pittsburgh, Pittsburgh, PA 15213

Igen Hongo
Department of Internal Medicine, Musashino Red Cross Hospital, Tokyo 180-8610, Japan

Michael G. Ison
Division of Infectious Diseases, Northwestern University Feinberg School of Medicine, Chicago, IL 60611

Frank J. Jenkins
Departments of Pathology and Infectious Diseases and Microbiology, University of Pittsburgh, Pittsburgh, PA 15213

Mary Louise Landry
Department of Laboratory Medicine, Yale University School of Medicine, P.O. Box 208035, New Haven, CT 06520-8035

Howard M. Lederman
Department of Medicine, The Johns Hopkins University School of Medicine, Baltimore, MD 21205

Elizabeth M. Marlowe
Southern California Permanente Medical Group, Regional Reference Laboratories, North Hollywood, CA 91605

William G. Merz
Department of Pathology, The Johns Hopkins Medical Institutions, Baltimore, MD 21287

Irving Nachamkin
Department of Pathology and Laboratory Medicine, University of Pennsylvania School of Medicine, Philadelphia, PA 19104-4823

Steven R. Nesheim
Department of Pediatrics, Emory University School of Medicine, 341 Ponce de Leon Avenue, Atlanta, GA 30308

Cathy A. Petti
Department of Pathology, University of Utah School of Medicine, Salt Lake City, UT 84312

Gary W. Procop
Department of Pathology, University of Miami Miller School of Medicine, Miami, FL 33136

Raymond R. Razonable
Division of Infectious Diseases, Department of Internal Medicine, Mayo Clinic and Mayo Clinic College of Medicine, Rochester, MN 55905

L. Barth Reller
Department of Pathology, Duke University School of Medicine, Durham, NC 27710

Paula A. Revell
Department of Pathology, Children's Medical Center Dallas, 1935 Motor St., Dallas, TX 75235

Glenn D. Roberts
Division of Clinical Microbiology, Mayo Clinic, Rochester, MN 55905

Beverly B. Rogers
Department of Pathology, Children's Medical Center Dallas, 1935 Motor St., Dallas, TX 75235

Eric S. Rosenberg
Department of Medicine, Massachusetts General Hospital, Boston, MA 02114

David T. Rowe
Graduate School of Public Health, University of Pittsburgh, Pittsburgh, PA 15213

Michael A. Saubolle
Division of Infectious Diseases, Laboratory Sciences of Arizona, University of Arizona College of Medicine, Phoenix, AZ 85006

Audrey N. Schuetz
Department of Pathology & Laboratory Medicine, Emory University School of Medicine, 1440 Clifton Road N.E., Atlanta, GA 30322

Cari R. Sloma
Division of Clinical Microbiology, Department of Laboratory Medicine and Pathology, Mayo Clinic and Mayo Clinic College of Medicine, Rochester, MN 55905

Yi-Wei Tang
Departments of Pathology and Medicine, Vanderbilt University School of Medicine, Nashville, TN 37232

Yun F. (Wayne) Wang
Department of Pathology & Laboratory Medicine, Emory University School of Medicine, Grady Memorial Hospital, P.O. Box 26248, 80 Jesse Hill Jr. Dr., Atlanta, GA 30303

Nancy G. Warren
Bureau of Laboratories, Pennsylvania Department of Health, 110 Pickering Way, Lionville, PA 19353

Susan Whittier
Clinical Microbiology Services, Columbia University Medical Center, New York-Presbyterian Hospital, 622 West 168th St., CHC 3-325, New York, NY 10032

Donna M. Wolk
Department of Pathology & Laboratory Medicine, University of Arizona, 1501 N. Campbell Ave., Tucson, AZ 85724-0001

Gail L. Woods
Department of Pathology and Laboratory Services, University of Arkansas for Medical Sciences, Mail Slot 502, 4301 W. Markham Street, Little Rock, AR 72205

Fann Wu
Clinical Microbiology Services, Columbia University Medical Center, New York-Presbyterian Hospital, 622 West 168th St., CHC 3-325, New York, NY 10032

FOREWORD

The subject matter of this book is focused on the implementation of diagnostic techniques for a special purpose, that of accurately and rapidly diagnosing infections of the immunocompromised host. In addition to the usual litany of pathogens that threaten all of us, patients with a compromised immune status are susceptible to normally harmless microbial cohabitants. Such infections may become invasive and life-threatening, in recent years affecting growing numbers of patients. This book provides a timely, up-to-date, and comprehensive summary of the ever-expanding array of technologies used to diagnose these infections and to monitor the effectiveness of specific therapies. For immunocompromised patients, the timing could not be better.

Many innovations in diagnostics over the past few decades have been fueled by the needs of specific patient populations. Immunocompromised patients are especially vulnerable to diagnostic delays because the time to intervention can have a dramatic impact on therapeutic efficacy. Only a few years ago, blood culture bottles were tested in "first morning batches" by Gram stain and culture, necessitating long delays in some cases. Differential counts were tallied by hand. In many pathology departments in the 1960s, immunoassays were performed in batches, perhaps once per week, within specialized laboratories built for containment of the radioisotopes used to label antibodies. Rabbit antiserum, sometimes collected from local farm animals, was radioiodinated with Bolton-Hunter reagent to generate the key diagnostic ingredient.

Fortunately, in clinical laboratories, as elsewhere in medicine, times have changed. Nonisotopic immunoassays were developed that could be automated, and more recently, random access immunoassay systems have been developed to eliminate batching requirements and improve turnaround time. Stat immunoassays for a variety of indications are now a reality, if not the standard of practice. Blood culture systems have been developed to allow continuous monitoring for the presence of microbial growth; these systems have largely replaced batch systems that require blind subculture. The fundamental impact of this technology is that it allows for delivery of "real-time" results. For immunocompromised patients, this need for speed is especially critical because infections in these patients often evolve quickly into life-threatening events. Earlier diagnoses can translate into earlier specific therapeutic interventions, which are more likely to result in favorable patient outcomes.

Nucleic acid amplification techniques occupy an increasingly important role in diagnosis and monitoring of infection in immunocompromised patients, and many chapters of this book are appropriately focused on detection of nucleic acid targets. Though the pathogens themselves come from entirely different phylogenetic domains, they are all similar in that they harbor genetic signatures in their genomes that can be used to identify them, quantify infectious burdens, determine virulence, and assess susceptibility or resistance to available drugs. Clumsy, contamination-prone techniques have largely been replaced with real-time detection technology performed in closed systems, and DNA sequencing and microarray technologies developed under the auspices of the human genome project are making steady inroads into clinical practice. Diagnosticians have taken great leaps forward in their level of overall sophistication and familiarity with this technology. Phylogenetic analysis and identification of bacteria, fungi, and viruses by direct DNA sequencing are quickly entering the mainstream and will require us to add a few new words, such as "bootstrapping" and "parsimonious," to our vocabulary. Microarray technology has yet to enter into routine diagnostic use for infectious disease applications, but it is only a matter of time until that happens.

As promising and important as it is, however, the practice of molecular diagnostics is currently more akin to that of the radioimmunoassay laboratory of the 1960s. Despite the speed of the underlying detection technologies, the requirements of specimen processing impose practical limits on turnaround time. As molecular methods evolve, the technology needs to keep up with requirements for optimal clinical management. For example, febrile patients who are immunocompromised would benefit greatly from rapid, on-demand testing of bacterial and fungal causes of sepsis, without the need to wait for initial blood culture results. Proof of concept of this approach has now been provided by at least one system, which is designed to detect and differentiate 23 pathogens by using broad-range PCR. As technologies like this become available, and as they evolve from batch mode to on-demand formats, they will have an ever greater impact on patient treatment and management. As real-time molecular diagnostic technology improves, so should the delivery of real-time patient results.

Few challenges are as urgent as determining definitively the cause of fever in an immunocompromised patient, given the sheer range of diagnostic possibilities. Fortunately, practitioners have filled this apparent diagnostic void with user-developed assays, and diagnostic companies themselves appear to be rising to the task. Hopefully, this trend will continue. Meeting the diagnostic challenges of immunocompromised patients will ultimately have broader implications; as described in this book, there is the potential to improve access to this technology for all patients, where and when they need it most.

David H. Persing
Cepheid, Sunnyvale, California

PREFACE

Over the past quarter century, health care of immunocompromised patients has grown progressively in importance. These individuals require high-intensity services and specialized care, often for a prolonged period of time. They are susceptible to a wide range of infectious diseases, which may manifest quite differently from those in an immunocompetent host. There are marked differences in how health care is delivered to such high-risk patients. Proper care depends on the etiology and degree of immune suppression as well as on underlying patient characteristics, such as demographics, nutritional status, and ongoing disease processes. Differences in clinical care include aspects of infection control practices, infectious disease prophylaxis, immune modulation, and pharmacologic therapy. In addition, the use and interpretation of laboratory tests, particularly tests for microorganisms, must be tailored carefully to fit these patients. Evidence-based diagnostic algorithms for the immunocompromised are evolving; however, many clinicians and laboratory professionals are challenged to best utilize the growing array of diagnostic tools at their disposal. Certainly there are books containing information on clinical testing; however, no standard laboratory reference focuses heavily on issues unique to the immunocompromised population. It is the goal of the authors to consolidate such discussions in a single, easily referenced text that can be used by clinical health care providers and laboratory professionals alike.

This book provides several approaches to the topic. The stage is set in the first section, wherein the essence of the problem is defined. That is, what are the causes of immune suppression, who are the populations at risk for infections, and to which infections are they prone? In Section II, the application of laboratory diagnostic methods is discussed, primarily in an organism-by-organism fashion, while in Section III, discussions are based on the organ system involved. Readers will find that these two approaches are complementary.

While in many cases the clinician may be more comfortable with an organ system approach, a focus on individual pathogens may be more useful in deciding upon screening strategies or follow-up of a known infection. Although laboratory professionals may turn most frequently to chapters on individual infectious agents, the systemic perspective will bring added value in making decisions on which new diagnostic methods to introduce in the laboratory. These sections will also be useful for a review of specimen-specific culture workup and exceptions to the rules, which may apply to immunocompromised patient units or clinics. In addition, many chapters include flow charts suggesting diagnostic pathways. We hope that these sections will provide a way to help to synthesize the material presented in the text into practical, user-friendly algorithms that can be applied to everyday patient care challenges. The concluding segment of the book pushes the envelope of current diagnostics, with a look at future trends in the diagnosis of infectious diseases in the immunocompromised patient. We hope that this section will enable health care facilities that treat these patients to plan for the future and to assess new technology from a global cost-benefit perspective as they attempt to wisely use scarce resources in the future.

This book is intended to have broad appeal to laboratory professionals, infectious disease physicians, oncologists, and other health care providers who play important roles in the health care of immunocompromised patients. The editorial board, as well as the contributors, comprise a diverse group of both clinical infectious disease practitioners and laboratory-based diagnosticians. We hope that this book will fill a void in many health care providers' libraries and contribute positively to the care of these increasingly complex patients. By sharing this information and working effectively together with a patient-focused approach, it is our hope that we can help to face the challenges of providing optimal diagnostic services to the immunocompromised patients in our care.

We extend our heartfelt thanks to all of the chapter authors, who devoted so much of their time and expertise to this project. Working with such a fine group of professionals has been our pleasure. We are also grateful for the support and patience of our families while we immersed ourselves in this project. We dedicate this work to all of them and to the immunocompromised patients whom we hope this book will serve.

Randall T. Hayden, Karen C. Carroll,
Yi-Wei Tang, and Donna M. Wolk

I. HOSTS AND INFECTIONS

Diagnostic Microbiology of the Immunocompromised Host
Edited by Randall T. Hayden, Karen C. Carroll, Yi-Wei Tang, and Donna M. Wolk

Chapter 1

Overview of Infections in the Immunocompromised Host

Lesia K. Dropulic and Howard M. Lederman

Host defense from infection depends upon a complex integrated system of physical barriers (e.g., skin, stomach acid, and mucociliary clearance), innate immunity (e.g., phagocytic cells, natural killer [NK] cells, and complement), and adaptive immunity (B and T lymphocytes). An individual may have deficiencies of one or more components of host defense, but no individual is defenseless. Because each functional compartment of the immune system plays a specialized role in host defense, defects in specific functions lead to increased susceptibility to specific pathogens. The key to understanding the susceptibility of a particular patient is to understand the specific host defense defects of that patient. This chapter briefly reviews the components of host defense and the types of infections that are most likely to occur with specific defects in those defense mechanisms. Following that review are descriptions of the infections that occur in patients with a variety of primary and secondary immunodeficiency disorders, with the intent of providing illustrative examples.

OVERVIEW OF HOST DEFENSE

Host defense depends upon physical barriers as well as the immune system. In conceptual terms, the components of the immune system can be divided into two compartments—innate and adaptive—with fundamentally different modes of action (Table 1). Innate host defense mechanisms are rapid (minutes to hours), depend upon patterned responses to pathogens (e.g., by phagocytic cells or complement), and do not improve with repeated exposure to one or many pathogens. In contrast, adaptive immune mechanisms are slow (days), depend upon very specific responses to individual antigens (e.g., by B and T lymphocytes), and improve with repeated exposure to an individual antigen. When they are successfully integrated and functioning together, physical barriers and the components of innate and adaptive immunity form a critical homeostatic mechanism necessary for the host's defense against infection and the generation of normal inflammatory responses (69, 70).

Physical Barriers

The initial defenses against infection are provided by physical and chemical barriers (98). These include the tight junctions between epithelial cells of the skin; a protective barrier of mucus that traps microorganisms on mucosal surfaces and then is swept out of the respiratory tract by cilia and from the gastrointestinal tract by peristalsis; lysozyme in saliva and tears; acid in the stomach; antimicrobial peptides, such as defensins; surfactant proteins that can opsonize microorganisms for easier phagocytosis; and mechanical flushing of the gastrointestinal and urinary tracts. In addition, microbial pathogens must compete for space and nutrients with the normal microbial flora on the skin and mucosal surfaces. Defects of physical barriers (e.g., disruption of the skin by burns or a vascular catheter, reduced mucociliary clearance because sedation is needed to keep a patient comfortable on mechanical ventilation, and obstruction of urinary drainage by a renal stone) are not the subject of this chapter, though they are common causes for increased susceptibility to infection.

Innate Immunity

The components of the innate immune system (neutrophils, macrophages, NK cells, and complement) recognize foreign antigens by receptors encoded by intact germ line genes (e.g., Toll-like receptors

Lesia K. Dropulic • Laboratory of Clinical Infectious Diseases, National Institutes of Health, Bethesda, MD 20892. **Howard M. Lederman** • Departments of Pediatrics, Medicine and Pathology, The Johns Hopkins University School of Medicine, Baltimore, MD 21287-3923.

Table 1. Components of the immune system

Feature	Innate immunity	Adaptive immunity
Cells	Neutrophils, monocytes/macrophages, NK cells	B lymphocytes, T lymphocytes
Receptors	Expressed by all cells of a particular type (e.g., macrophages); recognize broad classes of pathogens	Clonal distribution on individual cells; highly specific
Soluble factors	Complement, MBL, chemokines, cytokines (including IL-1 and TNF-α)	Antibody, cytokines (including IL-2, IL-4, IL-5, IL-6, and IL-10)
Change with repeated exposure to antigen	No	Yes (clonal expansion, memory lymphocytes)

[TLRs] and mannose-binding lectin [MBL]) (98, 282). These receptors bind to pathogen-associated molecular patterns (PAMPs) that are shared by many microorganisms (e.g., bacterial lipopolysaccharide [LPS]). For example, the macrophage mannose receptor binds specific sugar molecules found on the surfaces of many bacteria and viruses. Members of the TLR family of transmembrane receptors have specificities for a variety of PAMPs (138). Binding to these receptors triggers a signaling cascade, with induction of transcription factors and activation of proinflammatory genes. One particularly important role for TLRs is to trigger macrophage responses to bacterial LPS (186). Bacterial LPS in body fluids is bound by the circulating LPS-binding protein, and this complex then binds to CD14 on the macrophage surface. When the LPS–LPS-binding protein–CD14 complex binds to TLR-4, the transcription factor NF-κB is translocated into the nucleus, where it activates genes involved in host defense, such as tumor necrosis factor alpha (TNF-α) and inducible nitric oxide synthetase (26). The receptors for PAMPs are displayed nonclonally on cells of the innate immune system. That is, all neutrophils display the same set of PAMP receptors, whereas all NK cells display another set of PAMP receptors. Repeated exposure to an antigen does not alter the innate immune response to that antigen.

Phagocytes

Phagocytic cells ingest foreign antigens and microorganisms (2). Although many phagocytic cells are mobile and can move from the bloodstream through tissues to the site of microbial invasion or inflammation, other phagocytic cells are fixed in the sinusoids of the bloodstream and the lymphatic system, where they clear microorganisms and other particulate matter from the circulation. A variety of cells possess phagocytic activity, but neutrophils, monocytes, and macrophages are the most critical to the function of the immune system. Monocytes and macrophages can also present antigen to lymphoid cells and secrete a variety of proinflammatory substances (including cytokines and complement components). These cells thus play an important role in the generation of innate and adaptive immune responses, in addition to their role in phagocytosis.

To function properly, all phagocytic cells must attach to a substrate (adherence), move through tissues toward the site of microbial invasion (chemotaxis), attach to and ingest microbes (phagocytosis), and finally kill them (intracellular killing). The adherence of phagocytic cells is mediated by a family of cell surface glycoproteins (integrins), including CR3, LFA-1, and p150, and is enhanced by a number of soluble mediators (including C5a, thromboxane A_2, leukotrienes, and platelet-activating factor).

The directed movement of phagocytic cells toward a chemical stimulus is termed chemotaxis. Phagocytic cells sense chemical gradients across their length and then move in the direction of the higher concentration (i.e., the source of the chemotactic stimulus). A variety of substances act as chemoattractants (including C5a, which is produced by activation of the complement system; some bacterial peptides; prostaglandins; and monocyte- and lymphocyte-derived cytokines). Once phagocytic cells reach the site of infection, they ingest the microbes. The process is facilitated if the microbes have been coated (opsonized) with immunoglobulin G (IgG) antibody and/or the larger cleavage product of the third component of complement, C3b, as phagocytic cells have surface receptors for IgG and C3b.

The process of intracellular killing begins soon after the phagosome is internalized. Both primary (azurophilic) and secondary (specific) granules can fuse with the phagosome, and a number of antimicrobial substances are thereby introduced into the phagosome. These substances include lysozyme, lactoferrin, acid hydrolases, and cationic proteins. Perhaps the most important killing mechanism, however, is the myeloperoxidase-H_2O_2-halide system. After ingestion of microorganisms, molecular oxygen is reduced to superoxide by a series of reactions involving NADPH (reduced form) oxidase. The superoxide, in turn, undergoes further reactions, leading to the generation of reduced oxygen derivatives, such as hydrogen peroxide and hydroxyl radicals. Myeloperoxidase catalyzes the reaction of hydrogen peroxide with chloride to create hypochlorite ions. The net effect of these toxic derivatives of reduced molecular oxygen is to kill microorganisms within the phagocytic vacuole.

Complement system

The complement system is composed of about two dozen serum proteins that, when functioning in an ordered and integrated fashion, mediate a variety of defensive and inflammatory reactions (295, 296). The majority of the biologically significant effects of the complement system are mediated by the third component (C3) and the terminal components (C5 through C9). To subserve their biologic functions, however, C3 and C5 through C9 must first be activated via either the classical, the alternative, or the MBL complement pathway.

In the classical complement pathway, antigen-antibody complexes composed of either IgG or IgM activate the first component of complement (C1). Activation of the alternative complement pathway, in contrast, can occur in the absence of specific antibody if there is a "nonmammalian" cell surface. A third activation pathway, the MBL pathway, uses a molecule homologous to C1q to trigger the complement cascade. MBL binds to mannose residues on microbial surfaces but does not bind mannose on host cells because it is blocked by sialic acid.

Activation of any of these pathways leads to the proteolytic cleavage of C3 into C3a and C3b. The activation of C3 represents an amplification step because hundreds of C3 molecules can be cleaved by a single C3 convertase. C3a is released into the fluid phase, where it can act as an anaphylatoxin, releasing the tight junctions between vascular endothelial cells and thus facilitating the movement of inflammatory cells from the bloodstream to an infected tissue. C3b binds covalently to the surface of the activating cell or to the immunoglobulins of the activating immune complex, thereby acting as an opsonin or combining with either of the C3 convertases to create a C5 convertase. Activation of C5 creates a small cleavage product, C5a, which is released into the fluid phase, where it can act as an anaphylatoxin and a chemotaxin. C5b can combine with native C6 and thereby initiate the formation of a membrane attack complex (a multimolecular assembly of C5b, C6, C7, C8, and C9) which is inserted into cell membranes and is responsible for the cytolytic and bactericidal actions of complement.

NK cells

NK cells are derived from the common lymphoid progenitor cell (230). However, unlike other lymphocytes, NK cells have invariant receptors that are not expressed clonally. One type of receptor binds a variety of cell surface carbohydrates and is able to activate the NK cell. A second type of receptor binds major histocompatibility complex class I alleles and has inhibitory activity. NK cells can kill targets that express a net excess of activating versus inhibitory signals. This can occur, for example, if a virally infected host cell has decreased expression of major histocompatibility complex class I molecules. NK cells also have receptors for the Fc portion of IgG (FcγR), so they can bind to host cells expressing viral or tumor antigens to which IgG antibodies have attached. Once an NK cell has attached to a target, it can release cytotoxic granules that penetrate the target cell and induce programmed cell death (apoptosis). The cytotoxic activity of NK cells can be enhanced by prior exposure to interferons (IFNs) and the macrophage-derived cytokine interleukin-12 (IL-12).

Adaptive Immunity

The cells of the adaptive immune system (B and T lymphocytes) recognize antigen via receptors assembled from rearranged gene segments, and each lymphocyte expresses a unique antigen receptor (4). Repeated exposure to an antigen selects those cells with the highest-affinity receptors for that antigen and induces proliferation of that clonal population and differentiation into effector and long-lived memory cells. The net effect is to increase the kinetics and magnitude of the response to subsequent exposures of the same antigen.

B lymphocytes

Each B lymphocyte has a unique antigenic specificity, marked by the immunoglobulin receptor on its cell membrane. When antigen binds to the immunoglobulin (antibody) expressed on the surface of one of the B lymphocytes, that cell proliferates to form a population of cloned progeny cells with identical antibody specificity. These cells then differentiate into plasma cells that secrete immunoglobulins (IgM, IgG, IgA, IgE, or IgD). Most antigens are T-cell dependent, that is, optimal B-cell differentiation into plasma cells requires the presence of T-lymphocyte helper cells. There are a few antigens, however, including such clinically important ones as bacterial capsular polysaccharides, that are T-cell independent and able to trigger terminal B-cell differentiation even in the absence of T lymphocytes. In all cases, CD4 helper T lymphocytes (TH) are important modulators of B-cell function, influencing the degree, duration, and quality (affinity and class distribution) of the antibody response.

The five major classes of immunoglobulins are IgG, IgM, IgA, IgE, and IgD. Each class has unique structural and functional characteristics. Depending on the class, immunoglobulins function in host defense by opsonization of foreign microorganisms, activation of serum complement, neutralization of toxins and viruses, and inhibition of microbial attachment to mucosal surfaces. IgM is the first immunoglobulin produced in an immune response and is the most efficient activator of complement. IgG is the predominant serum immunoglobulin, is actively transported across the placenta, possesses opsonic activity, and activates complement. IgA, which is the major immunoglobulin secreted onto mucosal surfaces, is largely silent as an inflammatory mediator but can prevent microbial adherence and penetration across the mucosal surface and clears and disposes of antigens. IgE is a mediator of allergic disease. By means of interactions with mast cells and eosinophils, IgE can also play a role in host defense against parasitic infections. Most IgD is expressed on the surfaces of naïve B lymphocytes, though limited amounts are secreted. It has no known role in host defense.

T lymphocytes

T lymphocytes are the effectors of cell-mediated immunity. They also serve as important regulators of both the humoral and cell-mediated immune systems and modulate the activities of nonlymphoid cells, such as monocytes. Like B lymphocytes, each T lymphocyte has a unique antigenic specificity. The diverse effector and regulatory functions of T lymphocytes are carried out by distinct lymphocyte subpopulations. CD4 T lymphocytes carry out immunoregulatory functions by the release of cytokines, some of which stimulate B-lymphocyte (IL-2, IL-4, and IL-5) and T-lymphocyte (IL-2 and IL-4) proliferation and differentiation, activation of monocytes (IFN-γ), and proliferation of hematopoietic precursors of lymphoid and nonlymphoid cells (IL-3). Some lymphokines preferentially stimulate secretion of IgG1; others lead to the secretion of IgA and IgE. When CD4 TH lymphocytes proliferate, they differentiate into TH1 or TH2 cells. Although the factors determining whether a cell becomes a TH1 or TH2 cell have not been elucidated fully, the two cell types are known to have very different functions. The TH1 cell secretes cytokines (IL-2, IFN-γ, and TNF-α) that stimulate cell-mediated immune responses, such as activation of macrophage bactericidal function, delayed-type hypersensitivity (DTH), and cytotoxicity. The TH2 cell secretes cytokines (IL-4, IL-5, IL-6, and IL-10) that drive B-cell proliferation and differentiation, resulting in antibody synthesis. These TH subsets are not mutually exclusive, but most infectious pathogens induce a response that is predominantly TH1 or TH2. In addition, there is cross-regulation of TH1 and TH2 cells. The TH1 cytokine IFN-γ downregulates TH2 cells, whereas the TH2 cytokine IL-10 downregulates TH1 cells. Cytotoxic T cells (T_C) can kill target cells, such as virus-infected host cells, tumor cells, or the cells of a histoincompatible tissue graft. T_C cells reversibly bind to their targets by means of the T-cell antigen receptor as well as several other cell surface molecules.

Specific Immune Defects Predispose Patients to Specific Types of Infections

Because each functional compartment of the immune system plays a specialized role in host defense, infections with certain microorganisms are found characteristically in association with specific types of immunodeficiency (Table 2). For example, patients with abnormalities of cell-mediated immunity characteristically develop *Pneumocystis* pneumonia, disseminated fungal infections, mucocutaneous candidiasis, chronic or disseminated viral infections, and severe mycobacterial disease. Patients with defects of antibody or complement more often have infections with pyogenic encapsulated bacteria. Patients with phagocytic defects develop bacterial and fungal infections of the skin and reticuloendothelial system. These distinctions may be blurred, however, because the host's defense against any given microorganism depends on the successful integration of all components of the immune system. Thus, rare patients with antibody deficiencies can develop *Pneumocystis* pneumonia or chronic enteroviral meningitis, whereas patients with deficiencies of cell-mediated immunity can develop pyogenic bacterial infections. Recurrent infections at a single anatomic site should always prompt consideration of other predisposing conditions, such as ciliary dyskinesia, cystic fibrosis, or bronchial obstruction. The key to understanding the susceptibility of a particular compromised host is to understand the immune defects of that host. These defects are illustrated most easily by the primary immunodeficiency diseases, in which a single gene disorder causes one change in immune function (Table 3). Other disor-

Table 2. Patterns of illness associated with primary immunodeficiency diseases

Compartment of disorder	Illnesses	
	Infection	Other
Antibody	Sinopulmonary (pyogenic bacteria, viruses), gastrointestinal (enterovirus, *Giardia*)	Autoimmune disease (autoantibodies, inflammatory bowel disease)
Cell-mediated immunity	Wide range of microorganisms, including opportunistic pathogens; pneumonia (pyogenic bacteria, *Pneumocystis jirovecii*, viruses); gastrointestinal (viruses); skin, mucous membranes (fungi)	
Complement	Sepsis and other blood-borne infections (streptococci, pneumococci, neisseria)	Autoimmune disease (SLE, glomerulonephritis)
Phagocytes	Skin, reticuloendothelial system (staphylococcus, enteric bacteria, fungi, mycobacteria)	

ders which predispose the host to develop infection are often due to multiple factors. For example, cancer chemotherapy can cause neutropenia and mucositis, each of which will increase the host's susceptibility to infection.

PRIMARY IMMUNODEFICIENCY DISEASES

Antibody Disorders

XLA

X-linked agammaglobulinemia (XLA) is the prototypic disorder of humoral immunity that best illustrates the role of antibody in host defense. Male patients with this disease have no B lymphocytes and severe panhypogammaglobulinemia, but all other components of the immune system are normal. Boys with XLA are protected by transplacentally acquired maternal IgG for the first 3 to 4 months of life. Thereafter, chronic and recurrent infections are the predominant clinical manifestation of XLA. Otitis media, pneumonia, diarrhea, and sinusitis occur most often, usually in combination. *Streptococcus pneumoniae*, *Haemophilus influenzae*, and *Staphylococcus aureus* are the most frequently identified bacterial pathogens, but nontypeable, unencapsulated *H. influenzae*, *Salmonella*, *Pseudomonas*, and *Mycoplasma* infections occur with increased frequencies (306). Infections are not limited to mucosal surfaces, as bacterial meningitis, sepsis, and osteomyelitis occur in as many as 10% to 15% of untreated patients. Enterovirus infections are a particularly difficult clinical problem in patients with XLA. This group of viruses (coxsackieviruses, enteric cytopathogenic human orphan viruses [echoviruses], and polioviruses) tends to cause chronic diarrhea, hepatitis, pneumonitis, and meningoencephalitis in patients with XLA. The peculiar susceptibility to enteroviruses is perhaps best illustrated by the fact that

Table 3. Illustrative primary immunodeficiency diseases

Compartment of host defense	Disease(s)
Antibody	XLA
	CVID[a]
	Selective IgA deficiency
	Wiskott-Aldrich syndrome[a]
Antibody and cell-mediated immunity	SCID
	DiGeorge syndrome
Cell-mediated immunity	Chronic mucocutaneous candidiasis
Phagocytes	Congenital neutropenia
	Chronic granulomatous disease
	Congenital asplenia (Ivemark syndrome)
	Leukocyte adhesion deficiency
	Chediak-Higashi syndrome
Complement	Classical pathway (C1q, C1r, C1s, or C4, C2, or C3 deficiency)
	Alternative pathway (factor D, factor I, factor H, properdin)
	MBL pathway
	Terminal components (C5, C6, C7, C8, and C9)

[a]May have associated defects of cell-mediated immunity.

these children are at risk of developing chronic infection after receiving a live poliovirus vaccine or even being exposed to someone who was recently immunized (115). In an agammaglobulinemic host, viral replication can continue long enough for there to be reversion to wild-type virus, with the subsequent development of paralytic poliomyelitis. In some instances, enterovirus infections take the form of a dermatomyositis-like syndrome consisting of rash, edema of subcutaneous tissue, and muscle weakness (227). Enterovirus infections are often fatal for patients with XLA (184).

CVID

Common variable immunodeficiency (CVID) is a heterogeneous group of disorders that is characterized by hypogammaglobulinemia and impaired antibody responses. Additional immunologic abnormalities, such as T-cell dysfunction and autoimmune diseases, are expressed variably. Most patients do not manifest symptoms until after the first decade of life, but some patients present in early childhood or infancy. It has become increasingly apparent that the clinical phenotype of CVID can be the result of a wide variety of immunologic abnormalities. For example, genetic analyses have identified mutations of Btk (the gene causing XLA), SH2D1A (the gene causing X-linked lymphoproliferative syndrome), and ICOS (the "inducible stimulator" on activated T cells) among small numbers of individuals previously identified as having CVID (50). It is likely that such analyses will help to define subgroups of CVID patients who differ in presentation and outcome and, perhaps, lead to novel therapies. Because the one common abnormality of immune function in CVID is antibody deficiency, it is not surprising that the most frequent infections with CVID are similar to those seen with XLA (64, 224). Chronic or recurrent pneumonia, bronchitis, and/or sinusitis occurs in the majority of patients, and some eventually develop chronic pulmonary dysfunction. Most of the identified respiratory tract pathogens are encapsulated bacteria. In contrast to patients with XLA, disease of the gastrointestinal tract occurs with almost equal frequency to that of disease of the respiratory tract in patients with CVID. As many as 30% to 60% of patients with CVID have chronic diarrhea. An infectious agent is identified in approximately one-half of these patients, but many of the others have autoimmune/inflammatory bowel diseases. The most frequently documented gastrointestinal pathogen is *Giardia lamblia*. Bacterial overgrowth of the small bowel is an important cause of chronic diarrhea in patients with CVID; enteroviruses are less of a problem.

Selective IgA deficiency

Selective IgA deficiency is diagnosed by convention when a patient has a serum IgA level of <7 mg/dl with normal levels of other immunoglobulin classes, normal serum antibody responses, and normal cell-mediated immunity. The majority of patients with IgA deficiency lack both serum and secretory IgA, but rare cases occur in which there is a deficiency of secretory but not serum IgA. Unlike the other major serum immunoglobulin classes, IgA is largely silent as a mediator of inflammatory responses, but IgA provides an antimicrobial defense by inhibiting microbial adherence and neutralizing viruses and toxins. Some patients with selective IgA deficiency are more susceptible to infection, although disagreement exists about the relative risk of infection that IgA deficiency imposes on the host (63, 274). Among patients referred to tertiary care centers for evaluation of recurrent sinopulmonary infections, the incidence of IgA deficiency is significantly higher than that for the general population. However, many apparently asymptomatic IgA-deficient individuals have been identified by population-based screening. As might be expected by its role as the predominant secretory immunoglobulin, the most common infections in IgA-deficient patients occur on mucosal surfaces. Otitis media, sinusitis, bronchitis, pneumonia, and diarrhea are common; meningitis and bacterial sepsis are rare. The second major target for infections in IgA-deficient patients is the gastrointestinal tract. Chronic diarrhea is often idiopathic; *Giardia* is the most frequently identified microbial pathogen.

Disorders of Antibody- and Cell-Mediated Immunity

Severe combined immunodeficiency (SCID) causes an absence or near absence of humoral and cell-mediated adaptive immunity, but all components of the innate immune system are intact (40). This heterogenous group of disorders is almost always caused by defects intrinsic to the T lymphocyte (e.g., mutations in cytokine receptor genes) because of the essential role of T cells in controlling humoral as well as cell-mediated immune responses. For example, the absence of CD4 T lymphocytes interferes with the growth and differentiation of B lymphocytes as well as the growth and differentiation of T cells. Affected children have severe deficiencies of all T-lymphocyte subsets and have virtually no

T-lymphocyte function. They may or may not have normal numbers of B lymphocytes in the peripheral blood, but those B cells do not differentiate into plasma cells and the children do not make antibody responses to vaccines or infections. Infants will almost always become symptomatic within the first months of life. Unless the immunodeficiency is treated, most die from infections within the first year of life. These children are susceptible to virtually any microbial pathogen (272). Just as in patients with XLA, they are susceptible to infection by encapsulated bacteria and enteroviruses. However, they are also susceptible to a much wider array of viruses. Pathogens as diverse as adenovirus, rotavirus, cytomegalovirus (CMV), varicella-zoster virus (VZV), and respiratory syncytial virus (RSV) can cause chronic or fatal infections. Fungal infections (e.g., aspergillosis and candidiasis) are problematic because this group of patients lack the CD4 T-lymphocyte production of IFN-γ that is responsible for improving the intracellular killing of phagocytic cells. Patients with SCID can also be infected with opportunistic pathogens such as *Pneumocystis jirovecii*, *Mycobacterium avium-intracellulare*, and even *Mycobacterium bovis* bacillus Calmette-Guérin (BCG) from immunization. Curative treatment of most of these infections requires definitive treatment of the underlying immunodeficiency by bone marrow transplantation.

Disorders of T Cells

Chronic mucocutaneous candidiasis is one of very few disorders that affect only T cells. Patients with this immunodeficiency have a relatively selective defect in the cell-mediated immune defense against *Candida*. The etiology of this disease is not known. Patients have chronic candidal infections of the mucous membranes, nails, and skin but do not develop invasive candidiasis. Many, but not all of them, have clinically significant problems with other fungi (*Histoplasma* and *Aspergillus*), viruses (herpes simplex virus [HSV], VZV, RSV, and measles virus), and intracellular bacteria (*Mycobacterium avium-intracellulare*) (113).

Disorders of Phagocytes

Chronic granulomatous disease is a disorder of intracellular killing that is caused by defects in the NADPH oxidase-dependent respiratory burst system of phagocytic cells. Neutrophils and monocytes of affected individuals are able to follow chemotactic signals and ingest microbial pathogens. Once ingested, organisms such as the pneumococcus or group A streptococcus are killed efficiently because those bacteria produce hydrogen peroxide and thus compensate for the lack of respiratory burst. However, catalase-producing microbes are not killed. This leads to susceptibility to a restricted group of microorganisms, including *S. aureus*, *Burkholderia cepacia*, *Serratia marcescens* and other gram-negative rods, *Nocardia*, *Aspergillus*, and mycobacteria. The most frequent sites of infection are the lungs, lymph nodes, skin, perianal area, and gingivae (237, 305). Phagocytes with live intracellular organisms may travel to reticuloendothelial tissues, such as the liver and spleen, where micro or occasionally large abcesses occur. Typically, patients develop granulomas at the sites of infection as an increasing number of phagocytes and T lymphocytes are drawn to the areas of chronic infection. *Aspergillus* infections of the lung have historically had a dismal prognosis, with the infections spreading from the lung to overlying ribs to the vertebrae. Fortunately, outcomes have improved dramatically with the relatively recent use of IFN-γ to increase the killing capacity of phagocytes via a non-NADPH-dependent pathway combined with the use of nonnephrotoxic orally administered antifungals, such as itraconazole and voriconazole.

Congenital neutropenia can be caused by mutations in the elastase 2 gene and probably other genes (268). Patients present early in life with cellulitis, perirectal abcesses, stomatitis, and gingivitis. Pneumonia, sepsis, and meningitis can also occur (25). As expected, the risk of infection varies inversely with the neutrophil count, and the highest risk is with absolute neutrophil counts of $<500/mm^3$. Infections are caused by *S. aureus*; gram-negative rods, including *Klebsiella*, *Pseudomonas*, and *Escherichia coli*; and rarely, fungi. This group of organisms causes disease in these patients because of their prevalence on the skin and in the gastrointestinal tract, not because of the presence or absence of catalase.

Disorders of Complement

Diminished C3 activation via the classical pathway can be caused by an autosomal deficiency of C1q, C1r, C1s, C4, C2, or C3. Each of these disorders is associated with sepsis and other bloodstream infections (84, 238). The risk is highest for individuals with C3 deficiency because they are unable to mount complement effector function via the classical, the alternative, or the MBL pathway. The most common pathogens are *S. pneumoniae*, *H. influenzae*, gram-negative *Enterobacteriaceae*, *Neisseria meningitidis*, and staphylococci. Individuals with these deficiencies also have a propensity to develop immune complex-mediated diseases, such as systemic lupus erythematosus (SLE) and glomerulonephritis, at least in part because the inability to bind C3b to circulating IgG- and IgM-containing immune complexes impairs their clearance from the bloodstream.

Diminished activation of the terminal complement components/membrane attack complex can be caused by an autosomal deficiency of C5, C6, C7, C8, or C9. Individuals with any of these disorders have a markedly increased risk for neisserial infections, including meningococcemia, meningococcal meningitis, and disseminated gonococcal infections (84, 238). Despite the fact that C5a is an important chemoattractant, only a single C5a-deficient patient has been reported to have symptoms consistent with defective chemotaxis, i.e., recurrent infections of the skin and subcutaneous abscesses. The propensity to develop systemic neisserial infections is so great with these complement deficiencies that one in seven patients with nonepidemic invasive meningococcal infection will be found to have a terminal complement component deficiency. The chances increase to almost 1 in 3 for patients with more than one episode of invasive meningoccal disease.

Diminished activation of the MBL pathway, caused by MBL deficiency, increases the susceptibility of children under the age of 2 years to acute respiratory tract infections, as assessed in population-based studies (54, 153). No studies have yet reported information about the microbial pathogens seen in such children. Similar studies of adults have failed to show any correlation with risk for infection or death from infection.

Diminished activation of the alternative pathway can be caused by deficiencies of factor D, factor I, factor H, or properdin. Deficiency of the latter, a disease with X-linked inheritance, is the most common defect of the alternative pathway, but all of these disorders are very rare (or at least rarely diagnosed). Patients with properdin deficiency have a propensity to develop meningococcal meningitis and, to a lesser degree, invasive *S. pneumoniae* infections (84, 238). Patients with factor D deficiency present in childhood with systemic infections, usually caused by neisseria or *S. pneumoniae*. Those with factor I deficiency consume so much C3 that their presentation is identical to that of patients with C3 deficiency, as they develop invasive infections caused by *S. pneumoniae*, *H. influenzae*, *Enterobacteriaceae*, *N. meningitidis*, and staphylococci. Those with factor H deficiency appear to be most susceptible to autoimmune/chronic inflammatory diseases (especially hemolytic-uremic syndrome), but they are also susceptible to meningococcal infections.

SECONDARY IMMUNODEFICIENCIES

Secondary immunodeficiencies are those that are related to another illness or condition or occur as a result of treatment for such a condition. In this section, we review secondary immunodeficiencies and other compromises in host defenses that result because of treatment with a variety of agents (Table 4). The consequences of treatment for malignancy, of prevention and treatment of allograft rejection, and of the treatment of rheumatologic and autoimmune diseases are reviewed. The degree of immunodeficiency associated with various immunosuppressive agents used to treat a variety of conditions depends on the underlying condition, the doses of single agents, and drug combinations that may act synergistically.

Corticosteroids

Corticosteroids are used to treat a variety of diseases because of their anti-inflammatory and immunosuppressive properties (232, 250). They have many effects on innate and acquired immunity. Corticosteroids impair trafficking of neutrophils and monocytes to sites of inflammation and inhibit macrophage and neutrophil phagocytic and microbicidal function (35, 82). They inhibit the production of almost all known cytokines (16). Corticosteroids markedly reduce the numbers of circulating dendritic and T cells and affect antigen presentation by impairing the effector functions of macrophages and dendritic cells (17, 216, 234, 258). Their effects on the immune system are dose dependent. At doses of <2 mg/kg of body weight, T-lymphocyte numbers are slightly reduced ($CD4^+ > CD8^+$). Higher doses of >2 mg/kg result in suppression of lymphocyte activation and suppression of antibody production by B cells. Hence, corticosteroids predispose hosts to infection in a dose-dependent manner (95, 243, 275). The risk of infection is also determined by the underlying disorder and concomitant treatment with immunosuppressive agents. Viral (mainly herpesvirus), bacterial, and fungal (*Candida*) infections are encountered with greater frequencies—up to 40 times greater—in patients treated with corticosteroids (82). Pneumonia caused by *P. jirovecii* is the main opportunistic infection that occurs in patients treated with these agents. Reactivation of tuberculosis is also a potential complication of corticosteroid therapy.

Therapeutic Interventions for Neoplasia

In treating malignancy, the goal is to target mitotically active tumor cells as specifically as possible. However, in addition to the destruction of malignant

Table 4. Host immune deficits and infections associated with immunosuppressive and chemotherapeutic agents

Agent class	Effects on immune system	Associated pathogens or comments
Corticosteroids	Decrease chemotactic activity to sites of inflammation; inhibition of phagocytic, microbicidal, and T-cell functions	Herpesviruses, *Candida* sp., bacteria, *P. jirovecii*
Cytotoxic drugs (methotrexate, f-fluoruracil [antimetabolites], cyclophosphamide, chlorambucil, melphalan [alkylating agents], doxorubicin, daunorubicin, idarubicin, mitoxantrone [anthracyclines], vincristine, vinblastine, cisplatin, and bleomycin	Suppress bone marrow and significantly reduce counts at high doses; neutropenia; some cause lymphopenia, some cause significant mucositis	Bacteria (gram positive and negative), *Candida*, invasive mold infection with prolonged neutropenia
Purine analogs (e.g., fludarabine, cladribine, and pentostatin)	Neutropenia, lymphopenia, with or without hypogammaglobulinemia	Encapsulated, gram-positive, and gram-negative bacteria, HSV, VZV, CMV, mycobacteria, *Candida*, *Aspergillus*, *Cryptococcus*
Campath (alemtuzumab)	Profound depletion of B and T cells, NK cells, and monocytes; neutropenia	Bacteria, herpesviruses (HSV, VZV, EBV-associated PTLD), *P. jirovecii*, invasive fungi (*Candida*, *Aspergillus* species), *Histoplasma*, *Cryptococcus*, mycobacteria, *Toxoplasma*, *Nocardia*
Azathioprine	Inhibits B- and T-cell proliferation, decreased antibody production, myelosuppression	Bacterial infections (leukopenia), herpes zoster, exacerbation of viral hepatitis
Cyclosporine and tacrolimus	Inhibit production of IL-2 and other cytokines by CD4-positive T cells	Lower rates of infection in CSA-treated transplant patients; very low rates in patients with autoimmune diseases treated with CSA alone
MMF	Inhibits B- and T-cell proliferation; decreased antibody production; leukopenia	HSV, VZV, CMV, hepatitis C virus
Rapamycin (sirolimus)	Inhibits T-cell activation and proliferation; inhibits antibody production	Herpesviruses (HSV, VZV, possibly decreased incidence of CMV)
Polyclonal or monoclonal antilymphocyte serum and antithymocyte globulins (ALA, ATG, OKT3)	Deplete T lymphocytes, inhibits T-cell activation	Herpesviruses (CMV, HSV, EBV-associated PTLD), *P. jirovecii*
Anti-IL-2 receptor monoclonal antibodies (daclizumab, basiliximab)	Inhibit IL-2 binding to the IL-2 receptor on T cells; inhibit activation of T cells	Not associated with increased risk of infection when added to other immunosuppressive agents
Anti-CD20 monoclonal antibody (rituximab)	Lymphopenia, rarely neutropenia	HSV, VZV, CMV, hepatitis B and C viruses, JC virus
Anti-TNF-α inhibitors (infliximab, adalimumab, etanercept)	Decrease production of proinflammatory cytokines, chemokines, and endothelial adhesion molecules (decreased inflammatory response)	Mycobacteria, disseminated tuberculosis, *Aspergillus fumigatus*, *Histoplasma capsulatum*, *Cryptococcus neoformans*, *P. jirovecii*, *Listeria monocytogenes*, *Toxoplasma gondii*
Phenytoin	IgA deficiency or hypogammaglobulinemia	Bacteria (including encapsulated bacteria), respiratory viruses, enteroviruses, *Giardia*

cells, normal cells that are rapidly dividing will be affected by cytotoxic antineoplastic agents (chemotherapy). The primary nonmalignant cells affected include bone marrow cells and cells of the gastrointestinal mucosa. Therefore, the main chemotherapy-associated toxicities contributing to infectious risk are myelosuppression and mucositis (146). Neutropenia is usually an unavoidable consequence of the treatment of malignancy and significantly increases a patient's risk of infection (292). This risk increases with severity of neutropenia, with the highest risk of infection being associated with an absolute neutrophil count of <100 cells/mm^3 (31). In addition, chemotherapy results in chemotactic and phagocytic defects in neutrophils, further increasing the risk of severe infection (121, 189). The source of infection in the majority of patients is the patient's endogenous flora, which is enabled to translocate across mucocutaneous barriers secondary to chemotherapy-induced mucosal injury of the oral cavity and intestinal epithelium and due to indwelling vascular and urinary catheters (146). The course of mucositis after standard or high-dose chemotherapy parallels that of neutropenia. The onset of mucositis occurs at the nadir of the neutrophil count and resolves with count recovery.

The constellation of defects in host defense, neutropenia, mucositis, and indwelling catheters predispose the patient to bacterial, fungal, and viral (mainly HSV) infections. In prior years, bacterial infections were most often caused by gram-negative bacteria, such as *Pseudomonas* sp. (32). However, in recent years, a shift to gram-positive organisms has occurred and has been attributed to empiric antibiotic regimens targeted to gram-negative organisms, common use of long-term indwelling vascular catheters, and use of prophylactic antibiotics, such as fluoroquinolones, that are targeted against gram-negative organisms (307). The risk of invasive fungal infection with *Candida* species or molds, such as *Aspergillus*, increases with the severity and duration of neutropenia (58, 74).

Lymphocyte depletion can occur as a complication of cytotoxic antineoplastic therapy (169, 170). Some cancers, such as Hodgkin's lymphoma, are associated with lymphocyte dysfunction, but significant T-cell immunodeficiency is usually uncommon prior to initiation of cytotoxic therapy (118). Some agents, such as cyclophosphamide, administered as a single agent at a high-intensity dose or as part of a multiagent dose-intensive regimen, can cause profound depletion of the lymphocyte populations and predispose patients to opportunistic infections. Humoral immunity tends to be relatively spared from the effects of short courses of chemotherapy because of the long half-life of previously secreted IgG antibodies.

A variety of cytotoxic antineoplastic agents are used in combination to treat various malignancies. These agents are classified based on the mechanism by which they inhibit cell proliferation (Table 5). They all cause myelosuppression, and most cause some degree of mucositis.

The antimetabolite antineoplastic agents include methotrexate, fluorouracil, and gemcitabine. Methotrexate is an inhibitor of dihydrofolate reductase that interferes with the synthesis of purine nucleotides and, hence, with DNA synthesis, repair, and cellular replication. The major side effects of treatment with methotrexate include myelosuppression, causing significant neutropenia, and a dose-dependent, ulcerative mucositis (106, 212). High-dose methotrexate (>20 mg/kg) used in cancer therapy causes profound bone marrow suppression that also depresses primary and secondary cellular and humoral immune responses (256). Hence, opportunistic infections that arise in the setting of compromised T-cell function, such as those caused by *P. jirovecii*, CMV, *Cryptococcus*, *Histoplasma capsulatum*, *Nocardia* sp., and VZV, have been reported (12, 14, 140, 145, 157). Many of these patients were also receiving corticosteroids. Long-term treatment with methotrexate may place one at risk for Epstein-Barr virus (EBV)-associated lymphoma (139, 180). The risk of infection with low-dose methotrexate is not well established, but it is lower, given that the lymphocyte subsets and in vitro T-cell mitogen responses are unaffected. The other commonly used antimetabolites are fluorouracil and gemcitabine. These are antimetabolites of the pyrimidine analog type and are cell cycle specific (S phase) in inhibiting DNA synthesis. Both agents can cause significant myelosuppression and mucositis.

The alkylating agents cyclophosphamide, chlorambucil, and melphalan induce cytotoxic effects by chemically modifying nucleotides, cross-linking DNA or RNA, and inhibiting protein synthesis. Depending on the dose and duration of treatment, treatment with alkylating agents can result in significant bone marrow suppression, with a decline in neutrophil and T- and B-lymphocyte counts (169, 170). The tendency for alkylating agents to cause lymphopenia is enhanced by coadministration of corticosteroids. These cumulative negative effects on cellular host defense predispose recipients of alkylating agents to a variety of infections, including routine bacterial infections that cause pneumonia, sepsis, or urinary tract infection and opportunistic infections caused by *P. jirovecii*, fungi, *Nocardia*,

Table 5. Classes of cytotoxic antineoplastic agents

Class	Agents
Antimetabolites	Methotrexate, 5-fluorouracil, gemcitabine
Alkylating agents	Cyclophosphamide, chlorambucil, melphalan
Anthracyclines	Doxorubicin, daunorubicin, idarubicin, mitoxantrone
Vinca alkaloids	Vincristine, vinblastine, vinorelbine
Platinum compounds	Cisplatin, carboplatin
Glycopeptide antibiotic	Bleomycin
Podophylotoxin derivative	Etoposide
Taxanes	Paclitaxel, docetaxel

VZV, and *Mycobacterium tuberculosis* (100, 169, 226). Patients who are neutropenic or are treated concomitantly with high doses of glucocorticoids have an enhanced risk of infection (100, 226).

The anthracyclines, i.e., doxorubicin, daunorubicin, and idarubicin, are cytotoxic antineoplastic agents that are used to treat a variety of malignancies. They cause cytotoxicity by intercalating between DNA base pairs and by inhibiting topoisomerase II, resulting in inhibition of RNA and DNA synthesis. Neutropenia and mucositis are reported for a significant number of patients, depending on the agent used.

A variety of other classes of antineoplastic cytotoxic agents, including the vinca alkaloids, platinum compounds, taxanes, glycopeptide antibiotics, and topoisomerase inhibitors (Table 5), have similar effects to various degrees on the bone marrow and mucosa. Patients treated with these agents are primarily at risk for bacterial and candidal infections. This risk increases with the duration and depth of neutropenia and with the severity of mucositis (146, 291).

Purine analogs inhibit DNA synthesis and are used to treat a variety of hematologic malignancies. These agents induce severe immunosuppression, affecting multiple lineages of host defense, namely, T and B lymphocytes, neutrophils, and monocytes (57). After treatment with purine analogs, a profound T-cell lymphopenia, especially of CD4 cells, develops in 2 to 3 months and can persist for several years (57). Many patients develop neutropenia and a depletion of monocytes. Some patients may become hypogammaglobulinemic. Hence, a broad spectrum of infections is encountered in patients treated with purine analogs, including bacterial infections (staphylococci, streptococci, gram-negative rods, *Listeria*, *Nocardia*, *Legionella*, and mycobacteria), opportunistic viral infections (HSV, VZV, CMV, and EBV), and opportunistic fungal infections (*P. jirovecii*, *Candida*, and *Aspergillus*). Bacterial, fungal, and HSV infections occur early after treatment in the setting of neutropenia. Opportunistic infections, which are associated with depressed cell-mediated immunity, occur later after treatment. The type and stage of the underlying disorder, prior antineoplastic therapy, and concurrent treatment with steroids significantly influence the incidence of infectious complications. *Listeria*, *P. jirovecii*, and CMV infections occur more frequently in those treated concomitantly with corticosteroids.

Monoclonal antibody therapies

Rituximab is a monoclonal antibody directed against the CD20 antigen on B lymphocytes (171). Binding to the antigen results in complement- and/or cell-mediated cytotoxicity. Rituximab also binds to the Fc receptors, facilitating antibody-dependent cellular cytotoxicity. Rituximab is approved for the treatment of low-grade, follicular, or diffuse large-cell CD20-positive B-cell non-Hodgkin's lymphoma. In addition, this agent is approved for the treatment of rheumatoid arthritis, along with methotrexate. Treatment with rituximab causes lymphopenia (48%) and neutropenia (14%), with a median duration of 14 days. A variety of viral infections, either new or reactivated, some severe and potentially fatal, have been reported with rituximab use, including infections with HSV, VZV, CMV, parvovirus B19, hepatitis B and C viruses, and West Nile virus (3). These infections may be delayed, occurring up to a year after treatment. In December 2006, the Food and Drug Administration (FDA) issued a warning after two fatal cases of progressive multifocal leukoencephalopathy (PML) occurring in SLE patients treated with rituximab were reported (44, 221). Rituximab has been added to intense antirejection regimens used in the transplantation of organs (kidney and liver) into ABO blood group-incompatible and positive cross-match transplant recipients (101). A trend towards an increased incidence of a variety of infectious complications was demonstrated for patients who received rituximab as part of their immunosuppressive regimen compared to that for historical controls who did not (101). Persistent and severe hypogammaglobulinemia has been reported for rare patients who have been treated with rituximab (48, 124). These patients are at risk for the same infections (with encapsulated bacteria and some viruses) as those seen in patients with XLA or CVID.

Alemtuzumab (Campath) is a humanized monoclonal IgG1 antibody directed against the CD52 cell surface glycoprotein that is used as first- and second-line therapy for the treatment of B-cell chronic lymphocytic leukemia (CLL) and peripheral and cutaneous T-cell lymphomas. Its only FDA-labeled indication is for the treatment of CLL that is refractory to fludarabine (FDA approval of alemtuzumab use [http://www.cancer.gov/cancertopics/druginfo/fda-alemtuzumab]). Alemtuzumab has also been used for treatment of other malignancies (e.g., non-Hodgkin's lymphoma and T-cell prolymphocytic leukemia), as a conditioning agent in hematopoietic stem cell transplantation, for induction of immunosuppression or treatment of acute rejection in solid organ transplantation, for rheumatoid arthritis, and for the prevention of graft-versus-host-disease (GVHD) (179). The antibody binds to CD52 antigen on the surfaces of cells and causes cell lysis through complement activation and antibody-dependent cell-mediated toxicity (130). CD52 is also expressed on the surfaces of nonmalignant T and B lymphocytes, monocytes, macrophages, NK cells, some granulocytes, and normal bone marrow cells; therefore, cell destruction is not targeted to the malignant cell and significant impairment in cellular host defenses can occur. Profound and long-lasting depletion of mature B and T lymphocytes, NK cells, and monocytes occurs after treatment with alemtuzumab (165). Treated patients develop a profound lymphopenia by 1 to 2 weeks after initiation of treatment that may persist for over 1 year (231). Neutropenia (0.5×10^9 cells/liter) occurs in one-third of patients at around 4 weeks of therapy, but the cell number usually recovers in 2 to 3 weeks (144). As a consequence, the infections encountered are nonopportunistic and opportunistic. The incidence of infectious complications has been noted to range from 35 to 65%. However, the majority of studies reporting these data included patients with lymphoproliferative disorders who were pretreated with other agents, such as purine analogs, rituximab, and alkylating agents (280). This pretreatment and the advanced, refractory underlying malignancy further contributed to the net state of immunosuppression and enhanced risk of infection.

In a recent study reporting on infectious complications associated with alemtuzumab use for lymphoproliferative disorders, nonopportunistic bacterial infections causing sepsis, pneumonia, and catheter-related bacteremia were commonly encountered (179). Fifty-six percent of patients developed an opportunistic infection during the study period. Herpesvirus infections (HSV, VZV, and CMV) were the most common. CMV reactivation with resulting viremia is a well-described complication of therapy with alemtuzumab, with a reported incidence of as high as 50% (208, 308). Other opportunistic infections associated with alemtuzumab include adenovirus infection, PML, invasive pulmonary aspergillosis, disseminated histoplasmosis and cryptococcosis, pneumocystosis, tuberculosis, cerebral toxoplasmosis, and disseminated acanthamebiasis (179). CMV reactivation and invasive aspergillosis appear to be the most commonly reported opportunistic infections in the setting of lymphoproliferative disease. Prophylaxis for *P. jirovecii* pneumonia and herpes-related infections is recommended by experts and the drug manufacturer (280; Campath [alemtuzumab] package insert [Genzyme, Cambridge, MA]).

Clinical trials have revealed that alemtuzumab is efficacious for the prevention or treatment of acute allograft rejection in organ transplant recipients (22, 183, 298). A recent study of a large cohort of organ transplant recipients who received alemtuzumab for induction therapy or for the treatment of rejection reported a 10% incidence of opportunistic infections (220). Patients who received alemtuzumab for induction therapy were significantly less likely to develop an opportunistic infection than were patients who received it for the treatment of rejection (4.5% versus 21%). This is probably related to an enhanced net state of immunosuppression that results from the administration of other immunosuppressive agents in the setting of rejection. CMV disease and esophageal candidiasis were the most common opportunistic infections. Other infections included BK polyomavirus infection, EBV-associated posttransplant lymphoproliferative disorder (PTLD), invasive mold infections (aspergillosis, mucormycosis, and pseudoallescheriosis), nocardiosis, tuberculous and nontuberculous mycobacterial infections, and toxoplasmosis.

Immunosuppressive Therapy for Prevention and Treatment of Allograft Rejection

Maintenance immunosuppressive therapy is administered to organ transplant recipients to help prevent acute rejection. The maintenance regimen usually consists of a combination of immunosuppressive agents with different mechanisms of action. Currently, most transplant centers use a regimen consisting of prednisone, an antimetabolite (azathioprine or mycophenolate mofetil [MMF]), and a calcineurin inhibitor (cyclosporine or tacrolimus). The level of chronic immunosuppression is slowly decreased over time to lower the risk of infection and malignancy.

Antimetabolites

Azathioprine is a precursor of 6-mercaptopurine that inhibits purine biosynthesis and, hence, DNA, RNA, and protein synthesis. The effects of azathioprine include a decrease in circulating T and B lymphocytes, a decrease in immunoglobulin production, diminished IL-2 secretion, and myelosuppression (174). Leukopenia is the most serious side effect of azathioprine. Azathioprine is approved for the prevention of rejection in renal transplant recipients and for the treatment of rheumatoid arthritis. Infections reported in patients taking azathioprine are bacterial infections in the setting of leukopenia, herpes zoster, and exacerbation of hepatitis (65, 276).

Azathioprine has been used to prevent allograft rejection since the early 1980s. Several large trials comparing azathioprine to MMF have shown that MMF is superior to azathioprine in reducing the number of episodes of transplant rejection in heart, kidney, and liver transplant recipients (76, 270, 301). As a result, most transplant centers have switched to using MMF as part of their immunosuppressive regimen. MMF interferes with the de novo synthesis of purine nucleotides and, in this way, inhibits primarily T-cell proliferation. In addition, MMF inhibits B-cell proliferation and results in decreased antibody production (60, 123). Because of its potent inhibition of lymphocyte proliferation, treatment with MMF predisposes patients to infections associated with depressed cell-mediated immunity, such as infections with HSV, VZV, and CMV. A higher incidence of tissue-invasive CMV disease has been reported for renal and heart transplant recipients treated with MMF, particularly for those patients receiving >2 g of MMF per day (76, 247). However, clinical trials of liver and lung transplant recipients receiving MMF failed to show an increased incidence of CMV infection or disease (129, 217). Interestingly, mycophenolate exhibits an antimicrobial effect against *P. jirovecii* (213). Renal transplant patients taking MMF had no episodes of *Pneumocystis* pneumonia in a randomized trial comparing MMF to azathioprine for the prevention of acute rejection (270). Heart transplant patients receiving MMF had a higher rate of acute cholestatic hepatitis due to hepatitis C virus. No effect of MMF on bacterial infections in organ transplant recipients has been documented. Mycophenolate is also employed as a potential steroid-sparing agent in the treatment of a variety of autoimmune diseases.

Cyclosporine and tacrolimus

Organ allograft survival has improved significantly since the introduction of cyclosporine in the 1980s and tacrolimus in the 1990s. In addition, these agents are becoming increasingly popular for the treatment of a variety of rheumatic diseases. Cyclosporine is an 11-amino-acid cyclic peptide, and tacrolimus is a macrolide antibiotic. They bind to intracellular proteins called immunophilins—cyclosporine binds to cyclophilins and tacrolimus binds to FK binding proteins. The complex between drug and immunophilin inhibits calcineurin, a calcium- and calmodulin-dependent phosphatase. Hence, these agents are commonly referred to as calcineurin inhibitors. This inhibition of calcineurin results in prevention of translocation of a family of transcription factors, nuclear factor of activated T cells, into the nucleus. As a consequence, transcription of a variety of cytokine genes involved in T-cell activation is inhibited. The calcineurin inhibitors primarily affect TH cells, although some inhibition of T-suppressor and T_C cells may occur.

Over the past 2 decades, calcineurin inhibitors have become the cornerstone of immunosuppressive therapy in the organ transplant population. These agents are usually combined with corticosteroids and MMF. A large European randomized multicenter trial comparing the efficacies of tacrolimus plus low-dose corticosteroids and a conventional multidrug cyclosporine-based regimen (corticosteroids plus azathioprine) to prevent allograft rejection in liver transplant recipients revealed similar incidences of infection in patients receiving the tacrolimus- and cyclosporine-based regimens (206). The incidence of sepsis was approximately 20% for both groups, and the incidence of CMV infection ranged from 15 to 25%, with a lower incidence for the tacrolimus-treated patients. Despite an immunosuppressive effect that is estimated to be 36 to 100 times more potent than that of cyclosporine, tacrolimus has been associated with fewer CMV infections than cyclosporine-containing regimens have (206, 266). This is likely due to the fact that the incidence of rejection is lower with tacrolimus than with cyclosporine. Hence, the requirement for additional immunosuppression is lower (206, 279). It is likely that the increased incidence of CMV infection is not related to the particular calcineurin inhibitor but to the additional immunosuppressive agent, especially in T-cell-antibody-based therapies.

Other viral infections, such as EBV, hepatitis C virus, and polyomavirus infections, have been linked to treatment with calcineurin inhibitors. However, none of these infections are linked with an agent per se but likely arise as a result of the cumulative effect of immunosuppression. The more intense the immunosuppressive regimen, the more likely it is that a patient will acquire or reactivate one of these infections. As with CMV infection, T-cell-antibody therapy is the most significant component of the immunosuppressive regimen contributing to the risk of EBV-related PTLD (10, 42).

Patients on potent immunosuppressive regimens are at risk for fungal infections. The majority of these infections are caused by *Candida* and *Aspergillus* species. *Candida* infections often arise in the setting of neutropenia and compromised mucocutaneous barriers. Susceptibility to *Aspergillus* infections is influenced by the type and intensity of immunosuppressive regimen. High-dose steroids and OKT3 (muromonab-CD3) monoclonal antibody therapy are known to confer an increased risk for invasive aspergillosis (265). Interestingly, the calcineurin inhibitors possess in vitro activity against *Aspergillus* species (262). Because invasive aspergillosis continues to occur in patients treated with these agents, the immunosuppressant effects must predominate over the antifungal effects in vivo. There is evidence in animal models and in humans that calcineurin inhibitors may alter the pathogenesis of *Aspergillus* infection, with less dissemination (116, 262). Similar observations have been made for cryptococcal infection in organ transplant recipients (264).

Rapamycin

Rapamycin is a macrolide derived from the bacterium *Streptomyces hygroscopicus* that is used to prevent rejection in renal transplant patients. The drug binds intracellularly to the mammalian target of FKBP-12–rapamycin, mTOR, which regulates translation of mRNA required for cell division. A result of this interaction is the inhibition of T-lymphocyte activation and proliferation and an inhibition of antibody production. Patients treated with rapamycin are at increased risk for infections with intracellular pathogens such as herpesviruses. Conflicting data exist in terms of rapamycin predisposing patients to herpesvirus infection. Treatment with rapamycin does not appear to increase the risk of CMV infection in patients also treated with cyclosporine and corticosteroids (167). One study demonstrated a lower incidence of CMV infection in renal transplant recipients treated with rapamycin, MMF, and corticosteroids than in patients receiving cyclosporine to replace rapamycin in the same regimen (156). In a trial comparing rapamycin to azathioprine in a multidrug immunosuppressive regimen also containing cyclosporine and prednisone, the frequencies of infections due to sepsis, CMV, EBV, and VZV did not differ significantly in the first 12 months after transplantation (136). The incidence of CMV infection was similar in a multicenter study that randomized patients to receive rapamycin or cyclosporine in addition to receiving corticosteroids and azathioprine as part of a multidrug regimen. However, infection with HSV was reported more frequently for the rapamycin treatment group (102).

Interestingly, rapamycin exhibits potent in vitro fungicidal activity that is mediated by the inhibition of TOR1 in yeast species. Fungi inhibited include *Cryptococcus neoformans*, *Candida albicans*, *Aspergillus fumigatus*, *Aspergillus flavus*, and *Fusarium oxysporum*. It is not known whether the in vitro antifungal activity of rapamycin translates into a beneficial clinical effect.

Antilymphocyte antibody therapy

OKT3 is a murine IgG2a monoclonal antibody that binds the CD3-epsilon chain of the T-cell-receptor–CD3 complex on T cells. OKT3 has been used for induction immunosuppressive therapy and for the treatment of acute or steroid-resistant allograft rejection in transplant recipients. In vivo, OKT3 reacts with most peripheral blood T cells and T cells in tissues and causes a rapid and profound decrease in lymphocytes (34). T cells are not detectable between 2 and 7 days after administration but reappear rapidly and reach pretreatment levels within a week after termination of treatment. The antibody also causes T-cell-receptor modulation that interferes with T-cell activation. Patients treated with OKT3 are at significant risk for infectious complications, especially herpesvirus (HSV and CMV) infections that require functioning cytotoxic T cells for control of infection. In a prospective study that investigated risk factors for CMV disease in renal transplant recipients, treatment with OKT3 increased the risk of CMV disease fivefold in CMV-seropositive transplant patients (114). OKT3 administration is also associated with an increased risk of PTLD that, in most transplant patients, is EBV associated (9, 277). The impairment of T-cell cytotoxic function allows for the proliferation and transformation of EBV-infected B lymphocytes. The risk of transformation is highest when OKT3 is utilized for the treatment of rejection (43). Other infections related to depressed T-cell function induced by OKT3 include fungal infections, such as aspergillosis, cryptococcosis, and infections caused by *P. jirovecii*, *Listeria*, mycobacteria, *Nocardia*, and *Toxoplasma gondii* (OKT3 product information, p. 1–19; Orthoclone). Routine bacterial infections causing pneumonia and sepsis are also encountered.

Anti-thymocyte globulin (ATG) is a polyclonal antibody preparation of rabbit or equine origin that is used for the prevention or treatment of rejection in renal transplant recipients in conjunction with other immunosuppressive therapy. In addition, ATG has also been used in the field of hematologic malignancies to treat moderate or severe aplastic anemia, as part of conditioning regimens prior to bone marrow transplantation, or for the prevention of GVHD. The exact mechanism by which ATG causes immunosuppression is not known, but it is likely similar to the mechanism employed by OKT3. ATG acts on a variety of T-cell antigens, resulting in depletion of thymus-dependent lymphocytes and suppression of T-cell activation (110; ATGAM intravenous injection product information, 2005 [Pfizer]). Rabbit ATG also contains antibodies against NK cell markers as well as against CD20, a B-cell marker. Lymphopenia can persist for a year or more with rabbit ATG (37, 38). Severe infections can develop in patients treated with ATG, including infections caused by bacteria and organisms that depend on cell-mediated immunity for prevention or control of infection. For example, as with OKT3, ATG has been identified as a risk factor for CMV infection. However, in recent studies of ATG, a lower incidence of CMV infection was attributed to more effective antiviral prophylaxis (37).

Monoclonal antibodies that more specifically target the immune system have been developed. Basiliximab and daclizumab bind to the α chain of the IL-2 receptor (CD25). A reduction in allograft rejections has been demonstrated for kidney, heart, liver, lung, and kidney-pancreas transplant recipients treated with these agents (122, 228). Studies of rejection using daclizumab compared to placebo in multidrug immunosuppressive regimens found a similar or lower incidence of CMV infection or disease (204, 290). A significantly lower incidence of HSV infection in transplant recipients treated with basiliximab, cyclosporine, and corticosteroids than in patients treated with placebo, cyclosporine, and corticosteroids was attributed to the greater use of OKT3 and corticosteroids for rejection in the placebo group (137). Studies have revealed conflicting results regarding recurrence of hepatitis C virus infection in patients receiving daclizumab (111, 205). A difference in the incidences of bacterial and fungal infections in patients treated with these monoclonal antibodies has not been demonstrated (122).

Prevention and Treatment of GVHD in Bone Marrow Transplantation

The occurrence of GVHD posttransplantation remains the most important factor influencing the outcome following allogeneic blood and marrow transplantation (245). Acute GVHD is common in recipients with matched unrelated and haploidentical related donors. The agents reviewed above for the prevention and treatment of allograft rejection in organ transplant recipients are also used for the prevention and treatment of GVHD in allogeneic blood and marrow transplant recipients. The most common prophylactic regimen in use at many transplant centers is a combination of methotrexate and cyclosporine. If toxicities or drug interactions are encountered, alternative agents are employed. These include corticosteroids, tacrolimus, sirolimus, MMF, ATG, and daclizumab. Corticosteroids are the first-line agents for the treatment of acute GVHD. Second-line agents that include cyclosporine, tacrolimus, ATG, and MMF and that are generally less effective than high-dose steroids are utilized in steroid nonresponders. In addition to the enhanced susceptibility to infectious complications imposed by the immunosuppressive therapy to treat or prevent GVHD, GVHD itself also contributes to the risk of infection. GVHD of the skin and gut causes impairment of the mucocutaneous barrier. Chronic GVHD contributes to the immunocompromised state, as it contributes to the persistence of defects in cell-mediated and humoral immunity and reticuloendothelial system function in the postengraftment period (164, 249). Hence, patients with chronic GVHD remain at risk for a variety of opportunistic infections, including fungal (invasive aspergillosis or other mold infection) and viral (CMV, VZV, and EBV) infections and infections caused by *P. jirovecii* (55a). In addition, hypogammaglobulinemia that occurs in patients with GVHD predisposes these patients to infections with encapsulated bacteria, such as *H. influenzae* and *S. pneumoniae* (55a, 223).

Agents for the Treatment of Rheumatic Diseases

Many of the agents used for the treatment of malignancies and for the prevention of allograft rejection are also used for the treatment of a variety of rheumatic autoimmune diseases. Cyclophosphamide is the first-line immunosuppressive agent used for the treatment of severe systemic vasculitides and for severe immune complex-mediated manifestations of lupus. After disease control is achieved with cyclophosphamide, maintenance therapy with less toxic agents, such as methotrexate, azathioprine, MMF, and leflunomide, is often initiated.

Leflunomide

Leflunomide is an isoxazole derivative approved by the U.S. FDA for the treatment of rheumatoid arthritis (273). Leflunomide has anti-inflammatory and immunomodulatory properties. Its active metabolite inhibits the de novo synthesis of pyrimidine nucleotides, resulting in suppression of T-cell function (46). Limited data exist in the literature on the incidence of infections associated with leflunomide use. In the 24-month follow-up study of the double-blind placebo-controlled United States Leflunomide Trial in Rheumatoid Arthritis, the overall incidences of infections were not different in the active and placebo treatment groups, and no opportunistic infections were observed over the 24-month period (59). Since this study, several reports of infections occurring in patients with rheumatoid arthritis treated with leflunomide have been published (131). Common infections, such as lower respiratory tract infections and cellulitis, have been reported. However, opportunistic infections such as disseminated herpes zoster and reactivated tuberculosis are also reported (131, 281). It is important that the majority of the patients were treated with a combination disease-modifying antirheumatic drug regimen that, in addition to leflunomide, included prednisone, methotrexate, or both. A small case series study of pulmonary tuberculosis that occurred in patients with rheumatoid arthritis in Slovenia who were treated with leflunomide has been reported (281). These cases occurred in patients who received prior therapy with other disease-modifying agents or received methylprednisolone (≤8 mg daily) or methotrexate concomitantly. Therefore, with the evidence from the U.S. Leflunomide Trial and given that combination therapy is a common practice, at the present time direct causality of infection cannot be linked to leflunomide.

Anti-TNF therapy

TNF antagonists, such as etanercept, infliximab, and adalimumab, are used to treat moderate to severe rheumatoid arthritis, Crohn's disease, and other inflammatory syndromes. Treatment with these agents is associated with an increased risk of serious infection. Binding of TNF to the TNF receptor stimulates release of inflammatory cytokines and expression of chemokines and endothelial adhesion molecules (26, 289). Inhibition of these effects results in a decrease in migration of inflammatory cells to sites of infection and, hence, a decrease in granuloma maintenance and formation (6). Experiments of TNF blockade in animal models have revealed the importance of TNF in the control of infections caused by intracellular pathogens or those maintained in a latent state by cell-mediated immunity (7, 88, 120, 187). These pathogens include *M. tuberculosis*, *M. avium*, *M. bovis*, *Aspergillus fumigatus*, *P. jirovecii*, *Histoplasma capsulatum*, *Cryptococcus neoformans*, *C. albicans*, *Listeria monocytogenes*, and *Toxoplasma gondii* (141).

Infliximab has a broader spectrum of activity than etanercept and also predisposes patients to a higher risk of infection than etanercept does (294). This is likely linked to the mechanisms by which infliximab and etanercept inhibit TNF activity, i.e., they are a monoclonal antibody and a soluble TNF receptor fusion protein, respectively (75). In the FDA Adverse Event Reporting System, tuberculosis, histoplasmosis, coccidioidomycosis, and listeriosis occurred 2 to 10 times more often in patients treated with infliximab than in those treated with etanercept (294). *M. tuberculosis* infection remains the most commonly reported infection associated with anti-TNF inhibitor therapy and is more often associated with infliximab treatment (143). Tuberculosis in patients treated with TNF inhibitors is frequently atypical in presentation and appears as disseminated and/or extrapulmonary disease. A variety of fungal infections that are kept in check by granuloma formation have also been reported for these patients (Table 4), the most common of which are disseminated histoplasmosis, cryptococcosis, coccidioidomycosis, and aspergillosis (23, 158, 284, 297).

Other Drugs

Phenytoin

The commonly used anticonvulsant phenytoin causes a reversible decrease in the serum IgA level in approximately 20% of treated patients (36, 254). In 5% of cases, there is a severe deficiency of IgA, and in a smaller percentage, phenytoin can cause severe panhypogammaglobulinemia. These effects are usually but not always reversible. Similar effects have been reported, though much less often, with other anticonvulsants, including valproic acid (229). Affected individuals have the same propensity to develop infection as those with IgA deficiency or hypogammaglobulinemia.

Solid Organ Transplantation

The population of patients receiving organ transplants to restore vital organ functions and to prolong life continues to grow. Immunosuppressive regimens that suppress T-cell immune function are employed to prevent organ rejection and to maintain long-term allograft function (188). The immunosuppressive regimens employed in all forms of organ transplantation are similar, with cyclosporine or tacrolimus providing the cornerstone of maintenance antirejection treatment, along with an antimetabolite, such as MMF, and possibly a low-dose corticosteroid. Hence, the types, patterns, and timetables of infections encountered are similar in all forms of organ transplantation. The risk of infection is determined primarily by the intensity of exposure to potential pathogens and the patient's net state of immunosuppression (86, 239). The main determinants of the net state of immunosuppression are the dose, duration, and temporal sequence of the immunosuppressive agents and the presence or absence of infection with immunomodulating viruses (CMV, EBV, hepatitis B or C virus, and human immunodeficiency virus [HIV]) (86). The infectious risks encountered by organ transplant recipients can be divided into the following three phases: the first month after transplantation, 1 to 6 months after transplantation, and more than 6 months after transplantation. These phases are determined by the degree of immunosuppression expected during these periods (Fig. 1) (7a).

First month after transplantation

During the first month after transplantation, the consequences of immunosuppressive therapy have not yet taken effect, and therefore there is usu-

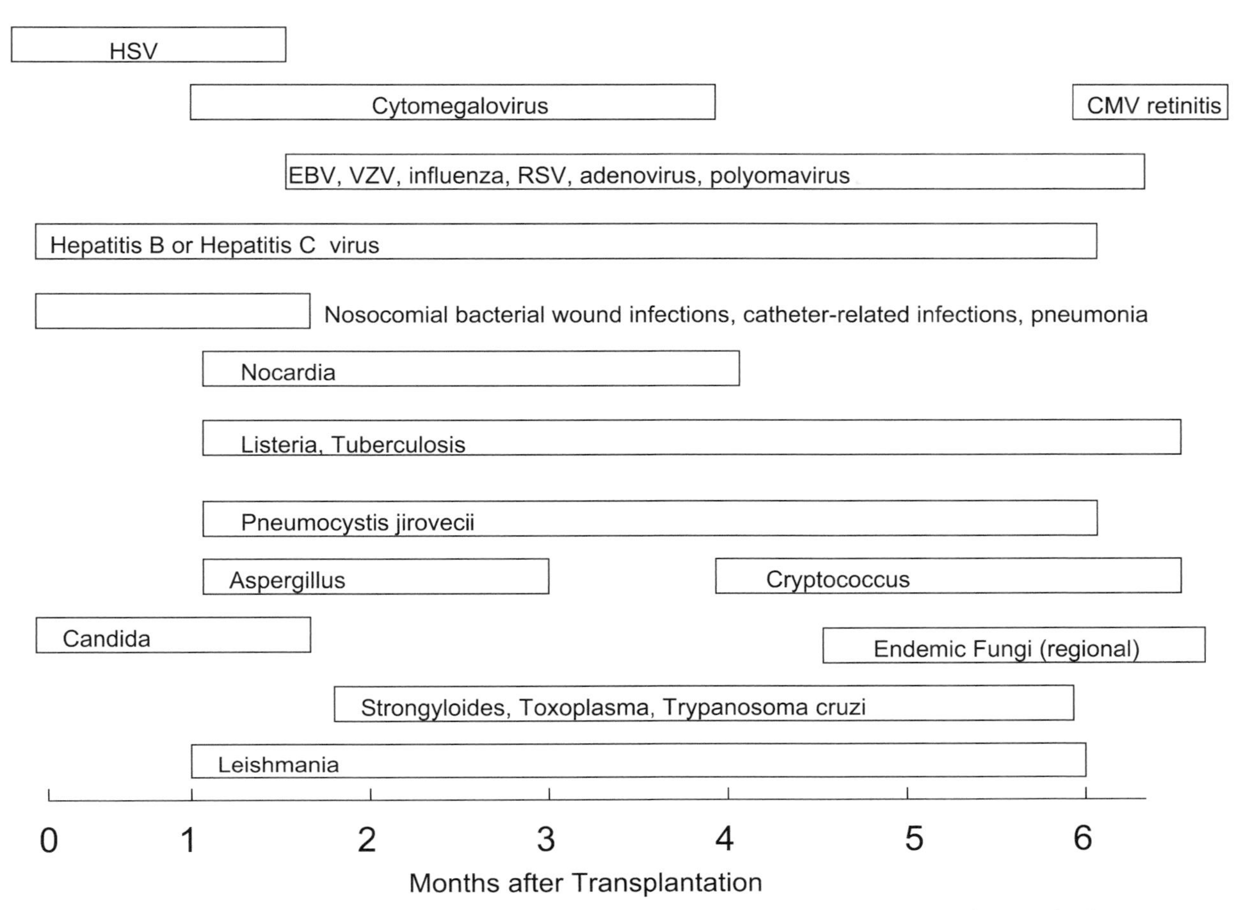

Figure 1. Timeline of infections after solid organ transplantation. Alterations in the timeline occur as a result of antimicrobial prophylaxis or in the presence of excessive immunosuppression or intense immunologic exposure to potential pathogens.

ally a notable absence of opportunistic infections during this time. Infections present in the allograft recipient prior to transplantation, infections transmitted via a contaminated allograft, and infections encountered in a postoperative setting are encountered. Undetected or incompletely treated infections occurring in the pretransplant setting may manifest or exacerbate in the first month after transplantation. Pneumonia and bloodstream infections related to vascular access devices are common in pretransplant recipients. Infections specific to different transplanted organs may occur, for example, peritonitis in a patient who received a liver transplant for end-stage liver disease. Infections transmitted via a contaminated allograft are rare but do occur and are frequent amongst lung transplant recipients (242). These allograft-transmitted infections are usually caused by a variety of bacteria from a donor who was not known to be bacteremic at the time of organ harvest (261). More unusual viral infections, such as rabies virus, lymphocytic choriomeningitis virus, and West Nile virus infections, have been transmitted via transplanted organs (55b, 85, 154, 271). Transmission of tuberculosis, *Trypanosoma cruzi*, and *Toxoplasma gondii* has been reported, with a higher incidence in areas where these are endemic (15, 18, 55, 200, 235). To prevent transmission of infection via the allograft, extensive screening of donors is performed and includes blood, urine, and sputum cultures; HIV, human T-cell leukemia virus, CMV, hepatitis C virus, hepatitis B virus, EBV, VZV, and toxoplasma (heart) serologies; purified protein derivative skin testing; and *Trypanosoma cruzi* serology for donors from areas where *T. cruzi* is endemic (7a).

The majority (>90%) of infections encountered in the first month after transplant are those related to the surgical procedure and are the same as those that occur in other postoperative patients. These infections result from the breakdown of normal mucocutaneous barriers and the presence of devitalized tissue and fluid collections at the operative site. Bacterial and candidal infections of surgical wounds, pneumonia, urinary tract infections, and infections related to vascular access devices, stents, and drainage catheters are very common. Any type of leak related to the surgical procedure that may form a biloma, urinoma, seroma, lymphocoele, or hematoma can readily become secondarily infected with bacteria or *Candida* in the posttransplant setting (244).

One to 6 months after transplantation

The period from 1 to 6 months after transplantation is the critical time after transplantation during which infections unique to these immunocompromised hosts most often arise (86, 239). During this period, the depth of immunosuppression is at its greatest and may be enhanced further by the presence of infection with immunomodulating viruses. CMV is a key player amongst these viruses and has been demonstrated to have multiple mechanisms by which it can induce further immunosuppression (192). Other herpesviruses, such as EBV, human herpesvirus 6 (HHV-6), and HHV-7, hepatitis B virus, hepatitis C virus, and HIV, if present, also are immunomodulating viruses (33, 103, 128, 207, 263).

The cellular immune system is the primary arm of host defense that is impaired, and therefore these patients acquire opportunistic infections that require an intact cell-mediated immune response to prevent or control the infection. CMV infection is the most important opportunistic infection occurring in the organ transplant population, given its direct and indirect effects on the organ transplant recipient (240). The direct effects include the clinical manifestations of CMV infection, such as viremia, which is often accompanied by the CMV syndrome or end organ disease, such as gastritis, colitis, pneumonitis, or hepatitis (222). The indirect effects of CMV infection include the virus' ability to modulate the immune system, contributing to the net state of immunosuppression, its contribution to oncogenesis, and its role in allograft injury and rejection (240). Other herpesviruses, such as VZV, can reactivate and present as shingles or disseminated infection (191). Parvovirus B19 can be acquired by the usual respiratory route or from the transplanted organ, or the virus can reactivate from a latent state (293). The most common manifestations of parvovirus B19 infection in renal transplant recipients are anemia with reticulocytopenia and pancytopenia. Intracellular bacterial pathogens, such as *Nocardia*, *Listeria*, *Legionella*, and *M. tuberculosis*, can cause infection during this period (200). In addition, patients are susceptible to infection with invasive molds, such as *Aspergillus*, and later in this period, to infection with *Cryptococcus* and with the geographically restricted, endemic mycoses caused by *Histoplasma capsulatum*, *Coccidioides immitis*, and *Blastomyces dermatitidis* (30, 91, 93, 259). Patients are at risk for infection with *P. jirovecii*, but this risk is minimized with antibiotic prophylaxis (87). The occurrence of parasitic infections (strongyloidiasis, toxoplasmosis, leishmaniasis, and trypanosomiasis) depends on prior exposure of the donor or recipient to the pathogens in areas where they are endemic (21).

The polyomaviruses BK virus and JC virus may reactivate during this time, causing nephropathy, particularly in renal transplant recipients, or PML in any organ transplant recipients (73, 257). Respiratory viral infections, such as those with influenza virus, RSV, parainfluenza virus, and adenovirus, can cause serious infection in an organ transplant recipient (126, 127, 162). Infections, often at this time caused by antibiotic-resistant bacteria (vancomycin-resistant enterococci [VRE], methicillin-resistant *S. aureus*, and gram-negative rods) and azole-resistant *Candida* species, linger on from the early postoperative period in the setting of persistent drains and catheters (160, 185, 252).

More than 6 months after transplantation

Susceptibility to infections in the period of more than 6 months after transplantation depends on the presence or absence of a well-functioning allograft and/or of chronic or progressive infection with CMV, EBV, or hepatitis B or C virus (86, 239). Patients with a well-functioning allograft on minimal maintenance immunosuppressive therapy are at risk for the usual community-acquired infections, primarily respiratory infections. Patients maintained on higher doses of immunosuppressive agents because of recurrent or chronic rejection remain at risk for opportunistic infections with the pathogens described above. Those patients chronically infected with the immunomodulating viruses remain at significant risk for secondary infections as well as for the virus-associated malignancies (263). Those infected with hepatitis C or B virus are at risk for hepatocellular carcinoma. Chronic EBV infection can result in PTLD, with its protean indolent and fulminant manifestations (159). Papillomavirus infection can result in squamous cell carcinoma of the skin that may be difficult to control and treat, resulting in metastases and significant morbidity and mortality (311).

Hematopoietic Stem Cell Transplantation

The implementation of bone marrow transplantation in the early 1970s revolutionized the treatment of hematologic malignancies (246). The successes of this treatment, however, have been counterbalanced by the significant morbidity and mortality associated with the transplantation of allogeneic or autologous bone marrow, peripheral blood stem cells, or umbilical cord blood cells. These negative consequences are the result of the toxicities associated with the preparative regimen given prior to infusion of the hematopoietic cells and the result of GVHD, the major risk of blood and marrow transplantation (39). The goal of the preparative regimen is twofold, i.e., to eradicate the disease for which the transplant is being performed and to prevent rejection of the graft. The primary toxicities associated with the preparative regimen, which usually consists of total body irradiation plus a chemotherapeutic agent, such as cyclophosphamide, are myelotoxicity and mucositis. Given the variable time frames during which these toxicities are manifested, the type of infection for which a particular patient is at risk is determined by the duration elapsed since transplant, using the following periods: the preengraftment period (<3 weeks since hematopoietic cell infusion), the immediate postengraftment period (3 weeks to 3 months), and the late postengraftment period (>3 months) (Fig. 2 and 3) (303).

Pre-engraftment period

The major risk factors for infection during the preengraftment period include mucocutaneous damage, neutropenia with a loss of neutrophil phagocytic ability, and organ dysfunction related to the preparative regimen. These defects predispose the patient primarily to bacterial and candidal infections. In the earlier era of bone marrow transplantation, bacterial infections caused by gram-negative rods, such as *Pseudomonas* sp. and *Enterobacteriaceae*, were most prominent, with bacteremia and pneumonia being the most common manifestations of infection (32). However, with the use of antibiotic prophylaxis and the employment of broad-spectrum antibiotics with gram-negative rod activity for neutropenic fever, the spectrum of bacterial infections has switched from gram-negative rods to gram-positive organisms, including coagulase-negative staphylococci, *S. aureus*, viridans group streptococci, *S. pneumoniae*, enterococci,

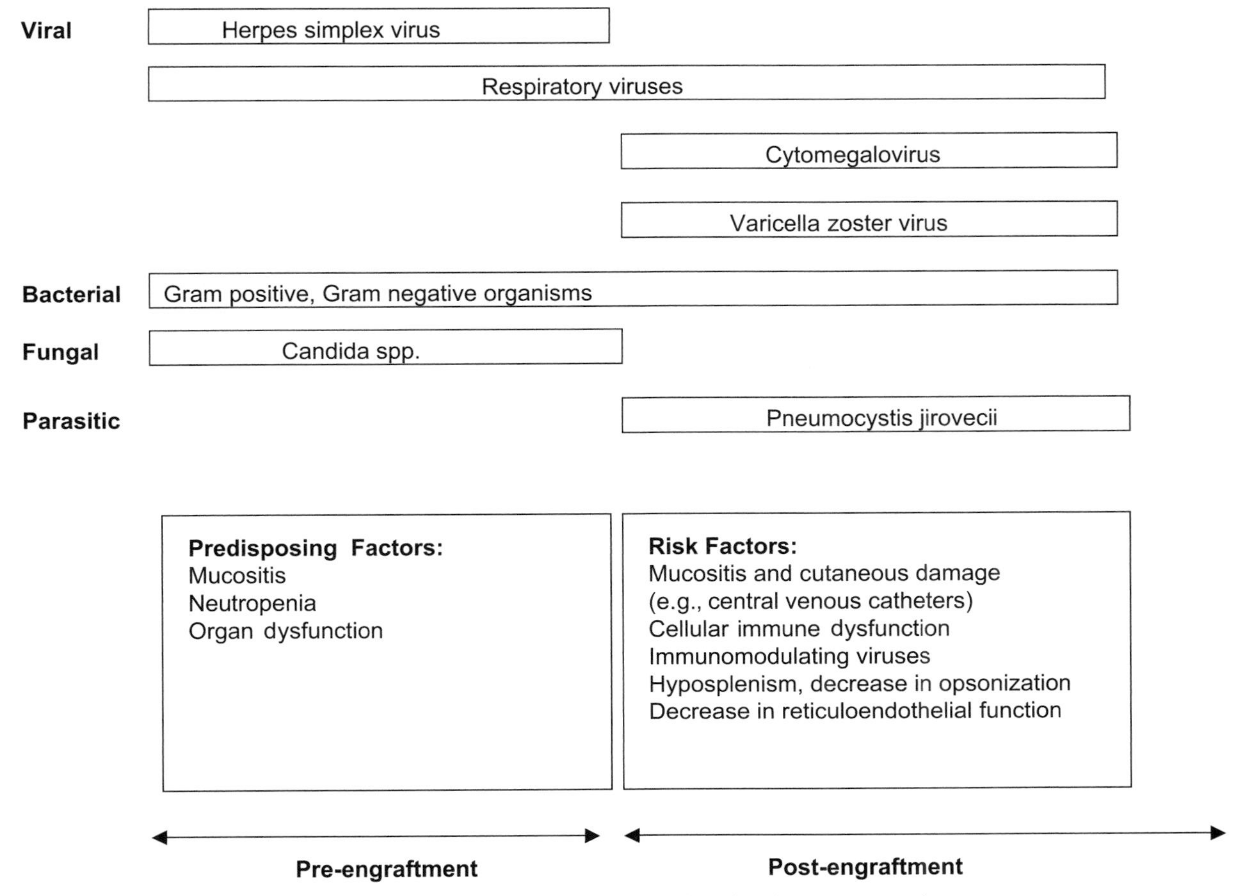

Figure 2. Timeline of infections after autologous blood and marrow transplantation.

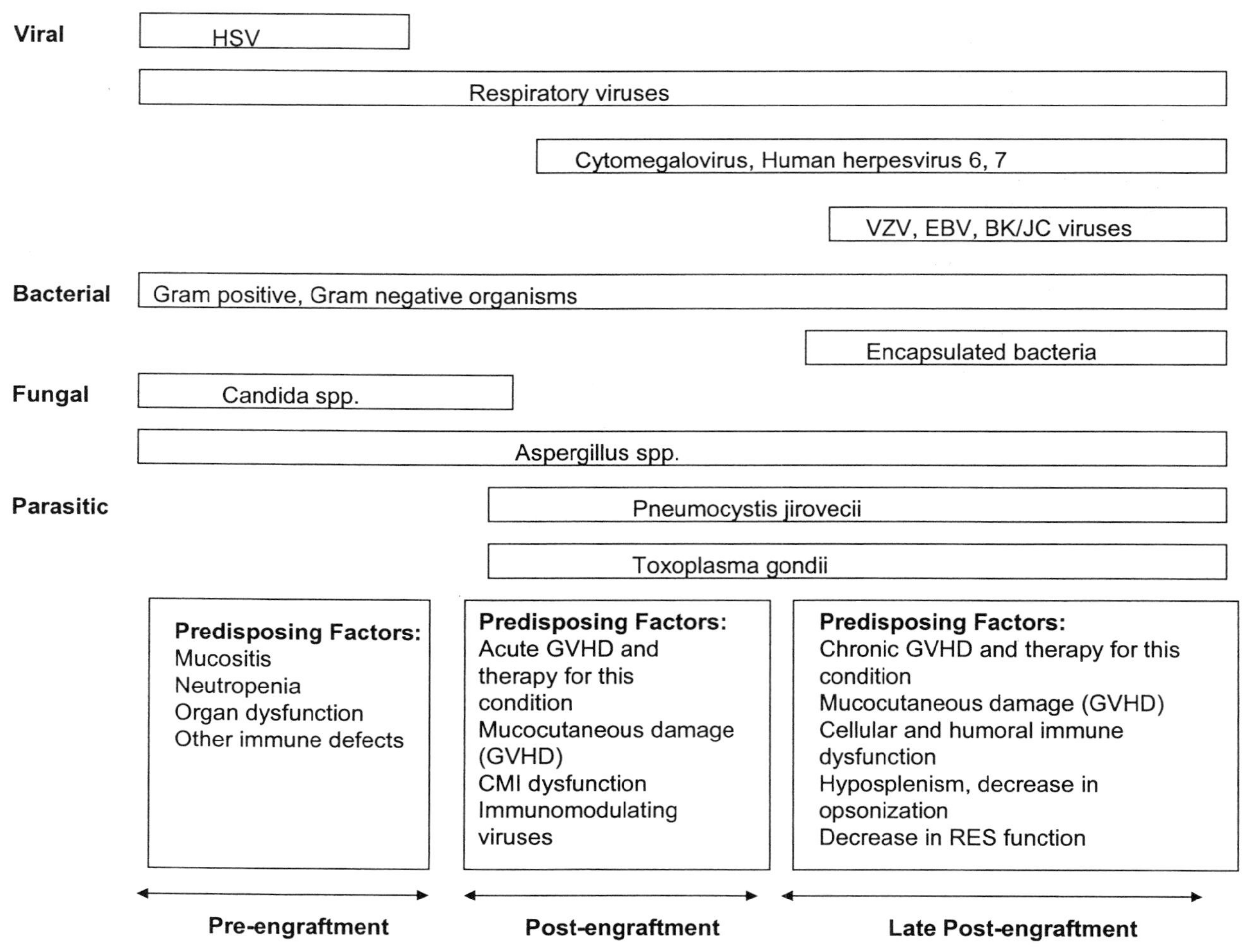

Figure 3. Timeline of infections after allogeneic blood and marrow transplantation. CMI, cell-mediated immunity; RES, reticuloendothelial system.

including VRE strains, and *Corynebacterium* species (49, 149, 307).

The defects predisposing patients to bacterial infection also predispose patients in the preengraftment period to infections with *Candida* species. In addition to neutropenia and mucocutaneous damage, broad-spectrum antibiotic use, organ dysfunction, and high-density yeast colonization are risk factors for invasive candidiasis (218). As with bacterial infections, in some centers there has been a shift in the spectrum of *Candida* species that cause invasive candidiasis due to the use of fluconazole prophylaxis (176, 178). The spectrum has switched from fluconazole-sensitive strains, such as *C. albicans*, *Candida parapsilosis*, and *Candida tropicalis*, to strains with partial or full resistance to fluconazole, such as *Candida glabrata* and *Candida krusei*.

Mold infections caused by *Aspergillus*, zygomycetes, or agents of hyalohyphomycosis (*Fusarium* and *Scedosporium*) or phaeohyphomycosis (*Curvularia*, *Alternaria*, and *Bipolaris*) are not as common during the preengraftment phase (177). However, these infections can occur in patients who have delayed engraftment, and hence prolonged neutropenia, in those whose disease process or its treatment has resulted in prolonged neutropenia, or in those who had a mold infection at some point prior to transplantation. In addition to delayed engraftment, allogeneic transplantation and positive pretransplant CMV serology have been identified as risk factors (19, 177). Mold infections manifest primarily in the lungs, sinuses, and skin.

HSV is the most common virus causing infection in the preengraftment period in the absence of antiviral prophylaxis. Virus reactivation occurs in >70% of seropositive patients, with comparable rates in allogeneic and autologous transplant recipients (302). The most common manifestation of herpes reactivation is severe mucositis, but end organ disease, such as esophagitis, tracheobronchitis, and pneumonitis, can occur. Prophylaxis with acyclovir has significantly reduced the incidence of HSV infections in transplant recipients (78).

Respiratory virus infections can play a prominent role in the preengraftment phase, and their incidence varies with the pattern of infection in the community and with community outbreaks. The most common respiratory viruses include RSV, the parainfluenza viruses, influenza A and B viruses, and rhinoviruses (127, 161). RSV, parainfluenza virus, and influenza A virus infections have been reported to cause outbreaks, severe pneumonitis, and fatal outcomes during epidemics in the community (62, 209, 269). The human metapneumoviruses, another group of pulmonary pathogens, are emerging pathogens in this patient population (77).

Immediate postengraftment period

The spectrum of infections encountered in the immediate postengraftment period (3 weeks to 3 months) is expanded because at this point after transplantation, significant cellular immune dysfunction comes into play as a result of the preparative regimen and myeloablation. The degree of genetic relatedness of the donor inversely correlates with the risk of acute GVHD and, hence, with the risk of infection (39, 303). Patients who receive autologous grafts are at significantly decreased risk of opportunistic infection compared to those who receive an allogeneic source of hematopoietic cells (Fig. 2). For patients who have received an allogeneic transplant, acute GVHD and therapy for this condition and the mucocutaneous damage caused by GVHD enhance the depth of immunosuppression.

Patients with HLA-matched related donors share more minor HLA antigens than do unrelated donors who are matched at all major HLA loci; hence, the risk of GVHD is substantially greater in the latter case. Haploidentical transplants in which the donor and recipient share half of the major HLA loci are at significant risk of serious viral and fungal infections, given the associated risk of GVHD (39). The risk of GVHD with mismatched donors can be reduced by T-cell depletion of the donor hematopoietic stem cell product. Despite a decreased incidence of GVHD, studies have revealed a significantly higher incidence of severe CMV or life-threatening or fatal *Aspergillus* infection in patients who received a T-cell-depleted marrow (288). The source of the hematopoietic cells, i.e., bone marrow versus peripheral blood versus umbilical cord, may influence the risk of infection (39). A recent study revealed that the risk of serious infections in children receiving umbilical cord blood grafts was comparable to that of children receiving unmanipulated marrow and lower than that of recipients of a T-cell-depleted stem cell source (287).

The bacterial and candidal infections encountered in the preengraftment period continue to occur in the postengraftment phase. In addition, opportunistic infections, such as with *Legionella* and *Listeria monocytogenes*, may occur. The incidence of *P. jirovecii* infection is very low ($<1\%$) in the setting of prophylaxis against this infection. Mold infections, especially those caused by *Aspergillus* spp., play a prominent role, and their incidence is associated with acute GVHD and prednisone therapy (doses of ≥ 1 mg/kg per day). Hence, the incidence of mold infections is significantly higher in allogeneic transplant recipients than in autologous recipients (5 to 30% versus 1 to 5%) (19, 286). As noted in the preengraftment section, less common mold infections, such as those caused by *Fusarium* or the *Zygomycetes*, are considered emerging pathogens.

Because of deficient CMV-specific T-cell immunity, CMV infection occurs during this period, with a median onset of 8 weeks after transplantation. The incidence of early CMV infection has decreased as a result of molecular monitoring for this infection and preemptive therapy. Late CMV infections (>3 months after transplantation) continue to occur in those patients with a persistent defect in CMV-specific T-cell immunity due to chronic GVHD and its treatment. In the absence of antiviral prophylaxis, CMV-seropositive allogeneic transplant recipients have a 70 to 80% risk of reactivating CMV (97). One-third of these patients will develop disease, most commonly pneumonitis. In contrast, only 40% of seropositive autologous hematopoietic cell recipients reactivate CMV, and $<5\%$ develop disease (304). Other herpesvirus infections, such as HHV-6 and HHV-7 infections, can manifest during this time. Respiratory virus infections continue to occur. Other less common viral infections, such as adenovirus infection, are encountered during this period. Reactivation of adenovirus occurs in $>80\%$ of autologous and allogeneic transplant recipients (79). Disease, such as pneumonitis, colitis, nephritis, or cystitis, occurs in $<2\%$ of these, with GVHD being a primary risk factor.

With the exception of toxoplasmosis, the occurrence of parasitic infections in this patient population requires specific exposures to particular parasitic pathogens, such as *Strongyloides*, *Leishmania*, *Trypanosoma*, or *Cryptosporidium*. Toxoplasmosis can reactivate in severely immunosuppressed, seropositive, usually allogeneic transplant patients and can cause localized or disseminated disease.

Mycobacterial infections are rare in the hematopoietic cell transplant population in areas where they are not endemic and occur more commonly in allogeneic than in autologous recipients. Infections caused by *M. tuberculosis* or the nontuberculous mycobacteria (NTM) (*Mycobacterium fortuitum*, *Mycobacterium chelonae*, *Mycobacterium abscessus*, and *M. avium* complex) arise due to reactivation or new exposure (61, 72). As expected, with *M. tuberculosis* infection, pneumonia is most common; however, extrapulmonary disease, such as catheter-related bloodstream infections, soft tissue infections, and bone and joint infections, are more common manifestations of NTM infections.

Late postengraftment period

Infections occurring in the last postengraftment period are typically seen in allogeneic transplant recipients who have chronic GVHD. By this time after transplantation, the immune systems of autologous recipients have begun to reconstitute, and most recipients do not experience opportunistic infections. Chronic GVHD and its treatment result in chronic mucocutaneous damage and immunodeficiency that continue to predispose patients with this condition to opportunistic infections in the late postengraftment period. The immune deficits that characterize chronic GVHD include cellular immunodeficiencies, humoral immune dysfunction, hyposplenism, decreased opsonization, and diminished reticuloendothelial cell function (164, 249). Because of this enhanced, prolonged immunodeficiency, many of the infections that occur in the preengraftment and immediate postengraftment periods continue to occur during this time. However, because of the additional defects in humoral immunity associated with chronic GVHD, patients are predisposed during this phase to serious infections caused by encapsulated bacteria, such as *S. pneumoniae*, *H. influenzae*, and *N. meningitidis*. Patients with chronic, severe GVHD have been found to have IgG subclass deficiencies in IgG2 and IgG4 that have been linked to severe pneumococcal disease, such as pneumonia, meningitis, and sepsis (108).

Patients with chronic GVHD may lose their immunity to several viruses that may play a prominent role during this period. VZV can reactivate and cause cutaneous, central nervous system, or disseminated disease. This infection occurs in up to 90% of children during the first year after transplantation (142). In adults, the infection usually occurs 6 to 9 months after transplantation and can present as a cutaneous, central nervous system, or disseminated infection (152). EBV is another virus to which patients with chronic GVHD lose immunity. The median time to onset of infection is 3 to 5 months after transplant. Infection is common, but disease is rare. The disease manifestations include a mononucleosis-like syndrome with fever and neutropenia, aplastic anemia, oral hairy leukoplakia, and PTLD. The patients at highest risk for PTLD include those patients who have received an allogeneic transplant from a matched unrelated, mismatched, or T-cell-depleted donor, have chronic GVHD, and have received antilymphocyte antibody therapy for GVHD prophylaxis (148). Patients may also lose B-cell-specific immunity to measles, mumps, rubella, parvovirus B19, and the polyomaviruses JC virus and BK virus (168). Examples of diseases attributed to infection of hematopoietic cell recipients with these viruses include parvovirus B19-induced severe anemia, late-onset hemorrhagic cystitis caused by BK virus, and PML caused by JC virus. Reactivation of hepatitis B or C virus occurs in up to 70% of patients, but rarely is disease associated with this reactivation (68, 163).

Nonmyeloablative conditioning regimens

In order to offer hematopoietic cell transplantation as a treatment option to patients who are elderly or have significant comorbidities, nonmyeloablative or reduced-intensity conditioning regimens have recently been introduced (39). These regimens are designed to reduce treatment-related toxicities, such as mucositis and neutropenia, but to preserve the graft-versus-leukemia effect. Many regimens include a purine analog, such as fludarabine, and an alkylating agent, such as cyclophosphamide, with or without an anti-T-cell monoclonal antibody, such as alemtuzumab. Despite the "reduced-intensity" regimen, the literature reveals that because of the potent immunosuppressing regimens utilized and because of the occurrence of GVHD after nonmyeloablative bone marrow transplant, the risk for opportunistic infections remains significant (203). Studies reporting the incidence of infectious complications in patients who have received a nonmyeloablative hematopoietic cell transplant are difficult to compare because of the variety of conditioning and GVHD prophylaxis regimens used. In addition, the studies have included patients with a wide variety of underlying illnesses, prior therapies, and antiprophylaxis regimens, all of which are determinants of the frequency and types of infections encountered (203). Studies of immune reconstitution after nonmyeloablative conditioning regimens revealed that peripheral T- and B-cell subsets were similar during the 12 months after transplantation or slightly depressed late after transplantation (>90 days) compared to those with myeloablative regimens (41, 175). Hence, these studies showed a similar rate of bacterial and CMV infections (41) or an increased rate of bacterial and fungal infections late posttransplant (>90 days) (175). Another study, in which fludarabine and cyclophosphamide composed the nonmyeloablative conditioning regimen, found similar frequencies of CMV and invasive fungal infections to infection frequencies in patients who received a myeloablative regimen (203). Overall, the data reveal that opportunistic infections, especially those caused by CMV and *Aspergillus* species, occur and are common after nonmyeloablative hematopoietic cell transplantation (39, 107, 135, 202). The frequency and type of infections are variable and are determined by underlying illness, prior therapies, and the conditioning regimen (203).

HIV Infection and AIDS

Infection with HIV type 1 (HIV-1) causes a chronic progressive immunodeficiency. Initially, it was thought that this immunodeficiency arose from the exclusive infection of $CD4^+$ T-cell lymphocytes, with subsequent cell death resulting in significant $CD4^+$ T-cell depletion. However, with time, it was discovered that HIV is capable of infecting a range of cells, including monocytes, follicular dendritic cells, epidermal Langerhans cells (83), alveolar macrophages, and cells within the central nervous system. In the early stages of infection, the high rate of viral replication in $CD4^+$ T cells results in a major increase in the daily production of $CD4^+$ cells that is balanced by a similar rate of $CD4^+$ cell destruction (117). Over time, a gradual decrease in cellular and humoral immune functions occurs. Accompanying this decline in $CD4^+$ T-cell counts, a wide array of functional abnormalities of the immune system have been characterized. These abnormalities include a decrease in responsiveness to mitogens and antigens, decreased proliferative responses to certain pathogens, decreased cytotoxic and NK cell activity, and decreased humoral immune responses. In concert, these abnormalities produce a state of general activation and disordered regulation of the cellular and humoral immune responses (198).

A unique example of dysregulation of the immune response in HIV-infected patients is the immune reconstitution inflammatory syndrome (IRIS) (198). This syndrome occurs in patients with AIDS (CD4 counts, often <100 cells/mm^3) who have a preexisting infectious process that worsens paradoxically after initiation of highly active antiretroviral therapy. The IRIS has been reported most frequently with VZV, *M. tuberculosis*, *M. avium* complex, CMV, and *Cryptococcus* infections (201, 267). The IRIS likely arises from an interaction of an exuberant, recovering immune response, following highly active antiretroviral therapy, with the residual antigenic burden and from host genetic susceptibility to such a process.

HIV-1 infection is divided into the following stages: primary infection with seroconversion, clinical latency, early symptomatic disease, and AIDS. The infectious complications that occur in patients infected with HIV-1 correlate with $CD4^+$ cell counts. After the resolution of primary infection, patients enter the phase of clinical latency, where they are primarily asymptomatic, if the peripheral $CD4^+$ T-cell count remains above 500/mm^3. As the CD4 count declines below 500/mm^3, patients begin to develop constitutional symptoms (109, 195). These symptoms are often accompanied by infections caused by reactivating herpesviruses, such as VZV, HSV, or EBV (oral hairy leukoplakia), or mucosal infection caused by *Candida* species (thrush or vaginitis). Additional infectious complications that occur when the CD4 count ranges from 200 to 500/mm^3 include pneumonia caused by pneumococci and other bacteria, reactivation of pulmonary tuberculosis, and cervical and anal dysplasia or cancer attributed to human papillomavirus infection. Other, less common syndromes that occur during this time and are associated with viral reactivation include EBV-associated B-cell lymphoma and HHV-8-associated Kaposi's sarcoma.

Late-stage disease and an AIDS-defining state occur when the CD4 count falls to <200/mm^3. The median time from the onset of this severe immunosuppression to an AIDS-defining diagnosis is 12 to 18 months for persons not receiving antiretroviral therapy. This phase is characterized by the development of opportunistic infections, tumors, wasting, and neurologic complications. The most common opportunistic infection encountered during this time, in the absence of antibiotic prophylaxis, is pneumonia caused by *P. jirovecii*. Disseminated endemic mycoses, such as histoplasmosis and coccidioidomycosis, and extrapulmonary tuberculosis are encountered. PML, caused by reactivation of the polyomavirus JC virus, is a less common diagnosis. With a further decline in the CD4 count, a broader array of opportunistic infections occurs. CD4 counts of below 100/mm^3 are associated with toxoplasmosis encephalitis, cryptococcal meningitis, candidal esophagitis, chronic diarrheal illnesses caused by *Cryptosporidium* or *Microsporidium*, and disseminated HSV infection. Disseminated *M. avium* complex and CMV infections are two of the main opportunistic infections usually encountered with severe immunosuppression, when the CD4 count falls below 50/mm^3.

Other Immunomodulating Viruses

EBV

EBV can cause a variety of immunologic perturbations. This virus infects human B lymphocytes and drives their proliferation until it is controlled by

T lymphocytes. (The activated T lymphocytes are the "atypical lymphocytes" seen in the peripheral blood of patients with acute EBV infection.) Despite the fact that numbers of peripheral blood lymphocytes increase, it is well known that patients with acute infectious mononucleosis have depressed cell-mediated immunity and can become anergic. The mechanism for this became apparent when it was discovered that the EBV gene BCRF1 shares extensive sequence homology with the human IL-10 gene and that EBV-infected B lymphocytes produce IL-10 (194). This cytokine has two roles in immunoregulation—it is a growth and differentiation factor for B lymphocytes, and it is a negative regulator of TH1 T-lymphocyte activity. These are the cells responsible for secreting cytokines such as IL-2 and IFN-γ that are needed for delayed hypersensitivity reactions and cell-mediated immune responses. Therefore, during and sometimes for months following an acute EBV infection, individuals may be anergic and have difficulty responding to other viral and intracellular bacterial (e.g., mycobacteria) pathogens (172). EBV can cause hypogammaglobulinemia and pancytopenia. These effects are usually, but not always, transient. Rarely among otherwise normal individuals but frequently among people with a rare immunodeficiency, the X-linked lymphoproliferative syndrome, EBV infection can cause chronic infection, pancytopenia, severe hypogammaglobulinemia, hemophagocytic syndrome, a lymphoproliferative syndrome, or B-cell lymphoma (92, 255). EBV infection is invariably fatal in people with X-linked lymphoproliferative syndrome.

Measles virus

Much of the morbidity and mortality from measles is due to secondary infections, particularly diarrhea, pneumonia, and reactivation of tuberculosis (190). Some of the predisposition to secondary infections is due to damage of mucosal barriers, but measles virus suppresses cell-mediated immune function for weeks following infection. During that period, people have decreased DTH skin test responses (278), in vitro NK cell function, and in vitro T-lymphoproliferative responses to mitogens. Similar changes, but of smaller magnitude, follow immunization with the measles vaccine. Measles virus appears to suppress the production of IL-12 by macrophages and dendritic cells, leading to a skewed activation of TH2 cells (310). As a result, infected individuals produce an excess of IL-4 and IL-10 but have decreased IFN-γ production for weeks.

Immunodeficiency Associated with Hematologic Malignancies

In addition to the effects of chemotherapy, cancers can predispose the host to develop infections if the cancer spreads to the bone marrow and/or lymph nodes and thereby reduces the number of normal hematopoietic cells. This is particularly true for lymphoid malignancies, such as leukemia and lymphoma.

CLL

CLL patients are at increased risk for infection because the leukemia cells can replace normal lymphocytes in the bone marrow and lymph nodes and because of adverse effects of the drugs used to treat the disease. Defects of humoral immunity are common, and the problem appears to be exacerbated by the use of rituximab (285). Sometimes humoral immune deficiency is accompanied by hypogammaglobulinemia, but some CLL patients have oligoclonal or monoclonal gammopathies that lead to the unusual combination of normal or elevated immunoglobulin levels with deficiency of antibody responses. Evaluation of humoral immune function in these patients requires immunofixation electrophoresis to test for gammopathy and measurement of antibody responses to T-cell-dependent and T-cell-independent vaccine antigens. Affected individuals are at increased risk for upper and lower respiratory tract infections caused by encapsulated bacteria (292). Prophylactic infusions of pooled human gamma globulin have been shown to be cost effective, though they may not alter the long-term prognosis (59a). Neutropenia and deficiencies of T-cell number and function can occur in CLL, though most likely as a consequence of chemotherapy. The types of infections can be predicted by the specific chemotherapeutic drugs that are being used in an individual patient.

Multiple myeloma

Multiple myeloma is a B-cell malignancy that causes effects similar to those seen with CLL (225). The malignant B-cell clone can displace other B cells, thus leading to humoral immune deficiency, even in the face of an elevated gamma globulin fraction. Myeloma cells secrete transforming growth factor beta, which suppresses the inflammatory responses of monocytes and macrophages and decreases the production of IgG and IgM antibodies by nonmalignant B cells. Myeloma cells can also secrete vascular endothelial growth factor, which has deleterious effects on dendritic cell differentiation and function, thus decreasing antigen presentation for adaptive immunity. Responses to pneumococcal vaccine are suboptimal. Patients with multiple myeloma have an increased risk for infection caused by pneumococci and other encapsulated bacteria. More widespread immunologic dysfunction is likely to be the result of chemotherapy, not a direct effect of the myeloma.

Metabolic Diseases

Diabetes mellitus

Both type 1 and type 2 diabetes mellitus can increase an individual's risk for infections because of reduced blood supply and denervation of peripheral tissues. It also appears that poor glycemic control is associated with impaired neutrophil function. Neutrophils from diabetics have decreased expression of adhesion molecules as well as impaired in vitro adhesion and chemotaxis (67). Phagocytosis appears to be normal, but bactericidal function is impaired. Adaptive immune function is normal. Finally, diabetics have increased nasal colonization with *S. aureus* and may have increased binding between *Candida* and epithelial cells of the oral mucosa and vagina. These defects lead to predictable patterns of infections. Diabetics are at increased risk for developing infections of the lower respiratory tract, urinary tract, skin, and mucous membranes (45, 134, 196). They are also at increased risk for recurrences of these infections. Infections caused by *S. aureus*, gram-negative rods, and *M. tuberculosis* occur at increased frequency, and infections caused by *S. pneumoniae* and influenza virus cause increased morbidity and mortality (155). Diabetics are at increased risk for developing cellulitis and necrotizing fasciitis, most often due to polymicrobial infections caused by combinations of gram-negative rods and anaerobes, but also due to *S. aureus* or group A streptococci (47). There are also several specific types of infections and specific pathogens that occur much more often in diabetics than in other people. These include rhinocerebral mucormycosis, necrotizing otitis externa associated with *Pseudomonas aeruginosa*, emphysematous cystitis, and emphysematous cholecystitis. The latter may be a secondary effect of diabetic gastrointestinal dysmotility.

Protein-losing enteropathy

Loss of protein across the gastrointestinal mucosa can occur in association with chronic inflammatory diseases (e.g., Crohn's colitis, celiac disease, and systemic lupus erythematosus). Though IgG levels may fall well below the normal range, affected individuals usually continue to produce normal antibody responses and do not have trouble with infections. There are two exceptions to this generalization. People who are treated with chronic immunosuppression may have reduced production of antibody in association with increased loss of immunoglobulin through the stool. This is rarely severe enough to cause a predisposition to develop infections. (The effects of immunosuppression on T-cell function are similar to those encountered by organ transplant recipients taking the same drugs at the same doses.) Protein-losing enteropathy can also be associated with lymphatic obstruction or intestinal lymphangiectasia, in which case there is a concomitant reduction in T-lymphocyte numbers (166). There can be a sufficient loss of T lymphocytes to cause anergy, thus compromising the diagnostic utility of DTH skin tests. In very rare cases, the loss of T cells can be so severe that it causes SCID.

Nephrotic syndrome

It would be expected that the consequences of protein loss through the urinary tract would be similar to those seen with protein loss through the gastrointestinal tract, that is, that most patients would have reduced levels of serum immunoglobulins with normal or near normal antibody production and no significant predisposition to infection. However, there is one glaring exception to this generalization, namely, the propensity of children and, less often, adults with nephrotic syndrome to develop primary or spontaneous pneumococcal peritonitis (99, 112). The explanation for this association is not known.

Asplenia

Individuals can have congenital asplenia as an isolated finding or in association with dextroposition of the heart, lungs, and abdominal viscera (Ivemark syndrome). The spleen can be removed surgically after trauma. There can be functional asplenia from splenic infarcts, most often caused by sickle cell disease but sometimes caused by portal hypertension or infiltration by malignant cells. Asplenia is associated with a reduction in serum IgM, impaired IgM antibody responses to polysaccharide vaccines (193), and a reduction in the capacity for phagocytosis of blood-borne microorganisms. The overall incidence of sepsis among asplenic individuals is 4.25% and the mortality rate is 2.52%, which represent approximately 200-fold increases above the risks among the general population (260). The risk of infection is greater among children less than 5 years old and in the first few years after surgical splenectomy. Infections are most often caused by encapsulated bacteria. The most frequent pathogen is *S. pneumoniae*, but *N. meningitidis*, *E. coli*, *H. influenzae* type b, and *S. aureus* are important pathogens, as are *Babesia* sp. and the malaria parasite.

Effects of Age

Infants and young children

The neonate and very young infant have increased susceptibility to infection for a variety of reasons (214). First, although the repertoire of the adaptive immune system is generated during fetal life, there is little clonal selection and terminal differentiation to generate effector T lymphocytes and antibody-secreting B lymphocytes until the child has been exposed to antigens. That is, adaptive immunity depends in large part upon previous exposure to pathogens via immunization or infection. (The mother provides some compensation for the lack of antibody, as maternal IgG antibodies are transported across the placenta and provide protection to the baby for up to 6 months.) In addition, T cells from the fetus and young infant have decreased cytokine production and moderate reductions in cytotoxic function compared to adult T cells. Neonates and young children are usually anergic to a panel of DTH skin tests, which limits their effectiveness as screening tests for deficiencies of cell-mediated immune function or mycobacterial infections (150, 199). The major defect of humoral immunity is that until children are 18 to 24 months old, they are generally unable to produce antibody in response to T-cell-independent antigens (including clinically important bacterial capsular polysaccharides, such as those produced by pathogenic pneumococci and *H. influenzae*) (20, 90). Before the development of polysaccharide-protein conjugate vaccines that activate B cells in a T-cell-dependent manner, children under 2 years of age were at high risk for developing invasive infections (sepsis and meningitis) caused by these organisms. It was most remarkable that even after recovering from sepsis and/or meningitis, such infants did not produce protective levels of IgG antibody to the infecting pathogen. Young infants and children are still at increased risk for developing infections with pneumococcal serotypes not contained in the polysaccharide-protein conjugate vaccine. The NK cells of neonates have reduced cytotoxic activity compared to that of adult cells (94). Multiple functions of phagocytic cells, including adhesion and chemotaxis, are reduced, but bacterial killing appears to be at or very near adult normal levels. The levels of complement components are generally reduced approximately 50% in comparison to those of normal adult cells (66). The effector functions of both phagocytes and the classical pathway of complement are further reduced by the absence of IgG antibodies to relevant pathogens. This is compounded in premature babies, who do not receive the full complement of maternal IgG antibodies, most of which are transferred during the last trimester of pregnancy.

Newborn babies are exposed to vaginal flora during delivery and to the same flora prior to delivery if there has been premature rupture of the amniotic sac. If the newborn is colonized by this flora, the combination of defects in the newborn makes it more susceptible to developing blood-borne infections that can be disseminated widely. In this host, infections by bacteria, including group B streptococci, *E. coli* and other *Enterobacteriaceae*, and *Listeria* (28), as well as by HSV (147) and enteroviruses (1) are particularly problematic.

Aging

It is difficult to clearly define the effects of aging on susceptibility to infection because of confounding comorbidities. Nevertheless, it is apparent that there are age-related deficiencies of adaptive immunity (51, 52, 53). Thymic involution leads to the decreased production of naïve T cells by age 40 years, and this is coupled with an age-related decline in the replicative potential of memory T cells. There is also a shift toward a TH2 T-cell response and a relative decrease in the CD4-to-CD8 ratio. Serum immunoglobulin levels do not decline, but both primary and secondary antibody responses are impaired relative to those of younger individuals. For example, it has been estimated that the influenza vaccine can prevent infection in 70 to 90% of people who are less than 65 years old (197), but in only 30 to 40% of people who are older (219). Similarly, the immunogenicity of pneumococcal polysaccharide vaccines is reduced in individuals over the age of 65 years compared to that in populations of younger subjects (236, 241). Innate immune function appears to be stable with aging. Based upon the diminished T- and B-cell function of the elderly, it should not be surprising that they have an increased incidence and severity of pneumococcal, RSV (81), and influenza virus pneumonia, group B streptococcal bacteremia and sepsis, tuberculosis, and herpes zoster.

Surgery or Trauma

Major trauma triggers a massive inflammatory response due to widespread activation of monocytes and macrophages by necrotic tissue (173). A consequence of this inflammatory state is depression of T- and B-cell function. Macrophages produce prostaglandins, such as prostaglandin E_2 (PGE_2), a powerful immunosuppressant. PGE_2 inhibits T-cell division, IL-2 production, and IL-2 receptor expression and affects the quality of antibody synthesis by B lymphocytes. In addition, PGE_2 induces a TH2 lymphocyte response that is immunosuppressive, in that immunosuppressive cytokines such as IL-4 and IL-10 are produced. This suppression of the adaptive immune response predisposes trauma patients to serious infections.

In patients who have experienced severe burns, the immune system undergoes a similar evolution in response to injury. Severe burns induce activation of an inflammatory cascade that contributes to the development of subsequent immunosuppression and increased susceptibility to infection and multiple organ system failure. The mechanisms by which burns cause immunosuppression have not been elucidated completely. However, recent studies reveal that macrophages likely play a major role in postburn immunosuppression, as their productive capacity for inflammatory mediators (e.g., prostaglandins, nitric oxide, TNF-α, IL-6, and other cytokines) is significantly increased at this time (253). A downstream consequence of this inflammatory state is that patients with extensive burns exhibit decreased T- and B-lymphocyte function and impaired function of circulating leukocytes and complement (27, 29). Despite recent advances in the care of burn patients, overwhelming infection remains the leading cause of death from serious burn injury. Approximately 30% of burn patients who die do so as a result of infection. In addition to immune system dysfunction, the burn destroys the physical barrier to infection, enabling tissue invasion by bacterial pathogens and possible systemic dissemination. High-density (up to 10^5 colonies of bacteria per gram of tissue) bacterial colonization of burn eschar occurs. Colonization with *P. aeruginosa* is most common, but colonization with other bacteria also occurs (e.g., *E. coli*, *Acinetobacter*, and *S. aureus*) (80). *P. aeruginosa* skin infection in burn victims remains a serious complication associated with a very high mortality despite aggressive antibiotic therapy (283).

Chronic Disease

SLE

Infections are a major cause of morbidity and mortality in patients with SLE. Infections account for the majority of deaths among SLE patients in developing countries and are the first or second most common cause of death in developed countries (5). Primary defects in innate and adaptive immunity that occur in SLE in addition to defects resulting from immunosuppressive therapy account for the high incidence of infections in this patient population. Immune defects in almost every component of the immune system have been reported. This impairment is not universal. The reported immune dysfunctions of the innate immune system include inherited complement deficiencies, decreased levels of complement proteins, reduced numbers of complement receptors, and abnormalities of chemotaxis, phagocytosis, and oxidative metabolism of polymorphonuclear leukocytes, monocytes, and macrophages. These immune abnormalities are more pronounced during periods of increased disease activity.

The adaptive immune system can also be compromised in patients with SLE. This appears to occur in the setting of enhanced disease activity and as a result of immunosuppressive therapy (95). During periods of exacerbation, patients with SLE have decreased levels of T cells, and TH cell responses to viral antigens, toxoids, and allogeneic cells are diminished (24). T-cell and other immune functions, as described above, are further impaired by corticosteroid therapy. Patients who are treated with cyclophosphamide and/or plasmapheresis because of serious disease manifestations unresponsive to corticosteroid therapy are at significant risk of developing fatal opportunistic infections.

Common bacteria, both gram positive and gram negative, are responsible for most infections in patients who have SLE (5). Some infections can have more severe manifestations if inherited deficiencies of complement or splenic dysfunction is present. Infections with encapsulated bacteria (*S. pneumoniae* and *N. meningitidis*) can cause meningitis and sepsis. *Salmonella* infections, such as bacteremia, occur in patients who have similar immune system deficiencies. Infection with *Listeria monocytogenes*, causing sepsis and meningitis, has been reported for patients receiving high doses of corticosteroids. Other opportunistic infections occur in patients who are treated with high-dose corticosteroids and/or additional immunosuppressive therapy, such as cyclophosphamide. Opportunistic infections with *Nocardia*, *M. tuberculosis*, atypical mycobacteria, VZV, CMV, *P. jirovecii*, *C. albicans*, *Cryptococcus neoformans*, *Aspergillus*, *Strongyloides stercoralis*, and *Toxoplasma gondii* have been reported for lupus patients.

Chronic active hepatitis and cirrhosis

Infections are a major complication and a major cause of death in advanced liver disease (133). The most common infections are bacterial and include spontaneous bacterial peritonitis, pneumonia, bacteremia, urinary tract infections, and endocarditis. Many immunologic abnormalities have been detected in patients with cirrhosis. Many of these studies have been performed on patients with alcohol-induced liver disease. Low levels of complement and decompensated alcoholic cirrhosis have been associated with an increased risk of infections and mortality (119). The liver is the primary site of C3 synthesis; therefore, in the presence of severe liver failure, opsonization of bacteria may be impaired. The reticuloendothelial system is an important filtering system for blood-borne pathogens. However, because of impaired macrophage activation and mobilization in the presence of cirrhosis, this filtering may be impaired (233). In addition, portal-systemic shunting occurs in cirrhosis, allowing portal blood to reach the systemic circulation without passing through the reticuloendothelial system. This is a suspected etiology of some bacteremias in patients with cirrhosis.

Patients with cirrhosis frequently demonstrate anergy and fail to respond to vaccination, suggesting delayed hypersensitivity and other impaired T-cell-dependent functions (251). Recent studies of vaccination with influenza vaccine demonstrated that patients with more advanced liver disease had significantly lower postimmunization levels of IFN-γ due to decreased lymphocyte responsiveness to specific antigen in advanced liver disease (56). This decreased lymphocyte responsiveness may be due to the effects of hepatitis C virus on dendritic cells, antigen-presenting cells that are essential for the development of an effective immune response (215). Hepatitis C virus binds to dendritic cells, replicates to a low level in these cells, and impairs their maturation. The result is an impaired ability of dendritic cells to stimulate alloreactive T cells. Similar findings have been reported for patients with chronic hepatitis B virus infection (13).

ESRD

Bacterial infections are the second most common cause of death in the end-stage renal disease (ESRD) population (248). Death rates from sepsis are 100- to 300-fold higher for ESRD patients than for the general population. The following immunologic abnormalities have been reported for patients with ESRD: decreased granulocyte and macrophage phagocytic function, reduced killing capacity of neutrophils, lower antibody titers, inability to maintain adequate antibody titers over time postvaccination, and impaired T-cell-mediated immunity (8, 71, 96, 181, 309). A recent study of patients with ESRD maintained on hemodialysis found a reduction of naïve and central memory T cells that may in part contribute to the increased predisposition to infection and the diminished response to vaccination in the ESRD population (309). This reduction in lymphocytes may result from apoptosis induced by uremia (182).

SCREENING FOR SUSPECTED IMMUNODEFICIENCY

Although immune system dysfunction can be suspected by the clinician after careful review of the history and physical exam, specific diagnoses of primary immunodeficiency are rarely evident without the use of the laboratory (89). Similarly, one may be able to suspect the most likely secondary immunomodulating consequences of a drug, infection, or other illness. However, the specific host defect and its severity may vary widely among different patients. Based upon information provided in this chapter, the types of infections, the drugs being used, and other symptoms should help to focus the laboratory workup on specific parts of the immune system (Table 2). For example, patients with antibody deficiency typically have sinopulmonary infections caused by encapsulated bacteria and viruses as a prominent presenting feature. Deficiency of cell-mediated immunity predisposes individuals to develop infections caused by opportunistic pathogens, such as *P. jirovecii* and other fungi, bacteria of low virulence, and a variety of viruses. Abnormalities of phagocytic function should be suspected when patients have recurrent skin infections or visceral abscesses, whereas patients with complement deficiency most often present with bacterial sepsis or immune complex-mediated diseases. Screening tests should be guided by the clinical features of the patient, with the aim of identifying possible primary and secondary immunodeficiencies and defining the relevant defect(s) in host defense.

Examination of the Peripheral Blood Smear

A complete blood count, together with blood smear examination, is an inexpensive and readily available test that provides important diagnostic information relating to a number of immunodeficiency states (Table 6). Neutropenia most often occurs secondary to immunosuppressive drugs, infection, malnutrition, or autoimmunity but may be a primary problem (congenital or cyclic neutropenia). In contrast, persistent neutrophilia is characteristic of leukocyte adhesion molecule deficiency, and abnor-

Table 6. Laboratory tests to assess immune function

Compartment of disorder	Screening tests	Advanced tests
Antibody	Serum immunoglobulin levels (IgG, IgA, IgM), B-lymphocyte numbers	Antibody responses to immunization (recall and neo-antigens, T-cell-dependent and T-cell-independent antigens)
Cell-mediated immunity	Lymphocyte count, T-lymphocyte subset numbers (CD4, CD8), DTH tests	In vitro tests of lymphocyte proliferation and/or cytokine secretion in response to mitogens and antigens
Complement	Total hemolytic complement (CH_{50}), alternative complement function (AH_{50}), MBL level	Tests of individual complement levels and function
Phagocytes	White blood cell count and differential, nitroblue tetrazolium or dihydrorhodamine dye test	Measurement of chemotaxis, phagocytosis, and bactericidal capacity

mal cytoplasmic granules may be seen in the peripheral blood smears of patients with Chediak-Higaski syndrome.

The blood is predominantly a "T-cell organ," i.e., the majority (50 to 70%) of peripheral blood lymphocytes are T cells, whereas only 5 to 15% are B cells. Therefore, lymphopenia is often a presenting feature of T-cell or combined immunodeficiency disorders, such as SCID or DiGeorge syndrome (151). Pediatricians and specialists in infectious diseases are often so intent on looking for abnormal neutrophil counts and the presence of elevated numbers of young neutrophils (e.g., bands) that they fail to note this important abnormality when it is present.

Thrombocytopenia may occur as a secondary manifestation of immunodeficiency but is often a presenting manifestation of Wiskott-Aldrich syndrome (210). A unique finding in the latter group of patients is an abnormally small platelet volume, a measurement that is easily made by automated blood counters.

Examination of red blood cell morphology can yield clues about splenic function. Howell-Jolly bodies may be visible in peripheral blood in cases of splenic dysfunction or asplenia. However, the converse is not always true, and the absence of Howell-Jolly bodies does not guarantee that splenic function is normal.

Evaluation of Humoral Immunity

Measurement of serum immunoglobulin levels is an important screening test to detect immunodeficiency for the following three reasons: (i) more than 80% of patients with primary disorders of immunity will have abnormalities of serum immunoglobulins; (ii) immunoglobulin measurements yield indirect information about several disparate aspects of the immune system because immunoglobulin synthesis requires the coordinated function of B lymphocytes, T lymphocytes, and monocytes; and (iii) the measurement of serum immunoglobulin levels is readily available, highly reliable, and relatively inexpensive. The initial screening test for humoral immune function is the quantitative measurement of serum immunoglobulins (89). Neither serum protein electrophoresis nor immunoelectrophoresis is sufficiently sensitive or quantitative to be useful for this purpose. Instead, quantitative measurements of serum IgG, IgA, and IgM should be used to identify patients with panhypogammaglobulinemia as well as those with deficiencies of an individual class of immunoglobulins, such as selective IgA deficiency. Interpretation of results must be made in view of the marked variations in normal immunoglobulin levels with age. Therefore, age-related normal values must always be used for comparison.

A clue to immunodeficiency may be a low normal IgG level in an individual with recurrent infections. In such cases, it is critical to assess antibody function in addition to immunoglobulin levels. Antibody levels should be measured in response to T-cell-dependent (e.g., tetanus toxoid or influenza virus) and T-cell-independent (e.g., pneumococcal polysaccharide) vaccines. Two caveats for the latter group of vaccines are that pneumococcal polysaccharide-protein conjugate vaccines are not useful for this purpose because they are T cell dependent and that children under the age of 18 to 24 months generally cannot respond to T-cell-independent antigens, whether presented by immunization or natural infection. As an alternative, T-cell-independent antibody responses can be assessed by quantitating isoagglutinin titers, as the ABO blood group antigens are polysaccharides. However, the value of measuring anti-isoagglutinin antibodies is also limited for children of less than 2 years of age. It is also important to assess responses to new antigens as well as recall antigens, as some acquired (e.g., CVID) and secondary (rituximab therapy) immunodeficiency diseases may not affect previously generated memory cells. Meningococcal and hepatitis A vaccines are useful for this purpose, as many people have not had previous immunization or exposure to these antigens. Live vaccines (varicella-zoster, measles, mumps, rubella, or BCG) should never be used for diagnostic testing, as immunodeficient patients can sometimes develop serious and potentially life-threatening infections from vaccine strain microorganisms.

The role for IgG subclass measurements is controversial. There are four subclasses of IgG, and selective deficiencies of each of these have been described. However, the significance of an IgG subclass deficiency in the presence of normal antibody responses to protein and polysaccharide antigens is not known. Therefore, it is probably best to rely upon antibody measurements, not IgG subclass levels.

Evaluation of Cell-Mediated Immunity

Testing for defects of cell-mediated immunity is relatively difficult because of the lack of good screening tests. Lymphopenia is suggestive of T-lymphocyte deficiency because T lymphocytes comprise the majority (50 to 70%) of peripheral blood mononuclear cells. However, lymphopenia is not always present in patients with T-lymphocyte functional defects. Similarly, the lack of a thymus silhouette on chest X-ray is rarely helpful in the evaluation of T-lymphocyte disorders because the thymus may involute following stress and give the appearance of thymic hypoplasia.

Indirect information about T-cell function may be obtained by flow cytometric enumeration of peripheral blood T lymphocytes with appropriate monoclonal antibodies (anti-CD3 for total T cells, anti-CD4 for TH cells, and anti-CD8 for T_C cells). Patients with SCID and DiGeorge syndrome generally have decreased numbers of both $CD4^+$ and $CD8^+$ T lymphocytes. In contrast, patients infected with HIV have decreased T-lymphocyte numbers because there are decreased numbers of $CD4^+$ lymphocytes.

DTH skin testing with a panel of antigens has historically been used to screen for defects of cell-mediated immunity in older children and adults (54a), with the presence of one or more positive delayed-type skin tests being considered indicative of intact cell-mediated immunity. However, there are significant limitations to this testing, as follows. (i) Prior exposure to antigen is a prerequisite. (ii) Immunocompetent patients may have transient depression of DTH with acute viral infections, such as infectious mononucleosis, with poor nutrition, or with stress. In fact, the incidence of anergy among hospitalized patients is significantly higher than what one might suppose based upon their admitting diagnoses and the lack of infections with opportunistic pathogens. (iii) A positive skin test to some antigens does not ensure that the patient has normal cell-mediated immunity to all antigens (e.g., patients with chronic mucocutaneous candidiasis have a limited defect in which cell-mediated immunity is generally intact except for the response to *Candida*). (iv) Immunocompetent children under the age of 12 months are frequently unresponsive to all of the antigens in the panel. (v) There are currently only three antigen preparations available in the United States that have been standardized and licensed for DTH testing (*Candida*, mumps, and purified protein derivative preparations) (211), and the last preparation is not useful for anergy testing, since most people should be expected to be negative. (vi) False-negative results can easily result from incorrect application of antigens, which is dependent on characteristics of the patient's skin as well as the skill of the operator. DTH skin tests are therefore generally not helpful for evaluation of suspected T-lymphocyte abnormalities, especially those that present early in life (e.g., SCID or DiGeorge syndrome). Instead, the preferred method is measurement of in vitro T-cell proliferation and/or cytokine production in response to mitogens and soluble antigens (89). When specimens are sent to an off-site reference laboratory for testing, they should always be accompanied by a specimen from a healthy individual to control for the effects of shipment on the results.

Evaluation of the Complement System

Most of the genetically determined deficiencies of complement can be detected with the total serum hemolytic complement (CH_{50}) assay (299). Since this assay depends on the functional integrity of the classical complement pathway (C1 through C9), a severe deficiency of any of these components leads to a marked reduction or absence of total hemolytic complement activity. Alternative pathway deficiencies (e.g., factor H, factor I, and properdin) are extremely rare; they may be suspected if the total serum hemolytic complement is in the low range of normal and the serum C3 level is low. The screening test for alternative pathway abnormalities is the AH_{50} assay. The screening test for abnormalities of the mannose-binding pathway is measurement of the MBL level. The final identification of the specific complement component that is deficient usually rests on both functional and immunochemical tests, and highly specific assays have been developed for each individual complement component. These tests are all commercially available at reference laboratories.

Evaluation of Phagocytic Cells

Evaluation of phagocytic cells usually entails assessment of both their number and their function. Disorders such as congenital agranulocytosis and cyclic neutropenia, which are characterized by a deficiency in phagocytic cell number, can easily be detected by using a white blood cell count and differential count. Beyond that, assessment of phagocytic cell function is relatively specialized because it depends upon a variety of in vitro assays, including measurement of directed cell motility (chemotaxis), ingestion (phagocytosis), and intracellular killing (bactericidal activity) (237). The most common of the phagocyte function disorders, chronic granulomatous disease, can be identified by the nitroblue tetrazolium dye test or the dihydrorhodamine dye flow cytometry assay (132), either of which measures the oxidative metabolic response of neutrophils and monocytes. This is another test for which patient specimens sent to an off-site reference laboratory for testing should always be accompanied by a specimen from a healthy individual.

SUMMARY

Immunocompromised hosts have an increased susceptibility to infections for a wide variety of reasons. As a result, it is not correct, nor is it particularly useful, for the clinician to consider all immunocompromised hosts to have the same susceptibility to infection. This chapter should lay the groundwork for understanding why an individual patient may have an increased susceptibility to specific types of pathogens. Some of the defects in host defense are due to the effects of the environment (e.g., exposure to immunosuppressive drugs or viruses), and other defects are due to genetic traits (e.g., primary immunodeficiency diseases). We are likely to learn much more about both of these processes as data from the human genome project enable us to identify genetic variants (polymorphisms) that make some individuals more susceptible to a toxic effect of a drug or to have some variation (but not complete defect) in a host defense mechanism.

The identification of specific host defense defects in an individual patient should lead to a more efficient strategy for identification and management of specific pathogens. Subsequent chapters of this book elaborate on the specific pathogens and syndromes peculiar to the expanding group of immunocompromised hosts.

REFERENCES

1. **Abzug, M. J., M. J. Levin, and H. A. Rotbart.** 1993. Profile of enterovirus disease in the first two weeks of life. *Pediatr. Infect. Dis. J.* **12:**820–824.
2. **Aderem, A.** 2003. Phagocytosis and the inflammatory response. *J. Infect. Dis.* **187**(Suppl. 2):S340–S345.
3. **Aksoy, S., H. Harputluoglu, S. Kilickap, D. S. Dede, O. Dizdar, K. Altundag, and K. Barista.** 2007. Rituximab-related viral infections in lymphoma patients. *Leukoc. Lymphoma* **48:** 1307–1312.
4. **Alam, R., and M. Gorska.** 2003. Lymphocytes. *J. Allergy Clin. Immunol.* **111**(Suppl. 2):S476–S485.
5. **Alarcon, G. S.** 2006. Infections in systemic connective tissue diseases: systemic lupus erythematosus, scleroderma, and polymyositis/dermatomyositis. *Infect. Dis. Clin. N. Am.* **20:** 849–875.
6. **Algood, H. M., P. L. Lin, and J. L. Flynn.** 2005. Tumor necrosis factor and chemokine interactions in the formation and maintenance of granulomas in tuberculosis. *Clin. Infect. Dis.* **41**(Suppl 3):S189–S193.
7. **Allendoerfer, R., and G. S. Deepe, Jr.** 1998. Blockade of endogenous TNF-alpha exacerbates primary and secondary pulmonary histoplasmosis by different mechanisms. *J. Immunol.* **160:**6072–6082.

7a. **American Journal of Transplantation.** 2004. Guidelines for the prevention and management of infectious complications of solid organ transplantation. *Am. J. Transplant.* **4**(Suppl. 10):6–20.

8. **Anding, K., P. Gross, J. M. Rost, D. Allgaier, and E. Jacobs.** 2003. The influence of uraemia and haemodialysis on neutrophil phagocytosis and antimicrobial killing. *Nephrol. Dial. Transplant.* **18:**2067–2073.
9. **Andreone, P., G. Annagiulia, S. Lorenzini, M. Biselli, C. Cursaro, S. Pileri, and M. Bernardi.** 2003. Post-transplantation lymphoproliferative disorders. *Arch. Intern. Med.* **163:**1997–2004.
10. **Andoniou, C. E., D. M. Andrews, and M. A. Degli-Esposti.** 2006. Natural killer cells in viral infection: more than just killers. *Immunol. Rev.* **214:**239–250.
11. Reference deleted.
12. **Antonelli, M. A., L. W. Moreland, and J. E. Brick.** 1991. Herpes zoster in patients with rheumatoid arthritis treated with weekly, low-dose methotrexate. *Am. J. Med.* **90:**295–298.
13. **Arima, S., S. M. Akbar, and K. E. Michitaka.** 2003. Impaired function of antigen-presenting dendritic cells in patients with chronic hepatitis B: localization of HBV DNA and HBV RNA in blood dendritic cells by in situ hybridization. *J. Mol. Med.* **11:**169–174.
14. **Arunkumar, P., T. Crook, and J. Ballard.** 2004. Disseminated histoplasmosis presenting as pancytopenia in a methotrexate-treated patient. *Am. J. Hematol.* **77:**86–87.
15. **Assi, M. A., J. E. Rosenblatt, and W. F. Marshall.** 2007. Donor-transmitted toxoplasmosis in liver transplant recipients: a case report and literature review. *Transpl. Infect. Dis.* **9:**132–136.
16. **Auphan, N., J. A. DiDonato, C. Rosette, A. Helmberg, and M. Karin.** 1995. Immunosuppression by glucocorticoids: inhibition of NF-κβ activation through induction of I kappa biosynthesis. *Science* **270:**286–290.
17. **Balow, J. E., and A. S. Rosenthal.** 1973. Glucocorticoid suppression of macrophage migration inhibitory factor. *J. Exp. Med.* **137:**1031–1041.
18. **Barcán, L., C. Luna, L. Clara, A. Sinagra, A. Valledor, A. M. De Rissio, A. Gadano, M. M. García, E. de Santibañes, and A. Riarte.** 2005. Transmission of *T. cruzi* infection via liver transplantation to a nonreactive recipient for Chagas' disease. *Liver Transplant.* **11:**1112–1116.
19. **Barnes, P. D., and K. A. Marr.** 2007. Risks, diagnosis and outcomes of invasive fungal infections in haematopoietic stem cell transplant recipients. *Br. J. Haematol.* **139:**519–531.

20. **Barrett, D. J.** 1985. Human immune responses to polysaccharide antigens: an analysis of bacterial polysaccharide vaccines in infants. *Adv. Pediatr.* **32:**139–158.
21. **Barsoum, R. S.** 2004. Parasitic infections in organ transplantation. *Exp. Clin. Transplant.* **2:**258–267.
22. **Basu, A., M. Ramkumar, H. P. Tan, A. Khan, J. McCauley, A. Marcos, J. J. Fung, T. E. Starzl, and R. Shapiro.** 2005. Reversal of acute cellular rejection after renal transplantation with Campath H-1. *Transplant. Proc.* **37:**923–926.
23. **Bergstrom, L., D. E. Yocum, N. M. Ampel, I. Villanueva, J. Lisse, O. Gluck, J. Tesser, J. Posever, M. Miller, J. Araujo, D. M. Kageyama, M. Berry, L. Karl, and C. M.Yung.** 2004. Increased risk of coccidioidiomycosis in patients treated with tumor necrosis factor alpha antagonists. *Arthritis Rheum.* **50:**1959–1966.
24. **Bermas, B. L., M. Petri, D. Goldman, B. Mittleman, M. W. Miller, N. I. Stocks, C. S. Via, and G. M. Shearer.** 1994. T helper cell dysfunction in systemic lupus erythematosus (SLE): relation to disease activity. *J. Clin. Immunol.* **14:**169–177.
25. **Bernini, J. C.** 1996. Diagnosis and management of chronic neutropenia during childhood. *Pediatr. Clin. N. Am.* **43:**773–792.
26. **Beutler, B., and A. Cerami.** 1989. The biology of cachectin/TNF—a primary mediator of the host response. *Annu. Rev. Immunol.* **7:**625–655.
27. **Bhat, S., and S. Milner.** 2007. Antimicrobial peptides in burns and wounds. *Curr. Protein Pept. Sci.* **8:**506–520.
28. **Bizzarro, M. J., C. Raskind, R. S. Baltimore, and P. G. Gallagher.** 2005. Seventy-five years of neonatal sepsis at Yale: 1928-2003. *Pediatrics* **116:**595–602.
29. **Bjerknes, R., H. Vindenes, and O. D. Laerum.** 1990. Altered neutrophil functions in patients with large burns. *Blood Cells* **16:**127–141.
30. **Blair, J. E.** 2007. Coccidioidiomycosis in patients who have undergone transplantation. *Ann. N. Y. Acad. Sci.* **1111:**365–376.
31. **Bodey, G. P., M. Buckley, Y. S. Sathe, and E. J. Freireich.** 1966. Quantitative relationships between circulating leukocytes and infection in patients with acute leukemia. *Ann. Intern. Med.* **64:**328–340.
32. **Bodey, G. P., L. Jadeja, and L. Elting.** 1985. Pseudomonas bacteremia: retrospective analysis of 410 episodes. *Arch. Intern. Med.* 145:1621–1629.
33. **Boni, C., P. Fisicaro, C. Valdatta, B. Amadei, P. Di Vincenzo, T. Giuberti, D. Laccabue, A. Zerbini, A. Cavalli, G. Missale, A. Bertoletti, and C. Ferrari.** 2001. Characterization of hepatitis B virus (HBV)-specific T-cell dysfunction in chronic HBV infection. *J. Virol.* **81:**4215–4225.
34. **Bonnefoy-Berard, N., and J. P. Revillard.** 1996. Mechanisms of immunosuppression induced by antithymocyte globulins and OKT3. *J. Heart Lung Transplant.* **15:**435–442.
35. **Boumpas, D. T., G. P. Chrousos, R. L. Wilder, T. R. Cupps, and J. E. Balow.** 1993. Glucocorticoid therapy for immune-mediated diseases: basic and clinical correlates. *Ann. Intern. Med.* **119:**1198–1208.
36. **Braconier, J. H.** 1999. Reversible total IgA deficiency associated with phenytoin treatment. *Scand. J. Infect. Dis.* **31:**515–516.
37. **Brennan, D. C., J. A. Daller, K. D. Lake, D. Cibrik, and D. Del Castillo.** 2006. Rabbit antithymocyte globulin versus basiliximab in renal transplantation. *N. Engl. J. Med.* **355:**1967–1977.
38. **Brennan, D. C., K. Flavin, J. A. Lowell, T. K. Howard, S. Shenoy, S. Burgess, S. Dolan, J. M. Kano, M. Mahon, M. A. Schnitzler, R. Woodward, W. Irish, and G. G. Singer.** 1999. A randomized, double-blinded comparison of thymoglobulin versus Atgam for induction immunosuppressive therapy in adult renal transplant recipients. *Transplantation* **7:**1011–1018.
39. **Brown, J. M. Y.** 2005. The influence of the conditions of hematopoietic cell transplantation on infectious complications. *Curr. Opin. Infect. Dis.* **18:**346–351.
40. **Buckley, R. H.** 2004. The multiple causes of human SCID. *J. Clin. Investig.* **114:**1409–1411.
41. **Busca, A., E. Lovisone, S. Aliberti, F. Locatelli, A. Serra, P. Scaravaglio, P. Omede, G. Rossi, D. Cirillo, A. Barbui, V. Ghisetti, A. M. Dall'Omo, and M. Falda.** 2003. Immune reconstitution and early infectious complications following nonmyeloablative hematopoietic stem cell transplantation. *Hematology* **8:**303–311.
42. **Bustami, R. T., A. O. Ojo, R. A. Wolfe, R. M. Merion, W. M. Bennett, S. V. McDiarmid, A. B. Leichtman, P. J. Held, and F. K. Port.** 2004. Immunosuppression and the risk of post-transplant malignancy among cadaveric first kidney transplant recipients. *Am. J. Transplant.* **4:**87–93.
43. **Caillard, S., V. Dharnidharka, L. Agodoa, E. Bohen, and K. Abbott.** 2005. Post-transplant lymphoproliferative disorders after renal transplantation in the United States in the era of modern immunosuppression. *Transplantation* **80:**1233–1243.
44. **Calabrese, L. H., E. S. Molloy, D. Huang, and R. M. Ransohoff.** 2007. Progressive multifocal leukoencephalopathy in rheumatic diseases. *Arthritis Rheum.* **56:**2116–2128.
45. **Calvet, H. M., and T. T. Yoshikawa.** 2001. Infections in diabetes. *Infect. Dis. Clin. N. Am.* **15:**407–421.
46. **Cannon, G. W., and J. M. Kremer.** 2004. Leflunomide. *Rheum. Dis. Clin. N. Am.* **30:**295–309.
47. **Caputo, G. M., N. Joshi, and M. R. Weitekamp.** 1997. Foot infections in patients with diabetes. *Am. Fam. Physician* **56:**195–202.
48. **Castagnola, E., S. Dallorso, M. Faraci, G. Morreale, D. Di Martino, E. Cristina, L. Scarso, and E. Lanino.** 2003. Long-lasting hypogammaglobulinemia following rituximab administration for Epstein-Barr virus-related post-transplant lymphoproliferative disease preemptive therapy. *J. Hematother. Stem Cell Res.* **12:**9–10.
49. **Castagnola, E., V. Fontana, I. Caviglia, S. Caruso, M. Faraci, F. Fioredda, M. L. Garrè, C. Moroni, M. Conte, G. Losurdo, F. Scuderi, R. Bandettini, P. Tomà, C. Viscoli, and R. Haupt.** 2007. A prospective study on the epidemiology of febrile episodes during chemotherapy-induced neutropenia in children with cancer or after hemopoietic stem cell transplantation. *Clin. Infect. Dis.* **45:**1296–1304.
50. **Castigli, E., and R. S. Geha.** 2006. Molecular basis of common variable immunodeficiency. *J. Allergy Clin. Immunol.* **117:**740–746.
51. **Castle, S. C.** 2000. Impact of age-related immune dysfunction on risk of infections. *Z. Gerontol. Geriatr.* **33:**341–349.
52. **Castle, S. C.** 2000. Clinical relevance of age-related immune dysfunction. *Clin. Infect. Dis.* **31:**578–585.
53. **Castle, S. C., K. Uyemura, T. Fulop, and T. Makinodan.** 2007. Host resistance and immune responses in advanced age. *Clin. Geriatr. Med.* **23:**463–479.
54. **Cedzynski, M., J. Szemraj, A. S. Swierzko, L. Bak-Romaniszyn, M. Banasik, K. Zeman, and D. C. Kilpatrick.** 2004. Mannan-binding lectin insufficiency in children with recurrent infections of the respiratory system. *Clin. Exp. Immunol.* **136:**304–311.

54a. **Centers for Disease Control and Prevention.** 1997. Anergy skin testing and tuberculosis preventive therapy for HIV-infected persons: revised recommendations. *MMWR Recommend. Rep.* **46**(RR-15):1–10.

55. **Centers for Disease Control and Prevention.** 2002. Chagas disease after organ transplantation—United States, 2001. *MMWR Morb. Mortal. Wkly. Rep.* **51**:210–212.
55a. **Centers for Disease Control and Prevention.** 2000. Guidelines for preventing opportunistic infections among hematopoietic stem cell transplant recipients. Recommendations of the CDC, the Infectious Diseases Society of America, and the American Society of Blood and Marrow Transplantation. *MMWR Morb. Mortal. Wkly. Rep.* **49** (RR10):1–128.
55b. **Centers for Disease Control and Prevention.** 2005. West Nile virus infections in organ transplant recipients—New York and Pennsylvania, August-September, 2005. *MMWR Morb. Mortal. Wkly. Rep.* **54**:1021–1023.
56. **Cheong, H.-J., J.-Y. Song, J.-W. Park, J.-E. Yeon, K.-S. Byun, C.-H. Lee, H.-I. Cho, T.-G. Kim, and W.-J. Kim.** 2006. Humoral and cellular immune responses to influenza vaccine in patients with advanced cirrhosis. *Vaccine* **24**:2417–2422.
57. **Cheson, B. D.** 1995. Infectious and immunosuppressive complications of purine analog therapy. *J. Clin. Oncol.* **13**:2431–2448.
58. **Chung, J. W., S. O. Lee, S. H. Choi, J. H. Woo, J. Ryu, Y. S. Kim, and N. J. Kim.** 2006. Risk factors and outcome for breakthrough candidaemia in patients with cancer. *Mycoses* **49**:114–118.
59. **Cohen, S., G. W. Cannon, M. Schiff, A. Weaver, R. Fox, N. Olsen, D. Furst, J. Sharp, L. Moreland, J. Caldwell, J. Kaine, and V. Strand.** 2001. Two-year, blinded, randomized, controlled trial of treatment of active rheumatoid arthritis with leflunomide compared with methotrexate. *Arthritis Rheum.* **44**:1984–1992.
59a. **Cooperative Group for the Study of Immunoglobulin in Chronic Lymphocytic Leukemia.** 1988. Intravenous immunoglobulin for the prevention of infection in chronic lymphocytic leukemia: a randomized, controlled clinical trial. *N. Engl. J. Med.* **319**:902–907.
60. **Corales, R., J. Chua, S. Mawhorter, J. B. Young, R. Starling, J. W. Tomford, P. McCarthy, W. E. Braun, N. Smedira, R. Hobbs, G. Haas, D. Pelegrin, M. Majercik, K. Hoercher, D. Cook, and R. K. Avery.** 2000. Significant post-transplant hypogammaglobulinemia in six heart transplant recipients: an emerging clinical phenomenon? *Transpl. Infect. Dis.* **2**: 133–139.
61. **Cordonnier, C., R. Martino, P. Trabasso, T. K. Held, H. Akan, M. S. Ward, K. Fabian, A. J. Ullmann, N. Wulffraat, P. Ljungman, E. P. Alessandrino, J. Pretnar, J. Gmür, R. Varela, A. Vitek, S. Sica, M. Rovira, and European Blood and Marrow Transplant Group Infectious Diseases Working Party.** 2004. Mycobacterial infection: a difficult and late diagnosis in stem cell transplant recipients. *Clin. Infect. Dis.* **38**:1229–1236.
62. **Cortez, K. J., D. D. Erdman, T. C. Peret, V. J. Gill, R. Childs, A. J. Barrett, and J. E. Bennett.** 2001. Outbreak of human parainfluenza virus 3 infections in a hematopoietic stem cell transplant population. *J. Infect. Dis.* **184**:1093–1097.
63. **Cunningham-Rundles, C.** 2001. Physiology of IgA and IgA deficiency. *J. Clin. Immunol.* **21**:303–309.
64. **Cunningham-Rundles, C., and C. Bodian.** 1999. Common variable immunodeficiency: clinical and immunological features of 248 patients. *Clin. Immunol.* **92**:34–48.
65. **David-Neto, E., J. A. da Fonseca, F. J. de Paula, W. C. Nahas, E. Sabbaga, and L. E. Ianhez.** 1999. Is azathioprine harmful to chronic viral hepatitis in renal transplantation? A long-term study on azathioprine withdrawal. *Transplant. Proc.* **31**:1149–1150.
66. **Davis, C. A., E. H. Vallota, and J. Forristal.** 1979. Serum complement levels in infancy: age related changes. *Pediatr. Res.* **13**:1043–1046.
67. **Delamaire, M., D. Maugendre, M. Moreno, M. C. Le Goff, H. Allannic, and B. Genetet.** 1997. Impaired leucocyte functions in diabetic patients. *Diabetes Med.* **14**:29–34.
68. **de Latour, R. P., V. Levy, T. Asselah, P. Marcellin, C. Scieux, L. Ades, R. Traineau, A. Devergie, P. Ribaud, H. Esperou, E. Gluckman, D. Valla, and G. Socie.** 2004. Long-term outcome of hepatitis C infection after bone marrow transplantation. *Blood* **103**:1618–1624.
69. **Delves, P. J., and I. M. Roitt.** 2000. The immune system. First of two parts. *N. Engl. J. Med.* **343**:37–49.
70. **Delves, P. J., and I. M. Roitt.** 2000. The immune system. Second of two parts. *N. Engl. J. Med.* **343**:108–117.
71. **Dinits-Pensy, M., G. N. Forrest, A. S. Cross, and M. K. Hise.** 2005. The use of vaccines in adult patients with renal disease. *Am. J. Kidney Dis.* **46**:997–1011.
72. **Doucette, K., and J. A. Fishman.** 2004. Nontuberculous mycobacterial infection in hematopoietic stem cell and solid organ transplant recipients. *Clin. Infect. Dis.* **38**:1428–1439.
73. **Drachenberg, C. B., H. H. Hirsch, J. C. Papadimitriou, R. Gosert, R. K. Wali, R. Munivenkatappa, J. Nogueira, C. B. Cangro, A. Haririan, S. Mendley, and E. Ramos.** 2007. Polyomavirus BK versus JC replication and nephropathy in renal transplant recipients: a prospective evaluation. *Transplantation* **84**:323–330.
74. **Dvorak, C. C., W. J. Steinbach, J. M. Y. Brown, and R. Agarwal.** 2005. Risk factors and outcomes of invasive fungal infections in pediatric patients undergoing allogeneic hematopoietic cell transplantation. *Bone Marrow Transplant.* **36**:621–629.
75. **Ehlers, S.** 2005. Tumor necrosis factor and its blockade in gramulomatous infections: differential modes of action of infliximab and etanercept? *Clin. Infect. Dis.* **41**(Suppl. 3): S199–S203.
76. **Eisen, H. J., J. Kobashigawa, A. Keogh, R. Bourge, D. Renlund, R. Mentzer, E. Aldreman, H. Valantine, G. Dureau, D. Mancini, R. Mamelok, R. Gordon, W. Wang, M. Mehra, M. R. Constanzo, M. Hummel, J. Johnson.** 2005. Three-year results of a randomized, double-blind, controlled trial of mycophenolate mofetil versus azathioprine in cardiac transplant recipients. *J. Heart Lung Transplant.* **24**:517–525.
77. **Englund, J. A., M. Boeckh, J. Kuypers, W. G. Nichols, R. C. Hackman, R. A. Morrow, D. N. Fredricks, and L. Corey.** 2006. Brief communication: fatal human metapneumovirus infection in stem-cell transplant recipients. *Ann. Intern. Med.* **144**:344–349.
78. **Epstein, J. B., A. Ransier, C. H. Sherlock, J. J. Spinelli, and D. Reece.** 1996. Acyclovir prophylaxis of oral herpes during bone marrow transplantation. *Eur. J. Cancer B* **32B**:158–162.
79. **Erard, V., M. L. Huang, J. Ferrenberg, L. Nguy, T. L. Stevens-Ayers, R. C. Hackman, L. Corey, and M. Boeckh.** 2007. Quantitative real-time polymerase chain reaction for detection of adenovirus after T cell-replete hematopoietic cell transplantation: viral load as a marker for invasive disease. *Clin. Infect. Dis.* **45**:958–965.
80. **Estahbanati, H. K., P. P. Kashani, and F. Ghanaatpisheh.** 2002. Frequency of Pseudomonas aeruginosa serotypes in burn wound infections and their resistance to antibiotics. *Burns* **28**:340–348.
81. **Falsey, A. R., P. A. Hennessey, M. A. Formica, C. Cox, and E. E. Walsh.** 2005. Respiratory syncytial virus infection in elderly and high-risk adults. *N. Engl. J. Med.* **352**:1749–1759.

82. **Fauci, A. S., D. C. Dale, and J. E. Balow.** 1976. Glucocorticoid therapy: mechanisms of action and clinical considerations. *Ann. Intern. Med. 84*:304–315.
83. **Fauci, A. S., G. Pantaleo, S. Stanley, and D. Weissman.** 1996. Immunopathogenic mechanisms of HIV infection. *Ann. Intern. Med.* **124**:654–663.
84. **Figueroa, J. E., and P. Densen.** 1991. Infectious diseases associated with complement deficiencies. *Clin. Microbiol. Rev.* **4**:359–395.
85. **Fischer, S. A., M. B. Graham, M. J. Kuehnert, C. N. Kotton, A. Srinivasan, F. M. Marty, J. A. Comer, J. Guarner, C. D. Paddock, D. L. DeMeo, W. J. Shieh, B. R. Erickson, U. Bandy, A. DeMaria, Jr., J. P. Davis, F. L. Delmonico, B. Pavlin, A. Likos, M. J. Vincent, T. K. Sealy, C. S. Goldsmith, D. B. Jernigan, P. E. Rollin, M. M. Packard, M. Patel, C. Rowland, R. F. Helfand, S. T. Nichol, J. A. Fishman, T. Ksiazek, S. R. Zaki, et al.** 2006. Transmission of lymphocytic choriomeningitis virus by organ transplantation. *N. Engl. J. Med.* **354**:2235–2249.
86. **Fishman, J. A., and R. H. Rubin.** 1998. Infection in organ transplant recipients. *N. Engl. J. Med.* **338**:1741–1751.
87. **Fishman, J. A.** 2001. Prevention of infection caused by Pneumocystis carinii in transplant recipients. *Clin. Infect. Dis.* **33**:1397–1405.
88. **Flynn, J. L., M. M. Goldstein, J. Chan, K. J. Triebold, K. Pfeffer, C. J. Lowenstein, R. Schreiber, T. W. Mak, and B. R. Bloom.** 1995. Tumor necrosis factor-alpha is required in the protective immune response against Mycobacterium tuberculosis in mice. *Immunity* **2**:561–572.
89. **Folds, J. D., and J. L. Schmitz.** 2003. Clinical and laboratory assessment of immunity. *J. Allergy Clin. Immunol.* **111** (Suppl. 2):S702–S711.
90. **Fothergill, L. D., and J. Wright.** 1933. Influenzal meningitis: the relation of age incidence to the bactericidal power of blood against the causal organism. *J. Immunol.* **24**:273–284.
91. **Freifeld, A. G., P. C. Iwen, B. L. Lesiak, R. K. Gilroy, R. B. Stevens, and A. C. Kalil.** 2005. Histoplasmosis in solid organ transplant recipients at a large Midwestern university transplant center. *Transpl. Infect. Dis.* **7**:109–115.
92. **Gaspar, H. B., R. Sharifi, K. C. Gilmour, and A. J. Thrasher.** 2002. X-linked lymphoproliferative disease: clinical, diagnostic and molecular perspective. *Br. J. Haematol.* **119**:585–595.
93. **Gauthier, G. M., N. Safdar, B. S. Klein, and D. R. Andes.** 2007. Blastomycosis in solid organ transplant recipients. *Transpl. Infect. Dis.* **9**:310–317.
94. **Georgeson, G. D., B. J. Szony, K. Streitman, A. Kovács, L. Kovács, and A. László.** 2001. Natural killer cell cytotoxicity is deficient in newborns with sepsis and recurrent infections. *Eur. J. Pediatr.* **160**:478–482.
95. **Ginzler, E., H. Diamond, D. Kaplan, M. Weiner, M. Schlesinger, and M. Seleznick.** 1978. Computer analysis of factors influencing frequency of infection in systemic lupus erythematosus. *Arthritis Rheum.* **21**:37–44.
96. **Girndt, M., M. Sester, U. Sester, H. Kaul, and H. Köhler.** 2001. Molecular aspects of T and B cell function in uremia. *Kidney Int.* **59**(Suppl. 78):S206–S211.
97. **Gluckman, E., R. Traineau, A. Devergie, H. Esperou-Bourdeau, and I. Hirsch.** 1992. Prevention and treatment of CMV infection after allogeneic bone marrow transplant. *Ann. Hematol.* **64**(Suppl. A):158–161.
98. **Goodarzi, H., J. Trowbridge, and R. L. Gallo.** 2007. Innate immunity: a cutaneous perspective. *Clin. Rev. Allergy Immunol.* **33**:15-26.
99. **Gorensek, M. J., M. H. Lebel, and J. D. Nelson.** 1988. Peritonitis in children with nephrotic syndrome. *Pediatrics* **81**:849–856.
100. **Gourley, M. F., H. A. Austin III, D. Scott, C. H. Yarboro, E. M. Vaughan, J. Muir, D. T. Boumpas, J. H. Klippel, J. E. Balow, and A. D. Steinberg.** 1996. Methylprednisolone and cyclophosphamide, alone or in combination, in patients with lupus nephritis. A randomized, controlled trial. *Ann. Intern. Med.* **125**:549–557.
101. **Grim, S. A., T. Pham, J. Thielke, H. Sankary, J. Oberholzer, E. Benedetti, and N. M. Clark.** 2007. Infectious complications associated with rituximab use for ABO-incompatible and positive cross-match renal transplant recipients. *Clin. Transplant.* **21**:628–632.
102. **Groth, C. G., L. Bäckman, J. M. Morales, R. Calne, H. Kreis, P. Lang, J. L. Touraine, K. Claesson, J. M. Campistol, D. Durand, L. Wramner, C. Brattström, B. Charpentier, et al.** 1999. Sirolimus (rapamycin)-based therapy in human renal transplantation: similar efficacy and different toxicity compared with cyclosporine. *Transplantation* **67**:1036–1042.
103. **Guerreiro-Cacais, A. O., L. Li, D. Donati, M. T. Bejarano, A. Morgan, M. G. Masucci, L. Hutt-Fletcher, and V. Levitsky.** 2004. Capacity of Epstein-Barr virus to infect monocytes and inhibit their development into dendritic cells is affected by the cell type supporting virus replication. *J. Gen. Virol.* **85**:2767–2778.
104. Reference deleted.
105. Reference deleted.
106. **Gutierrez-Ureña, S., J. F. Molina, C. O. García, M. L. Cuéllar, and L. R. Espinoza.** 1996. Pancytopenia secondary to methotrexate therapy in rheumatoid arthritis. *Arthritis Rheum.* **39**:272–276.
107. **Hagen, E. A., H. Stern, D. Porter, K. Duffy, K. Foley, S. Luger, S. J. Schuster, E. A. Stadtmauer, and M. G. Schuster.** 2003. High rate of invasive fungal infections following nonmyeloablative allogeneic transplantation. *Clin. Infect. Dis.* **36**:9–15.
108. **Hammarström, V., K. Pauksen, H. Svensson, B. Lönnqvist, B. Simonsson, O. Ringdén, and P. Ljungman.** 2000. Serum immunoglobulin levels in relation to levels of specific antibodies in allogeneic and autologous bone marrow transplant recipients. *Transplantation* **69**:1582–1586.
109. **Hanson, D. L., S. Y. Chu, M. Farizo, J. W. Ward, et al.** 1995. Distribution of CD4+ lymphocytes at diagnosis of the acquired immunodeficiency syndrome-defining and other human immunodeficiency virus-related illnesses. *Arch. Intern. Med.* **155**:1537–1542.
110. **Hardinger, K. L.** 2006. Rabbit antithymocyte globulin induction therapy in adult renal transplantation. *Pharmacotherapy* **26**:1771–1783.
111. **Heffron, T. G., G. A. Smallwood, M. E. de Vera, L. Davis, E. Martinez, and A. C. Stieber.** 2001. Daclizumab induction in liver transplant recipients. *Transplant. Proc.* **33**:1527.
112. **Hemsley, C., and S. J. Eykyn.** 1998. Pneumococcal peritonitis in previously healthy adults: case report and review. *Clin. Infect. Dis.* **27**:376–379.
113. **Herrod, H. G.** 1990. Chronic mucocutaneous candidiasis in childhood and complications of non-Candida infection: a report of the Pediatric Immunodeficiency Collaborative Study Group. *J. Pediatr.* **116**:377–382.
114. **Hibberd, P. L., N. E. Tolkoff-Rubin, A. B. Cosimi, R. T. Schooley, D. Isaacson, M. Doran, A. Delvecchio, F. L. Delmonico, H. Auchincloss, Jr., and R. H. Rubin.** 1992. Symptomatic cytomegalovirus disease in the cytomegalovirus antibody seropositive renal transplant recipient treated with OKT3. *Transplantation* **53**:68–72.
115. **Hidalgo, S., M. García Erro, D. Cisterna, and M. C. Freire.** 2003. Paralytic poliomyelitis caused by a vaccine-derived

polio virus in an antibody-deficient Argentinean child. *Pediatr. Infect. Dis. J.* **22**:570–572.

116. **High, K. P.** 1994. The antimicrobial activities of cyclosporine, FK506, and rapamycin. *Transplantation* **57**:1689–1700.
117. **Ho, D. D., A. U. Neumann, and A. S. Perelson.** 1995. Rapid turnover of plasma virions and CD4 lymphocytes in HIV-1 infection. *Nature 373*:123–126.
118. **Holm, G., H. Mellstedt, M. Björkholm, B. Johansson, D. Killander, R. Sundblad, and G. Söderberg.** 1976. Lymphocyte abnormalities in untreated patients with Hodgkin's disease. *Cancer* **37**:751–762.
119. **Homann, C., K. Varming, K. Hogasen, T. E. Mollnes, N. Graudal, A. C. Thomsen, and P. Garred.** 1997. Acquired C3 deficiency in alcoholic cirrhosis predisposing to infection and increased mortality. *Gut* **40**:544–549.
120. **Huffnagle, G. B., G. B. Toews, M. D. Burdick, M. B. Boyd, K. S. McAllister, R. A. McDonald, S. L. Kunkel, and R. M. Strieter.** 1996. Afferent phase production of TNF-alpha is required for the development of protective T cell immunity to Cryptococcus neoformans. *J. Immunol.* **157**:4529–4536.
121. **Humphreys, J. M., R. E. Stringer, C. A. Hart, and S. W. Edwards.** 1993. Effect of cytotoxic drugs on mature neutrophil function in the presence and absence of granulocyte-macrophage colony stimulating factor. *Br. J. Haematol.* **84**:316–321.
122. **Husain, S., and N. Singh.** 2002. The impact of novel immunosuppressive agents on infections in organ transplant recipients and interactions of these agents with antimicrobials. *Clin. Infect. Dis.* **35**:53–61.
123. **Hutchison, P., M. Jose, R. C. Atkins, and S. R. Holdsworth.** 2004. Ex vivo lymphocyte proliferative function is severely inhibited in renal transplant patients on mycophenolate mofetil treatment. *Transplant. Immunol.* **13**:55–61.
124. **Imashuku, S., T. Teramura, A. Morimoto, M. Naya, and H. Kuroda.** 2004. Prolonged hypogammaglobulinemia following rituximab treatment for post transplant Epstein-Barr virus-associated lymphoproliferative disease. *Bone Marrow Transplant.* **33**:129–130.
125. Reference deleted.
126. **Ison, M. G.** 2007. Respiratory viral infections in transplant recipients. *Antivir. Ther.* **12**:627–638.
127. **Ison, M. G., and F. G. Hayden.** 2002. Viral infections in immunocompromised patients: what's new with respiratory viruses? *Curr. Opin. Infect. Dis.* **15**:355–367.
128. **Jabs, W. J., H. J. Wagner, S. Maurmann, H. Hennig, and B. Kreft.** 2002. Inhibition of macrophage inflammatory protein-1 alpha production by Epstein-Barr virus. *Blood* **99**:1512–1516.
129. **Jain, A. B., I. Hamad, J. Rakela, F. Dodson, D. Kramer, J. Demetris, J. McMichael, T. E. Starzl, and J. J. Fung.** 1998. A prospective randomized trial of tacrolimus and prednisone versus tacrolimus, prednisone, and mycophenolate mofetil in primary adult liver transplant recipients: an interim report. *Transplantation* **66**:1395–1398.
130. **James, L. C., G. Hale, H. Waldmann, and A. C. Bloomer.** 1999. 1.9A structure of the therapeutic antibody Campath-1H fab in complex with a synthetic peptide antigen. *J. Mol. Biol.* **289**:293–301.
131. **Jenks, K. A., L. K. Stamp, J. L. O'Donnell, R. L. Savage, and P. T. Chapman.** 2007. Leflunomide-associated infections in rheumatoid arthritis. *J. Rheumatol.* **34**:2201–2203.
132. **Jirapongsananuruk, O., H. L. Malech, D. B. Kuhns, J. E. Niemela, M. R. Brown, M. Anderson-Cohen, and T. A. Fleisher.** 2003. Diagnostic paradigm for evaluation of male patients with chronic granulomatous disease, based on the dihydrorhodamine 123 assay. *J. Allergy Clin. Immunol.* **111**:374–379.
133. **Johnson, D. H., and B. A. Cunha.** 2001. Infections in cirrhosis. *Infect. Dis. Clin. N. Am.* **15**:363–371.
134. **Joshi, N., G. M. Caputo, M. R. Weitekamp, and A. W. Karchmer.** 1999. Infections in patients with diabetes mellitus. *N. Engl. J. Med. 341*:1906–1912.
135. **Junghanss, G., M. Boeckh, R. A. Carter, B. M. Sandmaier, M. B. Maris, D. G. Maloney, T. Chauncey, P. A. McSweeney, M. T. Little, L. Corey, and R. Storb.** 2002. Incidence and outcome of cytomegalovirus infections following nonmyeloablative compared with myeloablative allogeneic stem cell transplantation, a matched control study. *Blood* **99**:1978–1985.
136. **Kahan, B. D.** 2000. Efficacy of sirolimus compared with azathioprine for reduction of acute renal allograft rejection: a randomised multicentre study. *Lancet* **356**:194–202.
137. **Kahan, B. D., P. R. Rajagopalan, and M. Hall.** 1999. Reduction of the occurrence of acute cellular rejection among renal allograft recipients treated with basiliximab, a chimeric anti-interleukin-2-receptor monoclonal antibody. *Transplantation* **67**:276–284.
138. **Kaisho, T., and S. Akira.** 2006. Toll-like receptor function and signaling. *J. Allergy Clin. Immunol.* **117**:979–987.
139. **Kamel, O. W., M. van de Rijn, L. M. Weiss, G. J. Del Zoppo, P. K. Hench, B. A. Robbins, P. G. Montgomery, R. A. Warnke, and R. F. Dorfman.** 1993. Brief report: reversible lymphomas associated with Epstein-Barr virus occurring during methotrexate therapy for rheumatoid arthritis and dermatomyositis. *N. Engl. J. Med.* **328**:1317–1321.
140. **Kanik, K. S., and J. M. Cash.** 1997. Does methotrexate increase the risk of infection or malignancy? *Rheum. Dis. Clin. N. Am.* **23**:955–967.
141. **Kaur, N., and T. C. Mahl.** 2007. Pneumocystis jiroveci (carinii) pneumonia after infliximab therapy: a review of 84 cases. *Dig. Dis. Sci.* **52**:1481–1484.
142. **Kawasaki, H., J. Takayama, and M. Ohira.** 1996. Herpes zoster infection after bone marrow transplantation in children. *J. Pediatr.* **128**:353–356.
143. **Keane, J., S. Gershon, R. P. Wise, E. Mirabile-Levens, J. Kasznica, W. D. Schwieterman, J. N. Siegel, and M. M. Braun.** 2001. Tuberculosis associated with infliximab, a tumor necrosis factor α-neutralizing agent. *N. Engl. J. Med.* **345**:1098–1104.
144. **Keating, M., S. Coutre, K. Rai, A. Osterborg, S. Faderl, B. Kennedy, T. Kipps, G. Bodey, J. C. Byrd, S. Rosen, C. Dearden, M. J. Dyer, and P. Hillmen.** 2004. Management guidelines for use of alemtuzumab in B-cell chronic lymphocytic leukemia. *Clin. Lymphoma* **4**:220–227.
145. **Keegan, J. M., and J. W. Byrd.** 1988. Nocardiosis associated with low dose methotrexate for rheumatoid arthritis. *J. Rheumatol.* **15**:1585–1586.
146. **Khan, S. A., and J. R. Wingard.** 2001. Infection and mucosal injury in cancer treatment. *J. Natl. Cancer Inst. Monogr.* **29**:31–36.
147. **Kimberlin, D. W.** 2007. Herpes simplex virus infections of the newborn. *Semin. Perinatol.* **31**:19–25.
148. **Kinch, A., G. Oberg, J. Arvidson, K. I. Falk, A. Linde, and K. Pauksens.** 2007. Post-transplant lymphoproliferative disease and other Epstein-Barr virus diseases in allogeneic haematopoietic stem cell transplantation after introduction of monitoring of viral load by polymerase chain reaction. *Scand. J. Infect. Dis.* **39**:235–244.
149. **Klastersky, J., L. Ameye, J. Maertens, A. Georgala, F. Muanza, M. Aoun, A. Ferrant, B. Rapoport, K. Rolston, and M. Paesmans.** 2007. Bacteraemia in febrile neutropenic

cancer patients. *Int. J. Antimicrob. Agents* **30**(Suppl. 1): S51–S59.

150. **Kniker, W. T., B. M. Lesourd, J. L. McBryde, and R. N. Corriel.** 1985. Cell-mediated immunity assessed by multitest CMI skin testing in infants and preschool children. *Am. J. Dis. Child.* **139**:840–845.
151. **Kobrynski, L. J., and K. E. Sullivan.** 2007. Velocardiofacial syndrome, DiGeorge syndrome: the chromosome 22q11.2 deletion syndromes. *Lancet* **370**:1443–1452.
152. **Koc, Y., K. B. Miller, D. P. Schenkein, J. Griffith, M. Akhtar, J. DesJardin, and D. R. Snydman.** 2000. Varicella zoster virus infections following allogeneic bone marrow transplantation: frequency, risk factors, and clinical outcome. *Biol. Blood Marrow Transplant.* **6**:44–49.
153. **Koch, A., M. Melbye, P. Sørensen, P. Homøe, H. O. Madsen, K. Mølbak, C. H. Hansen, L. H. Andersen, G. W. Hahn, and P. Garred.** 2001. Acute respiratory tract infections and mannose-binding lectin insufficiency during early childhood. *JAMA* **285**:1316–1321.
154. **Kotton, C. N.** 2007. Zoonoses in solid-organ and hematopoietic stem cell transplant recipients. *Clin. Infect. Dis.* **44**:857–866.
155. **Koziel, H., and M. J. Koziel.** 1995. Pulmonary complication of diabetes mellitus: pneumonia. *Infect. Dis. Clin. N. Am.* **9**:65–96.
156. **Kreis, H., J. M. Cisterne, W. Land, L. Wramner, J. P. Squifflet, D. Abramowicz, J. M. Campistol, J. M. Morales, J. M. Grinyo, G. Mourad, F. C. Berthoux, C. Brattström, Y. Lebranchu, and P. Vialtel.** 2000. Sirolimus in association with mycophenolate mofetil induction for the prevention of acute graft rejection in renal allograft recipients. *Transplantation* **69**:1252–1260.
157. **Lang, B., W. Riegel, T. Peters, and H. H. Peter.** 1991. Low dose methotrexate therapy for rheumatoid arthritis complicated by pancytopenia and Pneumocystis carinii pneumonia. *J. Rheumatol.* **18**:1257–1259.
158. **Lee, J.-H., N. R. Slifman, S. K. Gershon, E. T. Edwards, W. D. Schwieterman, J. N. Siegel, R. P. Wise, S. L. Brown, J. N. Udall, Jr., and M. M. Braun.** 2002. Life-threatening histoplasmosis complicating immunotherapy with tumor necrosis factor α antagonists infliximab and etanercept. *Arthritis Rheum.* **46**:2565–2570.
159. **Lim, W. H., G. R. Russ, and P. T. H. Coates.** 2006. Review of Epstein-Barr virus and post-transplant lymphoproliferative disorder post-solid organ transplantation. *Nephrology* **11**:355–356.
160. **Linares, L., C. Cervera, F. Cofán, M. J. Ricart, N. Esforzado, V. Torregrosa, F. Oppenheimer, J. M. Campistol, F. Marco, and A. Moreno.** 2007. Epidemiology and outcomes of multiple antibiotic-resistant bacterial infection in renal transplantation. *Transplant. Proc.* **39**:2222–2224.
161. **Ljungman, P., K. N. Ward, B. N. Crooks, A. Parker, R. Martino, P. J. Shaw, L. Brinch, M. Brune, R. De La Camara, A. Dekker, K. Pauksen, N. Russell, A. P. Schwarer, and C. Cordonnier.** 2001. Respiratory virus infections after stem cell transplantation: a prospective study from the Infectious Diseases Working Party of the European Group for Blood and Marrow Transplantation. *Bone Marrow Transplant.* **28**:479–484.
162. **López-Medrano, F., J. M. Aguado, M. Lizasoain, D. Folgueira, R. S. Juan, C. Díaz-Pedroche, C. Lumbreras, J. M. Morales, J. F. Delgado, and E. Moreno-González.** 2007. Clinical implications of respiratory virus infections in solid organ transplant recipients: a prospective study. *Transplantation* **84**:851–856.
163. **Lubel, J. S., A. G. Testro, and P. W. Angus.** 2007. Hepatitis B virus reactivation following immunosuppressive therapy: guidelines for prevention and management. *Intern. Med. J.* 37:705–712.
164. **Lum, L. G., M. C. Seigneuret, R. F. Storb, R. P. Witherspoon, and E. D. Thomas.** 1981. In vitro regulation of immunoglobulin synthesis after marrow transplantation. I. T-cell and B-cell in patients with and without chronic graft-versus-host-diseases. *Blood 58:*431–439.
165. **Lundin, J., A. Porwit-MacDonald, E. D. Rossmann, C. Karlsson, P. Edman, M. R. Rezvany, E. Kimby, A. Osterborg, and H. Mellstedt.** 2004. Cellular immune reconstitution after subcutaneous alemtuzumab (anti-CD52 monocloncal antibody, CAMPATH-1H) treatment of first-line therapy for B-cell chronic lymphocytic leukemia. *Leukemia* **18**:484–490.
166. **Lynn, J., A. K. Knight, M. Kamoun, and A. I. Levinson.** 2004. A 55-year-old man with hypogammaglobulinemia, lymphopenia, and unrelenting cutaneous warts. *J. Allergy Clin. Immunol. 114:*409–414.
167. **MacDonald, A. S., et al.** 2001. A worldwide, phase III, randomized, controlled, safety and efficacy study of a sirolimus/cyclosporine regimen for prevention of acute rejection in recipients of primary mismatched renal allografts. *Transplantation* **71**:271–280.
168. **Machado, C. M., F. B. Gonçalves, C. S. Pannuti, F. L. Dulley, and V. A. de Souza.** 2002. Measles in bone marrow transplant recipients during an outbreak in São Paulo, Brazil. *Blood* **99**:83–87.
169. **Mackall, C.** 2000. T-cell immunodeficiency following cytotoxic antineoplastic therapy: a review. *Stem Cells* **18**:10–18.
170. **Mackall, C. L., T. A. Fleisher, M. R. Brown, I. T. Magrath, A. T. Shad, M. E. Horowitz, L. H. Wexler, M. A. Adde, L. L. McClure, and R. E. Gress.** 1994. Lymphocyte depletion during treatment with intensive chemotherapy for cancer. *Blood* **84**:2221–2228.
171. **Maloney, D. G., B. Smith, and A. Rose.** 2002. Rituximab: mechanism of action and resistance. *Semin. Oncol.* **29** (1 Suppl. 2):2–9.
172. **Mangi, R. J., J. C. Niederman, J. E. Kelleher, Jr., J. M. Dwyer, A. S. Evans, and F. S. Kantor.** 1974. Depression of cell-mediated immunity during acute infectious mononucleosis. *N. Engl. J. Med.* **291**:1149–1153.
173. **Mannick, J. A., M. L. Rodrick, and J. A. Lederer.** 2001. The immunologic response to injury. *J. Am. Coll. Surg.* **193**: 237–244.
174. **Marder, W., and W. J. McCune.** 2000. Advances in immunosuppressive therapy. *Semin. Respir. Crit. Care Med.* **28**:398–417.
175. **Maris, M., M. Boeckh, B. Storer, M. Dawson, K. White, M. Keng, B. Sandmaier, D. Maloney, R. Storb, and J. Storek.** 2003. Immunologic recovery after hematopoietic cell transplantation with nonmyeloablative conditioning. *Exp. Hematol.* **31**:941–952.
176. **Marr, K. A.** 2004. Invasive candida infections: the changing epidemiology. *Oncology* **18**:9–14.
177. **Marr, K. A., and R. A. Bowden.** 1999. Fungal infections in patients undergoing blood and marrow transplantation. *Transpl. Infect. Dis.* **1**:237–246.
178. **Marr, K. A., K. Seidel, T. C. White, and R. A. Bowden.** 2000. Candidemia in allogeneic blood and marrow transplant recipients: evolution of risk factors after the adoption of prophylactic fluconazole. *J. Infect. Dis.* **181**:309–316.
179. **Martin, S. I., F. M. Marty, K. Fiumara, S. P. Treon, J. G. Gribben, and L. R. Baden.** 2006. Infectious complications associated with alemtuzumab use for lymphoproliferative disorders. *Clin. Infect. Dis.* **43**:16–24.

180. **Maruani, A., E. Wierzbicka, M. C. Machet, M. Abdallah-Lotf, A. de Muret, and L. Machet.** 2007. Reversal of multifocal cutaneous lymphoproliferative disease associated with Epstein-Barr virus after withdrawal of methotrexate therapy for rheumatoid arthritis. *J. Am. Acad. Dermatol.* **57**(Suppl. 5):S69–S71.
181. **Massry, S., and M. Smogorzewski.** 2001. Dysfunction of polymorphonuclear leukocytes in uremia: role of parathyroid hormone. *Kidney Int.* **59**(Suppl. 78):S195–S196.
182. **Matsumoto, Y., T. Shinzato, I. Amano, I. Takai, Y. Kimura, H. Morita, M. Miwa, K. Nakane, Y. Yoshikai, and K. Maeda.** 1995. Relationship between susceptibility to apoptosis and Fas expression in peripheral blood T cells from uremic patients: a possible mechanism for lymphopenia in chronic renal failure. *Biochem. Biophys. Res. Commun.* **215**:98–105.
183. **McCurry, K. R., A. Iacono, A. Zeevi, S. Yousem, A. Girnita, S. Husain, D. Zaldonis, B. Johnson, B. G. Hattler, and T. E. Starzl.** 2005. Early outcomes in human lung transplantation with thymoglobulin or Campath-1H for recipient pretreatment followed by posttransplant tacrolimus near-monotherapy. *J. Thorac. Cardiovasc. Surg.* **130**:528–537.
184. **McKinney, R. E., Jr., S. L. Katz, and C. M. Wilfert.** 1987. Chronic enteroviral meningoencephalitis in agammaglobulinemic patients. *Rev. Infect. Dis.* **9**:334–356.
185. **McNeil, S. A., P. N. Malani, C. E. Chenoweth, R. J. Fontana, J. C. Magee, J. D. Punch, M. L. Mackin, and C. A. Kauffman.** 2006. Vancomycin-resistant enterococcal colonization and infection in liver transplant candidates and recipients: a prospective surveillance study. *Clin. Infect. Dis.* **42**:195–203.
186. **Medzhitov, R., and C. Janeway, Jr.** 2000. The Toll receptor family and microbial recognition. *Trends Microbiol.* **8**:452–456.
187. **Mehrad, B., R. M. Strieter, and T. J. Standiford.** 1999. Role of TNF-alpha in pulmonary host defense in murine invasive aspergillosis. *J. Immunol.* **162**:1633–1640.
188. **Meier-Kriesche, H. U., S. Li, R. W. Gruessner, J. J. Fung, R. T. Bustami, M. L. Barr, and A. B. Leichtman.** 2006. Immunosuppression: evolution in practice and trends, 1994–2004. *Am. J. Transplant.* **6**:1111–1131.
189. **Mendonca, M. A. O., F. Q. Cunha, E. F. C. Murta, and B. M. Tavares-Murta.** 2006. Failure of neutrophil chemotactic function in breast cancer patients treated with chemotherapy. *Cancer Chemother. Pharmacol.* **57**:663–670.
190. **Miller, D. L.** 1964. Frequency of complications of measles. *Br. Med. J.* **2**:75–78.
191. **Miller, G. G., and J. S. Dummer.** 2007. Herpes simplex and varicella zoster viruses: forgotten but not gone. *Am. J. Transplant.* **7**:741–747.
192. **Mocarski, E. S., Jr.** 2002. Immunomodulation by cytomegaloviruses: manipulative strategies beyond evasion. *Trends Microbiol.* **10**:332–339.
193. **Molrine, D. C., G. R. Siber, Y. Samra, D. S. Shevy, K. MacDonald, R. Cieri, and D. M. Ambrosino.** 1999. Normal IgG and impaired IgM responses to polysaccharide vaccines in asplenic patients. *J. Infect. Dis.* **179**:513–517.
194. **Moore, K. W., P. Vieira, D. F. Fiorentino, M. L. Trounstine, T. A. Khan, and T. R. Mosmann.** 1990. Homology of cytokine synthesis inhibitory factor (IL-10) to the Epstein-Barr virus gene BCRFI. *Science* **248**:1230–1234.
195. **Moore, R. D., and R. E. Chaisson.** 1996. Natural history of opportunistic disease in an HIV-infected urban clinical cohort. *Ann. Intern. Med.* **124**:633–642.
196. **Muller, L. M., K. J. Gorter, E. Hak, W. L. Goudzwaard, F. G. Schellevis, A. I. Hoepelman, and G. E. Rutten.** 2005. Increased risk of common infections in patients with type 1 and type 2 diabetes mellitus. *Clin. Infect. Dis.* **41**:281–288.
197. **Mullooly, J. P., M. D. Bennett, M. C. Hornbrook, W. H. Barker, W. W. Williams, P. A. Patriarca, and P. H. Rhodes.** 1994. Influenza vaccination programs for elderly persons: cost-effectiveness in a health maintenance organization. *Ann. Intern. Med.* **121**:947–952.
198. **Munier, M. L., and A. D. Kelleher.** 2007. Acutely dysregulated, chronically disabled by the enemy within: T-cell responses to HIV-1 infection. *Immunol. Cell Biol.* **85**:6–15.
199. **Munoz, A. I., and D. Limbert.** 1977. Skin reactivity to Candida and streptokinase-streptodornase antigens in normal pediatric subjects: influence of age and acute illness. *J. Pediatr.* **91**:565–568.
200. **Munoz, P., C. Rodriguez, and E. Bouza.** 2005. Mycobacterium tuberculosis infection in recipients of solid organ transplants. *Clin. Infect. Dis.* **40**:581–587.
201. **Murdoch, D. M., W. D. Venter, A. Van Rie, and C. Feldman.** 2007. Immune reconstitution inflammatory syndrome (IRIS): review of common infectious manifestations and treatment options. *AIDS Res. Ther.* **8**:4–9.
202. **Nachbaur, D., C. Larcher, B. Kircher, G. Eibl, W. Nussbaumer, E. Gunsilius, M. Haun, K. Grunewald, and G. Gastl.** 2003. Risk for cytomegalovirus infection following reduced intensity allogeneic stem cell transplantation. *Ann. Hematol.* **82**:621–627.
203. **Narreddy, S., S. Mellon-Reppen, M. H. Abidi, J. L. Klein, E. Peres, L. K. Heilbrun, D. Smith, G. Alangaden, and P. H. Chandrasekar.** 2007. Non-bacterial infections in allogeneic non-myeloablative stem cell transplant recipients. *Transpl. Infect. Dis.* **9**:3–10.
204. **Nashan, B., S. Light, I. R. Hardie, A. Lin, and J. R. Johnson.** 2000. Reduction of acute renal allograft rejection by daclizumab. *Lancet* **356**:194–202.
205. **Nelson, D. R., C. Soldevila-Pico, A. Reed, M. F. Abdelmalek, A. W. Hemming, W. J. Van der Werf, R. Howard, and G. L. Davis.** 2001. Anti-interleukin-2 receptor therapy in combination with mycophenolate mofetil is associated with more severe hepatitis C recurrence after liver transplantation. *Liver Transplant.* **7**:1064–1070.
206. **Neuhaus, P., R. Pichlmayr, and R. Williams.** 1994. Randomised trial comparing tacrolimus (FK506) and cyclosporine in prevention of liver allograft rejection. *Lancet* **344**:423–428.
207. **Neuman, M. G., J. P. Benhamou, I. M. Malkiewicz, R. Akremi, N. H. Shear, T. Asselah, A. Ibrahim, N. Boyer, M. Martinot-Peignoux, P. Jacobson-Brown, G. G. Katz, V. Le Breton, G. Le Guludec, A. Suneja, and P. Marcellin.** 2001. Cytokines as predictors for sustained response and as markers for immunomodulation in patients with chronic hepatitis C. *Clin. Biochem.* **34**:173–182.
208. **Ng, A. P., L. Worth, L. Chen, J. F. Seymour, H. M. Prince, M. Slavin, and K. Thursky.** 2005. Cytomegalovirus DNAemia and disease: incidence, natural history and management in setting other than allogeneic stem cell transplantation. *Haematologica* **90**:1672–1679.
209. **Nichols, W. G., K. A. Guthrie, L. Corey, and M. Boeckh.** 2004. Influenza infections after hematopoietic stem cell transplantation: risk factors, mortality, and the effect of antiviral therapy. *Clin. Infect. Dis.* **39**:1300–1306.
210. **Ochs, H. D., and A. J. Thrasher.** 2006. The Wiskott-Aldrich syndrome. *J. Allergy Clin. Immunol.* **117**:725–738.
211. **Ohri, L. K., J. M. Manley, A. Chatterjee, and N. E. Cornish.** 2004. Pediatric case series evaluating a standardized Candida albicans skin test product. *Ann. Pharmacother.* **38**:973–977.

212. **Ortiz, Z., B. Shea, M. E. Suarez-Almazor, D. Moher, G. A. Wells, and P. Tugwell.** 1998. The efficacy of folic acid and folinic acid in reducing methotrexate gastrointestinal toxicity in rheumatoid arthritis. A meta-analysis of randomized controlled trials. *J. Rheumatol.* **25**:36–43.
213. **Oz, H. S., and W. T. Hughes.** 1997. Novel anti-Pneumocystis carinii effects of the immunosuppressant mycophenolate mofetil in contrast to the provocative effects of tacrolimus, sirolimus, and dexamethasone. *J. Infect. Dis.* **175**:901–904.
214. **Pabst, H. F., and H. W. Kreth.** 1980. Ontogeny of the immune response as a basis of childhood disease. *J. Pediatr.* **97**:519–534.
215. **Pachiadakis, I., G. Pollara, B. M. Chain, and N. V. Naoumov.** 2005. Is hepatitis C infection of dendritic cells a mechanism of facilitating viral persistence. *Lancet Infect. Dis.* **5**:298–304.
216. **Paliogianni, F., S. S. Ahuja, J. P. Balow, J. E. Balow, and D. T. Boumpas.** 1993. Novel mechanism for inhibition of human T cells by glucocorticoids: glucocorticoids inhibit signal transduction through IL-2 receptor. *J. Immunol.* **151**:4081–4089.
217. **Palmer, S. M., M. A. Baz, L. Sanders, A. P. Miralles, C. M. Lawrence, J. B. Rea, D. S. Zander, L. J. Edwards, E. D. Staples, V. F. Tapson, and R. D. Davis.** 2001. Results of a randomized, prospective, multicenter trial of mycophenolate mofetil versus azathioprine in the prevention of acute lung allograft rejection. *Transplantation* **71**:1772–1776.
218. **Pappas, P. G.** 2006. Invasive candidiasis. *Infect. Dis. Clin. N. Am.* **20**:485–506.
219. **Patriarca, P. A., J. A. Weber, R. A. Parker, W. N. Hall, A. P. Kendal, D. J. Bregman, and L. B. Schonberger.** 1985. Efficacy of influenza vaccine in nursing homes: reduction in illness and complications during an influenza A (H3N2) epidemic. *JAMA* **253**:1136–1139.
220. **Peleg, A. Y., S. Husain, E. J. Kwak, F. P. Silveira, M. Ndirangu, J. Tran, K. A. Shutt, R. Shapiro, N. Thai, K. Abu-Elmagd, K. R. McCurry, A. Marcos, and D. L. Paterson.** 2007. Opportunistic infections in 547 organ transplant recipients receiving alemtuzumab, a humanized monoclonal CD-52 antibody. *Clin. Infect. Dis.* **44**:204–212.
221. **Pelosini, M., D. Focosi, F. Rita, S. Galimberti, F. Caracciolo, E. Benedetti, F. Papineschi, and M. Petrini.** 2008. Progressive multifocal leukoencephalopathy: report of three cases in HIV-negative hematological patients and review of the literature. *Ann. Hematol.* **87**:405–412.
222. **Pereyra, F., and R. H. Rubin.** 2004. Prevention and treatment of cytomegalovirus infection in solid organ transplant recipients. *Curr. Opin. Infect. Dis.* **17**:357–361.
223. **Perreault, C., M. Giasson, M. Gyger, R. Belanger, M. David, Y. Bonny, J. Boileau, R. Barcelo, and J. P. Moguin.** 1985. Serum immunoglobulin levels following allogeneic bone marrow transplantation. *Blut* **51**:137–142.
224. **Plebani, A., A. Soresina, R. Rondelli, G. M. Amato, C. Azzari, F. Cardinale, G. Cazzola, R. Consolini, D. De Mattia, G. Dell'Erba, M. Duse, M. Fiorini, S. Martino, B. Martire, M. Masi, V. Monafo, V. Moschese, L. D. Notarangelo, P. Orlandi, P. Panei, A. Pession, M. C. Pietrogrande, C. Pignata, I. Quinti, V. Ragno, P. Rossi, A. Sciotto, A. Stabile, et al.** 2002. Clinical, immunological, and molecular analysis in a large cohort of patients with X-linked agammaglobulinemia: an Italian multicenter study. *Clin. Immunol.* **104**:221–230.
225. **Pratt, G., O. Goodyear, and P. Moss.** 2007. Immunodeficiency and immunotherapy in multiple myeloma. *Br. J. Haematol.* **138**:563–579.
226. **Pryor, B. D., S. G. Bologna, and L. E. Kahl.** 1996. Risk factors for serious infection during treatment with cyclophosphamide and high-dose corticosteroids for systemic lupus erythematosus. *Arthritis Rheum.* **39**:1475–1482.
227. **Quartier, P., S. Foray, J. L. Casanova, I. Hau-Rainsard, S. Blanche, and A. Fischer.** 1987. Enteroviral meningoencephalitis in X-linked agammaglobulinemia: intensive immunoglobulin therapy and sequential viral detection in cerebrospinal fluid by polymerase chain reaction. *Rev. Infect. Dis.* **9**:334–356.
228. **Ramirez, C. B., and I. R. Marino.** 2007. The role of basiliximab induction therapy in organ transplantation. *Expert Opin. Biol. Ther.* **7**:137–148.
229. **Ranua, J., K. Luoma, A. Auvinen, J. Peltola, A. M. Haapala, J. Raitanen, and J. Isojärvi.** 2005. Serum IgA, IgG, and IgM concentrations in patients with epilepsy and matched controls: a cohort-based cross-sectional study. *Epilepsy Behav.* **6**:191–195.
230. **Raulet, D. H.** 2004. Interplay of natural killer cells and their receptors with the adaptive immune response. *Nat. Immunol.* **5**:996–1002.
231. **Rawstron, A. C., B. Kennedy, P. Moreton, A. J. Dickinson, M. J. Cullen, S. J. Richards, A. S. Jack, and P. Hillmen.** 2004. Early prediction of outcome and response to alemtuzumab therapy in chronic lymphocytic leukemia. *Blood* **103**:2027–2031.
232. **Rhen, T., and J. A. Cidlowski.** 2005. Antiinflammatory action of glucocorticoids—new mechanisms for old drugs. *N. Engl. J. Med.* **353**:1711–1723.
233. **Rimola, A., R. Soto, F. Bory, V. Arroyo, C. Piera, and J. Rodes.** 1984. Reticuloendothelial cell system phagocytic activity in cirrhosis and its relation to bacterial infection and prognosis. *Hepatology* **4**:53–58.
234. **Rinehart, J. J., S. P. Balcerzak, A. L. Sagone, and A. F. LoBuglio.** 1974. Effects of glucocorticoids on human monocyte function. *J. Clin. Investig.* **54**:1337–1343.
235. **Rogers, N. M., C. A. Peh, R. Faull, M. Pannell, J. Cooper, and G. R. Russ.** 2008. Transmission of toxoplasmosis in two renal allograft recipients receiving an organ from the same donor. *Transpl. Infect Dis.* **10**:71–74.
236. **Romero-Steiner, S., D. M. Musher, M. S. Cetron, L. B. Pais, J. E. Groover, A. E. Fiore, B. D. Plikaytis, and G. M. Carlone.** 1999. Reduction in functional antibody activity against Streptococcus pneumoniae in vaccinated elderly individuals highly correlates with decreased IgG antibody avidity. *Clin. Infect. Dis.* **29**:281–288.
237. **Rosenzweig, S. D., and S. M. Holland.** 2004. Phagocyte immunodeficiencies and their infections. *J. Allergy Clin. Immunol.* **113**:620–626.
238. **Ross, S. C., and P. Densen.** 1984. Complement deficiency states and infection: epidemiology, pathogenesis and consequences of neisserial and other infections in an immune deficiency. *Medicine* (Baltimore) *63*:243–273.
239. **Rubin, R. H.** 2002. Infection in the organ transplant recipient, p. 573–679. *In* R. H. Rubin and L. S. Young (ed.), *Clinical Approach to Infection in the Compromised Host*, 4th ed. Kluwer Academic/Plenum Publishers, New York, NY.
240. **Rubin, R. H.** 2007. The pathogenesis and clinical management of cytomegalovirus infection in the organ transplant recipient: the end of the 'silo hypothesis.' *Curr. Opin. Infect. Dis.* **20**:399–407.
241. **Rubins, J. B., A. K. Puri, J. Loch, D. Charboneau, R. MacDonald, N. Opstad, and E. N. Janoff.** 1998. Magnitude, duration, quality, and function of pneumococcal vaccine responses in elderly adults. *J. Infect. Dis.* **178**:431–440.
242. **Ruiz, I., J. Gavalda, V. Montforte, O. Len, A. Roman, C. Bravo, A. Ferrer, L. Tenorio, F. Roman, J. Maestre,**

I. Molina, F. Morell, and A. Pahissa. 2006. Donor-to-host transmission of bacterial and fungal infections in lung transplantation. *Am. J. Transplant.* **6**:178–182.

243. **Saag, K. G., R. Koehnke, J. R. Caldwell, R. Brasington, L. F. Burmeister, B. Zimmerman, J. A. Kohler, and D. E. Furst.** 1994. Low dose long-term glucocorticoid therapy in rheumatoid arthritis: an analysis of serious adverse events. *Am. J. Med.* 96:115–123.
244. **Safdar, N., A. Said, M. R. Lucey, S. J. Knechtle, A. D'Alessandro, A. Musat, J. Pirsch, J. McDermott, M. Kalayoglu, and D. G. Maki.** 2004. Infected bilomas in liver transplant recipients: clinical features, optimal management, and risk factors for mortality. *Clin. Infect. Dis.* **39**: 517–525.
245. **Sakoda, Y., D. Hashimoto, S. Asakura, K. Takeuchi, M. Harada, M. Tanimoto, and T. Teshima.** 2007. Donor-derived thymic-dependent T cells cause chronic graft-versus-host disease. *Blood* **109**:1756–1764.
246. **Santos, G. W.** 1971. Application of marrow grafts in human disease. *Am. J. Pathol.* **65**:658–668.
247. **Sarmiento, J. M., D. H. Dockrell, T. R. Schwab, S. R. Munn, and C. V. Paya.** 2000. Mycophenolate mofetil increases cytomegalovirus invasive organ disease in renal transplant patients. *Clin. Transplant.* **14**:136–138.
248. **Sarnak, M. J., and B. L. Jaber.** 2000. Mortality caused by sepsis in patients with end-stage renal disease compared with the general population. *Kidney Int.* **58**:1758–1764.
249. **Saxon, A., R. E. McIntyre, R. H. Stevens, and R. P. Gale.** 1981. Lymphocyte dysfunction in chronic graft versus host disease. *Blood* **58**:746–751.
250. **Scheinman, R. I., P. C. Cogswell, A. Lofquist, and A. S. Balwin, Jr.** 1995. Role of transcriptional activation of I kappa B alpha in mediation of immunosuppression by glucocorticoids. *Science* **270**:283–286.
251. **Schirren, C. A., M. C. Jung, R. Zachoval, H. Diepolder, R. Hoffmann, G. Riethmuller, and G. R. Pape.** 1997. Analysis of T cell activation pathways in patients with liver cirrhosis, impaired delayed hypersensitivity and other T cell-dependent functions. *Clin. Exp. Immunol.* **108**:144–150.
252. **Schneider, C. R., J. F. Buell, M. Gearhart, M. Thomas, M. J. Hanaway, S. M. Rudich, and E. S. Woodle.** 2005. Methicillin-resistant Staphylococcus aureus infection in liver transplantation: a matched controlled study. *Transplant. Proc.* **37**:1243–1244.
253. **Schwacha, M. G.** 2003. Macrophages and post-burn dysfunction. *Burns* **29**:1–14.
254. **Seager, J., D. L. Jamison, J. Wilson, A. R. Hayward, and J. F. Soothill.** 1975. IgA deficiency, epilepsy, and phenytoin treatment. *Lancet* **ii**:632–635.
255. **Seemayer, T. A., T. G. Gross, R. M. Egeler, S. J. Pirruccello, J. R. Davis, C. M. Kelly, M. Okano, A. Lanyi, and J. Sumegi.** 1995. X-linked lymphoproliferative disease: twenty-five years after the discovery. *Pediatr. Res.* **38**:471–478.
256. **Segal, B. H., and M. C. Sneller.** 1997. Infectious complications of immunosuppressive therapy in patients with rheumatic diseases. *Rheum. Dis. Clin. N. Am.* **23**:219–237.
257. **Shitrit, D., N. Lev, A. Bar-Gil-Shitrit, and M. R. Kramer.** 2005. Progressive multifocal leukoencephalopathy in transplant recipients. *Transpl. Int.* **11**:658–665.
258. **Shodell, M., K. Shah, and F. P. Siegal.** 2003. Circulating human plasmacytoid dendritic cells are highly sensitive to corticosteroid administration. *Lupus* **12**:222–230.
259. **Silveira, F. P., and S. Husain.** 2007. Fungal infections in solid organ transplantation. *Med. Mycol.* **45**:305–320.
260. **Singer, D. B.** 1973. Postsplenectomy sepsis. *Perspect. Pediatr. Pathol.* **1**:285–311.
261. **Singh, N.** 2002. Impact of donor bacteremia on outcome in organ transplant recipients. *Liver Transpl.* **8**:975–976.
262. **Singh, N.** 2005. Infectious complications in organ transplant recipients with the use of calcineurin-inhibitor agent-based immunosuppressive regimens. *Curr. Opin. Infect. Dis.* **18**:342–345.
263. **Singh, N.** 2005. Interactions between viruses in transplant recipients. *Clin. Infect. Dis.* **40**:430–436.
264. **Singh, N., B. D. Alexander, O. Lortholary, F. Dromer, K. L. Gupta, G. T. John, R. del Busto, G. B. Klintmalm, J. Somani, G. M. Lyon, K. Pursell, V. Stosor, P. Munoz, A. P. Limaye, A. C. Kalil, T. L. Pruett, J. Garcia-Diaz, A. Humar, S. Houston, A. A. House, D. Wray, S. Orloff, L. A. Dowdy, R. A. Fisher, J. Heitman, M. W. Wagener, S. Husain, and the Cryptococcal Collaborative Transplant Study Group.** 2007. Cryptococcus neoformans in organ transplant recipients: impact of calcineurin-inhibitor agents on mortality. *J. Infect. Dis.* **195**:756–764.
265. **Singh, N., R. K. Avery, P. Munoz, T. L. Pruett, B. Alexander, R. Jacobs, J. G. Tollemar, E. A. Dominguez, C. M. Yu, D. L. Paterson, S. Husain, S. Kusne, and P. Linden.** 2003. Trends in risk profiles for and mortality associated with invasive aspergillosis among liver transplant recipients. *Clin. Infect. Dis.* **36**:46–52.
266. **Singh, N., L. Mieles, V. L. Yu, and T. E. Starzl.** 1994. Decreased incidence of viral infections in liver transplant recipients: possible effects of FK506. *Dig. Dis. Sci.* **39**: 15–18.
267. **Singh, N., and J. Perfect.** 2007. Immune reconstitution syndrome associated with opportunistic mycoses. *Lancet Infect Dis.* **7**:395–401.
268. **Skokowa, J., M. Germeshausen, C. Zeidler, and K. Welte.** 2007. Severe congenital neutropenia: inheritance and pathophysiology. *Curr. Opin. Hematol.* **14**:22–28.
269. **Small, T. N., A. Casson, S. F. Malak, F. Boulad, T. E. Kiehn, J. Stiles, H. M. Ushay, and K. A. Sepkowitz.** 2002. Respiratory syncytial virus infection following hematopoietic stem cell transplantation. *Bone Marrow Transplant.* **29**:321–327.
270. **Sollinger, H. W., et al.** 1995. Mycophenolate mofetil for the prevention of acute rejection in primary cadaveric renal allograft recipients. *Transplantation* **60**:225–232.
271. **Srinivasan, A., E. C. Burton, M. J. Kuehnert, C. Rupprecht, W. L. Sutker, T. G. Ksiazek, C. D. Paddock, J. Guarner, W. J. Shieh, C. Goldsmith, C. A. Hanlon, J. Zoretic, B. Fischbach, M. Niezgoda, W. H. El-Feky, L. Orciari, E. Q. Sanchez, A. Likos, G. B. Klintmalm, D. Cardo, J. LeDuc, M. E. Chamberland, D. B. Jernigan, and S. R. Zaki.** 2005. Transmission of rabies virus from an organ donor to four transplant recipients. *N. Engl. J. Med.* **352**:1103–1111.
272. **Stephan, J. L., V. Vlekova, F. Le Deist, S. Blanche, J. Donadieu, G. De Saint-Basile, A. Durandy, C. Griscelli, and A. Fischer.** 1993. Severe combined immunodeficiency: a retrospective single-center study of clinical presentation and outcome in 117 patients. *J. Pediatr.* **123**:564–572.
273. **Strand, V., S. Cohen, M. Schiff, A. Weaver, R. Fleischmann, G. Cannon, R. Fox, L. Moreland, N. Olsen, D. Furst, J. Caldwell, J. Kaine, J. Sharp, F. Hurley, and I. Loew-Friedrich.** 1999. Treatment of active rheumatoid arthritis with leflunomide compared with placebo and methotrexate. *Arch. Intern. Med.* **159**:2542–2550.
274. **Strober, W., and M. C. Sneller.** 1991. IgA deficiency. *Ann. Allergy* **66**:363–375.
275. **Stuck, A. E., C. E. Minder, and F. J. Frey.** 1989. Risk of infectious complications in patients taking glucocorticoids. *Rev. Infect. Dis.* **11**:954–963.

276. **Sullivan, K. M., R. P. Witherspoon, R. Storb, P. Weiden, N. Flournoy, S. Dahlberg, H. J. Deeg, J. E. Sanders, K. C. Doney, F. R. Appelbaum, R. McCuffin, G. B. McDonald, J. Meyers, M. M. Schubert, J. Gauvreau, H. M. Shulman, G. E. Sale, C. Anasetti, T. P. Loughran, S. Strom, J. Nims, and E. D. Thomas.** 1988. Prednisone and azathioprine compared with prednisone and placebo for treatment of chronic graft-v-host disease: prognostic influence of prolonged thrombocytopenia after allogeneic marrow transplantation. *Blood* **72:**546–554.
277. **Swinnen, L. J., M. R. Costanzo-Nordin, S. G. Fisher, E. J. O'Sullivan, M. R. Johnson, A. L. Heroux, G. J. Dizikes, R. Pifarre, and R. I. Fisher.** 1990. Increased incidence of lymphoproliferative disorder after immunosuppression with monoclonal antibody OKT3 in cardiac-transplant recipients. *N. Engl. J. Med.* **323:**1723–1728.
278. **Tamashiro, V. G., H. H. Perez, and D. E. Griffin.** 1987. Prospective study of the magnitude and duration of changes in tuberculin reactivity during uncomplicated and complicated measles. *Pediatr. Infect. Dis. J.* **6:**451–454.
279. **The U.S. Multicenter FK506 Liver Study Group.** 1994. A comparison of tacrolimus FK506 and cyclosporine for immunosuppression in liver transplantation. *N. Engl. J. Med.* **331:**1110–1115.
280. **Thursky, K. A., L. J. Worth, J. F. Seymour, H. M. Prince, and M. A. Slavin.** 2006. Spectrum of infection, risk and recommendations for prophylaxis and screening among patients with lymphoproliferative disorders treated with alemtuzumab. *Br. J. Haemotol.* **132:**3–12.
281. **Tomsic, M.** 2006. Leflunimide-associated tuberculosis. *Rheumatology* **45:**228–229.
282. **Tosi, M. F.** 2005. Innate immune responses to infection. *J. Allergy Clin. Immunol.* **116:**241–249.
283. **Tredgett, E. E., H. A. Shankowsky, R. Rennie, R. E. Burrell, and S. Logsetty.** 2004. Pseudomonas infections in the thermally injured patient. *Burns* **30:**3–26.
284. **True, D. G., M. Penmetcha, and S. J. Peckham.** 2002. Disseminated cryptococcal infection in rheumatoid arthritis treated with methotrexate and infliximab. *J. Rheumatol.* **29:**1561–1563.
285. **Tsiodras, S., G. Samonis, M. J. Keating, and D. P. Kontoyiannis.** 2000. Infection and immunity in chronic lymphocytic leukemia. *Mayo Clin. Proc.* **75:**1039–1054.
286. **Upton, A., and K. A. Marr.** 2006. Emergence of opportunistic mould infections in the hematopoietic stem cell transplant patient. *Curr. Infect. Dis. Rep.* **8:**434–441.
287. **van Burik, J. A., and C. G. Brunstein.** 2007. Infectious complications following unrelated cord blood transplantation. *Vox Sang.* **92:**289–296.
288. **van Burik, J. A., S. L. Carter, A. G. Freifeld, K. P. High, K. T. Godder, G. A. Papanicolaou, A. M. Mendizabal, J. E. Wagner, S. Yanovich, and N. A. Kernan.** 2007. Higher risk of cytomegalovirus and aspergillus infections in recipients of T cell-depleted unrelated bone marrow: analysis of infectious complications in patients treated with T cell depletion versus immunosuppressive therapy to prevent graft-versus-host disease. *Biol. Blood Marrow Transplant.* **13:**1487–1498.
289. **Vassalli, P.** 1992. The pathophysiology of tumor necrosis factors. *Annu. Rev. Immunol.* **10:**411–452.
290. **Vincenti, F., R. Kirkman, S. Light, G. Bumgardner, M. Pescovitz, P. Halloran, J. Neylan, A. Wilkinson, H. Ekberg, R. Gaston, L. Backman, and J. Burdick.** 1998. Interleukin-2-receptor blockade with daclizumab to prevent acute rejection in renal transplantation. *N. Engl. J. Med.* **338:**161–165.
291. **Viscoli, C., O. Varnier, and M. Machetti.** 2005. Infections in patients with febrile neutropenia: epidemiology, microbiology, and risk stratification. *Clin. Infect. Dis.* **40**(Suppl. 4):S240–S245.
292. **Wadhwa, P. D., and V. A. Morrison.** 2006. Infectious complications of chronic lymphocytic leukemia. *Semin. Oncol.* **33:**240–249.
293. **Waldman, M., and J. B. Kopp.** 2007. Parvovirus-B19-associated complications in renal transplant recipients. *Nat. Clin. Pract. Nephrol.* **3:**540–550.
294. **Wallis, R. S., M. Broder, J. Wong, A. Lee, and L. Hoq.** 2005. Reactivation of latent granulomatous infections by infliximab. *Clin. Infect. Dis.* **41**(Suppl. 3):S194–S198.
295. **Walport, M. J.** 2001. Complement. First of two parts. *N. Engl. J. Med.* **344:**1058–1066.
296. **Walport, M. J.** 2001. Complement. Second of two parts. *N. Engl. J. Med.* **344:**1140–1144.
297. **Warris, A., A. Bjørneklett, and P. Gaustad.** 2001. Invasive pulmonary aspergillosis associated with infliximab therapy. *N. Engl. J. Med.* **344:**1099–1100.
298. **Watson, C. J., J. A. Bradley, P. J. Friend, J. Firth, C. J. Taylor, J. R. Bradley, K. G. Smith, S. Thiru, N. V. Jamieson, G. Hale, H. Waldmann, and R. Calne.** 2005. Alemtuzumab (CAMPATH 1H) induction therapy in cadaveric kidney transplantation—efficacy and safety at five years. *Am. J. Transplant.* **5:**1347–1353.
299. **Wen, L., J. P. Atkinson, and P. C. Giclas.** 2004. Clinical and laboratory evaluation of complement deficiency. *J. Allergy Clin. Immunol.* **113:**585–593.
300. Reference deleted.
301. **Wiesner, R., J. Rabkin, G. Klintmalm, S. McDiarmid, A. Langnas, J. Punch, P. McMaster, M. Klayoglu, G. Levy, R. Freeman, H. Bismuth, P. Neuhaus, R. Mamelok, and W. Wang.** 2001. A randomized double-blind comparative study of mycophenolate mofetil and azathioprine, in combination with cyclosporine and corticosteroids in primary liver transplant recipients. *Liver Transplant.* **7:**442–450.
302. **Wingard, J. R.** 1993. Infections in allogeneic bone marrow transplant recipients. *Semin. Oncol.* **20:**80–87.
303. **Wingard, J. R.** 1999. Opportunistic infections after blood and marrow transplantation. *Transpl. Infect. Dis.* **1:** 3–20.
304. **Wingard, J. R., D. Y. Chen, W. H. Burns, D. J. Fuller, H. G. Braine, A. M. Yeager, H. Kaiser, P. J. Burke, M. L. Graham, and G. W. Santos.** 1988. Cytomegalovirus infection after autologous bone marrow transplantation with comparison to infection after allogeneic bone marrow transplantation. *Blood* **71:**1432–1437.
305. **Winkelstein, J. A., M. C. Marino, R. B. Johnston, Jr., J. Boyle, J. Curnutte, J. I. Gallin, H. L. Malech, S. M. Holland, H. Ochs, P. Quie, R. H. Buckley, C. B. Foster, S. J. Chanock, and H. Dickler.** 2000. Chronic granulomatous disease. Report on a national registry of 368 patients. *Medicine* (Baltimore) **79:**155–169.
306. **Winkelstein, J. A., M. C. Marino, H. M. Lederman, S. M. Jones, K. Sullivan, A. W. Burks, M. E. Conley, C. Cunningham-Rundles, and H. D. Ochs.** 2006. X-linked agammaglobulinemia: report on a United States registry of 201 patients. *Medicine* (Baltimore) **85:**193–202.
307. **Wisplinghoff, H., H. Seifert, R. P. Wenzel, and M. B. Edmond.** 2003. Current trends in the epidemiology of nosocomial bloodstream infections in patients with hematologic malignancies and solid neoplasms in hospitals in the United States. *Clin. Infect. Dis.* **36:**1103–1110.

308. **Worth, L. J., and K. A. Thursky.** 2006. Cytomegalovirus reactivation in patients with chronic lymphocytic leukemia treated with alemtuzumab: prophylaxis vs. pre-emptive strategies for prevention. *Leuk. Lymphoma* **47:**2435–2436.
309. **Yoon, J.-W., S. Gollapudi, M. V. Pahl, and N. D. Vaziri.** 2006. Naïve and central memory T-cell lymphopenia in end-stage renal disease. *Kidney Int.* **70:**371–376.
310. **Zilliox, M. J., W. J. Moss, and D. E. Griffin.** 2007. Gene expression changes in peripheral blood mononuclear cells during measles virus infection. *Clin. Vaccine Immunol.* **14:**918–923.
311. **Zur Hausen, H.** 2000. Papillomaviruses causing cancer: evasion from host-cell control in early events in carcinogenesis. *J. Natl. Cancer Inst.* **92:**690–698.

II. LABORATORY DIAGNOSIS: SPECIFIC ETIOLOGIC AGENTS

Diagnostic Microbiology of the Immunocompromised Host
Edited by Randall T. Hayden, Karen C. Carroll, Yi-Wei Tang, and Donna M. Wolk

Chapter 2

Human Immunodeficiency Virus

YUN F. (WAYNE) WANG, MOLLY E. EATON, AUDREY N. SCHUETZ, AND STEVEN R. NESHEIM

Infection with the virus which would later be identified as human immunodeficiency virus type 1 (HIV-1) was first reported in the United States in 1981 (10, 11, 37). HIV-1 was subsequently isolated in 1983, and HIV-2 was isolated in 1986. Both HIV-1 and HIV-2 are retroviruses that cause worldwide infection and AIDS. Since initial reports, more than 25 million persons have succumbed to the infection and more than 50 million are currently living with HIV/AIDS (107). HIV-1 infection has become the fourth most common cause of death worldwide and the most common cause of death in Africa, surpassing respiratory infections, malaria, and diarrheal diseases (107). Research has elucidated much about the molecular and cellular biology of this retrovirus. Successful antiretroviral (ARV) treatment has made HIV/AIDS a manageable chronic disease in developed countries; however, cure and vaccine prevention remain elusive.

Since AIDS was first reported 26 years ago, more scientific effort at both the national and international levels has been focused on HIV than on any other virus in modern history. There is still no vaccine available for prevention of HIV infection, but multiple ARV drugs are available for combination therapy to control the infection. Many laboratory technologies, including conventional or rapid immunological and molecular methods, have been developed and used for diagnosis of HIV infection, surveillance, epidemiology, and monitoring of therapy (19, 116). It is thus important to summarize the available technologies and methods in the clinical or diagnostic laboratory setting, particularly those tests approved by the U.S. Food and Drug Administration (FDA) or approved for CE marking, which is a mandatory European mark for products. Tests for initial diagnosis of infection include enzyme immunoassay (EIA) or enzyme-linked immunosorbent assay (ELISA), Western blotting (WB) and other confirmatory assays, rapid immunoassay, quantitative viral RNA detection assays (commonly known as viral load [VL] assays), and qualitative proviral DNA assays. Tests for determining prognosis and for monitoring of ARV therapy include VL assays, genotyping, and phenotyping assays. Many of these tests are reviewed only briefly here due to space limitations.

BACKGROUND/CLINICAL INFORMATION ON HIV/AIDS

HIV-1 infection occurs following exposure to infected secretions, most commonly blood or semen, as well as through mother-to-child transmission. At the target mucosal site, the virus infects local T lymphocytes and binds to the surface receptors on Langerhans cells. These dendritic cells then carry HIV-1 to encounter the T-cell-rich environment of the local lymph nodes. Here the dendritic cells produce costimulatory molecules and present HIV-1 to activated $CD4^+$ T lymphocytes that have been primed for productive infection by HIV-1. Transport to local lymph nodes can occur within hours of mucosal exposure. Viremia can be detected within 4 to 9 days following exposure, as shown in Fig. 1 (77, 103).

Clinically, acute HIV-1 infection presents as a nonspecific viral syndrome in up to 90% of people who seroconvert (49, 99). Symptoms develop 2 to 4 weeks after exposure. The clinical infection is characterized by fever, lymphadenopathy, pharyngitis, and occasionally, exudates, macular or papular rash, and malaise. Mucosal ulceration seen in a minority of patients can be an important clue to diagnosis. Ulceration can occur in the oropharynx or esophagus or on the external genitalia. Neurological manifestations are common, ranging from headaches or

Yun F. (Wayne) Wang and Audrey N. Schuetz • Department of Pathology & Laboratory Medicine, Emory University School of Medicine, Atlanta, GA 30303. **Molly E. Eaton** • Department of Medicine, Emory University School of Medicine, Atlanta, GA 30308. **Steven R. Nesheim** • Department of Pediatrics, Emory University School of Medicine, Atlanta, GA 30308.

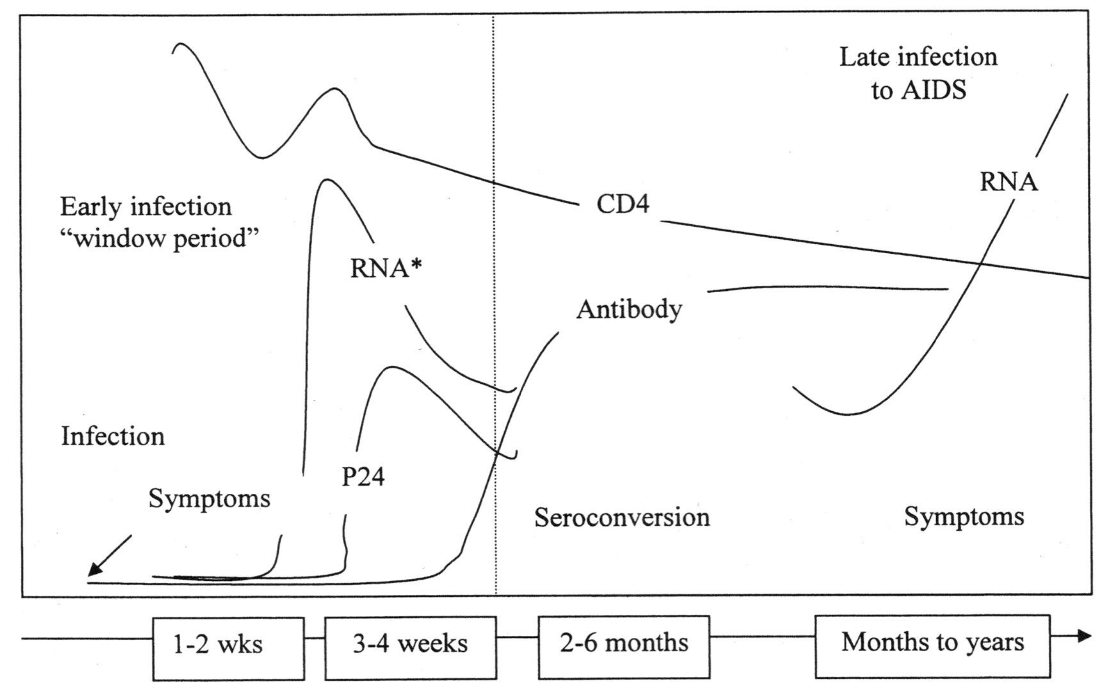

Figure 1. Schematic of HIV infection. RNA*, HIV-1 RNA or proviral DNA in the newborn. (Modified with permission from references 17, 19, and 82.)

aseptic meningitis to more severe meningoencephalitis or brachial neuritis. Rhabdomyolysis, acute pancreatitis, and cold agglutinin hemolysis are rare manifestations of acute HIV-1 infection. The median duration of symptoms is 14 days, and patients often seek medical care. In one study, 94% of patients sought medical care, but HIV was considered as a potential diagnosis for only 26% of these patients (99). During acute HIV-1 infection, viral replication usually explodes, unchecked by the immune system. $CD4^+$ T lymphocytes are depleted by this uncontrolled viremia. Over time, CD4 counts can fall from the normal range (600 to 1,200 cells/mm^3) to <200 cells/mm^3. This condition defines the onset of AIDS. Opportunistic infections such as oral thrush, esophageal candidiasis, and *Pneumocystis jirovecii* pneumonia have been reported during this early stage of infection (111).

Immune responses to HIV-1 develop gradually after initial infection. HIV-1 antibodies are often absent when patients first present with fever, pharyngitis, and lymphadenopathy. Serological tests become positive 22 to 27 days after the acute infection (5). High levels of HIV-1 exist and can be measured to diagnose HIV-1 infection. HIV-1 RNA levels are usually very high, often >100,000 copies/ml. Measurement of p24 antigen is less sensitive than PCR-based tests for HIV RNA in documenting infection and is impractical in most clinical settings (21).

With the development of HIV-specific cytotoxic T lymphocytes and, possibly, neutralizing antibody responses, most patients develop some degree of control of the HIV infection after the first few weeks. The high-level viremia drops to a lower set point that varies between individuals. The level of the new viral set point determines the subsequent rate of progression to AIDS (71). This process likely reflects the degree of immune control of the virus. Symptoms of acute infection resolve, and the $CD4^+$ T-lymphocyte count increases, although usually not back to baseline. HIV-1 and HIV-2 have similar lymphotropisms and modes of transmission (96).

During the next stage of HIV, patients are usually asymptomatic. However, this is not a truly latent stage due to ongoing viral replication. $CD4^+$ T lymphocytes are gradually depleted, at an average rate of decline of 70 cells/mm^3/year (71). This decline is extremely variable: in some patients, $CD4^+$ T-cell counts decline to <200 cells/mm^3 in as few as 2 years after acute HIV infection, while in others the counts remain in the normal range indefinitely. How HIV-1 actually leads to such marked $CD4^+$ T-cell depletion is still a controversial issue that remains to be elucidated completely.

Damage to the host immune system is indicated by a reduction in CD41 T-helper cells and an increase in HIV VL. In most untreated cases, CD41 T-cell counts eventually fall enough that patients begin to develop opportunistic infections. Initially, more virulent pathogens, such as *Mycobacterium tuberculosis* or agents of bacterial pneumonia, are observed. However, as T-cell counts decrease further, patients may develop oral thrush or seborrheic dermatitis. In addition, *P. jirovecii* pneumonia can develop when $CD4^+$ T-cell counts fall below 200 cells/mm^3 (71). As T cells decrease to <50 cells/mm^3, the patient is at risk of multiple opportunistic infections. Infections in patients with AIDS include bacterial diseases, such as bacterial respiratory and enteric diseases, syphilis, and bartonellosis; disseminated mycobacterial infections, such as *Mycobacterium tuberculosis* and *Mycobacterium avium* complex infections; fungal infections, such as *Pneumocystis jirovecii* pneumonia, candidiasis, cryptococcosis, histoplasmosis, coccidioidomycosis, and aspergillosis; parasitic infections, such as central nervous system toxoplasmosis, cryptosporidiosis, and microsporidiosis; and viral infections, such as cytomegalovirus, herpes simplex virus, varicella-zoster virus, human herpesvirus 8, human papillomavirus, hepatitis C virus (HCV), and hepatitis B virus (HBV) infections. With impaired immune surveillance, malignancies such as non-Hodgkin's and Hodgkin's lymphomas and Kaposi's sarcoma can develop. Untreated patients eventually succumb to an opportunistic infection or to malignancy (11, 71, 107)

DIAGNOSTIC APPROACHES

General Diagnostic Approaches for HIV Infection

HIV-1 and HIV-2 belong to the *Lentivirus* genus in the family *Retroviridae* and can infect humans. They are both enveloped RNA viruses with the following genes: *gag*, encoding the internal core or capsid proteins; *pol*, encoding reverse transcriptase (RT), integrase, and other enzymes; and *env*, encoding surface envelope glycoproteins that permit viral binding and entry. HIV-2 infection is found primarily in West Africa and is less prevalent worldwide than HIV-1 infection (68, 96). All subsequent comments refer to HIV-1 infection, unless otherwise specified.

HIV infection is diagnosed by testing for host and viral markers, including HIV-specific antibody, p24 antigen, viral RNA, proviral DNA, and CD4 counts. Understanding the course of infection is important for early detection before seroconversion and for monitoring the progression of infection and treatment (Fig. 1). In addition, understanding global HIV epidemiology, including the distributions of HIV types (HIV-1 and HIV-2) and HIV-1 groups and subtypes, is useful for development and validation of any new test for HIV infection (24, 50, 60, 68, 100). Recent reviews have been published on epidemiological surveillance and improvements in diagnosis (68, 79). New HIV groups and variants, such as group M non-B strains and group O strains, are targeted in newer versions of assays (Table 1). As shown in Table 1, there are three groups of HIV-1, namely, the M (major), O (outlier), and N (new; also called non-M and non-O) groups. Group M includes clades or subtypes A to D, F to H, J, and K, with B being the predominant subtype in the United States and Western Europe. Group M non-B subtypes as well as circulating recombinant forms (CRFs) are prevalent worldwide. Group M subtypes E and I are classified as recombinants (i.e., CRF-A/E) and are no longer used. HIV-1 group O is identified mainly in Cameroon and west-central Africa. Although rare, group O strains have been reported in Europe and the United States (95). Group N is extremely rare and, to date, has been found only in Cameroon and Senegal (95). HIV-2, which is closely related to HIV-1 but is rare worldwide, causes a disease that is indistinguishable from AIDS. Serological cross-reactivity between HIV-1 and HIV-2 varies highly from sample to sample.

As shown in Fig. 1, viral RNA in blood (viremia) can be detected within 10 to 14 days of infection (43, 70, 71). HIV p24 antigen, a viral protein, can be detected later than RNA can. The time interval from initial infection to the appearance of HIV-specific antibody is usually 3 to 4 weeks and is known as the "window period." However, it can take as long as 2 to 6 months for antibody to be detected. After seroconversion, there is a relatively long period of asymptomatic infection prior to late-stage AIDS. The time from infection to development of AIDS can take 10 years or longer (19).

The Centers for Disease Control and Prevention (CDC) recently published revised recommendations for HIV testing, based on the following rationale (12): (i) many HIV-infected persons access health care but are not tested for HIV until they are symptomatic, (ii) effective treatment is available, (iii) awareness of HIV infection leads to substantial reductions in high-risk sexual behavior, and (iv) HIV testing, including rapid tests, is available. The recommendations for adults and adolescents include the following: (i) routine, voluntary HIV screening should be conducted for all persons aged 13 to 64 in all health care settings (not based on risk); (ii) HIV screening should be repeated at least annually for persons with known risk; (iii) HIV consent should be included in the general consent for care (thus removing the need for separate signed informed

Table 1. Global epidemiology of HIV variants

HIV type	Origin (date of isolation)	Group	Subtype or clade	Geographic region[a]	Prevalence[a]
HIV-1	Chimpanzee in Central Africa (1983)	M (main, major)	A	E. Africa, WC Africa, Russia	High
			B[b]	America, W Europe, Australia, SE Asia	High
			C	SE Africa, India	High
			D	E Africa, W Africa	High
			F	WC Africa	Rare
			G	WC Africa	Rare
			H	WC Africa	Rare
			J	WC Africa	Rare
			K	WC Africa	Rare
			Circulating recombinant forms		High
			CRF01-AE	SE Asia	High
			CRF02-AG	W Africa	High
			CRF03-AB	Russia	Rare
			Unique recombinant forms		
		N (new, non-M, non-O)[b]		WC Africa	Low
		O (outlier)[b]		WC Africa	Rare
HIV-2	Monkeys in West Africa (1986)		A-E[b]	W Africa	Low

[a]S, south; E, east; SE, Southeast; W, west; WC, west central; C, central. Group M is prevalent throughout the world. The estimated infected population worldwide is 40 million, which includes 25.4 million people in sub-Saharan Africa and 7.1 million people in Southeast Asia.
[b]Important subtypes needed or already included in EIA test. Modified from references 50 and 68.

consent for HIV testing or HIV screening), with the option to decline (the so-called "opt-out" strategy); and (iv) test results should be communicated in the same manner as that for other diagnostic/screening tests. In addition, it is suggested that the consent for routine prenatal care include HIV testing with notification and the option to decline; this approach allows for universal opt-out HIV screening of pregnant women (12). A second or repeat HIV test is recommended for pregnant women during the third trimester in areas of high seroprevalence or for individuals who are known to be at risk of HIV infection. Rapid testing is recommended if the mother's status is unknown at labor and delivery. Based on rapid test results, ARV therapy can be initiated in a timely manner (92).

Since the diagnosis of HIV infection in infants is complicated by transplacental passage of maternal immunoglobulin G (IgG) antibodies, nearly all infants of HIV-infected women have positive antibody tests in the neonatal period. Positive antibody tests for uninfected HIV-exposed infants can persist until 18 months of age. There is no official recommendation for routine performance of antibody tests on HIV-exposed infants. However, repeatedly negative antibody results after 6 months of age or one negative antibody result after 12 months of age is presumptive evidence that an HIV-exposed infant is not infected (85). The median time from seroreversion to a negative antibody test is 9 to 10 months. Therefore, molecular testing such as HIV proviral DNA detection is useful in determining infection of HIV-exposed infants.

Laboratory Diagnosis of HIV Infection

Methods for detection of HIV used for initial diagnosis include antibody tests such as EIA, ELISA, rapid immunoassay, and WB. Other, non-antibody tests include p24 antigen detection, viral culture, and viral RNA and proviral DNA detection.

A suggested diagnostic algorithm using available cost-effective tests is shown in Fig. 2. EIA is the standard screening test. For routine clinical diagnosis of HIV infection, positive screening results must be confirmed with WB to ensure specificity. In addition, molecular methods are now used routinely to minimize the window period for diagnosis of acute or early infection. In the following discussion, methodologies are compared in terms of sensitivity, specificity, predictive value/clinical correlative data, ease or practicality of use, turnaround time, cost, and assay limitations.

EIA

Since the first test to detect serum HIV-1 antibodies was developed and licensed in 1985, the EIA method has been the primary test used for diagnosis,

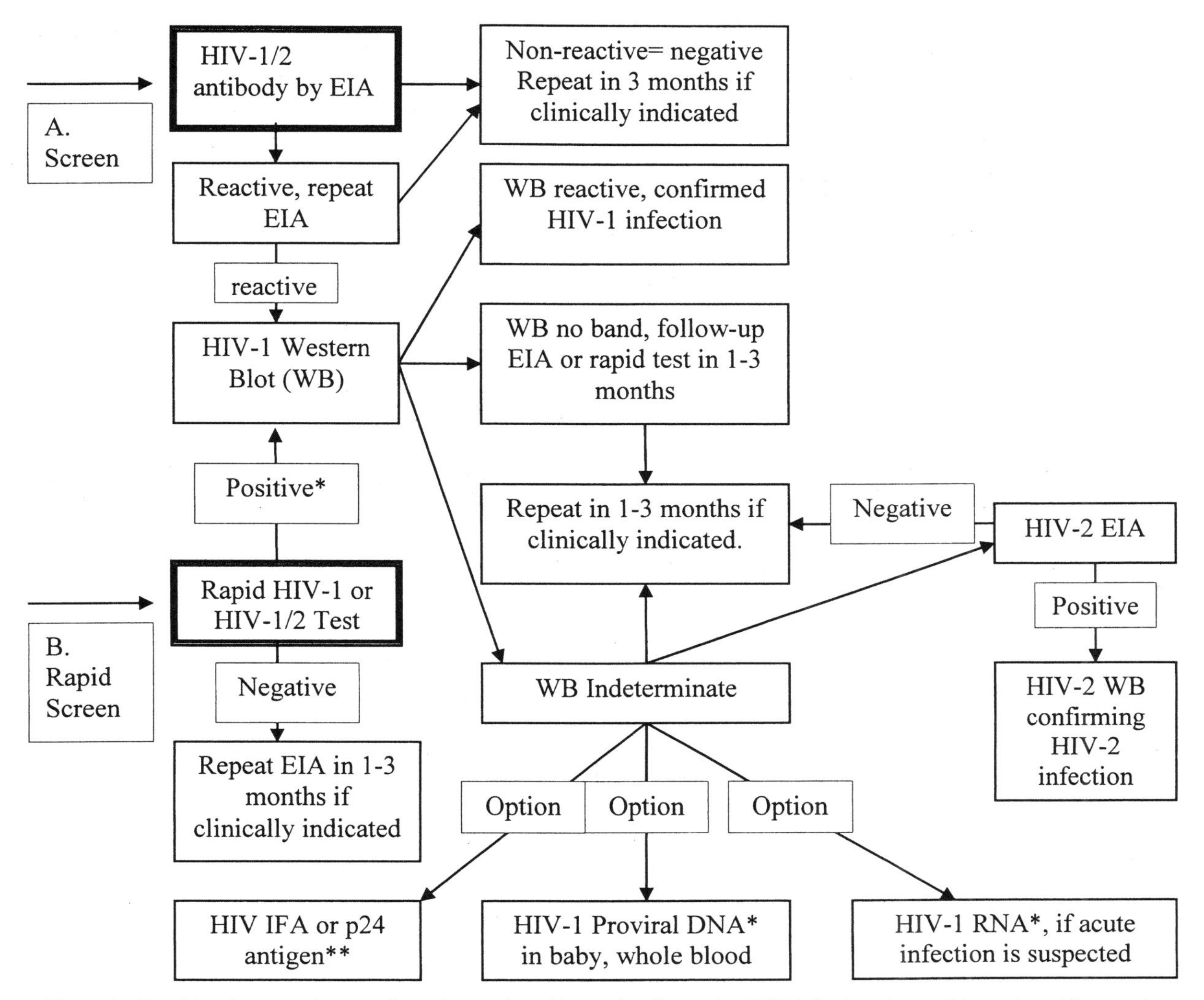

Figure 2. Algorithm for screening, confirmation, and rapid tests for diagnosing HIV infection. *, possibly repeat with second different rapid assay or even third different assay for rapid confirmation. **, if positive, confirmed HIV-1 infection; if negative, repeat serology in 1 to 3 months.

screening, and surveillance of HIV infection. General antibody detection methodologies have been reviewed recently (112). If antibodies to HIV-1 or HIV-2 antigens are present in serum, they react to the antigens coated on microplate wells and form antigen-antibody complexes. These complexes can be detected by color changes produced from the reaction between the substrate and a conjugate-coated secondary antibody. The antibodies detected are primarily IgGs. Many assays approved by the FDA have a diagnostic window of 19 to 35 days for detection of recent seroconversion or acute infection (116). EIA is useful for batch testing and thus is less expensive. However, the initial laboratory investment of testing instruments, particularly for automated systems, may be costly. Human serum, plasma, or dried blood spots can be used for EIA testing.

Most commercially available EIA test kits used in the United States for detecting the presence of HIV-1 and/or HIV-2 antibodies are so-called second- or third-generation tests (Table 2). Recombinant antigens and synthetic peptides are used in these tests. High sensitivities (99.3 to 99.7%) and specificities (99.7% or higher) can be expected with these tests, with early detection of seroconverters (19, 116).

One example of a third-generation test is the Genetic Systems HIV-1/HIV-2 Plus O EIA (Bio-Rad). It detects the broadest range of antibodies to HIV-1 (groups M and O) and HIV-2, thus minimizing the possibility of false-negative results in detection of HIV-1 non-B subtypes or HIV-2. This assay utilizes a minimal sample volume (around 75 µl) on a microplate coated with HIV-1 gp160 recombinant DNA protein, HIV-1 p24 recombinant DNA protein, a

Table 2. Methods of screening and confirmation, including rapid methods

Assay type	Name of test	Test comments	Sample type[b]	FDA and CLIA approved[a]	Company
EIA for antibody	HIV-1/HIV-2 (rRNA gene)	Recombinant (core and Env protein), specific	Serum, plasma (2nd generation)	Yes	Abbott Diagnostic
	Genetic Systems HIV-1/ HIV-2 Plus O	2.5 h, recombinant (gp160, p24) and synthetic peptide (gp36 and group O) specific	Serum, plasma (3rd generation)	Yes	Bio-Rad
	Genetic Systems HIV-1/ HIV-2 peptide	Synthetic peptide specific	Serum, plasma (2nd generation)	Yes	Bio-Rad
	Vironostika HIV-1 Plus O	Viral lysate, envelope, and group O peptide specific	Serum, plasma, dried blood spots (3rd generation)	Yes	bioMerieux
	Vironostika HIV-1	Viral lysate specific	Serum, plasma, or oral fluid	Yes	bioMerieux
	Genetic Systems HIV-2	Viral lysate specific	Serum, plasma	Yes	Bio-Rad
EIA for antigen	HIV-1 p24 Ag	Antibody to HIV-1 p24	Serum, plasma	Yes	Beckman Coulter (or Trinity Biotech)
IFA	Fluorognost HIV-1		Serum, plasma	Yes	Sanochemia
Agglutination	Capillus HIV-1/HIV-2	Direct latex aggregation assay	Serum		Trinity Biotech
WB	Calypte Cambridge Biotech	Uses viral lysate	Serum, plasma	Yes	Maxim Biomedical
	Genetic Systems HIV-1	Uses viral lysate	Serum, plasma	Yes	Bio-Rad
	New LAV Blot 1		Serum, plasma		Bio-Rad
	HIV-1 for oral fluid		Oral fluid	Yes	OraSure
Rapid	OraQuick Advance HIV-1/2	20-40 min, uses peptides gp41 and gp36	Whole blood (finger stick or venipuncture), oral fluid, plasma*	Yes W	OraSure/Abbott
	Uni-Gold Recombigen HIV-1	10-12 min, uses recombinant protein	Whole blood, serum,* or plasma*	Yes W	Trinity Biotech
	Reveal G-3 HIV-1	Read immediately, uses conserved peptide	Serum,* plasma*	Yes	MedMira
	MultiSpot HIV-1/HIV-2[c]	Read immediately or up to 24 h, uses peptide gp36 or gp41, recombinant gp41	Serum* or plasma*	Yes	Bio-Rad
	Clearview HIV-1/2 STA-PAK	15-20 min, uses gp36, gp41, and gp120	Whole blood, serum,* or plasma*	Yes W	Chembio/Inverness
	Clearview Complete HIV-1/2		Whole blood,* serum,* or plasma*	Yes	Chembio/Inverness

[a]W, CLIA-waived category.
[b]*, moderate complexity CLIA category.
[c]Can differentiate HIV-1 from HIV-2.

synthetic polypeptide mimicking an artificial HIV-1 group O epitope, and HIV-2 gp36 polypeptide. The test is easily adapted to automated (i.e., Evolis system) and semiautomated microplate technology. Results can be available in 2.5 hours. Diversified samples from different continents have been used for comparative evaluation of commercial EIAs (2). Recently, another EIA, the Advia Centaur HIV-1/O/2 enhanced EIA (Simens, Tarrytown, NY), received FDA approval. This test uses a microparticle chemiluminometric immunoassay for detection of antibodies to HIV-1, including group O strains, and/or HIV-2 in serum or plasma (106).

HIV test kits which simultaneously detect HIV antibodies and HIV-1 p24 antigen are called fourth-generation assays and are used in many countries outside the United States (31, 117). The diagnostic window for antibody detection during acute HIV infection is always a concern, i.e., there is the possibility of false-negative EIA results (7, 49). In addition to purified synthetic peptides and recombinant proteins representing immunodominant domains of the HIV-1 and HIV-2 proteins, monoclonal antibodies directed against the HIV-1 p24 antigen are included for detection of the preseroconversion stage of the infection. Examples of fourth-generation assays include the AxSYM HIV Ag/Ab Combo or Architect HIV Ag/Ab Combo (Abbott), Vironostika Uni-Form II Ag/Ab or Vidas HIV Duo Ultra (bioMerieux), HIV Ag/Ab Combination (Murex), Genscreen Plus HIV Ag/Ab (Bio-Rad), and Enzygnost HIV Integral Ag/Ab (Dade Behring) tests (67, 98, 109, 114, 116). Fourth-generation tests can reduce the diagnostic window of HIV infection by 2 to 5 days (detecting infection 2 to 5 days earlier) compared to that with third-generation EIA tests and thus are slightly more sensitive than antibody-only second- and third-generation tests (115, 117).

WB

Since many available EIA test kits are designed for blood donor testing and target a low-prevalence population, there are more issues with false-positive than with false-negative EIA results in routine clinical testing. WB is routinely used for confirmation of positive EIA or rapid test results due to its higher specificity (up to 100%) than that of EIAs (Fig. 2) (82, 116). There are only a limited number of HIV-1 WB assays available on the market, namely, the FDA-approved Cambridge Biotech product (Maxim; formerly from Calypte Biomedical) and Genetic Systems HIV-1 WB product (Bio-Rad). The Genetic Systems HIV-1 WB test is a qualitative assay for detecting and identifying antibodies to HIV-1 in human serum, plasma, or dried blood spots. Both kits use interpretive guidelines developed by the Association of Public Health Laboratories/CDC. The guidelines define as a positive result the presence of any two of the three essential bands, i.e., bands for p24, glycoprotein 41 (gp41), and gp120/160. The p24 band represents the core shell or capsid, while the gp41 band represents transmembrane glycoprotein. The gp120 and gp160 bands correspond to outer envelope or surface glycoproteins and the envelope protein precursor. The presence of gp120 and/or gp160 is considered a single reaction for the purpose of interpreting the WB results. The presence of a band(s) not meeting the criteria for a positive result is defined as an "indeterminate" result. These bands include those for p17/18, p31/32, p51, p55, and p65/66. If no band is present, the test is defined as negative.

WB assays can have a same-day turnaround time. For automation, a WB processing system with an automated strip processor (Bio-Rad) features pipetting robotics in a compact setting. Cross-reactivity between the HIV-2 antibody and the HIV-1 antigen(s) in the HIV-1 WB kit has been observed (our unpublished data).

Although there is no FDA-approved WB assay for HIV-2, an HIV-2 WB can be used to confirm a positive HIV-2 EIA result. In general, HIV-2 antibody will react to form bands such as those for p17, p26, p31, gp36, p51, p55, gp58, p65/66, gp105, gp125, and gp140. Because the prevalence of HIV-2 is extremely low in the United States, the sensitivity and specificity of HIV-2 EIA or WB are not well known.

Another HIV-1 confirmation method uses an HIV-1 multiplex method on a Luminex platform. This platform is a multiplex, fluorescent bead-based system and is designed to perform multiple assays simultaneously from a single sample in one well (unpublished data).

Rapid tests for HIV detection

Conventional EIA and WB for HIV antibody detection are still used widely in routine clinical settings. However, due to the turnaround time of at least 1 day required by these methods, many simplified and rapid HIV tests have been developed. These tests offer a number of advantages in circumstances that require a near-the-patient test, known as a point-of-care test. For regulation purposes, the Clinical Laboratory Improvement Amendment (CLIA) provides waivers for certain tests that need no additional equipment and require minimal laboratory experience. Rapid tests can be performed in 20 minutes, in most cases, and require no special laboratory equipment or extensive personnel training (19, 33, 38, 39). Although the control reagent still needs to be refrigerated, storage requirements for rapid test kits are less stringent than those for EIA kits, and thus, these tests can be performed outside the laboratory. Results of a rapid test for HIV-1 and/or HIV-2 can be available in less than 1 hour and can eliminate the need for return visits.

Among many rapid tests developed, several have received FDA approval. The FDA has required, as a condition of approval, that the lower confidence limit for estimated test sensitivity and specificity be 98% or greater. The SUDS single-use diagnostic system (Abbott/Murex) was the first FDA-approved rapid test. The SUDS test was later withdrawn from the market because newer rapid assays have apparently improved performance characteristics and require less expertise to perform. The OraQuick Advance HIV-1 test by OraSure (Bethlehem, PA), which is based on lateral-flow technology, became the first FDA-approved test to receive a CLIA waiver for rapid HIV point-of-care use. This test can provide rapid test results using either finger-stick samples, whole blood or plasma specimens from venipuncture, or oral swabs. Several tests were approved subsequently as either CLIA-waived assays or assays of moderate complexity (Table 2).

Technologies used for rapid testing include several approaches, as follows. (i) For lateral-flow technology, immobilized recombinant HIV antigens and synthetic peptides (such as gp41 and gp120 for HIV-1 and gp36 for HIV-2) are linked with colloidal gold on a nitrocellulose strip. HIV-1 or HIV-2 antibodies bind to antigen to form a distinct reactive line. Kits using this technology include Determine (Abbott Diagnostics, Abbott Park, IL), Vikia HIV-1/2 (bioMerieux, France), OraQuick Advance HIV-1/2 (OraSure Technologies), and UniGold Recombigen HIV (Trinity Biotech). (ii) Flowthrough immunoconcentration technologies are used for Multispot HIV-1/HIV-2 (Bio-Rad) and Reveal G-3 HIV-1 (Med-Mira) tests. The Multispot HIV-1/HIV-2 test is the first single-use test in the United States to offer both detection (screening) and differentiation of HIV-1 and HIV-2 antibodies. Other approaches include (iii) agglutination, in which recombinant HIV-1 and -2 proteins are bound to latex beads; (iv) immunodot or -spot assays; and (v) magnetic bead assays.

The sensitivities and specificities of these tests range from 98.4% to 100% (19, 33, 38, 39). Compared to traditional batch EIAs, rapid tests are quick and easy to perform but cost more. In addition, positive rapid tests, like positive EIAs, should be confirmed by WB. Rapid tests with the mentioned features and flexible temperature requirements are expected to be utilized in a wide range of situations.

Rapid tests have been validated and used in settings where pregnant women present to the delivery ward without prenatal care records for HIV (4). The rapid testing approach is particularly appealing for the identification of HIV-infected pregnant women, since a positive test would allow them to receive proper treatment and prophylaxis to prevent perinatal mother-to-child transmission. The test can be performed on women in late-stage pregnancy or in labor. In either setting, the rapid test could serve as the vital first test for women of unknown HIV status (4). Rapid HIV tests are useful when source patient identification is necessary during a needle-stick incidence among health care workers. Rapid tests can also be performed on patients presenting to the emergency room, where follow-up might be difficult. In addition, with relatively higher specificities than those of EIAs, rapid tests are approved for diagnostic purposes but not for blood donor screening. Thus, rapid tests can be utilized as supplemental tests to rule out possible false-positive EIA results on samples which are positive by EIA but negative by WB (our unpublished data).

EIA for p24 antigen

The HIV-1 p24 antigen test was developed to decrease the window period before the appearance of HIV antibodies. However, it is not commonly used due to the formation of immune complexes between p24 antigen and its antibodies after seroconversion and its relatively low sensitivity, ranging from 10% to 95% (6). HIV-1 p24 antigen testing is not as sensitive as nucleic acid amplification testing (NAAT) and has largely been replaced by NAAT (21). As discussed earlier, p24 antigen is now incorporated into fourth-generation EIA tests.

NAAT for viral RNA

PCR is a key technology in molecular diagnostics and can be considered the prototypical example of a NAAT. PCR enables detection of HIV RNA directly and with a high degree of sensitivity, making it possible to detect infections earlier in the infection cycle than with immunoassays. One of the NAAT methods other than PCR is transcription-mediated amplification (TMA). TMA technology is utilized in the Aptima HIV-1 RNA qualitative assay and the Procleix HIV-1/HCV assay, also allowing HIV detection prior to the appearance of antibodies (Gen-Probe, San Diego, CA; Chiron, Emeryville, CA). These qualitative assays approved for blood bank donor plasma screening can detect a smaller amount of viral RNA than quantitative tests can, i.e., <100 copies of HIV-1 RNA per ml by TMA or down to 25 copies/ml by PCR. Differences in detection rates between PCR and TMA technologies are very small and may be observed only in samples with very-low-level viremia (8, 62).

NAAT technologies can be utilized in the diagnosis of acute or primary infection when viral RNA can be detected earlier than the antibody or p24 antigen. An RNA test with a sensitivity of 50 copies/ml could detect HIV infection approximately 7 days before a p24 antigen test (from 16 days to 11 days) and 12 days before a third-generation anti-HIV antibody test (31).

Qualitative RNA NAAT assays have been developed and are commercially available for blood donor screening in blood banks to ensure the highest safety level possible. The addition of NAAT to an HIV testing algorithm in blood bank screening has been shown to significantly increase the identification of cases of infection without impairing the performance of diagnostic testing (45, 87).

NAAT for proviral DNA and diagnosis of HIV in infants

It is important to establish the infection status of an HIV-exposed infant at an early point because the effectiveness of highly active ARV therapy (HAART) at an early age has been demonstrated. Parental anxiety can be allayed earlier as well. Prophylaxis for *Pneumocystis jirovecii* pneumonia, usually with trimethoprim-sulfamethoxazole, is recommended for HIV-exposed infants until the diagnosis of HIV is excluded. Since there is transplacental passage of maternal antibodies, guidelines for diagnosing HIV in infants recommend the use of primary viral detection tests (virologic tests) in this context (120).

Diagnosis of HIV infection in HIV-exposed infants presenting at birth. For diagnosis of HIV infection in HIV-exposed infants who present at birth, the public health guidelines recommend that a virologic test be performed on children before 48 hours of age, between 1 and 2 months of age, and again between 3 and 6 months of age (1, 120). In recent years, official guidelines for making the diagnosis of HIV infection in exposed infants have recommended the use of PCR to detect HIV-1 proviral DNA (120). Due to the concern of transplacental passage of HIV-1 viral RNA, proviral DNA detection is felt to be a better indicator of retroviral infection in the infant host. Thus, although HIV viral RNA testing may have a slightly better sensitivity than that of proviral DNA testing for the group of infants of 3 weeks of age or younger, false-positive RNA results were found for a small number of infants who were not HIV infected (23, 76). Even though a PCR assay for the detection of proviral DNA is not FDA cleared, proviral DNA testing is used to establish the diagnosis of infection in infants born to HIV-1-infected mothers. The testing intervals suggested by the American Academy of Pediatrics are slightly different (<48 hours, 1 to 2 months, and 2 to 4 months of age). The American Academy of Pediatrics does recommend that there be a negative viral test at 4 months to rule out HIV infection (1).

For diagnosis of HIV infection in infants, the sensitivity of qualitative proviral DNA PCR is >95%, with an even higher specificity (26, 27, 58). A meta-analysis of studies of DNA PCR reported a median sensitivity and specificity of 91.6% and 100%, respectively (80). Reports have shown consistently that the sensitivity increases from 31 to 50% in the first days of life to >95% in approximately 3 weeks. The ability to make the diagnosis is not improved by using an additional test such as the p24 antigen test (75).

The sensitivity of DNA PCR appears to be unaffected by maternal or neonatal use of zidovudine (ZDV) in developed countries (27). However, in a study of perinatal ZDV prophylaxis conducted in Thailand (88), there was an apparent effect of duration of maternal ZDV on the time it took for infected infants to have a positive DNA PCR result. In this study, different durations of maternal and neonatal ZDV were used. Infants with a short course of ZDV had positive DNA PCR results earlier (11 days) than did infants with a long course of ZDV (43 days) (88).

Diagnosis of HIV infection in HIV-exposed infants presenting at 18 months. For HIV-exposed infants presenting at 18 months, antibody testing is recommended, regardless of results from prior virologic testing (1, 120).

Miscellaneous tests for HIV

Other EIA methods, used for testing oral fluid and urine samples, are also available. The Vironostika HIV-1 Microelisa oral fluid system (bioMerieux, Durham, NC) is an EIA for the detection of antibody to HIV-1 in human oral fluid (Table 2). In addition, the OraSure HIV-1 WB is an FDA-cleared confirmatory test for use with an OraSure HIV-1 oral fluid sample. Indirect immunofluorescence assay (IFA) is an alternative confirmatory assay to WB. Fluorognost HIV-1 (Sanochemia) is currently the only commercially available IFA for confirming the presence of human antibodies to HIV-1. IFA is not used widely in the United States but can be used in settings where WB is not available. The agglutination format uses proteins bound to polystyrene latex beads to form the basis of a direct latex aggregation assay for the detection of antibodies to HIV. Agglutination antibody assays, such as Capillus HIV-1/HIV-2 (Trinity Biotech) or Serodia (Fujirubio Inc.), are designed to be initial screening tests in low-volume testing facilities, as these tests require no sophisticated equipment and can be used as a rapid qualitative assay for the detection of antibodies to HIV-1 and HIV-2 in whole blood, serum, or plasma. These tests can be used as a supplemental assay in test algorithms. However, the test has been shown to have unexpectedly low sensitivity in a South African population of HIV-infected children (16).

Viral culture or peripheral blood lymphocyte coculture. Viral culture or peripheral blood lymphocyte coculture can also be performed for HIV confirmation. The culture method, which requires experience, a controlled facility, a long turnaround time (around 28 days), and thus a high cost, has been used in clinical trials but is not routinely available in clinical settings (14, 25, 119). Another unconventional EIA method to discriminate "early" from "long-standing" HIV infection is the sensitive/less-sensitive dual-EIA testing strategy, otherwise known as "detuned assay." This assay is based on the concept that HIV antibody titers rise during the early months of infection, and the detection of low levels of antibody indicates recent infection (<6 months postseroconversion). Thus, a positive EIA result by a highly sensitive test combined with a negative EIA result by a less sensitive method (by dilution, for example) would indicate a recent seroconversion. This sensitive/less-sensitive testing algorithm or strategy for diagnosing early infection is relatively complicated and has not been adopted in the clinical setting. This approach has been used to detect recent seroconverters and to estimate HIV incidence rates and is thus not covered in this chapter (18, 47, 69, 81, 82, 90, 102).

Result Interpretation for Laboratory Diagnosis of HIV Infection

Result interpretation

As shown in Fig. 2, repeat of HIV-1/2 EIA in duplicate before WB is needed for samples with reactive EIA results. If the repeat EIA is negative, a negative result is reported. If the repeat HIV-1/2 EIA is reactive, HIV-1 WB is performed for confirmation. If an indeterminate result is observed by HIV-1 WB, follow-up with HIV-2 testing is necessary to rule out HIV-2 infection. Different algorithms for HIV screening have been proposed, particularly to confirm infection. If HIV-1 WB is not readily available for confirmation, a combination of EIA and a rapid test, EIA and NAAT, EIA and another EIA, or a rapid test and another rapid test might be used, or even a third rapid test might be used if the first two do not agree. These approaches may be performed in resource-limited settings and are not FDA cleared or approved for use in blood banking or clinical diagnostics.

When HIV EIA is positive but WB is indeterminate, early seroconversion against HIV-1 is possible, but HIV-2 infection should be ruled out (Fig. 2). Another possible cause for an indeterminate WB result includes the receipt of HIV trial vaccines or antibodies to cellular components that might cross-react to HIV-1/2 antibodies in certain populations, such as multiparous women, multitransfused patients, patients receiving multihemodialysis, or patients with autoimmune diseases. Because increasing numbers of clinical trials for HIV vaccine immunogenicity and efficacy are being conducted globally, vaccines that contain multiple viral components can produce positive results in current HIV antibody detection tests, including EIA and WB. It may be a challenge to differentiate vaccine-induced from virus-induced anti-HIV antibodies (53, 104). NAAT is recommended in such cases.

NAAT can be useful in suspected false-negative EIA cases. Patients with a diminished immune response after an immunosuppressive therapy or an insufficient host antibody response seen in advanced HIV disease can give false-negative EIA results.

In conclusion, rapid tests can be useful in obstetric wards during labor, in cases of health care worker occupational exposure, in urgent care clinics, in emergency departments, in developing countries, and even in primary care offices (52). Since the false-negative EIA rate can be around 0.14 to 0.17% (64), evaluation of rapid tests must include test accuracy, ease of performance, a long shelf life, temperature, and reasonable cost. Rapid testing is also considered for test algorithms (Fig. 2B).

Laboratory testing for acute or primary HIV infection

In cases of suspected acute or primary HIV-1 infection, initial random blips of viremia and then exponential ramp-up viremia before seroconversion could be detected by using a NAAT method in conjunction with an HIV antibody test (see Fig. 1 in reference 17 and references 31, 32, and 64). PCR tests such as the Cobas AmpliScreen HIV-1/HCV (Roche) test can reduce the window period of detection for HIV-1 from 22 days to approximately 13 to 15 days. Small numbers of seroconverters (6.7%) have been found to have spontaneous control of HIV viremia to undetectable levels and have slower CD4 count declines (65). NAATs, including HIV-1 RNA quantitative assays discussed below and HIV-1 proviral qualitative DNA assay, are useful for diagnosis of acute or primary infection.

PROGNOSTICATION

Treatment is targeted at reducing viral replication, thereby allowing the $CD4^+$ T lymphocytes to recover. At least three ARV medications, targeting two or more phases of the viral life cycle, are required in most patients to achieve and maintain viral suppression at low levels (108). Combination therapy against HIV is known as HAART. The FDA-approved ARV drugs for HIV are classified into the following four groups: nucleoside/nucleotide reverse transcriptase inhibitors (NRTIs/NtRTIs), nonnucleoside reverse transcriptase inhibitors (NNRTIs), protease inhibitors (PIs), and entry/fusion inhibitors (see Table 4). When a resistance mutation is present, ARV medications may no longer be effective at suppressing viral replication. Therefore, periodic monitoring of VL is essential, as recommended by many panels, such as the International AIDS Society and the U.S. Department of Health and Human Services (44, 54, 108, 121). Either HIV RNA PCR or branched-chain DNA (bDNA) tests can be used to monitor the VL. If VL fails to fall adequately or if it rebounds to >1,000 copies/ml, tests for ARV resistance are recommended. Genotypic and phenotypic testing is available, but the latter is much more expensive and is usually reserved for patients with prior viral resistance.

RNA VL Assays

VL testing for prognostic purposes was introduced into the diagnostic laboratory 10 years ago to monitor and optimize ARV therapy. The accurate monitoring of VL levels is essential to measure the effectiveness of therapeutic regimens and to assess responses to therapy and potential disease progression, i.e., drug resistance evidenced by treatment failure. Decisions regarding initiation or changes in ARV therapy are guided by monitoring the plasma VL, the $CD4^{+}$ T-cell count, and the patient's clinical condition. The goal of ARV therapy is to reduce the HIV level in plasma to below detectable levels. For quantitative assays, <3-fold variation (0.5 $\log_{10}$ copies) is considered intra-assay or biological variability, but a >10-fold (1 $\log_{10}$ copies) change is considered clinically significant (66). One month after the patient starts taking medications, VL should fall at least 1 log. By 4 to 6 months into therapy, VL should have fallen below the detection limit of the test, usually to <50 to 75 copies/ml. Changes in VL and immunological response (CD4) together are good predictors of clinical outcomes after the initiation of HAART (44, 54).

Sensitive measurement of VL with a broad dynamic range of detection (from small to large numbers of copies/ml) and an enhanced ability to quantitate HIV-1 group M subtypes A to G are two major requirements for quantitative assay for HIV RNA. While HIV-1 subtype B continues to predominate in Western countries, studies now confirm that the incidence of HIV-1 non-B subtypes is increasing all over the world. The ability of a test to detect a broader range of these genetically diverse viruses is therefore crucial to HIV patient care on a global basis.

Current VL testing

Unlike antibody or antigen, viral RNA is relatively unstable and thus requires the blood samples (plasma) to be prepared within 4 to 6 hours of collection and stored at −20°C or −70°C. Among several HIV-1 VL tests which have been developed (Table 3), the following three are FDA approved for routine clinical testing: Amplicor 1.5 PCR manual or Cobas by Roche (Indianapolis, IN), bDNA Versant 3.0 assay 340 by Siemens Medical Solutions Diagnostics (formerly Bayer Diagnostics, Tarrytown, NY), and NucliSens, a nucleic acid sequence-based amplification (NASBA) assay by bioMerieux (Durham, NC). Two more assays, the Amplicor Taqman assay by Roche and the m2000 real-time assay by Abbott Molecular, are also now FDA approved.

The first FDA-approved HIV VL test, approved in 1996, was the Amplicor HIV-1 Monitor test (Roche), which measures VLs at levels as low as 400 HIV-1 RNA copies/ml. The Amplicor Ultra Sensitive test, approved in 1999, uses a slightly different sample-processing protocol and measures VLs down to 50 HIV-1 RNA copies/ml. The updated Amplicor, version 1.5, test, which can detect and quantify HIV non-B subtypes (group M subtypes A to G), received FDA approval in 2002. This version offers a detection range of 50 to 750,000 copies/ml of plasma, with a detection rate of >95% at 50 copies/ml, by using an ultrasensitive method. Suppression of HIV RNA to below 50 copies/ml (undetectable) is associated with more complete and durable viral suppression (3, 46). In addition, semiautomated versions of the Cobas as well as the AmpliPrep/Cobas for Amplicor HIV-1 Monitor test 1.5, which uses Cobas for PCR and detection and AmpliPrep for sample preparation, have been used (3).

The bDNA test known as the VERSANT HIV-1 RNA 3.0 assay (Siemens) has good reproducibility, since no amplification variation is expected due to its signal amplification technology (113). This test has a good precision across a wide reporting range, from 75 to 500,000 copies/ml, can distinguish threefold (0.5 $\log_{10}$) changes across the assay range, and quantifies subtypes A to G. The specificity of the bDNA test is reported by the manufacturer as 95 to 98%. A small portion of false-positive results may reflect assay chemistry. The bDNA test can be used to quantitate the VL down to 75 copies/ml (3, 30, 36, 78).

The NucliSens HIV-1 QT assay (bioMerieux) uses the following three key technologies: silica-based extraction of nucleic acids, NASBA for HIV RNA amplification, and electrochemiluminescence ECLdetection for quantification of the amplified RNA. NASBA technology is an isothermal process that uses three enzymes (avian myeloblastosis virus RT, RNase H, and T7 RNA polymerase) and target-specific oligonucleotides. The isothermal process runs at 41°C, generating single-stranded RNA as the

Table 3. NAAT methods for HIV-1

Test	Product and technology	Technology[a]	FDA approved	Sample (type and vol [ml])	Sensitivity (copies/ml) (range)	Company
RNA quantitation/ viral load	VERSANT HIV-1 3.0 bDNA (340)	bDNA, signal amplification, *pol* region, subtypes A to G	Yes	EDTA (1.0–2.0)	<75 (75–500,000)	Siemens (Bayer)
	NucliSens HIV-1 QT	NASBA, target amplification, *gag* region, isothermal, subtypes A to G	Yes	ACD, EDTA, or heparin (1.0)	<176 (80–3,470,000)	bioMerieux
	Amplicor HIV-1 Monitor 1.5	PCR, target amplification, *gag* region, subtypes A to F	Yes	ACD or EDTA (0.2)	<400 (400–750,000)	Roche Molecular
	Amplicor UltraSensitive (US) 1.5			0.5 (manual or Cobas)	<50 (50–100,000)	
	AmpliPrep Cobas	PCR	Yes	Automated	Same as above	Roche Molecular
	AmpliPrep Cobas TaqMan HIV-1	TaqMan real-time PCR *gag* region, subtypes A to F	Yes	Automated (up to 48 tests), EDTA (0.85)	<48 (48–10,000,000)	Roche Molecular
	RealTime HIV-1 *m*2000rt	Real-time PCR *pol* integrase region, subtypes A to G, CRF AE, AG, and group O	Yes	Automated (up to 96 tests), ACD, EDTA (1.0, 0.5, or 0.2)	<40 (40–10,000,000) (40, 75, 150 option)	Abbott Molecular
	EasyQ	NASBA/real-time detection		Automated		bioMerieux
	VERSANT HIV-1 3.0 bDNA 440	bDNA		Automated		Siemens (Bayer)
RNA qualitative test	Procleix HIV-1/ HCV by TMA	TMA, target amplification	Yes[b]	Automated		Gen-Probe, Siemens
	APTMA HIV-1 by TMA	LTR and Pol region	Yes[b]			Gen-Probe
	AmpliScreen by PCR	PCR, *gag* region	Yes[b]	Automated		Roche Molecular
DNA qualitative test	Amplicor by PCR	PCR, proviral DNA	No	Whole blood (0.1–0.5)		Roche Molecular

[a]NASBA, nucleic acid sequence-based amplification, bDNA, branched DNA; TMA, transcript-mediated amplification.
[b]Approved blood bank testing for plasma donors.

end product. The linear range of the NucliSens assay is 51 to 5,390,000 copies/ml.

Current VL testing, such as the Amplicor PCR assay, requires multistep detection of amplified products at the end of the PCR. "Real-time" PCR, however, has become a new standard of PCR technology because the amplified DNA can be detected during the PCR process (55). Real-time PCR will continue automatically with no user intervention. This feature decreases the total reaction time and reduces the possibility of cross-contamination that might occur during the traditional detection process. Real-time PCR appears to improve productivity and to provide consistent and reproducible VL results (35, 40, 96, 97).

Newer versions of VL tests with improved automation of sample preparation and RNA amplification, detection, and throughput (work flow or number of results reported per test run) have recently been cleared by the FDA, including AmpliPrep/Cobas by Roche and m2000 assay by Abbott Molecular Diagnostics (Des Plaines, IL), which are automated sample preparatory systems. Other tests, such as bDNA 440 (Siemens) and easyMAG for EasyQ by bioMerieux, are under FDA review. The instrument platforms for the AmpliPrep/Cobas TaqMan HIV-1 real-time PCR tests include an AmpliPrep instrument for automated sample preparation and either the Cobas TaqMan analyzer or Cobas TaqMan 48 analyzer for automated and high-throughput amplification and detection.

Real-time HIV-1 assay in the *m*2000 system (Abbott Molecular) consists of two instruments, an automated sample preparation instrument and an integrated real-time PCR amplification and detection instrument. The *m*2000 system, which was recently FDA cleared, is designed to detect more HIV-1 groups, including group O, and to provide more automation and throughput than previous tests.

The NucliSens EasyQ system (bioMerieux) is an automated system which combines a single-step NASBA test and real-time molecular beacon detection (using a type of DNA hybridization probe that fluoresces upon hybridization). The system uses molecular beacons that have a stem-loop structure and contain a fluorophore and a quencher group. In its normal state, the stem keeps the fluorophore and the quencher together, which prevents emission or fluorescence. In the presence of a sequence complementary to the loop sequence, the probe unfolds upon hybridization. As a result, the quencher can no longer absorb photons emitted by the fluorophore and the probe starts to fluoresce. A fluorescent reader, EasyQ Analyze, can detect this emission. Accurate results are ensured because the assay runs in a closed-tube format, minimizing risk for amplicon contamination.

NAAT RNA testing for perinatal diagnosis

In addition to HIV proviral DNA PCR for diagnostic purposes mentioned above, the usefulness of quantitative RNA PCR in diagnosing infants has been investigated (20, 23, 61, 76, 93, 122). Sensitivity of RNA PCR testing has been, in general, comparable to that of proviral DNA PCR, particularly after 6 to 8 weeks of age (20, 23, 61, 122). Specificity of <100% at ages up to 8 weeks have been reported by some non-clinical-trial series (23, 76, and 93). False-positive RNA PCR results have occurred in nonperinatal contexts (91). Certainly in the first 2 months of life, there is no clear advantage of RNA PCR over DNA PCR, but there may be situations where RNA PCR is the preferred test, particularly where systems are not able to maintain both tests in their repertoire. The sensitivity and specificity of RNA PCR in populations of HIV-exposed infants with a high proportion of maternal HAART treatment during pregnancy have not been demonstrated. Although HIV proviral DNA PCR is the preferred test for diagnosis, in the situation when a positive DNA PCR has already been obtained and verification is indicated, RNA PCR can accurately verify the positive test and can serve as baseline before beginning HAART.

Summary of VL testing

The current HIV-1 VL assays offer reliable testing with excellent sensitivity and specificity. However, consolidation and automation of VL testing are necessary. According to surveys by the College of American Pathologists, there are significant and consistent variations in absolute VL values determined using the different test kits. It is important to monitor patients over time, using the same assay, and to interpret changes of <3-fold (0.5 log) with caution. Guidelines for using HIV-1 RNA VL testing are available online (www.hivatis.org or www.iasusa.org). In clinical practice, $\log_{10}$-transformed data in addition to results in copies/ml or IU/ml should be used for reporting results. VL testing can potentially offer accurate quantitation of HIV-1, with excellent sensitivity and specificity, for broad groups (including O and N) and subtypes, and possibly in conjunction with HCV and HBV testing, especially for individuals with coinfection. Unlike current PCR assays, which require ultrasensitive in addition to regular assay in order to cover low concentrations, linear quantification from extremely low concentrations up to very high VLs, as demonstrated by two recently approved real-time PCR assays (*m*2000 by Abbott Molecular and AmpliPrep/Cobas TagMan by Roche), will help to optimize the management of antiviral therapy. Full automation, which has flexible configurations to streamline the work flow from sample input to result output, will conserve labor time and possibly laboratory space. In addition, user-defined applications on a single platform could be a good option for molecular laboratories in the near future.

Susceptibility Testing for ARV Resistance by Genotyping Methods

There are many drugs and drug combinations used to treat HIV infection. FDA-approved drug classes for HIV include (i) NRTIs/NtRTIs, (ii) NNRTIs, (iii) PIs, and (iv) entry/fusion inhibitors (Table 4). At least three ARV medications, targeting two or more sites in the viral life cycle, are required for most patients to achieve and maintain viral suppression at low levels (108).

In the presence of ARV medications, mutations in the viral genome can allow resistance to develop. Genotyping and phenotyping assays are used to monitor resistance to ARV therapy (22, 48, 63). In addition, with increasing availability of ARV therapy, resistant strains of HIV can be transmitted primarily; thus, for newly diagnosed patients, resistance testing is recommended before initiation of ARV therapy. In ARV-treated patients whose medications are discontinued, resistance mutations, especially those affecting the NNRTIs and PIs, have been found to persist for up to 2 years (13, 42, 110, 121). Finally, genotyping methods used to monitor therapy

Table 4. The FDA-approved ARV drugs for HIV

Group of ARV drugs	Abbreviation of group	Name of drug	Abbreviation of drug
Nucleoside/nucleotide reverse transcriptase inhibitors	NRTI/NtRTI	Abacavir	ABC
		Didanosine	ddI
		Emtricitabine	FTC
		Lamivudine	3TC
		Stavudine	d4T
		Tenofovir	TDF
		Zalcitabine	ddC
		Zidovudine	AZT or ZDV
			ZDV-3TC
			ZDV-3TC-ABC
			3TC-ABC and TDF-FTC
Nonnucleoside reverse transcriptase inhibitors	NNRTI	Delavirdine	DLV
		Efavirenz	EFV
		Nevirapine	NVP
Protease inhibitors	PI	Amprenavir	APV
		Atazanavir	ATV
		Darunavir	DRV
		Fosamprenavir	fosAPV
		Indinavir	IDV
		Lopinavir/ritonavir	LPV
		Nelfinavir	NFV
		Ritonavir	RTV
		Saquinavir	SQV
		Tipranavir	TPV
Entry/fusion inhibitor	FI	Enfuvirtide	T-20

can also be utilized for epidemiologic analysis, such as analysis of HIV types.

Resistance tests such as genotyping or phenotyping are recommended in the case of virologic failure during ARV treatment. ARV resistance tests determine the role of resistance in drug failure and maximize the number of active drugs in the new regimen. The tests can also be used to determine whether resistant virus was transmitted in acute or chronic HIV infection (108). Resistant HIV-1 can be transmitted and is now detected in over 15% of newly infected (naïve) patients, with higher levels in some areas. Due to the increasing prevalence of resistance in treatment-naïve patients, HIV genotype tests are now recommended for all patients (including infants) prior to the start of therapy (108).

Common genotyping method

HIV-1 genotype testing uses a DNA sequencing-based or line probe hybridization-based method to identify known resistance-associated genetic mutations in the reverse transcriptase (RT) and protease (PR) genes, with prediction of resistance based on interpretation algorithms (41, 42, 118). While genotypic testing is an indirect measure of drug susceptibility, it is the primary methodology for the analysis of HIV drug resistance. Two main commercially available tests (Table 5) in the United States are Trugene (Bayer or Siemens) and the ViroSeq HIV-1 genotyping system (Abbott Molecular with Celera Diagnostics).

For better results, plasma samples collected with a VL of ≥1,000 copies/ml should be used for genotyping (84). Viral RNA is extracted from plasma and amplified through RT-PCR. The products undergo dye terminator sequencing technology. In the ViroSeq HIV-1 genotyping system, the entire PR gene (codons 1 to 99) and the first 320 codons of the RT gene (codons 1 to 321) are sequenced. ViroSeq system software is then used to compare the sequences with wild-type viral sequences (HXB2) and against a database of known resistance mutations (15, 73).

Trugene HIV-1 genotyping is a full sequencing system, by Visible Genetics (now Siemens), which utilizes RT-PCR and sequencing technology (CLIP) to sequence the entire PR gene (codons 10 to 99) and the RT gene (codons 40 to 247), with a sensitivity of ≥1,000 copies/ml of viral RNA. Currently, the report and interpretation are based on HIV-1 mutations that are known to be associated with drug resistance. A rule-based algorithm developed by a consensus panel is used for interpretation of the entire pattern of mutations (59). This panel uses information available in the peer-reviewed literature, from clini-

Table 5. Genotyping and phenotyping methods for HIV-1

Technology	Name of available test	FDA approved	Sensitivity and mutation detection	Company
Genotyping[a]	Trugene HIV-1	Yes	Full sequencing (>1,000 copies/ml) sequencing of RT (codons 40 to 247) and PR (codons 1 to 99)	Siemens (Visible Genetics/Bayer)
	ViroSeq HIV-1	Yes	Full sequence (>1,000 copies/ml) sequencing of RT (codons 1 to 320) and PR (codons 1 to 99)	Abbott Molecular/Celera
	INNO-LiPA HIV-1		Reverse hybridization with line probe, point mutation detection in RT (seven codons) or PR (eight codons)	Siemens (Innogenetics/Bayer)
	GeneSeq		>500 copies/ml	Monogram
Database	Stanford HIV-1		RT and PR database	Stanford (http://hivdb.stanford.edu)
	Los Alamos		RT and PR database	Los Alamos National Lab (http://hiv-web.lanl.gov)
Phenotyping	Phenosense		RT and PR region (>1,000–2,000 copies/ml)	Monogram (previously Virologic, S. San Francisco, CA)
	Antivirogram		RT and PR region (>1,000–2,000 copies/ml)	Virco (Mechelen, Belgium)
Virtual phenotyping	Virtual Phenotype			Monogram
	Virtual Phenotype			Virco

[a]Some genotyping tests offered by commercial laboratories, including Monogram and Virco, in conjunction with their phenotyping tests are not listed here.

cal drug trials, and from the pharmaceutical industry for interpretating ARV levels of resistance. The levels consist of "no evidence of resistance," "possible resistance," and "resistance." The report lists rules and types of data used for each drug to make the resistance interpretation (59).

Other genotyping methods

DNA line probe assays determine point mutations in the viral genome by nucleic acid sequencing or hybridization techniques, using DNA probes on nitrocellulose strips. The Inno-LiPA HIV-1 PR and Inno-LiPA HIV-1 RT tests by Innogenetics (Siemens) use reverse hybridization techniques in multiparameter assays that simultaneously detect wild-type amino acids and mutations in the HIV-1 RT and PR genes (Table 5). Oligonucleotide probes specific for codons 41, 69, 70, 74, 184, 214, and 215 (RT) and codons 30, 46, 48, 50, 54, 82, 84, and 90 (PR) are coated as discrete lines on a nitrocellulose membrane in a strip format. Amplified biotinylated DNA material, if present, hybridizes with the specific probes.

In addition, several reference laboratories have developed "in-house" genotyping methods, including the GenSeq HIV test by Monogram. The GenSeq HIV test is available as a single assay or in the combination assay with phenotyping mentioned below. Samples with VL of >500 copies/ml can be tested. HIV-1 subtype determination is performed as a component of some genotyping processes.

Susceptibility Testing for ARV by Phenotyping Methods

Conventional phenotyping

The phenotype or phenotyping assay tests the viability of a synthetic version of the patient's HIV grown in the presence of different concentrations of ARV drugs and measures drug activity against the virus by comparing the drug susceptibility of patient virus to that of a laboratory reference strain of drug-sensitive (wild-type) virus. It thus directly and quantitatively measures the viral susceptibility to ARV drugs (28, 124). The degree of susceptibility is defined by using susceptibility cutoffs associated with expected clinical outcomes with ARV agents (118). Phenotypic changes in susceptibility have been associated with failure of HIV-1 triple combination therapy (83). Phenotypic testing results enhance the ability to predict sustained long-term suppression of VL (9).

The PhenoSense HIV (Monogram [previously Virologic]) assay directly measures the susceptibility of a patient's virus to currently available ARV drugs. The test involves several steps. PR and RT sequences are amplified from a patient's blood sample by PCR. The RT and PR segments plus the luciferase (indicator) gene are inserted into a resistance test vector, a proviral clone. The luciferase indicator is inserted into a deleted portion of the HIV-1 envelope gene. The resistance test vector is introduced into cell culture, along with a murine leukemia virus envelope vector, for assembly of the resulting virus particles. Virus particles collected during the transfection step are used to infect the target cells. A single round of viral replication results in the production of luciferase. Drug susceptibility is measured by comparing luciferase activities produced in the presence and absence of drugs: a low level of luciferase activity is shown in susceptible viruses, while a high level of luciferase activity is shown in viruses with reduced susceptibility. The drug susceptibility curve is used to calculate the concentration of drug required to inhibit viral replication by 50% (IC_{50}). Reduced drug susceptibility is indicated by a shift in the patient inhibition curve toward higher drug concentrations. The change is defined as the ratio of the patient IC_{50} to the IC_{50} of a drug-sensitive reference virus strain that contains PR and RT sequences derived from the NL4-3 strain of HIV-1. For a 1.0-fold change, patient virus must exhibit the same degree of drug susceptibility as the drug-sensitive virus. If the change is <1.0-fold, the patient virus is "more susceptible" than the drug-sensitive reference virus. A >1.0-fold change shows that the patient virus is "less susceptible" than the drug-sensitive reference virus (86). HIV drug resistance/susceptibility occurs along a continuum, as each mutation or pattern of mutations has a range of susceptibility to a drug. Phenotype testing can determine where a patient's virus falls within a range. Another phenotyping method is called Antivirogram (Virco, Bridgewater, NJ). Like the case for the PhenoSense assay, recombinant virus is produced by combining the PR-RT region from the sampled virus with a cloned provirus from which this region of the genome has been removed. The result is a replication-competent virus with PR-RT genes that represent the sampled virus population that can then be tested for susceptibility to the various drugs. Antivirogram uses the entire PR gene and the RT gene, out to codon 400 (instead of codon 320), from the patient's virus. Genotypes have been reported to match these Antivirogram phenotypes (89, 123). A less known phenotyping method is called Phenoscript (Viralliance), but it is not discussed here.

Recent phenotyping developments, including virtual phenotyping

A recent development in phenotyping technology involves an assay which assesses resistance to enfuvirtide, also known as T-20 (Fuzeon), the first approved member of the entry inhibitor class of ARV drugs. Enfuvirtide interacts with glycoprotein gp41 of the HIV envelope and interferes with the entry of HIV-1 into cells by inhibiting the merging of the virus with the cellular membrane. Since inhibition by enfuvirtide blocks HIV before it enters the human immune cell and does not act at the genetic level, genotyping tests cannot provide sufficient resistance information. The PhenoSense Entry test (Monogram) also assesses resistance to entry inhibitors that interact with gp120 and is used in their clinical development.

Another phenotyping-like method, virtual phenotyping, is marketed as a service that enhances the existing genotyping result by applying an algorithm derived from a database of genotype-phenotype matches. After a conventional genotyping method, virtual phenotyping provides a report of the estimation of the phenotype by averaging viruses with similar genotypes from a large database of historical genotype-phenotype results. Prediction of resistance is made based on an interpretation algorithm in which a patient virus falls within a range based on the genotyping test, without the high cost of a phenotyping test (34).

Summary of Genotyping and Phenotyping

Genotyping is a method to determine mutations in relevant parts of the viral genome, indirectly or qualitatively (yes or no) measuring the viral susceptibility to ARV drugs. This approach is relatively less expensive than phenotyping but more expensive than VL testing. Because genotyping does not directly measure the net effect of multiple mutations, the test has several limitations. The exact significance of some mutations remains to be confirmed. Interactions between mutations need to be clarified for estimation of the clinical impact of a given set of mutations. Resistant strains may not be detected if they comprise <10 to 20% of the viral population. Genotypic databases and algorithms for newer drugs need to be updated according to the resistances identified.

Phenotyping is an in vitro method to evaluate susceptibility to ARV drugs. Phenotyping uses clinical cutoffs associated with treatment outcome data and estimates the net effect of multiple mutations directly. Data from this assay are relatively simpler to interpret, and the report format of IC_{50} or IC_{90} is familiar to clinicians. However, phenotyping requires accurate clinical cutoff values for prediction of response. The test also requires a longer turnaround time, as it is technically more complex than genotyping. More data are needed for evidence of clinical utility for phenotyping than for genotyping (74). As with phenotypes, where clinical cutoffs relevant to each system are important for interpreting test data, interpreting virtual phenotypic data requires knowledge of clinical cutoffs in susceptibility for each drug.

Viruses with resistance mutations may become minor species or quasispecies in the absence of selective drug pressure, and the assay is unreliable if VL is low; thus, resistance testing is not recommended after discontinuation of drugs or if plasma VL is <1,000 copies/ml. Since studies of drug resistance have focused primarily on the known subtype B drug resistance mutations, the impact of HIV-1 subtype and ARV therapy or PR and RT genotype on the global epidemiology of non-B subtypes needs to be assessed (51).

SUMMARY

Much progress has been made in the laboratory diagnosis, therapy, and monitoring of HIV infection. Methods for diagnosis include EIA, WB, rapid testing, and qualitative and quantitative nucleic acid amplification; prognostic methods, such as genotyping and phenotyping, have been developed and implemented in clinical settings (Tables 2, 3, and 5). The abilities of newer-generation EIA and NAAT to detect HIV infection have been evaluated and will be incorporated into the test algorithm for detection of HIV infection during the acute or window period of infection and for the detection of non-B HIV subtypes (Fig. 1 and Table 1). Recommendations and guidelines for diagnostic and prognostic tests of HIV have been updated accordingly (Fig. 2 and 3).

Nevertheless, since HIV causes infections worldwide, it is necessary to develop more rapid, simplified, and cost-effective technologies for developing countries or resource-limited settings, where HIV is more prevalent, and for developed countries as well. Relatively complicated and costly technologies currently used in developed countries, such as VL testing, CD4 counts, genotyping, and even phenotyping, could be modified for use in both developed and developing countries. The potential use of

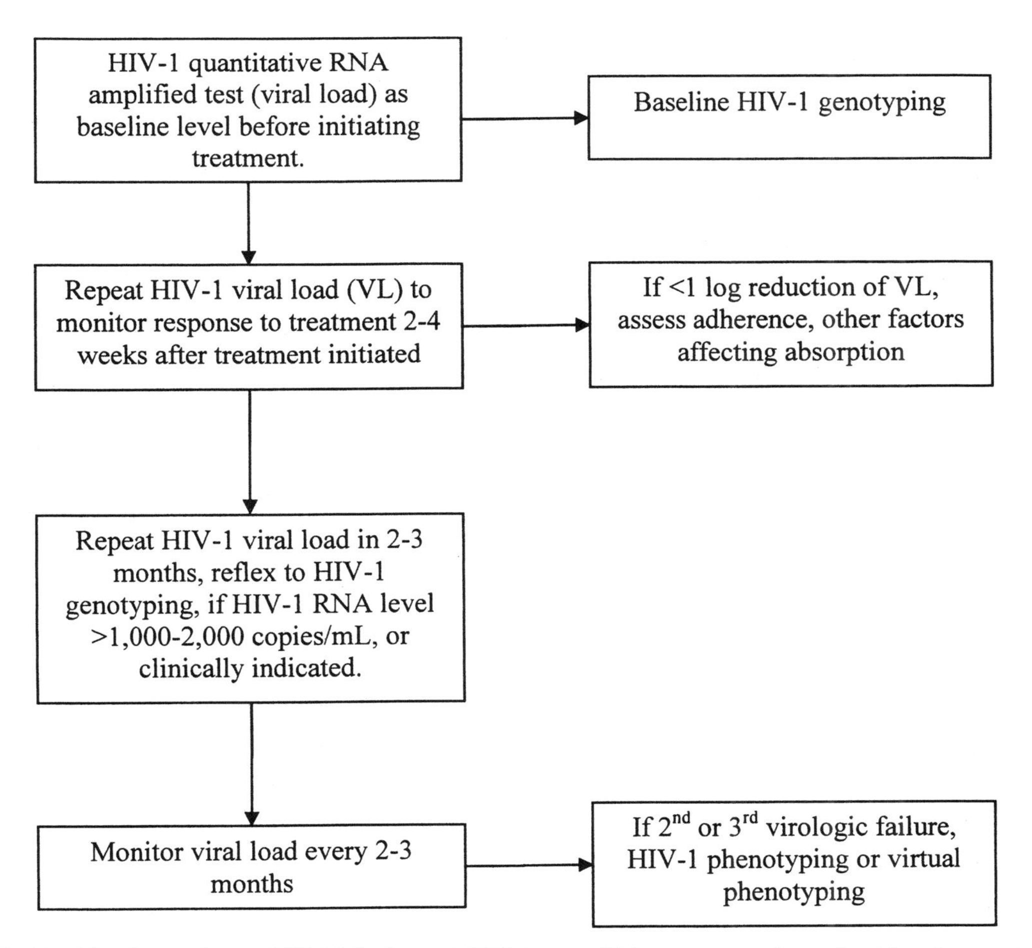

Figure 3. Algorithm for monitoring HIV-1 infection and ARV therapy. CD4 measurement is not listed here because it is not HIV specific.

dried blood spots as an alternative specimen for HIV serologic and nucleic acid testing has been studied. Rapid tests, real-time quantitative PCR, and multiplex assays for quantitating mixed drug-resistant populations have been developed and studied for broader application in HIV testing (57, 72, 94, 101, 105).

In summary, there have been many achievements in diagnostic microbiology for patients who are immunocompromised because of HIV infection. Newer immunological and molecular technologies will continue to lead the way in laboratory diagnosis of infection caused by HIV.

Acknowledgments. We thank Astrid Suantio of Emory University for assistance and proofreading as well as Chunfu Yang of the Centers for Disease Control for manuscript review.

REFERENCES

1. **American Academy of Pediatrics.** 2006. Human immunodeficiency virus infection, p. 378–401. *In Red Book: Report of the Committee on Infectious Diseases*, 27th ed. American Academy of Pediatrics, Elk Grove Village, IL.
2. **Bealert, G., G. Vercauteren, K. Fransen, M. Mangelschots, M. De Rooy, S. Garcia-Ribas, and G. van der Groen.** 2002. Comparative evaluation of eight commercial enzyme linked immunoabsorbent assays and 14 simple assays for detection of antibodies to HIV. *J. Virol. Methods* **105**:197–206.
3. **Berger, A., L. Scherzed, M. Sturmer, W. Preiser, H. W. Doerr, and H. F. Rabenau.** 2005. Comparative evaluation of the Cobas Amplicor HIV-1 Monitor ultrasensitive test, the new Cobas AmpliPrep/Cobas Amplicor HIV-1 Monitor ultrasensitive test and the Versant HIV RNA 3.0 assays for quantitation of HIV-1 RNA in plasma samples. *J. Clin. Virol.* **33**:43–51.
4. **Bulterys, M., D. J. Jamieson, M. J. O'Sullivan, M. H. Cohen, R. Maupin, S. Nesheim, M. P. Webber, R. Van Dyke, J. Wiener, B. M. Branson, et al.** 2004. Rapid HIV-1 testing during labor: a multicenter study. *JAMA* **292**:219–223.
5. **Busch, M.P., L. L. Lee, G. A. Satten, D. R. Henrard, H. Farzadegan, K. E. Nelson, S. Read, R. Y. Dodd, and L. R. Peterson.** 1995. Time course of detection of viral and serologic markers preceding human immunodeficiency virus type 1 seroconversion: implications for screening of blood and tissue donors. *Transfusion* **35**:91–97.
6. **Busch, M. P., and G. A. Satten.** 1997. Time course of viremia and antibody seroconversion following human immunodeficiency virus exposure. *Am. J. Med.* **102**:117–124.
7. **Busch, M. P., S. H. Kleinman, B. Jackson, S. L. Stramer, I. Hewlett, and S. Preston.** 2000. Committee report. Nucleic acid amplification testing of blood donors for transfusion-transmitted infectious diseases: report of the Interorganizational Task Force on Nucleic Acid Amplification Testing of Blood Donors. *Transfusion* **40**:143–159.
8. **Busch, M. P., S. A. Glynn, D. J. Wright, D. Hirschkorn, M. E. Laycock, J. McAuley, Y. Tu, C. Giachetti, J. Gallarda, J. Heitman, S. H. Kleinman, et al.** 2005. Relative sensitivities of licensed nucleic acid amplification tests for detection of viremia in early human immunodeficiency virus and hepatitis C virus infection. *Transfusion* **45**:1853–1863.
9. **Call, S. A., M. S. Saag, A. O. Westfall, J. L. Raper, S. V. Pham, J. M. Tolson, N. S. Hellmann, G. A. Cloud, and V. A. Johnson.** 2001. Phenotypic drug susceptibility testing predicts long-term virologic suppression better than treatment history in patients with human immunodeficiency virus infection. *J. Infect. Dis.* **183**:401–408.
10. **Centers for Disease Control (CDC).** 1981. Pneumocystis pneumonia—Los Angeles. *MMWR Morb. Mortal. Wkly. Rep.* **30**:250–252.
11. **Centers for Disease Control (CDC).** 1981. Kaposi's sarcoma and Pneumocystis pneumonia among homosexual men—New York City and California. *MMWR Morb. Mortal. Wkly. Rep.* **30**:305–308.
12. **Centers for Disease Control (CDC).** 2006. Revised recommendations for HIV testing of adults, adolescents, and pregnant women in health-care settings. *MMWR Morb. Mortal. Wkly. Rep.* **55**:1–16.
13. **Chearskul, P., C. Rongkavilit, H. Al-Tatari, and B. Asmar.** 2006. New antiretroviral drugs in clinical use. *Indian J. Pediatr.* **73**:335–341.
14. **Chun, T. W., D. C. Nickle, J. S. Justement, D. Large, A. Semerjian, M. E. Curlin, M. A. O'Shea, C. W. Hallahan, M. Daucher, D. J. Ward, S. Moir, J. I. Mullins, C. Kovacs, and A. S. Fauci.** 2005. HIV-infected individuals receiving effective antiviral therapy for extended periods of time continually replenish their viral reservoir. *J. Clin. Investig.* **115**:3250–3255.
15. **Church, J. D., D. Jones, T. Flys, D. Hoover, N. Marlowe, S. Chen, C. Shi, J. R. Eshleman, L. A. Guay, J. B. Jackson, N. Kumwenda, T. E. Taha, and S. H. Eshleman.** 2006. Sensitivity of the ViroSeq HIV-1 genotyping system for detection of the K103N resistance mutation in HIV-1 subtypes A, C, and D. *J. Mol. Diagn.* **8**:430–432.
16. **Claassen, M., G. U. van Zyl, S. N. J. Korsman, L. Smit, M. F. Cotton, and W. Preiser.** 2006. Pitfalls with rapid HIV antibody testing in HIV-infected children in the Western Cape, South Africa. *J. Clin. Virol.* **37**:68–71.
17. **Cohen, M. S., J. A. Anderson, and R. Swanstrom.** 2005. Acute HIV infection: implications for HIV spread, disease progression, and vaccine development. *HIV Mol. Immun.* **2005**:21–31.
18. **Constantine, N. T., A. M. Sill, N. Jack, K. Kreisel, J. Edwards, T. Cafarella, H. Smith, C. Bartholomew, F. R. Cleghorn, and W. A. Blattner.** 2003. Improved classification of recent HIV-1 infection by employing a two-stage sensitive/less-sensitive test strategy. *J. Acquir. Immune Defic. Syndr.* **32**:94–103.
19. **Constantine, N. T., and H. Zink.** 2005. HIV testing technologies after two decades of evolution. *Indian J. Med. Res.* **121**:519–538.
20. **Cunningham, C. K., T. T. Charbonneau, and K. Song.** 1999. Comparison of human immunodeficiency virus 1 DNA polymerase chain reaction and qualitative and quantitative RNA polymerase chain reaction in human immunodeficiency virus 1-exposed infants. *Pediatr. Infect. Dis. J.* **18**:30–35.
21. **Daar, E., S. Little, J. Pitt, J. Santangelo, P. Ho, N. Harawa, P. Kempt, J. Giorgi, J. Bai, P. Gaut, D. Richman, S. Mandel, and S. Nichols.** 2001. Diagnosis of primary HIV-1 infection. *Ann. Intern. Med.* **134**:25–29.
22. **DeGruttola, V., L. Dix, R. D'Aquila, D. Holder, A. Phillips, M. Ait-Khaled, J. Baxter, P. Clevenbergh, S. Hammer, R. Harrigan, D. Katzenstein, R. Lanier, M. Miller, M. Para, S. Yerly, A. Zolopa, J. Murray, A. Patick, V. Miller, S. Castillo, L. Pedneault, and J. Mellors.** 2000. The relation between baseline HIV drug resistance and response to antiretroviral therapy: re-analysis of retrospective and prospective studies using a standardized data analysis plan. *Antivir. Ther.* **5**:41–48.
23. **Delamare, C., M. Burgard, M. J. Mayaux, et al.** 1997. HIV-1 RNA detection in plasma for the diagnosis of infection in

neonates. *J. Acquir. Immune Defic. Syndr. Hum. Retrovirol.* **15**:121–125.

24. **Delwart, E., M. C. Kuhns, and M. P. Busch.** 2006. Surveillance of the genetic variation in incident HIV, HCV, and HBV infections in blood and plasma donors: implications for blood safety, diagnostics, treatment, and molecular epidemiology. *J. Med. Virol.* **78**:S30–S35.
25. **Dezube, B. J., J. Proper, J. Zhang, V. J. Choy, W. Weeden, J. Morrissey, E. M. Burns, J. D. Dixon, C. O'Loughlin, L. A. Williams, P. J. Pickering, C. S. Crumpacker, and F. B. Gelder.** 2003. A passive immunotherapy, (PE)HRG214, in patients infected with human immunodeficiency virus: a phase I study. *J. Infect. Dis.* **187**:500–503.
26. **Dunn, D. T., C. D. Brandt, A. Krivine, S. A. Cassol, P. Roques, W. Borkowsky, A. De Rossi, E. Denamur, A. Ehrnst, and C. Loveday.** 1995. The sensitivity of HIV-1 DNA polymerase chain reaction in the neonatal period and the relative contributions of intra-uterine and intra-partum transmission. *AIDS* **9**:F7–F11.
27. **Dunn, D. T., R. J. Simonds, M. Bulterys, L. A. Kalish, J. Moyer, Jr., A. de Maria, C. Kind, C. Rudin, E. Denamur, A. Krivine, C. Loveday, and M. L. Newell.** 2000. Interventions to prevent vertical transmission of HIV-1: effect on viral detection rate in early infant samples. *AIDS* **14**:1421–1428.
28. **Dunne, A. L., F. M. Mitchell, S. K. Coherly, N. S. Hellmann, J. Hoy, A. Mijch, C. J. Petropoulos, J. Mills, and S. M. Crowe.** 2001. Comparison of genotyping and phenotyping methods for determining susceptibility of HIV-1 to antiretroviral drugs. *AIDS* **15**:1471–1475.
29. **Erali, M., S. Page, L. G. Reimer, and D. R. Hillyard.** 2001. Human immunodeficiency virus type 1 drug resistance testing: a comparison of three sequence-based methods. *J. Clin. Microbiol.* **39**:2157–2165.
30. **Erice, A., D. Brambilla, J. Bremer, J. B. Jackson, R. Kokka, B. Yen-Lieberman, and R. W. Coombs.** 2000. Performance characteristics of the QUANTIPLEX HIV-1 RNA 3.0 assay for detection and quantitation of human immunodeficiency virus type 1 RNA in plasma. *J. Clin. Microbiol.* **38**:2837–2845.
31. **Fiebig, E. W., D. J. Wright, B. D. Rawal, P. E. Garrett, R. T. Schumacher, L. Peddada, C. Heldebrant, R. Smith, A. Conrad, S. Kleinman, and M. P. Busch.** 2003. Dynamics of HIV viremia and antibody seroconversion in plasma donors: implications for diagnosis and staging of primary HIV infection. *AIDS* **17**:1871–1879.
32. **Fiebig, E. W., C. M. Heldebrant, R. I. Smith, A. J. Conrad, E. L. Delwart, and M. P. Busch.** 2005. Intermittent low-level viremia in very early primary HIV-1 infection. *J. Acquir. Immune Defic. Syndr.* **39**:133–137.
33. **Franco-Paredes, C., I. Tellez, and C. del Rio.** 2006. Rapid HIV testing: a review of the literature and implications for the clinician. *Curr. HIV/AIDS Rep.* **3**:169–175.
34. **Gallego, O., L. Martin-Carbonero, J. Aguero, C. de Mendoza, A. Corral, and V. Soriano.** 2004. Correlation between rules-based interpretation and virtual phenotype interpretation of HIV-1 genotypes for predicting drug resistance in HIV-infected individuals. *J. Virol. Methods* **121**:115–118.
35. **Gibellini, D., F. Vitone, E. Gori, M. La Placa, and M. C. Re.** 2004. Quantitative detection of human immunodeficiency virus type 1 (HIV-1) viral load by SYBR green real-time RT-PCR technique in HIV-1 seropositive patients. *J. Virol. Methods* **115**:183–189.
36. **Gleaves, C. A., J. Welle, M. Campbell, T. Elbeik, V. Ng, P. E. Taylor, K. Kuramoto, S. Aceituno, E. Lewalski, B. Joppa, L. Sawyer, C. Schaper, D. McNairn, and T. Quinn.** 2002. Multicenter evaluation of the Bayer VERSANT HIV-1 RNA 3.0 assay: analytical and clinical performance. *J. Clin. Virol.* **25**:205–216.
37. **Gottlieb, M. S.** 2001. AIDS—past and future. *N. Engl. J. Med.* **344**:1788–1791.
38. **Granade, T. C.** 2005. Use of rapid HIV antibody testing for controlling the HIV pandemic. *Expert Rev. Anti Infect. Ther.* **3**:957–969.
39. **Greenwald, J. L., G. R. Burstein, J. Pincus, and B. Branson.** 2006. A rapid review of rapid HIV antibody tests. *Curr. Infect. Dis. Rep.* **8**:125–131.
40. **Gueudin, M., J.-C. Platier, F. Damond, P. Rouques, P. Mauclere, and F. Simon.** 2003. Plasma viral RNA assay in HIV-1 group O infection by real-time PCR. *J. Virol. Methods* **113**:43–49.
41. **Hanna, G. J., and R. T. D'Aquila.** 2001. Clinical use of genotypic and phenotypic drug resistance testing to monitor antiretroviral chemotherapy. *Clin. Infect. Dis.* **32**:774–782.
42. **Hirsch, M. S., F. Brun-Vezinet, B. Clotet, B. Conway, D. R. Kuritzkes, R. T. D'Aquila, L. M. Demeter, S. M. Hammer, V. A. Johnson, C. Loveday, J. W. Mellors, D. M. Jacobsen, and D. D. Richman.** 2003. Antiretroviral drug resistance testing in adults infected with human immunodeficiency virus type 1: 2003 recommendations of an international AIDS society—USA panel. *Clin. Infect. Dis.* **37**:113–128.
43. **Ho, D. D., T. Moudgil, and M. Alam.** 1989. Quantitation of human immunodeficiency virus type 1 infection. *N. Engl. J. Med.* **132**:1621–1625.
44. **Hughes, M. D., V. A. Johnson, M. S. Hirsch, J. W. Bremer, T. Elbeik, A. Erice, D. R. Kuritzkes, W. A. Scott, S. A. Spector, N. Basgoz, M. A. Fischl, and R. T. D'Aquila for the ACTG 241 Protocol Virology Substudy Team.** 1997. Monitoring plasma HIV-1 RNA levels in addition to CD4+ lymphocyte count improves assessment of antiretroviral therapeutic response. *Ann. Intern. Med.* **126**:929–938.
45. **Jackson, J. B., K. Smith, C. Knott, A. Korpela, A. Simmons, E. Piwowar-Manning, S. McDonough, L. Mimms, and J. M. Vargo.** 2002. Sensitivity of the Procleix HIV-1/HCV assay for detection of human immunodeficiency virus type 1 and hepatitis C virus RNA in a high-risk population. *J. Clin. Microbiol.* **40**:2387–2391.
46. **Jackson, J. B., E. Piwowar-Manning, L. Johnson-Lewis, R. Bassett, L. M. Demeter, and D. Brambilla.** 2004. Comparison of versions 1.0 and 1.5 of the UltraSensitive AMPLICOR HIV-1 MONITOR test for subjects with low viral load. *J. Clin. Microbiol.* **42**:2774–2776.
47. **Janssen, R. S., G. A. Satten, S. L. Stramer, B. D. Rawal, T. R. O'Brien, B. J. Weiblen, F. M. Hecht, N. Jack, F. R. Cleghorn, J. O. Kahn, M. A. Chesney, and M. P. Busch.** 1998. New testing strategy to detect early HIV-1 infection for use in incidence estimates and for clinical and prevention purposes. *JAMA* **280**:42–48.
48. **Johnson, V. A., F. Brun-Vezinet, B. Clotet, D. R. Kuritzkes, D. Pillay, J. M. Schapiro, and D. D. Richman.** 2006. Update of the drug resistance mutations in HIV-1: fall 2006. *Top. HIV Med.* **14**:125–130.
49. **Kahn, J. O., and B. D. Walker.** 1996. Acute human immunodeficiency type 1 infection. *N. Engl. J. Med.* **339**:33–39.
50. **Kandathil, A. J., S. Ramalingam, R. Kannangai, S. David, and G. Sridharan.** 2005. Molecular epidemiology of HIV. *Indian J. Med. Res.* **121**:333–344.
51. **Kantor, R., D. A. Katzenstein, B. Efron, A. P. Carvalho, B. Wynhoven, P. Cane, J. Clarke, S. Sirivichayakul, M. A. Soares, J. Snoeck, C. Pillay, H. Rudich, R. Rodrigues, A. Holguin, K. Ariyoshi, M. B. Bouzas, P. Cahn, W. Sugiura, V. Soriano, L. F. Brigido, Z. Grossman, L. Morris, A.-M. Vandamme,**

A. Tanuri, P. Phanuphak, J. N. Weber, D. Pillay, P. R. Harrigan, R. Camacho, J. M. Schapiro, and R. W. Shafer. 2005. Impact of HIV-1 subtype and antiretroviral therapy on protease and reverse transcriptase genotype: results of a global collaboration. *PLoS Med.* **2:**e113. http://www.plosmedicine.org.

52. **Keenan, P. A., J. M. Keenan, and B. M. Branson.** 2005. Rapid HIV testing: wait time reduced from days to minutes. *Postgrad. Med.* **117:**47–52.
53. **Khurana, S., J. Needham, S. Park, B. Mathieson, M. P. Busch, G. Nemo, P. Nyambi, S. Zolla-Pazner, S. Laal, J. Mulenga, E. Chomba, E. Hunter, S. Allen, J. McIntyre, I. Hewlett, S. Lee, S. Tang, E. Cowan, C. Beyrer, M. Altfeld, X. G. Yu, A. Tounkara, O. Koita, A. Kamali, N. Nguyen, B. S. Graham, D. Todd, P. Mugenyi, O. Anzala, E. Sanders, N. Ketter, P. Fast, and H. Golding.** 2006. Novel approach for differential diagnosis of HIV infections in the face of vaccine-generated antibodies: utility for detection of diverse HIV-1 subtypes. *J. Acquir. Immune Defic. Syndr.* **43:**304–312.
54. **Kitchen, C. M., S. G. Kitchen, J. A. Dubin, and M. S. Gottlieb.** 2001. Initial virological and immunologic response to highly active antiretroviral therapy predicts long-term clinical outcome. *Clin. Infect. Dis.* **33:**466–472.
55. **Klein, D.** 2002. Quantification using real-time PCR technology: applications and limitations. *Trends Mol. Med.* **8:**257–260.
56. **Kline, M., D. Lewis, and B. Hollinger.** 1994. A comparative study of HIV culture, polymerase chain reaction and anti-HIV immunoglobulin A antibody detection in the diagnosis during early infancy of vertically acquired human immunodeficiency virus infection. *Pediatr. Infect. Dis. J.* **13:**90–94.
57. **Knuchel, M. C., Z. Tomasik, R. F. Speck, R. Lüthy, and J. Schüpbach.** 2006. Ultrasensitive quantitative HIV-1 p24 antigen assay adapted to dried plasma spots to improve treatment monitoring in low-resource settings. *J. Clin. Virol.* **36:**64–67.
58. **Kovacs, A., J. Xu, and S. Rasheed.** 1995. Comparison of a rapid nonisotopic polymerase chain reaction assay with four commonly used methods for the early diagnosis of human immunodeficiency virus type 1 infection in neonates and children. *Pediatr. Infect. Dis. J.* **14:**948–954.
59. **Kuritzkes, D. R., R. M. Grant, P. Feorino, M. Griswold, M. Hoover, R. Young, S. Day, R. M. Lloyd, Jr., C. Reid, G. F. Morgan, and D. L. Winslow.** 2003. Performance characteristics of the TRUGENE HIV-1 genotyping kit and the open gene DNA sequencing system. *J. Clin. Microbiol.* **41:**1594–1599.
60. **Lal, R. B., S. Chakrabarti, and C. Yang.** 2005. Impact of genetic diversity of HIV-1 on diagnosis, antiretroviral therapy and vaccine development. *Indian J. Med. Res.* **121:**287–314.
61. **Lambert, J. S., D. R. Harris, and E. R. Stiehm.** 2003. Performance characteristics of HIV-1 culture and HIV-1 DNA and RNA amplification assays for early diagnosis of perinatal HIV-1 infection. *J. Acquir. Immune Defic. Syndr.* **34:**512–519.
62. **Lelie, P. N., H. A. J. van Drimmelen, H. T. M. Cuypers, S. J. Best, S. L. Stramer, C. Hyland, J.-P. Allain, P. Moncharmont, C. Defer, M. Nübling, A. Glauser, M. da Silva Cardoso, J.-F. Viret, M. H. Lankinen, L. Grillner, U. Wirthmüller, J. Coste, V. Schottstedt, B. Masecar, and E. M. Dax.** 2002. Sensitivity of HCV RNA and HIV RNA blood screening assays. *Transfusion* **42:**527.
63. **Little, S. J., S. Holte, J.-P. Routy, E. S. Daar, M. Markowitz, A. C. Collier, R. A. Koup, J. W. Mellors, E. Connick, B. Conway, M. Kilby, L. Wang, J. M. Whitcomb, N. S. Hellmann, and D. D. Richman.** 2002. Antiretroviral-drug resistance among patients recently infected with HIV. *N. Engl. J. Med.* **347:**385–394.
64. **Ly, T. D., C. Edlinger, and A. Vabret.** 2000. Contribution of combined detection assays of p24 antigen and anti human immunodeficiency virus (HIV) antibodies in diagnosis of primary HIV infection by routine testing. *J. Clin. Microbiol.* **38:**2459–2461.
65. **Madec, Y., F. Boufassa, K. Porter, and L. Meyer on behalf of the CASCADE Collaboration.** 2005. Spontaneous control of viral load and CD4 cell count progression among HIV-1 seroconverters. *AIDS* **19:**2001–2007.
66. **Martin, D.** 2000. Appropriate laboratory monitoring of HIV. *S. Afr. Med. J.* **90:**33–36.
67. **Martinez-Martinez, P., E. Martin del Barrio, J. De Benito, and R. Landinez.** 1999. New lineal immunoenzymatic assay for simultaneous detection of p24 antigen and HIV antibodies. *Eur. J. Clin. Microbiol. Infect. Dis.* **18:**591–594.
68. **McCutchan, F. E.** 2006. Global epidemiology of HIV. *J. Med. Virol.* **78:**S7–S12.
69. **McFarland, W., M. P. Busch, T. A. Kellogg, B. D. Rawal, G. A. Satten, M. H. Katz, J. Dilley, and R. S. Janssen.** 1999. Detection of early HIV infection and estimation of incidence using a sensitive/less-sensitive enzyme immunoassay testing strategy at anonymous counseling and testing sites in San Francisco. *J. Acquir. Immune Defic. Syndr.* **22:**484–489.
70. **Mellors, J. W., C. R. Rinaldo, P. Gupta, R. M. White, J. A. Todd, and L. A. Kingsley.** 1996. Prognosis in HIV-1 infection predicted by the quantity of virus in plasma. *Science* **272:**1167–1170.
71. **Mellors, J. W., A. Munoz, J. V. Giorgi, J. B. Margolick, C. J. Tassoni, P. Gupta, L. A. Kingsley, J. A. Todd, A. J. Saah, R. Detels, J. P. Phair, and C. R. Rinaldo, Jr.** 1997. Plasma viral load and CD4+ lymphocytes as prognostic markers of HIV-1 infection. *Ann. Intern. Med.* **126:**946–954.
72. **Moser, M. J., M. Ruckstuhl, C. A. Larsen, A. J. Swearingen, M. Kozlowski, L. Bassit, P. L. Sharma, R. F. Schinazi, and J. R. Prudent.** 2005. Quantifying mixed populations of drug-resistant human immunodeficiency virus type 1. *Antimicrob. Agents Chemother.* **49:**3334–3340.
73. **Mracna, M., G. Becker-Pergola, J. Dileanis, L. A. Guay, S. Cunningham, J. B. Jackson, and S. H. Eshleman.** 2001. Performance of Applied Biosystems ViroSeq HIV-1 genotyping system for sequence-based analysis of non-subtype B human immunodeficiency virus type 1 from Uganda. *J. Clin. Microbiol.* **39:**4323–4327.
74. **Munoz, M., R. Carmona, L. Perez-Alvarez, G. Cilla, M. D. Suarez, E. Delgado, G. Contreras, J. Corral, M. J. de Goicoetxea, L. Medrano, M. J. Lezaun, and R. Najera.** 2005. Analysis of discrepancies in the interpretation of antiretroviral drug resistance results in HIV-1 infected patients of Basque Country, Spain. *J. Clin. Virol.* **33:**224–229.
75. **Nesheim, S., F. Lee, M. L. Kalish, C. Y. Ou, M. Sawyer, L. Meadows, V. Grimes, R. J. Simonds, and A. Nahmias.** 1997. Diagnosis of perinatal human immunodeficiency virus infection by polymerase chain reaction and p24 antigen detection after immune complex dissociation in an urban community hospital. *J. Infect. Dis.* **175:**1333–1336.
76. **Nesheim, S., P. Palumbo, K. Sullivan, F. Lee, P. Vink, E. Abrams, and M. Bulterys.** 2003. Quantitative RNA testing for diagnosis of HIV-infected infants. *J. Acquir. Immune Defic. Syndr.* **32:**192–195.
77. **Niu, M. T., J. A. Jermano, P. Reichelderfer, and S. M. Schnittman.** 1993. Summary of the National Institutes of Health workshop on primary human immunodeficiency virus type 1 infection. *AIDS Res. Hum. Retrovir.* **9:**913–924.

78. **Nolte, F. S., J. Boysza, C. Thurmond, W. S. Clark, and J. L. Lennox.** 1998. Clinical comparison of an enhanced-sensitivity branched-DNA assay and reverse transcription-PCR for quantitation of HIV type 1 RNA in plasma. *J. Clin. Microbiol.* **36:**716–720.

79. **Osmanov, S., C. Pattou, N. Walker, B. Schwardlander, J. Esparza, et al.** 2002. Estimated global distribution and regional spread of HIV-1 genetic subtypes in the year 2000. *J. Acquir. Immune Defic. Syndr.* **29:**184–190.

80. **Owens, D. K., M. Holodniy, T. W. McDonald, J. Scott, and S. Sonnad.** 1996. A meta-analytic evaluation of the polymerase chain reaction for the diagnosis of HIV infection in infants. *JAMA* **275:**1342–1348.

81. **Parekh, B. S., M. S. Kennedy, T. Dobbs, C. P. Pau, R. Byers, T. Green, D. J. Hu, S. Vanichseni, N. L. Young, K. Choopanya, T. D. Mastro, and J. S. McDougal.** 2002. Quantitative detection of increasing HIV type 1 antibodies after seroconversion: a simple assay for detecting recent HIV infection and estimating incidence. *AIDS Res. Hum. Retrovir.* **18:**295–307.

82. **Parekh, B. S., and J. S. McDougal.** 2005. Application of laboratory methods for estimation of HIV-1 incidence. *Indian J. Med. Res.* **121:**510–518.

83. **Parkin, N. T., Y. S. Lie, N. S. Hellmann, M. Markowitz, S. Bonhoeffer, D. D. Ho, and C. J. Petropoulos.** 1999. Phenotypic changes in drug susceptibility associated with failure of HIV-1 triple combination therapy. *J. Infect. Dis.* **180:**865–870.

84. **Parkin, N. T., S. G. Deeks, M. T. Wrin, J. Yap, R. M. Grant, K. H. Lee, D. Heeren, N. S. Hellmann, and C. J. Petropoulos.** 2000. Loss of antiretroviral drug susceptibility at low viral load during early virological failure in treatment-experienced patients. *AIDS* **14:**2877–2887.

85. **Perinatal HIV Guidelines Working Group.** 12 October 2006, posting date. Public health service task force recommendations for use of antiretroviral drugs in pregnant HIV-1 infected women for maternal health and interventions to reduce perinatal HIV-1 transmission in the United States. http://aidsinfo.nih.gov/ContentFiles/PerinatalGL.pdf.

86. **Petropoulos, C. J., N. T. Parkin, K. L. Limoli, Y. S. Lie, T. Wrin, W. Huang, H. Tian, D. Smith, G. A. Winslow, D. J. Capon, and J. M. Whitecomb.** 2000. A novel phenotypic drug susceptibility assay for human immunodeficiency virus type 1. *Antimicrob. Agents Chemother.* **44:**920–928.

87. **Pilcher, C. D., S. A. Fiscus, T. Q. Nguyen, E. Foust, L. Wolf, D. Williams, R. Ashby, J. O. O'Dowd, J. T. McPherson, B. Stalzer, L. Hightow, W. C. Miller, J. J. Eron, Jr., M. S. Cohen, and P. A. Leone.** 2005. Detection of acute infections during HIV testing in North Carolina. *N. Engl. J. Med.* **352:**1873–1883.

88. **Prasitwattanaseree, S., M. Lallemant, D. Costagliola, G. Jourdain, and J. Y. Mary.** 2004. Influence of mother and infant zidovudine treatment duration on the age at which HIV infection can be detected by polymerase chain reaction in infants. *Antivir. Ther.* **9:**179–185.

89. **Qari, S. H., R. Respess, H. Weinstock, E. M. Beltrami, K. Hertogs, B. A. Larder, C. J. Petropoulos, N. Hellmann, and W. Heneine.** 2002. Comparative analysis of two commercial phenotypic assays for drug susceptibility testing of human immunodeficiency virus type 1. *J. Clin. Microbiol.* **40:** 31–35.

90. **Rawal, B. D., A. Degula, L. Lebedeva, R. S. Janssen, F. M. Hecht, H. W. Sheppard, and M. P. Busch.** 2003. Development of a new less-sensitive enzyme immunoassay for detection of early HIV-1 infection. *J. Acquir. Immune Defic. Syndr.* **33:**349–355.

91. **Rich, J. D., N. Merriman, E. Mylonakis, T. C. Greenough, T. P. Flanigan, B. J. Mady, and C. C. Carpenter.** 1999. Misdiagnosis of HIV infection by HIV-1 plasma viral load testing: a case series. *Ann. Intern. Med.* **130:**37–39.

92. **Rothman, R. E.** 2004. Current centers for disease control and prevention guidelines for HIV counseling, testing, and referral: critical role of and a call to action for emergency physicians. *Ann. Emerg. Med.* **44:**31–42.

93. **Rouet, F., C. Montcho, C. Rouzioux, V. Leroy, P. Msellati, J. B. Kottan, B. You, I. Viho, F. Dabis, et al.** 2001. Early diagnosis of pediatric HIV-1 infection among African breastfed children using quantitative plasma HIV RNA assay. *AIDS* **15:**1849–1856.

94. **Rouet, F., D. K. Ekouevi, M.-L. Chaix, M. Burgard, A. Inwoley, T. D'Aquin Tony, C. Danel, X. Anglaret, V. Leroy, P. Msellati, F. Dabis, and C. Rouzioux.** 2005. Transfer and evaluation of an automated, low-cost real-time reverse transcription-PCR test for diagnosis and monitoring of human immunodeficiency virus type 1 infection in a West African resource-limited setting. *J. Clin. Microbiol.* **43:** 2709–2717.

95. **Rox, J. M., A. M. Eis-Hubinger, J. Muller, M. Vogel, R. Kaiser, P. Hanfland, and M. Daumer.** 2004. First human immunodeficiency virus-1 group O infection in a European blood donor. *Vox Sang.* **87:**44–45.

96. **Ruelle, J., B. K. Mukadi, M. Schutten, and P. Goubau.** 2004. Quantitative real-time PCR on LightCycler for detection of human immunodeficiency virus type 2 (HIV-2). *J. Virol. Methods* **117:**67–74.

97. **Saha, B. K., B. Tian, and R. P. Bucy.** 2001. Quantitation of HIV-1 by real-time PCR with a unique fluorogenic probe. *J. Virol. Methods* **93:**33–42.

98. **Saville, R. D., N. T. Constantine, N. Jack, C. Bartholomew, J. Edwards, P. Gomez, and W. A. Blattner.** 2001. Fourth-generation enzyme-linked immunosorbent assay for the simultaneous detection of human immunodeficiency virus antigen and antibody. *J. Clin. Microbiol.* **39:**2518–2524.

99. **Schacker, T., A. C. Collier, J. Hughes, T. Shea, and L. Corey.** 1996. Clinical and epidemiologic features of primary HIV infection. *Ann. Intern. Med.* **125:**257–264.

100. **Schochetman, G., and M. C. Kuhns.** 2006. HIV variations and hepatitis B surface antigen mutants: diagnostic challenges for immunoassays. *J. Med. Virol.* **78:**S3–S6.

101. **Sherman, G. G., G. Stevens, S. A. Jones, P. Horsfield, and W. S. Stevens.** 2005. Dried blood spots improve access to HIV diagnosis and care for infants in low-resource settings. *J. Acquir. Immune. Defic. Syndr.* **38:**615–617.

102. **Soroka, S. D., T. C. Granade, D. Candal, and B. S. Parekh.** 2005. Modification of rapid human immunodeficiency virus (HIV) antibody assay protocols for detecting recent HIV seroconversion. *Clin. Diagn. Lab. Immunol.* **12:**918–921.

103. **Spira, A. I., P. A. Marx, B. K. Patterson, J. Mahoney, R. A. Koup, S. M. Wolinsky, and D. D. Ho.** 1996. Cellular targets of infection and route of viral dissemination after an intravaginal inoculation of simian immunodeficiency virus into rhesus macaques. *J. Exp. Med.* **183:**215–225.

104. **Suthon, V., R. Archawin, C. Chanchai, L. John, K. Wichuda, P. Wiroj, T. Hansa, S. Pathom, S. N. Pongnuwat, P. Silaporn, and I. Wimala.** 2002. Impact of HIV vaccination on laboratory diagnosis: case reports. *J. Infect. Dis.* **2:**19.

105. **Svarovskaia, E. S., M. J. Moser, A. S. Bae, J. R. Prudent, M. D. Miller, and K. Borroto-Esoda.** 2006. MultiCode-RTx real-time PCR system for detection of subpopulations of K65R human immunodeficiency virus type 1 reverse transcriptase mutant viruses in clinical samples. *J. Clin. Microbiol.* **44:**4237–4241.

106. **Taylor, P., G. Pickard, A. Gammie, and M. Atkins.** 2004. Comparison of the ADVIA Centaur and Abbott AxSYM immunoassay systems for a routine diagnostic virology laboratory. *J. Clin. Virol.* **30**(Suppl. 1):S11–S15.
107. **UNAIDS.** 2006. Report on the global AIDS epidemic: executive summary. A UNAIDS 10th anniversary special edition. UNAIDS, Geneva, Switzerland. http://www.unaids.org.
108. **U.S. Department of Health and Human Services Panel on Antiretroviral Guidelines for Adults and Adolescents.** 29 January 2008, posting date. Guidelines for the use of antiretroviral agents in HIV-1-infected adults and adolescents. U.S. Department of Health and Human Services, Washington, DC. http://aidsinfo.nih.gov/ContentFiles/AdultandAdolescentGL.pdf.
109. **Van Binsbergen, J., A. Siebelink, A. Jacobs, W. Keuer, F. Bruynis, M. van de Graaf, J. van der Heijden, D. Kambel, and J. Toonen.** 1999. Improved performance of seroconversion with a 4th generation HIV antigen/antibody assay. *J. Virol. Methods* **82**:77–84.
110. **Vandamme, A. M., A. Sönnerborg, M. Ait-Khaled, J. Albert, B. Asjo, L. Bacheler, D. Banhegyi, C. Boucher, F. Brun-Vézinet, R. Camacho, P. Clevenbergh, N. Clumeck, N. Dedes, A. De Luca, H. W. Doerr, J.-L. Faudon, G. Gatti, J. Gerstoft, W. W. Hall, A. Hatzakis, N. Hellmann, A. Horban, J. D. Lundgren, D. Kempf, M. Miller, V. Miller, T. W. Myers, C. Nielsen, M. Opravil, L. Palmisano, C. F. Perno, A. Phillips, D. Pillay, T. Pumarola, L. Ruiz, M. Salminen, J. Schapiro, B. Schmidt, J.-C. Schmit, R. Schuurman, E. Shulse, V. Soriano, S. Staszewski, S. Vella, M. Youle, R. Ziermann, and L. Perrin.** 2004. Updated European recommendations for the clinical use of HIV drug resistance testing. *Antivir. Ther.* **9**:829–848.
111. **Vanhems, P., C. Dassa, J. Lambert, D. A. Cooper, L. Perrin, J. Vizzard, B. Hirschel, S. K. Loes, A. Carr, and R. Allard.** 1999. Comprehensive classification of symptoms and signs reported among 218 patients with acute HIV-1 infection. *J. Acquir. Immune Defic. Syndr. Hum. Retrovir.* **21**:99–106.
112. **Wang, Y. F.** 2006. Advanced antibody detection, p. 42–62. *In* Y. W. Tang and C. W. Stratton (ed.), *Advanced Techniques in Diagnostic Microbiology*. Springer, New York, NY.
113. **Wang, Y. F.** 2006. Signal amplification techniques, p. 228–242. *In* Y. W. Tang and C. W. Stratton (ed.), *Advanced Techniques in Diagnostic Microbiology*. Springer, New York, NY.
114. **Weber, B., A. Berger, H. Rabenau, and H. W. Doerr.** 2002. Evaluation of a new combined antigen and antibody human immunodeficiency virus screening assay, VIDAS HIV DUO Ultra. *J. Clin. Microbiol.* **40**:1420–1426.
115. **Weber, B., R. Thorstensson, S. Tanprasert, U. Schmitt, and W. Melchior.** 2003. Reduction of the diagnostic window in three cases of human immunodeficiency-1 subtype E primary infection with fourth-generation HIV screening assays. *Vox Sang.* **85**:73–79.
116. **Weber, B.** 2006. Screening of HIV infection: role of molecular and immunological assays. *Expert Rev. Mol. Diagn.* **6**:399–411.
117. **Weber, B., B. Orazi, A. Raineri, R. Thorstensson, P. Burgisser, A. Muhlbacher, C. Areal, A. Eiras, R. Villaescusa, R. Camacho, I. Diogo, H. J. Roth, I. Zahn, J. Bartel, V. Bossi, F. Piro, K. Atamasirikul, P. Permpikul, L. Webber, and S. Singh.** 2006. Multicenter evaluation of a new 4th generation HIV screening assay Elecsys HIV combi. *Clin. Lab.* **52**:463–473.
118. **Wilson, J. W., and P. Bean.** 2000. A physician's primer to antiretroviral drug resistance testing. *AIDS Read.* **10**:469–473, 476–478.
119. **Wong, J. K., M. Hezareh, H. F. Gunthard, D. V. Havlir, C. C. Ignacio, C. A. Spina, and D. D. Richman.** 1997. Recovery of replication-competent HIV despite prolonged suppression of plasma viremia. *Science* **278**:1291–1295.
120. **Working Group on Antiretroviral Therapy and Medical Management of HIV-Infected Children.** 28 February 2008, posting date. Guidelines for the use of antiretroviral agents in pediatric HIV infection. http://aidsinfo.nih.gov/ContentFiles/PediatricGuidelines.pdf.
121. **Yeni, P. G., S. M. Hammer, C. C. J. Carpenter, D. A. Cooper, M. A. Fischl, J. M. Gatell, B. G. Gazzard, M. S. Hirsch, D. M. Jacobsen, D. A. Katzenstein, J. S. G. Montaner, D. D. Richman, M. S. Saag, M. Schechter, R. T. Schooley, M. A. Thompson, S. Vella, and P. A. Volberding.** 2002. Antiretroviral treatment for adult HIV infection in 2002. Updated recommendations of the International AIDS Society—-USA panel. *JAMA* **288**:222–235.
122. **Young, N. L., N. Shaffer, and T. Chaowanachan.** 2000. Early diagnosis of HIV-1-infected infants in Thailand using RNA and DNA PCR assays sensitive to non-B subtypes. *J. Acquir. Immune. Defic. Syndr.* **24**:401–407.
123. **Zhang, J., S. Y. Rhee, J. Taylor, and R. W. Shafer.** 2005. Comparison of the precision and sensitivity of the Antivirogram and PhenoSense HIV drug susceptibility assays. *J. Acquir. Immune. Defic. Syndr.* **38**:439–444.
124. **Zhang, M., and J. Versalovic.** 2002. HIV update. Diagnostic tests and markers of disease progression and response to therapy. *Am. J. Clin. Pathol.* **118**:S26–S32.

Diagnostic Microbiology of the Immunocompromised Host
Edited by Randall T. Hayden, Karen C. Carroll, Yi-Wei Tang, and Donna M. Wolk

Chapter 3

Cytomegalovirus

CARI R. SLOMA, THOMAS E. GRYS, AND RAYMUND R. RAZONABLE

VIROLOGY AND PATHOGENESIS

Cytomegalovirus (CMV), one of the largest viruses known to cause human disease, is the fifth member of the human herpesvirus family. CMV is 150 to 200 nm in diameter and has an icosahedral shape. CMV has four fundamental structural elements, as follows: an outer lipid envelope, tegument, a nucleocapsid, and an internal nucleoprotein core that contains its genome. These components are essential for the biology of the virus and, as discussed in this chapter, are also major targets for diagnostic and therapeutic modalities. The viral envelope of CMV contains lipoproteins and at least 33 structural proteins, including the glycoproteins that are involved in viral entry into cells. The tegument is composed of structural proteins, including the pp65 antigen (70), which is a major target for CMV detection. The CMV genome is a 64-nm linear double-stranded DNA molecule that contains nonoverlapping open reading frames for over 230 proteins. One of the proteins is CMV DNA polymerase, which plays an integral role in viral replication and serves as the main target for antiviral therapy. The CMV genome also encodes proteins that downregulate the immune system, thereby contributing to the immunosuppressive properties of CMV. For example, two CMV proteins—US2 and US11—target major histocompatibility complex class I heavy-chain molecules for proteasomal degradation (4). Consequently, major histocompatibility complex class I and CMV glycopeptides cannot form complexes on the cell surface to trigger recognition by $CD8^+$ T cells. Hence, impairment in antigen presentation occurs, which has been postulated as a mechanism for viral persistence and latency.

EPIDEMIOLOGY AND CLINICAL MANIFESTATIONS

Initially isolated from the salivary gland of an infant with cytomegalic inclusion disease (75), CMV can cause disease in individuals with healthy (42) and impaired immune function. Epidemiologic studies have demonstrated the ubiquitous nature of CMV. In urban U.S. cities, a CMV seroprevalence rate of 60 to 70% has been reported; this rate may reach as high as 100% in some developing countries.

CMV is acquired most commonly during childhood to early adulthood (59). With a healthy immune system, primary infection is usually manifested as a benign illness. It presents as asymptomatic infection, nonspecific febrile illness, or an infectious mononucleosis-like syndrome characterized by fever, lymphadenopathy, and lymphocytosis (59). After a self-limited course, CMV establishes latency and persists during the lifetime of the host. CMV exists in leukocytes, endothelial cells, renal epithelial cells, and salivary glands, among other sites (59). These cellular sites of latency serve as a reservoir for transmission to susceptible individuals and as sites of reactivation during critical illness and periods of immunocompromise, such as organ transplantation, pharmacologic immunosuppression, or use of chemotherapeutic agents (59).

The clinical presentation of CMV infection is highly influenced by the immune fitness of the host (59). Patients are at high risk of severe and sometimes fatal CMV infection if they suffer from a deficiency in their immune function. The three major groups of patients at highest risk of CMV disease are (i) patients with human immunodeficiency virus (HIV) and AIDS, (ii) recipients of solid organ transplants (SOT)

Cari R. Sloma and Thomas E. Grys • Division of Clinical Microbiology, Department of Laboratory Medicine and Pathology, Mayo Clinic and the Mayo Clinic College of Medicine, Rochester, MN 55905. **Raymund R. Razonable** • Division of Infectious Diseases, Department of Internal Medicine, and The William J. von Liebig Transplant Center, Mayo Clinic and the Mayo Clinic College of Medicine, Rochester, MN 55905.

or hematopoietic stem cell transplants (HSCT), and (iii) the immunologically underdeveloped fetus (Table 1).

CMV in Patients with AIDS

CMV disease in AIDS patients is most commonly manifested as sight-threatening retinitis (Table 1). The onset of CMV retinitis typically occurs when the $CD4^+$ T-cell count falls below 50 cells/mm^3. CMV causes a complete-thickness infection of retinal cells, which, if left untreated, results in a subacute progressive retinal destruction that leads to irreversible blindness. Less commonly, CMV may involve other organ systems, including the central nervous system (CNS), causing polyradiculopathy and ventriculoencephalitis. CMV may cause pneumonitis, which manifests with fever, cough, and dyspnea, and often it is diagnosed during coinfections with *Pneumocystis jirovecii*, *Aspergillus fumigatus*, or other pathogens. Gastrointestinal involvement by CMV includes esophagitis, gastritis, ileitis, colitis, pancreatitis, and hepatitis.

CMV in SOT Recipients

Allostimulation, allograft rejection, and intense pharmacologic immunosuppression combine to provide an environment that permits CMV reactivation after SOT. Depending on the immunologic status of the transplant recipient, CMV can cause either a primary or reactivation disease. Primary infection occurs when a CMV-seronegative (CMV R^-) individual receives an allograft from a CMV-seropositive donor (CMV D^+). This CMV "donor-positive–recipient-negative" (D^+/R^-) mismatch constitutes the highest-risk scenario for CMV disease. Reactivation CMV disease may occur in SOT recipients with prior CMV infection (CMV R^+) before transplantation. If antiviral prophylaxis is not administered, CMV disease occurs during the first 3 months after SOT. The clinical illness due to CMV can be manifested either as a CMV syndrome or as tissue-invasive disease (Table 1). CMV syndrome is characterized by fever and myelosuppression. In roughly half of cases, CMV could invade an organ system, hence the term tissue-invasive CMV disease. The most commonly involved organ is the gastrointestinal tract, accounting for over 70% of tissue-invasive CMV disease cases. Other organs may be involved, and CMV disease may manifest as hepatitis, pneumonitis, myocarditis, polyradiculopathy, meningitis, nephritis, pancreatitis, or retinitis. Unlike the case for patients with AIDS, retinitis occurs very rarely in transplant recipients.

CMV in HSCT Recipients

Allogeneic stimulation and intense pharmacologic immunosuppression to treat or prevent graft-versus-host disease increase the risk of CMV disease after HSCT. Since the severity of CMV disease is related to the degree of immunosuppression, allogeneic HSCT recipients are at higher risk of CMV disease than are autologous HSCT recipients. Like the case with SOT recipients, CMV in HSCT patients can be either a primary or reactivation disease. The clinical manifestations are also similar to those of SOT patients and can be classified as CMV syndrome or tissue-invasive disease. Allogeneic HSCT recipients are particularly at very high risk of CMV pneumonia. On chest radiographs, CMV pneumonitis is typically

Table 1. Risk factors and clinical manifestations of CMV disease in immunocompromised patients

Patient population	Major risk factors	Clinical disease
Patients with AIDS	CD4 count of < 50 cells/mm^3	CMV retinitis is the most common manifestation; other organ systems may be involved to cause polyradiculopathy, hepatitis, pneumonitis, or gastrointestinal disease
SOT patients	Lack of preexisting CMV-specific immunity, immunosuppressive drugs, donor transmission of CMV, allograft rejection	CMV syndrome of fever and myelosuppression; tissue-invasive disease, which is most commonly gastrointestinal CMV disease—Transplanted allografts are particularly at risk; numerous indirect effects, such as acute and chronic allograft rejection, increased risk of opportunistic infections, and mortality
HSCT patients	Lack of preexisting CMV-specific immunity in donors and recipients, immunosuppressive drugs, graft-versus-host disease	CMV syndrome of fever and myelosuppression; tissue-invasive disease, which is most commonly gastrointestinal CMV disease; CMV pneumonitis can be severe and fatal; numerous indirect effects, such as increased risk of opportunistic infections and mortality
Newborns	Immature immune system	Congenital CMV disease, which is manifested as cytomegalic inclusion disease and characterized by jaundice, petechial rash, microcephaly, hepatosplenomegaly, chorioretinitis, cerebral calcifications, hearing defects, lethargy, and seizures

manifested as diffuse interstitial infiltrates, although a nodular pattern may be observed. CMV pneumonia has an acute onset and rapid clinical course and could be fatal. Other organ systems may be involved, including the gastrointestinal tract, liver, and CNS.

CMV in Newborns and Infants

CMV infection in the susceptible pregnant woman can lead to an intrauterine fetal infection and congenital CMV disease. In most cases, the illness occurs in infants born to mothers with primary CMV infection during pregnancy. The risk is highest if infection occurs during the first half of pregnancy, although it may occur at any stage. The illness in immunologically immature newborns is cytomegalic inclusion disease, which is characterized by jaundice, petechiae, hepatosplenomegaly, microcephaly, motor disability, chorioretinitis, cerebral calcifications, and multiple organ involvement. Shortly after birth, infants manifest with lethargy and seizures. Extraneural abnormalities such as hearing deficits may occur. Congenital CMV infection is often fatal, and the infant may die within days to weeks after birth.

CMV in Other Immunocompromised Hosts

Other immunocompromised patients, such as patients with leukemia, lymphoma, or other malignancies or those on immunosuppressive treatment, may develop CMV disease. For example, CMV disease is common among patients receiving alemtuzumab for the treatment of lymphoma (22). The manifestations of CMV disease in this population are similar to those observed in transplant patients.

LABORATORY METHODS FOR DETECTION OF CMV

A variety of methods are used for the laboratory diagnosis of CMV (Table 2). Generally, the methods can be classified into nonmolecular and molecular tests (70). The nonmolecular techniques include (i) the isolation or growth of virus from blood, urine, or other body fluids (viral culture); (ii) the demonstration of CMV-specific immunoglobulin G (IgG) and IgM (serology); (iii) the detection of virion components in clinical samples, such as in the pp65 antigenemia assay; and (iv) the demonstration of characteristic nuclear inclusion-bearing cells (histopathology). Viral nucleic acid amplification and detection through PCR constitute the major molecular method for CMV detection, although non-PCR methods are also available.

Viral Culture

The detection of CMV by culture techniques, either through conventional tube cell culture or shell vial assay, was the primary laboratory method for the diagnosis of CMV for many decades (70). Clinical specimens such as blood, urine, respiratory secretions, cerebrospinal fluid (CSF), or other body fluids are inoculated into conventional tube cell cultures, such as cultures of MRC-5 human embryonic lung fibroblasts. Thereafter, the presence of cytopathic effect (CPE), which is indicated by large, rounded infected cells that contain cytoplasmic "ground glass"-appearing inclusions, is observed (Fig. 1). When CPE is observed, the identity of the specific viral isolate is then confirmed by immunofluorescence with the use of specific antisera (such as antisera against CMV) (Color Plate 1). Cell culture assays are highly predictive of CMV disease and are relatively specific, especially if a monoclonal antibody reagent is used (27, 60, 73). The major drawback of viral culture is its poor sensitivity. In a study of 47 liver biopsy specimens obtained from patients with histopathologically proven CMV inclusion bodies, the sensitivity of cell culture was only 52% (58). The slow growth of CMV in human fibroblast cultures is another major limitation of culture. Conventional culture techniques take at least a week and may take up to 4 weeks for CPE to be observed. Even after CPE is observed, a confirmatory test is required. This long turnaround time limits the clinical utility of culture for rapid diagnostic and therapeutic decision-making processes.

The shell vial assay, which utilizes low-speed centrifugation and uses monoclonal antibodies directed against early antigens of replicating CMV, accelerates the diagnostic process (60, 70, 73). Some examples of techniques that detect CMV in shell vial culture systems are direct and indirect fluorescent monoclonal antibody staining and in situ hybridization with a biotinylated DNA probe or a horseradish peroxidase-labeled probe directly linked to a DNA molecule (14, 53). In comparative studies, the shell vial assay not only had a more rapid turnaround than that of conventional cultures but was also more sensitive (27). Nonetheless, compared to antigenemia and molecular assays (discussed below), the culture-based systems generally suffer from a relatively poor sensitivity. The labor-intensive procedure, slow turnaround time, and poor sensitivity all combine to limit the use of viral culture in contemporary clinical practice.

Table 2. Laboratory methods for diagnosis of CMV infection[a]

Method	Principle	Sample processing	Turn-around time (h)[b]	Results and clinical utility	Advantages	Disadvantages
Nonmolecular methods						
Serology	Detection of antibody against CMV (IgG, IgM)	Requires serum samples	6	CMV IgG indicates past CMV infection, CMV IgM implies acute or recent infection	Prognostication of patients prior to transplant, screening post-transplantation for evidence of infection, diagnosis of acute congenital CMV disease	May require paired acute- and convalescent-phase sera for complete interpretation, not helpful for immunocompromised patients who have attenuated and delayed antibody production
Histopathology	Histologic detection of CMV-infected cells	Requires tissue specimens microscopy	24–48	Detection of CMV-infected cells indicates the presence of active tissue-invasive disease	Confirmatory test for tissue-invasive disease, highly specific	Need for invasive method to obtain tissue specimen
Virus cultures						
Tube culture	Viral growth and CPE	Cell culture facility, light microscopy	1–4 wk	Detection of characteristic CPE indicates presence of virus	Specific for CMV infection, the viral isolate can be tested for phenotypic susceptibility	Prolonged processing time is not clinically useful in real-time, poor sensitivity, requires viable CMV, very slow CPE
Shell vial assay	Viral growth	Cell culture facility, immunofluorescence detection	16–48	Infectious foci detected by monoclonal antibody directed to immediate-early antigen of CMV	Specific for CMV infection, more sensitive and rapid than conventional tube cultures	Relatively low sensitivity compared to molecular methods, rapid decrease of CMV activity in clinical specimens
Antigenemia	Detection of pp65 antigen	Recovery of PMN within 4 to 6 hours, cytospin, light microscopy or immunofluorescence	6	Number of CMV-infected cells per total (e.g, 5×10^4) cells evaluated, early detection of CMV replication	Rapid diagnosis of CMV infection, quantification used as guide for preemptive therapy	Subjective interpretation of results, requires rapid processing, not useful for leukopenic patients, lack of standardization
Molecular methods (Nucleic acid detection)						
Cobas Amplicor CMV Monitor	PCR amplification and detection of CMV DNA	Plasma sample recommended, Cobas Amplicor instrument	4	Reported as CMV copies per ml of plasma (lowest limit, 400 copies/ml), rapid detection of CMV infection, monitors CMV DNA decline, surrogate marker for antiviral drug resistance	Highly sensitive for CMV infection, highly specific for CMV infection, monitors response to therapy, short turn-around time	No widely accepted threshold for predicting CMV disease, cannot be compared directly with other PCR platforms
LightCycler	PCR amplification and detection of CMV DNA, detection by FRET mechanism	Various blood compartments, LightCycler instrument, nucleic acid extraction	1–2 (45 min for PCR amplification and detection)	Viral genomic copies per PCR input	Real-time PCR assay; short turn-around time, highly sensitive and specific, monitors response to therapy, surrogate marker for antiviral drug resistance	No widely accepted threshold for predicting CMV disease, cannot be compared directly with other PCR platforms, not standardized across centers, home brew assay

TaqMan-based PCR assays	PCR amplification and detection of CMV DNA	Various blood compartments, PCR instrument, nucleic acid extraction	1–2	Viral genomic copies per PCR input	Real-time PCR; short turnaround time, highly sensitive and specific, monitors response to therapy, surrogate marker for antiviral drug resistance	No widely accepted threshold for predicting CMV disease, cannot be compared directly with other PCR platforms, not standardized across centers, home brew assay
NucliSens pp67	mRNA detection	Whole blood is recommended, sample needs to be processed within 24 h (or kept in lysis buffer at −80°C indefinitely)	6	Qualitative assay, early detection of CMV replication	Highly specific for viral replication because it measures the replicative intermediate, clinical utility for preemptive therapy, clinical utility for monitoring response to treatment	Qualitative assay, less sensitive than DNA assays
Digene assay	DNA-RNA hybrid	Whole blood, delayed processing possible	6	Number of CMV copies per ml (lowest limit of detection 7×10^2 copies per ml of whole blood)	Highly specific for viral replication, rapid procedure, simple processing of samples	

[a]The examples noted in this table are representative of multiple assays and are not intended to be comprehensive. PMN, polymorphonuclear cells; FRET, fluorescence resonance energy transfer.
[b]Unless indicated otherwise.

Serology

Detection of CMV IgG and IgM antibody responses has been used to diagnose acute or previous CMV infection (70). A few examples of serologic assays to detect CMV antibody are the Vidas CMV IgG and IgM enzyme-linked fluorescence immunoassay and the Abbott CMV Total AB enzyme immunoassay. In principle, a serologic assay utilizes latex particles or beads that are coated with CMV antigens recognized by the CMV antibody present in a patient's serum or plasma. The bound patient antibodies are detected by anti-human immunoglobulins conjugated with horseradish peroxidase. Clinically, the presence of CMV IgM or a ratio of ≥2 for paired serum CMV IgG titers is indicative of acute or recent CMV infection. Detecting CMV IgM in umbilical cord blood suggests acute intrauterine CMV infection. However, the presence of CMV IgM may not always be indicative of acute primary infection, since IgM may persist for months in some individuals and since it is also produced during CMV reactivation. The acuity of CMV infection may be differentiated by the use of IgG avidity testing, which distinguishes primary CMV infection from reactivation. In using this test, patients with low-avidity IgG are believed to have acute CMV infection. Low-avidity IgG persists for approximately 17 weeks following acute infection, and its full maturation takes approximately 25 weeks after the onset of clinical symptoms (45).

Overall, there are important limitations to the clinical application of serology for the diagnosis of acute CMV infection in immunocompromised individuals. First, there is a time lag between the onset of clinical illness and the appearance of CMV IgM, and as a result, the diagnosis and treatment of CMV disease may be delayed if the diagnosis is based solely on the detection of CMV IgM. Second, some individuals may have CMV IgM antibody that persists long after the resolution of acute clinical illness, and thus, detection of CMV IgM does not necessarily indicate active infection. Third, some individuals, such as immunocompromised individuals, fail to develop a serologic response or may have a markedly delayed or attenuated response (70). These variables have contributed to the nonreliability of CMV IgM and IgG serology testing for the diagnosis of acute CMV infection in immunocompromised individuals (70). Probably the most important indication for serology among immunocompromised individuals is in the assessment of risk and susceptibility. Specifically, CMV serology is useful in assessing CMV exposure during pretransplantation evaluation of transplant candidates in order to determine the risk of either primary or reactivation CMV infection

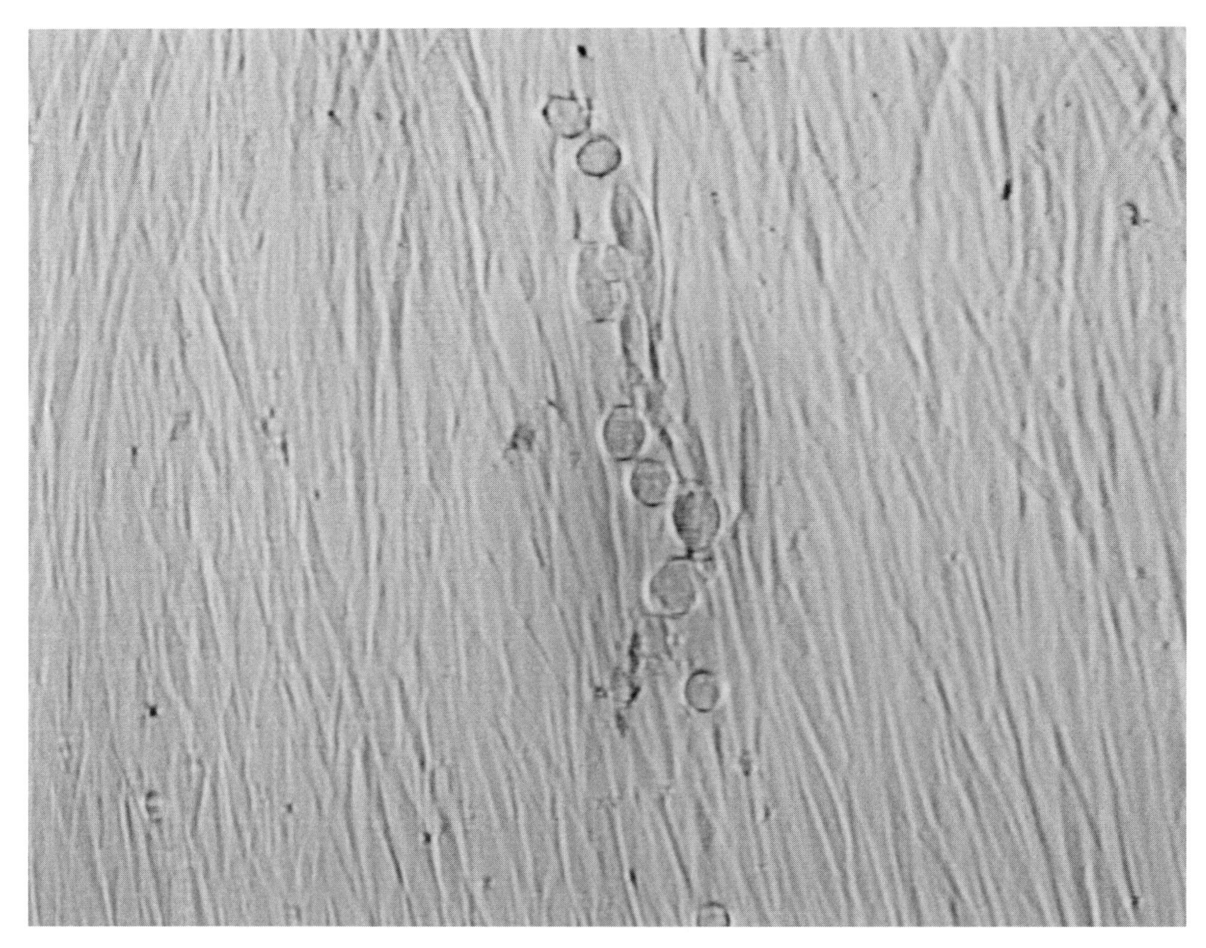

Figure 1. CMV-induced CPE. The image shows an unstained preparation. Original magnification, ×100. (Courtesy of Annette Ehrich.)

after transplantation. Likewise, serology is useful in assessing the CMV status of an individual with AIDS, as it will guide clinicians in prognosticating an individual's risk of developing CMV retinitis or other CMV-related diseases.

Antigenemia

The direct detection of CMV antigens in neutrophils by using monoclonal antibodies against the CMV matrix protein pp65 has proven very useful in the diagnosis of CMV in immunocompromised individuals (70). In this assay, cytospin preparations of a patient's peripheral blood mononuclear cells are formalin fixed and permeabilized, and the presence of CMV lower matrix protein pp65 is demonstrated by immunofluorescence, immunoperoxidase, and other antigen detection methods (70). Various antigen detection methods are commercially available and have comparable sensitivities, specificities, and performance characteristics (76). One example is the demonstration of CMV pp65-positive cells by immunoperoxidase staining with monoclonal antibody. Generally, the number of positive cells is counted and reported for every 50,000 leukocytes on the cytocentrifuge preparation (44). Quantitative assessment of antigen-positive cells assists in risk stratification and in monitoring therapeutic responses. Across many transplant centers, the pp65 antigenemia assay is used to determine the risk of subsequent disease in transplant patients, to diagnose acute CMV infection in patients with AIDS, and to monitor response to treatment (70). Studies have shown conclusively that the pp65 antigenemia assay is much more sensitive than shell vial and tube culture systems (77). Others have demonstrated that the performance of the pp65 antigenemia assay is comparable or equivalent to those of some molecular methods (discussed below), such as the Cobas Amplicor CMV Monitor (63) or Hybrid Capture CMV DNA assay, with both assays demonstrating 95% sensitivity and 94% specificity (52).

The major limitation of pp65 antigenemia assays is the lack of standardization across laboratories and the subjective and operator-dependent nature of the assay (70). A comparative evaluation of pp65 antigenemia assays among four laboratory sites, however, did not demonstrate significant differences in assay performance (76). Another limitation of this assay is the need to have a sufficient number of neutrophils for the test to be performed. This may not always be possible for some individuals, such as leukopenic patients undergoing myeloablative and cytotoxic chemotherapy, including the large population of HSCT recipients. Since the assay detects virus-associated cells, the presence of free viruses in biological fluids such as plasma is not detected. Some experts argue that the presence of virus in a cell-free environment such as plasma is more indicative of active viral replication. Another limitation of the pp65 antigenemia assay is the need to process the clinical specimens shortly after collection (i.e., within 6 to 8 hours), thereby limiting its utility in major referral laboratories (which process samples shipped from distant sites). Hence, a long hands-on time, the subjective nature of quantification, the need for an adequate number of cells, and the immediate sample processing time combine to make CMV pp65 antigenemia assay a less useful test for many referral laboratories that perform large-scale testing of clinical samples (1).

Molecular Methods for CMV Detection

The limitations of nonmolecular diagnostic tests have paved the way for the increasing clinical application of various molecular methods for the diagnosis of CMV. In principle, the molecular assays are based on the detection and/or amplification of CMV nucleic acid. For over a decade now, there has been an increasing volume of scientific data to support the use of molecular assays in the diagnosis of CMV. Generally, molecular methods have higher sensitivities than do nonmolecular methods. Among the various molecular assays, CMV PCR is the most widely used methodology. The basic principle of CMV PCR is to generate a huge number of target CMV gene sequence copies that can easily be detected and, in the case of quantitative assays, easily be quantified. The amount of CMV DNA theoretically doubles for every PCR cycle, resulting in an exponential increase in the quantity of amplified DNA. The amplification process, combined with the fact that the sample does not require viable virus, greatly increases the sensitivity of the PCR assay compared to that of culture systems. In contrast to serology, molecular assays have the ability to detect infection at early stages, even prior to serologic conversion. The major drawbacks to the various molecular assays are the potential for contamination and the lack of standardization for the various assays. Different molecular assays amplify different targets and use different types of samples, thereby limiting direct comparison of results. For example, the commercially available quantitative Cobas Amplicor CMV Monitor test (Roche) targets and amplifies the UL54 region of CMV (DNA polymerase gene), while others, such as in-house-developed (home brew) assays, may amplify a different gene sequence, such as the UL83 gene (64). Several of these molecular assay methods are described below.

Conventional CMV PCR assay

The conventional CMV PCR assay has long been used for the rapid diagnosis of CMV disease (54) and has been integrated into clinical practice as a surveillance measure that could assist in the early diagnosis of CMV. As a result of the amplification process, the conventional CMV PCR assays are highly sensitive. In one report, the sensitivity of a qualitative PCR assay was 100%, and its negative predictive value was 100% (54). As a result, conventional PCR has been used in guiding preemptive therapeutic strategies in order to prevent the progression of CMV infection to overt clinical disease (54). However, despite its high sensitivity, the conventional CMV PCR assay may not be rapid enough for timely diagnosis, and because of its qualitative nature, it may not necessarily differentiate active from latent infections (71). The PCR method requires a high level of technical expertise and additional time to detect the amplified DNA. The potential risk of contamination should be minimized by ensuring the physical separation of spaces that are dedicated to reagent preparation, nucleic acid extraction, and the performance of PCR amplification and detection.

Quantitative PCR assays

Quantitative PCR assays, such as the Cobas Amplicor CMV Monitor assay (Roche), represent a major improvement compared to conventional qualitative PCR assays, as they afford the ability to quantify the amounts of CMV in clinical samples. Studies with the transplant population and with CMV-infected newborns have demonstrated the high degree of correlation between CMV load and the risk of CMV disease. In a comparative study of the Cobas Amplicor CMV Monitor assay and an in-house (home brew) PCR assay, there was a high concordance between the two PCR assays (47). Among patients who developed CMV disease, the CMV loads were higher and persisted longer than those for patients who did not develop CMV disease (47). One study demonstrated the good correlation between the Cobas Amplicor CMV Monitor assay and the nonmolecular CMV pp65 antigenemia assay for transplant patients (1), while another study demonstrated the superiority of PCR assay compared to the antigenemia assay for infants with congenital CMV infection (43). The major advantage of quantitative PCR assays is the ability to sensitively measure the amount of virus in a clinical sample. Because of this ability, these assays have assisted in the prognostication of CMV disease in immunocompromised individuals and in assessing the responses to treatment. However, because there are different gene targets among the various quantitative PCR assays, it is difficult to directly compare viral load outputs. To illustrate, in a head-to-head comparative analysis between a home brew quantitative CMV PCR assay based on the LightCycler format (described below) and the Cobas Amplicor CMV Monitor assay, the absolute viral loads were significantly different, although highly correlated (65, 66).

Real-time CMV PCR assays

Real-time CMV PCR assays, which allow for simultaneous amplification and detection of viral targets, represent a major advance that revolutionized the diagnosis of CMV (21). Different real-time PCR assays generally follow very similar principles and methodologies. In most real-time PCR assays (and in contrast to PCR using conventional thermocyclers), the processes of amplification and detection occur simultaneously in a closed system so as to reduce contamination. As a result of the shorter thermocycling time and simultaneous amplicon detection, real-time PCR systems have a short turnaround time. In contrast to other assays, such as viral cultures, which take several days to complete, the results of the real-time PCR assays can be reported within 1 to 2 days of specimen receipt. Real-time PCR performed using a LightCycler instrument (Roche), for example, can be completed in 30 to 40 minutes (65). This short turnaround time is especially useful for high-risk patients, as a rapid diagnostic and therapeutic decision-making process is important in the clinical care of patients (71).

In addition to the LightCycler (Roche) (65) platform, other automated real-time PCR technologies that are currently available include the ABI Prism (Applied Biosystems), iCycler (Biorad), SmartCycler (Cepheid), and Mx4000 (Stratagene) systems, among others (48). Different targets have been used in various real-time PCR assays, including commercialized and institutionally developed homebrewed methods (21), and they include UL54 (65), US17 (38, 81), UL65 (62), and UL83 (25), among others (21). Some experts have recommended the use of analyte-specific reagents or assay kits which have passed through good manufacturing practices (21). Real-time nucleic acid detection assays may utilize one of the following detection probes: 5′ nuclease (TaqMan probes), molecular beacons, and fluorescence resonance energy transfer hybridization probes.

The recommended clinical sample and volume of sample vary among assays, and this further contributes to the variability of the results obtained from each system. The range of samples that have been used include plasma, serum, peripheral blood leukocytes, peripheral blood mononuclear cells, and whole blood (66). Which of these components of blood is ideal for PCR testing has been the subject of numerous investigations (66). Based on the collective interpretation of these studies, it appears that whole blood may provide a more sensitive and convenient (i.e., easier to process) specimen for CMV detection than do other blood components (24, 66). Several manufacturers have developed automated nucleic acid extraction instruments to facilitate the diagnostic process, such as the MagnaPure (for use with LightCycler systems) and ABI Prism 6700 (for use with ABI Prism systems) instruments.

Several investigators have utilized real-time PCR in the diagnosis and management of CMV in immunocompromised individuals (28, 48, 55, 56, 65-67, 70). However, differences in amplification, detection, components, and reagents make direct comparison among real-time PCR assays difficult. Real-time PCR assays using the LightCycler and TaqMan platforms have been compared with the nonmolecular CMV pp65 antigenemia assay in numerous studies, and while many of these studies suggest that real-time PCR assays are superior regarding early diagnosis of CMV in SOT recipients (33), there is also a high degree of correlation between pp65 antigenemia and molecular assays (48).

NASBA

Nucleic acid sequence-based amplification (NASBA), an isothermal reaction that uses CMV RNA (instead of DNA) as a substrate for reverse transcriptase (RT), is another molecular method for CMV detection. It is based on the premise that the presence of RNA is more indicative of active viral replication. One example of this molecular diagnostic approach is the Nuclisens pp67 assay (bioMerieux). This method has been evaluated with immunocompromised individuals, including a comparative study with a CMV DNA PCR assay for 33 allogeneic HSCT patients (32). Overall, there were high rates of agreement between PCR and NASBA in a per-sample (85.3%) and per-patient (87.9%) analysis (32). There was no difference between PCR and NASBA assays in the time to the first positive test result, but the time to the first negative test result upon initiation of antiviral therapy was significantly shorter for the NASBA assay (32). In another study that compared a real-time PCR assay and NASBA, the PCR assay was highly sensitive, but with modest specificity (95% and 50%, respectively), while the converse was true for pp67 NASBA, which had poor sensitivity (20%) and high specificity (93%) (55). For thoracic organ transplant recipients, the low sensitivity (26 to 56%) of NASBA compared to pp65 antigenemia assay limited its potential for guiding preemptive therapy (46). When applied to bronchoalveolar lavage specimens, NASBA may occasionally identify CMV in specimens that were negative by shell vial culture assay (i.e., NASBA is more sensitive than culture), but it does not have sufficient sensitivity or positive predictive value to be employed routinely for the management of CMV disease (41, 46). On the other hand, others have found this assay to be useful for CMV surveillance and in guiding preemptive therapy (6, 26). These contradictory results indicate that differences in patient populations and clinical practices affect the clinical utility of NASBA (and other CMV tests).

Hybrid capture assay

Hybrid capture assay is another molecular method that utilizes an RNA or DNA probe that hybridizes to a CMV DNA or RNA target, respectively. The hybrids are then captured on a solid phase coated with capture antibodies specific for RNA-DNA hybrids. The clinical utility of quantitative hybrid capture assay for CMV detection was demonstrated with bronchoalveolar lavage and blood samples from lung transplant recipients (11). In a comparative study, this assay had performance characteristics comparable to those of antigenemia assay, at least in terms of sensitivity and specificity (52).

Histopathology

For the diagnosis of tissue-invasive disease, examination of biopsy specimens for the presence of CMV is essential. The specimens are obtained by the use of fine-needle aspiration or through open surgical approaches. The most common specimens in the evaluation of tissue-invasive CMV disease are lung, gastrointestinal, and liver tissues. The choice of tissue specimen is dictated by the clinical presentation of the patient. For example, when a patient presents with diarrhea and abdominal pain, upper gastrointestinal endoscopy and colonoscopy are performed to detect the presence of characteristic mucosal ulcers, hyperemia, and inflammation associated with gastrointestinal CMV disease. At the same time, gastrointestinal tissue is obtained to demonstrate the presence of tissue-invasive CMV disease. The histologic criteria for the diagnosis of tissue-invasive CMV disease vary widely, from the demonstration of viral inclusion disease to the demonstration of CMV-specific antigens or DNA in tissue specimens by in situ techniques. Biopsy specimens may be stained with hematoxylin-eosin stain to evaluate histological alterations and to demonstrate the presence of giant cells with typical intracellular viral inclusions (40) (Color Plate 2). Immunocytochemical studies may be performed to detect CMV antigens in tissues specimens (Color Plate 3). For example, the presence of viral antigens in biopsy specimens may be demonstrated by indirect immunoperoxidase staining with monoclonal antibodies against CMV antigens, such as the viral matrix protein pp65 (44). To confirm the presence of CMV in tissue, CMV DNA may be demonstrated by in situ hybridization, such as that with the use of a biotinylated DNA probe (14, 44, 80).

GOALS AND ALGORITHMS OF LABORATORY TESTING FOR CMV INFECTION

The various assays for CMV detection discussed above can be used in the different aspects of CMV management of the immunocompromised patient. Specifically, these assays may be utilized for surveillance and screening, for prognostication and assessment of risk, for rapid and accurate diagnosis, for evaluation of therapeutic response, for assessment of the risk of relapse, and for the detection of antiviral drug resistance. Illustrations on the utility of the various assays in the management of CMV disease in immunocompromised hosts are depicted in Fig. 2 to 4.

Screening, Surveillance, and Prevention

One of the most common indications for laboratory testing of CMV is the assessment of disease risk. Depending on the clinical situation, serologic, antigenemia, and molecular testing may be utilized. A typical situation for a transplant recipient is depicted in Fig. 2. CMV serology is used to determine the risk of primary infection in pregnant women or of reactivation in AIDS patients. In the field of transplantation, CMV serology is performed on blood samples from the prospective organ donor and recipient to determine the risk of primary infection or reactivation disease. The knowledge of these risks influences the type of preventive efforts for each patient population. Specifically, a CMV D^+/R^- mismatch SOT recipient has the highest risk of primary CMV disease and therefore would benefit from antiviral prophylaxis. Some centers have also used serology to determine ongoing risk of primary CMV disease after transplantation, since seroconversion would suggest protection from subsequent CMV disease (37).

Serial surveillance for CMV replication with the use of antigenemia or PCR assay is a common practice during the early period after transplantation. This diagnostic approach is an integral component in the strategy of preemptive therapy against CMV disease (Fig. 3). Using this strategy, blood samples from transplant patients are collected on a weekly basis during the first 8 to 12 weeks after transplantation and tested for the presence of CMV, either by pp65 antigenemia or molecular assay. If CMV above a certain threshold is detected, the risk of subsequent CMV disease is high and therefore antiviral treatment is initiated. The success of this approach is therefore dependent on the diagnostic test, which should be highly sensitive and specific and have good predictive characteristics. In this context, both the PCR and antigenemia assays have been tested and compared. Currently, however, the choice of which assay (PCR versus antigenemia) to use is much debated (10, 32, 50, 55, 65, 76). Both antigenemia and molecular methods have demonstrated clinical utility in detecting CMV and guiding preemptive therapy (55, 61). It is generally recommended that each center optimize its diagnostic approach according to its clinical practices and patient population. The main advantages of the molecular assays are the sensitive detection of CMV with quantitative results, short turnaround time, and the ability to perform the assay on stored specimens with nonviable virus. The pp65 antigenemia assay, on the other hand, also provides a rapid measure of viral replication, but it is limited by subjectivity and requires immediate specimen processing and thus would be logistically difficult for samples that need shipping (such as in major referral laboratories).

Prognostication and Risk Assessment

For immunocompromised hosts, CMV pp65 antigenemia, CMV PCR, and other molecular tests have been used to predict the development of CMV

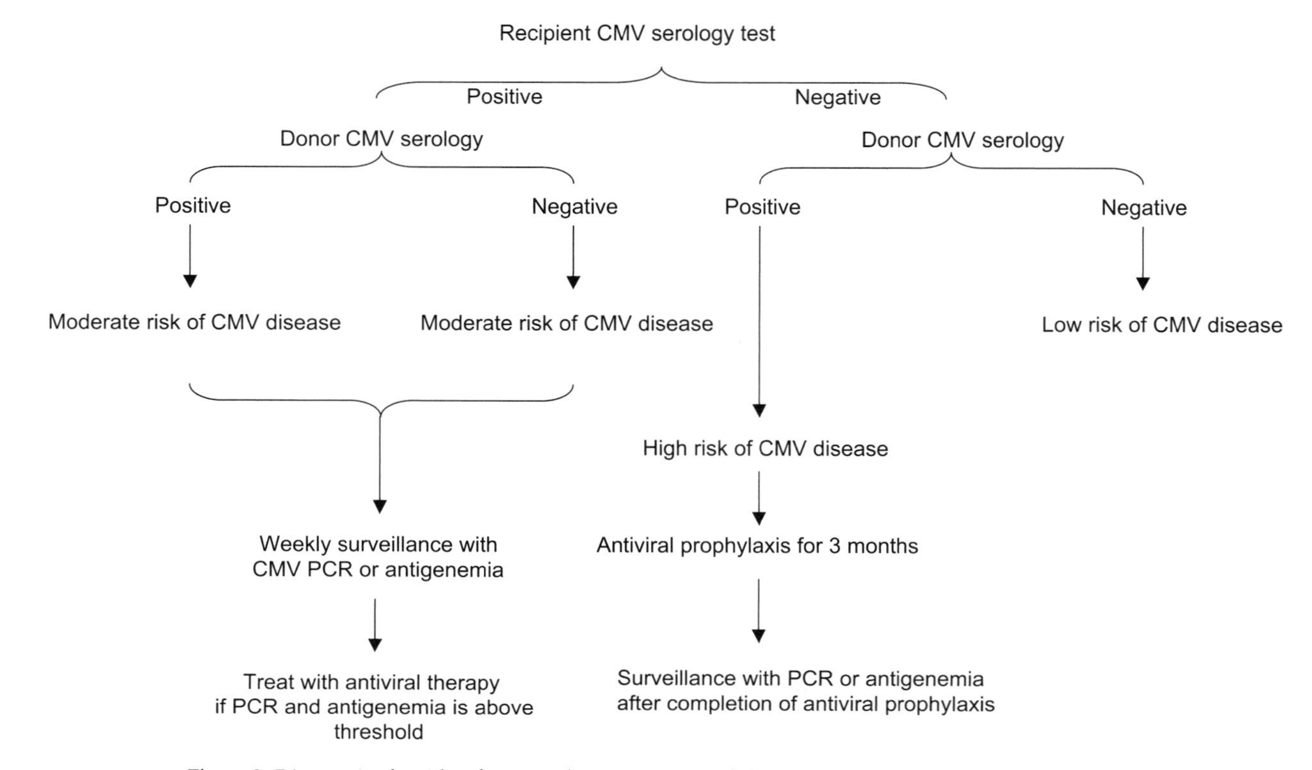

Figure 2. Diagnostic algorithm for transplant recipients and their donors with the use of CMV serology.

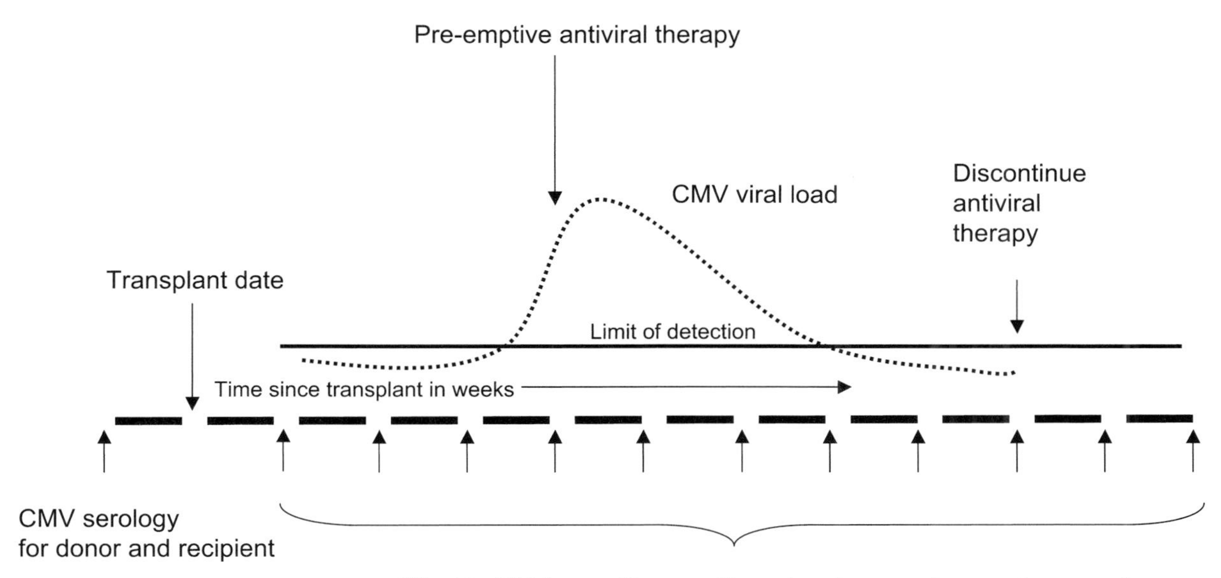

Figure 3. Illustrative algorithm utilizing various diagnostic assays for a transplant recipient prior to transplantation and during the period following transplantation.

disease. Among transplant patients, it is well described that a higher degree of CMV replication (as measured by the viral load) translates to a higher risk of progressing to clinical disease. However, a clinically relevant CMV threshold that highly predicts the development of CMV disease is not yet defined. This difficulty in defining a widely applicable viral load threshold is due to the variability of the different assays, clinical samples, and patient populations and their immunosuppressive regimens. In this regard, the kinetics of CMV replication for an individual patient (i.e., trends in viral load) may be a better and more clinically useful measure for predicting CMV disease than the absolute viral load number (19, 65, 71). Molecular methods such as PCR have also been used in predicting subsequent CMV retinitis among AIDS patients (23).

Diagnosis of CMV Infection

The diagnosis of CMV disease in the immunocompromised host requires the presence of clinical signs and symptoms and the demonstration of CMV in clinical specimens (Fig. 4). Traditionally, viral culture was the standard assay for the laboratory diagnosis of CMV infection. However, the poor sensitivity and long turnaround time of viral culture systems limited their clinical utility and led to the adaptation of more rapid and sensitive tests, such as the pp65 antigenemia and molecular assays (70). For instance, molecular assays such as PCR are now considered the methods of choice for the detection of CMV in CSF of patients with CMV encephalitis or polyradiculopathy (7, 30). Congenital CMV infection can also be identified by testing neonatal dried blood on filter paper (known as Guthrie cards) for the presence of CMV DNA by PCR (2). In one study, this test was found to be 100% sensitive and 99% specific for the diagnosis of congenital CMV infection in symptomatic and asymptomatic babies (2, 3). On the other hand, CMV serology is not usually used for the diagnosis of acute CMV infection in immunocompromised hosts. A negative CMV serology assay does not completely rule out primary CMV infection. For example, CMV IgM is not detectable in 10 to 30% of cord blood sera from infants who demonstrate infection in the first week of life. Likewise, approximately 25% of pregnant women with primary CMV infection do not have detectable CMV IgM within 2 months after infection.

Monitoring Therapeutic Response

The quantitative results generated by pp65 antigenemia assay and the various molecular tests have allowed for their use as a monitor of the response to treatment, to individualize and assess the duration of antiviral treatment, and to determine the risk of

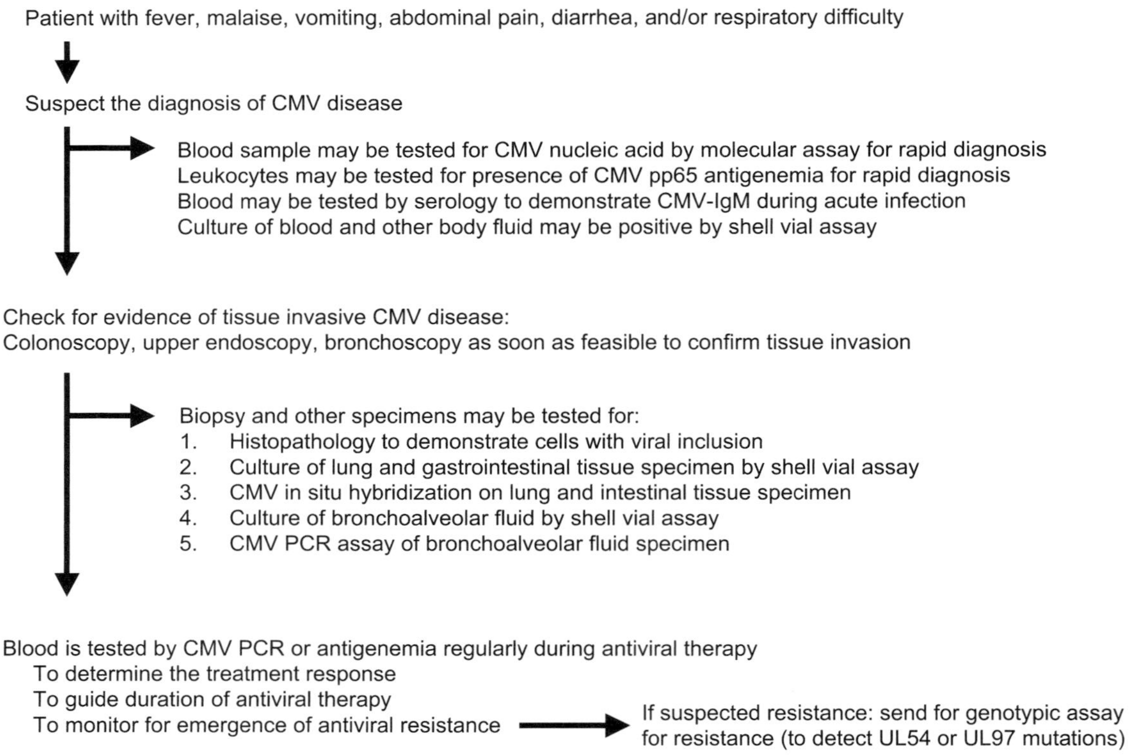

Patient with fever, malaise, vomiting, abdominal pain, diarrhea, and/or respiratory difficulty

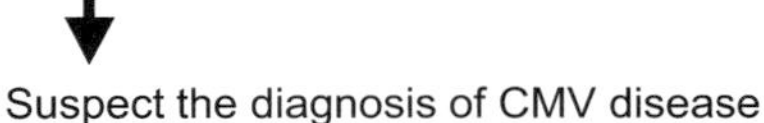

Suspect the diagnosis of CMV disease

→ Blood sample may be tested for CMV nucleic acid by molecular assay for rapid diagnosis
Leukocytes may be tested for presence of CMV pp65 antigenemia for rapid diagnosis
Blood may be tested by serology to demonstrate CMV-IgM during acute infection
Culture of blood and other body fluid may be positive by shell vial assay

Check for evidence of tissue invasive CMV disease:
Colonoscopy, upper endoscopy, bronchoscopy as soon as feasible to confirm tissue invasion

→ Biopsy and other specimens may be tested for:

1. Histopathology to demonstrate cells with viral inclusion
2. Culture of lung and gastrointestinal tissue specimen by shell vial assay
3. CMV in situ hybridization on lung and intestinal tissue specimen
4. Culture of bronchoalveolar fluid by shell vial assay
5. CMV PCR assay of bronchoalveolar fluid specimen

Blood is tested by CMV PCR or antigenemia regularly during antiviral therapy
To determine the treatment response
To guide duration of antiviral therapy
To monitor for emergence of antiviral resistance → If suspected resistance: send for genotypic assay for resistance (to detect UL54 or UL97 mutations)

Figure 4. Algorithm for the diagnosis of CMV disease in an immunocompromised patient.

disease relapse. For example, the presence of detectable virus at the end of antiviral treatment indicates a higher risk of CMV disease relapse (35, 36, 65, 67, 74). Certain results of these molecular tests could also be an indication of the development of a viral strain with drug resistance (70); a rising viral load during antiviral treatment suggests drug-resistant CMV (70).

CLINICAL APPLICATION AND INTERPRETATION OF DATA

CMV in Patients with AIDS

CMV retinitis

CMV retinitis is diagnosed based on findings of characteristic white fluffy retinal infiltrates and hemorrhages on fundoscopic examination. Based on this ophthalmologic finding, one can initiate anti-CMV therapy. If one needs to confirm the diagnosis, CMV can be demonstrated in aqueous humor through either viral culture or CMV PCR. In a study that compared CMV DNA loads in the aqueous humors of patients with active and inactive retinitis by the use of the Cobas Amplicor CMV Monitor assay, CMV DNA was detected in 37 of 42 eyes (71%) with active CMV retinitis but not in the eyes with inactive CMV retinitis. Following treatment with intravitreal ganciclovir, a decline in CMV DNA was observed in 29 of 37 eyes (78%) with active CMV retinitis. Hence, CMV DNA detection in aqueous humor can be useful in differentiating active from inactive CMV retinitis and in monitoring therapeutic response (82). Performing CMV PCR with a more accessible bodily fluid, such as the peripheral blood, may provide a less invasive means of diagnosis. The detection by PCR of CMV DNA in the blood of patients with AIDS may predict, several months ahead of time, the development of CMV retinitis (23). The positive predictive value of CMV PCR was 60% (23). Hence, serial monitoring of CMV DNA in the blood of patients at risk of CMV retinitis, such as those with low $CD4^+$ T-cell counts, could guide preemptive therapy (23).

CMV polyradiculopathy and ventriculoencephalitis

Molecular diagnostic assays offer a very sensitive and specific method for detecting CMV in CSF. These molecular tests are now considered standard for the diagnosis of CMV in the CNS (7, 15, 30). CMV polyradiculopathy is suspected in patients with low back pain, urinary retention, progressive bilateral leg weakness, and CSF characterized by high protein, low glucose, and pleocytosis. Quantitation of CMV DNA in CSF may be helpful in confirming the diagnosis, in evaluating disease severity, and for monitoring therapy (7, 30). The sensitivity of PCR is over 90% for diagnosis of neurological CMV disease (7, 30). However, despite the high sensitivities of molecular assays, the analytical performances of the various molecular tests differ. In a three-way comparative study using CSF specimens, nested PCR, Cobas Amplicor CMV Monitor assay, and NucliSens CMV pp67 NASBA had sensitivities of 93.3%, 86.6%, and 93.3%, respectively, and specificities of 93.7%, 93.7%, and 87.5%, respectively, in the diagnosis of CMV encephalitis in HIV-infected patients (5). In another study, the sensitivities of PCR, NucliSens CMV pp67 NASBA, and viral culture were 100%, 84.6%, and 18%, respectively, while their specificities were 90.5%, 100%, and 100%, respectively (83). The corresponding positive predictive values were 68.4%, 100%, and 100%, respectively, while the negative predictive values were 100%, 97.0%, and 82.7%, respectively (83).

Gastrointestinal CMV disease

The gastrointestinal tract is the most common extraocular site of CMV infection in AIDS patients (12). The clinical presentation of gastrointestinal CMV disease, which depends on the site of infection, could be painful esophageal ulcers or extensive enterocolitis. While CMV can be detected in the blood of patients in many of these cases, this is not always the case, and thus the diagnosis requires the demonstration of characteristic lesions on endoscopy and CMV inclusion bodies on histopathology. The tissue specimen obtained during endoscopy can be tested for CMV by the use of viral culture, PCR, or in situ hybridization.

CMV pneumonitis

The diagnosis of CMV pneumonia is difficult to establish in AIDS patients because of the high incidence of prior CMV infection, high prevalence of asymptomatic viral shedding, and frequent presence of other pulmonary pathogens. Isolation of CMV by culture or detection of CMV DNA from bronchoalveolar lavage fluid or lung tissue does not usually differentiate asymptomatic infection from active pneumonitis. CMV is isolated from pulmonary secretions or lung tissue for approximately 50% of HIV-infected patients who undergo bronchoscopic examination. Very high levels of CMV in the bronchoalveolar lavage fluid may, however, indicate true invasive infection (11). While there is no pathognomonic clinical finding to indicate CMV pneumonitis, the diagnosis should be suspected in a patient with fever, cough, dyspnea, and pulmonary infiltrates. Characteristic intracellular inclusions are observed in the lung tissue of patients with CMV pneumonitis. Demonstration of CMV in tissue by viral culture, antigen detection, or nucleic acid testing supports the diagnosis.

CMV in HSCT Patients

CMV syndrome and tissue-invasive disease

Molecular methods for CMV detection and pp65 antigenemia assays are currently the standard methods for the laboratory diagnosis of CMV in HSCT recipients. However, CMV pp65 antigenemia assay may not be optimal during the prolonged neutropenic phase of these patients. After HSCT, patients with CMV syndrome present with fever, myalgias, and some degree of myelosuppression. In some, the manifestations of tissue-invasive CMV disease may occur. The clinical diagnosis of CMV disease is confirmed when the symptoms are accompanied by the detection of CMV nucleic acid or pp65 antigenemia (11). Generally, the higher the level of CMV viral load or antigenemia, the greater the severity of disease. The clinical utility of viral culture for diagnosis is modest because of its low sensitivity, lack of quantitation, and long turnaround time. Likewise, CMV serology is not usually used for diagnosis of CMV disease in HSCT patients.

Preemptive therapy

One of the most common clinical applications of the antigenemia and molecular diagnostic tests is in the strategy of preemptive therapy, an approach of CMV prevention whereby an antiviral drug is administered upon the detection of CMV. Blood samples are collected on a weekly basis and tested for the presence of CMV, either by pp65 antigenemia assay or a molecular diagnostic test. For this to work, blood specimens should be processed immediately so that the results can be provided to the clinician in a timely manner. The key to preemptive therapy is the availability of a rapid diagnostic test with high sensitivity and predictive qualities. The advantages of preemptive therapy over the other major approach for CMV disease prevention (i.e., antiviral prophylaxis) are the reduction in the number of patients receiving antiviral drugs and, hence, reductions in the risk of antiviral resistance, adverse drug effects, and drug cost. For this approach to be effective, the diagnostic surveillance should be performed frequently, which for most centers is once or even twice weekly. In one trial, PCR-guided preemptive therapy with ganciclovir resulted in a significant reduction in the incidence of CMV disease among HSCT patients (17).

The successful implementation of preemptive anti-CMV therapy is therefore dependent on the availability of a sensitive and predictive diagnostic assay. In this regard, real-time molecular assays may offer an advantage over nonmolecular methods. PCR assays have been demonstrated by many investigators to be more sensitive than virus culture and even pp65 antigenemia assay. In a study of 45 HSCT recipients, the clinical values of virus culture, a CMV PCR assay, and RT-PCR assay (for late viral mRNA detection) were compared in the context of CMV disease. The sensitivities were 36% for urine culture, 43% for leukocyte culture, 100% for PCR, and 71% for RT-PCR; the corresponding specificities were 74% for urine culture, 84% for leukocyte culture, 65% for PCR, and 94% for RT-PCR. Collectively, these data indicate that CMV PCR is the most sensitive test, while RT-PCR is the most predictive of CMV disease (29). In contrast, virus culture of urine and leukocyte components had poor sensitivity, thereby limiting the use of culture in the preemptive therapy approach.

Monitoring response to treatment

Molecular techniques have an advantage over nonmolecular techniques (such as virus culture) in their ability to generate quantitative results. In a study of HSCT (and SOT) patients, there was a demonstrable decline in blood CMV DNA level during intravenous ganciclovir treatment (67). The half-life of decline in CMV load (termed viral decay) was estimated to be 1 to 5 days (19, 36, 67, 74). The slope of decline correlated with therapeutic efficacy (67). A longer duration of CMV DNA decline has been associated with CMV relapse (67). In addition, patients with detectable viral loads at the end of treatment are at higher risk of relapse (36, 67, 74). In general, most clinicians monitor for CMV DNA decline during antiviral treatment and demonstrate serial (at least two) weekly negativity for CMV DNA before discontinuing antiviral treatment. In some cases, viral DNA monitoring can serve as an early indirect measure for antiviral drug resistance (e.g., when there is failure of viral load decline during antiviral therapy).

CMV in SOT Patients

CMV syndrome and tissue-invasive disease

The molecular methods for detection of CMV and pp65 antigenemia assays are currently the standard methods for the diagnosis of CMV infection and disease after SOT. Like that in HSCT patients, CMV disease in SOT patients occurs either as a viral syndrome or as tissue-invasive disease. The diagnosis of CMV disease is made when clinical symptoms of fever, myalgias, and chills are accompanied by the detection of CMV nucleic acid or antigen. SOT patients may develop tissue-invasive CMV disease with clinical signs and symptoms related to dysfunction of the organ system affected. While gastrointestinal involvement is the most common manifestation of tissue-invasive disease, the transplanted organ also seems to be more predisposed to develop tissue invasion by CMV. Hence, liver transplant recipients are predisposed to develop CMV hepatitis, lung transplant recipients are predisposed to develop CMV pneumonitis, heart transplant recipients are predisposed to develop CMV carditis, kidney recipients are predisposed to develop CMV nephritis, and pancreas transplant recipients are predisposed to develop CMV pancreatitis. Among various tissue-invasive CMV disease states, CMV interstitial pneumonia is one of the most life-threatening because of the rapid clinical course, which is characterized by fever, cough, dyspnea, and hypoxia. Rapid diagnosis and prompt initiation of antiviral drug treatment are therefore essential. While demonstrating CMV in biopsy specimens is necessary for confirming the diagnosis, the use of rapid molecular tests, such as PCR assays and antigenemia assay, on a more accessible sample, such as the peripheral blood, may facilitate prompt diagnosis. Other appropriate specimens that can be used include bronchial and respiratory specimens. In a recent study that evaluated the hybrid capture assay, the CMV load in bronchoalveolar lavage specimens was more predictive of CMV pneumonitis for some patients than the CMV load in peripheral blood was (11).

Gastrointestinal CMV disease is the most common end organ CMV disease among SOT patients, and its diagnosis is suggested clinically by abdominal pain, nausea, vomiting, and diarrhea. Endoscopic examination shows characteristic mucosal ulcerations, which should be biopsied and tested for CMV. The diagnosis is supported by the detection of CMV in the peripheral blood and confirmed by the demonstration of CMV through in situ hybridization, culture, or PCR with tissue specimens. SOT patients with CMV hepatitis manifest prolonged fever, elevated bilirubin levels, and elevated liver enzyme levels. The diagnosis of CMV hepatitis in liver transplant recipients may be difficult because the clinical presentation may be indistinguishable from graft rejection. The presence of CMV in blood may indicate CMV hepatitis, although the coexistence of CMV hepatitis and allograft rejection is not uncommon.

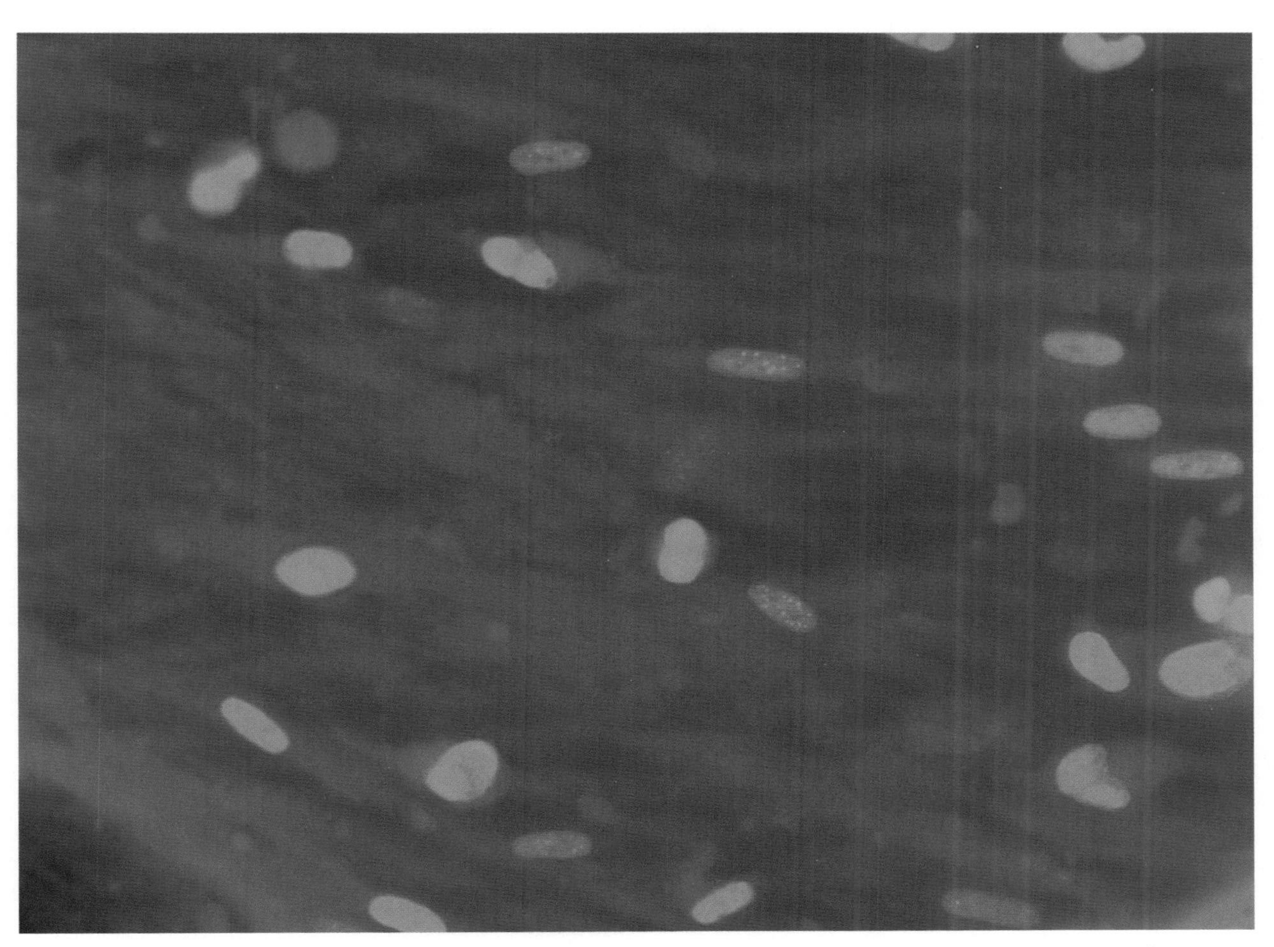

Color Plate 1. Detection of CMV antigens in infected MRC-5 cells. Following shell vial culture, cells are stained with fluorescently labeled antibodies which detect CMV immediate-early antigen. Original magnification, ×200.

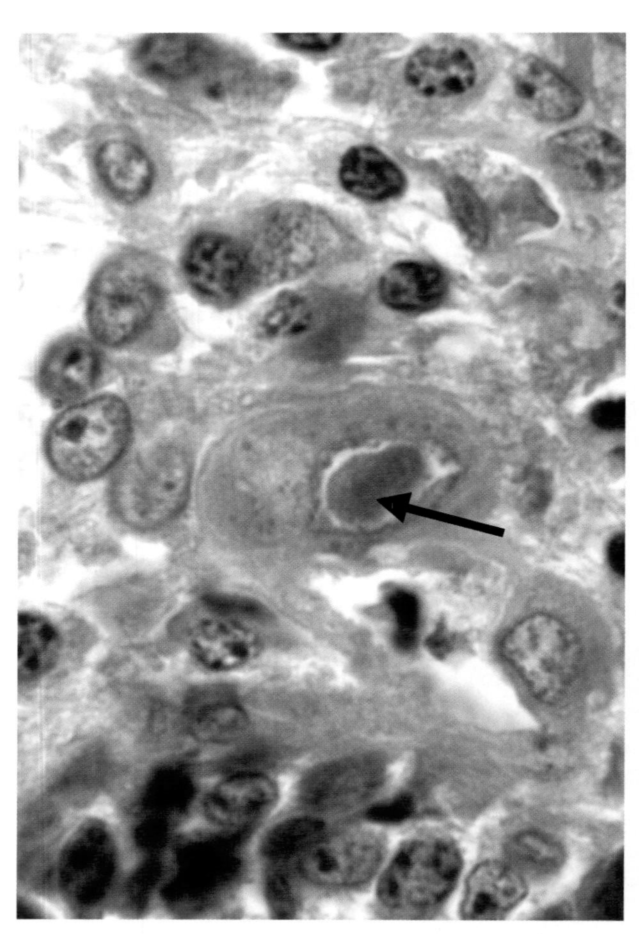

Color Plate 2. Biopsy specimen from a patient with CMV colitis, showing a classic "owl-eye" intranuclear inclusion (arrow) and intracytoplasmic inclusions. The dense intranuclear inclusion with surrounding halo is formed when the mass of viral particles shrinks away from the nuclear membrane during fixation. While herpes simplex virus intranuclear inclusions can have a similar appearance, CMV is the only member of the family *Herpesviridae* that contains both intranuclear and intracytoplasmic inclusions. The section was stained with hematoxylin and eosin stain. Original magnification, ×1,000 (oil immersion). (Courtesy of Bobbi Pritt.)

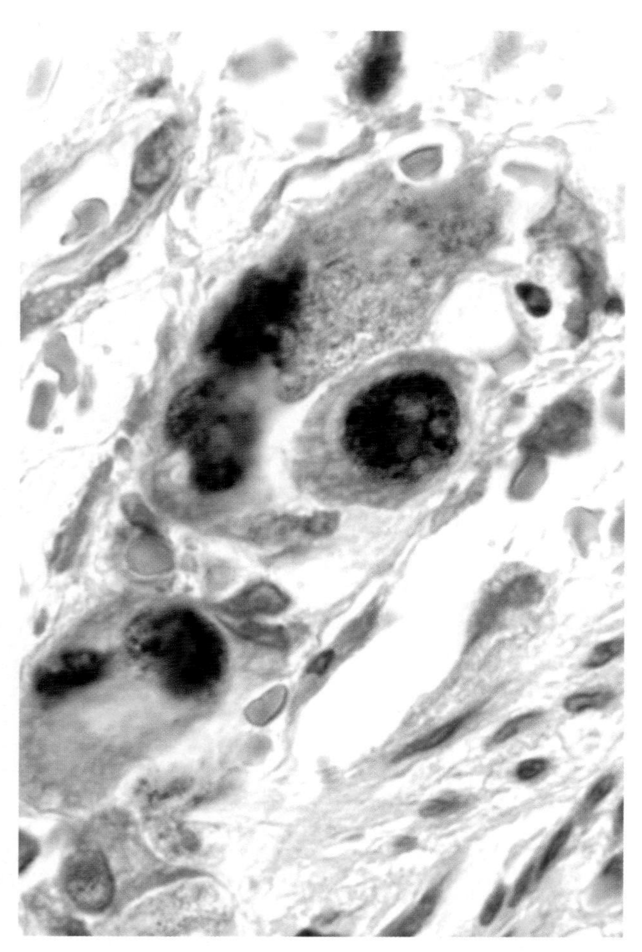

Color Plate 3. Immunoperoxidase staining of CMV antigens in a biopsy from a patient with CMV colitis. Viral inclusions stain brown (blue counterstain). Original magnification, ×1,000 (oil immersion). (Courtesy of Bobbi Pritt.)

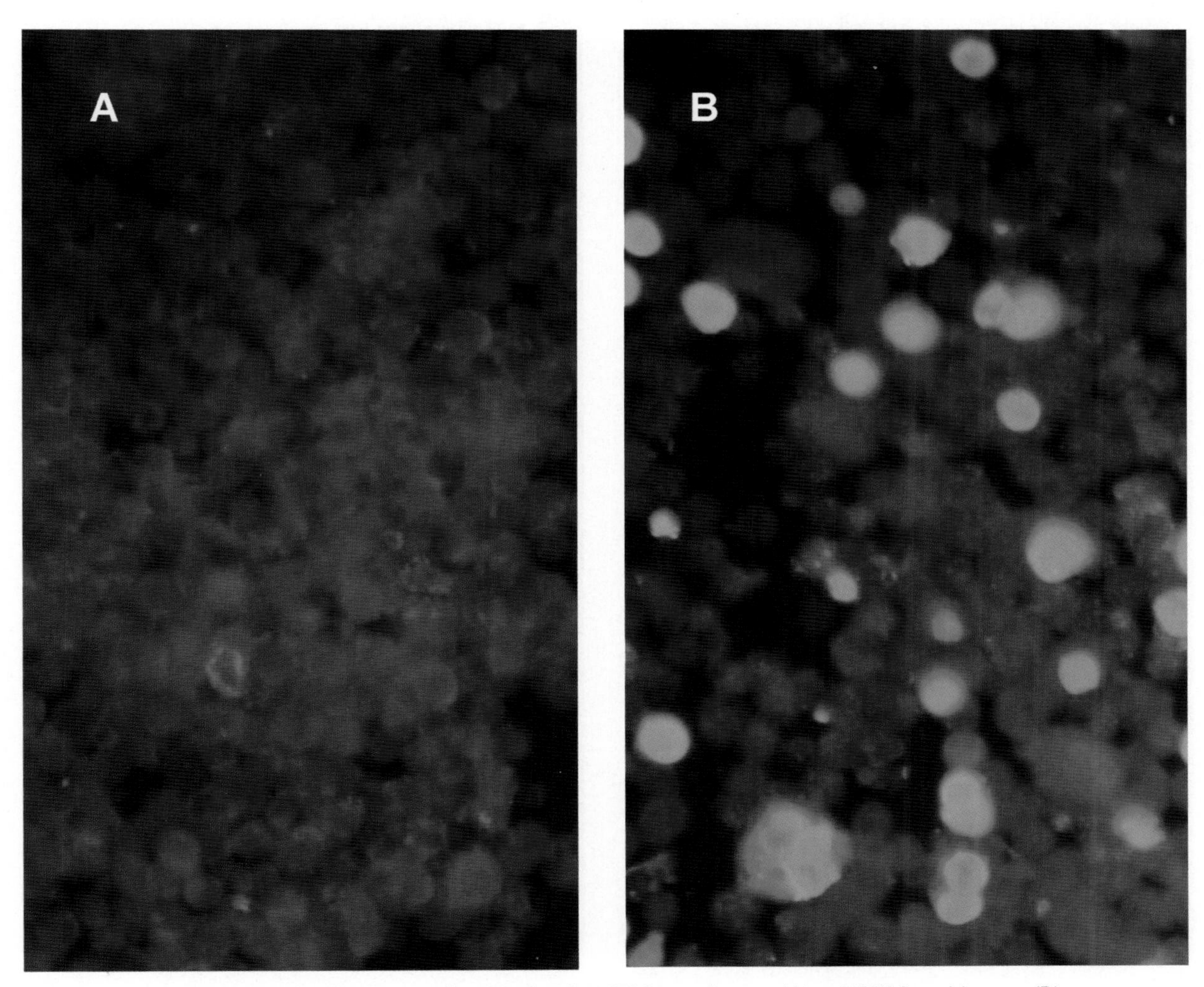

Color Plate 4. Induced BCBL-1 cells stained with HHV-8-negative sera (A) or HHV-8-positive sera (B).

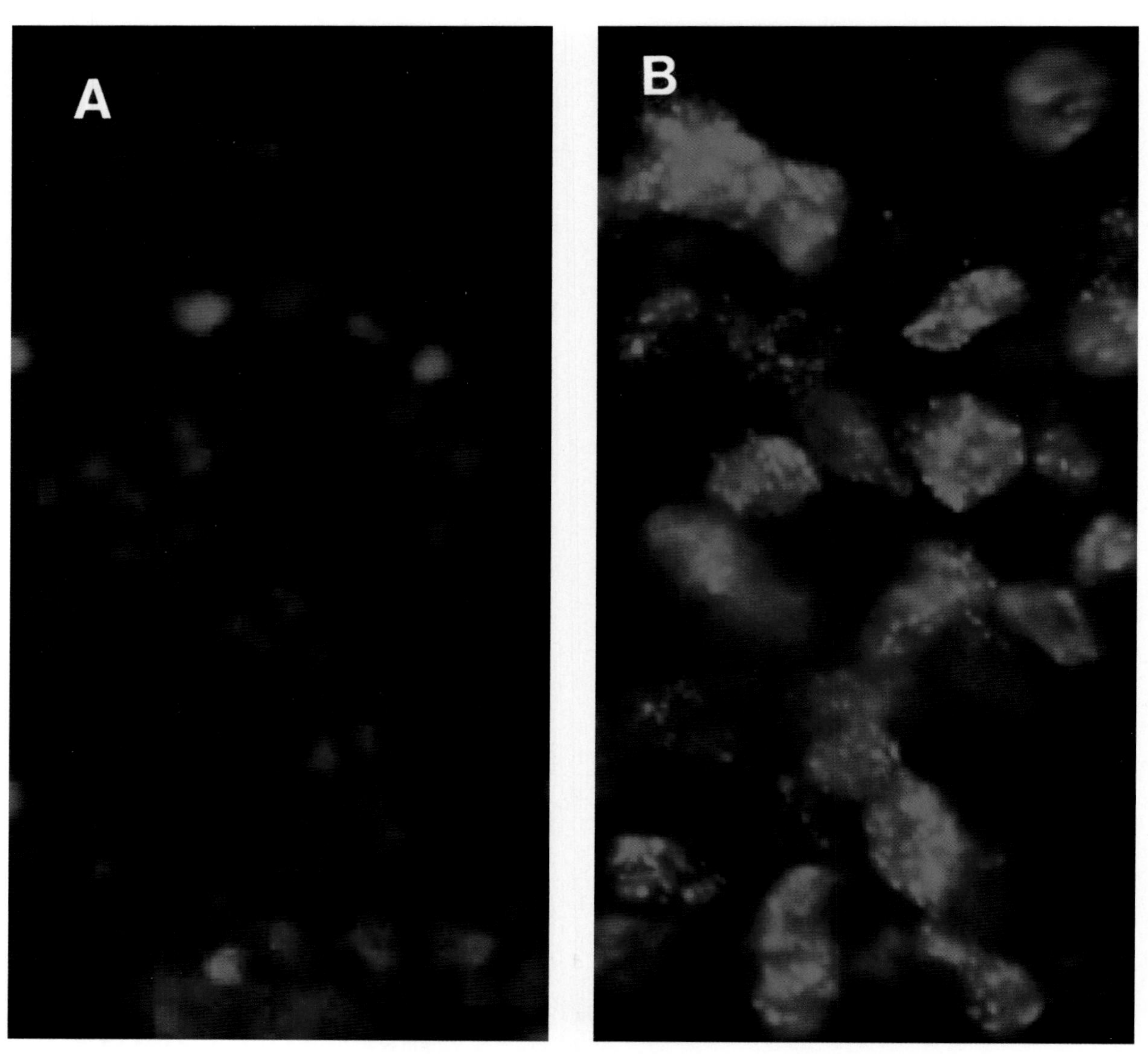

Color Plate 5. Untreated BCBL-1 cells stained with HHV-8-negative sera (A) or anti-LANA-1 antibody (B).

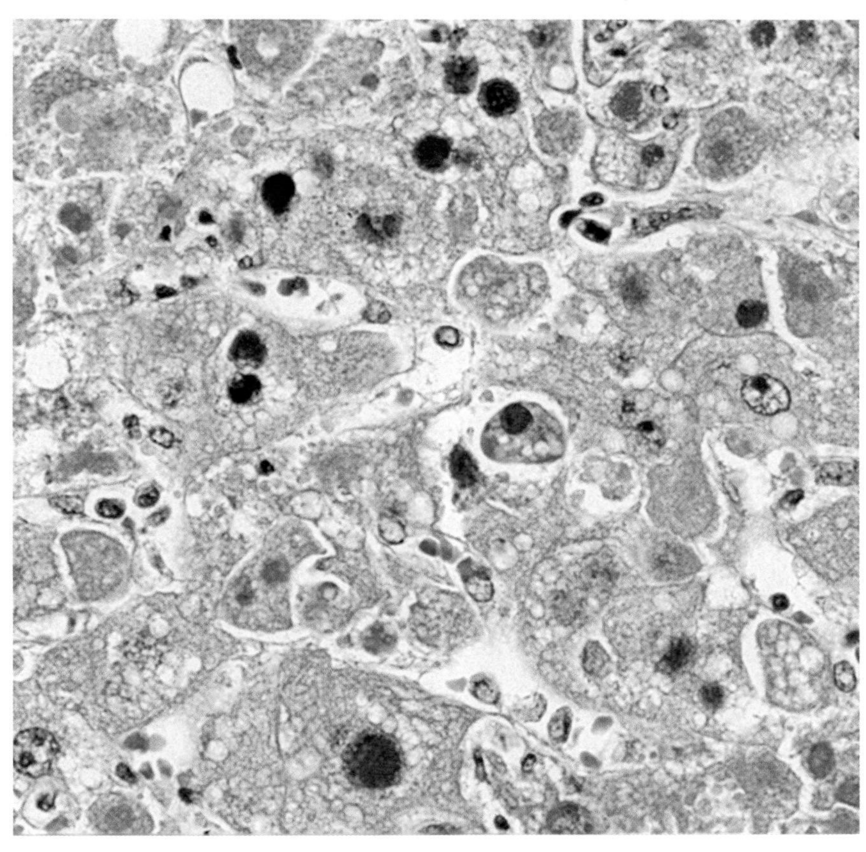

Color Plate 6. Hematoxylin- and eosin-stained section of liver showing reactive changes and intranuclear inclusions of adenovirus. (Photograph courtesy of Armed Forces Institute of Pathology, Washington, DC.)

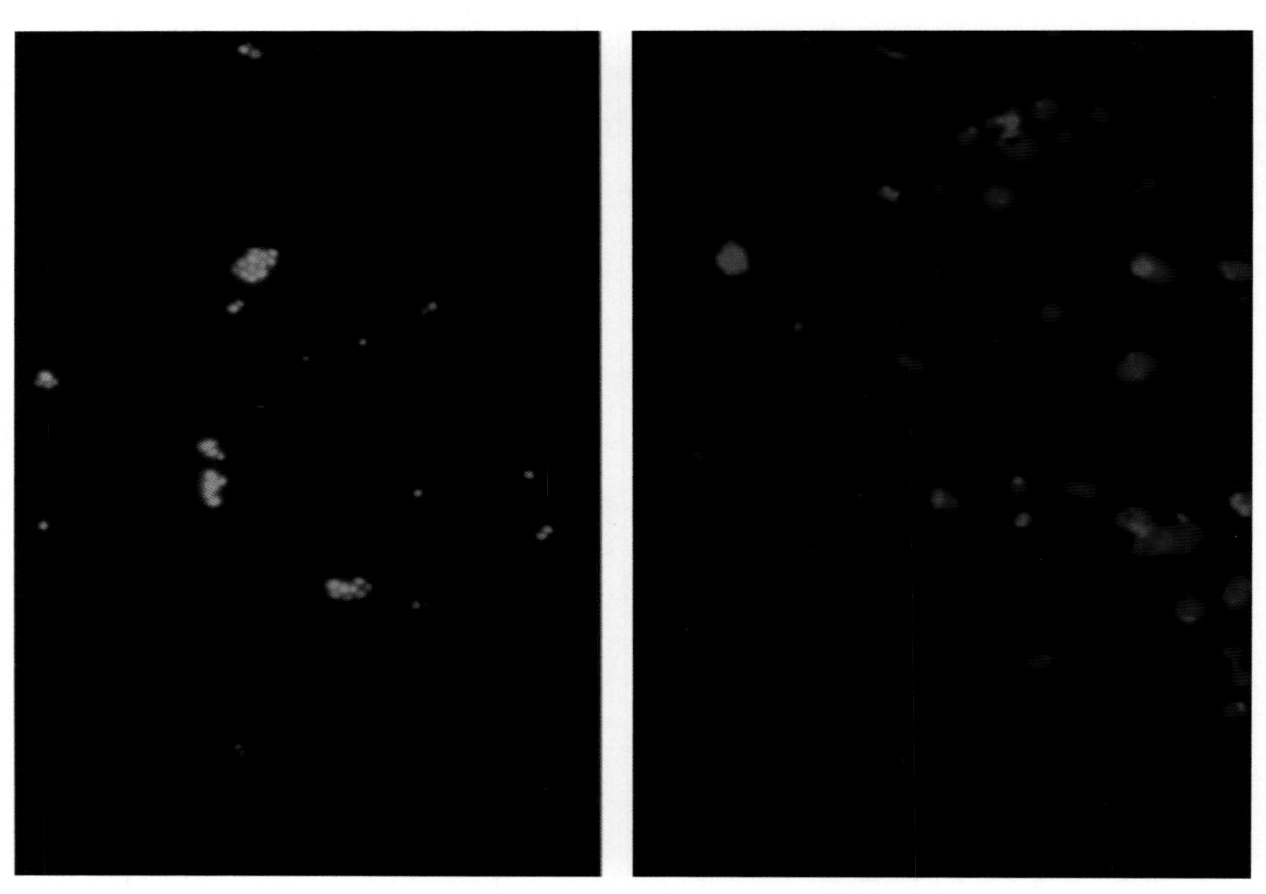

Color Plate 7. Fluorescent yeast cells from blood culture bottles with a Gram stain positive for yeast following testing with species-specific PNA FISH probes. (Left) Green fluorescent cells after reaction with *C. albicans*-specific PNA FISH probe. (Right) Red fluorescent cells after reaction with *C. glabrata*-specific probe.

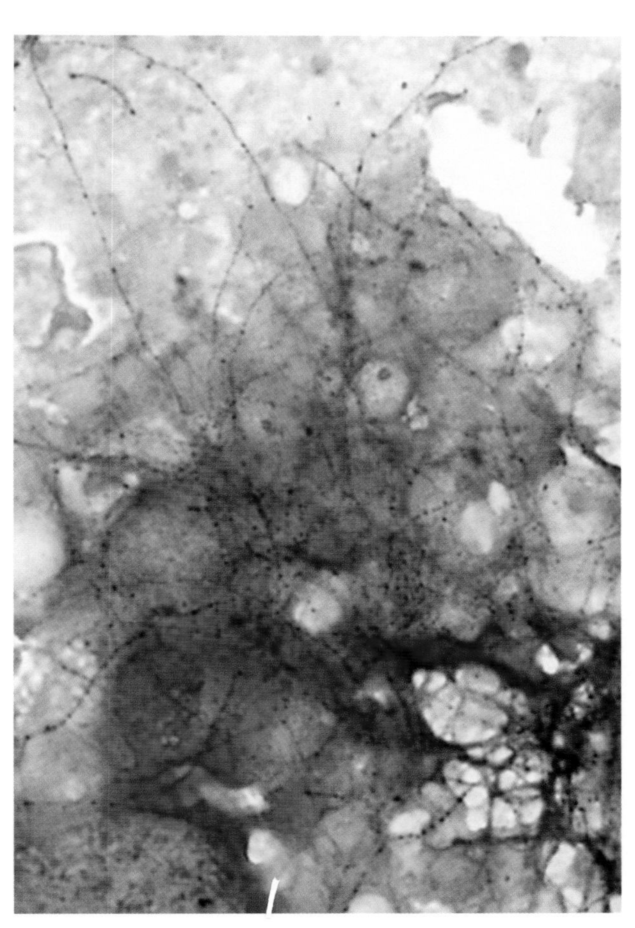

Color Plate 8. Gram-positive, filamentous, branching bacilli characteristic of the *Nocardia* species, as seen directly in a bronchoalveolar lavage specimen. Magnification, ×1,200.

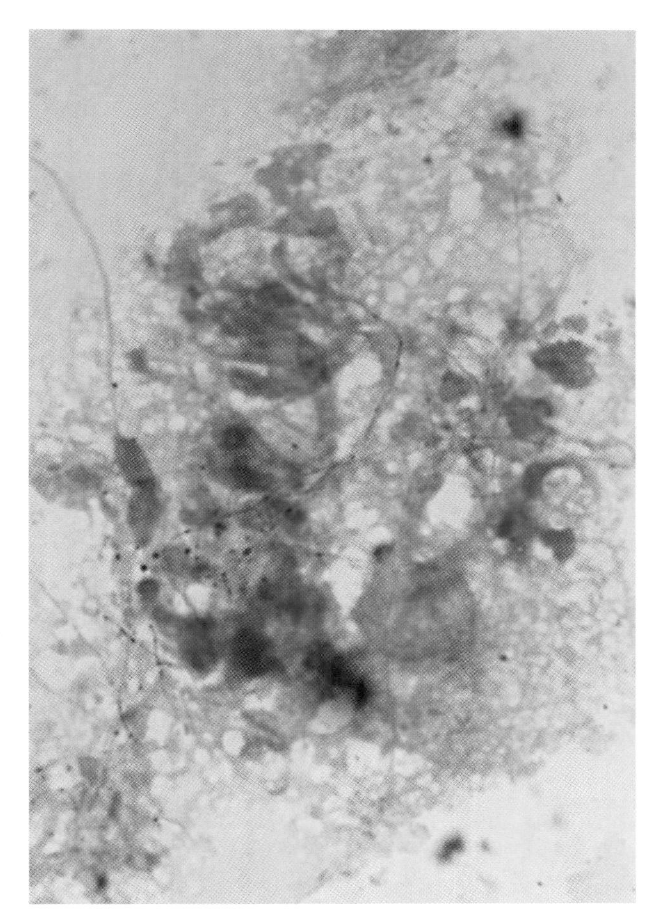

Color Plate 9. MAFS-positive, filamentous, branching bacilli characteristic of the *Nocardia* species, as seen directly in a bronchoalveolar lavage specimen. Magnification, ×1,200.

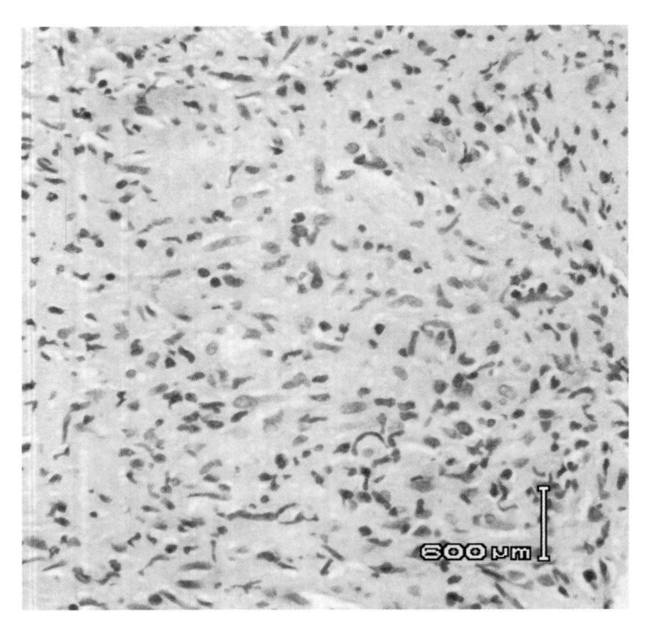

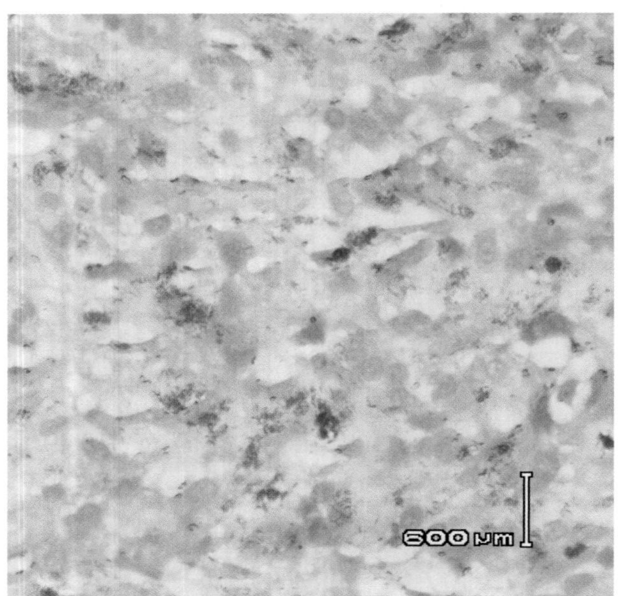

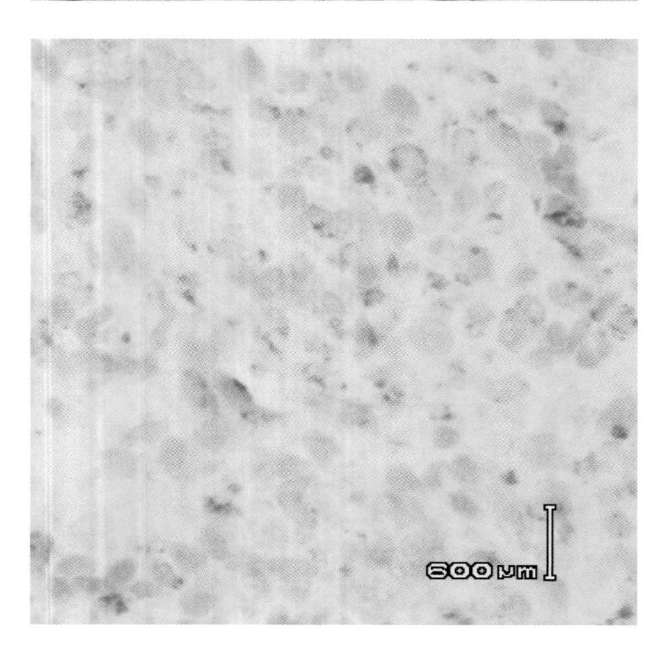

Color Plate 10. Mycobacterial spindle cell pseudotumor in a woman with previous AIDS and *M. avium* infection. (A) Spindle cell proliferation (hematoxylin and eosin stain). Magnification, ×200. (B) Acid-fast bacilli (Ziehl-Neelsen stain). Magnification, ×400. (C) Macrophage origin of the spindle cells (immunohistochemical stain with marker CD68). Magnification, ×400. Courtesy of Michael Deavers, M.D.

Hence, liver biopsy is the only reliable way to distinguish rejection from CMV hepatitis. It is important to distinguish these two clinical entities, since rejection is treated by increasing immunosuppression whereas CMV is treated by antiviral treatment and reduction in immunosuppression. For the diagnosis of CMV hepatitis, CMV can be demonstrated in liver tissue specimens by PCR, viral culture, histology, or in situ hybridization. In most cases of tissue-invasive CMV disease, detection of CMV DNA or RNA in blood by the use of PCR is helpful. However, there are cases where CMV may not be detected in blood or is detected only transiently and in small quantities. These compartmentalized cases can be diagnosed by biopsy and examination of the tissue specimen for the presence of characteristic CMV inclusion bodies and by detection of CMV by in situ hybridization and other molecular techniques, such as PCR.

Preemptive therapy

Antigenemia and molecular tests have been utilized for early detection of CMV and as a trigger for initiation of preemptive therapy after SOT (78, 79). Peripheral blood samples are collected on a weekly basis and tested for the presence of CMV, either by pp65 antigenemia assay or a molecular diagnostic test. For this to work, the blood should be processed immediately so that the results can be provided to the clinician in a timely manner. In a study of 69 liver transplant patients, the administration of ganciclovir upon detection of CMV by a highly sensitive qualitative CMV PCR assay resulted in reductions of CMV viremia, from 21% to 3%, and of CMV disease, from 12% to 0% (61). In another study of 49 liver transplant recipients, there were no cases of CMV disease when patients were monitored weekly with quantitative CMV PCR and treatment was initiated upon the detection of CMV DNA in the blood (13). The use of NASBA for guiding preemptive therapy has also been evaluated; the detection of CMV late gene expression was more predictive than that of CMV immediate-early and early gene expression, with a sensitivity of 97% and a specificity of 93% for CMV infection (57).

Monitoring response to antiviral treatment

In a study of mostly SOT recipients, serial measurements of CMV DNA demonstrated the rapid decline of CMV levels in the various blood compartments during antiviral drug treatment (65, 66). In SOT recipients, the half-life of decline was estimated to be approximately 1 to 5 days (65, 66), and patients with a slower decline in viral load were more likely to have a higher risk of CMV relapse. In a study of 52 SOT patients with CMV disease, the time to viral clearance was longer among patients with recurrent CMV disease (35, 36). The presence of detectable virus at the end of therapy was associated with CMV disease recurrence (74). Viral kinetics followed a logarithmic decay curve; the viral load half-life was 8.8 days among patients with recurrence versus 3.17 days among patients without recurrence (35, 36). The rate of decline has been suggested as a method to assess the efficacy of response (18, 19, 35, 36, 65, 71, 74); a longer duration of CMV DNA decline has been associated with a higher risk of CMV disease relapse (18, 19, 35, 36, 65, 71, 74). In addition, patients with detectable viral loads at the end of treatment are at higher risk of relapse (18, 19, 35, 36, 65, 71, 74). Hence, in most centers, at least two weekly negative PCR results are required prior to cessation of antiviral treatment.

Congenital CMV Infection

CMV acquired in utero may cause serious congenital CMV disease. Congenital CMV infection is seen among infants born to mothers with primary CMV infection during pregnancy. These primary infections are diagnosed in the mothers by detection of CMV IgM antibody. Clinical manifestations of infants with cytomegalic inclusion disease include jaundice, hepatosplenomegaly, petechial rash, and multiple organ involvement. CNS findings of microcephaly, motor disability, chorioretinitis, and cerebral calcifications are present. Shortly after birth, lethargy, respiratory distress, and seizures ensue; the newborn may die within a few days or weeks. The diagnosis of congenital infection is often demonstrated by viruria within the first week of life, and testing of urine for CMV DNA is often used. The presence of CMV IgM in umbilical cord serum also suggests congenital infection. More recently, congenital CMV infection has also been identified by testing by PCR for the presence of CMV DNA in neonatal dried blood on filter paper (known as Guthrie cards) (2). In one study, this test was found to be 100% sensitive and 99% specific for the diagnosis of congenital CMV infection in symptomatic and asymptomatic babies (2, 3).

Urinary CMV PCR has become a reliable, rapid, and convenient method for detecting CMV in newborns and may serve as a screening tool for congenital CMV infection (72). The severity of clinical symptoms appears to be correlated with the levels of virus in blood and urine (31). In a cohort of 76 infants with congenital CMV infection, infants with clinical abnormalities at birth had higher CMV DNA levels in the blood than did infants with no symptoms (9). Infants with congenital CMV infection who have CMV DNA in the blood (DNAemic) during the early period after birth are more likely to have indicators of active CMV disease, including elevated levels of alanine aminotransferase, petechial rash, and organomegaly (10).

Hearing deficits represent the most common extraneural defects associated with congenital CMV infection. Molecular diagnostics, such as viral load measurement, have been used to determine the relationship between viral replication and hearing loss. A cohort of 76 infants with congenital CMV infection was identified by means of newborn virologic screening, and peripheral blood samples were obtained for CMV DNA quantitation using a real-time quantitative PCR assay. The infants with hearing loss had significantly greater amounts of CMV DNA in the blood during infancy than those for infants with normal hearing (9). Another study investigated the viral load-hearing loss relationship in a cohort of 50 patients, wherein infants who were DNAemic during the early period after birth were more likely to have hearing loss and other indicators of active CMV disease (10). The utility of this molecular test is particularly useful since hearing loss may be decreased by ganciclovir administration.

THERAPEUTIC CONSIDERATIONS

Therapeutic Options

Ganciclovir, foscarnet, and cidofovir are the three drugs that are used for treatment of CMV disease (59). All of these drugs halt CMV replication by inhibiting CMV DNA polymerase. The most commonly used drug is ganciclovir, which is activated by the process of triphosphorylation by cellular and viral kinases. Foscarnet and cidofovir are considered second-line agents because of their associated toxicities and are used mainly when there is clinical and virologic failure with ganciclovir (59). For the purpose of treatment, ganciclovir should be administered intravenously. Oral ganciclovir should not be used for treatment of established disease since the oral drug has poor bioavailability and the systemic ganciclovir levels that are required to halt CMV replication are not achieved (59, 69). Valganciclovir, the valine ester of ganciclovir, is available in oral formulation, has high bioavailability, and attains high systemic levels, so it is used for the treatment of CMV retinitis in patients with AIDS (59, 69). Valganciclovir is also used in mildly symptomatic cases of CMV syndrome and tissue-invasive disease in organ and tissue transplant recipients (59, 69). The major toxicity of ganciclovir is myelosuppression, and this usually resolves upon the discontinuation of the drug. The major limiting toxicity of foscarnet and cidofovir is nephrotoxicity, and this has relegated them to use as alternative agents. Electrolyte abnormalities are common with foscarnet use.

Impact of Diagnostic Assays on Therapy

The major advantages afforded by diagnostic assays, particularly the pp65 antigenemia and molecular assays, have been the high sensitivity and the rapidity of diagnosis (70). This has allowed for the prompt initiation of antiviral therapy. In the transplant population, it is possible to perform surveillance for CMV by using these highly sensitive and quantitative assays and to initiate antiviral therapy even prior to the onset of clinical disease. As a result, use of these assays has likely resulted in the reduction of the severity of CMV disease.

Another major advantage of diagnostic assays is the ability to monitor response to treatment. While CMV disease was traditionally treated with a predefined duration of treatment, such as 2 or 3 weeks for CMV syndrome or tissue-invasive disease, respectively, current diagnostic methodologies have allowed for individualized therapies for patients. For example, in the transplant population, it is now a standard practice to continue therapy with induction doses of antiviral drugs at least until CMV is no longer detected in the peripheral blood (19, 35, 36, 70, 74). This practice has improved clinical outcomes by reducing the incidence of relapse and recurrence. Previous studies have demonstrated that the presence of viral load at the end of therapy is associated with disease relapse (19, 35, 36, 70, 74). The serial monitoring of CMV during therapy can also serve as a guide for the emergence of antiviral drug resistance. For example, viral load that does not decline is an early indicator that the isolate is resistant to the antiviral drug being used.

SUSCEPTIBILITY TESTING

The increasing use of antiviral drugs for prevention and treatment of CMV infection and disease has resulted in an emerging problem of antiviral drug-resistant CMV. Antiviral therapy selects for CMV gene mutations that confer antiviral drug resistance. The two general methods for testing the susceptibilities of viral isolates to antiviral drugs are genotypic and phenotypic methods.

The standard phenotypic method for detection of antiviral drug resistance is the plaque reduction assay (39). This method requires a lengthy viral propagation process in order to obtain sufficient infectivity. Cell monolayers are inoculated with a number of infected cells in a medium containing various concentrations of antiviral drugs, such as ganciclovir. After a period of incubation, the 50% inhibitory concentration, which is the concentration of drug producing a 50% reduction in the number of plaques, is determined. For ganciclovir, a 50% inhibitory concentration of ≤6 μM indicates sensitivity, while a value of >6 μM indicates resistance to ganciclovir. Drug-resistant CMV may harbor genotypic mutations that confer replication inefficiency, thereby slowing viral infectivity (16, 51). The overall slow growth of cell-associated clinical isolates and the subjectivity of the assay limit its value for real-time therapeutic decisions. Other phenotypic assays, which require less time, include detection of the CMV immediate-early antigen by flow cytometry and detection of viral DNA by DNA hybridization. Nonetheless, these assays still require weeks of cell culture to produce the initial CMV inoculum for the assay. Hence, phenotypic assays do not generally offer a short turnaround time for real-time guidance in the clinical management of patients (39).

The other major method for susceptibility testing is genotypic testing. Genetic mutations that confer resistance to one or all of the three antiviral drugs have been described. In general, compared to phenotypic testing, the genotypic assay provides a rapid measure that could guide the clinical decision-making process. Genotypic methods use primers for PCR that are designed to amplify the region containing the drug resistance mutations. The PCR products are subsequently sequenced and analyzed using an automated DNA sequencer to identify mutations associated with drug resistance. Currently, CMV strains that are resistant to ganciclovir, foscarnet, and cidofovir exhibit mutations in the UL97 (for ganciclovir resistance) and UL54 (for resistance to ganciclovir, foscarnet, and cidofovir) genes (20). The CMV UL97 gene is an important viral target for genotypic assays because all documented mutations conferring ganciclovir resistance have been found at one of three sites within the coding region for the C-terminal half of the phosphotransferase. Most commonly, CMV UL97 mutations associated with ganciclovir resistance are observed at codons 460, 594, and 595 (8). Based on a limited number of comparative studies, genotypic assays have a higher degree of sensitivity than do phenotypic studies, especially during the early phases of resistance. In one study, genotypically detected ganciclovir resistance always preceded phenotypic detection (16). Nonetheless, phenotypic assays remain useful to corroborate novel genotypic mutations associated with drug resistance (51). Hence, as new drugs are developed and used clinically, the utility of phenotypic assays will remain because they have the ability to relate drug resistance directly to the biological functions of the virus (51).

Among patients with AIDS, the presence of drug-resistant CMV has been associated with poor outcome. Among 197 patients who received ganciclovir therapy, 18 (9.1%) patients developed genotypic resistance to ganciclovir. The presence of ganciclovir-resistant CMV was associated with a 4.17- to 5.61-fold increase in the odds of retinitis progression and with greater loss of visual acuity (39). In a study of 87 patients with AIDS and CMV retinitis, sequence analysis of vitreous specimens showed that 15% of the patients had either a ganciclovir resistance-conferring mutation or a polymorphism in the CMV UL97 gene (34). In a study of 148 AIDS patients with CMV retinitis, the cumulative percentages of patients with UL97 mutant viruses at 3, 6, 12, and 18 months of antiviral treatment were 2.2%, 6.5%, 12.8%, and 15.3%, respectively (8). In a study of 23 CMV isolates from 10 immunocompromised patients, there were 8 isolates with a UL54 mutation that were correlated with resistance to ganciclovir. Four of these isolates were cross-resistant to cidofovir (20).

The incidence of antiviral drug resistance in transplant recipients is low but increasing (49, 68). In this patient population, the most common marker of drug-resistant CMV is a "nonresponding" viral load (i.e., the viral load does not decline to undetectable levels during antiviral therapy). Reassuringly, this is still an uncommon clinical problem in the transplant population, but it is anticipated to increase in the face of increasing antiviral drug exposure during prolonged antiviral prophylaxis.

SUMMARY

CMV is an important pathogen that causes severe disease in immunocompromised hosts, including transplant patients, patients with AIDS, and immunologically immature newborns. Advances in diagnostic modalities over the past decade have revolutionized the clinical management of CMV disease in these patients. Currently, the clinical diagnosis of CMV can be confirmed rapidly in the laboratory, thereby facilitating the early initiation of treatment in an effort to curtail the morbid effects of CMV disease. Current methods have even allowed for the early detection of CMV so that therapy can be instituted even prior to the onset of clinical disease. Monitoring responses to therapy has also been one of the major advances in diagnostic microbiology over the past years. Collectively, numerous diagnostic tests exist for the detection of CMV, from traditional methods of culture and serology to the more rapid and quantitative methods of antigenemia and PCR testing, with each method maintaining its own utility and clinical application.

REFERENCES

1. **Allice, T., M. Enrietto, F. Pittaluga, S. Varetto, A. Franchello, G. Marchiaro, and V. Ghisetti.** 2006. Quantitation of cytomegalovirus DNA by real-time polymerase chain reaction in peripheral blood specimens of patients with solid organ transplants: comparison with end-point PCR and pp65 antigen test. *J. Med. Virol.* **78:**915–922.
2. **Barbi, M., S. Binda, and S. Caroppo.** 2006. Diagnosis of congenital CMV infection via dried blood spots. *Rev. Med. Virol.* **16:**385–392.
3. **Barbi, M., S. Binda, V. Primache, S. Caroppo, P. Dido, P. Guidotti, C. Corbetta, and D. Melotti.** 2000. Cytomegalovirus DNA detection in Guthrie cards: a powerful tool for diagnosing congenital infection. *J. Clin. Virol.* **17:**159–165.
4. **Barel, M. T., G. C. Hassink, S. V. Voorden, and E. J. Wiertz.** 2005. Human cytomegalovirus-encoded US2 and US11 target unassembled MHC class I heavy chains for degradation. Mol. Immunol.
5. **Bestetti, A., C. Pierotti, M. Terreni, A. Zappa, L. Vago, A. Lazzarin, and P. Cinque.** 2001. Comparison of three nucleic acid amplification assays of cerebrospinal fluid for diagnosis of cytomegalovirus encephalitis. *J. Clin. Microbiol.* **39:**1148–1151.
6. **Blok, M. J., I. Lautenschlager, V. J. Goossens, J. M. Middeldorp, C. Vink, K. Hockerstedt, and C. A. Bruggeman.** 2000. Diagnostic implications of human cytomegalovirus immediate early-1 and pp67 mRNA detection in whole-blood samples from liver transplant patients using nucleic acid sequence-based amplification. *J. Clin. Microbiol.* **38:**4485–4491.
7. **Boivin, G.** 2004. Diagnosis of herpesvirus infections of the central nervous system. *Herpes 11 Suppl. 2:*48A–56A.
8. **Boivin, G., S. Chou, M. R. Quirk, A. Erice, and M. C. Jordan.** 1996. Detection of ganciclovir resistance mutations quantitation of cytomegalovirus (CMV) DNA in leukocytes of patients with fatal disseminated CMV disease. *J. Infect. Dis.* **173:**523–528.
9. **Boppana, S. B., K. B. Fowler, R. F. Pass, L. B. Rivera, R. D. Bradford, F. D. Lakeman, and W. J. Britt.** 2005. Congenital cytomegalovirus infection: association between virus burden in infancy and hearing loss. *J. Pediatr.* **146:**817–823.
10. **Bradford, R. D., G. Cloud, A. D. Lakeman, S. Boppana, D. W. Kimberlin, R. Jacobs, G. Demmler, P. Sanchez, W. Britt, S. J. Soong, and R. J. Whitley.** 2005. Detection of cytomegalovirus (CMV) DNA by polymerase chain reaction is associated with hearing loss in newborns with symptomatic congenital CMV infection involving the central nervous system. *J. Infect. Dis.* **191:**227–233.
11. **Chemaly, R. F., B. Yen-Lieberman, J. Chapman, A. Reilly, B. N. Bekele, S. M. Gordon, G. W. Procop, N. Shrestha, C. M. Isada, M. Decamp, and R. K. Avery.** 2005. Clinical utility of cytomegalovirus viral load in bronchoalveolar lavage in lung transplant recipients. *Am. J. Transplant.* **5:**544–548.
12. **Cheung, T. W., and S. A. Teich.** 1999. Cytomegalovirus infection in patients with HIV infection. *Mt. Sinai J. Med.* **66:**113–124.
13. **Daly, J. S., A. Kopasz, R. Anandakrishnan, T. Robins, S. Mehta, M. Halvorsen, and E. Katz.** 2002. Preemptive strategy for ganciclovir administration against cytomegalovirus in liver transplantation recipients. *Am. J. Transplant* **2:**955–958.
14. **Denijn, M., H. J. Schuurman, K. C. Jacobse, and R. A. De Weger.** 1992. In situ hybridization: a valuable tool in diagnostic pathology. Apmis. **100:**669–681.
15. **Drago, L., A. Lombardi, E. De Vecchi, G. Giuliani, R. Bartolone, and M. R. Gismondo.** 2004. Comparison of nested PCR and real time PCR of Herpesvirus infections of central nervous system in HIV patients. *BMC Infect. Dis.* **4:**55.
16. **Eckle, T., G. Jahn, and K. Hamprecht.** 2004. The influence of mixed HCMV UL97 wildtype and mutant strains on ganciclovir susceptibility in a cell associated plaque reduction assay. *J. Clin. Virol.* **30:**50–56.
17. **Einsele, H., H. Hebart, C. Kauffmann-Schneider, C. Sinzger, G. Jahn, P. Bader, T. Klingebiel, K. Dietz, J. Loffler, C. Bokemeyer, C. A. Muller, and L. Kanz.** 2000. Risk factors for treatment failures in patients receiving PCR-based preemptive therapy for CMV infection. *Bone Marrow Transplant* **25:**757–763.
18. **Emery, V. C.** 1999. Viral dynamics during active cytomegalovirus infection and pathology. *Intervirology* **42:**405–411.
19. **Emery, V. C., C. A. Sabin, A. V. Cope, D. Gor, A. F. Hassan-Walker, and P. D. Griffiths.** 2000. Application of viral-load kinetics to identify patients who develop cytomegalovirus disease after transplantation. *Lancet* **355:**2032–2036.
20. **Erice, A., C. Gil-Roda, J. L. Perez, H. H. Balfour, Jr., K. J. Sannerud, M. N. Hanson, G. Boivin, and S. Chou.** 1997. Antiviral susceptibilities and analysis of UL97 and DNA polymerase sequences of clinical cytomegalovirus isolates from immunocompromised patients. *J. Infect. Dis.* **175:**1087–1092.
21. **Espy, M. J., J. R. Uhl, L. M. Sloan, S. P. Buckwalter, M. F. Jones, E. A. Vetter, J. D. Yao, N. L. Wengenack, J. E. Rosenblatt, F. R. Cockerill, 3rd, and T. F. Smith.** 2006. Real-time PCR in clinical microbiology: applications for routine laboratory testing. *Clin. Microbiol. Rev.* **19:**165–256.
22. **Faderl, S., D. A. Thomas, S. O'Brien, G. Garcia-Manero, H. M. Kantarjian, F. J. Giles, C. Koller, A. Ferrajoli, S. Verstovsek, B. Pro, M. Andreeff, M. Beran, J. Cortes, W. Wierda, N. Tran, and M. J. Keating.** 2003. Experience with alemtuzumab plus rituximab in patients with relapsed and refractory lymphoid malignancies. *Blood* **101:**3413–3415.
23. **Feinberg, J.** 1996. The status of CMV viral load testing. *AIDS Clin. Care* **8:**81.
24. **Garrigue, I., S. Boucher, L. Couzi, A. Caumont, C. Dromer, M. Neau-Cransac, R. Tabrizi, M. H. Schrive, H. Fleury, and M. E. Lafon.** 2006. Whole blood real-time quantitative PCR for cytomegalovirus infection follow-up in transplant recipients. *J. Clin. Virol.* **36:**72–75.
25. **Gault, E., Y. Michel, A. Dehee, C. Belabani, J. C. Nicolas, and A. Garbarg-Chenon.** 2001. Quantification of human cytomegalovirus DNA by real-time PCR. *J. Clin. Microbiol.* **39:**772–775.
26. **Gerna, G., F. Baldanti, D. Lilleri, M. Parea, M. Torsellini, B. Castiglioni, P. Vitulo, C. Pellegrini, M. Vigano, P. Grossi,**

and M. G. Revello. 2003. Human cytomegalovirus pp67 mRNAemia versus pp65 antigenemia for guiding preemptive therapy in heart and lung transplant recipients: a prospective, randomized, controlled, open-label trial. *Transplantation* **75:**1012–1019.

27. **Gleaves, C. A., T. F. Smith, E. A. Shuster, and G. R. Pearson.** 1985. Comparison of standard tube and shell vial cell culture techniques for the detection of cytomegalovirus in clinical specimens. *J. Clin. Microbiol.* **21:**217–221.
28. **Gourlain, K., D. Salmon, E. Gault, C. Leport, C. Katlama, S. Matheron, D. Costagliola, M. C. Mazeron, and A. M. Fillet.** 2003. Quantitation of cytomegalovirus (CMV) DNA by real-time PCR for occurrence of CMV disease in HIV-infected patients receiving highly active antiretroviral therapy. *J. Med. Virol.* **69:**401–407.
29. **Gozlan, J., J. P. Laporte, S. Lesage, M. Labopin, A. Najman, N. C. Gorin, and J. C. Petit.** 1996. Monitoring of cytomegalovirus infection and disease in bone marrow recipients by reverse transcription-PCR and comparison with PCR and blood and urine cultures. *J. Clin. Microbiol.* **34:**2085–2088.
30. **Griffiths, P.** 2004. Cytomegalovirus infection of the central nervous system. *Herpes 11 Suppl.* **2:**95A–104A.
31. **Halwachs-Baumann, G., B. Genser, S. Pailer, H. Engele, H. Rosegger, A. Schalk, H. H. Kessler, and M. Truschnig-Wilders.** 2002. Human cytomegalovirus load in various body fluids of congenitally infected newborns. *J. Clin. Virol.* **25 Suppl 3:**S81–S87.
32. **Hebart, H., T. Rudolph, J. Loeffler, J. Middeldorp, T. Ljubicic, G. Jahn, and H. Einsele.** 2002. Evaluation of the NucliSens CMV pp67 assay for detection and monitoring of human cytomegalovirus infection after allogeneic stem cell transplantation. *Bone Marrow Transplant* **30:**181–187.
33. **Hernando, S., L. Folgueira, C. Lumbreras, R. San Juan, S. Maldonado, C. Prieto, M. J. Babiano, J. Delgado, A. Andres, E. Moreno, J. M. Aguado, and J. R. Otero.** 2005. Comparison of cytomegalovirus viral load measure by real-time PCR with pp65 antigenemia for the diagnosis of cytomegalovirus disease in solid organ transplant patients. *Transplant Proc.* **37:**4094–4096.
34. **Hu, H., D. A. Jabs, M. S. Forman, B. K. Martin, J. P. Dunn, D. V. Weinberg, and J. L. Davis.** 2002. Comparison of cytomegalovirus (CMV) UL97 gene sequences in the blood and vitreous of patients with acquired immunodeficiency syndrome and CMV retinitis. *J. Infect. Dis.* **185:**861–867.
35. **Humar, A., D. Gregson, A. M. Caliendo, A. McGeer, G. Malkan, M. Krajden, P. Corey, P. Greig, S. Walmsley, G. Levy, and T. Mazzulli.** 1999. Clinical utility of quantitative cytomegalovirus viral load determination for predicting cytomegalovirus disease in liver transplant recipients. *Transplantation* **68:**1305–1311.
36. **Humar, A., D. Kumar, G. Boivin, and A. M. Caliendo.** 2002. Cytomegalovirus (CMV) virus load kinetics to predict recurrent disease in solid-organ transplant patients with CMV disease. *J. Infect. Dis.* **186:**829–833.
37. **Humar, A., T. Mazzulli, G. Moussa, R. R. Razonable, C. V. Paya, M. D. Pescovitz, E. Covington, and E. Alecock.** 2005. Clinical utility of cytomegalovirus (CMV) serology testing in high-risk CMV D+/R- transplant recipients. *Am. J. Transplant.* **5:**1065–1070.
38. **Ikewaki, J., E. Ohtsuka, R. Kawano, M. Ogata, H. Kikuchi, and M. Nasu.** 2003. Real-time PCR assay compared to nested PCR and antigenemia assays for detecting cytomegalovirus reactivation in adult T-cell leukemia-lymphoma patients. *J. Clin. Microbiol.* **41:**4382–4387.
39. **Jabs, D. A., B. K. Martin, M. S. Forman, L. Hubbard, J. P. Dunn, J. H. Kempen, J. L. Davis, and D. V. Weinberg.** 2003. Cytomegalovirus resistance to ganciclovir and clinical outcomes of patients with cytomegalovirus retinitis. *Am. J. Ophthalmol.* **135:**26–34.
40. **Kasprzak, A., M. Zabel, J. Wysocki, W. Biczysko, J. Surdyk-Zasada, A. Olejnik, J. Gorczynski, and D. Jurczyszyn.** 2000. Detection of DNA, mRNA and early antigen of the human cytomegalovirus using the immunomax technique in autopsy material of children with intrauterine infection. *Virchows Arch.* **437:**482–490.
41. **Keightley, M. C., C. Rinaldo, A. Bullotta, J. Dauber, and K. St George.** 2006. Clinical utility of CMV early and late transcript detection with NASBA in bronchoalveolar lavages. *J. Clin. Virol.* **37:**258–264.
42. **Klemola, E., and L. Kaariainen.** 1965. Cytomegalovirus as a possible cause of a disease resembling infectious mononucleosis. *Br. Med. J.* **5470:**1099–1102.
43. **Lanari, M., T. Lazzarotto, V. Venturi, I. Papa, L. Gabrielli, B. Guerra, M. P. Landini, and G. Faldella.** 2006. Neonatal cytomegalovirus blood load and risk of sequelae in symptomatic and asymptomatic congenitally infected newborns. *Pediatrics* **117:**e76–83.
44. **Lautenschlager, I., L. Halme, K. Hockerstedt, L. Krogerus, and E. Taskinen.** 2006. Cytomegalovirus infection of the liver transplant: virological, histological, immunological, and clinical observations. *Transpl. Infect. Dis.* **8:**21–30.
45. **Lazzarotto, T., P. Spezzacatena, P. Pradelli, D. A. Abate, S. Varani, and M. P. Landini.** 1997. Avidity of immunoglobulin G directed against human cytomegalovirus during primary and secondary infections in immunocompetent and immunocompromised subjects. *Clin. Diagn. Lab. Immunol.* **4:**469–473.
46. **Lehto, J. T., K. Lemstrom, M. Halme, M. Lappalainen, J. Lommi, J. Sipponen, A. Harjula, P. Tukiainen, and P. K. Koskinen.** 2005. A prospective study comparing cytomegalovirus antigenemia, DNAemia and RNAemia tests in guiding pre-emptive therapy in thoracic organ transplant recipients. *Transpl. Int.* **18:**1318–1327.
47. **Lengerke, C., T. Ljubicic, C. Meisner, J. Loeffler, C. Sinzger, H. Einsele, and H. Hebart.** 2006. Evaluation of the COBAS Amplicor HCMV Monitor for early detection and monitoring of human cytomegalovirus infection after allogeneic stem cell transplantation. *Bone Marrow Transplant* **38:**53–60.
48. **Li, H., J. S. Dummer, W. R. Estes, S. Meng, P. F. Wright, and Y. W. Tang.** 2003. Measurement of human cytomegalovirus loads by quantitative real-time PCR for monitoring clinical intervention in transplant recipients. *J. Clin. Microbiol.* **41:**187–191.
49. **Limaye, A. P., L. Corey, D. M. Koelle, C. L. Davis, and M. Boeckh.** 2000. Emergence of ganciclovir-resistant cytomegalovirus disease among recipients of solid-organ transplants. *Lancet* **356:**645–649.
50. **Loginov, R., K. Hockerstedt, and I. Lautenschlager.** 2003. Detection of CMV-DNA in peripheral blood leukocytes of liver transplant patients after ganciclovir treatment. *Arch. Virol.* **148:**1269–1274.
51. **Lurain, N. S., A. Weinberg, C. S. Crumpacker, and S. Chou.** 2001. Sequencing of cytomegalovirus UL97 gene for genotypic antiviral resistance testing. *Antimicrob. Agents Chemother.* **45:**2775–2780.
52. **Mazzulli, T., L. W. Drew, B. Yen-Lieberman, D. Jekic-McMullen, D. J. Kohn, C. Isada, G. Moussa, R. Chua, and S. Walmsley.** 1999. Multicenter comparison of the digene hybrid capture CMV DNA assay (version 2.0), the pp65 antigenemia assay, and cell culture for detection of cytomegalovirus viremia. *J. Clin. Microbiol.* **37:**958–963.
53. **McClintock, J. T., S. R. Thaker, M. Mosher, D. Jones, M. Forman, P. Charache, K. Wright, J. Keiser, and F. E. Taub.**

1989. Comparison of in situ hybridization and monoclonal antibodies for early detection of cytomegalovirus in cell culture. *J. Clin. Microbiol.* **27**:1554–1559.

54. **Mendez, J., M. Espy, T. F. Smith, J. Wilson, R. Wiesner, and C. V. Paya.** 1998. Clinical significance of viral load in the diagnosis of cytomegalovirus disease after liver transplantation. *Transplantation* **65**:1477–1481.
55. **Mengoli, C., R. Cusinato, M. A. Biasolo, S. Cesaro, C. Parolin, and G. Palu.** 2004. Assessment of CMV load in solid organ transplant recipients by pp65 antigenemia and real-time quantitative DNA PCR assay: correlation with pp67 RNA detection. *J. Med. Virol.* **74**:78–84.
56. **Meyer-Koenig, U., M. Weidmann, G. Kirste, and F. T. Hufert.** 2004. Cytomegalovirus infection in organ-transplant recipients: diagnostic value of pp65 antigen test, qualitative polymerase chain reaction (PCR) and quantitative Taqman PCR. *Transplantation* **77**:1692–1698.
57. **Oldenburg, N., K. M. Lam, M. A. Khan, B. Top, N. M. Tacken, A. McKie, G. W. Mikhail, J. M. Middeldorp, A. Wright, N. R. Banner, and M. Yacoub.** 2000. Evaluation of human cytomegalovirus gene expression in thoracic organ transplant recipients using nucleic acid sequence-based amplification. *Transplantation* **70**:1209–1215.
58. **Paya, C. V., K. E. Holley, R. H. Wiesner, K. Balasubramaniam, T. F. Smith, M. J. Espy, J. Ludwig, K. P. Batts, P. E. Hermans, and R. A. Krom.** 1990. Early diagnosis of cytomegalovirus hepatitis in liver transplant recipients: role of immunostaining, DNA hybridization and culture of hepatic tissue. *Hepatology* **12**:119–126.
59. **Paya, C. V., and R. R. Razonable.** 2003. Cytomegalovirus infection after solid organ transplantation, p. 298-325. *In* R. Bowden, P. Ljungman, and C. Paya (ed.), Transplant Infections, vol. 1. Lippincott Williams and Wilkins, Philadelphia.
60. **Paya, C. V., T. F. Smith, J. Ludwig, and P. E. Hermans.** 1989. Rapid shell vial culture and tissue histology compared with serology for the rapid diagnosis of cytomegalovirus infection in liver transplantation. *Mayo Clin. Proc.* **64**:670–675.
61. **Paya, C. V., J. A. Wilson, M. J. Espy, I. G. Sia, M. J. DeBernardi, T. F. Smith, R. Patel, G. Jenkins, W. S. Harmsen, D. J. Vanness, and R. H. Wiesner.** 2002. Preemptive use of oral ganciclovir to prevent cytomegalovirus infection in liver transplant patients: a randomized, placebo-controlled trial. *J. Infect. Dis.* **185**:854–860.
62. **Persson, L., H. Dahl, A. Linde, P. Engervall, T. Vikerfors, and U. Tidefelt.** 2003. Human cytomegalovirus, human herpesvirus-6 and human herpesvirus-7 in neutropenic patients with fever of unknown origin. *Clin. Microbiol. Infect.* **9**: 640–644.
63. **Piiparinen, H., I. Helantera, M. Lappalainen, J. Suni, P. Koskinen, C. Gronhagen-Riska, and I. Lautenschlager.** 2005. Quantitative PCR in the diagnosis of CMV infection and in the monitoring of viral load during the antiviral treatment in renal transplant patients. *J. Med. Virol.* **76**:367–372.
64. **Pumannova, M., K. Roubalova, A. Vitek, and J. Sajdova.** 2006. Comparison of quantitative competitive polymerase chain reaction-enzyme-linked immunosorbent assay with LightCycler-based polymerase chain reaction for measuring cytomegalovirus DNA in patients after hematopoietic stem cell transplantation. *Diagn. Microbiol. Infect. Dis.* **54**:115–120.
65. **Razonable, R. R., R. A. Brown, M. J. Espy, A. Rivero, W. Kremers, J. Wilson, C. Groettum, T. F. Smith, and C. V. Paya.** 2001. Comparative quantitation of cytomegalovirus (CMV) DNA in solid organ transplant recipients with CMV infection by using two high-throughput automated systems. *J. Clin. Microbiol.* **39**:4472–4476.
66. **Razonable, R. R., R. A. Brown, J. Wilson, C. Groettum, W. Kremers, M. Espy, T. F. Smith, and C. V. Paya.** 2002. The clinical use of various blood compartments for cytomegalovirus (CMV) DNA quantitation in transplant recipients with CMV disease. *Transplantation* **73**:968–973.
67. **Razonable, R. R., and V. C. Emery.** 2004. Management of CMV infection and disease in transplant patients. 27-29 February 2004. *Herpes* **11**:77–86.
68. **Razonable, R. R., and C. V. Paya.** 2003. Herpesvirus infections in transplant recipients: current challenges in the clinical management of cytomegalovirus and Epstein-Barr virus infections. *Herpes* **10**:60–65.
69. **Razonable, R. R., and C. V. Paya.** 2004. Valganciclovir for the prevention and treatment of cytomegalovirus disease in immunocompromised hosts. *Expert Rev. Anti. Infect. Ther.* **2**:27–41.
70. **Razonable, R. R., C. V. Paya, and T. F. Smith.** 2002. Role of the laboratory in diagnosis and management of cytomegalovirus infection in hematopoietic stem cell and solid-organ transplant recipients. *J. Clin. Microbiol.* **40**:746–752.
71. **Razonable, R. R., H. van Cruijsen, R. A. Brown, J. A. Wilson, W. S. Harmsen, R. H. Wiesner, T. F. Smith, and C. V. Paya.** 2003. Dynamics of cytomegalovirus replication during preemptive therapy with oral ganciclovir. *J. Infect. Dis.* **187**:1801–1808.
72. **Schlesinger, Y., D. Halle, A. I. Eidelman, D. Reich, D. Dayan, B. Rudensky, D. Raveh, D. Branski, M. Kaplan, V. Shefer, and D. Miron.** 2003. Urine polymerase chain reaction as a screening tool for the detection of congenital cytomegalovirus infection. *Arch. Dis. Child Fetal Neonatal Ed.* **88**:F371–F374.
73. **Shuster, E. A., J. S. Beneke, G. E. Tegtmeier, G. R. Pearson, C. A. Gleaves, A. D. Wold, and T. F. Smith.** 1985. Monoclonal antibody for rapid laboratory detection of cytomegalovirus infections: characterization and diagnostic application. *Mayo Clin. Proc.* **60**:577–585.
74. **Sia, I. G., J. A. Wilson, C. M. Groettum, M. J. Espy, T. F. Smith, and C. V. Paya.** 2000. Cytomegalovirus (CMV) DNA load predicts relapsing CMV infection after solid organ transplantation. *J. Infect. Dis.* **181**:717–720.
75. **Smith, M. G.** 1956. Propagation in tissue cultures of a cytopathogenic virus from human salivary gland virus (SGV) disease. *Proc. Soc. Exp. Biol. Med.* **92**:424–430.
76. **St George, K., M. J. Boyd, S. M. Lipson, D. Ferguson, G. F. Cartmell, L. H. Falk, C. R. Rinaldo, and M. L. Landry.** 2000. A multisite trial comparing two cytomegalovirus (CMV) pp65 antigenemia test kits, biotest CMV brite and Bartels/ Argene CMV antigenemia. *J. Clin. Microbiol.* **38**:1430–1433.
77. **St George, K., and C. R. Rinaldo, Jr.** 1999. Comparison of cytomegalovirus antigenemia and culture assays in patients on and off antiviral therapy. *J. Med. Virol.* **59**:91–7.
78. **Tong, C. Y., L. E. Cuevas, H. Williams, and A. Bakran.** 2000. Prediction and diagnosis of cytomegalovirus disease in renal transplant recipients using qualitative and quantitative polymerase chain reaction. *Transplantation* **69**:985–991.
79. **Turgeon, N., J. A. Fishman, N. Basgoz, N. E. Tolkoff-Rubin, M. Doran, A. B. Cosimi, and R. H. Rubin.** 1998. Effect of oral acyclovir or ganciclovir therapy after preemptive intravenous ganciclovir therapy to prevent cytomegalovirus disease in cytomegalovirus seropositive renal and liver transplant recipients receiving antilymphocyte antibody therapy. *Transplantation* **66**:1780–1786.
80. **Wood, N. B., M. Sheikholeslami, M. Pool, and J. S. Coon.** 1994. PCR production of a digoxigenin-labeled probe for the detection of human cytomegalovirus in tissue sections. *Diagn. Mol. Pathol.* **3**:200–208.
81. **Yakushiji, K., H. Gondo, K. Kamezaki, K. Shigematsu, S. Hayashi, M. Kuroiwa, S. Taniguchi, Y. Ohno, K. Takase,**

A. Numata, K. Aoki, K. Kato, K. Nagafuji, K. Shimoda, T. Okamura, N. Kinukawa, N. Kasuga, M. Sata, and M. Harada. 2002. Monitoring of cytomegalovirus reactivation after allogeneic stem cell transplantation: comparison of an antigenemia assay and quantitative real-time polymerase chain reaction. *Bone Marrow Transplant* **29:**599–606.

82. **Yamamoto, N., T. Wakabayashi, K. Murakami, and S. Hommura.** 2003. Detection of CMV DNA in the aqueous humor of AIDS patients with CMV retinitis by AMPLICOR CMV test. *Ophthalmologica.* **217:**45–48.

83. **Zhang, F., S. Tetali, X. P. Wang, M. H. Kaplan, F. V. Cromme, and C. C. Ginocchio.** 2000. Detection of human cytomegalovirus pp67 late gene transcripts in cerebrospinal fluid of human immunodeficiency virus type 1-infected patients by nucleic acid sequence-based amplification. *J. Clin. Microbiol.* **38:**1920–1925.

Diagnostic Microbiology of the Immunocompromised Host
Edited by Randall T. Hayden, Karen C. Carroll, Yi-Wei Tang, and Donna M. Wolk

Chapter 4

Epstein-Barr Virus and Human Herpesvirus 8

DAVID T. ROWE, FRANK J. JENKINS, AND HEATHER R. HENSLER

Epstein-Barr virus (EBV or human herpesvirus 4 [HHV-4]) and Kaposi's sarcoma-associated virus (KSHV or HHV-8) are distantly related gammaherpesviruses, both of which have associations with cancer in humans. These tumor associations involve immunosuppression as a contributing factor, and hence monitoring of their infections in immunosuppressed patients has taken on increased significance in recent years. Neither of these viruses is easily culturable or associated with profound antigenemia, which limits diagnostic detection options. Serological assays have been the principle method for detection of infections in otherwise healthy subjects, but these assays depend on robust host immune responses. Developments in nucleic acid detection techniques have brought new diagnostic capabilities to bear, revealing new associations and activities of these viruses in the setting of immunosuppression. Here we review these disease associations and the techniques that are currently being used to diagnose EBV and KSHV infections.

EBV

Background and Clinical Information

EBV is a member of the *Gammaherpesvirinae* subfamily of human herpesviruses and has a tropism for epithelial cells and B lymphocytes (169). Primary infection usually occurs in infants or young children and causes either no symptoms or a mild fever and lymphadenopathy that is indistinguishable from many other brief, inconsequential illnesses of childhood. When primary infection with EBV occurs during adolescence, the combination of high fever, lymphadenopathy, and pharyngitis may lead to a diagnosis of infectious mononucleosis (IM) in approximately half of cases (234). Splenomegaly or hepatomegaly may develop, but heart and/or central nervous system involvement is rare, and in immunocompetent patients IM is almost never fatal. A recent study provides the most up-to-date description of the clinical and laboratory presentation of IM in young adults (69). The symptoms of IM usually resolve in 1 to 2 months, but the virus establishes a lifelong persistence within the pool of resting memory B cells. During acute infection, 10% or more of circulating B cells are infected, and most of these have a memory phenotype (78, 92, 100, 174, 175). Latently infected cells in healthy carriers are all memory B cells and number between $1/10^5$ and $1/10^6$ of the circulating B-cell population (137, 213). Asymptomatic reactivation with virus particle production is common, and infectious virus may periodically be recovered from the saliva of reactivating individuals.

In the immunocompetent host, cell-mediated immunity is responsible for controlling EBV infection (95, 171). Acute infection with EBV elicits a pronounced cellular immune response, which results in the production of large numbers of specific $CD8^+$ cytotoxic T cells (CTLs). By limiting-dilution analysis, these CTLs have been estimated to range from 1/100 to 1/500 of the circulating T cells, but more recent analyses with fluorochrome-labeled HLA tetramers provided numerical estimates that are even higher (77). Through their release of (Th1-type) cytokines, these EBV-specific CTLs are responsible for most of the symptoms of IM (208, 235). Healthy carriers of a latent infection maintain circulating CTL precursors (CTLp) at frequencies in the range of 1/400 to 1/42,000 (21, 102a, 114, 171). This is a remarkably large pool for memory CTLp, but it is apparently needed to maintain surveillance against virus reactivation. Although the frequency of EBV-specific CTLp is high, their range of specificity for EBV target antigens is narrowly focused. One or two peptide epitopes,

David T. Rowe • Graduate School of Public Health, University of Pittsburgh, Pittsburgh, PA 15213.
Frank J. Jenkins • Departments of Pathology and Infectious Diseases and Microbiology, University of Pittsburgh, Pittsburgh, PA 15213.
Heather R. Hensler • Department of Infectious Diseases and Microbiology, University of Pittsburgh, Pittsburgh, PA 15213.

drawn primarily from the Epstein-Barr nuclear antigen 3 proteins (EBNA3A, EBNA3B, and EBNA3C), dominate their specificities (96, 143, 171, 207). Epitope choice in immune responses to latently infected B cells is highly HLA allele specific, with each HLA allele presenting only a particular set of epitopes. For reasons that are not understood, HLA-B alleles appear to be more efficient than HLA-A alleles at presenting EBV antigens to CTLs (39, 171, 208).

Infections in the Setting of Immune Dysregulation

Because of the requirement for persistent immunosurveillance against latent EBV, diseases with clearly recognized underlying immunologic dysregulation, such as rheumatoid arthritis (RA), autoimmune thyroiditis, multiple sclerosis, and chronic fatigue syndrome, are associated with rises in circulating virus loads and elevations of anti-EBV antibody titers (17, 79, 144, 156, 196, 227). In addition, EBV has been implicated in the etiology of some autoimmune diseases (89). Numerous reports have documented a potential association between EBV and RA. Anti-EBV antibody titers are higher than normal, circulating peripheral viral loads are elevated, and T-cell responses are impaired in RA patients (12). Molecular mimicry between EBV antigens and self antigens has been implicated. The glycine/alanine repeat sequence in EBNA1 resembles that of synovial proteins, and the EBV gB envelope glycoprotein also contains a shared epitope sequence (11). A temporal link between primary EBV infection and the onset of type 1 diabetes has been reported in a small number of cases, with the evidence again suggesting that molecular mimicry between an EBV antigen and, in this instance, a region of HLA-DQw8β may be involved (182).

The changes in immune responses and viral loads that seem to coincide with, and therefore reinforce, these correlations are actually rather difficult to interpret. Even factors as benign as psychological or physical stress in otherwise healthy individuals can adversely impact EBV-specific T-cell immunosurveillance and lead to the development of similar signs of EBV reactivation (64, 198). In addition, age-associated changes in the immune system, such as thymic involution, changes in lymphocyte counts, clonal expansions, weaker responses to mitogens and antigens, and proliferative senescence, result in decreased immunosurveillance and increased incidence and severity of not just EBV but all types of infections (73). An immune risk phenotype (IRP) among the elderly is characterized by an inverted CD4-to-CD8 ratio, poor T-cell proliferative responses to mitogens, increased $CD8^+$, $CD28^-$, and $CD57^+$ cell counts, low B-cell counts, and seropositivity for another herpesvirus, cytomegalovirus (CMV). The IRP was initially identified to be predictive of 2-year mortality in a longitudinal study of octogenarians by use of a cluster analysis approach (160). In general, herpesvirus-specific immune responses decrease in the elderly, although the characteristic phenotypes of the memory populations are largely retained. A decrease in EBV-specific CTLp actually depends upon CMV serostatus, because CMV is causally implicated in the alterations to T-cell immunity that make up the IRP. Aging is associated with a profound accumulation of CMV-specific CTLp that have a marked oligoclonality and decreased effector function. In the CMV-infected elderly, there is a corresponding increase in the clonality of EBV-specific CTLp, a decrease in CTLp frequency, and a significant decrease in the fraction of cells able to secrete gamma interferon (IFN- γ) after stimulation with specific antigenic peptides (157). These alterations are considered to represent immunosenescence, a deterioration of the immune system that is associated with a general increase in morbidity and mortality of infectious diseases.

Infections in Genetically Immunocompromised Individuals

B-cell deficiencies

XLA. Individuals with X-linked (Bruton's) agammaglobulinemia (XLA) carry defects in Bruton's tyrosine kinase (Btk), which mediates activation and proliferation signals from the B-cell antigen receptor to the nucleus (224). Because these signals are absent, B-cell development is arrested in the bone marrow and XLA patients produce no B lymphocytes and make no antibodies. Despite the absence of humoral immunity, XLA patients have functional T-cell immunity and, with immunoglobulin supplements, lead basically normal lives. Of interest here is that there is no detectable T-cell-mediated immune response to EBV antigens and no evidence of EBV infection (53). From analyses of XLA patients, it has been concluded that persistent latent infection and the development of specific immune responses absolutely depend on the host possessing a susceptible population of B cells. Predictably, eradication of an infected person's B-cell population, such as occurs prior to bone marrow transplantation, will cure a persistent latent EBV infection (67).

X-linked hyper-IgM syndrome. Hyper-immunoglobulin M (hyper-IgM) syndrome is caused by a group of rare genetic disorders, all of which involve defective immunoglobulin class switch recombination. The result is a perturbed B-cell population lacking a classical memory B-cell pool (where latent EBV normally resides) and the presence of high concentrations of serum IgM, with a profound lack of other immunoglobulin isotypes (47). In these patients, EBV infection takes place and produces no obviously different symptomology. One study of this phenomenon demonstrated a short-term virus persistence in B cells, but with highly variable longitudinal detection. The investigators concluded that a stable long-term latent infection did not appear to develop due to the absence of germinal-center reactions and a classical memory B-cell population (37).

T-cell deficiencies

XLP. In X-linked lymphoproliferative syndrome (XLP) patients, a number of mutations have been identified in the SH2D1A gene, which encodes a protein called SLAM-associated protein (SAP). The XLP-associated genetic defects usually cause a complete absence of the SAP molecule. SAP is a mediator of signaling through the SLAM (signaling lymphocytic activation molecule) family of receptors on the surfaces of T cells and NK cells (50). Disruption of SAP in transgenic mice results in hyperproliferation of IFN-γ-producing T cells and aberrant Th2-cell polarization after infection with lymphocytic choriomeningitis virus, features that recapitulate some of the immunological alterations found in patients with XLP. Dysgammaglobulinemia is also observed in SAP-deficient mice and can develop in patients with XLP. IM in XLP patients is fatal within weeks. There is a large polyclonal proliferation of T and B cells, leading to marked liver necrosis and bone marrow failure. The destruction of the liver and bone marrow often leads to fulminant hepatitis and virus-associated hemophagocytic syndrome (50). The EBV-induced lymphoproliferative syndrome is thought to result from an inability to mount appropriate CD4 T-cell helper responses along with extended proliferation of IFN-γ-producing CD8 T cells. Exactly why patients with XLP cannot control EBV infections while they appear to be able to adequately handle most other viruses is not clear.

CVID. Common variable immunodeficiency (CVID) is a heterogeneous group of immunologic disorders characterized by low serum immunoglobulin concentrations, defective specific antibody production, and increased susceptibility to bacterial infections (38). CVID is also associated with various degrees of T-cell dysfunction, and the severity of these T-cell abnormalities is associated with the degree of loss of $CD4^+$ naive T cells from the lymphocyte population (63). Despite this, CD8 T-cell-mediated responses to EBV and control of EBV infections appear to be essentially normal in CVID patients. The number of tetramer-positive CD8 T cells and the ability of these cells to kill large-cell lymphoma targets are not significantly different from those of healthy age-matched seropositive controls. Approximately one-half of CVID patients have an exaggerated CD8 T-cell response to CMV (another common persistent herpesvirus), and it has been proposed that the response to this herpesvirus may actually contribute to the T-cell abnormalities of CVID and to the previously unexplained severe enteropathy that occurs in about 5% of patients (165).

CAEBV (NK/T-cell lymphoproliferative disease). Chronic active EBV (CAEBV) is a rare disease with a high morbidity and mortality and a varied clinical presentation. As suggested by its name, most patients initially present with a mononucleosis-like illness that fails to resolve (98). The best-studied cases come from Japan, where a majority of patients had EBV infections of T or NK cells, while only 2 of the 82 documented CAEBV cases exhibited B-cell infections (97). Most patients with CAEBV exhibit oligo- or monoclonality in the infected cell populations, and chromosomal alterations are observed in 50% of cases. There are two distinct clinical classifications of CAEBV that depend on whether T cells or NK cells are the predominant infected cell type; the T-cell-type infections have a poorer prognosis (202). In terms of clinicopathogenesis, CAEBV may be most similar to opportunistic B-cell lymphoproliferative disorders in that there is a progression from polyclonal disease through oligoclonal lymphoproliferation to monoclonal lymphoma. Patients cannot control EBV-infected cells because CTL activity is impaired. They have very small numbers of EBV-specific CTLp in the peripheral blood that perform very poorly in functional CTL assays in vitro (103, 201). Both the mechanism of EBV infection of T and NK cells and the genetic and/or environmental basis of the deficient immune responses remain unknown. Immunotherapeutic approaches to treatment have been proposed, and successful treatment of CAEBV by allogeneic stem cell transplantation has been reported (65, 83, 152, 199, 206, 240).

Infections in Acquired Immunodeficiency

LPDs

Virus-induced B-cell proliferation underlies the pathogenesis of most EBV-associated lymphoproliferative disorders (LPDs). It only occurs in the setting of suppressed T-cell immunity that is unable to control primary or persistent EBV infection (27, 66). Posttransplant LPDs (PTLDs) have an overall incidence of 1 to 2%, but individual risk of developing a PTLD depends upon pretransplant EBV serostatus, the type of organ transplanted, and the immunosuppressive regimen (35, 44, 154, 203). PTLDs occur in approximately 1% of renal (36, 109, 146, 155, 164, 233), 2% of liver (146, 155, 233), 6% of pancreas (42, 116, 200), 2 to 10% of heart (33, 109, 146, 155, 164, 204, 233), 4 to 10% of heart/lung and lung (6, 7, 109, 146, 164, 236), 20% of small bowel (101), 23% of multivisceral, and 2% of bone marrow (40, 43, 101, 214, 217) transplant recipients. Recent estimates indicate that 87% of PTLDs are of B-cell origin, 12.5% are of T-cell origin, and 0.5% are of null cell origin. The World Health Organization (WHO) classification includes the following three histologic categories of PTLD: hyperplastic, polymorphic, and monomorphic (or lymphomatous) (85). As with CAEBV, these categories correspond to stages in the evolution of disease from polyclonal lymphoproliferation to monoclonal lymphoma (147).

There are EBV-specific CTLp cells in the circulation of immunosuppressed transplant recipients during primary EBV infections (52). These cells have the normal activated/memory phenotype of responders, but their frequency is at least an order of magnitude lower than that expected for an immunocompetent response. This suggests that immunosuppression is directly responsible for reducing the magnitude of the response, while not eliminating it entirely. Not surprisingly, the incidence of PTLD directly correlates with the intensity and duration of the immunoablative therapy, and low-dose immunosuppression reduces the incidence of PTLD in solid organ transplant recipients (29, 58, 131). Reduction in immunosuppression (the frontline therapeutic approach to PTLD) increases the CTLp frequency. Measuring the EBV-specific T-cell response in immunocompromised patients with primary EBV infection has been suggested as a predictive marker for the risk for PTLD (192).

Classic HL

Hodgkin's lymphoma (HL) is a major lymphoma subtype in the Western world, accounting for nearly one-third of all lymphomas (145). The different WHO subtypes, i.e., nodular sclerosis (NS), mixed-cellularity (MC), lymphocyte-depleted (LD), lymphocyte-rich (LR), and nodular lymphocyte-predominant (NLP) types, are based on the characteristics of the reactive infiltrates and the morphology of Hodgkin/Reed-Sternberg (H/RS) tumor cells. The NLP type of HL is separated from the other classical HL types on the basis of different biological and clinical features (85). In addition to geographic and ethnic differences in incidence and presentation, there are familial, socioeconomic, age, and environmental risk associations. Most of these associations correlate with enhanced hygienic practices that result in reduced exposure to common infectious agents. This has led to the "delayed exposure" model, which postulates the involvement of aberrant host responses to unnaturally delayed primary infections by common agents (71). Accordingly, there is an increased risk for HL among patients with a history of IM (i.e., a delayed primary EBV infection). But the relationship between EBV and HL is more involved than just aberrant immune responses in the general sense. In situ hybridization and immunohistochemical analyses have demonstrated that EBV can be found in H/RS cells in one-half of all HLs and that, when it is present, EBV latent genes are being expressed. For the two most common HL subtypes, EBV is more likely to be associated with the MC subtype (70% of cases) than with the NS subtype (20% of cases), and in developing countries and among underprivileged patients in industrial countries, >90% of HL cases are associated with EBV (Table 1).

The genetic analysis of H/RS cells in lymphoma biopsy specimens suggests that germinal-center or post-germinal-center B cells are their precursors. They are most likely derived from a B-cell population with low antigen affinity or nonfunctional immunoglobulin receptors (i.e., affinity maturation failures) that should have been eliminated in the germinal-center reaction but have somehow survived. In contrast to those of follicular lymphomas, which have a similar cellular origin and retain a GC gene expression phenotype, H/RS cell expression profiles more closely resemble those of activated B cells. EBV gene expression, which is restricted to EBNA1, LMP1, and LMP2, may be involved in regulating B-cell-specific gene transcription patterns that promote and maintain the H/RS cell phenotype (150). In situ H/RS cells are surrounded by reactive infiltrates composed of T cells, histiocytes, eosinophils, and plasma cells. It is unclear why these immune cells cannot eliminate the H/RS cells, but HL-infiltrating lymphocytes appear to be anergic to stimulation with mitogens and antigens and produce poor Th1 and Th2 cytokine responses (123). The prevailing notion is that H/RS cells attract lymphocytes with T-regulatory cell phenotypes and, through an interplay of cytokines amongst the cells, induce a profoundly immunosuppressive microenvironment that serves to protect the tumor cells.

Table 1. EBV associations with disease in immunocompromised hosts

Disease	EBV association (%)	Latency type	EBNA1	EBNA2	LMP1	LMP2
IM	90	III	+	+	+	+
CAEBV	100	III	+	+	+	+
XLP	100?	III	+	+	+	+
LPDs						
PTLD	85	II/III	+	+/−	+	+
AIDS-associated LPDs	100	II/III	+	+/−	+	+
HL						
NS	20	II	+	−	+	+
MC	70	II	+	−	+	+
LD	50	II	+	−	+	+
LR	80	II	+	−	+	+
NLP	?					
AIDS-associated HL	90	II	+	−	+	+
Burkitt's lymphoma						
Endemic	95	I	+	−	−	−
Sporadic	25	I	+	−	−	−
AIDS-associated Burkitt's lymphoma	40	I/II	+	−	+/−	+/−
NK/T-cell lymphoma	100	II	+	−	+/−	−

Immunodeficiency-associated HL

HL has been reported for patients with primary immune disorders, such as autoimmune lymphoproliferative syndrome, Wiskott-Aldrich syndrome, and ataxia telangiectasia. HL also develops in the setting of immunosuppression after transplantation. An increased incidence of HL after allogeneic bone marrow transplantation has been reported, with an observed-to-expected case ratio of 6.2:1 (34). In polymorphic PTLDs, H/RS-like cells appear in the admixture of inflammatory cells (HL-like PTLD) infiltrating the lesion. EBV-infected H/RS cells are present in almost all cases of posttransplant HL (36). HL is one of the most common neoplasms that develop in human immunodeficiency virus (HIV)-infected individuals. Most of these HLs present with a histologic picture of either the MC or the LD subtype, and nearly all cases are associated with EBV infection (21).

Burkitt's lymphoma

Burkitt's lymphoma is a mature aggressive B-cell lymphoma that has geographically defined endemic and sporadic types and cases associated with progressive EBV disease in the setting of immunodeficiency or immunosuppression (181, 241). Malaria is a recognized cofactor in endemic Burkitt's lymphoma, through a mechanism that probably involves chronic parasite-related stimulation of B cells along with T-cell immunosuppression. EBV antibody titers, EBV viral loads, and numbers of circulating EBV-infected cells in the peripheral blood were higher in children from an area of Africa with holoendemic malaria than in children from an area with sporadic infection (139, 213).

The WHO consensus histologic classification of classic or atypical Burkitt's lymphoma is $CD20^+$ $BCL6^+$ $CD10^+$ $BCL2^-$ $CD5^-$ Ki-67^+ (≥95%), with a Ch8-Ch14 translocation (85). Nevertheless, the distinction of sporadic Burkitt's lymphoma from the much more common diffuse large-B-cell lymphomas is not reliable using the current triad of morphology, immunophenotypes, and cytogenetic abnormalities. Two recent expression array studies have been used to develop a molecular signature for Burkitt's lymphoma diagnosis that correlates with clinical outcome (41, 82). The key conclusion of these studies was that this new molecular definition of lymphomas needs to be adopted as a prerequisite for planning clinical trials for the treatment of mature aggressive B-cell lymphomas.

NK/T-cell lymphomas

NK/T-cell lymphomas are reported mainly from Asia and South America, where there is a consistent presence of clonal episomal EBV DNA in the tumor cells (5, 85, 104, 105, 190). Clonality implies that EBV is involved in the etiology and has important clinical implications. The tumor cells consistently express EBNA1 and LMP2, whereas expression of LMP1 is more variable (90, 241). They present clinically as lethal midline granulomas and histologically as pleomorphic malignant tumors variably associated with angiocentricity, angioinvasion, and necrosis

(90). NK lymphoma cells are typically positive for CD2, CD3[cepsilon], CD56, and cytotoxic effector molecules (perforin, granzyme B, and TIA-1). In situ hybridization for Epstein-Barr-encoded RNAs (EBERs) (see "In situ techniques," below) distinguishes NK lymphoma cells in histopathologic sections. The peripheral EBV DNA load is high (10^5 to 10^{10} copies/cm^3) at diagnosis and correlates with clinical staging, implying that it is a useful surrogate tumor biomarker in the initial evaluation of NK-cell lymphomas (9).

AIDS-related NHLs

Non-Hodgkin's lymphomas (NHLs) in patients with AIDS are heterogeneous and include systemic lymphomas, primary lymphomas of the central nervous system, primary effusion lymphomas (PELs), and plasmablastic lymphomas of the oral cavity (28, 237). EBV associations are highest among the immunoblastic lymphomas (95 to 100% EBV positive) and the PELs (90 to 100% EBV positive). These high association rates suggest that there is a pathogenic role for the virus in these tumors, and they are regarded as EBV-driven lymphoproliferations with similarities to PTLDs. Lower association rates (40% or less) for Burkitt's lymphomas, centroblastic lymphomas, and plasmacytic lymphomas imply that they arise by mechanisms that need not involve EBV infection.

Diagnostic Approaches

Virus culture techniques are not amenable to diagnosing EBV infection, principally because infection of B cells does not lead to a titratable cytopathic effect. Laboratory stocks of virus are titrated by a limiting-dilution immortalization assay that requires freshly isolated human B lymphocytes and at least 4 to 6 weeks of incubation (134). In addition, the immortalizing titers are dependent upon laboratory handling and culture conditions that are not standardized. A speedier assay, the EBNA-inducing assay, is an immunofluorescence assay (IFA) for EBNA1 expression 3 to 6 days after infection of a susceptible EBV-negative cell line, such as BJAB or DG75 (178). The relationship between the EBNA-inducing titer and the immortalizing titer has not been determined rigorously. Plasma virus titration assays (and antigenemia assays) are not suitable for EBV diagnosis because very little virus production occurs in infected cells in the circulation (78). The virus remains largely cell associated even as mononucleosis symptoms peak. Therefore, diagnostic approaches to EBV infection rely mainly on detecting serologic responses to classical EBV antigens and, more recently, on nucleic acid amplification and quantitation techniques.

EBV serology

EBV serology is based on detecting humoral responses to the following three classes of antigen: viral capsid antigen (VCA), early antigen (EA), and EBNA (170). Major epitopes within each class have been linked to the products of specific viral genes, and this knowledge has led to the production of a number of enzyme-linked immunosorbent assays (ELISAs) employing native antigens, purified recombinant proteins, and synthetic peptides (25, 26, 60, 136, 166, 184, 212, 221, 226). In one comprehensive study, commercially available EBV enzyme immunoassays (EIAs) differed considerably in performance compared to a "gold standard" EBNA IFA. Although the sensitivities are generally good, with only small differences between the different assays, there was a greater variation in specificity, which varied between 100% (Enzygnost's EIA) and 86% (Biotest's EIA) (25). The different performances of the individual parameters can be explained by the various antigens used. Native antigens, rather than synthetic peptides, are more reliable for EBV serology testing (25, 60). EIAs offer the benefit of automated high-throughput analyses and objective results in routine diagnostics, whereas IFA is more time-consuming and takes experienced personnel for fluorescence image reading. EIA kits have proved to be reliable and convenient alternatives to the indirect IFA for routine diagnostics.

The diagnosis of primary EBV infection is based upon an EBV-specific test for IgG and IgM antibodies to VCAs and IgG antibodies to the EBNAs, especially EBNA1, as the minimal requirement. After primary infection, IgM antibodies against EA and VCA are detected first, while IgG antibody titers against these antigens peak within 1 to 2 months. During convalescence, an IgG anti-EBNA response emerges, the IgG anti-VCA titer becomes very low, and anti-EA becomes nondetectable. A typical seroconverter will have a low-titer (1:20 to 1:160 by ELISA) anti-EBNA IgG response directed against epitopes in the EBNA1 protein and a similarly persistent low anti-VCA response dominated by an epitope in the BFRF3 gene (187, 221). Differentiating between a primary infection and the reactivation of latent virus would be valuable in some circumstances, but current serological methods based on detection of IgM antibodies to VCA and EBNA are unreliable (173). Because the anti-EA response is transient, the presence of anti-EA antibodies is suggestive of a recent or ongoing active infection. The

IgG avidity state (a measure of affinity maturation) may also have the potential to distinguish a primary from a reactivated infection. In one study, avidity measurement improved the sensitivity of serologic diagnosis of primary EBV infection from 93% to 100%. The specificity of IgM anti-VCA testing alone was poor (49% false-positive results) but improved to 97% by the demonstration of high-avidity IgG anti-VCA. The combination of negative IgG anti-EBNA and low-avidity IgG anti-VCA had a sensitivity and specificity of 100% (173).

A single test kit for IgM and IgG antibodies to VCA, EA, and EBNA does not exist, but most commercial kits are suitable for routine diagnostics as long as their limitations are appreciated. The ideal test kit for detection of EBV-specific antibodies would have both a high sensitivity for analysis of specimens from patients with a primary infection and a high specificity for analysis of cases with atypical symptoms that could be due to EBV infection. Clinical labs should combine different EIA kits for VCA, EA, and EBNA, and selecting kits based on native antigens is recommended.

Many attempts have been made to draw correlations between antibody responses against EBV and various diseases, syndromes, and cancers. For most genetically immunocompromised patients, none of the EBV serological tests provide either a sensitive or specific diagnosis. One possible exception might be made for patients with CAEBV. As a characteristic of the syndrome, these patients usually develop persistent exceptionally high anti-VCA and anti-EA titers (153, 216). IgM and IgA anti-VCA titers also appear to persist, and a portion of CAEBV patients fail to develop anti-EBNA responses. However, the high antibody titers against VCA and EA and the lack of anti-EBNA are not alone diagnostic of CAEBV. It is essential to demonstrate the presence of EBV in affected tissues and/or peripheral blood (97).

Perhaps not surprisingly, the development of LPDs posttransplant comes with no clear pattern of anti-EBV responses, making serological testing useful only for determining pretransplant serostatus. Anti-EBNA titers often decrease and anti-VCA titers rise slightly following transplantation in seropositive recipients, but there is no striking change in antibody titers preceding a clinical diagnosis of PTLD (129, 164, 172, 177). Because of immunosuppression, there is a rather limited humoral response to primary EBV infection in seronegative transplant recipients, which is dominated by anti-VCA antibodies, with little or no anti-EA and no anti-EBNA (222). The pattern is similar for AIDS-related immunoblastic lymphomas, where HIV-infected individuals show slightly elevated anti-EBV antibody titers but no serological correlation between antibody status and the development of AIDS-related lymphomas.

The development of Hodgkin's disease may in some patients be preceded by enhanced activation of EBV. For HL, individuals with a history of mononucleosis and elevated EBNA and VCA antibodies are at greater risk of disease (141). The relative risks of HL associated with elevated levels of IgG and IgA antibodies against capsid antigen were found to be 2.6 and 3.7, respectively. For anti-EBNA, the relative risk was 4.0, and for anti-EA it was 2.6 (141). There is no strong correlation between EBV serology and the presence of EBV in the H/RS cells of HL patients (2, 49, 113, 131). Antibodies to LMP1 (a viral protein that is often highly expressed in H/RS cells) were detected in 30% of EBV-seropositive Hodgkin's disease patients. The strongest antibody responses to LMP1 were among patients with EBV-negative H/RS cells, suggesting that the response might be associated with immune surveillance (130). LMP1 epitopes are not typically part of the humoral response to EBV infection, and the anti-LMP1 response detected was not part of a diagnostically distinctive anti-EBV antibody profile.

Many of the malignancies with an EBV association (NK/T-cell lymphoma, hairy cell leukemia, and gastric carcinoma) show elevations of anti-EA responses reminiscent of viral reactivation and may be the consequence of reduced immune surveillance (31, 88, 151, 188, 205).

Viral load measurements by PCR

Direct viral load measurement using DNA amplification technologies (particularly PCR) is gaining wider use in clinical laboratories to assist in diagnosing and monitoring virus-associated diseases (120). A large number of studies have been reported on the use of PCR in detecting and monitoring EBV infection (68, 70, 76, 99, 102, 133, 149, 179, 218, 229). The best assays use targets in highly conserved open reading frames (ORFs) (e.g., LMP1, LMP2, EBNA1, gp350, and gB ORFs), with few or no false-negative results due to primer or probe sequence variation. The approaches use slightly different sources and processing techniques for the nucleic acid and employ different quantitation standards. For the highest sensitivity, it is best to use either whole blood or isolated peripheral blood mononuclear cells as the source of virus DNA, since the virus is almost always cell associated and rarely found at comparable levels in cell-free compartments (such as plasma or cerebrospinal fluid) (228). Normalization (a separate quantitative PCR determination of the amount of amplifiable human DNA present) is a useful and desirable tool for smoothing out differences in specimen preparation and loading as well as providing an independent control for amplification inhibition (239). The assays have different readouts, using units such as copies/ml, copies/μg, or copies/10^6 cells. It is therefore not possible to generalize what an accurate enumeration of the viral load should be. In our experience, real-time PCR values below 4,000 copies/μg of DNA extracted from whole blood are considered low loads, PCR values above 4,000 copies/μg are considered high loads, and PCR values above 30,000 copies/μg are considered very high loads (228).

In primary EBV infection, the viral load is very high in the peripheral blood and in oral secretions. The level of the load is associated with severity of illness, and there is a rapid clearance of virus (mainly in the form of virus-infected cells) from the blood by a first-order process with a median half-life of 3 days (13, 34). High levels of shedding persist for months afterwards in the absence of symptoms. In the post-primary infection setting, viral load and anti-VCA IgG titers (but not anti-EA or anti-EBNA titers) are directly correlated in immunocompetent individuals, HL patients, transplant recipients, and HIV-infected individuals (15, 177, 196). At least one report suggests that quantitative EBV DNA PCR could lead to an increase of >16% in the number of positive diagnoses of primary infections by confirming a positive IgM anti-VCA result (115).

In genetically immunodeficient patients with XLP or CAEBV, the virus is directly involved in the lymphoproliferative syndrome and large numbers of infected lymphocytes are in the circulation. The level of viral load is very high. Because of this, monitoring for decreases in the circulating level of the load can be a useful adjunct to the resolution of clinical features in determining the effectiveness of therapeutic interventions (50, 234).

EBV load in HL. Patients with EBV-associated HL have elevations of EBV-positive B cells in the peripheral blood (94). These cells have a memory B-cell phenotype, and their relationship to H/RS cells is not clear because it is unlikely that H/RS cells are derived from the memory B-cell compartment. Despite this observation, other studies indicate no disease association with an elevated EBV load in peripheral blood mononuclear cells for patients with active EBV-positive HL compared to those with EBV-negative HL or to HL patients in long-term remission. However, free plasma EBV DNA is detectable in EBV-positive (and not EBV-negative) HL prior to therapy and disappears following treatment, suggesting that plasma-associated loads may be a noninvasive biomarker for EBV-positive HL and that serial monitoring could be used to monitor the response to therapy (57, 230).

EBV load in PTLD. Because EBV viral loads in the peripheral blood are elevated at the time of diagnosis of PTLD, EBV DNA load monitoring is a promising tool for the identification of patients at risk for PTLD development (10). In quantitative PCR assays, there is consistently a difference of 3 to 4 orders of magnitude between the normal latent viral load of a healthy carrier and the load detected in a PTLD patient (179). Monitoring viral load in the peripheral blood by quantitative PCR provides early detection of infection, an opportunity for preemptive therapeutic intervention, and a means of gauging the effectiveness of therapeutic strategies aimed at treating EBV infections and lymphoproliferative diseases (68, 110, 128, 180). Determining a threshold value for EBV load that triggers therapeutic intervention has been a problem, since other factors, such as pretransplant serology, type of organ transplanted, type and level of immunosuppression, time after transplantation, and other infectious complications, affect the load in the absence of PTLD (10). Serial monitoring for increases in viral load might be a better indicator of impending PTLD than observing a static load at an arbitrary threshold value (112, 197). For identification of patients at risk for PTLD in the absence of monitoring or when persistent high viral loads have been detected, combining an EBV DNA load measurement with a measure of the level of EBV-specific CTL responses has been suggested (110, 191). Low cellular immune responses combined with high EBV DNA loads in patients with primary EBV infection are associated with an increased risk for PTLD development (192).

EBV load in AIDS. The observation of persistently elevated EBV loads after iatrogenic immunosuppression in the setting of transplantation is recapitulated in HIV infection. In the first year after HIV seroconversion, there is an approximately fivefold increase in latent EBV load in the peripheral blood, but after this initial rise, there is no further alteration. The absolute number of effector CTLs also increases, suggesting that there is a response to the increased burden and that the latent infection is being controlled. Significantly, it has been shown that there is a lack of predictive value for elevated EBV load in relation to the occurrence of AIDS-related EBV-associated immunoblastic NHLs and primary central nervous system lymphomas (54, 163, 219). However, there is a significant decrease of EBV load in patients treated with chemotherapy, and a strong association between a favorable NHL outcome and the drop in the peripheral EBV load has been observed (19, 80). Detecting EBV in cerebrospinal fluid by EBV PCR also has a low sensitivity, specificity, and positive predictive value for primary central nervous system lymphomas (84).

In situ techniques

In situ techniques are used on tissue biopsies, primarily to determine whether or not EBV is present in the cells responsible for the lesion. Two in situ hybridization techniques are used, utilizing EBV-specific DNA probes to detect the presence of genomic DNA and oligonucleotide probes to detect EBER RNA (32, 87). The latter is by far the most commonly used technique, and probe kits are available from several sources. They rely on detection of one or both of two small (~200 bases) RNA molecules that are made by RNA polymerase III from the most actively transcribed region of the EBV genome. The great advantage of in situ hybridization for EBERs is that accumulation of these RNAs has been estimated to exceed 10^7 copies per cell. Thus, they are relatively easy to detect and can provide unequivocal evidence of the location of EBV within tissue sections in <2 hours. EBV has been detected in tumors of various lineages, but infected nonneoplastic B cells may also occupy the background of EBV-negative tumors. The assignment of cell type by double-labeling techniques to simultaneously detect virus and cell lineage provides an improved view of the relationship of the infected cells to the lesion.

Attempts are being made to augment the diagnostic utility of the detection of viral presence (EBER probes) with in situ characterizations of the pattern of viral gene expression in the infected cells. The following three main categories of latency-type infections are recognized: latency I (EBNA1 positive [EBNA1$^+$] LMP1$^-$ EBNA2$^-$), latency II (EBNA1$^+$ LMP1$^+$ EBNA2$^-$), and latency III (EBNA1$^+$ LMP1$^+$ EBNA2$^+$). IFA kits for EBNA1, EBNA2, and LMP1 are commonly used. A note of caution regarding the widely distributed and popular EBNA1 monoclonal antibody 2B4 is necessary. 2B4 also binds to MAGE-4, a testis cancer antigen expressed in a variety of tumor cells, including cells from breast carcinomas, seminomas, and EBV-negative cases of HL (75). The reports (too many to enumerate) documenting an EBV-disease association on the basis of reactivity with this antibody must be considered unreliable (142). Expression profiles for latent gene products can also be noncanonical (e.g., as in H/RS cells), especially when an expanded set of latency-associated proteins (including the EBNA3s, EBNA-LP, and LMP2, for which kits are not available) is considered, and the pattern may not be homogeneous and applicable to all EBV-positive cells within a section. There are also kits for lytic cycle antigens ZEBRA (BZLF1 Epstein-Barr replication activator), EA-D (EA diffuse cytoplasmic staining), and EA-R (EA restricted nuclear staining) and for the so-called M antigen (gp350/220 major envelope glycoprotein). Any of these can be used to detect lytic virus production. Temporally, ZEBRA is expressed first in cells that have entered virus particle production and has become the most commonly probed lytic cycle product even though it is not a virion component. Available kits rely on the BZ-1 antibody, which was made against a bacterial fusion protein and recognizes an epitope in the carboxy-terminal DNA binding domain of ZEBRA (242). It is important to verify that virion components (e.g., the gp350/220 major capsid glycoprotein) are also expressed, as studies relying on ZEBRA expression alone are unable to distinguish an abortive from a fully productive lytic cycle.

The current state of EBV gene expression profiling (with the exception of EBER expression profiling) is that it is used primarily as an investigative, not diagnostic, tool (10). Clearly, certain lymphoproliferative syndromes and lymphomas have characteristic patterns of EBV gene expression (Table 1). The challenge has been to show that strong correlations exist amongst the EBV expression patterns, the infected cell phenotypes, and pathogenic stages of disease and that they have clinical diagnostic significance. For instance, since as many as 14% of PTLD biopsies may be EBER negative, EBV positivity has not been adopted as an absolute criterion for the diagnosis of PTLD. Yet the distinction is worth noting, as truly EBV-negative PTLDs are suspected to be a distinct clinical entity (148). EBV-negative PTLDs have a later onset (a median of 50 months posttransplant versus 10 months) than do EBV-positive PTLDs and have been reported to be more aggressive (87, 148). Similarly, the presence of EBV in H/RS cells is associated with better survival in young patients and poorer survival in older patients with NS-type lymphomas, completely independent of other factors (93).

HHV-8

Background, Clinical Information, and Epidemiology

Similar to EBV, HHV-8 is a gammaherpesvirus that has the ability to cause cancer when the host is subjected to an immunocompromised state. HHV-8, like every known herpesvirus, establishes both lytic and latent infections. During a lytic infection, the entire genome is transcribed, resulting in virus replication. During latent infection, a small subset of viral genes (termed the latent genes) is expressed and there is no viral replication. HHV-8 infection varies geographically, with rates ranging from 5 to 10% in the U.S. general population to over 40% in some areas of Africa (48, 86, 108, 189, 193). In the United States, the infection rate also varies among different subpopulations. Infection is rarely seen in children (18) and has been reported to be, on average, 3.5% among blood donors (161). Among transplant recipients, it has been found to be as high as 20% (86), while among men who have sex with men it is often over 40% (125).

In immunocompetent hosts, HHV-8 primary infection appears to be mostly asymptomatic or, in some cases, associated with a flu-like syndrome (231) or febrile maculopapular skin rash (4), and thus individuals undergoing a primary infection are likely to be unaware of the infection. Among immunocompromised hosts, such as solid organ transplant recipients, a primary infection has been associated with severe disease, including rash, fever, anemia, adenopathy, and in some cases, death (118, 122, 167).

HHV-8 is the causative agent of Kaposi's sarcoma (KS) as well as a rare B-cell pleural effusion and some forms of multicentric Castleman's disease. KS is a tumor originating from endothelial cells that occurs in four forms (209). The classic form is found in older men of Mediterranean, Eastern European, or Jewish descent, while an endemic form is seen in sub-Saharan Africa and affects predominantly young adults and young children. An epidemic or AIDS-associated form is the most common cancer seen in HIV-infected homosexual men, and an iatrogenic form is seen in solid organ transplant recipients. HHV-8 infection in healthy persons does not result in cancer, with the exception of African endemic KS, which affects normally healthy persons.

KS is the most common of the posttransplant HHV-8-associated cancers, occurring in 0.5 to 5% of all cases in the United States (121), but it has been reported to represent as high as 47.7% of all posttransplant cancers in other areas (140). KS arises most often after kidney transplantation and less often in cases of liver, heart, and lung transplantation (3), but it can arise as a result of immunosuppression related to the transplantation of any organ type. However, incidences of PEL (46), EBV-negative PTLD (91, 126), and bone marrow failure and complications (119), and possibly a case of multicentric Castleman's disease, have been associated with HHV-8 posttransplant (211). KS often results from either a secondary infection or reactivation of a latent infection of the recipient, but there are also a number of reports suggesting that the seronegative recipient developed KS upon receiving an organ from a seropositive donor (118, 122, 158). Further studies have elegantly shown that these KS cells were indeed donor derived (14).

In contrast to AIDS-KS and classical Mediterranean KS, which almost exclusively affects males, the incidence of posttransplant KS in the United States is only slightly higher for males than females (127) but may vary by geographical region elsewhere (140). Also, like that of classical KS, the risk of posttransplant KS is correlated with increasing age (127).

Posttransplant KS is found most commonly in the skin, with a smaller proportion of cases also having visceral involvement (20 to 47%, depending on geographic area of origin) (127, 140, 238). While posttransplant KS often previously resulted in organ rejection and/or death of the patient, current therapy regimens have an extremely positive outlook (see "Therapy," below).

Therapy

Current treatment for posttransplant KS typically involves a decrease or complete cessation of immunosuppressive agents, which unfortunately may lead to transplant rejection. Sirolimus (also known as rapamycin or Rapamune), which inhibits the interleukin-2 pathway by binding to the FK-binding protein 12 and preventing its interaction with mTOR (mammalian target of rapamycin), appears to be a more ideal drug, allowing for therapeutic levels of immunosuppression concurrent with regression of KS in a number of patients who are converted to the regimen (72, 183, 225). Additional therapies have been used with limited efficacy, including anti-CD20 monoclonal antibody treatment (210), chemotherapy agents (159), and antiviral agents (117, 223), but in many cases, the only successful treatment is cessation of immunosuppression or conversion to sirolimus. A suitable specific antiviral agent has not been discovered for treatment of KS.

Risk Stratification

Currently, there are no tests available for determining susceptibility to posttransplant HHV-8-related cancers distinct from simple determination of HHV-8 infection. A recent report has suggested that persons with a polymorphism in the interleukin-6 promoter region (G174C) have a higher risk of developing KS after renal transplantation (62). Additionally, other polymorphisms in immune-related genes have been shown to be associated with classical KS (23, 45, 55, 61, 111). However, it has yet to be determined whether any of these genes will serve as a strong tool for genotypic risk profiling. Currently, the strongest association for development of KS remains transplantation of an organ from an HHV-8-seropositive donor into a seronegative recipient.

Diagnostic Approaches

Diagnoses of KS cancers are most often performed histologically. Demonstration of HHV-8 nucleic acids or proteins in tumor cells is usually performed to support the pathology diagnosis. The most common method is immunohistochemistry, using antibodies directed against the latency-associated nuclear antigen (LANA; encoded by ORF73). All KS lesions express LANA in the tumor cells. Since a primary infection with HHV-8 is most often asymptomatic or subclinical and therefore does not result in overt disease, a clinical diagnosis of a primary HHV-8 infection is rarely given. However, given that transplantation of a solid organ from a seropositive donor into a seronegative recipient results in an increased risk of morbidity and mortality, determination of HHV-8 serostatus may become more commonplace in future clinical settings. Diagnostic approaches for the determination of HHV-8 infection have centered on the detection of virus-specific antibodies, viral nucleic acid, or viral proteins in patient samples.

Detection of viral antibodies

The most common approach to diagnosis of HHV-8 infection is the detection of virus-specific antibodies. This is due to inconsistent detection of viral DNA or proteins in circulating blood cells of infected individuals. A variety of serological assays have been developed to detect latent or lytic cycle viral proteins. The assays can be divided into three groups. The first group consists of IFAs which use B-cell lines from PEL patients. These cells are infected with HHV-8, and the virus maintains a latent infection in 97 to 98% of cells (168). The remaining 2% to 3% of the cells undergo spontaneous reactivation and express lytic cycle proteins. Examples of the cell lines used in these assays are BCBL-1 (168), BC-1 (132), BC-3 (8), and BCP-1 (20). Treatment of the PEL cell lines with phorbol esters and/or butyric acid results in the induction of the lytic cycle in 10% to 20% of the cells. These induced cells can be used in IFA to detect antibodies against lytic cycle proteins. Since only 10% to 20% of the cells express the lytic cycle proteins, the use of induced PEL cells provides the necessary mixture of positive and negative cells. Positively stained cells (expressing lytic cycle proteins) show predominantly whole-cell staining (Color Plate 4). IFAs using uninduced PEL cell lines detect the presence of antibodies directed against latent cycle proteins. In these assays, positively stained cells typically show punctate nuclear fluorescence representing the staining of LANA-1, encoded by ORF73 (Color Plate 5). For this assay, the uninduced cells should be mixed with an HHV-8-negative B-cell line in order to provide both positive and negative cells within the same microscopic field.

The second set of HHV-8 serological assays consists of ELISAs using either individual recombinant viral proteins or lysates from induced cells or purified virus particles (for examples, see references 16, 30, 51, 106, 124, 161, 186, 194, and 215). These assays can detect antibodies directed against latent or lytic cycle proteins, depending on the target protein used to construct the assay. The viral proteins most often used in this set of assays are LANA-1, for detection of antibodies directed against latent proteins, and the structural proteins ORF65 and K8.1 (a viral glycoprotein), for detection of antibodies directed against lytic cycle proteins.

The third set of HHV-8 serological assays is based on immunoblots of viral proteins separated by polyacrylamide gel electrophoresis. These assays can use either purified proteins or lysates from infected cells (232, 243).

As mentioned earlier, routine clinical diagnostics are not currently performed for HHV-8 infections. In research settings, however, the field is divided between those labs that utilize IFAs and those that use ELISAs.

Serological assays designed to detect viral antibodies should be both sensitive and specific. Determinations of sensitivity and specificity are dependent upon the availability of a set of carefully evaluated sera consisting of known positive and negative samples. For the majority of reports testing individual assays, HHV-8-positive samples have consisted of serum samples from patients with KS. The negative samples have been more problematic, since primary infections with HHV-8 have been reported to be asymptomatic or subclinical (231) and, as described below, seropositivity cannot be confirmed by detection of viral DNA, since most seropositive non-KS patients do not have detectable HHV-8 DNA in blood or tissue samples. Therefore, the determination of true negative samples is difficult and open to debate. In some cases, investigators have assumed that blood donors or healthy adult controls are seronegative and have used these samples to define the cutoff limit for negative controls. This approach can result in false-negative results since some of these samples may be seropositive. In other cases, laboratories have compared individual assays by using sera from different groups, such as HIV-infected subjects with and without KS and healthy HIV-seronegative subjects. These types of analyses allow for determinations of sensitivity but do not permit specificity to be evaluated accurately. Until the issue of true HHV-8-seronegative samples can be resolved, the debate regarding the specificities of different assays will continue.

Studies comparing these different assays have reported various results. There is a consensus that the IFA using induced PEL cells is the most sensitive assay compared to IFAs using uninduced PEL cells, individual ELISAs, or immunoblot assays (161, 185, 194). This is not particularly surprising because this assay has the greatest number of viral proteins, and therefore protein epitopes, available for binding to anti-HHV-8 antibodies. In support of this, laboratories using ELISAs that detect individual viral proteins often define a seropositive cell as one in which there is a positive reaction to one or more of the ELISAs. Thus, by this definition, different serum samples will react with different viral proteins and sensitivity is increased as the number of viral proteins present in the assay increases. Most studies have used serum samples from KS patients as their positive controls. KS patients have very high HHV-8 antibody titers, and therefore all serological assays should readily identify sera from KS patients as seropositive. Assays that fail to detect sera from KS subjects as seropositive are clearly not sensitive. The inability to detect HHV-8 antibodies in a KS patient is analogous to an inability to detect EBV antibodies in patients with mononucleosis. The largest disadvantages to the lytic IFA are that it is time-consuming and the results can be somewhat subjective. Therefore, this assay is best used in situations where the number of readers is minimized in order to reduce variance between the readers. As described earlier, the specificities of the different assays are under debate, since true seronegative samples are not clearly defined.

Detection of viral DNA

Viral DNA detection methods have focused on using PCR for both quantitative and qualitative assays. For detection of viral DNA, i.e., to demonstrate the presence of virus in a particular sample, qualitative PCR is as useful as quantitative PCR. While studies have indicated that the viral DNA load in circulating lymphocytes increases prior to the onset of clinical KS (16, 135), unlike the case for EBV, there are no known viral loads that are predictive for disease development. The major concern with using detection of viral DNA as a marker of viral infection is an apparent lack of sensitivity and indecision on which tissue type is best suited for detection of viral DNA. In the absence of an HHV-8-related cancer, the most common samples used for detection of viral DNA are circulating lymphocytes, saliva, and plasma. Since HHV-8 is a gammaherpesvirus, it is reasonable to think that peripheral B cells would represent a reservoir of virus from which viral DNA could be detected. However, most studies indicate that viral DNA can be detected in some but not all seropositive individuals (1, 107, 138). The most common source for detecting viral DNA appears to be saliva, inasmuch as studies that have looked at both B cells and saliva from the same patients have reported the detection of HHV-8 among saliva samples more often than among B cells (reviewed in reference 162). In one particular study, the researchers were able to identify and quantitate viral transcripts in clinical samples, including saliva, rectal tissue, and peripheral blood lymphocytes (195). However, these samples were from HIV-positive patients, some of whom had KS, and the assay was able to detect viral DNA in only some, but not all, samples, which varied by tissue type among the patients tested. This is in agreement with an additional study that found 83% of KS, PEL, or multicentric Castleman's disease cases to be DNA positive in serum, while 96% of the same samples were seropositive. In cases of HIV-positive/HHV-8 disease-negative patients, this assay found 3% to be DNA positive (versus 38% who were seropositive), while 0% of healthy blood donors were DNA positive (versus 5% who were seropositive) (22). A more recent study was able to find viral DNA in 100% of KS biopsy samples but did not attempt it in any other tissue types (74). When viral DNA loads were assayed for persons who were known to be HHV-8 seropositive but did not have KS, only 16.5% had detectable HHV-8 in their peripheral blood (24).

A new trend in the field of viral detection methods is to construct assays that are able to detect multiple pathogens in a single assay. For example, one group has devised a nested PCR assay that is able to detect all eight herpesviruses at once, with a lower detection limit of 10 to 20 copies, depending on the virus (81). Similarly, another group has developed a multiplex PCR array for detection of EBV, CMV, and HHV-8. This particular assay was able to detect HHV-8 and CMV down to 1 copy but was less effective at EBV detection (limit of detection, 1,000 copies) (56). Since both of these assays are new, they have been assessed only with known infected cell lines or clinical samples and have yet to be evaluated for diagnostic cases.

Thus, it appears that while the ability to quantitate viral load is an attractive idea, current strategies are too insensitive to be very useful for widespread screening of asymptomatic carriers. Additionally, while real-time PCR is currently the "gold standard" for DNA quantitation, the specialized equipment required is quite expensive and may not be available in smaller markets.

In summary, serological assays are the best method for determining HHV-8 infection, and assays designed to detect lytic cycle proteins, especially lytic IFAs, are the most sensitive assays currently available.

REFERENCES

1. **Albrecht, D., T. Meyer, T. Lorenzen, A. Stoehr, R. Arndt, and A. Plettenberg.** 2004. Epidemiology of HHV-8 infection in HIV-positive patients with and without Kaposi sarcoma: diagnostic relevance of serology and PCR. *J. Clin. Virol.* **30:** 145–149.
2. **Alexander, F., C. P. Daniel, A. A. Armstrong, D. A. Clark, D. E. Onions, R. A. Cartwright, and R. F. Jarrett.** 1995. Case clustering, Epstein-Barr virus Reed-Sternberg cell status and herpes virus serology in Hodgkin's disease: results of a case-control study. *Eur. J. Cancer* **31:**1479–1486.
3. **Andreoni, M., D. Goletti, P. Pezzotti, A. Pozzetto, P. Monini, L. Sarmati, F. Farchi, G. Tisone, A. Piazza, F. Pisani, M. Angelico, P. Leone, F. Citterio, B. Ensoli, and G. Rezza.** 2001. Prevalence, incidence and correlates of HHV-8/KSHV infection and Kaposi's sarcoma in renal and liver transplant recipients. *J. Infect.* **43:**195–199.
4. **Andreoni, M., L. Sarmati, E. Nicastri, G. El Sawaf, M. El Zalabani, I. Uccella, R. Bugarini, S. G. Parisi, and G. Rezza.** 2002. Primary human herpesvirus 8 infection in immunocompetent children. *JAMA* **287:**1295–1300.
5. **Arber, D. A. M. D., L. M. M. D. Weiss, P. F. M. D. Albujar, Y.-Y. B. S. Chen, and E. S. M. D. Jaffe.** 1993. Nasal lymphomas in Peru: high incidence of T-cell immunophenotype and Epstein-Barr virus infection. *Am. J. Surg. Pathol.* **17:** 392–399.
6. **Aris, R. M., D. M. Maia, I. P. Neuringer, K. Gott, S. Kiley, K. Gertis, and J. Handy.** 1996. Post-transplantation lymphoproliferative disorder in the Epstein-Barr virus-native lung transplant recipient. *Am. J. Respir. Crit. Care Med.* **154:** 1712.
7. **Armitage, J. M., R. L. Kormos, R. S. Stuart, F. J. Fricker, B. P. Griffith, M. Nalesnik, R. L. Hardesty, and J. S. Dummer.** 1991. Posttransplant lymphoproliferative disease in thoracic organ transplant patients: ten years of cyclosporine-based immunosuppression. *J. Heart Lung Transplant.* **10:**877.
8. **Arvanitakis, L., E. A. Mesri, R. G. Nador, J. W. Said, A. S. Asch, D. M. Knowles, and E. Cesarman.** 1996. Establishment and characterization of a primary effusion (body cavity-based) lymphoma cell line (BC-3) harboring Kaposi's sarcoma-associated herpesvirus (KSHV/HHV-8) in the absence of Epstein-Barr virus. *Blood* **88:**2648–2654.
9. **Au, W.-Y., A. Pang, C. Choy, C.-S. Chim, and Y.-L. Kwong.** 2004. Quantification of circulating Epstein-Barr virus (EBV) DNA in the diagnosis and monitoring of natural killer cell and EBV-positive lymphomas in immunocompetent patients. *Blood* **104:**243–249.
10. **Bakker, N. A., G. W. van Imhoff, E. A. M. Verschuuren, and W. J. van Son.** 2007. Presentation and early detection of post-transplant lymphoproliferative disorder after solid organ transplantation. *Transpl. Int.* **20:**207–218.
11. **Balandraud, N., J. Roudier, and C. Roudier.** 2004. Epstein-Barr virus and rheumatoid arthritis. *Autoimmun. Rev.* **3:** 362.
12. **Balandraud, N. I. A., H. Sovran, B. Mugnier, D. Reviron, and J. Roudier.** 2003. Epstein-Barr virus load in the peripheral blood of patients with rheumatoid arthritis, accurate quantification using real time polymerase chain reaction. *Arthritis Rheum.* **48:**1223–1228.
13. **Balfour, H. H., Jr., C. J. Holman, K. M. Hokanson, M. M. Lelonek, J. E. Giesbrecht, D. R. White, D. O. Schmeling, C.-H. Webb, W. Cavert, D. H. Wang, and R. C. Brundage.** 2005. A prospective clinical study of Epstein-Barr virus and host interactions during acute infectious mononucleosis. *J. Infect. Dis.* **192:**1505–1512.
14. **Barozzi, P., M. Luppi, F. Facchetti, C. Mecucci, M. Alu, R. Sarid, V. Rasini, L. Ravazzini, E. Rossi, S. Festa, B. Crescenzi, D. G. Wolf, T. F. Schulz, and G. Torelli.** 2003. Post-transplant Kaposi sarcoma originates from the seeding of donor-derived progenitors. *Nat. Med.* **9:**554–561.
15. **Besson, C., C. Amiel, C. Le-Pendeven, P. Brice, C. Ferme, P. Carde, O. Hermine, M. Raphael, L. Abel, and J.-C. Nicolas.** 2006. Positive correlation between Epstein-Barr virus viral load and antiviral capsid immunoglobulin G titers determined for Hodgkin's lymphoma patients and their relatives. *J. Clin. Microbiol.* **44:**47–50.
16. **Biggar, R. J., E. A. Engels, D. Whitby, D. H. Kedes, and J. J. Goedert.** 2003. Antibody reactivity to latent and lytic antigens to human herpesvirus-8 in longitudinally followed homosexual men. *J. Infect. Dis.* **187:**12–18.
17. **Blaschke, S., G. Schwarz, D. Moneke, L. Binder, G. Muller, and M. Reuss-Borst.** 2000. Epstein-Barr virus infection in peripheral blood mononuclear cells, synovial fluid cells, and synovial membranes of patients with rheumatoid arthritis. *J. Rheumatol.* **27:**866–873.
18. **Blauvelt, A., S. Sei, P. M. Cook, T. F. Schulz, and K. T. Jeang.** 1997. Human herpesvirus 8 infection occurs following adolescence in the United States. *J. Infect. Dis.* **176:**771–774.
19. **Bonnet, F., A. C. Jouvencel, M. Parrens, M. J. Leon, E. Cotto, I. Garrigue, P. Morlat, J. Beylot, H. Fleury, and M. E. Lafon.** 2006. A longitudinal and prospective study of Epstein-Barr virus load in AIDS-related non-Hodgkin lymphoma. *J. Clin. Virol.* **36:**258–263.
20. **Boshoff, C., S. J. Gao, L. E. Healy, S. Matthews, A. J. Thomas, L. Coignet, R. A. Warnke, J. A. Strauchen, E. Matutes, O. W. Kamel, P. S. Moore, R. A. Weiss, and Y. Chang.** 1998. Establishing a KSHV+ cell line (BCP-1) from peripheral blood and characterizing its growth in Nod/SCID mice. *Blood* **91:**1671–1679.
21. **Bourgault, I., A. Gomez, E. Gomard, and J. P. Levy.** 1991. Limiting-dilution analysis of the HLA restriction of anti-Epstein-Barr virus-specific cytotoxic T lymphocytes. *Clin. Exp. Immunol.* **84:**501–507.
22. **Broccolo, F., G. Locatelli, L. Sarmati, S. Piergiovanni, F. Veglia, M. Andreoni, S. Butto, B. Ensoli, P. Lusso, and M. S. Malnati.** 2002. Calibrated real-time PCR assay for quantitation of human herpesvirus 8 DNA in biological fluids. *J. Clin. Microbiol.* **40:**4652–4658.

23. **Brown, E. E., D. Fallin, I. Ruczinski, A. Hutchinson, B. Staats, F. Vitale, C. Lauria, D. Serraino, G. Rezza, G. Mbisa, D. Whitby, A. Messina, J. J. Goedert, and S. J. Chanock.** 2006. Associations of classic Kaposi sarcoma with common variants in genes that modulate host immunity. *Cancer Epidemiol. Biomarkers Prev.* **15:**926–934.
24. **Brown, E. E., D. Whitby, F. Vitale, P. C. Fei, C. Del Carpio, V. Marshall, A. J. Alberg, D. Serraino, A. Messina, L. Gafa, and J. J. Goedert.** 2005. Correlates of human herpesvirus-8 DNA detection among adults in Italy without Kaposi sarcoma. *Int. J. Epidemiol.* **34:**1110–1117.
25. **Bruu, A. L., R. Hjetland, E. Holter, L. Mortensen, O. Natas, W. Petterson, A. G. Skar, T. Skarpaas, T. Tjade, and B. Asjo.** 2000. Evaluation of 12 commercial tests for detection of Epstein-Barr virus-specific and heterophilic antibodies. *Clin. Diagn. Lab. Immunol.* **7:**451–456.
26. **Buisson, M., B. Fleurent, M. Mak, P. Morand, L. Chan, A. Ng, M. Guan, D. Chin, and J. M. Seigneurin.** 1999. Novel immunoblot assay using four recombinant antigens for diagnosis of Epstein-Barr virus primary infection and reactivation. *J. Clin. Microbiol.* **37:**2709–2714.
27. **Capello, D., D. Rossi, and G. Gaidano.** 2005. Post-transplant lymphoproliferative disorders: molecular basis of disease histogenesis and pathogenesis. *Hematol. Oncol.* **23:**61–67.
28. **Carbone, A., and A. Gloghini.** 2005. AIDS-related lymphomas: from pathogenesis to pathology. *Br. J. Haematol.* **130:**662–670.
29. **Carpentier L., B. Tapiero, F. Alvarez, C. Viau, and C. Alfieri.** 2003. Epstein-Barr virus (EBV) early-antigen serologic testing in conjunction with peripheral blood EBV DNA load as a marker for risk of posttransplantation lymphoproliferative disease. *J. Infect. Dis.* **188:**1853–1864.
30. **Casper, C., E. Krantz, H. Taylor, J. Dalessio, D. Carrell, A. Wald, L. Corey, and R. Ashley.** 2002. Assessment of a combined testing strategy for detection of antibodies to human herpesvirus 8 (HHV-8) in persons with Kaposi's sarcoma, persons with asymptomatic HHV-8 infection, and persons at low risk for HHV-8 infection. *J. Clin. Microbiol.* **40:**3822–3825.
31. **Chan, K. H., Y. L. Gu, F. Ng, P. S. Ng, W. H. Seto, J. S. Sham, D. Chua, W. Wei, Y. L. Chen, W. Luk, Y. S. Zong, and M. H. Ng.** 2003. EBV specific antibody-based and DNA-based assays in serologic diagnosis of nasopharyngeal carcinoma. *Int. J. Cancer* **105:**706–709.
32. **Chang, K., Y. Y. Chen, D. Shibata, and L. M. Weiss.** 1992. Description of an in situ hybridization methodology for detection of Epstein-Barr virus RNA in paraffin-embedded tissues, with a survey of normal and neoplastic tissues. *Diagn. Mol. Pathol.* **1:**246–255.
33. **Chen, J. M. B., M. L. Barr, A. Chadburn, G. Frizzera, F. A. Schenkel, R. R. Sciacca, D. S. Reison, L. J. Addonizio, E. A. Rose, and D. M. Knowles.** 1993. Management of lymphoproliferative disorders after cardiac transplantation. *Ann. Thorac. Surg.* **56:**527–532.
34. **Claudia, C., S. Bauer, et al.** 2005. Serum Epstein-Barr virus DNA load in primary Epstein-Barr virus infection. *J. Med. Virol.* **75:**54–58.
35. **Cockfield, S. M.** 2001. Identifying the patient at risk for post-transplant lymphoproliferative disorder. *Transpl. Infect. Dis.* **3:**70–78.
36. **Cockfield, S. M., J. K. Preiksaitis, L. D. Jewell, and A. A. Parfrey.** 1993. Post-transplant lymphoproliferative disorder in renal allograft recipients: clinical experience and risk factor analysis in a single center. *Transplantation* **56:**88–96.
37. **Conacher, M., R. Callard, K. McAulay, H. Chapel, D. Webster, D. Kumararatne, A. Chandra, G. Spickett, P. A. Hopwood, and D. H. Crawford.** 2005. Epstein-Barr virus can establish infection in the absence of a classical memory B-cell population. *J. Virol.* **79:**11128–11134.
38. **Conley, M. E., L. D. Notarangelo, and A. Etzioni.** 1999. Diagnostic criteria for primary immunodeficiencies: representing PAGID (Pan-American Group for Immunodeficiency) and ESID (European Society for Immunodeficiencies). *Clin. Immunol.* **93:**190-197.
39. **Crotzer, V. L., R. E. Christian, J. M. Brooks, J. Shabanowitz, R. E. Settlage, J. A. Marto, F. M. White, A. B. Rickinson, D. F. Hunt, and V. H. Engelhard.** 2000. Immunodominance among EBV-derived epitopes restricted by HLA-B27 does not correlate with epitope abundance in EBV-transformed B-Lymphoblastoid cell lines. *J. Immunol.* **164:**6120–6129.
40. **Curtis, R. E., L. B. Travis, P. A. Rowlings, G. Socie, D. W. Kingma, P. M. Banks, E. S. Jaffe, G. E. Sale, M. M. Horowitz, R. P. Witherspoon, D. A. Shriner, D. J. Weisdorf, H.-J. Kolb, K. M. Sullivan, K. A. Sobocinski, R. P. Gale, R. N. Hoover, J. F. Fraumeni, Jr., and H. J. Deeg.** 1999. Risk of lymphoproliferative disorders after bone marrow transplantation: a multi-institutional study. *Blood* **94:**2208–2216.
41. **Dave, S. S., K. Fu, G. W. Wright, L. T. Lam, P. Kluin, E.-J. Boerma, T. C. Greiner, D. D. Weisenburger, A. Rosenwald, G. Ott, H.-K. Muller-Hermelink, R. D. Gascoyne, J. Delabie, L. M. Rimsza, R. M. Braziel, T. M. Grogan, E. Campo, E. S. Jaffe, B. J. Dave, W. Sanger, M. Bast, J. M. Vose, J. O. Armitage, J. M. Connors, E. B. Smeland, S. Kvaloy, H. Holte, R. I. Fisher, T. P. Miller, E. Montserrat, W. H. Wilson, M. Bahl, H. Zhao, L. Yang, J. Powell, R. Simon, W. C. Chan, L. M. Staudt, et al.** 2006. Molecular diagnosis of Burkitt's lymphoma. *N. Engl. J. Med.* **354:**2431–2442.
42. **Davis, C. L., K. L. Harrison, J. P. McVicar, P. J. Forg, M. P. Bronner, and C. L. Marsh.** 1995. Antiviral prophylaxis and the Epstein Barr virus-related post-transplant lymphoproliferative disorder. *Clin. Transplant.* **9:**53–61.
43. **Deeg, H. J., J. Sanders, and P. Martin.** 1984. Secondary malignancies after marrow transplantation. *Exp. Hematol.* **12:**660.
44. **Domingo-Domenech, E., S. de Sanjosé, E. Gonzalez-Barca, V. Romagosa, A. Domingo-Claros, S. Gil-Vernet, J. Figueras, N. Manito, B. Oton, J. Petit, A. Granena, and A. Fernandez de Sevilla.** 2001. Post-transplant lymphomas: a 20-year epidemiologic, clinical and pathologic study in a single center. *Haematologica* **86:**715–721.
45. **Dorak, M. T., L. J. Yee, J. Tang, W. Shao, E. S. Lobashevsky, L. P. Jacobson, and R. A. Kaslow.** 2005. HLA-B,-DRB1/3/4/5, and -DQB1 gene polymorphisms in human immunodeficiency virus-related Kaposi's sarcoma. *J. Med. Virol.* **76:**302–310.
46. **Dotti, G., R. Fiocchi, T. Motta, B. Facchinetti, B. Chiodini, G. M. Borleri, G. Gavazzeni, T. Barbui, and A. Rambaldi.** 1999. Primary effusion lymphoma after heart transplantation: a new entity associated with human herpesvirus-8. *Leukemia* **13:**664–670.
47. **Durandy, A., P. Revy, K. Imai, and A. Fischer.** 2005. Hyper-immunoglobulin M syndromes caused by intrinsic B-lymphocyte defects. *Immunol. Rev.* **203:**67–79.
48. **Eltom, M. A., S. M. Mbulaiteye, A. J. Dada, D. Whitby, and R. J. Biggar.** 2002. Transmission of human herpesvirus 8 by sexual activity among adults in Lagos, Nigeria. *AIDS* **16:**2473–2478.
49. **Enblad, G., K. Sandvej, E. Lennette, C. Sundstrom, G. Klein, B. Glimelius, and G. Pallesen.** 1997. Lack of correlation between EBV serology and presence of EBV in the Hodgkin and Reed-Sternberg cells of patients with Hodgkin's disease. *Int. J. Cancer* **72:**394–397.

50. **Engel, P., M. J. Eck, and C. Terhorst.** 2003. The SAP and SLAM families in immune responses and X-linked lymphoproliferative disease. *Nat. Rev. Immunol.* **3:**813.
51. **Engels, E. A., D. Whitby, P. B. Goebel, A. Stossel, D. Waters, A. Pintus, L. Contu, R. J. Biggar, and J. J. Goedert.** 2000. Identifying human herpesvirus 8 infection: performance characteristics of serologic assays. *J. Acquir. Immune Defic. Syndr.* **23:**346–354.
52. **Falco, D. A., R. R. Nepomuceno, S. M. Krams, P. P. Lee, M. M. Davis, O. Salvatierra, S. R. Alexander, C. O. Esquivel, K. L. Cox, L. R. Frankel, and O. M. Martinez.** 2002. Identification of Epstein-Barr virus-specific CD8+ T lymphocytes in the circulation of pediatric transplant recipients. *Transplantation* **74:**501–510.
53. **Faulkner, G. C., S. R. Burrows, R. Khanna, D. J. Moss, A. G. Bird, and D. H. Crawford.** 1999. X-linked agammaglobulinemia patients are not infected with Epstein-Barr virus: implications for the biology of the virus. *J. Virol.* **73:**1555–1564.
54. **Fellner, M. D., K. Durand, R. M. Correa, L. Redini, C. Yampolsky, A. Colobraro, G. Sevlever, A. R. Teyssie, J. Benetucci, and M. A. Picconi.** 2007. Circulating Epstein-Barr virus (EBV) in HIV-infected patients and its relation with primary brain lymphoma. *Int. J. Infect. Dis.* **11:**172.
55. **Foster, C. B., T. Lehrnbecher, S. Samuels, S. Stein, F. Mol, J. A. Metcalf, K. Wyvill, S. M. Steinberg, J. Kovacs, A. Blauvelt, R. Yarchoan, and S. J. Chanock.** 2000. An IL6 promoter polymorphism is associated with a lifetime risk of development of Kaposi sarcoma in men infected with human immunodeficiency virus. *Blood* **96:**2562–2567.
56. **Fujimuro, M., K. Nakaso, K. Nakashima, H. Sadanari, I. Hisanori, Y. Teishikata, S. D. Hayward, and H. Yokosawa.** 2006. Multiplex PCR-based DNA array for simultaneous detection of three human herpesviruses, EVB, CMV and KSHV. *Exp. Mol. Pathol.* **80:**124–131.
57. **Gandhi, M. K., E. Lambley, J. Burrows, U. Dua, S. Elliott, P. J. Shaw, H. M. Prince, M. Wolf, K. Clarke, C. Underhill, T. Mills, P. Mollee, D. Gill, P. Marlton, J. F. Seymour, and R. Khanna.** 2006. Plasma Epstein-Barr virus (EBV) DNA is a biomarker for EBV-positive Hodgkin's lymphoma. *Clin. Cancer Res.* **12:**460–464.
58. **Ganschow, R., T. Schulz, T. Meyer, D. C. Broering, and M. Burdelski.** 2004. Low-dose immunosuppression reduces the incidence of post-transplant lymphoproliferative disease in pediatric liver graft recipients. *J. Pediatr. Gastroenterol. Nutr.* **38:**198–203.
59. Reference deleted.
60. **Gartner, B. C., R. D. Hess, D. Bandt, A. Kruse, A. Rethwilm, K. Roemer, and N. Mueller-Lantzsch.** 2003. Evaluation of four commercially available Epstein-Barr virus enzyme immunoassays with an immunofluorescence assay as the reference method. *Clin. Diagn. Lab. Immunol.* **10:**78–82.
61. **Gaya, A., A. Esteve, J. Casabona, J. J. McCarthy, J. Martorell, T. F. Schulz, and D. Whitby.** 2004. Amino acid residue at position 13 in HLA-DR beta chain plays a critical role in the development of Kaposi's sarcoma in AIDS patients. *AIDS* **18:**199–204.
62. **Gazouli, M., G. Zavos, I. Papaconstantinou, J. C. Lukas, A. Zografidis, J. Boletis, and A. Kostakis.** 2004. The interleukin-6[hyphen]174 promoter polymorphism is associated with a risk of development of Kaposi's sarcoma in renal transplant recipients. *Anticancer Res.* **24:**1311–1314.
63. **Giovannetti, A., M. Pierdominici, F. Mazzetta, M. Marziali, C. Renzi, A. M. Mileo, M. De Felice, B. Mora, A. Esposito, R. Carello, A. Pizzuti, M. G. Paggi, R. Paganelli, W. Malorni, and F. Aiuti.** 2007. Unravelling the complexity of T cell abnormalities in common variable immunodeficiency. *J. Immunol.* **178:**3932–3943.
64. **Glaser R., S. B. Friedman, J. Smyth, R. Ader, P. Bijur, P. Brunell, N. Cohen, L. R. Krilov, S. T. Lifrak, A. Stone, and P. Toffler.** 1999. The differential impact of training stress and final examination stress on herpesvirus latency at the United States Military Academy at West Point. *Brain Behav. Res.* **13:**240–251.
65. **Gottschalk, S., O. L. Edwards, U. Sili, M. H. Huls, T. Goltsova, A. R. Davis, H. E. Heslop, and C. M. Rooney.** 2003. Generating CTLs against the subdominant Epstein-Barr virus LMP1 antigen for the adoptive immunotherapy of EBV-associated malignancies. *Blood 101:*1905–1912.
66. **Gottschalk, S., C. M. Rooney, and H. E. Heslop.** 2005. Post-transplant lymphoproliferative disorders. *Annu. Rev. Med.* **56:**29–44.
67. **Gratama, J., M. Oosterveer, F. Zwaan, J. Lepoutre, G. Klein, and I. Ernberg.** 1988. Eradication of Epstein-Barr virus by allogeneic marrow transplantation: implication for the site of viral latency. *Proc. Natl. Acad. Sci. USA* **85:**8693–8699.
68. **Green, M., T. V. Cacciarelli, G. Mazariegos, L. Sigurdsson, L. Qu, D. T. Rowe, and G. Reyes.** 1998. Serial measurement of EBV viral load in peripheral blood in pediatric liver transplant recipients during treatment for post transplant lymphoproliferative disease. *Transplantation* **66:**1641–1644.
69. **Grotto, I., D. Mimouni, M. Huerta, M. Mimouni, D. Cohen, G. Robin, S. Pitlik, and M. Green.** 2003. Clinical and laboratory presentation of EBV positive infectious mononucleosis in young adults. *Epidemiol. Infect.* **131:**683–689.
70. **Gulley, M., H. Fan, and S. H. Elmore.** 2006. Validation of Roche LightCycler Epstein-Barr virus quantification reagents in a clinical laboratory setting. *J. Mol. Diagn.* **8:**589–597.
71. **Gutensohn, N., and P. Cole.** 1980. Epidemiology of Hodgkin's disease. *Semin. Oncol.* **7:**92–102.
72. **Gutierrez-Dalmau, A., A. Sanchez-Fructuoso, A. Sanz-Guajardo, A. Mazuecos, A. Franco, M. C. Rial, P. Iranzo, J. V. Torregrosa, F. Oppenheimer, and J. M. Campistol.** 2005. Efficacy of conversion to sirolimus in posttransplantation Kaposi's sarcoma. *Transpl. Proc.* **37:**3836–3838.
73. **Hadrup, S. R., J. Strindhall, T. Kollgaard, T. Seremet, B. Johansson, G. Pawelec, P. thor Straten, and A. Wikby.** 2006. Longitudinal studies of clonally expanded CD8 T cells reveal a repertoire shrinkage predicting mortality and an increased number of dysfunctional cytomegalovirus-specific T cells in the very elderly. *J. Immunol.* **176:**2645–2653.
74. **Hammock, L., A. Reisenauer, W. Wang, C. Cohen, G. Birdsong, and A. L. Folpe.** 2005. Latency-associated nuclear antigen expression and human herpesvirus-8 polymerase chain reaction in the evaluation of Kaposi sarcoma and other vascular tumors in HIV-positive patients. *Mod. Pathol.* **18:**463–468.
75. **Hennard, C., T. Pfuhl, M. Buettner, K. F. Becker, T. Knöfel, J. Middeldorp, E. Kremmer, G. Niedobitek, and F. A. Grässer.** 2006. The antibody 2B4 directed against the Epstein-Barr virus (EBV)-encoded nuclear antigen 1 (EBNA1) detects MAGE-4: implications for studies on the EBV association of human cancers. *J. Pathol.* **209:**430–435.
76. **Hill, C., S. Harris, E. Culler, J. Zimring, F. S. Nolte, and A. M. Caliendo.** 2006. Performance characteristics of two real-time PCR assays for the quantification of Epstein-Barr virus DNA. *Am. J. Clin. Pathol.* **125:**665–671.
77. **Hislop, A. D., N. E. Annels, N. H. Gudgeon, A. M. Leese, and A. B. Rickinson.** 2002. Epitope-specific evolution of human CD8+ T cell responses from primary to persistent phases of Epstein-Barr virus infection. *J. Exp. Med.* **195:**893–905.

78. **Hochberg, D., T. Souza, M. Catalina, J. L. Sullivan, K. Luzuriaga, and D. A. Thorley-Lawson.** 2004. Acute infection with Epstein-Barr virus targets and overwhelms the peripheral memory B-cell compartment with resting, latently infected cells. *J. Virol.* **78:**5194–5204.

79. **Hollsberg, P., H. J. Hansen, and S. Haahr.** 2003. Altered CD8+ T cell responses to selected Epstein-Barr virus immunodominant epitopes in patients with multiple sclerosis. *Clin. Exp. Immunol.* **132:**137–143.

80. **Hongxin, F., et al.** 2005. Epstein-Barr viral load as a marker of lymphoma in AIDS patients. *J. Med. Virol.* **75:**59–69.

81. **Hudnall, S. D., T. Chen, and S. K. Tyring.** 2004. Species identification of all eight human herpesviruses with a single nested PCR assay. *J. Virol. Methods* **116:**19–26.

82. **Hummel, M., S. Bentink, H. Berger, W. Klapper, S. Wessendorf, T. F. E. Barth, H.-W. Bernd, S. B. Cogliatti, J. Dierlamm, A. C. Feller, M.-L. Hansmann, E. Haralambieva, L. Harder, D. Hasenclever, M. Kuhn, D. Lenze, P. Lichter, J. I. Martin-Subero, P. Moller, H.-K. Muller-Hermelink, G. Ott, R. M. Parwaresch, C. Pott, A. Rosenwald, M. Rosolowski, C. Schwaenen, B. Sturzenhofecker, M. Szczepanowski, H. Trautmann, H.-H. Wacker, R. Spang, M. Loeffler, L. Trumper, H. Stein, R. Siebert, et al.** 2006. A biologic definition of Burkitt's lymphoma from transcriptional and genomic profiling. *N. Engl. J. Med.* **354:**2419–2430.

83. **Imoto, S., T. Murayama, H. Gomyo, I. Mizuno, T. Sugimoto, T. Nakagawa, and T. Koizumi.** 2000. Long-term molecular remission induced by donor lymphocyte infusions for recurrent acute myeloblastic leukemia after allogeneic bone marrow transplantation. *Bone Marrow Transplant.* **26:**809–810.

84. **Ivers, L., A. Y. Kim, and P. E. Sax.** 2004. Predictive value of polymerase chain reaction of cerebrospinal fluid for detection of Epstein-Barr virus to establish the diagnosis of HIV-related primary central nervous system lymphoma. *Clin. Infect. Dis.* **38:**1629–1632.

85. **Jaffe, E., N. L. Harris, H. Stein, and J. W. Vardiman (ed.).** 2001. World Health Organization classification of tumours: pathology and genetics, tumours of haematopoietic and lymphoid tissues. World Health Organization, Geneva, Switzerland.

86. **Jenkins, F. J., L. J. Hoffman, and A. Liegey-Dougall.** 2002. Reactivation of and primary infection with human herpesvirus 8 among solid-organ transplant recipients. *J. Infect. Dis.* **185:**1238–1243.

87. **Johnson, L. R. M. D., M. A. M. D. Nalesnik, and S. H. M. D. Swerdlow.** 2006. Impact of Epstein-Barr virus in monomorphic B-cell posttransplant lymphoproliferative disorders: a histogenetic study. *Am. J. Surg. Pathol.* **30:**1604–1612.

88. **Jumbou, O., C. Mollat, J. M. N'Guyen, S. Billaudel, P. Litoux, and B. Dreno.** 1997. Increased anti-Epstein-Barr virus antibodies in epidermotropic cutaneous T-cell lymphoma: a study of 64 patients. *Br. J. Dermatol.* **136:**212–216.

89. **Jun, H.-S., and Y. Ji-Won.** 2003. A new look at viruses in type 1 diabetes. *Diabetes Metab. Res. Rev.* **19:**8–31.

90. **Kanavaros, P., P. C. De Bruin, J. Briere, C. J. Meijer, and P. Gaulard.** 1995. Epstein-Barr virus (EBV) in extranodal T-cell non-Hodgkin's lymphomas (T-NHL). Identification of nasal T-NHL as a distinct clinicopathological entity associated with EBV. *Leuk. Lymphoma* **18:**27–34.

91. **Kapelushnik, J., S. Ariad, D. Benharroch, D. Landau, A. Moser, G. Delsol, and P. Brousset.** 2001. Post renal transplantation human herpesvirus 8-associated lymphoproliferative disorder and Kaposi's sarcoma. *Br. J. Haematol.* **113:** 425–428.

92. **Katsuki, T., Y. Hinuma, T. Saito, J. Yamamoto, Y. Hirashima, H. Sudoh, M. Deguchi, and M. Motokawa.** 1979. Simultaneous presence of EBNA-positive and colony-forming cells in peripheral blood of patients with infectious mononucleosis. *Int. J. Cancer* **23:**746–750.

93. **Keegan, T. H. M., S. L. Glaser, C. A. Clarke, M. L. Gulley, F. E. Craig, J. A. DiGiuseppe, R. F. Dorfman, R. B. Mann, and R. F. Ambinder.** 2005. Epstein-Barr virus as a marker of survival after Hodgkin's lymphoma: a population-based study. *J. Clin. Oncol.* **23:**7604–7613.

94. **Khan, G., A. Lake, L. Shield, J. Freeland, L. Andrew, F. E. Alexander, R. Jackson, P. R. A. Taylor, E. A. B. McCruden, and R. F. Jarrett.** 2005. Phenotype and frequency of Epstein-Barr virus-infected cells in pretreatment blood samples from patients with Hodgkin lymphoma. *Br. J. Haematol.* **129:** 511–519.

95. **Khanna, R., and S. R. Burrows.** 2000. Role of cytotoxic T lymphocytes in Epstein-Barr virus-associated diseases. *Annu. Rev. Microbiol.* **54:**19–48.

96. **Khanna, R., S. R. Burrows, A. Neisig, J. Neefjes, D. J. Moss, and S. L. Silins.** 1997. Hierarchy of Epstein-Barr virus-specific cytotoxic T-cell responses in individuals carrying different subtypes of an HLA allele: implications for epitope-based antiviral vaccines. *J. Virol.* **71:**7429–7435.

97. **Kimura, H.** 2006. Pathogenesis of chronic active Epstein-Barr virus infection: is this an infectious disease, lymphoproliferative disorder, or immunodeficiency? *Rev. Med. Virol.* **16:**251–261.

98. **Kimura, H., Y. Hoshino, H. Kanegane, I. Tsuge, T. Okamura, K. Kawa, and T. Morishima.** 2001. Clinical and virologic characteristics of chronic active Epstein-Barr virus infection. *Blood* **98:**280–286.

99. **Kimura, H., M. Morita, Y. Yabuta, K. Kuzushima, K. Kato, S. Kojima, T. Matsuyama, and T. Morishima.** 1999. Quantitative analysis of Epstein-Barr virus load by using a real-time PCR assay. *J. Clin. Microbiol.* **37:**132–136.

100. **Klein, G., E. Svedmyr, M. Jondal, and P. O. Persson.** 1976. EBV-determined nuclear antigen (EBNA)-positive cells in the peripheral blood of infectious mononucleosis patients. *Int. J. Cancer* **17:**21–26.

101. **Kocoshis, S. A.** 1994. Small bowel transplantation in infants and children. *Pediatr. Gastroenterol.* **23:**727.

102. **Kozic, S., A. Vince, J. I. Bes, O. D. Rode, S. Z. Lepej, M. Poljak, M. Bozic, and H. H. Kessler.** 2006. Evaluation of a commercial real-time PCR assay for quantitation of Epstein-Barr virus DNA in different groups of patients. *J. Virol. Methods* **135:**263.

102a. **Kusunoki, Y., H. Huang, Y. Fukuda, K. Ozaki, M. Saito, Y. Hirai, and M. Akiyama.** 1993. A positive correlation between the precursor frequency of cytotoxic lymphocytes to autologous Epstein-Barr virus-transformed B cells and antibody titer level against Epstein-Barr virus-associated nuclear antigen in healthy seropositive individuals. *Microbiol. Immunol.* **37:** 461–469.

103. **Kuzushima, K., N. Hayashi, A. Kudoh, Y. Akatsuka, K. Tsujimura, Y. Morishima, and T. Tsurumi.** 2003. Tetramer-assisted identification and characterization of epitopes recognized by HLA A*2402-restricted Epstein-Barr virus-specific CD8+ T cells. *Blood* **101:**1460–1468.

104. **Kwong, Y.** 2005. Natural killer-cell malignancies: diagnosis and treatment. *Leukemia* **19:**2186–2194.

105. **Kwong, Y. L., Chan, R. Liang, A. K. S. Chiang, C. S. Chim, T. K. Chan, D. Todd, and F. C. S. Ho.** 1997. CD56+ NK lymphomas: clinicopathological features and prognosis. *Br. J. Haematol.* **97:**821–829.

106. **Lam, L. L., C. P. Pau, S. C. Dollard, P. E. Pellett, and T. J. Spira.** 2002. Highly sensitive assay for human herpesvirus 8

antibodies that uses a multiple antigenic peptide derived from open reading frame K8.1. *J. Clin. Microbiol.* **40**:325–329.

107. **Laney, A. S., S. C. Dollard, H. W. Jaffe, M. K. Offermann, T. J. Spira, C. J. Gunthel, P. E. Pellett, and M. J. Cannon.** 2004. Repeated measures study of human herpesvirus 8 (HHV-8) DNA and antibodies in men seropositive for both HHV-8 and HIV. *AIDS* **18**:1819–1826.
108. **Lavreys, L., B. Chohan, R. Ashley, B. A. Richardson, L. Corey, K. Mandaliya, J. O. Ndinya-Achola, and J. K. Kreiss.** 2003. Human herpesvirus 8: seroprevalence and correlates in prostitutes in Mombasa, Kenya. *J. Infect. Dis.* **187**:359–363.
109. **Leblond, V., L. Sutton, R. Dorent, F. Davi, M. O. Bitker, J. Gabarre, F. Charlotte, J. J. Ghoussoub, C. Fourcade, and A. Fischer.** 1995. Lymphoproliferative disorders after organ transplantation: a report of 24 cases observed in a single center. *J. Clin. Oncol.* **13**:961–968.
110. **Lee, T. C., B. Savoldo, C. M. Rooney, H. E. Heslop, A. P. Gee, Y. Caldwell, N. R. Barshes, J. D. Scott, L. J. Bristow, C. A. O'Mahony, and J. A. Goss.** 2005. Quantitative EBV viral loads and immunosuppression alterations can decrease PTLD incidence in pediatric liver transplant recipients. *Am. J. Transplant.* **5**:2222–2228.
111. **Lehrnbecher, T. L., C. B. Foster, S. Zhu, D. Venzon, S. M. Steinberg, K. Wyvill, J. A. Metcalf, S. S. Cohen, J. Kovacs, R. Yarchoan, A. Blauvelt, and S. J. Chanock.** 2000. Variant genotypes of FcgammaRIIIA influence the development of Kaposi's sarcoma in HIV-infected men. *Blood* **95**:2386–2390.
112. **Leung, E., B. K. Shenton, K. Green, G. Jackson, F. K. Gould, C. Yap, and D. Talbot.** 2004. Dynamic EBV gene loads in renal, hepatic, and cardiothoracic transplant recipients as determined by real-time PCR light cycler. *Transpl. Infect. Dis.* **6**:156–164.
113. **Levine, P., G. Pallesen, P. Ebbesen, N. Harris, A. S. Evans, and N. Mueller.** 1994. Evaluation of Epstein-Barr virus antibody patterns and detection of viral markers in the biopsies of patients with Hodgkin's disease. *Int. J. Cancer* **59**:48–50.
114. **Lucas, K. G., T. N. Small, G. Heller, B. Dupont, and R. J. O'Reilly.** 1996. The development of cellular immunity to Epstein-Barr virus after allogeneic bone marrow transplantation. *Blood* **87**:2594–2603.
115. **Luderer, R., M. Kok, H. G. M. Niesters, R. Schuurman, O. de Weerdt, and S. F. T. Thijsen.** 2005. Real-time Epstein-Barr virus PCR for the diagnosis of primary EBV infections and EBV reactivation. *Mol. Diagn.* **9**:195–200.
116. **Lumbreras, C., I. Fernandez, J. Velosa, S. Munn, S. Sterioff, and C. V. Paya.** 1995. Infectious complications following pancreatic transplantation: incidence, microbiological and clinical characteristics, and outcome. *Clin. Infect. Dis.* **20**:514.
117. **Luppi, M., P. Barozzi, V. Rasini, and G. Torelli.** 2002. HHV-8 infection in the transplantation setting: a concern only for solid organ transplant patients? *Leuk. Lymphoma* **43**:517–522.
118. **Luppi, M., P. Barozzi, G. Santagostino, R. Trovato, T. F. Schulz, R. Marasca, D. Bottalico, L. Bignardi, and G. Torelli.** 2000. Molecular evidence of organ-related transmission of Kaposi sarcoma-associated herpesvirus or human herpesvirus-8 in transplant patients. *Blood* **96**:3279–3281.
119. **Luppi, M., P. Barozzi, T. F. Schulz, G. Setti, K. Staskus, R. Trovato, F. Narni, A. Donelli, A. Maiorana, R. Marasca, S. Sandrini, and G. Torelli.** 2000. Bone marrow failure associated with human herpesvirus 8 infection after transplantation. *N. Engl. J. Med.* **343**:1378–1385.
120. **Mackay, I. M., K. E. Arden, and A. Nitsche.** 2002. Real-time PCR in virology. *Nucleic Acids Res.* **30**:1292–1305.
121. **Marcelin, A. G., V. Calvez, and E. Dussaix.** 2007. KSHV after an organ transplant: should we screen? *Curr. Top. Microbiol. Immunol.* **312**:245–262.
122. **Marcelin, A. G., A. M. Roque-Afonso, M. Hurtova, N. Dupin, M. Tulliez, M. Sebagh, Z. A. Arkoub, C. Guettier, D. Samuel, V. Calvez, and E. Dussaix.** 2004. Fatal disseminated Kaposi's sarcoma following human herpesvirus 8 primary infections in liver-transplant recipients. *Liver Transplant.* **10**:295–300.
123. **Marshall, N. A., L. E. Christie, L. R. Munro, D. J. Culligan, P. W. Johnston, R. N. Barker, and M. A. Vickers.** 2004. Immunosuppressive regulatory T cells are abundant in the reactive lymphocytes of Hodgkin lymphoma. *Blood* **103**:1755–1762.
124. **Martin, J. N., Z. Amad, C. Cossen, P. K. Lam, D. H. Kedes, K. A. Page-Shafer, D. H. Osmond, and B. Forghani.** 2000. Use of epidemiologically well-defined subjects and existing immunofluorescence assays to calibrate a new enzyme immunoassay for human herpesvirus 8 antibodies. *J. Clin. Microbiol.* **38**:696–701.
125. **Martin, J. N., D. E. Ganem, D. H. Osmond, K. A. Page-Shafer, D. Macrae, and D. H. Kedes.** 1998. Sexual transmission and the natural history of human herpesvirus 8 infection. *N. Engl. J. Med.* **338**:948–954.
126. **Matsushima, A. Y., J. A. Strauchen, G. Lee, E. Scigliano, E. E. Hale, M. T. Weisse, D. Burstein, O. Kamel, P. S. Moore, and Y. Chang.** 1999. Posttransplantation plasmacytic proliferations related to Kaposi's sarcoma-associated herpesvirus. *Am. J. Surg. Pathol.* **23**:1393–1400.
127. **Mbulaiteye, S. M., and E. A. Engels.** 2006. Kaposi's sarcoma risk among transplant recipients in the United States (1993–2003). *Int. J. Cancer* **119**:2685–2691.
128. **McDiarmid, S., S. Jordan, and G. S. Lee.** 1999. Prevention and preemptive therapy of posttransplant lymphoproliferative disease in pediatric liver recipients. *Transplantation* **66**:1604–1611.
129. **McKnight, J. L., H. Cen, S. A. Riddler, M. C. Breinig, P. A. Williams, M. Ho, and P. S. Joseph.** 1994. EBV gene expression, EBNA antibody responses and EBV+ peripheral blood lymphocytes in post-transplant lymphoproliferative disease. *Leuk. Lymphoma* **15**:9–16.
130. **Meij P., M. B. Vervoort, E. Bloemena, T. E. Schouten, C. Schwartz, S. Grufferman, R. F. Ambinder, and J. M. Middeldorp.** 2002. Antibody responses to Epstein-Barr virus-encoded latent membrane protein-1 (LMP1) and expression of LMP1 in juvenile Hodgkin's disease. *J. Med. Virol.* **68**:370–377.
131. **Meijer, E., A. W. Dekker, A. J. L. Weersink, M. Rozenberg-Arska, and L. F. Verdonck.** 2002. Prevention and treatment of Epstein-Barr virus-associated lymphoproliferative disorders in recipients of bone marrow and solid organ transplants. *Br. J. Haematol.* **119**:596–607.
132. **Mesri, E. A., E. Cesarman, L. Arvanitakis, S. Rafii, M. A. Moore, D. N. Posnett, D. M. Knowles, and A. S. Asch.** 1996. Human herpesvirus-8/Kaposi's sarcoma-associated herpesvirus is a new transmissible virus that infects B cells. *J. Exp. Med.* **183**:2385–2390.
133. **Middledorp, J. M.** 1999. Monitoring of Epstein-Barr virus DNA load in the peripheral blood by quantitative competitive PCR. *J. Clin. Microbiol.* **37**:2852–2857.
134. **Miller, I. G.** 1990. Epstein-Barr virus, p. 2397–2436. *In* B. N. Fields et al. (ed.), *Fields Virology,* 2nd ed. Lippincott-Raven Publishers, Philadelphia, PA.
135. **Min, J., and D. A. Katzenstein.** 1999. Detection of Kaposi's sarcoma-associated herpesvirus in peripheral blood cells in human immunodeficiency virus infection: association with Kaposi's sarcoma, CD4 cell count, and HIV RNA levels. *AIDS Res. Hum. Retrovir.* **15**:51–55.

136. **Mitchell, J. L., C. M. Doyle, M. V. Land, and P. L. Devine.** 1998. Comparison of commercial ELISA for detection of antibodies to the viral capsid antigen (VCA) of Epstein-Barr virus (EBV). *Dis. Markers* **13:**245–249.
137. **Miyashita, E., B. Yang, G. Babcock, and D. Thorley-Lawson.** 1997. Identification of the site of Epstein-Barr virus persistence in vivo as a resting B cell. *J. Virol.* **71:**4882–4891.
138. **Moore, P. S., L. A. Kingsley, S. D. Holmberg, T. Spira, P. Gupta, D. R. Hoover, J. P. Parry, L. J. Conley, H. W. Jaffe, and Y. Chang.** 1996. Kaposi's sarcoma-associated herpesvirus infection prior to onset of Kaposi's sarcoma. *AIDS* **10:**175–180.
139. **Moormann, A., K. Chelimo, O. P. Sumba, M. L. Lutzke, R. Ploutz-Snyder, D. Newton, J. Kazura, and R. Rochford.** 2005. Exposure to holoendemic malaria results in elevated Epstein-Barr virus loads in children. *J. Infect. Dis.* **191:** 1233–1238.
140. **Moosa, M. R.** 2005. Kaposi's sarcoma in kidney transplant recipients: a 23-year experience. *QJM* **98:**205–214.
141. **Mueller, N., A. Evans, N. L. Harris, G. W. Comstock, E. Jellum, K. Magnus, N. Orentreich, B. F. Polk, and J. Vogelman.** 1989. Hodgkin's disease and Epstein-Barr virus. Altered antibody pattern before diagnosis. *N. Engl. J. Med.* **320:**689–695.
142. **Murray, P. G.** 2006. Epstein-Barr virus in breast cancer: artefact or aetiological agent? *J. Pathol.* **209:**427–429.
143. **Murray, R. J., M. G. Kurilla, J. M. Brooks, W. A. Thomas, M. Rowe, E. Kieff, and A. B. Rickinson.** 1992. Identification of target antigens for the human cytotoxic T cell response to Epstein-Barr virus (EBV): implications for the immune control of EBV-positive malignancies. *J. Exp. Med.* **176:**157–168.
144. **Myhr, K. M., T. Riise, E. Barrett-Connor, H. Myrmel, C. Vedeler, M. Gronning, M. B. Kalvenes, and H. Nyland.** 1998. Altered antibody pattern to Epstein-Barr virus but not to other herpesviruses in multiple sclerosis: a population based case-control study from western Norway. *J. Neurol. Neurosurg. Psych.* **64:**539–542.
145. **Nakatsuka, S., and K. Aozasab.** 2006. Epidemiology and pathologic features of Hodgkin lymphoma. *Int. J. Hematol.* **83:**391–397.
146. **Nalesnik, M., R. Jaffe, T. E. Starzl, A. J. Demetris, K. Porter, J. A. Burnham, L. Makowka, M. Ho, and J. Locker.** 1988. The pathology of post-transplant lymphoproliferative disease occurring in the setting of cyclosporine A-prednisone immunosuppression. *J. Pathol.* **133:**173–183.
147. **Nalesnik, M. A.** 2001. The diverse pathology of post-transplant lymphoproliferative disorders: the importance of a standardized approach. *Transpl. Infect. Dis.* **3:**88–96.
148. **Nelson, B., M. Nalesnik, D. Bahler, J. Locker, J. Fung, and S. Swerdlow.** 2000. Epstein-Barr virus-negative post-transplant lymphoproliferative disorders: a distinct entity? *Am. J. Surg. Pathol.* **24:**375–385.
149. **Niesters, H. G., J. van Esser, E. Fries, K. C. Wolhers, J. Corneilissen, and A. D. Osterhaus.** 2000. Development of a real-time quantitative assay for detection of Epstein-Barr virus. *J. Clin. Microbiol.* **38:**712–715.
150. **Nishikori, M., and T. Uchiyama.** 2006. Molecular pathogenesis of Hodgkin lymphoma. *Int. J. Hematol.* **83:**398–403.
151. **Nordstrom, M., L. Hardell, A. Linde, L. Schloss, and A. Nasman.** 1999. Elevated antibody levels to Epstein-Barr virus antigens in patients with hairy cell leukemia compared to controls in relation to exposure to pesticides, organic solvents, animals, and exhausts. *Oncol. Res.* **11:**539–544.
152. **Okamura, T., Y. Hatsukawa, H. Arai, M. Inoue, and K. Kawa.** 2000. Blood stem-cell transplantation for chronic active Epstein-Barr virus with lymphoproliferation. *Lancet* **356:**223.
153. **Okano, M., S. Matsumoto, T. Osato, Y. Sakiyama, G. M. Thiele, and D. T. Purtilo.** 1991. Severe chronic active Epstein-Barr virus infection syndrome. *Clin. Microbiol. Rev.* **4:**129–135.
154. **Opelz, G., and B. Dohler.** 2004. Lymphomas after solid organ transplantation: a collaborative transplant study report. *Am. J. Transplant.* **4:**222–230.
155. **Opelz, G., and R. Henderson.** 1993. Incidence of non-Hodgkin lymphoma in kidney and heart transplant recipients. *Lancet* **342:**1514.
156. **O'Sullivan, C. E., R. Peng, K. S. Cole, R. C. Montelaro, T. Sturgeon, H. B. Jenson, and P. D. Ling.** 2002. Epstein-Barr virus and human immunodeficiency virus serological responses and viral burdens in HIV-infected patients treated with HAART. *J. Med. Virol.* **67:**320–326.
157. **Ouyang, Q., W. M. Wagner, S. Walter, C. A. Muller, A. Wikby, G. Aubert, T. Klatt, S. Stevanovic, T. Dodi, and G. Pawelec.** 2003. An age-related increase in the number of CD8+ T cells carrying receptors for an immunodominant Epstein-Barr virus (EBV) epitope is counteracted by a decreased frequency of their antigen-specific responsiveness. *Mech. Ageing Dev.* **124:**477.
158. **Parravicini, C., S. J. Olsen, M. Capra, F. Poli, G. Sirchia, S. J. Gao, E. Berti, A. Nocera, E. Rossi, G. Bestetti, M. Pizzuto, M. Galli, M. Moroni, P. S. Moore, and M. Corbellino.** 1997. Risk of Kaposi's sarcoma-associated herpes virus transmission from donor allografts among Italian posttransplant Kaposi's sarcoma patients. *Blood* **90:**2826–2829.
159. **Patel, N., M. Salifu, N. Sumrani, D. Distant, J. Hong, M. Markell, and A. S. Braverman.** 2002. Successful treatment of post-renal transplant Kaposi's sarcoma with paclitaxel. *Am. J. Transplant.* **2:**877–879.
160. **Pawelec, G., A. Akbar, C. Caruso, R. Effros, B. Grubeck-Loebenstein, and A. Wikby.** 2004. Is immunosenescence infectious? *Trends Immunol.* **25:**406.
161. **Pellett, P. E., D. J. Wright, E. A. Engels, D. V. Ablashi, S. C. Dollard, B. Forghani, S. A. Glynn, J. J. Goedert, F. J. Jenkins, T. H. Lee, F. Neipel, D. S. Todd, D. Whitby, G. J. Nemo, and M. P. Busch.** 2003. Multicenter comparison of serologic assays and estimation of human herpesvirus 8 seroprevalence among US blood donors. *Transfusion* **43:** 1260–1268.
162. **Pica, F., and A. Volpi.** 2007. Transmission of human herpesvirus 8: an update. *Curr. Opin. Infect. Dis.* **20:**152–156.
163. **Piriou, E. R., K. van Dort, N. M. Nanlohy, F. Miedema, M. H. van Oers, and D. van Baarle.** 2004. Altered EBV viral load setpoint after HIV seroconversion is in accordance with lack of predictive value of EBV load for the occurrence of AIDS-related non-Hodgkin lymphoma. *J. Immunol.* **172:** 6931–6937.
164. **Preiksaitis, J. D., F. Diaz-Mitoma, F. Mirzayans, S. Roberts, and D. L. J. Tyrrell.** 1992. Quantitative oropharyngeal Epstein-Barr virus shedding in renal and cardiac transplant recipients: relationship to immunosuppressive therapy, serologic responses, and the risk of posttransplant lymphoproliferative disorder. *J. Infect. Dis.* **166:**986–990.
165. **Raeiszadeh, M., J. Kopycinski, S. J. Paston, T. Diss, M. Lowdell, G. A. D. Hardy, A. D. Hislop, S. Workman, A. Dodi, V. Emery, and A. D. Webster.** 2006. The T cell response to persistent herpes virus infections in common variable immunodeficiency. *Clin. Exp. Immunol.* **146:**234–242.

166. **Rea, T. D., R. L. Ashley, J. E. Russo, and D. S. Buchwald.** 2002. A systematic study of Epstein-Barr virus serologic assays following acute infection. *Am. J. Clin. Pathol.* **117:**3715–3724.
167. **Regamey, N., M. Tamm, M. Wernli, A. Witschi, G. Thiel, G. Cathomas, and P. Erb.** 1998. Transmission of human herpesvirus 8 infection from renal-transplant donors to recipients. *N. Engl. J. Med.* **339:**1358–1363.
168. **Renne, R., W. Zhong, B. Herndier, M. McGrath, N. Abbey, D. Kedes, and D. Ganem.** 1996. Lytic growth of Kaposi's sarcoma-associated herpesvirus (human herpesvirus 8) in culture. *Nat. Med.* **2:**342–346.
169. **Rickinson, A.** 2001. Epstein-Barr virus. *Virus Res.* **82:**109.
170. **Rickinson, A. B., and E. Kieff.** 1996. Epstein-Barr virus, p. 2397–2436. *In* D. Knipe, et al. (ed.), *Fields Virology,* 3rd ed. Lippincott-Raven Publishers, Philadelphia, PA.
171. **Rickinson, A. B., and D. J. Moss.** 1997. Human cytotoxic T lymphocyte responses to Epstein-Barr virus infection. *Annu. Rev. Immunol.* **15:**405–431.
172. **Riddler, S. A., M. C. Breinig, and J. L. McKnight.** 1994. Increased levels of circulating Epstein-Barr virus (EBV)-infected lymphocytes and decreased EBV nuclear antigen antibody responses are associated with the development of posttransplant lymphoproliferative disease in solid-organ transplant recipients. *Blood* **84:**972–984.
173. **Robertson P., S. Beynon, R. Whybin, C. Brennan, U. Vollmer-Conna, I. Hickie, and A. Lloyd.** 2003. Measurement of EBV-IgG anti-VCA avidity aids the early and reliable diagnosis of primary EBV infection. *J. Med. Virol.* **70:**617–623.
174. **Robinson, J., D. Smith, and J. Niederman.** 1980. Plasmacytic differentiation of circulating Epstein-Barr virus-infected B lymphocytes during acute infectious mononucleosis. *J. Exp. Med.* **153:**235–244.
175. **Robinson, J., D. Smith, and J. Niederman.** 1980. Mitotic EBNA-positive lymphocytes in peripheral blood during infectious mononucleosis. *Nature* **287:**334–335.
176. Reference deleted.
177. **Rogers, B. B., C. Conlin, C. F. Timmons, D. B. Dawson, K. Krisher, and W. S. Andrews.** 1997. Epstein-Barr virus PCR correlated with viral histology and serology in pediatric liver transplant recipients. *Pediatr. Pathol. Lab. Med.* **17:**391–400.
178. **Rooney, C., J. G. Howe, S. H. Speck, and G. Miller.** 1989. Influence of Burkitt's lymphoma and primary B cells on latent gene expression by the nonimmortalizing P3J-HR-1 strain of Epstein-Barr virus. *J. Virol.* **63:**1531–1539.
179. **Rowe, D. T., L. Qu, J. Reyes, N. Jabbour, E. Yunis, P. Putnam, S. Todo, and M. Green.** 1997. Use of quantitative competitive PCR to measure Epstein-Barr virus genome load in the peripheral blood of pediatric transplant patients with lymphoproliferative disorders. *J. Clin. Microbiol.* **35:**1612–1615.
180. **Rowe, D. T., S. Webber, E. M. Schauer, J. Reyes, and M. Green.** 2001. Epstein-Barr virus load monitoring: its role in the prevention and management of post-transplant lymphoproliferative disease. *Transpl. Infect. Dis.* **3:**79–87.
181. **Ruth, F. J.** 2006. Viruses and lymphoma/leukaemia. *J. Pathol.* **208:**176–186.
182. **Sairenji, T., M. Daibata, C. H. Sorli, H. Qvistb äck, R. E. Humphreys, J. Ludvigsson, J. Palmer, M. Landin-Olsson, G. Sundkvist, B. Michelsen, A. Lernmark, and T. Dyrberg.** 1991. Relating homology between the Epstein-Barr virus BOLF1 molecule and HLA-DQw8 β chain to recent onset type 1 (insulin-dependent) diabetes mellitus. *Diabetologia* **34:**33–39.
183. **Sanchez-Fructuoso, A., J. Conesa, I. Perez Flores, N. Ridao, N. Calvo, D. Prats, A. Rodriguez, and A. Barrientos.** 2006. Conversion to sirolimus in renal transplant patients with tumors. *Transplant. Proc.* **38:**2451–2452.
184. **Schaade, L., M. Kleines, and M. Hausler.** 2001. Application of virus-specific immunoglobulin M (IgM), IgG, and IgA antibody detection with a polyantigenic enzyme-linked immunosorbent assay for diagnosis of Epstein-Barr virus infections in childhood. *J. Clin. Microbiol.* **39:**3902–3905.
185. **Schatz, O., P. Monini, R. Bugarini, F. Neipel, T. F. Schulz, M. Andreoni, P. Erb, M. Eggers, J. Haas, S. Butto, M. Lukwiya, J. R. Bogner, S. Yaguboglu, J. Sheldon, L. Sarmati, F. D. Goebel, R. Hintermaier, G. Enders, N. Regamey, M. Wernli, M. Sturzl, G. Rezza, and B. Ensoli.** 2001. Kaposi's sarcoma-associated herpesvirus serology in Europe and Uganda: multicentre study with multiple and novel assays. *J. Med. Virol.* **65:**123–132.
186. **Sergerie, Y., Y. Abed, J. Roy, and G. Boivin.** 2004. Comparative evaluation of three serological methods for detection of human herpesvirus 8-specific antibodies in Canadian allogeneic stem cell transplant recipients. *J. Clin. Microbiol.* **42:**2663–2667.
187. **Shedd, D., A. Angeloni, J. Niederman, and G. Miller.** 1995. Detection of human serum antibodies to the BFRF3 Epstein-Barr virus capsid component by means of a DNA-binding assay. *J. Infect. Dis.* **172:**1367–1370.
188. **Shinkura, R., N. Yamamoto, C. Koriyama, Y. Shinmura, Y. Eizuru, and M. Tokunaga.** 2000. Epstein-Barr virus-specific antibodies in Epstein-Barr virus-positive and -negative gastric carcinoma cases in Japan. *J. Med. Virol.* **60:**411–416.
189. **Sitas, F., H. Carrara, V. Beral, R. Newton, G. Reeves, D. Bull, U. Jentsch, R. Pacella-Norman, D. Bourboulia, D. Whitby, C. Boshoff, and R. Weiss.** 1999. Antibodies against human herpesvirus 8 in black South African patients with cancer. *N. Engl. J. Med.* **340:**1863–1871.
190. **Siu, L., J. K. Chan, and Y. L. Kwong.** 2002. Natural killer cell malignancies: clinicopathologic and molecular features. *Histol. Histopathol.* **17:**539–554.
191. **Smets, F., and E. M. Sokal.** 2002. Epstein-Barr virus-related lymphoproliferation in children after liver transplant: role of immunity, diagnosis, and management. *Pediatr. Transplant.* **6:**280–287.
192. **Smets, F., D. Latinne, H. Bazin, R. Reding, J. B. Otte, J. P. Buts, and E. M. Sokal.** 2002. Ratio between Epstein-Barr viral load and anti-Epstein-Barr virus specific T-cell response as a predictive marker of posttransplant lymphoproliferative disease. *Transplantation* **73:**1603–1610.
193. **Smith, N. A., C. A. Sabin, R. Gopal, D. Bourboulia, W. Labbet, C. Boshoff, D. Barlow, B. Band, B. S. Peters, A. de Ruiter, D. W. Brown, R. A. Weiss, J. M. Best, and D. Whitby.** 1999. Serologic evidence of human herpesvirus 8 transmission by homosexual but not heterosexual sex. *J. Infect. Dis.* **180:**600–606.
194. **Spira, T. J., L. Lam, S. C. Dollard, Y. X. Meng, C. P. Pau, J. B. Black, D. Burns, B. Cooper, M. Hamid, J. Huong, K. Kite-Powell, and P. E. Pellett.** 2000. Comparison of serologic assays and PCR for diagnosis of human herpesvirus 8 infection. *J. Clin. Microbiol.* **38:**2174–2180.
195. **Stamey, F. R., M. M. Patel, B. P. Holloway, and P. E. Pellett.** 2001. Quantitative, fluorogenic probe PCR assay for detection of human herpesvirus 8 DNA in clinical specimens. *J. Clin. Microbiol.* **39:**3537–3540.
196. **Stevens, S. J. C., B. S. N. A. Blank, P. H. M. B. Smits, P. L. C. Meenhorst, and J. M. Middeldorp.** 2002. High Epstein-Barr virus (EBV) DNA loads in HIV-infected pa-

tients: correlation with antiretroviral therapy and quantitative EBV serology. *AIDS* **16**:993–1001.

197. **Stevens, S. J. C., E. A. M. Verschuuren, S. A. W. M. Verkuijlen, A. J. C. V. D. Brule, C. J. L. M. Meijer, and J. M. Middeldorp.** 2002. Role of Epstein-Barr virus DNA load monitoring in prevention and early detection of post-transplant lymphoproliferative disease. *Leuk. Lymphoma* **43**:831–840.
198. **Stowe, R. P., D. L. Pierson, D. L. Feeback, and A. D. Barrett.** 2000. Stress-induced reactivation of Epstein-Barr virus in astronauts. *Neuroimmunomodulation* **8**:51–58.
199. **Straathof, K., C. M. Bollard, C. M. Rooney, and H. E. Heslop.** 2003. Immunotherapy for Epstein-Barr virus-associated cancers in children. *Oncologist* **8**:83–98.
200. **Stratta, R. J., R. J. Taylor, J. S. Bynon, J. A. Lowell, M. S. Cattral, K. Frisbie, S. Miller, S. J. Radio, and D. C. Brennan.** 1994. Viral prophylaxis in combined pancreas-kidney transplant recipients. *Transplantation* **57**:506–511.
201. **Sugaya, N., H. Kimura, S. Hara, Y. Hoshino, S. Kojima, T. Morishima, T. Tsurumi, and K. Kuzushima.** 2004. Quantitative analysis of Epstein-Barr virus (EBV)-specific CD8+ T cells in patients with chronic active EBV infection. *J. Infect. Dis.* **190**:985–988.
202. **Suzuki, K., K. Ohshima, K. Karube, J. Suzumiya, J. Ohga, S. Ishihara, K. Tamura, and M. Kikuchi.** 2004. Clinicopathological states of Epstein-Barr virus-associated T/NK-cell lymphoproliferative disorders (severe chronic active EBV infection) of children and young adults. *Int. J. Oncol.* **24**:1165–1174.
203. **Swinnen, L.** 2000. Diagnosis and treatment of transplant-related lymphoma. *Ann. Oncol.* **11**:45–48.
204. **Swinnen, L. J., M. R. Costanzo-Nordin, S. G. Fisher, E. J. O'Sullivan, M. R. Johnson, A. L. Heroux, G. J. Dizikes, R. Pifarre, and R. I. Fisher.** 1990. Increased incidence of lymphoproliferative disorder after immunosuppression with the monoclonal antibody OKT3 in cardiac-transplant recipients. *N. Engl. J. Med.* **323**:1723–1728.
205. **Szkaradkiewicz, A., A. Kruk-Zagajewska, M. Wal, A. Jopek, M. Wierzbicka, and A. Kuch.** 2002. Epstein-Barr virus and human papillomavirus infections and oropharyngeal squamous cell carcinomas. *Clin. Exp. Med.* **2**:137–141.
206. **Taeko Uehara, C., et al.** 2004. Successful control of Epstein-Barr virus (EBV)-infected cells by allogeneic nonmyeloablative stem cell transplantation in a patient with the lethal form of chronic active EBV infection. *Am. J. Hematol.* **76**:368–372.
207. **Tamaki H., B. Beaulieu, M. Somasunrdaran, and J. Sullivan.** 1995. Major histocompatibility complex class I restricted CTL responses to EBV in children. *J. Infect. Dis.* **172**:739–752.
208. **Tan, L. C., N. Gudgeon, N. E. Annels, P. Hansasuta, C. A. O'Callaghan, S. Rowland-Jones, A. J. McMichael, A. B. Rickinson, and M. F. C. Callan.** 1999. A re-evaluation of the frequency of CD8+ T cells specific for EBV in healthy virus carriers. *J. Immunol.* **162**:1827–1835.
209. **Tappero, J. W., M. A. Conant, S. F. Wolfe, and T. G. Berger.** 1993. Kaposi's sarcoma. Epidemiology, pathogenesis, histology, clinical spectrum, staging criteria and therapy. *J. Am. Acad. Dermatol.* **28**:371–395.
210. **Thaunat, O., M. F. Mamzer-Bruneel, F. Agbalika, F. Valensi, M. Venditto, C. Lebbe, C. Frances, R. Kania, L. Chatenoud, C. Antoine, S. Kaveri, H. Kreis, and E. Morelon.** 2006. Severe human herpesvirus-8 primary infection in a renal transplant patient successfully treated with anti-CD20 monoclonal antibody. *Blood* **107**:3009–3010.
211. **Theate, I., L. Michaux, J. P. Squifflet, A. Martin, and M. Raphael.** 2003. Human herpesvirus 8 and Epstein-Barr virus-related monotypic large B-cell lymphoproliferative disorder coexisting with mixed variant of Castleman's disease in a lymph node of a renal transplant recipient. *Clin. Transplant.* **17**:451–454.
212. **Thiele, G. M., and M. Okano.** 1993. Diagnosis of Epstein-Barr virus infections in the clinical laboratory. *Clin. Microbiol. News* **15**:41–46.
213. **Thorley-Lawson, D. A., and A. Gross.** 2004. Persistence of the Epstein-Barr virus and the origins of associated lymphomas. *N. Engl. J. Med.* **350**:1328–1337.
214. **Todo, S., A. Tzakis, K. Abu-Elmagd, J. Reyes, H. Furukawa, B. Nour, J. Fung, A. Demetris, and T. E. Starzl.** 1995. Abdominal multivisceral transplantation. *Transplantation* **59**:234–239.
215. **Topino, S., L. Vincenzi, I. Mezzaroma, E. Nicastri, M. Andreoni, and M. C. Sirianni.** 2001. Correlation between enzyme-linked immunosorbent assay and immunofluorescence assay with lytic antigens for detection of antibodies to human herpesvirus 8. *Clin. Diagn. Lab. Immunol.* **8**:203–205.
216. **Tosato, G., S. Straus, W. Henle, S. E. Pike, and R. M. Blaese.** 1985. Characteristic T cell dysfunction in patients with chronic active Epstein-Barr virus infection (chronic infectious mononucleosis). *J. Immunol.* **134**:3082–3088.
217. **Tsao, L.** 2007. The clinicopathologic spectrum of posttransplantation lymphoproliferative disorders. *Arch. Pathol. Lab. Med.* **131**:1209–1218.
218. **Vajro P., S. Lucariello, and F. Migliaro.** 2000. Predictive value of Epstein-Barr virus genome copy number and BZLF1 expression in blood lymphocytes of transplant recipients at risk for lymphoproliferative disease. *J. Infect. Dis.* **170**:2050–2054.
219. **van Baarle, D., K. C. Wolthers, E. Hovenkamp, H. Niesters, A. Osterhaus, F. Miedema, and M. van Oers.** 2002. Absolute level of Epstein-Barr virus DNA in human immunodeficiency virus type 1 infection is not predictive of AIDS-related non-Hodgkin lymphoma. *J. Infect. Dis.* **186**:405–409.
220. Reference deleted.
221. **van Grunsven, W. M. J., W. J. M. Spaan, and J. M. Middledorp.** 1994. Localization and diagnostic application of immunodominant domains of the BFRF3-encoded Epstein-Barr virus capsid protein. *J. Infect. Dis.* **170**:13–19.
222. **Verschuuren, E., W. van der Bij, W. de Boer, W. Timens, J. Middeldorp, et al.** 2003. Quantitative Epstein-Barr virus (EBV) serology in lung transplant recipients with primary EBV infection and/or post-transplant lymphoproliferative disease. *J. Med. Virol.* **69**:258–266.
223. **Verucchi, G., L. Calza, F. Trevisani, A. Zambruni, M. Tadolini, R. Giuliani, R. Manfredi, P. Andreone, F. Chiodo, and M. Bernardi.** 2005. Human herpesvirus-8-related Kaposi's sarcoma after liver transplantation successfully treated with cidofovir and liposomal daunorubicin. *Transpl. Infect. Dis.* **7**:34–37.
224. **Vihinen, M., P. T. Mattsson, and C. I. Smith.** 2000. Bruton tyrosine kinase (BTK) in X-linked agammaglobulinemia (XLA). *Front. Biosci.* **5**:17–28.
225. **Volkow, P., J. W. Zinser, and R. Correa-Rotter.** 2007. Molecularly targeted therapy for Kaposi's sarcoma in a kidney transplant patient: case report, "what worked and what did not." *BMC Nephrol.* **8**:6.
226. **Votava, M., D. Bartosova, A. Krchnakova, K. Crhova, and L. Kubinova.** 1996. Diagnostic importance of heterophile antibodies and immunoglobulins IgA, IgE, IgM and low-

avidity IgG against Epstein-Barr virus capsid antigen in children. *J. Med. Virol.* **69**:258–266.

227. **Vrbikova, J., I. Janatkova, V. Zamrazil, F. Tomiska, and T. Fucikova.** 1996. Epstein-Barr virus serology in patients with autoimmune thyroiditis. *Exp. Clin. Endocrinol. Diabetes* **104**:89–92.
228. **Wadowsky, R. M., S. Laus, M. Green, S. A. Webber, and D. Rowe.** 2003. Measurement of Epstein-Barr virus DNA loads in whole blood and plasma by TaqMan PCR and in peripheral blood lymphocytes by competitive PCR. *J. Clin. Microbiol.* **41**:5245–5249.
229. **Wagner, H. J., G. Bein, A. Bitsch, and H. Kirchner.** 1992. Detection and quantitation of latently infected B lymphocytes in EBV seropositive healthy individuals by polymerase chain reaction. *J. Clin. Microbiol.* **30**:2826–2831.
230. **Wagner, H. J., F. Schlager, A. Claviez, and P. Bucsky.** 2001. Detection of Epstein-Barr virus DNA in peripheral blood of paediatric patients with Hodgkin's disease by real-time polymerase chain reaction. *Eur. J. Cancer* **37**:1853.
231. **Wang, Q. J., F. J. Jenkins, L. P. Jacobson, L. A. Kingsley, R. D. Day, Z. W. Zhang, Y. X. Meng, P. E. Pellett, K. G. Kousoulas, A. Baghian, and C. R. Rinaldo, Jr.** 2001. Primary human herpesvirus 8 infection generates a broadly specific CD8(+) T-cell response to viral lytic cycle proteins. *Blood* **97**:2366–2373.
232. **Wang, Y. F., S. B. Lee, L. C. Cheng, M. H. Tai, and I. J. Su.** 2002. Detection of serum antibodies to three different recombinant antigens of human herpesvirus 8 by immunoblotting: seroprevalence studies in Taiwan. *Clin. Chim. Acta* **320**:37–42.
233. **Wilkinson, A. H., J. L. Smith, L. G. Hunsicker, J. Tobacman, D. P. Kapelanski, M. Johnson, F. H. Wright, D. M. Behrendt, and R. J. Corry.** 1989. Increased frequency of posttransplant lymphomas in patients treated with cyclosporine, azathioprine, and prednisone. *Transplantation* **47**:293–296.
234. **Williams, H., and D. H. Crawford.** 2006. Epstein-Barr virus: the impact of scientific advances on clinical practice. *Blood* **107**:862–869.
235. **Williams, H., K. Macsween, K. McAulay, C. Higgins, N. Harrison, A. Swerdlow, K. Britton, and D. Crawford.** 2004. Analysis of immune activation and clinical events in acute infectious mononucleosis. *J. Infect. Dis.* **190**:63–71.
236. **Wood, B. L., D. Sabath, V. C. Broudy, and G. Raghu.** 1996. The recipient origin of posttransplant lymphoproliferative disorders in pulmonary transplant patients. *Cancer* **78**: 2223–2227.
237. **Wood, C., and W. Harrington, Jr.** 2005. AIDS and associated malignancies. *Cell Res.* **15**:947.
238. **Woodle, E. S., M. Hanaway, J. Buell, T. Gross, M. R. First, J. Trofe, and T. Beebe.** 2001. Kaposi sarcoma: an analysis of the US and international experiences from the Israel Penn International Transplant Tumor Registry. *Transplant. Proc.* **33**:3660–3661.
239. **Xu, S., M. Green, L. Kingsley, S. Webber, and D. Rowe.** 2006. A comparison of quantitative-competitive and real-time PCR assays using an identical target sequence to detect Epstein-Barr virus viral load in the peripheral blood. *J. Virol. Methods* **137**:205.
240. **Yagita, M., H. Iwakura, T. Kishimoto, T. Okamura, A. Kunitomi, R. Tabata, Y. Konaka, and K. Kawa.** 2001. Successful allogeneic stem cell transplantation from an unrelated donor for aggressive Epstein-Barr virus-associated clonal T-cell proliferation with hemophagocytosis. *Int. J. Hematol.* **74**:451–454.
241. **Young, L., and A. B. Rickinson.** 2004. Epstein-Barr virus: 40 years on. *Nat. Rev. Cancer* **4**:757–768.
242. **Young, L. S., R. Lau, M. Rowe, G. Niedobitek, G. Packham, F. Shanahan, D. T. Rowe, D. Greenspan, J. S. Greenspan, and A. B. Rickinson.** 1991. Differentiation-associated expression of the Epstein-Barr virus BZLF1 transactivator protein in oral hairy leukoplakia. *J. Virol.* **65**: 2868–2874.
243. **Zhu, L., R. Wang, A. Sweat, E. Goldstein, R. Horvat, and B. Chandran.** 1999. Comparison of human sera reactivities in immunoblots with recombinant human herpesvirus (HHV)-8 proteins associated with the latent (ORF73) and lytic (ORFs 65, K8.1A, and K8.1B) replicative cycles and in immunofluorescence assays with HHV-8-infected BCBL-1 cells. *Virology* **256**:381–392.

Diagnostic Microbiology of the Immunocompromised Host
Edited by Randall T. Hayden, Karen C. Carroll, Yi-Wei Tang, and Donna M. Wolk

Chapter 5

Herpes Simplex Virus, Varicella-Zoster Virus, Human Herpesvirus 6, and Human Herpesvirus 7

Paula A. Revell, James H. Clark III, and Beverly B. Rogers

The four viruses discussed in this chapter are all members of the family *Herpesviridae*. The ubiquitous nature of these viruses, as well as their ability to establish a lifelong latent infection, makes them particularly important in the immunocompromised population. To better describe the specific details of the pathogenesis, epidemiology, and diagnostic testing of each of these viruses, each organism is discussed separately in the following four sections: herpes simplex virus (HSV), varicella-zoster virus (VZV), human herpesvirus 6 (HHV-6), and HHV-7.

HSV

The Organism

HSV is a member of the alphaherpesviruses, which consist of HSV-1, HSV-2, and VZV. The HSV virion is composed of a linear double-stranded DNA genome; an internal core; an icosahedral capsid; the tegument, which is the amorphous material that surrounds the capsid; and a lipid envelope containing viral glycoproteins on the surface. All members of the *Herpesviridae* family are morphologically similar, with HSV-1 and HSV-2 being the most closely related. Despite this common ancestry and morphology, HSV-1 and HSV-2 are serologically and genetically distinct. The HSV-1 and HSV-2 genomes share only 50% sequence homology, making them easily distinguishable by DNA sequence analysis. HSV has the ability to establish latency in an infected host following primary infection, with the potential for reactivation.

Pathogenesis

Infection in the immunocompetent patient is often subclinical. However, HSV-1 may result in gingivostomatitis and ulcerating lesions. When clinically apparent, HSV-2 infection typically presents as recurrent painful vesicular ulcerative lesions in the genital area. HSV-1 can inoculate genital areas by contact with oral secretions and is responsible for >25% of genital herpes infections. Rarely, HSV-1 and -2 can result in more serious diseases, such as encephalitis and neonatal infections. Initial infection with HSV occurs when the virus comes in contact with mucosal surfaces, where it then replicates. After replication in either the sensory or autonomic nerve ending, the virus is transported intra-axonally to the nerve cell bodies, where it establishes latency. HSV-1 infection typically establishes latency within the trigeminal ganglia, whereas HSV-2 establishes latency in the sacral nerve root ganglia. The virus periodically reactivates and travels through the axon to a peripheral site, either oral or genital, resulting in the release of infectious virus. Reactivation can be triggered by various stimuli, including stress, fever, ultraviolet light, and immunosuppression, and reactivation can result in lesion formation at the peripheral site (18). The transmission of HSV occurs by direct contact of mucosal surfaces with virally infected secretions. Transmission occurs intermittently in asymptomatic infected individuals (18). Individuals with defects in cell-mediated immunity are particularly susceptible to developing symptomatic HSV disease. The immunocompromised states covered in this section include patients with human immunodeficiency virus type 1 (HIV-1) infection and bone marrow and solid organ transplant recipients.

In solid organ and bone marrow transplant patients, HSV primary infection or reactivation may result in more frequent and severe infections. The majority of cases of HSV reactivation occur in the initial few weeks after transplantation, at the time of maximum pharmacologic immunosuppression (90). Disease reactivation has been reported to occur in

Paula A. Revell, James H. Clark III, and Beverly B. Rogers • The University of Texas Southwestern Medical Center and Children's Medical Center Dallas, Dallas, TX 75235.

approximately 60% of solid organ transplant patients not receiving prophylaxis; approximately 50% of those will develop symptomatic lesions (29). HSV infection can be acquired by the transplantation of an infected organ, resulting in primary infection in the recipient. In the majority of cases, the consequences of primary infection are more severe than those of reactivation. In solid organ and bone marrow transplant patients, HSV infection may result in more invasive lesions that take longer to heal and may increase the potential for dissemination to visceral organs. Dissemination may involve the lungs, gastrointestinal tract, liver, and skin (67, 74, 97). HSV pneumonia is most common in lung and heart-lung transplant recipients, but it has been seen in all types of solid organ transplant recipients (40, 78, 92). HSV pneumonia has been associated with significant mortality, at approximately 75%, for solid organ transplant patients (78). In renal transplant patients, HSV infections represented 23.8% of viral illnesses detected in a recent study group (47). Solid organ transplant patients with unexplained elevated transaminase levels should be evaluated for HSV infection (28). Encephalitis is rare after solid organ transplantation (78). Among bone marrow transplant patients, reactivation of HSV occurs in up to 70% of patients who are seropositive before transplantation. Most oncologists use routine prophylaxis in this patient group because of the difficulty in differentiating HSV infection from chemotherapy mucositis (18). In most bone marrow transplant patients, HSV infections are confined to the oropharynx; however, infections can extend to other areas (98).

HIV-1-infected patients have an increased risk of the presence of HSV-2 (95). Several studies have demonstrated that HSV infection is a common coinfection in the HIV-1-infected patient population (87). Epidemiological evidence suggests that HSV infection increases the efficiency of HIV-1 transmission (45, 66). HIV-infected patients have more severe and chronic HSV lesions and have increased rates of asymptomatic genital shedding of HSV-2 (95). HIV patients coinfected with HSV-2 have lesions that are more persistent than lesions in immunocompetent individuals. This is more pronounced in patients with $CD4^+$ counts of $<100 \times 10^6$ cells/liter (3, 94).

Epidemiology

HSVs are present worldwide, with humans being the only known natural reservoir. HSV-1 is transmitted through oral secretions, usually during childhood by nonsexual contact. In the United States, the seroprevalence of HSV-1 was 57.7% in the period 1999–2004 (106). HSV-2, on the other hand, is transmitted through sexual contact in almost all cases; the seroprevalence of HSV-2 was 17.0% in the period 1999–2004 (106). Worldwide, 85% of the population is seropositive for HSV-1 (108).

Diagnostic Testing

Laboratory tests available for the diagnosis of HSV include direct smear, viral culture, direct fluorescent-antibody staining (DFA), serologic testing, and nucleic acid amplification assays.

Direct smear

Direct examination of herpetic skin lesions via Tzanck smear can demonstrate changes consistent with HSV infection; the lesions will demonstrate intranuclear inclusions and multinucleated squamous epithelial cells (18). This detection method is not sensitive, with only 60% of culture-positive lesions being positive by Tzanck smear (73). It is also important that HSV cannot be distinguished from VZV by direct smear.

Viral culture

Viral culture for HSV has been considered the gold standard for virus detection. However, it is important that the sensitivity of culture depends on the cell type used (112), the quality of viral transport (50), and the stage of HSV disease in the patient from whom the specimen is collected. Culture is most sensitive when the specimen tested is cellular material from the base of a fresh lesion; culture is not reliable for older, crusted lesions. Importantly, recurrent infections often result in decreased viral shedding; as a result, culture is less sensitive in this setting. HSV replicates in several cell culture lines, including human diploid fibroblasts (MRC-5 and WI-38), mink lung cells, rhabdomyosarcoma cells, and human epidermoid carcinoma lines (HEp-2 and A549). As mentioned above, the sensitivity of viral culture depends on the cell line used; for example, mink lung cells are more sensitive than MRC-5 cells. Due to this differential sensitivity of cell lines, in many laboratories two different cell types are used to detect HSV. Cytopathic effect (CPE) is usually apparent after 1 to 3 days of incubation, with >95% of positive cultures producing CPE by 5 days. When CPE is detected, the samples can then be stained with specific monoclonal antibodies to distinguish HSV-1 from HSV-2, which can be useful for epidemiologic studies. Cultures need to be held for up to 6 days in order to detect all culture-positive specimens.

Methods that involve modifications to traditional viral culture have also been developed to decrease the time needed for detection of HSV, including spin-amplification shell vial culture and the enzyme-linked virus-inducible system (ELVIS). Shell vial culture involves inoculation of the specimen onto a monolayer of cells by centrifugation, followed by detection of viral proteins prior to development of CPE (31). The sensitivity of this method is equal to that of traditional culture, with a much improved turnaround time. ELVIS is a system based on BHK cells with a beta-galactosidase reporter gene driven by the HSV-1 UL39 promoter. The UL39 promoter is not transactivated by other viruses. HSV-infected cells express this reporter gene, which results in a color change that is visible by light microscopy (93). Like shell vial culture, detection of HSV with ELVIS has a sensitivity comparable to that of standard viral culture, but with a decreased turnaround time. All culture-based detection methods have the same limitation in requiring live virus for detection; as such, appropriate viral transport becomes crucial. HSV is heat labile, and delays in transport or failure to maintain an appropriate temperature results in reduced sensitivity.

Direct antigen detection

Direct detection of HSV viral antigen in patient specimens can be done using DFA. Typically, a specimen is fixed to a slide by use of acetone, followed by washes with antibody preparations against HSV. Both HSV-1 and HSV-2 can be detected in the same well if the antibodies are conjugated to distinct fluorophores. DFA can be up to 90% as sensitive as culture if the appropriate specimen is used. Specimens that include cells from the base of a young vesicular lesion result in the highest sensitivity, and the sensitivity goes down as the lesions begin to heal (60). DFA has a much shorter turnaround time than culture or shell vial culture; a result can be reported within 1.5 hours of receipt in the laboratory. In addition, DFA does not require live virus, making variations in transport conditions less significant.

Serology

Serologic testing for HSV is limited primarily to epidemiologic purposes. Importantly, because the majority of immunocompromised patients have recurrent infections with HSV, the role of serology in these patients is of limited diagnostic value. Serologic diagnosis is also hampered by the common structural proteins of HSV-1 and HSV-2. Only HSV serologic assays directed against the glycoprotein G envelope can be used to differentiate HSV-1, HSV-2, and VZV infections due to heterotypic antibody responses that may be present (18). There are commercially available immunoblot assays that can be used to reliably differentiate HSV-1 immunoglobulin G (IgG) antibodies from HSV-2 IgG antibodies. No commercially available method is capable of differentiating HSV-1 and HSV-2 IgM antibodies (100).

Molecular detection

The use of HSV DNA testing has increased the sensitivity of virus detection three to four times over that of culture (18). PCR-based methods are the most sensitive methods for detection of HSV. Multiple studies have demonstrated increased sensitivity for detection of HSV in samples from genital and dermal lesions, corneal scrapings, blood, and cerebrospinal fluid (CSF) (5, 16, 30, 71, 101). For patients coinfected with HSV-1 and HSV-2, PCR testing is capable of detecting and differentiating these viruses. This is possible because HSV-1 and HSV-2 have different melting temperatures due to DNA sequence differences (30). Prior to the development of PCR technology, viral culture of HSV from brain tissue was the gold standard for detection of central nervous system (CNS) HSV infection. Studies have demonstrated that for the detection of HSV in CSF, PCR is the diagnostic method of choice (41). The sensitivity and specificity of PCR testing of the CSF for HSV have been reported to be up to 100.0% and 98.6%, respectively. Due to the dramatic increase in sensitivity and decrease in time to detection, PCR has replaced viral culture as the standard method of detection for HSV in the CSF (2, 5). Yet another example of the utility of the sensitive PCR-based detection method may be in detection of disseminated disease. Although it occurs rarely, HSV has been described as causing disseminated infection with multiorgan involvement, particularly in recipients of solid organ transplants. Despite the availability of effective antiviral therapies, the outcomes of disseminated infections are often poor. This is in part due to the difficulty in diagnosing the infection in a timely manner. In one review of cases of HSV hepatitis, only 23% of cases were diagnosed premortem (52). In cases of disseminated HSV infection, early viral detection is critical for initiation of appropriate therapy. Detection of HSV nucleic acid in the blood of patients with suspected disseminated disease may facilitate early detection and result in initiation of appropriate antiviral therapy (48, 76).

In cases where HSV nucleic acid can be present without clinical disease, such as in respiratory secretions, quantitative PCR may offer a solution to determining the clinical significance of detecting viral nucleic acid. It is possible that a rise in viral load over time may indicate active infection, although further study is required to evaluate the clinical utility of this method (88). Importantly, detection of HSV DNA does not require live virus; therefore, specimen quality is not jeopardized by delays in transport time or fluctuations in transport temperature like it would be for viral culture. The benefits of HSV DNA testing include increased sample stability, a short turnaround time, and increased sensitivity (30).

Therapy

Several antiviral agents are available for the management of HSV infections, including acyclovir, valacyclovir, famciclovir, and foscarnet. Acyclovir is most widely used both for treatment and for prophylaxis. In solid organ and bone marrow transplant patients, the widespread use of prophylaxis to prevent cytomegalovirus (CMV) infections is effective in preventing HSV in most cases. HSV prophylactic regimens using oral acyclovir have reduced symptomatic HSV infections in kidney and heart transplant recipients as well (90). Acyclovir in low doses for the first 3 to 4 weeks prevents most cases (78). In bone marrow transplant patients who are HSV seropositive, a higher rate of HSV reactivation and acyclovir-resistant strains are seen. HSV-seropositive bone marrow transplant patients are usually given prophylaxis until the mucositis has decreased (77). For solid organ and bone marrow transplant patients with symptomatic disease, intravenous antivirals are indicated (90). HIV-1-infected patients treated with antivirals have a decrease in the frequency and severity of HSV recurrences as well as in the frequency and severity of asymptomatic reactivation (95). Acyclovir-resistant strains have been reported and are seen most commonly in the immunocompromised patient population. Transplant recipients and AIDS patients who fail to respond to therapy for HSV are potentially infected with resistant virus. For these patients, antivirals such as foscarnet or ganciclovir can be used, although these drugs have higher rates of toxicity than acyclovir (18). The Clinical and Laboratory Standards Institute (CLSI) has established a standard for antiviral susceptibility testing of HSV (96). This standard utilizes a plaque reduction assay to determine a decrease in susceptibility to an antiviral agent in vitro. For HSV, in vitro resistance has been correlated with clinical resistance (4, 63).

VZV

The Organism

Like all herpesviruses, VZV consists of a linear, double-stranded DNA genome surrounded by an icosahedral nucleocapsid. The capsid is surrounded by a proteinaceous tegument, which is enclosed in a lipid envelope that is composed of both host cell membranes and viral glycoproteins. The 125-kilobase VZV genome is the smallest of the human herpesvirus genomes and encodes 71 known genes. There is very little nucleotide variation in the VZV genome; however, there are epidemiologically distinct VZV strains that can be distinguished by sequencing or restriction endonuclease patterns. This ability to distinguish certain strains is useful because most cases of VZV due to wild-type virus in the United States, Europe, and Australia can be distinguished from the Oka strain, which was a clinical strain isolated from a Japanese patient and used to produce the live attenuated varicella vaccine (68). Most Japanese isolates cannot easily be distinguished from the Oka strain.

Pathogenesis

VZV is the causative agent of two distinct clinical manifestations, varicella and herpes zoster.

Primary infection results in varicella, or chicken pox, usually a relatively benign disease of childhood in the nonvaccinated population. This disease is characterized by a vesicular rash, mild fever, and malaise. Cutaneous lesions typically occur in groups, often on the trunk, scalp, and face, and several stages of maturity are present at one time.

The virus is acquired via inhalation of aerosolized particles; once internalized, the virus infects the mucous membranes of the respiratory tract and then spreads to the regional lymph nodes. Viral replication in the lymph nodes is followed by a period of subclinical viremia, during which the virus replicates in the visceral organs; this incubation period lasts about 2 weeks. A second viremic phase occurs 4 to 5 days prior to the outbreak of the disseminated rash. The vesicles contain high titers of infectious virus, and the contagious period ends when all lesions have crusted over. Upon resolution of varicella, the virus establishes latency in the dorsal and trigeminal ganglia, where it remains throughout the lifetime of the host.

Varicella tends to be a milder disease in young, otherwise healthy children than in adolescents and adults. While this disease is typically mild and self-limiting, complications can occur. The most common complication is bacterial superinfection; however, more serious invasive complications can be life-threatening. The more serious complications include pneumonia (more common in adults) and CNS involvement (ranging from benign cerebellar ataxia to meningoencephalitis and meningitis) (6, 34, 72). Severe disease more commonly occurs in immunocompromised patients, primarily in individuals with defects in cellular immunity. In this patient population, the rash can persist for weeks accompanied by high fever, and these patients are at increased risk for disseminated disease (21, 99).

Herpes zoster, or shingles, is a painful, localized vesicular rash that results from reactivation of the virus from the latently infected neuronal cells. Zoster lesions are characteristically grouped in a dermatomal distribution; the sensory nerve ganglion in which the virus was latent typically innervates the dermatome affected by the rash. The rash can be preceded by and associated with acute hypersensitivity and pain. In otherwise healthy patients, the affected area typically heals in 2 weeks, but lesions can persist for up to 6 weeks. The most common complication of zoster is postherpetic neuralgia, defined as pain that remains after the rash has resolved. As is the case for varicella, zoster can be a significant problem in the immunocompromised population. In addition to the risk of a prolonged and more severe local disease, these patients are at risk for developing disseminated zoster. Disseminated disease can manifest as cutaneous disease, with skin lesions appearing outside the primary dermatome, and with more severe systemic involvement, such as pneumonia, encephalitis, and hepatitis (25, 33, 38, 39). Immunocompromised patients can also pose a unique diagnostic dilemma because severe disease is not always associated with the appearance of a preceding rash (38, 99). It is crucial for this patient population to have rapid diagnostic tests available in order to initiate appropriate antiviral therapy.

Epidemiology

VZV, like its fellow alphaherpesvirus HSV, has a worldwide distribution. This virus is strictly a human pathogen, and person-to-person spread is the sole source of transmission. In temperate climates, prior to the widespread implementation of the vaccine, varicella was primarily a childhood disease, with peak incidence in the late winter and early spring. This is in contrast to the case in tropical climates, where the disease is acquired later in life. VZV is highly contagious and unique among the herpesviruses in that both varicella and disseminated zoster require airborne isolation precautions due to its transmissibility through the respiratory route (44). The ease of transmission makes diagnosis of both the infected patient and the infected health care provider critical due to the extreme risk for rapid, extensive transmission in a susceptible population. Herpes zoster does not exhibit the same seasonal pattern as varicella, but it does provide a source for VZV transmission because the virus can be transmitted through contact with vesicular contents. Contact with VZV from a patient with zoster can lead to varicella in an immune-naïve or immunocompromised individual.

The introduction of a live attenuated varicella vaccine has greatly impacted the epidemiology of VZV. The vaccine was licensed for use in healthy children in the United States in 1995. Since the vaccine was introduced, the number of varicella cases reported to surveillance sites has been reduced substantially, as has the rate of mortality associated with varicella. According to the 2005 National Immunization Survey, the coverage rate for the varicella vaccine among children aged 19 to 35 months is 87.9% (8a). Worldwide impact is difficult to assess at this point due to a lack of routine vaccination in many countries (70, 81). Despite the success of the vaccine, there are a substantial number of individuals who remain at risk for severe disease due to VZV primary infection or reactivation; these include severely immunocompromised patients, for whom the live attenuated vaccine is contraindicated (84). This group includes patients with severe cellular immune deficiencies, such as AIDS patients and allogeneic stem cell transplant recipients. Vaccination of severely immunocompromised patients can result in disease due to the vaccine strain (56, 64).

Diagnostic Testing

There are two main aspects of VZV diagnostics: one is identification of VZV during active infection, and the other is determination of a person's immune status toward VZV.

Diagnosis based on clinical presentation alone is adequate for typical cases of varicella or zoster. Atypical presentations in either previously healthy or immunocompromised patients can require laboratory-based diagnosis. Techniques available for diagnosis include viral culture, DFA, and PCR. The staining of tissue scrapings from the base of a lesion, or Tzanck smear, can be used as a rapid method to establish the presence of HSV or VZV in a lesion. However, the Tzanck smear is not specific for VZV and is not a useful diagnostic tool for atypical VZV infections (73, 83).

Serology

Serologic assays for VZV are not typically used for diagnosis of infection but are critical in determining the patient's immune status. This is particularly important in the transplant population. Pretransplant serostatus can aid in determining posttransplant infection risk. Patients with previous VZV exposure are at increased risk for posttransplant reactivation, and patients who are seronegative for VZV pretransplant are at risk for primary infection after transplantation. The fluorescent-antibody membrane antigen (FAMA) assay detects binding of antibodies in sera to unfixed VZV-infected cells. This assay is very sensitive but highly complex and is not available in the clinical setting. Enzyme-linked immunosorbent assay (ELISA) is the most widely used assay to determine VZV IgG levels. ELISA can be used to assess a patient's exposure history or to evaluate the patient's immune status following vaccination. A correlation between protective immunity and the presence of anti-VZV glycoprotein antibody levels has been shown and was used as the primary end point for determining immunogenicity in clinical trials with the VZV vaccines (65). It is important that ELISA is not as sensitive as FAMA assay; therefore, a positive ELISA result confirms exposure, but a negative ELISA result may not reliably determine susceptibility. In addition, there have been reported cases of varicella infection in individuals with low levels of VZV antibodies detected by FAMA assay (36). These cases indicate that a positive antibody titer in the immunocompromised population does not always indicate protection against disease. It has been suggested that a reliable history of clinical varicella infection is the only true indication of VZV immunity in the immunocompromised patient population (35).

Viral culture

Culture can be used for diagnosis of active VZV infection; however, it lacks sensitivity compared to PCR or DFA, in part due to the extremely labile nature of the virus and the transient nature of viral infectivity. More so than HSV, VZV is quite labile on transport, and thus it often does not grow well in cell culture. The cell types that support growth are human lung carcinoma (A-549) and human lung fibroblast (MRC-5) cells; 5 to 7 days of growth are required before detection of CPE. Time to positivity can be decreased with shell vial culture. Shell vial culture combines culture and fluorescent-antibody staining, and thus virus can be detected before the formation of CPE, with a limited loss of sensitivity. The optimal specimen for viral culture is fluid from an early vesicle, along with cells from the base of the lesion. Late lesions do not contain adequate infectious virus for viral culture, making the window of time for appropriate specimen collection limited.

Direct antigen detection

DFA can be more sensitive than culture for detecting VZV and has a short turnaround time. In this case, the optimal specimen requires cellular material from the base of a lesion. Vesicular fluid may contain cell-free virus, but due to the lack of cellularity, it is not appropriate for detection by DFA. The material is placed on a microscope slide, fixed with acetone, and then stained with either a monoclonal or polyclonal fluorescently linked antibody. DFA has been shown to have a sensitivity of 97.5% compared to culture, which had a sensitivity of 49.4% for 79 cases of clinically diagnosed VZV infection (14). In another study using a different DFA stain and procedure, DFA had a sensitivity of 87% when culture was defined as the gold standard (10). An additional benefit to DFA is that staining of the same specimen for HSV and VZV can determine the viral etiology of an atypically presenting skin lesion.

Molecular detection

In the immunocompromised patient, it is possible that the typical vesicular presentation may not occur or that systemic symptoms may mimic other disease entities, such as graft-versus-host disease (99). Due to the unusual presentation and critical need for a rapid diagnosis, culture and DFA have less diagnostic value in these cases, and PCR has become the mainstay of diagnosis. PCR assays are designed to detect the presence of viral DNA in the specimen; these assays can be qualitative or quantitative. PCR is extremely sensitive and has been used to confirm severe, atypical, or fatal cases. VZV DNA has been detected in crusted lesions, CSF, ocular fluid, lungs, liver, brain, whole blood, plasma, and peripheral blood mononuclear cells (23, 54). Qualitative PCR for VZV DNA in the CSF has been particularly useful in the diagnosis of VZV-associated CNS disease in both immunocompetent and immunocompromised patients (5, 37). It is possible that quantitative assays might be useful in establishing disease severity of CNS disease, but further studies are needed. In addition, studies have been done to evaluate the efficacy of PCR testing for evaluating the response to therapy of VZV CNS disease (12).

Like the case for CMV, VZV DNA can be detected and quantitated in the blood of patients with active disease. Studies have shown that no VZV DNA is detected in the blood of healthy controls with a documented history of VZV infection or vaccination (49, 55). In contrast, high levels of VZV DNA have been detected in patients with active disease (49, 57). One important difference between VZV and CMV or Epstein-Barr virus (EBV) is that there seems to be a nearly simultaneous rise in VZV DNA in the blood and onset of clinical symptoms; this may make it challenging to monitor DNA levels in order to preemptively treat patients, as is common with CMV (51). The highest DNA levels have been reported for patients with disseminated disease (with the presence or absence of lesions) and primary varicella. There are contrasting reports on the ability to detect VZV DNA in the blood or peripheral blood mononuclear cells of patients with active, limited zoster. Although multiple studies have shown the value in using quantitative PCR testing to detect VZV DNA in the blood or plasma in diagnosing VZV-related disease in the immunocompromised patient, there are no standard assays available. Most labs have developed inhouse assays, and there is variation in everything from specimen type, i.e., blood versus plasma, to target DNA sequence. This variation makes comparing results from institution to institution difficult. These challenges make uniform statements about sensitivity, specificity, positive predictive value, and negative predictive value impossible. There is a consensus agreement, however, that PCR testing makes diagnosing atypical presentations of VZV possible in a rapid time frame, with sensitivities that are superior to those of DFA or culture for specimens other than cells harvested from the base of a fresh lesion.

Therapy

Antiviral drugs can effectively treat VZV infections, but the window of time for implementation of antiviral therapy is small. Initiation of treatment within the first 24 hours maximizes efficacy. This emphasizes the importance of a rapid and accurate diagnosis. Intravenous acyclovir is recommended for immunocompromised patients; famciclovir and valacyclovir are both available for treatment of adults, but no pediatric formulation is available. These drugs all require phosphorylation by the viral tyrosine kinase for activity. Importantly, tyrosine kinase-negative VZV strains have been identified, and infections caused by these strains should be treated with foscarnet. To date, there are no standardized clinical assays to determine susceptibilities to acyclovir.

HHV-6

As the sixth herpesvirus, HHV-6 is an important pathogen with an as yet ill-defined spectrum of disorders. Even one of the simplest and most well-accepted clinical associations, febrile seizure, has been called into question based on the prevalence of HHV-6 in the population (46). The immunocompromised patient presents an even more difficult diagnostic dilemma, with difficulty in viral burden testing and interpretation compounded not only by latency but also by the fact that the virus may integrate into the human genome and actually be transferred from donor to recipient through transplantation (13).

The Organism

The viral genome is linear, double-stranded DNA of approximately 160 kilobases (22). Two variants, HHV-6A and HHV-6B, have approximately 90% sequence homology and, while causing similar symptoms, may vary in prevalence and age of symptom onset. Conserved genes within various strains of HHV-6B include a portion of the DR region and the IE-A gene. One important feature of the HHV-6 genome is that it has been found to be incorporated into the human genome (13, 20, 105). Repeat sequences within the viral genome reveal similarity with human telomeric sequences; transfer of HHV-6 genetic material from incorporated DNA from a parent to a child has been described (20). There has also been a report of incorporation of almost the entire viral genome of HHV-6A into a stem cell line, resulting in detection of the viral genome in the recipient when the stem cells engrafted and proliferated (13).

Pathogenesis

HHV-6 is thought to be transmitted through the saliva, as high titers have been identified in salivary secretions (110). In addition, the viral genome has been identified by in situ hybridization of oral mucosal epithelium overlying tonsillar tissue, suggesting that this may be a reservoir for the virus (82). A survey of postmortem tissue for herpesvirus genomes resulted in identification of HHV-6 in locations similar to those for EBV, most commonly the salivary gland and gastrointestinal tract (11). Vertical transmission has also been proposed, with 1 to 2% of umbilical cord blood specimens containing the HHV-6 genome (8). Documentation is not available to identify the course of infection once the virus enters the host, but it is known that the virus replicates most efficiently in T lymphocytes of the $CD4^+$

immunophenotype. HHV-6 has been shown to infect a variety of hematopoietic and nonhematopoietic cells, with particular interest applied to astrocytes, oligodendroglia, and microglia in the brain. Latency can be established by this virus, and this predisposes it to reactivation in the immunosuppressed host (22).

Coinfection with other herpesviruses may potentiate the pathologic effects of HHV-6; conversely, HHV-6 is thought to enhance the activity of certain viral pathogens. CMV and EBV, common pathogens in transplant patients, are frequently detected in concert with HHV-6; in one report, HHV-6 was found to increase prior to or in association with elevations of CMV (43, 61). HHV-6 also interacts with the immune system and HIV-1, although the net effect of its properties is not clear. The literature is conflicting in this regard, with some authors suggesting that HHV-6 potentiates the immunosuppressive effect and some indicating that it is, in fact, protective (19, 22).

Epidemiology

Acute HHV-6 infection causes roseola infantum, or sixth disease, in the immunocompetent child, with infection occurring most commonly between 9 and 21 months of age in industrialized countries, although intrauterine and neonatal acquisition is described for approximately 2% of patients (111). The prevalence of HHV-6 infection, determined using antibody positivity in healthy blood donors, shows >90% positivity for most countries across the world, including developing countries, with the lowest rates of positivity found in Poland and Belgium (58). Fever, fussiness, and rhinorrhea are the most common presenting symptoms, with the characteristic rash of roseola occurring in 23% of cases. Fever is seen in 57% of children with primary HHV-6 infection, and febrile seizures are known to be associated with HHV-6 infection; however, in a prospective study evaluating the acquisition of HHV-6 with seizures, none of 130 children who acquired HHV-6 developed seizures. Multiple review articles cite an association of HHV-6 with febrile seizures, and prevailing thought cites a correlation due to the prevalence of HHV-6 in children with febrile seizures (8, 22, 109). However, the only case-control study of HHV-6 and febrile seizures reported a 43% prevalence of HHV-6 in patients with febrile seizures, which was almost identical to the prevalence in febrile patients without seizures, who were matched for clinical parameters (46). This raises the question of whether HHV-6 infection is an "innocent bystander" in children with febrile seizures.

The natural history of HHV-6 infection includes primary infection, latency, and reactivation (8). Following primary infection in healthy children and adults, the virus replicates in the salivary glands and is secreted into the saliva. Although this stage of HHV-6 infection is not associated with disease, it is important for the primary mode of viral transmission. The virus remains latent in lymphocytes and monocytes and can also persist in other cell types throughout various tissues. HHV-6 reactivation occurs rarely but is associated with the most serious clinical disease. Due to the fact that most individuals are infected in infancy or childhood, any immunosuppressed individual is at risk for reactivated infection. The immunocompromised states covered in this section include patients who have undergone bone marrow and solid organ transplantation.

The majority of the literature surrounding the role of HHV-6 in the transplant population regards patients with bone marrow transplantation. Patients who receive allogeneic stem cell transplants are at a greater risk of developing symptomatic infection than are patients who receive autologous transplants (89, 107). HHV-6 reactivation typically occurs within a year of receiving the transplant, with most studies reporting reactivation occurring around 21 days following transplantation (7, 102). The development of a rash is one of the most common presenting symptoms of HHV-6 reactivation. One study reported HHV-6 reactivation in 66% of patients presenting with a rash in the first 100 days following transplantation, whereas only 20% of patients without HHV-6 reactivation developed rash (102). HHV-6 also affects the hematopoietic system, with platelet engraftment delayed in patients presenting with HHV-6 reactivation within 28 days of transplantation (80). However, another study failed to identify an effect on engraftment (102).

The most severe disease associated with HHV-6 infection following bone marrow transplantation is encephalitis. Symptoms at presentation include confusion, coma, seizures, headache, and speech disturbance (89). Depression has been described in one case (24). Most patients respond to antiviral therapy, and symptoms resolve in conjunction with resolution of viral DNA in the CSF. Histologic evaluation of the brain at autopsy reveals necrosis of both gray and white matter, with prominent demyelination (27).

Solid organ transplant recipients are also at risk for developing symptomatic HHV-6 reactivation. A prospective analysis of HHV-6 primary infection in seven patients who were seronegative for HHV-6 prior to solid organ transplantation revealed seroconversion in four of the seven patients, with a viral load detected in only one patient, who developed cholestatic hepatitis (9). One patient with encephalitis following liver transplantation presented with confusion, headache, and involuntary movements (79). HHV-6 DNA was detected in the CSF, and symptoms resolved following antiviral therapy. Another patient with a cardiac transplant presented with fever, confusion, decreasing responsiveness, and diplopia, in conjunction with HHV-6 detected in CSF and plasma (75).

While most immunosuppressed patients develop reactivated HHV-6 infections, there are reports of primary infection following transplantation occurring in children (86). Two patients who were seronegative prior to bone marrow transplantation developed fever and a rash associated with the acquisition of HHV-6. One of these patients, who was 2 1/2 years old at the time of infection, had a self-limited course; the other patient, who was 3 months of age, died shortly after infection, with signs of multiorgan failure.

HHV-6 has been implicated to have a causal relationship to disease in patients with pneumonitis and hepatitis. Leach et al. stated, in 1992, "further improvements in diagnostic testing, and results from carefully designed clinical and virologic studies, will eventually provide a more complete understanding of the clinical consequences of infection with this presumably ancient but recently discovered herpesvirus" (62). In another article, published in the *New England Journal of Medicine* in 1993, HHV-6 was associated with idiopathic interstitial pneumonitis in bone marrow transplant recipients. This association was based on identifying increased HHV-6 DNA levels in open lung biopsies from patients with a clinical presentation of interstitial pneumonitis compared to those in controls (15). However, a cause-and-effect relationship has not been defined. The same authors, in 1996, identified HHV-6A and HHV-6B DNAs in lung tissue from bone marrow transplant recipients as well as previously healthy individuals who suffered sudden death (17). The strain of HHV-6 most commonly associated with infection in both the immunocompetent and immunosuppressed patient is HHV-6B (22). HHV-6A has been associated with childhood febrile illness in a specific geographic location and in a case of posttransplant encephalitis presenting as depression (8, 59).

Diagnostic Testing

The methods available to detect HHV-6 include viral culture, serology, immunohistochemistry, antigen testing (antigenemia), and PCR amplification.

Viral culture

Culture is not readily available in the clinical laboratory due to the requirement for lymphocyte coculture, and the diagnosis in the immunocompetent patient is typically made based on clinical presentation.

Serology

If serologic evaluation is desired, specific IgG or IgM antibodies to HHV-6 are the best means of confirming a diagnosis. The use of paired sera to demonstrate seroconversion is the appropriate method to identify acute infection by using the IgG antibody test. Cross-reactivity has been described between HHV-6 and HHV-7 serologic testing, but this will vary with the method and manufacturer (104).

Molecular detection

PCR amplification of viral DNA from whole blood has been used to diagnose viremia in the immunocompetent patient, with one study using real-time PCR to detect virus from saliva and plasma (110). Children younger than 2 years of age with fever were evaluated, and in children with a detectable viral burden, the median plasma viral load was 7,530 copies/ml, with a range of 1,298 to 21,393 copies/ml. HHV-6 was detected in the saliva of 28% of patients who were nonviremic, presumably reflective of viral shedding following past infection; viral DNA was detected in the saliva of 65% of those who were viremic. For these reasons, a clinical diagnosis or serologic evaluation may be more useful for the immunocompetent host.

PCR assay is the most common assay employed to detect viral burden in the immunosuppressed patient, although antigen detection has also been employed (7, 9, 75, 85). The samples most commonly analyzed include CSF and whole blood. Interpretation of PCR testing can be difficult because finding virus in a sample does not necessarily indicate disease, particularly in the immunosuppressed population. Viral burden testing using quantitative PCR analysis is more useful in this regard, but a discrete value is not as useful as a viral burden which is increasing or decreasing. As with viral burden testing using quantitative PCR for other herpesviruses, such as EBV and CMV, a change in the viral burden is most useful if seen as a trend (consistent over three or more time points). There is no specific number indicating an increase or decrease that is uniformly reflective of clinical disease. With that aside, the literature suggests that a viral load of $>10^3$ targets/10^6 peripheral blood mononuclear cells in stem cell transplant patients is associated with fever, rash, pneumonitis, or partial myelosuppression (7). However, a single viral burden analysis should be followed up to assess trends to assist in determining significance. In addition, identification of viral genomes in tissues by PCR testing has not been associated with disease and should be interpreted with caution. It is likely that the interpretation of PCR testing for HHV-6 will follow the same course as that for EBV, with low viral burdens typically unassociated with disease and high viral burdens more likely to reflect disease.

As described below, markedly high HHV-6 viral burdens may be seen with viral integration into the chromosomes, without signs of infection, or high viral burdens may reflect true infection. This further complicates the interpretation of a positive viral burden in blood. A positive PCR test result for HHV-6 in CSF carries greater significance than a single level in the blood, although even the CSF can be positive due to viral integration or contamination from peripheral blood during the spinal tap. Likely, the best axiom is to interpret HHV-6 test results with caution and, as always, in the clinical context of the patient.

One study reported viral burdens in patients with encephalitis following stem cell transplantation (85). In that report, there were two patients with detectable HHV-6 who had encephalitis and no other infectious or noninfectious reason. The viral burdens for these two patients were 64,600 and 135,700 genome equivalents/ml of plasma. Patients without disease attributable to HHV-6 had viral burdens in the plasma of between 0 and 14,050 genome equivalents/ml, with the majority having $<$1,000 genome equivalents/ml.

HHV-6 reactivation in immunosuppressed patients has also been linked to reactivation of CMV and HHV-7. Whether reactivation of one of these viruses potentiates the reactivation of another cannot be differentiated from a propensity to reactivate secondary to the effects of immunosuppressive therapy. There are reports of HHV-6 and HHV-7 antigenemia associated with CMV disease following liver transplantation (43, 61) as well as stem cell transplantation (85).

There is one additional caveat related to HHV-6 testing that makes the interpretation of increasing viral burdens especially treacherous, and that is the fact that HHV-6 can incorporate into the human genome. When this happens, there can be elevations of HHV-6 simply related to cell replication, as reported following bone marrow transplantation (13). One patient developed a persistently elevated HHV-6 viral burden following stem cell transplantation, with levels of HHV-6 rising to 10^6 genome equivalents/ml of blood 10 days following transplantation, and this level of HHV-6 persisted in the peripheral blood for months. The donor had an HHV-6 viral load of 8×10^6 genome equivalents/ml of whole blood prior to transplant, and the recipient was negative. The authors attributed clinical signs such as fever to other causes, and there was no history of encephalitis. Fluorescence in situ hybridization for HHV-6A was positive with donor lymphocytes, with integration at 17p13.3. The elevation of HHV-6 viral burden was attributed to engraftment. Other reports of persistently high levels of HHV-6 have also been reported for patients with lymphoma, multiple sclerosis, and chronic fatigue syndrome (69). It has been indicated that persistent viral burdens of $>10^7$ genome equivalents/ml are suggestive of chromosomal integration and should be interpreted with caution. Ward et al. (105) noted that viral infection by HHV-6 typically occurs with HHV-6B, so if HHV-6A is detected, there should be a suspicion that the viral burden may result from integration. In general, it is suggested that if the HHV-6 viral burden remains persistently elevated despite therapy, there may be integration.

Immunohistochemistry and in situ hybridization

Immunohistochemistry and in situ hybridization are direct methods of detecting viral protein and genomes, respectively. Whether these methods are superior to quantitative PCR testing of blood/plasma or CSF to aid in the diagnosis of clinical infection remains to be determined. While PCR analysis can be performed on tissue, such as lymph nodes, obtaining a true quantitative result can be difficult due to the heterogeneity of tissue and the lack of a true common denominator. Similarly, liquids, such as bronchoalveolar lavage fluid, are variable in the quantity of liquid, based on the amount of liquid infused into the bronchiolar tree. PCR-positive results are almost universal in evaluating lymph nodes, as demonstrated by Roush et al. (82). This study compared staining for HHV-6 by using immunohistochemistry and in situ hybridization to PCR analysis of tonsillar epithelia of nonimmunosuppressed individuals undergoing tonsillectomy in an attempt to determine if lymphoid tissue was a site of latency in the immunocompetent population (82). All tonsils tested were positive for HHV-6 by PCR, and of those tested, HHV-6 was localized to the tonsillar epithelium by in situ hybridization and immunohistochemistry. It was suggested that the tonsillar epithelium was a site of viral latency, given the uniformly positive PCR results and the strong staining by in situ hybridization of the tonsillar epithelium.

HHV-6 has also been localized to other tissues histologically, and its correlation with PCR positivity varies with the study. While PCR testing is generally felt to be the most sensitive method for detecting HHV-6, one study revealed immunohistochemical detection of HHV-6 in reactive astrocytes of the CNS associated with waning viral burden in the CSF (32).

Therapy

Therapy for HHV-6 is the same as that for CMV, with in vitro sensitivity demonstrated to ganciclovir, foscarnet, and cidofovir (91). When to treat patients, rather than what to use, is the bigger question relating to this virus. An immunocompetent patient developed HHV-6-associated encephalomyelitis and was treated successfully with cidofovir (26). Immunosuppressed patients have received similar therapy, with some responding and others progressing (75, 89, 107). HHV-6 resistance has rarely been described and is hypothesized to result from resistance mechanisms similar to those of CMV (22).

HHV-7

The Organism and Pathogenesis

HHV-7 is a betaherpesvirus and thus shares genomic homology with HHV-6 and CMV. First isolated in 1990, the virus infects $CD4^+$ T cells and remains latent following infection, similar to other herpesviruses. Compared to HHV-6, for which the clinical spectrum of disease is certain in only a few instances, the role of HHV-7 is even less clear. HHV-7 is associated with febrile illnesses and reactivates following immunosuppression, but whether the organism causes disease in the immunosuppressed host is unclear (11).

Epidemiology

Infection occurs early in life, but typically later than that with HHV-6 (42). It is associated with febrile illness in childhood and may present with symptoms similar to those of exanthema subitum (22, 103). There have not been reports of congenital infection like that seen with HHV-6, with some authors proposing that this is due to the fact that HHV-7 does not have the propensity to integrate into the human genome that HHV-6 does (103). The prevalence of HHV-7 in transplant patients varies from 0% to 46%. While there is no well-recognized clinical syndrome associated with HHV-7 infection or reactivation in the immunosuppressed population, one study reported an association of HHV-7 reactivation with the occurrence of graft-versus-host disease, fever, fatigue, and vomiting in children following bone marrow transplantation (53). However, other infections also occurred commensurate with HHV-7 reactivation, so it was unclear if HHV-7 actually caused disease or simply reactivated concurrent with symptoms related to other infections. There is also an indication that HHV-7 may be associated with reactivation of HHV-6 (22).

Laboratory Testing

Serologic evaluation may be used to diagnose HHV-7 infections in the immunocompetent patient, with paired acute- and convalescent-phase IgG titers most commonly being used. Antigenemia and PCR analyses have been used for the immunosuppressed population, although without a defined disease state, interpretation of the significance of the results is unclear. Issues related to interpretation of laboratory tests in the face of viral integration do not exist with HHV-7.

Therapy

As with HHV-6, in vitro sensitivity of HHV-7 has been demonstrated for ganciclovir, foscarnet, and cidofovir (91).

SUMMARY

The four viruses discussed in this chapter have both similarities and dissimilarities. HSV and VZV are both neurotropic and may reactivate at a time distant from the primary infection, and infection can occur with or without vesicular lesions. In the immunosuppressed population, disseminated disease can be fatal, and treatments are similar. Both can be cultured in the clinical laboratory, although of the two, VZV is extremely labile and slower growing, making diagnostic modalities other than culture of particular value (Table 1). The CPEs of HSV and VZV are identical; therefore, differentiation of these two viruses by cytologic or histologic evaluation is not possible without the use of immunohistochemical testing. Molecular methods of detection for each of these viruses are sensitive, specific, and available with a relatively short turnaround time.

Although they are herpesviruses, HHV-6 and HHV-7 are dissimilar from HSV and VZV. Infections by HHV-6 and HHV-7 do not produce vesicular

Table 1. Diagnostic methods for detection of the four viruses discussed in this chapter[a]

Detection method	Usefulness for: HSV	VZV	HHV-6	HHV-7	Comments
Direct smear	Tzanck smear	Tzanck smear	NA	NA	Tzanck smear lacks sensitivity and specificity
Viral culture	Relatively sensitive Sensitivity depends on cell type Long turnaround time (TAT) Modifications available to decrease TAT (shell vial culture) Does not distinguish HSV-1 and HSV-2 without additional staining Can perform susceptibility testing on cultured virus	Lacks sensitivity compared to PCR or DFA Long TAT Modifications available to decrease TAT (shell vial culture)	NA	NA	Less sensitive for recurrent infections due to decreased viral shedding Viral stability critical Sensitivity depends on quality transport and live virus in specimen
Direct antigen detection (DFA)	Nearly as sensitive as culture Stage of lesion important to sensitivity Short TAT Specific	More sensitive than culture Short TAT Stage of lesion important to sensitivity Specific	NA	NA	
Serology	Primarily for epidemiology	Useful in determining immune status	Can document primary infection in children	Useful in immunocompetent patients	
Molecular detection	Most sensitive detection method for all specimen types Can detect virus in healing lesions Can detect virus in atypical settings and in specimens in which culture and DFA are insensitive, such as blood and CSF Can type strains easily	Most sensitive detection method for all specimen types Critical in diagnosis of atypical presentation of VZV Quantitative PCR may be useful in establishing disease severity Standard assays not currently available	Diagnostic method of choice Viral nucleic acid can be detected in the absence of disease Viral integration without infection can lead to high viral burden Quantitative PCR may aid in determining significance	Interpretation of significance not clear	Does not require live virus Increased specimen stability
Immunohistochemistry	Used as a confirmatory test for histologic inclusions	Not commonly available in clinical laboratories	Not commonly available in clinical laboratories	Not available	

[a]NA, not applicable.

lesions; these infections present as disseminated disease with fever, an erythematous rash, and possibly lymphocytosis and may present with neurologic symptoms if CNS involvement is prominent. Neither HHV-6 nor HHV-7 is cultured in the clinical laboratory by standard culture techniques, and thus detection of the viruses relies primarily on molecular methods (Table 1). Viral DNA remains latent following primary infection and may reactivate without disease in the immunosuppressed population; in addition, viral nucleic acid can integrate into the host genome, resulting in high levels of viral nucleic acid in the absence of disease. As a result of the many cases where HHV-6 nucleic acid can be detected in the absence of disease, positive PCR results must be interpreted with caution. Finally, quantitative PCR analysis may prove useful in differentiating when the presence of HHV-6 nucleic acid is significant and related to a disease state.

REFERENCES

1. Reference deleted.
2. **Allawi, H. T., H. Li, T. Sander, A. Aslanukov, V. I. Lyamichev, A. Blackman, S. Elagin, and Y. W. Tang.** 2006. Invader plus method detects herpes simplex virus in cerebrospinal fluid and simultaneously differentiates types 1 and 2. *J. Clin. Microbiol.* **44:**3443–3447.
3. **Aoki, F. Y.** 2001. Management of genital herpes in HIV-infected patients. *Herpes* **8:**41–45.
4. **Bean, B., C. Fletcher, J. Englund, S. N. Lehrman, and M. N. Ellis.** 1987. Progressive mucocutaneous herpes simplex infection due to acyclovir-resistant virus in an immunocompromised patient: correlation of viral susceptibilities and plasma levels with response to therapy. *Diagn. Microbiol. Infect. Dis.* **7:**199–204.
5. **Boivin, G.** 2004. Diagnosis of herpesvirus infections of the central nervous system. *Herpes* **11**(Suppl. 2):48A–56A.
6. **Bonhoeffer, J., G. Baer, B. Muehleisen, C. Aebi, D. Nadal, U. B. Schaad, and U. Heininger.** 2005. Prospective surveillance of hospitalisations associated with varicella-zoster virus infections in children and adolescents. *Eur. J. Pediatr.* **164:** 366–370.
7. **Boutolleau, D., C. Fernandez, E. Andre, B. M. Imbert-Marcille, N. Milpied, H. Agut, and A. Gautheret-Dejean.** 2003. Human herpesvirus (HHV)-6 and HHV-7: two closely related viruses with different infection profiles in stem cell transplantation recipients. *J. Infect. Dis.* **187:**179–186.
8. **Campadelli-Fiume, G., P. Mirandola, and L. Menotti.** 1999. Human herpesvirus 6: an emerging pathogen. *Emerg. Infect. Dis.* **5:**353–366.

8a. **Centers for Disease Control and Prevention.** 2006. National, state, and urban area vaccination coverage among children aged 19–35 months—United States, 2005. *Morb. Mortal. Wkly. Rep.* **55:**988–993.

9. **Cervera, C., M. A. Marcos, L. Linares, E. Roig, N. Benito, T. Pumarola, and A. Moreno.** 2006. A prospective survey of human herpesvirus-6 primary infection in solid organ transplant recipients. *Transplantation* **82:**979–982.
10. **Chan, E. L., K. Brandt, and G. B. Horsman.** 2001. Comparison of Chemicon SimulFluor direct fluorescent antibody staining with cell culture and shell vial direct immunoperoxidase staining for detection of herpes simplex virus and with cytospin direct immunofluorescence staining for detection of varicella-zoster virus. *Clin. Diagn. Lab. Immunol.* **8:**909–912.
11. **Chen, T., and S. D. Hudnall.** 2006. Anatomical mapping of human herpesvirus reservoirs of infection. *Mod. Pathol.* **19:**726–737.
12. **Cinque, P., S. Bossolasco, L. Vago, C. Fornara, S. Lipari, S. Racca, A. Lazzarin, and A. Linde.** 1997. Varicella-zoster virus (VZV) DNA in cerebrospinal fluid of patients infected with human immunodeficiency virus: VZV disease of the central nervous system or subclinical reactivation of VZV infection? *Clin. Infect. Dis.* **25:**634–639.
13. **Clark, D. A., E. P. Nacheva, H. N. Leong, D. Brazma, Y. T. Li, E. H. Tsao, H. C. Buyck, C. E. Atkinson, H. M. Lawson, M. N. Potter, and P. D. Griffiths.** 2006. Transmission of integrated human herpesvirus 6 through stem cell transplantation: implications for laboratory diagnosis. *J. Infect. Dis.* **193:**912–916.
14. **Coffin, S. E., and R. L. Hodinka.** 1995. Utility of direct immunofluorescence and virus culture for detection of varicella-zoster virus in skin lesions. *J. Clin. Microbiol.* **33:**2792–2795.
15. **Cone, R. W., R. C. Hackman, M. L. Huang, R. A. Bowden, J. D. Meyers, M. Metcalf, J. Zeh, R. Ashley, and L. Corey.** 1993. Human herpesvirus 6 in lung tissue from patients with pneumonitis after bone marrow transplantation. *N. Engl. J. Med.* **329:**156–161.
16. **Cone, R. W., A. C. Hobson, J. Palmer, M. Remington, and L. Corey.** 1991. Extended duration of herpes simplex virus DNA in genital lesions detected by the polymerase chain reaction. *J. Infect. Dis.* **164:**757–760.
17. **Cone, R. W., M. L. Huang, R. C. Hackman, and L. Corey.** 1996. Coinfection with human herpesvirus 6 variants A and B in lung tissue. *J. Clin. Microbiol.* **34:**877–881.
18. **Corey, L.** 2005. Herpes simplex virus, p. 1762–1774. *In* G. L. Mandell, J. E. Bennett, and R. Dolin (ed.), *Principles and Practice of Infectious Diseases,* 6th ed., vol. 2. Churchill Livingston, Philadelphia, PA.
19. **Csoma, E., T. Deli, J. Konya, L. Csernoch, Z. Beck, and L. Gergely.** 2006. Human herpesvirus 6A decreases the susceptibility of macrophages to R5 variants of human immunodeficiency virus 1: possible role of RANTES and IL-8. *Virus Res.* **121:**161–168.
20. **Daibata, M., T. Taguchi, Y. Nemoto, H. Taguchi, and I. Miyoshi.** 1999. Inheritance of chromosomally integrated human herpesvirus 6 DNA. *Blood* **94:**1545–1549.
21. **David, D. S., B. R. Tegtmeier, M. R. O'Donnell, I. B. Paz, and T. M. McCarty.** 1998. Visceral varicella-zoster after bone marrow transplantation: report of a case series and review of the literature. *Am. J. Gastroenterol.* **93:**810–813.
22. **De Bolle, L., L. Naesens, and E. De Clercq.** 2005. Update on human herpesvirus 6 biology, clinical features, and therapy. *Clin. Microbiol. Rev.* **18:**217–245.
23. **de Jong, M. D., J. F. Weel, M. H. van Oers, R. Boom, and P. M. Wertheim-van Dillen.** 2001. Molecular diagnosis of visceral herpes zoster. *Lancet* **357:**2101–2102.
24. **de Labarthe, A., A. Gauthert-Dejean, P. Bossi, J. P. Vernant, and N. Dhedin.** 2005. HHV-6 variant A meningoencephalitis after allogeneic hematopoietic stem cell transplantation diagnosed by quantitative real-time polymerase chain reaction. *Transplantation* **80:**539.
25. **De La Blanchardiere, A., F. Rozenberg, E. Caumes, O. Picard, F. Lionnet, J. Livartowski, J. Coste, D. Sicard, P. Lebon, and D. Salmon-Ceron.** 2000. Neurological complications of varicella-zoster virus infection in adults with human immunodeficiency virus infection. *Scand. J. Infect. Dis.* **32:**263–269.

26. **Denes, E., L. Magy, K. Pradeau, S. Alain, P. Weinbreck, and S. Ranger-Rogez.** 2004. Successful treatment of human herpesvirus 6 encephalomyelitis in immunocompetent patient. *Emerg. Infect. Dis.* **10:**729–731.
27. **Drobyski, W. R., K. K. Knox, D. Majewski, and D. R. Carrigan.** 1994. Brief report: fatal encephalitis due to variant B human herpesvirus-6 infection in a bone marrow-transplant recipient. *N. Engl. J. Med.* **330:**1356–1360.
28. **Duckro, A. N., B. E. Sha, S. Jakate, M. K. Hayden, D. M. Simon, S. N. Saltzberg, S. Arai, and H. A. Kessler.** 2006. Herpes simplex virus hepatitis: expanding the spectrum of disease. *Transpl. Infect. Dis.* **8:**171–176.
29. **Dummer, J. S.** 2005. Infections in solid organ transplant recipients, p. 3501–3512. *In* G. L. Mandell, J. E. Bennett, and R. Dolin (ed.), *Principles and Practice of Infectious Diseases,* 6th ed., vol. 2. Churchill Livingston, Philadelphia, PA.
30. **Espy, M. J., J. R. Uhl, L. M. Sloan, S. P. Buckwalter, M. F. Jones, E. A. Vetter, J. D. Yao, N. L. Wengenack, J. E. Rosenblatt, F. R. Cockerill III, and T. F. Smith.** 2006. Real-time PCR in clinical microbiology: applications for routine laboratory testing. *Clin. Microbiol. Rev.* **19:**165–256.
31. **Espy, M. J., A. D. Wold, D. J. Jespersen, M. F. Jones, and T. F. Smith.** 1991. Comparison of shell vials and conventional tubes seeded with rhabdomyosarcoma and MRC-5 cells for the rapid detection of herpes simplex virus. *J. Clin. Microbiol.* **29:**2701–2703.
32. **Fotheringham, J., N. Akhyani, A. Vortmeyer, D. Donati, E. Williams, U. Oh, M. Bishop, J. Barrett, J. Gea-Banacloche, and S. Jacobson.** 2007. Detection of active human herpesvirus-6 infection in the brain: correlation with polymerase chain reaction detection in cerebrospinal fluid. *J. Infect. Dis.* **195:**450–454.
33. **Fox, R. J., S. L. Galetta, R. Mahalingam, M. Wellish, B. Forghani, and D. H. Gilden.** 2001. Acute, chronic, and recurrent varicella zoster virus neuropathy without zoster rash. *Neurology* **57:**351–354.
34. **Galil, K., C. Brown, F. Lin, and J. Seward.** 2002. Hospitalizations for varicella in the United States, 1988 to 1999. *Pediatr. Infect. Dis. J.* **21:**931–935.
35. **Gershon, A. A., J. Chen, P. LaRussa, and S. P. Steinberg.** 2007. Varicella-zoster virus, p. 1537–1548. *In* E. J. B. Patrick, R. Murray, M. L. Landry, M. A. Pfaller, and J. H. Jorgensen (ed.), *Manual of Clinical Microbiology,* 9th ed., vol. 2. ASM Press, Washington, DC.
36. **Gershon, A. A., S. P. Steinberg, et al.** 1990. Live attenuated varicella vaccine: protection in healthy adults compared with leukemic children. *J. Infect. Dis.* **161:**661–666.
37. **Gilden, D.** 2004. Varicella zoster virus and central nervous system syndromes. *Herpes* **11**(Suppl. 2)**:**89A–94A.
38. **Gilden, D. H., R. R. Wright, S. A. Schneck, J. M. Gwaltney, Jr., and R. Mahalingam.** 1994. Zoster sine herpete, a clinical variant. *Ann. Neurol.* **35:**530–533.
39. **Gnann, J. W., Jr.** 2002. Varicella-zoster virus: atypical presentations and unusual complications. *J. Infect. Dis.* **186**(Suppl. 1)**:**S91–S98.
40. **Green, M., and R. K. Avery.** 2004. Herpes simplex virus (HSV)-1 and -2, and varicella zoster virus (VZV). *Am. J. Transplant.* **4:**69–71.
41. **Grover, D., W. Newsholme, N. Brink, H. Manji, and R. Miller.** 2004. Herpes simplex virus infection of the central nervous system in human immunodeficiency virus-type 1-infected patients. *Int. J. STD AIDS* **15:**597–600.
42. **Hall, C. B., M. T. Caserta, K. C. Schnabel, M. P. McDermott, G. K. Lofthus, J. A. Carnahan, L. M. Gilbert, and S. Dewhurst.** 2006. Characteristics and acquisition of human herpesvirus (HHV) 7 infections in relation to infection with HHV-6. *J. Infect. Dis.* **193:**1063–1069.
43. **Harma, M., K. Hockerstedt, O. Lyytikainen, and I. Lautenschlager.** 2006. HHV-6 and HHV-7 antigenemia related to CMV infection after liver transplantation. *J. Med. Virol.* **78:**800–805.
44. **Heininger, U., and J. F. Seward.** 2006. Varicella. Lancet **368:**1365–1376.
45. **Hook, E. W., III, R. O. Cannon, A. J. Nahmias, F. F. Lee, C. H. Campbell, Jr., D. Glasser, and T. C. Quinn.** 1992. Herpes simplex virus infection as a risk factor for human immunodeficiency virus infection in heterosexuals. *J. Infect. Dis.* **165:**251–255.
46. **Hukin, J., K. Farrell, L. M. MacWilliam, M. Colbourne, E. Waida, R. Tan, L. Mroz, and E. Thomas.** 1998. Case-control study of primary human herpesvirus 6 infection in children with febrile seizures. *Pediatrics* **101:**E3.
47. **Hwang, E. A., M. J. Kang, S. Y. Han, S. B. Park, and H. C. Kim.** 2004. Viral infection following kidney transplantation: long-term follow-up in a single center. *Transplant. Proc.* **36:**2118–2119.
48. **Ichai, P., A. M. Afonso, M. Sebagh, M. E. Gonzalez, L. Codes, D. Azoulay, F. Saliba, V. Karam, E. Dussaix, C. Guettier, D. Castaing, and D. Samuel.** 2005. Herpes simplex virus-associated acute liver failure: a difficult diagnosis with a poor prognosis. *Liver Transpl.* **11:**1550–1555.
49. **Ishizaki, Y., J. Tezuka, S. Ohga, A. Nomura, N. Suga, R. Kuromaru, K. Kusuhara, Y. Mizuno, N. Kasuga, and T. Hara.** 2003. Quantification of circulating varicella zoster virus-DNA for the early diagnosis of visceral varicella. *J. Infect.* **47:**133–138.
50. **Jensen, C., and F. B. Johnson.** 1994. Comparison of various transport media for viability maintenance of herpes simplex virus, respiratory syncytial virus, and adenovirus. *Diagn. Microbiol. Infect. Dis.* **19:**137–142.
51. **Kalpoe, J. S., A. C. Kroes, S. Verkerk, E. C. Claas, R. M. Barge, and M. F. Beersma.** 2006. Clinical relevance of quantitative varicella-zoster virus (VZV) DNA detection in plasma after stem cell transplantation. *Bone Marrow Transplant.* **38:**41–46.
52. **Kaufman, B., S. A. Gandhi, E. Louie, R. Rizzi, and P. Illei.** 1997. Herpes simplex virus hepatitis: case report and review. *Clin. Infect. Dis.* **24:**334–338.
53. **Khanani, M., A. Al-Ahmari, R. Tellier, U. Allen, S. Richardson, J. J. Doyle, and A. Gassas.** 2007. Human herpesvirus 7 in pediatric hematopoietic stem cell transplantation. *Pediatr. Blood Cancer* **48:**567–570.
54. **Kido, S., T. Ozaki, H. Asada, K. Higashi, K. Kondo, Y. Hayakawa, T. Morishima, M. Takahashi, and K. Yamanishi.** 1991. Detection of varicella-zoster virus (VZV) DNA in clinical samples from patients with VZV by the polymerase chain reaction. *J. Clin. Microbiol.* **29:**76–79.
55. **Kimura, H., S. Kido, T. Ozaki, N. Tanaka, Y. Ito, R. K. Williams, and T. Morishima.** 2000. Comparison of quantitations of viral load in varicella and zoster. *J. Clin. Microbiol.* **38:**2447–2449.
56. **Kramer, J. M., P. LaRussa, W. C. Tsai, P. Carney, S. M. Leber, S. Gahagan, S. Steinberg, and R. A. Blackwood.** 2001. Disseminated vaccine strain varicella as the acquired immunodeficiency syndrome-defining illness in a previously undiagnosed child. *Pediatrics* **108:**E39.
57. **Kronenberg, A., W. Bossart, R. P. Wuthrich, C. Cao, S. Lautenschlager, N. D. Wiegand, B. Mullhaupt, G. Noll, N. J. Mueller, and R. F. Speck.** 2005. Retrospective analysis of varicella zoster virus (VZV) copy DNA numbers in plasma of immunocompetent patients with herpes zoster, of immunocompromised patients with disseminated VZV disease, and of asymptomatic solid organ transplant recipients. *Transpl. Infect. Dis.* **7:**116–121.

58. **Krueger, G. R., B. Koch, N. Leyssens, Z. Berneman, J. Rojo, C. Horwitz, T. Sloots, M. Margalith, J. D. Conradie, S. Imai, I. Urasinski, M. de Bruyere, V. Ferrer Argote, and J. Krueger.** 1998. Comparison of seroprevalences of human herpesvirus-6 and -7 in healthy blood donors from nine countries. *Vox Sang.* **75:**193–197.
59. **Labarthe, M. C.** 2005. Food quality and safety. *Rev. Infirm.* **2005:**24–25.
60. **Lafferty, W. E., S. Krofft, M. Remington, R. Giddings, C. Winter, A. Cent, and L. Corey.** 1987. Diagnosis of herpes simplex virus by direct immunofluorescence and viral isolation from samples of external genital lesions in a high-prevalence population. *J. Clin. Microbiol.* **25:**323–326.
61. **Lautenschlager, I., M. Lappalainen, K. Linnavuori, J. Suni, and K. Hockerstedt.** 2002. CMV infection is usually associated with concurrent HHV-6 and HHV-7 antigenemia in liver transplant patients. *J. Clin. Virol.* **25**(Suppl. 2):S57–S61.
62. **Leach, C. T., C. V. Sumaya, and N. A. Brown.** 1992. Human herpesvirus-6: clinical implications of a recently discovered, ubiquitous agent. *J. Pediatr.* **121:**173–181.
63. **Lehrman, S. N., J. M. Douglas, L. Corey, and D. W. Barry.** 1986. Recurrent genital herpes and suppressive oral acyclovir therapy. Relation between clinical outcome and in-vitro drug sensitivity. *Ann. Intern. Med.* **104:**786–790.
64. **Levy, O., J. S. Orange, P. Hibberd, S. Steinberg, P. LaRussa, A. Weinberg, S. B. Wilson, A. Shaulov, G. Fleisher, R. S. Geha, F. A. Bonilla, and M. Exley.** 2003. Disseminated varicella infection due to the vaccine strain of varicella-zoster virus, in a patient with a novel deficiency in natural killer T cells. *J. Infect. Dis.* **188:**948–953.
65. **Li, S., I. S. Chan, H. Matthews, J. F. Heyse, C. Y. Chan, B. J. Kuter, K. M. Kaplan, S. J. Vessey, and J. C. Sadoff.** 2002. Inverse relationship between six week postvaccination varicella antibody response to vaccine and likelihood of long term breakthrough infection. *Pediatr. Infect. Dis. J.* **21:**337–342.
66. **Lingappa, J. R., and C. Celum.** 2007. Clinical and therapeutic issues for herpes simplex virus-2 and HIV co-infection. *Drugs* **67:**155–174.
67. **Longerich, T., C. Eisenbach, R. Penzel, T. Kremer, C. Flechtenmacher, B. Helmke, J. Encke, T. Kraus, and P. Schirmacher.** 2005. Recurrent herpes simplex virus hepatitis after liver retransplantation despite acyclovir therapy. *Liver Transpl.* **11:**1289–1294.
68. **Loparev, V. N., T. Argaw, P. R. Krause, M. Takayama, and D. S. Schmid.** 2000. Improved identification and differentiation of varicella-zoster virus (VZV) wild-type strains and an attenuated varicella vaccine strain using a VZV open reading frame 62-based PCR. *J. Clin. Microbiol.* **38:**3156–3160.
69. **Luppi, M., P. Barozzi, R. Bosco, D. Vallerini, L. Potenza, F. Forghieri, and G. Torelli.** 2006. Human herpesvirus 6 latency characterized by high viral load: chromosomal integration in many, but not all, cells. *J. Infect. Dis.* **194:**1020–1021.
70. **Macartney, K. K., P. Beutels, P. McIntyre, and M. A. Burgess.** 2005. Varicella vaccination in Australia. *J. Paediatr. Child Health* **41:**544–552.
71. **Madhavan, H. N., K. Priya, J. Malathi, and P. R. Joseph.** 2003. Laboratory methods in the detection of herpes simplex virus (HSV) in keratitis—a 9-year study including polymerase chain reaction (PCR) during last 4 years. *Indian J. Pathol. Microbiol.* **46:**109–112.
72. **Meyer, P. A., J. F. Seward, A. O. Jumaan, and M. Wharton.** 2000. Varicella mortality: trends before vaccine licensure in the United States, 1970–1994. *J. Infect. Dis.* **182:**383–390.
73. **Nahass, G. T., B. A. Goldstein, W. Y. Zhu, U. Serfling, N. S. Penneys, and C. L. Leonardi.** 1992. Comparison of Tzanck smear, viral culture, and DNA diagnostic methods in detection of herpes simplex and varicella-zoster infection. *JAMA* **268:**2541–2544.
74. **Naik, H. R., and P. H. Chandrasekar.** 1996. Herpes simplex virus (HSV) colitis in a bone marrow transplant recipient. *Bone Marrow Transplant.* **17:**285–286.
75. **Nash, P. J., R. K. Avery, W. H. Tang, R. C. Starling, A. J. Taege, and M. H. Yamani.** 2004. Encephalitis owing to human herpesvirus-6 after cardiac transplant. *Am. J. Transplant.* **4:**1200–1203.
76. **Nebbia, G., F. M. Mattes, M. Ramaswamy, A. Quaglia, G. Verghese, P. D. Griffiths, A. Burroughs, and A. M. Geretti.** 2006. Primary herpes simplex virus type-2 infection as a cause of liver failure after liver transplantation. *Transpl. Infect. Dis.* **8:**229–232.
77. **Nichols, W. G., L. Corey, T. Gooley, C. Davis, and M. Boeckh.** 2002. High risk of death due to bacterial and fungal infection among cytomegalovirus (CMV)-seronegative recipients of stem cell transplants from seropositive donors: evidence for indirect effects of primary CMV infection. *J. Infect. Dis.* **185:**273–282.
78. **Patel, R., and C. V. Paya.** 1997. Infections in solid-organ transplant recipients. *Clin. Microbiol. Rev.* **10:**86–124.
79. **Paterson, D. L., N. Singh, T. Gayowski, D. R. Carrigan, and I. R. Marino.** 1999. Encephalopathy associated with human herpesvirus 6 in a liver transplant recipient. *Liver Transpl. Surg.* **5:**454–455.
80. **Radonic, A., O. Oswald, S. Thulke, N. Brockhaus, A. Nitsche, W. Siegert, and J. Schetelig.** 2005. Infections with human herpesvirus 6 variant B delay platelet engraftment after allogeneic haematopoietic stem cell transplantation. *Br. J. Haematol.* **131:**480–482.
81. **Rentier, B., and A. A. Gershon.** 2004. Consensus: varicella vaccination of healthy children—a challenge for Europe. *Pediatr. Infect. Dis. J.* **23:**379–389.
82. **Roush, K. S., R. K. Domiati-Saad, L. R. Margraf, K. Krisher, R. H. Scheuermann, B. B. Rogers, and D. B. Dawson.** 2001. Prevalence and cellular reservoir of latent human herpesvirus 6 in tonsillar lymphoid tissue. *Am. J. Clin. Pathol.* **116:**648–654.
83. **Sadick, N. S., P. D. Swenson, R. L. Kaufman, and M. H. Kaplan.** 1987. Comparison of detection of varicella-zoster virus by the Tzanck smear, direct immunofluorescence with a monoclonal antibody, and virus isolation. *J. Am. Acad. Dermatol.* **17:**64–69.
84. **Sartori, A. M.** 2004. A review of the varicella vaccine in immunocompromised individuals. *Int. J. Infect. Dis.* **8:**259–270.
85. **Sassenscheidt, J., J. Rohayem, T. Illmer, and D. Bandt.** 2006. Detection of beta-herpesviruses in allogenic stem cell recipients by quantitative real-time PCR. *J. Virol. Methods* **138:**40–48.
86. **Savolainen, H., I. Lautenschlager, H. Piiparinen, U. Saarinen-Pihkala, L. Hovi, and K. Vettenranta.** 2005. Human herpesvirus-6 and -7 in pediatric stem cell transplantation. *Pediatr. Blood Cancer* **45:**820–825.
87. **Schacker, T.** 2001. The role of HSV in the transmission and progression of HIV. *Herpes* **8:**46–49.
88. **Simoons-Smit, A. M., E. M. Kraan, A. Beishuizen, R. J. Strack van Schijndel, and C. M. Vandenbroucke-Grauls.** 2006. Herpes simplex virus type 1 and respiratory disease in critically-ill patients: real pathogen or innocent bystander? *Clin. Microbiol. Infect.* **12:**1050–1059.
89. **Singh, N., and D. L. Paterson.** 2000. Encephalitis caused by human herpesvirus-6 in transplant recipients: relevance of a novel neurotropic virus. *Transplantation* **69:**2474–2479.

90. **Slifkin, M., S. Doron, and D. R. Snydman.** 2004. Viral prophylaxis in organ transplant patients. *Drugs* **64:**2763–2792.
91. **Smith, J. M., and R. A. McDonald.** 2006. Emerging viral infections in transplantation. *Pediatr. Transplant.* **10:**838–843.
92. **Smyth, R. L., T. W. Higenbottam, J. P. Scott, T. G. Wreghitt, S. Stewart, C. A. Clelland, J. P. McGoldrick, and J. Wallwork.** 1990. Herpes simplex virus infection in heart-lung transplant recipients. *Transplantation* **49:**735–739.
93. **Stabell, E. C., S. R. O'Rourke, G. A. Storch, and P. D. Olivo.** 1993. Evaluation of a genetically engineered cell line and a histochemical beta-galactosidase assay to detect herpes simplex virus in clinical specimens. *J. Clin. Microbiol.* **31:**2796–2798.
94. **Stewart, J. A., S. E. Reef, P. E. Pellett, L. Corey, and R. J. Whitley.** 1995. Herpesvirus infections in persons infected with human immunodeficiency virus. *Clin. Infect. Dis.* **21**(Suppl. 1)**:**S114–S120.
95. **Strick, L. B., A. Wald, and C. Celum.** 2006. Management of herpes simplex virus type 2 infection in HIV type 1-infected persons. *Clin. Infect. Dis.* **43:**347–356.
96. **Swierkosz, E. M., R. L. Hodinka, B. M. Moore, S. Sacks, D. R. Scholl, and D. K. Wright.** 2004. Antiviral susceptibility testing: herpes simplex virus by plaque reduction assay; approved standard. *In* CLSI (ed.), document M33-A. CLSI, Wayne, PA.
97. **Tan, H. H., and C. L. Goh.** 2006. Viral infections affecting the skin in organ transplant recipients: epidemiology and current management strategies. *Am. J. Clin. Dermatol.* **7:**13–29.
98. **Van Burik, J., and D. Weisdorf.** 2005. Infections in recipients of hematopoietic stem cell transplantation. *In* G. L. Mandell, J. E. Bennett, and R. Dolin (ed.), *Principles and Practice of Infectious Diseases,* 6th ed., vol. 2. Churchill Livingston, Philadelphia, PA.
99. **Vinzio, S., B. Lioure, and B. Goichot.** 2006. Varicella in immunocompromised patients. *Lancet* **368:**2208.
100. **Wald, A., and R. Ashley-Morrow.** 2002. Serological testing for herpes simplex virus (HSV)-1 and HSV-2 infection. *Clin. Infect. Dis.* **35:**S173–S182.
101. **Wald, A., M. L. Huang, D. Carrell, S. Selke, and L. Corey.** 2003. Polymerase chain reaction for detection of herpes simplex virus (HSV) DNA on mucosal surfaces: comparison with HSV isolation in cell culture. *J. Infect. Dis.* **188:**1345–1351.
102. **Wang, L. R., L. J. Dong, M. J. Zhang, and D. P. Lu.** 2006. The impact of human herpesvirus 6B reactivation on early complications following allogeneic hematopoietic stem cell transplantation. *Biol. Blood Marrow Transplant.* **12:**1031–1037.
103. **Ward, K. N.** 2005. Human herpesviruses-6 and -7 infections. *Curr. Opin. Infect. Dis.* **18:**247–252.
104. **Ward, K. N.** 2005. The natural history and laboratory diagnosis of human herpesviruses-6 and -7 infections in the immunocompetent. *J. Clin. Virol.* **32:**183–193.
105. **Ward, K. N., H. N. Leong, A. D. Thiruchelvam, C. E. Atkinson, and D. A. Clark.** 2007. Human herpesvirus 6 DNA levels in cerebrospinal fluid due to primary infection differ from those in chromosomal viral integration and have implications for the diagnosis of encephalitis. *J. Clin. Microbiol.* **45:**1298–1304.
106. **Xu, F., M. R. Sternberg, B. J. Kottiri, G. M. McQuillan, F. K. Lee, A. J. Nahmias, S. M. Berman, and L. E. Markowitz.** 2006. Trends in herpes simplex virus type 1 and type 2 seroprevalence in the United States. *JAMA* **296:**964–973.
107. **Yamane, A., T. Mori, S. Suzuki, A. Mihara, R. Yamazaki, Y. Aisa, T. Nakazato, T. Shimizu, Y. Ikeda, and S. Okamoto.** 2007. Risk factors for developing human herpesvirus 6 (HHV-6) reactivation after allogeneic hematopoietic stem cell transplantation and its association with central nervous system disorders. *Biol. Blood Marrow Transplant.* **13:**100–106.
108. **Yeung-Yue, K. A., M. H. Brentjens, P. C. Lee, and S. K. Tyring.** 2002. Herpes simplex viruses 1 and 2. *Dermatol. Clin.* **20:**249–266.
109. **Zerr, D. M.** 2006. Human herpesvirus 6: a clinical update. *Herpes* **13:**20–24.
110. **Zerr, D. M., L. M. Frenkel, M. L. Huang, M. Rhoads, L. Nguy, M. A. Del Beccaro, and L. Corey.** 2006. Polymerase chain reaction diagnosis of primary human herpesvirus-6 infection in the acute care setting. *J. Pediatr.* **149:**480–485.
111. **Zerr, D. M., A. S. Meier, S. S. Selke, L. M. Frenkel, M. L. Huang, A. Wald, M. P. Rhoads, L. Nguy, R. Bornemann, R. A. Morrow, and L. Corey.** 2005. A population-based study of primary human herpesvirus 6 infection. *N. Engl. J. Med.* **352:**768–776.
112. **Zhao, L. S., M. L. Landry, E. S. Balkovic, and G. D. Hsiung.** 1987. Impact of cell culture sensitivity and virus concentration on rapid detection of herpes simplex virus by cytopathic effects and immunoperoxidase staining. *J. Clin. Microbiol.* **25:**1401–1405.

Diagnostic Microbiology of the Immunocompromised Host
Edited by Randall T. Hayden, Karen C. Carroll, Yi-Wei Tang, and Donna M. Wolk

Chapter 6

Adenovirus

Michael G. Ison and Randall T. Hayden

Adenoviruses are nonenveloped, double-stranded DNA viruses associated with a wide range of clinical syndromes in humans (45, 86). To date, 51 immunologically distinct types of adenoviruses have been described and further classified into one of six (A to F) species based on hemagglutinin properties, DNA homology, oncogenic potential in rodents, and clinical disease (Table 1) (45, 55).

Adenoviruses cause mostly self-limited respiratory, gastrointestinal (GI), or conjunctival disease throughout the year, without significant seasonal variation. Transmission can occur via inhalation of aerosolized droplets, direct conjunctival inoculation, fecal-oral spread, or exposure to infected tissue or blood (86). The incubation period is dependent on the viral serotype and the mechanism of transmission and can range from 2 days to 2 weeks (86). Viral shedding in immunocompetent patients occurs for approximately 1 to 3 days from the throat of adults with common colds; 3 to 5 days from the nose, throat, stool, or eyes of patients with pharyngoconjunctival fever; 2 weeks from eye cultures for keratoconjunctivitis; and 3 to 6 weeks from the throat or stool of children with respiratory or generalized illness. Although detailed information on viral shedding is limited for immunocompromised patients, shedding is typically prolonged. Irrespective of the primary site of infection, stool cultures from patients with most clinical syndromes are commonly positive over the first 2 weeks of illness (86). Latency may occur, with various groups documenting intermittent detection for months or even years, particularly when highly sensitive detection techniques, such as nucleic acid amplification, are used. Proposed sites of viral latency include the pharynx (tonsils and adenoids), intestine, urinary tract, and lymphocytes, although some of these remain controversial (37, 47).

Adenovirus infections are most common among children, people living in close quarters or closed populations, such as college students and military recruits (85), and immunocompromised patients (45). Although disease among immunocompetent patients is almost always self-limited, adenoviruses cause a wider spectrum of clinical disease in immunocompromised patients, with more end organ involvement, disseminated disease, and higher mortality (45). Among immunocompromised patients, infection is most commonly described for transplant recipients; adenovirus infection in such patients is the primary focus of this chapter.

HSCT RECIPIENTS

In the stem cell transplant population, the incidence of disease due to adenovirus ranges from 3 to 47% (4, 10, 13, 23, 30, 36, 39, 54, 60, 93). Available data suggest that adenoviral infections are more frequent in allogeneic stem cell transplant recipients than in those receiving autologous grafts (8.5 to 30% versus 2 to 12%) (4, 10, 23, 39), in children than in adults (31 to 47% versus 13.6%) (12, 23, 39), in patients who receive T-cell-depleted grafts (45% versus 11%) (13), and in patients with acute graft-versus-host disease (23, 54, 84, 93). Most retrospective studies have documented the onset of adenovirus disease primarily during the first 100 days following hematopoietic stem cell transplantation (HSCT) (median, 36 to 90 days) (8, 13, 30, 39, 45, 54, 69, 84, 92), although adenovirus disease after 100 days has also been documented clearly, especially for adults (13, 23).

In HSCT recipients, adenovirus is commonly associated with upper and/or lower respiratory tract

Michael G. Ison • Divisions of Infectious Diseases and Organ Transplantation, Transplant & Immunocompromised Host Infectious Diseases Service, Northwestern University Feinberg School of Medicine, Chicago, IL 60611. **Randall T. Hayden** • Department of Pathology, St. Jude Children's Research Hospital, Memphis, TN 38105.

Table 1. Infections associated with adenovirus species and serotype

Species	Serotype	Major site of infection
A	12, 18, 31	GI tract
B	3, 7, 11, 14, 16, 21, 34, 35	Respiratory, urinary tracts
C	1, 2, 5, 6	Respiratory tract
D	8-10, 13, 15, 17, 19, 20, 22–30, 32, 33, 36–39, 42–49	Eye, GI tract
E	4	Respiratory tract
F	40, 41	GI tract

infection, GI disease, hepatitis, and cystitis (45). Respiratory tract disease ranges from mild upper tract involvement, typically with nonspecific cold-like symptoms, to severe pneumonia (14, 23, 28, 93). GI disease ranges from mild diarrhea to hemorrhagic colitis; severe hepatitis has also been described (14, 23, 93). Adenovirus can cause an interstitial nephritis rarely but more commonly is associated with hemorrhagic cystitis. Detection of adenovirus in the urine of a patient with hemorrhagic cystitis is diagnostic of adenovirus-induced hemorrhagic cystitis (2, 14, 28, 31). Unlike other forms of adenoviral disease, hemorrhagic cystitis can typically be treated with local therapy and rarely progresses to disseminated infection (10 to 20%) (45).

The outcomes of infections can be severe. Adenovirus infections may be associated with graft failure or delayed engraftment. Additionally, coinfection with cytomegalovirus, *Aspergillus*, or bacteria frequently occurs and may contribute to the poor outcomes associated with adenovirus disease (45). If adenovirus infection is left untreated, the mortality for HSCT patients approaches 26% for all symptomatic patients, while pneumonia and disseminated disease portend more ominous outcomes (50% and 80% mortality, respectively) (45).

There are limited options for therapy of adenovirus infection. There have been no prospective studies to date, and the optimal timing for therapeutic intervention during the course of illness is unclear (45). Cidofovir appears to have the best in vitro and in vivo efficacy; unfortunately, significant toxicity has so far limited its application to wider patient populations (58). Lipid ester preparations of cidofovir appear to have increased in vitro efficacy against adenovirus and may have less toxicity (45). Ribavirin has in vitro activity against serogroup C viruses but does not appear to have significant activity in humans (52). Other approved agents, including vidarabine, ddC, and ganciclovir, may have activity, but their efficacy in managing adenoviral infections remains uncertain (45). Donor lymphocyte infusions may also have a role in the management of adenoviral infections, although experience with this method is limited (22, 40).

SOT RECIPIENTS

Adenovirus infection has been reported for a wide variety of solid organ transplant (SOT) recipient populations, including those receiving heart, lung, liver, intestinal, and renal transplants (35). Incidence data for adenoviral disease in SOT recipients are even more limited than those for HSCT recipients. Among SOT recipients, risk factors for adenovirus infection include renal or hepatic transplant, pediatric age group, T-cell depletion, and serologic mismatch (transplant from an adenovirus-seropositive donor to a seronegative recipient) (45). Invasive disease occurs in up to 10% of patients (64, 67). While adenovirus nucleic acid is frequently detected in blood from asymptomatic adult SOT recipients, unlike the case with HSCT patients, this does not appear to predict progression to invasive disease (42).

Among SOT recipients, hemorrhagic cystitis, nephritis, pneumonia, hepatitis, enterocolitis, and disseminated disease have been described (11, 48, 64, 65, 67, 88, 95). With the exception of hemorrhagic cystitis (the most common form of symptomatic disease in renal transplant recipients), the transplanted organ is typically the site of infection; pneumonia is most frequent in lung transplant recipients (9, 25, 83); hepatitis is most frequent in liver transplant recipients (64, 67), and enterocolitis, which may mimic rejection, is most frequent in small bowel recipients (73, 74).

DIAGNOSTIC APPROACHES

The diagnostic approach to patients with adenovirus disease is complex. The optimal diagnostic strategy depends on the specific indication, i.e., screening asymptomatic patients versus diagnosing end organ disease. It is imperative to consider adenovirus in the differential diagnosis of any clinically compatible condition. Failure to do so will likely delay the diagnosis, as demonstrated repeatedly in the literature (45). Although there are limited treatment options, outcomes appear to be maximized when antiviral therapy is started before extensive disseminated disease is recognized (45). Throat swabs, nasal washes, conjunctival swabs or scrapings, anal swabs, stool, urine, and blood may be collected for testing (86). Sample types collected and tested should correspond to clinically evident sites of infection (for example, stool for GI disease, urine for genitourinary disease, etc.). Serial monitoring of peripheral blood by quantitative PCR should be considered for all clinical syndromes in immunocompromised patients (45).

Various diagnostic techniques, including detection of antibodies, antigen detection methods, culture, electron microscopy, PCR, in situ hybridization, immunohistochemistry, and histopathology, have been described. Some of these are not widely used on a routine basis. Serology, for example, has limited practical value due to both a poor humoral immune response in many high-risk patients and a high population seroprevalence for adenovirus seen in the absence of active infection. Viral subtyping and serotyping can be accomplished through immunologic or nucleic acid sequence-based methods. These techniques are currently in use predominantly for epidemiologic studies and other research applications. The review below focuses primarily on culture, antigen detection, and molecular diagnostic methods.

Culture

Traditional culture methods have been described elsewhere in detail (37, 86, 103). Most adenoviruses can best be grown efficiently on human embryonic kidney (HEK) cells, although continuous cell lines, such as A549, KB, HeLa, HEp-2, and MRC-5, are more commonly used (49). Notably, adenovirus type 40 (AdV40) and AdV41 grow best on Graham 293 (HEK) cells and therefore may be missed in traditional cell cultures (97, 98). For maximum yield, initial cultures of primary clinical specimens should be set up on at least two optimal cell lines (86). The cytopathic effect associated with adenovirus is classically described as grape-like, with irregular aggregates of enlarged, rounded-up cells showing characteristic refractile intranuclear inclusions. Cytopathic effect develops slowly and typically begins at the periphery of the cell monolayer (Fig. 1) (103). Changes may be observed for up to 28 days following inoculation, with most appearing in the first week of culture (33). The time to detection

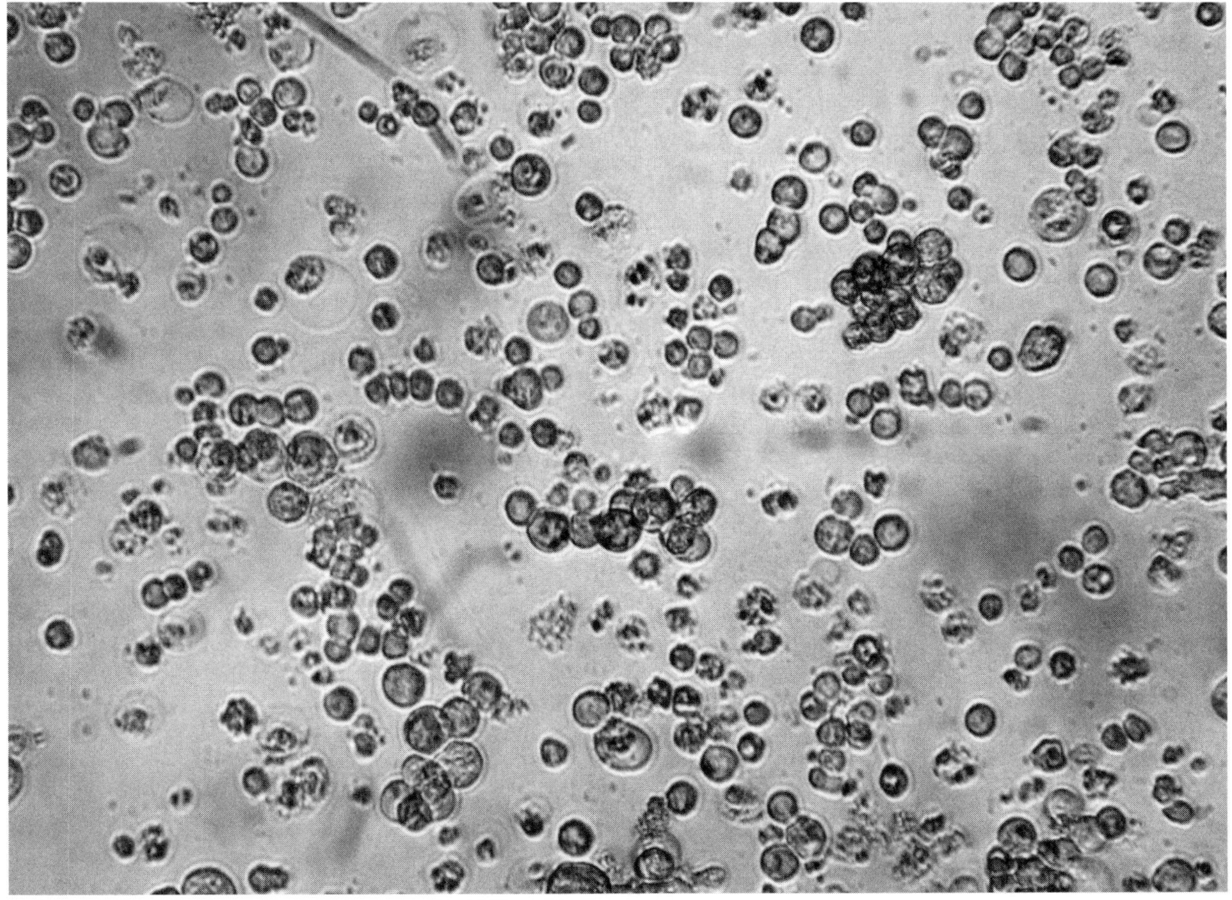

Figure 1. Typical adenovirus cytopathic effect in hybrid cell line composed of A549 and mink lung (Mv1Lu) cells. (Photograph courtesy of M. Bankowski.)

is dependent on multiple factors, including viral load inoculated, subgroup or serotype of adenovirus, cell lines in use, and culture conditions (86). Once cytopathic effect is noted, identification can be confirmed by immunofluorescence, enzyme immunoassay, or latex agglutination (3, 86).

Shell vial cultures can enhance the speed of adenovirus detection in clinical specimens (76). In general, a cell line monolayer is grown on a coverslip and inoculated using centrifugation (21). The slides are incubated for 1 to 2 days and then stained with monoclonal antibodies to the adenovirus hexon protein (3, 76, 86). Cell lines used are similar to those for conventional cell culture, with the recent advent of mixed cell lines used primarily to target multiple viruses simultaneously in respiratory tract specimens (105). Shell vial technology offers an improved turnaround time with potential cost savings (5), but it typically has a lower sensitivity than that of conventional cell culture methods and may be more susceptible to degradation with certain specimen types; as a result, more false-negative cultures may arise (102).

Antigen Detection

Monoclonal and polyclonal antibodies directed against the group-specific hexon antigen can be used for the direct detection of all adenoviruses and are commercially available (3). Signals may be detected for up to 3 weeks postinfection, with the greatest sensitivity achieved within the first 5 days of illness. Sensitivities for immunofluorescence assays performed on primary specimens have ranged from 28 to 75%, while other antigen detection methods have had sensitivities that ranged from 43 to 89% (34, 56, 76, 78, 79, 99, 100). Sensitivity is greatest for children and is markedly lower for adults (86). In general, commercially available assays (Table 2) are not widely used clinically, and their true analytical and clinical sensitivities are poorly studied, particularly among immunocompromised patients; available data suggest that antigen detection is not sufficiently sensitive for clinical care in this population (77).

Fluorescent-antibody (FA) testing and immunohistochemistry (see section below on morphologic methods) are used clinically to diagnose respiratory and tissue-invasive adenoviral disease, respectively (50, 83, 110). FA is widely used, both for primary, direct detection (usually from respiratory specimens) and for confirmatory identification of viral isolates grown using both conventional and shell vial methods. Although FA testing may be limited to adenovirus, the virus is typically one of several viruses tested as part of either a screening pool or a panel consisting of antibodies targeting several of the most common viral respiratory pathogens. FA testing offers a rapid turnaround time for the detection of respiratory viruses but may have limited sensitivity (83), particularly in the detection of dual infections (50, 83, 110).

Electron Microscopy

Electron microscopy may be used to identify adenovirus, particularly in stool. In general, although enteric adenoviruses occur in stool in consid-

Table 2. FDA-approved rapid diagnostic tests for adenovirus

Test	Company	Serotype specificity	Sensitivity[a] (%) (reference)	Specimen type(s)[b]	Notes
Imagen Adenovirus	Dako Diagnostics	Pan-AdV	86-100	Any clinical specimen or CC	Immunofluorescence vs hexon antigen
Ideia Adenovirus	Dako Diagnostics	Pan-AdV	90-98	Stool or CC	Genus-specific MAb vs hexon antigen
Premier Adenoclone	Meridian Bioscience	Pan-AdV	85-100[e]	Stool or CC[f]	
Premier Adenoclone type 40/41	Meridian Bioscience	AdV40 and -41	85-100(68)	Stool	
Adenolex	Orion Diagnostica	Pan-AdV	95(27)	Stool	Latex agglutination test
RPS AdenoDetector	Rapid Pathogen Screening Inc.	Pan-AdV	88-89 (89)	Conjuctival fluid/scrapings	Combined specimen collection and diagnostic in single device
SAS Adeno Test[d]	SA Scientific	Pan-AdV	73-95	Eye swab, stool, CC, NPS, TS, NW, NPA, FS	

[a]Values are per product package inserts except as indicated.
[b]CC, cell culture; NPS, nasopharyngeal swab; TS, throat swab; NW, nasal wash; NPA, nasopharyngeal aspirate; FS, fecal swab.
[c]Also detects influenza A and B viruses, parainfluenza virus types 1 to 3, and respiratory syncytial virus.
[d]Use of oxybuprocaine to anesthetize the eye may result in a high false-positive rate (38).
[e]Much lower sensitivities have been reported for ocular specimens (3, 62, 106, 107).
[f]Has been studied with other specimen types.

erably larger quantities than do respiratory adenovirus isolates, the presence of adenovirus in the stool is not, in and of itself, diagnostic of GI disease. Adenoviruses are nonenveloped icosahedral viruses (20 triangular surfaces and 12 vertices). They form typical diagnostic intranuclear inclusions which appear as "paracrystalline" arrays in an infected cell. Individual mature virions are 65 to 80 nm in diameter, and the entire virus is about 1.5×10^8 Da (Fig. 2) (86). The sensitivity of detection may be enhanced through the use of immunoelectron microscopy. While in the past such methods were a principle means of identifying adenovirus in clinical specimens (particularly in GI specimens), their routine use has diminished in recent years.

Histopathology

Although there is little call for such methods in the immunocompetent host, histopathology can be very useful in the detection of invasive adenovirus in those with impaired immunity. Routine examination of hematoxylin- and eosin-stained slides, most commonly from sites such as the lung, liver, or GI tract, may show various degrees of inflammation and reactive and regenerative changes in the presence of adenovirus infection. In addition, characteristic intranuclear inclusions are often seen, which can be very suggestive of the diagnosis (16) (Color Plate 6). Immunohistochemistry (24, 73, 88) and in situ hybridization (6, 63, 96) techniques for detection of adenovirus have been described elsewhere. When used as an adjunct to morphologic examination, either method may help to definitively identify adenovirus infection in tissue sections, improving the sensitivity of detection, while also localizing disease and providing evidence for causality (63). Since adenoviral enteritis may mimic rejection, the use of these methods is of particular value in situations such as small bowel transplantation (7, 26, 44, 51, 59, 72-74, 101).

Nucleic Acid Detection

Adenovirus can be detected by a number of molecular methods. In situ hybridization (6) techniques are briefly discussed above as adjuncts to morphologic detection in tissue. Outside this limited application, direct hybridization techniques have not gained widespread use. Since the transactivating regions of

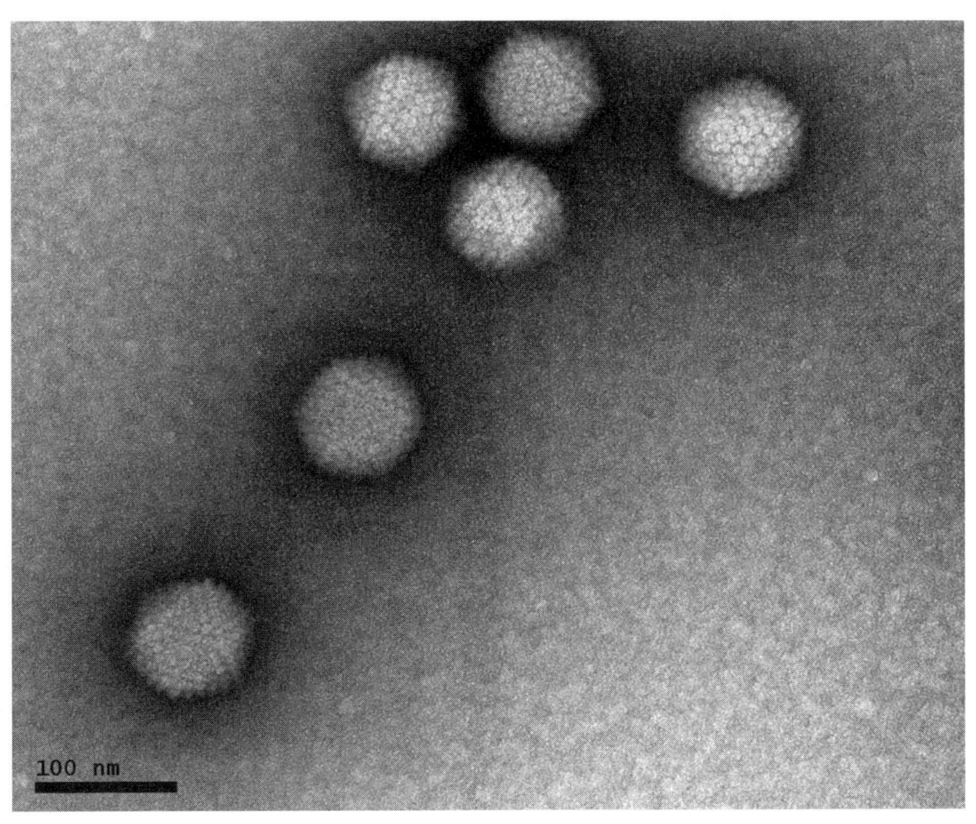

Figure 2. Transmission electron microscopy of AdV41 negatively stained with uranyl acetate. (Photograph courtesy of Karin Boucke, University of Zürich.)

the E1A and N-terminal regions of the hexon gene are well conserved between serotypes, PCR assays with primers directed at these regions are commonly used (87, 91). Unfortunately, there is still significant genetic heterogeneity in the adenovirus genome (>80% sequence dissimilarity between some species) (86). This presents challenges in designing a single robust assay to detect and quantify all types. Numerous traditional and quantitative real-time PCR techniques have been described, often using nested primers or multiplex methods to detect many or all virus types (15, 18, 20, 32, 42, 46, 52, 58, 63, 66, 71, 75, 77, 78, 80, 90, 104). Although false-negative results have been described (43), these PCR methods generally have the advantage of high sensitivity and, in some cases, the ability to provide quantitative data or to differentiate specific subspecies (15, 18, 20, 32, 42, 45, 52, 58, 63, 66, 71, 75, 77, 78, 80, 90). PCR applied to whole blood has emerged as a significant screening method with a proven impact on pediatric HSCT recipients (14). The specific blood compartment used for testing has varied widely, with different groups using whole blood, peripheral blood mononuclear cells, or plasma. There are few direct comparative data available, but a recent study showed similar sensitivities among the three sample types, with peripheral blood mononuclear cell testing yielding reduced viral loads compared to the other two compartments (J. Perlman, personal communication). Even less has been published regarding other (nonblood) specimens, with the notable exception of PCR for detection of adenovirus in respiratory tract specimens (71, 78, 80, 81, 83, 104). The use of PCR and other molecular amplification assays has increased markedly in recent years. As noted elsewhere in this chapter, the evaluation of PCR results must be interpreted with caution, and there is much yet to be achieved in aspects of testing related to standardizing methods and quantitative controls. In addition, some assays that claim to detect and quantify all serotypes of adenovirus may not in fact do so. It is important, particularly in the case of such a genetically heterogeneous group of viruses, that the performance characteristics of each method be verified thoroughly against all expected targets. Other factors, including collection, preservation, and preparation of samples prior to testing, may also significantly affect results.

Serotyping

Although not currently in wide clinical use, specific subgroups and serotypes of adenovirus can be used to predict oncogenic potential, clinical disease, and susceptibility to antiviral agents (Table 1). Species F viruses, for example, almost exclusively cause GI disease, have specific growth characteristics, and are resistant to ribavirin (70, 86). Classification of identified viruses into individual species (A to F) and serotypes (1 to 51) can be accomplished by traditional serotyping, restriction endonuclease digestion of viral DNA, or molecular techniques. Serotyping by neutralization or hemagglutination inhibition assays using reference sera is losing favor (33, 82). Classical methods are labor- and time-intensive and require specialized reagents that are in short supply; as a result, only a few state labs and the CDC commonly perform traditional serotyping. Type-specific monoclonal antibodies have been developed for a few adenovirus serotypes, but they are not widely available (108). Restriction fragment length polymorphism analysis is also labor-intensive and may be difficult to interpret for some viruses (1). Numerous PCR assays have been developed based on species- and type-specific primers targeting the hexon, fiber, and virus-associated RNA coding regions; unfortunately, reference sequences are not readily available for all recognized adenoviruses (86). A more adaptable method focused on PCR and sequencing of hexon gene hypervariable regions 1 to 6 (HVR_{1-6}) and HVR_7 has been developed as well (61).

Antiviral Susceptibility Testing

Few labs are able to perform antiviral susceptibility testing. There are no standard methods that have been agreed upon universally. An in vitro susceptibility assay has been developed in Europe and approximates the methods used in prior susceptibility testing (71). The use of susceptibility testing in the context of clinical care has not been studied.

RESULT INTERPRETATION

Interpreting the results of adenovirus testing requires careful correlation of results with the clinical situation to rule out other potential pathogens or comorbid conditions that may confound efforts to determine disease causality. As described above, shedding of virus from some sites may be prolonged and may represent either persistent active infection or asymptomatic latent virus. Despite a growing body of published work on the subject, interpretation of positive test results can be difficult, particularly for highly sensitive nucleic acid amplification techniques (42, 45). In general, while a positive test for adenovirus in the appropriate clinical setting may be considered diagnostic of adenovirus infection, pathologic evidence of invasive disease is still required to definitively prove causality. Unfortunately, patients with altered immune systems often have multiple concomitant problems, so coinfections must also be sought. Management of immunocompromised patients with adenovirus infections can be complex and should be done in consultation with an expert in the field (14, 45, 57) and with communication among multiple laboratory subspecialties. In summary, definitive diagnosis typically combines pathologic evidence of invasive adenoviral disease with positive culture or PCR from the site of infection. The interpretation of PCR-based data has become an important issue in the clinical care of immunocompromised patients with suspected systemic adenoviral infection and will constitute the balance of this discussion.

Quantitative viral load measurements can contribute to the diagnosis of infection and act as surrogates that correlate with the clinical response to therapy (52, 58). The lack of assay standardization remains an important limitation, as most molecular testing for adenovirus is performed with user-developed assays. Furthermore, no universal quantitative standards exist to allow normalization between different tests. As such, caution must be exercised in comparing values from different laboratories, who often use different techniques on different specimen types. These issues should be considered not only in interpreting individual patient results but also in reviewing the literature. In part as a result of this lack of standardization, there are currently no well-defined diagnostic or treatment thresholds. In the appropriate clinical setting, such as fever and evidence of end organ disease, a high viral load from appropriately collected samples is highly suggestive of invasive adenoviral infection. In general, larger viral loads have a stronger association with disease (19). However, it is critical that interpretation of actual values depends on the assay used and the blood compartment or other sample type tested. Perhaps more importantly, the trend of viral loads over time provides insight into the risk of disease—persistent or rising viral replication is suggestive of progressive or disseminated disease (14).

PCR, particularly with blood samples, may be an effective screening modality to identify asymptomatic patients at risk for progressive adenovirus-associated disease. However, the value of such surveillance has been demonstrated clearly only among pediatric HSCT recipients (14, 29, 109). In this population, adenoviremia, but not detectable virus from the urine, throat, and stool, has been associated with progression to invasive disease and with increased mortality (19, 60). In most patients, detectable adenoviremia preceded the onset of clinical symptoms (60). Persistent adenoviremia in pediatric HSCT recipients has been linked to disease progression, especially in the presence of high viral load, lymphopenia, or continued immunosuppression (14). Routine screening of peripheral blood by PCR and of end organ sites of infection (stool, throat, and urine) by culture or PCR has been suggested for pediatric HSCT patients (14). Patients would be treated with antiviral therapy in the setting of two consecutive PCR-positive blood samples or any positive end organ result together with either severe lymphopenia or an inability to reduce immunosuppression.

Adenoviremia does not appear to be predictive of disease or complications among adult SOT recipients. In prospective studies, adenoviremia was detected in 6.5%, 6.7%, 8.3%, and 22.5% of adult kidney, heart, liver, and lung recipients, respectively (41, 42). Few of the patients had symptoms at the time that viremia was detected; viral loads were, in general, low, and none of the patients developed end organ disease. There was no compromise of pulmonary function among lung transplant recipients (41, 42). While there are currently no published studies showing prospective data on screening among adult stem cell transplant recipients (45), research in this area is ongoing.

An important use of PCR is the monitoring of therapeutic responses (45, 52, 53, 57, 58). Often, patients with severe adenovirus infections have comorbidities, complicating the determination of response to therapy based on clinical signs and symptoms. Dynamic trends in viral load measurements can help to overcome such obstacles. Two recent studies have provided insight regarding the optimal management of adenovirus infection among HSCT recipients and have demonstrated the clinical course of disease during treatment. In both instances, serial samples of blood were collected for quantitative viral load testing (52, 58). In the first study, four patients with adenovirus infection were treated with ribavirin (52); three of the patients had C species viruses, for which ribavirin has been shown to be active in vitro (70). None of the patients had a decline in viral load, and all showed continued progressive clinical disease (52). In the second study, eight patients with adenovirus disease with adenoviremia were treated with cidofovir (58), which appears to have in vitro activity against all tested adenoviruses (70). In this case series, five of the eight patients had reductions in viral load, which were associated with a clinical response; the remaining three had no improvement in viral load and had progressive clinical disease (58). A lack of reduction of viral load following the first two doses of cidofovir was predictive of a progressive clinical course (58). Based on these findings, some experts suggest that all patients with adenoviremia who are on antiviral therapy should have serial viral load testing to monitor the response to therapy.

PROGNOSTICATION

As described above, detection of adenovirus DNA in the peripheral blood of pediatric HSCT recipients may predict the onset of adenovirus disease (Fig. 3) (45). Similarly, sequential quantitative monitoring of adenoviremia may help to indicate the likelihood of therapeutic response; most patients who will respond to therapy will do so within the first two doses of antiviral agent (58). Viral load may also be useful for determining prognosis, as higher values ($>1 \times 10^6$ copies/ml in one study) have been associated with a greater likelihood of death among pediatric transplant recipients (15, 90).

Detection of the adenoviral genome in myocardial biopsy specimens posttransplantation may also be predictive of adverse clinical outcomes, including coronary vasculopathy and graft loss (odds ratio, 4.7 for comparison to adenovirus-negative patients; 95% confidence interval, 1.3 to 17.1), among cardiac transplant recipients (94). A more recent study found that 11% of donor hearts had evidence of latent adenovirus by PCR, suggesting that donor-derived infections are common (17). Still lacking is a prospective study to determine whether adenovirus infection of the donor heart prior to transplantation is associated with graft loss or other preventable outcomes.

APPLICATION OF DIAGNOSTIC PROCEDURES

One must consider adenovirus when immunocompromised patients present with any clinical syndrome that could be compatible with adenoviral infection. Once disease is suspected, cultures of appropriate specimens should be obtained (Fig. 4) and blood should be sent for adenovirus PCR. PCR testing of appropriate specimen types for the presenting symptoms should be considered. There is no consensus on when to start antiviral therapy, although data suggest that institution of therapy earlier in the disease course may prevent dissemination.

CONCLUSION

Adenovirus is associated with a wide spectrum of clinical disease in immunosuppressed patients, with sig-

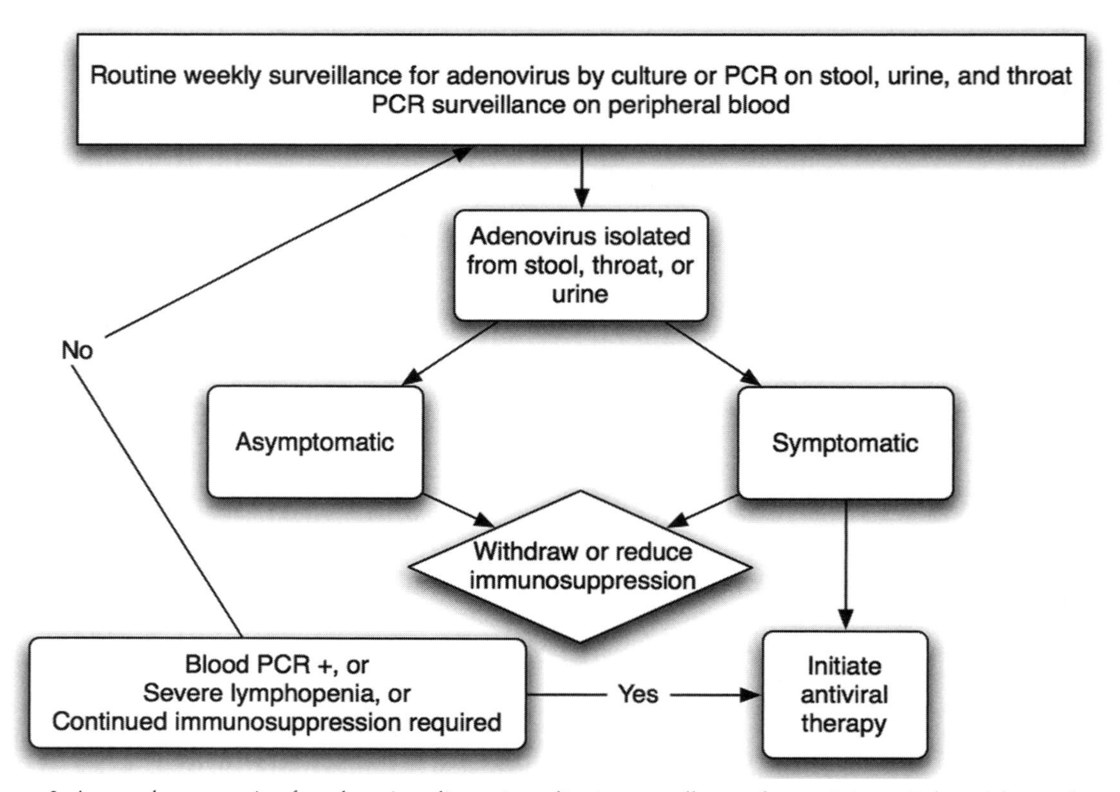

Figure 3. Approach to screening for adenovirus disease in pediatric stem cell transplant recipients. (Adapted from reference 14 with permission of the publisher.)

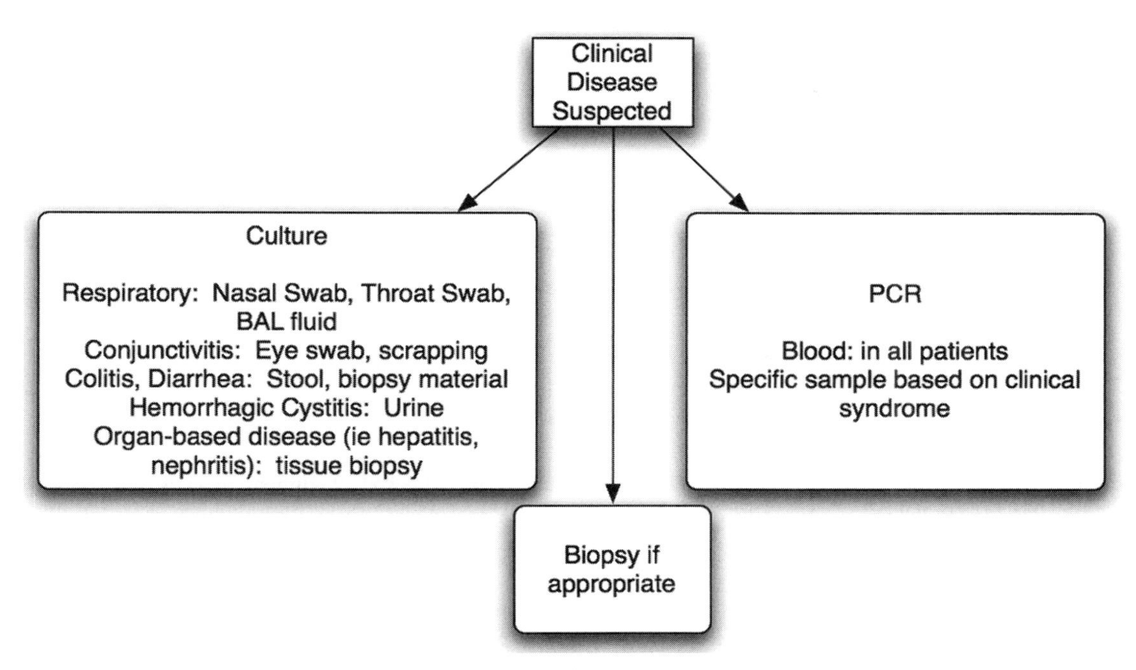

Figure 4. Approach to diagnosis of suspected adenovirus disease (45).

nificant attributable morbidity and mortality. Nucleic acid testing, in particular, is emerging as a primary diagnostic modality, primarily as a surveillance tool to enable preemptive antiviral therapy, and as a critical means of guiding patient management and treatment with antiviral agents. Management of patients with proven adenovirus disease has not been studied prospectively, and the care of such individuals, where possible, should be done in consultation with an expert in infectious diseases of immunocompromised hosts.

REFERENCES

1. **Adrian, T., G. Wadell, J. C. Hierholzer, and R. Wigand.** 1986. DNA restriction analysis of adenovirus prototypes 1 to 41. *Arch. Virol.* **91:**277–290.
2. **Akiyama, H., T. Kurosu, C. Sakashita, T. Inoue, S. Mori, K. Ohashi, S. Tanikawa, H. Sakamaki, Y. Onozawa, Q. Chen, H. Zheng, and T. Kitamura.** 2001. Adenovirus is a key pathogen in hemorrhagic cystitis associated with bone marrow transplantation. *Clin. Infect. Dis.* **32:**1325–1330.
3. **August, M. J., and A. L. Warford.** 1987. Evaluation of a commercial monoclonal antibody for detection of adenovirus antigen. *J. Clin. Microbiol.* **25:**2233–2235.
4. **Baldwin, A., H. Kingman, M. Darville, A. B. Foot, D. Grier, J. M. Cornish, N. Goulden, A. Oakhill, D. H. Pamphilon, C. G. Steward, and D. I. Marks.** 2000. Outcome and clinical course of 100 patients with adenovirus infection following bone marrow transplantation. *Bone Marrow Transplant.* **26:**1333–1338.
5. **Barenfanger, J., C. Drake, N. Leon, T. Mueller, and T. Troutt.** 2000. Clinical and financial benefits of rapid detection of respiratory viruses: an outcomes study. *J. Clin. Microbiol.* **38:**2824–2828.
6. **Bateman, E. D., S. Hayashi, K. Kuwano, T. A. Wilke, and J. C. Hogg.** 1995. Latent adenoviral infection in follicular bronchiectasis. *Am. J. Respir. Crit. Care Med.* **151:**170–176.
7. **Berho, M., M. Torroella, A. Viciana, D. Weppler, J. Thompson, J. Nery, A. Tzakis, and P. Ruiz.** 1998. Adenovirus enterocolitis in human small bowel transplants. *Pediatr. Transplant.* **2:**277–282.
8. **Bordigoni, P., A. S. Carret, V. Venard, F. Witz, and A. Le Faou.** 2001. Treatment of adenovirus infections in patients undergoing allogeneic hematopoietic stem cell transplantation. *Clin. Infect. Dis.* **32:**1290–1297.
9. **Bridges, N. D., T. L. Spray, M. H. Collins, N. E. Bowles, and J. A. Towbin.** 1998. Adenovirus infection in the lung results in graft failure after lung transplantation. *J. Thorac. Cardiovasc. Surg.* **116:**617–623.
10. **Bruno, B., T. Gooley, R. C. Hackman, C. Davis, L. Corey, and M. Boeckh.** 2003. Adenovirus infection in hematopoietic stem cell transplantation: effect of ganciclovir and impact on survival. *Biol. Blood Marrow Transplant.* **9:**341–352.
11. **Carter, B. A., S. J. Karpen, R. E. Quiros-Tejeira, I. F. Chang, B. S. Clark, G. J. Demmler, H. E. Heslop, J. D. Scott, P. Seu, and J. A. Goss.** 2002. Intravenous cidofovir therapy for disseminated adenovirus in a pediatric liver transplant recipient. *Transplantation* **74:**1050–1052.
12. **Chakrabarti, S., K. E. Collingham, C. D. Fegan, D. Pillay, and D. W. Milligan.** 2000. Adenovirus infections following haematopoietic cell transplantation: is there a role for adoptive immunotherapy? *Bone Marrow Transplant.* **26:**305–307.
13. **Chakrabarti, S., V. Mautner, H. Osman, K. E. Collingham, C. D. Fegan, P. E. Klapper, P. A. Moss, and D. W. Milligan.** 2002. Adenovirus infections following allogeneic stem cell transplantation: incidence and outcome in relation to graft manipulation, immunosuppression, and immune recovery. *Blood* **100:**1619–1627.
14. **Chakrabarti, S., D. W. Milligan, P. A. Moss, and V. Mautner.** 2004. Adenovirus infections in stem cell transplant recipients: recent developments in understanding of pathogenesis, diagnosis and management. *Leuk. Lymphoma* **45:**873–885.

15. **Claas, E. C., M. W. Schilham, C. S. de Brouwer, P. Hubacek, M. Echavarria, A. C. Lankester, M. J. van Tol, and A. C. Kroes.** 2005. Internally controlled real-time PCR monitoring of adenovirus DNA load in serum or plasma of transplant recipients. *J. Clin. Microbiol.* **43:**1738–1744.
16. **Conner, D. H., F. W. Chandler, D. A. Schwartz, H. J. Manz, and E. E. Lack.** 1997. *Pathology of Infectious Diseases.* Appleton & Lange, Stamford, CT.
17. **Donoso Mantke, O., R. Meyer, S. Prosch, A. Nitsche, K. Leitmeyer, R. Kallies, and M. Niedrig.** 2005. High prevalence of cardiotropic viruses in myocardial tissue from explanted hearts of heart transplant recipients and heart donors: a 3-year retrospective study from a German patients' pool. *J. Heart Lung Transplant.* **24:**1632–1638.
18. **Ebner, K., M. Suda, F. Watzinger, and T. Lion.** 2005. Molecular detection and quantitative analysis of the entire spectrum of human adenoviruses by a two-reaction real-time PCR assay. *J. Clin. Microbiol.* **43:**3049–3053.
19. **Echavarria, M., M. Forman, M. J. van Tol, J. M. Vossen, P. Charache, and A. C. Kroes.** 2001. Prediction of severe disseminated adenovirus infection by serum PCR. *Lancet* **358:** 384–385.
20. **Echavarria, M. S., S. C. Ray, R. Ambinder, J. S. Dumler, and P. Charache.** 1999. PCR detection of adenovirus in a bone marrow transplant recipient: hemorrhagic cystitis as a presenting manifestation of disseminated disease. *J. Clin. Microbiol.* **37:**686–689.
21. **Espy, M. J., J. C. Hierholzer, and T. F. Smith.** 1987. The effect of centrifugation on the rapid detection of adenovirus in shell vials. *Am. J. Clin. Pathol.* **88:**358–360.
22. **Feuchtinger, T., S. Matthes-Martin, C. Richard, T. Lion, M. Fuhrer, K. Hamprecht, R. Handgretinger, C. Peters, F. R. Schuster, R. Beck, M. Schumm, R. Lotfi, G. Jahn, and P. Lang.** 2006. Safe adoptive transfer of virus-specific T-cell immunity for the treatment of systemic adenovirus infection after allogeneic stem cell transplantation. *Br. J. Haematol.* **134:**64–76.
23. **Flomenberg, P., J. Babbitt, W. R. Drobyski, R. C. Ash, D. R. Carrigan, G. V. Sedmak, T. McAuliffe, B. Camitta, M. M. Horowitz, N. Bunin, et al.** 1994. Increasing incidence of adenovirus disease in bone marrow transplant recipients. *J. Infect. Dis.* **169:**775–781.
24. **Friedrichs, N., A. M. Eis-Hubinger, A. Heim, E. Platen, H. Zhou, and R. Buettner.** 2003. Acute adenoviral infection of a graft by serotype 35 following renal transplantation. *Pathol. Res. Pract.* **199:**565–570.
25. **Garbino, J., M. W. Gerbase, W. Wunderli, L. Kolarova, L. P. Nicod, T. Rochat, and L. Kaiser.** 2004. Respiratory viruses and severe lower respiratory tract complications in hospitalized patients. *Chest* **125:**1033–1039.
26. **Gavel, G., S. Marven, M. J. Evans, J. Walker, and A. J. Vora.** 2003. Obliterative enteritis complicating graft versus host disease. *Bone Marrow Transplant.* **32:**1097–1098.
27. **Grandien, M., C. A. Pettersson, L. Svensson, and I. Uhnoo.** 1987. Latex agglutination test for adenovirus diagnosis in diarrheal disease. *J. Med. Virol.* **23:**311–316.
28. **Green, M., P. Ljungman, and M. G. Michaels.** 2003. Adenovirus, parvovirus B19, papilloma virus, and polyomaviruses after hemopoietic stem cell or solid organ transplantation, p. 412–438. *In* R. A. Bowden, P. Ljungman, and C. V. Paya (ed.), *Transplant Infections.* Lippincott Williams & Wilkins, Philadelphia, PA.
29. **Gu, Z., S. W. Belzer, C. S. Gibson, M. J. Bankowski, and R. T. Hayden.** 2003. Multiplexed, real-time PCR for quantitative detection of human adenovirus. *J. Clin. Microbiol.* **41:**4636–4641.
30. **Hale, G. A., H. E. Heslop, R. A. Krance, M. A. Brenner, D. Jayawardene, D. K. Srivastava, and C. C. Patrick.** 1999. Adenovirus infection after pediatric bone marrow transplantation. *Bone Marrow Transplant.* **23:**277–282.
31. **Hatakeyama, N., N. Suzuki, T. Kudoh, T. Hori, N. Mizue, and H. Tsutsumi.** 2003. Successful cidofovir treatment of adenovirus-associated hemorrhagic cystitis and renal dysfunction after allogenic bone marrow transplant. *Pediatr. Infect. Dis. J.* **22:**928–929.
32. **Heim, A., C. Ebnet, G. Harste, and P. Pring-Akerblom.** 2003. Rapid and quantitative detection of human adenovirus DNA by real-time PCR. *J. Med. Virol.* **70:**228–239.
33. **Heirholzer, J.** 1995. Adenoviruses, p. 169–188. *In* E. H. Lennette, D. A. Lennette, and E. T. Lennette (ed.), *Diagnostic Procedures for Viral, Rickettsial, and Chlamydial Infections.* American Public Health Association, Washington, DC.
34. **Hierholzer, J. C., K. H. Johansson, L. J. Anderson, C. J. Tsou, and P. E. Halonen.** 1987. Comparison of monoclonal time-resolved fluoroimmunoassay with monoclonal capture-biotinylated detector enzyme immunoassay for adenovirus antigen detection. *J. Clin. Microbiol.* **25:**1662–1667.
35. **Hoffman, J. A.** 2006. Adenoviral disease in pediatric solid organ transplant recipients. *Pediatr. Transplant.* **10:**17–25.
36. **Hoffman, J. A., A. J. Shah, L. A. Ross, and N. Kapoor.** 2001. Adenoviral infections and a prospective trial of cidofovir in pediatric hematopoietic stem cell transplantation. *Biol. Blood Marrow Transplant.* **7:**388–394.
37. **Horwitz, M. S.** 2001. Adenoviruses, p. 2313–2316. *In* D. M. Knipe, P. M. Howley, D. E. Griffin, R. A. Lamb, M. A. Martin, B. Roizman, and S. E. Straus (ed.), *Fields Virology.* Lippincott Williams & Wilkins, Philadelphia, PA.
38. **Hoshino, T., T. Takanashi, M. Okada, and S. Uchida.** 2002. Oxybuprocaine induces a false-positive response in immunochromatographic SAS Adeno test. *Ophthalmology* **109:** 808–809.
39. **Howard, D. S., I. G. Phillips, D. E. Reece, R. K. Munn, J. Henslee-Downey, M. Pittard, M. Barker, and C. Pomeroy.** 1999. Adenovirus infections in hematopoietic stem cell transplant recipients. *Clin. Infect. Dis.* **29:**1494–1501.
40. **Hromas, R., K. Cornetta, E. Srour, C. Blanke, and E. R. Broun.** 1994. Donor leukocyte infusion as therapy of life-threatening adenoviral infections after T-cell-depleted bone marrow transplantation. *Blood* **84:**1689–1690.
41. **Humar, A., K. Doucette, D. Kumar, X. L. Pang, D. Lien, K. Jackson, and J. Preiksaitis.** 2006. Assessment of adenovirus infection in adult lung transplant recipients using molecular surveillance. *J. Heart Lung Transplant.* **25:**1441–1446.
42. **Humar, A., D. Kumar, T. Mazzulli, R. R. Razonable, G. Moussa, C. V. Paya, E. Covington, E. Alecock, and M. D. Pescovitz.** 2005. A surveillance study of adenovirus infection in adult solid organ transplant recipients. *Am. J. Transplant.* **5:**2555–2559.
43. **Ishida, H., N. Shin-Nakai, T. Yoshihara, T. Teramura, T. Imamura, A. Mukouyama, A. Morimoto, and S. Imashuku.** 2005. Invasive adenoviral infection in a recipient of unrelated bone marrow transplantation: problems with diagnostic PCR. *Pediatr. Blood Cancer* **44:**416–419.
44. **Ishii, T., G. V. Mazariegos, J. Bueno, S. Ohwada, and J. Reyes.** 2003. Exfoliative rejection after intestinal transplantation in children. *Pediatr. Transplant.* **7:**185–191.
45. **Ison, M. G.** 2006. Adenovirus infections in transplant recipients. *Clin. Infect. Dis.* **43:**331–339.
46. **Kohdera, U., M. Kino, and M. Ito.** 2006. Detection of adenovirus DNA in throat swabs and blood by SYBR green real-

time PCR assay in patients with adenovirus-associated tonsillitis. *Jpn. J. Infect. Dis.* **59**:394–396.
47. **Kojaoghlanian, T., P. Flomenberg, and M. S. Horwitz.** 2003. The impact of adenovirus infection on the immunocompromised host. *Rev. Med. Virol.* **13**:155–171.
48. **Koneru, B., R. Jaffe, C. O. Esquivel, R. Kunz, S. Todo, S. Iwatsuki, and T. E. Starzl.** 1987. Adenoviral infections in pediatric liver transplant recipients. *JAMA* **258**:489–492.
49. **Krisher, K. K., and M. A. Menegus.** 1987. Evaluation of three types of cell culture for recovery of adenovirus from clinical specimens. *J. Clin. Microbiol.* **25**:1323–1324.
50. **Landry, M. L., and D. Ferguson.** 2000. SimulFluor respiratory screen for rapid detection of multiple respiratory viruses in clinical specimens by immunofluorescence staining. *J. Clin. Microbiol.* **38**:708–711.
51. **Landry, M. L., C. K. Fong, K. Neddermann, L. Solomon, and G. D. Hsiung.** 1987. Disseminated adenovirus infection in an immunocompromised host. Pitfalls in diagnosis. *Am. J. Med.* **83**:555–559.
52. **Lankester, A. C., B. Heemskerk, E. C. Claas, M. W. Schilham, M. F. Beersma, R. G. Bredius, M. J. van Tol, and A. C. Kroes.** 2004. Effect of ribavirin on the plasma viral DNA load in patients with disseminating adenovirus infection. *Clin. Infect. Dis.* **38**:1521–1525.
53. **Lankester, A. C., M. J. van Tol, E. C. Claas, J. M. Vossen, and A. C. Kroes.** 2002. Quantification of adenovirus DNA in plasma for management of infection in stem cell graft recipients. *Clin. Infect. Dis.* **34**:864–867.
54. **La Rosa, A. M., R. E. Champlin, N. Mirza, J. Gajewski, S. Giralt, K. V. Rolston, I. Raad, K. Jacobson, D. Kontoyiannis, L. Elting, and E. Whimbey.** 2001. Adenovirus infections in adult recipients of blood and marrow transplants. *Clin. Infect. Dis.* **32**:871–876.
55. **Leen, A. M., C. M. Bollard, G. D. Myers, and C. M. Rooney.** 2006. Adenoviral infections in hematopoietic stem cell transplantation. *Biol. Blood Marrow Transplant.* **12**:243–251.
56. **Lehtomaki, K., I. Julkunen, K. Sandelin, J. Salonen, M. Virtanen, M. Ranki, and T. Hovi.** 1986. Rapid diagnosis of respiratory adenovirus infections in young adult men. *J. Clin. Microbiol.* **24**:108–111.
57. **Lenaerts, L., and L. Naesens.** 2006. Antiviral therapy for adenovirus infections. *Antivir. Res.* **71**:172–180.
58. **Leruez-Ville, M., V. Minard, F. Lacaille, A. Buzyn, E. Abachin, S. Blanche, F. Freymuth, and C. Rouzioux.** 2004. Real-time blood plasma polymerase chain reaction for management of disseminated adenovirus infection. *Clin. Infect. Dis.* **38**:45–52.
59. **Levy, M. F., J. S. Crippin, O. Abbasoglu, M. Vodapally, R. M. Goldstein, B. S. Husberg, T. A. Gonwa, and G. B. Klintmalm.** 1996. Adenovirus infection of the human intestinal allograft: a case report. *Transplant. Proc.* **28**:2786–2787.
60. **Lion, T., R. Baumgartinger, F. Watzinger, S. Matthes-Martin, M. Suda, S. Preuner, B. Futterknecht, A. Lawitschka, C. Peters, U. Potschger, and H. Gadner.** 2003. Molecular monitoring of adenovirus in peripheral blood after allogeneic bone marrow transplantation permits early diagnosis of disseminated disease. *Blood* **102**:1114–1120.
61. **Lu, X., and D. D. Erdman.** 2006. Molecular typing of human adenoviruses by PCR and sequencing of a partial region of the hexon gene. *Arch. Virol.* **151**:1587–1602.
62. **Mahafzah, A. M., and M. L. Landry.** 1989. Evaluation of immunofluorescent reagents, centrifugation, and conventional cultures for the diagnosis of adenovirus infection. *Diagn. Microbiol. Infect. Dis.* **12**:407–411.
63. **Matsuse, T., H. Matsui, C. Y. Shu, T. Nagase, T. Wakabayashi, S. Mori, S. Inoue, Y. Fukuchi, and H. Orimo.** 1994. Adenovirus pulmonary infections identified by PCR and in situ hybridisation in bone marrow transplant recipients. *J. Clin. Pathol.* **47**:973–977.
64. **McGrath, D., M. E. Falagas, R. Freeman, R. Rohrer, R. Fairchild, C. Colbach, and D. R. Snydman.** 1998. Adenovirus infection in adult orthotopic liver transplant recipients: incidence and clinical significance. *J. Infect. Dis.* **177**:459–462.
65. **McLaughlin, G. E., S. Delis, L. Kashimawo, G. P. Cantwell, N. Mittal, R. E. Cirocco, P. Ruiz, T. Kato, and A. G. Tzakis.** 2003. Adenovirus infection in pediatric liver and intestinal transplant recipients: utility of DNA detection by PCR. *Am. J. Transplant.* **3**:224–228.
66. **Metzger-Boddien, C., and J. Kehle.** 2005. Development and evaluation of a sensitive PCR-ELISA for detection of adenoviruses in feces. *Intervirology* **48**:297–300.
67. **Michaels, M. G., M. Green, E. R. Wald, and T. E. Starzl.** 1992. Adenovirus infection in pediatric liver transplant recipients. *J. Infect. Dis.* **165**:170–174.
68. **Moore, P. L., A. D. Steele, and J. J. Alexander.** 2000. Relevance of commercial diagnostic tests to detection of enteric adenovirus infections in South Africa. *J. Clin. Microbiol.* **38**:1661–1663.
69. **Morfin, F., A. Boucher, F. Najioullah, Y. Bertrand, N. Bleyzac, F. Poitevin-Later, F. Bienvenu, V. Simonet, C. Galambrun, N. Philippe, M. Aymard, D. Thouvenot, and G. Souillet.** 2004. Cytomegalovirus and adenovirus infections and diseases among 75 paediatric unrelated allogeneic bone marrow transplant recipients. *J. Med. Virol.* **72**:257–262.
70. **Morfin, F., S. Dupuis-Girod, S. Mundweiler, D. Falcon, D. Carrington, P. Sedlacek, M. Bierings, P. Cetkovsky, A. C. Kroes, M. J. van Tol, and D. Thouvenot.** 2005. In vitro susceptibility of adenovirus to antiviral drugs is species-dependent. *Antivir. Ther.* **10**:225–229.
71. **Morris, D. J., R. J. Cooper, T. Barr, and A. S. Bailey.** 1996. Polymerase chain reaction for rapid diagnosis of respiratory adenovirus infection. *J. Infect.* **32**:113–117.
72. **Ozolek, J. A., K. Cieply, J. Walpusk, and S. Ranganathan.** 2006. Adenovirus infection within stromal cells in a pediatric small bowel allograft. *Pediatr. Dev. Pathol.* **9**:321–327.
73. **Parizhskaya, M., J. Walpusk, G. Mazariegos, and R. Jaffe.** 2001. Enteric adenovirus infection in pediatric small bowel transplant recipients. *Pediatr. Dev. Pathol.* **4**:122–128.
74. **Pinchoff, R. J., S. S. Kaufman, M. S. Magid, D. D. Erdman, G. E. Gondolesi, M. H. Mendelson, K. Tane, S. G. Jenkins, T. M. Fishbein, and B. C. Herold.** 2003. Adenovirus infection in pediatric small bowel transplantation recipients. *Transplantation* **76**:183–189.
75. **Pring-Åkerblom, P., F. E. J. Trijssenaar, T. Adrian, and H. Hoyer.** 1999. Multiplex polymerase chain reaction for subgenus-specific detection of human adenoviruses in clinical samples. *J. Med. Virol.* **58**:87–92.
76. **Rabalais, G. P., G. G. Stout, K. L. Ladd, and K. M. Cost.** 1992. Rapid diagnosis of respiratory viral infections by using a shell vial assay and monoclonal antibody pool. *J. Clin. Microbiol.* **30**:1505–1508.
77. **Raboni, S. M., M. M. Siqueira, S. R. Portes, and R. Pasquini.** 2003. Comparison of PCR, enzyme immunoassay and conventional culture for adenovirus detection in bone marrow transplant patients with hemorrhagic cystitis. *J. Clin. Virol.* **27**:270–275.
78. **Raty, R., M. Kleemola, K. Melen, M. Stenvik, and I. Julkunen.** 1999. Efficacy of PCR and other diagnostic methods for the detection of respiratory adenoviral infections. *J. Med. Virol.* **59**:66–72.
79. **Ray, C. G., and L. L. Minnich.** 1987. Efficiency of immunofluorescence for rapid detection of common respiratory viruses. *J. Clin. Microbiol.* **25**:355–357.

80. **Rector, A., N. Azzi, C. Liesnard, K. Zlateva, D. Van Beers, R. Snoeck, and M. Van Ranst.** 2002. Use of polymerase chain reaction for diagnosis of disseminated adenovirus infection. *Pediatr. Infect. Dis. J.* **21:**1176–1178.
81. **Roghmann, M., K. Ball, D. Erdman, J. Lovchik, L. J. Anderson, and R. Edelman.** 2003. Active surveillance for respiratory virus infections in adults who have undergone bone marrow and peripheral blood stem cell transplantation. *Bone Marrow Transplant.* **32:**1085–1088.
82. **Rosen, L.** 1960. A hemagglutination-inhibition technique for typing adenoviruses. *Am. J. Hyg.* **71:**120–128.
83. **Rovida, F., E. Percivalle, M. Zavattoni, M. Torsellini, A. Sarasini, G. Campanini, S. Paolucci, F. Baldanti, M. G. Revello, and G. Gerna.** 2005. Monoclonal antibodies versus reverse transcription-PCR for detection of respiratory viruses in a patient population with respiratory tract infections admitted to hospital. *J. Med. Virol.* **75:**336–347.
84. **Runde, V., S. Ross, R. Trenschel, E. Lagemann, O. Basu, K. Renzing-Kohler, U. W. Schaefer, M. Roggendorf, and E. Holler.** 2001. Adenoviral infection after allogeneic stem cell transplantation (SCT): report on 130 patients from a single SCT unit involved in a prospective multi center surveillance study. *Bone Marrow Transplant.* **28:**51–57.
85. **Russell, K. L., M. P. Broderick, S. E. Franklin, L. B. Blyn, N. E. Freed, E. Moradi, D. J. Ecker, P. E. Kammerer, M. A. Osuna, A. E. Kajon, C. B. Morn, and M. A. Ryan.** 2006. Transmission dynamics and prospective environmental sampling of adenovirus in a military recruit setting. *J. Infect. Dis.* **194:**877–885.
86. **Ruuskanen, O., O. Meurman, and G. Akusjärvi.** 2002. Adenoviruses, p. 515–535. *In* D. D. Richman, R. J. Whitley, and F. G. Hayden (ed.), *Clinical Virology*. ASM Press, Washington, DC.
87. **Rux, J. J., P. R. Kuser, and R. M. Burnett.** 2003. Structural and phylogenetic analysis of adenovirus hexons by use of high-resolution X-ray crystallographic, molecular modeling, and sequence-based methods. *J. Virol.* **77:**9553–9566.
88. **Saad, R. S., A. J. Demetris, R. G. Lee, S. Kusne, and P. S. Randhawa.** 1997. Adenovirus hepatitis in the adult allograft liver. *Transplantation* **64:**1483–1485.
89. **Sambursky, R., S. Tauber, F. Schirra, K. Kozich, R. Davidson, and E. J. Cohen.** 2006. The RPS adeno detector for diagnosing adenoviral conjunctivitis. *Ophthalmology* **113:**1758–1764.
90. **Schilham, M. W., E. C. Claas, W. van Zaane, B. Heemskerk, J. M. Vossen, A. C. Lankester, R. E. Toes, M. Echavarria, A. C. Kroes, and M. J. van Tol.** 2002. High levels of adenovirus DNA in serum correlate with fatal outcome of adenovirus infection in children after allogeneic stem-cell transplantation. *Clin. Infect. Dis.* **35:**526–532.
91. **Scott-Taylor, T. H., and G. W. Hammond.** 1992. Conserved sequences of the adenovirus genome for detection of all human adenovirus types by hybridization. *J. Clin. Microbiol.* **30:**1703–1710.
92. **Seidemann, K., A. Heim, E. D. Pfister, H. Koditz, A. Beilken, A. Sander, M. Melter, K. W. Sykora, M. Sasse, and A. Wessel.** 2004. Monitoring of adenovirus infection in pediatric transplant recipients by quantitative PCR: report of six cases and review of the literature. *Am. J. Transplant.* **4:**2102–2108.
93. **Shields, A. F., R. C. Hackman, K. H. Fife, L. Corey, and J. D. Meyers.** 1985. Adenovirus infections in patients undergoing bone-marrow transplantation. *N. Engl. J. Med.* **312:**529–533.
94. **Shirali, G. S., J. Ni, R. E. Chinnock, J. K. Johnston, G. L. Rosenthal, N. E. Bowles, and J. A. Towbin.** 2001. Association of viral genome with graft loss in children after cardiac transplantation. *N. Engl. J. Med.* **344:**1498–1503.
95. **Singhal, S., D. A. Muir, D. A. Ratcliffe, J. A. Shirley, P. A. Cane, J. G. Hastings, D. Pillay, and D. J. Mutimer.** 1999. Respiratory viruses in adult liver transplant recipients. *Transplantation* **68:**981–984.
96. **Strickler, J. G., T. P. Singleton, C. M. Copenhaver, A. Erice, and D. C. Snover.** 1992. Adenovirus in the gastrointestinal tracts of immunosuppressed patients. *Am. J. Clin. Pathol.* **97:**555–558.
97. **Takiff, H. E., and S. E. Straus.** 1982. Early replicative block prevents the efficient growth of fastidious diarrhea-associated adenoviruses in cell culture. *J. Med. Virol.* **9:**93–100.
98. **Takiff, H. E., S. E. Straus, and C. F. Garon.** 1981. Propagation and in vitro studies of previously noncultivable enteral adenoviruses in 293 cells. *Lancet* **ii:**832–834.
99. **Takimoto, S., M. Grandien, M. A. Ishida, M. S. Pereira, T. M. Paiva, T. Ishimaru, E. M. Makita, and C. H. Martinez.** 1991. Comparison of enzyme-linked immunosorbent assay, indirect immunofluorescence assay, and virus isolation for detection of respiratory viruses in nasopharyngeal secretions. *J. Clin. Microbiol.* **29:**470–474.
100. **Trabelsi, A., B. Pozzetto, A. D. Mbida, F. Grattard, A. Ros, and O. G. Gaudin.** 1992. Evaluation of four methods for rapid detection of adenovirus. *Eur. J. Clin. Microbiol. Infect. Dis.* **11:**535–539.
101. **Turner, D., S. Martin, B. Y. Ngan, D. Grant, and P. M. Sherman.** 2006. Anastomotic ulceration following small bowel transplantation. *Am. J. Transplant.* **6:**236–240.
102. **Van Doornum, G. J., and J. C. De Jong.** 1998. Rapid shell vial culture technique for detection of enteroviruses and adenoviruses in fecal specimens: comparison with conventional virus isolation method. *J. Clin. Microbiol.* **36:**2865–2868.
103. **Wadell, G., A. Allard, and J. C. Heirholzer.** 1999. Adenoviruses, p. 970–982. *In* P. R. Murray, E. J. Baron, M. A. Pfaller, F. C. Tenover, and R. H. Yolken (ed.), *Manual of Clinical Microbiology*. ASM Press, Washington, DC.
104. **Watanabe, M., U. Kohdera, M. Kino, T. Haruta, S. Nukuzuma, T. Suga, K. Akiyoshi, M. Ito, S. Suga, and Y. Komada.** 2005. Detection of adenovirus DNA in clinical samples by SYBR green real-time polymerase chain reaction assay. *Pediatr. Int.* **47:**286–291.
105. **Weinberg, A., L. Brewster, J. Clark, and E. Simoes.** 2004. Evaluation of R-Mix shell vials for the diagnosis of viral respiratory tract infections. *J. Clin. Virol.* **30:**100–105.
106. **Wiley, L., D. Springer, R. P. Kowalski, R. Arffa, M. I. Roat, R. A. Thoft, and Y. J. Gordon.** 1988. Rapid diagnostic test for ocular adenovirus. *Ophthalmology* **95:**431–433.
107. **Wiley, L. A., L. A. Roba, R. P. Kowalski, E. G. Romanowski, and Y. J. Gordon.** 1996. A 5-year evaluation of the adenoclone test for the rapid diagnosis of adenovirus from conjunctival swabs. *Cornea* **15:**363–367.
108. **Wood, S. R., I. R. Sharp, E. O. Caul, I. Paul, A. S. Bailey, M. Hawkins, S. Pugh, J. Treharne, and S. Stevenson.** 1997. Rapid detection and serotyping of adenovirus by direct immunofluorescence. *J. Med. Virol.* **51:**198–201.
109. **Yusuf, U., G. A. Hale, J. Carr, Z. Gu, E. Benaim, P. Woodard, K. A. Kasow, E. M. Horwitz, W. Leung, D. K. Srivastava, R. Handgretinger, and R. T. Hayden.** 2006. Cidofovir for the treatment of adenoviral infection in pediatric hematopoietic stem cell transplant patients. *Transplantation* **81:**1398–1404.
110. **Zavattoni, M., E. Percivalle, E. Cattaneo, M. G. Revello, M. Torsellini, and G. Gerna.** 2003. Optimized detection of respiratory viruses in nasopharyngeal secretions. *New Microbiol.* **26:**133–140.

Diagnostic Microbiology of the Immunocompromised Host
Edited by Randall T. Hayden, Karen C. Carroll, Yi-Wei Tang, and Donna M. Wolk

Chapter 7

RNA Respiratory Viruses

Michael G. Ison and Eric S. Rosenberg

RNA respiratory viruses, including influenza virus, respiratory syncytial virus (RSV), parainfluenza virus (PIV), human metapneumovirus (hMPV), human rhinovirus (HRV), and coronavirus, are increasingly being recognized as causing significant morbidity, graft failure, and death among immunocompromised patient populations (94, 96). There are very few prospective studies of respiratory viral infections in the immunocompromised host. Most retrospective studies have relied on diagnoses made while patients were hospitalized. As a result, these studies likely underestimated the incidence of respiratory viral infections and overestimated the severity of these infections. The more recent prospective studies, which have used more contemporary diagnostic strategies, have shown that most immunocompromised patients have approximately two respiratory viral infections per year (63, 138, 146, 203). These studies have shown the impact of infections with rhinovirus, a virus that was previously felt to be a rare and insignificant pathogen, in this patient population (104, 138, 203).

The incidence and seasonality of respiratory viral infections in immunocompromised patients mimic those for the general population (94, 96). Transplant recipients often have atypical presentations of respiratory infection; this is particularly true of lung transplant recipients, in whom the only manifestation of disease may be changes in pulmonary function testing (96). A variety of clinical syndromes are associated with these viruses, including laryngitis, influenza-like illness, cold-like illness, bronchitis, and croup; no one virus is exclusively associated with one clinical syndrome (94, 96). Viral shedding tends to be prolonged in transplant recipients and may occur without significant symptoms (158). This prolonged shedding increases the risk of emergence of antiviral resistance during therapy (100). Respiratory viral infections are associated with increased risk of death, development of viral pneumonia, and coinfections, particularly bacterial pneumonia and invasive aspergillosis (94, 96, 137). Lastly, respiratory viral infections are associated with both acute and chronic rejection. The risk of chronic rejection is most clearly demonstrated when there is lower tract involvement among lung transplant recipients (23, 33, 112, 204). Whether this chronic rejection is the result of direct infection of the lung or systemic inflammation remains to be determined.

Clinically, it is often difficult to differentiate one respiratory virus from another. However, with the availability of effective antiviral therapy, particularly for influenza, and effective infection control methods, diagnosing the correct virus is clinically important. Diagnostic methods include serology, culture, antigen detection, and nucleic acid testing (NAT). In general, it is important to cast a wide net and perform a thorough screening for respiratory viruses during the initial evaluation (Table 1). Immunocompromised patients with upper respiratory tract symptoms should have upper airway samples collected and tested (Fig. 1). More rapid testing methods are preferred, although their sensitivity may be limited among transplant recipients. In general, serology is not useful for management of infections in transplant recipients since serologic responses are not universal in this population and convalescent-phase titers are often needed to document infection. Failure to determine the identity of a pathogen, particularly in patients with lower airway involvement, should prompt collection of lower airway samples, using methods such as bronchoalveolar lavage.

Michael G. Ison • Divisions of Infectious Diseases and Organ Transplantation, Transplant & Immunocompromised Host Infectious Diseases Service, Northwestern University Feinberg School of Medicine, Chicago, IL 60611. **Eric S. Rosenberg** • Departments of Medicine and Pathology, Transplant Infectious Diseases and Compromised Host Program, Massachusetts General Hospital, Boston, MA 02114.

Table 1. Epidemiology, diagnosis, and therapy for respiratory viral infections in transplant recipients

Virus	Prevalence (%)		Diagnostic method	Therapy
	Stem cell transplants	Solid organ transplants		
Influenza virus	11-18	10	Culture, DFA, EIA, PCR	Neuraminidase inhibitors[a] (oseltamivir and zanamivir)
RSV	35-49	38	Culture, DFA, EIA, PCR	Antibody-based therapy (IVIG, RSV Ig, palivizumab), aerosolized ribavirin
Parainfluenza virus	9-30	43	Culture, DFA, PCR	Ribavirin[b]
Rhinovirus	18-24		Culture, DFA, PCR	Pleconaril
Multiple pathogens		14	Culture, DFA, PCR	

[a]M2 inhibitors (amantadine and rimantadine) were previously active against influenza virus; with the rapid emergence of resistance to these agents within the United States, they are no longer recommended for the prevention or treatment of influenza (29).
[b]Ribavirin is associated with a mortality benefit for solid organ transplant recipients only; its use is not associated with a mortality benefit or a reduction of viral titers in stem cell transplant recipients (94).

All of the respiratory viruses are highly contagious, and outbreaks in transplant wards are well described (94, 96). In general, once a respiratory viral infection is considered, the patient should be placed in a private room and appropriate isolation precautions should be used. Once a virus is isolated, infection control should be contacted immediately to begin appropriate infection control practices. Lastly, isolation should be continued until all viral testing returns to negative to avoid exposure to others on the floor.

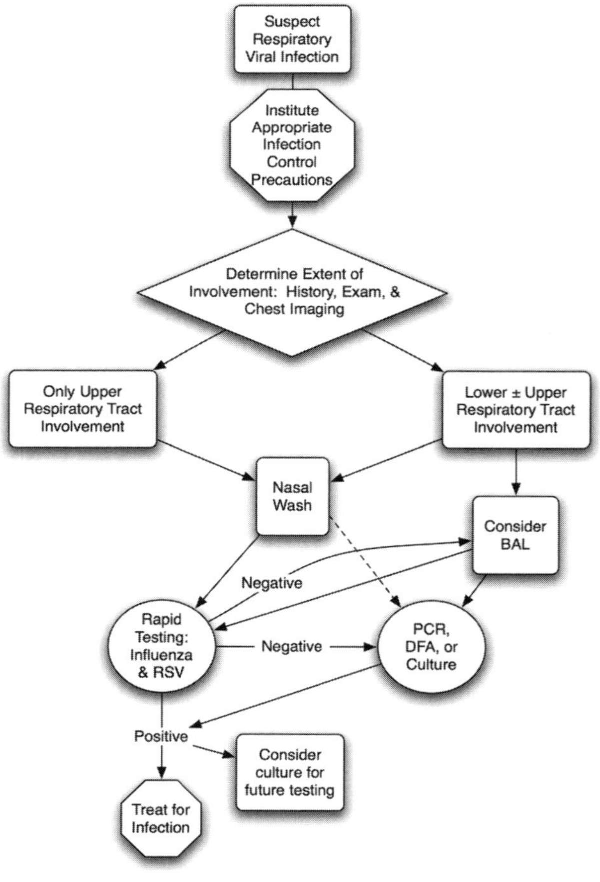

Figure 1. Approach to diagnosis of suspected viral respiratory tract infection.

INFLUENZA VIRUS

Background and Clinical Information

Influenza viruses are orthomyxoviruses and consist of types A, B, and C. Influenza A and B viruses are classically associated with influenza-like illnesses, while influenza C virus is more commonly associated with milder, cold-like syndromes (159). Rarely, influenza C virus has been associated with more severe disease (159). These enveloped, single-stranded RNA viruses contain eight gene segments (seven for influenza C virus) (159). The segmented nature of the genome allows reassortment and facilitates antigenic variation. Surface glycoprotein spikes possess either hemagglutinin (HA) or neuraminidase (NA) activity; identification of the specific HA and NA helps to further classify type A viruses. Influenza viruses are named by type, location of isolation, isolate number, year of recovery, and for influenza A viruses, subtype (e.g., A/Charlottesville/03/2004 [H3N2]) (83).

Although influenza is a common cause of morbidity in the community, resulting in over 130,000 excess hospitalizations and 36,000 excess deaths per year (193, 194), immunocompromised patients appear to suffer greater morbidity and mortality (94, 96). In general, the attack rate of influenza likely depends on various factors, including patient age (higher in children and the elderly), likelihood of exposure (community versus nosocomial; contact with children), level of specific immunity (from prior infections and immunizations), degree of immunocompromise, and nature of the epidemic (magnitude and antigenic type). Prevalence estimates for transplant recipients have ranged from 2 to 29%, compared to 5 to 20% for the general population (10, 47, 81, 94, 96, 136, 206). In transplant recipients, viral replication may be prolonged, especially in the absence of therapy, underscoring the need for rapid diagnosis and treatment. One study found a mean duration of viral shedding of 6.7 days for autogenic hematopoietic stem cell transplant (HSCT) recipients without treatment, 11.1 days for allogeneic HSCT recipients without treatment, 9.7 days for any HSCT recipients who received M2 inhibitors (amantadine and rimantadine), and 7.5 days when NA inhibitors were used (158). HSCT recipients have higher rates of infectious complications (94, 96, 137) and progression to viral pneumonia (96) than do other immunocompromised patients. Acute and, especially in lung transplant recipients, chronic rejection occurs following influenza virus infection, particularly with lower tract involvement (23, 33, 112, 204). Mortality has historically been high (attributable mortality of 17 to 43%) (52, 94, 96), although it appears to be lower with the use of antiviral therapy (94, 96, 103, 128, 129). Risk factors for more severe disease include being an allogeneic HSCT recipient (odds ratio [OR], 5.26), pediatric age group, infection early posttransplant, and lymphopenia (OR, 6.5) (94, 96).

Influenza Prevention

Prevention of influenza depends on either vaccination or antiviral therapy (187). Since the influenza virus surface glycoproteins change frequently, causing antigenic drift and shift, vaccines are reformulated annually (187). Unfortunately, influenza vaccination in transplant recipients appears to be less effective at inducing protective immunity than that in the general population (94, 96, 110). As a result, some immunocompromised patients, particularly stem cell transplant recipients during the first 6 to 12 months posttransplant, are not effectively protected with vaccination alone (94). As such, there is interest in the use of antiviral agents to prevent influenza. NA inhibitors have been documented to be 70 to 93% effective at preventing influenza in immunocompetent adults and children (151). High-level protection was noted in high-risk patients who received antiviral therapy with NA inhibitors (NAIs) in addition to influenza vaccination (172), suggesting that a combination of vaccine plus antiviral therapy may be efficacious.

Influenza Therapy

Once patients are infected, antiviral therapy appears to be associated with reduced morbidity and mortality (94). Worldwide emerging resistance to M2 inhibitors is limiting therapeutic options to NAIs and possibly ribavirin (29, 187). Several studies have documented the safety of using NAIs in immunocompromised patients and have shown that these drugs are associated with more rapid clearance of virus, reduced symptomatology, reduced progression to pneumonia, and reduced mortality (103, 128, 129, 158). The dose and duration of oseltamivir and zanamivir treatment are based on data from previously healthy ambulatory adults and children (151); the optimal dose and duration of therapy for immunocompromised patients have not been established. As discussed below, some recommend that therapy should be continued until viral replication has ceased; almost universally, this is longer than the approved 5-day course (94, 96). Additionally, there may be benefits to starting therapy well after symptom onset, to using higher-than-approved doses, and possibly to using multidrug therapy; further studies of these interventions are needed (94, 96).

Antiviral Resistance

One major issue related to the use of antiviral agents in this population is the emergence of resistant variants (31a, 69, 95, 142, 143, 219). Emergence of M2 inhibitor resistance occurs frequently in both immunocompetent and immunocompromised patients and results in cross-resistance among all drugs in the class (31a, 29, 53). NAI resistance can occur as the result of mutations in either the NA or HA gene and typically does not result in cross-resistance among all NAIs (69, 95, 108, 143). Further work is needed to identify risk factors and strategies to prevent the emergence of resistant variants, to develop more rapid methods of detection of resistant variants, and to more efficiently manage patients with infections caused by resistant variants (100).

Diagnostic Approaches

Specimen collection

Diagnosis of influenza may be achieved by culture, antigen detection, NA detection, and NAT. Although specimens from the nose, throat, nasopharynx, and lung may be submitted, nasopharyngeal washes tend to provide higher viral burdens than do throat specimens (83). This may not be true with avian strains, where shedding is increased in the throat compared to that of human influenza virus (192a). Growth is maximized by inoculation of fresh specimens into tissue culture; freezing at −20°C can result in reduced culture yield. Short-term storage can be achieved with refrigeration at

4°C for up to 4 days; longer-term storage requires storage at or below −70°C to preserve infectivity (83).

Tissue culture

Virus can be cultured using tissue culture techniques on primary rhesus monkey kidney, cynomolgus monkey kidney, Madin-Darby canine kidney (MDCK), LLC-MK2, Vero, mink lung, or MRC-5 cells (83). Yields are improved with the addition of trypsin to the serum-free medium and by incubation at 33°C with rolling (83). Virus can be detected by typical cytopathic effect (83) or by hemadsorption using guinea pig or turkey erythrocytes (101). Shell vial culture techniques combined with antigen detection can give even faster results, within 24 to 48 hours (83). Individual isolates can be identified by HA inhibition testing for current strains, by immunofluorescence, or by enzyme immunoassays (EIAs) (83).

Antigen detection

Numerous antigen tests have been developed as point-of-care tests and are FDA cleared for the rapid detection of influenza virus; many are used by central laboratories as well (165). These tests vary in their methods, acceptable sample types, testing characteristics, sensitivity, specificity, and Clinical Laboratory Improvement Amendment (CLIA) waiver status (Table 2). Most detect influenza A and B viruses and some differentiate A- from B-type viruses. All tests are reasonably easy to perform and provide results in less than 30 minutes. Their use has been associated with selection of more appropriate antimicrobial therapy and infection control (161, 178, 186). Despite these potential advantages, rapid antigen testing has limited sensitivity; many experts believe that rapid influenza virus antigen testing in routine clinical care has a lower sensitivity than what is reported in package inserts (99). There are limited data about the sensitivity of rapid antigen testing of immunocompromised patients (160, 165, 185). It is important that not all kits have been tested prospectively for the ability to detect avian strains of influenza virus and may result in false-negative results in the setting of a pandemic. In general, these assays have higher sensitivities with samples from children than from adults; as with culture, nasal or nasopharyngeal washes and aspirates have superior sensitivity compared to swabs. Pelleting of cells from combined nasopharyngeal and throat swabs increases sensitivity (124).

Indirect immunofluorescence assay and direct immunofluorescence assay (DFA) can be used on pa-

Table 2. FDA-approved rapid diagnostic tests for influenza

Test	Method	Virus type(s) detected	CLIA waived	Sensitivity[d] (%)	Specificity[d] (%)	Specimen type(s)[c]
Directigen Flu A (Becton Dickinson)	Membrane filter EIA	A	No	67-96	88-100	
Directigen Flu A+B (Becton Dickinson)	Membrane filter EIA	A and B[a]	No	71-100	90-100	NPW, NPA, NPS, NS, TS, BAL
Directigen EZ Flu A+B (Becton Dickinson)	Lateral-flow EIA	A and B[a]	No	69-86	86-100	NPW, NPA, TS
NOW Influenza A (Binax)	EIA	A	Yes	78-82	92-94	NPW, NPA, NPS
NOW Influenza B (Binax)	EIA	B	Yes	58-71	97	NPW, NPA, NPS
NOW Influenza A&B (Binax)	EIA	A and B[a]	Yes	93-100	93-100	NPW, NPA, NPS
Flu OIA (Biostar)	Optical surface EIA	A and B[b]	No	62-88	52-80	NPA, NPS, TS, sputum
Flu OIA A/B (Biostar)	Optical surface EIA	A and B[a]	No	62-88	52-80	NPA, NPS, TS, sputum
OSOM Influenza A&B (Genzyme)	Membrane filter EIA	A and B[a]		56-74	92-98	NS
QuickVue influenza test (Quidel)	Lateral-flow EIA	A and B[b]	Yes	73-81	96-99	NS, NPW, NPA
QuickVue Influenza A+B test (Quidel)	Lateral-flow EIA	A and B[a]	Yes	62-94	89-99	NS, NPS, NPA, NPW
XPECT Flu A&B (Remel)	Membrane filter EIA	A and B[a]	No	92-98	100	
SAS FluAlert (SA Scientific)	EIA	A and B[a]	No	76-90.5	97.9-100	NPW, NPA
TRU FLU (Meridian Bioscience)	Lateral flow EIA	A and B[a]	No	50-87.2	89.3-100	NPW, NPA, NPS, NS
ZstatFlu (ZymeTx)	NA enzyme activity	A and B[b]	Yes	57-65	98-100	TS

[a]This test is able to distinguish influenza A from influenza B viruses.
[b]This test can detect both A and B viruses but does not differentiate between the two types.
[c]NPW, nasopharyngeal wash; NPA, nasopharyngeal aspirate; NPS, nasopharyngeal swab; NS, nasal swab; TS, throat swab; BAL, bronchoalveolar lavage.
[d]Values are per product package inserts.

tient specimens and cell cultures (88). Because of the changing antigenicity of influenza virus, detection of conserved proteins, such as the nucleoprotein (NP), is preferred (83). The sensitivity of DFA in some studies approaches that of PCR (117, 179). Since DFA is typically used to test for multiple pathogens, it is discussed in further detail in the section on multipathogen testing.

NAT

PCR-based testing for influenza virus has been refined over the years and is being used more commonly in routine clinical laboratory diagnostics (18, 34, 109, 117, 177, 179, 189, 208, 210). For most populations, the sensitivity of PCR exceeds that of other testing methods; available data suggest a 7 to 20% increase in the number of detected viruses for PCR compared to culture (85, 150, 189). Quantitative methods have been developed recently and have shown a correlation with clinical improvement associated with antiviral treatment (26, 175, 180). In the future, viral load measurement may replace quantitative cultures in clinical care. One limitation of NAT is that nucleic acids may be detected after viral replication has ceased (85). PCR testing is not yet available in an FDA-cleared format for respiratory viral diagnosis, although analyte-specific or research-use-only reagents are commercially available and are being used more widely. Most of the commercially available assays for influenza virus provide results on multiple respiratory pathogens in one test (see below). The best-studied assay is the Hexaplex system (Prodesse, Inc., Wukesha, WI). This system uses primers for the matrix (M) gene of influenza A virus and the nonstructural gene of influenza B virus. Fifty-three percent of symptomatic patients had positive PCR results using this system alone, compared to tissue culture, in one study (126). Other systems use primers directed against the HA, NA, and M genes. Several companies have developed assays that are able to identify influenza A virus by PCR and to accurately determine the specific HA and NA subtypes (CombiMatrix influenza A virus detection system), potentially allowing differentiation of emerging strains of virus, such as avian influenza virus.

Assessment of Antiviral Resistance

Antiviral resistance is an emerging issue among immunocompromised patients infected with influenza virus, as discussed above (53, 69, 95), although resistance testing is not yet widely available outside the research setting. M2 inhibitor resistance results from mutations in the M2 gene that result in either reduced binding of M2 inhibitors or enlargement of the M2 protein pore diameter; both changes allow protons to pass despite binding of the inhibitor (14, 78, 90). NAI resistance can result from changes in either (or both) the HA or NA gene. In general, resistance to M2 inhibitors may be detected by cell culture (28) or M2 gene sequencing (20, 28, 78), while NAI resistance can be determined by plaque assays, NA inhibition assays (69, 95), and NA and HA gene sequencing (69, 142, 143, 219). In the case of NAI resistance testing, neither plaque assay nor NA inhibition assay can determine if the alteration in sensitivity is the result of changes in the NA or the HA protein; only sequencing of the individual genes will identify which protein is likely responsible for the altered sensitivity. It should be noted that cell culture-based plaque assays have traditionally been unreliable methods to assess the susceptibility of NAIs, as individual viruses differ in their susceptibilities, and susceptibility in cell culture does not always correlate with clinical response or other markers of susceptibility (70, 71, 215). If available for clinical use, resistance testing may be considered for patients with persistent viral replication beyond the expected period (i.e., 7 to 10 days) while on antiviral therapy or if a clinical relapse occurs.

Result Interpretation

In general, when influenza is being considered, nasal swab or nasopharyngeal wash specimens should be obtained and tested using direct methods, including either rapid influenza virus antigen assay, fluorescent-antibody (FA) detection, or NAT (Fig. 1). A positive result should be considered diagnostic of influenza, and antiviral therapy should be considered for all immunocompromised patients, irrespective of the severity of illness or duration of symptoms prior to diagnosis. A negative test should result in a wider workup for the cause of the illness. Additional testing, such as NAT, DFA, or culture, should screen for multiple viruses, as it is difficult to distinguish influenza from other causes of respiratory viral disease. If there are lower tract signs or symptoms, a bronchoscopy should be considered early in the course, typically within the first 24 hours if a pathogenic cause is not documented. Bronchoalveolar lavage fluid should be sent for NAT, DFA, or culture. Although NAT is usually the most sensitive test, it may not be readily available at all centers; if possible, samples should be sent out for PCR testing if it cannot be done locally.

Alternative testing, with DFA or culture, based on local availability, should be performed if PCR results will not be available within 24 to 36 hours.

Therapeutic Monitoring

There is a paucity of data on how best to monitor therapy. Viral replication is prolonged among transplant recipients, and the FDA-approved 5-day course of NAI may not be sufficient (94, 96). Many experts recommend continuing antiviral therapy until replication is eradicated. Animal model data suggest that a rebound in viral titers may occur if antivirals are stopped prematurely, which may be detrimental and may contribute to the emergence of resistance (100). Using cultures to monitor the response to therapy can provide quantitative data which may suggest a reduction in viral replication over time; however, such methods are not widely available for clinical use. Cultivation of virus also allows isolates to be preserved for future resistance testing (94, 96). There are limited data regarding the use of rapid antigen testing to monitor viral clearance. For immunocompetent patients, one rapid kit (QuickVue; Quidel, San Diego, CA) has been shown to be an acceptable alternative to culture in monitoring responses to therapy (19). NAT can also be used to monitor responses to therapy, although a positive signal may persist for up to several weeks, after replication has ceased (85).

RSV

Background and Clinical Information

RSV is a member of the family *Paramyxoviridae* and the genus *Pneumovirus*. These viruses are enveloped single-stranded RNA viruses that code for 11 proteins. RSV is named for its ability to cause cell fusion in tissue culture cell lines (173). There are two key antigenic proteins of the virus, namely, the G surface glycopeptide, which determines if the virus is a member of the A or B subgroup, and the F (fusion) glycopeptide, which is conserved between the two subgroups (173).

RSV is a major cause of morbidity and mortality in immunocompromised recipients and consistently represents a common cause of respiratory viral infections in this population. RSV causes disease predominantly during the midwinter and is frequently superimposed on the seasonality of influenza (157, 220). Like other respiratory viruses, RSV is associated with prolonged replication, increased risk of infectious and noninfectious complications, and increased mortality (OR, 1.6) among immunosuppressed patients (94, 96, 157). There appears to be an increased risk of progression to lower tract disease among patients who have received lymphocyte-depleting therapies or who have lymphopenia, a history of prior lung disease, or onset of upper respiratory tract infection prior to HSCT or prior to engraftment (94, 96, 157). The risk of proceeding with HSCT in a patient diagnosed with RSV prior to conditioning is controversial. Patients who proceed with transplant frequently need oxygen therapy, and progressive disease may occur (6); there appears to be little risk of progressive malignancy when the transplant is delayed (168). Based on this information, most centers defer stem cell transplantation until RSV has been cleared.

The cornerstone of prevention of RSV is strict infection control practices (32a). Use of passive immunoprophylaxis with immunoglobulin (intravenous immunoglobulin [IVIG], RSV Ig, or palivizumab) might reduce the frequency and severity of RSV in immunocompromised patients; its utility in high-risk children has been shown (5, 7, 8). Although studies of humans are limited (37), RSV Ig and palivizumab were shown to result in lower virus titers after infection in an immunocompromised mouse model of RSV infection (163, 164).

There are also limited data on the management of RSV infection in transplant recipients. Aerosolized ribavirin may provide some clinical and virologic efficacy in stem cell transplant recipients (24, 94, 96). The addition of intravenous antibodies (IVIG, RSV Ig, or palivizumab) appears to have the greatest benefit in reducing mortality (94, 96). The use of donor lymphocyte infusions has rarely been attempted but may improve survival (51, 107).

Diagnostic Approaches

Specimen collection

RSV infection can be diagnosed by culture, antigen detection, and NAT. In general, nasopharyngeal washes are superior to nasopharyngeal swabs and are therefore the preferred specimen type (173). Since RSV is extremely labile (infectivity is lost within 1 hour at 37°C), specimens should be placed immediately on ice and transported urgently to the laboratory, where they should be set up for culture as quickly as possible (173). Slow freezing to −20°C results in loss of viability (21), and therefore specimens should be either inoculated freshly (preferred) or flash frozen to <−70°C using alcohol and dry ice baths (173). Use of sucrose or glycerol in the medium may enhance recovery after freezing (173).

Tissue culture

RSV is readily cultured on HEp-2 cell lines, although human embryonic lung diploid fibroblast, monkey kidney, human amniotic, and human kidney cell lines can support growth as well (173); mixed cell lines can also be used, as discussed below. Cultures typically demonstrate syncytium formation within 3 to 5 days of inoculation (173). Cytopathic effect may be confirmed by detection of viral antigen using FA methods (106). Shell vial techniques have also been developed and may yield positive results as early as 10 hours after inoculation (173). Over time, cell lines change their ability to form syncytia, and sensitivity needs to be monitored carefully (173).

Antigen detection

There are a number of FDA-approved rapid antigen detection kits for RSV (Table 3). For immunosuppressed adults, rapid RSV testing was found to be only 15 to 50% sensitive for nasal washes compared to culture; use of endotracheal and bronchoalveolar lavage fluid increased the sensitivity to 71 to 89% (54, 211). Other, more recent studies have shown low sensitivity compared to NAT for immunosuppressed patients (179). Rapid RSV antigen tests commonly provide false-positive results with bloody samples and those with thick, tenacious mucus (173).

Indirect immunofluorescence assay and DFA can be used on patient specimens and cell cultures (173). Since these assays typically test for multiple pathogens, they are discussed in further detail below.

NAT

Several NAT methods for detecting RSV have been developed, but most are not widely available clinically (18, 34, 45, 56, 112, 115, 117, 146, 162, 179, 210). Most assays use primers directed toward the N, F, and G genes. In general, there are three types of assays, those that detect RSV, those that detect RSV and differentiate between subgroup A and B viruses, and those that detect just subgroup A or B viruses. Differentiating the A and B subgroups has limited clinical significance, and sensitivity is slightly reduced (56.7% versus 48.3%) for methods that differentiate subgroups (62). Likewise, real-time PCR methods have been shown to be 60% more sensitive than enzyme-linked immunosorbent assay (ELISA)-based antigen detection (144). NAT is also more sensitive than culture; in one study, 13 cases of RSV infection were identified by PCR, but only 4 were positive by culture; all culture-positive specimens were also PCR positive (202). Additionally, there are several multiplex methods (discussed below) that detect and differentiate RSV from other respiratory viral agents.

Result Interpretation

Patients with suspected RSV infection should initially have a nasal or nasopharyngeal wash sent for direct testing using rapid antigen detection, FA testing, or NAT. If rapid testing is done first and yields a negative result, PCR, DFA, or culture should be done. Many laboratories utilize a combination of these methods. A positive result with any of the

Table 3. FDA approved rapid diagnostic tests for RSV

Test	Method	CLIA waived	Sensitivity[b] (%)	Specificity[b] (%)	Specimen type(s)
Directigen RSV (Becton Dickinson)	EIA	No	93-97	90-97	NPW, NPA, NPS
Directigen EZ RSV (Becton Dickinson)	Lateral-flow EIA	No	67-87	86-95	NPW, NPA, NPS
NOW RSV (Binax)	EIA	Yes	74-98[c]	92-100[c]	NW, NPS
OIA RSV (Biostar)	Optical surface EIA	No	67-87	83-96	NW, NPS
Fisher Sure-Vue RSV (Fisher Scientific)		Yes	96	94	NPW, NPA, NPS
Quicklab RSV (Integrated Biotechnology)	EIA	Yes	93-100	87-98	NW, NPS
Immuno*Card* STAT! RSV Plus (Meridian Bioscience)	Lateral-flow EIA	Yes	91[d]	80[d]	NPS, NPA
QuickVue RSV test (Quidel)	Lateral-flow EIA	No	92-99	92	NPA, NPS
XPECT RSV (Remel)	Membrane-filter EIA	Yes	75-96[c]	94-98[c]	NA
SAS RSVAlert (SAS)	EIA	Yes	91[e]	NA	NPW, NPA, NS
Clearview RSV (Wampole Laboratories)	Lateral-flow EIA	Yes	93-100	88-97	NPS, NPW, NPA

[a]NPW, nasopharyngeal wash; NPA, nasopharyngeal aspirate; NPS, nasopharyngeal swab; NS, nasal swab.
[b]Values are per product package inserts, unless noted otherwise. NA, not available.
[c]From reference 27.
[d]From reference 79.
[e]From reference 115.

above methods for an immunocompromised host suggests that therapy should be considered. As with all viruses, an aggressive attempt to identify copathogens should also be made. If therapy is initiated, it should be continued until viral replication has ceased. Traditionally, this has been monitored by culture, although quantitative PCR methods may eventually be used in this way. As with other viruses, the presence of viral nucleic acid may be detected after replication has ceased.

PIV

Background and Clinical Information

PIV is a member of the family *Paramyxoviridae* and the genus *Paramycovirus*. These viruses are enveloped single-stranded RNA viruses (173). There are four recognized strains that affect humans; PIV 3 is the most prevalent among transplant recipients, and PIV 4 is typically associated with mild cold-like symptoms, although more severe disease has been recognized (87, 145). PIV causes a range of diseases, including colds, croup, bronchiolitis, and viral pneumonia. PIV disease occurs throughout the year (94, 96, 156, 205), and risk factors for progressive disease include young age, presence of graft-versus-host disease, antilymphocyte therapy, and steroid use (94, 96, 155). Lower tract disease (hazard ratio [HR] 3.4), the need for ventilatory support (HR 3.3), and the presence of copathogens (HR 2.8) have been associated strongly with increased mortality (155).

Much like the case for RSV infection, strict infection control practices are the cornerstone in the prevention of PIV infection in transplant patients (94, 96). Several live attenuated PIV vaccines are being developed, but their use in the immunosuppressed patient may not be feasible (98, 99). Therapeutic options for the management of PIV are also limited. There are no prospective studies of antiviral agents against this infection in transplant recipients. Among heart and lung transplant recipients, intravenous ribavirin may provide a mortality benefit (35, 80, 140, 211). Among stem cell transplant recipients, neither IVIG nor intravenous ribavirin has been associated with a reduction of viral shedding or mortality (50, 96, 155).

Diagnostic Approaches

Specimen collection

PIV infection can be diagnosed by culture, antigen detection, or NAT. PIV is stable in viral transport medium at 4°C for up to 5 days. Freezing to −20°C does decrease infectivity of the virus, but long-term storage can be achieved easily by adding sucrose or glycerol to the holding medium and freezing samples to <−70°C (173).

Tissue culture

PIV is a stable virus that is readily culturable on LLC-MK2 as well as primary monkey kidney cells, using standard and shell vial techniques; mixed cell cultures can also be used and are discussed below (173). Hemadsorption-inhibition, hemagglutination inhibition, or immunofluorescence is used to identify PIV in cultures (173).

Antigen detection

Unlike the case for RSV and influenza virus, no rapid antigen test kits are available for detection of PIV. There are several monoclonal antibody (MAb) systems that allow for identification and differentiation of the different PIV serotypes in primary patient samples (see the section on multiple pathogen testing for full details) and in cell cultures (173).

NAT

Several NAT methods have also been developed to detect PIV (18, 46, 112, 117, 146, 177, 179, 210). Typically, these are directed toward the HA-NA (HN) gene, although exact primers have differed between assays. Different PCR methods and probes have not been compared to one another for selection of a preferred method. The sensitivity and specificity of NAT have consistently been superior to those of culture and are comparable to those of FA-based testing (126, 173). In one study, 50% of pediatric patients with PIV infection had positive PCR results only, and none were detected by DFA alone (117).

Result Interpretation

Since there are no rapid antigen tests for PIV, appropriate upper or lower respiratory tract samples should be sent for PCR, DFA, or culture. All three methods have been shown to have high degrees of specificity, although their sensitivities vary. If therapeutic options become available, positive results should result in infectious disease consultation to determine the appropriateness of therapy. If initial testing is negative for PIV, testing for a wider range of pathogens should be considered to make a definitive diagnosis. Likewise, sampling of deeper respiratory compartments, if not already done, should be considered. As with all viruses, an aggressive attempt to identify copathogens should be made.

hMPV

Background and Clinical Information

hMPV is a recently described enveloped, single-stranded RNA virus (197). hMPV is increasingly recognized as a significant pathogen in both immunocompetent and immunocompromised patients; it appears to have a similar epidemiology and clinical course to those of RSV infection (25, 31, 40, 43, 61, 75, 89, 125, 132, 138, 170, 198, 199, 212). One study noted that up to 25% of lung transplant recipients had respiratory infections caused by hMPV, highlighting its importance (122). Fatal infections have been noted (25, 31, 125, 170, 212), and the presence of copathogens, particularly RSV, appears to predispose patients to more severe disease (65, 182). Few studies have investigated therapeutic options for hMPV, and none have studied preventative strategies. In vitro, ribavirin, $NMSO_3$, and pooled Ig, but not palivizumab, inhibit hMPV (216-218).

Diagnostic Approaches

Tissue culture

hMPV can be diagnosed by culture, antigen detection (FA testing), or PCR. Because of the novelty of the virus, optimal diagnostic techniques are still being studied. Initial studies showed that tertiary monkey kidney and LLC-MK2 cell lines supported growth, while most other cell lines supported either poor or no growth of the virus (25, 197). In these culture methods, there was no hemagglutination, growth required a mean of 17 days to demonstrate cytopathic effect, and trypsin was required for efficient growth (196). Shell vial cultures, which have the advantage of decreased time to positive result, are currently being used experimentally in some centers (121, 171).

Antigen detection

Antigen detection is an easy way to detect hMPV from primary clinical specimens and cultures (22, 49, 93, 117, 171). However, the necessary reagents are not widely used in most centers and have only recently become available commercially. FA assays applied directly to patient specimens have excellent sensitivity (49, 121).

NAT

PCR remains slightly superior to other methods for hMPV detection (49, 117, 171, 179). From the limited data, the greatest yields result from assays with primers directed at the nucleoprotein (N) and polymerase (L) genes; most assays have a lower limit of detection of ~100 copies (38, 127). There are insufficient data to suggest the superiority of one molecular method to another.

Result Interpretation

PCR is currently the preferred diagnostic method for diagnosing hMPV infection; FA and EIA testing has become available, but there are limited data on the sensitivities of these methods in comparison to PCR. The presence of hMPV with an appropriate clinical picture is diagnostic of infection. In situ hybridization studies have correlated the presence of hMPV by PCR with clinical disease in lung transplant recipients (14, 37, 89, 92, 172). In these cases, histological examination of lung tissues demonstrated acute and organizing injury, with smudge cell formation that was unique to hMPV (191). Coinfection is frequent, so copathogens (viral, bacterial, and fungal) should be sought in all cases. Data are too limited regarding quantitative PCR measurements to determine if these can be used for clinical prognostication or for determination of response to therapy.

RHINOVIRUSES

Background and Clinical Information

HRVs are members of the family *Picornaviridae* and are single-stranded RNA viruses surrounded by a protein capsid (72). The capsid must be in its hydrated form to be infectious. There are over 100 serotypes, which are divided into the following three groups based on their cellular receptor: the major group, which binds intercellular cell adhesion molecule 1; the minor group, which binds the low-density lipoprotein receptor; and HRV87, which binds sialoprotein (66, 86, 195). HRVs can be differentiated from other picornaviruses by their acid lability and their buoyant density in cesium chloride (72).

Clinically, HRV is the most common cause of colds in adults and children and peaks in activity in the early fall and the late spring (73, 148, 149). From experimental human data, the onset of symptoms occurs early after inoculation, typically within 12 hours, and correlates with a rise in viral load

(76). Prolonged shedding, sometimes in the absence of symptoms, progression to the lower tract, and complications may occur in immunocompromised patients (64, 97, 104). The clinical importance of HRV, particularly when it is shed asymptomatically, in transplant recipients is only now being understood (64, 104, 177, 203). With the available data, though, HRV is likely the most frequent cause of respiratory viral infections in immunocompromised patients, causing a wide range of illnesses, from mild colds to life-threatening pneumonias (77, 97, 104, 138, 177, 203).

There are no proven preventative interventions other than careful hand hygiene. Likewise, there are currently limited therapeutic options. Pleconaril was studied extensively in healthy adults with rhinoviral colds, with findings that the drug was well tolerated and led to a median 1-day reduction in duration of symptoms, in addition to more rapid improvement in symptom scores and clearance of virus from nasal mucus (82). This drug was not approved by the FDA because of the potential enhanced clearance of ethinyl estradiol (82). Since commonly used immunosuppressants are metabolized by the CyP-450 3A enzymes, this compound would likely have a significant impact on drug levels (192). The compound has been reformulated into an intranasal formulation, which is currently undergoing clinical evaluation.

Diagnostic Approaches

Tissue culture

Two cell lines, i.e., human diploid embryonic lung cells (such as WI-38 and MRC5) and HeLa cells (72), appear to best support the growth of HRV. Although the virus can grow in a wide range of environments, the best growth is achieved at 33 to 34°C with gentle motion (72, 123, 167). Yields are greatest from upper tract specimens, among which nasal or nasopharyngeal washes have the best recovery (11, 72). HRV can also be isolated from the lower tract (97, 104). Classic cytopathic effect develops in 8 to 48 hours and is dependent on the pH, temperature, loading dose of virus, cell line used, and serotype of virus used (72).

Antigen detection

FA and immunoperoxidase methods for detecting rhinoviral antigen have been used in clinical studies (3, 4, 15, 42, 72, 111). Low concentrations of antigen in respiratory secretions and limited commercial availability of reagents have limited the clinical utility of these assays (72).

NAT

PCR offers improved sensitivity and ease in detecting multiple pathogens with one or similar tests (18, 68, 91, 92, 162). In one study, 18% of HRV-infected patients were detected by PCR alone (13). As such, several PCR protocols have been developed for the detection of HRV. Several methods have been applied to immunocompromised patient populations and have demonstrated clinical efficacy (12, 17, 18, 34, 39, 68, 91, 92, 97, 102, 104, 166, 174, 177, 181, 188, 203). Different nucleic acid amplification techniques have not been compared directly to determine which method or primer sets are preferred, although available data suggest that real-time PCR is more sensitive than conventional PCR techniques (39).

Result Interpretation

Detection of virus by culture, DFA, or PCR is highly suggestive for patients with an appropriate clinical presentation. Quantitative cultures have been shown to correlate with clinical improvement when antiviral therapy is used, although this is not widely available clinically. Given the variable course of HRV infection in immunocompromised patients and the limited therapeutic options, careful monitoring and screening for coinfecting or complicating pathogens should be conducted (64, 97, 104, 203).

CORONAVIRUSES

Background and Clinical Information

There are currently five clinically significant human coronaviruses recognized, namely, OC43, 229E, NL63, HKU1, and severe acute respiratory syndrome coronavirus (SARS-CoV) (141, 169, 176, 200, 201, 213, 214). These viruses are all positive-sense membrane-bound RNA viruses (141) and typically cause mild cold symptoms (135). Severe lower tract disease has been described, and at least five cases of severe infection among immunocompromised patients, with one death, have been reported (57-59, 179).

SARS-CoV, a rare coronavirus, caused a short-lived pandemic in 2003, affecting >8,000 individuals (114, 118, 169). Three transplant recipients (liver, lung, and HSCT recipients) were recognized to be infected with SARS; two had progressive disease leading to death, while a third, treated with ribavirin and prednisolone, survived (114, 118, 169). The SARS-CoV loads in lung and heart tissues from one patient were at least 10,000-fold greater than those in most immunocompetent SARS patients (113).

No therapy (with the possible exception of interferon) has been proven effective in the management of coronavirus infections. Several strategies have been tried in patients with SARS, but none has proven sufficiently effective to recommend use, although there are suggestive data to support the use of interferon. Coronavirus 3C-protease inhibitors are currently under development and hold promise as therapeutic options for the future (44, 67, 74).

Diagnostic Approaches

Tissue culture

Coronaviruses require specialized cell lines (organ cultures of human embryonic trachea) and usually require subpassage, making routine culture very difficult (105, 141). Because of the difficulty in growing coronaviruses and the danger associated with growing SARS-CoV, other diagnostic methods are preferred. If SARS is suspected, the local health authorities should be consulted immediately and further workup of the specimen should be performed by the Centers for Disease Control and Prevention or other public health facilities.

Antigen detection

Immunofluorescence assays have been developed for OC43, 229E, NL63, and HKU1 and are commercially available for OC43 and 229E, including an ELISA for detection of antigen from nasal swabs and secretions (130, 131, 183, 184). In studies comparing DFA to PCR (147), the major difference in the two methods was the lower limit of detection for the 229E and OC43 viruses, favoring PCR (183).

NAT

Molecular diagnosis of coronaviruses appears to have the greatest sensitivity and is the diagnostic method of choice for diagnosing SARS-CoV (84, 152-154, 200, 213). Pancoronavirus PCR methods that increase efficiency in detecting the different clinically significant coronaviruses have been developed but are still available mainly on a research basis (147). Electron microscopy demonstrates round, membrane-bound, pleomorphic particles measuring 100 to 150 nm and covered with widely spaced projections (peplomers) (41, 141).

Result Interpretation

Detection of any of the human coronaviruses in the appropriate clinical setting is highly suggestive of infection. Prolonged shedding and asymptomatic shedding have been described (141). As with all RNA respiratory tract viruses, it is imperative that coinfecting viruses, bacteria, and fungi are ruled out. Although lower tract disease has been described, if the coronavirus is detected from the lower respiratory tract, even in the appropriate clinical setting, alternative causes of infection must be ruled out because of the infrequency of this infection and the lack of available antiviral therapy (63, 97). In the case of SARS, the virus may be detected at a number of clinical sites and may be difficult to detect in the respiratory tract during the early phases of disease (169). The local health department should be contacted as soon as a patient is suspected of having SARS infection, as evaluation and management have significant public health implications. Consultation with appropriate health authorities is recommended to guide isolation, diagnosis, and management (32).

MULTIPATHOGEN TESTING

It is often clinically difficult to differentiate among the potential causative pathogens in immunocompromised patients presenting with respiratory tract symptoms. This is particularly problematic among lung transplant recipients, whose sole symptom may be a decline in pulmonary function testing. Given this clinical limitation, a broad screening approach for potential pathogens initially, using a rapid, sensitive, and specific test, is preferable. A number of different testing methods, mostly based on tissue culture, antigen detection, or PCR, have been developed to diagnose a wide array of potential respiratory pathogens.

Tissue Culture

Tissue culture is uniquely suited to detecting multiple pathogens. Unfortunately, single cell lines may not effectively grow all of the respiratory viruses, and thus, multiple lines are frequently used. Mixed cell lines have recently become used widely, allowing culture of a broad array of respiratory viruses by use of a reduced number of inoculated tubes or shell vials. In the past, several tubes had to be set up to maximize recovery of individual viruses. Several different mixtures of cells (e.g., R-Mix [Diagnostic Hybrids, Athens, OH]) are commercially available. In general, these mixed cell lines allow greater ease in setting up and monitoring viral cultures, with similar sensitivities to those of individual cell lines (16, 60, 190, 209). Unfortunately, despite their ease of use and relative cost efficiency, these mixtures have variable sensitivity to different

Table 4. Sensitivity of R-Mix cell cultures in detecting specific respiratory viruses[b]

Virus	Sensitivity[a](%)	Avg time to positivity (days)
Influenza A virus	98.8	1.9
Influenza B virus	94.7	3.1
PIV 1	86.7	3.8
PIV 2	100	1.7
PIV 3	83.3	3.7
RSV	86.7	2.1
Adenovirus	68.6	3.5

[a]Compared to traditional culture techniques.
[b]Adapted from the *Journal of Clinical Microbiology* (48) with permission.

viruses as well as variable time to positivity (Table 4) (48, 190, 209). Some studies have found similar sensitivities to those of single cell lines. Likewise, viruses that do not grow well in cell culture (i.e., hMPV and coronavirus) cannot be isolated with these systems. Despite these limitations, many laboratories are moving to this technology, and some viruses may be grown in laboratories where resources previously limited the number of cell lines used. If these rapid culture methods fail to determine a pathogen in an immunosuppressed patient, alternative methods should be considered.

Antigen Detection

Several MAb pools are available for the simultaneous detection of multiple viral pathogens. These can be applied to culture supernatants or directly to patient specimens. Techniques for direct staining of respiratory cells from samples collected directly from the patient are well described elsewhere (221). Several commercially available reagents that screen for RSV, influenza A and B viruses, PIV 1 to 3, and adenovirus have been tested widely in immunocompromised and nonimmunocompromised patient populations. The assay contains two components, namely, one that is expressed in bright apple green when any of the viruses are present and a second that fluoresces yellow-gold in the presence of RSV (120). Individual fluorescein-conjugated MAbs are commercially available to differentiate between the different viruses. Although the sensitivity of this assay was excellent compared to those of culture and PCR, in a study using one of the available pools in an immunocompromised patient population, PCR was more sensitive in detecting dual infections (179); detailed sensitivity data are presented in Table 5.

NAT

Several PCR-based products are currently available or are being developed to screen for respiratory viral pathogens. Several platforms have been tested in transplant populations and appear to have excellent sensitivity (18, 112, 177), although sensitivities vary and are dependent on the platform used and the particular virus detected. For some viruses, such as coronavirus and hMPV, PCR is clearly the diagnostic method of choice. In most cases, PCR-based diagnostics are more sensitive than other diagnostic modalities and can screen for a broad range of respiratory pathogens from a single sample (18, 34, 68, 133, 146, 162, 177, 179, 182, 183, 189, 208, 210); multivirus NATs are available commercially from several reference laboratories, but there are limited head-to-head comparative data for individual methods. The most-studied individual system for immunosuppressed patients is the Hexaplex system (Prodesse, Inc., Waukesha, WI); this multiplex method was superior to immunofluorescence assay and/or culture for all viruses tested (126). This system was recently expanded to include hMPV (Hexaplex Plus). These techniques are being expanded further with approaches which are more efficient at detecting novel viruses and viruses that differ from those de-

Table 5. Detection of respiratory viruses by MAbs versus reverse transcription-PCR[a]

Virus	MAb result	No. of patients with reverse transcription-PCR result	
		Positive	Negative
Influenza A virus	Positive	0	0
	Negative	1	196
Influenza B virus	Positive	4	0
	Negative	0	193
Parainfluenza virus	Positive	6	0
	Negative	0	191
RSV	Positive	48	3
	Negative	3	143
Adenovirus	Positive	1	0
	Negative	0	196

[a]Adapted from *Journal of Medical Virology* (179) 75:336–347, 2005, with permission from Wiley-Liss, Inc., a subsidiary of John Wiley & Sons, Inc.

tected with traditional primer sets (119, 207). Several novel techniques, such as those that use Luminex's xMAP technology (Luminex Corp., Austin, TX) (30, 134, 139) and automated fluorescent capillary electrophoresis with GeneScan software analysis (55), may simplify multiplexed NAT and help to broaden its availability. These protocols can give very specific data about individual viruses (i.e., specific HA or NA types) and can easily be modified to include newly discovered pathogens. Recently, the xTAG Respiratory Viral Panel has been 501(k) cleared by the FDA and CE marked for sale in Europe and is commercially available in these areas. As more commercial methods become available, centers may have increased flexibility. Unfortunately, data on sensitivity and specificity using these systems are currently limited.

CONCLUSION

RNA respiratory viruses cause a broad range of clinically significant upper and lower respiratory tract diseases among immunocompromised patients. Since therapy is different for each virus and since it is often difficult to clinically differentiate the viral causes of disease, accurate laboratory diagnosis is essential. Rapid antigen testing, particularly for influenza virus, is useful because of its speed and specificity, but its sensitivity appears to be low among immunocompromised patients. NATs often provide the highest yields, although few clinical centers currently have the ability to perform these tests. With increasing numbers of commercial tests, increased automation, and improved ease of testing, NAT will become more widely used in the coming years. DFA techniques are currently more widely available, in many cases may approach the sensitivity of PCR, and can screen for several viruses at one time. DFA should be considered an alternative to PCR if this technology is not readily available. Traditionally, diagnostics for respiratory pathogens have been limited by methodological problems and limited treatment options. The past several years have seen a tremendous growth in diagnostic techniques, particularly using nucleic acid amplification methods and point-of-care testing. Similarly, more antivirals are under development, making it critical to continue efforts to improve testing and to allow appropriate application of these new compounds.

REFERENCES

1. Reference deleted.
2. Reference deleted.
3. **al-Mulla, W., A. el Mekki, and W. al-Nakib.** 1994. Rapid culture-amplified immunofluorescent test for the detection of human rhinoviruses in clinical samples: evidence of a common epitope in culture. *J. Med. Virol.* **42:**182–187.
4. **al-Nakib, W., C. J. Dearden, and D. A. Tyrrell.** 1989. Evaluation of a new enzyme-linked immunosorbent assay (ELISA) in the diagnosis of rhinovirus infection. *J. Med. Virol.* **29:**268–272.
5. **American Academy of Pediatrics Committee on Infectious Diseases and Committee of Fetus and Newborn.** 1998. Prevention of respiratory syncytial virus infections: indications for the use of palivizumab and update on the use of RSV-IGIV. *Pediatrics* **102:**1211–1216.
6. **Anaissie, E. J., T. H. Mahfouz, T. Aslan, A. Pouli, R. Desikan, A. Fassas, and B. Barlogie.** 2004. The natural history of respiratory syncytial virus infection in cancer and transplant patients: implications for management. *Blood* **103:**1611–1617.
7. **Anonymous.** 1999. Palivizumab (Synagis) for prevention of RSV infection. *Med. Lett. Drugs Ther.* **41:**3–4.
8. **Anonymous.** 2003. Revised indications for the use of palivizumab and respiratory syncytial virus immune globulin intravenous for the prevention of respiratory syncytial virus infections. *Pediatrics* **112:**1442–1446.
9. Reference deleted.
10. **Apalsch, A. M., M. Green, J. Ledesma-Medina, B. Nour, and E. R. Wald.** 1995. Parainfluenza and influenza virus infections in pediatric organ transplant recipients. *Clin. Infect. Dis.* **20:**394–399.
11. **Arruda, E., C. E. Crump, B. S. Rollins, A. Ohlin, and F. G. Hayden.** 1996. Comparative susceptibilities of human embryonic fibroblasts and HeLa cells for isolation of human rhinoviruses. *J. Clin. Microbiol.* **34:**1277–1279.
12. **Arruda, E., and F. G. Hayden.** 1993. Detection of human rhinovirus RNA in nasal washings by PCR. *Mol. Cell. Probes* **7:**373–379.
13. **Arruda, E., A. Pitkaranta, T. J. J. Witek, C. A. Doyle, and F. G. Hayden.** 1997. Frequency and natural history of rhinovirus infections in adults during autumn. *J. Clin. Microbiol.* **35:**2864–2868.
14. **Astrahan, P., I. Kass, M. A. Cooper, and I. T. Arkin.** 2004. A novel method of resistance for influenza against a channel-blocking antiviral drug. *Proteins* **55:**251–257.
15. **Barclay, W. S., and W. Al-Nakib.** 1987. An ELISA for the detection of rhinovirus specific antibody in serum and nasal secretion. *J. Virol. Methods* **15:**53–64.
16. **Barenfanger, J., C. Drake, T. Mueller, T. Troutt, J. O'Brien, and K. Guttman.** 2001. R-Mix cells are faster, at least as sensitive and marginally more costly than conventional cell lines for the detection of respiratory viruses. *J. Clin. Virol.* **22:**101–110.
17. **Bates, P. J., G. Sanderson, S. T. Holgate, and S. L. Johnston.** 1997. A comparison of RT-PCR, in-situ hybridisation and in-situ RT-PCR for the detection of rhinovirus infection in paraffin sections. *J. Virol. Methods* **67:**153–160.
18. **Bellau-Pujol, S., A. Vabret, L. Legrand, J. Dina, S. Gouarin, J. Petitjean-Lecherbonnier, B. Pozzetto, C. Ginevra, and F. Freymuth.** 2005. Development of three multiplex RT-PCR assays for the detection of 12 respiratory RNA viruses. *J. Virol. Methods* **126:**53–63.
19. **Bellei, N., D. Benfica, A. H. Perosa, R. Carlucci, M. Barros, and C. Granato.** 2003. Evaluation of a rapid test (QuickVue) compared with the shell vial assay for detection of influenza virus clearance after antiviral treatment. *J. Virol. Methods* **109:**85–88.
20. **Belshe, R. B., M. H. Smith, C. B. Hall, R. Betts, and A. J. Hay.** 1988. Genetic basis of resistance to rimantadine emerging during treatment of influenza virus infection. *J. Virol.* **62:**1508–1512.

21. **Belshe, R. B., L. P. Van Voris, and M. A. Mufson.** 1982. Parenteral administration of live respiratory syncytial virus vaccine: results of a field trial. *J. Infect. Dis.* **145**:311–319.
22. **Biacchesi, S., M. H. Skiadopoulos, L. Yang, B. R. Murphy, P. L. Collins, and U. J. Buchholz.** 2005. Rapid human metapneumovirus microneutralization assay based on green fluorescent protein expression. *J. Virol. Methods* **128**:192–197.
23. **Billings, J. L., M. I. Hertz, K. Savik, and C. H. Wendt.** 2002. Respiratory viruses and chronic rejection in lung transplant recipients. *J. Heart Lung Transplant.* **21**:559–566.
24. **Boeckh, M., J. Englund, Y. Li, C. Miller, A. Cross, H. Fernandez, J. Kuypers, H. Kim, J. Gnann, R. J. Whitley, et al.** 2007. Randomized controlled multicenter trial of aerosolized ribavirin for respiratory syncytial virus upper respiratory tract infections in hematopoietic cell transplant recipients. *Clin. Infect. Dis.* **44**:245–249.
25. **Boivin, G., Y. Abed, G. Pelletier, L. Ruel, D. Moisan, S. Cote, T. C. Peret, D. D. Erdman, and L. J. Anderson.** 2002. Virological features and clinical manifestations associated with human metapneumovirus: a new paramyxovirus responsible for acute respiratory-tract infections in all age groups. *J. Infect. Dis.* **186**:1330–1334.
26. **Boivin, G., Z. Coulombe, and C. Wat.** 2003. Quantification of the influenza virus load by real-time polymerase chain reaction in nasopharyngeal swabs of patients treated with oseltamivir. *J. Infect. Dis.* **188**:578–580.
27. **Borek, A. P., S. H. Clemens, V. K. Gaskins, D. Z. Aird, and A. Valsamakis.** 2006. Respiratory syncytial virus detection by Remel Xpect, Binax Now RSV, direct immunofluorescence staining, and tissue culture. *J. Clin. Microbiol.* **44**:1105–1107.
28. **Bright, R. A., M. J. Medina, X. Xu, G. Perez-Oronoz, T. R. Wallis, X. M. Davis, L. Povinelli, N. J. Cox, and A. I. Klimov.** 2005. Incidence of adamantane resistance among influenza A (H3N2) viruses isolated worldwide from 1994 to 2005: a cause for concern. *Lancet* **366**:1175–1181.
29. **Bright, R. A., D. K. Shay, B. Shu, N. J. Cox, and A. I. Klimov.** 2006. Adamantane resistance among influenza A viruses isolated early during the 2005–2006 influenza season in the United States. *JAMA* **295**:891–894.
30. **Brunstein, J., and E. Thomas.** 2006. Direct screening of clinical specimens for multiple respiratory pathogens using the Genaco respiratory panels 1 and 2. *Diagn. Mol. Pathol.* **15**:169–173.
31. **Cane, P. A., B. G. van den Hoogen, S. Chakrabarti, C. D. Fegan, and A. D. Osterhaus.** 2003. Human metapneumovirus in a haematopoietic stem cell transplant recipient with fatal lower respiratory tract disease. *Bone Marrow Transplant.* **31**:309–310.
31a. **Centers for Disease Control and Prevention.** 2006. High levels of adamantane resistance among influenza A (H3N2) viruses and interim guidelines for use of antiviral agents—United States, 2005–06 influenza season. *MMWR Morb. Mortal. Wkly. Rep.* **55**:44–46.
32. **Centers for Disease Control and Prevention.** 3 May 2005, posting date. Public health guidance for community-level preparedness and response to severe acute respiratory syndrome (SARS). Centers for Disease Control and Prevention, Atlanta, GA. http://www.cdc.gov/ncidod/sars/guidance/index.htm.
32a. **Centers for Disease Control and Prevention.** 2000. Guidelines for preventing opportunistic infections among hematopoietic stem cell transplant recipients. *MMWR Morb. Mortal. Wkly. Rep.* **49**:1–128.
33. **Chakinala, M. M., and M. J. Walter.** 2004. Community acquired respiratory viral infections after lung transplantation: clinical features and long-term consequences. *Semin. Thorac. Cardiovasc. Surg.* **16**:342–349.
34. **Christensen, M. S., L. P. Nielsen, and H. Hasle.** 2005. Few but severe viral infections in children with cancer: a prospective RT-PCR and PCR-based 12-month study. *Pediatr. Blood Cancer* **45**:945–951.
35. **Cobian, L., S. Houston, J. Greene, and J. T. Sinnott.** 1995. Parainfluenza virus respiratory infection after heart transplantation: successful treatment with ribavirin. *Clin. Infect. Dis.* **21**:1040–1041.
36. Reference deleted.
37. **Cortez, K., B. R. Murphy, K. N. Almeida, J. Beeler, R. A. Levandowski, V. J. Gill, R. W. Childs, A. J. Barrett, M. Smolskis, and J. E. Bennett.** 2002. Immune-globulin prophylaxis of respiratory syncytial virus infection in patients undergoing stem-cell transplantation. *J. Infect. Dis.* **186**:834–838.
38. **Cote, S., Y. Abed, and G. Boivin.** 2003. Comparative evaluation of real-time PCR assays for detection of the human metapneumovirus. *J. Clin. Microbiol.* **41**:3631–3635.
39. **Dagher, H., H. Donninger, P. Hutchinson, R. Ghildyal, and P. Bardin.** 2004. Rhinovirus detection: comparison of real-time and conventional PCR. *J. Virol. Methods* **117**:113–121.
40. **Dare, R., S. Sanghavi, A. Bullotta, M. C. Keightley, K. St. George, R. M. Wadowsky, D. Paterson, K. R. McCurry, T. A. Reinhart, S. Husain, and C. R. Rinaldo.** 2006. Diagnosis of human metapneumovirus infection in immunosuppressed lung transplant recipients and children evaluated for pertussis. *J. Clin. Microbiol.* **45**:548–552.
41. **Davies, H. A., and M. R. Macnaughton.** 1979. Comparison of the morphology of three coronaviruses. *Arch. Virol.* **59**:25–33.
42. **Dearden, C. J., and W. Al-Nakib.** 1987. Direct detection of rhinoviruses by an enzyme-linked immunosorbent assay. *J. Med. Virol.* **23**:179–189.
43. **Debiaggi, M., F. Canducci, M. Sampaolo, M. C. Marinozzi, M. Parea, C. Terulla, A. A. Colombo, E. P. Alessandrino, L. Z. Bragotti, M. Arghittu, A. Goglio, R. Migliavacca, E. Romero, and M. Clementi.** 2006. Persistent symptomless human metapneumovirus infection in hematopoietic stem cell transplant recipients. *J. Infect. Dis.* **194**:474–478.
44. **De Clercq, E.** 2006. Potential antivirals and antiviral strategies against SARS coronavirus infections. *Expert Rev. Anti Infect. Ther.* **4**:291–302.
45. **Dewhurst-Maridor, G., V. Simonet, J. E. Bornand, L. P. Nicod, and J. C. Pache.** 2004. Development of a quantitative TaqMan RT-PCR for respiratory syncytial virus. *J. Virol. Methods* **120**:41–49.
46. **Dingle, K. E., D. Crook, and K. Jeffery.** 2004. Stable and noncompetitive RNA internal control for routine clinical diagnostic reverse transcription-PCR. *J. Clin. Microbiol.* **42**:1003–1011.
47. **Duchini, A., R. M. Hendry, D. C. Redfield, and P. J. Pockros.** 2000. Influenza infection in patients before and after liver transplantation. *Liver Transpl.* **6**:531–542.
48. **Dunn, J. J., R. D. Woolstenhulme, J. Langer, and K. C. Carroll.** 2004. Sensitivity of respiratory virus culture when screening with R-Mix fresh cells. *J. Clin. Microbiol.* **42**:79–82.
49. **Ebihara, T., R. Endo, X. Ma, N. Ishiguro, and H. Kikuta.** 2005. Detection of human metapneumovirus antigens in nasopharyngeal secretions by an immunofluorescent-antibody test. *J. Clin. Microbiol.* **43**:1138–1141.
50. **Elizaga, J., E. Olavarria, J. Apperley, J. Goldman, and K. Ward.** 2001. Parainfluenza virus 3 infection after stem cell transplant: relevance to outcome of rapid diagnosis and ribavirin treatment. *Clin. Infect. Dis.* **32**:413–418.

51. **El Saleeby, C. M., J. Suzich, M. E. Conley, and J. P. DeVincenzo.** 2004. Quantitative effects of palivizumab and donor-derived T cells on chronic respiratory syncytial virus infection, lung disease, and fusion glycoprotein amino acid sequences in a patient before and after bone marrow transplantation. *Clin. Infect. Dis.* **39:**e17–e20.
52. **Englund, J. A.** 2001. Diagnosis and epidemiology of community-acquired respiratory virus infections in the immunocompromised host. *Biol. Blood Marrow Transplant.* 7(Suppl.): 2S–4S.
53. **Englund, J. A., R. E. Champlin, P. R. Wyde, H. Kantarjian, R. L. Atmar, J. Tarrand, H. Yousuf, H. Regnery, A. I. Klimov, N. J. Cox, and E. Whimbey.** 1998. Common emergence of amantadine- and rimantadine-resistant influenza A viruses in symptomatic immunocompromised adults. *Clin. Infect. Dis.* **26:**1418–1424.
54. **Englund, J. A., P. A. Piedra, A. Jewell, K. Patel, B. B. Baxter, and E. Whimbey.** 1996. Rapid diagnosis of respiratory syncytial virus infections in immunocompromised adults. *J. Clin. Microbiol.* **34:**1649–1653.
55. **Erdman, D. D., G. A. Weinberg, K. M. Edwards, F. J. Walker, B. C. Anderson, J. Winter, M. González, and L. J. Anderson.** 2003. GeneScan reverse transcription-PCR assay for detection of six common respiratory viruses in young children hospitalized with acute respiratory illness. *J. Clin. Microbiol.* **41:**4298–4303.
56. **Falsey, A. R., M. C. Criddle, and E. E. Walsh.** 2006. Detection of respiratory syncytial virus and human metapneumovirus by reverse transcription polymerase chain reaction in adults with and without respiratory illness. *J. Clin. Virol.* **35:**46–50.
57. **Falsey, A. R., R. M. McCann, W. J. Hall, M. M. Criddle, M. A. Formica, D. Wycoff, and J. E. Kolassa.** 1997. The "common cold" in frail older persons: impact of rhinovirus and coronavirus in a senior daycare center. *J. Am. Geriatr. Soc.* **45:** 706–711.
58. **Falsey, A. R., E. E. Walsh, and F. G. Hayden.** 2002. Rhinovirus and coronavirus infection-associated hospitalizations among older adults. *J. Infect. Dis.* **185:**1338–1341.
59. **Folz, R. J., and M. A. Elkordy.** 1999. Coronavirus pneumonia following autologous bone marrow transplantation for breast cancer. *Chest* **115:**901–905.
60. **Fong, C. K., M. K. Lee, and B. P. Griffith.** 2000. Evaluation of R-Mix FreshCells in shell vials for detection of respiratory viruses. *J. Clin. Microbiol.* **38:**4660–4662.
61. **Franquet, T., S. Rodriguez, R. Martino, T. Salinas, A. Gimenez, and A. Hidalgo.** 2005. Human metapneumovirus infection in hematopoietic stem cell transplant recipients: high-resolution computed tomography findings. *J. Comput. Assist. Tomogr.* **29:**223–227.
62. **Freymuth, F., G. Eugene, A. Vabret, J. Petitjean, E. Gennetay, J. Brouard, J. F. Duhamel, and B. Guillois.** 1995. Detection of respiratory syncytial virus by reverse transcription-PCR and hybridization with a DNA enzyme immunoassay. *J. Clin. Microbiol.* **33:**3352–3355.
63. **Garbino, J., S. Crespo, J. D. Aubert, T. Rochat, B. Ninet, C. Deffernez, W. Wunderli, J. C. Pache, P. M. Soccal, and L. Kaiser.** 2006. A prospective hospital-based study of the clinical impact of non-severe acute respiratory syndrome (non-SARS)-related human coronavirus infection. *Clin. Infect. Dis.* **43:**1009–1015.
64. **Ghosh, S., R. Champlin, R. Couch, J. Englund, I. Raad, S. Malik, M. Luna, and E. Whimbey.** 1999. Rhinovirus infections in myelosuppressed adult blood and marrow transplant recipients. *Clin. Infect. Dis.* **29:**528–532.
65. **Greensill, J., P. S. McNamara, W. Dove, B. Flanagan, R. L. Smyth, and C. A. Hart.** 2003. Human metapneumovirus in severe respiratory syncytial virus bronchiolitis. *Emerg. Infect. Dis.* **9:**372–375.
66. **Greve, J. M., G. Davis, A. M. Meyer, C. P. Forte, S. C. Yost, C. W. Marlor, M. E. Kamarck, and A. McClelland.** 1989. The major human rhinovirus receptor is ICAM-1. *Cell* **56:**839–847.
67. **Groneberg, D. A., S. M. Poutanen, D. E. Low, H. Lode, T. Welte, and P. Zabel.** 2005. Treatment and vaccines for severe acute respiratory syndrome. *Lancet Infect. Dis.* **5:**147–155.
68. **Gruteke, P., A. S. Glas, M. Dierdorp, W. B. Vreede, J. W. Pilon, and S. M. Bruisten.** 2004. Practical implementation of a multiplex PCR for acute respiratory tract infections in children. *J. Clin. Microbiol.* **42:**5596–5603.
69. **Gubareva, L. V.** 2004. Molecular mechanisms of influenza virus resistance to neuraminidase inhibitors. *Virus Res.* **103:**199–203.
70. **Gubareva, L. V., M. S. Nedyalkova, D. V. Novikov, K. G. Murti, E. Hoffmann, and F. G. Hayden.** 2002. A release-competent influenza A virus mutant lacking the coding capacity for the neuraminidase active site. *J. Gen. Virol.* **83:** 2683–2692.
71. **Gubareva, L. V., R. G. Webster, and F. G. Hayden.** 2002. Detection of influenza virus resistance to neuraminidase inhibitors by an enzyme inhibition assay. *Antivir. Res.* **53:**47–61.
72. **Gwaltney, J. M., Jr., and B. A. Heinz.** 2002. Rhinovirus, p. 995–1018. *In* D. D. Richman, R. J. Whitley, and F. G. Hayden (ed.), *Clinical Virology,* 2nd ed. ASM Press, Washington, DC.
73. **Gwaltney, J. M., Jr., J. O. Hendley, G. Simon, and W. S. Jordan, Jr.** 1966. Rhinovirus infections in an industrial population. I. The occurrence of illness. *N. Engl. J. Med.* **275:**1261–1268.
74. **Haagmans, B. L., and A. D. Osterhaus.** 2006. Coronaviruses and their therapy. *Antivir. Res.* **71:**397–403.
75. **Hamelin, M. E., Y. Abed, and G. Boivin.** 2004. Human metapneumovirus: a new player among respiratory viruses. *Clin. Infect. Dis.* **38:**983–990.
76. **Harris, J. M., II, and J. M. Gwaltney, Jr.** 1996. Incubation periods of experimental rhinovirus infection and illness. *Clin. Infect. Dis.* **23:**1287–1290.
77. **Hassan, I. A., R. Chopra, R. Swindell, and K. J. Mutton.** 2003. Respiratory viral infections after bone marrow/peripheral stem-cell transplantation: the Christie hospital experience. *Bone Marrow Transplant.* **32:**73–77.
78. **Hay, A. J., M. C. Zambon, A. J. Wolstenholme, J. J. Skehel, and M. H. Smith.** 1986. Molecular basis of resistance of influenza A viruses to amantadine. *J. Antimicrob. Chemother.* **18**(Suppl. B):19–29.
79. **Hayashi, K., T. Uchiyama, M. Iwat, K. Sano, M. Yanai, K. Kumasaka, and Y. Inamo.** 2005. The clinical usefulness of a newly rapid diagnosis kit, detection of respiratory syncytial virus. *Kansenshogaku Zasshi* **79:**276–283.
80. **Hayden, F. G.** 1996. Combination antiviral therapy for respiratory virus infections. *Antivir. Res.* **29:**45–48.
81. **Hayden, F. G.** 1997. Prevention and treatment of influenza in immunocompromised patients. *Am. J. Med.* **102:**55–60.
82. **Hayden, F. G., D. T. Herrington, T. L. Coats, K. Kim, E. C. Cooper, S. A. Villano, S. Liu, S. Hudson, D. C. Pevear, M. Collett, and M. McKinlay.** 2003. Efficacy and safety of oral pleconaril for treatment of colds due to picornaviruses in adults: results of 2 double-blind, randomized, placebo-controlled trials. *Clin. Infect. Dis.* **36:**1523–1532.
83. **Hayden, F. G., and P. Palese.** 2002. Influenza virus, p. 891–920. *In* D. D. Richman, R. J. Whitley, and F. G. Hayden (ed.), *Clinical Virology*, 2nd ed. ASM Press, Washington, DC.

84. **Hays, J. P., and S. H. Myint.** 1998. PCR sequencing of the spike genes of geographically and chronologically distinct human coronaviruses 229E. *J. Virol. Methods* **75:**179–193.
85. **Herrmann, B., C. Larsson, and B. W. Zweygberg.** 2001. Simultaneous detection and typing of influenza viruses A and B by a nested reverse transcription-PCR: comparison to virus isolation and antigen detection by immunofluorescence and optical immunoassay (FLU OIA). *J. Clin. Microbiol.* **39:**134–138.
86. **Hofer, F., M. Gruenberger, H. Kowalski, H. Machat, M. Huettinger, E. Kuechler, and D. Blaas.** 1994. Members of the low density lipoprotein receptor family mediate cell entry of a minor-group common cold virus. *Proc. Natl. Acad. Sci. USA* **91:**1839–1842.
87. **Hohenthal, U., J. Nikoskelainen, R. Vainionpaa, R. Peltonen, M. Routamaa, M. Itala, and P. Kotilainen.** 2002. Human parainfluenza type 4 virus (hPIV4) infection provoking many questions regarding the role of this virus as the causative agent of respiratory tract infections in the post-transplant setting. *Bone Marrow Transplant.* **29:**629–630.
88. **Hopkins, P. M., M. L. Plit, I. W. Carter, P. N. Chhajed, M. A. Malouf, and A. R. Glanville.** 2003. Indirect fluorescent antibody testing of nasopharyngeal swabs for influenza diagnosis in lung transplant recipients. *J. Heart Lung Transplant.* **22:**161–168.
89. **Huck, B., M. Egger, H. Bertz, G. Peyerl-Hoffman, W. V. Kern, D. Neumann-Haefelin, and V. Falcone.** 2006. Human metapneumovirus infection in a hematopoietic stem cell transplant recipient with relapsed multiple myeloma and rapidly progressing lung cancer. *J. Clin. Microbiol.* **44:**2300–2303.
90. **Hurt, A. C., P. Selleck, N. Komadina, R. Shaw, L. Brown, and I. G. Barr.** 2006. Susceptibility of highly pathogenic A(H5N1) avian influenza viruses to the neuraminidase inhibitors and adamantanes. *Antivir. Res.* **73:**228–231.
91. **Hyypia, T., T. Puhakka, O. Ruuskanen, M. Makela, A. Arola, and P. Arstila.** 1998. Molecular diagnosis of human rhinovirus infections: comparison with virus isolation. *J. Clin. Microbiol.* **36:**2081–2083.
92. **Ishibashi, T., H. Monobe, Y. Nomura, M. Shinogami, and J. Yano.** 2003. Multiplex nested reverse transcription-polymerase chain reaction for respiratory viruses in acute otitis media. *Ann. Otol. Rhinol. Laryngol.* **112:**252–257.
93. **Ishiguro, N., T. Ebihara, R. Endo, X. Ma, R. Shirotsuki, S. Ochiai, H. Ishiko, and H. Kikuta.** 2005. Immunofluorescence assay for detection of human metapneumovirus-specific antibodies by use of baculovirus-expressed fusion protein. *Clin. Diagn. Lab. Immunol.* **12:**202–205.
94. **Ison, M. G.** 2005. Respiratory viral infections in transplant recipients. *Curr. Opin. Organ Transplant.* **10:**312–319.
95. **Ison, M. G., L. V. Gubareva, R. L. Atmar, J. Treanor, and F. G. Hayden.** 2006. Recovery of drug-resistant influenza virus from immunocompromised patients: a case series. *J. Infect. Dis.* **193:**760–764.
96. **Ison, M. G., and F. G. Hayden.** 2002. Viral infections in immunocompromised patients: what's new with respiratory viruses? *Curr. Opin. Infect. Dis.* **15:**355–367.
97. **Ison, M. G., F. G. Hayden, L. Kaiser, L. Corey, and M. Boeckh.** 2003. Rhinovirus infections in hematopoietic stem cell transplant recipients with pneumonia. *Clin. Infect. Dis.* **36:**1139–1143.
98. **Ison, M. G., S. L. Johnston, P. Openshaw, B. Murphy, and F. Hayden.** 2004. Current research on respiratory viral infections: fifth international symposium. *Antivir. Res.* **62:**75–110.
99. **Ison, M. G., J. Mills, P. Openshaw, M. Zambon, A. Osterhaus, and F. Hayden.** 2002. Current research on respiratory viral infections: fourth international symposium. *Antivir. Res.* **55:**227–278.
100. **Ison, M. G., V. P. Mishin, T. J. Braciale, F. G. Hayden, and L. V. Gubareva.** 2006. Comparative activities of oseltamivir and A-322278 in immunocompetent and immunocompromised murine models of influenza virus infection. *J. Infect. Dis.* **193:**765–772.
101. **Ito, T., Y. Suzuki, L. Mitnaul, A. Vines, H. Kida, and Y. Kawaoka.** 1997. Receptor specificity of influenza A viruses correlates with the agglutination of erythrocytes from different animal species. *Virology* **227:**493–499.
102. **Jenison, R., M. Rihanek, and B. Polisky.** 2001. Use of a thin film biosensor for rapid visual detection of PCR products in a multiplex format. *Biosens. Bioelectron.* **16:**757–763.
103. **Johny, A. A., A. Clark, N. Price, D. Carrington, A. Oakhill, and D. I. Marks.** 2002. The use of zanamivir to treat influenza A and B infection after allogeneic stem cell transplantation. *Bone Marrow Transplant.* **29:**113–115.
104. **Kaiser, L., J. D. Aubert, J. C. Pache, C. Deffernez, T. Rochat, J. Garbino, W. Wunderli, P. Meylan, S. Yerly, L. Perrin, I. Letovanec, L. Nicod, C. Tapparel, and P. M. Soccal.** 2006. Chronic rhinoviral infection in lung transplant recipients. *Am. J. Respir. Crit. Care Med.* **174:**1392–1399.
105. **Kapikian, A. Z., H. D. James, Jr., S. J. Kelly, J. H. Dees, H. C. Turner, K. McIntosh, H. W. Kim, R. H. Parrott, M. M. Vincent, and R. M. Chanock.** 1969. Isolation from man of "avian infectious bronchitis virus-like" viruses (coronaviruses) similar to 229E virus, with some epidemiological observations. *J. Infect. Dis.* **119:**282–290.
106. **Kisch, A. L., K. M. Johnson, and R. M. Chanock.** 1962. Immunofluorescence with respiratory syncytial virus. *Virology* **16:**177–189.
107. **Kishi, Y., M. Kami, Y. Oki, Y. Kazuyama, M. Kawabata, S. Miyakoshi, S. Morinaga, R. Suzuki, S. Mori, and Y. Muto.** 2000. Donor lymphocyte infusion for treatment of life-threatening respiratory syncytial virus infection following bone marrow transplantation. *Bone Marrow Transplant.* **26:**573–576.
108. **Kiso, M., K. Mitamura, Y. Sakai-Tagawa, K. Shiraishi, C. Kawakami, K. Kimura, F. G. Hayden, N. Sugaya, and Y. Kawaoka.** 2004. Resistant influenza A viruses in children treated with oseltamivir: descriptive study. *Lancet* **364:**759–765.
109. **Klimov, A. I., E. Rocha, F. G. Hayden, P. A. Shult, L. F. Roumillat, and N. J. Cox.** 1995. Prolonged shedding of amantadine-resistant influenza A viruses by immunodeficient patients: detection by polymerase chain reaction-restriction analysis. *J. Infect. Dis.* **172:**1352–1355.
110. **Kobashigawa, J. A., L. Warner-Stevenson, B. L. Johnson, J. D. Moriguchi, N. Kawata, D. C. Drinkwater, and H. Laks.** 1993. Influenza vaccine does not cause rejection after cardiac transplantation. *Transplant. Proc.* **25:**2738–2739.
111. **Kremser, L., T. Konecsni, D. Blaas, and E. Kenndler.** 2004. Fluorescence labeling of human rhinovirus capsid and analysis by capillary electrophoresis. *Anal. Chem.* **76:**4175–4181.
112. **Kumar, D., D. Erdman, S. Keshavjee, T. Peret, R. Tellier, D. Hadjiliadis, G. Johnson, M. Ayers, D. Siegal, and A. Humar.** 2005. Clinical impact of community-acquired respiratory viruses on bronchiolitis obliterans after lung transplant. *Am. J. Transplant.* **5:**2031–2036.
113. **Kumar, D., and A. Humar.** 2005. Emerging viral infections in transplant recipients. *Curr. Opin. Infect. Dis.* **18:**337–341.
114. **Kumar, D., R. Tellier, R. Draker, G. Levy, and A. Humar.** 2003. Severe acute respiratory syndrome (SARS) in a liver transplant recipient and guidelines for donor SARS screening. *Am. J. Transplant.* **3:**977–981.

115. **Kuroiwa, Y., K. Nagai, L. Okita, S. Ukae, T. Mori, T. Hotsubo, and H. Tsutsumi.** 2004. Comparison of an immunochromatography test with multiplex reverse transcription-PCR for rapid diagnosis of respiratory syncytial virus infections. *J. Clin. Microbiol.* **42:**4812–4814.
116. Reference deleted.
117. **Kuypers, J., N. Wright, J. Ferrenberg, M. L. Huang, A. Cent, L. Corey, and R. Morrow.** 2006. Comparison of real-time PCR assays with fluorescent-antibody assays for diagnosis of respiratory virus infections in children. *J. Clin. Microbiol.* **44:**2382–2388.
118. **Lam, M. F., G. C. Ooi, B. Lam, J. C. Ho, W. H. Seto, P. L. Ho, P. C. Wong, R. Liang, W. K. Lam, and K. W. Tsang.** 2004. An indolent case of severe acute respiratory syndrome. *Am. J. Respir. Crit. Care Med.* **169:**125–128.
119. **Lamson, D., N. Renwick, V. Kapoor, Z. Liu, G. Palacios, J. Ju, A. Dean, K. St. George, T. Briese, and W. I. Lipkin.** 2006. MassTag polymerase-chain-reaction detection of respiratory pathogens, including a new rhinovirus genotype, that caused influenza-like illness in New York State during 2004–2005. *J. Infect. Dis.* **194:**1398–1402.
120. **Landry, M. L., and D. Ferguson.** 2000. SimulFluor respiratory screen for rapid detection of multiple respiratory viruses in clinical specimens by immunofluorescence staining. *J. Clin. Microbiol.* **38:**708–711.
121. **Landry, M. L., D. Ferguson, S. Cohen, T. C. Peret, and D. D. Erdman.** 2005. Detection of human metapneumovirus in clinical samples by immunofluorescence staining of shell vial centrifugation cultures prepared from three different cell lines. *J. Clin. Microbiol.* **43:**1950–1952.
122. **Larcher, C., C. Geltner, H. Fischer, D. Nachbaur, L. C. Muller, and H. P. Huemer.** 2005. Human metapneumovirus infection in lung transplant recipients: clinical presentation and epidemiology. *J. Heart Lung Transplant.* **24:**1891–1901.
123. **Lennette, E. H.** 1992. Laboratory diagnosis of viral infections. Marcel Dekker, Inc., New York, NY.
124. **Leonardi, G. P., H. Leib, G. S. Birkhead, C. Smith, P. Costello, and W. Conron.** 1994. Comparison of rapid detection methods for influenza A virus and their value in health care management of institutionalized geriatric patients. *J. Clin. Microbiol.* **32:**70–74.
125. **Levin, M. D., and G. J. van Doornum.** 2004. An immunocompromised host with bilateral pulmonary infiltrates. *Neth. J. Med.* **62:**197–210.
126. **Liolios, L., A. Jenney, D. Spelman, T. Kotsimbos, M. Catton, and S. Wesselingh.** 2001. Comparison of a multiplex reverse transcription-PCR-enzyme hybridization assay with conventional viral culture and immunofluorescence techniques for the detection of seven viral respiratory pathogens. *J. Clin. Microbiol.* **39:**2779–2783.
127. **Lopez-Huertas, M. R., I. Casas, B. Acosta-Herrera, M. L. Garcia, M. T. Coiras, and P. Perez-Brena.** 2005. Two RT-PCR based assays to detect human metapneumovirus in nasopharyngeal aspirates. *J. Virol. Methods* **129:**1–7.
128. **Machado, C. M., L. S. Boas, A. V. Mendes, I. F. da Rocha, D. Sturaro, F. L. Dulley, and C. S. Pannuti.** 2004. Use of oseltamivir to control influenza complications after bone marrow transplantation. *Bone Marrow Transplant.* **34:**111–114.
129. **Machado, C. M., L. S. Boas, A. V. Mendes, M. F. Santos, I. F. da Rocha, D. Sturaro, F. L. Dulley, and C. S. Pannuti.** 2003. Low mortality rates related to respiratory virus infections after bone marrow transplantation. *Bone Marrow Transplant.* **31:**695–700.
130. **Macnaughton, M. R.** 1982. Occurrence and frequency of coronavirus infections in humans as determined by enzyme-linked immunosorbent assay. *Infect. Immun.* **38:**419–423.
131. **Macnaughton, M. R., D. Flowers, and D. Isaacs.** 1983. Diagnosis of human coronavirus infections in children using enzyme-linked immunosorbent assay. *J. Med. Virol.* **11:**319–325.
132. **Madhi, S. A., H. Ludewick, Y. Abed, K. P. Klugman, and G. Boivin.** 2003. Human metapneumovirus-associated lower respiratory tract infections among hospitalized human immunodeficiency virus type 1 (HIV-1)-infected and HIV-1-uninfected African infants. *Clin. Infect. Dis.* **37:**1705–1710.
133. **Magnard, C., M. Valette, M. Aymard, and B. Lina.** 1999. Comparison of two nested PCR, cell culture, and antigen detection for the diagnosis of upper respiratory tract infections due to influenza viruses. *J. Med. Virol.* **59:**215–220.
134. **Mahony, J., S. Chong, F. Merante, S. Yaghoubian, T. Sinha, C. Lisle, and R. Janeczko.** 2007. Development of a respiratory virus panel test for detection of twenty human respiratory viruses by use of multiplex PCR and a fluid microbead-based assay. *J. Clin. Microbiol.* **45:**2965–2970.
135. **Makela, M. J., T. Puhakka, O. Ruuskanen, M. Leinonen, P. Saikku, M. Kimpimaki, S. Blomqvist, T. Hyypia, and P. Arstila.** 1998. Viruses and bacteria in the etiology of the common cold. *J. Clin. Microbiol.* **36:**539–542.
136. **Malavaud, S., B. Malavaud, K. Sandres, D. Durand, N. Marty, J. Icart, and L. Rostaing.** 2001. Nosocomial outbreak of influenza virus A (H3N2) infection in a solid organ transplant department. *Transplantation* **72:**535–537.
137. **Marr, K. A., R. A. Carter, M. Boeckh, P. Martin, and L. Corey.** 2002. Invasive aspergillosis in allogeneic stem cell transplant recipients: changes in epidemiology and risk factors. *Blood* **100:**4358–4366.
138. **Martino, R., R. P. Porras, N. Rabella, J. V. Williams, E. Ramila, N. Margall, R. Labeaga, J. E. Crowe, Jr., P. Coll, and J. Sierra.** 2005. Prospective study of the incidence, clinical features, and outcome of symptomatic upper and lower respiratory tract infections by respiratory viruses in adult recipients of hematopoietic stem cell transplants for hematologic malignancies. *Biol. Blood Marrow Transplant.* **11:**781–796.
139. **Martins, T. B.** 2002. Development of internal controls for the Luminex instrument as part of a multiplex seven-analyte viral respiratory antibody profile. *Clin. Diagn. Lab. Immunol.* **9:**41–45.
140. **McCurdy, L. H., A. Milstone, and S. Dummer.** 2003. Clinical features and outcomes of paramyxoviral infection in lung transplant recipients treated with ribavirin. *J. Heart Lung Transplant.* **22:**745–753.
141. **McIntosh, K.** 2002. Coronaviruses, p. 1087–1096. *In* D. D. Richman, R. J. Whitley, and F. G. Hayden (ed.), *Clinical Virology,* 2nd ed. ASM Press, Washington, DC.
142. **McKimm-Breschkin, J., T. Trivedi, A. Hampson, A. Hay, A. Klimov, M. Tashiro, F. Hayden, and M. Zambon.** 2003. Neuraminidase sequence analysis and susceptibilities of influenza virus clinical isolates to zanamivir and oseltamivir. *Antimicrob. Agents Chemother.* **47:**2264–2272.
143. **McKimm-Breschkin, J. L.** 2005. Management of influenza virus infections with neuraminidase inhibitors: detection, incidence, and implications of drug resistance. *Treat. Respir. Med.* **4:**107–116.
144. **Mentel, R., U. Wegner, R. Bruns, and L. Gütler.** 2003. Real-time PCR to improve the diagnosis of respiratory syncytial virus infection. *J. Med. Microbiol.* **52:**893–896.
145. **Miall, F., A. Rye, M. Fraser, A. Hunter, and J. A. Snowden.** 2002. Human parainfluenza type 4 infection: a case report highlighting pathogenicity and difficulties in rapid diagnosis in the post-transplant setting. *Bone Marrow Transplant.* **29:**541–542.

146. **Milstone, A. P., L. M. Brumble, J. Barnes, W. Estes, J. E. Loyd, R. N. Pierson III, and S. Dummer.** 2006. A single-season prospective study of respiratory viral infections in lung transplant recipients. *Eur. Respir. J.* **28:**131–137.
147. **Moes, E., L. Vijgen, E. Keyaerts, K. Zlateva, S. Li, P. Maes, K. Pyrc, B. Berkhout, L. van der Hoek, and M. Van Ranst.** 2005. A novel pancoronavirus RT-PCR assay: frequent detection of human coronavirus NL63 in children hospitalized with respiratory tract infections in Belgium. *BMC Infect. Dis.* **5:**6.
148. **Monto, A. S.** 1994. Studies of the community and family: acute respiratory illness and infection. *Epidemiol. Rev.* **16:**351–373.
149. **Monto, A. S., E. R. Bryan, and S. Ohmit.** 1987. Rhinovirus infections in Tecumseh, Michigan: frequency of illness and number of serotypes. *J. Infect. Dis.* **156:**43–49.
150. **Moore, C., S. Hibbitts, N. Owen, S. A. Corden, G. Harrison, J. Fox, C. Gelder, and D. Westmoreland.** 2004. Development and evaluation of a real-time nucleic acid sequence based amplification assay for rapid detection of influenza A. *J. Med. Virol.* **74:**619–628.
151. **Moscona, A.** 2005. Neuraminidase inhibitors for influenza. *N. Engl. J. Med.* **353:**1363–1373.
152. **Myint, S., D. Harmsen, T. Raabe, and S. G. Siddell.** 1990. Characterization of a nucleic acid probe for the diagnosis of human coronavirus 229E infections. *J. Med. Virol.* **31:**165–172.
153. **Myint, S., S. Johnston, G. Sanderson, and H. Simpson.** 1994. Evaluation of nested polymerase chain methods for the detection of human coronaviruses 229E and OC43. *Mol. Cell. Probes* **8:**357–364.
154. **Ng, E. K., and Y. M. Lo.** 2006. Molecular diagnosis of severe acute respiratory syndrome. *Methods Mol. Biol.* **336:** 163–175.
155. **Nichols, W. G., L. Corey, T. Gooley, C. Davis, and M. Boeckh.** 2001. Parainfluenza virus infections after hematopoietic stem cell transplantation: risk factors, response to antiviral therapy, and effect on transplant outcome. *Blood* **98:**573–578.
156. **Nichols, W. G., D. D. Erdman, A. Han, C. Zukerman, L. Corey, and M. Boeckh.** 2004. Prolonged outbreak of human parainfluenza virus 3 infection in a stem cell transplant outpatient department: insights from molecular epidemiologic analysis. *Biol. Blood Marrow Transplant.* **10:**58–64.
157. **Nichols, W. G., T. Gooley, and M. Boeckh.** 2001. Community-acquired respiratory syncytial virus and parainfluenza virus infections after hematopoietic stem cell transplantation: the Fred Hutchinson Cancer Research Center experience. *Biol. Blood Marrow Transplant.* **7**(Suppl.):11S–15S.
158. **Nichols, W. G., K. A. Guthrie, L. Corey, and M. Boeckh.** 2004. Influenza infections after hematopoietic stem cell transplantation: risk factors, mortality, and the effect of antiviral therapy. *Clin. Infect. Dis.* **39:**1300–1306.
159. **Nicholson, K. G., J. M. Wood, and M. Zambon.** 2003. Influenza. *Lancet* **362:**1733–1745.
160. **Noyola, D. E., B. Clark, F. T. O'Donnell, R. L. Atmar, J. Greer, and G. J. Demmler.** 2000. Comparison of a new neuraminidase detection assay with an enzyme immunoassay, immunofluorescence, and culture for rapid detection of influenza A and B viruses in nasal wash specimens. *J. Clin. Microbiol.* **38:**1161–1165.
161. **Noyola, D. E., and G. J. Demmler.** 2000. Effect of rapid diagnosis on management of influenza A infections. *Pediatr. Infect. Dis. J.* **19:**303–307.
162. **Ong, G. M., D. E. Wyatt, H. J. O'Neill, C. McCaughey, and P. V. Coyle.** 2001. A comparison of nested polymerase chain reaction and immunofluorescence for the diagnosis of respiratory infections in children with bronchiolitis, and the implications for a cohorting strategy. *J. Hosp. Infect.* **49:**122–128.
163. **Ottolini, M. G., S. R. Curtis, A. Mathews, S. R. Ottolini, and G. A. Prince.** 2002. Palivizumab is highly effective in suppressing respiratory syncytial virus in an immunosuppressed animal model. *Bone Marrow Transplant.* **29:**117–120.
164. **Ottolini, M. G., D. D. Porter, V. G. Hemming, M. N. Zimmerman, N. M. Schwab, and G. A. Prince.** 1999. Effectiveness of RSVIG prophylaxis and therapy of respiratory syncytial virus in an immunosuppressed animal model. *Bone Marrow Transplant.* **24:**41–45.
165. **Pachucki, C. T.** 2005. Rapid tests for influenza. *Curr. Infect. Dis. Rep.* **7:**187–192.
166. **Papadopoulos, N. G., J. Hunter, G. Sanderson, J. Meyer, and S. L. Johnston.** 1999. Rhinovirus identification by BglI digestion of picornavirus RT-PCR amplicons. *J. Virol. Methods* **80:**179–185.
167. **Papadopoulos, N. G., G. Sanderson, J. Hunter, and S. L. Johnston.** 1999. Rhinoviruses replicate effectively at lower airway temperatures. *J. Med. Virol.* **58:**100–104.
168. **Peck, A. J., L. Corey, and M. Boeckh.** 2004. Pretransplantation respiratory syncytial virus infection: impact of a strategy to delay transplantation. *Clin. Infect. Dis.* **39:**673–680.
169. **Peiris, J. S., Y. Guan, and K. Y. Yuen.** 2004. Severe acute respiratory syndrome. *Nat. Med.* **10:**S88–S97.
170. **Pelletier, G., P. Déry, Y. Abed, and G. Boivin.** 2002. Respiratory tract reinfections by the new human metapneumovirus in an immunocompromised child. *Emerg. Infect. Dis.* **8:**976–979.
171. **Percivalle, E., A. Sarasini, L. Visai, M. G. Revello, and G. Gerna.** 2005. Rapid detection of human metapneumovirus strains in nasopharyngeal aspirates and shell vial cultures by monoclonal antibodies. *J. Clin. Microbiol.* **43:**3443–3446.
172. **Peters, P. H., Jr., S. Gravenstein, P. Norwood, V. De Bock, A. Van Couter, M. Gibbens, T. A. von Planta, and P. Ward.** 2001. Long-term use of oseltamivir for the prophylaxis of influenza in a vaccinated frail older population. *J. Am. Geriatr. Soc.* **49:**1025–1031.
173. **Piedra, P. A., J. A. Englund, and W. P. Glezen.** 2002. Respiratory syncytial virus and parainfluenza viruses, p. 763–790. *In* D. D. Richman, R. J. Whitley, and F. G. Hayden (ed.), *Clinical Virology,* 2nd ed. ASM Press, Washington, DC.
174. **Pitkaranta, A., T. Puhakka, M. J. Makela, O. Ruuskanen, O. Carpen, and A. Vaheri.** 2003. Detection of rhinovirus RNA in middle turbinate of patients with common colds by in situ hybridization. *J. Med. Virol.* **70:**319–323.
175. **Puhakka, T., H. Lehti, R. Vainionpaa, V. Jormanainen, M. Pulkkinen, S. Sharp, C. Kerr, M. Dempsey, C. J. Ring, C. Ward, and M. Tisdale.** 2003. Zanamivir: a significant reduction in viral load during treatment in military conscripts with influenza. *Scand. J. Infect. Dis.* **35:**52–58.
176. **Pyrc, K., B. Berkhout, and L. van der Hoek.** 2007. The novel human coronaviruses NL63 and HKU1. *J. Virol.* **81:**3051–3057.
177. **Roghmann, M., K. Ball, D. Erdman, J. Lovchik, L. J. Anderson, and R. Edelman.** 2003. Active surveillance for respiratory virus infections in adults who have undergone bone marrow and peripheral blood stem cell transplantation. *Bone Marrow Transplant.* **32:**1085–1088.
178. **Rothberg, M. B., D. Fisher, B. Kelly, and D. N. Rose.** 2005. Management of influenza symptoms in healthy children:

cost-effectiveness of rapid testing and antiviral therapy. *Arch. Pediatr. Adolesc. Med.* **159:**1055–1062.

179. **Rovida, F., E. Percivalle, M. Zavattoni, M. Torsellini, A. Sarasini, G. Campanini, S. Paolucci, F. Baldanti, M. G. Revello, and G. Gerna.** 2005. Monoclonal antibodies versus reverse transcription-PCR for detection of respiratory viruses in a patient population with respiratory tract infections admitted to hospital. *J. Med. Virol.* **75:**336–347.
180. **Sauerbrei, A., A. Ulbricht, and P. Wutzler.** 2003. Semi-quantitative detection of viral RNA in influenza A virus-infected mice for evaluation of antiviral compounds. *Antivir. Res.* **58:**81–87.
181. **Scheltinga, S. A., K. E. Templeton, M. F. Beersma, and E. C. Claas.** 2005. Diagnosis of human metapneumovirus and rhinovirus in patients with respiratory tract infections by an internally controlled multiplex real-time RNA PCR. *J. Clin. Virol.* **33:**306–311.
182. **Semple, M. G., A. Cowell, W. Dove, J. Greensill, P. S. McNamara, C. Halfhide, P. Shears, R. L. Smyth, and C. A. Hart.** 2005. Dual infection of infants by human metapneumovirus and human respiratory syncytial virus is strongly associated with severe bronchiolitis. *J. Infect. Dis.* **191:**382–386.
183. **Sizun, J., N. Arbour, and P. J. Talbot.** 1998. Comparison of immunofluorescence with monoclonal antibodies and RT-PCR for the detection of human coronaviruses 229E and OC43 in cell culture. *J. Virol. Methods* **72:**145–152.
184. **Sizun, J., D. Soupre, M. C. Legrand, J. D. Giroux, S. Rubio, J. M. Cauvin, C. Chastel, D. Alix, and L. de Parscau.** 1995. Neonatal nosocomial respiratory infection with coronavirus: a prospective study in a neonatal intensive care unit. *Acta Paediatr.* **84:**617–620.
185. **Smit, M., K. A. Beynon, D. R. Murdoch, and L. C. Jennings.** 2007. Comparison of the NOW Influenza A & B, NOW Flu A, NOW Flu B, and Directigen Flu A+B assays, and immunofluorescence with viral culture for the detection of influenza A and B viruses. *Diagn. Microbiol. Infect. Dis.* **57:**67–70.
186. **Smith, K. J., and M. S. Roberts.** 2002. Cost-effectiveness of newer treatment strategies for influenza. *Am. J. Med.* **113:**300–307.
187. **Smith, N. M., J. S. Bresee, D. K. Shay, T. M. Uyeki, N. J. Cox, and R. A. Strikas.** 2006. Prevention and control of influenza: recommendations of the Advisory Committee on Immunization Practices (ACIP). *MMWR Recommend. Rep.* **55:**1–42.
188. **Steininger, C., S. W. Aberle, and T. Popow-Kraupp.** 2001. Early detection of acute rhinovirus infections by a rapid reverse transcription-PCR assay. *J. Clin. Microbiol.* **39:**129–133.
189. **Steininger, C., M. Kundi, S. W. Aberle, J. H. Aberle, and T. Popow-Kraupp.** 2002. Effectiveness of reverse transcription-PCR, virus isolation, and enzyme-linked immunosorbent assay for diagnosis of influenza A virus infection in different age groups. *J. Clin. Microbiol.* **40:**2051–2056.
190. **St. George, K., N. M. Patel, R. A. Hartwig, D. R. Scholl, J. A. Jollick, Jr., L. M. Kauffmann, M. R. Evans, and C. R. Rinaldo, Jr.** 2002. Rapid and sensitive detection of respiratory virus infections for directed antiviral treatment using R-Mix cultures. *J. Clin. Virol.* **24:**107–115.
191. **Sumino, K. C., E. Agapov, R. A. Pierce, E. P. Trulock, J. D. Pfeifer, J. H. Ritter, M. Gaudreault-Keener, G. A. Storch, and M. J. Holtzman.** 2005. Detection of severe human metapneumovirus infection by real-time polymerase chain reaction and histopathological assessment. *J. Infect. Dis.* **192:**1052–1060.
192. **Thervet, E., C. Legendre, P. Beaune, and D. Anglicheau.** 2005. Cytochrome P450 3A polymorphisms and immunosuppressive drugs. *Pharmacogenomics* **6:**37–47.

192a. **The Writing Committee of the World Health Organization (WHO) Consultation on Human Influenza A/H5.** 2005. Avian influenza A (H5N1) infection in humans. *N. Engl. J. Med.* **353:**1374–1385.

193. **Thompson, W. W., D. K. Shay, E. Weintraub, L. Brammer, C. B. Bridges, N. J. Cox, and K. Fukuda.** 2004. Influenza-associated hospitalizations in the United States. *JAMA* **292:**1333–1340.
194. **Thompson, W. W., D. K. Shay, E. Weintraub, L. Brammer, N. Cox, L. J. Anderson, and K. Fukuda.** 2003. Mortality associated with influenza and respiratory syncytial virus in the United States. *JAMA* **289:**179–186.
195. **Uncapher, C. R., C. M. DeWitt, and R. J. Colonno.** 1991. The major and minor group receptor families contain all but one human rhinovirus serotype. *Virology* **180:**814–817.
196. **van Burik, J. A.** 2006. Human metapneumovirus: important but not currently diagnosable. *Ann. Intern. Med.* **144:**374–375.
197. **van den Hoogen, B. G., J. C. de Jong, J. Groen, T. Kuiken, R. de Groot, R. A. Fouchier, and A. D. Osterhaus.** 2001. A newly discovered human pneumovirus isolated from young children with respiratory tract disease. *Nat. Med.* **7:**719–724.
198. **van den Hoogen, B. G., D. M. Osterhaus, and R. A. Fouchier.** 2004. Clinical impact and diagnosis of human metapneumovirus infection. *Pediatr. Infect. Dis. J.* **23:**S25–S32.
199. **van den Hoogen, B. G., G. J. van Doornum, J. C. Fockens, J. J. Cornelissen, W. E. Beyer, R. de Groot, A. D. Osterhaus, and R. A. Fouchier.** 2003. Prevalence and clinical symptoms of human metapneumovirus infection in hospitalized patients. *J. Infect. Dis.* **188:**1571–1577.
200. **van der Hoek, L., K. Pyrc, and B. Berkhout.** 2006. Human coronavirus NL63, a new respiratory virus. *FEMS Microbiol. Rev.* **30:**760–773.
201. **van der Hoek, L., K. Pyrc, M. F. Jebbink, W. Vermeulen-Oost, R. J. Berkhout, K. C. Wolthers, P. M. Wertheim-van Dillen, J. Kaandorp, J. Spaargaren, and B. Berkhout.** 2004. Identification of a new human coronavirus. *Nat. Med.* **10:**368–373.
202. **van Elden, L. J., A. M. van Loon, A. van der Beek, K. A. Hendriksen, A. I. Hoepelman, M. G. van Kraaij, P. Schipper, and M. Nijhuis.** 2003. Applicability of a real-time quantitative PCR assay for diagnosis of respiratory syncytial virus infection in immunocompromised adults. *J. Clin. Microbiol.* **41:**4378–4381.
203. **van Kraaij, M. G., L. J. van Elden, A. M. van Loon, K. A. Hendriksen, L. Laterveer, A. W. Dekker, and M. Nijhuis.** 2005. Frequent detection of respiratory viruses in adult recipients of stem cell transplants with the use of real-time polymerase chain reaction, compared with viral culture. *Clin. Infect. Dis.* **40:**662–669.
204. **Vilchez, R. A., J. Dauber, and S. Kusne.** 2003. Infectious etiology of bronchiolitis obliterans: the respiratory viruses connection—myth or reality? *Am. J. Transplant.* **3:**245–249.
205. **Vilchez, R. A., J. Dauber, K. McCurry, A. Iacono, and S. Kusne.** 2003. Parainfluenza virus infection in adult lung transplant recipients: an emergent clinical syndrome with implications on allograft function. *Am. J. Transplant.* **3:** 116–120.
206. **Vilchez, R. A., K. McCurry, J. Dauber, A. Lacono, B. Griffith, J. Fung, and S. Kusne.** 2002. Influenza virus infection in adult solid organ transplant recipients. *Am. J. Transplant.* **2:**287–291.

207. **Wang, D., L. Coscoy, M. Zylberberg, P. C. Avila, H. A. Boushey, D. Ganem, and J. L. DeRisi.** 2002. Microarray-based detection and genotyping of viral pathogens. *Proc. Natl. Acad. Sci. USA* **99:**15687–15692.
208. **Watzinger, F., M. Suda, S. Preuner, R. Baumgartinger, K. Ebner, L. Baskova, H. G. Niesters, A. Lawitschka, and T. Lion.** 2004. Real-time quantitative PCR assays for detection and monitoring of pathogenic human viruses in immunosuppressed pediatric patients. *J. Clin. Microbiol.* **42:** 5189–5198.
209. **Weinberg, A., L. Brewster, J. Clark, and E. Simoes.** 2004. Evaluation of R-Mix shell vials for the diagnosis of viral respiratory tract infections. *J. Clin. Virol.* **30:**100–105.
210. **Weinberg, A., M. R. Zamora, S. Li, F. Torres, and T. N. Hodges.** 2002. The value of polymerase chain reaction for the diagnosis of viral respiratory tract infections in lung transplant recipients. *J. Clin. Virol.* **25:**171–175.
211. **Wendt, C. H., J. M. Fox, and M. I. Hertz.** 1995. Paramyxovirus infection in lung transplant recipients. *J. Heart Lung Transplant.* **14:**479–485.
212. **Williams, J. V., R. Martino, N. Rabella, M. Otegui, R. Parody, J. M. Heck, and J. E. Crowe, Jr.** 2005. A prospective study comparing human metapneumovirus with other respiratory viruses in adults with hematologic malignancies and respiratory tract infections. *J. Infect. Dis.* **192:**1061–1065.
213. **Woo, P. C., S. K. Lau, C. M. Chu, K. H. Chan, H. W. Tsoi, Y. Huang, B. H. Wong, R. W. Poon, J. J. Cai, W. K. Luk, L. L. Poon, S. S. Wong, Y. Guan, J. S. Peiris, and K. Y. Yuen.** 2005. Characterization and complete genome sequence of a novel coronavirus, coronavirus HKU1, from patients with pneumonia. *J. Virol.* **79:**884–895.
214. **Woo, P. C., S. K. Lau, H. W. Tsoi, Y. Huang, R. W. Poon, C. M. Chu, R. A. Lee, W. K. Luk, G. K. Wong, B. H. Wong, V. C. Cheng, B. S. Tang, A. K. Wu, R. W. Yung, H. Chen, Y. Guan, K. H. Chan, and K. Y. Yuen.** 2005. Clinical and molecular epidemiological features of coronavirus HKU1-associated community-acquired pneumonia. *J. Infect. Dis.* **192:**1898–1907.
215. **Woods, J. M., R. C. Bethell, J. A. Coates, N. Healy, S. A. Hiscox, B. A. Pearson, D. M. Ryan, J. Ticehurst, J. Tilling, S. M. Walcott, et al.** 1993. 4-Guanidino-2,4-dideoxy-2,3-dehydro-*N*-acetylneuraminic acid is a highly effective inhibitor both of the sialidase (neuraminidase) and of growth of a wide range of influenza A and B viruses in vitro. *Antimicrob. Agents Chemother.* **37:**1473–1479.
216. **Wyde, P. R., S. N. Chetty, A. M. Jewell, G. Boivin, and P. A. Piedra.** 2003. Comparison of the inhibition of human metapneumovirus and respiratory syncytial virus by ribavirin and immune serum globulin in vitro. *Antivir. Res.* **60:**51–59.
217. **Wyde, P. R., S. N. Chetty, A. M. Jewell, S. L. Schoonover, and P. A. Piedra.** 2005. Development of a cotton rat-human metapneumovirus (hMPV) model for identifying and evaluating potential hMPV antivirals and vaccines. *Antivir. Res.* **66:**57–66.
218. **Wyde, P. R., E. H. Moylett, S. N. Chetty, A. Jewell, T. L. Bowlin, and P. A. Piedra.** 2004. Comparison of the inhibition of human metapneumovirus and respiratory syncytial virus by NMSO3 in tissue culture assays. *Antivir. Res.* **63:** 51–59.
219. **Zambon, M., and F. G. Hayden.** 2001. Position statement: global neuraminidase inhibitor susceptibility network. *Antivir. Res.* **49:**147–156.
220. **Zambon, M. C., J. D. Stockton, J. P. Clewley, and D. M. Fleming.** 2001. Contribution of influenza and respiratory syncytial virus to community cases of influenza-like illness: an observational study. *Lancet* **358:**1410–1416.
221. **Zavattoni, M., E. Percivalle, E. Cattaneo, M. G. Revello, M. Torsellini, and G. Gerna.** 2003. Optimized detection of respiratory viruses in nasopharyngeal secretions. *New Microbiol.* **26:**133–140.

Diagnostic Microbiology of the Immunocompromised Host
Edited by Randall T. Hayden, Karen C. Carroll, Yi-Wei Tang, and Donna M. Wolk

Chapter 8

Enteroviruses

JAMES J. DUNN

CLASSIFICATION AND BIOLOGY

The genus *Enterovirus* is currently comprised of 68 distinct serotypes within the family *Picornaviridae* (175). Taking into account molecular and biologic characteristics, the human enteroviruses (EVs) are divided into four species, namely, human EVs A, B, C, and D (Table 1). With the use of molecular sequencing techniques, a number of new EVs have recently been identified (130, 132, 133). Genetic characterization of two echovirus serotypes, 22 and 23, has resulted in their reclassification into a separate new genus, *Parechovirus*, and they are termed human parechoviruses 1 and 2, respectively (175).

EVs are small (approximately 300 Å in diameter) nonenveloped viruses. The genome is a positive-sense, single-stranded RNA molecule of approximately 7,400 nucleotides that is covalently bound at the 5′ end to a virally encoded protein (VPg) and polyadenylated at the 3′ end. A single open reading frame encoding approximately 2,185 amino acids is flanked 5′ and 3′ by nontranslated regions of approximately 750 and 100 nucleotides, respectively. The coding region is subdivided into the following three regions (5′ to 3′): the P1 region encodes the structural proteins that comprise the virion capsid (VP4, VP2, VP3, and VP1), and the P2 and P3 regions encode the nonstructural proteins, such as the RNA-dependent RNA polymerase, proteases, and other proteins necessary for intracellular replication. The entire polyprotein is cleaved co- and posttranslationally by the virally encoded proteases to generate the 11 gene products (202). The most conserved regions of the genome are the 5′-nontranslated region and those coding for the VPg protein and the RNA polymerase. The regions coding for the structural proteins are less conserved, and there is considerable variation within the regions that code for epitopes that bind neutralizing antibodies (114, 137).

EPIDEMIOLOGY AND CLINICAL MANIFESTATIONS

EV infections occur worldwide, although rates of infections may vary noticeably by location and are seasonal in nature. Individuals living in temperate climates experience considerably higher rates of infection in the summer and fall months. In the United States from 1970 to 2005, EV infections occurring from June through October accounted for approximately 78% of reports with a known month of specimen collection (86). In tropical and subtropical areas, transmission occurs throughout the year. In the United States, an estimated 10 million to 15 million symptomatic nonpolio EV infections occur each year (179).

The predominant EV serotypes typically produce endemic disease with various periodicities and in patterns that vary regionally, a feature likely due to the availability of susceptible host populations within a community (179). Typically, on an annual basis, nonpolio EV infections are caused by only a few dominant serotypes. The 15 most common EV serotypes reported to the National Enterovirus Surveillance System, a voluntary, passive surveillance system, between 1970 and 2005 accounted for 83.5% of reports with a known serotype (86). Antibodies to the more commonly circulating EV serotypes are seen in 30 to 80% of individuals by adulthood (121). Although EV infections occur in all age groups, it has been shown that infants and young children have the highest rates of infection and disease (86).

Since the introduction of poliovirus vaccines in 1955, endemic poliovirus infections have largely been controlled, and the region of the Americas has been certified as eradicated of wild-type polioviruses since 1994 (31). However, certain underdeveloped parts of the world continue to have endemic poliovirus infections, causing significant morbidity and

James J. Dunn • Department of Pathology, Cook Children's Medical Center, Fort Worth, TX 76104.

Table 1. Classification of the genus *Enterovirus*

Virus group and type
Human enterovirus A
Coxsackievirus types A2 to A8, A10, A12, A14, and A16
Enterovirus types 71 and 76
Human enterovirus B
Coxsackievirus A9
Coxsackievirus types B1 to B6
Echovirus types 1 to 7, 9, 11 to 21, 24 to 27, and 29 to 33
Enterovirus types 69, 73 to 75, 77, and 78
Human enterovirus C
Coxsackievirus types A1, A11, A13, A17, A19 to A22, and A24
Poliovirus types 1 to 3
Human enterovirus D
Enterovirus types 68 and 70

mortality (33). Sporadic cases and outbreaks of paralytic poliomyelitis due to strains contained in the live, attenuated oral poliovirus vaccine (OPV) continue to occur in many parts of the world (34). The incidence of vaccine-associated paralytic poliomyelitis (VAPP) is approximately 1 case per 2.9 million doses of vaccine distributed, and it is associated with the first dose more often than with subsequent doses. Since the policy change to exclusive use of inactivated poliovirus vaccine (IPV) in 2000, no cases of VAPP have occurred in the United States (7).

In immunocompetent hosts, nonpoliovirus EVs are responsible for numerous types of infections that result in a broad spectrum of clinical manifestations, from asymptomatic or mild self-limiting disease to severe, disseminated disease that is sometimes fatal. The contributions of different EV serotypes to human infection may vary markedly depending on the age of the host, type of disease, and implicated serotype.

Human EV infections are acquired by contact with the virus, which is shed in the feces or upper respiratory tract of direct and indirect contacts (138). Studies of wild-type and attenuated polioviruses in primates showed that only small concentrations of infectious virus are needed to cause infection. The incubation period for enteroviral infections may vary according to the clinical syndrome but is generally 7 to 14 days (range, 3 to 35 days). Following primary infection, EV can be shed in the upper respiratory tract for 1 to 3 weeks and in the feces for several weeks to months. Reinfection with the same serotype can occur despite previous immunity. However, fecal shedding occurs with lower virus concentrations, with the time of shedding usually reduced to less than 3 weeks; oropharyngeal shedding is rare, and viremia does not occur.

Protective immunity to EV infection is serotype specific. In the well-studied poliovirus model, serum neutralizing antibody develops about 1 week after infection and persists for life. Although both humoral and cell-mediated immune response mechanisms occur following EV infection, it is the presence of serotype-specific antibody that is important in limiting disease and eradicating virus. Supporting evidence for this is illustrated by the pathogenesis of EV infections in individuals producing little or no antibody (71, 109, 164). Severe, chronic, and often fatal EV infections occur in these patients. Passive immunity conferred by immunoglobulin preparations has been used to effectively prevent or treat EV infections in these antibody-deficient individuals (146, 172). Transplacental transfer of maternal antibody also effectively prevents EV disease (2, 119, 120).

While the majority of EV infections are asymptomatic, the most common symptomatic manifestation of infection is an acute, nonfocal febrile illness affecting mainly infants of less than 1 year of age (26, 81, 153, 161). EVs may cause upper respiratory tract symptoms alone or with fever, but demonstrating the clinical significance of infection has been complicated by the occasional isolation of these viruses from healthy individuals with no apparent clinical evidence of disease (43). Involvement of the lower respiratory tract may cause exacerbations of asthma, chronic bronchitis, laryngotracheobronchitis, and less commonly, bronchiolitis and pneumonia.

Herpangina, characterized by painful vesicular lesions on the soft palate, uvula, tonsils, and posterior pharynx, is seen most commonly in children of 1 to 7 years of age and is often accompanied by fever, sore throat, dysphagia, and malaise (144). Hand-foot-and-mouth disease is characterized by vesicular stomatitis and cutaneous lesions on the distal ends of the extremities and is commonly associated with coxsackievirus A16 and EV 71. Outbreaks of hand-foot-and-mouth disease caused by EV 71 have been associated with serious central nervous system (CNS) disease, including acute motor neuron disease and brain stem encephalitis (110, 135). Several EV serotypes, particularly EV 70 and coxsackievirus A24, are known to cause acute conjunctivitis, keratoconjunctivitis, and hemorrhagic conjunctivitis (121).

The foremost EV CNS infections include aseptic meningitis, encephalitis, poliomyelitis, and poliomyelitis-like illness (163). Nonpolio EV infections have also been implicated in cases of Guillain-Barré syndrome, acute transverse myelitis, and cerebellar ataxia (113, 176, 183). A number of myotropic diseases have been ascribed to the EVs. These include pleurodynia (Bornholm disease), acute myocarditis, chronic dilated cardiomyopathy, and myositis (163).

EV infections in the newborn period are common. EVs may be transmitted to the fetus during delivery in 30 to 50% of mothers with seroconversion in pregnancy (120, 167). EV infection of the placenta has also been described (55, 62, 166). In population-based studies, enteroviral infections were found in 4 to 26% of hospitalized febrile infants younger than 3 months of age being evaluated for sepsis (26, 153, 157). In the summer and fall months, EV infections in infants of less than 3 months of age accounted for 13 to 65% of hospital admissions (81, 161). Data reported to the National Enterovirus Surveillance System from 1983 to 2003 revealed that the most commonly isolated serotypes from infants of less than 1 month of age were echoviruses 6, 9, 11, and 30, coxsackieviruses B1 to B5, and coxsackievirus A9 (87) (Table 2). These 10 serotypes accounted for 73.3% of all detections of EVs of known serotype among neonates. Of reported neonatal infections with known outcomes, 11.5% resulted in death, and infection with coxsackievirus B4 was associated with a significantly higher risk of mortality than infection with other EV serotypes (87). In other reports, case fatality rates for neonatal EV infections ranged between 0% and 83% (1, 4, 119).

The number of immunocompromised individuals has grown extensively. With the global epidemic of human immunodeficiency virus (HIV) and available antiviral medications, more intensive and successful cancer chemotherapy regimens, and the availability of new immunosuppressive agents, the number of individuals surviving HIV infection, hematologic malignancies, and solid organ and hematopoietic stem cell transplants has continued to increase. However, these immunocompromised patient populations are at increased risk of infection, and in fact, the leading cause of death in all of these groups is infection.

EVs have occasionally been implicated in serious infections in immunocompromised patients. In a 5-year study of pediatric patients with malignancy conducted in Greece, EV infections were documented for 55 patients (125). Of 77 children with acute leukemia, 43 (56%) developed documented enteroviral infection, whereas 5 of 37 (13%) patients with lymphoma and 7 of 236 (3%) patients with solid tumors developed EV infection. Severe manifestations, including encephalitis and myocarditis, were noted in 20% of those infected. In a prospective study of febrile episodes in 66 pediatric oncology patients, EVs were identified in two patients, with one from a respiratory infection concomitantly with rhinovirus and varicella-zoster virus and the other from stool (45). EVs were identified in 4.7% of 105 viral infections identified in pediatric patients being treated for leukemia (204). Outbreaks of EV infection in oncology and stem cell transplant units have also been described (123, 190).

Table 2. Clinical syndromes associated with EV infection in immunocompromised hosts and reported EV serotypes

Patient population	Clinical manifestation[a]	EV serotype(s)
Patients with congenital B-cell deficiency	Chronic meningoencephalitis	Echovirus 2, 3, 5, 6, 7, 9, 11, 13-15, 17-19, 21, 24, 25-27, 29, 30, 33; coxsackievirus A4, A11, A15, B2, B3
	Dermatomyositis-like syndrome	Echovirus 2, 3, 5, 7, 9, 11, 17, 19, 24, 30, 33; coxsackievirus B3
	VAPP	Poliovirus 1, 2, 3
Transplant recipients	URI, LRTI, ARDS	Echovirus 6, 11; coxsackievirus A9, B (untyped)
	Meningoencephalitis	Echovirus 11, 19; coxsackievirus A9
	Myocarditis/pericarditis	Echovirus 11; coxsackievirus B3
	Acute flaccid paralysis	Echovirus 19
	Gastroenteritis	Echovirus (untyped); coxsackievirus A1
Patients with malignancy	Meningoencephalitis	Echovirus 11, 13, 20; coxsackievirus B3 to B5
	URI, LRTI	Echovirus 11; coxsackievirus B3
	Myocarditis	Coxsackievirus B3 to B5
Neonates	Meningoencephalitis	Echovirus 4, 6, 7, 9, 11, 17, 19, 30; coxsackievirus B1 to B5
	Pneumonia	Echovirus 6, 7, 9, 11, 17, 19, 20; coxsackievirus B3
	Hepatitis	Echovirus 4, 6, 7, 9, 11, 14, 17, 19, 20, 21 Coxsackievirus B1 to B4
	Myocarditis	Echovirus 7, 11; coxsackievirus B1 to B5
	DIC	Echovirus 4, 6, 11, 17; coxsackievirus B1 to B4
HIV-positive patients	URI, LRTI	Echovirus 33
	Meningoencephalitis	Echovirus 6
	VAPP	Poliovirus 2

[a]VAPP, vaccine-associated paralytic poliomyelitis; URI, upper respiratory tract infection; LRTI, lower respiratory tract infection; ARDS, acute respiratory distress syndrome; DIC, disseminated intravascular coagulation.

In a series examining the prevalence of respiratory viruses in immunocompromised patients, EVs were recovered from bronchoalveolar lavage (BAL) samples from 6 of 36 symptomatic patients, including 2 with HIV infection, 3 who had undergone bone marrow transplants, and 1 heart transplant recipient (149). Over an 8-year period, EVs were isolated from 38% of BAL samples from patients undergoing chemotherapy or from stem cell transplant recipients with suspected lower respiratory tract infections (68). An estimated 3 to 21% of transplant recipients excrete EV in their stool (35, 192). Chakrabarti et al. (36) identified symptomatic EV upper respiratory tract infections in 7 of 64 adult patients who had received T-cell-depleted allogeneic stem cell transplants.

Patients with congenital isolated B-cell deficiency or combined immunodeficiencies are at considerable risk of infection. Individuals with congenital immunodeficiency have approximately 3,000 times the risk of VAPP as do normal children (181). A U.S. registry of 201 patients with X-linked agammaglobulinemia (XLA), a primary immunodeficiency resulting in the nearly total absence of antibody production, identified enteroviral infections in 10 individuals, the vast majority of whom had chronic meningoencephalitis (203). Many EV serotypes can be involved, including wild-type and vaccine-associated polioviruses (Table 2). Echovirus 11 has been the most commonly identified EV serotype in antibody-deficient patients with chronic meningoencephalitis (71, 109, 117). Cases of EV meningoencephalitis have been reported as a complication for patients with X-linked hyper-immunoglobulin M (hyper-IgM) (XHIM) syndrome (49, 96).

Few reports exist of EV infections in patients with HIV infection and immunosuppression, but sporadic asymptomatic shedding (for up to 6 months) of EV has been described for up to 28% of individuals (69, 101, 115, 139, 149). EVs have been identified in up to 3% of HIV-positive patients with signs and symptoms of respiratory tract illness (88, 112). Prolonged excretion of vaccine-associated polioviruses is uncommon, although cases of paralytic poliomyelitis in undiagnosed HIV infection have been described (42, 78, 101).

EV INFECTIONS IN SPECIFIC IMMUNOCOMPROMISED HOST POPULATIONS

Primary B-Cell-Associated Immunodeficiencies

Among hereditary B-cell-associated immunodeficiency syndromes, XLA, common variable and severe combined immunodeficiency syndromes, and XHIM syndrome have been associated with persistent progressive EV infection. Antibody-deficient individuals infected with EV may develop chronic meningitis or meningoencephalitis lasting many years, often with a fatal outcome (71, 109, 164) (Table 2). However, with the availability of intravenous immunoglobulin (IVIG) preparations and antipicornaviral drugs and the early recognition of EV infection, fewer of these patients appear to be progressing to the classical description of chronic EV meningoencephalitis of agammaglobulinemia (CEMA).

XLA is characterized by a block in B-cell development at the pre-B-cell stage, typically resulting in markedly decreased levels or an absence of both mature B cells and immunoglobulin production in affected male subjects. T-cell and myeloid functions are intrinsically normal, although neutropenia under conditions of stress has been described. The gene encoding the cytoplasmic Bruton's tyrosine kinase (BTK) is defective in XLA. Mutations of the BTK gene have also been reported for male subjects with milder clinical forms and incomplete B-cell defects. The diagnosis of XLA is based on the presence of lymphoid hypoplasia, markedly reduced serum levels of all three major classes of immunoglobulins, failure to make antibody to antigenic stimulation, and almost complete absence of B lymphocytes in the peripheral blood (134). Because IgG is actively transported across the placenta, affected newborns have normal levels of serum IgG at birth and few, if any, symptoms. As the maternally derived antibodies are catabolized, serum IgG gradually decreases, and since immunoglobulin production is impaired in XLA, severe hypogammaglobulinemia and increased susceptibility to infection develop. Thus, many patients with XLA are asymptomatic for the first few months of life and begin to have recurrent infections between 4 and 12 months of age. A smaller percentage of patients with XLA will present after the first year of life (203).

Bacterial infections are the most common clinical manifestation of XLA, both before the diagnosis is made and, to a lesser degree, after therapy with immunoglobulin has been initiated (203). Most commonly, the infections are localized in the respiratory tract, causing otitis, sinusitis, bronchitis, and pneumonia. Resistance to viral infections is generally intact in patients with XLA, except for that to EVs. Because antibodies are important in neutralizing these viruses during their passage through the bloodstream, patients with XLA lack an important mechanism of resistance to the EVs. Historically, when OPV was in common use, some patients with XLA would develop VAPP after immunization (71, 73).

CEMA is a disease marked by prolonged enteroviral infection of the CNS. Patients may present with slowly progressive neurological symptoms, including ataxia, loss of cognitive skills, deafness, dysarthria, and paresthesias. Consistent with gradual onset, early signs of CNS involvement are pleocytosis, moderately elevated protein levels, and low glucose in the cerebrospinal fluid (CSF). A few patients have a more acute presentation, with fever, headache, and seizures. In time, the majority of patients with persistent CNS infection manifest neurological symptoms, such as lethargy, weakness, and headache (109). Once CEMA is recognized, its progression can generally be slowed by the administration of immunoglobulin preparations, although most patients continue to have episodic relapses. Some patients develop a dermatomyositis-like syndrome with "woody" edema of their extremities, erythematous rash, and biopsy evidence of fasciitis and myositis. EVs can occasionally be isolated from skin, muscle, or liver biopsies (109, 111). Prior to the early, consistent use of IVIG preparations and, more recently, antipicornaviral drugs, neurologic manifestations of EV infection associated with congenital B-cell immunodeficiency included paralysis/paresis, seizures, long-term cognitive impairment, developmental regression, and coma (71).

At the onset of disease, viral cultures from CSF may be negative and remain so for long periods. Molecular methods may show that culture-negative CSF samples contain EV, even in the presence of normal cell counts and protein levels (147, 199). EV may persist in the CNS, where levels of passively transferred antibody are low. The CNS is then the site from which the virus may spread secondarily to other organs and tissues. The most commonly isolated EVs are echoviruses (primarily echovirus 11), followed by polioviruses and coxsackieviruses (71, 109) (Table 2). A change in the infecting EV serotype has been documented during the course of illness (109). It may be that the original strain was replaced by one causing a second infection or that recombination occurred between two viruses.

XHIM syndrome is a rare cause of immunodeficiency characterized by normal or elevated levels of IgM and low levels of IgG, IgA, and IgE. Clinical manifestations include neutropenia, cholangitis, tumors, and frequent infections with bacteria, *Pneumocystis jirovecii*, *Candida* spp., and cryptosporidium. XHIM syndrome is caused by mutations of the gene encoding CD40 ligand. This protein is expressed on the surfaces of activated $CD4^+$ T cells and interacts with CD40, a receptor constitutively expressed on B cells. This interaction leads to B-cell activation and switching of immunoglobulin classes. Failure to express functional CD40 ligand interferes with normal B-cell function, resulting in abnormal antibody responses to T-cell-dependent antigens, a failure to switch from IgM to other immunoglobulin isotypes, a lack of immunologic memory, and an inability to form germinal centers in lymph nodes. Meningitis, encephalitis, and encephalopathy have been reported as complications in up to 12% of patients with XHIM syndrome (96). In one series of patients with XHIM syndrome, three had EV meningoencephalitis despite IVIG treatment yielding therapeutic serum immunoglobulin levels (49).

On rare occasions, reverse transcription-PCR (RT-PCR) results may intermittently be negative during the course of CEMA (147, 158, 159, 199). Virus is isolated only rarely from the stools of patients with CEMA, except during the disseminated phases of infection or in those infected with poliovirus strains. A negative PCR result for CSF does not necessarily exclude the diagnosis, and at least another sample of CSF should be tested if enteroviral infection is suspected.

Transplant Recipients

Opportunistic viral infections occur frequently in patients undergoing bone marrow transplantation and can be a major cause of morbidity and mortality. The main group of viruses responsible for infection after transplantation are the herpesviruses, which may either be transmitted by the graft or cause a clinically manifest reactivation of infection. Certainly, given the trend for more aggressive immunosuppressive therapy both pre- and posttransplant by use of B-cell-depleting antibodies, the incidence and severity of EV infections may increase among transplant recipients. Moreover, manipulation of hematopoietic stem cell grafts (e.g., CD34 selection) can further delay already impaired B-cell reconstitution and place patients at risk for enteroviral infections during their immune recovery phases. Ideally, patients with prior EV infections should have documented clearance of disease prior to proceeding to transplant.

Several reports of fatal disseminated EV infections have been reported for bone marrow transplant recipients (10, 18, 59, 162, 169, 184) (Table 2). Initial signs of infection were apparent from 12 to 53 days posttransplant and included clinical syndromes such as encephalitis, myocarditis, pericarditis, and pneumonia. Most fatal cases among bone marrow transplant recipients were those with unrelated donors. The EV serotypes identified in these cases were predominantly echoviruses (types 6, 11,

13, and 19) and coxsackieviruses (types A9 and B3) (Table 2). Reports of infectious gastroenteritis caused by coxsackieviruses among bone marrow transplant recipients have also been associated with significant mortality (190, 210). Recovery from severe EV infection following bone marrow or solid organ transplantation has been described, even in cases with extensive CNS involvement (57, 162, 205).

Starlin et al. (178) reported the only published case of acute flaccid paralysis due to a nonpoliovirus EV (echovirus 19) in a solid organ transplant recipient. This patient, who had cystic fibrosis, had undergone two bilateral lung transplants and a renal transplant and was on an intensive immunosuppressive regimen. Symptoms began as a viral syndrome with low-grade fever, sore throat, and malaise and progressed to severe headache, photophobia, fever, vomiting, and acute flaccid paralysis of the left leg. The patient was treated with IVIG and pleconaril, with some objective clinical improvement in neurological function. Subsequent cultures for EV were negative, although the patient succumbed to bacterial pneumonia and fungal sepsis (178).

Respiratory viral infections are increasingly being recognized as the cause of complications in bone marrow transplant recipients. While considerable experience has been gained regarding the role of infections with viruses such as respiratory syncytial virus, influenza virus, parainfluenza virus, and adenoviruses, there are relatively few data available regarding the role of EV in respiratory tract infections in transplant recipients (79, 200). Not unlike those in immunocompetent hosts, respiratory illnesses due to EV in transplant recipients are confined mostly to the upper respiratory tract and result in mild to moderate respiratory disease, with clinical improvement within 1 to 2 weeks (36, 72, 194). Upper respiratory symptoms attributable to EV infection can be seen as early as 25 days after transplant but may not appear for more than a year (108). On the other hand, lower respiratory tract EV infection may be more severe in transplant recipients. Of one cardiac transplant and seven stem cell transplant recipients with radiographic evidence of lower respiratory tract disease due to EV, five developed acute respiratory distress syndrome (ARDS), three required mechanical ventilation, and three died (68, 149). In the fatal cases, the median onset of respiratory illness was 120 days after transplant (range, 37 to 160 days), and all patients either did not have engraftment or were being treated for relapse of leukemia/lymphoma. A fatal outcome after ARDS and subsequent respiratory failure due to EV infection have also been described for a child after stem cell transplantation (184).

Malignancy

Viral pneumonia is a frequent complication in patients undergoing intensive chemotherapy for treatment of hematologic malignancies. Herpesviruses and community-associated respiratory viruses, such as influenza virus, respiratory syncytial virus, parainfluenza virus, and adenovirus, have been recognized as a significant cause of pneumonia in immunocompromised hosts (79, 200). The significance of EVs in lower respiratory tract disease was examined in one study in which EVs were isolated from three and five BAL specimens from patients with hematologic malignancies and recipients of stem cell transplants, respectively, with suspected lower respiratory tract infections (67). All patients with EV infection presented with fever at diagnosis, and 75% had pneumonia. Five patients subsequently died from pneumonia, although three of them had concomitant viral and/or bacterial infections. A case of hand-foot-and-mouth disease in a patient undergoing chemotherapy for Ewing's sarcoma has also been described (56).

EV meningoencephalitis and disseminated infection have been described for patients undergoing chemotherapy for hematologic malignancies or solid tumors (63, 124, 125, 173, 198) (Table 2). In these patients, manifestations of EV infection included meningoencephalitis lasting at least 1 month, myositis, pneumonia, myocarditis, and hemophagocytic syndrome. The implicated EVs were coxsackieviruses (B3 to B5) and echoviruses (types 11 and 20). Hypogammaglobulinemia and/or leukopenia was reported for the vast majority of these patients. While most patients survived infection with nearly complete resolution of neurological function, several fatalities attributed to disseminated EV infection have been reported for pediatric patients with hematologic malignancies (125).

An outbreak of EV infection in a 2-month period (November and December) among five pediatric patients with acute lymphoblastic leukemia and immunosuppression resulted in severe encephalitis in three patients and in two deaths, one with chronic encephalitis and one with acute brain stem encephalitis (123). The index patient was admitted with vomiting and upper respiratory tract illness and subsequently developed parotitis, headaches, and seizures. The three patients who recovered were in the maintenance phase of chemotherapy, while the two who died were in the induction phase.

Treatment of hematological malignancies with the chimeric anti-CD20 monoclonal antibody (MAb) rituximab induces rapid and long-lasting depletion of circulating B cells. Rituximab administration is associated with profound antibody deficiency of usually 6 to 12 months in duration. The occurrence of EV meningoencephalitis following rituximab therapy has been reported for a child with immune thrombocytopenia and two adults with relapsed B-cell lymphomas (136, 148). These patients presented 6 to 11 months following rituximab therapy, with neurological symptoms including aphasia, headaches, paresthesias, sensorimotor deafness, and ataxia. Initial CSF studies revealed moderate to marked pleocytosis (predominantly lymphocytes) and elevated protein levels, and EV was detected by RT-PCR or culture. Brain magnetic resonance imaging studies of one patient revealed diffuse white matter abnormalities, while evidence of myelitis was present in another (148). Following treatment with IVIG (two patients had combined pleconaril treatment), subsequent CSF samples from two patients were negative for EV by RT-PCR. Only one patient recovered completely, one underwent stem cell transplantation with recurrence of EV meningoencephalitis and persistent deafness, and continued neurologic deterioration was noted in the third patient, who subsequently died. Echovirus 13 was isolated from multiple CSF specimens from the patient with recurrent EV infection, and all strains displayed >99% nucleotide sequence similarity (12). Notably, the initial identification of echovirus 13 in this patient coincided with an outbreak of EV meningitis with this serotype in the region.

Neonates

EV infection is one of the most common causes of fever in the neonatal stage and early infancy, especially in the summer and fall months (26, 81, 153, 161). These infections are often seen in term neonates who have no perinatal risk factors for sepsis and who have been discharged in seemingly good health after birth. The manifestations of neonatal EV disease range from inapparent infection to overwhelming severe, life-threatening disease. The majority of neonates experience asymptomatic infection or benign illness without sequelae. Severe illness develops in a minority of infected newborns, resulting in high rates of mortality and morbidity because multiple organ systems, such as the CNS, lungs, liver, and heart, are commonly involved (Table 2). Severe infant disease is correlated with prematurity, male sex, and onset of maternal EV disease in the 2 weeks before delivery. Often, the clinical presentation in neonates mimics that of fulminant bacterial sepsis, so a viral etiology is not considered. Because of the potential for substantial morbidity and mortality, it is important to perform viral diagnostic testing for this group of patients, preferably early in the course of illness.

Transmission of EV from mother to infant is relatively common and may occur transplacentally or through contact with maternal secretions during vaginal delivery, blood, or upper respiratory tract secretions (120). Transmission at the time of birth appears to be the most common route of exposure, occurring in 30 to 50% of mothers with seroconversion in pregnancy (120, 167). Maternal viral illness within 2 weeks prior to delivery, with symptoms such as fever, respiratory tract infection, or abdominal pain, has been reported for 59 to 68% of infected neonates (185). Premature infants are particularly vulnerable to infection due to the limited transplacental transfer of maternal immunoglobulins, low IgM and IgA production, deficiency of complement components, and impaired cellular function (15). It is likely that transplacentally acquired maternal antibody has an important role in modifying perinatal infection in the newborn (118). This passively acquired specific antibody is lacking in infants whose mothers become infected within the last few days of pregnancy. EVs have been isolated from amniotic fluid obtained during the third trimester via amniocentesis, despite the presence of intact membranes, suggesting that the virus is transmitted transplacentally (44, 180). Infection of the placenta has been demonstrated in cases of fetal or neonatal death or severe neonatal morbidity (55, 62, 166).

Clinical manifestations of EV infection can develop as early as day 1 of life and throughout the neonatal period. Early signs and symptoms of newborn infection may be nonspecific. In those with mild disease, fever and other symptoms typically regress within a week. Severe disease may occasionally follow a biphasic pattern, with a short period of recovery after mild manifestations followed by progressively worsening illness. Severe disease is frequently seen when neonates develop EV infection within the first 2 weeks of life. Major systemic manifestations, such as hepatic necrosis, myocarditis, encephalitis, pneumonia, necrotizing enterocolitis, and disseminated intravascular coagulation, may develop, mimicking overwhelming bacterial infection. CNS involvement may or may not be associated with signs of meningeal inflammation. Encephalitis may develop, with seizures, focal neurologic abnormalities, and depressed consciousness. Manifestations of myocarditis include cardiomegaly, congestive heart failure, arrhythmias, and myocardial infarction. Hepatitis is often associated with thrombocytopenia and coagulopathy. The most significant factors associated with hepatic necrosis with coagulopathy are prematurity, maternal history of illness, early age of onset, high white blood cell count, and low hemoglobin level (99). When death occurs, it is typically due to hepatic failure with coagulopathy and/or myocarditis. Significant risk factors associated with fatality from hepatic necrosis with coagulopathy are a total bilirubin level of >14.3 mg/dl and concurrent myocarditis (99).

Most survivors of severe neonatal disease do not have long-term cardiac or hepatic dysfunction. However, there are reports of residual myocardial dysfunction, chronic calcific myocarditis with chronic heart failure and dysrhythmias, and ventricular aneurysm (40, 170). Some studies suggest that those with CNS involvement may have long-term sequelae, such as delayed speech and language development, intellectual deficits, motor abnormalities, seizure disorders, ocular defects, or microcephaly.

Neonates are also at increased risk of nosocomial EV infections, as evidenced by reports of epidemic spread and sporadic transmission occurring in hospital nurseries (19, 28, 39, 41, 80, 141, 151, 201). Generally, nosocomial EV infections are associated with lower rates of morbidity and mortality than those of vertically acquired infections (119). The most common risk factors for nosocomial acquisition of EV infection among neonates in the nursery include prematurity, low birth weight, intensive care, and the use of nasopharyngeal or oropharyngeal instrumentation (90).

HIV/AIDS

HIV-associated immune system dysfunction is characterized by a slowly progressive and complex impairment of immune functions, in particular antigen presentation and functions mediated by $CD4^+$ T lymphocytes. The main effect is an impairment of cellular immunity, although at some stages of disease, humoral immunity can subside. Severe EV disease is observed only sporadically in patients with only cell-mediated immune defects, as illustrated by the absence of EV from lists of reported opportunistic infections in AIDS patients.

In the course of HIV infection, symptomatic respiratory tract infections due to EV are uncommon (50, 88, 112, 149). While gastrointestinal disorders frequently affect HIV-positive patients, prospective studies on excretion of EV have found little or no shedding and no significant association with gastrointestinal illness (69, 101, 115, 139). While coxsackieviruses are a common cause of myocarditis and dilated cardiomyopathy in the general population, a matched case control study among 24 HIV-infected children with cardiac impairment and 24 HIV-infected control subjects without cardiac impairment did not identify differences in coxsackievirus infections or a causal role for coxsackieviruses in HIV-associated cardiomyopathy (82).

A variety of neurological syndromes of uncertain etiology can occur during the course of HIV infection, but there are scant reports of CNS infection caused by EV (52, 189). VAPP with Sabin poliovirus type 2 has been reported for one HIV-positive child in Africa and one in Romania (42, 78). The child from Africa presented approximately 2 weeks after receiving the second dose of OPV, with flaccid paralysis of the right leg that persisted for several years (42). The Romanian child had completed a four-dose schedule of OPV approximately 1 year prior to presentation with flaccid paralysis of the right leg (78).

A protracted outbreak of EV 71 infections has been described among eight HIV-infected children residing at two orphanages in Kenya in 1999, during which time large EV 71 pandemics were occurring in Southeast Asia and Australia (37). Clinical findings included myalgias, excoriation and lip swelling, and dermatitis. No life-threatening complications were noted.

VAPP

Persons with immunodeficiency, either hereditary or acquired, are at risk for infection with the live, attenuated vaccine-derived polioviruses (VDPVs) contained in the OPV. The primary identified underlying conditions have included B-cell-associated immunodeficiencies (XLA, common variable immunodeficiency, transient hypogammaglobulinemia of infancy, IgG subclass deficiency) or T- and B-cell immunodeficiency (SCID). Individuals with these conditions also represent a potential reservoir for reintroduction of poliovirus because of chronic excretion of vaccine strains. After administration of OPV to immunocompetent persons, the vaccine viruses are excreted for a limited period of time, typically less than 2 to 3 months. However, immunodeficient individuals may excrete vaccine strains for 10 years or longer (85, 105).

Although OPV is not recommended for immunodeficient patients, it is often administered inadvertently because certain primary immunodeficiencies are not recognized until later in life. Selective pressures in the human intestine can cause partial reversion of the attenuated OPV strains to neurovirulence, resulting in VAPP, which is clinically identical to the paralysis caused by wild-type virus, affecting both vaccine recipients and their contacts. The risk of VAPP is highest after the first dose of OPV, estimated as 1 case per 750,000 children vaccinated, and is 3,000 times higher for immunocompromised persons. Some patients with B-cell-associated immunodeficiency who excrete VDPVs may experience onset of poliomyelitis several years after the implicated OPV dose was administered (85). Long-term excretors of VDPVs are more likely to die than are patients with VDPVs who do not have a known long-term persistence (85).

Some patients who received immunoglobulin replacement therapy and excreted virus for prolonged periods of time spontaneously stopped shedding VDPVs (17, 107, 116). Four patients with SCID and persistent, nonparalytic, vaccine-related poliovirus infection ceased shedding polioviruses after successful bone marrow transplantation with reconstitution of humoral immunity, as evidenced by increased levels of serum neutralizing antibodies to the respective poliovirus serotypes (104). In some patients, persistent shedding of VDPVs cannot be eliminated, despite the use of multiple therapeutic interventions (105).

TREATMENT

Since the primary mechanism of clearance of EV by the host is humoral immunity, those patients with impaired antibody production because of congenital or acquired immunodeficiency are particularly susceptible to infection. Similarly, neonates are at high risk for severe enteroviral disease because of a relative deficiency of antibodies. Immunoglobulin preparations have been used prophylactically and therapeutically against EV disease. In neonates with disseminated disease, some clinical success has been achieved with the use of maternal serum or plasma or commercially prepared immunoglobulin preparations against a number of EV serotypes (2, 141, 186). However, some infants may have progressive disease and die despite such therapy (22, 208). With the early recognition of immunodeficiency and treatment with IVIG, the incidence and severity of chronic, progressive EV meningoencephalitis in patients with congenital or acquired antibody deficiencies have declined (172).

Results of IVIG therapy in those patients with established CNS infection have been mixed. Despite ongoing IVIG therapy, some patients may have evidence of viral persistence in the CNS, as demonstrated by RT-PCR and progressive neurological deterioration (158, 164). The failure of IVIG therapy in these patients may be related to the low concentrations of antibodies achieved in CSF (61). Additionally, IVIG only replaces IgG and is unable to correct the secretory immune defect. Moreover, the use of nonselected lots of IVIG does not provide uniformly high levels of serotype-specific antibody (60). There is some evidence that the disease can be stabilized by infusing immunoglobulin-containing antibodies to the relevant EV into the CSF, usually through an Ommaya ventricular reservoir (54).

The antiviral drug pleconaril (Schering-Plough, Kenilworth, NJ) has been used in trials of healthy hosts with common EV infections as well as antibody-deficient individuals with potentially life-threatening infections. Pleconaril integrates into the capsid canyon hydrophobic pocket, preventing viral replication by inhibiting viral uncoating and blocking viral attachment to host cell receptors. It is absorbed well from the gastrointestinal tract and has a long half-life. Effective antiviral concentrations are attained in both serum and CSF (155).

A small series of patients with humoral immune deficits who received pleconaril included 17 patients with CEMA, 6 neonates with disseminated EV infection, 2 patients with severe EV infection following bone marrow transplantation, and 2 patients with VAPP (162). Twelve of the patients with CEMA showed clinical improvement after pleconaril treatment, and viral clearance in the CSF was noted for six of seven patients who had subsequent CSF samples taken. Five newborns with overwhelming infection survived with only mild sequelae, and both bone marrow transplant recipients had clinical improvement. Other reports have shown similar efficacies of pleconaril in newborns with fulminant infection and patients with CEMA (11, 16, 168). However, in a few patients who failed to respond, the infecting EV strain was formally shown to be resistant. Moreover, a survey of 215 clinical EV isolates has shown that about 10% may be resistant to the levels of pleconaril likely to be achieved in vivo (i.e., about 2 μg/ml) (143).

LABORATORY DIAGNOSTIC TECHNIQUES

A specific diagnosis of EV infection requires detection of the virus itself, of EV-specific antibodies, or of the viral genome in patient samples. Because of the myriad clinical manifestations of EV infections, confirmation of the diagnosis can be important in reducing hospitalization, antibiotic use, and additional diagnostic testing often performed to exclude or treat other conditions. The development of antiviral therapy for EV infections also demands rapid and accurate diagnostic tests. Available methods to identify EV infection include nucleic acid detection, culture, serology, and antigen detection.

Nucleic Acid Detection

Nucleic acid amplification methods, such as RT-PCR and nucleic acid sequence-based amplification, have replaced culture-based techniques as the gold standard for detection of EV, given the significant increase in sensitivity and the rapid turnaround time. This is particularly notable for testing of CSF specimens for the diagnosis of aseptic meningitis, where isolation of EV by conventional cell culture requires mean incubation times of 3.7 to 8.2 days and is up to 75% less sensitive than RT-PCR (5, 23, 29, 154, 160). The rapid diagnosis (as few as 3 to 5 hours) of EV meningitis afforded by RT-PCR allows for significant health care-associated cost savings, such as decreased length of hospitalization and antibiotic or antiviral usage (128, 152, 154, 161). For some patients, CSF may be positive for EV by culture or RT-PCR despite normal cell counts and chemistries. RT-PCR is also more sensitive than culture for identification of EV in respiratory tract secretions, urine, and blood (3, 160, 161). It has been used to detect enteroviral RNA in cardiac tissue from patients with myocarditis and other heart diseases and in liver tissue from neonates with fulminant infection (9, 38, 46, 97, 174).

For neonates and young infants, testing for EV in serum or plasma by RT-PCR is often a more sensitive indicator of active infection than testing by RT-PCR with CSF specimens (3, 129, 195). In one study, serum RT-PCR was the only positive test for nearly one-quarter of infants of <90 days of age with symptoms of sepsis (153). Combined testing of CSF and serum for infants and young children can increase the diagnostic yield by up to 20% (8, 153). For infants and children, testing of urine or throat swab specimens by RT-PCR is more sensitive than culture but less likely to be positive than testing by RT-PCR with CSF and/or blood (3, 160, 161). In pediatric cases of acute myocarditis with or without viral pneumonia, there was a strong correlation between EVs detected in tracheal aspirates and myocardial biopsies by RT-PCR (6). In adults with EV meningitis, the yield of CSF testing by RT-PCR more than 2 days after the onset of clinical symptoms is often less than that of culture or RT-PCR of stool specimens (94). Additionally, testing of serum specimens by RT-PCR for adults with aseptic meningitis is infrequently positive (8, 94).

Most EV RT-PCR assays have been designed to detect genomic sequences within the 5′-nontranslated region, where short spans of nucleotides with high sequence identity exist among the majority of serotypes. Some EV serotypes, notably parechoviruses 1 and 2, may not be detected by all RT-PCR assays that use this target region. The recent availability of real-time PCR methods has further facilitated the rapid and sensitive detection of EV in the CSF and other types of specimens (47, 64, 83, 95, 122, 129, 150, 196). The performance characteristics of several real-time PCR assays are listed in Table 3. Although cross-amplification in EV RT-PCR assays with bacteria, fungi, and most nonrelated viruses has not been seen, high sequence identity with rhinoviruses within the 5′-nontranslated region has resulted in the amplification of certain serotypes, depending on the primers and/or probe sequences utilized (27, 84, 95). This cross-reactivity is notable for testing of respiratory samples but not for CSF, since rhinoviruses have not been isolated from the CNS.

While most laboratories develop and verify nucleic acid amplification assays for EV in-house, reagents for EV RNA amplification for conventional and real-time PCR platforms are commercially available as analyte-specific reagents or research-use-only systems (Argene Biosoft, Varilhes, France; bioMerieux, Durham, NC; Cepheid, Sunnyvale, CA; Chemicon, Temecula, CA; and Nanogen, San Diego, CA). Currently, the only FDA-approved test for molecular detection of EV in CSF is the Xpert EV assay performed on the GeneXpert system (Cepheid). This is a self-contained, random-access, fully automated system in which sample preparation, amplification, and real-time multiplex detection are integrated in a single module for detection of EVs in approximately 2.5 hours (93, 106). The sensitivity of the Xpert EV

Table 3. Performance characteristics of selected real-time PCR assays for detection of EV RNA

Assay reference	Sample type(s)	Sensitivity (%)	Comment
47	CSF/throat swabs	100	Detection of enteroviruses and parechoviruses 1 and 2; 16% more viruses were detected by PCR than by cell culture
64	CSF	98	31% more viruses were detected by PCR than by cell culture
83	CSF/stool/nasopharyngeal swab	100	Fifteen of 20 samples negative by cell culture were positive by PCR
93,106	CSF	94-97	FDA-approved Xpert EV assay on GeneXpert system
122	CSF	97	Sensitivity of culture was 35%
129	Stool/CSF/BAL fluid	100	42% more viruses were detected by PCR than by cell culture
150	CSF	89	Twenty-four of 27 samples positive by real-time PCR, compared to conventional PCR
196	CSF	100	45% more viruses were detected by PCR than by cell culture

assay for detection of EV in CSF ranges from 94 to 97%, with a specificity of 100%, a positive predictive value of 100%, and a negative predictive value of 98 to 99% (93, 106). The Xpert EV assay and the Enterovirus Consensus product (Argene) have CE marking in Europe.

The correlation between EV disease severity and blood viral load has been assessed using an in-house real-time RT-PCR on serial samples from infants of <2 months of age (209). For all patients, the highest viral RNA load was detected in the initial blood specimen, with a subsequent gradual decrease over time. Those patients with severe, disseminated disease had higher initial viral loads (>23,000 copies/μl) and viremia that persisted for up to 2 months, while those with mild disease had initial viral loads of <2,000 copies/μl, with viremia persisting no longer than 10 days. A rapid and significant decrease in viral load was seen for two patients with severe disease following administration of IVIG.

In situ hybridization (ISH) techniques have been utilized to demonstrate the presence of both positive- and negative-sense EV RNA, primarily in patients with heart disease (46, 98). It has also been described for use with muscle biopsy specimens and placental material (55, 211). ISH is typically less sensitive than immunohistochemistry (IHC), likely due to the fact that RNA can be lost or degraded during fixation or histological processing (98).

Culture

Isolation of EV in cell culture has historically been the mainstay for diagnosis, particularly since most strains grow readily. However, no single cell line allows for propagation of all EV serotypes. In addition, some EV serotypes fail to grow in cell culture. This is particularly true for a number of the group A coxsackieviruses, for which isolation by inoculation in suckling mice has historically been used. In the clinical laboratory, the principal cell lines used for recovery of EV include primary monkey kidney, A549, buffalo green monkey kidney (BGMK), human embryonic lung fibroblast, and RD cells, which are derived from a human rhabdomyosarcoma. Polioviruses, group B coxsackieviruses, and echoviruses grow efficiently in monkey kidney cells, and many group A coxsackieviruses can be isolated using human embryonic lung fibroblast and RD cells (92, 163). Typically, detection of cytopathic effect (CPE) in these cell lines requires five or more days of incubation.

The sensitivity of cell culture is highly dependent on adequate collection, handling, and processing of specimens. Recovery of virus may be affected for these reasons as well as by antibody neutralization in situ or because of the intrinsic insensitivity of the cell lines used. Recovery of virus in cell culture is optimized by sampling multiple sites. For example, viral culture specimens with the highest yields in neonatal EV infections are rectum or stool (88 to 93% positive), CSF (62 to 83% positive), and nasopharynx or throat (52 to 67% positive) specimens (1, 2, 127). Cultures of serum and urine have lower yields (24 to 74%) than do cultures from mucosal sites; however, serum specimens may grow virus more rapidly than do specimens from other body fluids/sites. Serum cultures are more likely to be positive with echoviruses, low serum neutralizing antibody titers, and an onset of illness within the first 5 days of life. In infants and young children with EV infection, most can be found to be shedding virus in stool by culture (70 to 80%), whereas only 44 to 66% of CSF, 48 to 68% of throat or nasopharynx, 42 to 58% of urine, and 37 to 39% of blood samples will yield an EV in cell culture (127, 160, 161, 195). For adults with aseptic meningitis, culture of stool specimens typically has a high yield (94). For immunocompromised patients with CNS infection receiving IVIG, the rates of recovery of EV from CSF by culture are quite low compared to those for RT-PCR (159).

Centrifugation-enhanced inoculation and immunofluorescence or immunoperoxidase staining prior to development of CPE in shell vial cell cultures have been effective at decreasing the time to detection of EV from various sources (20, 91, 100, 142, 171, 193). By this method, EVs that may replicate poorly are also detected. Fifty-seven percent to 94% of EV-containing cultures are positive by this method within 72 hours, whereas only 23% to 51% of tube cultures are CPE positive in the same time frame (91, 100, 142, 193). However, because these centrifugation cultures are routinely stained blindly only for a specific virus or viruses at a designated time interval rather than evaluated for CPE, only the viruses sought will be detected, and unanticipated viruses will be missed. In fact, for persons with immunosuppression, there is a heightened potential for coinfection with more than one virus in respiratory specimens. Although the frequency of mixed infections is difficult to estimate, it has been shown that approximately 10% of respiratory samples from immunocompromised patients contain more than one virus (197).

Transgenic and cocultured cell lines can increase the sensitivity of recovery of EV and eliminate the need for multiple shell vials. BGMK cells stably transfected with human decay accelerating factor, a cofactor involved in the binding and entry of some echoviruses, coxsackieviruses, and EV 70, have been cocultured with CaCo-2 cells, a human colon adenocarcinoma cell line, or A549 cells (76) (Super E-Mix; Diagnostic Hybrids, Athens, OH). These mixed cell cultures have been shown to be more sensitive for recovery of EV from clinical samples than conventional tube or shell vial cultures with single cell lines (23, 76).

From an epidemiologic standpoint, viral culture has been important to isolate and identify the serotype of a virus causing disease. Traditionally, EV serotypes have been determined by using intersecting pools of lyophilized antisera adopted by Lim and Benyesh-Melnick (LBM pools) (156). However, the pools are in limited supply (World Health Organization) and are limited by the number of serotypes that can be identified (130, 132, 133). Alternatives to the LBM pools include serotype-specific EV MAb and pan-reactive antibody preparations that are commercially available for cell culture confirmation by immunofluorescence (Chemicon, Temecula, CA; Dako, Carpinteria, CA; and Diagnostic Hybrids, Athens, OH). Cross-reactions of pan-reactive reagents with other viruses, such as hepatitis A virus, reovirus, astrovirus, adenovirus, and rhinovirus, have been demonstrated (91, 206). Additionally, some preparations may not react with strains of EV 71 or parechoviruses 1 and 2 (165, 193). Recently, molecular serotyping methods have been developed to identify individual EV serotypes. These include VP1 gene sequence-based typing and serotype-specific PCR assays (21, 75, 89, 126, 131).

Serotyping of EV is useful for continued public health surveillance and, in certain circumstances, for clinical diagnostic purposes, since there are differences noted in the predominant clinical manifestations and outcomes associated with different EV serotypes and various sensitivities to antipicornavirus medications currently under development (143, 155). In addition, in those areas where use of OPV occurs, it is important to distinguish between isolates of the vaccine strain polioviruses and the nonpoliovirus EVs. Following vaccination, isolates can be recovered from the throat for 1 to 2 weeks and from stool for several weeks to months.

Serology

Serology has limited clinical value in routine diagnosis of EV infections because of the time needed to obtain acute- and convalescent-phase samples, the large number of serotypes that exist, and the lack of a single common antigen for use in serologic assays. Documenting a rise in serotype-specific IgG antibody titer of ≥4-fold from the acute to the convalescent phase of infection may be useful in epidemiologic investigations, such as studies of outbreaks of a specific EV serotype. EV serotype-specific antibodies can be determined using microneutralization, complement fixation, or hemagglutination inhibition. Generally, these assays have limited usefulness because they are relatively insensitive, poorly standardized, and labor-intensive and they lack specificity (74, 182, 188). Enzyme immunoassays (EIAs) for group- and serotype-specific determination of IgA, IgM, and IgG are also often limited by their sensitivity and specificity (48, 66, 182, 188). For culture-confirmed cases of EV infection, the sensitivities and specificities of IgM detection by EIA range from 34% to 77% and 88% to 94%, respectively (58, 182, 188).

Antigen Detection

IHC detection of EV has been described primarily for the diagnosis of myocardial infections but has also been used for other tissues, such as the spleen, lung, kidney, intestine, bone marrow, and pancreas (46, 97, 125, 213). The most commonly used MAb for IHC, 5-D8/1 (Dako Ltd., Cambridge, UK), is a group-specific IgG2a antibody that recognizes an epitope of EV capsid protein VP1 that has been localized to residues 40 to 48 of the N terminus, where the protein does not form an exposed domain of the virus capsid (165). Sequence alignment shows that this is a highly conserved region among serotypes, and experiments have shown immunoreactivity with at least 39 EV serotypes, including parechoviruses 1 and 2 (191). The epitope is recognized by MAb 5-D8/1 in formalin-fixed, paraffin-embedded (FFPE) sections after heating but not after proteinase digestion, suggesting that it is a linear or nonconformational epitope (212). No cross-reactions with this MAb preparation have been noted.

IHC for detection of EVs is typically more sensitive than ISH techniques, likely due to RNA loss or degradation during fixation or histological processing (98). However, despite potential issues with RNA in FFPE sections compared to RT-PCR of tissue sections, reported sensitivities of IHC range from 57% to 100% (9, 46, 97). Although the manufacturer's indications for use of anti-EV antibodies are limited to culture confirmation and identification, fluorescent-antibody techniques using serotype-specific MAbs on tissue sections have been reported (174). Detection of EV antigen directly in stool samples by EIA has been reported to be as sensitive as traditional cell culture but only 58% sensitive compared to RT-PCR (187).

Antiviral Susceptibility Testing

While not routinely used in clinical practice, in vitro susceptibility testing of pleconaril against a number of EV serotypes, using a well-characterized cell culture protection assay, has been reported (143). Serial dilutions of drug were added to 96-well cell culture plates infected with the EV serotype of interest, and after 3 days, the monolayers were examined for CPE. The concentration of drug that protected 50% of the cell monolayer from virus-induced CPE was defined as the 50% inhibitory concentration. Of 215 clinical isolates representing the 15 most common EV serotypes and 15 prototypic EV strains, all but 2 were inhibited by pleconaril, with 50% inhibitory concentrations of ≤3.4 μM. One prototypic strain of coxsackievirus B3 and one clinical isolate of echovirus 24 were insensitive to the drug, likely due to amino acid changes in the drug-binding pocket within the capsid (70, 143).

Test Selection, Application, and Interpretation

EV can be detected in multiple specimen types, such as blood, CSF, urine, or tissues, depending on the clinical syndrome. Identification of EV in these sites is associated with a high likelihood of causality for the disease state. However, because some individuals may be colonized and shed EV in the oropharynx or stool for weeks to months, detection of virus from these sites may make it difficult to establish causality for the illness in question. In fact, most individuals shedding EV in the stool at any given time are asymptomatic. On the other hand, nearly all patients with EV meningitis will have detectable virus in stool. OPV strains are also shed in stool for weeks after administration and will be detected in cell culture.

If EV infection is suspected in an immunocompromised patient, a rapid diagnostic method is desirable, not only so treatment can be instituted as early as possible but also to reduce unneeded diagnostic evaluations and treatments. Nucleic acid amplification assays are appropriate for CSF, serum, plasma, stool, urine, respiratory, and tissue specimens. Cell culture is appropriate for CSF, stool, respiratory, and tissue specimens, and ISH/IHC is appropriate for tissue specimens. While cell culture allows for identification of a broad range of viruses, the time required for isolation and identification of EV may be too long to be of clinical utility. Nucleic acid amplification assays typically afford greater sensitivity along with a more rapid time to result. The specificities of nucleic acid amplification assays are generally excellent. The notable exception to this is testing of respiratory specimens, where rhinoviruses may also be present. It behooves the testing laboratory to determine the validity of their EV assay for respiratory samples by assessing potential cross-amplification with many of the more common rhinovirus serotypes.

RT-PCR is particularly useful for documenting EV infection in those patients with B-cell-associated immunodeficiencies, where recovery of viruses in cell culture is inadequate, particularly if IVIG has been administered. During the course of CEMA, RT-PCR testing of CSF may intermittently be negative, but this result should not exclude the diagnosis and another sample of CSF should be tested. For those patients with immunodeficiency and EV infection, the sequential recovery of virus over time has provided an opportunity to study the genomic changes that EV strains undergo during prolonged replication in a human host (13, 25, 51). Viral load testing by real-time RT-PCR may prove to be a useful tool for some patient populations (e.g., neonates) for early diagnosis and prediction of disease severity, although assays for this type of testing are not yet commercially available (209). IHC may be useful in detecting EV in FFPE tissue samples, particularly if the RNA has been compromised, thus making detection by RT-PCR or ISH unreliable. For neonates, sampling from multiple sites increases the diagnostic yield of EV. The identification of EV from a mucosal site in a neonate along with EV-compatible illness and the absence of other infectious or noninfectious causes are strong indications that EV is the likely etiologic agent.

Interpretation of serologic results with paired serum samples may be complicated by the high prevalence of EV antibodies in the general population and the fact that antibody may already be present in the acute sample due to long incubation and prodromal periods with EV illness. Interpretation of EIA results is complicated by the heterotypic responses of many individuals caused by other EV infections and therefore cannot be considered strictly serotype specific (58, 182). The performance characteristics of the serologic assays and the fact that a large percentage of immunocompromised individuals will have diminished antibody production preclude the routine use of this type of testing for diagnostic purposes for this patient population.

EV vaccines are available only for the three poliovirus serotypes, and they provide no protection against the nonpoliovirus EV serotypes. Recent policy changes in the United States have resulted in a return to the use of IPV in an attempt to reduce or eliminate the few remaining vaccine-associated cases of poliomyelitis that occur annually in this country (31). Live viral vaccines such as the OPV are contraindicated for immunocompromised patients (30). Only IPV should be administered. In addition, it is recommended that family members, close contacts, and health care workers of immunodeficient individuals be given only the IPV (32).

Because stem cell transplant recipients lose immunity to poliovirus early after transplantation, a revaccination schedule with three doses of the IPV is recommended by both the European Group for Blood and Marrow Transplantation and the Centers for Disease Control and Prevention (CDC) (32, 103). The European Group for Blood and Marrow Transplantation recommendations call for revaccination starting 6 to 12 months after transplantation, with subsequent doses 1 to 3 months apart, while the CDC recommends a schedule of individual doses 12, 14, and 24 months after transplantation. Revaccination after stem cell transplantation results in protective immunity in most patients, although a few patients may become seronegative over the course of several years (53, 102, 140). Younger patients and those with graft-versus-host disease are more likely to lose immunity after revaccination. However, for vaccines that typically elicit a good immune response, such as IPV, routine pre- and postvaccination testing of neutralizing antibody titers is not necessary (103).

Ideally, solid organ transplant patients should be immunized prior to transplantation, since the response to many vaccines may be diminished in the setting of organ failure. Although there are no data regarding timing of vaccination after transplantation, it is recommended that vaccination series be started no sooner than 6 to 12 months posttransplantation, since there is an increased risk of graft dysfunction and high-dose immunosuppressive treatment may inhibit an adequate immune response prior to this time (24, 177). Hypogammaglobulinemia has been described for heart, lung, and kidney transplant recipients and has been associated with the development of recurrent infections (65, 145, 207). In adult renal transplant patients evaluated at least 1 year after transplantation, few had protective levels of antibody to all three poliovirus serotypes (77). Four weeks after revaccination with a single booster of IPV, approximately 92% of patients demonstrated protective antibody levels to all three serotypes. Similarly, booster immunization with IPV in children after liver transplantation produced similar serologic responses to those noted in controls (14).

REFERENCES

1. **Abzug, M. J., M. J. Levin, and H. A. Rotbart.** 1993. Profile of enterovirus disease in the first two weeks of life. *Pediatr. Infect. Dis. J.* **12:**820–824.
2. **Abzug, M. J., H. L. Keyserling, M. L. Lee, M. J. Levin, and H. A. Rotbart.** 1995. Neonatal enterovirus infection: virology, serology, and effects of intravenous immune globulin. *Clin. Infect. Dis.* **20:**1201–1206.
3. **Abzug, M. J., M. Loeffelholz, and H. A. Rotbart.** 1995. Diagnosis of neonatal enterovirus infection by polymerase chain reaction. *J. Pediatr.* **126:**447–450.
4. **Abzug, M. J.** 2001. Prognosis for neonates with enterovirus hepatitis and coagulopathy. *Pediatr. Infect. Dis. J.* **20:**758–763.
5. **Ahmed, A., F. Brito, C. Goto, S. M. Hickey, K. D. Olsen, M. Trujillo, and G. H. McCracken.** 1997. Clinical utility of the polymerase chain reaction for diagnosis of enteroviral meningitis in infancy. *J. Pediatr.* **131:**393–397.
6. **Akhtar, N., J. Ni, D. Stromberg, G. L. Rosenthal, N. E. Bowles, and J. A. Towbin.** 1999. Tracheal aspirate as a substrate for polymerase chain reaction detection of viral genome in childhood pneumonia and myocarditis. *Circulation* **99:**2011–2018.
7. **Alexander, L. N., J. F. Seward, T. A. Santibanez, M. A. Pallansch, O. M. Kew, D. R. Prevots, P. M. Strebel, J. Cono, M. Wharton, W. A. Orenstein, and R. W. Sutter.** 2004. Vaccine policy changes and epidemiology of poliomyelitis in the United States. *JAMA* **292:**1696–1701.
8. **Andréoletti, L., N. Blassel-Damman, A. Dewilde, L. Vallée, R. Cremer, D. Hober, and P. Wattré.** 1998. Comparison of use of cerebrospinal fluid, serum, and throat swab specimens in diagnosis of enteroviral acute neurological infection by a rapid RNA detection PCR assay. *J. Clin. Microbiol.* **36:**589–591.
9. **Andréoletti, L., T. Bourlet, D. Moukassa, L. Rey, D. Hot, Y. Li, V. Lambert, B. Gosselin, J. Mosnier, C. Stankowiak, and P. Wattré.** 2000. Enteroviruses can persist with or without active viral replication in cardiac tissue of patients with end-stage ischemic or dilated cardiomyopathy. *J. Infect. Dis.* **182:**1222–1227.
10. **Aquino, V. M., R. A. Farah, M. C. Lee, and E. S. Sandler.** 1996. Disseminated coxsackie A9 infection complicating bone marrow transplantation. *Pediatr. Infect. Dis. J.* **15:**1053–1054.
11. **Aradottir, E., E. M. Alonso, and S. T. Shulman.** 2001. Severe neonatal enteroviral hepatitis treated with pleconaril. *Pediatr. Infect. Dis. J.* **20:**457–459.
12. **Archimbaud, C., J. Bailly, M. Chambon, O. Tournilhac, P. Travade, and H. Peigue-Lafeuille.** 2003. Molecular evidence of persistent echovirus 13 meningoencephalitis in a patient with relapsed lymphoma after an outbreak of meningitis in 2000. *J. Clin. Microbiol.* **41:**4605–4610.
13. **Bailly, J., M. Chambon, C. Henquell, J. Icart, and H. Peigue-Lafeuille.** 2000. Genomic variations in echovirus 30 persistent isolates recovered from a chronically infected immunodeficient child and comparison with the reference strain. *J. Clin. Microbiol.* **38:**552–557.
14. **Balloni, A., B. M. Assael, L. Ghio, C. Pedrazzi, G. Nebbia, B. Gridelli, E. Melada, A. Panuccio, M. Foti, M. Barbi, and C. Luraschi.** 1999. Immunity to poliomyelitis, diphtheria and tetanus in pediatric patients before and after renal or liver transplantation. *Vaccine* **17:**2507–2511.
15. **Ballow, M., K. L. Cates, J. C. Rowe, C. Goetz, and C. Desbonnet.** 1986. Development of the immune system in very low birth weight (less than 1500 g) premature infants: concentrations of plasma immunoglobulins and patterns of infection. *Pediatr. Res.* **20:**899–904.
16. **Bauer, S., G. Gottesman, L. Sirota, I. Litmanovitz, S. Ashkenazi, and I. Levi.** 2002. Severe coxsackie virus B infection in preterm newborns treated with pleconaril. *Eur. J. Pediatr.* **161:**491–493.

17. **Bellmunt, A., G. May, R. Zell, P. Pring-Åkerblom, W. Verhagen, and A. Heim.** 1999. Evolution of poliovirus type 1 during 5.5 years of prolonged enteral replication in an immunodeficient patient. *Virology* 265:178–184.
18. **Biggs, D. D., B. C. Toorkey, D. R. Carrigan, G. A. Hanson, and R. C. Ash.** 1990. Disseminated echovirus infection complicating bone marrow transplantation. *Am. J. Med.* **88:**421–425.
19. **Birenbaum, E., R. Handsher, J. Kuint, R. Dagan, B. Raichman, E. Mendelson, and N. Linder.** 1997. Echovirus type 22 outbreak associated with gastro-intestinal disease in a neonatal intensive care unit. *Am. J. Perinatol.* **14:**469–473.
20. **Bourlet, T., J. Gharbi, S. Omar, M. Aouni, and B. Pozzetto.** 1998. Comparison of a rapid culture method combining an immunoperoxidase test and a group specific anti-VP1 monoclonal antibody with conventional virus isolation for routine detection of enteroviruses in stools. *J. Med. Virol.* **54:**204–209.
21. **Brown, B. A., D. R. Kilpatrick, M. S. Oberste, and M. A. Pallansch.** 2000. Serotype-specific identification of enterovirus 71 by PCR. *J. Clin. Virol.* **16:**107–112.
22. **Bryant, P. A., D. Tingay, P. A. Dargaville, M. Starr, and N. Curtis.** 2004. Neonatal coxsackie B virus infection—a treatable disease? *Eur. J. Pediatr.* **163:**223–228.
23. **Buck, G. E., M. Wiesemann, and L. Stewart.** 2002. Comparison of mixed cell culture containing genetically engineered BGMK and CaCo-2 cells (Super E-Mix) with RT-PCR and conventional cell culture for the diagnosis of enteroviral meningitis. *J. Clin. Virol.* **25**(Suppl. 1):S13–S18.
24. **Burroughs, M. and A. Moscona.** 2000. Immunization of pediatric solid organ transplant candidates and recipients. *Clin. Infect. Dis.* **30:**857–869.
25. **Buttinelli, G., V. Donati, S. Fiore, J. Marturano, A. Plebani, P. Balestri, A. R. Soresina, R. Vivarelli, F. Delpeyroux, J. Martin, and L. Fiore.** 2003. Nucleotide variation in Sabin type 2 poliovirus from an immunodeficient patient with poliomyelitis. *J. Gen. Virol.* **84:**1215–1221.
26. **Byington, C. L., E. W. Taggart, K. C. Carroll, and D. R. Hillyard.** 1999. A polymerase chain reaction-based epidemiologic investigation of the incidence of nonpolio enteroviral infections in febrile and afebrile infants 90 days and younger. *Pediatrics* **103:**e27.
27. **Capaul, S. E., and M. Gorgievski-Hrisoho.** 2005. Detection of enterovirus RNA in cerebrospinal fluid using Nuclisens EasyQ enterovirus assay. *J. Clin. Virol.* **32:**236–240.
28. **Carolane, D. J., A. M. Long, P. A. McKeever, S. J. Hobbs, and A. P. Roome.** 1985. Prevention of spread of echovirus 6 in a special care baby unit. *Arch. Dis. Child.* **60:**674–676.
29. **Carroll, K. C., B. Taggart, J. Robison, C. Byington, and D. Hillyard.** 2000. Evaluation of the Roche AMPLICOR enterovirus PCR assay in the diagnosis of enteroviral central nervous system infections. *J. Clin. Virol.* **19:**149–156.
30. **Centers for Disease Control and Prevention.** 1993. Recommendations of the Advisory Committee on Immunization Practices (ACIP): use of vaccines and immune globulins in persons with altered immunocompetence. *MMWR Morb. Mortal. Wkly. Rep.* **42**(RR-4):1–18.
31. **Centers for Disease Control and Prevention.** 2000. Poliomyelitis prevention in the United States. Updated recommendations of the Advisory Committee on Immunization Practices (ACIP). *MMWR Morb. Mortal. Wkly. Rep.* **49**(RR-5):1–22.
32. **Centers for Disease Control and Prevention.** 2000. Guidelines for preventing opportunistic infections among hematopoietic stem cell transplant recipients. Recommendations of CDC, the Infectious Disease Society of America, and the American Society of Blood and Marrow Transplantation. *Morb. Mortal. Wkly. Rep.* **49**(RR-10):1–125.
33. **Centers for Disease Control and Prevention.** 2006. Progress toward interruption of wild poliovirus transmission—worldwide, January 2005-March 2006. *MMWR Morb. Mortal. Wkly. Rep.* **55:**458–462.
34. **Centers for Disease Control and Prevention.** 2006. Update on vaccine-derived polioviruses. *MMWR Morb. Mortal. Wkly. Rep.* **55:**1093–1097.
35. **Chakrabarti, S., K. E. Collingham, R. H. Stevens, D. Pillay, C. D. Fegan, and D. W. Milligan.** 2000. Isolation of viruses from stools in stem cell transplant recipients: a prospective surveillance study. *Bone Marrow Transplant.* **25:**277–282.
36. **Chakrabarti, S., H. Osman, K. E. Collingham, C. D. Fegan, and D. W. Milligan.** 2004. Enterovirus infections following T-cell depleted allogeneic transplants in adults. *Bone Marrow Transplant.* **33:**425–430.
37. **Chakraborty, R., M. Iturriza-Gómara, R. Musoke, T. Palakudy, A. D'Agostino, and J. Gray.** 2004. An epidemic of enterovirus 71 infection among HIV-1-infected orphans in Nairobi. *AIDS* **18:**1968–1970.
38. **Chambon, M., C. Delage, J. Bailly, J. Gaulme, P. Dechelotte, C. Henquell, C. Jallat, and H. Peigue-Lafeuille.** 1997. Fatal hepatic necrosis in a neonate with echovirus 20 infection: use of the polymerase chain reaction to detect enterovirus in liver tissue. *Clin. Infect. Dis.* **24:**523–524.
39. **Chambon, M., J. L. Bailly, A. Béguet, C. Henquell, C. Archimbaud, J. Gaulme, A. Labbé, G. Malpuech, and H. Peigue-Lafeuille.** 1999. An outbreak due to echovirus type 30 in a neonatal unit in France in 1997: usefulness of PCR diagnosis. *J. Hosp. Infect.* **43:**63–68.
40. **Chan, S. H., and K. S. Lun.** 2001. Ventricular aneurysm complicating neonatal coxsackievirus B4 myocarditis. *Pediatr. Cardiol.* **22:**247–249.
41. **Chen, J., N. Chiu, J. Chang, F. Huang, K. Wu, and T. Lin.** 2005. A neonatal echovirus 11 outbreak in an obstetric clinic. *J. Microbiol. Immunol. Infect.* **38:**332–337.
42. **Chitsike, I., and R. van Furth.** 1999. Paralytic poliomyelitis associated with live oral poliomyelitis vaccine in child with HIV infection in Zimbabwe: case report. *BMJ* **318:**841–843.
43. **Chonmaitree, T., and L. Mann.** 1995. Respiratory infections, p. 255–270. *In* H. A. Rotbart (ed.), *Human Enterovirus Infections.* ASM Press, Washington, DC.
44. **Chow, K., C. Lee, T. Lin, W. Shen, J. Wang, C. Peng, and C. Lee.** 2000. Congenital enterovirus 71 infection: a case study with virology and immunohistochemistry. *Clin. Infect. Dis.* **31:**509–512.
45. **Christensen, M. S., L. P. Nielson, and H. Hasle.** 2005. Few but severe viral infections in children with cancer: a prospective RT-PCR and PCR-based 12-month study. *Pediatr. Blood Cancer* **45:**945–951.
46. **Cioc, A. M., and G. J. Nuovo.** 2001. Histologic and in situ viral findings in the myocardium in cases of sudden, unexpected death. *Mod. Pathol.* **15:**914–922.
47. **Corless, C. E., M. Guiver, R. Borrow, V. Edwards-Jones, A. J. Fox, E. B. Kaczmarski, and K. J. Mutton.** 2002. Development and evaluation of a real-time RT-PCR for the detection of enterovirus and parechovirus RNA in CSF and throat swab samples. *J. Med. Virol.* **67:**555–562.
48. **Craig, M. E., P. Robertson, N. J. Howard, M. Silink, and W. D. Rawlinson.** 2003. Diagnosis of enterovirus infection by genus-specific PCR and enzyme-linked immunosorbent assays. *J. Clin. Microbiol.* **41:**841–844.
49. **Cunningham, C. K., C. A. Bonville, H. D. Ochs, K. Seyama, P. A. John, H. A. Rotbart, and L. B. Weiner.** 1999.

Enteroviral meningoencephalitis as a complication of X-linked hyper IgM syndrome. *J. Pediatr.* **134:**584–588.
50. **Dechamps, C., H. H. Peigue-Lafeuille, H. Laveran, J. Beytout, H. Roger, and D. Beytout.** 1988. Four cases of vesicular lesions in adults caused by enterovirus infections. *J. Clin. Microbiol.* **26:**2182–2183.
51. **Dunn, J. J., J. R. Romero, R. Wasserman, and H. A. Rotbart.** 2000. Stable enterovirus 5′ nontranslated region over a 7-year period in a patient with agammaglobulinemia and chronic infection. *J. Infect. Dis.* **182:**298–301.
52. **Dyer, J. R., R. H. Edis, and M. A. H. French.** 1998. Enterovirus associated neurological disease in an HIV-1 infected man. *J. Neurovirol.* **4:**569–571.
53. **Engelhard, D., R. Handsher, E. Naparstek, I. Hardan, N. Strauss, M. Aker, R. Or, H. Baciu, and S. Slavin.** 1991. Immune response to polio vaccination in bone marrow transplant recipients. *Bone Marrow Transplant.* **8:**295–300.
54. **Erlendsson, K., T. Swartz, and J. M. Dwyer.** 1985. Successful reversal of echovirus encephalitis in X-linked hypogammaglobulinemia by intraventricular administration of immunoglobulin. *N. Engl. J. Med.* **312:**352–353.
55. **Euscher, E., J. Davis, I. Holzman, and G. J. Nuovo.** 2001. Coxsackie virus infection of the placenta associated with neurodevelopmental delays in the newborn. *Obstet. Gynecol.* **98:**1019–1026.
56. **Faulkner, C. F., A. M. Godbolt, B. DeAmbrosis, and J. Triscott.** 2003. Hand, foot and mouth disease in an immunocompromised adult treated with acyclovir. *Australas. J. Dermatol.* **44:**203–206.
57. **Fischmeister, G., P. Wiesbauer, H. Holzmann, C. Peters, M. Eibl, and H. Gadner.** 2000. Enteroviral meningoencephalitis in immunocompromised children after matched unrelated donor-bone marrow transplantation. *Pediatr. Hematol. Oncol.* **17:**393–399.
58. **Frisk, G., E. Nilsson, A. Ehrnst, and H. Diderholm.** 1989. Enterovirus IgM detection: specificity of mu-antibody-capture radioimmunoassay using virions and procapsids of coxsackie B virus. *J. Virol. Methods* **24:**191–202.
59. **Galama, J. M. D., N. de Leeuw, S. Wittebol, H. Peters, and W. J. G. Melchers.** 1996. Prolonged enteroviral infection in a patient who developed pericarditis and heart failure after bone marrow transplantation. *Clin. Infect. Dis.* **22:**1004–1008.
60. **Galama, J. M. D.** 1997. Enteroviral infections in the immunocompromised host. *Rev. Med. Microbiol.* **8:**33–40.
61. **Galama, J. M. D., M. Gielen, and C. M. R. Weemaes.** 2000. Enterovirus antibody titers after IVIG replacement in agammaglobulinemic children. *Clin. Microbiol. Infect.* **6:**629–632.
62. **Garcia, A. G., N. G. Basso, M. E. Fonseca, and H. N. Outani.** 1990. Congenital echovirus infection—morphological and virological study of fetal and placental tissue. *J. Pathol.* **160:**123–127.
63. **Geller, T. J., and D. Condie.** 1995. A case of protracted coxsackie virus meningoencephalitis in a marginally immunodeficient child treated successfully with intravenous immunoglobulin. *J. Neurol. Sci.* **129:**131–133.
64. **Ginocchio, C. G., F. Zhang, A. Malhotra, R. Manji, P. Sillekens, H. Foolen, M. Overdyk, and M. Peeters.** 2005. Development, technical performance, and clinical evaluation of a NucliSens basic kit application for detection of enterovirus RNA in cerebrospinal fluid. *J. Clin. Microbiol.* **43:**2616–2623.
65. **Goldfarb, N. S., R. K. Avery, M. Goormastic, A. C. Mehta, R. Schilz, N. Smedira, L. Pien, M. T. Haug, S. M. Gordon, L. K. Hague, J. M. Dresing, T. Evans-Walker, and J. R. Maurer.** 2001. Hypogammaglobulinemia in lung transplant recipients. *Transplantation* **71:**242–246.
66. **Goldwater, P. N.** 1995. Immunoglobulin M capture immunoassay in investigation of coxsackievirus B5 and B6 outbreaks in South Australia. *J. Clin. Microbiol.* **33:**1628–1631.
67. **Gonzalez, Y., R. Martino, N. Rabella, R. Labega, I. Badell, and J. Sierra.** 1999. Community respiratory virus infections in patients with hematologic malignancies. *Haematologica* **84:**820–823.
68. **Gonzalez, Y., R. Martino, I. Badell, N. Pardo, A. Sureda, S. Brunet, J. Sierra, and N. Rabella.** 1999. Pulmonary enterovirus infections in stem cell transplant recipients. *Bone Marrow Transplant.* **23:**511–513.
69. **Gouandjika-Vasilache, I., C. Akoua-Koffi, E. Begaud, and A. Dosseh.** 2005. No evidence of prolonged enterovirus excretion in HIV-seropositive patients. *Trop. Med. Int. Health* **10:**743–747.
70. **Groarke, J. M., and D. C. Pevear.** 1999. Attenuated virulence of pleconaril-resistant coxsackievirus B3 variants. *J. Infect. Dis.* **179:**1538–1541.
71. **Halliday, E., J. Winkelstein, and A. D. B. Webster.** 2003. Enteroviral infections in primary immunodeficiency (PID): a survey of morbidity and mortality. *J. Infect.* **46:**1–8.
72. **Hassan, I. A., R. Chopra, R. Swindell, and K. J. Mutton.** 2003. Respiratory viral infections after bone marrow/peripheral stem-cell transplantation: the Christie hospital experience. *Bone Marrow Transplant.* **32:**73–77.
73. **Hidalgo, S., M. G. Erro, D. Cisterna, and M. C. Freire.** 2003. Paralytic poliomyelitis caused by a vaccine-derived polio virus in an antibody-deficient Argentinean child. *Pediatr. Infect. Dis. J.* **22:**570–572.
74. **Hierholzer, J. C., P. G. Bingham, R. A. Coombs, Y. O. Stone, and M. H. Hatch.** 1984. Quantitation of enterovirus 70 antibody by microneutralization test and comparison with standard neutralization, hemagglutination inhibition, and complement fixation tests with different virus strains. *J. Clin. Microbiol.* **19:**826–830.
75. **Huang, Q. S., J. M. Carr, W. A. Nix, M. S. Oberste, D. R. Kilpatrick, M. A. Pallansch, M. C. Croxson, J. A. Lindeman, M. G. Baker, and K. Grimwood.** 2003. An echovirus type 33 winter outbreak in New Zealand. *Clin. Infect. Dis.* **37:**650–657.
76. **Huang, Y. T., P. Tam, H. Yan, and Y. Sun.** 2002. Engineered BGMK cells for sensitive and rapid detection of enteroviruses. *J. Clin. Microbiol.* **40:**366–371.
77. **Huzly, D., S. Neifer, P. Reinke, K. Schröder, C. Schönfeld, T. Hofman, and U. Bienzle.** 1997. Routine immunizations in adult renal transplant recipients. *Transplantation* **63:**839–845.
78. **Ion-Nedelcu, N., A. Dobrescu, P. M. Strebel, and R. W. Sutter.** 1994. Vaccine-associated paralytic poliomyelitis and HIV infection. *Lancet* **343:**51–52.
79. **Ison, M. G., and F. G. Hayden.** 2002. Viral infections in immunocompromised patients: what's new with respiratory viruses? *Curr. Opin. Infect. Dis.* **15:**355–367.
80. **Jankovic, B., S. Pasic, B. Kanjuh, K. Bukumirovic, G. Cvetanovic, N. Todorovic, and S. Djuricic.** 1999. Severe neonatal echovirus 17 infection during a nursery outbreak. *Pediatr. Infect. Dis. J.* **18:**393–394.
81. **Jenista, J. A., K. R. Powell, and M. A. Menegus.** 1984. Epidemiology of neonatal enterovirus infection. *J. Pediatr.* **104:**685–690.
82. **Jenson, H. B., C. J. Gauntt, K. A. Easley, J. Pitt, S. E. Lipshultz, K. McIntosh, and W. T. Shearer.** 2002. Evaluation of coxsackievirus infection in children with human immunodeficiency virus type 1-associated cardiomyopathy. *J. Infect. Dis.* **185:**1798–1802.

83. **Kares, S., M. Lönnrot, P. Vuorinen, S. Oikarinen, S. Taurianen, and H. Hyöty.** 2004. Real-time PCR for rapid diagnosis of entero- and rhinovirus infections using LightCycler. *J. Clin. Virol.* **29:**99–104.

84. **Kessler, H. H., B. Santner, H. Rabenau, A. Berger, A. Vince, C. Lewinski, B. Weber, K. Pierer, D. Stuenzner, E. Marth, and H. W. Doerr.** 1997. Rapid diagnosis of enterovirus infection by a new one-step reverse transcription-PCR assay. *J. Clin. Microbiol.* **35:**976–977.

85. **Khetsuriani, N., D. R. Prevots, L. Quick, M. E. Elder, M. Pallansch, O. Kew, and R. W. Sutter.** 2003. Persistence of vaccine-derived polioviruses among immunodeficient persons with vaccine-associated paralytic poliomyelitis. *J. Infect. Dis.* **188:**1845–1852.

86. **Khetsuriani, N., A. LaMonte-Fowlkes, M. S. Oberste, and M. A. Pallansch.** 2006. Enterovirus surveillance—United States, 1970–2005. *MMWR Morb. Mortal. Wkly. Rep. Surveill. Summ.* **55:**1–20.

87. **Khetsuriani, N., A. LaMonte, M. S. Oberste, and M. Pallansch.** 2006. Neonatal enterovirus infections reported to the national enterovirus surveillance system in the United States, 1983–2003. *Pediatr. Infect. Dis. J.* **25:**889–893.

88. **Khoo, S. H., M. Hajia, C. C. Storey, P. E. Klapper, E. G. Wilkins, D. W. Denning, E. M. Dunbar, G. Corbitt, and B. K. Mandal.** 1998. Influenza-like episodes in HIV-positive patients: the role of viral and 'atypical' infections. *AIDS* **12:**751–757.

89. **Kilpatrick, D. R., J. Quay, M. A. Pallansch, and M. S. Oberste.** 2001. Type-specific detection of echovirus 30 isolates using degenerate reverse transcriptase PCR primers. *J. Clin. Microbiol.* **39:**1299–1302.

90. **Kinney, J. S., E. McCray, J. E. Kaplan, D. E. Low, G. W. Hammond, G. Harding, P. F. Pinsky, M. J. Davi, S. F. Kovnats, P. Riben, W. J. Martone, L. B. Schonberger, and L. J. Anderson.** 1986. Risk factors associated with echovirus 11 infection in a hospital nursery. *Pediatr. Infect. Dis.* **5:**192–197.

91. **Klespies, S. L., D. E. Cebula, C. L. Kelley, D. Galehouse, and C. C. Maurer.** 1996. Detection of enteroviruses from clinical specimens by spin amplification shell vial culture and monoclonal antibody assay. *J. Clin. Microbiol.* **34:**1465–1467.

92. **Kok, T. W., T. Pryor, and L. Payne.** 1998. Comparison of rhabdomyosarcoma, buffalo green monkey kidney epithelial, A549 (human lung epithelial) cells and human embryonic lung fibroblasts for isolation of enteroviruses from clinical samples. *J. Clin. Virol.* **11:**61–65.

93. **Kost, C. B., B. Rogers, M. S. Oberste, C. Robinson, B. L. Eaves, K. Leos, S. Danielson, M. Satya, F. Weir, and F. S. Nolte.** 2007. Multicenter beta trial of the GeneXpert enterovirus assay. *J. Clin. Microbiol.* **45:**1081–1086.

94. **Kupila, L., T. Vuorinen, R. Vainionpää, R. J. Marttila, and P. Kotilainen.** 2005. Diagnosis of enteroviral meningitis by use of polymerase chain reaction of cerebrospinal fluid, stool, and serum specimens. *Clin. Infect. Dis.* **40:**982–987.

95. **Lai, K. K., L. Cook, S. Wendt, L. Corey, and K. R. Jerome.** 2003. Evaluation of real-time PCR versus PCR with liquid-phase hybridization for detection of enterovirus RNA in cerebrospinal fluid. *J. Clin. Microbiol.* **41:**3133–3141.

96. **Levy, J., T. Espanol-Boren, C. Thomas, A. Fischer, P. Tovo, P. Bordigoni, I. Redneck, A. Fasth, M. Baer, L. Gomez, E. A. Sanders, M. D. Tabone, D. Plantaz, A. Etzioni, V. Monafo, M. Abinun, L. Hammarstrom, T. Abrahamsen, A. Jones, A. Finn, T. Klemola, E. DeVries, O. Sanal, M. C. Peitsch, and L. D. Notarangelo.** 1997. Clinical spectrum of X-linked hyper-IgM syndrome. *J. Pediatr.* **131:**47–54.

97. **Li, Y., T. Bourlet, L. Andreoletti, J. Mosnier, T. Peng, Y. Yang, L. C. Archard, B. Pozzetto, and H. Zhang.** 2000. Enteroviral capsid protein VP1 is present in myocardial tissues from some patients with myocarditis or dilated cardiomyopathy. *Circulation* **101:**231–234.

98. **Li, Y., Z. Pan, Y. Ji, T. Peng, L. C. Archard, and H. Zhang.** 2002. Enterovirus replication in valvular tissue from patients with chronic rheumatic heart disease. *Eur. Heart J.* **23:**567–573.

99. **Lin, T., H. Kao, S. Hsieh, Y. Huang, C. Chiu, Y. Chou, P. Yang, R. Lin, K. Tsao, K. Hsu, and L. Chang.** 2003. Neonatal enterovirus infections: emphasis on risk factors of severe and fatal infections. *Pediatr. Infect. Dis. J.* **22:**889–894.

100. **Lipson, S. M., K. David, F. Shaikh, and L. Qian.** 2001. Detection of precytopathic effect of enteroviruses in clinical specimens by centrifugation-enhanced antigen detection. *J. Clin. Microbiol.* **39:**2755–2759.

101. **Liste, M. B., I. Natera, J. A. Suarez, F. H. Pujol, F. Liprandi, and J. E. Ludert.** 2000. Enteric virus infections and diarrhea in healthy and human immunodeficiency virus-infected children. *J. Clin. Microbiol.* **38:**2873–2877.

102. **Ljungman, P., J. Aschan, B. Gustafsson, I. Lewensohn-Fuchs, J. Winiarski, and O. Ringdén.** 2004. Long-term immunity to poliovirus after vaccination of allogeneic stem cell transplant recipients. *Bone Marrow Transplant.* **34:**1067–1069.

103. **Ljungman, P., D. Engelhard, R. De la Cámara, H. Einsele, A. Locasciulli, R. Martino, P. Ribaud, K. Ward, and C. Cordonnier.** 2005. Vaccination of stem cell transplant recipients: recommendations of the Infectious Diseases Working Party of the EBMT. *Bone Marrow Transplant.* **35:**737–746.

104. **Lopez, C., W. D. Biggar, B. H. Park, and R. A. Good.** 1974. Nonparalytic poliovirus infections in patients with severe combined immunodeficiency diseases. *J. Pediatr.* **84:**497–502.

105. **MacLennan, C., G. Dunn, A. P. Hulssoon, D. S. Kumalaratne, J. Martin, P. O'Leary, R. A. Thompson, H. Osman, P. Wood, P. Minor, D. J. Wood, and D. Pillay.** 2004. Failure to clear persistent vaccine-derived neurovirulent poliovirus infection in an immunodeficient man. *Lancet* **363:**1509–1513.

106. **Marlowe, E. M., S. M. Novak, J. J. Dunn, A. Smith, J. Cumpio, E. Makalintal, D. Barnes, and R. J. Burchette.** 2008. Performance of the GeneXpert enterovirus assay for detection of enteroviral RNA in cerebrospinal fluid. *J. Clin. Virol.*, in press. doi:10.1016/j.jcv.2008.04.006 (e-pub ahead of print).

107. **Martín, J., G. Dunn, R. Hull, V. Patel, and P. D. Minor.** 2000. Evolution of the Sabin strain of type 3 poliovirus in an immunodeficient patient during the entire 637-day period of virus excretion. *J. Virol.* **74:**3001–3010.

108. **Martino, R., R. P. Porras, N. Rabella, J. V. Williams, E. Ramila, N. Margall, R. Labeaga, J. E. Crowe, P. Coll, and J. Sierra.** 2005. Prospective study of the incidence, clinical features and outcome of symptomatic upper and lower respiratory tract infections by respiratory viruses in adult recipients of hematopoietic stem cell transplants for hematologic malignancies. *Biol. Blood Marrow Transplant.* **11:**781–796.

109. **McKinney, R. E., S. L. Katz, and C. M. Wilfert.** 1987. Chronic enteroviral meningoencephalitis in agammaglobulinemic patients. *Rev. Infect. Dis.* **9:**334–356.

110. **McMinn, P., I. Stratov, L. Nagarajan, and S. Davis.** 2001. Neurological manifestations of enterovirus 71 infection in children during an outbreak of hand, foot, and mouth disease in Western Australia. *Clin. Infect. Dis.* **32:**236–242.

111. **Mease, P. J., H. D. Ochs, and R. J. Wedgwood.** 1981. Successful treatment of echovirus meningoencephalitis and

myositis-fasciitis with intravenous immune globulin therapy in a patient with X-linked agammaglobulinemia. *N. Engl. J. Med.* **304:**1278–1281.

112. **Miller, R. F., C. Loveday, J. Holton, Y. Sharvell, G. Patel, and N. S. Brink.** 1996. Community-based respiratory viral infections in HIV positive patients with lower respiratory tract disease: a prospective bronchoscopic study. *Genitourin. Med.* **72:**9–11.
113. **Minami, K., Y. Tsuda, H. Maeda, T. Yanagawa, G. Izumi, and N. Yoshikawa.** 2004. Acute transverse myelitis caused by coxsackie virus B5 infection. *J. Paediatr. Child. Health* **40:**66–68.
114. **Minor, P. D., M. Ferguson, D. M. Evans, J. W. Almond, and J. P. Icenogle.** 1986. Antigenic structure of poliovirus serotypes 1, 2, and 3. *J. Gen. Virol.* **67:**1283–1291.
115. **Minosse, C., M. S. Zaniratti, S. Calcaterra, F. Carletti, M. Muscillo, M. Pisciotta, L. Pillitteri, A. Corpolongo, F. N. Lauria, P. Narciso, G. Anzidei, and M. R. Capobianchi.** 2005. Application of a molecular panel to demonstrate enterotropic virus shedding by healthy and human immunodeficiency virus-infected patients. *J. Clin. Microbiol.* **43:**1979–1981.
116. **Misbah, S. A., P. A. Lawrence, J. B. Kurtz, and H. M. Chapel.** 1991. Prolonged faecal excretion of poliovirus in a nurse with common variable hypogammaglobulinemia. *Postgrad. Med. J.* **67:**301–303.
117. **Misbah, S. A., G. P. Spickett, P. C. J. Ryba, J. M. Hockaday, J. S. Kroll, C. Sherwood, J. B. Kurtz, E. R. Moxon, and H. M. Chapel.** 1992. Chronic enteroviral meningoencephalitis in agammaglobulinemia: case report and literature review. *J. Clin. Immunol.* **12:**266–270.
118. **Modlin, J. F., B. F. Polk, P. Horton, P. Etkind, E. Crane, and A. Spiliotes.** 1981. Perinatal echovirus infection: risk of transmission during a community outbreak. *N. Engl. J. Med.* **305:**368–371.
119. **Modlin, J. F.** 1986. Perinatal echovirus infection: insights from a literature review of 61 cases of serious infection and 16 outbreaks in nurseries. *Rev. Infect. Dis.* **8:**918–926.
120. **Modlin, J. F.** 1988. Perinatal echovirus and group B coxsackievirus infections. *Clin. Perinatol.* **15:**233–245.
121. **Modlin, J. F.** 2003. Enteroviruses: coxsackieviruses, echoviruses, and newer enteroviruses, p. 1179–1187. *In* S. S. Long, L. K. Pickering, and C. G. Prober (ed.), *Principles and Practices of Pediatric Infectious Diseases,* 2nd ed. Churchill Livingstone, Philadelphia, PA.
122. **Monpoeho, S., M. Costa-Burel, M. Costa-Mattioli, B. Besse, J. J. Chomel, S. Billaudel, and V. Ferré.** 2002. Application of a real-time polymerase chain reaction with internal positive control for detection and quantification of enterovirus in cerebrospinal fluid. *Eur. J. Microbiol. Infect. Dis.* **21:**532–536.
123. **Moschovi, M. A., M. Theodoridou, V. Papaevangelou, S. Youroukos, A. Nitsa, P. Matsiota, and F. Tzortzatou-Stathopoulou.** 2002. Outbreak of enteroviral infection in a pediatric hematology-oncology unit. *Clin. Infect. Dis.* **34:**1269–1272.
124. **Moschovi, M. A., P. Sterpi, S. Youroukos, and F. Tzortzatou-Stathopoulou.** 2002. Encephalitis and myocarditis in a child with acute lymphoblastic leukemia: role of coxsackievirus B5? *Pediatr. Hematol. Oncol.* **19:**205–210.
125. **Moschovi, M. A., K. Katsibardi, M. Theodoridou, A. G. Michos, A. Tsakris, N. Spanakis, and F. Tzortzatou-Stathopoulou.** 2007. Enteroviral infections in children with malignant disease: a 5-year study in a single institution. *J. Infect.* **54:**387–392.
126. **Mullins, J. A., N. Khetsuriani, W. A. Nix, M. S. Oberste, A. LaMonte, D. R. Kilpatrick, J. Dunn, J. Langer, P. McMinn, Q. S. Huang, K. Grimwood, C. Huang, and M. A. Pallansch.** 2004. Emergence of echovirus type 13 as a prominent enterovirus. *Clin. Infect. Dis.* **38:**70–77.
127. **Nielsen, L. P., J. F. Modlin, and H. A. Rotbart.** 1996. Detection of enteroviruses by polymerase chain reaction in urine samples of patients with aseptic meningitis. *Pediatr. Infect. Dis. J.* **15:**625–627.
128. **Nigrovic, L. E., and V. W. Chiang.** 2000. Cost analysis of enteroviral polymerase chain reaction in infants with fever and cerebrospinal fluid pleocytosis. *Arch. Pediatr. Adolesc. Med.* **154:**817–821.
129. **Nijhuis, M., N. van Maarseveen, R. Schuurman, S. Verkuijlen, M. De Vos, K. Hendrikson, and A. M. van Loon.** 2002. Rapid and sensitive routine detection of all members of the genus enterovirus in different clinical specimens by real-time PCR. *J. Clin. Microbiol.* **40:**3666–3670.
130. **Norder, H., L. Bjerregaard, L. Magnius, B. Lina, M. Aymard, and J. J. Chomel.** 2003. Sequencing of 'untypable' enteroviruses reveals two new types, EV-77 and EV-78, within human enterovirus type B and substitutions in the BC loop of the VP1 protein for known types. *J. Gen. Virol.* **84:**827–836.
131. **Oberste, M. S., W. A. Nix, K. Maher, and M. A. Pallansch.** 2003. Improved molecular identification of enteroviruses by RT-PCR and amplicon sequencing. *J. Clin. Virol.* **26:**375–377.
132. **Oberste, M. S., S. M. Michele, K. Maher, D. Schnurr, D. Cisterna, N. Junttila, M. Uddin, J. J. Chomel, C. S. Lau, W. Ridha, S. al Busaidy, H. Norder, L. O. Magnius, and M. A. Pallansch.** 2004. Molecular identification and characterization of two proposed new enterovirus serotypes, EV74 and EV75. *J. Gen. Virol.* **85:**3205–3212.
133. **Oberste, M. S., K. Maher, S. M. Michele, G. Belliot, M. Uddin, and M. A. Pallansch.** 2005. Enteroviruses 76, 89, 90 and 91 represent a novel group within the species human enterovirus A. *J. Gen. Virol.* **86:**445–451.
134. **Ochs, H. D., and C. I. E. Smith.** 1996. X-linked agammaglobulinemia. A clinical and molecular analysis. *Medicine* (Baltimore) **75:**287–299.
135. **Ooi, M. H., S. C. Wong, Y. Podin, W. Akin, S. del Sel, A. Mohan, C. H. Chieng, D. Perera, D. Clear, D. Wong, E. Blake, J. Cardosa, and T. Solomon.** 2007. Human enterovirus 71 disease in Sarawak, Malaysia: a prospective clinical, virological, and molecular epidemiological study. *Clin. Infect. Dis.* **44:**646–656.
136. **Padate, B. P., and J. Keidan.** 2006. Enteroviral meningoencephalitis in a patient with non-Hodgkin's lymphoma treated previously with rituximab. *Clin. Lab. Haematol.* **28:**69–71.
137. **Page, G. S., A. G. Mosser, J. M. Hogle, D. J. Filman, R. R. Reuckert, and M. Chow.** 1988. Three-dimensional structure of poliovirus serotype 1 neutralizing determinants. *J. Virol.* **63:**1781–1794.
138. **Pallansch, M., and R. Roos.** 2007. Enteroviruses, polioviruses, coxsackieviruses, echoviruses, and new enteroviruses, p. 839–893. *In* D. M. Knipe et al. (ed.), *Fields Virology,* 5th ed. Lippincott Williams & Wilkins, Philadelphia, PA.
139. **Papaventsis, D., P. Markoulatos, N. Mangafas, M. Lazanas, and S. Levidiotou-Stefanou.** 2004. Enteroviral infection in Greek AIDS patients. *Mol. Diagn.* **8:**11–16.
140. **Parkkali, T., T. Ruutu, M. Stenvik, T. Kuronen, H. Kayhty, T. Hovi, R. M. Olander, L. Volin, and P. Ruutu.** 1996. Loss of protective immunity to polio, diphtheria and Haemophilus influenzae type b after allogeneic bone marrow transplantation. *APMIS* **104:**383–388.

141. **Pasic, S., B. Jankovic, M. Abinun, and B. Kanjuh.** 1997. Intravenous immunoglobulin prophylaxis in an echovirus 6 and echovirus 4 nursery outbreak. *Pediatr. Infect. Dis. J.* **16:** 718–720.
142. **Pérez-Ruiz, M., J. M. Navarro-Marí, E. Palacios del Valle, and M. Rosa-Fraile.** 2003. Human rhabdomyosarcoma cells for rapid detection of enteroviruses by shell-vial assay. *J. Med. Microbiol.* **52:**789–791.
143. **Pevear, D. C., T. M. Tull, M. E. Seipel, and J. M. Groarke.** 1999. Activity of pleconaril against enteroviruses. *Antimicrob. Agents Chemother.* **43:**2109–2115.
144. **Pichichero, M. E., S. McLinn, H. A. Rotbart, M. A. Menegus, M. Cascino, and B. E. Reidenberg.** 1998. Clinical and economic impact of enterovirus illness in private practice. *Pediatrics* **102:**1126–1134.
145. **Pollock, C. A., J. F. Mahony, L. S. Ibels, R. J. Caterson, D. A. Waugh, J. V. Wells, and A. G. Sheil.** 1989. Immunoglobulin abnormalities in renal transplant recipients. *Transplantation* **47:**952–956.
146. **Quartier, P., M. Debré, J. De Blic, R. De Sauverzac, N. Sayegh, N. Jabado, E. Haddad, S. Blanche, J. Casanova, C. I. E. Smith, F. Le Deist, G. de Saint Basile, and A. Fischer.** 1999. Early and prolonged intravenous immunoglobulin replacement therapy in childhood agammaglobulinemia: a retrospective survey of 31 patients. *J. Pediatr.* **134:**589–596.
147. **Quartier, P., S. Foray, J. Casanova, I. Hau-Rainsard, S. Blanche, and A. Fischer.** 2000. Enteroviral meningoencephalitis in X-linked agammaglobulinemia: intensive immunoglobulin therapy and sequential viral detection in cerebrospinal fluid by polymerase chain reaction. *Pediatr. Infect. Dis. J.* **19:**1106–1108.
148. **Quartier, P., O. Tournilhac, C. Archimbaud, L. Lazaro, C. Chaleteix, P. Millet, H. Peigue-Lafeuille, S. Blanche, A. Fischer, J. Casanova, P. Travade, and M. Tardieu.** 2003. Enteroviral meningoencephalitis after anti-CD20 (rituximab) treatment. *Clin. Infect. Dis.* **36:**e47–e49.
149. **Rabella, N., P. Rodriguez, R. Labeaga, M. Otegui, M. Mercader, M. Gurgui, and G. Prats.** 1999. Conventional respiratory viruses recovered from immunocompromised patients: clinical considerations. *Clin. Infect. Dis.* **28:**1043–1048.
150. **Rabenau, H. F., A. M. Clarici, G. Muhlbauer, A. Berger, A. Vince, S. Muller, E. Daghofer, B. I. Santner, E. Marth, and H. H. Kessler.** 2002. Rapid detection of enterovirus infection by automated RNA extraction and real-time fluorescence PCR. *J. Clin. Virol.* **25:**155–164.
151. **Rabkin, C. S., E. E. Telzak, M. Ho, J. Goldstein, Y. Bolton, M. Pallansch, L. Anderson, E. Kilchevsky, S. Solomon, and W. J. Martone.** 1988. Outbreak of echovirus 11 infection in hospitalized neonates. *Pediatr. Infect. Dis. J.* **7:**186–190.
152. **Ramers, C., G. Billman, M. Hartin, S. Ho, and M. H. Sawyer.** 2000. Impact of a diagnostic cerebrospinal fluid enterovirus polymerase chain reaction test on patient management. *JAMA* **283:**2680–2685.
153. **Rittichier, K. R., P. A. Bryan, K. E. Bassett, E. W. Taggart, F. R. Enriquez, D. R. Hillyard, and C. L. Byington.** 2005. Diagnosis and outcomes of enterovirus infections in young infants. *Pediatr. Infect. Dis. J.* **24:**546–550.
154. **Robinson, C. C., M. Willis, A. Meagher, K. E. Gieseker, H. Rotbart, and M. P. Glodé.** 2002. Impact of rapid polymerase chain reaction results on management of pediatric patients with enteroviral meningitis. *Pediatr. Infect. Dis. J.* **21:**283–286.
155. **Romero, J. R.** 2001. Pleconaril: a novel antipicornaviral drug. *Investig. Drugs* **10:**369–379.
156. **Romero, J. R., and H. A. Rotbart.** 2003. Enteroviruses, p. 1427–1438. *In* P. R. Murray, E. J. Baron, J. H. Jorgensen, M. A. Pfaller, and R. H. Yolken (ed.), *Manual of Clinical Microbiology,* 8th ed. ASM Press, Washington, DC.
157. **Rosenlew, M., M. Stenvik, M. Roivainen, A. L. Jarvenpaa, and T. Hovi.** 1999. A population-based prospective survey of newborn infants with suspected systemic infection: occurrence of sporadic enterovirus and adenovirus infections. *J. Clin. Virol.* **12:**211–219.
158. **Rotbart, H. A., J. P. Kinsella, and R. L. Wasserman.** 1990. Persistent enterovirus infection in culture-negative meningoencephalitis: demonstration by enzymatic RNA amplification. *J. Infect. Dis.* **161:**787–791.
159. **Rotbart, H. A.** 1995. Enteroviral infections of the central nervous system. *Clin. Infect. Dis.* **20:**971–981.
160. **Rotbart, H. A., A. Ahmed, S. Hickey, R. Dagan, G. H. McCracken, R. J. Whitley, J. F. Modlin, M. Cascino, J. F. O'Connell, M. A. Menegus, and D. Blum.** 1997. Diagnosis of enterovirus infection by polymerase chain reaction of multiple specimen types. *Pediatr. Infect. Dis. J.* **16:**409–411.
161. **Rotbart, H. A., G. H. McCracken, R. J. Whitley, J. F. Modlin, M. Cascino, S. Shah, and D. Blum.** 1999. Clinical significance of enteroviruses in serious summer febrile illnesses in children. *Pediatr. Infect. Dis. J.* **18:**869–874.
162. **Rotbart, H. A., and A. D. Webster.** 2001. Treatment of potentially life-threatening enterovirus infections with pleconaril. *Clin. Infect. Dis.* **32:**228–235.
163. **Rotbart, H. A.** 2002. Enteroviruses, p. 971–994. *In* D. D. Richman, R. J. Whitley, and F. G. Hayden (ed.), *Clinical Virology,* 2nd ed. ASM Press, Washington, DC.
164. **Rudge, P., A. D. B. Webster, T. Revesz, T. Warner, T. Espanol, C. Cunningham-Rundles, and N. Hyman.** 1996. Encephalomyelitis in primary hypogammaglobulinemia. *Brain* **119:**1–15.
165. **Samuelson, A., M. Forsgren, and M. Sällberg.** 1995. Characterization of the recognition site and diagnostic potential of an enterovirus group-reactive monoclonal antibody. *Clin. Diagn. Lab. Immunol.* **2:**385–386.
166. **Satosar, A., N. C. Ramirez, D. Bartholomew, J. Davis, and G. J. Nuovo.** 2004. Histologic correlates of viral and bacterial infection of the placenta associated with severe morbidity and mortality in the newborn. *Hum. Pathol.* **35:**536–545.
167. **Sauerbrei, A., B. Glück, K. Jung, H. Bittrich, and P. Wutzler.** 2000. Congenital skin lesions caused by intrauterine infection with coxsackievirus B3. *Infection* **28:**326–328.
168. **Schmugge, M., R. Lauener, W. Bossart, R. A. Seger, and T. Güngör.** 1999. Chronic enteroviral meningo-encephalitis in X-linked agammaglobulinaemia: favourable response to anti-enteroviral treatment. *Eur. J. Pediatr.* **158:**1010–1011.
169. **Schwarer, A. P., S. S. Opat, A. L. Watson, D. Spelman, F. Firkin, and N. Lee.** 1997. Disseminated echovirus infection after allogeneic bone marrow transplantation. *Pathology* **29:**424–425.
170. **Shah, S. S., W. E. Hellenbrand, and P. G. Gallagher.** 1998. Atrial flutter complicating neonatal coxsackievirus B2 myocarditis. *Pediatr. Cardiol.* **19:**185–186.
171. **She, R. C., G. Crist, E. Billetdeaux, J. Langer, and C. A. Petti.** 2006. Comparison of multiple shell vial cell lines for isolation of enteroviruses: a national perspective. *J. Clin. Virol.* **37:**151–155.
172. **Skull, S., and A. Kemp.** 1996. Treatment of hypogammaglobulinemia with intravenous immunoglobulin, 1973–93. *Arch. Dis. Child.* **74:**527–530.
173. **Smith, J. K., D. S. Chi, J. Guarderas, P. Brown, A. Verghese, and S. L. Berk.** 1989. Disseminated echovirus infection in a patient with multiple myeloma and a functional defect in

complement. Treatment with intravenous immunoglobulin. *Arch. Intern. Med.* **149**:1455–1457.
174. **Spanakis, N., E. N. Manolis, A. Tsakris, S. Tsiodra, T. Panagiotopoulos, G. Saroglou, and N. J. Legakis.** 2005. Coxsackievirus B3 sequences in the myocardium of fatal cases in a cluster of acute myocarditis in Greece. *J. Clin. Pathol.* **58**:357–360.
175. **Stanway, G., F. Brown, P. Christian, T. Hovi, T. Hyypiä, A. M. Q. King, N. J. Knowles, S. M. Lemon, P. D. Minor, M. A. Pallansch, A. C. Palmenberg, and T. Skern.** 2005. *Picornaviridae*, p. 757–778. *In* C. M. Fauquet, M. A. Mayo, J. Maniloff, U. Desselberger, and L. A. Ball (ed.), *Virus Taxonomy: Eighth Report of the International Committee on the Taxonomy of Viruses*. Elsevier Academic Press, London, United Kingdom.
176. **Starakis, I., M. Marangos, S. Giali, and H. Bassaris.** 2005. Acute transverse myelitis due to coxsackie virus. *J. Clin. Neurosci.* **12**:296–298.
177. **Stark, K., M. Günther, C. Schönfeld, S. G. Tullius, and U. Bienzle.** 2002. Immunisations in solid-organ transplant recipients. *Lancet* **359**:957–965.
178. **Starlin, R., N. Reed, B. Leeman, J. Black, E. Trulock, and L. M. Mundy.** 2001. Acute flaccid paralysis syndrome associated with echovirus 19, managed with pleconaril and intravenous immunoglobulin. *Clin. Infect. Dis.* **33**:730–732.
179. **Strikas, R. A., L. J. Anderson, and R. A. Parker.** 1986. Temporal and geographic patterns of isolates of nonpolio enteroviruses in the United States, 1970–1983. *J. Infect. Dis.* **153**:346–351.
180. **Strong, B. S., and S. A. Young.** 1995. Intrauterine coxsackie virus group B type 1 infection: viral cultivation from amniotic fluid in the third trimester. *Am. J. Perinatol.* **12**:78–79.
181. **Sutter, R. W., and D. R. Prevots.** 1994. Vaccine-associated paralytic poliomyelitis among immunodeficient persons. *Infect. Med.* 11:426–438.
182. **Swanink, C. M. A., L. Veenstra, Y. A. G. M. Poort, J. A. Kaan, and J. M. D. Galama.** 1993. Coxsackievirus B1-based antibody-capture enzyme-linked immunosorbent assay for detection of immunoglobulin G (IgG), IgM, and IgA with broad specificity for enteroviruses. *J. Clin. Microbiol.* **31**:3240–3246.
183. **Takahashi, S., A. Miyamoto, J. Oki, H. Azuma, and A. Okuna.** 1995. Acute transverse myelitis caused by echovirus type 18 infection. *Eur. J. Pediatr.* **154**:378–380.
184. **Tan, P., M. R. Verneris, and L. R. Charnas.** 2005. Outcome of CNS and pulmonary enteroviral infections after hematopoietic cell transplantation. *Pediatr. Blood Cancer* **45**:74–75.
185. **Tang, J. W., J. W. A. Bendig, and I. Ossuetta.** 2005. Vertical transmission of human echovirus 11 at the time of Bornholm disease in late pregnancy. *Pediatr. Infect. Dis. J.* **24**:88–89.
186. **Tarcan, A., N. Özbek, and B. Gürakan.** 2001. Bone marrow failure with concurrent enteroviral infection in a newborn. *Pediatr. Infect. Dis. J.* **20**:719–721.
187. **Terletskaia-Ladwig, E., C. Metzger, G. Schalasta, and G. Enders.** 2000. A new enzyme immunoassay for the detection of enteroviruses in faecal specimens. *J. Med. Virol.* **60**:439–445.
188. **Terletskaia-Ladwig, E., C. Metzger, G. Schalasta, and G. Enders.** 2000. Evaluation of enterovirus serological tests IgM-EIA and complement fixation in patients with meningitis, confirmed by detection of enteroviral RNA by RT-PCR in cerebrospinal fluid. *J. Med. Virol.* **61**:221–227.
189. **Tormey, V. J., J. R. Buscombe, M. A. Johnson, A. P. Thomson, and A. D. Webster.** 2003. SPECT scans for monitoring response to pleconaril therapy in chronic enteroviral meningoencephalitis. *J. Infect.* **46**:138–140.
190. **Townsend, T. R., E. A. Bolyard, R. H. Yolken, W. E. Beschorner, C. A. Bishop, W. H. Burns, G. W. Santos, and R. Saral.** 1982. Outbreak of coxsackie A1 gastroenteritis: a complication of bone-marrow transplantation. *Lancet* **i**: 820–823.
191. **Trabelsi, A., F. Grattard, M. Nejmeddine, M. Aouni, T. Bourlet, and B. Pozzetto.** 1995. Evaluation of an enterovirus group-specific anti-VP1 monoclonal antibody, 5-D8/1, in comparison with neutralization and PCR for rapid identification of enteroviruses in cell culture. *J. Clin. Microbiol.* **33**:2454–2457.
192. **Troussard, X., F. Bauduer, E. Gallet, F. Freymuth, P. Boutard, J. J. Ballet, O. Reman, and M. Leporrier.** 1993. Virus recovery from stools of patients undergoing bone marrow transplantation. *Bone Marrow Transplant.* **12**:573–576.
193. **Van Doornum, G. J. J., and J. C. de Long.** 1998. Rapid shell vial culture technique for detection of enteroviruses and adenoviruses in fecal specimens: comparison with conventional virus isolation. *J. Clin. Microbiol.* **36**:2865–2868.
194. **Van Kraaij, M. G. J., L. J. R. van Elden, A. M. van Loon, K. A. W. Hendriksen, L. Laterveer, A. W. Dekker, and M. Nijhuis.** 2005. Frequent detection of respiratory viruses in adult recipients of stem cell transplants with the use of real-time polymerase chain reaction, compared with viral culture. *Clin. Infect. Dis.* **40**:662–669.
195. **Verboon-Maciolek, M. A., M. Nijhuis, A. M. van Loon, N. van Maarssenveen, H. van Wieringen, M. A. Pekelharing-Berghuis, T. G. Krediet, L. J. Gerards, R. J. A. Diepersloot, and S. F. T. Thijsen.** 2003. Diagnosis of enterovirus infection in the first 2 months of life by real-time polymerase chain reaction. *Clin. Infect. Dis.* **37**:1–6.
196. **Verstrepen, W. A., P. Bruynseels, and A. H. Mertens.** 2002. Evaluation of a rapid real-time PCR assay for detection of enterovirus RNA in cerebrospinal fluid specimens. *J. Clin. Virol.* **25**(Suppl. 1):S39–S43.
197. **Waner, J. L.** 1994. Mixed viral infections: detection and management. *Clin. Microbiol. Rev.* **7**:143–151.
198. **Watanabe, M., C. T. Cho, R. C. Trueworthy, and K. L. Saving.** 1986. Prolonged echoviral meningitis in a cancer patient with normal serum immunoglobulins. *Med. Pediatr. Oncol.* **14**:342–344.
199. **Webster, A. D. B., H. A. Rotbart, T. Warner, P. Rudge, and N. Hyman.** 1993. Diagnosis of enterovirus brain disease in hypogammaglobulinemic patients by polymerase chain reaction. *Clin. Infect. Dis.* **17**:657–661.
200. **Wendt, C. H.** 1997. Community respiratory viruses: organ transplant recipients. *Am. J. Med.* **102**:31–36.
201. **Wilson, C. W., D. K. Stevenson, and A. M. Arvin.** 1989. A concurrent epidemic of respiratory syncytial virus and echovirus 7 in an intensive care nursery. *Pediatr. Infect. Dis. J.* **8**:24–29.
202. **Wimmer, E., C. U. Hellen, and X. Cao.** 1993. Genetics of poliovirus. *Ann. Rev. Genet.* **27**:353–436.
203. **Winkelstein, J. A., M. C. Marino, H. M. Lederman, S. M. Jones, K. Sullivan, A. W. Burks, M. E. Conley, C. Cunningham-Rundles, and H. D. Ochs.** 2006. X-linked agammaglobulinemia: report on a United States registry of 201 patients. *Medicine* (Baltimore) **85**:193–202.
204. **Wood, D. J., and G. Corbitt.** 1985. Viral infections in childhood leukemia. *J. Infect. Dis.* **152**:266–273.
205. **Wreghitt, T. G., C. E. D. Taylor, J. E. Banatvala, J. Bryan, and J. Wallwork.** 1986. Concurrent cytomegalovirus and coxsackie B virus infections in a heart-lung transplant recipient. *J. Infect.* **13**:51–54.

206. **Yagi, S., D. Schnurr, and J. Lin.** 1992. Spectrum of monoclonal antibodies to coxsackievirus B3 includes type- and group-specific antibodies. *J. Clin. Microbiol.* **30:**2498–2501.

207. **Yamani, M. H., R. K. Avery, S. D. Mawhorter, J. B. Young, N. B. Ratliff, R. E. Hobbs, P. M. McCarthy, N. G. Smedira, M. Goormastic, D. Pelegrin, and R. C. Starling.** 2001. Hypogammaglobulinemia following cardiac transplantation: a link between rejection and infection. *J. Heart Lung Transplant.* **20:**425–430.

208. **Yen, H., R. Lien, R. Fu, and L. Chang.** 2003. Hepatic failure in a newborn with maternal peripartum exposure to echovirus 6 and enterovirus 71. *Eur. J. Pediatr.* **162:**648–649.

209. **Yen, M., K. Tsao, Y. Huang, C. Huang, Y. Huang, R. Lin, M. Chang, C. Huang, D. Yan, and T. Lin.** 2007. Viral load in blood is correlated with disease severity of neonatal coxsackievirus B3 infection: early diagnosis and predicting disease severity is possible in severe neonatal enterovirus infection. *Clin. Infect. Dis.* **44:**e78–e81.

210. **Yolken, R. H., C. A. Bishop, T. R. Townsend, E. A. Bolyard, J. Bartlett, G. W. Santos, and R. Saral.** 1982. Infectious gastroenteritis in bone-marrow-transplant recipients. *N. Engl. J. Med.* **306:**1010–1012.

211. **Yousef, G. E., D. A. Isenberg, and J. F. Mowbray.** 1990. Detection of enterovirus specific RNA sequences in muscle biopsy specimens from patients with adult onset myositis. *Ann. Rheum. Dis.* **49:**310–315.

212. **Zhang, H., Y. Li, T. Peng, M. Aasa, L. Zhang, Y. Yang, and L. C. Archard.** 2000. Localization of enteroviral antigen in myocardium and other tissues from patients with heart muscle disease by an improved immunohistochemical technique. *J. Histochem. Cytochem.* **48:**579–584.

213. **Zhang, H., Y. Li, D. R. McClean, P. J. Richardson, R. Florio, M. Sheppard, K. Morrison, N. Latif, M. J. Dunn, and L. C. Archard.** 2004. Detection of enterovirus capsid protein VP1 in myocardium from cases of myocarditis or dilated cardiomyopathy by immunohistochemistry: further evidence of enterovirus persistence in myocytes. *Med. Microbiol. Immunol.* **193:**109–114.

Diagnostic Microbiology of the Immunocompromised Host
Edited by Randall T. Hayden, Karen C. Carroll, Yi-Wei Tang, and Donna M. Wolk

Chapter 9

Parvovirus

Marie Louise Landry

DESCRIPTION OF THE AGENT AND PATHOPHYSIOLOGY OF INFECTION

In 1975, Yvonne Cossart first identified parvovirus particles by electron microscopy while evaluating tests for hepatitis B virus (34). This new virus, detected in serum number 19 in plate B, was thus named parvovirus B19. Subsequently, fifth disease or erythema infectiosum (EI), polyarthralgias/polyarthritis, fetal hydrops, transient aplastic crisis, and chronic red cell aplasia have been linked to this virus.

Until the discovery of *Human bocavirus* (2), *Parvovirus B19* was the only member of the large *Parvoviridae* family to be associated unequivocally with human disease. Human bocavirus appears to be a common cause of respiratory infection in the healthy host (5, 10, 70); however, its impact in the compromised host has not yet been defined. Therefore, this chapter is focused on parvovirus B19.

Members of the *Parvoviridae* family are named for their small size, with "parvum" being Latin for "small." The family, which includes both animal and insect pathogens, is divided into two subfamilies based on infection of vertebrate or invertebrate cells. The vertebrate *Parvovirinae* subfamily consists of three genera, namely, *Parvovirus* (autonomous replication), *Dependovirus* (replication requiring a helper virus), and *Erythrovirus* (replication in erythroid precursor cells).

Parvovirus B19, of the genus *Erythrovirus*, consists of a highly stable nonenveloped icosahedral virion of 22 to 24 nm in diameter, with a single-stranded 5.6-kilobase DNA genome. The stability of parvovirus allows infectious virus to persist despite standard heat treatments of blood products and to be transmitted in products such as albumin, immunoglobulin, and pooled factor VIII and factor IX concentrates. The genome encodes a major nonstructural protein (NS1) and two structural proteins, viral protein 1 (VP1) and VP2. The function of NS1 is not well defined, but it appears to induce cytotoxicity and growth arrest in infected cells (20, 69). VP2 is the predominant protein, comprising 95% of the virus capsid. VP1 is the same as VP2 except for an additional 226 amino acids at its amino terminus (111). VP1 makes up only 5% of the capsid, has its unique region external to the viral capsid itself, and is thought to be the main target of neutralizing antibodies (111). Sequence analysis reveals that NS1 is highly conserved, while VP1 and VP2 show greater variation (38). Despite variations in VP1 and VP2, the antigens are commonly and successfully used in serologic tests (24, 53, 59).

Several novel variants of parvovirus B19 have been discovered (85-87), leading to a proposal that the *Erythrovirus* genus should include the following genotypes: genotype 1 (parvovirus B19), genotype 2 (strains A6 and K71), and genotype 3 (strains V9 and D91.1) (97). While the prevalence and significance of these variants remain to be determined, sporadic infections in both healthy and immunocompromised hosts have been reported in Europe (76) and the United Kingdom (32), especially for older patients. The V9 variant (genotype 3) appears to be especially common in Africa and in France but was not detected in samples from the United States in recent studies (89, 97).

Parvovirus B19 infection is thought to occur via the respiratory route, and the virus is presumed to replicate in nasopharyngeal lymphatic tissue (4). Approximately 1 week after inhalation of infected droplets, viremia is observed, which leads to infection of erythroid precursor cells in bone marrow and

Marie Louise Landry • Department of Laboratory Medicine, Yale University School of Medicine, P.O. Box 208035, New Haven, CT 06520-8035.

in fetal liver. The virus binds to P antigen or globoside (Gb4) on the cell surface (19), and entry into the cell is facilitated by the coreceptor $\alpha5\beta1$ integrin (109). Erythroid progenitor cells express high levels of both P antigen and $\alpha5\beta1$ integrin. Once infected, these cells produce infectious virus and are destroyed, leading to a drop in circulating reticulocytes (71), followed by a drop in hematocrit. Although infectious virus is not produced in leukocyte and platelet precursors, these cells can be infected and express the cytotoxic NS1 protein. A transient decline in white cell and platelet counts can also occur. Virus titers in the blood reach levels of up to 10^{13} particles per milliliter (17, 53, 111). With the appearance of immunoglobulin M (IgM) antibodies, the virus titer in the peripheral blood begins to fall. Shortly thereafter, IgG is detectable in the blood and the symptoms associated with EI, namely, erythematous rash and arthralgia, appear. During acute infection, parvovirus B19 has been detected in the nasopharynx, blood, bone marrow, liver, skin, cerebrospinal fluid, and synovium. In addition, B19 is known to infect erythroblasts, megakaryoblasts, granulocytes, macrophages, follicular dendritic cells, B and T lymphocytes, and endothelial cells (53, 57).

It is not clear whether parvovirus B19 is eliminated from the host or remains in an inactive state capable of reactivation. Parvovirus B19 DNA has been detected in bone marrow samples in the absence of disease and can be detected at low levels in peripheral blood of healthy hosts for months and even years after primary infection and resolution of symptoms (26-28, 51, 53, 68). Virus has also been found to persist in synovium, myocardium, and skin (22, 57, 103). The cells harboring the virus and the mechanism of persistence remain uncertain. It is possible that parvovirus B19 may integrate into the human genome, as occurs with other parvoviruses, such as minute virus of mice and dependoviruses.

CLINICAL AND EPIDEMIOLOGIC FEATURES OF PARVOVIRUS INFECTIONS

Immunocompetent Patients

Acquisition of parvovirus B19 infection begins in childhood and continues throughout life. By young adulthood, up to 40 to 60% of people have been infected, and in old age, ~90% of people are antibody positive. Infection with parvovirus occurs year-round but may peak in late winter to early summer. Every 3 to 4 years, epidemics of increased activity occur. Although acute infections can be manifested as EI, infection is asymptomatic in up to 50% of children. A prodromal nonspecific illness consisting of fever, chills, headache, malaise, and myalgias, coinciding with B19 viremia, can occur, followed by a typical "slapped cheek" facial rash and a lacy, reticular erythematous rash on the trunk and extremities coincident with the immune response. The typical EI rash often waxes and wanes over a period of days.

In adults, arthralgias predominate, particularly in women, and the typical EI rash is much less common. Symmetric painful, swollen joints, especially wrists, knees, and hands, can last for weeks to months and can be confused with Lyme disease or rheumatoid arthritis. Indeed, some features of B19 infection resemble those of autoimmune diseases, and autoantibodies such as antinuclear antibody and rheumatoid factor are commonly found (111). Prolonged fatigue has been associated with elevated levels of gamma interferon and tumor necrosis factor, and some cases meet the criteria for chronic fatigue syndrome (63-67).

In healthy individuals with a normal red cell lifespan of 120 days, the drop in hematocrit is modest and not clinically significant. However, in anemic individuals, especially those with high reticulocyte counts, such as in sickle cell anemia or hereditary spherocytosis, parvovirus B19 infection leads to a dramatic fall in hematocrit and an acute "transient aplastic crisis" requiring red cell transfusion. These individuals, who can have a red cell lifespan as short as 15 to 20 days, present earlier in the course of infection than typical fifth disease and thus may not yet have detectable antibodies. Once antibodies develop, the virus is neutralized, reticulocytosis reappears, and the hematocrit returns to baseline. Transient declines in white blood cell or platelet counts may also be seen but are rarely clinically significant.

When infection occurs during pregnancy, maternal viremia can lead to transplacental transmission of virus. Fetal infection leads to interruption of erythropoiesis and fetal anemia; in a minority of cases, parvovirus-associated myocarditis and severe anemia can lead to heart failure, fetal hyrops, and/or fetal demise.

Other rashes can occur, in a generalized or localized distribution. These include petechial rashes, Henoch-Schönlein purpura, papular-purpuric gloves-and-socks syndrome, Gianotti-Crosti syndrome, desquamation, erythema multiforme, and erythema nodosum (7, 30, 44, 62, 80, 102, 107).

Thrombocytopenia can be mediated by bone marrow suppression, NS1 toxicity to megakaryocytes, and antiplatelet antibodies. Virus-associated hemophagocytic syndrome has also been associated with B19, as well as other viruses, and occurs primarily in previously healthy hosts (8, 78).

Parvovirus B19 has also been linked to myocarditis, a variety of neurologic syndromes, uveitis, hepatitis, renal syndromes, vasculitis, and chronic fatigue syndrome (8, 9, 14, 66, 84, 88). The mechanisms responsible, including which cell types are infected and whether disease is mediated by immune complexes, NS1 toxicity, cytokine up-regulation, or autoantibodies, are not clear. HLA and cytokine gene polymorphisms have been linked to chronic arthritis and chronic fatigue syndrome following acute B19 infection, and it is possible that inherited variations in antigen presentation and cytokine responses are responsible for symptoms (63).

Immunocompromised Patients

Disease spectrum

Many immunocompromised hosts have preexisting antibody to B19 or may be able to mount a sufficient immune response to primary infection. However, in others, neutralizing antibodies are not produced (75). Consequently, active parvovirus B19 infection of erythroid precursors is not terminated and viremia persists. The most common consequence of persistent active infection is chronic pure red cell aplasia (PRCA), resulting in chronic or recurrent anemia with reticulocytopenia (43, 74). Bone marrow involvement can also manifest as neutropenia, agranulocytosis, pancytopenia, thrombocytopenia, and hemophagocytosis (25, 39, 45, 58, 71). Nonspecific symptoms of fever (71, 90, 95), myalgias, and malaise can occur. Immunomediated symptoms, such as the rash and arthralgias of EI, are not observed unless antibody develops or intravenous immunoglobulin (IVIG) is administered. However, atypical rashes, some due to endothelial cell infection, can be seen (13, 39, 55). Myocarditis, glomerulopathy, and central nervous system vasculitis have also been reported (9, 13, 52, 81, 82, 96).

Host factors and subgroups

Persistent parvovirus B19 infection can occur in patients with congenital, acquired, and iatrogenic immunodeficient states, including those with acute and chronic leukemias (37, 39, 92), patients on chemotherapy for cancer (48, 72, 91, 99, 101), hematopoietic stem cell and solid organ transplant recipients (36, 45), patients undergoing immunosuppressive therapy for autoimmune disease (49, 71), patients with congenital immunodeficiencies such as Nezelof's syndrome (73, 106), and patients with AIDS (42). The common denominator is an inability to produce sufficient neutralizing antibodies to eliminate the virus.

Although B19 is responsible for a minority of PRCA cases (46), B19-associated PRCA tends to be more severe and is more likely to follow primary B19 infection. B19 should be considered for patients with oncologic or hematologic malignancies who are undergoing either induction or maintenance chemotherapy with unexplained or severe anemia and reticulocytopenia (79). Rituximab therapy, which inhibits B cells and leads to a reduction in immunoglobulins, has been linked to persistent B19 infection in lymphoma patients (58). Persistent B19 infection can be the first manifestation of congenital immunodeficiencies, such as X-linked hyper-IgM syndrome (98) and Nezelof's syndrome (73).

Numerous cases of PRCA due to B19 have been reported for solid organ transplant recipients, especially those receiving renal transplants, and they occur weeks to years after transplantation (45). Regimens that include the potent antirejection drugs mycophenolate mofetil and tacrolimus may carry a greater risk of PRCA than those using cyclosporine (36, 110), although cyclosporine has also been implicated. Persistent B19 infection in renal transplant recipients has led to collapsing glomerulonephritis and allograft loss (45). In a lung transplant patient with cystic fibrosis, total lymphoid irradiation for bronchiolitis obliterans was considered a contributing factor to persistent B19 infection (62).

Parvovirus B19 has been linked to engraftment failure after peripheral blood stem cell transplantation (104) as well as to PRCA (31). However, problems are infrequent in hematopoietic stem cell transplant recipients due to the common practice of administering periodic IVIG (95).

PRCA due to B19 has been described for AIDS patients with low CD4 cell counts, predominantly homosexual men. In contrast, AIDS patients with hemophilia generally have not had problems with persistent B19 infection (38, 47), probably because they have a very high prevalence of B19 neutralizing antibodies from repeated exposure to B19-contaminated blood products and clotting factors.

In immunocompromised hosts, B19 infection can be acquired via the respiratory route, from blood products, or from endogenous reactivation. Exposure to other immunocompromised hosts, who shed very high titers for prolonged periods, has been implicated in outbreaks in cancer and transplant patients (39).

Morbidity and mortality

Prognostic factors. If immunosuppressive agents can be stopped, reduced, or changed, the host may be able to mount an effective antibody response and control or terminate active infection. In AIDS cases, highly active antiretroviral therapy can result in increased CD4 cell counts, restoration of the immune system, and resolution of the anemia (29, 108). Likewise, once cancer chemotherapy has been completed, the immune system hopefully rebounds and is able to control the infection. Some patients may have a delayed antibody response but still recover with some transfusion support, even with no other change in regimen (3). However, if the underlying immune deficiency is severe and cannot be corrected or worsens, spontaneous resolution is less likely.

Therapeutic considerations. If immune reconstitution is not anticipated, passive administration of B19 antibodies will lead to virus neutralization, resumption of reticulocytosis, and a rise in hematocrit. In the majority of reported cases, only one course of IVIG has been needed for long-term remission (36). Since treatment with IVIG reduces virus shedding a millionfold, IVIG should reduce transmission to others (42). However, IVIG also has side effects, including allergic reactions, transfusion-transmitted diseases, and a decline in renal function. It is also expensive and can be in limited supply.

GOALS OF LABORATORY TESTING

Screening and Prevention

No vaccine or antiviral is available for parvovirus B19 infection. It is not practical to prevent community acquisition of B19 infection, since transmission via respiratory droplets usually occurs from persons without recognized symptoms. In addition, B19 can be transmitted in plasma, bone marrow, stem cells, erythrocyte and platelet concentrates, clotting factors, immunoglobulin, serum albumin, and other products from human blood. The incidence of positive blood products is highest at times of peak B19 community transmission (33, 100). In one study, 1% of blood products administered to patients with hematologic malignancies tested positive for B19 DNA, with the risk being highest for allogeneic peripheral blood progenitor cells (17.6%) and pooled products (2%) (76).

The FDA and the European Pharmacopoeia have determined that 10^4 IU/ml is the maximum level of B19 DNA acceptable in plasma pools and their products (17), since recipients of product lots containing $<10^4$ IU/ml B19 DNA did not become infected. Manufacturers now perform B19 PCR on plasma minipools in order to remove plasmas with high viral loads. To aid in screening, a WHO international standard, consisting of a positive donation diluted in pooled plasma, has been established for B19 nucleic acid amplification techniques (93). Following a collaborative study, it was assigned a concentration of 10^6 IU/ml.

Notably, current B19 PCR assays usually either miss or underquantitate B19 variants; fortunately, the prevalence of genotypes 2 and 3 in blood products is likely low (12, 16, 56). Clearly, better methods for pathogen inactivation, filtration, and detection in blood products are needed.

Diagnosis and Prognostication

Since PRCA can have other causes, laboratory diagnosis of parvovirus B19 infection is essential for immunocompromised hosts. Of the available methods, quantitative B19 DNA PCR is of greatest value in establishing a diagnosis and in following the response to treatment for the compromised host.

Treatment and Monitoring

In cases of PRCA, hematocrit should be monitored and red cells transfused as needed. In individuals treated with IVIG, PRCA may recur once passive antibody wanes. Therefore, if the hematocrit begins to fall again, the B19 viral load should be determined and a second course of IVIG administered, if indicated. B19 DNA can also be monitored quantitatively and prospectively as an early indicator of relapse (17, 49). However, since B19 DNA in serum and tissues can be detected by PCR for months to years at lower levels in the absence of disease (25-28, 41, 111), a positive PCR result alone should not lead to therapy. Rather significant increases in viral load (>1 $\log_{10}$), in concert with a drop in hematocrit, should be indications for concern.

LABORATORY TESTING OPTIONS, APPLICATIONS, AND INTERPRETATION

Laboratory testing options and applications for diagnosis of parvovirus B19 infection are summarized in Table 1.

Serology

Detection of IgM and IgG antibodies is the mainstay of parvovirus B19 diagnosis in the immunocompetent host. In general, IgM to B19 appears 7 to 10 days after infection, followed within a few days by IgG. While IgM remains positive for 2 to 4 months, IgG remains positive for life. In contrast, immunocompromised hosts may not develop antibodies, or IgM can develop but remain positive for months or years as an indicator of persistent infection, without development of IgG (90).

A variety of tests to detect parvovirus B19 antibodies are available (Table 2), and an international standard for B19 IgG assays has been developed and tested in collaborative studies (40). However, the only FDA-approved assay uses baculovirus-expressed conformational VP2 antigen in an enzyme-linked immunosorbent assay (ELISA) format (Biotrin International, Dublin, Ireland). Class (Mu) capture ELISA using baculovirus-expressed VP2 conformational antigen has been shown to generate far fewer equivocal and inaccurate results than other test formats (24, 59, 60, 89). Bacterium-based expression systems, such as *Escherichia coli* systems, produce B19 linear epitopes more suitable for immunoblots than for ELISA (60, 89). Although B19 genetic variability affects detection of DNA by PCR, VP2, the major capsid antigen, appears to be conserved among known genotypes. Thus, routine serology using VP2 antigen detects antibodies to all three B19 genotypes (53). Since ELISA does not assess neutralizing ability, a positive IgG result by ELISA does not exclude active B19 disease in the compromised host. IgG can also reflect passive transmission of antibody from blood products or IVIG therapy. VP1-specific immunoblot assays and neutralization assays of virus in cell culture are generally confined to the research setting (73, 75). Antibodies to NS1 have been linked to persistent infection in some studies but not in others (33).

Bone Marrow Examination

Bone marrow examination can provide the first clue to an unrecognized B19 infection (79). The hallmark is erythroid hypoplasia, as erythrocyte precursors

Table 1. Parvovirus B19 diagnostic methods in order of diagnostic utility

Method	Use	Advantages	Limitations
IgM antibody detection by ELISA	Diagnosis of acute or recent infection, mainstay of diagnosis in healthy hosts	Widely available; less expensive than molecular tests; detects antibodies to all known genotypes; class capture ELISA provides best results	Antibody may not develop in immunocompromised hosts; baculovirus-derived conformational VP2 antigens provide more sensitive results; IgM assays are prone to false-positive results; indirect ELISA is less sensitive and specific than class capture ELISA for IgM
IgG antibody detection by ELISA	Diagnosis of current or past infection	Development of IgG is usually associated with neutralization of virus, fall in virus titers, and resolution of acute infection; detects seroconversion; presence of IgG generally correlates with immunity to reinfection; detects antibodies to all known genotypes	In compromised hosts, IgG may not develop; IgG may be poorly neutralizing; ELISA does not assess neutralizing ability; neutralization and immunoblot tests are confined to research laboratories; baculovirus-derived conformational VP2 antigens provide more sensitive results; IgG can reflect passive transfer of antibody from blood products, not a host immune response
DNA amplification (e.g., PCR)	Detection and quantification of virus in serum, plasma, bone marrow, or biopsy tissues	Main diagnostic test for use in hosts who may not produce antibody; real-time PCR allows wider availability and quantification; two quantitative real-time commercial tests are available; viral DNA can be quantitated and monitored for response to therapy and for risk of recurrence; international B19 quantitation standard is available; available in reference laboratories and larger centers	Protocols in different laboratories give varying results; quantification may not be standardized among labs; may not detect or accurately quantify all genotypes unless additional primer sets and probes are used; real-time probes are more likely to miss mutations due to short length; PCR can remain positive at low levels for months to years after treatment or resolution of disease, even in healthy hosts; cross-contamination leads to false-positive results
Cytopathology in bone marrow	Detects giant pronormoblasts, viral inclusions, and red cell aplasia	Can be first clue to parvovirus B19 infection; standard technique available in most hospitals	Giant pronormoblasts or inclusions may not be present or may be missed; atypical findings are reported for compromised hosts; PRCA is not specific to parvovirus B19
Antigen detection	Detection of viral antigen in bone marrow, tissues	Immunohistochemistry localizes viral antigen to specific cells	Not as sensitive as in situ hybridization; not widely available; not standardized
DNA hybridization	Detection of DNA in serum, bone marrow, tissues	Positive result in serum has strong predictive value for disease or recent infection; can be quantitated; in situ hybridization can localize infected cells; detects all three genotypes	Less sensitive than DNA amplification methods; not widely available; not standardized
Virus isolation	Isolation of infectious virus	Can assay infectious virus, develop neutralization tests, and study mechanisms of replication	Requires specialized cultures (human erythroid progenitor cells and specialized cell lines); available only in research laboratories

Table 2. Parvovirus B19 serologic assays

Manufacturer	Method[c]	Antigen and source[a]	FDA approval
Biotrin International (Dublin, Ireland)	Indirect IgG EIA and class capture IgM EIA	Baculovirus-expressed VP2	Yes
Denka Seiken (Tokyo, Japan)	Indirect IgG and IgM EIA	Baculovirus-expressed VP1 and VP2	No
Medac Diagnosticka (Wedel, Germany)	Indirect IgG EIA and class capture IgM EIA	Baculovirus-expressed VP1 and VP2	No
IBL (Hamburg, Germany)	Indirect IgG and IgM EIA	*E. coli*-expressed VP1	No
Mikrogen (Martinsreid, Germany)	Indirect IgG and IgM EIA	*E. coli*-expressed VP1, baculovirus-expressed VP2	No
Biotrin International (Dublin, Ireland)	Indirect immunofluorescence (IFA) IgG and IgM[b]	Baculovirus-expressed VP1	No
Home brew	Immunoblot	*E. coli*-expressed VP1, VP2, and/or NS1	No

[a] VP2 comprises >95% of capsid antigen and appears to be conserved among genotypes; baculovirus-derived conformational VP2 antigen provides the most accurate EIA results.
[b] Pretreatment with adsorbent reagent is needed to prevent interference from rheumatoid factors.
[c] EIA, enzyme immunoassay.

are destroyed by B19 infection. Giant pronormoblasts, early erythroid cells in which B19 is actively replicating, are considered pathognomonic for B19 infection, but they may be absent or missed. Typical giant pronormoblasts contain large eosinophilic nuclear inclusion bodies, cytoplasmic vacuolization, and occasional "dog-ear" projections (105). Atypical bone marrow findings have also been reported for compromised hosts, including near-normal erythroid precursors and inclusions throughout the erythroid spectrum (35).

Nucleic Acid Detection

In situ DNA hybridization is used for tissue localization of B19 DNA in formalin-fixed tissues, including bone marrow and tissue biopsies, but is usually confined to the research setting (23). Direct hybridization, usually in the dot blot format, has a detection limit of $\sim 10^5$ genome copies/ml, can readily be quantified, and with the use of a full-length probe, detects all known variants (17, 18). However, it is rarely performed by clinical laboratories and has been replaced by more sensitive genome amplification methods, such as PCR.

PCR-ELISA and nested PCR assays are labor-intensive and susceptible to contamination and have largely been replaced by more automated, rapid, and contained real-time methods (60, 61, 77, 83, 89). Real-time technology has allowed wider clinical use and made quantification simpler (61). Assays have targeted conserved regions of either the NS1 (12) or VP1/VP2 (1, 21, 61) genes. Home brew real-time PCR assays have been reported by a number of laboratories (1, 21, 49, 61, 94), and two commercial B19 quantitative assays using the LightCycler instrument are available as analyte-specific reagents in the United States (Table 3). In published reports, these commercial assays have equaled or exceeded the sensitivity of nested PCR for genotype 1, but the Roche assay failed to detect genotypes 2 and 3 and the Artus assay underquantitated the genotype 3 variant D91.1 by 3 $\log_{10}$ (12, 16, 49, 56). Even some strains of genotype 1 may be missed (16). This is a particular concern because genome variability may be more common in persistent infection (54). Due to the short probe lengths used in real-time assays, the ability to detect variants may be reduced unless multiple primers and probes are incorporated (89). Several real-time consensus assays for detection and differentiation of B19 variants have been reported (Table 3) (11, 12, 94).

An internal control to detect inhibition is recommended and is provided with the commercial assays. In one study, inhibition occurred in 2 of 165 serum samples but 3 of 5 bone marrow samples extracted on a MagNA Pure instrument (16).

Culture

Isolation of B19 requires specialized cultures, such as bone marrow erythroid progenitor cells or fetal liver cells. While useful for assaying virus infectivity, developing neutralization tests, and studying viral replication, these methods are confined to research laboratories (17, 50).

Antigen Detection

Detection of B19 antigens in bone marrow or tissue samples by immunostaining using monoclonal antibodies may be available in some settings, but it is not as sensitive as in situ hybridization (23).

ORGAN SYSTEM- AND SAMPLE TYPE-SPECIFIC INTERPRETATION OF DATA

Prior to development of neutralizing antibodies, viral loads of 10^{11} to 10^{13} B19 genome

Table 3. Examples of parvovirus B19 real-time PCR assays

Kit (manufacturer)	Method[b]	Target	Ability to detect variants	Reference
Lightcycler parvovirus B19 quantification kit[a] (Roche Diagnostics)	Lightcycler, FRET	Proprietary	Highly sensitive for genotype 1, not suitable for genotype 2 or 3	12, 56
RealArt Parvo B19 PCR[a] (Artus/Qiagen)	Lightcycler, ABI Prism and Rotor-Gene kits	Proprietary	Detects all three genotypes, but with lower sensitivity for genotype strain 3 D91.1	12, 56
B19 assay (home brew)	ABI 7700, TaqMan	VP1/VP2	Tested only against genotype 1	61
Consensus assay (home brew)	Lightcycler, FRET	VP1	Detects genotypes 1 and 3	94
	ABI 7700, TaqMan	NS1	Detects genotypes 1, 2, and 3	76
	Lightcycler, TaqMan	NS1	Detects genotypes 1, 2, and 3	11, 12

[a]Available as analyte-specific reagents in United States.
[b]FRET, fluorescence resonance energy transfer.

copies/ml of serum are commonly seen, even in healthy persons without symptoms. Levels fall as antibodies develop, but low levels of B19 DNA can be detected by PCR in serum, bone marrow, joints, liver, kidney, and skin for months to years after symptoms have resolved (26-28, 70). As PCR assays become more sensitive, more low-level positive results of questionable significance will be detected. Quantitation of viral load may help with interpretation of clinical significance for the individual patient, but no data are currently available to correlate specific viral loads with disease. Rising titers in compromised hosts merit close monitoring of the reticulocyte count and hematocrit, assessment of immunosuppressive or antiretroviral regimens, and the need for IVIG.

The frequency and clinical relevance of missing or underquantitating genotypes 2 and 3 are unclear and need further study in various geographic locations with various patient groups (89).

In detecting B19 DNA in blood samples, serum or plasma is commonly used. In monitoring viral load over time, the same sample type should be used for consistency. It is not clear whether detection of B19 by PCR in tissue extracts reflects viremia, infection of endothelial cells lining blood vessels, or tissue replication. In situ hybridization, if available, can help with tissue localization. In some cases of encephalitis and meningitis, B19 DNA has been detected in cerebrospinal fluid by PCR (8, 9, 21), but the site of replication has not been determined.

THERAPY

Therapeutic Options

No antiviral therapy is available. IVIG can be administered, especially when anemia is severe and improvement in the immune system in the short term is not anticipated. The optimal dose and duration of IVIG have not been determined, and published reports have varied widely (36, 39, 45). A regimen of 0.4 g/kg of body weight/day for 5 to 10 days has been suggested (111). Reticuloycte counts rise within a week of IVIG therapy. For patients who relapse, a second course may be needed. In rare circumstances, monthly therapy of 0.4 g/kg/day for 1 day per month has been used. Preparations that are not sucrose based are now available and should be used when possible in renal transplants and other patients with impaired renal function. Serum creatinine should be monitored and IVIG decreased or withheld as needed. Erythropoietin has been used in a few cases, but some investigators report worsening disease, presumably due to an increase in target cells for B19 replication (6, 15).

Impact of Diagnostic and Screening Assays on Therapy

Although detection of high titers of B19 DNA by PCR confirms the etiologic role of B19 in the patient's anemia, data are not available correlating specific viral loads with clinical disease. A low or negative B19 viral load in the serum and an absence of supporting pathology in the bone marrow should call into question the diagnosis of B19-induced PRCA.

Use of Different Diagnostic Modalities To Monitor Therapeutic Response

If IVIG is not administered, but rather a change in immunosuppressive or antiretroviral drugs is initiated, quantitative PCR should be helpful in assessing the effectiveness of the strategy. The viral load should be documented once the patient is stable. Should the hematocrit start to fall again, a repeat viral load test is warranted and a significant rise should be expected. Monitoring B19 viral load as an early predictor of relapse is now feasible (49), but there are no data yet to correlate viral load with disease or to support the cost-effectiveness of this approach. Documenting IgG seroconversion in patients who have not been treated with IVIG, or after IVIG has waned, is recommended.

SUMMARY

Parvovirus B19 is an infrequent but serious and treatable cause of chronic anemia in immunocompromised hosts. Patients with congenital, acquired, and iatrogenic immunodeficiencies are susceptible, and the common denominator is an inability to produce sufficient neutralizing antibodies to control the virus. The anemia persists until neutralizing antibody is either produced by the host or passively administered. B19 should be suspected in compromised hosts with unexplained or severe anemia and reticulocytopenia or when bone marrow examination shows giant pronormoblasts or an absence of red cell precursors. Diagnosis is established by detection of B19 DNA in serum in the absence of IgG antibody to B19. However, in some cases, IgG antibody is detected by ELISA but is not a neutralizing antibody. Anti-B19 IgM may or may not be present. Therapy includes any or all of the following: red cell transfusion, adjustment in medications to restore or improve the patient's immune system, and administration of IVIG. Following treatment, patients should be monitored closely, especially if immunosuppression is unchanged or increased. Should the hematocrit trend downward and parvovirus DNA trend upward, the therapeutic options above should be revisited. In a few instances, monthly maintenance IVIG may be indicated. B19 variants, though infrequent, can be missed or underquantitated, especially by probe-based real-time PCR methods.

REFERENCES

1. **Aberham, C., C. Pendl, P. Gross, G. Zerlauth, and M. Gessner.** 2001. A quantitative, internally controlled real-time PCR assay for the detection of parvovirus B19 DNA. *J. Virol. Methods* **92:**183–191.
2. **Allander, T., M. T. Tammi, M. Eriksson, A. Bjerkner, A. Tiveljung-Lindell, and B. Andersson.** 2005. Cloning of a human parvovirus by molecular screening of respiratory tract samples. *Proc. Natl. Acad. Sci. USA* **102:**12891–12896.
3. **Amiot, L., T. Langanay, B. Drenou, B. Lelong, P. Y. Le Prise, Y. Logeais, R. Colimon, and R. Fauchet.** 1998. Spontaneous recovery from severe parvovirus B19 pure red cell aplasia, in a heart transplant recipient, as demonstrated by marrow culture. *Hematol. Cell Ther.* **40:**71–73.
4. **Anderson, M. J., P. G. Higgins, L. R. Davis, J. S. Willman, S. E. Jones, I. M. Kidd, J. R. Pattison, and D. A. Tyrrell.** 1985. Experimental parvoviral infection in humans. *J. Infect. Dis.* **152:**257–265.
5. **Arnold, J. C., K. K. Singh, S. A. Spector, and M. H. Sawyer.** 2006. Human bocavirus: prevalence and clinical spectrum at a children's hospital. *Clin. Infect. Dis.* **43:**283–288.
6. **Arzouk, N., R. Snanoudj, A. Beauchamp-Nicoud, G. Mourad, B. Charpentier, G. Tchernia, and A. Durrbach.** 2006. Parvovirus B19-induced anemia in renal transplantation: a role for rHuEPO in resistance to classical treatment. *Transpl. Int.* **19:**166–169.
7. **Bakhshi, S., S. A. Sarnaik, C. Becker, W. W. Shurney, M. Nigro, and S. Savasan.** 2002. Acute encephalopathy with parvovirus B19 infection in sickle cell disease. *Arch. Dis. Child.* **87:**541–542.
8. **Barah, F., P. J. Vallely, M. L. Chiswick, G. M. Cleator, and J. R. Kerr.** 2001. Association of human parvovirus B19 infection with acute meningoencephalitis. *Lancet* **358:**729–730.
9. **Barah, F., P. J. Vallely, G. M. Cleator, and J. R. Kerr.** 2003. Neurological manifestations of human parvovirus B19 infection. *Rev. Med. Virol.* **13:**185–199.
10. **Bastien, N., K. Brandt, K. Dust, D. Ward, and Y. Li.** 2006. Human bocavirus infection, Canada. *Emerg. Infect. Dis.* **12:**848–850.
11. **Baylis, S. A., J. F. Fryer, and P. Grabarczyk.** 2007. Effects of probe binding mutations in an assay designed to detect parvovirus B19: implications for the quantitation of different virus genotypes. *J. Virol. Methods* **139:**97–99.
12. **Baylis, S. A., N. Shah, and P. D. Minor.** 2004. Evaluation of different assays for the detection of parvovirus B19 DNA in human plasma. *J. Virol. Methods* **121:**7–16.
13. **Bilge, I., B. Sadikoglu, S. Emre, A. Sirin, K. Aydin, and B. Tatli.** 2005. Central nervous system vasculitis secondary to parvovirus B19 infection in a pediatric renal transplant patient. *Pediatr. Nephrol.* **20:**529–533.
14. **Bock, C. T., K. Klingel, S. Aberle, A. Duechting, A. Lupescu, F. Lang, and R. Kandolf.** 2005. Human parvovirus B19: a new emerging pathogen of inflammatory cardiomyopathy. *J. Vet. Med. B* **52:**340–343.
15. **Borkowski, J., M. Amrikachi, and S. D. Hudnall.** 2000. Fulminant parvovirus infection following erythropoietin treatment in a patient with acquired immunodeficiency syndrome. *Arch. Pathol. Lab. Med.* **124:**441–445.
16. **Braham, S., J. Gandhi, S. Beard, and B. Cohen.** 2004. Evaluation of the Roche LightCycler parvovirus B19 quantification kit for the diagnosis of parvovirus B19 infections. *J. Clin. Virol.* **31:**5–10.
17. **Brown, K. E.** 2004. Detection and quantitation of parvovirus B19. *J. Clin. Virol.* **31:**1–4.
18. **Brown, K. E.** 2004. Variants of B19. *Dev. Biol.* (Basel) **118:**71–77.
19. **Brown, K. E., S. M. Anderson, and N. S. Young.** 1993. Erythrocyte P antigen: cellular receptor for B19 parvovirus. *Science* **262:**114–117.
20. **Brown, K. E., N. S. Young, and J. M. Liu.** 1994. Molecular, cellular and clinical aspects of parvovirus B19 infection. *Crit. Rev. Oncol. Hematol.* **16:**1–31.
21. **Buller, R. S., and G. Storch.** 2004. Evaluation of a real-time PCR assay using the LightCycler system for detection of parvovirus B19 DNA. *J. Clin. Microbiol.* **42:**3326–3328.
22. **Bultmann, B. D., K. Klingel, K. Sotlar, C. T. Bock, H. A. Baba, M. Sauter, and R. Kandolf.** 2003. Fatal parvovirus B19-associated myocarditis clinically mimicking ischemic heart disease: an endothelial cell-mediated disease. *Hum. Pathol.* **34:**92–95.

23. **Bultmann, B. D., K. Klingel, K. Sotlar, C. T. Bock, and R. Kandolf.** 2003. Parvovirus B19: a pathogen responsible for more than hematologic disorders. *Virchows Arch.* **442:**8–17.
24. **Butchko, A. R., and J. A. Jordan.** 2004. Comparison of three commercially available serologic assays used to detect human parvovirus B19-specific immunoglobulin M (IgM) and IgG antibodies in sera of pregnant women. *J. Clin. Microbiol.* **42:**3191–3195.
25. **Calvet, A., M. O. Pujol, M. Bertocchi, O. Bastien, P. Boissonnat, and J. F. Mornex.** 1999. Parvovirus B19 infection in thoracic organ transplant recipients. *J. Clin. Virol.* **13:**37–42.
26. **Cassinotti, P., G. Burtonboy, M. Fopp, and G. Siegl.** 1997. Evidence for persistence of human parvovirus B19 DNA in bone marrow. *J. Med. Virol.* **53:**229–232.
27. **Cassinotti, P., D. Schultze, P. Schlageter, S. Chevili, and G. Siegl.** 1993. Persistent human parvovirus B19 infection following an acute infection with meningitis in an immunocompetent patient. *Eur. J. Clin. Microbiol. Infect. Dis.* **12:** 701–704.
28. **Cassinotti, P., and G. Siegl.** 2000. Quantitative evidence for persistence of human parvovirus B19 DNA in an immunocompetent individual. *Eur. J. Clin. Microbiol. Infect. Dis.* **19:**886–887.
29. **Chen, M. Y., C. C. Hung, C. T. Fang, and S. M. Hsieh.** 2001. Reconstituted immunity against persistent parvovirus B19 infection in a patient with acquired immunodeficiency syndrome after highly active antiretroviral therapy. *Clin. Infect. Dis.* **32:**1361–1365.
30. **Chinsky, J. M., and R. R. Kalyani.** 2006. Fever and petechial rash associated with parvovirus B19 infection. *Clin. Pediatr.* (Philadelphia) **45:**275–280.
31. **Cohen, B. J., S. Beard, W. A. Knowles, J. S. Ellis, D. Joske, J. M. Goldman, P. Hewitt, and K. N. Ward.** 1997. Chronic anemia due to parvovirus B19 infection in a bone marrow transplant patient after platelet transfusion. *Transfusion* **37:** 947–952.
32. **Cohen, B. J., J. Gandhi, and J. P. Clewley.** 2006. Genetic variants of parvovirus B19 identified in the United Kingdom: implications for diagnostic testing. *J. Clin. Virol.* **36:**152–155.
33. **Corcoran, A., and S. Doyle.** 2004. Advances in the biology, diagnosis and host-pathogen interactions of parvovirus B19. *J. Med. Microbiol.* **53:**459–475.
34. **Cossart, Y. E., A. M. Field, B. Cant, and D. Widdows.** 1975. Parvovirus-like particles in human sera. *Lancet* **i:**72–73.
35. **Crook, T. W., B. B. Rogers, R. D. McFarland, S. H. Kroft, P. Muretto, J. A. Hernandez, M. J. Latimer, and R. W. McKenna.** 2000. Unusual bone marrow manifestations of parvovirus B19 infection in immunocompromised patients. *Hum. Pathol.* **31:**161–168.
36. **Eid, A. J., R. A. Brown, R. Patel, and R. R. Razonable.** 2006. Parvovirus B19 infection after transplantation: a review of 98 cases. *Clin. Infect. Dis.* **43:**40–48.
37. **El-Mahallawy, H. A., T. Mansour, S. E. El-Din, M. Hafez, and S. Abd-el-Latif.** 2004. Parvovirus B19 infection as a cause of anemia in pediatric acute lymphoblastic leukemia patients during maintenance chemotherapy. *J. Pediatr. Hematol. Oncol.* **26:**403–406.
38. **Erdman, D. D., E. L. Durigon, and B. P. Holloway.** 1994. Detection of human parvovirus B19 DNA PCR products by RNA probe hybridization enzyme immunoassay. *J. Clin. Microbiol.* **32:**2295–2298.
39. **Fattet, S., P. Cassinotti, and M. B. Popovic.** 2004. Persistent human parvovirus B19 infection in children under maintenance chemotherapy for acute lymphocytic leukemia. *J. Pediatr. Hematol. Oncol.* **26:**497–503.
40. **Ferguson, M., and A. Heath.** 2004. Report of a collaborative study to calibrate the second international standard for parvovirus B19 antibody. *Biologicals* **32:**207–212.
41. **Flunker, G., A. Peters, S. Wiersbitzky, S. Modrow, and W. Seidel.** 1998. Persistent parvovirus B19 infections in immunocompromised children. *Med. Microbiol. Immunol.* (Berlin) **186:**189–194.
42. **Frickhofen, N., J. L. Abkowitz, M. Safford, J. M. Berry, J. Antunez-de-Mayolo, A. Astrow, R. Cohen, I. Halperin, L. King, D. Mintzer, et al.** 1990. Persistent B19 parvovirus infection in patients infected with human immunodeficiency virus type 1 (HIV-1): a treatable cause of anemia in AIDS. *Ann. Intern. Med.* **113:**926–933.
43. **Frickhofen, N., Z. J. Chen, N. S. Young, B. J. Cohen, H. Heimpel, and J. L. Abkowitz.** 1994. Parvovirus B19 as a cause of acquired chronic pure red cell aplasia. *Br. J. Haematol.* **87:**818–824.
44. **Gallinella, G., M. Zerbini, M. Musiani, S. Venturoli, G. Gentilomi, and E. Manaresi.** 1997. Quantitation of parvovirus B19 DNA sequences by competitive PCR: differential hybridization of the amplicons and immunoenzymatic detection on microplate. *Mol. Cell. Probes* **11:**127–133.
45. **Geetha, D., J. B. Zachary, H. M. Baldado, J. D. Kronz, and E. S. Kraus.** 2000. Pure red cell aplasia caused by parvovirus B19 infection in solid organ transplant recipients: a case report and review of literature. *Clin. Transplant.* **14:**586–591.
46. **Gilsanz, F., J. Garcia Vela, J. A. Vargas, J. Ibanez, F. Ona, J. Lopez, and M. Roggendorf.** 1995. Acquired pure red cell aplasia: a study of six cases. *Ann. Hematol.* **71:**181–183.
47. **Goedert, J. J., D. D. Erdman, B. A. Konkle, T. J. Torok, M. M. Lederman, D. Kleinert, T. Mandalaki, C. M. Kessler, L. J. Anderson, and N. L. Luban.** 1997. Parvovirus B19 quiescence during the course of human immunodeficiency virus infection in persons with hemophilia. *Am. J. Hematol.* **56:**248–251.
48. **Graeve, J. L., P. A. de Alarcon, and S. J. Naides.** 1989. Parvovirus B19 infection in patients receiving cancer chemotherapy: the expanding spectrum of disease. *Am. J. Pediatr. Hematol. Oncol.* **11:**441–444.
49. **Harder, T. C., M. Hufnagel, K. Zahn, K. Beutel, H. J. Schmitt, U. Ullmann, and P. Rautenberg.** 2001. New LightCycler PCR for rapid and sensitive quantification of parvovirus B19 DNA guides therapeutic decision-making in relapsing infections. *J. Clin. Microbiol.* **39:**4413–4419.
50. **Heegaard, E. D., and K. E. Brown.** 2002. Human parvovirus B19. *Clin. Microbiol. Rev.* **15:**485–505.
51. **Heegaard, E. D., B. L. Petersen, C. J. Heilmann, and A. Hornsleth.** 2002. Prevalence of parvovirus B19 and parvovirus V9 DNA and antibodies in paired bone marrow and serum samples from healthy individuals. *J. Clin. Microbiol.* **40:**933–936.
52. **Heegaard, E. D., N. A. Peterslund, and A. Hornsleth.** 1995. Parvovirus B19 infection associated with encephalitis in a patient suffering from malignant lymphoma. *Scand. J. Infect. Dis.* **27:**631–633.
53. **Heegaard, E. D., K. Qvortrup, and J. Christensen.** 2002. Baculovirus expression of erythrovirus V9 capsids and screening by ELISA: serologic cross-reactivity with erythrovirus B19. *J. Med. Virol.* **66:**246–252.
54. **Hemauer, A., A. von Poblotzki, A. Gigler, P. Cassinotti, G. Siegl, H. Wolf, and S. Modrow.** 1996. Sequence variability among different parvovirus B19 isolates. *J. Gen. Virol.* **77:** 1781–1785.
55. **Hofbauer, G. F., A. Boehler, R. Speich, G. Burg, and F. O. Nestle.** 2002. Painless erythema of the hands associated with

non-Hodgkin's lymphoma in a lung transplant recipient. *J. Am. Acad. Dermatol.* **46:**S159–S160.

56. **Hokynar, K., P. Norja, H. Laitinen, P. Palomaki, A. Garbarg-Chenon, A. Ranki, K. Hedman, and M. Soderlund-Venermo.** 2004. Detection and differentiation of human parvovirus variants by commercial quantitative real-time PCR tests. *J. Clin. Microbiol.* **42:**2013–2019.
57. **Hokynar, K., M. Soderlund-Venermo, M. Pesonen, A. Ranki, O. Kiviluoto, E. K. Partio, and K. Hedman.** 2002. A new parvovirus genotype persistent in human skin. *Virology* **302:** 224–228.
58. **Isobe, Y., K. Sugimoto, Y. Shiraki, M. Nishitani, K. Koike, and K. Oshimi.** 2004. Successful high-titer immunoglobulin therapy for persistent parvovirus B19 infection in a lymphoma patient treated with rituximab-combined chemotherapy. *Am. J. Hematol.* **77:**370–373.
59. **Jordan, J. A.** 2000. Comparison of a baculovirus-based VP2 enzyme immunoassay (EIA) to an *Escherichia coli*-based VP1 EIA for detection of human parvovirus B19 immunoglobulin M and immunoglobulin G in sera of pregnant women. *J. Clin. Microbiol.* **38:**1472–1475.
60. **Jordan, J. A.** 2001. Diagnosing human parvovirus B19 infection: guidelines for test selection. *Mol. Diagn.* **6:**307–312.
61. **Jordan, J. A., S. J. Faas, E. R. Braun, and M. Trucco.** 1996. Exonuclease-released fluorescence detection of human parvovirus B19 DNA. *Mol. Diagn.* **1:**321–328.
62. **Kariyawasam, H. H., K. M. Gyi, M. E. Hodson, and B. J. Cohen.** 2000. Anaemia in lung transplant patient caused by parvovirus B19. *Thorax* **55:**619–620.
63. **Kerr, J. R.** 2005. Pathogenesis of parvovirus B19 infection: host gene variability, and possible means and effects of virus persistence. *J. Vet. Med. B* **52:**335–339.
64. **Kerr, J. R., F. Barah, M. L. Chiswick, G. V. McDonnell, J. Smith, M. D. Chapman, J. B. Bingham, P. Kelleher, and M. N. Sheppard.** 2002. Evidence for the role of demyelination, HLA-DR alleles, and cytokines in the pathogenesis of parvovirus B19 meningoencephalitis and its sequelae. *J. Neurol. Neurosurg. Psych.* **73:**739–746.
65. **Kerr, J. R., F. Barah, D. L. Mattey, I. Laing, S. J. Hopkins, I. V. Hutchinson, and D. A. Tyrrell.** 2001. Circulating tumour necrosis factor-alpha and interferon-gamma are detectable during acute and convalescent parvovirus B19 infection and are associated with prolonged and chronic fatigue. *J. Gen. Virol.* **82:**3011–3019.
66. **Kerr, J. R., J. Bracewell, I. Laing, D. L. Mattey, R. M. Bernstein, I. N. Bruce, and D. A. Tyrrell.** 2002. Chronic fatigue syndrome and arthralgia following parvovirus B19 infection. *J. Rheumatol.* **29:**595–602.
67. **Kerr, J. R., V. S. Cunniffe, P. Kelleher, R. M. Bernstein, and I. N. Bruce.** 2003. Successful intravenous immunoglobulin therapy in 3 cases of parvovirus B19-associated chronic fatigue syndrome. *Clin. Infect. Dis.* **36:**e100–e106.
68. **Kerr, J. R., M. D. Curran, J. E. Moore, and P. G. Murphy.** 1995. Parvovirus B19 infection—persistence and genetic variation. *Scand. J. Infect. Dis.* **27:**551–557.
69. **Kerr, J. R., and S. Modrow.** 2006. Human and primate parvovirus infections and associated disease, p. 385–416. *In* J. R. Kerr, S. G. Cotmore, M. E. Bloom, R. M. Linden, and C. R. Parrish (ed.), *Parvoviruses*. Edward Arnold Limited, London, United Kingdom.
70. **Kesebir, D., M. Vazquez, C. Weibel, E. D. Shapiro, D. Ferguson, M. L. Landry, and J. S. Kahn.** 2006. Human bocavirus infection in young children in the United States: molecular epidemiological profile and clinical characteristics of a newly emerging respiratory virus. *J. Infect. Dis.* **194:**1276–1282.
71. **Koch, W. C., G. Massey, C. E. Russell, and S. P. Adler.** 1990. Manifestations and treatment of human parvovirus B19 infection in immunocompromised patients. *J. Pediatr.* **116:** 355–359.
72. **Kuo, S. H., L. I. Lin, C. J. Chang, Y. R. Liu, K. S. Lin, and A. L. Cheng.** 2002. Increased risk of parvovirus B19 infection in young adult cancer patients receiving multiple courses of chemotherapy. *J. Clin. Microbiol.* **40:**3909–3912.
73. **Kurtzman, G., N. Frickhofen, J. Kimball, D. W. Jenkins, A. W. Nienhuis, and N. S. Young.** 1989. Pure red-cell aplasia of 10 years' duration due to persistent parvovirus B19 infection and its cure with immunoglobulin therapy. *N. Engl. J. Med.* **321:**519–523.
74. **Kurtzman, G. J., B. Cohen, P. Meyers, A. Amunullah, and N. S. Young.** 1988. Persistent B19 parvovirus infection as a cause of severe chronic anaemia in children with acute lymphocytic leukaemia. *Lancet* **ii:**1159–1162.
75. **Kurtzman, G. J., B. J. Cohen, A. M. Field, R. Oseas, R. M. Blaese, and N. S. Young.** 1989. Immune response to B19 parvovirus and an antibody defect in persistent viral infection. *J. Clin. Investig.* **84:**1114–1123.
76. **Liefeldt, L., A. Plentz, B. Klempa, O. Kershaw, A. S. Endres, U. Raab, H. H. Neumayer, H. Meisel, and S. Modrow.** 2005. Recurrent high level parvovirus B19/genotype 2 viremia in a renal transplant recipient analyzed by real-time PCR for simultaneous detection of genotypes 1 to 3. *J. Med. Virol.* **75:**161–169.
77. **Manaresi, E., G. Gallinella, E. Zuffi, F. Bonvicini, M. Zerbini, and M. Musiani.** 2002. Diagnosis and quantitative evaluation of parvovirus B19 infections by real-time PCR in the clinical laboratory. *J. Med. Virol.* **67:**275–281.
78. **Matsumoto, Y., D. Naniwa, S. Banno, and Y. Sugiura.** 1998. The efficacy of therapeutic plasmapheresis for the treatment of fatal hemophagocytic syndrome: two case reports. *Ther. Apher.* **2:**300–304.
79. **McNall, R. Y., D. R. Head, C. H. Pui, and B. I. Razzouk.** 2001. Parvovirus B19 infection in a child with acute lymphoblastic leukemia during induction therapy. *J. Pediatr. Hematol. Oncol.* **23:**309–311.
80. **McNeely, M., J. Friedman, and E. Pope.** 2005. Generalized petechial eruption induced by parvovirus B19 infection. *J. Am. Acad. Dermatol.* **52:**S109–S113.
81. **Moudgil, A., C. C. Nast, A. Bagga, L. Wei, A. Nurmamet, A. H. Cohen, S. C. Jordan, and M. Toyoda.** 2001. Association of parvovirus B19 infection with idiopathic collapsing glomerulopathy. *Kidney Int.* **59:**2126–2133.
82. **Moudgil, A., H. Shidban, C. C. Nast, A. Bagga, S. Aswad, S. L. Graham, R. Mendez, and S. C. Jordan.** 1997. Parvovirus B19 infection-related complications in renal transplant recipients: treatment with intravenous immunoglobulin. *Transplantation* **64:**1847–1850.
83. **Musiani, M., A. Azzi, M. Zerbini, D. Gibellini, S. Venturoli, K. Zakrzewska, M. C. Re, G. Gentilomi, G. Gallinella, and M. La Placa.** 1993. Nested polymerase chain reaction assay for the detection of B19 parvovirus DNA in human immunodeficiency virus patients. *J. Med. Virol.* **40:**157–160.
84. **Nakazawa, T., N. Tomosugi, K. Sakamoto, M. Asaka, T. Yuri, I. Ishikawa, and S. Kitagawa.** 2000. Acute glomerulonephritis after human parvovirus B19 infection. *Am. J. Kidney Dis.* **35:**E31.
85. **Nguyen, Q. T., C. Sifer, V. Schneider, X. Allaume, A. Servant, F. Bernaudin, V. Auguste, and A. Garbarg-Chenon.** 1999. Novel human erythrovirus associated with transient aplastic anemia. *J. Clin. Microbiol.* **37:**2483–2487.
86. **Nguyen, Q. T., C. Sifer, V. Schneider, F. Bernaudin, V. Auguste, and A. Garbarg-Chenon.** 1998. Detection of an eryth-

rovirus sequence distinct from B19 in a child with acute anaemia. *Lancet* **352**:1524.

87. **Nguyen, Q. T., S. Wong, E. D. Heegaard, and K. E. Brown.** 2002. Identification and characterization of a second novel human erythrovirus variant, A6. *Virology* **301**:374–380.
88. **Nigro, G., V. Bastianon, V. Colloridi, F. Ventriglia, P. Gallo, G. D'Amati, W. C. Koch, and S. P. Adler.** 2000. Human parvovirus B19 infection in infancy associated with acute and chronic lymphocytic myocarditis and high cytokine levels: report of 3 cases and review. *Clin. Infect. Dis.* **31**:65–69.
89. **Peterlana, D., A. Puccetti, R. Corrocher, and C. Lunardi.** 2006. Serologic and molecular detection of human parvovirus B19 infection. *Clin. Chim. Acta* **372**:14–23.
90. **Plentz, A., J. Hahn, E. Holler, W. Jilg, and S. Modrow.** 2004. Long-term parvovirus B19 viraemia associated with pure red cell aplasia after allogeneic bone marrow transplantation. *J. Clin. Virol.* **31**:16–19.
91. **Rao, S. P., S. T. Miller, and B. J. Cohen.** 1994. B19 parvovirus infection in children with malignant solid tumors receiving chemotherapy. *Med. Pediatr. Oncol.* **22**:255–257.
92. **Rao, S. P., S. T. Miller, and B. J. Cohen.** 1990. Severe anemia due to B19 parvovirus infection in children with acute leukemia in remission. *Am. J. Pediatr. Hematol. Oncol.* **12**:194–197.
93. **Saldanha, J., N. Lelie, M. W. Yu, and A. Heath.** 2002. Establishment of the first World Health Organization international standard for human parvovirus B19 DNA nucleic acid amplification techniques. *Vox Sang.* **82**:24–31.
94. **Schalasta, G., M. Schmid, T. Lachmund, and G. Enders.** 2004. LightCycler consensus PCR for rapid and differential detection of human erythrovirus B19 and V9 isolates. *J. Med. Virol.* **73**:54–59.
95. **Schleuning, M., G. Jager, E. Holler, W. Hill, C. Thomssen, C. Denzlinger, T. Lorenz, G. Ledderose, W. Wilmanns, and H. J. Kolb.** 1999. Human parvovirus B19-associated disease in bone marrow transplantation. *Infection* **27**:114–117.
96. **Schowengerdt, K. O., J. Ni, S. W. Denfield, R. J. Gajarski, N. E. Bowles, G. Rosenthal, D. L. Kearney, J. K. Price, B. B. Rogers, G. M. Schauer, R. E. Chinnock, and J. A. Towbin.** 1997. Association of parvovirus B19 genome in children with myocarditis and cardiac allograft rejection: diagnosis using the polymerase chain reaction. *Circulation* **96**:3549–3554.
97. **Servant, A., S. Laperche, F. Lallemand, V. Marinho, G. De Saint Maur, J. F. Meritet, and A. Garbarg-Chenon.** 2002. Genetic diversity within human erythroviruses: identification of three genotypes. *J. Virol.* **76**:9124–9134.
98. **Seyama, K., R. Kobayashi, H. Hasle, A. J. Apter, J. C. Rutledge, D. Rosen, and H. D. Ochs.** 1998. Parvovirus B19-induced anemia as the presenting manifestation of X-linked hyper-IgM syndrome. *J. Infect. Dis.* **178**:318–324.
99. **Shaw, P. J., T. Eden, and B. J. Cohen.** 1993. Parvovirus B19 as a cause of chronic anemia in rhabdomyosarcoma. *Cancer* **72**:945–949.
100. **Siegl, G., and P. Cassinotti.** 1998. Presence and significance of parvovirus B19 in blood and blood products. *Biologicals* **26**:89–94.
101. **Smith, M. A., N. R. Shah, J. S. Lobel, P. J. Cera, G. W. Gary, and L. J. Anderson.** 1988. Severe anemia caused by human parvovirus in a leukemia patient on maintenance chemotherapy. *Clin. Pediatr.* (Philadelphia) **27**:383–386.
102. **Smith, P. T., M. L. Landry, H. Carey, J. Krasnoff, and E. Cooney.** 1998. Papular-purpuric gloves and socks syndrome associated with acute parvovirus B19 infection: case report and review. *Clin. Infect. Dis.* **27**:164–168.
103. **Soderlund-Venermo, M., K. Hokynar, J. Nieminen, H. Rautakorpi, and K. Hedman.** 2002. Persistence of human parvovirus B19 in human tissues. *Pathol. Biol.* (Paris) **50**:307–316.
104. **Solano, C., O. Juan, C. Gimeno, and J. Garcia-Conde.** 1996. Engraftment failure associated with peripheral blood stem cell transplantation after B19 parvovirus infection. *Blood* **88**:1515–1517.
105. **Takahashi, T., K. Ozawa, K. Takahashi, S. Asano, and F. Takaku.** 1990. Susceptibility of human erythropoietic cells to B19 parvovirus in vitro increases with differentiation. *Blood* **75**:603–610.
106. **Tang, M. L., A. S. Kemp, and L. D. Moaven.** 1994. Parvovirus B19-associated red blood cell aplasia in combined immunodeficiency with normal immunoglobulins. *Pediatr. Infect. Dis. J.* **13**:539–542.
107. **Vafaie, J., and R. A. Schwartz.** 2004. Parvovirus B19 infections. *Int. J. Dermatol.* **43**:747–749.
108. **Ware, A. J., and T. Moore.** 2001. Resolution of chronic parvovirus B19-induced anemia, by use of highly active antiretroviral therapy, in a patient with acquired immunodeficiency syndrome. *Clin. Infect. Dis.* **32**:E122–E123.
109. **Weigel-Kelley, K. A., M. C. Yoder, and A. Srivastava.** 2003. Alpha5beta1 integrin as a cellular coreceptor for human parvovirus B19: requirement of functional activation of beta1 integrin for viral entry. *Blood* **102**:3927–3933.
110. **Wong, T. Y., P. K. Chan, C. B. Leung, C. C. Szeto, J. S. Tam, and P. K. Li.** 1999. Parvovirus B19 infection causing red cell aplasia in renal transplantation on tacrolimus. *Am. J. Kidney Dis.* **34**:1132–1136.
111. **Young, N. S., and K. E. Brown.** 2004. Parvovirus B19. *N. Engl. J. Med.* **350**:586–597.

Diagnostic Microbiology of the Immunocompromised Host
Edited by Randall T. Hayden, Karen C. Carroll, Yi-Wei Tang, and Donna M. Wolk

Chapter 10

Filamentous Fungi

GARY W. PROCOP, RANDALL T. HAYDEN, AND GLENN D. ROBERTS

The term "emerging" is an overused and misused word in the infectious disease and microbiology literature. A standard definition demands that truly emerging infections have "newly appeared in the population, or have existed but are rapidly increasing in incidence or geographic range" (137). Elsewhere, these are defined as "new, re-emerging or drug resistant infections whose incidence in humans has increased within the past two decades or whose incidence threatens to increase in the near future" (104). Therefore, many infections by fungi may be considered truly emerging infections by these standards.

A variety of factors are associated with the emergence of infections, including changes in microorganisms that may enhance virulence or enable resistance to antimicrobial agents. Alternatively, the primary change may be in the host or the population affected. For example, it was primary host factor differences that were responsible for the decimation of native populations in the Americas when viral diseases were introduced from Europe. It is our opinion that host factors are primarily responsible for the increased incidence of fungal infections. These are usually linked in some way to the numerous advances in the types of therapeutics and new surgical procedures that have been developed over the past 50 years. These interventions extend the lives of patients with malignancies or end organ failure who in the past would have otherwise died. However, many of these treatments result in an increased risk of fungal infections through breeches in nonspecific barriers (e.g., skin), alteration in the patient's normal microbiota, or alteration in host immune function.

Patients who may not be considered immunosuppressed in the classic sense but who have severe illnesses, such as uncontrolled diabetes mellitus, or who require complicated surgical procedures, have extended hospitalization in intensive care units, or have long-term indwelling catheters (intravascular, peritoneal, or urinary) are also at risk for fungal infections. Patients with impaired cellular immunity, whether hereditary, acquired, or iatrogenic, are at increased risk of developing fungal infections (182). Patients who undergo solid organ or bone marrow/stem cell transplantation are at heightened risk for invasive fungal infections. Although all of these are "transplant recipients," these individuals are each unique, and each patient must be considered individually with respect to risk for acquiring a fungal infection. The variables that affect the risk for fungal infections in these patients include the complexity of the surgical procedures, the immunosuppressive regimen used, and other factors (e.g., the presence of graft-versus-host disease) (160, 216, 234). Similarly, immunosuppressive agents, such as corticosteroids and other, newer agents, which have proven useful for the control of autoimmune and other inflammatory diseases, also increase the risk for fungal infections (33, 41, 154). Finally, patients with inherited immunodeficiency syndromes or AIDS are also at elevated risk for fungal infections; the degree to which these patients are at risk and the type of fungi that are most likely to cause infection are different, depending on the degree of immunosuppression and the specific type of immunodeficiency present (182).

ETIOLOGIC AGENTS OF DISEASE

Any fungus may cause disease in the profoundly immunocompromised host. However, certain fungi are more virulent for humans and more likely to be

Gary W. Procop • Department of Pathology, University of Miami Miller School of Medicine, Miami, FL 33136.
Randall T. Hayden • Department of Pathology, St. Jude Children's Research Hospital, Memphis, TN 38105.
Glenn D. Roberts • Division of Clinical Microbiology, Mayo Clinic, Rochester, MN 55905.

the cause of a systemic mycotic infection. In this chapter, the most common causes of invasive mold infections are discussed in greater detail, with a brief mention of particular molds that are less frequent causes of infection. The hyaline septate molds are covered first, with emphasis on *Aspergillus* species, *Fusarium* species, and *Pseudallescheria boydii*. Two dimorphic fungal pathogens, *Histoplasma capsulatum* and *Penicillium marneffei*, are also covered in this section, as well as selected less commonly encountered hyaline septate molds. Next, the zygomycetes are discussed, and *Rhizopus* and *Mucor* species are included because they are the genera most frequently encountered. The dematiaceous fungi are briefly discussed; a description of all the dematiaceous molds that have been described as human pathogens is beyond the scope of this text.

Hyalohyphomycoses (Common Causes)

Invasive fungal disease caused by molds that produce hyaline or colorless, septate hyphae may generically be termed an invasive hyalohyphomycosis (234). The most frequent causes of invasive hyalohyphomycosis are *Aspergillus* species, most commonly *Aspergillus fumigatus*, *Fusarium* species, and *Pseudallescheria boydii*. A brief description of the less common causes of invasive hyalohyphomycoses is also provided. The appropriate identification of the cause of an invasive hyalohyphomycosis has prognostic and therapeutic significance, since accurate identification influences the selection of the appropriate antifungal agent and, ultimately, the outcome of the infection.

The morphologic features of these molds are often indistinguishable from one another in the direct examination of tissues and fluid aspirates as well as in cytologic and histopathologic preparations. It is not acceptable to characterize all hyaline septate hyphae that occur in tissue sections (i.e., histopathology) or body fluids as "*Aspergillus* species." Molds such as *Fusarium* species, *P. boydii*, and others may produce identical morphologic features in tissue; this is important to recognize since *Fusarium* may be resistant to amphotericin B and *P. boydii* is intrinsically resistant to this antifungal agent (164). Culture or advanced molecular methods are necessary to differentiate these fungi so that appropriate therapy may be initiated.

Aspergillus species

Aspergillus species are the most common causes of invasive hyalohyphomycosis (7, 64, 66, 111, 118, 162, 189). Invasive aspergillosis has significantly increased in incidence over the past several decades because of an increase in the number of immunosuppressed patients (i.e., the number of successful transplantations continues to increase), as well as the type/degree of immunosuppression. Invasive aspergillosis commonly occurs in patients with profound neutropenia, lung and heart transplant recipients, and those who develop graft-versus-host disease (26, 35, 82, 89, 150, 158, 232). Aspergillosis may also infect patients with burns and vascular grafts (4, 78, 138, 180, 223).

Of the numerous species of *Aspergillus* that exist, only a few are common human pathogens. Of these, *Aspergillus fumigatus* is the species most frequently recovered from patients with invasive aspergillosis, and some suggest it accounts for >90% of invasive mold infections (120). However, regional or local differences may exist (i.e., in some hospitals *Aspergillus* species other than *A. fumigatus* may be the species most commonly encountered). Any of the *Aspergillus* species, however, may cause disease, so mycology laboratories should become competent in the identification of the most commonly occurring species. The mycology section of the clinical microbiology laboratory should be able to identify *A. fumigatus*, *A. flavus*, *A. niger*, and *A. terreus*. Other *Aspergillus* species that we encourage laboratories to gain expertise in identifying and reporting include *A. glaucus*, *A. ustus*, *A. nidulans*, *A. versicolor*, and *A. clavatus*.

The aspergillum (fruiting head) is the asexual reproductive structure, is most important for definitively identifying *Aspergillus* isolates to the genus level, and is often helpful in further identification to the species level. Definitive species identification can be based on the presence or absence of other features, such as aleuroconidia, hülle cells, and cleistothecia, as well as on colony morphology. A number of excellent laboratory guides are available to facilitate the identification of *Aspergillus* species.

Although most *Aspergillus* species are susceptible to amphotericin B, *Aspergillus terreus* is an important exception (135, 221). The definitive identification of *Aspergillus* to the species level, or at least to exclude the possibility of an infection with *Aspergillus terreus*, is therefore important to guide therapy.

Fusarium

Fusarium species have emerged as significant fungal pathogens in immunocompromised patients. The patient populations at risk for invasive fusariosis are the same as those at risk for infections by other fungal pathogens. The individuals at highest risk for fusariosis include those with neutropenia secondary to hematopoietic malignancy or due to the use of therapeutic agents used to treat malignancies; patients who have received high doses of corticosteroids and other immunosuppressive agents are also at risk (7, 21, 58, 68, 73, 75, 98, 118, 122, 123, 140, 149, 184). *Fusarium* species have been reported to be the second most common cause of invasive hyalohyphomycosis in some studies (7, 48, 118). This fungus has also been reported to cause infections in immunocompetent patients with trauma or severe burns and in those who receive peritoneal dialysis and have a catheter-associated infection (1, 75, 103, 242).

Fusarium species produce hyaline septate hyphae like those of *Aspergillus*. Asexual reproduction is accomplished through the production of single-celled microconidia and crescent-shaped multicellular macroconidia. The microconidia produced by *Fusarium* are oval and nondistinct and resemble those produced by *Acremonium* species, another hyaline septate mold that may infect immunocompromised patients. The multicellular macroconidia characteristic of *Fusarium* usually contain more than two cells per conidium (the exception is *Fusarium dimerum*). Close inspection of the multicelled macroconidia of *Fusarium* demonstrates the presence of a foot cell, which differentiates this organism from *Cylindrocarpon* (204).

Fusarium species, like *Aspergillus* species, produce hyaline septate hyphae that have acute-angle branching in infected tissues. Sometimes, however, the hyphae from a *Fusarium* species may be arranged more loosely (i.e., not as tightly packed or associated one with another as in many cases of invasive aspergillosis). This is a clue that the infecting fungus may be something other than an *Aspergillus* species. The hyphae of *Fusarium* may also appear moniliform, wherein the diameter of the hypha varies from one area to another. *Fusarium* may produce adventitious conidia in tissue that may be analogous to the microconidia seen in culture (111). It is thought that these structures are responsible, at least in part, for dissemination throughout the body; it is well known that *Fusarium*, unlike *Aspergillus*, may be recovered in blood cultures (7, 9, 10, 21, 31, 149). Rarely, the characteristic macroconidia of *Fusarium* may be seen upon histologic examination of wounds or burns when the growing fungus is exposed to the air.

Fusarium species are among the fungi that are most resistant to antifungal agents (48, 234). Many strains exhibit resistance to amphotericin B, although some patients have responded to high-dose amphotericin B or the lipid formulation treatment (8, 38, 50, 62, 132, 177, 178, 230, 233, 235, 236, 247). Fluconazole and itraconazole are inactive against *Fusarium* species, but some of the newer antifungals, such as voriconazole and posaconazole, hold promise (51, 235). Three factors are important for recovery of patients with invasive fusariosis. These are (i) the use of the most active antifungal agent available; (ii) surgical excision of infected tissues, as indicated; and (iii) the restoration of a normal immune response, which may be accelerated with the use of granulocyte colony-stimulating factor or granulocyte transfusions.

Pseudallescheria boydii (*Scedosporium apiospermum*)

Pseudallescheria boydii is the teleomorphic or sexual form of *Scedosporium apiospermum*. This is another septate mold that is a well-known cause of invasive disease of immunocompromised patients (15, 94, 112, 161, 213, 219). Whether this is a hyaline septate or lightly pigmented dematiaceous mold is an argument that will be left for the taxonomists; here we consider it with the hyaline septate molds. In addition to causing invasive pseudallescheriosis, this fungus is also a cause of white grain mycetoma. It is the most common cause of this type of mycetoma in North America. *Pseudallescheria boydii*, like *Aspergillus*, may also produce a fungus ball in the sinuses, bronchiectatic airways, or old lung cavities caused by tuberculosis (219).

The hyphae present in infected tissues of patients with invasive pseudallescheriosis are hyaline and septate and are indistinguishable from those of *Aspergillus* or *Fusarium* in histologic sections (112). It is important, however, to establish the diagnosis of pseudallescheriosis, since this fungus is innately resistant to amphotericin B, unlike most *Aspergillus* species.

Pseudallescheria boydii and its asexual anamorphic form, *Scedosporium apiospermum*, produce pear-shaped (i.e., pyriform) terminal annelloconidia from simple conidiophores. These conidia are distinctly flattened at the base. Rarely, the conidiophores are aggregated into the so-called *Graphium* anamorph. *Pseudallescheria boydii* is distinguished from *S. apiospermum* by the production of darkly pigmented cleistothecia (i.e., the sexual reproductive structure that contains ascospores).

Scedosporium apiospermum is differentiated from *Scedosporium prolificans* (previously called *Scedosporium inflatum*) by colony morphology (*S. prolificans* is unmistakably dematiaceous) and by the absence of a swollen or inflated conidiophore. Differentiation between these two organisms is important, since *S. prolificans* is multidrug resistant.

Dimorphic molds

The thermally dimorphic fungi are those that produce a yeast or spherule phase in tissue but grow as a mold in culture at 25 to 30°C. The incidences of infections caused by *Histoplasma capsulatum* and *Penicillium marneffei* have increased largely because of an increase of infections in individuals with compromised immune systems.

Histoplasma capsulatum. Patients with immunocompromising conditions, particularly those that affect the T-helper cells, are at increased risk for histoplasmosis. These patients more commonly develop disseminated disease rather than a localized pulmonary infection. Disseminated histoplasmosis affects patients with advanced human immunodeficiency virus (HIV) infection, those receiving high-dose corticosteroids, and more recently, those who have received tumor necrosis factor-alpha blockade (106, 144, 248, 252). The inflammatory response in these patients is variable and depends upon the degree and type of immunosuppression. Responses range from poorly formed granulomas to nonspecific necrosis. The yeasts may be seen clustered within phagocytes and occasionally are seen in peripheral blood smears. The examination of a bone marrow biopsy or aspirate often confirms the diagnosis. Rarely, hyaline membrane disease may be the principle finding for immunocompromised patients with disseminated histoplasmosis. The yeasts in such instances are inconspicuous by hematoxylin and eosin staining and require a Gomori methenamine silver stain or another fungal stain for detection.

Penicillium marneffei. *Penicillium marneffei* is a dimorphic *Penicillium* species that is endemic to Thailand and Southeast Asia, where it causes disseminated penicilliosis in patients with AIDS (202, 209-211, 234). The disseminated disease caused by this fungus is similar in many aspects to that caused by *Histoplasma capsulatum*.

This fungus grows as a yeast-like organism in the bodies of infected patients and in the laboratory at 35 to 37°C. It is designated "yeast-like" because it replicates asexually by binary fission rather than by budding. At 25°C or room temperature, mold colonies that in many ways resemble typically encountered *Penicillium* strains are produced. However, the colonies of *P. marneffei* produce a soluble red pigment that diffuses into the agar. Although the presence of this pigment is not specific for *P. marneffei*, it is a clue that should prompt the laboratorian to consider the possibility of this organism. The confirmation of a suspect isolate as *P. marneffei* may be achieved through the in vitro conversion of the mold form to the yeast-like form or through molecular techniques.

Less common causes of hyalohyphomycosis

It is well known that a wide variety of molds, even those that in the past were considered only laboratory contaminants, may cause disease in the immunocompromised patient. Infections caused by these unusually encountered molds often appear in the literature as case reports (20, 27, 46, 53, 59, 63, 81, 88, 113, 124, 139, 146, 147, 154, 175, 192, 212, 215, 237, 246, 250). Given the immunologic status of many patients today, it is important that the clinical mycologist does not immediately disregard an unusually encountered mold as a likely contaminant or nonpathogen. Prompt correlation with clinical, radiologic, and pathologic findings is more important than ever and is usually necessary to establish the validity of these organisms as true pathogens. Conversely, the lack of supportive evidence of infection will help to establish some as contaminants. Infections caused by these "uncommon" pathogens usually occur in patients who are immunocompromised or have a serious underlying disease. This common thread reiterates the opportunistic nature of these molds and the importance of a normally functioning immune system in averting these infections in everyday life. Even among these uncommon pathogens, there are certain molds that are seen with greater frequencies than the others and that warrant further discussion.

Acremonium. *Acremonium* is well recognized as a human pathogen, although it is infrequently encountered (234). It causes an invasive hyalohyphomycosis that may disseminate in immunocompromised patients (54). In addition, it is a cause of white grain mycetoma, corneal infection, fungal endocarditis, fungal sinusitis, bone and joint infections, pneumonia, and peritonitis (56, 69, 72, 97, 133, 181, 205).

Hyaline septate hyphae that have acute-angle branching are produced in infected tissues. This is therefore another fungus that may be mistaken for an *Aspergillus* sp. in tissue sections or in direct examination. *Acremonium* produces small conidia from simple, straight conidiophores similar to those producing microconidia of *Fusarium* (111). It is these small conidia, which have been shown to be produced in vivo, that are thought to be responsible for the dissemination of this mold. *Acremonium strictum* is the species most commonly associated with human disease (188). *Acremonium* is relatively resistant to several common antifungal agents; antifungal activity has been shown with amphotericin B but not with fluconazole (205, 234).

Paecilomyces. Although *Paecilomyces* species may cause disease in the appropriate setting, in many instances they represent either contaminants or transient flora. *Paecilomyces* produces a "penicillus" and therefore superficially resembles a *Penicillium* species in the microscopic examination. However, this genus can be differentiated successfully from *Penicillium* species through close examination of the conidiophores. *Paecilomyces* species produce small, delicate, tapered phialides, in contrast to the larger, blunt-ended phialides of *Penicillium*. Two species of *Paecilomyces*, *Paecilomyces varioti* and *Paecilomyces lilacinus*, have been associated with human disease more frequently than other species have.

Paecilomyces species have been reported to cause disseminated infection, fungemia, pneumonia, skin infection, eye infections, and catheter-related infection (18, 30, 32, 44, 77, 107, 111, 143, 155, 163, 193, 214, 245). Outbreaks have been associated with this mold; one was due to a contaminated solution that was used during intraocular lens implantation, and the other was associated with contaminated hand lotion (155, 163).

The identification of *P. varioti* and *P. lilacinus* to the species level is recommended, since these species are more likely to be associated with human disease and there is some evidence that there may be important differences in the antifungal susceptibility profiles of these fungi. *Paecilomyces varioti* has been reported to be susceptible to amphotericin B, whereas *P. lilacinus* is resistant. Voriconazole and other new broad-spectrum triazoles may be useful for the treatment of *P. lilacinus* infections (5, 56, 244).

Scopulariopsis. There are two species of *Scopulariopsis* that may cause disease in the immunocompromised host. These are the hyaline variant, *Scopulariopsis brevicaulis*, and the dematiaceous variant, *Scopulariopsis brumptii*. Although these were once thought of as either merely contaminants or causes of onychomycosis, they subsequently have been shown to be capable of causing serious fungal disease in humans, including pulmonary infections, sinusitis, and disseminated disease (190, 234). *Microascus* is the teleomorphic form of *Scopulariopsis*; this form has also been recovered from immunocompromised patients when cultures are allowed to incubate for two or more weeks at 30°C (13, 134, 222).

Zygomycosis (Mucormycosis)

There are two orders in the class Zygomycetes, namely, Mucorales and Entomophthorales. The Mucorales are of primary importance throughout the world, whereas the Entomophthorales cause disease primarily in Africa. The members of the Mucorales are the primary causes of zygomycosis in immunocompromised patients and are the focus of this discussion. These fungi are important causes of rapidly progressive disease. The most notable disease caused by the zygomycetes is rhinocerebral zygomycosis, a rapidly progressive and often fatal disease. Infection in this disease often begins in the paranasal sinus and/or palate, subsequently extending into the eye and brain. Rhinocerebral zygomycosis occurs most commonly in patients with severe or uncontrolled diabetes mellitus with ketoacidosis. The zygomycetes also cause invasive pulmonary infections in immunosuppressed patients and are known pathogens in the lungs of patients who have undergone pulmonary transplantation. Zygomycetes cause progressive nodular to cavitary lesions of the lung. The incidence of pulmonary disease caused by these fungi has increased significantly. These fungi may also cause wound and soft tissue infections and may invade the gastrointestinal tract. They thrive in necrotic tissues and are known to cause infection in patients with severe burns. Dissemination following localized infection may occur. Skin and soft tissue infections may be seen in the setting of profound immunosuppression or as complications of surgery. An increase in the incidence of zygomycete infections may be associated with the sole use of voriconazole for antifungal prophylaxis (220). Although voriconazole is an excellent new broad-spectrum triazole, it is not useful for treating zygomycosis (220).

The zygomycetes often fail to grow in culture. This is thought to be due to the pauciseptate nature of the hyphae of these molds. A disruption of the hyphal strand by too aggressive tissue processing will kill the fungal cells of these molds, and the culture will be negative for growth. Therefore, mincing versus grinding of tissue is preferred by some, particularly if the possibility of zygomycosis is raised. However, when zygomycetes grow, the colonies are readily apparent and the colony rapidly fills the petri plate. The presence of a wooly, rapidly growing mold is the first clue that a zygomycete may be present. Initial growth may often be detected the day following inoculation. The identification is then achieved using the microscopic morphologic features of the mold. The features of the sporangiophores (e.g., branched versus not branched) and the presence, absence, and location of rhizoids, among other features, are used for identification. The genera most commonly isolated are *Rhizopus* and *Mucor*, but several other members of the Mucorales have been reported to cause human disease. These include *Absidia*, *Apophysomyces*, and *Cunninghamella*, among others.

The members of the Mucorales are usually readily distinguished from most hyaline septate molds in the direct examination or in histologic sections, but occasionally, the hyphae of *Aspergillus flavus* may resemble those of the zygomycetes. Additionally, antifungal therapy may alter the morphology of the hyaline septate molds; globose forms may be seen, which may cause these fungi to be confused with the zygomycetes. The zygomycetes produce broad and twisted (i.e., ribbon-like) hyphae, with rare septations that may not be seen in a limited sampling. The hyphae are classically described as having 90° branching rather than 45° branching, such as that seen with *Aspergillus*. This, however, may be difficult to determine in tissue due to bending and folding of the hyphal strand. Of all the molds that may be seen in histologic sections, the zygomycetes stain the poorest with Gomori methenamine silver but are usually apparent in routine hematoxylin- and eosin-stained sections. Like *Aspergillus* species, these molds are angioinvasive and associated with tissue infarcts (55, 171). Perineural invasion has also been described and hypothesized as a potential conduit of spread (55).

Dematiaceous Molds

The dematiaceous molds are a heterogeneous group of fungi that contain melanin or melanin-like pigments in their cell walls. It is this pigment that makes them appear brown to black. These fungi cause black grain mycetoma, chromoblastomycosis, and phaeohyphomycosis (22). They also cause brain abscesses and phaeohyphomycotic cysts, which some consider variants of phaeohyphomycosis. Phaeohyphomycosis is the type of infection that more commonly occurs in the immunocompromised host.

An exhaustive coverage of the causative agents of phaeohyphomycosis is beyond the scope of this chapter, but a brief introduction to these fungi is provided. These fungi are often initially separated into slowly and rapidly growing molds. The slowly growing types often have a velvety colony morphology, whereas the rapid growers are commonly more wooly in appearance. Interestingly, some of the slowly growing variants, those belonging to the genus *Exophiala*, often produces yeast-like cells in early cultures. One of the most commonly isolated species, *Exophiala jeanselmei*, was reported to cause a phaeohyphomycotic cyst and invasive phaeohyphomycosis in a renal transplant recipient (34). Another species, *Exophiala dermatitidis*, has been documented to cause intravascular line-associated infections and is one of the neurotropic fungi that may disseminate to the central nervous system. Other neurotropic dematiaceous molds include *Cladophialophora bantiana*, *Ochrochonis* (*Dactylaria*) *gallopavum*, and *Ramichloridium mackenziei* (19). A publication by Revankar et al. describes 101 cases of cerebral phaeohyphomycosis, whereas Kantarcioglu et al. published a more limited number of cases caused by dematiaceous fungi (90, 179).

There are a number of more rapidly growing dematiaceous molds that should be considered potential pathogens in the appropriate host (Table 1). *Curvularia* species, *Bipolaris* species, and *Scedosporium prolificans* are a few of the more common rapidly growing dematiaceous molds that may be encountered and have been shown to cause serious infections in the immunocompromised host (3, 16, 22, 47, 64, 100, 157, 167, 224, 234). *Scedosporium prolificans* is particularly important to recognize, since it is multidrug resistant (164, 203).

The treatment of phaeohyphomycosis is often problematic because of innate resistance of many of the etiologic agents to antifungal agents (164). A combination of antifungal therapy and surgical excision is often needed for a complete cure. The triazoles may have superior activity for the treatment of infections caused by dematiaceous molds (65, 126, 127, 191). Additional studies are needed, but posaconazole has been shown to be effective in the treatment of patients with serious infections caused by dematiaceous fungi and could emerge as the drug of choice for this group of infections (145, 170).

EPIDEMIOLOGY

Apart from *Histoplasma capsulatum* and *Penicillium marneffei*, which are endemic to particular regions, the fungi that cause infections in the immunocompromised host are quite ubiquitous. In most instances, it is not possible to identify the source of the fungus that has caused an infection in an immunocompromised host, since these fungi are natural saprophytic molds that are widely distributed in nature. However, certain risk factors have been associated with infection. For example, we recently

Table 1. Filamentous fungi that medical mycology laboratories should be able to identify

Fungus	Comments
Hyaline septate molds	
More common molds	
Aspergillus species	*Aspergillus* species that must be recognized and reported include *A. fumigatus*, *A. niger*, *A. flavus*, and *A. terreus* It is important to recognize and report *A. terreus*, since this species is innately resistant to amphotericin B Other species that we encourage laboratories to gain expertise to identify and report include *A. glaucus*, *A. ustus*, *A. nidulans*, *A. versicolor*, and *A. clavatus*
Fusarium species	Species-level identification is not yet necessary, but this may change as we learn more about species-level differences with respect to antifungal susceptibility profiles These molds are often multidrug resistant
Pseudallescheria boydii (*Scedosporium apiospermum*)	This mold is innately resistant to amphotericin B There has been confusion regarding how to report this mold; foremost, the report should be consistent and understood by the primary health care provider. Since *P. boydii* is a heterothallic mold, all strains theoretically have the ability to form cleistothecia under the appropriate conditions
Dimorphic molds (*Histoplasma capsulatum*, *Blastomyces dermatitidis*, *Coccidioides immitis*, *Paracoccidioides brasiliensis*, *Penicillium marneffei*)	The systemic dimorphic fungi are always pathogens and should be reported promptly The correlation of the yeast or yeast-like form seen in the direct examination of the clinical specimen or in histologic or cytologic examination with the mold form in culture is an excellent means to presumptively identify these pathogens
Less common molds	
Acremonium species, *Paecilomyces*, *Scopulariopsis*, *Geotrichum*, dermatophytes, *Penicillium* species other than *P. marneffei*, *Scytalidium*	The other less commonly encountered molds may be pathogens, contaminants, or transient fungal flora; clinicopathologic correlation is required to elucidate the nature of these isolates Although dermatophytes overall are common, subcutaneous disease (i.e., a pseudomycetoma) that may present in the immunocompromised host is rarely seen Although the frequency of recovery of *Penicillium* species other than *P. marneffei* is relatively high, the vast majority of these are not associated with clinical disease
Zygomycetes	
More common molds	
Rhizopus species, *Mucor* species	Identification of these to the species level is not necessary Zygomycetes are not susceptible to voriconazole but are susceptible to posaconazole
Less common molds	
Rhizomucor, *Absidia*, *Syncephalastrum*, *Apophysomyces*, *Saksenaea*	Although infrequently encountered, these may cause rapidly progressive, fatal disease in the immunocompromised patient
Dematiaceous molds	
More common molds	
Curvularia, *Bipolaris*, *Scedosporium prolificans*, *Alternaria*	*Scedosporium prolificans* is multidrug resistant; infections with this mold are particularly difficult to treat *Alternaria* is most commonly encountered as a contaminant or transient fungal flora; however, in rare circumstances, it may be associated with an infection
Less common molds	
Exophiala species, *Cladophialophora*, *Phialophora*, *Fonsecaea*, *Ochroconis* (*Phoma*, *Chaetomium* [occasional contaminants], *Sporothrix schenckii*	The less commonly encountered dematiaceous molds may be pathogens, contaminants, or transient fungal flora; clinicopathologic correlation is required to elucidate the nature of these isolates

described a skin and soft tissue infection in a gardener in South Florida. His infection was caused by *Mycoleptodiscus indicus*, a plant pathogen of the cycad, of which he had many. He certainly contracted his infection through contact with infected plants or contaminated soil.

The individuals at risk for acquiring fungal infections who are considered in the scope of this text may be separated into those with particular, usually definable, defects in their immune systems (e.g., those with HIV infection) and those who have a more general suppression of their immune system because of a severe underlying disease (e.g., a patient in the intensive care unit who has severe burns), chemotherapy, or organ transplantation. Specific defects in immunity may be congenital, acquired, or iatrogenic. The most profound changes influencing the host immune response over the past few decades have been secondary to acquired immunodeficiency due to the spread of HIV and because of iatrogenic immunosuppression. Fungal infections also affect those individuals with congenital immunosuppression. Other people who may develop serious fungal infections have functionally suppressed immune systems secondary to extensive burns, complicated surgical procedures, trauma, diabetes mellitus, or long-term indwelling catheters. These subgroups are discussed briefly.

HIV-Associated Fungal Disease

The HIV epidemic was heralded by an increased incidence of a number of opportunistic infections and unusual neoplasms. Patients with HIV infection become infected by the same types of fungi that cause infections of patients with T-cell depletion or defects due to other causes. These are predominantly the yeasts *Candida albicans* and *Cryptococcus neoformans* and the unusual fungus *Pneumocystis jirovecii*, which are covered elsewhere. Interestingly, there has not been a comparable increase in infections caused by some of the more important hyalohyphomycetes (e.g., *Aspergillus* and *Fusarium*) or the zygomycetes in this group of patients.

The two molds that are important causes of infections in patients with advanced HIV infection/AIDS are the endemic dimorphic fungal pathogens *Histoplasma capsulatum* and *Penicillium marneffei*. *Histoplasma capsulatum* is likely the most widespread of the endemic fungi. It has been reported from at least 55 countries throughout the world and in North America; it is endemic to the Ohio and Mississippi river valleys (241). Conversely, *Penicillium marneffei* is a geographically limited fungus that is found in Thailand and Southeast Asia; in these areas, the incidence of penicilliosis parallels that of HIV infection (202, 209). Infection usually presents as disseminated disease with fungemia and involvement of multiple organ systems. The clinical manifestations of histoplasmosis and penicilliosis have been reviewed and compared (42, 136, 209, 241).

Cancer/Neutropenia/Stem Cell Transplantation/Graft-versus-Host Disease

Although patients with cancer or neutropenia, stem cell transplant recipients, and those with graft-versus-host disease make up a heterogeneous group, many patients have a hematologic malignancy and require chemotherapy. The bone marrow changes in patients undergoing treatment for malignancies are diverse, and the status of their immunity depends to a large extent upon the particular neoplastic process and the type of therapy required. The bone marrow changes range from mild suppression of hematopoiesis to complete bone marrow ablation. Patients with significant bone marrow suppression and, particularly, neutropenia have a significantly elevated risk for fungal infections (83, 108, 118, 120, 196). Those who receive a hematopoietic stem cell (bone marrow) transplant and develop graft-versus-host disease are also at risk for fungal infections (21). In addition to the immunosuppression due to therapy, these patients are often critically ill, which results in further immunosuppression, and may have some type of long-term indwelling catheter that may become infected.

Aspergillus and *Candida* species are the types of fungi that most commonly infect this patient population (118, 120, 232, 234). However, differences among institutions may influence the types of patients seen. The type of antifungal prophylaxis used and the types of fungi that are prevalent in the environment also influence the types of infecting fungi isolated. A variety of filamentous fungi may cause infections in these patients. Some of the other commonly encountered fungi include *Fusarium* species, *P. boydii*, and zygomycetes, particularly *Rhizopus* and *Mucor* species; occasionally, one may encounter infections caused by the dematiaceous fungi in this population (13, 48, 96, 148, 234). It is important to recognize that virtually any fungus, even those with low intrinsic virulence, may cause disease in patients with profound neutropenia.

Solid Organ Transplantation

Solid organ transplant recipients, too, are a very heterogeneous group and are consolidated here for this basic overview. A variety of factors increase the risk of a solid organ transplant recipient contracting fungal infection. The risk associated with surgery arises from the complexity of the surgical procedure and complications that may arise in the surgical intensive care unit during the postoperative period (197). In addition to a complicated surgery, these patients may have extended stays in an intensive care unit, may have long-term indwelling catheters, and often receive prophylactic broad-spectrum antibacterial antibiotics that alter their normal microbiota. All of these factors, combined with the immunosuppressive agents given to control rejection of the transplanted organ, place these patients at high risk for opportunistic fungal infections (39, 99, 195, 216).

There are great differences in the immunosuppressive regimens used to control transplant rejection. These are given at the discretion of the transplant physician and are selected based on the experience of the attending physician, the type of organ that has been transplanted, and the individual patient's medical profile. The mechanisms of action of the various immunosuppressive agents used are beyond the scope of this text, but many of these increase the patient's risk of developing a fungal infection to some extent (11, 76, 99, 152). A thorough knowledge of these mechanisms and associated risks for opportunistic infections is requisite for the clinician to be able to appropriately manage the health care of a transplant recipient.

Similarly, the organisms most commonly associated with infection in patients with solid organ transplants are *Aspergillus* and *Candida* species (25, 26, 29, 35, 39, 82, 89, 108, 120, 150, 158, 196, 200, 201, 234). Infections with *Candida* and other yeasts or yeast-like fungi are discussed elsewhere. Invasive aspergillosis usually manifests with lung involvement (198, 218). The most feared complication is dissemination of the infection, which can go to any organ system but is most frequently fatal when it involves the brain. Solid organ transplant recipients may also develop infections with a variety of other fungi, including *Fusarium* and other hyaline molds, members of the zygomycetes, and occasionally the dematiaceous fungi (83, 85-87, 171, 184, 199).

Corticosteroids and Other Causes of Immunosuppression

Corticosteroids have proven extremely useful for a wide variety of conditions, but these leave the host susceptible to fungal infections (80, 109, 176). Infections seen in patients receiving long-term corticosteroid therapy include invasive aspergillosis, invasive hyalohyphomycosis caused by other hyaline septate molds, such as *P. boydii* and *Fusarium*, and cryptococcosis (covered elsewhere), amongst others (99, 234). Corticosteroid therapy has also been shown to be an important risk factor for the development of zygomycosis (96, 105, 148, 159, 171).

Newer therapeutic immunomodulating drugs, such as infliximab, a tumor necrosis factor alpha inhibitor, have been developed and are useful for controlling diseases such as severe rheumatoid arthritis, fistulizing inflammatory bowel disease, and other disorders (23, 110, 169, 185). Although it is useful in controlling these diseases, this type of inflammatory blockade has been associated with disseminated (miliary) tuberculosis, disseminated histoplasmosis, aspergillosis, and pneumocystosis (142, 238). The degree of risk and types of infection to which even the newer immunomodulating agents predispose patients will have to be determined through careful monitoring.

The Critically Ill Patient: Intensive Care Unit/Burns/Surgery/Trauma

Persons without specifically identifiable defects in their cellular or humoral immune systems may also develop fungal infections. The empiric "coverage" of patients in the intensive care unit and during surgery with broad-spectrum antibacterial agents is common and necessary to avert bacterial infections. However, an imbalance of the normal microbiota (flora) often results secondary to treatment with broad-spectrum antibacterial agents. This imbalance creates an environment that increases the risk of fungal infection (6, 187). In addition, these patients often have a breech in their innate host defenses, such as their skin or mucous membranes. Like the normal microbiota, the skin and mucous membranes are innate barriers of defense against invasion by microorganisms. Breeches in these barriers are associated with bacterial and fungal infections. Patients with severe burns, who have large areas of their epidermis denuded, are at high risk for both bacterial and fungal infections (14, 45, 48, 103, 173). A variety of filamentous molds have been described to cause infections in patients with severe burns. These include *Fusarium* and members of the zygomycetes (75, 242). Not surprisingly, fungal infections may also begin at sites of trauma or surgery (12). Traumatic wounds that are heavily contaminated with environmental debris are most likely to become infected (17, 75, 174).

Patients in the intensive care unit are often severely ill and sometimes must be cared for over extended periods. It is well described that the critically ill patient is at heightened risk for the development of fungal infections (6, 49, 116). These patients are thought to have nonspecific immunosuppression due to the severity of their illness. For example, it is well known that patients with uncontrolled severe diabetes mellitus and ketoacidosis are at high risk for developing often fatal rhinocerebral zygomycosis (80, 171). Many patients who are severely ill and/or in intensive care units have long-term intravascular catheters (187). The entrance sites for catheters, like the breeches in the mucosa described above, represent foci in the skin into which fungi may be introduced. These catheters may become colonized by fungi, and patients may develop fungemia and have infection of the surrounding soft tissue (37, 183, 234). Hematogenous spread from these sites may result in dissemination, with deep-seated fungal abscesses and/or fungal endocarditis. Similarly, peritoneal dialysis catheters are also important and common sites where fungi may be introduced into a normally sterile site and cause infection.

SPECTRUM OF DISEASE, INCLUDING END ORGAN INVOLVEMENT

Any of a number of opportunistic fungal infections may be seen in a severely immunocompromised host. Certain fungi are usually associated with a particular immunologic deficit and often infect a particular organ. For example (although covered in more detail elsewhere), cryptococcosis commonly presents as meningitis, whereas pneumocystosis causes pneumonia. The filamentous fungi need a portal of entry to gain access to the host. This is usually the lung, so pulmonary infections are most commonly seen.

These may spread via the bloodstream to any organ. In some instances, skin lesions (e.g., *Fusarium* infections) may be the presenting manifestation of a deep-seated infection. Dissemination to the brain is the most feared complication, since swelling in this limited space may cause herniation and death.

The paranasal sinuses are another portal of entry, and invasive disease may follow fungal sinusitis or a fungus ball. Fungi may cause wound infections following trauma and/or surgery. They may also gain access through disrupted mucosal barriers and at catheterization sites. Locally invasive disease and/or disseminated disease to any organ may follow, depending upon the ability to control the infection through surgery, reconstitution of the immune system, and antifungal therapy.

MORBIDITY, MORTALITY, AND PROGNOSTIC FACTORS

There is significant morbidity and mortality associated with fungal infections in the immunocompromised host (52, 115, 234). In many parts of the world, there has been a significant decrease in the incidence of fungal diseases associated with advanced HIV infection, largely because of the availability and use of highly active antiretroviral therapy. Similarly, advances in antifungal prophylaxis, most notably with newer antifungal agents, have helped, to some extent, to prevent fungal infections in other immunocompromised hosts. However, in resource-poor countries, medications to treat HIV infection and the newer antifungal agents may not be available. The outcomes of fungal diseases in the immunocompromised host have been improved somewhat by using newer, more effective antifungal drugs for prophylaxis and therapy, but many improvements are still needed.

The factors that are traditionally associated with the outcome of invasive fungal infections in the immunocompromised host concern host factors, the fungus, and the antifungal agents given. However, the time to diagnosis and diagnostic accuracy are also factors that likely influence the outcome in these patients and are variables that will be analyzed critically in future studies. Diagnostic modalities that offer identification of fungi to a taxonomic level that predicts likely antifungal susceptibility will prove particularly attractive. For example, although an assay that detects the presence of *Aspergillus fumigatus* may prove useful, one that detects all *Aspergillus* species and differentiates the amphotericin B-resistant species *A. terreus* from the other *Aspergillus* species would be optimal.

The restoration of the host's immune system to normal or as close to normal as possible is critical. This is particularly true for severe neutropenia (251). Invasive fungal infections are often associated with vascular invasion and tissue infarcts. Thrombosis of the affected blood vessels and associated tissue necrosis limit the ability to adequately deliver antifungal drugs to infected tissues. Therefore, surgical resection of infected tissues is often needed in conjunction with appropriate antifungal therapy. The biology of the fungus itself also has prognostic significance. The rapid growth of zygomycetes makes them particularly lethal, especially when infections occur in the paranasal sinuses and palate. In addition, highly resistant molds, such as *Scedosporium prolificans*, are not likely to be cured by antifungal therapy alone. The appropriate selection of the antifungal agent used based on the antifungal resistance pattern of the mold causing the infection is an important prognostic factor. For example, in the past, when all patients with invasive aspergillosis were treated with amphotericin B, a poor outcome was demonstrable for patients infected with *Aspergillus terreus* (i.e., the species innately resistant to amphotericin B) compared with the outcome for those infected by another *Aspergillus* species (102). The correct identification of the fungus causing the infection is the first step in the selection of the appropriate antifungal agent, since certain fungi are known to be resistant to certain antifungal agents.

THERAPEUTIC CONSIDERATIONS AND ANTIFUNGAL SUSCEPTIBILITY ISSUES

The development of resistance to antimicrobial agents is distinctly different in the fungi from that in the bacteria. Antifungal profiles can often be assumed based on the identity of the fungus present. However, as with any microorganism, the overuse of any antimicrobial agent will naturally select for resistant clones in the population. Fortunately, in this era of the immunocompromised host, the past decade has witnessed the release of many new, highly effective antifungal agents. However, as with any therapeutic agent, there are certain microorganisms that are susceptible and others that are resistant to each antifungal agent. The important known resistance profiles among the more commonly encountered filamentous fungi are presented in Table 1. Although there are situations where antifungal susceptibility testing may be warranted, the limitations of such testing (particularly the paucity of data correlating in vitro results with clinical therapeutic efficacy) should be recognized.

It is important to recognize that antifungal susceptibility testing of molds is considerably more difficult to perform than susceptibility testing of yeasts or bacteria. Our recommendation is that antifungal susceptibility testing of filamentous fungi should not be attempted by laboratories that plan to provide this testing on a limited basis. It is best for most laboratories that do not have a dedicated interest in antifungal susceptibility testing for molds or a sufficient quantity to maintain proficiency to send the isolates to a qualified reference laboratory for testing. Clinical Laboratory Standards Institute (CLSI) document 38-A addresses the standard broth dilution methods for determining antifungal resistance in the filamentous fungi (165). Alternatively, some authors have shown the utility of other methods, such as the Etest, for determining antifungal resistance in the filamentous fungi (28, 101, 166). If non-CLSI methods are verified, then these should be standardized against results obtained using CLSI standard methods.

GOALS OF LABORATORY TESTING

Pathology and laboratory medicine play a critical role in the diagnosis and management of patients with immunocompromising conditions. Newer laboratory tests are being used to screen patients who are at high risk for invasive mold infections so that preemptive therapy may be used to prevent the development of invasive disease. Once a disease becomes established, the laboratorian has the responsibility to determine the etiologic agent of infection and to aid in determining the extent of disease, both of which have prognostic value. Finally, accurate identification and, if needed, an antifungal susceptibility profile may be provided to guide therapy.

Screening and Prevention

Prophylactic therapy is commonly used to help prevent fungal infections in patient populations who are at considerable risk. Prophylactic therapy is necessary when diagnostic and monitoring tools are unavailable or insufficient to predict which patients are developing or likely to develop disease. Although useful in many regards, one concern about prophylactic versus preemptive therapy is that the dosage of the drug used in prophylaxis is often considerably less than that used to treat an established infection. Subtherapeutic levels of a drug may be nonlethal to the organisms present, promoting antimicrobial resistance. In addition, when prophylactic therapy is used, many patients who do not have and would likely not develop infections may be treated; this has implications with respect to drug side effects and cost.

Parallels can be drawn with the early days of transplantation, wherein prophylactic rather than preemptive therapy was given for cytomegalovirus (CMV) disease. This was necessary since the laboratory was unable to provide information necessary to enable the use of preemptive therapy. However, with advances in molecular diagnostics (i.e., CMV viral load monitoring), preemptive therapy has become the standard of care in many institutions. The CMV viral load of patients at risk is routinely monitored, and those with rising viral loads are preemptively treated to curtail incipient CMV disease. This may be what the future holds for early detection and treatment of invasive fungal infections. However, well-designed clinical trials are necessary to determine which treatment approach (i.e., preemptive versus prophylactic) is superior. Whether we will screen for and monitor fungal antigens, fungal DNA, or human determinants of disease remains to be determined (see "Antigen Detection Assays" and "Nucleic Acid Amplification-Based Assays" below).

Diagnosis and Prognostication

The definitive diagnosis of an invasive mold infection is not possible without the laboratory. However, there are many clinical and radiologic findings (e.g., the air crescent sign) that can be highly suggestive of an invasive fungal infection, particularly in the appropriate clinical setting. The definitive diagnosis of an invasive fungal infection is usually achieved with the finding of a positive culture and/or the demonstration of fungal elements in tissue or body fluids by morphologic examination. More recently, the demonstration of an elevated or rising amount of a fungus-associated antigen (e.g., galactomannan) has been used to support the diagnosis of an invasive fungal infection. Similarly, the presence of fungal DNA could also be used to support such a diagnosis, but nucleic acid-based tests for this purpose are not yet commercially available, and extensive studies will be required before these tests are used routinely.

The diagnosis of disseminated histoplasmosis may necessitate the use of many diagnostic tests, so guidance is given here. If the possibility of disseminated histoplasmosis is considered, then blood or bone marrow biopsy culture, histologic examination, and histoplasma urinary antigen testing are warranted. A blood culture should be collected specifically for *Histoplasma*, and this should be obtained by using an Isolator lysis-centrifugation tube (Remel, Lenexa, KS). Although lytic blood culture bottles may also recover the organism, the sensitivities of these systems have never been proven to be equivalent to that of the Isolator system for the recovery of *H. capsulatum* (231). If disseminated histoplasmosis is suspected, then a *Histoplasma* urinary antigen test (MiraVista Laboratories, Indianapolis, IN) should also be performed. This assay is more rapid than culture, since *H. capsulatum* is a slow-growing mold, and is very sensitive in the setting of disseminated disease. However, false-positive test results may occur in patients with disease caused by one of the other endemic fungi or because of the presence of cross-reacting substances (e.g., the original *Histoplasma* antigen test demonstrated cross-reactivity in specimens from patients who had received rabbit antithymocyte globulin) (40). Real-time PCR assays for *H. capsulatum* have been described and are highly specific, and they have been shown to be positive with blood or bronchoalveolar lavage samples or with formalin-fixed, paraffin-embedded tissues from anecdotal cases (70, 121). These assays, although perhaps useful in the future or in particular cases, have yet to be examined in a well-designed clinical trial.

Treatment and Monitoring

Patients with invasive fungal infections are treated by use of antimicrobial agents that particularly target elements or enzymatic processes found in the fungal cell but not in human cells. The development of such therapeutics is more challenging for fungi, which are eukaryotic (like humans), than for the prokaryotic bacteria, for which there are more substantial differences. Although it is challenging, we are in an era wherein many new antifungal agents, some representing new classes of drugs (e.g., echinocandins), have been released (Table 2). These hold great promise for curing many patients with serious fungal infections. These agents, however, have discrete spectra of activity that have been and continue to be defined. The accurate identification of the etiologic agent of a mycotic disease is more important than ever, since certain antifungal drugs are ineffective or less effective against certain fungi.

The monitoring of patients with invasive fungal infections has remained largely a clinical and radiologic function. The vital signs, temperature, and any other clinical signs that may be associated with a particular infection are monitored to assess improvement of the individual patient to the antifungal therapy. Radiologic studies are critical to document the sizes of lesions so they may be monitored for signs of progression (i.e., enlargement) or resolution. The traditional laboratory values that are monitored are indirect. For example, the complete blood count may be the most important parameter to monitor for bone marrow transplant recipients to document their recovery from profound neutropenia. Similarly, the CD4 T-cell subsets and viral load may be the most important indicators to monitor for patients with HIV infection. More recently, we have gained the ability to monitor antigens released from the fungal cell, whether these are relatively specific antigens (e.g., galactomannan of *Aspergillus* or *Histoplasma* antigen) or more generic fungal antigens (e.g., beta-glucan). The monitoring of fungus-related antigens gives health care providers a new opportunity to monitor the response to therapy and to identify potential relapses. Antigen levels have been shown to decline with effective therapy. Significant rises in the level of antigen following an initial decline could be due to a relapse (see "Antigen Detec-

Table 2. Strengths and limitations of selected antifungal agents

Agent and class	Strengths	Limitations	Comments
Polyene			
Amphotericin B	Broad spectrum of antifungal activity	Resistant fungi include *Trichosporon* species, *Pseudallescheria boydii*, many strains of *Fusarium*, and *Aspergillus terreus* Renal toxicity	Acts by binding to ergosterol, a sterol in fungal cell membranes, which results in osmotic leakage Toxicity is significantly reduced with lipid formulations
Azoles			
Fluconazole	Useful primarily for yeasts, with notable exceptions (see Chapter 11)	Not useful for most molds (some exceptions exist for some of the dimorphic fungi)	Azoles work through the inhibition of ergosterol synthesis by inhibiting the enzyme 14-α-demethylase
Itraconazole	Expanded activity to cover more yeasts and several filamentous fungi, including dimorphic molds	Does not adequately cover *Fusarium* species or the zygomycetes	MICs are not as low as with the newer triazoles
Voriconazole	Expanded coverage compared with itraconazole Covers fluconazole-resistant yeast, dimorphic molds, and many filamentous fungi, including many *Fusarium* strains	Does not cover the zygomycetes Some *Fusarium* strains remain resistant	Visual disturbances are an important side effect; these are time limited but may cause significant concern for patients unless they are warned
Posaconazole	Expanded antifungal coverage, even compared with voriconazole; this includes activity against zygomycetes and dematiaceous fungi	Some *Fusarium* strains remain resistant	Only oral formulations are currently available, but absorption is good
Echinocandins			
Caspofungin	Good activity against *Candida* species and *Aspergillus* species	Limited activity against *Fusarium* and *Rhizopus*; importantly, no activity against *Cryptococcus neoformans*	The echinocandins are inhibitors of glucan synthesis, an important component of the fungal cell wall
Micafungin	Similar to caspofungin, but moderate activity against certain dematiaceous fungi	No activity against *Fusarium solani*, *Pseudallescheria boydii*, or the zygomycetes	Although it has in vitro activity against the mycelial form of the dimorphic fungi, it is inactive against the yeast forms/spherules produced in vivo
Anidulafungin	Similar to the above regarding the *Candida* species, *C. neoformans*, and *Aspergillus* species	Limited activity against *Fusarium*, *Acremonium*, *Rhizopus*, and several other fungi	

tion Assays"). The monitoring of fungal DNA has not yet been demonstrated to be useful in this manner (i.e., similar to how CMV or HIV viral load testing is used), but this is another potential means for monitoring patients at risk for fungal infections.

LABORATORY TESTING

Culture

Direct examination and culture are the standard microbiologic methods of assessing tissues and body fluids for the presence of fungi. The direct examination is a critical part of this process, as it quickly provides information to the clinician upon which therapeutic decisions can be made. This is usually performed using a combination of potassium hydroxide (KOH) and calcofluor white. The KOH digests the human tissues, whereas the calcofluor white binds to the fungal cell walls, making them more easily visualized with fluorescence microscopy. Although the sensitivity of the direct examination is lower than that of culture, it still has value, particularly when fungal elements are seen (i.e., the positive predictive value is high when the direct examination is performed by an experienced microscopist, whereas the negative predictive value is not high).

Clinical specimens, apart from most blood cultures, are plated directly onto a variety of fungal media, but many fungi, including molds, may also be recovered from standard bacteriologic media. An extensive discussion regarding the types of media is beyond the scope of this text, but two points deserve further discussion. Media with antibacterial antibiotics should be included in the selection of media used for specimens that are likely to be contaminated with bacterial flora (e.g., respiratory specimens). In addition, the inclusion of a cycloheximide-containing medium should be considered to inhibit the overgrowth of culture plates by contaminating environmental fungi, such as basidiomycetes. If a cycloheximide-containing medium is used, then it is important to also include another standard fungal medium that does not contain cycloheximide, since some molds that are inhibited by cycloheximide may be true pathogens in the immunocompromised host. For example, *Aspergillus* species and members of the zygomycetes are common pathogens in the immunocompromised host and are inhibited by cycloheximide.

In considering the results of assays that test for fungal pathogens, one must consider a number of factors, whether one is interpreting culture results or data from a molecular test (see below). Foremost, one must consider the identity of the fungus isolated. Is the isolated mold a common pathogen, such as *A. fumigatus*, or an infrequent pathogen (e.g., *Chaetomium*) that would cause disease only in the most severely immunosuppressed patient? One should also consider the number of cultures or culture plates that are growing the organism. The greater the number of plates that are positive, the more likely it is that the isolate is a true pathogen. It is also important to determine if fungal elements were also seen in the direct examination of the specimen (i.e., the calcofluor-KOH preparation) and, if possible, to correlate the culture results with histopathologic and/or cytologic findings. One should also bear in mind the types of normal fungal flora that may be present, even transiently, in certain clinical specimens. The presence of a single colony of a mold in a respiratory specimen may simply represent the presence of transient fungal flora or a plate contaminant; however, it could also (rarely) be evidence of an invasive fungal infection in an immunocompromised host. Herein lies the importance of the correlation of laboratory, radiologic, pathologic, and clinical findings. The isolation of a known pathogen, particularly in adequate quantity and/or from a normally sterile site, usually represents a true infection.

Every mycology laboratory should be able to definitively identify certain filamentous fungi (Table 1). Smaller laboratories may not maintain the materials, such as genetic probes, necessary to achieve the timely identification of the dimorphic fungi. In such instances, the laboratorian should be able to recognize key morphologic features that raise the possibility of these fungi, and the isolate should promptly be sent to a reference laboratory for identification.

The first examination of the colony gives one an impression of the growth rate of the fungus. Although this is true and there are general impressions that can be made, there are a variety of factors that affect growth rate and time to recovery, such as antifungal therapy and the quantity of fungus present in the specimen. For example, zygomycetes are very rapidly growing molds, whereas the fungi that cause chromoblastomycosis (i.e., *Fonsecea*, *Chladophilaphora*, and *Phialophora*) are slowly growing molds. Examination of the surface of the colony may give an indication of the color of the spores produced, whereas an examination of the reverse of the colony (i.e., the hyphal mat) often affords the separation of the mold into dematiaceous versus hyaline categories. Microscopic examination usually confirms the suspicion that the mold under examination is a zygomycete, a hyaline septate mold, or a dematiaceous mold. Microscopic examination reveals the spore morphology, the conidiophore, and the method of conidiogenesis. These features, when studied with identification keys and atlases, should afford the identification of the most commonly encountered filamentous fungi. Occasionally, ancillary testing is necessary to differentiate morphologically similar fungi. For example, the two most commonly encountered *Exophiala* species, *E. jeanselmei* and *E. dermatitidis*, are morphologically identical. However, *E. jeanselmei* has the ability to utilize nitrate and has inhibited growth at 42°C, whereas *E. dermatitidis* cannot utilize nitrate but grows at 42°C.

Anatomic Pathology

The anatomic pathologist often plays a critical role in the diagnosis of fungal infections. The evaluation of excised tissues and body fluids is similar in many regards to the direct examination methods discussed above. The histopathologist, however, has the opportunity to see the infecting organism in situ. The presence of mold forms in tissue, particularly in association with tissue destruction and inflammation (sometimes lacking in this patient group), can help to differentiate true infection from colonization or contamination when a culture is positive, particularly for an uncommonly encountered fungus. Such determinations are not always possible based on the results of culture. For example, a single colony of an unusual mold that is recovered in culture could represent either a plate contaminant or the agent of disease. In addition, the type of inflammatory response (if present), as well as the morphology of the fungus present, may be useful for appropriately characterizing the infecting agent. It is not possible in most instances to characterize the mold present in a tissue or cytologic preparation to the same degree as that from a culture, but some categorization is possible. The pathologist usually can categorize an invasive mold infection as a hyalohyphomycosis (invasive disease caused by a hyaline septate mold), a zygomycosis (disease caused by a zygomycete), or a phaeohyphomycosis (invasive disease caused by a dematiaceous mold).

Unfortunately, many pathologists characterize all hyalohyphomycoses as aspergillosis, based on the presence of hyaline septate hyphae with acute-angle branching. This characterization is unacceptable and dangerous, since some may take it to imply susceptibility to amphotericin B. *Fusarium*, *P. boydii*, and other hyaline septate molds have morphologic features that significantly overlap with those of *Aspergillus* species. One can definitively identify an *Aspergillus* species by morphologic means alone only if an aspergillum structure is found. This finding is more common in fungus balls and would be seen in invasive disease only if an air space, such as a bronchus, were involved.

Zygomycetes are classically characterized by the presence of broad, "ribbon-like" hyphae that contain few to no discernible septations. Occasionally, isolates of *Aspergillus flavus* may demonstrate a morphology that is difficult to distinguish from that of a zygomycete in tissue sections. Dematiaceous molds causing phaeohyphomycosis appear as pigmented, septate hyphae. The pigment may not be pronounced but can be demonstrated using the Fontana-Masson stain or, sometimes, by examining unstained preparations. *Exophiala* species causing phaeohyphomycotic cysts appear more as yeast forms and pseudohypha-like forms but are dematiaceous. The dimorphic molds *Histoplasma capsulatum* and *Penicillium marneffei* appear as small budding yeast cells and small yeast-like structures with internal septations, respectively. Further differentiation of molds in histologic sections and cytology preparations is possible using adjunctive tools, such as immunohistochemistry, in situ hybridization, and PCR.

Antigen Detection Assays

The detection of fungal antigens has proven useful to identify and/or detect several types of fungi and holds promise for future diagnostics. One of the early uses of antigen detection was for the identification of cultured fungi suspected of being dimorphic fungi. Specific antibodies were used to detect particular fungal antigens in an immunodiffusion-type format (i.e., the exoantigen test). Although highly effective, this method has largely been replaced by identification using genetic probes and other molecular methods. However, these tests remain a viable alternative for laboratories that do not have access to molecular methods. These tests are highly reliable with appropriate training and experience. It is important for the user to be able to recognize true lines of identity and to differentiate these from nonidentity and nonspecific precipitants. Likely the most effective use of antigen testing to date has been the development of the antigen detection assays used to detect *Cryptococcus neoformans* in the cerebrospinal fluid (see Chapter 11).

The detection of *Histoplasma capsulatum* and *Blastomyces dermatitidis* antigens in serum and urine specimens is useful for the identification of disseminated disease and acute pulmonary disease, which are seen particularly in immunocompromised hosts. Unfortunately, these tests may be misused by some clinicians, affecting the predictive values of the assays. With the appropriate patient population, the sensitivity is very good and the specificity is acceptable, with the caveat that there is known cross-reactivity amongst the dimorphic fungi. The *Histoplasma* urinary antigen test is particularly useful for immunocompromised patients suspected of having disseminated disease.

Antigen detection is an important and relatively new tool in the detection of invasive aspergillosis in the immunocompromised patient. An enzyme immunoassay-based assay for detection of galactomannan has been marketed in Europe for several years and has more recently become available on the U.S. market. This assay targets a heat-stable heteropolysaccharide cell wall constituent of *Aspergillus* species, using a β-1,5-galactofuranose-specific monoclonal antibody in a sandwich enzyme immunoassay microtiter plate format (206, 207). Studies of galactomannan expression and of assay performance characteristics have demonstrated a wide range of sensitivities, specificities, and clinical predictive values, with sensitivities ranging from 29 to 100% and specificities typically >85%, depending on underlying illnesses of the patients being tested, patient age, treatment with mold-active antifungal agents, laboratory technique, and interpretative criteria for determining positive results, among other variables (79, 128, 131, 225, 239). The assay is marketed for detection of galactomannan in serum samples, and the majority of published data relate to this specimen type.

The sensitivity of galactomannan testing in the adult allogeneic hematopoietic stem cell transplant population appears most encouraging, with results of a recent meta-analysis demonstrating an overall sensitivity of 82% and a specificity of 86% (167).

Performance characteristics also appear favorable among patients with hematologic malignancies. However, limited studies among solid organ transplant recipients demonstrate a much lower sensitivity (22% in the meta-analysis just cited), although specificity appears to be maintained. The relative lack of sensitivity may have to do with reduced degrees of immunosuppression and more effective clearance of circulating antigen in the latter patient group and in other populations, such as those with chronic granulomatous disease and Job's syndrome (79, 226). Detection of galactomannan may also be reduced in patients receiving concomitant mold-active antifungal therapy (119).

The specificity of the assay, while generally reported as 85 to 95%, may be reduced due to a number of potential causes. The most commonly cited reasons for false-positive results with this test appear to involve the use of piperacillin-tazobactam and other β-lactams, such as amoxicillin-clavulanate, which have been shown to sometimes produce falsely elevated galactomannan levels (2, 125). Such false-positive results may be difficult to recognize. It is important, as always, to correlate positive findings with the patient's clinical picture. Positive findings that are temporally correlated with the start of treatment with one of these antimicrobials should be viewed with particular caution. In addition, while galactomannan appears to be present at especially high levels in *Aspergillus* and *Penicillium* species, other fungal agents have been shown to produce lesser amounts of the antigen and may cross-react in the assay (43, 61, 91, 240). While many of these organisms are more uncommon causes of human disease than *Aspergillus*, the possibility of confounding infection with other fungal agents should always be a clinical consideration.

Specificity has also been suggested as a concern in the pediatric population, which has been less well studied than adults with respect to galactomannan expression and the clinical predictive value of antigen testing (72, 194, 208). Initial reports seemed to indicate a potentially higher rate of false-positive results among the pediatric group. However, these findings have not been replicated by all investigators, and others have found sensitivities and specificities similar to those seen for adults (71). Neonates remain a population potentially at higher risk for false-positive results, possibly due to increased gastrointestinal permeability, gastrointestinal colonization with *Bifidobacterium* sp., and dietary sources of galactomannan (60, 130).

The clinical use of galactomannan testing has been evaluated primarily as a screening tool. When performed at least twice a week, testing of sera from highly immunocompromised patients (particularly hematopoietic stem cell transplant recipients and patients with hematologic malignancies) has been shown to have a high degree of sensitivity for invasive aspergillosis and to provide diagnostic evidence of infection prior to clinical or radiographic evidence of disease (114, 208). In addition, there is evidence that serial monitoring of galactomannan can be correlated with the response to therapy and may provide important prognostic information for such patients (24, 225, 249). The use of this assay for single-time-point diagnostic purposes rather than for screening of patients has not been well studied. Therefore, one should cautiously interpret isolated testing results and should preferentially look for increases in circulating antigen over time in order to improve the clinical predictive value.

The use of this methodology with nonserum specimen types has also been studied (95). The most well-characterized nonserum specimen is bronchoalveolar lavage fluid; a number of studies have shown promise in the use of galactomannan testing as a diagnostic tool for patients undergoing bronchoalveolar lavage. Sensitivities have been reported to range from 47 to 100%, and specificities are even more variable. As with serum testing, performance characteristics appear to vary depending on patient population, transplant type, and underlying disease (36, 84, 141, 186, 227). Data from any single patient population remain limited. Unlike the case with serum-based testing, airway colonization may result in false-positive results, with one study showing a particularly high rate of false-positive results among lung transplant recipients (36). Detection of galactomannan in the cerebrospinal fluid has also been suggested as a means of diagnosing cerebral aspergillosis. A few studies have shown promise in this area, although the number of cases studied remains small (228, 229). Data are insufficient for other specimen types to assess the potential value of galactomannan detection. It is important to remember that testing of cerebrospinal fluid or bronchoalveolar lavage specimens should be considered an off-label use of the method, to be performed only with substantial in-house validation; results from these specimens should be interpreted with an added degree of caution.

Another fungal cell wall constituent, (1,3)-β-D-glucan (BG), can also be used to diagnose invasive aspergillosis as well as other invasive fungal infections. Detection of BG is typically based on activation of the horseshoe crab coagulation cascade, in a manner similar to that used for detection of endotoxin (93, 153). This methodology has been marketed primarily in Japan for a number of years, with early studies appearing primarily in the Japanese literature. More recently, BG testing has become available more widely, with a U.S. FDA-cleared assay now available. BG is not specific for aspergillosis and can be used as a marker of fungal infection by a wide variety of pathogens. Importantly, this list excludes species of *Cryptococcus* and zygomycetes, which appear to express very low levels of this antigen. Although BG assays have not been widely available for as long as tests targeting galactomannan, the literature suggests similar ranges of sensitivity and specificity for invasive aspergillosis in highly immunocompromised patient populations, with published sensitivities of 55 to 100% and specificities of 72 to 100% (92, 131, 151, 153, 168). Few data are available for patients outside the hematopoietic stem cell transplant and hematologic malignancy groups, so it is difficult to say whether sensitivity is diminished in a manner similar to that of galactomannan tests. Like that of galactomannan testing, it has been suggested that the sensitivity of BG testing may be reduced by mold-active antifungal therapy (156).

Specificity of BG testing may also be diminished in a number of cases. Hemodialysis with cellulose membranes, intravenous treatment with immunoglobulin, albumin, coagulation factors, and other plasma protein fractions may result in false-positive results (131). Antimicrobials may also play a role in false-positive results, with one recent report associating amoxicillin-clavulanic acid with detection of BG (129). Importantly, early studies have shown possible cross-reactivity in the presence of bacteremia (168). In limited studies, upper airway or urinary colonization did not seem to give false-positive results (151). No studies have targeted the pediatric population in assessing the performance characteristics of the BG assay, and minimal data exist on evaluating this test for use on specimens other than plasma.

Nucleic Acid Amplification-Based Assays

The precise place where nucleic acid testing will fit in the routine evaluation of the immunocompromised host remains to be determined, but it holds promise and will likely be an alternative technology that will be competitive or possibly complementary with antigen testing. A variety of nucleic acid amplification technologies have been designed to detect the most important filamentous fungal pathogens. *Aspergillus*-specific PCR assays predominate and have been used to detect this pathogen in different clinical specimens, such as bronchoalveolar lavage fluid and blood. Importantly, the United Kingdom-Ireland Fungal PCR Consensus Group has taken a systematic approach to the evaluation of different nucleic acid extraction techniques and different assays for the detection of fungal pathogens (243). Such studies are necessary to critically evaluate procedures in a nonbiased manner so that the most efficient assay may be used for routine diagnostics (172).

The optimal type of specimen also must be defined for particular disease states. For example, although we have anecdotal experience with the detection of *Histoplasma capsulatum* from the blood of patients with disseminated histoplasmosis, it is not clear whether serum, plasma, whole blood, or the buffy coat would prove the best specimen for such an assay. There are likely significant differences between the body fluids that may be used for antigen testing and those that are suitable for DNA-based testing. For example, Tang et al. found that *Histoplasma* PCR of the urine is insensitive compared with antigen testing of the urine (214a). Differences such as these are likely based largely on the physiologic differences in the clearance of proteins and nucleic acids. In short, extensive studies need to be performed to define optimal specimens, the best extraction methods, and the optimal assays for the detection of fungi in clinical specimens.

Although genus-specific PCR assays are useful, the major limitation of such assays resides in the fact that patients at risk for fungal infection may be infected by a fungal pathogen other than the one targeted by the assays (i.e., a patient dying of a *Fusarium* infection would have an appropriately negative *Aspergillus*-specific PCR). Another approach that may prove useful in such instances is a broad-range PCR followed by postamplification analysis. A broad-range nucleic acid amplification reaction is one that utilizes primers that target a conserved region of a gene that is in turn conserved to a certain degree throughout the fungal kingdom. An amplification product would therefore be generated if any fungus were present. The type of fungus present could subsequently be determined by postamplification techniques, such as DNA sequencing, reverse hybridization, or the use of a microarray. The method of postamplification analysis used would depend upon the type of specimen and the likelihood of more than one type of fungus being present. The most significant drawback of this type of approach, particularly when DNA sequencing is used for postamplification analysis, is the problem of specimen or test system contamination by ubiquitous fungal spores.

Serology

Serologic studies have not proven particularly useful for the identification of fungal infections in the immunocompromised host. The presence of H and M bands by immunodiffusion for *Histoplasma capsulatum* is suggestive of active infection. However, because immunocompromised patients may not be able to generate an adequate humoral response, a negative result does not exclude the possibility of disease. In addition, serologic assays are not available for many of the fungi that may infect the patients at risk. There are serologic assays for *Aspergillus*, but these have not proven useful for the diagnosis of acute infections in the immunocompromised host. They are more useful for the identification of patients with allergic aspergillosis. They are likely not useful for the profoundly immunocompromised host because of the inability or truncated ability of the host to mount an adequate humoral response and/or the rapidly progressive nature of the infection.

SUMMARY

The prevalence of invasive fungal infections is likely greater now than ever before in history. This is largely because of the overwhelming success of solid and bone marrow/stem cell transplantation programs and new therapeutics to treat diseases such as Crohn's disease and rheumatoid arthritis. Unfortunately, these treatments leave the patient with a compromised immune system. In addition, although highly active antiretroviral therapy has been very effective in slowing the progression of HIV infection in developed countries, it is not a cure for the disease, and many patients in resource-poor parts of the world do not have access to these medications. These patients, who have great variations in the degree of their immunosuppression, are at increased risk for fungal infections. Although *Aspergillus* and *Candida* species are the most commonly encountered pathogens, any fungus may cause disease in a sufficiently immunocompromised patient. Fortunately, a variety of new, highly active antifungal agents have been developed and released. The availability of more treatment options increases the need for early diagnostics in order to maximize the advantage of the newly achieved potential for therapy. In addition, like any therapeutic agent, these newer agents have finite spectra of activity and do not cover all fungi. Therefore, it is important that the mycologist be able to accurately identify most fungal pathogens. Finally, because even molds that were previously thought to be only saprophytic may cause infections, it is important to correlate the findings in the mycology laboratory with those of the anatomic pathologist and the clinician.

REFERENCES

1. **Abramowsky, C. R., D. Quinn, W. D. Bradford, and N. F. Conant.** 1974. Systemic infection by fusarium in a burned child. The emergence of a saprophytic strain. *J. Pediatr.* **84:** 561–564.
2. **Adam, O., A. Auperin, F. Wilquin, J.-H. Bourhis, B. Gachot, and E. Chachaty.** 2004. Treatment with piperacillin-tazobactam and false-Positive Aspergillus galactomannan antigen test results for patients with hematological malignancies. *Clin. Infect. Dis.* **38:**917–920.
3. **Adam, R. D., M. L. Paquin, E. A. Petersen, M. A. Saubolle, M. G. Rinaldi, J. G. Corcoran, J. N. Galgiani, and R. E. Sobonya.** 1986. Phaeohyphomycosis caused by the fungal genera Bipolaris and Exserohilum. A report of 9 cases and review of the literature. *Medicine* (Baltimore) **65:**203–217.
4. **Aguado, J. M., R. Valle, R. Arjona, J. C. Ferreres, and J. A. Gutierrez.** 1992. Aortic bypass graft infection due to Aspergillus: report of a case and review. *Clin. Infect. Dis.* **14:** 916–921.
5. **Aguilar C., I. Pujol, J. Sala, and J. Guarro.** 1998. Antifungal susceptibilities of *Paecilomyces* species. *Antimicrob. Agents Chemother.* **42:**1601–1604.
6. **Alvarez-Lerma, F., J. Nolla-Salas, C. Leon, M. Palomar, R. Jorda, N. Carrasco, F. Bobillo, and E. S. Group.** 2003. Candiduria in critically ill patients admitted to intensive care medical units. *Intensive Care Med.* **29:**1069–1076.
7. **Anaissie, E., H. Kantarijian, J. Ro, R. Hopfer, K. Rolston, V. Fainstein, and G. Bodey.** 1988. The emerging role of *Fusarium* infection in patients with cancer. *Medicine* **67:**77–83.
8. **Anaissie, E. J., R. Hachem, C. Legrand, P. Legenne, P. Nelson, and G. P. Bodey.** 1992. Lack of activity of amphotericin B in systemic murine fusarial infection. *J. Infect. Dis.* **165:** 1155–1157.
9. **Anaissie E. J., P. Nelson, M. Beremand, D. Kontoyiannis, and M. Rinaldi.** 1992. Fusarium-caused hyalohyphomycosis: an overview. *Curr. Top. Med. Mycol.* **4:**231–249.
10. **Anaissie, E. J., and M. G. Rinaldi.** 1990. *Fusarium* and the immunocompromised host: liaisons dangeureuses. *New York State J. Med.* **90:**586–587.
11. **Arend, S. M., R. G. Westendorp, F. P. Kroon, J. W. van't Wout, J. P. Vandenbroucke, A. van Es, et al.** 1996. Rejection treatment and cytomegalovirus infection as risk factors for *Pneumocystis carinii* pneumonia in renal transplant recipients. *Clin. Infect. Dis.* **22:**920–925.
12. **Ayliffe, G. A.** 1991. Role of the environment of the operating suite in surgical wound infection. *Rev. Infect. Dis.* **13**(Suppl. 10):S800–S804.
13. **Baddley, J. W., T. P. Stroud, D. Salzman, and P. G. Pappas.** 2001. Invasive mold infections in allogeneic bone marrow transplant recipients. *Clin. Infect. Dis.* **32:**1319–1324.
14. **Becker, W. K., W. G. Cioffi, Jr., A. T. McManus, S. H. Kim, W. F. McManus, A. D. Mason, and B. A. Pruitt, Jr.** 1991. Fungal burn wound infection. A 10-year experience. *Arch. Surg.* **126:**44–48.
15. **Berenguer, J., J. Diaz-Mediavilla, D. Urra, and P. Munoz.** 1990. Central nervous system infection caused by *Pseudallescheria boydii*: case report and review. *Rev. Infect. Dis.* **60:**531–537.
16. **Bhat, S. V., D. L. Paterson, M. G. Rinaldi, and P. J. Veldkamp.** 2007. Scedosporium prolificans brain abscess in a patient with chronic granulomatous disease: successful combination therapy with voriconazole and terbinafine. *Scand. J. Infect. Dis.* **39:**87–90.
17. **Bisno, A.** 1984. Cutaneous infections: microbiologic and epidemiologic considerations. *Am. J. Med.* **76:**172–179.
18. **Blackwell, V., K. Ahmed, C. O'Docherty, and R. J. Hay.** 2000. Cutaneous hyalophyphomycosis caused by *Paecilomyces lilacinus* in a renal transplant patient. *Br. J. Dermatol.* **143:**873–875.

19. **Boggild, A. K., S. M. Poutanen, S. Mohan, and M. A. Ostrowski.** 2006. Disseminated phaeohyphomycosis due to Ochroconis gallopavum in the setting of advanced HIV infection. *Med. Mycol.* **44:**777–782.
20. **Bolignano, G., and G. Criseo.** 2003. Disseminated nosocomial fungal infection by *Aureobasidium pullulans* var. *melanigenum*: a case report. *J. Clin. Microbiol.* **41:**4483–4485.
21. **Boutati, E. I., and E. J. Anaissie.** 1997. *Fusarium*, a significant emerging pathogen in patients with hematologic malignancy: ten years' experience at a center and implications for management. *Blood* **90:**999–1008.
22. **Brandt, M. E., and D. W. Warnock.** 2003. Epidemiology, clinical manifestations, and therapy of infections caused by dematiaceous fungi. *J. Chemother.* **15**(Suppl. 2):36–47.
23. **Braun, J., F. de Keyser, J. Brandt, H. Mielants, J. Sieper, and E. Veys.** 2001. New treatment options in spondyloarthropathies: increasing evidence for significant efficacy of antitumor necrosis factor therapy. *Curr. Opin. Rheumatol.* **13:** 245–249.
24. **Bretagne, S., A. Marmorat-Khuong, M. Kuentz, J. P. Latge, E. Bart-Delabesse, and C. Cordonnier.** 1997. Serum Aspergillus galactomannan antigen testing by sandwich ELISA: practical use in neutropenic patients. *J. Infect.* **35:**7–15.
25. **Briegel, J., H. Forst, B. Spill, A. Haas, B. Grabein, M. Haller, et al.** 1995. Risk factors for systemic fungal infections in liver transplant recipients. *Eur. J. Clin. Microbiol. Infect. Dis.* **14:**375–382.
26. **Cahill, B. C., J. R. Hibbs, K. Savik, B. A. Juni, and B. M. Dosland.** 1997. *Aspergillus* airway colonization and invasive disease after lung transplantation. *Chest* **112:**1160–1164.
27. **Carey, J., R. D'Amico, D. A. Sutton, and M. G. Rinaldi.** 2003. *Paecilomyces lilacinus* vaginitis in an immuno-competent patient. *Emerg. Infect. Dis.* **9:**1155–1158.
28. **Carrillo-Munoz, A. J., G. Quindos, M. Ruesga, O. del Valle, J. Peman, E. Canton, J. M. Hernandez-Molina, and P. Santos.** 2006. In vitro antifungal susceptibility testing of filamentous fungi with Sensititre Yeast One. *Mycoses* **49:**293–297.
29. **Castaldo, P., et al.** 1991. Clinical spectrum of fungal infections after orthotopic liver transplantation. *Arch. Surg.* **126:**149–156.
30. **Castro, L. G., A. Salebian, and M. N. Sotto.** 1990. Hyalohyphomycosis by *Paecilomyces lilacinus* in a renal transplant patient and a review of human *Paecilomyces* species infections. *J. Med. Vet. Mycol.* **28:**15–26.
31. **Caux, F., S. Aractingi, H. Baurmann, P. Reygagne, H. Dombret, S. Romand, and L. Dubertret.** 1993. *Fusarium solani* cutaneous infection in a neutropenic patient. *Dermatology* **186:**232–235.
32. **Chan-Tack, K. M., C. L. Thio, N. S. Miller, C. L. Karp, C. Ho, and W. G. Merz.** 1999. *Paecilomyces lilacinus* fungemia in adult marrow transplant recipient. *Med. Mycol.* **37:**57–60.
33. **Christianson, J. C., W. Engber, and D. Andes.** 2003. Primary cutaneous cryptococcosis in immunocompetent and immunocompromised hosts. *Med. Mycol.* **41:**177–188.
34. **Chua, J. D., S. M. Gordon, J. Banbury, G. S. Hall, and G. W. Procop.** 2001. Relapsing Exophiala jeanselmei phaeohyphomycosis in a lung-transplant patient. *Transpl. Infect. Dis.* **3:**235–238.
35. **Cisneros, J. M., et al.** 1998. Pneumonia after heart transplantation: a multiinstitutional study. *Clin. Infect. Dis.* **27:**324–331.
36. **Clancy, C. J., R. A. Jaber, H. L. Leather, J. R. Wingard, B. Staley, L. J. Wheat, C. L. Cline, K. H. Rand, D. Schain, M. Baz, and M. H. Nguyen.** 2007. Bronchoalveolar lavage galactomannan in diagnosis of invasive pulmonary aspergillosis among solid-organ transplant recipients. *J. Clin. Microbiol.* **45:**1759–1765.
37. **Clarke, D. E., and T. A. Raffin.** 1990. Infectious complications of indwelling long-term central venous catheters. *Chest* **97:**966–972.
38. **Cofrancesco, E., C. Boschetti, M. A. Viviani, C. Bargiggia, A. M. Tortorano, M. Cortellaro, and C. Zanussi.** 1992. Efficacy of liposomal amphotericin B (AmBisome) in the eradication of *Fusarium* infection in a leukaemic patient. *Haematologica* **77:**280–283.
39. **Collins, L. A., et al.** 1994. Risk factors for invasive fungal infections complicating orthotopic liver transplantation. *J. Infect. Dis.* **170:**644–652.
40. **Connolly, P., M. Durkin, A. LeMonte, J. Witt III, E. Hackett, L. Egan, et al.** 2007. Rapid diagnosis of systemic and invasive mycoses. *Clin. Microbiol. Newsl.* **29:**1–6.
41. **Cornet, M., H. Mallat, D. Somme, E. Guerot, G. Kac, J. L. Mainardi, P. Fornes, L. Gutmann, and V. Lavarde.** 2003. Fulminant invasive pulmonary aspergillosis in immunocompetent patients—a two-case report. *Clin. Microbiol. Infect.* **9:** 1224–1227.
42. **Couppie, P., C. Aznar, B. Carme, and M. Nacher.** 2006. American histoplasmosis in developing countries with a special focus on patients with HIV: diagnosis, treatment, and prognosis. *Curr. Opin. Infect. Dis.* **19:**443–449.
43. **Cummings, J. R., G. R. Jamison, J. W. Boudreaux, M. J. Howles, T. J. Walsh, and R. T. Hayden.** 2007. Cross-reactivity of non-Aspergillus fungal species in the Aspergillus galactomannan enzyme immunoassay. *Diagn. Microbiol. Infect. Dis.* **59:**113–115.
44. **Das, A., M. E., L. A. Ross, H. L. Monforte, M. V. Horn, G. L. Lam, and W. H. Mason.** 2000. *Paecilomyces variotii* in a pediatric patient with lung transplantation. *Pediatr. Transplant.* **4:**328–332.
45. **Dean, D. A., and K. W. Burchard.** 1996. Fungal infection in surgical patients. *Am. J. Surg.* **171:**374–382.
46. **Decostere, A., K. Hermans, T. De Baere, F. Pasmans, and F. Haesebrouck.** 2003. First report on *Cryptococcus laurentii* associated with feather loss in a glossy starling (*Lamprotornis chalybaeus*). *Avian Pathol.* **32:**309–311.
47. **de la Monte, S. M., and G. M. Hutchins.** 1985. Disseminated Curvularia infection. *Arch. Pathol. Lab. Med.* **109:**872–874.
48. **Dignani, M. C., and E. Anaissie.** 2004. Human fusariosis. *Clin. Microbiol. Infect.* **10:**67–75.
49. **Eggimann, P., J. Garbino, and D. Pittet.** 2003. Epidemiology of Candida species infections in critically ill non-immunosuppressed patients. *Lancet Infect. Dis.* **3:**685–702.
50. **Ellis, M. E., H. Clink, D. Younge, and B. Hainau.** 1994. Successful combined surgical and medical treatment of fusarium infection after bone marrow transplantation. *Scand. J. Infect. Dis.* **26:**225–228.
51. **Espinel-Ingroff, A.** 1998. Comparison of in-vitro activities of new triazole SCH 56592 and the echinocandins MK-0991 (L-743,872) and LY303366 against opportunistic filamentous and dimorphic fungi and yeasts. *J. Clin. Microbiol.* **36:**2950–2956.
52. **Fantry, L.** 2000. Gastrointestinal infections in the immunocompromised host. *Curr. Opin. Gastroenterol.* **16:**45.
53. **Fica, A., M. C. Diaz, M. Luppi, R. Olivares, L. Saez, M. Baboor, and P. Vasquez.** 2003. Unsuccessful treatment with voriconazole of a brain abscess due to *Cladophialophora bantiana*. *Scand. J. Infect. Dis.* **35:**892–893.
54. **Fincher, R. M., J. F. Fisher, R. D. Lovell, C. L. Newman, A. Espinel-Ingroff, and H. J. Shadomy.** 1991. Infection due to the fungus *Acremonium* (*Cephalosporium*). *Medicine* **70:** 398–409.

55. **Frater, J. L., G. S. Hall, and G. W. Procop.** 2001. Histologic features of zygomycosis: emphasis on perineural invasion and fungal morphology. *Arch. Pathol. Lab. Med.* **125:**375–378.
56. **Fung-Tomc, J. C., E. Huczko, B. Minassian, and D. P. Bonner.** 1998. In vitro activity of a new oral triazole, BMS-207147 (ER-3-346). *Antimicrob. Agents Chemother.* **42:**313–318.
57. Reference deleted.
58. **Gamis, A. S., T. Gudnason, G. S. Giebink, and N. K. Ramsay.** 1991. Disseminated infection with *Fusarium* in recipients of bone marrow transplants. *Rev. Infect. Dis.* **13:**1077–1088.
59. **Gan, G. G., A. Kamarulzaman, K. Y. Goh, K. P. Ng, S. L. Na, and T. S. Soo-Hoo.** 2002. Non-sporulating Chrysosporium: an opportunistic fungal infection in a neutropenic patient. *Med. J. Malaysia* **57:**118–122.
60. **Gangneux, J. P., D. Lavarde, S. Bretagne, C. Guiguen, and V. Gandemer.** 2002. Transient Aspergillus antigenaemia: think of milk. *Lancet* **359:**1251.
61. **Giacchino, M., N. Chiapello, S. Bezzio, F. Fagioli, P. Saracco, A. Alfarano, V. Martini, G. Cimino, P. Martino, and C. Girmenia.** 2006. *Aspergillus* galactomannan enzyme-linked immunosorbent assay cross-reactivity caused by invasive *Geotrichum capitatum*. *J.Clin.Microbiol.* **44:**3432–3434.
62. **Goldblum, D., B. E. Frueh, S. Zimmerli, and M. Bohnke.** 2000. Treatment of postkeratitis Fusarium endophthalmitis with amphotericin B lipid complex. *Cornea* **19:**853–856.
63. **Greig, J., M. Harkness, P. Taylor, C. Hashmi, S. Liang, and J. Kwan.** 2003. Peritonitis due to the dermatiaceous mold *Exophiala dermatitidis* complicating continuous ambulatory peritoneal dialysis. *Clin. Microbiol. Infect.* **9:**713–715.
64. **Groll, A. H., and T. J. Walsh.** 2001. Uncommon opportunistic fungi: new nosocomial threats. *Clin. Microbiol. Infect.* **7:**8–24.
65. **Groll, A., and T. J. Walsh.** 1998. Pharmacology of antifungal compounds. *Adv. Pharmacol.* **44:**343–350.
66. **Groll, A. H., P. M. Shah, C. Mentzel, M. Schneider, G. Just-Nuebling, and K. Huebner.** 1996. Trends in the postmortem epidemiology of invasive fungal infections at a university hospital. *J. Infect.* **33:**23–32.
67. Reference deleted.
68. **Guarro, J., and J. Gené.** 1995. Opportunistic fusarial infections in humans. *Eur. J. Clin. Microbiol. Infect. Dis.* **14:**741–754.
69. **Guarro, J., W. Gams, I. Pujol, and J. Genéé.** 1997. *Acremonium* species: new emerging fungal opportunists—in vitro antifungal susceptibilities and review. *Clin. Infect. Dis.* **25:** 1222–1229.
70. **Guiot, H. M., J. Bertran-Pasarell, L. M. Tormos, C. Gonzalez-Keelan, G. W. Procop, J. Fradera, C. Sanchez-Sergenton, and W. Mendez.** 2007. Ileal perforation and reactive hemophagocytic syndrome in a patient with disseminated histoplasmosis: the role of the real-time polymerase chain reaction in the diagnosis and successful treatment with amphotericin B lipid complex. *Diagn. Microbiol. Infect. Dis.* **57:**429–433.
71. **Hayden, R. T., S. Pounds, R. Schaufele, T. Sein, and T. J. Walsh.** Expression of galactomannan antigenemia in pediatric oncology patients with invasive aspergillosis. *Pediatr. Infect. Dis. J.*, in press.
72. **Hebrecht, R., et al.** 2002. *Acremonium strictum* pulmonary infecion in a leukemic patient successfully treated with posaconazole after failure of amphotericin B. *Eur. J. Clin. Microbiol. Infect. Dis.* **21:**814–817.
73. **Heenequin, C., et al.** 1997. Invasive *Fusarium* infections: a retrospective survey of 31 cases. *J. Med. Vet. Mycol.* **33:**107–114.
74. Reference deleted.
75. **Hiemenz, J. W., B. Kennedy, and K. Kwon-Chung.** 1990. Invasive fusariosis associated with an injury with a stringray barb. *J. Med. Vet. Mycol.* **28:**209–213.
76. **Hiestand, P. C., et al.** 1992. The new cyclosporine derivative, SDZ IMM 125: in vitro and in vivo pharmacological effects. *Transplant. Proc.* **19:**31–38.
77. **Hilmarsdottir, I., S. B. Thorsteinsson, P. Asmundsson, M. Bodvarsson, and M. Arnadottir.** 2000. Cutaneous infection caused by *Paecilomyces lilicanus* in a renal transplant patient: treatment with voriconazole. *Scand. J. Infect. Dis.* **32:**331–332.
78. **Holzheimer, R. G., and H. Dralle.** 2002. Management of mycoses in surgical patients—review of the literature. *Eur. J. Med. Res.* **7:**200–226.
79. **Hope, W. W., T. J. Walsh, and D. W. Denning.** 2005. Laboratory diagnosis of invasive aspergillosis. *Lancet Infect. Dis.* **5:**609–622.
80. **Hopkins, M. A., and D. M. Treloar.** 1997. Mucormycosis in diabetes. *Am. J. Crit. Care* **6:**363–367.
81. **Horre, R., B. Jovanic, G. Marklein, G. Schumacher, N. Friedrichs, T. Neuhaus, G. S. de Hoog, W. H. Becker, S. M. Choi, and K. P. Schaal.** 2003. Fatal pulmonary scedosporiosis. *Mycoses* **46:**418–421.
82. **Horvath, J., S. Dummer, J. Lloyd, B. Walker, W. H. Merrill, and W. H. Frist.** 1993. Infection in the transplanted and native lung after single lung transplantation. *Chest* **104:**681–685.
83. **Husain, S., et al.** 2003. Opportunistic mycelial fungal infections in organ transplant recipients: emerging importance of non-*Aspergillus* mycelial fungi. *Clin. Infect. Dis.* **37:**221–229.
84. **Husain, S., D. L. Paterson, S. M. Studer, M. Crespo, J. Pilewski, M. Durkin, J. L. Wheat, B. Johnson, L. McLaughlin, C. Bentsen, K. R. McCurry, and N. Singh.** 2007. Aspergillus galactomannan antigen in the bronchoalveolar lavage fluid for the diagnosis of invasive aspergillosis in lung transplant recipients. *Transplantation* **83:**1330–1336.
85. **Husain, S., M. M. Wagener, and N. Singh.** 2001. *Cryptococcus neoformans* infection in organ transplant recipients: variables influencing clinical characteristics and outcome. *Emerg. Infect. Dis.* **7:**1–7.
86. **Jabbour, N., J. Reyes, S. Kusne, M. Martin, and J. Fung.** 1996. Cryptococcal meningitis after liver transplantation. *Transplantation* **61:**146–167.
87. **John, G. T., et al.** 1994. Cryptococcosis in renal allograft recipients. *Transplantation* **58:**855–856.
88. **Kaliamurthy, J., C. M. Kalavathy, M. D. Ramalingam, D. A. Prasanth, C. A. Jesudasan, and P. A. Thomas.** 2004. Keratitis due to a coelomycetous fungus: case reports and review of the literature. *Cornea* **23:**3–12.
89. **Kanj, S. S., et al.** 1996. Fungal infections in lung and heart-lung transplant recipients: report of 9 cases and review of the literature. *Medicine* **75:**142–156.
90. **Kantarcioglu, A. S., and G. S. de Hoog.** 2004. Infections of the central nervous system by melanized fungi: a review of cases presented between 1999 and 2004. *Mycoses* **47:**4–13.
91. **Kappe, R., and A. Schulze-Berge.** 1993. New cause for false-positive results with the Pastorex Aspergillus antigen latex agglutination test. *J. Clin. Microbiol.* **31:**2489–2490.
92. **Kawazu, M., Y. Kanda, Y. Nannya, K. Aoki, M. Kurokawa, S. Chiba, T. Motokura, H. Hirai, and S. Ogawa.** 2004. Prospective comparison of the diagnostic potential of real-time PCR, double-sandwich enzyme-linked immunosorbent assay for galactomannan, and a (1→3)-beta-D-glucan test in weekly screening for invasive aspergillosis in patients with hematological disorders. *J. Clin. Microbiol.* **42:**2733–2741.

93. **Kedzierska, A.** 2007. (1→3)-Beta-D-glucan—a new marker for the early serodiagnosis of deep-seated fungal infections in humans. *Pol. J. Microbiol.* **56:**3–9.
94. **Kershaw, P., R. Freeman, D. Templeton, P. C. DeGirolami, U. DeGirolami, D. Tarsy, S. Hoffmann, G. Eliopoulos, and A. W. Karchmer.** 1990. *Pseudallescheria boydii* infection of the central nervous system. *Arch. Neurol.* **47:**468–472.
95. **Klont, R. R., M. A. Mennink-Kersten, and P. E. Verweij.** 2004. Utility of Aspergillus antigen detection in specimens other than serum specimens. *Clin. Infect. Dis.* **39:**1467–1474.
96. **Kontoyiannis, D. P., V. C. Wessel, G. P. Bodey, and K. V. Rolston.** 2000. Zygomycosis in the 1990s in a tertiary care cancer center. *Clin. Infect. Dis.* **30:**851–856.
97. **Kouvousis, N., G. Lazaros, E. Christoforatou, S. Deftereos, D. Petropoulou-Milona, M. Lelekis, and A. Zacharoulis.** 2002. Pacemaker pocket infection due to *Acremonium* species. *Pacing Clin. Electrophysiol.* **25:**378–379.
98. **Krcmery, V., Jr., et al.** 1997. Fungaemia due to *Fusarium* spp. in cancer patients. *J. Hosp. Infect.* **36:**223–228.
99. **Kusne, S., et al.** 1988. Infections after liver transplantation, an analysis of 101 consecutive cases. *Medicine* **67:**132–143.
100. **Lamaris, G. A., G. Chamilos, R. E. Lewis, A. Safdar, I. I. Raad, and D. P. Kontoyiannis.** 2006. Scedosporium infection in a tertiary care cancer center: a review of 25 cases from 1989–2006. *Clin. Infect. Dis.* **43:**1580–1584.
101. **Lass-Florl, C., M. Cuenca-Estrella, D. W. Denning, and J. L. Rodriguez-Tudela.** 2006. Antifungal susceptibility testing in Aspergillus spp. according to EUCAST methodology. *Med. Mycol.* **44**(Suppl.):319–325.
102. **Lass-Florl, C., K. Griff, A. Mayr, A. Petzer, G. Gastl, H. Bonatti, M. Freund, G. Kropshofer, M. P. Dierich, and D. Nachbaur.** 2005. Epidemiology and outcome of infections due to Aspergillus terreus: 10-year single centre experience. *Br. J. Haematol.* **131:**201–207.
103. **Latenser, B. A.** 2003. Fusarium infections in burn patients: a case report and review of the literature. *J. Burn Care Rehab.* **24:**285–288.
104. **Lederberg, J., R. Shope, and S. Oaks.** 1992. *Emerging Infections: Microbial Threats to Health in the United States.* National Academy Press, Washington, DC.
105. **Lee, F. Y., S. B. Mossad, and K. A. Adal.** 1999. Pulmonary mucormycosis: the last 30 years. *Arch. Intern. Med.* **159:**1301–1309.
106. **Lee, J. H., N. R. Slifman, S. K. Gershon, E. T. Edwards, W. D. Schwieterman, J. N. Siegel, R. P. Wise, S. L. Brown, J. N. Udall, Jr., and M. M. Braun.** 2002. Life-threatening histoplasmosis complicating immunotherapy with tumor necrosis factor alpha antagonists infliximab and etanercept. *Arthritis Rheum.* **46:**2565–2570.
107. **Lee, J., W. W. Yew, C. S. Chiu, P. C. Wong, C. F. Wong, and E. P. Wang.** 2002. Delayed sternotomy wound infection due to *Paecilomyces variotii* in a lung transplant recipient. *J. Heart Lung Transplant.* **21:**1131–1134.
108. **Lin, S. J., J. Schranz, and S. M. Teutsch.** 2001. Aspergillosis case-fatality rate: systematic review of the literature. *Clin. Infect. Dis.* **32:**358–366.
109. **Lionakis, M. S., and D. P. Kontoyiannis.** 2003. Glucocorticoids and invasive fungal infections. *Lancet* **362:**1828–1838.
110. **Lipsky, P. E., D. M. van der Heijde, E. W. St. Clair, D. E. Furst, F. C. Breedveld, J. R. Kalden, J. S. Smolen, M. Weisman, P. Emery, M. Feldmann, G. R. Harriman, R. N. Maini, et al.** 2000. Infliximab and methotrexate in the treatment of rheumatoid arthritis. *N. Engl. J. Med.* **343:**1594–1602.
111. **Liu, K., et al.** 2002. Morphologic criteria for the preliminary indentification of *Fusarium*, *Paecilomyces*, and *Acremonium* species by histopathology. *Am. J. Clin. Pathol.* **109:**45–54.
112. **Lopez, F. A., R. S. Crowley, L. Wastila, H. A. Valantine, and J. S. Remington.** 1998. *Scedosporium apiospermum* (*Pseudallescheria boydii*) infection in a heart transplant recipient: a case of mistaken identity. *J. Heart Lung Transplant.* **17:**321–324.
113. **Madariaga, M. G., A. Tenorio, and L. Proia.** 2003. *Trichosporon inkin* peritonitis treated with caspofungin. *J. Clin. Microbiol.* **41:**5827–5829.
114. **Maertens, J., J. Van Eldere, J. Verhaegen, E. Verbeken, J. Verschakelen, and M. Boogaerts.** 2002. Use of circulating galactomannan screening for early diagnosis of invasive aspergillosis in allogeneic stem cell transplant recipients. *J. Infect. Dis.* **186:**1297–1306.
115. **Mahfouz, T., and E. Anaissie.** 2003. Prevention of fungal infections in the immunocompromised host. *Curr. Opin. Investig. Drugs* **4:**974–990.
116. **Marchetti, O., J. Bille, U. Fluckiger, P. Eggimann, C. Ruef, J. Garbino, T. Calandra, M. P. Glauser, M. G. Tauber, D. Pittet, et al.** 2004. Epidemiology of candidemia in Swiss tertiary care hospitals: secular trends, 1991–2000. *Clin. Infect. Dis.* **38:**311–320.
117. Reference deleted.
118. **Marr, K. A., R. A. Carter, F. Crippa, A. Wald, and L. Corey.** 2002. Epidemiology and outcome of mould infections in hematopoietic stem cell transplant recipients. *Clin. Infect. Dis.* **34:**909–917.
119. **Marr, K. A., M. Laverdiere, A. Gugel, and W. Leisenring.** 2005. Antifungal therapy decreases sensitivity of the Aspergillus galactomannan enzyme immunoassay. *Clin. Infect. Dis.* **40:**1762–1769.
120. **Marr, K. A., T. Patterson, and D. Denning.** 2002. Aspergillosis: pathogenesis, clinical manifestations, and therapy. *Infect. Dis. Clin. N. Am.* **16:**875–894.
121. **Martagon-Villamil, J., N. Shrestha, M. Sholtis, C. M. Isada, G. S. Hall, T. Bryne, B. A. Lodge, L. B. Reller, and G. W. Procop.** 2003. Identification of *Histoplasma capsulatum* from culture extracts by real-time PCR. *J. Clin. Microbiol.* **41:**1295–1298.
122. **Martino, P., R. Gastaldi, R. Raccah, and C. Girmenia.** 1994. Clinical patterns of *Fusarium* infections in immunocompromised patients. *J. Infect.* **1:**7–15.
123. **Martino, P., et al.** 1990. *Blastoschizomyces capitatus*: an emerging cause of invasive fungal disease in leukemic patients. *Rev. Infect. Dis.* **12:**570–582.
124. **Marty, F. M., D. H. Barouch, E. P. Coakley, and L. R. Baden.** 2003. Disseminated trichosporonosis caused by *Trichosporon loubieri*. *J. Clin. Microbiol.* **41:**5317–5320.
125. **Mattei, D., D. Rapezzi, N. Mordini, F. Cuda, N. C. Lo, M. Musso, A. Arnelli, S. Cagnassi, and A. Gallamini.** 2004. False-positive *Aspergillus* galactomannan enzyme-linked immunosorbent assay results in vivo during amoxicillin-clavulanic acid treatment. *J. Clin. Microbiol.* **42:**5362–5363.
126. **McGinnis, M. R., and L. Pasarell.** 1998. In vitro evaluation of terbinafine and itraconazole against dematiaceous fungi. *Med. Mycol.* **36:**243–246.
127. **McGinnis, M. R., and L. Pasarell.** 1998. In vitro testing of susceptibilities of filamentous ascomycetes to voriconazole, itraconazole, and amphotericin B, with consideration of phylogenetic implications. *J. Clin. Microbiol.* **36:**2353–2355.
128. **Mennink-Kersten, M. A., J. P. Donnelly, and P. E. Verweij.** 2004. Detection of circulating galactomannan for the diagnosis and management of invasive aspergillosis. *Lancet Infect. Dis.* **4:**349–357.

129. **Mennink-Kersten, M. A., A. Warris, and P. E. Verweij.** 2006. 1,3-Beta-D-glucan in patients receiving intravenous amoxicillin-clavulanic acid. *N. Engl. J. Med.* **354:**2834–2835.

130. **Mennink-Kersten, M. A., D. Ruegebrink, R. R. Klont, A. Warris, F. Gavini, H. J. Op den Camp, and P. E. Verweij.** 2005. Bifidobacterial lipoglycan as a new cause for false-positive Platelia Aspergillus enzyme-linked immunosorbent assay reactivity. *J. Clin. Microbiol.* **43:**3925–3931.

131. **Mennink-Kersten, M., and P. E. Verweij.** 2007. Non-culture-based diagnostics for opportunistic fungi. *Infect. Dis. Clin. N. Am.* **20:**711–727.

132. **Merz, W. G., J. E. Karp, M. Hoagland, M. Jett-Goheen, J. M. Junkins, and A. F. Hood.** 1988. Diagnosis and successful treatment of fusariosis in the compromised host. *J. Infect. Dis.* **158:**1046–1055.

133. **Miro, O., J. Ferrando, V. Lecha, and J. M. Campistol.** 1994. Subcutaneous abscesses caused by *Acremonium falciforme* in a kidney transplant recipient. *Med. Clin.* (Barcelona) **102:**316.

134. **Mohammedi, I., M. A. Piens, C. Audigier-Valette, J. C. Gantier, L. Argaud, O. Martin, and D. Robert.** 2004. Fatal Microascus trigonosporus (anamorph Scopulariopsis) pneumonia in a bone marrow transplant recipient. *Eur. J. Clin. Microbiol. Infect. Dis.* **23:**215–217.

135. **Moore, C. B., N. Sayers, J. Mosquera, J. Slaven, and D. W. Denning.** 2000. Antifungal drug resistance in Aspergillus. *J. Infect.* **41:**203–220.

136. **Mootsikapun, P., and S. Srikulbutr.** 2006. Histoplasmosis and penicilliosis: comparison of clinical features, laboratory findings and outcome. *Int. J. Infect. Dis.* **10:**66–71.

137. **Morse, S.** 1996. Factors in the emergence of infectious diseases. *Emerg. Infect. Dis.* **7:**7–15.

138. **Mousa, H. A.** 1999. Fungal infection of burn wounds in patients with open and occlusive treatment methods. *East Mediterr. Health J.* **5:**333–336.

139. **Murayama, N., R. Takimoto, M. Kawai, M. Hiruma, K. Takamori, and K. Nishimura.** 2003. A case of subcutaneous phaeohyphomycotic cyst due to Exophiala jeanselmei complicated with systemic lupus erythematosus. *Mycoses* **46:**145–148.

140. **Musa, M. O., A. Al Eisa, M. Halim, E. Sahovic, M. Gyger, N. Chaudhri, F. Al Mohareb, P. Seth, M. Aslam, and M. Aljurf.** 2000. The sprectrum of Fusarium infection in immunocompromised patients with hematological malignancies and in non-immunocompromised patients. *Br. J. Hematol.* **108:**544–548.

141. **Musher, B., D. Fredricks, W. Leisenring, S. A. Balajee, C. Smith, and K. A. Marr.** 2004. *Aspergillus* galactomannan enzyme immunoassay and quantitative PCR for diagnosis of invasive aspergillosis with bronchoalveolar lavage fluid. *J. Clin. Microbiol.* **42:**5517–5522.

142. **Myers, A., J. Clark, and H. Foster.** 2002. Tuberculosis and treatment with infliximab. *N. Engl. J. Med.* **346:**623–626.

143. **Naidu, J., and S. M. Singh.** 1992. Hyalohyphomycosis caused by *Paecilomyces variotii*: a case report, animal pathogenicity and in vitro sensitivity. *Antonie Van Leeuwenhoek* **62:**225–230.

144. **Nakelchik, M., and J. E. Mangino.** 2002. Reactivation of histoplasmosis after treatment with infliximab. *Am. J. Med.* **112:**78.

145. **Negroni, R., S. H. Helou, N. Petri, A. M. Robles, A. Arechavala, and M. H. Bianchi.** 2004. Case study. Posaconazole treatment of disseminated phaeohyphomycosis due to *Exophiala spinifera*. *Clin. Infect. Dis.* **38:**e15–e20.

146. **Nir-Paz, R., H. Elinav, G. E. Pierard, D. Walker, A. Maly, M. Shapiro, R. C. Barton, and I. Polacheck.** 2003. Deep infection by *Trichophyton rubrum* in an immunocompromised patient. *J. Clin. Microbiol.* **41:**5298–4301.

147. **Nobrega, J. P., S. Rosemberg, A. M. Adami, E. M. Heins-Vaccari, S. Lacaz Cda, and T. de Brito.** 2003. Fonsecaea pedrosoi cerebral phaeohyphomycosis ("chromoblastomycosis"): first human culture-proven case reported in Brazil. *Rev. Inst. Med. Trop. Sao Paulo* **45:**217–220.

148. **Nosari, A., et al.** 2000. Mucormycosis in hematologic malignancies: an emerging fungal infection. *Haematologica* **85:**1068–1071.

149. **Nucci, M., and and E. Anaissie.** 2002. Cutaneous infection by *Fusarium* species in healthy and immunocompromised hosts: implication for diagnosis and management. *Clin. Infect. Dis.* **35:**909–920.

150. **Nunley, D. R., O. N., W. F. Grgurich, A. T. Iacono, P. A. Williams, R. J. Keenan, et al.** 1998. Pulmonary aspergillosis in cystic fibrosis lung transplant recipients. *Chest* **114:**1321–1329.

151. **Obayashi, T., et al.** 1995. Plasma (1→3)-beta-D-glucan measurement in diagnosis of invasive deep mycosis and fungal febrile episodes. *Lancet* **345:**17–20.

152. **Ochiai, T. A., et al.** 1987. Effect of a new immunosuppressive agent, FK506 on heterotropic cardiac allotransplantation in the rat. *Transplant. Proc.* **19:**1284–1286.

153. **Odabasi, Z., G. Mattiuzzi, E. Estey, H. Kantarjian, F. Saeki, R. J. Ridge, P. A. Ketchum, M. A. Finkelman, J. H. Rex, and L. Ostrosky-Zeichner.** 2004. Beta-D-glucan as a diagnostic adjunct for invasive fungal infections: validation, cutoff development, and performance in patients with acute myelogenous leukemia and myelodysplastic syndrome. *Clin. Infect. Dis.* **39:**199–205.

154. **Ohira, S., K. Isoda, H. Hamanaka, K. Takahashi, K. Nishimoto, and H. Mizutani.** 2002. Case report. Phaeohyphomycosis caused by Phialophora verrucosa developed in a patient with non-HIV acquired immunodeficiency syndrome. *Mycoses* **45:**50–54.

155. **Orth, B., R. Frei, P. H. Itin, M. G. Rinaldi, B. Speck, A. Gratwohl, and A. F. Widmer.** 1996. Outbreak of invasive mycosis caused by *Paecilomyces lilacinus* from a contaminated skin lotion. *Ann. Intern. Med.* **125:**799–806.

156. **Ostrosky-Zeichner, L., B. D. Alexander, D. H. Kett, J. Vazquez, P. G. Pappas, F. Saeki, P. A. Ketchum, J. Wingard, R. Schiff, H. Tamura, M. A. Finkelman, and J. H. Rex.** 2005. Multicenter clinical evaluation of the (1→3) beta-D-glucan assay as an aid to diagnosis of fungal infections in humans. *Clin. Infect. Dis.* **41:**654–659.

157. **Panackal, A. A., and K. A. Marr.** 2004. Scedosporium/Pseudallescheria infections. *Semin. Respir. Crit. Care Med.* **25:**171–181.

158. **Paradowski, L.** 1997. Saprophytic fungal infections and lung transplantation—revisited. *J. Heart Lung Transplant.* **16:**524–531.

159. **Parfrey, N.** 1986. Improved diagnosis and prognosis of mucormycosis. A clinicopathologic study of 33 cases. *Medicine* **65:**113–123.

160. **Pasqualotto, A. C., and D. W. Denning.** 2006. Post-operative aspergillosis. *Clin. Microbiol. Infect.* **12:**1060–1076.

161. **Patterson, T. F., V. T. Andriole, M. J. Zervos, D. Therasse, and C. A. Kauffman.** 1990. The epidemiology of pseudallescheriasis complicating transplantation: nosocomial and community acquired infection. *Mycoses* **33:**297–302.

162. **Perfect, J. R., and W. A. Schell.** 1996. The new fungal opportunists are coming. *Clin. Infect. Dis.* **22:**S112–S118.

163. **Pettit, T. H., R. J. Olson, R. Y. Foos, and W. J. Martin.** 1980. Fungal endophthalmitis following intraocular lens implantation. A surgical epidemic. *Arch. Ophthalmol.* **98:**1025–1039.

164. **Pfaller, M. A., and D. J. Diekema.** 2004. Rare and emerging opportunistic fungal pathogens: concern for resistance beyond *Candida albicans* and *Aspergillus fumigatus*. *J. Clin. Microbiol.* **42:**4419–4431.
165. **Pfaller, M. A., et al.** *Reference Method for Broth Dilution Antifungal Susceptibility Testing. CLSI Document 38-A22.* CLSI, Wayne, PA.
166. **Pfaller, M. A., F. Marco, S. A. Messer, and R. N. Jones.** 1998. In vitro activity of two echinocandin derivatives, LY303366 and MK-0991 (L-743,792), against clinical isolates of *Aspergillus*, *Fusarium*, *Rhizopus*, and other filamentous fungi. *Diagn. Microbiol. Infect. Dis.* **30:**251–255.
167. **Pfeiffer, C. D., J. P. Fine, and N. Safdar.** 2006. Diagnosis of invasive aspergillosis using a galactomannan assay: a meta-analysis. *Clin. Infect. Dis.* **42:**1417–1427.
168. **Pickering, J., H. Sant, C. Bowles, W. Roberts, and G. Woods.** 2005. Evaluation of a (1→3)-beta-D-glucan assay for diagnosis of invasive fungal infections. *J. Clin. Microbiol.* **43:**5957–5962.
169. **Pisetsky, D. S.** 2000. Tumor necrosis factor blockers in rheumatoid arthritis. *N. Engl. J. Med.* **342:**810–811.
170. **Pitisuttithum, P., R. Negroni, J. R. Graybill, B. Bustamante, P. Pappas, S. Chapman, R. S. Hare, and C. J. Hardalo.** 2005. Activity of posaconazole in the treatment of central nervous system fungal infections. *J. Antimicrob. Chemother.* **56:**745–755.
171. **Prabhu, R. M., and R. Patel.** 2004. Mucormycosis and entomophthoramycosis: a review of the clnical manifestations, diagnosis and treatment. *Clin. Microbiol. Infect.* **10:**31–47.
172. **Procop, G. W.** 2006. Evaluation of molecular diagnostic assays for fungal infections. *J. Mol. Diagn.* **8:**297–298.
173. **Pruitt, B. A., Jr., A. T. McManus, S. H. Kim, and C. W. Goodwin.** 1998. Burn wound infections: current status. *World J. Surg.* **22:**135–145.
174. **Queiroz-Telles, F., M. R. McGinnis, I. Salkin, and J. R. Graybill.** 2003. Subcutaneous mycoses. *Infect. Dis. Clin. N. Am.* **17:**59–85.
175. **Rajendran, C., B. K. Khaitan, R. Mittal, M. Ramam, M. Bhardwaj, and K. K. Datta.** 2003. Phaeohyphomycosis caused by Exophiala spinifera in India. *Med. Mycol.* **41:** 437–441.
176. **Rangel-Guerra, R. A., H. R. Martinez, C. Saenz, F. Bosques-Padilla, and I. Estrada-Bellmann.** 1996. Rhinocerebral and systemic mucormycosis. Clinical experience with 36 cases. *J. Neurol. Sci.* **143:**19–30.
177. **Reis, A., R. Sundmacher, K. Tintelnot, H. Agostini, H. E. Jensen, and C. Althaus.** 2000. Successful treatment of ocular invasive mould infection (fusariosis) with the new antifungal agent voriconazole. *Br. J. Ophthalmol.* **84:**932–933.
178. **Reuben, A., E. Anaissie, P. E. Nelson, R. Hashem, C. Legrand, D. H. Ho, and G. P. Bodey.** 1989. Antifungal susceptibility of 44 clinical isolates of *Fusarium* species determined by using a broth microdilution method. *Antimicrob. Agents Chemother.* **33:**1647–1649.
179. **Revankar, S. G., D. A. Sutton, and M. G. Rinaldi.** 2004. Primary central nervous system phaeohyphomycosis: a review of 101 cases. *Clin. Infect. Dis.* **38:**206–216.
180. **Rodgers, G. L., J. Mortensen, M. C. Fisher, A. Lo, A. Cresswell, and S. S. Long.** 2000. Predictors of infectious complications after burn injuries in children. *Pediatr. Infect. Dis. J.* **19:**990–995.
181. **Roilides, E. S., et al.** 1995. *Acremonium* fungemia in two immunocompromised children. *Pediatr. Infect. Dis. J.* **14:** 548–550.
182. **Romani, L.** 2004. Immunity to fungal infections. *Nat. Rev. Immunol.* **4:**1–23.
183. **Rotstein, C., L. Brock, and R. S. Roberts.** 1995. The incidence of first Hickman catheter-related infection and predictors of catheter removal in cancer patients. *Infect. Control Hosp. Epidemiol.* **16:**451–458.
184. **Sampathkumar, P., and C. V. Paya.** 2001. Fusarium infection after solid-organ transplantation. *Clin. Infect. Dis.* **32:**1237–1240.
185. **Sandborn, W. J., and S. B. Hanauer.** 1999. Antitumor necrosis factor therapy for inflammatory bowel disease: a review of agents, pharmacology, clinical results, and safety. *Inflamm. Bowel Dis.* **5:**119–133.
186. **Sanguinetti, M., B. Posteraro, L. Pagano, G. Pagliari, L. Fianchi, L. Mele, M. La Sorda, A. Franco, and G. Fadda.** 2003. Comparison of real-time PCR, conventional PCR, and galactomannan antigen detection by enzyme-linked immunosorbent assay using bronchoalveolar lavage fluid samples from hematology patients for diagnosis of invasive pulmonary aspergillosis. *J. Clin. Microbiol.* **41:**3922–3925.
187. **Schelenz, S., and W. R. Gransden.** 2003. Candidaemia in a London teaching hospital: analysis of 128 cases over a 7-year period. *Mycoses* **46:**390–396.
188. **Schell, W. A., and J. R. Perfect.** 1996. Fatal disseminated *Acremonium strictum* infection in a neutropenic host. *J. Clin. Microbiol.* **34:**1333–1336.
189. **Schwartz, J.** 1982. The diagnosis of deep mycosis by morphological methods. *Hum. Pathol.* **13:**519–533.
190. **Sellier, P., J. J. Monsuez, C. Lacroix, C. Feray, J. Evans, C. Minozzi, F. Vayre, P. Del Giudice, M. Feuilhade, C. Pinel, D. Vittecoq, and J. Passeron.** 2000. Recurrent subcutaneous infection due to *Scopulariopsis brevicaulis* in a liver transplant recipient. *Clin. Infect. Dis.* **30:**820–823.
191. **Sharkey, P. K., et al.** 1990. Itraconazole treatment of phaeohyphomycosis. *J. Am. Acad. Dermatol.* **23:**577–586.
192. **Shenoy, R., U. A. Shenoy, and Z. H. al Mahrooqui.** 2003. Keratomycosis due to Trichophyton mentagrophytes. *Mycoses* **46:**157–158.
193. **Shing, M. M., M. Ip, C. K. Li, K. W. Chik, and P. M. Yuen.** 1996. *Paecilomyces variotii* fungemia in a bone marrow transplant patient. *Bone Marrow Transplant.* **17:**281–283.
194. **Siemann, M., M. Koch-Dorfler, and M. Gaude.** 1998. False-positive results in premature infants with the Platelia Aspergillus sandwich enzyme-linked immunosorbent assay. *Mycoses* **41:**373–377.
195. **Singh, N.** 2000. Antifungal prophylaxis in organ transplant recipients: seeking clarity amidst controversy. *Clin. Infect. Dis.* **31:**545–553.
196. **Singh, N.** 2002. The changing face of invasive aspergillosis in liver transplant recipients. *Liver Transplant.* **8:**1071–1072.
197. **Singh, N.** 2003. Fungal infections in the recipients of solid organ transplantation. *Infect. Dis. Clin. N. Am.* **17:**113–134.
198. **Singh, N., et al.** 1997. Invasive aspergillosis in liver transplant recipients in the 1990s. *Transplantation* **64:**716–720.
199. **Singh, N., F. Y. Chang, T. Gayowski, and I. R. Marino.** 1997. Infections due to dematiaceous fungi in organ transplant recipients. *Clin. Infect. Dis.* **24:**369–374.
200. **Singh, N., T. Gayowski, M. M. Wagener, H. Doyle, and I. R. Marino.** 1997. Invasive fungal infections in liver transplant recipients receiving tacrolimus as primary immunosuppressive agent. *Clin. Infect. Dis.* **24:**179–184.
201. **Singh, N., and S. Husain.** 2003. *Aspergillus* infections after lung transplantation: clinical differences in type of transplant and implications for management. *J. Heart Lung Transplant.* **22:**258–266.
202. **Sirisanthana, T., and K. Supparatpinyo.** 1998. Epidemiology and management of penicilliosis in human immunodeficiency virus-infected patients. *Int. J. Infect.* Dis. **3:**48–53.

203. **Steinbach, W. J., and J. R. Perfect.** 2003. Scedosporium species infections and treatments. *J. Chemother.* **15**(Suppl. 2):16–27.
204. **St.-Germain, G., et al.** 1996. Fusarium, p. 122–123. *In* G. St.-Germain et al. (ed.), *Identifying Filamentous Fungi.* Blackwell Science, Blemont, CA.
205. **Strabelli, T. M., et al.** 1990. *Acremonium* infection after heart transplant. *Rev. Soc. Bras. Med. Trop.* **23**:233.
206. **Stynen, D., A. Goris, J. Sarfati, and J. P. Latge.** 1995. A new sensitive sandwich enzyme-linked immunosorbent assay to detect galactofuran in patients with invasive aspergillosis. *J. Clin. Microbiol.* **33**:497–500.
207. **Stynen, D., J. Sarfati, A. Goris, M. C. Prevost, M. Lesourd, H. Kamphuis, V. Darras, and J. P. Latge.** 1992. Rat monoclonal antibodies against *Aspergillus* galactomannan. *Infect. Immun.* **60**:2237–2245.
208. **Sulahian, A., F. Boutboul, P. Ribaud, T. Leblanc, C. Lacroix, and F. Derouin.** 2001. Value of antigen detection using an enzyme immunoassay in the diagnosis and prediction of invasive aspergillosis in two adult and pediatric hematology units during a 4-year prospective study. *Cancer* **91**:311–318.
209. **Supparatpinyo, K., C. Khamwan, V. Baosoung, K. E. Nelson, and T. Sirisanthana.** 1994. Disseminated *Penicillium marneffei* infection in Southeast Asia. *Lancet* **344**:110–113.
210. **Supparatpinyo, K., K. E. Nelson, W. G. Merz, B. J. Breslin, C. R. Cooper, Jr., C. Kamwan, and T. Sirisanthana.** 1993. Response to antifungal therapy by human immunodeficiency virus-infected patients with disseminated *Penicillium marneffei* infections and in vitro susceptibilities of isolates from clinical specimens. *Antimicrob. Agents Chemother.* **37**:2407–2411.
211. **Supparatpinyo, K., J. Perriens, K. E. Nelson, and T. Sirisanthana.** 1998. A controlled trial of itraconazole to prevent relapse of *Penicillium marneffei* infection in patients infected with the human immunodeficiency virus. *N. Engl. J. Med.* **339**:1739–1743.
212. **Swoboda-Kopec, E., M. M. Wroblewska, A. Rokosz, and M. Luczak.** 2003. Mixed bloodstream infection with Staphylococcus aureus and Penicillium chrysogenum in an immunocompromised patient: case report and review of the literature. *Clin. Microbiol. Infect.* **9**:1116–1117.
213. **Tadros, T. S., K. A. Workowski, R. J. Siegel, S. Hunter, and D. A. Schwartz.** 1998. Pathology of hyalohyphomycosis caused by *Scedosporium apiospermum* (*Pseudallescheria boydii*): an emerging mycosis. *Hum. Pathol.* **29**:1266–1272.
214. **Tan, T. Q., A. K. Ogden, J. Tillman, G. J. Demmler, and M. G. Rinaldi.** 1992. *Paecilomyces lilacinus* catheter-related fungemia in immunocompromised pediatric patient. *J. Clin. Microbiol.* **30**:2479–2483.
214a. **Tang, Y. W., L. Haijing, M. M. Durkin, S. E. Sefers, S. Meng, P. A. Connolly, C. W. Sutton, and L. J. Wheat.** 2006. Urine polymerase chain reaction is not as sensitive as urine antigen for the diagnosis of disseminated histoplasmosis. *Diagn. Microbiol. Infect. Dis.* **4**:283–287.
215. **Teixeira, A. B., P. Trabasso, M. L. Moretti-Branchini, F. H. Aoki, A. C. Vigorito, M. Miyaji, Y. Mikami, M. Takada, and A. Z. Schreiber.** 2003. Phaeohyphomycosis caused by Chaetomium globosum in an allogeneic bone marrow transplant recipient. *Mycopathologia* **156**:309–312.
216. **Tollemar, J., B. G. Ericzon, L. Barkholt, J. Andersson, O. Ringden, and C. G. Groth.** 1990. Risk factors for deep *Candida* infections in liver transplant recipients. *Transplant. Proc.* **22**:1826–1827.
217. Reference deleted.
218. **Torre-Cisnero, J., O. L. Lopez, S. Kusne, A. J. Martinez, and T. E. Starzl.** 1993. CNS aspergillosis in organ transplantation: a clinicopathologic study. *J. Neurol. Neurosurg. Psych.* **56**:188–193.
219. **Travis, L. B., G. D. Roberts, and W. R. Wilson.** 1985. Clinical significance of *Pseudallescheria boydii*: a review of 10 years' experience. *Mayo Clinic Proc.* **60**:531–537.
220. **Trifilio, S. M., C. L. Bennett, P. R. Yarnold, J. M. McKoy, J. Parada, J. Mehta, G. Chamilos, F. Palella, L. Kennedy, K. Mullane, M. S. Tallman, A. Evens, M. H. Scheetz, W. Blum, and D. P. Kontoyiannis.** 2007. Breakthrough zygomycosis after voriconazole administration among patients with hematologic malignancies who receive hematopoietic stem-cell transplants or intensive chemotherapy. *Bone Marrow Transplant.* **39**:425–429.
221. **Tritz, D., and G. L. Woods.** 1993. Fatal disseminated infection with *Aspergillus terreus* in the immunocompromised patient. *Clin. Infect. Dis.* **16**:118–122.
222. **Ustun, C., G. Huls, M. Stewart, and K. A. Marr.** 2006. Resistant Microascus cirrosus pneumonia can be treated with a combination of surgery, multiple anti-fungal agents and a growth factor. *Mycopathologia* **162**:299–302.
223. **van Burik, J. A., R. Colven, and D. H. Spach.** 1998. Cutaneous aspergillosis. *J. Clin. Microbiol.* **36**:3115–3121.
224. **Vartivarian, S. E., E. J. Anaissie, and G. P. Bodey.** 1993. Emerging fungal pathogens in immunocompromised patients: classification, diagnosis, and management. *Clin. Infect. Dis.* **17**(Suppl. 2):S487–S491.
225. **Verdaguer, V., T. J. Walsh, W. Hope, and K. J. Cortez.** 2007. Galactomannan antigen detection in the diagnosis of invasive aspergillosis. *Expert Rev. Mol. Diagn.* **7**:21–32.
226. **Verweij, P. E., C. M. Weemaes, J. H. Curfs, S. Bretagne, and J. F. Meis.** 2000. Failure to detect circulating *Aspergillus* markers in a patient with chronic granulomatous disease and invasive aspergillosis. *J. Clin. Microbiol.* **38**:3900–3901.
227. **Verweij, P. E., J. P. Latge, A. J. Rijs, W. J. Melchers, B. E. de Pauw, J. A. Hoogkamp-Korstanje, and J. F. Meis.** 1995. Comparison of antigen detection and PCR assay using bronchoalveolar lavage fluid for diagnosing invasive pulmonary aspergillosis in patients receiving treatment for hematological malignancies. *J. Clin. Microbiol.* **33**:3150–3153.
228. **Verweij, P. E., K. Brinkman, H. P. Kremer, B. J. Kullberg, and J. F. Meis.** 1999. *Aspergillus* meningitis: diagnosis by non-culture-based microbiological methods and management. *J. Clin. Microbiol.* **37**:1186–1189.
229. **Viscoli, C., M. Machetti, P. Gazzola, A. De Maria, D. Paola, M. T. Van Lint, F. Gualandi, M. Truini, and A. Bacigalupo.** 2002. *Aspergillus* galactomannan antigen in the cerebrospinal fluid of bone marrow transplant recipients with probable cerebral aspergillosis. *J. Clin. Microbiol.* **40**:1496–1499.
230. **Viviani, M. A., E. Cofrancesco, C. Boschetti, A. M. Tortorano, and M. Cortellaro.** 1991. Eradication of *Fusarium* infection in a leukopenic patient treated with liposomal amphotericin B. *Mycoses* **34**:255–256.
231. **Waite, R. T., and G. L. Woods.** 1998. Evaluation of BACTEC MYCO/F lytic medium for recovery of mycobacteria and fungi from blood. *J. Clin. Microbiol.* **36**:1176–1179.
232. **Wald, A., et al.** 1997. Epidemiology of *Aspergillus* infections in a large cohort of patients undergoing bone marrow transplantation. *J. Infect. Dis.* **175**:1459–1466.
233. **Walsh, T. J., R. W. Finberg, C. Arndt, J. Hiemenz, C. Schwartz, D. Bodensteiner, P. Pappas, N. Seibel, R. N. Greenberg, S. Dummer, M. Schuster, and J. S. Holcenberg.** 1999. Liposomal amphotericin B for empirical therapy in

patients with persistent fever and neutropenia. *N. Engl. J. Med.* **340**:764–771.

234. **Walsh, T. J., A. Groll, J. Hiemenz, R. Fleming, E. Roilides, and E. Anaissie.** 2004. Infections due to emerging and uncommon medically important fungal pathogens. *Clin. Microbiol. Infect.* **10**:48–66.
235. **Walsh, T. J., J. L. Goodman, P. Pappas, I. Bekersky, D. N. Buell, M. Roden, J. Barrett, and E. J. Anaissie.** 2001. Safety, tolerance, and pharmacokinetics of high-dose liposomal amphotericin B (AmBisome) in patients infected with *Aspergillus* species and other filamentous fungi: a maximum tolerated dose study. *Antimicrob. Agents Chemother.* **45**:3487–3496.
236. **Walsh, T. J., J. W. Hiemenz, N. L. Seibel, J. R. Perfect, G. Horwith, L. Lee, J. L. Silber, M. J. DiNubile, A. Reboli, E. Bow, J. Lister, and E. J. Anaissie.** 1998. Amphotericin B lipid complex in patients with invasive fungal infections: analysis of safety and efficacy in 556 cases. *Clin. Infect. Dis.* **26**:1383–1396.
237. **Wang, T. K., W. Chiu, S. Chim, T. M. Chan, S. S. Wong, and P. L. Ho.** 2003. Disseminated Ochroconis gallopavum infection in a renal transplant recipient: the first reported case and a review of the literature. *Clin. Nephrol.* **60**:415–423.
238. **Warris, A., A. Bjorneklett, and P. Gaustad.** 2001. Invasive pulmonary aspergillosis associated with infliximab therapy. *N. Engl. J. Med.* **344**:1099–1100.
239. **Wheat, J.** 2005. Galactomannan antigenemia detection for diagnosis of invasive aspergillosis. *Clin. Microbiol. Newsl.* **27**:51.
240. **Wheat, J., E. Hackett, M. Durkin, P. Connolly, R. Petraitiene, T. Walsh, K. Knox, and C. Hage.** 2007. Histoplasmosis-associated cross-reactivity in the BioRad Platelia Aspergillus enzyme immunoassay. *Clin. Vaccine Immunol.* **14**:638–640.
241. **Wheat, L. J., and C. A. Kauffman.** 2003. Histoplasmosis. *Infect. Dis. Clin. N. Am.* **17**:1–19.
242. **Wheeler, M. S., M. R. McGinnis, W. A. Schell, and D. H. Walker.** 1981. Fusarium infection in burned patients. *Am. J. Clin. Pathol.* **75**:304–311.
243. **White, P. L., R. Barton, M. Guiver, C. J. Linton, S. Wilson, M. Smith, B. L. Gomez, M. J. Carr, P. T. Kimmitt, S. Seaton, K. Rajakumar, T. Holyoake, C. C. Kibbler, E. Johnson, R. P. Hobson, B. Jones, and R. A. Barnes.** 2006. A consensus on fungal polymerase chain reaction diagnosis?: a United Kingdom-Ireland evaluation of polymerase chain reaction methods for detection of systemic fungal infections. *J. Mol. Diagn.* **8**:376–384.
244. **Wildfeurer, A., H. P. Seidl, I. Paule, and A. Haberreiter.** 1997. *In vitro* activity of voriconazole against yeasts, moulds, and dermatophytes in comparison with fluconazole, amphotericin B and griseofulvin. *Arzneimittelforschung* **47**:1257–1263.
245. **Williamson, P. R., K. J. Kwon-Chung, and J. I. Gallin.** 1992. Successful treatment of *Paecilomyces variotii* in a patient with chronic granulomatous disease and a review of *Paecilomyces* species infections. *Clin. Infect. Dis.* **14**:1023–1026.
246. **Willinger, B., G. Kopetzky, F. Harm, P. Apfalter, A. Makristathis, A. Berer, A. Bankier, and S. Winkler.** 2004. Disseminated infection with *Nattrassia mangiferae* in an immunosuppressed patient. *J. Clin. Microbiol.* **42**:478–480.
247. **Wolff, M. A., and R. Ramphal.** 1995. Use of amphotericin B lipid complex for treatment of disseminated cutaneous *Fusarium* infection in a neutropenic patient. *Clin. Infect. Dis.* **20**:1568–1569.
248. **Wood, K. L., C. A. Hage, K. S. Knox, M. B. Kleiman, A. Sannuti, R. B. Day, L. J. Wheat, and H. L. Twigg III.** 2003. Histoplasmosis after treatment with anti-tumor necrosis factor-alpha therapy. *Am. J. Respir. Crit. Care Med.* **167**:1279–1282.
249. **Woods, G., M. H. Miceli, M. L. Grazziutti, W. Zhao, B. Barlogie, and E. Anaissie.** 2007. Serum Aspergillus galactomannan antigen values strongly correlate with outcome of invasive aspergillosis: a study of 56 patients with hematologic cancer. *Cancer* **110**:830–834.
250. **Yehia, M., M. Thomas, H. Pilmore, W. Van Der Merwe, and I. Dittmer.** 2004. Subcutaneous black fungus (phaeohyphomycosis) infection in renal transplant recipients: three cases. *Transplantation* **77**:140–142.
251. **Yoo, J. H., S. M. Choi, D. G. Lee, J. H. Choi, W. S. Shin, W. S. Min, and C. C. Kim.** 2005. Prognostic factors influencing infection-related mortality in patients with acute leukemia in Korea. *J. Korean Med. Sci.* **20**:31–35.
252. **Zhang, Z., H. Correa, and R. E. Begue.** 2002. Tuberculosis and treatment with infliximab. *N. Engl. J. Med.* **346**:623–626.

Diagnostic Microbiology of the Immunocompromised Host
Edited by Randall T. Hayden, Karen C. Carroll, Yi-Wei Tang, and Donna M. Wolk

Chapter 11

Yeasts

JUSTIN A. BISHOP AND WILLIAM G. MERZ

Yeasts are fungi that reproduce vegetatively (asexually) by either budding or fission, yielding single-celled progeny genetically identical to the mother cell. In their vegetative stage, most yeast species are haploid, although important diploid and aneuploid species do exist. The number of known yeast genera and species continues to increase. A quantitative measure of this increase can be estimated by comparing the numbers of genera and species described in serial editions of the textbook entitled *The Yeasts: A Taxonomic Study*, edited by Cletus Kurtzman and J. W. Fell. There were 60 genus and 500 species descriptions in the third edition (103) and 100 genus and >1,000 species descriptions in the fourth edition (104), and approximately 150 genera and 2,000 species will be included in the fifth edition, which is currently in revision (Cletus Kurtzman, personal communication). This increase is probably multifactorial and is due at least in part to the advent of molecular taxonomic analysis. Many phenotypically defined species are now being found to represent complexes of phenotypically related but molecularly distinct species. The number of yeast species documented to cause invasive infections in humans has also increased. Species recently recognized to cause human infections are opportunistic pathogens that typically cause disease in patients with defects in mucosal barriers, cell-mediated immunity, or phagocyte numbers and/or function.

Sexual and molecular taxonomic studies support the observation that yeast morphology has evolved in disparate portions of the fungal phylogenetic tree, suggesting that it is competitive with other microorganisms in nature. Yeasts have key genetic mechanisms to ensure that they can adapt successfully to changing environments, including the generation of strains with a new genetic makeup from heterothallic sexual reproduction, selection of mutations during mitotic divisions, and phenotypic switching systems. Yeast species that are pathogenic to humans may lack a known sexual cycle or may be the anamorph, or asexual, stage of many ascomycetous or basidiomycetous yeast species. Interestingly, there are some black yeast species, and even a few zygomycetous fungi (*Mucor rouxii*, for example) can be induced into a yeast morphology (73). Table 1 provides *Candida* anamorphs and their teleomorph (sexual-stage) genera. This list, which includes more than 11 teleomorphic genera with one or more anamorphic *Candida* spp., dramatically emphasizes the fact that the genus *Candida* is a very heterogeneous group of yeasts that share similar phenotypic characteristics. Therefore, it is not surprising that there are differences in susceptibility to antifungals and differences in detection by diagnostic methodologies.

ASSOCIATION OF SPECIFIC HOST IMMUNE DEFECTS AND TYPES OF INFECTION

Of the yeast genera and species covered in this chapter, *Cryptococcus neoformans* is most capable of causing infections in the absence of known immunocompromising diseases, immunosuppressive drugs, etc., although the majority of severe infections occur in hosts with impaired cell-mediated immunity. *Candida* and other yeasts documented to cause human infections are considered opportunistic pathogens; host-yeast interactions and outcomes are depicted in Fig. 1. Exceptions include diseases such as vulvovaginal candidiasis, diaper rash, and denture stomatitis, which may occur in immunocompetent hosts (59). Today, the opportunistic yeasts are important as established and emerging causes of both

Justin A. Bishop and William G. Merz • Department of Pathology, The Johns Hopkins Medical Institutions, Baltimore, MD 21287.

Table 1. Ascomycetous yeasts reported to cause human disease

Teleomorphic species	Anamorphic species
Clavispora lusitaniae	*Candida lusitaniae*
Debaromyces hansenii	*Candida famata*
Dipodascus capitatus	*Blastoschizomyces (Geotrichum) capitatus*
Galactomyces candidum	*Geotrichum candidum*
Hanseniospora uvarum	*Kloeckera apiculata*
Hanseniospora valbyensis	*Kloeckera africana*
Issatchenkia orientalis	*Candida krusei*
Kluyveromyces lactis var. *lactis*	*Candida spherica*
Kluyveromyces marxianus	*Candida kefyr*
Pichia anomala	*Candida pelliculosa*
Pichia fabianii	*Candida fabianii*
Pichia ohmeri	*Candida guilliermondii*
Pichia norvegensis	*Candida norvegensis*
Saccharomyces cerevisiae	*Candida robusta*
Stephanoascus ciferrii	*Canddia ciferrii*
Yarrowia lipolytica	*Candida lipolytica*
None	*Candida boidinii*
	Candida haemulonii
	Candida inconspicua
	Candida maltosa
	Candida rugosa

hospital- and community-acquired invasive infection. As medical care has become more sophisticated and increasing numbers of people are living longer with a multitude of immunosuppressive conditions, such as human immunodeficiency virus infection/AIDS (HIV/AIDS), organ transplantation, cancer, and critical illness, the incidence of invasive opportunistic yeast infections has increased, as has the recognition of the wide array of species capable of causing invasive disease in humans.

INFECTIONS DUE TO *CANDIDA* SPP.

Of the yeasts that are pathogenic to humans, *Candida* spp. are the most common, and they cause a wide array of diseases, from superficial dermatitis to chronic mucocutaneous disease to disseminated and deeply invasive infections, particularly among patients with significant immunosuppression. Interactions with *Candida* spp. that have evolved the ability to colonize our mucosal and cutaneous barri-

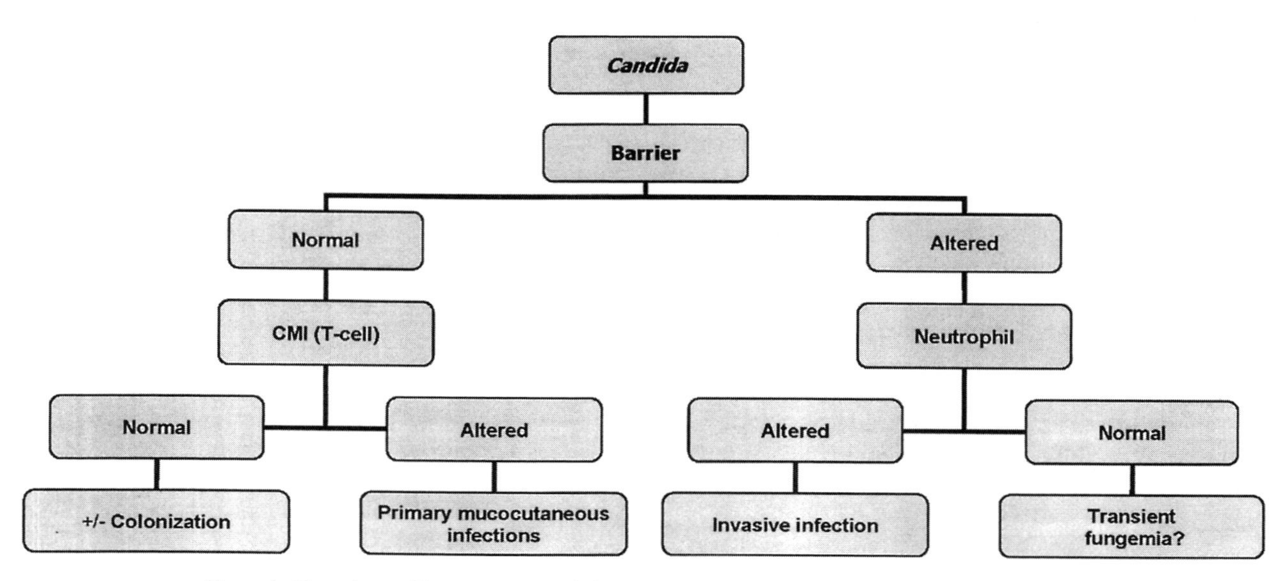

Figure 1. Flow chart of host-yeast (*Candida*) interactions and types of infection or colonization.

ers are shown in Fig. 1. Primary mucocutaneous infections occur in hosts with altered cell-mediated immunity, whereas invasive infections are more common in hosts with altered neutrophils.

The incidence of invasive *Candida* infections increased dramatically during the 1980s to 1990s. In a large epidemiological study of nosocomial bloodstream infections (BSIs) occurring between 1995 and 2002 in 49 U.S. hospitals, *Candida* spp. were the fourth most common cause of nosocomial BSIs overall and the third most common cause among intensive care unit (ICU) patients (233). In a population-based active surveillance project conducted in Connecticut and in Baltimore City and County between 1998 and 2000, the average annual incidence of candidemia was found to be 10 cases per 100,000 people, and 28% of these candidemias had their onset in the outpatient setting (74). An analysis of data from this population-based surveillance also revealed the substantial mortality and costs associated with candidemia. Mortality attributable to candidemia was 19 to 24%, with case patients having 5.3- to 8.5-fold greater odds of dying than matched controls (139). Lengths of stay were significantly longer for patients with candidemia than for those without, and the cost of candidemia treatment significantly increased total hospital charges (139).

Worldwide, *Candida albicans* continues to be the most prevalent cause of invasive candidiasis (IC) (170), but data from the ARTEMIS DISK surveillance program show that the proportion of IC cases due to *C. albicans* has decreased, from 73.3% in the period 1997–1998 to 62.3% in 2003 (170). Furthermore, the number of species reported to cause IC increased during that time (170). In some regions, the non-*C. albicans* species as a group have been shown to account for more than half of all cases of candidemia and IC (74). The emergence of the most common non-*C.albicans* species, *Candida glabrata*, *Candida tropicalis*, *Candida parapsilosis*, and *Candida krusei*, is likely due to a multitude of factors, including host factors (for example, degree of immunosuppression, age, and comorbid conditions) and the use of antifungal agents. Studies performed in the United States during the 1990s showed that significant changes in the epidemiology of candidemia, and particularly increases in the prevalence of *C. glabrata*, were occurring during that time (150). Trick and colleagues reported that the incidence of BSI due to *C. glabrata* increased among U.S. ICU patients from 1989 to 1999, while the incidence of BSI due to *C. albicans* actually decreased (214). Although *C. glabrata* is regarded as being less virulent than *C. albicans*, the organism's reduced azole susceptibility and tendency to cause disease in patients with a greater severity of illness (170) create management challenges for clinicians. The prevalence of *C. glabrata* varies widely according to geographic area; for example, *C. glabrata* is a less common pathogen in Latin America, where *C. tropicalis* and *C. parapsilosis* are the second and third most common causes of candidemia (170).

C. parapsilosis tends to be associated with vascular catheters and has caused numerous outbreaks of candidemia in the hospital setting (13, 48, 224). Although most isolates of *C. parapsilosis* remain fluconazole susceptible, this species has inherently higher echinocandin MICs (166, 167, 168). The clinical significance of this finding remains to be determined. *C. tropicalis*, which also remains generally fluconazole susceptible, has been associated with invasive infections in neutropenic hosts, patients with hematologic malignancies, and recipients of hematopoietic stem cell transplants in particular (152). *C. krusei*, which remains an uncommon pathogen overall and typically accounts for <5% of all cases of IC, is intrinsically resistant to fluconazole. Like *C. tropicalis*, *C. krusei* also tends to infect patients with neutropenia and hematologic malignancies. Perhaps surprisingly, despite increasingly widespread use of fluconazole and earlier reports of emergence of *C. krusei* in this setting (124, 231), larger epidemiologic studies have not shown an increase in the prevalence of this species (170).

Several uncommon *Candida* species have now been described to cause invasive infection in humans. Some, like *Candida lusitaniae*, have been recognized human pathogens for some time, while others, such as *Candida kefyr* (196), *Candida pelliculosa* (77), and *Candida haemulonii* (93), are more recent additions to the list. In addition, the use of molecular identification techniques has revealed that in some cases isolates identified as one species by traditional methods are actually members of a different species. For example, oral *Candida* isolates from HIV-infected patients were initially thought to be *C. albicans*, but DNA sequencing showed that they were in fact a new, distinct species, *Candida dubliniensis* (211). Clinical isolates phenotypically similar but not identical to *C. glabrata* were determined to be a new species, *Candida bracarensis* (41). Isolates thought to be *C. parapsilosis* were determined by sequencing to be two closely related but distinct species, *Candida metapsilosis* and *Candida orthopsilosis* (212). Whether recognition of these species has clinical relevance remains to be determined. As molecular techniques for identification become used more commonly in the clinical microbiology laboratory, it is highly likely that the list of potentially pathogenic *Candida* species will extend even further.

INFECTIONS DUE TO *CRYPTOCOCCUS* SPP.

Cryptococcosis has been a recognized disease of humans since the late 1800s-early 1900s (165). *Cryptococcus neoformans* var. *grubii* and *C. neoformans* var. *neoformans* are capable of causing disease in both immunocompetent and immunocompromised hosts, but it was not until the acceleration of the HIV/AIDS epidemic in the 1980s that they became a commonly diagnosed cause of meningitis in the United States. After inhalation of the infectious particle (and it remains controversial whether this is the yeast cell or basidiospore) (33), *C. neoformans* can infect virtually any organ system, although the organism has a propensity for spreading to the central nervous system and causing meningitis or meningoencephalitis, particularly in patients with HIV/AIDS (33). Other major sites of infection are the lungs and skin. Although the incidence of HIV-associated cryptococcosis in the United States has declined since the introduction of highly active antiretroviral therapy (129), cryptococcosis remains an important cause of morbidity and mortality among immunosuppressed patients worldwide. The incidence continues to be high in parts of the developing world with high HIV prevalence, where cryptococcosis is a major cause of death (121, 147). *C. neoformans* is the third most common cause of invasive fungal disease among organ transplant recipients (218). From a review of published reports, Husain and colleagues found that cryptococcosis was diagnosed in 2.8% of solid organ transplant recipients, with an overall mortality of 42% (87). Cryptococcosis has also been described to occur as a complication of tumor necrosis factor antagonist therapy (71, 202).

Cryptococcus gattii is the second most common *Cryptococcus* species associated with human disease and historically has caused infections in both immunocompetent and immunocompromised persons residing in subtropical areas, such as Australia, New Zealand, and parts of Asia and Africa, where the organism is endemic (33). It was typically found in the environment in association with eucalyptus trees (54). However, beginning in 1999, an outbreak of *C. gattii* infection was reported from Vancouver Island, British Columbia, an area that had never before been recognized to harbor the organism. Environmental strains of *C. gattii* in this outbreak were indistinguishable from most patient strains and were found in new environmental niches, in association with fir and oak trees (33). The *C. gattii* outbreak was determined to have been caused by a hypervirulent strain and has been characterized by a large proportion of cases among immunocompetent individuals as well as by high degrees of morbidity and mortality and a preponderance of pulmonary disease (84, 94). A case of cryptococcosis due to the Vancouver Island strain of *C. gattii* acquired in Washington State was also recently reported (216).

Other *Cryptococcus* species are also increasingly being recognized as causes of invasive disease among immunocompromised hosts. The *Cryptococcus laurentii* complex has been reported to cause invasive disease among patients with cancer (9, 102), HIV/AIDS (99), and primary immunodeficiency (203) and among hematopoietic stem cell transplant recipients (100). BSI due to *C. laurentii* has also been reported for a premature neonate (36). Interestingly, this pathogen has caused breakthrough fungemia in patients receiving azole antifungal prophylaxis (101). *Cryptococcus albidus* has been reported to cause numerous human infections, such as pulmonary infections (27), infections in persons with AIDS (99, 114), disseminated disease in a renal transplant recipient (109), cutaneous infection in a child who had received tumor necrosis factor antagonist therapy (83), and keratitis after corneal transplantation (44).

INFECTIONS DUE TO NON-*CANDIDA*, NON-*CRYPTOCOCCUS* YEASTS

Invasive infections due to yeasts other than *Candida* and *Cryptococcus* tend to mimic IC. They occur in similar patient populations, frequently in hospital settings, and patients present with similar, nonspecific clinical signs and symptoms. Risk factors for the development of infections are similar to those for IC, including receipt of broad-spectrum antibacterial antibiotics, presence of indwelling vascular catheters, receipt of parenteral nutrition, high severity of illness, prolonged hospitalization, particularly in ICU settings, and the presence of underlying diseases, such as hematologic malignancies.

Rapid identification of these pathogens to the species level is essential for clinical care and for epidemiological reasons. Today the antifungal armamentarium is far more substantial than it was even a few years ago, and the "one-size-fits-all" approach to administering antifungal treatment is no longer valid (if it ever was). Accurate identification of emerging pathogens aids in the selection of appropriate antifungal therapy and allows for identification of risk factors and predictors of outcome. If clinical laboratories do not identify invasive yeasts to the species level, there will be no way to determine which organisms are emerging pathogens in vulnerable patient populations. Most frequently, diagnosis of invasive infection is established by recovery of the organism from blood cultures, although the sensitivity of blood cultures does not approach 90 to 100%. Invasive procedures to obtain tissue for culture and histopathological examination may be necessary in some instances. Such procedures may be associated with significant risk in some patient populations, such as those with hematological malignancies, underscoring the need for rapid, sensitive, and specific diagnostics that can be performed on blood or other easily sampled body fluids.

The non-*Candida*, non-*Cryptococcus* yeast genera and species presented herein have been limited to those that cause human infections with relative frequency. They have been divided into ascomycetous and basidiomycetous yeasts, since this division reflects phylogeny, and susceptibility to antifungal agents may sometimes be inferred based on phylogeny. Characteristics used to designate placement into the Ascomycota or Basidiomycota include the recognition of a sexual cycle, with production of basidia/basidiospores or asci/ascospores, diazonium blue B staining, guanine/cytosine ratios, DNA homologous association rates, and sequencing of regions of rRNA genes (105).

Ascomycetous Yeasts

Clavispora, *Debaryomyces*, *Dipodascus*, *Hanseniospora*, *Issatchenkia*, *Kluyveromyces*, *Pichia*, *Saccharomyces*, and *Yarrowia* are teleomorphic, ascomycetous genera that have species documented to cause infections in humans. Many of these teleomorph species have *Candida* anamorphs (Table 1).

Blastoschizomyces

Blastoschizomyces capitatus is an opportunistic fungus that has features in common with other genera. The anamorph is *Blastoschizomyces capitatus*, and the teleomorph is *Dipodascus capitatus* (105). Girmenia et al. (66) performed a retrospective, multicenter study and also a literature review on invasive infections caused by this fungus. The study was a 20-year review of cases of invasive infections in patients with hematologic diseases conducted at 15 centers throughout Italy. Thirty-five cases were found in the multicenter study, and 92% of them were in patients with hematologic diseases (84% in patients with acute leukemia). Ninety-four percent occurred in patients with polymorphonuclear leukocyte counts of <100 cells/μl. Twenty-nine of 35 (82.9%) patients had proven infection, and 26 of these 29 patients had positive blood cultures (9 without and 17 with evidence of tissue invasion). A total of 31 patients received antifungal treatment, and 20 died. The literature review yielded 99 cases with similar characteristics (66).

B. capitatus may morphologically resemble *Trichosporon* spp., but there are various and specific characteristics that separate them (79). *B. capitatus* is urease negative, produces hyphae (true hyphae and/or pseudohyphae) and annellospores that resemble blastospores, and is a nonfermenter. Carbohydrate assimilation patterns are helpful, and the yeast grows at 42°C and in the presence of cycloheximide.

Pichia

Of the species in the ascomycetous genus *Pichia*, *Pichia anomala* (formerly *Hansenula anomala*) is the most prominent one documented to cause human infections. The anamorph of this species is *Candida pelliculosa*, which is found in the environment and sometimes may colonize humans and even cause superficial mucosal infection similar to other *Candida* infections. Cases of fungemia (144), urinary tract infections (178), endocarditis (151), and pneumonia (223) have also been described. There have been several outbreaks caused by this species in hospitalized pediatric (30), pediatric nursery (4), and pediatric ICU (145, 163, 213) patients and in adults (91). Again, these infections mimic candidiasis in risk factors, signs, and symptoms. Treatment has been successful with amphotericin B, with or without flucytosine or fluconazole, and with removal of catheters where possible. *P. anomala* is a fermenter that usually produces pseudohyphae but not true hyphae. Hat-shaped ascospores (four per cell) can be seen with some clinical strains cultured on appropriate media. Commercially available kits or systems are useful and should provide an identification of this species.

Pichia ohmeri (the teleomorph of *Candida guilliermondii*) is the second most common *Pichia* species that causes human infections. Documented infections include fungemia (16, 76, 115, 160), endocarditis (180), phlebitis (201), peritonitis (37), and urinary tract infections (177). In addition, *Pichia fabianii* has been documented to cause fungemia in a neonate (19) and to cause prostatitis (50).

Saccharomyces

Cases of invasive *Saccharomyces* infection in the world's literature through 30 May 2005 were reviewed by Enache-Angoulvant and Hennequin (55). Ninety-two cases were collected, and 72 of the 92 (78%) patients had *Saccharomyces* recovered from blood specimens. Risk factors included intravascular catheters and antibiotic therapies. Approximately two-thirds of the patients had favorable outcomes; removing the catheter and using amphotericin B or fluconazole proved to be an effective strategy. Interestingly, while approximately one-half of the infections were due to *Saccharomyces cerevisiae*, the remaining infections were due to the species found in probiotic preparations, i.e., *Saccharomyces boulardii*. An outbreak of three cases of *S. boulardii* fungemia among patients receiving a probiotic product occurred in an ICU over a 2- to 3-week period. The strains of *S. boulardii* isolated from the three patients and the probiotic product were indistinguishable (143).

Basidiomycetous Yeasts

Compared to ascomycetes, yeasts of the phylum Basidiomycota are generally less frequently implicated in human disease. However, it is important to recognize a yeast isolate as a basidiomycete because these organisms, due to differences in cell wall composition, are generally resistant to the echinocandin class of antifungal agents.

Malassezia

Malassezia spp. are basidiomycetous yeasts that require an exogenous source of long-chain fatty acids for growth. The one exception, *Malassezia pachydermatis*, grows better in the presence of lipids but does not require lipids for growth. The taxonomy of this genus has undergone major revisions due to molecular studies. There are at least 11 species (8, 69, 79, 82, 207, 208, 209, 210) that have been described. Currently, most laboratories do not identify *Malassezia* isolates to the species level; they default to *Malassezia furfur* or *M. pachydermatis* based on an absolute or partial requirement for an exogenous source of fatty acids. The microscopic appearance of the typical *Malassezia* yeast cell usually can be recognized by clinical laboratory members. The cells may be oval, round, or milk bottle-shaped, with monopolar budding with a collarette or bud scar from multiple bud production. Assays that can be used to identify many of the species include growth on various concentrations of Tween 20, 40, and 80 or on Cremophor EL (an emulsifying agent that contains castor oil) and the ability to grow at 37°C. In addition, the catalase reaction and the splitting of esculin are also helpful (127); however, molecular assays may be required for identification.

Malassezia spp. often can be found as part of the normal flora of the skin or as causes of cutaneous infections and, less often, can cause fungemia with or without deep tissue invasion. Pityriasis (tinea) versicolor and folliculitis are common cutaneous diseases due to *Malassezia* spp. that occur in immunocompetent as well as immunocompromised hosts. Tinea versicolor presents with hypo- or hyperpigmented lesions on the upper trunk and is usually diagnosed clinically and confirmed when needed by direct examination of skin scrapings. Folliculitis typically affects the upper trunk and usually occurs in patients with underlying conditions who are receiving antibiotics. Biopsy may be required for diagnosis (6, 10, 81).

Malassezia infections occur more often in warmer climates. *M. furfur* was historically reported as the major species causing infection; however, the role of this species as a human pathogen needs to be redefined, since there is evidence that *Malassezia sympodialis* and *Malassezia globosa* are more common etiologies than *M. furfur* (8, 148). *Malassezia* spp. do cause fungemias that are usually catheter associated, with or without tissue invasion. Fungemia due to *Malassezia* was first documented for neonates receiving intralipids, although it may also occur in other immunosuppressed hosts with indwelling catheters. *M. furfur* is the common etiology, but *M. pachydermatis* also may cause this infection, probably in association with an animal source, as documented for a neonatal ICU outbreak (32). Cases of fungemia also should be considered for patients not receiving intralipids (12). Treatment with amphotericin B or azoles (fluconazole) is usually effective, based on clinical and in vitro results (65).

Trichosporon

Trichosporon spp. are basidiomycetous yeasts that have significant variation in colony morphology, from extremely dry (mold-like) to a mucoid texture, and that produce extensive hyphae, blastospores, and arthrospores. Over the last 30 years, there have been major changes in the spectrum of diseases caused by *Trichosporon* spp. and in the major revisions in the genus based on molecular studies. Clinically, trichosporonosis has evolved from the classic white piedra with soft white hair nodules caused by a single species, *Trichosporon beigelii*, to life-threatening, disseminated infections in immunocompromised patients. Major taxonomic revisions based on molecular studies have defined 19 species (186), and their incidence, clinical significance, and appropriate antifungal regimens are being collected.

Trichosporon spp. produce a wide range of superficial infections, including white piedra and secondary nail lesions in immunocompetent hosts and serious infections in immunosuppressed hosts. *T. beigelii* or *Trichosporon capitatum* was implicated as the etiology of serious infections prior to taxonomic revisions. Today, however, with better species delineation, other *Trichosporon* spp. (not *T. beigelii*) are more common etiologic agents and are responsible for the morbidity and mortality in high-risk patient populations. These infections are more often caused by *Trichosporon asahii*, followed by *Trichosporon mucoides*, *Trichosporon inkin*, and two newer species, *Trichosporon louberi* and *Trichosporon mycotoxinivorans* (79, 98, 118, 149, 186). Risk factors include vascular catheters, hematologic malignancies, and neutropenia. The majority of cases are diagnosed by positive blood cultures. Biopsy and autopsy specimens reveal hyphae and, usually, arthrospores and blastospores upon extensive examinations. Individuals with *Trichosporon* infections may occasionally have a positive cryptococcal antigen test (122, 123). When visceral organs are involved, they usually are the lungs, brain, liver, spleen, skin, and eyes.

The formulation of specific antifungal regimens requires recovery and identification of the yeast at least to the genus level (*Trichosporon*). There are cases of infection which are apparently refractory to all systemic antifungals. *Trichosporon* sp. infections may not respond to amphotericin B; therefore, azoles or combination therapy may be necessary for adequate anti-*Trichosporon* activity. Return of normal phagocytic cells/activity is extremely important in this infection. *Trichosporon* spp. produce hyphae, arthrospores, and blastospores (although the latter may be difficult to visualize). They are nonfermenters; most are urease positive, and they are phenoloxidase negative and identified by carbohydrate assimilation patterns (API 20C) or molecular assays.

Rhodotorula

Rhodotorula spp. are anamorphs of basidiomycetous yeasts, including *Rhodosporidium* species. The genus is characterized by the production of salmon-red colonies due to carotenoid pigment production. *Rhodotorula glutinus*, *Rhodotorula minuta*, and *Rhodotorula mucilaginosa* are important in relation to human colonization and/or infection (105). Infections are opportunistic and most commonly are catheter-associated fungemia and peritonitis. Removal of indwelling catheters and treatment with an antifungal are usually effective. Amphotericin B has been the most common antifungal used. *Rhodotorula* spp. are characterized by carotenoid pigment production, occasional hyphal production, a lack of the ability to ferment carbohydrates, and the lack of assimilation of inositol. Most species associated with humans are urease positive, and some may produce mucoid colonies.

Sporobolomyces

Sporobolomyces spp. are the anamorphs of the basidiomycetous yeast genus *Sporidium*. They produce pink to red colonies due to carotenoid pigment production. *Sporobolomyces salmonicolor*, *Sporobolomyces holsaticus*, and *Sporobolomyces roseus* have been reported to cause allergic manifestations and infections, including dermatitis and endophthalmitis (141, 142, 174). Since the number of documented infections is low, optimal antifungal protocols are lacking. Amphotericin B and the newer azoles appear to be effective in minimal cases, and in vitro susceptibility data are supportive of these treatments.

DIAGNOSIS OF YEAST INFECTIONS IN IMMUNOCOMPROMISED HOSTS

Advances in Culture (Growth)-Dependent Systems for Diagnosis of Yeast Infections

Culture methods remain the most frequently utilized laboratory tools for the diagnosis of invasive yeast infections. Traditional means of identification of species in culture have relied on morphological characteristics of the yeast in culture. However, a number of advancements have been made in culture-based identification of yeasts. These include the use of chromogenic media, rapid biochemical assays, commercial biochemical panels, in vitro susceptibility testing, and molecular identification techniques (Table 2).

Chromogenic media

The high prevalence of yeast infections has led to the development of differential media that can distinguish different species of *Candida* and also detect infections with a mixed yeast population. Many of these media were developed and are primarily available in Europe. These include Oxoid Chromogenic Candida agar (Oxoid Ltd., Basingstoke, UK) (11), CPS ID2 medium (bioMerieux, France), UriSelect3 medium (Sanofi Diagnostics Pasteur, France) (29), Rainbow Agar UTI medium (Biolog) (29), Chromogenic UTI medium (Oxoid) (29), Candida ID (bioMerieux) (158, 229), Albicans ID (bioMerieux) (158), and Candi Select (Sanofi) (195). However, the chromogenic medium that has been

Table 2. Culture-dependent diagnostic assays

Test	Description	Example(s)	Commercial availability in the United States
Chromogenic media	Agars with chromogenic substrates that yield colony colors specific to some *Candida* spp.	CHROMagar Candida	+
Rapid trehalose	Bench test assessing hydrolysis of trehalose for presumptive identification of *C. glabrata*	Rapid trehalose broth	+
Biochemical panels	Panels of miniature biochemical assays that determine an isolate's pattern of substrate assimilation; the patterns are listed in databases, with corresponding identifications	RapID Yeast Plus	+
		API 20C Aux	+
		Vitek 2	+
In vitro susceptibility testing	Assays that use various concentrations of a panel of antifungal agents to determine MICs on isolated yeast	Sensititre YeastOne	+
		Etest	+
Molecular identification	Various nucleic acid-based techniques for identifying yeasts grown in culture; these assays depend on either nucleic acid amplification, hybridization, or both	PNA FISH	+/− (see text)
		Microarrays	+/−
		PCR-based assay	+/−[a]
		Sequencing	+/−[a]
		Gen-Probe	−

[a]No FDA-cleared platforms.

used most widely, particularly in the United States, is CHROMagar Candida (Becton Dickinson, Sparks, MD; and Hardy Diagnostics, Santa Maria, CA).

Introduced in the early 1990s, CHROMagar Candida is a chromogenic medium that is selective for most fungi. In addition, it is a differential agar, providing a presumptive identification, by color reactions and colony morphology, of *C. albicans*, *C. krusei*, and *C. tropicalis* (15, 155, 171, 175). These color reactions are produced by the hydrolysis of chromogenic substrates (175). *C. albicans* is identified by the growth of green colonies, *C. tropicalis* yields steel blue colonies, and *C. krusei* produces rough, matted, rose-colored colonies (146, 155). It has also been suggested that *C. glabrata* may be identified by producing pink or pink-purple colonies (17, 171, 175) and that *C. dubliniensis* may be identified by producing a darker shade of green than *C. albicans* (89, 156). This medium can yield presumptive species-level identifications 24 to 48 hours sooner than those by standard mycological identification procedures (171). In addition, the chromogenic substrate in CHROMagar Candida facilitates both the detection and identification of different yeast species in mixed cultures (155, 171).

CHROMagar Candida has been evaluated extensively in laboratories worldwide. The accuracy of green colony growth for *C. albicans*, in particular, has been well studied, with sensitivities ranging from 82% to 88.6% and specificities ranging from 90 to 100% (15, 17, 20, 28, 40, 62, 88, 155, 175, 190). The sensitivities and specificities for *C. krusei* and *C. tropicalis* are reported less frequently but seem to be similarly high, at 94% or higher (17, 190). In general, a high degree of agreement has been observed in comparing identifications made with CHROMagar Candida and those made with traditional mycological methods (22, 171). The use of CHROMagar Candida is also economical; Iyampillai and colleagues (88) reported that the cost of identification of yeast in India by CHROMagar Candida was lower than that of traditional methods. Although CHROMagar Candida may detect yeast growth not detected by traditional media, it also may not detect some yeasts that are detected by general-use fungal media, including some strains of common species such as *C. albicans*, *C. glabrata*, and *C. parapsilosis* (146).

Although some authors have reported high accuracy in the identification of *C. glabrata* with CHROMagar Candida, many different yeast species can produce pink colonies on this medium, and therefore this appearance should not be interpreted as presumptive *C. glabrata* (62, 85, 155, 156, 175, 190, 191). The apparent success that some investigators have reported in identifying *C. glabrata* may result from the high prevalence of *C. glabrata* in clinical specimens, particularly surveillance cultures (156, 171), compared to these other species. In addition, although some authors have reported being able to differentiate *C. albicans* and *C. dubliniensis* by growth on CHROMagar Candida, this has not been verified consistently; in particular, older cultures of *C. dubliniensis* seem to lose the ability to produce "darker" green colonies in culture (156). This darker growth may, however, be a useful way to determine that further studies may be needed to identify the isolate. Finally, *C. parapsilosis* produces a wide variety of colony colors and morphologies (85).

In conclusion, chromogenic agars, particularly CHROMagar Candida, have become an important addition to the mycology laboratory. They appear to contribute to the rapid identification of *C. albicans*, *C. krusei*, *C. tropicalis*, and *C. glabrata* and to aid in the detection of multiple yeast species in a clinical specimen in a cost-effective manner.

Rapid trehalose assays

Biochemical tests that may be performed rapidly have been investigated as "bench tests" that may aid in identification of yeasts in culture. The most studied assay is that based on the rapid hydrolysis of trehalose by *C. glabrata*. An assay for maltose is sometimes added to increase specificity. The rapid hydrolysis of these carbohydrates is usually specific to *C. glabrata* among commonly isolated yeast species. In the United States, because *C. glabrata* is usually the second most commonly identified *Candida* species and because it often rapidly acquires resistance to fluconazole, rapid identification of this yeast is indeed a desirable goal. Results of rapid trehalose tests, however, have varied. Generally, they have good performances when colonies are recovered from chromogenic agars; the number of false-positive results increases when colonies are grown on blood agar or Sabouraud agar (63, 173, 229).

Commercial panels

While traditional culture identification methods depend on morphology as well as carbohydrate assimilation patterns, they are often tedious to perform. As a result, a number of rapid biochemical identification systems have been introduced, and these are commercially available. Examples of the commonly used commercial systems are RapID Yeast Plus (Remel Inc., Lenexa, KS), API 20C Aux (bioMerieux-Vitek, Hazelwood, MO), and the Vitek 2 biochemical card (bioMerieux-Vitek, Marcy l'Etoile, France). Many studies have evaluated the use of these systems in clinical laboratories, usually reporting accuracies of >90% (63, 126, 189, 204). Studies directly comparing the methods have generally found relatively high rates of agreement between the various systems (56, 217). The API 20C Aux system has emerged, however, as the reference method in the United States (63). The reader is directed to a review by Freydiere et al. for further discussion of these systems (63).

In vitro susceptibility testing

Compared to the situation with antibacterial agents, the utility of antifungal susceptibility testing of yeast in the laboratory is less clearly defined. The reference method recommended by the Clinical and Laboratory Standards Institute (CLSI; formerly NCCLS) for in vitro susceptibility testing of *Candida* spp. uses broth microdilution (38a). This method was developed by a consensus process (184). Commercial systems have become available for determining MICs of various antifungal agents for yeast species. Sensititre YeastOne test panels (Trek Diagnostic Systems, Inc., Westlake, OH) are one available method for susceptibility testing of yeasts (3). Currently, the panels include ketoconazole, fluconazole, itraconazole, voriconazole, flucytosine, amphotericin B, and caspofungin. The 2008 panels will include the three echinocandins and posaconazole, while ketoconazole will be omitted. Etest (AB Biodisk, Solna, Sweden) is the other major commercially available method. Etest is an agar-based testing method that uses strips with a predefined concentration gradient to determine the MICs of various agents for *Candida* spp. (3). Both the Etest and Sensititre systems show good agreement with CLSI M27-A results as well as excellent intralaboratory reproducibility (135).

As mentioned previously, *Candida* spp. often have predictable susceptibility patterns, hence the importance of accurate identification. *C. albicans* is usually susceptible to a wide range of antifungal agents, including fluconazole, itraconazole, amphotericin B, echinocandins, and the newer triazoles (e.g., voriconazole and posaconazole) (164); *C. tropicalis* is also generally susceptible to most agents, but with higher amphotericin B MICs and occasional fluconazole resistance. *C. parapsilosis* is typically susceptible to the azole antifungals and amphotericin B but has high echinocandin MICs, a finding of unclear clinical significance at this time. *C. krusei* is resistant to fluconazole and has high amphotericin B MICs. *C. glabrata* is the least predictable species; many *C. glabrata* isolates do remain susceptible to azoles, but resistance to fluconazole is commonly seen and resistance to newer triazoles also occurs (164). Predictions, however, cannot be generalized; there are differences in resistance in different geographic global regions, differences among institutions, and even differences among services within a single institute.

In vitro testing, as performed with bacteria, is not carried out routinely with yeasts. Testing of species that may be resistant is one approach, but there is often a delay of days if testing is performed only after identifications are known. Reflexive testing of the initial yeast recovered from blood cultures or other selected sterile sites, in advance of identification, seems prudent. Additionally, breakthrough strains should be tested (164). Finally, in vitro susceptibility testing should be performed on rarely recovered species for which there are few or no published susceptibility data. Smaller institutes should employ a reliable reference laboratory to carry out this testing.

Interpretation of the results is of paramount importance. Clearly, standardization of in vitro susceptibility testing methods and interpretive MIC breakpoints are essential. The CLSI has established that isolates of *Candida* spp. with fluconazole MICs of ≤8 μg/ml should be considered susceptible, while those with MICs of ≥64 μg/ml should be considered resistant (164). If results fall between these breakpoints (16 to 32 μg/ml), the isolate may be considered "susceptible-dose dependent," i.e., the susceptibility of that isolate is dependent on achieving the maximal possible blood level of fluconazole (184). In addition, the utility of in vitro susceptibility testing as a tool in optimizing fluconazole therapy for *C. albicans* infections has been validated (7). Breakpoints for itraconazole (susceptible, MICs of ≤0.125 μg/ml; susceptible-dose dependent, MICs of 0.25 to 0.5 μg/ml; and resistant, MICs of >1 μg/ml), voriconazole (susceptible, MICs of <1 μg/ml; susceptible-dose dependent, MICs of 1 to 2 μg/ml; and resistant, MICs of ≥4 μg/ml), and flucytosine (susceptible, MICs of ≤4 mg/ml; susceptible-dose dependent, MICs of 8 to 16 mg/ml; and resistant, MICs of ≥32 mg/ml) have also been established (7). However, breakpoints for other yeast genera and for newer antifungals are not yet available, and physicians have to infer interpretations of the assessment of the MIC and the range of concentrations tested.

Molecular identification of yeasts grown in culture

There is considerable overlap between the topics of detection of fungal nucleic acid in clinical specimens, as discussed below, and the identification by molecular means of yeasts grown in culture. The advantages of using such methods are similar. Molecular identifications have the potential to be much more rapid than those obtained with traditional, phenotypic techniques (53). More importantly, the identifications are, ideally, more accurate (53).

FISH. Hybridization of DNA with species-specific probes holds promise for the identification of yeasts grown in culture. In particular, probes consisting of fluorescein-labeled peptide nucleic acid (PNA), a DNA mimic with a polyamide backbone, have recently been used to rapidly identify yeasts in culture. PNA has the advantages of being hydrophobic (thus allowing the probes to diffuse well through yeast cell walls), and it is more stable than DNA or RNA under a broad spectrum of conditions (157). A major advantage of PNA-based fluorescence in situ hybridization (FISH) (AdvanDx, Inc., Woburn, MA) over standard phenotypic identification methods is the short amount of time (2.5 hours) from a positive blood culture to species-level identification. In fact, there are data to support an even more rapid, 1.25-hour identification protocol (W. G. Merz, unpublished data).

Oliveira et al. used this modality to differentiate between *C. albicans* and *C. dubliniensis*, with 100% accuracy (157). In a multicenter study, Wilson and colleagues used PNA FISH with probes specific for *C. albicans* to identify this yeast directly from 244 positive blood culture bottles. They reported a sensitivity of 99% and a specificity of 100% (230). Alexander et al. (2) examined the cost-effectiveness of using PNA FISH to differentiate between *C. albicans* and non-*C. albicans Candida* spp. In their decision analysis model, they found that although the assay was more expensive than germ tube testing ($82.72 compared to $2.83), the use of this assay had the potential to save the hospital $1,837 per patient. The savings were due primarily to the decreased use of costly echinocandin antifungal agents (2). Forrest et al. (61) found that the *C. albicans* PNA FISH assay was 100% sensitive and specific. Use of the assay was associated with an overall cost savings of $1,729 per patient (61). Reller and colleagues examined the use in 238 clinical isolates of PNA FISH probes for the five most commonly isolated *Candida* spp., i.e., *C. albicans*, *C. glabrata*, *C. krusei*, *C. parapsilosis*, and *C. tropicalis* (183). In addition to 99% sensitivity and 99% specificity, the tests were performed consistently in 2.5 to 3 hours, compared to days with standard phenotypic techniques (183). More recently, Shepard et al. proved that *C. glabrata* as well as *C. albicans* can be detected separately or simultaneously in blood bottles by use of red and green fluorescent probes (Color Plate 7) (198). Identifications are, of course, limited by the specific probes available. In addition, the assay is limited by the number of fluorescent color reactions available; currently, a maximum of three different species may be identified at one time. Finally, the assay generally needs a certain number of yeast cells (1×10^5) to have a high likelihood of positive results. In all, however, PNA FISH technology seems to be a sensitive, specific, and cost-effective molecular assay that allows for rapid species-level identification of *Candida* species directly from positive blood culture bottles. It should be noted that this assay is FDA cleared only for yeasts isolated from blood. However, validation has been performed on yeasts from sterile body sites, with similarly promising results (183).

PCR-based assays. Most nucleic acid-based identification systems utilize PCR as an initial step, to amplify the fungal DNA. An appropriate target must first be chosen for this step. In general, most methods use panfungal primer sets to recognize highly conserved regions among pathogenic yeasts. The ribosomal DNA (rDNA) cluster is the most commonly utilized target (34). The advantage of these targets is that an amplicon may be produced from all *Candida* spp. under the same conditions (53). Following the amplification step, the amplicon must be analyzed. This was originally done by gel electrophoresis, with or without restriction enzyme digestion, followed by ethidium bromide staining (53, 137, 179). Southern blotting was later utilized to increase sensitivity (53, 90, 182). However, Southern blot analysis is time-consuming and often requires the use of radioactive materials, making it an impractical technology for most clinical laboratories (34).

Newer techniques used to analyze amplification products include probe hybridization, length polymorphism analysis of the internal transcribed spacer (ITS) regions of the rRNA gene, restriction fragment length polymorphism (RFLP) analysis, probe hybridization, flow cytometry, and DNA sequencing.

Probe hybridization. One form of probe hybridization, PCR-enzyme immunoassay (PCR-EIA), is a cost-effective method of analyzing amplicons (53). It is generally user-friendly and may be automated, making it a feasible option for many clinical laboratories (53). Shin et al. (53) used PCR with universal fungal primers, Elie and colleagues (52) used a similar assay but expanded the panel of *Candida* probes from 5 to 18, and Coignard et al. (39) used PCR-EIA to resolve discrepancies in the phenotypic identifications of the Centers for Disease Control and Prevention and referring hospitals. Results were encouraging, with accuracies of identification in the range of 98 to 100%. Other assays utilizing nucleic acid probe hybridization following PCR amplification have been investigated (188, 236), also with encouraging data. Commercialization, however, has not occurred.

Microarrays. Most recently, oligonucleotide microarrays have been utilized in the rapid identification of yeast species in culture. Microarrays can simultaneously analyze hundreds of targets and have the potential to identify a wide spectrum of yeasts with high sensitivity and specificity (108). Leaw and colleagues constructed a microarray of oligonucleotides specific to the ITS1 and ITS2 regions of 77 yeast species. Following amplification with a pair of fungus-specific primers, the digoxigenin-labeled amplicon was hybridized with a panel of oligonucleotide probes immobilized on a nylon membrane. The assay was performed on 452 yeast isolates, yielding a sensitivity of 100% and a specificity of 97%. The detection limit was 10 pg of genomic DNA per assay (108).

SSCP analysis. Single-strand conformational polymorphism (SSCP) analysis has been used as a method to identify yeast species in culture. In SSCP analysis, the migration of amplicons is a function not only of the amplicon size but also of the amplicon conformation (34, 220). As a result, SSCP analysis can distinguish amplicons that differ by as little as one base pair (34). Walsh et al. applied SSCP analysis to the identification of *Candida* and *Cryptococcus* spp. as well as to various filamentous fungi (220). This method has not been utilized widely in clinical laboratories, however.

Length polymorphism analysis of ITS region. Another PCR-based method of yeast identification is analysis of sequence length polymorphisms in the ITS2 region of rRNA genes. Turenne et al. reported that when the rRNA ITS2 is amplified and its length is determined by fragment analysis, the length of the ITS2 region is characteristic for most species and is almost identical for all strains within a species (215). Chen et al. (35) and Chang (31), using the length of the ITS2 region, have published data supporting correct identification of 92 to 97% of isolates. de Baere et al. (43) also utilized fragment length analysis of the amplified ITS2 region. After amplification of the rRNA ITS2 fragment, the length of the amplicon was determined with a capillary electrophoresis system. The authors found that 22 of 26 species could be identified unambiguously. *Trichosporon ateroides* and *T. inkin* could not be differentiated, and neither could *Trichosporon ovoides* and *T. mucoides*. Importantly, the identifications could be made in approximately 3.5 hours (43).

RFLP analysis. Esteve-Zarzoso et al. proposed a method for yeast identification based on analysis of RFLPs of the 5.8S rRNA gene and ITS1 and -2 (57). De Llanos Frutos et al. applied this methodology to analyze 75 species of *Candida* (45). Mirhendi et al. utilized this technology in the clinical laboratory (128). They used fungus-specific universal primer pairs to amplify the ITS regions of 137 clinical yeast isolates. They then digested the amplicons with MspI, a restriction enzyme. They determined that the RFLP pattern for each species tested was unique and allowed 100% accuracy in the identification of six *Candida* spp. from clinical specimens (128).

Flow cytometry. Flow cytometry is another method used to analyze amplified nucleic acid products. This may be done in two ways. One method uses species-specific DNA oligonucleotides that are attached to beads and hybridized to amplicons of the D1/D2 domain of the LSU (26S) rRNA gene, which are formed by initiation with biotinylated primers. Only amplicons with a sequence sufficiently complementary to the species-specific primer will hybridize to the bead and therefore be detected by a flow cytometer that was especially developed for this procedure by Luminex Corp. (Austin, TX). In the second method, species-specific primers are hybridized to complementary sites on the D1/D2 amplicons and, during the subsequent extension reaction, incorporate biotin-dCTP. All extension primers are hybridized to beads, but only extended products are biotinylated and detected by the Luminex flow cytometer (161). The former method takes less than 5 hours to complete from culture to identification, while the latter method takes 8 to 10 hours (161). Diaz and Fell used this method to identify various species of *Trichosporon* (47). Page et al. used this method to analyze 88 clinical specimens; 79 were identified correctly, compared to 81 correct identifications with phenotypic methods (162). No misidentifications were made; however, nine isolates were not identified by flow cytometry because oligonucleotide probes were not available for those species. Importantly, the assay took only 5 hours, compared to up to 5 days for phenotypic identification (162).

Real-time PCR. Most recently, "real-time" PCR methods have been developed for the detection and identification of yeast isolates from clinical specimens. Advantages of these systems include quantitative results and the lack of postamplification manipulation (53). Shin and colleagues used a TaqMan real-time PCR assay to analyze 81 blood culture bottles, including 61 with positive growth of *Candida*. The probes detected and correctly identified the yeast species in 58 of 61 specimens (95%) and gave no false-positive results (200). Guiver et al. used a TaqMan real-time PCR assay to identify *C. albicans*, *C. glabrata*, *C. kefyr*, *C. krusei*, and *C. parapsilosis* with species-specific primer and probe sets. The assay was shown to be 100% specific, with no cross-reactions (70). Metwally et al. used three different TaqMan real-time PCR assays to identify six *Candida* spp. directly from blood culture bottles positive for yeast. Specifically, the investigators sought to differentiate *C. glabrata* and *C. krusei*, species that are often resistant to fluconazole, from species typically susceptible to this agent (*C. albicans*, *C. dubliniensis*, *C. parapsilosis*, and *C. tropicalis*). The assay for the fluconazole-susceptible group utilized universal fungal primers and a fluconazole-sensitive species-specific probe to amplify a conserved portion of the 18S rRNA sequences, whereas the sequences of probes for the *C. glabrata* and *C. krusei* assays were specific to the variable sequences of the ITS2 region. The assays were applied to 43 blood culture bottles, including 33 that were culture positive for yeast; 100% sensitivity and specificity were seen in the identifications made by the PCR assays compared with phenotypic assays. Furthermore, the assays took less than 3 hours to reach a species identification (125). Finally, Selvarangan et al. used a real-time PCR assay to study 61 blood culture bottles which were positive for yeast. Identification directly from positive blood culture bottles was achieved by rapid-cycle nucleic acid amplification of the ITS2 DNA, fluorescent-probe hybridization to specific sequences in the ITS2 amplicons, and melting curve analysis. The authors reported 100% agreement with identifications by traditional methods, and the assay took approximately 2 hours to complete (194).

Automated PCR. Most recently, Wise et al. used a commercially available automated system, DiversiLab (Bacterial Barcodes, Inc., Athens, GA), to sequence clinical *Candida* isolates by using repetitive extragenic palindromic sequence-based PCR (rep-PCR). This method uses primers that target noncoding repetitive sequences interspersed throughout the fungal genome. Wise and colleagues used this system to examine 156 clinical *Candida* isolates originally identified biochemically and morphologically as *C. albicans*, *C. glabrata*, *C. parapsilosis*, *C. tropicalis*, *C. krusei*, and *C. lusitaniae*. Their initial results showed that the DiversiLab system was 95% concordant with biochemical/morphological identification. However, ITS sequencing of the discrepant samples showed that nine were misidentified by traditional methods. For these isolates, the ITS sequencing identification was in agreement with that of the DiversiLab system, yielding an actual correlation of 99%. The DiversiLab rep-PCR system is commercially available and has shown, in limited studies, promise for use in the identification of medically important yeasts (232).

Sequencing. Sequence analysis is one of the most promising techniques for the identification of fungi, including yeasts, following nucleic acid amplification with PCR (34). Following amplification using universal fungal primers, amplicon sequences are determined and compared with sequences of the same fragment found in public databases (34). In fact, the sequence of the D1/D2 domain of the LSU (26S) rRNA gene has become accepted as the main tool for yeast taxonomy (45). Kurtzman and Robnett used this methodology to analyze approximately 500 species of ascomycetous yeasts, reporting sequences and phylogenic relationships (105). Fell et al. and Scorzetti et al. showed that D1/D2 sequencing could identify most basidiomycetous yeast species as well (58, 193). Linton et al. described their experience of sequencing the D1/D2 region of the LSU gene at the United Kingdom Mycology Reference Laboratory over a 2-year period. Of a total of 3,033 clinical isolates encompassing 50 different yeast species, >90% could be identified by using a commercial biochemical identification kit. One hundred fifty-three isolates (5%) consisting of 47 species could not be identified and were therefore subjected to molecular identification. All 47 species requiring molecular identification were identified unambiguously with D1/D2 sequencing (110). Hall et al. used the commercially available MicroSeq D2 LSU rDNA sequencing kit (Applied Biosystems, Foster City, CA) to analyze 131 yeast isolates identified from clinical specimens. They identified 94% of the specimens correctly; most discrepancies were due to a lack of sequence inclusion in the MicroSeq database (75).

Leaw and colleagues sequenced the ITS1 and ITS2 regions to identify 299 reference strains and 73 clinical isolates. The rates of correct identification using ITS1 and ITS2 sequencing were 97% and 99.7%, respectively. Only one strain (*Rhodotorula glutinis*) could not be identified by ITS2 sequence analysis (107). Similarly, Ciardo et al. used ITS sequence analysis to identify 113 yeast strains; 98% of all strains were identified correctly to the species level, compared to 87% with a rapid phenotypic assay (38).

Sequencing methods are not without potential obstacles. Hall and colleagues demonstrated that sequencing costs $29.50 more than an API 20C Aux biochemical kit (75). This cost, in addition to the technical expertise required to apply these methods, is prohibitive to most laboratories. In addition, the identifications reached by sequencing are only as good as the sequence libraries available. For example, Hall et al. showed that the MicroSeq library was somewhat limited, not including *C. dubliniensis* at the time of their study (75). In addition, publicly maintained databases such as GenBank must be used with caution, as they are not refereed and therefore may contain errors (34, 53). Despite these potential limitations, sequencing may have an important role as a means to provide identifications for organisms that cannot be identified by other, more routinely used methods.

Molecular identification of *Cryptococcus* spp.

Studies of molecular identification methods for yeast infections have concentrated on the genus *Candida*. Molecular identification of *C. neoformans* has not been investigated thoroughly, perhaps because culture and antigen detection are highly reliable. However, a DNA probe specific to *C. neoformans* (Gen-Probe Corp., San Diego, CA) may be applied to yeast in culture to identify *C. neoformans* with 100% sensitivity and specificity (86). This product, however, was recently taken off the market. Finally, Mitchell et al. utilized nested PCR for the identification of *C. neoformans* in 40 clinical cerebrospinal fluid (CSF) samples, using the ITS regions. They correctly identified all 21 culture-proven cases of cryptococcal meningitis (131).

Drawbacks to Culture-Based Systems for Identification

Culture systems are not without drawbacks. Generally, the time from the collection of the specimen to the identification of the yeast species is at least 24 hours, and often more. As a result, the patient often has quite an advanced stage of infection before targeted treatment can be initiated. In addition, these methods often suffer from low sensitivity. Blood cultures, still considered the most reliable marker of IC, are estimated to be positive in only 58% of cases (34). In addition, although commercially available biochemical identification systems are quite accurate with commonly encountered yeasts, their identifications are less precise with less frequently isolated species (75). Indeed, as mentioned above, these less common yeast species may be more resistant to traditional antifungal agents.

Non-Culture-Based Methods

Assays that can be performed directly on the clinical specimen have the potential to guide initial, focused therapy. Historically, these methods were host response assessments and direct examination. Because yeast species other than *C. albicans* and *C. neoformans* are increasingly encountered and because non-*C. albicans Candida* spp. are frequently resistant to fluconazole and other azole drugs, identifying yeasts to the species level is critically important. The diagnosis of IC and other invasive yeast infections is difficult to make, as the signs and symptoms are relatively nonspecific (34). These reasons have provided an incentive to develop new technologies for the detection of yeast (Table 3).

Table 3. Non-culture-dependent diagnostic assays

Test	Description	Example(s)	Commercial availability in the United States
Direct microscopic examination	Microscopic evaluation of organisms obtained directly from clinical specimens	Gram stain	+
		Calcofluor white	+
Cryptococcal antigen detection	Detection of antigen from the polysaccharide capsule of *C. neoformans* in various clinical specimens	Latex agglutination	+
		EIA	+
Mannan detection	Detection of mannan, a polysaccharide released from the outer cell wall of many *Candida* spp. during invasive infection	Pastorex Candida (latex agglutination)	+
		Platelia Candida (EIA)	+
Unknown antigen detection	Agglutination of anti-*Candida* antibody-coated latex particles in the presence of an unknown antigen	Cand-Tec (latex agglutination)	−
Cytoplasmic antigen detection	Detection of proteins found in the cytoplasm of many pathologic yeast species; presumably, the antigens are more detectable during an active infection	Enolase	−
		Heat shock protein	−
BDG detection	BDG is detected spectrophotometrically in the sera of patients with invasive yeast infections	Fungitell	+
Metabolite detection	Detection of D-arabinitol, a metabolite produced by many yeast species, by an enzymatic chromogenic assay	D-Arabinitol	−
Nucleic acid detection	Detection of yeast genetic material directly from clinical specimens	Traditional PCR	+[a]
		Real-time PCR	+[a]
		NASBA	+[a]

[a]No FDA-cleared assays available.

Direct microscopic examination

Direct microscopic examination of clinical specimens remains an important approach to the diagnosis of yeast infections. These studies can be performed in a clinical laboratory or in an anatomic pathology service. Advantages include the following: (i) the time to results is relatively short; (ii) for those yeast species that produce hyphae, detection of the hyphae may help in assessing the clinical relevance of the yeast; and (iii) a positive result from a tissue specimen acquired by an invasive procedure under sterile conditions is usually diagnostic.

Limitations of direct examination include the following: (i) false-negative results occur, due to the small number of fungal elements at sites of infection, due to sampling differences, or simply due to the fact that direct examination is inherently less sensitive than culture; (ii) false-positive results may also occur due to artifacts that appear as budding yeast cells; (iii) examinations and interpretations need to be performed by well-trained individuals; and (iv) cultures remain necessary to determine species-level identification and in vitro susceptibility results.

Several stains can be used to aid in the detection of yeast and diagnosis of infections. Gram stain can be useful since most yeasts stain gram positive, with an exception being *C. neoformans*, since the capsule may prevent complete staining of the cell wall. Calcofluor white, a fluorescent stain that must be read on a UV microscope with a correct filter, can increase the sensitivity of detection of yeast in clinical specimens (72). Giemsa (Wright's) stain can be used to detect yeast in peripheral blood and bone marrow smears. In addition, yeast cells can usually be visualized without stains; the use of KOH is a standard approach in mycology laboratories. KOH lyses host cells, secretions, and other interfering material without affecting yeast cells. Training of readers and recognition of artifacts are crucial with KOH preps. A wet prep of urine can be used to detect yeast cells. The capsule of *C. neoformans* can be visualized by use of an India ink preparation in the clinical laboratory or by use of mucicarmine, Alcian blue, or Halle's colloidal iron stains in an anatomic pathology laboratory (235). Specifics about the staining procedures are available in the *Manual of Clinical Microbiology* (79).

Antigen detection

During the 1970s and 1980s, many EIAs were developed, and the hope was to detect circulating antigens for all microbial pathogens, including medically important yeasts. This technology never reached that goal but has contributed significantly to our diagnostic capabilities, in a few cases even replacing culture as a primary diagnostic modality.

Cryptococcal polysaccharide capsule antigen. Cryptococcal antigen detection is one of the best antigen tests for the diagnosis of an infectious disease. The capsular antigen reaches excess rapidly and is very stable, and the test has excellent sensitivity and specificity. CSF, serum, bronchoalveolar lavage, and pleural fluid specimens can be assayed with this methodology. The sensitivity of testing of CSF for the diagnosis of meningitis is 90% or higher for immunocompromised patients (226), and antigenemia may be detected days before organisms may be detected by culture. A similar sensitivity is attainable by testing serum specimens from cases of disseminated infection in immunocompromised individuals. Serum testing is also important for patients with suspected central nervous system disease when CSF specimens are unattainable. For the diagnosis of cryptococcal pneumonia, testing of bronchoalveolar lavage and serum specimens yields sensitivities of 90% and 67%, respectively (14). False-negative results may occur for patients with minimal pathology or focal infections (usually pulmonary), presumably due to a small number of replicating cryptococcal cells. Semiquantitative data (titers) can be determined with CSF specimens; the use of these data is controversial, and the results must be interpreted with caution. Cultures may ultimately be required to address treatment efficacy and relapse issues.

Two different methodologies have been developed for detection of the cryptococcal antigen. The original format was a latex agglutination assay; subsequently, an EIA was developed. Kits for both formats are commercially available; they are reliable and provide important data for the diagnosis and management of cryptococcosis (64). Both formats may occasionally produce false-positive results with sera from patients with trichosporonosis, due to cross-reactivity. Agar (21) may also cause false-positive results, as may circulating proteins (rheumatoid factor, for example) with the latex agglutination assay format, requiring pronase pretreatment to increase both sensitivity and specificity (68).

Candida antigen detection

Mannan. Early attempts at antigen detection in the diagnosis of IC focused on mannan, a polysaccharide found in the cell walls of many *Candida* species (225). Mannan is released from the outer cell wall during invasive infection and enters the bloodstream (219, 237). A large body of evidence has accumulated for detection of this fungal macromolecule by latex agglutination assay and EIA. Early data were disappointing, with low sensitivities and/or specificities reported (46, 80, 169). The EIAs were generally more sensitive than the latex agglutination methods (237), but immune complexes and rapid clearance of mannan contributed to the poor sensitivities of these assays (219, 237). Dissociation of immune complexes by use of heat and the development of reliable monoclonal antibodies to react with specific defined epitopes improved the assay (219, 237). Like the case with other antigen detection assays, detection improved when multiple samples were tested (237). A double-sandwich assay that utilizes the monoclonal antibody EB-CA1 (Platelia Candida antigen test; Bio-Rad Laboratories, Marnes-la Coquette, France) (197) is commercially available. However, low reported sensitivities varying from 40% to 68% have hampered its clinical use. Some investigators have had encouraging results by combining the Platelia Candida antigen test with assays for detecting anti-mannan antibodies (197, 238). Despite the wealth of knowledge about mannan, the place of assays for its detection in the clinical setting remains unclear.

Latex agglutination of unknown antigens. The Cand-Tec test (Ramco Laboratories, Houston, TX), introduced in 1983, is an assay that was designed to detect unknown antigens by using a latex agglutination format (219). It is based on the agglutination of rabbit anti-*Candida* antibody-coated latex particles in the presence of antigen (130). Studies have reported widely variable sensitivities (0 to 100%) and specificities (29 to 100%). These inconsistent results, along with the unknown nature and function of the target antigen, have prevented the Cand-Tec assay from acceptance in the United States (237). However, a modified Cand-Tec test, the Cand-Tec microtiter system (Cand-Tec MT, Ramco, Japan), which is believed to offer a more objective means to evaluate results through quantitative colorimetric analysis, is available in Japan. One study showed improved sensitivity and specificity compared to the traditional Cand-Tec assay (130).

Cytoplasmic antigens: enolase and heat shock proteins. The utility of the detection of cytoplasmic antigens for the diagnosis of invasive *Candida* infections has been investigated. The rationale is that release of cytoplasmic antigens would result from degradation of yeast cells during deep infections as opposed to superficial, non-life-threatening mycoses or mucosal colonization (5, 53). Strockbine et al. reported the discovery of a 48-kDa immunodominant protein antigen of *C. albicans* by the use of Western blots (206). The protein was subsequently identified as the enzyme enolase. Walsh and colleagues performed a prospective multicenter trial investigating the recovery of enolase from the sera of 170 oncology patients at risk for IC and 140 healthy adults by using a double-sandwich liposomal immunoassay manufactured by Becton Dickinson (Directigen, Baltimore, MD). The results were encouraging, including a sensitivity of 76% and a positive predictive value of 63% for invasive yeast infections. In addition, the specificity and negative predictive value were >95%. Interestingly, it was noted that for some patients with antigenemia and histologic evidence of candidiasis, no positive blood cultures were recovered. In addition, in some cases, specificity was increased with additional samples being tested (221). Unfortunately, the assay was never accepted and was withdrawn by the company.

Matthews and Burnie investigated the detection of a 47-kDa heat shock protein in the sera of 105 patients with invasive yeast infections as well as 32 patients with superficial candidiasis or colonization and 125 patients with no evidence of fungal infection. They reported a sensitivity, specificity, positive predictive value, and negative predictive value of 76%, 94%, 89%, and 86%, respectively, for invasive infections caused by yeast. They noted false-positive results for hemolyzed samples and for one patient with candiduria. False-negative results were seen for non-*C. albicans* infections (*C. glabrata*, *T. beigelii*, and *C. parapsilosis*); limiting analysis to *C. albicans* only increased the sensitivity to 84% and the negative predictive value to 91%. Sensitivity was also increased markedly with additional samples taken (120). This assay, however, is not commercially available.

Detection of fungal structural macromolecule BDG by chemical methods

(1→3)-β-D-Glucan (BDG) is a major cell wall constituent in most pathogenic fungi; notable exceptions include most basidiomycetes (including *C. neoformans*) and the zygomycetes (53, 134, 153, 154). *C. albicans* has been shown to release glucan into the culture medium during growth (97); it is presumed that BDG is released into the tissues and the bloodstream during invasive fungal infections (IFIs), where it may be detected (140). BDG is not found in humans, bacteria, or viruses; consequently, its detection in the blood of patients has been investigated as a marker for IFI (53, 226, 227).

BDG as an indicator of IFI was first investigated in Japan. It was noted that tests which utilized the coagulation cascade of the horseshoe crab were not specific for bacterial endotoxin as originally believed; they were also sensitive for the detection of BDG (133). The *Limulus* amebocyte coagulation cascade was subsequently shown to consist of two pathways. One pathway consists of factor B and factor C, which react to endotoxin, while the second pathway consists of factor G, which was shown to respond to BDG (134, 153). BDG binds to the β-subunit of factor G and consequently activates the β-subunit, a serine protease, to activate the clotting cascade (154). Early studies measured BDG levels by subtracting the value of an endotoxin-specific *Limulus* assay from that of a nonspecific, traditional *Limulus* test to yield a "fungal index" (96, 132, 133, 134). Later, an assay was developed to measure BDG directly by adding a chromogenic substrate to factor G from *Tachypleus tridentatus* (the Japanese horseshoe crab), measuring BDG spectrophotometrically, and quantifying the value by comparison to a standard curve (154, 159, 172). The assay is commercially available in Japan as the Fungitec-G test (Seikagaku Corporation, Tokyo, Japan). Similar tests became available in the United States for research purposes as the Glucatell test (Associates of Cape Cod, Falmouth, MA) and commercially in 2004 as the Fungitell test (Associates of Cape Cod). The assay was FDA cleared in 2004 (119). These assays differ from the Fungitec-G assay in their use of factor G from a different species of horseshoe crab (*Limulus polyphemus*); due to a slightly lower affinity for factor G, a higher cutoff for a positive test is employed (154, 159). Both the Fungitec-G and Fungitell tests are able to produce results within 2 hours (53, 154).

Known causes of false-positive results include hemodialysis using cellulose membranes (which contain cross-reacting polysaccharides), intravenous administration of plasma components, such as intravenous immunoglobulin or albumin (which are filtered through cellulose membranes during their manufacture), and exposure to glucan-containing gauze or related materials (53, 154, 226). Known causes of false-negative results include infections caused by *C. neoformans* or zygomycetes (which have little or no BDG in their cell walls), aspergillomas (walled-off colonizations that usually do not allow fungal antigens to enter the bloodstream), and high concentrations of bilirubin or triglycerides (which are inhibitory to the assay) (154, 159, 172, 226). It has been suggested that certain antimicrobials may cause false-positive results, possibly because many of these agents are actually produced by fungi (119). Marty and associates found that some antimicrobials, including colistin, cefepime, and trimethoprim-sulfamethoxazole, did yield positive BDG assay results, but not at antimicrobial concentrations typically used therapeutically (119).

Many studies have been undertaken to investigate the performance of these assays in the detection of IFIs, including candidiasis and infections with other, noncryptococcal yeasts. Kohno and colleagues reported a sensitivity of 78% for proven candidemia in one of the first trials, using a "fungal index" of 60 pg/ml as the cutoff for a positive result (96). Mitsutake and coworkers also investigated the use of a "fungal index" and reported promising results, with a sensitivity, specificity, positive predictive value, and negative predictive value of 84%, 88%, 84%, and 88%, respectively, for IC (132). Trials investigating the Fungitec-G test produced similar results. Investigators reported sensitivities varying from 90 to 100%, specificities of 92 to 100%, positive predictive values of 64 to 100%, and negative predictive values of 88 to 100% for IC (97, 134, 140, 153). Notably, these studies varied in their settings (from single university hospitals to multicenter trials), patient populations investigated, and number of samples tested per patient. Predictably, cryptococcal infections, aspergillomas, and superficial yeast infections were consistently associated with negative BDG levels. Additional false-negative results included infections with *C. parapsilosis* and *C. tropicalis*. No detailed information about false-positive results was provided in these publications.

Clinical studies performed in the United States with the Fungitell or Glucatell assay have produced generally encouraging results and are summarized in Table 4. The sensitivities for proven invasive yeast infections ranged from 84% to 100%; the sensitivities were lower if "probable" infections were included or when all types of IFI (i.e., mould infections as well as yeast infections) were investigated. Sensitivity was increased, however, when multiple samples per patient were tested (154). Specificity for proven invasive yeast infection varied widely, from 53% to 93%. This is not unexpected, as the assay is not intended to be specific for yeast (i.e., ideally, most invasive mould infections are expected to yield a positive result). Positive predictive values for proven invasive yeast infections also ranged greatly, from 20 to 64%, while negative predictive values were consistently high (90 to 100%). Notably, Digby and associates utilized a cutoff of 20 pg/ml; the remaining studies either analyzed multiple cutoffs or used a cutoff of 60 pg/ml (the manufacturer recommends interpretation of levels of 60 to 79 pg/ml as "intermediate" and of 80 pg/ml or greater as positive). This likely contributed to the large number of false-positive results they reported, consequently yielding a specificity of only 54% for IFI. False-positive results have been seen for patients with bacterial infection (49, 172), patients on hemodialysis (159, 172), patients given intravenous immunoglobulins (172), and patients treated with beta-lactam antibacterial agents (172). As seen in the Japanese trials, false-negative results occurred with *C. parapsilosis* and *Candida rugosa* (159, 172).

In light of the evidence, detection of BDG does provide valuable, rapid results as an aid in the diagnosis of invasive yeast infections. The sensitivities and negative predictive values have been consistently high in various trials; clinicians using this test would be unlikely to miss a true invasive yeast infection, and a negative result could be interpreted with confidence as a true lack of such an infection. A relatively high rate of false-positive results has been noted by some authors, however. The predisposing factors that may lead to a false-positive result are fairly common in critical patients, so care must be taken in the interpretation of a positive result when the patient is on hemodialysis, for example.

D-Arabinitol detection

The diagnosis of invasive yeast infections may be aided by the detection of fungal metabolites. Metabolite detection seems very attractive because, theoretically, the concentration should increase in the host with growth and decrease with growth inhibition (237). The most widely studied of the yeast metabolites is D-arabinitol, a five-carbon acyclic polyol not produced by humans (185, 237). In the late 1970s, it was first discovered that *C. albicans* produces large amounts of D-arabinitol in culture and that this compound could be detected by gas liquid chromatography in 75% of patients with IC, with 97% specificity (95); it was later found that *C. tropicalis*, *C. parapsilosis*, *C. kefyr* (*pseudotropicalis*) (18), *C. lusitaniae*, and *C. guillermondii* (106) also produce this metabolite in large quantities in vitro. Interestingly, *C. glabrata*, *C. krusei*, and *C. neoformans* have not been shown to produce D-arabinitol in culture (18, 106).

Wong et al. hypothesized that D-arabinitol accumulates in patients and showed that determination of the ratio of D-arabinitol to serum creatinine (Cr) could compensate for renal dysfunction (234). Using this method, Gold et al. demonstrated 64% sensitivity and 97% specificity in the diagnosis of IC against 73 oncology controls, 21 of whom had renal failure (67). Several studies have been reported, including a study performed by Walsh and colleagues (222). That study was a multicenter trial with industry interest investigating the use of D-arabinitol/Cr from the sera of 678 patients by an enzymatic chromogenic assay. They demonstrated 62% sensitivity, 95% specificity, a 57% positive predictive value, and a 96% negative predictive value. In addition, these authors reported the early rise in D-arabinitol and the correlation of D-arabinitol/Cr values with clinical response to treatment (222).

To summarize, the detection of D-arabinitol has been investigated as a tool in the diagnosis of IC, and although there were promising results, commercialization has not occurred.

Table 4. American clinical trials investigating the BDG assay for diagnosis of invasive candidiasis[a]

Authors, yr (reference)	Assay	Site	Patient population	Cutoff(s) (pg/ml)	No. of samples taken/ patient	Sensitivity[b] (infection, %)	Specificity[b] (infection, %)	PPV[b] (infection, %)	NPV[b] (infection, %)	Cause of FP results	Cause of FN results
Digby et al., 2003 (49)	"G test" (Cape Cod)	Multicenter (three sites)	46 ICU patients with bacterial or fungal pulmonary or blood cultures, 9 control ICU patients, 8 healthy controls (fungal etiologies not specified)	20	1	IFI, 95	IFI, 53.7	IFI, 47.5	IFI, 95.7	Bacterial infection (blood or pulmonary)	Not stated
Odabasi et al., 2004 (154)	Glucatell (Cape Cod)	University hospital	30 nonneutropenic patients with culture-proven candidemia, 30 healthy adults	60	1	IC, 97	IC, 93	IC, 94	IC, 97	Not stated	Not stated
Odabasi et al., 2004 (154)	Glucatell (Cape Cod)	University hospital	283 neutropenic patients with newly diagnosed AML or MDS undergoing induction chemotherapy and receiving antifungal prophylaxis, including 16 with proven fungemia, 4 with probable fungal pneumonia, and 33 with possible fungal pneumonia	60	At least 2	Proven IFI, 100; proven invasive yeast, 100; proven and probable IFI, 100	Proven IFI, 88.4; proven invasive yeast; 88.7; proven and probable IFI, 89.7	Proven IFI, 34 proven invasive yeast; 29.8; proven and probable IFI, 42.6	Proven IFI, 100 proven invasive yeast; 100; proven and probable IFI, 100	Not stated	Not stated
Ostrosky-Zeichner et al., 2005 (159)	Fungitell (Cape Cod)	Multicenter (six sites)	142 proven and 21 probable cases of IFI, with various underlying diseases; 170 controls (healthy adults, adults with minor medical problems, or hospitalized nonmycotic patients)	40, 50, 60, 80, 100, 125, 150	1	Proven IFI, 73.9; proven IC, 81.3; proven and probable IFI, 69.9	Proven IFI, 83.8; proven IC, 78.3; proven and probable IFI, 87.1	Proven IFI, 77.2; Proven IC, 64; Proven and probable IFI, 83.8	Proven IFI, 81.2; proven IC, 89.8; proven and probable IFI, 75.1	HD	*C. parapsilosis*
Pickering et al., 2005 (172)	Fungitell (Cape Cod)	University hospital	15 patients with yeast in blood, 1 with yeast and GPC in blood, 14 with GPC in blood, 10 with GNR in blood, 16 with suspected histoplasmosis, 64 with suspected invasive aspergillosis, 36 healthy control blood donors	60, 80	1	Proven IFI, 94.4; proven and probable IFI, 96.3; proven IC, 94.1	Proven IFI, 52.9; proven and probable IFI, 70.6; proven IC, 52.5	Proven IFI, 20.7; proven and probable IFI, 63.4; proven IC, 19.5	Proven IFI, 98.6; proven and probable IFI, 97.3; proven IC, 98.6	GPC bacteremia, IVIG, HD, beta-lactams	*C. rugosa*

[a]PPV, positive predictive value; NPV, negative predictive value; FP, false positive; FN, false negative; IFI, invasive fungal infection; IC, invasive candidiasis; AML, acute myelogenous leukemia; MDS, myelodyplastic syndrome; HD, hemodialysis; GPC, gram-positive cocci; GNR, gram-negative rods; IVIG, intravenous immunoglobulin.
[b] When more than one cutoff was analyzed, the data for 60 pg/ml are presented

Yeast nucleic acid detection

Many recent studies have centered on the detection of nucleic acids from clinical specimens as a method of diagnosing invasive yeast infections. There are several reasons that this is an attractive methodology. First, these methods have the potential to reach species-level identifications very rapidly (53, 219). As opposed to culture methods that require viable organisms, nucleic acid may be detected from dead organisms as well (53). For this reason, nucleic acid amplification techniques have potential sensitivities superior to those of traditional culture techniques (53, 116, 237).

Sample preparation. In nucleic acid detection assays, the manner in which the samples are prepared can have a profound impact on sensitivity and specificity (53, 117, 219, 237). Clinical sources might include blood or plasma, urine, CSF, bronchoalveolar lavage fluid, or even intraocular fluid. Goals of sample preparation include the release of intracellular DNA, concentration of DNA targets, and elimination of contaminants or potential inhibitors without damaging target nucleic acid sequences. Although many methods have been attempted and reported as successful (24, 25, 92, 188), there is no standardization in these methods (117, 237).

Target selection. Various nucleic acid sequence targets have been selected for identification of yeast directly from clinical specimens. Single-copy and multicopy nuclear, mitochondrial, and RNA targets have been studied. Specifically, examples of targets have included genes for actin (92), chitin synthase (90), cytochrome P450 enzymes (25, 137), heat shock protein (42), ITS regions of rRNA (26, 205, 236), and 18S RNA (51, 187). The use of multicopy targets generally lowers detection thresholds and enhances sensitivity over that with single-copy targets (237). In particular, multicopy ribosomal genes are attractive, as they contain conserved sequences common to all fungi, which are useful for screening for any fungus, as well as variable domains and highly variable ITS regions that may be used for identification at the species level (237).

Detection and amplification methods. Nucleic acid hybridization and amplification are at the heart of molecular diagnostic methods for invasive yeast infections (237). Hybridization utilizes a single strand of synthesized nucleic acid complementary to a recognized nucleic acid sequence of the pathogen in question (237). ISH with rRNA-targeted, fluorescently or chemically labeled oligonucleotide probes for yeast may be performed on formalin-fixed tissue sections, a great advantage if fungal infection was not suspected and tissue was therefore not sent for culture. Some investigators have reported success with detecting and identifying *Candida* species in experimental animals by using ISH (111, 112). Hayden et al. reported excellent sensitivity and specificity with a *Candida*-specific oligonucleotide probe for a small series of tissue sections with fungal organisms visualized microscopically (78). Although this is a promising technique, the sensitivity of FISH is often lower than those of other molecular assays, and more studies are needed to determine its clinical utility (237).

PCR is the most utilized nucleic acid amplification technique (219, 237). In general, as mentioned above, multicopy gene targets have increased sensitivity over single-copy DNA sequences (181). Following amplification of targeted nucleic acid, the amplification product (amplicon) must be analyzed. In earlier studies, the PCR-generated product was usually analyzed by ethidium bromide-stained gel electrophoresis. Later, Southern blotting was utilized. However, both Southern blotting and gel electrophoresis are time-consuming and produce results that are somewhat subjective. As a result, PCR-EIA was developed and has been utilized (26, 39, 52, 199). In this method, detection of the amplification products by EIA provides a colorimetric or fluorescent result. Flahaut et al. reported 100% specificity as well as a sensitivity comparable to that of Southern blotting with PCR-EIA (60).

Most recently, quantitative real-time PCR methods have been described for the detection of *Candida* species in clinical specimens. The TaqMan PCR assay (Applied Biosystems, Foster City, CA) combines PCR, probe hybridization, and signal generation into one step (237). The assay is quantitative; the amount of reporter dye released is proportional to the amount of DNA amplified (116). Investigators have reported excellent sensitivity and specificity in using this method, with the time required to complete the assay decreased compared to that for traditional PCR methods (116, 200). The other major real-time PCR method relies on dual-probe chemistries, often together with LightCycler instrumentation (Roche Diagnostics, Mannheim, Germany). This method combines rapid thermocycling with glass capillaries with fluorescence detection of the amplicon. Like TaqMan assay, the assay is quantitative; the amount of fluorescence is proportional to the amount of DNA amplified, and the test may also be performed quickly, in 5 to 6 hours (192). Advantageously, both TaqMan and dual-probe chemistries require no postamplification manipulations, decreasing the likelihood of laboratory contamination (53, 237). Investigators have reported promising findings in the detection of *Candida* species in clinical specimens by use of the latter method (23, 176, 192).

Some authors have investigated the detection of RNA instead of DNA from clinical specimens for the diagnosis of invasive yeast infections (113). Nucleic acid sequence-based amplification (NASBA) is a non-PCR-based technique for the amplification of RNA that has been used for the detection of viruses and parasites as well as yeasts (113). NASBA specifically amplifies RNA sequences by using T7 RNA polymerase in an isothermal process resulting in amplification of a billionfold in approximately 2 hours (113). Unlike PCR, NASBA does not require thermal cycling, and amplification and detection take less time, on average, than that required for PCR (53, 113).

Clinical studies. Although many molecular assays (user-developed assays) have been designed for the detection of yeast nucleic acid from clinical specimens, most studies examining the utility of these tests have been performed on a small number of specimens, generally at single institutions (1, 23, 60, 90, 113, 137, 187, 192, 199). Consequently, it is difficult to draw conclusions about the usefulness of these assays, as relatively few studies have been performed using large numbers of patient specimens. For the following discussion, we arbitrarily chose studies that analyzed 50 clinical specimens or more for assessing clinical performances. These studies are summarized in Table 5.

Rand et al. (179) sought to identify *Candida* nucleic acid from the blood of 59 patients by using PCR targeting the 18S rRNA gene. They reported positive results for 11 of 25 patients with culture-proven candidemia, for a sensitivity of only 44%. Specificity was high, however, at 94%, and analytic sensitivity was reported to be 1 CFU/ml of blood (179). Einsele and colleagues (51) also investigated the detection of candidemia in 601 blood samples through amplification of the 18S rRNA gene. They reported 100% sensitivity, detecting all eight patients with culture-proven candidemia; the analytical sensitivity was 1 CFU/ml of blood. In addition, only two false-positive results were reported, for a specificity of 99% (51). Morace et al. (136) investigated the identification of candidemia in 72 patients with hematological malignancies, fever, and neutropenia who were not responding clinically to antibacterial agents. In a PCR-based assay using the *ERG11* gene as the amplification target, the investigators identified 13 of 14 patients with disseminated candidiasis (sensitivity of 93%); for comparison, culture identified only 4 of these patients. The analytical sensitivity was 5 CFU/ml of blood. The specificity and negative predictive value of the assay were both 98% (136).

White et al. (228) attempted to diagnose candidemia by using a real-time PCR assay for 113 pa-

Table 5. Clinical studies evaluating the detection of nucleic acid from clinical specimens in >50 patients[a]

Authors, yr (reference)	Method	Target	Analytic sensitivity (CFU/ml of blood)	Patient population	Clinical sensitivity (no. of positive results/no. of patients [%])	Clinical specificity (no. of positive results/no. of patients [%])
Rand et al., 1994 (179)	PCR	18S rRNA gene	1	In patients with candidemia	11/25 (44)	32/34 (94.1)
Einsele et al., 1997 (53)	PCR	18S rRNA gene	1	Patients with IC, colonized patients, febrile neutropenic patients, and healthy controls	8/8 (100)	453/456 (99)
Morace et al., 1999 (136)	PCR	*ERG11* gene	5	Patients with hematological malignancies, neutropenia, and suspected IC	13/14 IC cases (92.9), including 4/4 with positive blood cultures	
White et al., 2003 (228)	Real-time PCR	18S rRNA gene	1-5	Patients with suspected IFI	27/113 (PCR positive), 3/113 (blood culture positive)	Not given
Maaroufi et al., 2003 (116)	Real-time PCR	rDNA	5	Patients with proven or suspected candidemia	11/15 (73)	Not given
Moreira-Oliveira et al., 2005 (138)	PCR	5.8S rDNA	10 (cells/ml of blood)	Patients at high risk for IC	31/43 (72.1)	91.2

[a]IC, invasive candidiasis; IFI, invasive fungal infection.

tients "at risk" for IFI. The target sequence was within the 18S rRNA gene. The assay identified 27 patients with *Candida* DNA in their blood, compared with only 3 patients identified with standard blood cultures. Interestingly, only two of the three culture-positive patients were positive with the PCR assay. It is therefore unclear whether the assay detected true cases of candidiasis or if it yielded a large number of false-positive results. The analytic sensitivity of the assay was 1 to 5 CFU/ml of blood (228). Maaroufi et al. investigated a real-time PCR assay for the detection of *Candida* rDNA in the blood of 61 patients with proven or suspected candidemia. Their assay was able to detect 5 CFU/ml of blood. Two probes were utilized; one was specific to *C. albicans*, and the other was specific to five common *Candida* spp. The sensitivity for the *C. albicans*-specific probe was 100%, with a specificity of 97%. Meanwhile, the sensitivity of the broader *Candida* probe was 100%, with a specificity of 72%. Both probes had a 100% negative predictive value (116). Finally, Moreira-Oliveira et al. examined a PCR-based assay for the detection of candidemia in 225 patients at high risk for IC. The assay targeted a 5.8S rDNA sequence and was able to detect 10 cells/ml of blood. The investigators reported a sensitivity of 72% and a specificity of 91%, with a positive predictive value of 66% and a negative predictive value of 93% (138).

CONCLUSIONS

We and many others have shown that the spectrum of infections caused by yeasts is not static; there are clear epidemiologic trends in the patterns of yeast infections. A major contributing factor is the overall increase in the number of susceptible hosts. Truly a credit of medical science's progress, more patients are living longer with profound immunocompromising states. Another factor due to changing medical practice is the movement to treat more compromised patients in an outpatient environment, which probably exposes the compromised host to a wider array of unusual yeast species. Consequently, while in the past *C. albicans* and *C. neoformans* were, for practical purposes, the yeast species a clinician needed to be concerned with, other species have now taken the stage. Non-*C. albicans Candida* spp. (e.g., *C. glabrata*, *C. kefyr*, *C. lusitaniae*, *C. parapsilosis*, *C. tropicalis*, *C. guilliermondii*, etc.) and non-*Candida*/non-*Cryptococcus* genera (e.g., *Trichosporon*, *Pichia*, *Malessezia*, *Rhodotorula*, *Saccharomyces*, etc.) are now pathogens of which clinicians must be aware. Finally, the emergence of species with intrinsic antifungal drug resistance or the ability to acquire resistance rapidly is a disturbing trend that has a significant impact on clinicians treating immunocompromised patients. One may not simply assume that one agent (e.g., fluconazole) will be active for the treatment of infection due to any yeast species. Today, antifungal treatment decisions are often made based on species identification. Thus, the major challenge for the clinical laboratory is rapid, accurate identification to the species level. A looming problem with the widespread use of antifungal agents is that resistance may emerge in many species (164). Continued efforts to standardize antifungal susceptibility testing, evaluate the relationship between in vitro MICs and clinical outcomes, and establish breakpoints and interpretive guidelines are also important.

Diagnostic methods have not kept pace with the changing landscape of clinically important yeasts. For the reasons mentioned above, the rapid identification of pathogenic yeasts to the species level is extremely important to immunocompromised patients. However, in most clinical microbiology laboratories, the identification of yeasts is based, at least in part, on phenotypic assays, including colony and microscopic morphology, as well as some biochemical reactions, including carbohydrate assimilation patterns. Recent advances include the introduction of chromogenic agars and rapid identification systems which are culture dependent. Several institution-specific (user-developed) molecular assays have been reported and show promise and feasibility for rapid identification, but industry participation is important in the effort to simplify these assays and to make them available in a standardized format for clinical laboratories nationally and internationally. The DiversiLab rep-PCR assay and the use of PNA FISH for the rapid identification of the five most common *Candida* spp. fit these criteria. While investigations into the use of mannan, enolase, D-arabinitol, and other antigens have produced promising results, their use remains limited. Of the non-culture-dependent assays, only the BDG assay is commercially available for use in diagnostic algorithms.

Molecular amplification methods for direct detection in clinical specimens are encouraging. Unfortunately, assays used in clinical studies are not commercially available. Therefore, some of the clinical studies are more proof-of-principle than well-designed clinical studies. Indeed, multicenter protocols are needed.

Finally, the various diagnostic approaches analyzed in this chapter can be used to develop diagnostic algorithms. An initial algorithm that can apply to direct detection of yeast in sterile body fluids from a

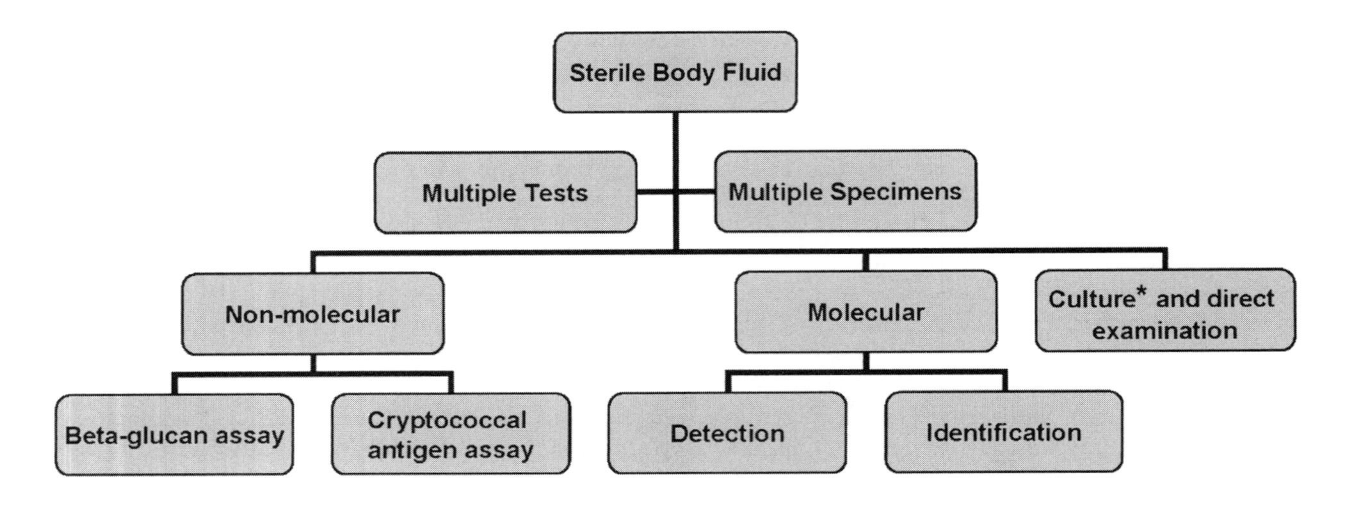

Figure 2. Algorithm for direct detection, with or without direct identification, of yeasts in clinical specimens.

patient with a clinically suspected invasive yeast infection or lower respiratory infection (when cryptococcosis is a possibility) is presented in Fig. 2. It must be emphasized that for every diagnostic modality presented, multiple samples need to be taken, preferably at different times, to obtain maximal sensitivity with high-risk populations. This could provide diagnoses before culture positivity or advanced clinical disease. In addition, there is no one completely reliable assay, and therefore a battery of tests is desirable. Of the nonmolecular, non-culture-based assays for yeast infections presented above, only BDG and cryptococcal capsule antigen assays are currently commercially available. A "positive" result for BDG in the appropriate context would support the presence of an IFI but would not be specific. The cryptococcal antigen tests, however, are organism specific. Nucleic acid detection assays, if available to the laboratory, can be useful as part of a battery of tests for diagnosis; however, these are not yet widely available or standardized. These assays may provide genus-specific (e.g., *Candida*) or species-specific identifications. Finally, direct examination and/or culture-based techniques remain the "gold standard" in the diagnosis of yeast infections.

An algorithm for a laboratory approach to culture-based yeast identification is provided in Fig. 3. When a yeast is recovered in the microbiology labo-

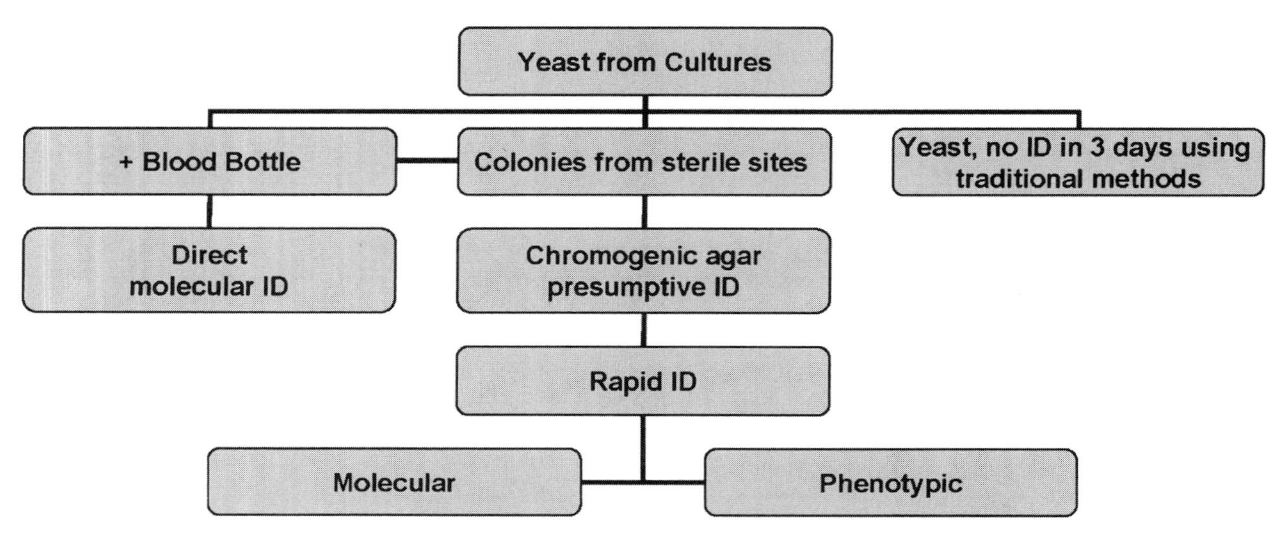

Figure 3. Algorithm for definitive identification of yeasts recovered from clinical specimens.

ratory, a multiassay strategy seems prudent. First, in the case of a positive blood culture bottle, identification could be made with direct molecular means. PNA FISH and real-time PCR are examples of assays that can yield species-level identifications rapidly, directly from blood culture bottles. Colonies recovered on a chromogenic medium and rapid phenotypic tests (e.g., trehalose assay) can provide a presumptive identification. These identifications can then be confirmed by commercial biochemical panels or molecular means, if available, to arrive at a final species identification. Examples of FDA-cleared molecular identification methods include *C. albicans* PNA FISH (from blood culture bottles) and Gen-Probe for *C. neoformans*. The rep-PCR (DiversiLab) method also has the potential to be very useful for rapid identification. If no identification can be reached within 3 days by using conventional means, the use of a molecular approach would be appropriate.

The hope is for the development of new, commercially available, FDA-cleared assays that can be incorporated into these algorithms, as well as integration of algorithms for in vitro susceptibility testing of pathogenic yeast species.

Acknowledgment. We thank Shelley Magill for her contributions to and thoughtful review of the manuscript.

REFERENCES

1. **Ahmad, S., Z. Khan, A. S. Mustafa, and Z. U. Khan.** 2002. Seminested PCR for diagnosis of candidemia: comparison with culture, antigen detection, and biochemical methods for species identification. *J. Clin. Microbiol.* **40:**2483–2489.
2. **Alexander, B. D., E. D. Ashley, L. B. Reller, and S. D. Reed.** 2006. Cost savings with implementation of PNA FISH testing for identification of *Candida albicans* in blood cultures. *Diagn. Microbiol. Infect. Dis.* **54:**277–282.
3. **Alexander, B. D., T. C. Byrne, K. L. Smith, K. E. Hanson, K. J. Anstrom, J. R. Perfect, and L. B. Reller.** 2007. Comparative evaluation of Etestin and Sensititre YeastOne panels against the clinical and laboratory standards institute M27-A2 reference broth microdilution method for testing *Candida* susceptibility to seven antifungal agents. *J. Clin. Microbiol.* **45:**698–706.
4. **Arago, P. A., I. C. Oshiro, E. I. Mannique, C. C. Gomes, L. L. Matsuo, C. Leone, M. L. Moretti-Branchini, A. S. Levin, and IRIS Study Group.** 2001. *Pichia anomala* outbreak in a nursery: exogenous source? *Pediatr. Infect. Dis. J.* **20:**843–848.
5. **Araj, G. F., R. L. Hopfer, S. Chesnut, V. Fainstein, and G. P. Bodey, Sr.** 1982. Diagnostic value of the enzyme-linked immunosorbent assay for detection of *Candida albicans* cytoplasmic antigen in sera of cancer patients. *J. Clin. Microbiol.* **16:**46–52.
6. **Archer-Dubon, C., E. Icaza-Chivez, R. Orozco-Topete, E. Reyes, R. Baez-Martinez, and S. Ponce de Leon.** 1999. An epidemic outbreak of *Malassezia* folliculitis in three adult patients in an intensive care unit: a previously unrecognized nosocomial infection. *Int. J. Dermatol.* **38:**453–456.
7. **Arikan, S.** 2007. Current status of antifungal susceptibility testing methods. *Med. Mycol.* **45:**569–587.
8. **Aspiroz, C., M. Ara, M. Varea, A. Rezusta, and C. Rubio.** 2002. Isolation of *Malassezia globosa* and *M. sympodialis* from patients with pityriasis versicolor in Spain. *Mycopathologia* **154:**111–117.
9. **Averbuch, D., T. Boekhoutt, R. Falk, D. Engelhard, M. Shapiro, C. Block, and I. Polacheck.** 2002. Fungemia in a cancer patient caused by fluconazole-resistant *Cryptococcus laurentii*. *Med. Mycol.* **40:**479–484.
10. **Back O., J. Faergemann, and R Hørnquist.** 1985. *Pityrosporum* folliculitis: a common disease of the young and middle-aged. *J. Am. Acad. Dermatol.* **12:**56–61.
11. **Baixench, M. T., A. Taillandier, and A. Paugam.** 2006. Clinical and experimental evaluation of a new chromogenic medium (OCCA, Oxoid) for direct identification of *Candida albicans*, *C. tropicalis*, and *C. krusei*. *Mycoses* **49:**311–315.
12. **Barber G. R., A. E. Brown, T. E. Kiehn, F. F. Edwards, and D. Armstrong.** 1993. Catheter-related *Malassezia furfur* fungemia in immunocompromised patients. *Am. J. Med.* **95:**365–370.
13. **Barchiesi, F., G. Caggiano, F. Di Francesco, M. T. Montagna, S. Barbuti, and G. Scalise.** 2004. Outbreak of fungemia due to *Candida parapsilosis* in a pediatric oncology unit. *Diagn. Microbiol. Infect. Dis.* **49:**269–271.
14. **Baughman, R. P., J. C. Rhodes, M. N. Dohn, H. Henderson, and P. T. Frame.** 1992. Detection of cryptococcal antigen in bronchoalveolar lavage fluid: a prospective study of diagnostic utility. *Am. Rev. Respir. Dis.* **145:**1226–1229.
15. **Baumgartner, C., A.-M. Freydiere, and Y. Gille.** 1996. Direct identification and recognition of yeast species from clinical material by using Albicans ID and CHROMagar Candida plates. *J. Clin. Microbiol.* **34:**454–456.
16. **Bergman, M. M., D. Gagnon, and G. V. Doern.** 1998. *Pichia ohmeri* fungemia. *Diagn. Microbiol. Infect. Dis.* **30:**229–231.
17. **Bernal, S., E. Martin Mazuelos, M. Garcia, A. I. Aller, M. A. Martinez, and M. J. Gutierrez.** 1996. Evaluation of CHROMagar Candida medium for the isolation and presumptive identification of species of *Candida* of clinical importance. *Diagn. Microbiol. Infect. Dis.* **24:**201–204.
18. **Bernard, E. M., K. J. Christiansen, S.-F. Tsang, T. E. Kiehn, and D. Armstrong.** 1981. Rate of arabinitol production by pathogenic yeast species. *J. Clin. Microbiol.* **14:**189–194.
19. **Bhally, H. S., C. Shields, N. Halsey, E. Cristofalo, and W. G. Merz.** 2006. Infection in a neonate caused by *Pichia fabianii*: importance of molecular identification. *Med. Mycol.* **44:**185–187.
20. **Birinci, A., L. Akkurt, C. Acuner, M. Unlu, and B. Durupinar.** 2004. Rapid identification of the *Candida* species from direct blood cultures by CHROMagar Candida. *J. Int. Med. Res.* **32:**484–487.
21. **Boom, W. H., D. J. Piper, K. L. Ruoff, and M. J. Ferro.** 1985. New cause for false-positive results with the cryptococcal antigen test by latex agglutination. *J. Clin. Microbiol.* **22:** 856–857.
22. **Bouchara, J. P., P. Declerck, B. Cimon, C. Planchenault, L. deGentile, and D. Chabasse.** 1996. Routine use of CHROMagar Candida medium for presumptive identification of *Candida* yeast species and detection of mixed fungal populations. *Clin. Microbiol. Infect.* **2:**202–208.
23. **Bu, R., R. K. Sathiapalan, M. M. Ibrahim, I. Al-Mohsen, E. Almodavar, M. I. Gutierrez, and K. Bhatia.** 2005. Monochrome LightCycler PCR assay for detection and quantification of five common species of *Candida* and *Aspergillus*. *J. Med. Microbiol.* **54:**243–248.

24. **Buchman, T. G., M. Rossier, W. G. Merz, and P. Charache.** 1990. Detection of surgical pathogens by in vitro DNA amplification. I. Rapid identification of *Candida albicans* by in vitro amplification of a fungus-specific gene. *Surgery* **108:** 338–347.
25. **Burgener-Kairuz, P., J.-P. Zuber, P. Jaunin, T. G. Buchman, J. Bille, and M. Rossier.** 1994. Rapid detection and identification of *Candida albicans* and *Torulopsis (Candida) glabrata* in clinical specimens by species-specific nested PCR amplification of a cytochrome P-450 lanosterol-(-demethylase (L1A1) gene fragment. *J. Clin. Microbiol.* **32:**1902–1907.
26. **Burnie, J. P., N. Golbang, and R. C. Matthews.** 1997. Semiquantitative polymerase chain reaction enzyme immunoassay for diagnosis of disseminated candidiasis. *Eur. J. Clin. Microbiol. Infect. Dis.* **19:**346–350.
27. **Burnik, C., N. D. Altintas, G. Ozkaya, T. Serter, Z. T. Selcuk, P. Firat, S. Arikan, M. Cuenca-Estrella, and A. Topeli.** 2007. Acute respiratory distress syndrome due to *Cryptococcus albidus* pneumonia: case report and review of the literature. *Med. Mycol.* **45:**469–473.
28. **Cardenes, C. D., A. J. Carrillo-Munoz, A. Arias, C. Rodriguez-Alvarez, A. Torres-Lana, A. Sierra, and M. P. Arevalo.** 2004. Comparative evaluation of four commercial tests for presumptive identification of *Candida albicans*. *J. Microbiol. Methods* **59:**293–297.
29. **Carricajo, A., S. Boiste, J. Thore, G. Aubert, Y. Gille, and A. M. Freydiere.** 1999. Comparative evaluation of five chromogenic media for detection, enumeration and identification of urinary tract pathogens. *Eur. J. Clin. Microbiol. Dis.* **18:**796–803.
30. **Chakrabarti, A., K. Singh, A. Narang, S. Singhi, R. Batra, K. L. Rao, P. Ray, S. Gopalan, S. Das, V. Gupta, A. K. Gupta, S. M. Bose, and M. M. McNeil.** 2001. Outbreak of *Pichia anomala* infection in the pediatric service of a tertiary-care center in northern India. *J. Clin. Microbiol.* **39:**1702–1706.
31. **Chang, H. C., S. N. Leaw, A. H. Huang, T. L. Wu, and T. C. Chang.** 2001. Rapid identification of yeasts in positive blood cultures by a multiplex PCR method. *J. Clin. Microbiol.* **39:**3466–3471.
32. **Chang H. J., H. L. Miller, N. Watkins, M. J. Arduino, D. A. Ashford, G. Midgely, S. M. Aquero, R. Pinto-Powell, C. F. von Reyn, W. Edwards, M. M. McNeil, and W. R. Jarvis.** 1998. An epidemic of *Malassezia pachydermatis* in an intensive care nursery associated with colonization of health care workers' pet dogs. *N. Engl. J. Med.* **338:**706–711.
33. **Chayakulkeeree, M., and J. R. Perfect.** 2006. Cryptococcosis. *Infect. Dis. Clin. N. Am.* **20:**507–544.
34. **Chen, S. C. A., C. L. Halliday, and W. Meyer.** 2002. A review of nucleic acid-based diagnostic tests for systemic mycoses with an emphasis on polymerase chain reaction-based assays. *Med. Mycol.* **40:**333–357.
35. **Chen, Y. C., J. D. Eisner, M. M. Kattar, S. L. Rassoulian-Barrett, K. LaFe, S. L. Yarfitz, A. P. Limaye, and B. T. Cookson.** 2000. Identification of medically important yeasts using PCR-based detection of DNA sequence polymorphisms in the internal transcribed spacer 2 region of the rRNA genes. *J. Clin. Microbiol.* **38:**2302–2310.
36. **Cheng, M. F., C. C. Chiou, Y. C. Liu, H. Z. Wang, and K. S. Hsieh.** 2001. *Cryptococcus laurentii* fungemia in a premature neonate. *J. Clin. Microbiol.* **39:**1608–1611.
37. **Choy, B. Y., S. S. Wong, T. M. Chan, and K. N. Lai.** 2000. *Pichia ohmeri* peritonitis in a patient on CAPD: response to treatment with amphotericin. *Perit. Dial. Int.* **20:**91.
38. **Ciardo, D. E., G. Schar, E. C. Bottger, M. Altwegg, and P. P. Bosshard.** 2006. Internal transcribed spacer sequencing versus biochemical profiling for identification of medically important yeasts. *J. Clin. Microbiol.* **44:**77–84.

38a. **CLSI.** 2002. CLSI document M27-A2. CLSI, Wayne, PA.

39. **Coignard, C., S. F. Hurst, L. E. Benjamin, M. E. Brandt, D. W. Warnock, and C. J. Morrison.** 2004. Resolution of discrepant results for *Candida* species identification by using DNA probes. *J. Clin. Microbiol.* **42:**858–861.
40. **Cooke, V. M., R. J. Miles, R. G. Price, G. Midgley, W. Khamri, and A. C. Richardson.** 2002. New chromogenic agar medium for the identification of *Candida* spp. *Appl. Environ. Microbiol.* **68:**3622–3627.
41. **Correia, A., P. Sampaio, S. James, and C. Pais.** 2006. *Candida bracarensis* sp. nov., a novel anamorphic yeast species phenotypically similar to *Candida glabrata*. *Int. J. Syst. Evol. Microbiol.* **56:**313–317.
42. **Crampin, A. C., and R. C. Matthews.** 1993. Amplification of the polymerase chain reaction to the diagnosis of candidosis by amplification of HSP 90 gene fragment. *J. Med. Microbiol.* **39:**233–238.
43. **de Baere, T., G. Claeys, D. Swinne, C. Massonet, G. Verschraegen, A. Muylaert, and M. Vaneechoutte.** 2002. Identification of cultured isolates of clinically important yeast species using fluorescent fragment length analysis of the amplified internally transcribed rRNA spacer 2 region. *BMC Microbiol.* **2:**21–28.
44. **de Castro, L. E. F., O. A. Sarraf, J. M. Lally, H. P. Sandoval, K. D. Solomon, and D. T. Vroman.** 2005. *Cryptococcus albidus* keratitis after corneal transplantation. *Cornea* **24:**882–883.
45. **de Llanos Frutos, R., M. T. Fernandez-Espinar, and A. Querol.** 2004. Identification of species of the genus *Candida* by analysis of the 5.8S rRNA gene and the two ribosomal internal transcribed spacers. *Antonie van Leeuwenhoek* **85:**175–185.
46. **de Repentigny, L.** 1992. Serodiagnosis of candidiasis, aspergillosis, and cryptococcosis. *Clin. Infect. Dis.* **14:**S11–S22.
47. **Diaz, M. R., and J. W. Fell.** 2004. High-throughput detection of pathogenic yeasts of the genus *Trichosporon*. *J. Clin. Microbiol.* **42:**3696–3706.
48. **Diekema, D. J., S. A. Messer, R. J. Hollis, R. P. Wenzel, and M. A. Pfaller.** 1997. An outbreak of *Candida parapsilosis* prosthetic valve endocarditis. *Diagn. Microbiol. Infect. Dis.* **29:**147–153.
49. **Digby, J., J. Kalbfleisch, A. Glenn, A. Larsen, W. Browder, and D. Williams.** 2003. Serum glucan levels are not specific for presence of fungal infections in intensive care unit patients. *Clin. Diagn. Lab. Immunol.* **10:**882–885.
50. **Dooley, D. P., M. L. Beckius, C. K. McAllister, and B. S. Jeffrey.** 1990. Prostatitis caused by *Hansenula fabianii*. *J. Infect. Dis.* **161:**1040–1041.
51. **Einsele, H., H. Hebart, G. Roller, J. Loffler, I. Rothenhofer, C. A. Muller, R. A Bowden, J. A. Van Burik, D Engelhard, L. Kanz, and U. Schumacher.** 1997. Detection and identification of fungal pathogens in blood by using molecular probes. *J. Clin. Microbiol.* **35:**1353–1360.
52. **Elie, C. M., T. J. Lott, E. Reiss, and C. J. Morrison.** 1998. Rapid identification of *Candida* species with species-specific DNA probes. *J. Clin. Microbiol.* **36:**3260–3265.
53. **Ellepola, A. N. B., and C. J. Morrison.** 2005. Laboratory diagnosis of invasive candidiasis. *J. Clin. Microbiol.* **43:**S65–S84.
54. **Ellis, D. H., and T. J. Pfeiffer.** 1990. Ecology, life cycle, and infectious propagule of *Cryptococcus neoformans*. *Lancet* **336:**923–925.
55. **Enache-Angoulvant, A., and C. Hennequin.** 2005. Invasive *Saccharomyces* infection: a comprehensive review. *Clin. Infect. Dis.* **41:**1559–1568.

56. **Espinel-Ingroff, A., L. Stockman, G. Roberts, D. Pincus, J. Pollack, and J. Marler.** 1998. Comparison of RapID Yeast Plus system with API 20C system for identification of common, new, and emerging yeast pathogens. *J. Clin. Microbiol.* **36:**883–886.
57. **Esteve-Zarzoso B., C. Belloch, F. Uruburu, and A. Querol.** 1999. Identification of yeasts by RFLP analysis of the 5.8S rRNA gene 184 and the two ribosomal internal transcribed spacers. *Int. J. Syst. Bacteriol.* **49:**329–337.
58. **Fell, J. W., T. Boekhout, A. Fonseca, G. Scorzetti, and A. Statzell-Tallman.** 2000. Biodiversity and systematics of basidiomycetous yeasts as determined by large-subunit rDNA D1/D2 domain sequence analysis. *Int. J. Syst. Evol. Microbiol.* **50:**1351–1371.
59. **Fidel, P.** 2007. History and update on host defenses against vaginal candidiasis. *Am. J. Reprod. Immunol.* **57:**2–12.
60. **Flahaut, M., D. Sanglard, M. Monod, J. Bille, and M. Rossier.** 1998. Rapid detection of *Candida albicans* in clinical samples by DNA amplification of common regions from *C. albicans*-secreted aspartic proteinase genes. *J. Clin. Microbiol.* **36:**395–401.
61. **Forrest, G. N., K. Mankes, M. A. Jabra-Rizk, E. Weekes, J. K. Johnson, D. P. Lincalis, and R. A. Venezia.** 2006. Peptide nucleic acid fluorescence in situ hybridization-based identification of *Candida albicans* and its impact on mortality and antifungal therapy costs. *J. Clin. Microbiol.* **44:**3381–3383.
62. **Freydiere, A. M., L. Buchaille, and Y. Gille.** 1997. Comparison of three commercial media for direct identification and discrimination of *Candida* species in clinical specimens. *Eur. J. Clin. Microbiol. Infect. Dis.* **16:**464–467.
63. **Freydiere, A.-M., R. Guinet, and P. Boiron.** 2001. Yeast identification in the clinical microbiology laboratory: phenotypic methods. *Med. Mycol.* **39:**9–33.
64. **Gade, W., S. W. Hinnefeld, L. S. Babcock, P. Gilligan, W. Kelly, K. Wait, D. Greer, M. Pinilla, and R. L. Kaplan.** 1991. Comparison of the Premier cryptococcal antigen enzyme immunoassay and the latex agglutination assay for detection of cryptococcal antigens. *J. Clin. Microbiol.* **29:**1616–1619.
65. **Garau, M., M. Pereito, Jr., and A. del Palacio.** 2003. In vitro susceptibilities of *Malassezia* species to a new triazole, albaconazole (UR-9825), and other antifungal compounds. *Antimicrob. Agents Chemother.* **47:**2342–2344.
66. **Girmenia, C., L. Pagano, B. Martino, D. D'Antonio, R. Fanci, G. Specchia, L. Melillo, M. Buelli, G. Pizzarelli, M. Venditti, P. Martino, and GIMEMA Infection Program.** 2005. Invasive infections caused by *Trichosporon* species and *Geotrichum capitatum* in patients with hematological malignancies: a retrospective multicenter study from Italy and review of the literature. *J. Clin. Microbiol.* **43:**1818–1828.
67. **Gold, J. W. M., B. Wong, E. M. Bernard, T. E. Kiehn, and D. Armstrong.** 1983. Serum arabinitol concentrations and arabinitol/creatinine ratios in invasive candidiasis. *J. Infect. Dis.* **147:**504–513.
68. **Gray, L. D., and G. D. Roberts.** 1988. Experience with the use of pronase to eliminate interference factors in the latex agglutination for cryptococcal antigen. *J. Clin. Microbiol.* **26:**2450–2451.
69. **Gueho, E., G. Midgley, and J. Guillot.** 1996. The genus *Malassezia* with description of four new species. *Antonie van Leeuwenhoek* **69:**337–355.
70. **Guiver, M., K. Levi, and B. A. Oppenheim.** 2001. Rapid identification of *Candida* species by TaqMan PCR. *J. Clin. Pathol.* **54:**362–366.
71. **Hage, C. A., K. L. Wood, H. T. Winer-Muram, S. J. Wilson, G. Sarosi, and K. S. Knox.** 2003. Pulmonary cryptococcosis after initiation of anti tumor necrosis factor-alpha therapy. *Chest* **124:**2395–2397.
72. **Hageage, G. J., and B. J. Harrington.** 1984. Use of calcofluor white in clinical mycology. *Lab. Med.* **15:**109–112.
73. **Haidle, C. W., and R. Storck.** 1966. Control of dimorphism in *Mucor rouxii. J. Bacteriol.* **92:**1236–1244.
74. **Hajjeh, R. A., A. N. Sofair, L. H. Harrison, G. M. Lyon, B. A. Arthington-Skaggs, S. A. Mirza, M. Phelan, J. Morgan, W. Lee-Yang, M. A. Ciblak, L. E. Benjamin, L. T. Sanza, S. Huie, S. F. Yeo, M. E. Brandt, and D. W. Warnock.** 2004. Incidence of bloodstream infections due to *Candida* species and in vitro susceptibilities of isolates collected from 1998 to 2000 in a population-based active surveillance program. *J. Clin. Microbiol.* **42:**1519–1527.
75. **Hall, L., S. Wohlfiel, and G. D. Roberts.** 2003. Experience with the MicroSeq D2 large-subunit ribosomal DNA sequencing kit for identification of commonly encountered, clinically important yeast species. *J. Clin. Microbiol.* **41:** 5099–5102.
76. **Han, X. Y., J. J. Tarrand, and E. Escudero.** 2004. Infections by the yeast *Kodomaee* (*Pichia*) *ohmeri*: two cases and literature review. *Eur. J. Clin. Microbiol. Infect. Dis.* **23:**127–130.
77. **Hanzen, J., and V. Krcmery.** 2002. Polyfungal candidaemia due to *Candida rugosa* and *Candida pelliculosa* in a haemodialyzed neonate. *Scand. J. Infect. Dis.* **34:**555.
78. **Hayden, R. T., X. Qian, G. W. Procop, G. D. Roberts, and R. V. Lloyd.** 2002. In situ hybridization for the identification of filamentous fungi in tissue section. *Diagn. Mol. Pathol.* **11:**119–126.
79. **Hazen, K. C., and S. A. Howell.** 2007. *Candida, Cryptococcus*, and other yeasts of medical importance, p. 1762–1788. *In* P. R. Murray, E. J. Baron, J. H. Jorgensen, M. L. Landry, and M. A. Pfaller (ed.), *Manual of Clinical Microbiology*, 9th ed. ASM Press, Washington, DC.
80. **Herent, P., D. Stynen, F. Hernando, J. Fruit, and D. Poulain.** 1992. Retrospective evaluation of two latex agglutination tests for detection of circulating antigens during invasive candidosis. *J. Clin. Microbiol.* **8:**2158–2164.
81. **Heymann, W. R., and D. J. Wolf.** 1986. *Malassezia* (*Pityrosporum*) folliculitis occurring during pregnancy. *Int. J. Dermatol.* **25:**49–51.
82. **Hirai, A., R. Kano, K. Makimura, E. R. Duarte, J. S. Hamdan, M. A. Lachance, H. Yamaguchi, and A. Hasegawa.** 2004. *Malassezia nana* sp. nov., a novel lipid dependent yeast species isolated from animals. *Int. J. Syst. Evol. Microbiol.* **54:**623–627.
83. **Hoang, J. K., and J. J. Burruss.** 2007. Localized cutaneous *Cryptococcus albidus* infection in a 14-year-old boy on etanercept therapy. *Pediatr. Dermatol.* **24:**285–288.
84. **Hoang, L. M., J. A. Maguire, P. Doyle, M. Fyfe, and D. L. Roscoe.** 2004. *Cryptococcus neoformans* infections at Vancouver Hospital and Health Sciences Centre (1997–2002): epidemiology, microbiology and histopathology. *J. Med. Microbiol.* **53:**935–940.
85. **Hospenthal, D. R., M. L. Beckius, K. L. Floyd, L. L. Horvath, and C. K. Murray.** 2006. Presumptive identification of *Candida* species other than *C. albicans*, *C. krusei*, and *C. tropicalis* with the chromogenic medium CHROMagar Candida. *Ann. Clin. Microbiol. Antimicrob.* **5:**1–5.
86. **Huffnagle, K. E., and R. M. Gander.** 1993. Evaluation of Gen-Probe's *Histoplasma capsulatum* and *Cryptococcus neoformans* AccuProbes. *J. Clin. Microbiol.* **31:**419–421.
87. **Husain, S., M. M. Wagener, and N. Singh.** 2001. *Cryptococcus neoformans* infection in organ transplant recipients: variables influencing clinical characteristics and outcome. *Emerg. Infect. Dis.* **7:**375–381.

88. **Iyampillai, T., S. Michael, E. Mathai, and M. S. Mathews.** 2004. Use of CHROMagar medium in the differentiation of *Candida* species: is it cost effective in developing countries? *Ann. Trop. Med. Parasitol.* **98:**279–282.
89. **Jabra-Rizk, M. A., T. M. Brenner, M. Romagnoli, A. A. M. A. Baqui, W. G. Merz, W. A. Falkler, Jr., and T. F. Meiller.** 2001. Evaluation of reformulated CHROMagar Candida. *J. Clin. Microbiol.* **39:**2015–2016.
90. **Jordan, J. A.** 1994. PCR identification of four medically important *Candida* species by using a single primer pair. *J. Clin. Microbiol.* **32:**2962–2967.
91. **Kalenic, S., M. Jandrlic, V. Vegar, N. Zuech, A. Sekulic, and E. Mlinaric-Missoni.** 2001. *Hansenula anomala* outbreak at a surgical intensive care unit: a search for risk factors. *Eur. J. Epidemiol.* **17:**491–496.
92. **Kan, V. L.** 1993. Polymerase chain reaction for the diagnosis of candidemia. *J. Infect. Dis.* **168:**779–783.
93. **Khan, Z. U., N. A. Al-Sweih, S. Amad, N. Al-Kazemi, S. Khan, L. Joseph, and R. Chandy.** 2007. Outbreak of fungemia among neonates caused by *Candida haemulonii* resistant to amphotericin B, itraconazole, and fluconazole. *J. Clin. Microbiol.* **45:**2025–2027.
94. **Kidd, S. E., F. Hagen, R. L. Tscharke, M. Huynh, K. H. Bartlett, M. Fyfe, L. Macdougall, T. Boekhout, K. J. Kwon-Chung, and W. Meyer.** 2004. A rare genotype of *Cryptococcus gattii* caused the cryptococcosis outbreak on Vancouver Island (British Columbia, Canada). *Proc. Nat. Acad. Sci. USA* **101:**17258–17263.
95. **Kiehn, T. E., E. M. Bernard, J. W. M. Gold, and D. Armstrong.** 1979. Candidiasis: detection by gas-liquid chromatography of D-arabinitol, a fungal metabolite, in human serum. *Science* **206:**577–580.
96. **Kohno, S., K. Mitsutake, S. Maesaki, A. Yasuoka, T. Miyazaki, M. Kaku, H. Koga, and K. Hara.** 1993. An evaluation of serodiagnostic tests in patients with candidemia: beta-glucan, mannan, Candida antigen by Cand-Tec and D-arabinitol. *Microbiol. Immunol.* **37:**207–212.
97. **Kondori, N., L. Edebo, and I. Mattsby-Baltzer.** 2004. Circulating β-(1-3)-glucan and immunoglobulin G subclass antibodies to *Candida albicans* cell wall antigens in patients with systemic candidiasis. *Clin. Diagn. Lab. Immunol.* **11:** 344–350.
98. **Kontoyiannis, D. P., H. A. Torres, M. Chagua, R. Hachem, J. J. Tarrand, G. P. Bodey, and I. I. Raad.** 2004. Trichosporonosis in a tertiary care cancer center: risk factors, changing spectrum and determinants of outcome. *Scand. J. Infect. Dis.* **36:**564–569.
99. **Kordossis, T., A. Avlami, A. Velegraki, I. Stefanou, G. Georgakopoulos, C. Papalambrou, and N. J. Legakis.** 1998. First report of *Cryptococcus laurentii* meningitis and a fatal case of *Cryptococcus albidus* cryptococcaemia in AIDS patients. *Med. Mycol.* **36:**335–339.
100. **Krcmery, V.** 1997. Nosocomial *Cryptococcus laurentii* fungemia in a bone marrow transplant patient after prophylaxis with ketoconazole successfully treated with oral fluconazole. *Infection* **25:**130.
101. **Krcmery, V., E. Oravcova, S. Spanik, M. Mrazova-Studena, J. Trupl, A. Kunova, K. Stopkova-Grey, E. Kukuckova, I. Krupova, A. Demitrovicova, and K. Kralovicova.** 1998. Nosocomial breakthrough fungaemia during antifungal prophylaxis or empirical antifungal therapy in 41 cancer patients receiving antineoplastic chemotherapy: analysis of aetiology risk factors and outcome. *J. Antimicrob. Chemother.* **41:**373–380.
102. **Kunova, A., and V. Krcmery.** 1999. Fungaemia due to thermophilic cryptococci: 3 cases of *Cryptococcus laurentii* bloodstream infections in cancer patients receiving antifungals. *Scand. J .Infect. Dis.* **31:**328.
103. **Kurtzman, C. P., and J. W. Fell (ed.).** 1984. *The Yeasts, a Taxonomic Study,* 3rd ed. Elsevier, New York, NY.
104. **Kurtzman, C. P., and J. W. Fell (ed.).** 1998. *The Yeasts, a Taxonomic Study,* 4th ed. Elsevier, New York, NY.
105. **Kurtzman, C. P., and C. J. Robnett.** 1998. Identification and phylogeny of ascomycetous yeasts from analysis of nuclear large subunit (26S) ribosomal DNA partial sequences. *Antonie van Leeuwenhoek* **73:**331–371.
106. **Larsson, L., C. Pehrson., T. Wiebe, and B. Christensson.** 1994. Gas chromatographic determination of D-arabinitol/L-arabinitol ratios in urine: a potential method for diagnosis of disseminated candidiasis. *J. Clin. Microbiol.* **32:**1855–1859.
107. **Leaw, S. N., H. C. Chang, H. F. Sun, R. Barton, J.-P. Boucara, and T. C. Chang.** 2006. Identification of medically important yeast species by sequence analysis of the internal transcribed spacer regions. *J. Clin. Microbiol.* **44:**693–699.
108. **Leaw, S. N., H. C. Chang, R. Barton, J.-P. Bouchara, and T. C. Chang.** 2007. Identification of medically important *Candida* and non-*Candida* yeast species by an oligonucleotide array. *J. Clin. Microbiol.* **45:**2220–2229.
109. **Lee, Y. A., H. J. Kim, T. W. Lee, M. J. Kim, M. H. Lee, J. H. Lee, and C. G. Ihm.** 2004. First report of *Cryptococcus albidus*-induced disseminated cryptococcosis in a renal transplant recipient. *Korean J. Intern. Med.* **19:**53–57.
110. **Linton, C. J., A. M. Borman, G. Cheung, A. D. Holmes, A. Szekely, M. D. Palmer, P. D. Bridge, C. K. Campbell, and E. M. Johnson.** 2007. Molecular identification of unusual pathogenic yeast isolates by large ribosomal subunit gene sequencing: 2 years of experience at the United Kingdom mycology reference laboratory. *J. Clin. Microbiol.* **45:**1152–1158.
111. **Lischewski, A., R. I. Amann, D. Harmsen, H. Merkert, J. Hacker, and J. Morschhauser.** 1996. Specific detection of *Candida albicans* and *Candida tropicalis* by fluorescent in situ hybridization with an 18S rRNA-targeted oligonucleotide probe. *Microbiology* **142:**2731–2740.
112. **Lischewski, A., M. Kretschmar, H. Hof, R. Amann, J. Hacker, and J. Morschhauser.** 1997. Detection and identification of *Candida* species in experimentally infected tissue and human blood by rRNA-specific fluorescent in situ hybridization. *J. Clin. Microbiol.* **35:**2943–2948.
113. **Loeffler, J., C. Dorn, H. Hebart, P. Cox, S. Magga, and H. Einsele.** 2003. Development and evaluation of the Nuclisens basic kit NASBA for the detection of RNA from *Candida* species frequently resistant to antifungal drugs. *Diagn. Microbiol. Infect. Dis.* **45:**217–220.
114. **Loison, J., P. Bouchara, E. Gueho, L. de Gentile, B. Cimon, J. M. Chennebault, and D. Chabasse.** 1996. First report of *Cryptococcus albidus* septicaemia in an HIV patient. *J. Infect.* **33:**139–140.
115. **Ma, J. S., P. Y. Chen, C. H. Chen, and C. S. Chi.** 2000. Neonatal fungemia caused by *Hansenula anomala*: a case report. *J. Microbiol. Immunol. Infect.* **33:**267–270.
116. **Maaroufi, Y., C. Heymans, J.-M. DeBruyne, V. Duchateau, H. Rodriguez-Villalobos, M. Aoun, and F. Crokaert.** 2003. Rapid detection of *Candida albicans* in clinical blood samples by using a TaqMan-based PCR assay. *J. Clin. Microbiol.* **41:**3293–3298.
117. **Maaroufi, Y., N. Ahariz, M. Husson and F. Crokaert.** 2004. Comparison of different methods of isolation of DNA of commonly encountered *Candida* species and its quantitation by using a real-time PCR-based assay. *J. Clin. Microbiol.* **42:**3159–3163.

118. **Marty, F. M., D. H. Barouch, E. P. Coakley, and L. R. Baden.** 2003. Disseminated trichosporonosis caused by *Trichosporon loubieri*. *J. Clin. Microbiol.* **41:**5317–5320.
119. **Marty, F. M., C. M. Lowry, S. J. Lempitski, D. W. Kubiak, M. A. Finkelman, and L. R. Baden.** 2006. Reactivity of (1→3)-β-D-glucan assay with commonly used intravenous antimicrobials. *Antimicrob. Agents Chemother.* **50:**3450–3453.
120. **Matthews, R., and J. Burnie.** 1988. Diagnosis of systemic candidiasis by an enzyme-linked immunobinding assay for a circulating immunodominant 47-kilodalton antigen. *J. Clin. Microbiol.* **26:**459–463.
121. **McCarthy, K. M., J. Morgan, K. A. Wannemuehler, S. A. Mirza, S. M. Gould, N. Mhlongo, P. Moeng, B. R. Maloba, H. H. Crewe-Brown, M. E. Brandt, and R. A. Hajjeh.** 2006. Population-based surveillance for cryptococcosis in an antiretroviral-naive South African province with a high HIV seroprevalence. *AIDS* **20:**2199–2206.
122. **McManus, E. J., and J. M. Jones.** 1985. Detection of a *Trichosporon beigelii* antigen cross-reactive with *Cryptococcus neoformans* capsular polysaccharide in serum from a patient with disseminated *Trichosporon* infection. *J. Clin. Microbiol.* **21:**681–685.
123. **Melcher, C. P., K. D. Reed, M. G. Rinaldi, J. W. Lee, P. A. Pizzo, and T. J. Walsh.** 1991. Demonstration of a cell wall antigen cross-reacting with cryptococcal polysaccharide in experimental disseminated trichosporonosis. *J. Clin. Microbiol.* **29:**192–196.
124. **Merz, W. G., J. E. Karp, D. Schron, and R. Saral.** 1986. Increased incidence of fungemia caused by *Candida krusei*. *J. Clin. Microbiol.* **24:**581–584.
125. **Metwally, L., G. Hogg, P. V. Coyle, R. J. Hay, S. Hedderwick, B. McCloskey, H. J. O'Neill, G. M. Ong, G. Thompson, C. H. Webb, and R. McMullan.** 2007. Rapid differentiation between fluconazole-sensitive and -resistant species of *Candida* directly from positive blood-culture bottles by real-time PCR. *J. Med. Microbiol.* **56:**964–970.
126. **Meurman, O., A. Koskensalo, and K. Rantakokko-Jalava.** 2006. Evaluation of Vitek 2 for identification of yeasts in the clinical laboratory. *Clin. Microbiol. Infect.* **12:**591–593.
127. **Midgley, G.** 2000. The lipophilic yeasts: state of the art and prospects. *Med. Mycol.* 38(Suppl. 1):9–16.
128. **Mirhendi, H., K. Makimura, M. Khoramizadeh, and H. Yamaguchi.** 2006. One enzyme PCR-RFLP assay for identification of six medically important *Candida* species. *Jpn. J. Med. Mycol.* **47:**225–229.
129. **Mirza, S. A., M. Phelan, D. Rimland, E. Graviss, R. Hamill, M. E. Brandt, T. Gardner, M. Sattah, G. P. de Leon, W. Baughman, and R. A. Hajjeh.** 2003. The changing epidemiology of cryptococcosis: an update from population-based active surveillance in 2 large metropolitan areas, 1992–2000. *Clin. Infect. Dis.* **36:**789–794.
130. **Misaki, H., H. Iwasaki, and T. Ueda.** 2003. A comparison of the specificity and sensitivity of two *Candida* antigen assay systems for the diagnosis of deep candidiasis in patients with hematologic diseases. *Med. Sci. Monit.* **9:**MT1–MT7.
131. **Mitchell, T. G., E. Z. Freedman, T. J. White, and J. W. Taylor.** 1994. Unique oligonucleotide primers in PCR for identification of *Cryptococcus neoformans*. *J. Clin. Microbiol.* **32:**253–255.
132. **Mitsutake, K., T. Miyazaki, T. Tashiro, Y. Yamamoto, H. Kakeya, T. Otsubo, S. Kawamura, M. A. Hossain, T. Noda, Y. Hirakata, and S. Kohno.** 1996. Enolase antigen, mannan antigen, Cand-Tec antigen, and _-glucan in patients with candidemia. *J. Clin. Microbiol.* **34:**1918–1921.
133. **Miyazaki, T. S. Kohno, H. Koga, M. Kaku, K. Mitsutake, S. Maesaki, A. Yasuoka, K. Hara, S. Tanaka, and H. Tamura.** 1992. G test, a new method for diagnosis of *Candida* infection: comparison with assays for beta-glucan and mannan antigen in a rabbit model of systemic candidiasis. *J. Clin. Lab. Anal.* **6:**315–318.
134. **Miyazaki, T., S. Kohno, K. Mitsutake, S. Maesaki, K.-I. Tanaka, N. Ishikawa, and K. Hara.** 1995. Plasma (1→3)-β-D-glucan and fungal antigenemia in patients with candidemia, aspergillosis, and cryptococcosis. *J. Clin. Microbiol.* **33:**3115–3118.
135. **Morace, G., G. Amato, F. Bistoni, G. Fadda, P. Marone, M. T. Montagna, S. Oliveri, L. Polonelli, R. Rigoli, I. Mancuso, S. La Face, L. Masucci, L. Romano, C. Napoli, D. Tato, M. G. Muscema, C. M. C. Belli, M. M. Piccirillo, S. Conti, S. Covan, F. Fanti, C. Cavanna, F. D'Alo, and L. Pitzurra.** 2002. Multicenter comparative evaluation of six commercial systems and the National Committee for Clinical Laboratory Standards M27-A broth microdilution method for fluconazole susceptibility testing of *Candida* species. *J. Clin. Microbiol.* **40:**2953–2958.
136. **Morace, G., L. Pagano, M. Sanguinetti, B. Posteraro, L. Mele, F. Equitani, G. D'Amore, G. Leone, and G. Fadda.** 1999. PCR-restriction enzyme analysis for detection of *Candida* DNA in blood from febrile patients with hematological malignancies. *J. Clin. Microbiol.* **37:**1871–1875.
137. **Morace, G., M. Sanquinetti, B. Posteraro, G. Lo Cascio, and G. Fadda.** 1997. Identification of various medically important *Candida* species in clinical specimens by PCR restriction enzyme analysis. *J. Clin. Microbiol.* **35:**667–672.
138. **Moreira-Oliveira, M. S., Y. Mikami, M. Miyaji, T. Imai, A. Z. Schreiber, and M. L. Moretti.** 2005. Diagnosis of candidemia by polymerase chain reaction and blood culture: prospective study in a high-risk population and identification of variables associated with development of candidemia. *Eur. J. Clin. Microbiol. Infect. Dis.* **24:**721–726.
139. **Morgan, J., M. I. Meltzer, B. D. Plikaytis, A. N. Sofair, S. Huie-White, S. Wilcox, L. H. Harrison, E. C. Seaberg, R. A. Hajjeh, and S. M. Teutsch.** 2005. Excess mortality, hospital stay, and cost due to candidemia: a case-control study using data from population-based candidemia surveillance. *Infect. Control Hosp. Epidemiol.* **26:**540–547.
140. **Mori, T., H. Ikemoto, M. Matsumura, M. Yoshida, K. Inada, S. Endo, A. Ito, S. Watanabe, H. Yamaguchi, M. Mitsuya, M. Kodama, T. Tani, T. Yokota, T. Kobayashi, J. Kambayashi, T. Nakamura, T. Masoaka, H. Teshima, T. Yoshinaga, S. Kohno, K. Hara, and S. Miyazaki.** 1997. Evaluation of plasma (1→3)-β-D-glucan measurement by the kinetic turbidimetric *Limulus* test, for the clinical diagnosis of mycotic infections. *Eur. J. Clin. Chem. Clin. Biochem.* **35:**553–560.
141. **Morris J. T., M. Beekius, and C. K. McAllister.** 1991. *Sporobolomyces* infection in an AIDS patient. *J. Infect. Dis.* **164:**623-624.
142. **Morrow, J. D.** 1994. Prosthetic cranioplasty infection due to *Sporobolomyces*. *J. Tenn. Med. Assoc.* **87:**466.
143. **Munoz, P., E. Bouza, and M. Cuenca-Estrella.** 2005. *Saccharomyces cerevisiae* fungemia: an emerging infectious disease. *Clin. Infect. Dis.* **40:**1625–1634.
144. **Munoz, P., M. E. Garcia Leoni, J. Berenguer, J. C. Bernaldo de Quiros, and E. Bouza.** 1989. Catheter-related fungemia by *Hansenula anomala*. *Arch. Intern. Med.* **149:**709–713.
145. **Murphy, N., C. R. Buchanan, V. Damjanovic, R. Whitaker, C. A. Hart, and R. W. Cooke.** 1986. Infection and colonization of neonates by *Hansenula anomala*. *Lancet* **i:**291–293.

146. **Murray, C. K., M. L. Beckius, J. A. Green, and D. R. Hospenthal.** 2005. Use of chromogenic medium in the isolation of yeasts from clinical specimens. *J. Med. Microbiol.* **54:**981–985.
147. **Mwaba, P., J. Mwansa, C. Chintu, J. Pobee, M. Scarborough, S. Portsmouth, and A. Zumla.** 2001. Clinical presentation, natural history, and cumulative death rates of 230 adults with primary cryptococcal meningitis in Zambian AIDS patients treated under local conditions. *Postgrad. Med. J.* 77:769–773.
148. **Nakabayashi A., Y. Sei, and J. Guillot.** 2000. Identification of *Malassezia* species isolated from patients with seborrhoeic dermatitis, atopic dermatidis, pityriasis versicolor and normal subjects. *Med. Mycol.* **38:**337–341.
149. **Nettles, R. E., L. S. Nichols, K. Bell-McGuinn, M. R. Pipeling, P. J. Scheel, Jr., and W. G. Merz.** 2003. Successful treatment of *Trichosporon mucoides* infection with fluconazole in a heart and kidney transplant recipient. *Clin. Infect. Dis.* **36:**e63–e66.
150. **Ngyuen, M. H., J. E. Peacock, Jr., A. J. Morris, D. C. Tanner, M. L. Nguyen, D. R. Snydman, M. M. Wagener, M. G. Rinaldi, and V. L. Yu.** 1996. The changing face of candidemia: emergence of non-*Candida albicans* species and antifungal resistance. *Am. J. Med.* **6:**617–623.
151. **Nohinek, B., C. S. Zu-Cheng, W. G. Barnes, L. Dall, and H. R. Gibbs.** 1987. Infective endocarditis of a bicuspid aortic valve caused by *Hansenula anomala*. *Am. J. Med.* **82:** 165–168.
152. **Nucci, M., and A. L. Colombo.** 2007. Candidemia due to *Candida tropicalis*: clinical, epidemiologic, and microbiologic characteristics of 188 episodes occurring in tertiary care hospitals. *Diagn. Microbiol. Infect. Dis.* **58:**77–82.
153. **Obayashi, T., M. Yoshida, T. Mori, H. Goto, A. Yasuoka, H. Iwasaki, H. Teshima, S. Kohno, A. Horiuchi, A. Ito, H. Yamaguchi, K. Shimada, and T. Kawai.** 1995. Plasma (1→3)-β-D-glucan measurement in diagnosis of invasive deep mycosis and fungal febrile episodes. *Lancet* **345:**17–20.
154. **Odabasi, Z., G. Mattiuzzi, E. Estey, H. Kantarjian, F. Saeki, R. J. Ridge, P. A. Ketchum, M. A. Finkelman, J. H. Rex, and L. Ostrosky-Zeichner.** 2005. (1→3)-β-D-glucan as a diagnostic adjunct for invasive fungal infections: validation, cutoff development, and performance in patients with acute myelogenous leukemia and myelodysplastic syndrome. *Clin. Infect. Dis.* **39:**199–205.
155. **Odds, F. C., and R. Bernaerts.** 1994. CHROMagar Candida, a new differential isolation medium for presumptive identification of clinically important *Candida* species. *J. Clin. Microbiol.* **32:**1923–1929.
156. **Odds, F. C., and A. Davidson.** 2000. "Room temperature" use of CHROMagar Candida. *Diagn. Microbiol. Infect. Dis.* **38:**147–150.
157. **Oliveira, K., G. Haase, C. Kurtzman, J. J. Hyldig-Nielson, and H. Stender.** 2001. Differentiation of *Candida albicans* and *Candida dubliniensis* by fluorescent in situ hybridization with peptide nucleic acid probes. *J. Clin. Microbiol.* **39:**4138–4141.
158. **Olver, W. J., J. Stafford, P. Cheetham, and T. C. Boswell.** 2002. Comparison of Candida ID medium with Sabouraud-chloramphenicol agar for the isolation of yeasts from clinical haematology surveillance specimens. *J. Med. Microbiol.* **51:**221–224.
159. **Ostrosky-Zeichner, L., B. D. Alexander, D. H. Kett, J. Vazquez, P. G. Pappas, F. Saeki, P. A. Ketchum, J. Wingard, S. Schiff, H. Tamura, M. A. Finkelman, and J. H. Rex.** 2005. Multicenter clinical evaluation of the (1→3) D-glucan assay as an aid to diagnosis of fungal infections in humans. *Clin. Infect. Dis.* **41:**654–659.
160. **Otag, F., N. Kuyucu, Z. Erturan, S. Sen, G. Emekdas, and T. Sugita.** 2005. An outbreak of *Pichia ohmeri* infection in the paediatric intensive care unit: case reports and review of the literature. *Mycoses* **48:**265–269.
161. **Page, B. T., and C. P. Kurtzman.** 2005. Rapid identification of *Candida* species and other clinically important yeast species by flow cytometry. *J. Clin. Microbiol.* **43:**4507–4514.
162. **Page, B. T., C. E. Shields, W. G. Merz, and C. P. Kurtzman.** 2006. Rapid identification of ascomycetous yeasts from clinical specimens by a molecular method based on flow cytometry and comparison with identifications from phenotypic assays. *J. Clin. Microbiol.* **44:**3167–3171.
163. **Pasqualotto, A. C., T. C. T. Sukiennik, L. C. Severo, C. S. de Amorim, and A. L. Colombo.** 2005. An outbreak of *Pichia anomala* fungemia in a Brazilian pediatric intensive care unit. *Infect. Control Hosp. Epidemiol.* **26:**553–558.
164. **Patterson, T. F.** 2002. Fungal susceptibility testing: where are we now? *Transpl. Infect. Dis.* **4:**38–45.
165. **Perfect, J. R., and G. M. Cox.** 2005. Cryptococcosis, p. 637–658. *In* W. G. Merz and R. J. Hay (ed.), *Topley and Wilson's Microbiology and Microbial Infections: Medical Mycology*. Hodder Publishing, London, United Kingdom.
166. **Pfaller, M. A., L. Boyken, R. J. Hollis, S. A. Messer, S. Tendolkar, and D. J. Diekema.** 2005. In vitro activities of anidulafungin against more than 2,500 clinical isolates of *Candida* spp., including 315 isolates resistant to fluconazole. *J. Clin. Microbiol.* **43:**5425–5427.
167. **Pfaller, M. A., L. Boyken, R. J. Hollis, S. A. Messer, S. Tendolkar, and D. J. Diekema.** 2006. Global surveillance of in vitro activity of micafungin against *Candida*: a comparison with caspofungin by CLSI-recommended methods. *J. Clin. Microbiol.* **44:**3533–3538.
168. **Pfaller, M. A., L. Boyken, R. J. Hollis, S. A. Messer, S. Tendolkar, and D. J. Diekema.** 2006. In vitro susceptibilities of *Candida* spp. to caspofungin: four years of global surveillance. *J. Clin. Microbiol.* **44:**760–763.
169. **Pfaller, M. A., I. Cabezudo, B. Buschelman, M. Bale, T. Howe, M. Vitug, H. J. Linton, and M. Densel.** 1993. Value of the Hybritech ICON *Candida* assay in the diagnosis of invasive candidiasis in high risk patients. *Diagn. Microbiol. Infect.* **16:**53–60.
170. **Pfaller, M. A., and J. Diekema.** 2007. Epidemiology of invasive candidiasis: a persistent public health problem. *Clin. Microbiol. Rev.* **20:**133–163.
171. **Pfaller, M. A., A. Houston, and S. Coffman.** 1996. Application of CHROMagar Candida for rapid screening of clinical specimens for *Candida albicans*, *Candida tropicalis*, *Candida krusei*, and *Candida* (*Torulopsis*) *glabrata*. *J. Clin. Microbiol.* **34:**58–61.
172. **Pickering, J. W., H. W. Sant, C. A. P. Bowles, W. L. Roberts, and G. L. Woods.** 2005. Evaluation of a (1→3)-β-D-glucan assay for diagnosis of invasive fungal infections. *J. Clin. Microbiol.* **43:**5957–5962.
173. **Piens, M. A., J. D. Perry, H. Raberin, F. Parant, and A. M. Freydiere.** 2003. Routine use of a one minute trehalase and maltase test for the identification of *Candida glabrata* in four laboratories. *J. Clin. Pathol.* **56:**687–689.
174. **Plazas, J., J. Portilla, V. Boix, and M. Parez-Mateo.** 1994. *Sporobolomyces salmonicolor* lymphadenitis in AIDS patient. Pathogen or passenger? *AIDS* **8:**387–388.
175. **Powell, H. L., C. A. Sand, and R. P. Rennie.** 1998. Evaluation of CHROMagar Candida for presumptive identification of clinically important *Candida* species. *Diagn. Microbiol. Infect. Dis.* **32:**201–204.

176. **Pryce, T. M., I. D. Kay, S. Palladino, and C. H. Heath.** 2003. Real-time automated polymerase chain reaction (PCR) to detect *Candida albicans* and *Aspergillus fumigatus* DNA in whole blood from high-risk patients. *Diagn. Microbiol. Infect. Dis.* **47:**487–496.
177. **Puerto, J. L., P. Garcia-Martos, and A. Saldarreaga.** 2002. First report of urinary tract infection due to *Pichia ohmeri*. *Eur. J. Clin. Microbiol. Infect. Dis.* **21:**630–631.
178. **Qadri, S. M. H., A. L. Dayel, M. J. Strampfer, and B. A. Cunha.** 1988. Urinary tract infection caused by *Hansenula anomala*. *Mycopathologia* **104:**99–101.
179. **Rand, K. H., H. Houck, and M. Wolff.** 1994. Detection of candidemia by polymerase chain reaction. *Mol. Cell. Probes* **8:**215–222.
180. **Reina, J. P., D. H. Larone, J. R. Sabetta, K. K. Krieger, and B. J. Hartman.** 2002. *Pichia ohmeri* prosthetic valve endocarditis and review of literature. *Scand. J. Infect. Dis.* **34:**140–141.
181. **Reiss, E., and C. J. Morrison.** 1993. Nonculture methods for diagnosis of disseminated candidiasis. *Clin. Microbiol. Rev.* **6:**311–323.
182. **Reiss, E., K. Tanaka, G. Bruker, V. Chazalet, D. Coleman, J. P. Debeaupuis, R. Hanazawa, J.-P. Latge, J. Lortholary, K. Makimura, C. J. Morrison, S. Y. Murayama, S. Naoe, S. Paris, J. Sarfari, K. Shibuya, D. Sullivan, K. Uchida, and H. Yamaguchi.** 1998. Molecular diagnosis and epidemiology of fungal infections. *Med. Mycol.* **36**(Suppl. 1):249–257.
183. **Reller, M. E., A. B. Mallonee, N. P. Kwiatkowski, and W. G. Merz.** 2007. Use of peptide nucleic acid fluorescence in situ hybridization for definitive, rapid identification of five common species of *Candida*. *J. Clin. Microbiol.* **45:** 3802–3803.
184. **Rex, J. H., M. A. Pfaller, J. N. Galgiani, M. S. Bartlett, A. Espinel-Ingroff, M. A. Ghannoum, M. Lancaster, F. C. Odds, M. G. Rinaldi, T. J. Walsh, and A. L. Barry.** 1997. Development of interpretive breakpoints for antifungal susceptibility testing: conceptual framework and analysis of in vitro-in vivo correlation data for fluconazole, itraconazole, and *Candida* infection. *Clin. Infect. Dis.* **24:**235–247.
185. **Roboz, J., and R. N. Katz.** 1992. Diagnosis of disseminated candidiasis based on serum D/L-arabinitol ratios using negative chemical ionization mass spectrometry. *J. Chromatogr.* **575:**281–286.
186. **Rodriguez-Tudela, J. L., and M. Cuenca-Estrella.** 2006. Molecular identification of *Trichosporon* species. *Clin. Infect. Dis.* **43:**1370.
187. **Sakai, T., K. Ikegami, E. Yoshinaga, R. Uesugi-Hayakawa, and A. Wakizaka.** 2000. Rapid, sensitive, and simple detection of *Candida* deep mycosis by amplification of 18S ribosomal RNA gene; comparison with assay of serum beta-D-glucan level in clinical samples. *Tohoku J. Exp. Med.* **190:** 119–128.
188. **Sandhu, G. S., B. C. Kline, L. Stockman, and G. D. Roberts.** 1995. Molecular probes for diagnosis of fungal infections. *J. Clin. Microbiol.* **33:**2913–2919.
189. **Sanguinetti, M., R. Porta, M. Sali, M. La Sorda, G. Pecorini, G. Fadda, and B. Posteraro.** 2007. Evaluation of Vitek 2 and RapID Yeast Plus systems for yeast species identification: experience at a large clinical microbiology laboratory. *J. Clin. Microbiol.* **45:**1343–1346.
190. **San Millan, R., L. Ribacoba, J. Ponton, and G. Quindos.** 1996. Evaluation of a commercial medium for identification of *Candida* species. *Eur. J. Clin. Microbiol. Infect. Dis.* **15:**153–158.
191. **Saunte, D. M., L. Klingspor, S. Jatal, J. Amau, and M. C. Arendrup.** 2005. Four cases of *Candida albicans* infections with isolates developing pink colonies on CHROMagar Candida plates. *Mycoses* **48:**378–381.
192. **Schabereiter-Gurtner, C., B. Selitsch, M. L. Rotter, A. M. Hirschl, and B. Willinger.** 2007. Development of novel real-time PCR assays for detection and differentiation of eleven medically important *Aspergillus* and *Candida* species in clinical specimens. *J. Clin. Microbiol.* **45:**906–914.
193. **Scorzetti, G., J. W. Fell, A. Fonseca, and A. Statzell-Tallman.** 2002. Systematics of basidiomycetous yeasts: a comparison of large subunit D1/D2 and internal transcribed spacer rDNA regions. *FEMS Yeast Res.* **2:**495–517.
194. **Selvarangan, R., U. Bui, A. P. Limaye, and B. T. Cookson.** 2003. Rapid identification of commonly encountered *Candida* species directly from blood culture bottles. *J. Clin. Microbiol.* **41:**5660–5664.
195. **Sendid, B., N. Francois, A. Standaert, E. Dehecq, F. Zerimech, D. Camus, and D. Poulain.** 2007. Prospective evaluation of the new chromogenic medium Candi*Select* 4 for differentiation and presumptive identification of the major pathogenic *Candida* species. *J. Med. Microbiol.* **56:**495–499.
196. **Sendid, B., C. Lacroix, and M. E. Bougnoux.** 2006. Is *Candida kefyr* an emerging pathogen in patients with oncohematological diseases? *Clin. Infect. Dis.* **43:**666–667.
197. **Sendid, B., M. Tabouret, J. L. Poirot, D. Mathieu, J. Fruit, and D. Poulain.** 1999. New enzyme immunoassay for sensitive detection of circulating *Candida albicans* mannan and antimannan antibodies: useful combined test for diagnosis of systemic candidiasis. *J. Clin. Microbiol.* **37:**1510–1517.
198. **Shepard, J. R., R. M. Addison, B. D. Alexander, P. Della-Latta, M. Gherna, G. Hasse, G. Hall, J. K. Johnson, W. G. Merz, H. Peltroche-Llacsahuanga, H. Stender, R.A. Venezia, D. Wilson, G.W. Procop, F. Wu, and M. J. Fiandaca.** 2008. Multicenter evaluation of the *Candida albicans/Candida glabrata* peptide nucleic acid fluorescent in situ hybridization method for simultaneous dual color identification of *Candida albicans* and *Candida glabrata* directly from blood culture bottles. *J. Clin. Microbiol.* **46:**50–55.
199. **Shin, J. H., F. S. Nolte, and C. J. Morrison.** 1997. Rapid identification of *Candida* species in blood cultures by a clinically useful PCR method. *J. Clin. Microbiol.* **35:**1454–1459.
200. **Shin, J. H., F. S. Nolte, B. P. Holloway, and C. J. Morrison.** 1999. Rapid identification of up to three *Candida* species in a single reaction tube by a 5′ exonuclease assay using fluorescent DNA probes. *J. Clin. Microbiol.* **37:**165–170.
201. **Shin, D. H., J. H. Park, J. H. Shin, S. P. Suh, D. W. Ryang, and S. J. Kim.** 2003. *Pichia ohmeri* fungaemia associated with phlebitis: successful treatment with amphotericin B. *J. Infect. Chemother.* **9:**88–89.
202. **Shrestha, R. K., J. K. Stoller, G. Honari, G. W. Procop, and S. M. Gordon.** 2004. Pneumonia due to *Cryptococcus neoformans* in a patient receiving infliximab: possible zoonotic transmission from a pet cockatiel. *Respir. Care* **49:**606–608.
203. **Simon, G., G. Simon, M. Erdos, and M. Laszlo.** 2005. Invasive *Cryptococcus laurentii* disease in a nine-year-old boy with X-linked hyper-immunoglobulin M syndrome. *Pediatr. Infect. Dis. J.* **24:**935–937.
204. **Smith, M. B., D. Dunklee, H. Vu, and G. L. Woods.** 1999. Comparative performance of the RapID Yeast Plus system and the API 20C AUX clinical yeast system. *J. Clin. Microbiol.* **37:**2697–2698.
205. **Spiess, B., W. Seifarth, M. Hummel, O. Frank, A. Fabarius, C. Zheng, H. Morz, R. Hehlmann, and D. Buchheidt.** 2007. DNA microarray-based detection and identification of

fungal pathogens in clinical samples from neutropenic patients. *J. Clin. Microbiol.* **45**:3743–3753.

206. **Strockbine, N. A., M. T. Largen, S. M. Zweibel, and H. R. Buckley.** 1984. Identification and molecular weight characterization of antigens from *Candida* albicans that are recognized by human sera. *Infect. Immun.* **43**:715–721.
207. **Sugita, T., M. Tajima, M. Amaya, R. Tsuboi, and A. Nishikawa.** 2004. Genotype analysis of *Malassezia restricta* as the major cutaneous flora in patients with atopic dermatitis and healthy subjects. *Microbiol. Immunol.* **48**: 755–759.
208. **Sugita, T., M. Tajima, M. Takashima, M. Amaya, M. Saito, R. Tsuboi, and A. Nishikawa.** 2004. A new yeast, *Malassezia yamatoensis*, isolated from a patient with seborrheic dermatitis, and its distribution in patients and healthy subjects. *Microbiol. Immunol.* **48**:579–583.
209. **Sugita, T., M. Takashima, M. Kodama, R. Tsuboi, and A. Nishikawa.** 2003. Description of a new yeast species, *Malassezia japonica*, and its detection in patients with atopic dermatitis and healthy subjects. *J. Clin. Microbiol.* **41**:4695–4699.
210. **Sugita, T., M. Takashima, T. Skinoda, H. Suto, T. Unno, R. Tsuboi, H. Ogawa, and A. Nishikawa.** 2002. A new yeast species, *Malassezia dermatis*, isolated from patients with atopic dermatitis. *J. Clin. Microbiol.* **40**:1363–1367.
211. **Sullivan, D. J., T. J. Westereng, K. A. Haynes, D. E. Bennett, and D. C. Coleman.** 1995. *Candida dubliniensis* sp. nov.: phenotypic and molecular characterization of a novel species associated with oral candidosis in HIV-infected individuals. *Microbiology* **141**:1507–1521.
212. **Tavanti, A., A. D. Davidson, N. A. Gow, M. C. Maiden, and F. C. Odds.** 2005. *Candida orthopsilosis* and *Candida metapsilosis* sp. nov. to replace *Candida parapsilosis* groups II and III. *J. Clin. Microbiol.* **43**:284–292.
213. **Thuler, L. C., S. Faivichenco, E. Velasco, C. A. Martins, C. R. Nascimento, and I. A. Castilho.** 1997. Fungemia caused by *Hansenula anomala*: an outbreak in a cancer hospital. *Mycoses* **40**:193–196.
214. **Trick, W. E., S. K. Fridkin, J. R. Edwards, R. A. Hajjeh, R. P. Gaynes, and National Nosocomial Infections Surveillance System Hospitals.** 2002. Secular trend of hospital-acquired candidemia among intensive care unit patients in the United States during 1989–1999. *Clin. Infect. Dis.* **35**:627–630.
215. **Turenne, C. Y., S. E. Sanche, D. J. Hoban, J. A. Karlowsky, and A. M. Kabani.** 1999. Rapid identification of fungi by using the ITS2 genetic region and an automated fluorescent capillary electrophoresis system. *J. Clin. Microbiol.* **37**: 1846–1851.
216. **Upton, A., J. A. Faser, S. E. Kidd, C. Bretz, K. H. Bartlett, J. Heitman, and K. A. Marr.** 2007. First contemporary case of human infection with *Cryptococcus gattii* in Puget Sound: evidence for spread of the Vancouver Island outbreak. *J. Clin. Microbiol.* **45**:3086–3088.
217. **Verweij, P. E., I. M. Breuker, A. J. Rijs, and J. F. Meis.** 1999. Comparative study of seven commercial yeast identification systems. *J. Clin. Pathol.* **52**:271–273.
218. **Vilchez, R. A., J. Fung, and S. Kusne.** 2002. Cryptococcosis in organ transplant recipients: an overview. *Am. J. Transplant.* **2**:575–580.
219. **Walsh, T. J., and S. J. Chanock.** 1998. Diagnosis of invasive fungal infections: advances in nonculture systems. *Curr. Clin. Top. Infect. Dis.* **18**:101–153.
220. **Walsh, T. J., A. Fracesconi, M. Kasai, and S. J. Chanock.** 1995. PCR and single-strand conformational polymorphism for recognition of medically important opportunistic fungi. *J. Clin. Microbiol.* **33**:3216–3220.
221. **Walsh, T. J., J. W. Hathorn, J. D. Sobel, W. G. Merz, V. Sanchez, S. M. Maret, H. R. Buckley, M. A. Pfaller, R. Schaufele, C. Sliva, E. Navarro, J. Lecciones, P. Chandrasekar, J. Lee, and P. A. Pizzo.** 1991. Detection of circulating enolase by immunoassay in patients with cancer and invasive candidiasis. *N. Engl. J. Med.* **324**:1026–1031.
222. **Walsh, T. J., W. G. Merz, J. W. Lee, R. Schaufele, T. Sein, P. O. Whitcomb, M. Ruddel, W. Burns, J. R. Wingard, A. C. Switchenko, T. Goodman, and P. A. Pizzo.** 1995. Diagnosis and therapeutic monitoring of invasive candidiasis by rapid enzymatic detection of serum D-arabinitol. *Am. J. Med.* **99**:164–172.
223. **Wang, C. J. K., and J. Schwarz.** 1958. The etiology of interstitial pneumonia. Identification as *Hansenula anomala* of yeast isolated from lung of infants. *Mycopathol. Mycol. Appl.* **9**:299–306.
224. **Warnock, D. W.** 2007. Trends in the epidemiology of invasive fungal infections. *Nippon Ishinkin Gakkai Zasshi* **48**:1–12.
225. **Weiner, M. H., and W. J. Yount.** 1976. Mannan antigenemia in the diagnosis of invasive *Candida* infections. *J. Clin. Investig.* **58**:1045–1053.
226. **Wheat, L. J.** 2006. Antigen detection, serology and molecular diagnosis of invasive mycoses in the immunocompromised host. *Transpl. Infect. Dis.* **8**:128–139.
227. **White, P. L., A. E. Archer, and R. A. Barnes.** 2005. Comparison of non-culture-based methods for detection of systemic fungal infections, with an emphasis on invasive *Candida* infections. *J. Clin. Microbiol.* **43**:2181–2187.
228. **White, P. L., A. Shetty, and R. A. Barnes.** 2003. Detection of seven *Candida* species using the Light-Cycler system. *J. Med. Microbiol.* **52**:229–238.
229. **Willinger, B., C. Hillowoth, B. Selitsch, and M. Manafi.** 2001. Performance of Candida ID, a new chromogenic medium for presumptive identification of *Candida* species, in comparison to CHROMagar Candida. *J. Clin. Microbiol.* **39**:3793–3795.
230. **Wilson, D. A., M. J. Joyce, L. S. Hall, L. B. Reller, G. D. Roberts, G. S. Hall, B. D. Alexander, and G. W. Procop.** 2005. Multicenter evaluation of a *Candida albicans* peptide nucleic acid fluorescent in situ hybridization probe for characterization of yeast isolates from blood cultures. *J. Clin. Microbiol.* **43**:2909–2912.
231. **Wingard, J. R., W. G. Merz, M. G. Rinaldi, T. R. Johnson, J. E. Karp, and R. Saral.** 1991. Increase in *Candida krusei* infection among patients with bone marrow transplantation and neutropenia treated prophylactically with fluconazole. *N. Engl. J. Med.* **325**:1274–1277.
232. **Wise, M. G., M. Healy, K. Reece, R. Smith, D. Walton, W. Dutch, A. Renwick, J. Huong, S. Young, J. Tarrand, and D. P. Kontoyiannis.** 2007. Species identification and strain differentiation of clinical *Candida* isolates using the DiversiLab system of automated repetitive sequence-based PCR. *J. Med. Microbiol.* **56**:778–787.
233. **Wisplinghoff, H., T. Bischoff, S. M. Tallent, H. Seifert, R. P. Wenzel, and M. B. Edmond.** 2004. Nosocomial bloodstream infections in US hospitals: analysis of 24,179 cases from a prospective nationwide surveillance study. *Clin. Infect. Dis.* **39**:309–317.
234. **Wong, B., E. M. Bernard, J. W. M. Gold, D. Fong, and D. Armstrong.** 1982. The arabinitol appearance rate in laboratory animals and humans: estimation from the arabinitol/creatinine ratio and relevance to the diagnosis of candidiasis. *J. Infect. Dis.* **146**:353–359.

235. **Woods, G. L., and D. H. Walker.** 1996. Detection of infection and identification of infectious agents by use of cytologic or histologic stains. *Clin. Microbiol. Rev.* **9:**382–404.

236. **Xiang, H., L. Xiong, X. Liu, and Z. Tu.** 2007. Rapid simultaneous detection and identification of six species of *Candida* using polymerase chain reaction and reverse line hybridization assay. *J. Microbiol. Methods* **69:**282–287.

237. **Yeo, S. F., and B. Wong.** 2002. Current status of nonculture methods for diagnosis of invasive fungal infections. *Clin. Microbiol. Rev.* **15:**465–484.

238. **Yera, H., B. Sendid, N. Francois, D. Camus, and D. Poulain.** 2001. Contribution of serologic tests and blood culture to the early diagnosis of systemic candidiasis. *Eur. J. Clin. Microbiol. Infect. Dis.* **20:**864–870.

Diagnostic Microbiology of the Immunocompromised Host
Edited by Randall T. Hayden, Karen C. Carroll, Yi-Wei Tang, and Donna M. Wolk

Chapter 12

Mycobacteria

NANCY G. WARREN AND GAIL L. WOODS

BRIEF DESCRIPTION OF THE PATHOGENS

Many mycobacteria can cause infections in the immunocompromised individual, but *Mycobacterium tuberculosis* and *Mycobacterium avium* typically are the most commonly reported. In general terms, the mycobacteria are divided into the *Mycobacterium tuberculosis* complex (MTBC) and the nontuberculous mycobacteria (NTM). As the incidence of tuberculosis has fallen and identification techniques have improved, infection with NTM has become recognized more readily. Previously, NTM were called "atypical" because they were thought to be unusual strains, but they now are recognized as legitimate species, and the term "atypical" is not appropriate.

The influence of human immunodeficiency virus (HIV) on infection with *M. tuberculosis* is extensive and affects the course of infection in several ways, including higher rates of reactivated disease, acute disease, extrapulmonary disease, skin test anergy, and malabsorption of antituberculosis drugs. Conversely, infection with *M. tuberculosis* may increase macrophage expression of HIV and progression to AIDS in untreated cases. These profound influences have led to an epidemic of tuberculosis within the HIV epidemic, especially in underserved countries with a high incidence of HIV infection (54).

NTM infections, once viewed primarily as sequelae to otherwise damaged lungs or traumatized skin and soft tissue, are able to opportunistically infect individuals with low $CD4^+$ cell counts. While detection of *M. tuberculosis* is considered evidence of infection, determining the clinical significance of the isolation of NTM is not as straightforward and should be based on clinical presentation, mycobacterial species, specimen source, quantification of the organisms in smear and culture, the number of positive cultures, and elimination of possible laboratory or environmental contamination.

MTBC

The MTBC is comprised of *M. tuberculosis* and several closely related mycobacteria (*M. bovis*, including bacillus Calmette-Guérin [BCG], which is used as a vaccine in many countries; *M. africanum*; *M. canetti*; and *M. microti*). These organisms are carried by airborne particles composed of droplet nuclei generated when patients with pulmonary tuberculosis cough. All of these species produce tuberculosis-like diseases in humans and animals and are generally transmissible from human to human; no environmental reservoir for MTBC exists.

M. tuberculosis is a slowly growing organism, requiring 2 weeks or more for buff-colored, dry, rough colonies to become visible on conventional laboratory media. Temperatures of 35 to 37°C are required for recovery of the organism, and growth is enhanced by the addition of 5 to 10% CO_2. Microscopically, *M. tuberculosis* is viewed as acid-fast, rod-shaped bacteria that may align in such as manner as to give the appearance of ropes or "serpentine cords." This distinctive characteristic is best viewed when smears are made from cultures and is not likely to be seen in the actual clinical specimen. Not all *M. tuberculosis* cultures produce cords, nor is cording specific for *M. tuberculosis*, but if present, the characteristic should be noted during the bacteriological workup. By conventional identification techniques, *M. tuberculosis* is niacin and nitrate positive; however, there are DNA probes that provide faster test results. Other members of the MTBC can be differentiated by various biochemical tests and several other methods (63). Table 1 shows some

Nancy G. Warren • Bureau of Laboratories, Pennsylvania Department of Health, 110 Pickering Way, Lionville, PA 19353.
Gail L. Woods • Department of Pathology and Laboratory Services, University of Arkansas for Medical Sciences, Mail Slot 502, 4301 W. Markham Street, Little Rock, AR 72205.

Table 1. Phenotypic characteristics of selected mycobacteria

Organism	Niacin	Nitrate	Probe available	Pigment	Acid-fast	Growth rate (days)
M. tuberculosis	+	+	Yes	Buff	+	> 7
M. bovis	−	−	Yes	Buff	+	>7
M. avium complex	−	−	Yes	Buff	+	>7
M. kansasii	−	+	Yes	Buff to yellow[a]	+	>7
M. haemophilum	−	−	No	Buff	+	>7
M. chelonae	−	−	No	Buff	Variable[b]	≤7

[a]Colonies are buff in the dark and turn yellow to orange when exposed to light (photochromogen).
[a]Bacilli may not stain with fluorochrome stains.

defining characteristics of the MTBC and several species of NTM.

NTM

The ecology of the NTM is very different from that of the MTBC. Many NTM are free-living saprophytes that have been detected in water, soil, dust, and foods. They grow and persist in these natural environments, often show resistance to disinfectants, and may be involved in nosocomial infections (23). Infection is acquired from the environment; person-to-person transmission does not occur. NTM are opportunistic pathogens, generally causing disease in immunocompromised patients, such as persons with AIDS.

MAC

The *M. avium* complex (MAC) consists of *M. avium*, *M. intracellulare*, and possibly a few additional less-defined species. In the immunocompromised host, the MAC is the most frequently recovered type of NTM, and in many laboratories, it is more commonly encountered than *M. tuberculosis*. MAC strains have been isolated repeatedly from many environmental sources and from drinking water distribution systems (23). These organisms are of low pathogenicity and often colonize the respiratory tracts of immunocompetent individuals. In the immunocompromised host, the presentation is quite different (discussed below).

MAC strains are slowly growing mycobacteria, often requiring more than 2 weeks to show visible growth on conventional solid culture media. In broth cultures, however, the MAC grows much more quickly; growth may be apparent in less than 1 week. Colonies are buff colored, smooth, and creamy, but occasionally, the growth may be so fine that it appears only as a thin film on the medium. Microscopically, the organisms are acid fast and vary in shape from coccobacillary forms to longer, thin rods. Biochemically, the MAC is niacin and nitrate negative; however, currently, identification is infrequently based on results of biochemical tests. More often, the MAC is identified by use of a DNA probe specific for the complex or by high-performance liquid chromatography (HPLC).

Mycobacterium kansasii

M. kansasii is a photochromogenic mycobacterium; colonies are buff when grown in the dark and turn bright lemon yellow to orange when exposed to light. It is intermittently found in water supplies and is the second most common cause of NTM disease in patients with HIV/AIDS. On solid media, the colonies appear after a week or more and may be rough or smooth in appearance. The bacilli of *M. kansasii* typically are longer and wider than those of *M. tuberculosis* and often have the appearance of a candy cane or shepherd's crook, but this feature is not reliable for definitive identification to the species level. Biochemically, *M. kansasii* is niacin negative and nitrate positive, but similar to the MAC, it is identified more often by its characteristic HPLC pattern or by using a DNA probe specific for the species.

Mycobacterium haemophilum

M. haemophilum is a "blood-loving" mycobacterium with several unique growth requirements that set it apart from other mycobacteria. Although first described in 1978, it may well have been noted before then, in cases where acid-fast bacilli (AFB) were visualized in tissue sections but cultures produced no growth. Most notably today, the organism causes skin and soft tissue infections in HIV-positive individuals. *M. haemophilum* is acid fast and grows as short rods that may clump together. Cultures require more than 2 weeks before growth is visible, and colonies are usually buff colored and rough in texture. Because of its special growth requirements, it is helpful if the physician or a pathologist routinely alerts the laboratory that *M.*

haemophilum is suspected so that culture incubation temperatures can be reduced to 30 to 32°C and growth media can be supplemented with hemin, hemoglobin, or ferric ammonium citrate. Alternatively, since this species cannot grow on media typically used to recover other mycobacteria, a supplemented medium may be included routinely for skin specimens and incubated at 30 to 32°C. This mycobacterium is relatively inert biochemically, so HPLC provides the most practical and rapid method of identifying this species (51).

Mycobacterium genavense

M. genavense is a slowly growing NTM that was first isolated from the blood of an AIDS patient and later described as causing cutaneous disease (25). Its environmental reservoir is not known, but it is a common cause of mycobacterial disease in psittacine birds. This mycobacterium is difficult to grow in culture; it prefers liquid media and requires extended incubation, and its growth is supported by mycobactin J and a pH of 6.2. Chromatographs of *M. genavense* show an HPLC lipid pattern that is identical to that of *M. simiae*. When AFB are detected on smears but cultures fail to yield positive growth, this species should be considered.

Rapidly growing mycobacteria

The rapidly growing mycobacteria most commonly associated with infection in humans are *M. chelonae*, *M. abscessus*, and *M. fortuitum* (10). These mycobacteria are commonly found in municipal water supplies and biofilms. They grow under a variety of environmental conditions and are resistant to many disinfectants and biocides. As a group, these mycobacteria are considered "rapid growers" because growth is detected on a solid medium in 7 days or less, one of the features that distinguish them from MTBC and the slowly growing NTM. There are many species of rapid growers; the taxonomy is evolving and their clinical significance is being defined. Rapid growers are coccoid to rod shaped and may show variability in acid fastness, especially with fluorochrome stains. Growth occurs on routine media not specifically designed for mycobacteria, such as blood agar, and often is better at 25 to 30°C than at 35 to 37°C (10). Colonies of *M. chelonae*, *M. abscessus*, and *M. fortuitum* are buff colored and may be smooth or rough. Biochemical reactions vary depending on the species but may not be sufficient to identify an isolate to the species level. Molecular tests may be necessary for species identification.

Other mycobacteria

Genotypic studies during the last decade have characterized a growing number of new mycobacterial species. Many of these species are difficult to distinguish from the more common mycobacteria based on phenotypic methods for identification, and several have been detected in immunocompromised individuals. During the 1990s, 42 new mycobacterial species were recognized. Most of them (33 species) were isolated from clinical specimens, and 9 were recovered from immunocompromised patients. Among the newly described mycobacteria are the species *M. celatum*, *M. hackensackense*, *M. conspicuum*, *M. interjectum*, *M. lentiflavum*, *M. canetti*, *M. genavense*, *M. sherrisii*, *M. triplex*, *M. mageritense*, *M. septicum*, and *M. parascrofulaceum* (62).

SPECTRUM OF CLINICAL DISEASE IN THE IMMUNOCOMPROMISED HOST

HIV/AIDS

Infection with HIV is considered to be the greatest risk factor for the development of tuberculosis (52). HIV infection increases the risk of progression to active tuberculosis among persons with latent *M. tuberculosis* infection, the risk of rapid progression to active tuberculosis in those newly infected with *M. tuberculosis*, and the risk of reinfection with *M. tuberculosis*. Moreover, tuberculosis is associated with an increase in the HIV viral load and with more rapid progression of HIV infection.

The clinical manifestations of tuberculosis in HIV-infected patients vary based on the degree of immune compromise. When the CD4 count is $>350/\mu l$, the usual presentation is "typical" pulmonary disease, i.e., upper lobe involvement, often with cavitation. In contrast, when the CD4 count is $<50/\mu l$, pulmonary disease generally involves the middle and lower lobes, cavitation usually is absent, and extrapulmonary tuberculosis with pleuritis, pericarditis, meningitis, and disseminated disease is more common. The clinical manifestations of extrapulmonary tuberculosis in persons with HIV resemble those in non-HIV-infected persons, but the survival rate, especially for tuberculous meningitis, is lower (60). Abscesses of the liver, pancreas, prostate, spleen, chest, abdominal wall, and other soft tissue may also occur.

Patients with HIV/AIDS are also at increased risk of disease due to NTM. The vast majority of NTM disease in the HIV-infected patient is due to the MAC; however, local or disseminated infections with many other species of NTM, including *M. kansasii*, *M. genavense*, *M. haemophilum*, *M. simiae*, *M. xenopi*,

M. marinum, *M. malmoense*, and various rapidly growing mycobacteria, have been reported (3, 6, 11, 18, 24, 27, 30). The incidence of disseminated MAC has declined considerably since the introduction of highly active antiretroviral therapy (HAART) and MAC prophylaxis. For example, at Johns Hopkins University, the proportion of persons with advanced HIV disease who developed disseminated MAC infection fell from 16% before 1996 to 4% after 1996, with a rate of <1% per year in 2004 (31). The clinical presentation of disseminated MAC infection also has changed over time. Prior to HAART, most patients had high fever, weight loss, night sweats, and severe anemia. Other features were abdominal pain, fatigue, chronic diarrhea, intra-abdominal lymphadenopathy, and hepatosplenomegaly. Common laboratory abnormalities included elevated liver function values, especially elevated alkaline phosphatase (≥2 times the upper limit of normal) and γ-glutamyl transpeptidase (≥3 times the upper limit of normal), and leukopenia. With HAART, severe anemia, significant weight loss, and elevated alkaline phosphatase have become less common. Concurrent pulmonary disease, characterized radiographically by alveolar infiltrates, nodules, cavitary lesions, pleural effusions, or lymphadenopathy, is infrequent (41).

The immune reconstitution syndrome (IRS), or paradoxical reaction, may occur in HIV-infected patients with either tuberculosis or NTM disease (most commonly disseminated MAC disease) after initiation of HAART. HAART reduces the HIV viral load, which is associated with both an early increase in memory CD4 cells and an increase in naïve CD4 cells about 4 to 6 weeks after the start of therapy. These changes allow enhanced in vitro proliferative responses to various pathogens, including mycobacteria, which very likely are responsible for the inflammatory lesions of IRS.

IRS due to *M. tuberculosis* is rare in the absence of recently documented tuberculosis. Findings include high fever, lymphadenopathy, large effusions, and expanding lesions. NTM IRS, in contrast, may be seen in HIV-infected patients with no previous history of mycobacterial disease, apparently representing an unmasking of subclinical disease in these cases. Symptoms typically occur 1 to 12 weeks after initiating HAART, with a median of 3 weeks in one large study (46). The most common clinical presentations are peripheral lymphadenitis, pulmonary-thoracic disease, and intra-abdominal disease. Localized involvement of the joints, spine, prostate, skin, or soft tissue and spontaneously resolving MAC bacteremia have also been described. Patients with abdominal involvement are more likely than patients with other manifestations to have previous disseminated MAC infection, lower CD4 counts, and positive blood cultures at NTM IRS diagnosis, and they have higher rates of hospitalization and relapse.

Transplantation

Tuberculosis is an important opportunistic infection in solid organ transplant recipients worldwide. Among such patients, the incidence of tuberculosis is estimated to be 20 to 74 times higher than that among the general population (43, 55). The reported incidence is 0.7 to 2.3% among adult liver transplant recipients, 2.5% among pediatric liver transplant recipients, 0.5 to 15% among kidney transplant recipients, 1 to 1.5% among heart transplant recipients, and 2 to 6.5% among lung transplant recipients (43). The incidence among bone marrow transplant recipients is considerably lower, at 0.23 to 0.79%. In a review of over 500 solid organ transplant recipients, the median time to diagnosis of tuberculosis after transplantation was 9 months (range, 0.5 to 144 months) overall, but it varied with the type of transplant (median time to onset, 11.5, 4, 4, and 3.5 months for kidney, liver, heart, and lung transplant recipients, respectively) (55). Significant predictors of early-onset tuberculosis (occurring within 1 year after transplantation) were nonkidney transplantation, allograft rejection within 6 months before the onset of tuberculosis, and the primary immunosuppressive regimen (more likely in patients receiving cyclosporine plus azathioprine or tacrolimus alone than in those receiving cyclosporine alone or azathioprine alone). The most frequent mode of acquisition of tuberculosis was reactivation of latent *M. tuberculosis* infection, although nosocomial acquisition and donor transmission have been described (55).

In the review of tuberculosis in solid organ transplant recipients by Singh and Paterson (55), 51% of patients had pulmonary tuberculosis, 16% had extrapulmonary involvement, and 33% had disseminated disease. The latter presentation was most common in those who had received OKT3. Fever was a frequent symptom, especially in patients with disseminated tuberculosis; night sweats and weight loss were also common. Radiographic findings for patients with pulmonary tuberculosis included focal or diffuse infiltrates, miliary disease, nodules, and pleural effusions; cavitation was uncommon (present in only 4% of cases). The mortality rate was 29% overall and 30%, 21%, 18%, and 17% among kidney, liver, heart, and lung transplant recipients, respectively. Significant predictors of death were disseminated disease, prior rejection, and receipt of OKT3 or anti-T-cell antibodies.

NTM are important opportunistic pathogens in hematopoietic stem cell transplant (HSCT) recipients and, to a lesser extent, in solid organ transplant recipients (21, 29, 48). The estimated incidence of NTM infection is 0.4 to 4.9% for HSCT recipients (50 to 600 times that in the general population) and 0.16 to 0.38%, 0.24 to 2.8%, and 0.46 to 2.3% for kidney, heart, and lung transplant recipients, respectively (21). The true incidence cannot be determined because NTM disease is not reportable.

In a large review of NTM infections in transplant recipients (21), the median time between transplantation and development of NTM disease was 4.2, 23.5, 30, 14.8, and 10 months for HSCT and kidney, heart, lung, and liver transplant recipients, respectively. For HSCT recipients, the incidence was the same for males and females, and graft-versus-host disease was present in nearly half of the cases. The most common manifestation of disease in these patients was central venous catheter-related infection, accounting for 37% of cases. Pulmonary, cutaneous, and disseminated diseases were also commonly reported. Rapidly growing mycobacteria (predominantly *Mycobacterium fortuitum*, *Mycobacterium abscessus*, and *M. chelonae*) were the most frequently isolated pathogens, accounting for 45% of isolates, and most often were associated with catheter-related infections. Other common pathogens were MAC strains, usually associated with pulmonary or disseminated disease, and *Mycobacterium haemophilum*, most commonly a cause of pulmonary or cutaneous disease.

Manifestations of NTM infection varied in solid organ transplant recipients. In kidney and heart transplant recipients, local or disseminated cutaneous disease was the most common presentation. Pleuropulmonary disease was the most frequent manifestation in lung transplant recipients, accounting for 54% of cases, and also was common in heart transplant recipients (26% of cases). Disseminated NTM disease also occurred frequently; it was the most common presentation in liver recipients, the second most common among kidney transplant recipients, and the third most common in heart and lung recipients. The most frequently isolated pathogens were rapidly growing mycobacteria, followed by *Mycobacterium kansasii* and *M. haemophilum*, for kidney transplant recipients; *M. kansasii*, MAC, and *M. haemophilum* for heart recipients; MAC, *M. abscessus*, and *M. haemophilum* for lung recipients; and MAC for liver transplant recipients (accounting for four of the eight cases).

Urothelial Bladder Cancer

Currently, the treatment of choice for noninvasive urothelial carcinomas and carcinoma in situ following transurethral resection is intravesical installation of BCG, a live attenuated strain of *Mycobacterium bovis* (33). When instilled into the bladder, the viable mycobacteria adhere to the urothelium, are ingested into urothelial cells, and elicit an inflammatory response, with release of cytokines and attraction of helper and killer T lymphocytes. This immune response is believed to be responsible for the destruction of malignant cells. Infectious complications of BCG therapy occur uncommonly. Granulomatous prostatitis was reported for 1% of patients in one large series (34) but is believed to occur more frequently. Usually, granulomatous prostatitis is asymptomatic, but patients may present with acute prostatitis or urinary retention, potentially resulting in induration of the prostate and elevation of the serum prostate-specific antigen. Granulomatous hepatitis and pneumonitis each occur in <1% of patients treated with BCG. They generally present with fever, malaise, and shortness of breath. The bone marrow also may be involved, resulting in cytopenias. The most serious and potentially fatal complication of BCG therapy is BCG sepsis, which developed in 0.4% of patients in the study reported by Lamm et al. (34). This results from systemic absorption of BCG following traumatic catheterization and/or absorption through the inflamed bladder wall in patients given BCG soon after tumor resection. The usual presentation is high fever, chills, hypotension, and mental confusion, with disseminated intravascular coagulation, respiratory failure, jaundice, and leukopenia. Urine and blood cultures generally are negative.

Treatment with TNF-α Antagonists

Tumor necrosis factor alpha (TNF-α) plays an important role in the pathogenesis of various inflammatory diseases (e.g., rheumatoid arthritis, ankylosing spondylitis, and psoriasis) as well as a critical role in the immune response to infection with many intracellular pathogens, such as *M. tuberculosis*. Treatment of certain inflammatory diseases with TNF-α antagonists improves symptoms and slows progression of the disease but also has the potential to increase the risk of infection with *M. tuberculosis* as well as other intracellular pathogens (2, 8, 19, 26, 32).

The infection most closely associated with the use of TNF-α antagonists is tuberculosis. The first case was reported in 1999 (38), in association with the TNF-α antagonist infliximab, and since then several investigators have explored the relationship. Among the first 117 cases of infliximab-associated tuberculosis reported to the Food and Drug Administration (FDA), the incident rate was 41 cases per 100,000 person-years, compared with the baseline rate of 6.2 cases of tuberculosis per 100,000 person-years among patients with rheumatoid arthritis who did not receive a TNF-α antagonist. Keane et al. (32) reported that not only do TNF-α antagonists increase the risk of tuberculosis, but they also are associated with a higher likelihood of complicated disease (56% of patients had extrapulmonary and 24% had disseminated tuberculosis). The increased risk of tuberculosis in patients receiving a TNF-α antagonist was also documented in a large study of 112,300 patients with rheumatoid arthritis (8). In that study, the adjusted rate ratio of tuberculosis with use of TNF-α antagonists was 1.5 per 1,000 person-years of follow-up, and the median time from the first prescription of TNF-α antagonist to presentation of tuberculosis was 17 weeks (range, 1 to 71 weeks) for infliximab and 79 weeks (range, 3 to 168 weeks) for the TNF-α antagonist etanercept. In contrast to the findings of Keane et al. (32), the latter study did not show a higher likelihood of complicated disease, although the proportion of pulmonary-only tuberculosis, versus extrapulmonary or miliary disease, was greater for those receiving a TNF-α antagonist than for controls.

Diabetes

Diabetes is a recognized risk factor for the development of pulmonary tuberculosis, both reactivation disease and recently transmitted infection (4, 44, 45, 47, 56). In a study conducted from 1987 to 1997 at Bellevue Hospital in New York City, Bashar et al. found a significant association between diabetes and multiple-drug-resistant tuberculosis (4). In contrast, Singla et al. reported a lower prevalence of drug-resistant disease in patients treated at Sahary Hospital in Saudi Arabia (56). However, they did find differences between patients with and without diabetes. In their study, patients with diabetes had a higher pretreatment bacillary load, based on the presence of numerous AFB in sputum specimens, and they had higher sputum conversion (from smear positive to smear negative) rates after 3 months of appropriate treatment. Radiological differences between patients with pulmonary tuberculosis who do and do not have diabetes have also been noted. Pérez-Guzmán et al. (45) reported that compared to patients without diabetes, those who had diabetes had a decreased frequency of upper lobe disease and an increased frequency of lower and upper lobe plus lower lung involvement. Moreover, patients with diabetes had higher frequencies of cavitary lesions, cavities in the lower lung fields, and multiple cavities. In fact, based on multiple logistic regression analysis, diabetes was the most important factor associated with tuberculosis involving lower lung fields and with lower lung cavities.

Other Conditions

Rheumatologic diseases themselves, especially systemic lupus erythematosus and perhaps rheumatoid arthritis, and traditional immunosuppressant therapies, particularly prednisone, used to treat them have been reported to be associated with an increased risk of tuberculosis (7, 8, 14). Patients with celiac disease also appear to be at increased risk of tuberculosis (37).

RECOMMENDED APPROACHES TO LABORATORY DIAGNOSTICS

Historically, mycobacteria have been detected and identified by using the conventional methods of AFB smear and culture. With the advent of newer, molecular methods, detection and identification capabilities have increased, time to results has decreased, and new species have been discovered.

Serology

An in vitro test recently received final approval from the FDA as an aid in diagnosing latent tuberculosis infection and tuberculosis disease. This enzyme-linked immunosorbent assay detects gamma interferon released in fresh heparinized whole blood of sensitized persons when the blood is incubated with mixtures of synthetic peptides simulating two proteins present in *M. tuberculosis*. Because these proteins are absent from all BCG vaccine strains and from commonly encountered NTM (except *M. kansasii*, *M. marinum*, and *M. szulgai*), this test is expected to be more specific for detecting latent infection with *M. tuberculosis* than tests that use tuberculin purified protein derivative as the antigen. Currently, trials are under way to explore the test's performance for more narrowly defined populations, but a specificity of 98.1% has been reported for BCG-vaccinated individuals, with a sensitivity of 89% for patients with culture-confirmed tuberculosis (42). Specific guidelines for using this test and its limitations have been published (39). In addition to interpretive limitations, practical limitations include the need to draw blood and to ensure its receipt in the laboratory within narrow time constraints.

Microscopy

All mycobacteria are acid fast, that is, they retain certain dyes even when subjected to strong acids and alcohols. The classic Ziehl-Neelsen (ZN) stain employs carbol fuchsin to stain the mycobacteria red. Other gram-positive and gram-negative bacteria and fungi do not stain red but take on the counterstain, which is usually blue. While the ZN stain is still used today, other fuchsin stains, such as Kinyoun, are generally preferred because they do not require the heating step needed in the ZN staining method.

Fluorescence microscopy, introduced over 30 years ago, offers the advantages of easier preparation and greater sensitivity than light microscopy. Bacilli are stained (fluoresce) yellow or yellow-orange against a dark background. This method does require a fluorescence microscope, which is more expensive than the light microscope used for fuchsin-based stains. Fluorescence microscopy reduces reading time by as much as 75% and is effective in detecting *M. tuberculosis* and NTM (66).

An algorithm for AFB testing is shown in Fig. 1. Smears are prepared by concentrating the clinical specimen, heat fixing it on a slide, staining the smear, and reading the prepared material to detect AFB. If present, the AFB are quantitated, using a scale of 1+ to 4+. No determination as to the exact species of mycobacteria present can be made. Because the sensitivity of microscopy is somewhat low (63) and mycobacteria can be recovered by culture when the smear results are negative, microscopy is considered a preliminary result, and culture is needed for confirmation. Results of smears should be available within 24 h of receipt of the specimen in the laboratory. Positive smears that have negative cultures can be a result of environmental contamination, misread artifacts, or dead organisms. Negative smears that have positive cultures can result from the presence of a few AFB, below the detection limit; poor quality of the specimen; or incomplete smear reading.

Direct Detection of MTBC in Clinical Specimens

Nucleic acid amplification tests (NAAT) can provide a rapid diagnosis of tuberculosis by detecting the MTBC directly in clinical specimens. Once digested and decontaminated sediments are prepared, NAAT results can be available within several hours (15). In the United States, two commercial NAAT kits are cleared by the FDA for testing of respiratory specimens. One, which utilizes transcription-mediated amplification, is FDA cleared for testing both AFB smear-positive and smear-negative specimens; the other, which uses PCR, is cleared for testing AFB smear-positive specimens only. While each kit varies in method, target, probe, and detection format, both are more sensitive than the AFB smear and perform best when used to test AFB smear-positive specimens. Neither kit is FDA cleared for testing of extrapulmonary specimens, but several studies have shown that they can be used successfully on such specimens (16). Commercial NAATs are expensive and require considerable technical expertise and environmental cleaning to prevent false-positive results by amplicon contamination. Additionally, a negative test result does not exclude the possibility of infection with the MTBC, especially for AFB smear-negative specimens. False-positive results have been seen with specimens containing NTM that have a close genetic affinity to the MTBC (61). NAAT results should be considered preliminary, as culture is still needed to confirm the identification and to provide material for drug susceptibility testing.

Culture

Specimens are digested and decontaminated using a sodium hydroxide and *N*-acetyl-L-cysteine method to prepare a concentrated sediment. This sediment can then be used to prepare smears for microscopy and to inoculate media for recovery of mycobacterial growth. Conventional methods use solid media, such as egg-based Lowenstein-Jensen medium, and recovery of *M. tuberculosis* can take 2 to 3 weeks or longer. More rapid growth has been achieved since the addition of Middlebrook agar to the workup. Middlebrook 7H10 or 7H11 agar increases the recovery of mycobacteria (both quantity and species) and decreases the time to growth detection by at least 1 week. Colony morphology is also easier to assess. The most rapid growth of mycobacteria is obtained by using liquid media. Today, there are several automated methods using broth cultures that substantially reduce recovery rates, allowing recovery of the MTBC in as little as 5 to 7 days. The broth can then easily be used for identification studies and drug susceptibility testing. The main disadvantage of these liquid systems is possible overgrowth by non-acid-fast organisms. In addition, the ability to assess colonial morphology is lost unless a subculture is prepared.

Identification of Mycobacteria

Conventional methods for identification to the species level involve a variety of biochemical and physiologic determinations. Cultures are examined for colony morphology on solid media, growth rate is determined, pigmentation is noted, and biochemical reactions are evaluated. Typically, conventional identification of a mycobacterium can take a minimum of several weeks and is laborious. Table 1 lists a few of the salient characteristics of some species. The niacin test is one that provides information to

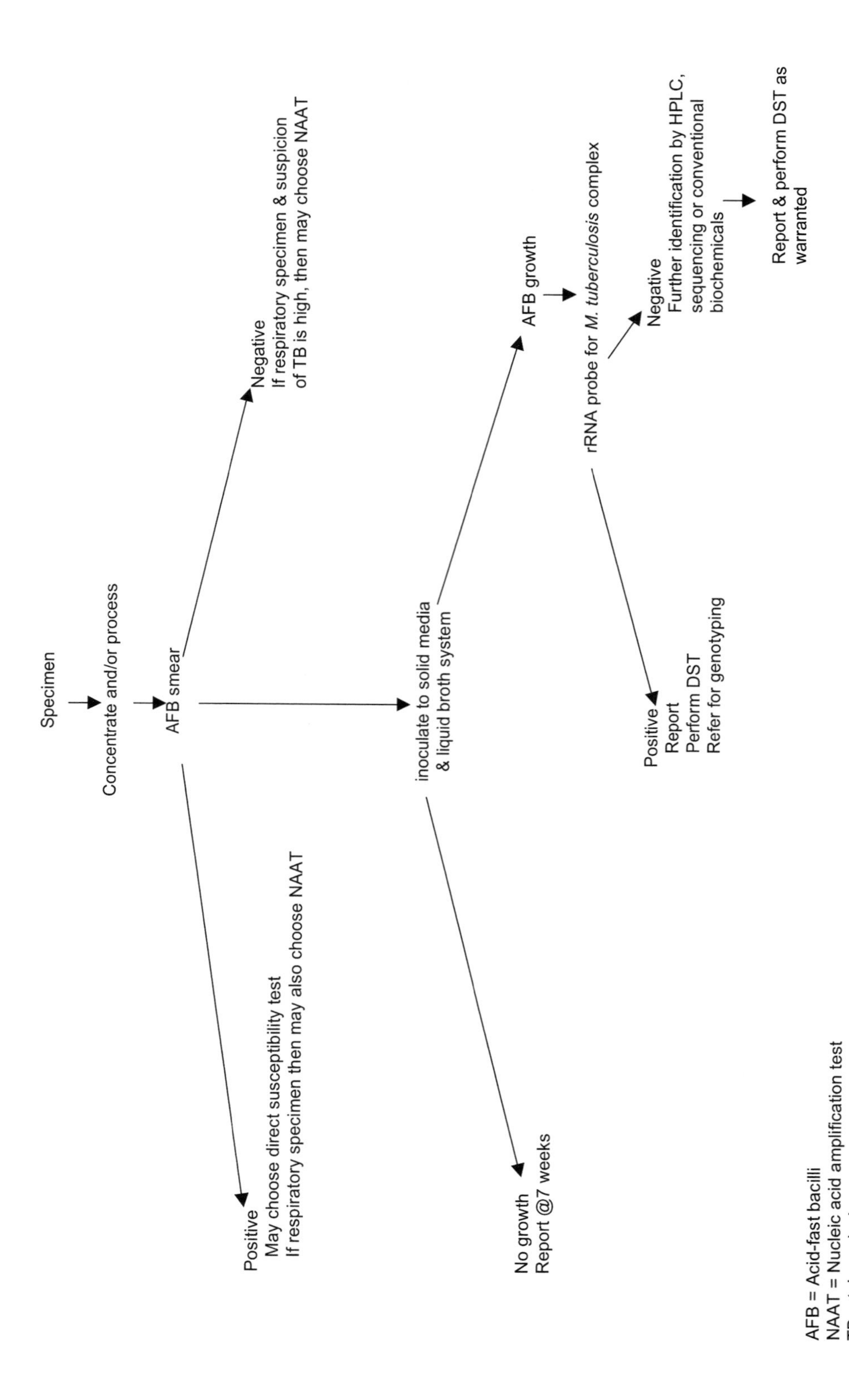

Figure 1. Algorithm for AFB testing.

separate *M. tuberculosis* from other mycobacteria: *M. tuberculosis* produces detectable amounts of niacin, while most NTM and *M. bovis* do not.

The NTM have been divided into four groups, based on growth rate and pigment production (Table 2). As more complete methods for species identification have become available, it is not unusual to see the "group" designation dropped and only the species name used.

DNA probes for the identification of several species of mycobacteria (MTBC, MAC, *M. avium*, *M. intracellulare*, *M. gordonae*, and *M. kansasii*) are available for use on isolates that have been recovered on solid media and in broth culture. These DNA probes are complementary to species-specific sequences of rRNA within the mycobacterial genome. Probes are a major component to achieving a shorter turnaround time for laboratory identification. Results of probe assays can be obtained in about 2 h. Although the number of available probes is limited, they do cover the majority of mycobacterial species that are encountered in clinical specimens. They do not separate members of the MTBC, and some rare isolates of *M. celatum* and a few other mycobacteria have produced false-positive reactions with the MTBC probe (61, 63).

Analysis of mycobacterial cell wall fatty acids is a newer technique that assists with identification of mycobacteria, especially NTM. Use of reverse-phase HPLC can easily recognize a wider spectrum of mycobacterial species and provide results in a much faster time frame than conventional methods. HPLC is not suitable for every clinical laboratory and is usually found in reference laboratories, as specialized equipment and technical expertise are required (13).

Several recent approaches to genotypic identification of mycobacteria have been developed in the last 10 years. Over 120 species of mycobacteria have now been recognized, many as a result of genotyping applications. The basis of genotypic taxonomy is related to the detection of highly conserved regions in the bacterial genome having hypervariable gene sequences in which species-specific deletions, insertions, or replacements of single nucleotides are present. Techniques that study the 16S portion of the rRNA are the mainstay of mycobacterial molecular methods, although other sequences, such as the *hsp65* gene, have been used (62). Sequencing systems generally have some limitations that include an inability to differentiate between species with similar sequences, such as between *M. kansasii* and *M. gastri* and between *M. chelonae* and *M. abscessus*. While these methods require technical expertise and specialized equipment, at least one method is available in kit form and has been used successfully to identify a number of clinically significant mycobacteria (58).

Genotyping

DNA fingerprinting of *M. tuberculosis* is a powerful tool in the control of tuberculosis. It has been applied to outbreak investigations and population-based studies. Genotyping provides information regarding the presence of endemic strains, the diversity of strains, and even detection of laboratory cross-contamination. The National Tuberculosis Genotyping and Surveillance Network, established by the CDC, currently accepts isolates from state and local public health laboratories for the characterization of *M. tuberculosis* strains. This capacity to differentiate *M. tuberculosis* strain patterns by DNA fingerprinting is helpful in the management and control of tuberculosis (40).

Approaches to Comprehensive Clinical Mycobacteriology

Laboratories play a key role in the diagnosis and control of tuberculosis and other mycobacterial diseases. The combination of the most sensitive and

Table 2. Phenotypic groupings of mycobacteria

Group	Time for growth on solid media (days)	Colony appearance	Examples
M. tuberculosis complex	>7	Rough, buff (light and dark)	*M. tuberculosis, M. bovis, M. africanum*
Photochromogen	>7	Smooth or rough, buff (dark), becoming lemon yellow or orange when exposed to light	*M. kansasii, M. marinum, M. simiae*
Scotochromogen	>7	Smooth or rough, yellow to orange (light and dark)	*M. gordonae*
Nonchromogen	>7	Smooth or rough, buff (light and dark), color may intensify with age	*M. avium, M. haemophilum, M. terrae*
Rapid growers	≤7	Smooth or rough, buff or orange (light and dark)	*M. abscessus, M. chelonae, M. fortuitum*

Table 3. Comparison of conventional and newer mycobacterial test methods

Test method using newer technology	Turnaround time	Comparable conventional test method	Turnaround time
Gamma interferon assay	24 h	Purified protein derivative skin test	48–72 h
Fluorescent stain for AFB	3 min each	Fuchsin stain for AFB	15 min each
Culture with liquid media	1–3 wk	Culture with solid media	1–6 wk
Identification—HPLC, DNA probes, sequencing	1–3 days	Identification—phenotypic patterns	3–8 wk
Nucleic acid amplification for direct detection of MTBC	2–6 h	No comparable method	NA[a]
Drug susceptibility test with liquid media	5–7 days	Drug susceptibility test with solid media	3 wk

[a]NA, not applicable.

rapid methods for detection, culture, identification, and susceptibility testing should be employed routinely. These methods include fluorochrome staining for mycobacteria; NAATs on clinical material; use of a broth-based culture recovery method; use of DNA probes, HPLC, or DNA sequencing for colony identification; and rapid drug susceptibility testing (Table 3). Laboratories should be encouraged to move away from routine conventional testing and to utilize current molecular and rapid methods in order to reduce turnaround times and support earlier disease diagnosis and control (59). Using newer methods, it is possible to have results of an AFB smear and a direct NAAT to confirm or exclude the presence of the MTBC within 24 h of specimen collection (Table 4).

INTERPRETATION OF TEST RESULTS

Smears for AFB

The reported sensitivity of the smear for AFB varies, ranging from approximately 20 to 80% (36). This variability is related to several factors, including the number of organisms present (approximately 10,000 organisms/ml of specimen must be present for a smear to be positive), type of stain performed, specimen type, infecting species, centrifugation speed, expertise of the person examining the smear, culture method, and patient population. For example, the likelihood that a smear will be AFB positive is higher if the infecting organism is *M. tuberculosis* than if it is an NTM, for sputum specimens than for other specimen types, and for patients attending a pulmonary clinic than for those at a primary care facility. Additionally, fluorochrome stains are more sensitive than carbol-fuchsin stains, and broth-based culture systems are more sensitive than solid media for isolation of mycobacteria.

In contrast to the sensitivity of the AFB stain for detection of mycobacteria, the specificity is very high. However, a positive smear is not diagnostic of tuberculosis, even for a sputum specimen. NAAT of smear-positive respiratory specimens will provide rapid differentiation of tuberculosis from NTM infection (unless inhibitors are present), thus allowing optimal utilization of tuberculosis control resources.

Specimens Positive for *M. tuberculosis*

In most cases, detection of *M. tuberculosis* in a specimen, either by culture and identification or by nucleic acid amplification, confirms the diagnosis of tuberculosis. However, false-positive results have been reported with both methods (5, 12), setting in motion tuberculosis control measures, which are both time-consuming for public health staff and

Table 4. Suggested turnaround times for mycobacteriology laboratory tests

Activity	Turnaround time
Process specimen and inoculate to media	Within 24 h of specimen collection
Read concentrated smear	Within 24 h of specimen receipt
Perform NAAT	Within 48 h of specimen receipt
Report negative culture	7 weeks
Report AFB detected by culture	Within 3 weeks of specimen receipt
Perform probe test for MTBC	Within 1 day of growth detection
Perform primary drug susceptibility test	Within 2 weeks of TB detection
Perform secondary drug susceptibility test	Usually 3 to 4 weeks after primary test
Perform genotyping	3 weeks after TB identification
Identify NTM	2 days to 6 weeks of growth detection

have potentially harmful repercussions for the patient, including drug toxicity and social ostracism. The median false-positive culture rate in reports involving ≥100 patients was 3.1% (interquartile range, 2.2 to 10.5%); however, when a major error in laboratory technique occurred, rates were much higher (12, 20, 35, 49). Several causes of false-positive cultures have been recognized. Clinical equipment (e.g., bronchoscopes) may become contaminated by mycobacteria and then inadequately high-level disinfected or sterilized, causing contamination of specimens from subsequent patients examined with the same instrument or transmission of tuberculosis (1, 9, 28). However, this is a relatively uncommon cause of false-positive cultures. Clerical errors in the laboratory are another possible, but infrequent, cause of false-positive results. The reason most commonly cited for false-positive cultures is laboratory cross-contamination. Outbreaks of false-positive cultures have been linked to defects in the exhaust system of a biological safety cabinet used for specimen processing (53). Cross-contamination due to equipment malfunction and subsequent carryover during instrument reading has been reported in the past (57) but was quickly corrected by the manufacturer. More often, cross-contamination occurs during batch processing. It could be an isolated event, due to a splash or a contaminated lid or pipette, or it may affect multiple specimens, typically via contamination of one of the reagents (e.g., the neutralizing buffer) used in batch processing.

The laboratory finding most characteristic of a false-positive result is the single positive culture (i.e., the false-positive culture was the only positive culture from that patient). A single positive culture, however, is nonspecific, because patients with a confirmed diagnosis of tuberculosis for whom no evidence of cross-contamination exists may have only one positive culture. Molecular strain typing of isolates should be performed to confirm or exclude cross-contamination (57).

Specimens Positive for NTM

NTM can be found throughout the environment (e.g., in fresh and salt water, potable water, soil, various foods, and cigarettes). Consequently, NTM transiently colonize the respiratory or gastrointestinal tract of many persons without causing disease. In such cases, determining the clinical significance of NTM recovered from respiratory specimens or feces becomes problematic.

Generally, NTM isolates from blood, other sterile body fluids, or tissues are considered clinically significant. One exception is *Mycobacterium gordonae* (nicknamed the tap water bacillus), a common laboratory contaminant that only rarely is a proven human pathogen (22, 65), regardless of the specimen type from which it was recovered. Single positive cultures that are AFB smear negative and/or contain small numbers of organisms (as suggested by only one medium showing growth) are unlikely to be clinically significant.

The American Thoracic Society has published the following criteria for diagnosing lung disease caused by NTM in symptomatic patients with reticulonodular or cavitary disease or a high-resolution computed tomography scan showing multifocal bronchiectasis, small nodules, or both (64):

- three AFB smear-negative, culture-positive sputum or bronchial wash specimens collected on three separate days within 12 months; or
- two positive sputum or bronchial wash cultures within 12 months, with one or both being AFB smear negative; or
- for patients unable to produce sputum, one positive bronchial wash culture with >2+ AFB on the smear and/or growth in both liquid and solid media; or
- in the absence of diagnostic sputum or bronchial wash studies, a transbronchial or lung biopsy yielding an NTM or histological examination of the biopsy showing granulomatous inflammation and/or AFB, with one or more respiratory cultures positive for an NTM.

Susceptibility Testing

Mycobacterial susceptibility testing should be performed according to the guidelines of the Clinical and Laboratory Standards Institute (CLSI) (17). Drugs recommended for testing the MTBC and the more commonly encountered NTM are listed in Table 5.

MTBC

Susceptibility testing should be performed on the initial isolate of MTBC from all patients and repeated if there is clinical evidence of failure to respond to appropriate therapy or if cultures do not convert to negative after 3 months of therapy. Results must be reported to the local public health department as soon as they are available. To ensure early detection of drug resistance, using a broth-based susceptibility test system with a shorter incubation time than that required for conventional agar proportion is recommended. Currently, in the United States, the following

Table 5. Primary and secondary drugs for susceptibility testing of mycobacteria

Mycobacterium species or group	Primary drugs	Secondary drugs
MTBC[a]	Isoniazid, rifampin, ethambutol, pyrazinamide	Capreomycin, ethionamide, amikacin, moxifloxacin, levofloxacin, *p*-aminosalicylic acid, rifabutin, streptomycin
MAC	Clarithromycin (or azithromycin)	Moxifloxacin, linezolid
M. kansasii	Rifampin	Ethambutol, clarithromycin, ciprofloxacin, linezolid, moxifloxacin, trimethoprim-sulfamethoxazole
M. marinum[b]	Rifampin, ethambutol, doxycycline (or minocycline), clarithromycin, trimethoprim-sulfamethoxazole, amikacin	NA[c]
Rapidly growing mycobacteria	Amikacin, cefoxitin, ciprofloxacin, clarithromycin, doxycycline (or minocycline), imipenem, linezolid, moxifloxacin, trimethoprim-sulfamethoxazole, tobramycin	NA[c]

[a]All secondary drugs plus ethambutol hydrochloride (10 (μ/ml) should be tested on isolates of MTBC that are resistant to rifampin or to any two primary agents.
[b]Routine susceptibility testing of *M. marinum* is not necessary. However, susceptibility testing should be performed if a patient fails therapy and cultures remain positive for *M. marinum* after several months of apparently adequate treatment (e.g., clarithromycin, doxycycline, minocycline, trimethoprim-sulfamethoxazole, or rifampin plus ethambutol).
[c]NA, not applicable.

three commercial broth systems have been cleared by the FDA for indirect susceptibility testing of the MTBC: Bactec 460TB, Bactec MGIT 960, and VersaTREK.

An indirect susceptibility test uses organisms grown in culture as the inoculum. In contrast, the direct susceptibility test uses a specimen that is smear positive for AFB, most commonly a processed sputum sample, as the inoculum. The direct test should be performed only on smear-positive specimens and only by the agar proportion method or by a commercial broth system that is FDA cleared for this purpose. The direct test has two major advantages. First, it is cost-effective because materials are much less expensive than those for the commercial broth systems. Second, it generally provides results within 3 weeks of receipt of the specimen in the laboratory, which is comparable to, if not faster than, the time required to grow the organism in culture and then perform indirect susceptibility testing using a commercial broth system. However, there are disadvantages to the direct test. The inoculum cannot be calculated accurately, potentially resulting in insufficient or excessive growth on drug-free control quadrants. Normal respiratory flora may survive digestion and decontamination, making results uninterpretable. Few laboratories have personnel who have sufficient experience with the direct procedure.

The complete panel of primary drugs for susceptibility testing of MTBC strains (Table 5) includes two concentrations of isoniazid and single concentrations of rifampin, ethambutol, and pyrazinamide. The first three drugs can be tested by both direct and indirect methods; pyrazinamide should be tested by the indirect method only. Testing the entire set of antituberculosis drugs, including both concentrations of isoniazid, provides the clinician with comprehensive data regarding the four-drug therapy currently recommended for treatment of most patients with tuberculosis in the United States. However, the commercial broth susceptibility systems are expensive, and laboratory directors must make decisions about cost-effective testing, such as using an abbreviated panel of drugs. For example, in laboratories where drug resistance is uncommon, it might be reasonable to consider testing a single, critical concentration of isoniazid, rifampin, and ethambutol initially. Optimally, before decisions concerning reducing the number of drugs/concentrations tested are made, input should be sought from pulmonary and/or infectious disease specialists and the local tuberculosis control officer.

Secondary drugs should be tested whenever an isolate of the MTBC is resistant to rifampin or to any two primary antituberculosis drugs. When such testing is indicated, all secondary drugs (Table 5) and a higher concentration of ethambutol (10 μg/ml by agar proportion) should be tested to prevent delays in reporting results. Cycloserine, which is an option therapeutically, should not be tested due to technical problems with the assay, causing unreliable results. Testing of all secondary drugs initially is especially important in using the agar proportion method, because results are not available for several weeks. Data from studies evaluating the MGIT 960 broth system for testing more commonly used secondary drugs suggest that the system provides results comparable to those with agar proportion, but with a much shorter turnaround time (50). However, because the MGIT 960 system is not FDA cleared for testing of secondary drugs, using the system for this purpose would require an extensive inhouse validation.

NTM

In general, susceptibility testing of NTM should be considered only for those isolates believed to be clinically significant (as discussed in the previous section). Susceptibility testing should be repeated after 3 months of therapy for patients with disseminated NTM disease and after 6 months of therapy for patients with chronic pulmonary disease if the patient fails to improve clinically or shows clinical deterioration and remains culture positive. In addition, only those species that exhibit variability in susceptibility to clinically useful antimicrobial agents and/or have a significant possibility of acquired mutational resistance to one or more of these agents need be tested. In general, neither of these criteria applies to *M. marinum*; therefore, routine susceptibility testing of initial isolates is not necessary. Drugs that have been used successfully as single agents for treating *M. marinum* infections are clarithromycin, doxycycline, minocycline, and trimethoprim-sulfamethoxazole; rifampin plus ethambutol has also been used. However, if a patient fails therapy and cultures remain positive for *M. marinum* after several months of apparently adequate treatment, susceptibility testing should be performed.

Broth dilution is considered the standard for susceptibility testing of NTM by the CLSI (17). Both the MIC and the interpretation (provided in CLSI document M24-A [17]) should be reported. Drugs recommended for testing of the more commonly encountered NTM are listed in Table 5. For MAC strains, in vitro MIC data can be obtained for drugs other than those listed in Table 5 (e.g., ethambutol, rifampin, rifabutin, amikacin, and streptomycin), but the data available are insufficient to establish correlation of those results with clinical response. For this reason, selecting breakpoints for these drugs to distinguish susceptible and resistant strains is not possible, and reporting results for agents other than those listed in Table 5 is not recommended. For slowly growing NTM other than those listed in Table 5, the primary and secondary drugs listed for *M. kansasii* should be tested, and the same interpretive criteria should be used. For the rapidly growing mycobacteria, MIC values and interpretations should be reported for all drugs listed in Table 5, with one exception. Clarithromycin data should not be reported for members of the *M. fortuitum* (*M. fortuitum*, *M. boenickei*, *M. houstonense*, *M. neworleansense*, *M. porcinum*, and *M. mageritense*) and *M. smegmatis* (*M. smegmatis*, *M. goodii*, and *M. wolinskyi*) groups. These species have an inducible chromosomal *erm* gene that confers macrolide resistance, even for isolates that have a clarithromycin MIC of <4 μg/ml and therefore appear to be susceptible. Moreover, the efficacy of the macrolides in the treatment of disease caused by these species has not been evaluated fully.

REFERENCES

1. **Agerton, T., S. Valway, B. Gore, C. Pozsik, B. Plikaytis, C. Woodley, and I. Ororato.** 1997. Transmission of a highly drug-resistant strain (strain W1) of *Mycobacterium tuberculosis*: community outbreak and nosocomial transmission via a contaminated bronchoscope. *JAMA* **278:**1073–1077.
2. **Askling, J., C. M. Fored, L. Brandt, E. Baecklund, L. Bertilsson, L. Cöster, P. Geborek, L. T. Jacobsson, S. Lindblad, J. Lysholm, S. Rantapää-Dahlqvist, T. Saxne, V. Romanus, L. Klareskog, and N. Feltelius.** 2006. Risk and case characteristics of tuberculosis in rheumatoid arthritis associated with tumor necrosis factor antagonists in Sweden. *Arthritis Rheum.* **52:**1986–1992.
3. **Barbar, T. W., D. E. Craven, and H. W. Farber.** 1991. *Mycobacterium gordonae*: a possible opportunistic respiratory tract pathogen in patients with advanced human immunodeficiency virus, type 1 infection. *Chest* **100:**716–720.
4. **Bashar, M., P. Alcabes, W. N. Rom, and R. Condos.** 2001. Increased incidence of multi-drug resistant tuberculosis in diabetic patients on the Bellevue chest service, 1987 to 1997. *Chest* **120:**1514–1519.
5. **Bhattacharya, M., S. Dietrich, L. Mosher, F. Saddiqui, B. E. Reisberg, W. S. Paul, and J. R. Warren.** 1998. Cross-contamination of specimens with *Mycobacterium tuberculosis*: clinical significance, causes, and prevention. *Am. J. Clin. Pathol.* **109:**324–330.
6. **Bloch, K. C., L. Zwerling, M. J. Pletcher, J. A. Hahn, J. L. Gerberding, S. M. Ostroff, D. J. Vugia, and A. I. Reingold.** 1998. Incidence and clinical implications of isolation of *Mycobacterium kansasii*: results of a 5 year, population-based study. *Ann. Intern. Med.* **129:**698–704.
7. **Bouza, E., J. G. Moya, and P. Munñoz.** 2001. Infections in systemic lupus erythematosus and rheumatoid arthritis. *Infect. Dis. Clin. N. Am.* **15:**335–361.
8. **Brassard, P., A. Kezouh, and S. Suissa.** 2006. Antirheumatic drugs and the risk of tuberculosis. *Clin. Infect. Dis.* **43:**717–722.
9. **Brown, N. M., E. A. Hellyar, J. E. Harvey, and D. S. Reeves.** 1993. Mycobacterial contamination of fibreoptic bronchoscopes. *Thorax* **48:**1283–1285.
10. **Brown-Elliott, B. A., and R. J. Wallace, Jr.** 2002. Clinical and taxonomic status of pathogenic nonpigmented or late-pigmenting rapidly growing mycobacteria. *Clin. Microbiol. Rev.* **15:**716–746.
11. **Brown-Elliot, B. A., and R. J. Wallace, Jr.** 2005. Infections caused by nontuberculous mycobacteria, p. 2909–2916. *In* G. L. Mandell, R. G. Douglass, Jr., and J. E. Bennett (ed.), *Principles and Practice of Infectious Diseases*, 6th ed. Elsevier Churchill Livingstone, Philadelphia, PA.

12. **Burman, W. J., and R. R. Reves.** 2000. Review of false-positive cultures for *Mycobacterium tuberculosis* and recommendations for avoiding unnecessary treatment. *Clin. Infect. Dis.* **31:**1390–1395.
13. **Butler, W. R., and L. S. Guthertz.** 2001. Mycolic acid analysis by high-performance liquid chromatography for identification of *Mycobacterium* species. *Clin. Microbiol. Rev.* **14:**704–726.
14. **Carmona, L., C. Hernández-García, C. Vadillo, E. Pato, A. Balsa, I. González-Álvaro, M. A. Belmonte, X. Tena, R. Sanmartí, and the EMECAR Study Group.** 2003. Increased risk of tuberculosis in patients with rheumatoid arthritis. *J. Rheumatol.* **30:**1436–1439.
15. **Centers for Disease Control and Prevention.** 2000. Update: nucleic acid amplification tests for tuberculosis. *MMWR Morb. Mortal. Wkly. Rep.* **49:**593–594.
16. **Clinical and Laboratory Standards Institute.** 2006. Laboratory detection and identification of mycobacteria; proposed guideline. CLSI document M48-P. Clinical and Laboratory Standards Institute, Wayne, PA.
17. **Clinical and Laboratory Standards Institute.** 2003. Susceptibility testing of Mycobacteria, Nocardiae, and other aerobic actinomycetes; approved standard. CSLI document M24-A. Clinical and Laboratory Standards Institute, Wayne, PA.
18. **Corbett, E. L, G. J. Churchyard, M. Hay, P. Herselman, T. Clayton, B. Williams, R. Hayes, D. Mulder, and K. M. De Cock.** 1999. The impact of HIV infection of *Mycobacterium kansasii* disease in South African gold miners. *Am. J. Respir. Crit. Care Med.* **160:**10–14.
19. **Crum, N. F., E. R. Lederman, and M. R. Wallace.** 2005. Infections associated with tumor necrosis factor-α antagonists. *Medicine* **84:**291–302.
20. **Doucet-Populaire, F., V. Lalande, E. Carpentier, A. Bourgoin, M. Dailloux, C. Bollet, A. Vachee, D. Moinard, J. Texier-Maugein, B. Carbonelle, and J. Grosset.** 1996. A blind study of the polymerase chain reaction for the detection of *Mycobacterium tuberculosis* DNA. *Tuber. Lung Dis.* **77:**358–362.
21. **Doucette, K., and J. A. Fishman.** 2004. Nontuberculous mycobacterial infection in hematopoietic stem cell and solid organ transplant recipients. *Clin. Infect. Dis.* **38:**1428–1439.
22. **Eckburg, P. B., E. O. Buadu, P. Stark, P. S. A. Sarinas, R. K. Chitkara, and W. G. Kuschner.** 2000. Clinical and chest radiographic findings among persons with sputum culture positive for *Mycobacterium gordonae. Chest* **117:**96–102.
23. **Falkinham, J. O., III.** 1996. Epidemiology of infection by nontuberculous mycobacteria. *Clin. Microbiol. Rev.* **9:**177–215.
24. **Field, S. K., and R. L. Cowie.** 2006. Lung disease due to the more common nontuberculous mycobacteria. *Chest* **129:** 1653–1672.
25. **Fournier, S., G. Piaious, and V. Vincent.** 1998. *Mycobacterium genavense* and cutaneous disease in AIDS. *Ann. Int. Med.* **128:**409–410.
26. **Gómez-Reino, J. J., L. Carmona, V. R. Valverde, E. M. Mola, and M. D. Montero on behalf of the BIOBADASER Group.** 2003. Treatment of rheumatoid arthritis with tumor necrosis factor inhibitors may predispose to significant increase in tuberculosis risk. *Arthritis Rheum.* **48:**2122–2127.
27. **Gordin, F. M., and R. C. Horsburgh.** 2005. *Mycobacterium avium* complex, p. 2897–2909. *In* G. L. Mandell, R. G. Douglass, and J. E. Bennett (ed.), *Principles and Practice of Infectious Disease,* 6th ed. Elsevier Churchill Livingstone, Philadelphia, PA.
28. **Gubler, J. G. H., M. Salfinger, and A. Graevenitz.** 1992. Pseudoepidemic of nontuberculous mycobacteria due to a contaminated bronchoscope cleaning machine: report of an outbreak and review of the literature. *Chest* **101:**1245–1251.
29. **Jie, T., A. J. Matas, K. J. Gillingham, D. E. R. Sutherland, D. L. Dunn, and A. Humar.** 2005. Mycobacterial infections after kidney transplant. *Transplant. Proc.* **37:**937–939.
30. **Juffermans, N. P., A. Verbon, S. A. Danner, E. J. Kuijper, and P. Speelman.** 1998. *Mycobacterium xenopi* in HIV-infected patients: an emerging pathogen. *AIDS* **12:**1661–1666.
31. **Karakousis, P. C., R. D. Moore, and R. E. Chaisson.** 2004. *Mycobacterium avium* complex in patients with HIV infection in the era of highly active antiretroviral therapy. *Lancet Infect. Dis.* **4:**557–565.
32. **Keane, J., S. Gershon, R. P. Wise, E. Mirabile-Levens, J. Kasznica, W. D. Schwieterman, J. N. Siegel, and M. M. Braun.** 2001. Tuberculosis associated with infliximab, a tumor necrosis factor α-neutralizing agent. *N. Engl. J. Med.* **345:**1098–1104.
33. **Koya, M. P., M. A. Simon, and M. S. Soloway.** 2006. Complications of intravesical therapy for urothelial cancer of the bladder. *J. Urol.* **175:**2004–2010.
34. **Lamm, D. L., A. P. M. Van Der Meijden, A. Morales, S. A. Brosman, W. J. Catalona, H. W. Herr, M. S. Soloway, A. Steg, and F. M. J. Debruyne.** 1992. Incidence and treatment of complications of bacillus Calmette-Guerin intravesical therapy in superficial bladder cancer. *J. Urol.* **147:**596–600.
35. **Lan, N. T. N., C. D. Wells, N. J. Binkin, J. E. Becerra, P. D. Linh, and N. V. Co.** 1999. Quality control of smear microscopy for acid-fast bacilli: the case for blinded re-reading. *Int. J. Tuberc. Lung Dis.* **3:**55–61.
36. **Lipsky, B. A., J. Gates, F. C. Tenover, and J. J. Plorde.** 1984. Factors affecting the clinical value of microscopy for acid-fast bacilli. *Rev. Infect. Dis.* **6:**214–222.
37. **Ludvigsson, J. F., J. Wahlstrom, J. Grunewald, A. Ekbom, and S. M. Montgomery.** 2006. Coeliac disease and risk of tuberculosis: a population based cohort study. *Thorax* **62:**23–28.
38. **Maini, R., E. W. St. Clair, F. Breedveld, D. Furst, J. Kalden, M. Weisman, J. Smolen, P. Emery, G. Harriman, M. Feldmann, and P. Lipsky for the ATTRACT Study Group.** 1999. Infliximab (chimeric anti-tumor necrosis factor α monoclonal antibody) versus placebo in rheumatoid arthritis patients receiving concomitant methotrexate: a randomized phase III trial. *Lancet* **354:**1932–1939.
39. **Mazurek, G. H., J. Jereb, P. LoBue, M. F. Iademarco, B. Metchock, and A. Vernon.** 2005. Guidelines for using the QuantiFERON-TB Gold test for detecting *Mycobacterium tuberculosis* infection, United States. *MMWR Morb. Mortal. Wkly. Rep.* **54:**49–55.
40. **McNabb, S. J. N., C. R. Braden, and T. R. Navin.** 2002. DNA fingerprinting of *Mycobacterium tuberculosis*: lessons learned and implications for the future. *Emerg. Infect. Dis.* **8:**1314–1319.
41. **Miguez-Burbano, M. J., M. Flores, D. Ashkin, A. Rodriguez, A. M. Granada, N. Quintero, and A. Pitchenik.** 2006. Nontuberculous mycobacteria disease as a cause of hospitalization in HIV-infected subjects. *Int. J. Infect Dis.* **10:**47–55.
42. **Mori, T., M. Sakatani, F. Yamagishi, T. Takashima, Y. Kawabe, K. Nagao, E. Shigeto, N. Harada, S. Mitarai, M. Okada, K. Suzuki, Y. Inoue, K. Tsuyuguchi, Y. Sasaki, G. H. Mazurek, and I. Tusyuguchi.** 2004. Specific detection of tuberculosis infection: an interferon-γ-based assay using new antigens. *Am. J. Respir. Crit. Care Med.* **170:**59–64.
43. **Munoâz, P., C. Rodríquez, and E. Bouza.** 2005. *Mycobacterium tuberculosis* infection in recipients of solid organ transplants. *Clin. Infect. Dis.* **40:**581–587.

44. **Pablos-Méndez, A., J. Blustein, and C. A. Knirsch.** 1997. The role of diabetes mellitus in the higher prevalence of tuberculosis among Hispanics. *Am. J. Public Health* **87**:574–579.
45. **Pérez-Guzmán, C., A. Torres-Cruz, H. Villarreal-Velarde, M. A. Salazar-Lezama, and M. H. Vargas.** 2001. Atypical radiological images of pulmonary tuberculosis in 192 diabetic patients: a comparative study. *Int. J. Tuberc. Lung Dis.* **5**: 455–461.
46. **Philips, P., S. Bonner, N. Gataric, T. Bai, P. Wilcox, R. Hogg, M. O'Shaughnessy, and J. Montaner.** 2005. Nontuberculous mycobacterial immune reconstitution syndrome in the HIV infected patients: spectrum of disease and long-term follow-up. *Clin. Infect. Dis.* **41**:1483–1497.
47. **Ponce-De-Leon, A., M. L. Garcia-Garcia, M. C. Garcia-Sancho, F. J. Gomez-Perez, J. L. Valdespino-Gomez, G. Olaiz-Fernandez, R. Rojas, L. Ferreyra-Reyes, B. Cano-Arellano, M. Bobadilla, P. M. Small, and J. Sifuentes-Osornio.** 2004. Tuberculosis and diabetes in southern Mexico. *Diabetes Care* **27**:1584–1590.
48. **Queipo, J. A., E. Broseta, M. Santos, J. Sánchez-Plumed, A. Budía, and F. Jiménez-Cruz.** 2003. Mycobacterial infection in a series of 1261 renal transplant recipients. *Clin. Microbiol. Infect.* **9**:518–525.
49. **Ramos, M. C., H. Soini, G. C. Roscanni, M. Jacques, M. C. Villares, and J. M. Musser.** 1999. Extensive cross-contamination of specimens with *Mycobacterium tuberculosis* in a reference laboratory. *J. Clin. Microbiol.* **37**:916–919.
50. **Rusch-Gerdes, S., G. E. Pfyffer, M. Casal, M. Chadwick, and S. Siddiqi.** 2006. Multicenter laboratory validation of the BACTEC MGIT 960 technique for testing susceptibilities of *Mycobacterium tuberculosis* to classical second-line drugs and newer antimicrobials. *J. Clin. Microbiol.* **44**:688–692.
51. **Saubolle, M. A., T. E. Kiehn, M. H. White, M. F. Rudinsky and D. Armstrong.** 1996. *Mycobacterium haemophilum*: microbiology and expanding clinical and geographic spectra of disease in humans. *Clin. Microbiol. Rev.* **9**:435–447.
52. **Schneider, E., M. Moore, and K. G. Castro.** 2005. Epidemiology of tuberculosis in the United States. *Clin. Chest Med.* **26**:183-195.
53. **Segal-Maurer, S., B. N. Kreisworth, J. M. Burns, S. Lavie, M. Lim, C. Urban, and J. J. Rahal.** 1998. *Mycobacterium tuberculosis* specimen contamination revisited: the role of laboratory environmental control in a pseudo-outbreak. *Infect. Control Hosp. Epidemiol.* **19**:101–105.
54. **Sepkowitz, K. A., J. Raffalli, L. Riley, T. E. Kiehn, and D. Armstrong.** 1995. Tuberculosis in the AIDS era. *Clin. Microbiol. Rev.* **8**:180–199.
55. **Singh, N., and D. L. Paterson.** 1998. *Mycobacterium tuberculosis* infection in solid organ transplant recipients: impact and implications for management. *Clin. Infect. Dis.* **27**: 1266–1277.
56. **Singla, R., N. Khan, N. Al-Sharif, M. O. Al-Sayegh, M. A. Shaikh, and M. M. Osman.** 2006. Influence of diabetes on manifestations and treatment outcome of pulmonary TB patients. *Int. J. Tuberc. Lung Dis.* **10**:74–79.
57. **Small, P. M., N. B. McClenny, S. P. Singh, G. K. Schoolnik, L. S. Tompkins, and P. A. Mickelsen.** 1993. Molecular strain typing of *Mycobacterium tuberculosis* to confirm cross-contamination in the mycobacteriology laboratory and modification of procedure to minimize occurrence of false-positive cultures. *J. Clin. Microbiol.* **31**:1677–1682.
58. **Suffys, P. N., A. da Silva Rocha, M. de Oliveira, C. E. D. Campos, A. M. W. Barreto, F. Portaels, L. Rigouts, G. Wouteers, G. Jannes, G. van Reybroeck, W. Mijs, and R. Vanderborght.** 2003. Rapid identification of mycobacteria to the species level using INNO-LiPA mycobacteria, a reverse hybridization assay. *J. Clin. Microbiol.* **39**:4477–4482.
59. **Tenover, F., J. Crawford, R. Huebner, L. Getter, C. R. Horsburgh, Jr., and R. C. Good.** 1993. The resurgence of tuberculosis: is your laboratory ready? *J. Clin. Microbiol.* **32**:767–770.
60. **Thwaites, G. E., D. B. Nguyen, H. D. Nguyen, T. Q. Hoang, T. T. O. Do, T. C. T. Nguyen, Q. H. Nguyen, T. T. Nguyen, N. H. Nguyen, T. N. L. Nguyen, N. L. Nguyen, H. D. Nguyen, N. T. Vu, H. H. Cao, T. H. C. Tran, P. M. Pham, T. D. Nguyen, S. Kasia, C. P. Simmons, N. J. White, T. H. Tran, and J. J. Farrar.** 2005. The influence of HIV infection on clinical presentation, response to treatment, and outcome in adults with tuberculous meningitis. *J. Infect. Dis.* **192**: 2134–2141.
61. **Tjhie, J. H. T., A. F. van Belle, M. Dessens-Droon, and D. van Soolingen.** 2001. Misidentification and diagnostic delay caused by a false-positive amplified *Mycobacterium tuberculosis* direct test in an immunocompetent patient with a *Mycobacterium celatum* infection. *J. Clin. Microbiol.* **39**: 2311–2312.
62. **Tortoli, E.** 2003. Impact of genotypic studies on mycobacterial taxonomy: the new mycobacteria of the 1990s. *Clin. Microbiol. Rev.* **16**:319–354.
63. **Vincent, V., B. A. Brown-Elliot, K. C. Jost, Jr., and R. J. Wallace, Jr.** 2003. *Mycobacterium:* phenotypic and genotypic identification, p. 560–584. *In* P. R. Murray, E. J. Baron, J. H. Jorgensen, M. A. Pfaller, and R. H. Yolken (ed.), *Manual of Clinical Microbiology,* 8th ed. ASM Press, Washington, DC.
64. **Wallace, R. J., Jr., J. Glassroth, D. E. Griffith, K. N. Oliver, J. L. Cook, and F. Gordin.** 1997. Diagnosis and treatment of disease caused by nontuberculous mycobacteria. *Am. J. Respir. Crit. Care Med.* **156**:S1–S25.
65. **Weinberger, M., S. L. Berg, I. M. Feuerstein, P. A. Pizzo, and F. G. Witebsky.** 1992. Disseminated infection with *Mycobacterium gordonae*: report of a case and critical review of the literature. *Clin. Infect. Dis.* **14**:1229–1239.
66. **Wright, P. W., R. J. Wallace, Jr., N. W. Wright, B. A. Brown, and D. E. Griffith.** 1998. Sensitivity of fluorochrome microscopy for detection of *Mycobacterium tuberculosis* versus nontuberculous mycobacteria. *J. Clin. Microbiol.* **36**:1046–1049.

Diagnostic Microbiology of the Immunocompromised Host
Edited by Randall T. Hayden, Karen C. Carroll, Yi-Wei Tang, and Donna M. Wolk

Chapter 13

Aerobic Actinomycetes

MICHAEL A. SAUBOLLE

The group of gram-positive bacillary organisms broadly known as "aerobic actinomycetes" consists of a vast array of taxonomically heterogeneous and divergent genera. The majority of genera containing human pathogens are placed in the suborder *Corynebacterinae*. The genus *Dermatophilus* alone is placed in the suborder *Micrococcineae* and is probably not closely related to the other aerobic actinomycetes. These organisms are characterized by the presence of often-branching filaments, with the ability to produce spores or to fragment (3, 7). Except for the *Mycobacterium tuberculosis* complex, *Mycobacterium leprae*, and the genus *Dermatophilus*, most species in this group are considered saprophytes; few are associated with human pathogenesis (3, 11, 14). This chapter deals only with those genera having the most impact on human health care, i.e., those that usually affect primarily patients with immunocompromising conditions. The mycobacterial genus is dealt with elsewhere in this monograph.

The etiologic prevalence of specific genera and species varies geographically within the United States and worldwide (14). Over 40 genera of aerobic actinomycetes have been described, but only a minority of these have been associated with human health (3, 11). The aerobic actinomycetes are found in a wide variety of natural and man-made environments but are rarely considered a part of the normal human flora; infections normally originate from exogenous sources. The association of these species with human disease is most frequently of an opportunistic nature, either through accidental means (i.e., traumatic percutaneous inoculation) or through involvement with an immunocompromised host (14). As a group, they cause a wide spectrum of disease; symptoms may differ greatly between the taxa, but there is a substantial amount of overlap.

Because of patient population changes, such as increasing age, increasing chronic debilitating conditions (e.g., chronic respiratory diseases such as chronic obstructive pulmonary disease), and decreasing immunocompetency (e.g., organ/tissue transplantation, cancer, and AIDS, with concomitant immunosuppression), the number of infections caused by aerobic actinomycetes is rising. Despite this rise, laboratory methods have been slow to maximize recovery and identification of these pathogens. The incidence of these infections may be greater than currently appreciated, as many cases may not be recognized. Traditional identification methods for separation of the many species in this group have found great difficulties. The development and application of molecular diagnostic methods for identification may provide a more objective understanding of the organisms' distribution, epidemiology, and pathogenicity in association with human disease (7).

NOMENCLATURE AND DESCRIPTION

Nocardia spp.

The nocardia cell wall is responsible for its staining attributes. The cell wall contains tuberculostearic acids made of shorter-chained (40- to 60-carbon) mycolates than those of the mycobacteria. They have a type IV cell wall, which is characterized by a peptidoglycan composed of *meso*-diaminopimelic acid (DAP), arabinose, and galactose (3, 14).

The taxonomy within the genus *Nocardia* is changing rapidly, as recent applications of modern taxonomic procedures, including molecular characterization and numerical taxonomic methods, have expanded our knowledge of the phylogenetic relatedness and taxonomic status of organisms in this genus (4, 7, 14). Traditionally, the nocardial species

Michael A. Saubolle • Division of Infectious Diseases, Laboratory Sciences of Arizona/Sonora Quest Laboratories, Banner Health, and Department of Medicine, University of Arizona College of Medicine, Phoenix, AZ 85006.

were identified by phenotypic means, such as microscopy, growth characteristics (color and aerial filamentation), and biochemical reactions, including resistance to lysozyme (separating the nocardia from other streptomycetes) and the hydrolysis of casein, tyrosine, xanthine, and hypoxanthine (14).

Historically, clinical laboratory methods were only able to separate the nocardia from other genera and to delineate the genus into three species, *Nocardia asteroides*, *Nocardia brasiliensis*, and *Nocardia otitidiscaviarum* (*caviae*). Subsequently, in 1988, Wallace and colleagues described six stable susceptibility patterns within *N. asteroides*. They were initially able to define two separate species within this group, namely, *Nocardia farcinica* (susceptibility pattern V) and *Nocardia nova* (susceptibility pattern III). As they studied these profiles, they found that more separable, stable species were discernable within the "species" *Nocardia asteroides* (4). The susceptibility profiles within the traditional species of *N. asteroides* and the newer species now allocated to those profiles are depicted in Table 1.

Although high-performance liquid chromatography and fatty acid analysis were used to characterize *Nocardia* species, newer molecular methods, such as PCR-restriction endonuclease analysis of the 16S RNA gene and the heat shock protein (*hsp*) gene and, most recently, sequencing (16S RNA) and pyrosequencing, have provided new tools for nocardia characterization. New methods have provided evidence that the *N. asteroides* species contains a multitude of distinguishable species. Similarly, other previously indistinguishable groups have been divided to yield separate species. The old *N. brasiliensis* species is now known to be made up of two separate species, *N. brasiliensis* and *Nocardia pseudobrasiliensis*. These two species have significant genetic and clinical differences; the former causes primarily skin and lymphocutaneous infections, while the latter is associated with pulmonary and disseminated disease (19).

Approximately 70 *Nocardia* species are presently recognized, and the genus remains heterogeneous, with taxonomy that continues to evolve (3, 4, 13). The nocardial species that have been associated with human infection are noted, together with their clinical presentations, in Table 2.

Rhodococcus spp.

The rhodococci were given their own genus status in the 1970s. Redefining of members within the genus *Rhodococcus* moved several species into a new genus, *Gordona* (now known as *Gordonia*), and one species into the genus *Tsukamurella* (7, 14). The rhodococcus cell wall contains 34- to 52-carbon mycolates, no mycobactins, and eight isoprene dehydrogenated menaquinones (14). Differences also exist between *Gordonia* and *Rhodococcus* in their chemotaxonomic characteristics and molecular 16S rRNA sequences. Of the 34 current species in the genus *Rhodococcus*, only *Rhodococcus equi* has repeatedly been associated with human disease (7, 14).

Gordonia spp.

The *Gordonia* genus was initially introduced as *Gordona* in 1988 to accommodate those members

Table 1. Newer species of *Nocardia* recently recognized and characterized within the stable susceptibility profiles of *N. asteroides* described by Wallace and colleagues[b]

Susceptibility group	New species name	Susceptibility profile[a]
I	*N. abscessus*	S to ampicillin, amoxicillin-clavulanate, ceftriaxone, linezolid, amikacin R to imipenem, ciprofloxacin, clarithromycin
II	*N. brevicatena-paucivorans* group (unnamed groups)	Same as group I, but S to ciprofloxacin and R to clarithromycin
III	*N. nova* complex (*N. nova, N. veterana, N. africana*)	S to ampicillin, erythromycin, clarithromycin, linezolid, ceftriaxone, imipenem, amikacin
IV	*N. transvalensis*	S to ciprofloxacin, ceftriaxone, linezolid, imipenem R to aminoglycosides, erythromycin, clarithromycin
V	*N. farcinica*	S to ciprofloxacin, linezolid, imipenem R to ampicillin, broad-spectrum cephalosporins, clarithromycin, aminoglycosides (except amikacin)
VI	*N. asteroides* complex, *N. cyriacigeorgica* (*N. asteroides* type IV unnamed)	S to ceftriaxone, amikacin, linezolid, imipenem R to ampicillin, amoxicillin-clavulanate, clarithromycin, ciprofloxacin

[a]S, susceptible; R, resistant.
[b]Compiled from reference 4.

Table 2. Species of *Nocardia* associated with human infection[a]

Species and isolation frequency	Infection		
	Primary skin	Pulmonary	Disseminated
Common species			
N. asteroides complex	+	+	+
N. cyriacigeorgica		+	+
N. farcinica		+	+
N. transvalensis		+	+
N. brasiliensis	+		
Less common species			
N. nova		+	+
N. abscessus		+	+
N. pseudobrasiliensis		+	+
N. otitidiscaviarum (old *N. caviae*)	+	+	+
Uncommon species			
N. africana		+	
N. paucivorans		+	+
N. veterana		+	
Rare species or those with prevalence unknown			
N. asiatica			
N. beijingensis			
N. higoensis			
N. veterana			
N. niigatensis			

[a]Compiled from reference 4.

of *Rhodococcus* which were aberrant in their molecular structure and had differences in their mycolates (48- to 66-carbon mycolates) as well as in their dehydrogenated menaquinones (nine isoprene units) (14). The genus name was later changed to *Gordonia*. Presently, more than 21 validly published species are in the genus. Approximately eight species have been implicated in human infections, albeit in very small numbers.

Tsukamurella spp.

Tsukamura, for whom the genus was named in 1980, was the first to study the *Tsukamurella* organism, which was isolated from sputum in 1971 (14). The primary species in this genus, *Tsukamurella paurometabola*, was originally called *Corynebacterium paurometabolum* (7). The species in the genus *Tsukamurella* contain 62 to 78 carbon atoms in their mycolates and menaquinones of the MK-9 type. Of the eight species recognized in this genus, all have been associated with very few infections in humans, with *Tsukamurella tyrosinosolvans* being the most common. New data indicate that the traditional type species of this genus, *T. paurometabola*, may actually be *Tsukamurella inchonensis* (7).

Other Aerobic Actinomycetes, Including the Genera *Actinomadura*, *Amycolata*, *Amycolatopsis*, *Dermatophilus*, *Dietzia*, *Nocardiopsis*, *Pseudonocardia*, and *Streptomyces*

Several genera in this group (*Actinomadura*, *Dermatophilus*, and *Streptomyces*) are primary pathogens in several infectious clinical syndromes. In the United States, they have rarely been associated with diseases in humans and rarely, if ever, involved specifically with immunocompromised patients. They have been limited geographically to tropical or subtropical areas. Others in this group are rarely described in human disease, and their in-depth characterization is not within the scope of this monograph. The reader is directed to several good reviews on the subject, including a recent chapter by Conville and Witebsky in the 2007 edition of the *Manual of Clinical Microbiology* (7).

Actinomadura spp.

The organisms within the genus *Actinomadura* were originally classified within the genus *Nocardia* but were given their own genus status later (14). They all share a chemotype III cell wall containing the sugar madurose (a characteristic shared only by

Dermatophilus). The genus contains at least 57 species, 2 of which (*Actinomadura madurae* and *Actinomadura pelletieri*) are commonly associated with human infections.

Amycolata spp., *Amycolatopsis* spp., and *Pseudonocardia* spp.

The members of the genera *Amycolata* and *Amycolatopsis* were placed into their own genera, moved from the *Nocardia* because they were gram positive but negative for modified acid-fast stain (MAFS). Like *Nocardia*, both genera have aerial mycelia. There are few isolates associated with human disease. Recently, 16S rRNA studies suggested that *Amycolata* be incorporated into the genus *Pseudonocardia* (7).

Nocardiopsis spp.

Members of the species *Actinomadura dassonvillei* were given their own genus status (*Nocardiopsis*) because they differed from other *Actinomadura* species both morphologically and chemotaxonomically (7, 8). The cell wall is made of DAP and is considered a chemotype III cell wall but lacks mycolates and madurose. Chains of arthrospores fragmenting from aerial and substrate hyphae are characteristically produced. Only 1 of the 27 species in this genus (*Nocardiopsis dassonvillei*) is considered pathogenic (7, 14).

Streptomyces spp.

The *Streptomyces* genus contains an extremely large assortment of mostly poorly defined species. With over 517 species, the nomenclature is in a state of flux and tentative for many of them (7, 14). The cell wall in this group is of chemotype I (L-DAP and glycine but no sugar). Only *Streptomyces somaliensis* and *Streptomyces anulatus* commonly cause human infections.

EPIDEMIOLOGY

The primary sources of human infection with aerobic actinomycetes are environmental. Even when reported as nosocomial infections, they have primarily been associated with point sources and the environment. In such settings, dusty conditions during construction or aerosolization during manipulation of infected material have been implicated in some outbreaks (7, 14).

Because of their ubiquitous presence in the environment, aerobic actinomycetes have on several occasions been reported to cause pseudo-outbreaks through point source contamination of specimens during processing or through cross-contamination of cultures being monitored using common needles (7).

In terms of overall numbers, infections are uncommon in the United States, but the frequency is dependent on geographic location. Species of *Nocardia* are the major group of aerobic actinomycetes that cause infections in humans, followed distantly by the rhodococci. *Gordonia* spp. and *Tsukamurella* spp. cause most of the remaining infections in immunocompromised patients. Although in some specific geographic areas mycetoma-causing species (*Actinomadura* and *Streptomyces*) are uncommon, the number of documented cases seems to be growing. Other genera of aerobic actinomycetes very rarely cause disease.

Nocardia spp.

Nocardia spp. are found extensively worldwide and are saprophytic, making up an important component of normal soil microflora and often associated with fresh and salt water. They may also be associated with dust and air as well as with decomposing plant and fecal material (3, 14).

At least 32 species of *Nocardia* have been associated with human disease (Table 2), but the geographic prevalence of each may be dramatically different throughout the world, with some being uncommon or rare. Seventeen species are more commonly implicated in human infection, while an additional 15 species have been associated with human infection but either are rarely isolated or their prevalence is not yet known; at least 21 other species have been described from the environment or from other animal species but have not yet been implicated in human infections (4).

The specific natural habitats of many nocardial species have not yet been ascertained (3, 8, 14, 15), but geographic distributions are known. *Nocardia cyriacigeorgica* seems to be distributed evenly throughout the United States, together with *Nocardia farcinica*, although the latter, debatably, is less prevalent. Distributions of other species may vary regionally. For example, *N. brasiliensis* is fairly commonly isolated in the United States but has a higher prevalence in the southwest and southeast regions and is most commonly associated with tropical environments. *Nocardia otitidiscaviarum* has infrequently been recovered from soil throughout the world.

The majority of nocardial infections in the United States are acquired through inhalation, followed by traumatic percutaneous inoculation, in far smaller numbers (3, 4, 14, 15). In the United States, the prevalence of nocardial infections seems to be greatest in association with the Southwest, where dispersal of infectious material may be facilitated by warm, dry, and windy (and thus dusty) conditions. Infections with *N. brasiliensis* and *N. otitidiscaviarum* are associated with percutaneous implantation via a foreign object. Little is known of the epidemiology of many of the newly recognized potentially pathogenic human species, such as *Nocardia africana*, *Nocardia paucivorans*, and *Nocardia veterana*. Nocardiosis is not thought to be transmitted from person to person and is rarely acquired nosocomially, although there have been reports of clusters of patients infected with identical strains of nocardia in which nosocomial acquisition was a probability (4, 6, 14, 15).

The most common species causing human disease include *N. cyriacigeorgica*, *N. farcinica*, *N. nova*, *Nocardia transvalensis*, *N. brasiliensis*, possibly *N. paucivorans*, and isolates identified as a *Nocardia* complex and often lumped into the "*Nocardia asteroides* complex." *Nocardia abscessus*, *N. otitidiscaviarum*, *N. pseudobrasiliensis*, and *N. africana* are recovered less frequently. *N. veterana* has been recovered only occasionally, while other species of *Nocardia* are rarely recovered in confirmed association with human disease. The prevalence of various species, however, again seems to be related to geographic location.

As noted above, the nocardias are primarily opportunistic pathogens with a predilection for immunocompromised patients, especially those on long-term corticosteroids for other respiratory conditions. Their primary clinical presentation is that of a respiratory infection, often of a chronic nature. Infection is characterized by acute inflammation, necrosis, and formation of abscesses; granulomas are not usually formed (14, 15).

In one evaluation of nocardiosis in patients over a 1-year period, 75% of cases were associated with underlying chronic lung conditions, followed by diabetes mellitus, hematologic and solid malignancies, transplantation, and AIDS (15). Fewer than 10% of patients had no definable immunocompromising conditions. Cystic fibrosis may also be a condition predisposing patients to infection or colonization with *Nocardia* (10).

The organisms may invade almost any organ system and present with a wide array of symptoms, none of which are pathognomonic. Additional manifestations of nocardiosis in the immunocompromised patient almost always occur via respiratory entry, followed by dissemination, and may include bacteremia, empyema, brain abscess, pericarditis, synovitis, peritonitis, and ocular infection. In reality, almost any organ site may be a target of dissemination. The brain has frequently been reported as a site of metastasis, especially with *N. farcinica*, which seems to have greater invasive capabilities than other species, as shown by animal studies (2). However, the central nervous system (CNS) was not a major source of *Nocardia* in at least one report from the arid Southwest United States (15). In that study, <2% of 433 evaluable cases of nocardiosis over a 5-year period had the CNS as a source of recovery. The lower respiratory tract yielded 76% of isolates, while wounds (soft tissue, lymph nodes, and surgical wounds, including deep lines) yielded 16% and blood yielded 3%. *Nocardia nova* seems to have the least invasive capability, as it is rarely recovered from infections outside the pulmonary tree. *Nocardia pseudobrasiliensis* is associated with lower respiratory tract infection and dissemination. Severe nocardial infections in the immunocompromised host are frequently fatal unless appropriate therapy is initiated.

Infection may also occur via transcutaneous and percutaneous inoculation or entry via soft tissue traumatic abrasions (thorns, cat scratch, or a lesion contamination by dirt) (14, 15).

Such infections almost always occur in immunocompetent hosts and are associated specifically with *N. brasiliensis*, *N. otitidiscaviarum*, *N. transvalensis*, and on occasion, other members of the *N. asteroides* complex (7, 15). Localized infections usually present as cellulitis, necrosis, or abscess formation or may spread lymphocutaneously in "sporotrichoid fashion." Corneal ulceration after trauma can also occur. In the Western hemisphere, *Nocardia brasiliensis* is also one of the most common causes of actinomycotic mycetoma, which is usually limited to the very southern-most areas of the United States and, especially, Mexico; it is rarely found in the northern parts of the United States.

Rhodococcus spp.

Species of *Rhodococcus* are distributed worldwide in the environment and are frequently isolated from soil. They have been isolated from feces of animals (especially herbivores) and grow well in soil contaminated by the latter. Other sources which have been implicated include freshwater, marine habitats, fish, animal urine, and the guts of some arthropods (7, 14).

Rhodococcus equi is the most common species associated with infections in humans as well as a wide range of animal hosts. It is playing an increasingly important role in pulmonary and disseminated disease (often with the presence of bacteremia) in immunocompromised humans. The organism has a predilection for causing pulmonary infection and bacteremia in patients with AIDS and those receiving organ transplants. Patients on corticosteroids and those with long-term presence of a foreign body (such as a catheter) are also at increased risk (4, 7). Other species of *Rhodococcus*, including *Rhodococcus aurianticus* (probable *T. inchonensis*), *Rhodococcus erythropolis*, *Rhodococcus fascians*, *Rhodococcus globerulus*, *Rhodococcus gordoniae*, and *Rhodococcus rhodochrous*, have very rarely been implicated in human disease.

In humans, *R. equi* isolates have plasmid types characteristic for specific animal hosts, suggesting zoonotic transmission; however, patients presenting with *R. equi* infection often have no history of direct exposure to animals. The presence of *R. equi* in sputum and central lines may indicate that inhalation or contamination of lines via air may be a primary source of infection (7).

Gordonia spp.

Species in the genus *Gordonia* are well distributed in nature and have been isolated from soil, sewage treatment plants, biofilters from waste gas treatment, and sputum (14). There have been only a few reports of disease in humans; many of these have questionable characterization of the isolates, and thus their identifications are in question (7). Patients from whom these isolates have been reported include those with immunocompromising conditions, often in the presence of foreign bodies (7). *Gordonia bronchialis* (from sternal wound infections and once from bacteremia) and *Gordonia terrae* (from catheter-related bacteremias and from CNS infections) have been reported most commonly (7). Other members of the *Gordonia* (including *Gordonia otitis*, *Gordonia polyisoprenivorans*, *Gordonia rubriperctincta*, and *Gordonia sputi*) have been reported only once each in cases of the following: bacteremia, pulmonary disease, and soft tissue infection (7, 14).

Tsukamurella spp.

Tsukamurella isolates have been retrieved from soil, sludge, and arthropods. Most isolates seem to prefer cooler temperatures. Isolates in this genus are rare opportunistic pathogens, with only 26 total cases reported in the literature (7). *Tsukamurella inchonensis*, *T. paurometabola* (probably also *T. inchonensis*), and *T. tyrosinosolvans* have been reported most commonly. Most of the reported cases have been in immunocompromised patients, in patients undergoing dialysis, and in those with indwelling catheters (15). The infections have been associated with bacteremia, pulmonary disease, skin infection, conjunctivitis, and knee prosthesis infection.

Other Aerobic Actinomycetes, Including the Genera *Actinomadura*, *Amycolata*, *Amycolatopsis*, *Dermatophilus*, *Dietzia*, *Nocardiopsis*, *Pseudonocardia*, and *Streptomyces*

Actinomadura spp.

Two members of the genus *Actinomadura* (*A. madurae* and *A. pelletieri*) most commonly cause disease in humans through evolution of a slow but progressive actinomycotic mycetoma, called "Madura foot." Normally, these infections are seen in nonimmunocompromised patients and do not spread via lymphatics (14). Mycetoma is common in tropical regions and is associated with walking shoeless; it is rarely found in the United States, except perhaps in some of the southeastern states. *Actinomadura madurae* has also been associated in a few cases of human disease through isolation from sputum (approximately 25 episodes), blood (1 episode), peritoneal fluid (1 episode), and a few other sites (7).

Amycolata, *Amycolatopsis*, and *Pseudonocardia* spp.

There have been no well-characterized or defined cases of infection with *Amycolata*, *Amycolatopsis*, or *Pseudonocardia* spp. in humans (7).

Dermatophilus spp.

Dermatophilus congolensis and *Dermatophilus chelonae* cause dermatitis in a wide variety of animals worldwide, including cattle, horses, goats, and sheep. *Dermatophilus congolensis* alone has been noted on only a few occasions as a possible cause of human infection, usually manifesting as a skin dermatitis (scaling; one episode) or tongue leukoplakia (one episode) (7).

Dietzia spp.

Only two cases of *Dietzia maris* infection have been noted in humans. One was a case of bacteremia in the presence of an intravascular catheter, and the other was an infection of a hip prosthesis. Only one of the two reported cases is plausible, since the isolate's identity was supported by DNA sequencing studies (7).

Nocardiopsis spp.

Infection with *Nocardiopsis* spp. in humans is very rare and always caused by *N. dassonvillei*. Only five noncutaneous infections (mycetoma, bacteremia, pneumonia, and conjunctivitis) and a few cutaneous infections have been reported to date (7).

Streptomyces spp.

The streptomycetes primarily cause localized, suppurative, mostly chronic mycetomas (see the *Actinomadura* section above) (15). *Streptomyces somaliensis* is the most common species in this genus to cause mycetoma. It prefers arid regions with sandy soil, and thus it is normally found primarily, but not exclusively, in Africa, Mexico, and portions of South America. Other species (*Streptomyces griseus* [only a few cases], *Streptomyces bikiniensis* and *Streptomyces thermovulgaris* [one case each associated with bacteremia], and *Streptomyces albus* [associated with mycetoma]) have rarely been documented to cause human infections (7). Most isolates found in the laboratory in cultures from patients presenting with infections other than mycetoma are almost always either contaminants or colonizers.

GOALS OF LABORATORY TESTING

For diagnosis of clinically significant infections caused by the aerobic actinomycetes, evaluation of appropriate specimens by direct microscopy and culture remains the primary goal of the laboratory. With appropriate attention, detection in smear and isolation on primary and/or selective media are usually not difficult (15). The laboratory should be able to differentiate quickly between probable saprophytic colonizers or contaminants and true pathogens. It should be able to help the clinician in interpretation of direct microscopic and culture results and to provide guidance in choosing both empiric and laboratory-guided therapeutic modalities. It is usually fairly simple to identify the presence of a potential pathogen in an overtly morbid infectious process and to identify it to the group or genus level, but it is much more difficult to identify it to the species level due to the vast number of organisms that have recently been implicated in human infection. A routine laboratory should be able to recognize when an organism must be identified to help in relevant patient care or for epidemiologic purposes. Laboratories must, when necessary, be ready to rapidly forward the isolate to a competent reference laboratory. Susceptibility studies are necessary to help guide therapy in some instances and should be performed or referred to other appropriate laboratories as necessary.

Screening for and prevention of infections by the aerobic actinomycetes in the immunocompromised host play no real role. Surveillance cultures are not of any help because most of the aerobic actinomycetes are ubiquitous in the environment and may easily contaminate specimens from nonsterile sites. Isolation from such sites does not imply pathogenesis and may confound or mislead the clinician.

Prophylactic or preemptive therapy of patients with prolonged immunocompromising states is not universally recommended. There are some authorities that support such suppressive therapy for nocardiosis, although there are no well-designed studies to show that such therapy has a positive effect on outcomes (16). Serologic screening or diagnosis is unreliable, and serologic tests are not commercially available for the aerobic actinomycetes. Skin tests may be of some value in diagnosis of actinomycotic mycetoma and have been reviewed elsewhere (9, 17, 18).

Specimen Collection

The collection and transportation of appropriate specimens are crucial to the diagnosis of infections with the aerobic actinomycetes. Respiratory specimens are most commonly submitted, except in more tropical regions of the world, where actinomycotic mycetomas prevail. Bronchoscopically collected lower respiratory secretions result in a higher recovery rate of true etiologies of infections than do expectorated sputum secretions (4). Other specimens, such as tissue and normally sterile body fluids (including cerebrospinal fluid, synovial fluid, peritoneal fluid, etc.), make up a significant portion of material submitted for evaluation; such specimens are normally associated with extrapulmonary dissemination and traumatically caused infection. Blood may also yield aerobic actinomycetes as etiologic agents, especially from immunocompromised patients and often from those with long-term catheters. Each clinical setting thus must be evaluated on its own characteristics in the decision of the most appropriate specimen choice.

Several specimen issues are notable. Refrigerated storage of specimens may adversely affect recovery rates. In addition, contaminated specimens are often pretreated in an attempt to decrease the contamination load prior to culture setup. Unfortunately, decontamination methods used for the mycobacteria (e.g., with *N*-acetyl-L-cysteine) are too harsh for the other aerobic actinomycetes and lower their recovery rates (14). To diminish the effect of the pretreatment process, the specimens should be exposed to the decontaminating reagents for shorter periods (15 minutes). Alternatively, Conville and Witebsky describe a procedure in which the sample is diluted 1:10 in a preparation of 0.2 M HCl–0.2 M KCl at a pH of 2.2 for 2 to 3 minutes before inoculation (7).

LABORATORY TESTING OPTIONS

Microscopy and Direct Visualization

Direct microscopic visualization can help to provide early clues to the identity of the etiologic agent and to its actual role in the disease process. An early diagnosis and therapy of nocardiosis have been associated with better clinical outcomes. A number of wet mount, Gram, and other special stains can be used to detect the aerobic actinomycetes either directly in specimens or after their decontamination and/or concentration (e.g., decontamination of specimens from highly contaminated areas or concentration of normally sterile specimens, such as synovial, peritoneal, or pleural fluids). Not only can stained or unstained smears and/or wet mounts of specimens show characteristic morphologies, but they can also demonstrate the presence or absence of specific cellular material, such as polymorphonuclear cells, mononuclear cells, macrophages, and squamous epithelial cells (indicating the presence of contaminated material). Evaluation of direct material can help to interpret the cellular response, presence or absence of normal flora, potential etiologic agent (in association with the appropriate cellular response), and elucidation of its clinical significance (15). Gross or microscopic examination may demonstrate granules in the drainage material from mycetomas and other chronic infections. These are usually associated with infections caused by *Actinomadura* and *Streptomyces* spp. but may also occasionally be found in chronic infections caused by *N. brasiliensis*. It is important to look for such granules in the right specimens and, if found, to wash the specimens in sterile saline and crush them between glass slides prior to microscopic examination.

Specialized stains, such as the MAFS (e.g., the Kinyoun stain), can further help to differentiate between some groups and genera of aerobic actinomycetes (15). Full-strength acid-fast stains should not be used on the nontuberculous aerobic actinomycetes except to separate the mycobacteria from the other genera in the group. The MAFS uses a 0.5% sulfuric acid decolorizer rather than the full concentration used in the acid-fast stain used on the mycobacteria.

A nonspecific acridine orange stain or a modified auramine-rhodamine acid stain may also be used to detect some organisms in specimens. Both methods, however, require use of a fluorescence microscope. Additionally, the Brown and Brenn stain and the Gram-Weigert stain will usually stain the aerobic actinomycetes, but the hematoxylin and eosin stain does not stain adequately (14).

Nocardia spp.

In Gram-stained preparations, nocardias are usually seen as gram-positive or beaded, filamentous, branching bacteria (Color Plate 8). They may, on occasion, fragment to coccobacillary forms. They are considered MAFS positive (Color Plate 9), but controversy surrounds the comparative efficacies of several staining techniques to demonstrate nocardias in specimens. The experience of at least some laboratories supports the Gram stain as the most sensitive microscopic method by which to visualize and recognize nocardias in clinical specimens (15). In one study of 50 individual patients in whom nocardias were visualized by direct Gram staining of specimens, only 51% had visually detectable MAFS-positive organisms; importantly, there were no gram-negative but MAFS-positive specimens. The MAFS is not as reliable as the Gram stain and should be used only to confirm the acid fastness of organisms detected by the Gram stain (15).

Rhodococcus spp.

Microscopically, rhodococci are usually diphtheroid-like, gram-positive coccobacilli without branching; they are often arranged in a zigzag fashion. They may appear rod-like in liquid media. The morphology may be cyclical, with bacillary forms reverting quickly to coccal forms as they age. Occasionally, rudimentary branching may be seen on filaments grown in liquid media. They are considered MAFS positive, but only a very small portion of the cellular population may exhibit this characteristic. Negative modified acid fastness is especially prevalent in populations grown on media containing tryptic soy agar with 5% sheep blood as well as on chocolate agar (7).

The rhodococci do not have a characteristic morphology and may readily be confused with diphtheroids. However, microscopic examination of bronchial secretions often shows the presence of short pleomorphic gram-positive bacilli within polymorphonuclear cells. Histopathology may also show microabscesses, pseudotumors, and granulomatous inflammation (14).

Gordonia spp.

Members of the genus *Gordonia* are seen as diphtheroid-like coccobacilli, with rudimentary or no branching.

Tsukamurella spp.

Microscopically, *Tsukamurella* organisms have been described as long, usually straight, thin rods that may show a slight curvature and no branching (7). They are MAFS positive but might show only slight staining.

Other aerobic actinomycetes, including the genera *Actinomadura*, *Amycolata*, *Amycolatopsis*, *Dermatophilus*, *Dietzia*, *Nocardiopsis*, *Pseudonocardia*, and *Streptomyces*

***Actinomadura* spp.** The colonies and organisms in the genus *Actinomadura* resemble those of *Streptomyces*. They are short, thin, branching rods which are MAFS negative. *Actinomadura madurae* often results in draining granules which are white to yellow, while *A. pelletieri* results in granules that are red to pink.

***Amycolata*, *Amycolatopsis*, and *Pseudonocardia* spp.** *Amycolata*, *Amycolatopsis*, and *Pseudonocardia* spp. are gram positive but MAFS negative.

***Dermatophilus* spp.** *Dermatophilus* organisms are gram positive but MAFS negative and may be coccobacillary. On culture, microscopic examination often reveals branching forms and the development of longitudinal as well as transverse septa along the filaments.

***Dietzia* spp.** *Dietzia* organisms almost never branch but show coccal and bacillary gram-positive cell forms and are MAFS positive (7).

***Nocardiopsis* spp.** *Nocardiopsis* organisms form long, gram-positive, branched filaments that fragment into zigzag chains of arthroconidium-like spores.

***Streptomyces* spp.** *Streptomyces* organisms are solidly gram-positive filamentous bacillary forms but are MAFS negative. Fragmentation of mycelia and beading may be found. Granules found with *S. somaliensis* are usually white to yellow.

Antigen Testing

Antigen testing or in vitro test systems for antigens of aerobic actinomycetes are not commercially available. Such studies play no role at this time in the diagnosis of infections caused by this group of organisms.

Culture

Most clinically relevant aerobic actinomycetes are able to grow on routine bacteriology or mycology media (14). Primary media such as 5% sheep blood agar, chocolate agar, buffered charcoal yeast extract agar, Mueller-Hinton agar, tryptic soy agar, brain heart infusion agar, and various broths will readily support the growth of clinical isolates.

Because other organisms present in contaminated specimens may readily overgrow the slower growing aerobic actinomycetes, it is imperative to add selective media to enhance the recovery of the latter in culture. Thus, for potentially contaminated specimens, primary culture media should be supplemented with selective media, such as selective buffered charcoal yeast extract agar (containing polymyxin B, anisomycin, and either vancomycin or cefamandole), modified Thayer-Martin agar, Lowenstein-Jensen agar, colistin-nalidixic acid agar, and Sabouraud dextrose agar. In addition, the nocardias grow well on fungal media containing cycloheximide. Chloramphenicol should not be used as a selective agent, as it can also inhibit the other aerobic actinomycetes. Selective media may also have an inhibitory effect on some species and should never be used alone without the primary media (3, 11, 14).

Various commercial blood culture systems can also support growth of the aerobic actinomycetes, with recovery usually occurring in between 3 and 19 days (7). Terminal subcultures are useful when these organisms are suspected. Use of fungal blood culture systems, such as Isolator tubes, may also be helpful.

Some support the notion that media should be incubated at both 30°C and 35°C. If the streptomycetes are suspected, then incubation at 25°C is also warranted. Cultures should be thought of in the same way as fungal cultures, with incubation extending for at least 2 weeks and perhaps longer, depending on the species suspected. Plates should be sealed to keep them from dehydrating. The cultures should be examined every 2 days for the first week and then at least twice a week for the remainder of the incubation period (14).

Once clinically significant isolates are recovered, they may be inoculated onto slide cultures to help with the identification process. Agar blocks (using agar with minimal nutrients, such as tap water agar) are covered with coverslips, incubated at 25°C for up to 2 to 3 weeks in a moist environment, and periodically examined with a dissecting microscope for aerial and substrate filaments (mycelia) and branching (7).

Nocardia spp.

Nocardias will normally appear within 2 to 7 days on most routine bacteriologic media, such as 5% sheep blood agar and chocolate agar. Although the majority of isolates will be detected within the first several days of incubation, media should be examined for up to 2 or 3 weeks. Use of a dissecting microscope may help in the recognition of the filamentous, white to yellow to orange colonies with aerial mycelia and delicate, dichotomously branched substrate mycelia characteristic of the genus *Nocardia* (14). Organisms can occasionally fragment into bacillary or coccal forms. Of the MAFS-positive species, nocardias are the only ones that have aerial mycelia.

Rhodococcus spp.

Rhodococci are frequently recovered from respiratory secretions, from material on infected catheter tips or lines, and from blood. On culture, the rhodococci produce short coccal forms which may lengthen with time. Colonies have variable morphologies and may be rough, smooth, or mucoid. Some, but not all, colonies may show red to yellow pigmentation with age. Except for colonies of a few nonpathogenic species, most colonies show no macroscopically visible aerial mycelia. Growth occurs optimally at 28 to 35°C but not at 45°C. The organisms are aerobic, catalase positive, and nonmotile.

Gordonia spp.

Colonies are commonly wrinkled and salmon colored to orange after several days of incubation.

Tsukamurella spp.

Colonies do not produce aerial mycelia and are usually small and smooth to rough, with pigmentation varying from whitish to orange. The organisms are catalase positive.

Other aerobic actinomycetes, including the genera *Actinomadura, Amycolata, Amycolatopsis, Dermatophilus, Dietzia, Nocardiopsis,* and *Streptomyces*

***Actinomadura* spp.** Colonies between isolates may vary, are frequently wrinkled, and have evidence of aerial mycelia (7).

***Amycolata, Amycolatopsis,* and *Pseudonocardia* spp.** Like those of *Nocardia*, colonies of *Amycolata*, *Amycolatopsis*, and *Pseudonocardia* spp. have aerial mycelia. Like the nocardias, they can occasionally fragment into bacillary, coccal, or square forms.

***Dermatophilus* spp.** The organisms are facultative anaerobes. The colonies produce aerial mycelia under increased CO_2 conditions and are heaped, opaque, and beta-hemolytic on sheep blood agar. The colonies may turn yellow to orange as they grow older (7). They have characteristic vegetative mycelia with long filaments that branch laterally as well as transversally and longitudinally.

***Nocardiopsis* spp.** The colonies are often wrinkled or folded. Aerial mycelia are produced but may be sparse in some isolates. The filaments associated with the agar fragment into coccal forms, while the aerial filaments break up into spore-like forms of various sizes (7).

***Streptomyces* spp.** Colonies may vary between isolates, are frequently wrinkled, and have evidence of aerial mycelia (7). Colonies are MAFS negative.

Identification

Initial visualization of phenotypic colony coloration and morphology, together with the presence of aerial hyphae, by use of a dissecting microscope often provides initial clues to the genus of the isolate. Presumptive identification can be achieved if a filamentous, branched isolate stains with MAFS but not with the traditional Kinyoun acid-fast stain. Resistance to 0.005% lysozyme differentiates *Nocardia* spp. from *Streptomyces* spp. *Tsukamurella* spp. are also MAFS positive and are resistant to lysozyme but do not produce aerial filaments. *Gordonia* spp. have variable resistance to lysozyme but have no aerial filaments.

Identification to the species level by use of phenotypic characteristics may be more tedious and problematic and, for some species, nearly impossible. Identification of the *Nocardia* spp. was historically based on hydrolysis of casein, tyrosine, xanthine, and hypoxanthine. Depending on the organism being identified, phenotypic characteristics that are useful in differentiating the more common clinically significant aerobic actinomycetes include the microscopic and colony morphologies and a number of biochemical or susceptibility profiles. Differing stable susceptibility profiles within *N. asteroides* showed that at least six unique species are identifiable; molecular and further phenotypic studies of the species confirmed their disparity and uniqueness. Molecular means are required to identify some of the new species.

Other phenotypic tests that are useful include use of acetamide and arylsulfatase, acid production from and utilization of carbohydrates as sole carbon sources, esculin hydrolysis, Middlebrook opacification, substrate decomposition, temperature studies, and urease studies (7).

Schematic flow charts, use of commercial test systems, and reviews of phenotypic characteristics have been described by others (3, 11, 14, 15). They do not address the large number of newer species that have recently been reported, and thus they cannot differentiate between many of the newer species recovered. To identify these would require a huge number of phenotypic studies and a large enough database with which to confirm identities. Evaluation of whole-cell hydrolysates, chromatographic procedures such as paper, gas, thin-layer, and high-performance liquid chromatography, and mass spectrometry for fatty acid as well as cell wall and mycolate analysis have been used to differentiate between some of the genera and species. These are not adequate for overall identification or amenable for use in the clinical laboratory.

Presently, molecular methods used to identify organisms to the species level include restriction endonuclease analysis of an amplified portion of the 16S rRNA gene, ribotyping, PCR using randomly amplified polymorphic DNA, repetitive-element sequence-based analysis, and the use of DNA probes (6, 8). These methods have a number of shortcomings, as they cannot be used to identify most of the aerobic actinomycete species and are not frequently used (6, 8). Gene sequencing methodologies such as sequencing of the 16S rRNA or DNA have recently become popular for identification of most aerobic actinomycetes. Gene sequencing of a 500-bp region of 16S rRNA has gained momentum for identification of many such isolates, but a longer region may be needed for better discrimination (6, 8).

CLINICAL SIGNIFICANCE

The presence of aerobic actinomycetes in the environment can lead to contamination and/or colonization of clinical specimens and can confuse the clinical relevance. The aerobic actinomycetes are rarely seen as contaminants in the laboratory, and each isolate has to be evaluated carefully regarding its clinical significance (5). The presence of organisms in normally sterile sites or on direct microscopic examination of potentially contaminated specimens (especially in association with a pyogenic cellular response), such as sputum, greatly increases the likelihood of an organism's role as an etiologic agent (4, 15). Immunocompromising conditions or corticosteroid use in patients also significantly increases the clinical relevance of isolates. Semiquantitation of isolates in specimens can further elucidate their significance. A colony or two on primary media or isolated from a single medium or from broth may not be considered as significant as the presence of large quantities of growth on multiple media and plates or tubes. It is extremely important that isolates be evaluated in conjunction with the patient's clinical presentation as well as any underlying conditions.

The organism's identity is also very important and is associated with clinical significance. For example, *Nocardia farcinica* is far less likely than a member of the streptomycetes to be a contaminant or colonizer. It is important to include all facets of the case in evaluating the significance of a specific isolate.

THERAPY

In general, therapeutic efficacy in patients is dependent on the presenting infection, susceptibilities of the recovered isolates, and the extent of the immunocompromising conditions of patients. In vitro susceptibilities of various isolates can vary dramatically, and susceptibility studies are indicated.

Nocardia

Sulfa-containing antimicrobials remain the drugs of choice for nocardiosis and improve survival if used alone for respiratory tract infections or in combination with other antimicrobials in dealing with severe or disseminated disease in immunocompromised patients (3-5). However, other antimicrobial agents are needed for patients intolerant to sulfonamides or in whom infection is refractory to sulfonamide therapy. Primary agents that have been used successfully, alone or in combination, include minocycline for less severe infections, such as skin and soft tissue or respiratory infections, and for more serious infections, amikacin, imipenem, linezolid, and ceftriaxone (normally in combinations). Linezolid should not be used for more than 4 weeks because of its high association with hematologic toxicity. Empiric therapy using combinations of a sulfa-containing agent and one or more of the latter antibiotics has been recommended for serious systemic disease (4, 5). For example, amikacin in combination with imipenem or ceftriaxone has been suggested for serious infection. Imipenem seems to be more active than meropenem against nocardias (4). Other potentially efficacious choices include the expanded-spectrum cephalosporins, amoxillin-clavulanate, ampicillin-sulbactam, newer macrolides, other aminoglycosides, and the fluoroquinolones. The latter agents should be avoided unless susceptibility data are available; susceptibility data should be sought with all significant isolates of *Nocardia*.

The duration of therapy is uncertain but should be protracted because of considerable relapse after shorter courses (5). Cutaneous disease requires at least 3 to 6 months of therapy, and serious infections require at least 6 weeks with maximal dosing followed by reduced dosing for between 6 and 12 months. Immunocompromised patients and those with CNS manifestations require at least a year of therapy.

Rhodococcus spp.

Isolates of *R. equi* are frequently susceptible in vitro to a number of agents, including erythromycin and other extended-spectrum macrolides, rifampin, fluoroquinolones, aminoglycosides, glycopeptides, and imipenem. Antimicrobial agents with some activity include trimethoprim-sulfamethoxazole, clindamycin, and chloramphenicol, but these should not be used until susceptibility studies show them to be active. Most isolates are resistant to the penicillins and cephalosporins, and resistance to beta-lactams has been noted to arise in initially susceptible strains during therapy. Resistance has also been noted to develop during therapy with trimethoprim-sulfamethoxazole, doxycycline, and rifampin.

Strains of *R. equi* may be resistant to intracellular killing, and thus an agent that penetrates into macrophages (e.g., azithromycin and fluoroquinolones) should be chosen in combination with other agents. In immunocompetent patients, single therapy with a macrolide or fluoroquinolone is adequate, while in immunocompromised patients combination therapy with two or more agents is required. Depending on the patient's immune status, the clinical presentation, and the susceptibility profile of the isolate recovered, one may choose amongst the macrolides and the fluoroquinolones in combinations with rifampin, vancomycin, and/or gentamicin. Rifabutin may be substituted for rifampin in patients with human immunodeficiency virus disease.

SUSCEPTIBILITY TESTING

The majority of aerobic actinomycetes have various susceptibilities to a wide variety of agents and require in vitro susceptibility testing to help guide therapy. A standard for susceptibility testing of the aerobic actinomycetes, using broth microdilution and cation-supplemented Mueller-Hinton broth, has been published by the Clinical Laboratory Standards Institute (CLSI) (12). The organisms in the standard include *Nocardia*, *Actinomadura*, *Rhodococcus*, *Gordonia*, *Tsukamurella*, and *Streptomyces* spp. The primary recommended antimicrobial agents for testing are amikacin, amoxicillin-clavunalate, ceftriaxone, ciprofloxacin, clarithromycin, imipenem, linezolid, minocycline, sulfamethoxazole or trimethoprim-sulfamethoxazole, and tobramycin. The secondary set of recommended agents includes cefepime, cefotaxime, doxycycline, moxifloxacin, and gentamicin. Interpretative guidelines and reporting formats are provided in the CLSI publication.

Nocardia spp.

Nocardia species can vary in their antimicrobial susceptibility patterns. Therapeutic efficacy in individual patients may depend on species identity and on in vitro susceptibility studies (12). Susceptibility testing should especially be considered in refractory cases. Besides the standard microdilution method published by CLSI, disk agar diffusion, agar dilution, gradient strip agar dilution (Etest), and Bactec radiometric methods have all also been used for susceptibility testing of the nocardias (1, 12, 20). Studies have shown inter- and intralaboratory agreement and reproducibility of above 90% between these methods (12, 20). Prospective clinical studies attempting to correlate results of susceptibility testing with patient therapy and outcomes have not, however, been performed systematically. Most of the available data are accumulated from anecdotal and individual cases.

SUMMARY

The aerobic actinomycetes have gained increased prominence as etiologic agents of significant disease, primarily in immunocompromised patients. The combination of increasing conditions impairing host resistance to invasion by environmental pathogens and the rapidly progressing technological capability to identify isolates recovered from such patients has allowed a wide expansion of the species known to be capable of causing disease in humans (albeit under special circumstances). A better understanding of the epidemiology, clinical course, and antimicrobial susceptibilities of aerobic actinomycetes is essential. Molecular identification and typing techniques will also bring a better understanding of the interrelatedness of the various genera and species. Evidence-based evaluation is paramount to rapid diagnosis and early choice in therapeutic modalities, aimed to achieve better outcomes in patients. Clinical laboratories must become aware of the role now played by the aerobic actinomycetes in disease and must work with clinical care providers to determine when identification, susceptibility testing, and therapeutic or surgical interventions are necessary.

REFERENCES

1. **Ambaye, A., P. Kohner, P. Wollan, K. Roberts, G. Roberts, and F. R. Cockerill.** 1997. Comparison of agar dilution, broth microdilution, disk diffusion, Etest, and BACTEC radiometric methods for antimicrobial susceptibility testing of clinical isolates of the *Nocardia asteroides* complex. *J. Clin. Microbiol.* **35:**847–852.
2. **Beaman, B., and L. Beaman.** 1994. *Nocardia* species: host-parasite relationships. *Clin. Microbiol. Rev.* **7:**213–264.
3. **Brown, J. M., and M. M. McNeill.** 2003. *Nocardia, Rhodococcus, Gordonia, Actinomadura, Streptomyces*, and other aerobic actinomycetes, p. 502–531. *In* P. R. Murray, E. J. Baron, M. A. Pfaller, F. C. Tenover, and R. H. Yolken (ed.), *Manual of Clinical Microbiology,* 8th ed. American Society for Microbiology, Washington, DC.
4. **Brown-Elliott, B., J. Brown, P. Conville, and R. J. Wallace.** 2006. Clinical and laboratory features of the *Nocardia* spp. based on current molecular taxonomy. *Clin. Microbiol. Rev.* **19:**259–282.
5. **Burgert, S. J.** 1999. Nocardiosis: a clinical review. *Infect. Dis. Clin. Pract.* **8:**27–32.
6. **Conville, P., S. Fischer, C. Cartwright, and F. Witebsky.** 2000. Identification of *Nocardia* species by restriction endonuclease analysis of an amplified portion of the 16S rRNA gene. *J. Clin. Microbiol.* **38:**158–164.
7. **Conville, P. A., and F. G. Witebsky.** 2007. *Nocardia, Rhodococcus, Gordonia, Actinomyces, Streptomyces*, and other aerobic actinomycetes, p. 515–542. *In* P. R. Murray, E. J. Baron, J. H. Jorgenson, M. L. Landry, and M. A. Pfaller (ed.), *Manual of Clinical Microbiology,* 9th ed. ASM Press, Washington, DC.
8. **Conville, P. S., and F. G. Witebsky.** 2007. Analysis of multiple differing copies of the 16S rRNA gene in five clinical isolates and three type strains of *Nocardia* species and implications for species assignment. *J. Clin. Microbiol.* **45:**1146–1151.
9. **Dravid, M. N., A. Venugopalan, R. S. Bharadwaj, and S. S. Nene.** 1991. Mycetoma due to Nocardia caviae—a case report. *Indian J. Pathol. Microbiol.* **34:**62–63.
10. **Lumb, R., H. Greville, J. Martin, N. Sangster, and M. Holmes.** 2002. Nocardia asteroides isolated from three patients with cystic fibrosis. *Eur. J. Clin. Microbiol. Infect. Dis.* **21:**230–233.
11. **McNeil, M., and J. Brown.** 1994. The medically important aerobic actinomycetes: epidemiology and microbiology. *Clin. Microbiol. Rev.* **7:**357–417.
12. **NCCLS.** 2003. Susceptibility testing of mycobacteria, nocardiae, and other aerobic actinomycetes. NCCLS document M24-A. NCCLS, Wayne, PA.
13. **Roth, A., S. Andrees, R. Kroppenstedt, D. Harmsen, and H. Mauch.** 2003. Phylogeny of the genus *Nocardia* based on reassessed 16S rRNA gene sequences reveals underspeciation and division of strains classified as *Nocardia asteroides* into three established species and two unnamed taxons. *J. Clin. Microbiol.* **41:**851–856.
14. **Saubolle, M. A.** 2002. Aerobic actinomycetes, p. 1201–1220. *In* K. D. McClatchey (ed.), *Clinical Laboratory Medicine,* 2nd ed. Lippincott Williams and Wilkins, Philadelphia, PA.
15. **Saubolle, M. A., and D. Sussland.** 2003. Nocardiosis: review of clinical and laboratory experience. *J. Clin. Microbiol.* **41:**4497–4501.
16. **Simpson, G., E. Stinson, M. Egger, and J. Remington.** 1981. Nocardial infections in the immunocompromised host: a detailed study in a defined population. *Rev. Infect. Dis.* **3:**492–507.
17. **Venugopal, P. L., and T. V. Venugopal.** 1990. Red grain mycetoma of the scalp due to Actinomadura pelletieri in Madurai. *Indian J. Pathol. Microbiol.* **33:**384–386.
18. **Venugopal, P. V., T. V. Venugopal, W. N. Laing, Y. al Humaidan, S. S. Namnyak, A. A. al Jama, and A. M. Elbashier.** 1990. Black grain mycetoma caused by Madurella grisea in Saudi Arabia. *Int. J. Dermatol.* **29:**434–435.
19. **Wallace, R. J., B. Brown, Z. Blacklock, R. Ulrich, K. Jost, J. Brown, M. McNeil, G. Onyi, V. Steingrube, and J. Gibson.** 1995. New *Nocardia* taxon among isolates of *Nocardia brasiliensis* associated with invasive disease. *J. Clin. Microbiol.* **33:**1528–1533.
20. **Woods, G. L., B. A. Brown-Elliott, E. P. Desmond, G. S. Hall, L. Heifets, G. E. Pfyffer, J. C. Ridderhof, M. R. Plaunt, R. J. Wallace, N. G. Warren, and F. G. Witebsky.** 2001. Susceptibility testing of mycobacteria, nocardiae, and other actinomycetes, vol. 20. NCCLS, Wayne, PA.

Diagnostic Microbiology of the Immunocompromised Host
Edited by Randall T. Hayden, Karen C. Carroll, Yi-Wei Tang, and Donna M. Wolk

Chapter 14

Parasites

Lynne S. Garcia

This chapter includes some of the opportunistic parasites that can cause disease in immunocompromised patients. Although most parasitic infections are known to cause more severe symptoms when a host's immune system is impaired, the representative organisms presented in this chapter have been identified as causing the most severe disease in this population group.

Humans have very effective defense mechanisms to protect themselves against parasites. The innate resistance defense system includes surface and mechanical barriers, pH, temperature, phagocytosis, and nonspecific inflammation. The induced defense system recognizes foreign organisms; the term for this system is adaptive immunity. The two major components of the adaptive immune response are humoral (antibody) and cellular (sensitized cells); B lymphocytes (B cells) are responsible for the humoral response, and T lymphocytes (T cells) are responsible for the cellular response. The human immune system is very complex, with different classes of antibody and various subsets of T cells that have different functions as immune effectors. Unfortunately, the immune system is not always protective and under certain circumstances can be directed toward the human body, not the parasite.

Parasitic infections in immunocompetent individuals are relatively common throughout the world, and symptoms in the normal host can range from none to mild to severe, even causing death. However, immunocompromised individuals tend to be more susceptible to infections with minimally pathogenic organisms. In addition to suffering numerous infectious episodes with bacterial, viral, and fungal organisms, many of these individuals contract parasitic infections. Immune system deficiencies can be attributed to congenital absence, abnormal development, malignancy, therapy with cytotoxic drugs, irradiation, or infections, such as with human immunodeficiency virus (HIV); even age and malnutrition can compromise the host's immune system. Of particular interest are patients with AIDS and those undergoing organ transplantation (8, 15).

Organ transplant recipients may acquire significant parasitic disease in the following three ways: transmission with the graft, de novo infection, or activation of dormant infection as a consequence of immunosuppression. Malaria, *Trypanosoma* spp., *Toxoplasma gondii*, and *Leishmania* spp. are the principal parasites that may be transmitted with bone marrow, kidney, or liver homografts, and microsporidia are the principle parasites transmitted with xenotransplants. De novo cases of malaria and kala-azar may occur in immunocompromised travelers visiting areas where these diseases are endemic, while immunocompromised natives are subject to superinfection with different strains of endemic parasites, reinfection with schistosomes, or rarely, primary infections such as *Acanthamoeba* sp. infection. The list of diseases that may be reactivated in the immunocompromised host includes giardiasis, balantidiasis, strongyloidiasis, capillariasis, malaria, Chagas' disease, and kala-azar. The broad clinical syndromes of parasitic infection in transplant recipients include prolonged pyrexia, lower gastrointestinal symptoms, bronchopneumonia, and meningoencephalitis. Specific syndromes include the hematologic manifestations of malaria, myocarditis in Chagas' disease, acute renal failure in malaria and leishmaniasis, and the typical skin lesions of Chagas' disease and cutaneous leishmaniasis. It is recommended that transplant clinicians have a high index of suspicion of parasitic infections as an important transmission threat as well as a potential cause of significant posttransplant morbidity.

Although any parasitic infection may be more severe in the immunocompromised host, certain organisms tend to produce greater pathologic sequelae in these patients, while other organisms occur with a

Lynne S. Garcia • LSG & Associates, Santa Monica, CA 90402-2908.

higher frequency in individuals with certain immune system deficiencies. The organisms discussed here are *Entamoeba histolytica*, free-living amebae, *Giardia lamblia*, *Toxoplasma gondii*, *Cryptosporidium* spp., *Cyclospora cayetanensis*, *Isospora belli*, *Sarcocystis* spp., microsporidia, *Leishmania* spp., *Strongyloides stercoralis*, and *Sarcoptes scabiei* (crusted scabies). It is important for the laboratorian and clinician to be aware of problems that these organisms can cause in immunocompromised patients and of the proper diagnostic techniques and their clinical relevance (Tables 1 and 2).

Table 1. Parasitic infections: clinical findings in immunocompetent and immunocompromised patients

Organism	Manifestations and comments	
	Immunocompetent patient	Immunocompromised patient
Entamoeba histolytica	Asymptomatic to chronic or acute colitis; extraintestinal disease may also occur (primary site; right upper quadrant of liver)	Diminished immune system may lead to extraintestinal disease; despite high prevalence, amebiasis is not a major source of morbidity within the HIV/AIDS pandemic; however, amebiasis may indeed be a risk for individuals living with HIV/AIDS
Free-living amebae	Patients tend to have eye infections with *Acanthamoeba* spp.; linked to poor contact lens care	PAM with *Naegleria fowleri*; GAE with *Acanthamoeba* spp. and *Balamuthia mandrillaris*
Giardia lamblia	Asymptomatic to malabsorption syndrome	Certain immunodeficiencies tend to predispose an individual to infection, including hypogammaglobulinemia, protein or caloric malnutrition, or previous gastrectomy
Toxoplasma gondii	Approximately 35 to 50% of individuals have antibody and organisms in tissue but are asymptomatic	Disease in compromised patients tends to involve the CNS, with various neurological symptoms, including TE
Cryptosporidium spp.	Self-limiting infection with diarrhea and abdominal pain	Due to autoinfective nature of life cycle, infection is not self-limiting and may produce diarrhetic fluid loss of over 10 liters/day; may be multisystem involvement; no known totally effective therapy
Cyclospora cayetanensis	Self-limiting infection with diarrhea (3 to 4 days), with relapses common	Diarrhea may persist for 12 weeks or more; biliary disease has also been reported for this group, particularly for those with AIDS
Isospora (Cystoisospora) belli	Self-limiting infection with mild diarrhea or no symptoms	May cause severe diarrhea, abdominal pain, possible death (rare case reports); nonrecognition of oocyst stage in feces may result in missed diagnosis; oocysts not seen when stool concentrated from polyvinyl alcohol fixative
Sarcocystis spp.	Self-limiting infection with diarrhea or mild symptoms	Symptoms may be more severe and last longer
Microsporidia (may be reclassified as fungi)	Less is known about infections in competent than in compromised patients; serologic evidence suggests that infections are probably more common than recognized	Can infect various parts of the body; diagnosis often depends on histologic examination of tissues; routine examination of clinical specimens (stool, urine, eye specimens, etc.) is very labor-intensive; testing is becoming more common; infections can lead to death
Leishmania spp.	Asymptomatic to mild disease	More serious manifestations of VL; some cutaneous species manifest visceral disease; difficult to treat and manage; definite coinfection with AIDS
Trypanosoma cruzi	Acute, indeterminate, and chronic stages; severest form in children under 5 years; tends to be a subacute form in children of more than 5 years and in adults; unilateral periorbital edema early; organism spreads to lymphatics; high fevers and signs of meningoencephalitis; chronic stage is characterized by cardiomyopathy, megacolon, megaesophagus; chronic infection may arise many years after acute infection	Major neurologic sequelae in HIV-infected patients; disease reactivation (severe multifocal or diffuse meningoencephalitis) is rare in non-AIDS patients; may also see protracted asymptomatic course; recrudescence occurs in transplantation patients; reactivation with cutaneous lesions is also seen; bone marrow transplants are also involved with reactivation; there is a link between HTLV-2 infection and *T. cruzi* seropositivity
Strongyloides stercoralis	Asymptomatic to mild abdominal complaints; can remain latent for 30 to 40+ years due to low-level infection maintained by internal autoinfective life cycle	Can result in disseminated disease (hyperinfection syndrome); abdominal pain, pneumonitis, sepsis-meningitis with gram-negative bacilli, eosinophilia; distinct link to some leukemias and lymphomas; can be fatal
Crusted (Norwegian) scabies (*Sarcoptes scabiei*)	Infections can range from asymptomatic to moderate itching	Severe infection with reduced itching response; hundreds of thousands of mites on body; infection very easily transferred to others; secondary infection very common

Table 2. Parasitic infections in the compromised host: diagnostic options[a,c]

Organism and disease	Size and stage	Specimen	Procedure[b]	Examination method	Laboratory findings	Comments
Entamoeba histolytica Intestinal disease Pathogenic *E. histolytica* can now be differentiated from nonpathogenic *E. dispar* by use of immunoassays; although morphologic characteristics are identical in cysts and trophozoites without RBCs in the cytoplasm, *E. histolytica* containing RBCs in the trophozoite cytoplasm can be identified morphologically as the true pathogen *E. histolytica*	Trophozoite, usually 15 to 20 μm (range, 12 to 60 μm); cyst, usually 12 to 15 μm (range, 10 to 20 μm) (sizes as seen in permanently stained smear [may appear 1 to 1.5 μm smaller than wet mount measurements])	Stool (fresh or preserved in fecal fixative, often containing PVA)	O&P examination (minimum of concentration and permanent stained smear)	Concentration (high, dry, ×400) Permanent stain (oil immersion, ×1,000)	Trophozoites or cysts	Series of three stools collected every other day recommended for O&P examination
		Stool (fresh or frozen)	Immunoassays	Depends on format (EIA, FA, membrane flow/rapid)	Presence or absence of antigen	Detection of *Entamoeba histolytica* or the *Entamoeba histolytica/E. dispar* group
		Sigmoidoscopic biopsy or scrapings	Permanent stained smear and routine histology	Permanent stain (oil immersion, ×1,000)	Trophozoites	Permanently stained smear mandatory for complete O&P examination
		Serum	Amebic serologic tests	IHA, IFA, CIE, ID, ELISA	Not recommended for intestinal disease; negative or possible low titer, indicating present or past exposure	At least two different serologic methods should be performed so results can be compared
			Radiographic studies	Barium	Intestinal abnormalities (may also be normal)	
Extraintestinal disease	Trophozoite stage only	Stool	O&P examination	See above	No organisms seen in about 50% of cases	There may be no evidence of GI disease before or during extraintestinal invasion
		Sigmoidoscopic biopsy or scrapings	Permanent stained smear and routine histology	See above	No organisms seen in about 50% of cases	
		Serum	Amebic serologic tests (antibody): IHA, EIA reagents available	See above	Positive titers may be lower in compromised patients	At least two different serologic methods should be performed so results can be compared
			Scans		Evidence of space-occupying lesion	
Free-living amebae *Naegleria fowleri*	Biphasic (amebic and flagellate forms); trophozoite, usual range, 8 to 15 μm (amebic form, with lobate pseudopodia); (cyst, not present in tissue, small, smooth, rounded	CSF	Centrifugation, wet/stained preparation; culture with bacterial overlay	Wet, low (×100) and high, dry (×400) Stained, oil immersion (×500, ×600 or 1,000) Culture, low (×100) and high, dry (×400), on material from culture plate surface	Trophozoites in specimens; trophozoites and cysts in culture	Since these organisms are normally free-living, they will survive shipment via mail to any laboratory for subsequent examination and/or culture; various culture temperatures are recommended in different situations

Continued on following page

Table 2. *Continued*

Organism and disease	Size and stage	Specimen	Procedure[b]	Examination method	Laboratory findings	Comments
Acanthamoeba spp.	Trophozoite, large, 15 to 25 µm (amebic form, with filiform pseudopodia); cyst, present in tissue, large with wrinkled double wall	CSF, eye specimens, cutaneous biopsies	Centrifugation, wet/stained preparation; culture with bacterial overlay; histology	Histology (routine microscopic exam)	Trophozoites and/or cysts in specimens and culture	Agar plates with bacterial overlay (culture) more sensitive than wet preparations
Balamuthia mandrillaris	Trophozoite, large, 12 to 60 µm (amebic form, with fingerlike pseudopodia); cyst, 6 to 30 µm, large with wrinkled double wall	CSF, eye specimens, cutaneous biopsies	Centrifugation, wet/stained preparation; cell culture only, will not grow on routine agar culture with bacterial overlay; histology	All of the above, including cell culture, if available	Trophozoites and/or cysts in specimens and cell culture	Becoming more widely recognized as causing same types of disease as *Acanthamoeba* spp.
Giardia lamblia	Trophozoite, usual range, 10 to 12 µm long by 5 to 10 µm wide; cyst, usual range, 11 to 14 µm (range, 7 to 10 µm wide)	Stool	O&P examination	Concentration (high, dry, ×400)	Trophozoites/cysts (may not always be detected)	The parasite may not be recovered from stool even after four or five examinations or different specimens; this does not mean that the diagnostic techniques are inadequate or the organisms are missed by inexperienced personnel; additional diagnostic techniques (including biopsy) may have to be used; organisms passed on a cyclical basis
				Permanent stain (oil immersion, ×1,000)		
			Immunoassays	Depends on format (EIA, FA, membrane flow/rapid)	Presence or absence of antigen or actual organisms	
		Duodenal aspirate	Wet preparations, permanent stained smears	See above	Trophozoites	
		Entero-Test capsule	Wet preparations, permanent stained smears	See above	Trophozoites	
		Biopsy	Routine histology and special stains	Generally trophozoites are seen	Possible upper GI tract changes and/or trophozoites	
		Serum	Serology	Experimental (not generally available)		
Toxoplasma gondii	Tachyzoites, 4 to 6 µm long by 2 to 3 µm wide; bradyzoites, individual tissue cysts may contain many hundreds of organisms; cysts, may range up to 50 to 75 µm	Serum	Serology	IHA, IFA, ELISA, Sabin-Feldman dye test	Results are often sufficient to confirm diagnosis; compromised patients may not exhibit positive serology	Patients with positive IgM serology results may have Epstein-Barr virus infections; there may be nonspecific stimulation of other antibody-producing cells; although organism isolation techniques using animals and/or

	Biopsy	Routine histology		Possible recovery of bradyzoites (organism may not be etiologic agent of disease)		tissue culture may be used, these results must be evaluated within the context of the total clinical picture
		Tissue culture or animal isolation		Possible recovery of tachyzoites (organism may not be etiologic agent of disease)		
	Spinal fluid	Organism isolation (direct, tissue culture, or animal isolation)		Presence of tachyzoites		Considered confirmatory
		Serology		Confirmatory if higher in CSF than in serum		Rare finding
Cryptosporidium species	Developmental stages within brush border of intestine; Cells, 1 to 2 µm; oocysts in stool, 4 to 6 µm	Stool and/or sputum (disseminated infections in compromised patient)	Modified acid-fast stains	Permanent stain (high, dry [×400], confirm with oil immersion ×1,000)	Oocysts visible	Regardless of staining technique used on stool, the more normal (formed) the stool specimen, the greater the chances that artifacts will present problems in identifying organisms and fewer organisms will be present; specimens representing classic diarrhea will allow easier recovery and identification of organisms; numbers of organisms will vary from day to day
			Immunoassays (more sensitive than routine modified acid-fast stains)	Depends on format (EIA, FA, membrane flow/rapid)	Presence or absence of antigen or actual oocysts	
		Intestinal mucosal biopsy	Routine histology	H&E stain	Developmental stages (1 µm) along brush border	
			EM	Routine EM	May see developmental stages or possibly oocysts	Very specific but not very sensitive approach
		Serum	Serology	Experimental only (not generally available)		Used for epidemiology
Cyclospora cayetanensis	Oocysts, 8 to 10 µm	Stool	Modified acid-fast stains	Permanent stain (high, dry [×400], confirm with oil immersion ×1,000)	Oocysts visible	Due to morphologic similarities with *Cryptosporidium* it is important to measure the oocysts; there is tremendous staining variability of

Continued on following page

Table 2. *Continued*

Organism and disease	Size and stage	Specimen	Procedure[b]	Examination method	Laboratory findings	Comments
Cyclospora cayetanensis (continued)			Wet prep exams; oocysts will autofluoresce (intensity tends to be about 1 to 3+)	Green using 450 to 490-nm DM excitation filter, blue using 365-nm DM excitation filter; wet prep (high, dry [×400])	Oocysts visible as green or blue circles (interior of oocyst does not fluoresce)	the oocysts with acid-fast stains (from clear to deep purple); oocysts look like they are wrinkled; life cycle not yet completely determined or described; oocysts in stool are unsporulated and do not contain any internal definition or structure
		Intestinal mucosal biopsy	Routine histology	H&E stain	Organisms resembling coccidia found in jejunal enterocytes	
			EM	Routine EM	May see developmental stages or possibly oocysts	
Sarcocystis suihominis, *S. bovihominis*	Oocyst is thin walled, contains two mature sporocysts, each with four sporozoites; sporocysts are 9 to 16 μm long by 7.5 to 12 μm wide Shapes and sizes of skeletal and cardiac muscle sarcocysts vary	Stool	O&P examination (concentration important); PVA may distort morphology	Concentration (low ×100 and high, dry [×400])	Oocysts or sporocysts	Not commonly seen; size can overlap that of *I. belli*
"*S. lindemanni*"		Tissue biopsy	Routine histology	H&E stain, high, dry (×400)	Developmental stages within tissue cells	Sarcocysts contain several hundred to thousands of trophozoites, each measuring 4 to 9 μm wide by 12 to 16 μm long; sarcocysts may be divided into compartments by septa (not seen in *Toxoplasma* cysts)
Microsporidia species *Nosema*, *Brachiola*, *Vittaforma*, *Encephalitozoon*, *Enterocytozoon*, *Pleistophora*, *Trachipleistophora*, *Microsporidium* spp.	Spores, 1 to 4 μm; dividing forms within epithelial cells	Tissue biopsy	Routine histology	Tissue Gram stain (Brown-Brenn), methenamine silver stain, acid-fast stain, PAS-positive granule, Giemsa stain, H&E stain	Spores, dividing forms	It may be difficult to find and identify the organisms using routine histological techniques; due to the difficulties in identification at the light microscopy level, EM methods have also been used; spores have been identified in stool, although current procedures are somewhat difficult to interpret and time-consuming to perform; may be reclassified as fungi
				Routine EM	Spores, dividing forms	
				Modified trichrome stain	Spores	

			EM	Routine EM (specific but insensitive)	Dividing forms	
These genera have been reported to infect humans; organisms that have not yet been classified to the genus level have been placed in the "catch-all" genus *Microsporidium*		Stool	Permanent stained smear	Modified trichrome stain	Spores	
			EM	Routine EM	Spores	
			Immunoassays	Not yet available commercially	Presence or absence of antigen or actual organisms	
		Serology	IFA, CF, microagglutination	Experimental (not generally available)	Titers	
Leishmania spp. (one of the most important opportunistic infections in HIV-infected patients); now present in Texas	Amastigotes in tissue measure from 1 to 3 μm; tend to appear larger when released from cells, such as in a touch or squash preparation; appear as dots within cells in routine histology preparations	Blood specimen (looking at buffy coat cells); tissue biopsy (bone marrow most common)	Buffy coat preparations, centrifugation, smear preparation; smears and culture (NNN) of bone marrow aspirates	Microscopic examination at high, dry, ×400 power and ×1,000 oil immersion	Presence of amastigotes in macrophages and/or tissues; presence of promastigotes (motile, flagellated) in culture	Multiple specimens may be required; cultures must not become contaminated or results may be falsely negative; controls must accompany all patient cultures
Trypanosoma cruzi (potential reactivation of Chagas' disease in HIV-infected patients); severe meningoencephalitis; now present in Texas	Trypomastigotes in blood (early infections); amastigotes found in tissues	Blood specimen (looking at buffy coat cell layer); tissue biopsy	Buffy coat preparations, centrifugation, smear preparation; smears and culture; lymph node and chagoma aspirates and biopsies	Microscopic examination at high, dry, ×400 power and ×1,000 oil immersion	Presence of amastigotes in tissues; presence of epimastigotes (motile, flagellated) in culture	Positive blood smear is key indicator of Chagas' disease reactivation in chronic patients; however, positive films may occur late during the reactivation process; recrudescence in organ transplantation patients is a concern, as is transplantation of an infected organ (including bone marrow transplants)
		Urine, serum	Immunoassays	Antigen detection	Not widely used in United States; blood banking reagents used in other parts of the world (e.g., Brazil)	
		Serum	Serology	Antibody detection	Used in other parts of the world in blood banking (e.g., Bolivia)	Recently introduced by Red Cross into U.S. blood banking protocols
		Sterile reduviid bugs	"Culture system" (xenodiagnosis)	Life cycle stages in hind gut of bugs	Used in parts of the world where Chagas' disease is endemic	Sterile bugs are allowed to feed on blood of individual suspected of having Chagas' disease; life cycle stages recovered from bug feces give confirmation of infection

Continued on following page

Table 2. *Continued*

Organism and disease	Size and stage	Specimen	Procedure[b]	Examination method	Laboratory findings	Comments
Strongyloides stercoralis (immunosuppression leads to hyperinfection and disseminated disease)	Larval stages, 180 to 380 μm by 14 to 20 μm, in feces; larvae (rhabditiform) have a short buccal cavity and a prominent, genital primordium; both noninfective and infective larvae, adults, and eggs may be found in duodenal specimens	Stool	O&P examination (concentration is important)	Direct smears and concentration (low [×100], confirm with high, dry [×400])	Rhabditiform larvae (infective filariform larvae may also be found, particularly in compromised patients)	Caution should be used in working with fresh specimens that may contain infective filariform larvae *Strongyloides* infections may be difficult to diagnose from stool specimens; examination of material from the duodenum may be necessary, and concentrations or cultures can also be used
		Duodenal aspirate	Baermann concentrate, Harada-Mori and petri dish cultures	See above	See above	
		Duodenal aspirate	Wet preparations (direct smears)	See above	Larvae (usually rhabditiform)	
		Entero-Test	Wet preparations (direct smears)	See above	Larvae (see above)	
Crusted (Norwegian) scabies (*Sarcoptes scabiei*)	Mites are microscopic and live in cutaneous burrows where females deposit eggs; mites range from 215 to 390 μm in length; eggs are 170 μm long by 92 μm wide; fecal pellets are yellow-brown	Skin scrapings	Use skin-scraping technique	Prepare slides of skin scraping material and examine microscopically, using low power (×100) and high power (×400)	Mites, eggs, or scybala (fecal pellets) may be present	Highly infectious; most common skin sites are interdigital spaces, backs of the hands, elbows, axillae, groin, breasts, umbilicus, penis, shoulder blades, small of the back, buttocks; may see kerotic excrescences on the body; itching may be absent; diagnosis of scabies should be considered for patients with advanced malignancies and associated pruritus

[a]PCR and other molecular methods are often developed in research laboratories; some of these methods are available if required for patient diagnosis. However, their availability tends to be somewhat limited, and on occasion, they may be available only outside the United States. It is strongly recommended that the Centers for Disease Control and Prevention be contacted for specific recommendations if molecular testing is suggested and commercial reagents or laboratory tests are not readily available.

[b]The use of immunoassay reagents for the direct and indirect detection of some of these parasites has become much more common. Reagents are available for *Entamoeba histolytica*, the *Entamoeba histolytica/E. dispar* group, *Giardia lamblia, Cryptosporidium* spp., and the microsporidia (not yet available, but reagents have been developed in research laboratories).

[c]CIE, counterimmunoelectrophoresis; EIA, enzyme immunoassay; ELISA, enzyme-linked immunosorbent assay; EM, electron microscopy; FA, fluorescent antibody; ID, immunodiffusion; IFA, indirect fluorescent antibody; IgG and IgM, immunoglobulins G and M; IHA, indirect hemagglutination; PVA, polyvinyl alcohol; GI, gastrointestinal; H&E, hematoxylin and eosin.

INTESTINAL AND TISSUE PROTOZOA

Entamoeba histolytica

Introduction

Entamoeba histolytica is the cause of amebiasis (Fig. 1). *E. histolytica* infection kills more than 100,000 people each year. The pathogenesis of this organism is related to its ability to lyse host cells and to cause tissue destruction. The trophozoites interact with the host through a series of steps, including adhesion to the target cell, phagocytosis, and cytopathic effect (22, 40, 116, 134-136). From the perspective of the host, *E. histolytica* induces both humoral and cellular immune responses, with cell-mediated immunity representing the major human host defense against this complement-resistant cytolytic protozoan.

In intestinal disease, the parasites are confined to the gastrointestinal tract, with no gross or microscopic invasion through the mucosal lining. Extraintestinal disease indicates organism invasion of the mucosal lining of the gastrointestinal tract. Parasites have passed through the mucosal lining, entered the bloodstream, and been carried to other body tissues,

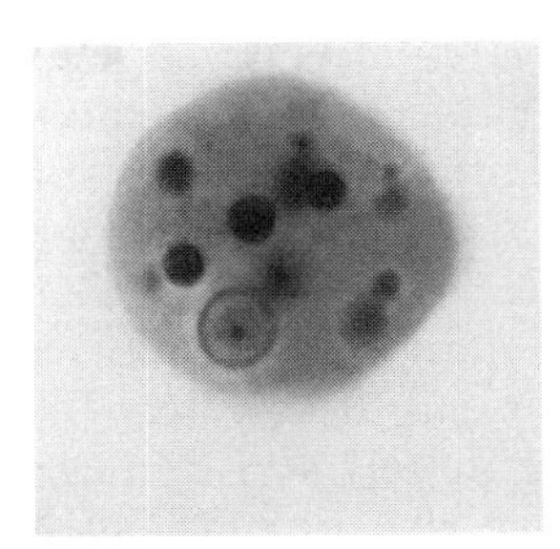

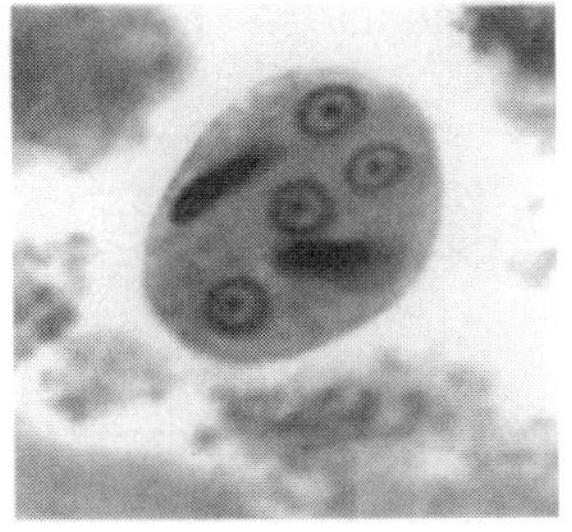

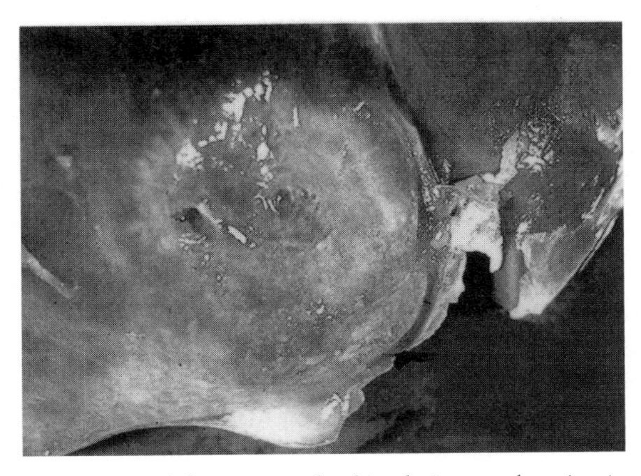

Figure 1. (Top left) *Entamoeba histolytica* trophozoite (true pathogen; note the ingested RBCs in the cytoplasm); (top right) *Entamoeba histolytica/E. dispar* cyst (unable to differentiate the true pathogen *E. histolytica* from the nonpathogen *E. dispar*); (bottom) gross specimen of amebic liver abscess. (Courtesy of Armed Forces Institute of Pathology.)

particularly the liver. For patients in whom the organisms have begun to invade the mucosal lining, interpretation of serologic tests may be difficult. In this situation, the terms "intestinal" and "extraintestinal" do not strictly apply. These differences are important to understand in discussing the clinical interpretation of serologic tests for amebiasis.

Only a small percentage of people infected with this organism will develop clinical symptoms. Depending on the geographic area, organism strain, and patient immune status, morbidity and mortality due to amebiasis vary. In the past, pathogenicity has been very controversial; historically, some believed that *E. histolytica* was really two separate species of *Entamoeba*, with *Entamoeba histolytica* being pathogenic and causing invasive disease and *Entamoeba dispar* being nonpathogenic and causing mild or asymptomatic infections.

Research emphasis has been placed on the molecular differences between pathogenic *E. histolytica* and nonpathogenic *E. dispar*. Since the molecules considered the most important for host tissue destruction (amebapore, galactose/*N*-acetyl galactosamine-inhibitable lectin, and cysteine proteases) can be found in both organisms (110, 111), differences in pathogenicity may be related to the composition and properties of the surface coat components (or pathogen-associated molecular patterns) and the ability of the innate immune response to recognize these components, thus eliminating the organisms. Targets of host immune system modulation appear to be both neutrophils and macrophages, which are rendered unable to abort the infection even if present at the site of the lesion. Evidence strongly supports the differentiation of the pathogenic species *E. histolytica* from the nonpathogenic species *E. dispar*. With publication of the *E. histolytica* genome, this key development has also led to increased research defining the complex multistage process of clinical amebiasis (2).

Organism description

Trophozoites vary in size from about 12 to 60 μm in diameter. Organisms from diarrheic or dysenteric stools are generally larger than those in a formed stool from an asymptomatic individual (Fig. 1). The motility is rapid, unidirectional, and sporadic, with pseudopods forming quickly in response to the conditions around the organism. Although this type of motility is often described, it is rare to diagnose amebiasis on the basis of motility seen in a direct wet mount. The cytoplasm is differentiated into a clear, outer ectoplasm and a more granular, inner endoplasm.

When the organism is examined on a permanent stained smear (trichrome or iron hematoxylin), the morphological characteristics of *E. histolytica/E. dispar* are more easily seen. The nucleus has evenly arranged chromatin on the nuclear membrane and a small, compact, centrally located karyosome. The cytoplasm is usually finely granular, with few ingested bacteria or debris in the vacuoles. In organisms isolated from a patient with dysentery, red blood cells (RBCs) may be visible in the cytoplasm, and this finding is diagnostic for *E. histolytica*. Most often, infection with *E. histolytica/E. dispar* is diagnosed on the basis of organism morphology without the presence of RBCs. Under these circumstances, the laboratory report would indicate the presence of the *Entamoeba histolytica/E. dispar* group rather than the true pathogen, *E. histolytica*.

The trophozoites may condense into a round mass (precyst), and a thin wall is secreted around the immature cyst. Within the cyst there may be a glycogen mass and highly refractile chromatoidal bars with smooth, rounded edges. As the cysts mature (metacyst), four nuclei are produced; on rare occasions, eight nuclei are seen. The cysts range in size from 10 to 20 μm. Often, as the cyst matures, the glycogen completely disappears; the chromatoidal bars may also be absent in the mature cyst. Cyst morphology does not differentiate between the true pathogen and nonpathogen; therefore, the report would be given as *Entamoeba histolytica/E. dispar* group. Cyst formation occurs only within the intestinal tract; once the stool has left the body, cyst formation does not occur. The one-, two-, and four-nucleated cysts are infective and represent the mode of transmission from one host to another (Fig. 1). After cyst ingestion, once the pH becomes neutral or slightly alkaline, the encysted organism becomes active, with the outcome being four separate trophozoites (small, metacystic trophozoites). These organisms develop into normal trophozoites when they become established in the large intestine.

Clinical aspects

In patients with intestinal disease, symptoms range from none to acute or chronic amebic colitis, both of which can mimic inflammatory bowel disease. Various factors influence the outcome of amebiasis, including the host immune system response. Although invasive *E. histolytica* organisms are usually found in tropical areas of the world, there are organisms within the United States that cause hepatic abscess in patients who have no history of travel outside the United States.

Asymptomatic infection. Individuals harboring *E. histolytica* may have a negative or low antibody titer and negative stools for occult blood but may be passing infective cysts. Although trophozoites may also be found, they will not contain any phagocytized RBCs and cannot be differentiated morphologically from *E. dispar*. Organisms isolated from asymptomatic individuals are generally *E. dispar*. Asymptomatic patients infrequently become symptomatic and may excrete cysts for a short period. However, this pattern can be found in patients infected with either nonpathogenic *E. dispar* or pathogenic *E. histolytica*.

Intestinal disease. Amebic ulcers most often develop in the cecum, appendix, or adjacent portion of the ascending colon; however, they can also be found in the sigmoidorectal area. Ulcers are often raised, with a small opening on the mucosal surface and a larger area of destruction below the surface, i.e., they are "flask-shaped." The mucosal lining usually appears normal between ulcers.

The incubation period varies from days to weeks, and in areas where *E. histolytica* is endemic, it is impossible to determine exactly when the original exposure occurred. Although the time frame normally ranges from 1 to 4 weeks, not every patient infected with *E. histolytica* will develop invasive amebiasis. The outcome will depend on the interaction between parasite virulence factors and the host response. Despite high prevalence in some areas, amebiasis is not a major source of morbidity within the HIV/AIDS pandemic; however, amebiasis may indeed be a risk for individuals living with HIV/AIDS (14). Organ transplant recipients are not at increased risk of having *E. histolytica* invasive disease. It is likely that the absence of intestinal colonization with *E. histolytica* in countries where transplants occur is an important factor (104).

There are four clinical forms of invasive intestinal amebiasis, all of which tend to be acute. They include dysentery or bloody diarrhea, fulminating colitis, amebic appendicitis, and ameboma of the colon. Dysentery and diarrhea account for 90% of cases of invasive intestinal amebiasis (40). Symptoms may range from none to those mimicking ulcerative colitis. Patients with colicky abdominal pain, frequent bowel movements, and tenesmus may present with a gradual onset of disease. With the onset of dysentery, bowel movements characterized by blood-tinged mucus are frequent (up to 10 per day). Although dysentery may last for months, it usually varies from severe to mild over that time and may lead to weight loss and prostration. In patients with severe cases, symptoms may begin very suddenly and include profuse diarrhea (over 10 stools per day), fever, dehydration, and electrolyte imbalances. In acute disease, the illness may mimic appendicitis, cholecystitis, intestinal obstruction, or diverticulitis.

Extraintestinal amebiasis. Blood flow draining the intestine and submucosa tends to return to the liver, most commonly the upper right lobe. Onset of extraintestinal amebiasis may be gradual or sudden, and the most consistent findings are upper right abdominal pain, with fever of 38 to 39°C. Weakness, weight loss, cough, and sweating are less common. There is often hepatomegaly with tenderness; however, liver function tests may be normal or slightly abnormal, with jaundice being very rare. There may be changes at the base of the right lung owing to the elevated diaphragm. The abscess can be visualized radiologically, sonically, or by radionuclear scan, and the majority of patients have a single abscess in the right lobe of the liver. The most common complication is rupture of the abscess into the pleural space. An abscess can also extend into the peritoneum and through the skin. Hematogenous spread to the brain, as well as to the lung, pericardium, and other sites, is possible. For patients in areas where infection is endemic or with a relevant travel history, amebic abscess should be suspected for spiking fever, weight loss, abdominal pain in the upper right quadrant or epigastrium, and tenderness in the liver area. The presence of leukocytosis, a high alkaline phosphatase level, and an elevated right diaphragm suggest a hepatic abscess. Pyogenic and amebic liver abscesses are the two most common hepatic abscesses. Pyogenic abscess severity depends on the bacterial source and the patient's underlying condition. Amebic abscess tends to be more prevalent in individuals with suppressed cell-mediated immunity, men, and younger individuals (46, 61).

E. histolytica has been seen in almost every soft organ and tissue of the body, including the lungs, heart, skin, and brain. Amebiasis of the lungs usually develops as an extension of a hepatic abscess that has ruptured through the diaphragm; however, it can also originate as a secondary site from the original infection in the intestine. The patient will experience expectoration of liver-colored sputum. In a very low percentage of cases, the pericardium can be involved; this is more likely to occur with abscess in the left lobe of the liver.

Amebiasis of the abdominal wall occurs due to rupture or open drainage of an internal intestinal or hepatic lesion (16). Amebiasis of the skin usually develops as a perianal extension of acute amebic colitis (16) and may also occur as a venereal infection of the penis after anal intercourse.

Amebic abscess in the brain is quite rare and usually arises from either hepatic or pulmonary involvement. Other rare ectopic sites include the spleen, adrenals, kidneys, ureters, urinary bladder, urethra, clitoris, and pericardium (46).

AIDS patients. In data from AIDS patients with diarrhea in India, the presence of various parasites in 56.4% (62/110 samples) of stool specimens indicates that their specific diagnosis is essential. Accurate diagnosis supports timely therapy to reduce the morbidity and mortality among these patients (67).

Invasive amebiasis appears to be an emerging parasitic disease in patients infected with HIV in areas where amebic infection is endemic. Medical, microbiological, and histopathologic records of 296 HIV-infected patients and serologic data from indirect hemagglutination assays (IHAs) of samples from 126 HIV-infected patients were reviewed to identify cases of invasive amebiasis. An IHA titer of 1:128 was considered positive. Eighteen of these HIV-infected patients (18/296 patients [6.1%]) were diagnosed with invasive amebiasis. Clinical manifestations included amebic colitis (13 patients), amebic liver abscess (9 patients), both colitis and abscess (4 patients), and pleural effusion (2 patients). Invasive amebiasis was the initial presentation of HIV infection for nine patients. Of the 18 patients diagnosed with invasive amebiasis, 13 had an IHA titer of ≥1:128. The sensitivity of IHA in the diagnosis of invasive amebiasis was 72.2%, and the specificity was 99.1%. The positive predictive value of IHA for invasive amebiasis in this patient population was 92.9%, and the negative predictive value was 95.5% (61). It appears that invasive amebiasis is becoming more important as one of the opportunistic parasitic infections in patients with HIV infection in areas of amebiasis endemicity. In this type of area of endemicity, the IHA should be considered a relevant diagnostic approach.

In a study from Mexico, prevalence estimates with PCR data showed that *E. histolytica* infection was more common in the HIV/AIDS group (25.32%) than in HIV-negative contacts (18.46%). *E. histolytica* and *E. dispar* infection was more frequent in HIV/AIDS patients (13.3%) than in HIV-negative contacts (0.7%). However, *E. histolytica* and/or *E. dispar* infection was highly prevalent in HIV/AIDS patients (34.1%) without evidence of recent or current invasive disease. Contacts of HIV/AIDS patients who were infected with *E. histolytica* were asymptomatic cyst passers. These data indicate that *E. histolytica* strains prevalent in the studied community appear to be of low pathogenic potential within the immunocompromised patient (92).

In another study, in Taiwan, 49 (5.2%) of 951 HIV-infected persons had 51 episodes of invasive amebiasis. High IHA titers were detected in 39 (6.2%) of 634 HIV-infected persons, compared with 10 (2.3%) of 429 uninfected controls with gastrointestinal symptoms and 0 of 178 uninfected healthy controls. Stool specimens from 40 (12.1%) of 332 HIV-infected persons and 2 (1.4%) of 144 uninfected healthy controls were positive for *E. histolytica* or *E. dispar* antigen. Ten (25.0%) of the 40 antigen-positive stool specimens from HIV-infected persons contained *E. histolytica*. Therefore, persons infected with HIV in Taiwan are at increased risk for invasive amebiasis and exhibit a relatively high frequency of elevated antibody titers and intestinal colonization with *E. histolytica* (52). Although amebiasis cutis is rare in HIV-positive patients, it should be considered in the differential diagnosis of perianal ulcers (16). Amebic osteomyelitis has also been reported for a child with AIDS (127).

Diagnosis

Intestinal disease. The diagnosis of amebiasis usually begins with an ova and parasite (O&P) examination. The O&P examination includes a direct saline wet mount, which is designed to allow the motility of the organisms to be seen; a concentration procedure, which provides a method to recover helminth eggs and larvae and protozoan cysts; and a permanent stained smear, which is the most important technique for diagnosis of the intestinal protozoa, including the trophozoites. Although motility can be seen on direct wet preparations, the material must be fresh, and a diagnosis of amebic infection should never be made solely from this type of examination. At the very least, the concentration procedure and permanent stained smear should be performed on every stool sample that is submitted to the laboratory for an O&P examination (46, 64).

Regardless of the patient's immune status, stools should be submitted to the laboratory on an every-other-day basis for a period of no more than 10 days. Although the recommended minimum number of stools to be examined is three, some laboratories are accepting two specimens. Multiple specimens are recommended because populations of intestinal protozoa tend to be cyclic. The specimen may be negative on collection day 1 but positive by collection day 3. It is also possible for only a nonpathogen to be found on collection day 1 but for pathogenic organisms to be recovered from the second or third specimen. If nonpathogenic protozoa are found, there is always the possibility of finding pathogenic organisms on subsequent examinations.

Presumptive identification of the organism as *Entamoeba histolytica*/*E. dispar* is based on the typical nuclear morphology (evenly arranged chromatin on the membrane and the presence of a small, central karyosome). The presence of RBCs within the trophozoite cytoplasm would provide definitive identification of true *E. histolytica* (rather than nonpathogenic *E. dispar*). In general, if the patient is symptomatic, one assumes that the causative agent is pathogenic *E. histolytica* (assuming that the cause is related to the presence of intestinal protozoa). Since organisms are shed sporadically, there is no relationship between the number of organisms seen and severity of disease and/or immune status of the patient.

The cysts of *E. histolytica*/*E. dispar* may contain chromatoidal bars with smooth, rounded edges. Even though the mature cyst contains four nuclei, these are not always visible, and the cyst can often be identified from the chromatoidal bars alone.

In a wet preparation, such as the direct smear or concentration sediment, the trophozoites and cysts of *E. histolytica* measure >12 and >10 μm, respectively. Because of artificial shrinkage during the preparation of permanent stained smears, the sizes are diminished somewhat, with the trophozoites measuring at least 10.5 to 11 μm and the cysts measuring at least 8.5 to 9 μm. Organisms measuring less than these limits and containing morphologic characteristics similar to those of *E. histolytica* would be identified as *Entamoeba hartmanni*, a nonpathogenic ameba.

It is important to remember that some human cells found in the stool can mimic *E. histolytica*. The macrophages or monocytes can look like the trophozoite form of *E. histolytica*, and polymorphonuclear leukocytes, when they have been in the stool for a while, can mimic the four-nucleus *E. histolytica* mature cyst. However, in a patient with diarrhea, the stool passage is too rapid for cyst formation.

Other tests include sigmoidoscopy, by which both scrapings and smears can be submitted for permanent stains. Biopsy specimens should also be submitted for histopathologic studies. Radiographic examination with barium is helpful; however, the presence of barium in the stool makes the O&P examination very difficult to perform. O&P examinations should be done before barium studies or at least 1 week to 10 days afterward. The sensitivities of serologic tests for antibody in patients with intestinal amebiasis vary, and a negative serologic test does not rule out the presence of amebic infection, particularly in the immunocompromised individual. Serologic tests for antibody are not recommended for the diagnosis of intestinal amebiasis.

Tests for antigen detection in the stool are much more relevant for intestinal disease, and these methods are rapid and technically simple. Reagents are currently available for antigen detection specifically for the true pathogen *Entamoeba histolytica* (141) and for the *Entamoeba histolytica/E. dispar* group. Fresh or frozen stool is required for these procedures; PCR and isoenzyme analyses are available for confirmation of *E. histolytica* infection but are not commonly available in the routine clinical laboratory.

If *E. histolytica* organisms are not seen in the stool and the final diagnosis is inflammatory bowel disease, the patient may receive corticosteroid therapy. Immunosuppression may predispose the patient to more severe disease if the true pathogen, *E. histolytica*, is present and not detected. Problems in differential diagnosis arise with inflammatory bowel disease grossly limited to the colon. In these cases, false-negative stool studies are increased by diarrhea and preparation for examinations, and both stool studies and serologies are compromised by steroids. If the clinical course of inflammatory bowel disease worsens, the clinician is justified in starting steroids even if evaluation for amebiasis is incomplete. If, however, the index of suspicion is high, concomitant treatment with metronidazole would be reasonable (74).

A single-round PCR assay has been developed for the detection and differential diagnosis of the three *Entamoeba* species found in humans, i.e., *Entamoeba moshkovskii*, *E. histolytica*, and *E. dispar*, all of which are morphologically identical as trophozoites and cysts (53). The only morphologic exception is the presence of RBCs within the trophozoite cytoplasm, a finding that could confirm the true pathogen, *E. histolytica*.

Extraintestinal disease. When organisms invade other tissues, such as the liver (extraintestinal or hepatic disease), different diagnostic approaches are necessary. Both fecal and sigmoidoscopic examinations for organisms are negative for approximately one-half of patients with extraintestinal amebiasis. Both radioactive and computed tomographic scans are available to assist the clinician in defining the presence of an abscess.

Lesions in the liver may range from less than a few centimeters to several inches in diameter. In either case, if the patient has a normal humoral immune system, serologic tests for antibody will generally be positive.

Serologic tests for extraintestinal amebiasis are very specific and quite sensitive. However, in certain geographic areas, sensitivity may be difficult to interpret for patients previously exposed to *E. histolytica*. In these individuals, a low to moderate titer may represent past infection rather than current disease. Also, patients who are immunocompromised may present with low or negative serologic titers. It is generally recommended that two different procedures be used so that the results can be compared. One approach is to use both qualitative screen and quantitative titer procedures. Commercial suppliers of kits are limited, and few laboratories routinely provide this type of testing. Serum for serologic tests should be sent to the state and local public health laboratories if the test is not routinely available.

Histology. Organisms must be differentiated from host cells, particularly histiocytes and ganglion cells. Periodic acid-Schiff (PAS) staining is often used to help locate the organisms. The organisms will appear bright pink with a green-blue background (depending on the counterstain used). Hematoxylin and eosin staining will also allow the typical morphology to be seen, thus allowing accurate identification. As a result of sectioning, some organisms will exhibit the evenly arranged nuclear chromatin with the central karyosome and some will no longer contain the nucleus.

Therapy

While carriers usually harbor nonpathogenic *E. dispar*, pathogenic *E. histolytica* may also be found in these individuals. At present, test methodologies which routinely differentiate between *E. histolytica* and *E. dispar* are not used routinely by all diagnostic laboratories. The diagnosis of *E. histolytica/E. dispar* infection is most often based on organism morphology. For this reason, in general, patients in the United States who harbor these organisms may be treated, regardless of the presence or absence of symptoms.

There are two classes of drugs used in the treatment of amebic infections: they are luminal amebicides, such as iodoquinol or diloxanide furoate, and tissue amebicides, such as metronidazole, chloroquine, or dehydroemetine. Because differences in drug efficacy exist, it is important that the laboratory report for the physician indicate whether cysts, trophozoites, or both are present in the stool specimen.

Epidemiology and control

Infections with *E. histolytica* are worldwide in distribution and are generally most prevalent in the tropics. In 1984, 500 million people were estimated to be infected with *E. histolytica*, 40 to 50 million of whom had extensive symptoms, including colitis or extraintestinal abscesses. Prevalence figures for the United States are generally thought to be <5%. Those with a higher incidence of amebiasis include recent immigrants and refugees from South and Central America and from Southeast Asia. Residents in southeastern and southwestern parts of the United States also tend to have more infections with intestinal parasites, as do other groups, including patients in mental institutions.

Epidemiologic evidence of sexual transmission of *E. histolytica* has grown significantly since the early 1970s, particularly in areas such as New York City and San Francisco. In San Francisco, the incidence of reported symptomatic intestinal amebiasis among homosexual men between 20 and 39 years of age has increased over 1,000% during the last 10 years. Although percentages vary, it appears that approximately 30% of urban homosexual men may be infected with *E. histolytica*, a sharp increase over the estimated rate of <5% seen in the general population within the United States. The key factor is not necessarily homosexuality but the frequency of sexual activity and the potential for fecal-oral contact.

There are certain urban areas (Mexico City, Mexico; Medellin, Colombia; and Durban, South Africa) where the incidence of invasive disease is considerably higher than that in the rest of the world. Contributing factors may include poor nutrition, tropical climate, decreased immunologic competence of the host, stress, altered bacterial flora in the colon, traumatic injuries to the colonic mucosa, alcoholism, and genetic factors.

Amebiasis can be considered a zoonotic waterborne infection, as well as being transmitted through human-to-human transmission. The asymptomatic cyst passer who is a food handler is generally thought to play the most important role in transmission.

A colonization-blocking vaccine could eliminate *E. histolytica* as a cause of human disease, particularly since humans serve as the only significant reservoir host. Continued research should lead to the development of a cost-effective oral combination "enteric pathogen" vaccine, capable of inducing protective mucosal immune responses to several important enteric diseases, including amebiasis (135).

Giardia lamblia

Introduction

Although host specificity, various body dimensions, and variations in structure have been used to designate species of *Giardia*, there is still debate over classification and nomenclature regarding this group of organisms. Within the United States, the term *Giardia lamblia* has been used commonly for many years, and it is used throughout this chapter to refer to those organisms found in humans and other mammals; many health care workers are accustomed to this name and continue to report the presence of the organism by using the term "*Giardia lamblia*." Other designations include *Giardia duodenalis* and *Giardia intestinalis*.

Giardia lamblia is a flagellate commonly found in many parts of the world. The incidence of giardiasis worldwide may be as high as 1 billion cases. Infections with this organism can be transmitted via the cyst form in both water and food contaminated with fecal material. Various conditions that have been associated with giardiasis in the immunocompromised patient include hypogammaglobulinemia, protein or caloric malnutrition, previous gastrectomy, histocompatibility antigen HLA YB-12, gastric achlorhydria, blood group A, differences in mucolytic proteins in immunoglobulin deficiencies, and reduced secretory immunoglobulin A (IgA) levels in the gut. Adults and children are mentioned in relation to giardiasis; however, it is important to remember that the elderly, in both tropical and temperate zones, are also susceptible. These patients tend to be somewhat immunodepressed due to age, and giardiasis may be overlooked in their differential diagnosis.

While the number of organisms in the crypts of the duodenal mucosa may be quite high, the parasites may not cause any pathology. The trophozoites feed on the mucous secretions and generally do not penetrate the mucosa. However, organisms have been seen in biopsy material from inside the intestinal mucosa, while others have been seen attached only to the epithelium. In symptomatic patients, there may be irritation of the mucosal lining, increased mucus secretion, and dehydration.

G. lamblia is able to undergo continuous antigenic variation of its major surface antigen, the variant surface protein. This phenomenon, as well as adhesion of the organism to the mucosa, loss of intestinal brush border surface area, villus flattening, inhibition of disaccharidase activities, and eventual overgrowth of the enteric bacterial flora, is involved in the pathophysiology of giardiasis (97).

Although some patients with HIV infection also have giardiasis, the infection does not appear to be more severe, regardless of the $CD4^+$ cell count (68). In some groups, the prevalence of *G. lamblia* is greater for HIV-infected subjects (151). Both humoral and cellular immune responses play a role in acquired immunity, and the mechanisms involved remain somewhat unclear. However, the immune response to intestinal parasites, including *G. lamblia*, might be a risk factor for HIV/AIDS and tuberculosis patients (3). It is well known that diagnosing intestinal parasites in HIV/AIDS patients is necessary, especially for those who report to be chronic alcoholics or are not on antiretroviral treatment (28, 52, 91, 131).

Organism description

Both the trophozoite and the cyst are included in the life cycle of *G. lamblia*. Trophozoites divide by means of longitudinal binary fission, producing two daughter trophozoites. The most common location of the organisms is in the crypts within the duodenum. The trophozoites are the intestinal dwelling stage and attach to the epithelium of the host villi by means of the ventral disc. The attachment is substantial and results in disc "impression prints" when the organism detaches from the surface of the epithelium. Trophozoites may remain attached or detach from the mucosal surface. Since the epithelial surface sloughs off the tip of the villus every 72 h, apparently the trophozoites detach at that time.

The trophozoite is usually described as being teardrop shaped from the front, with the posterior end being pointed (Fig. 2). If one examines the trophozoite from the side, it resembles the curved portion of a spoon. The concave portion is the area of the sucking disc. There are four pairs of flagella, two nuclei, two axonemes, and two slightly curved bodies, called the median bodies. The trophozoites usually measure 10 to 20 μm in length and 5 to 15 μm in width.

The cysts may be either round or oval, and they contain four nuclei, axonemes, and median bodies (Fig. 2). Often, some cysts appear to be shrunk or distorted, and one may see two halos, one around the cyst wall itself and one inside the cyst wall around the shrunken organism. The halo effect around the outside of the cyst is particularly visible on the permanent stained smear. Cysts normally measure 11 to 14 μm in length and 7 to 10 μm in width.

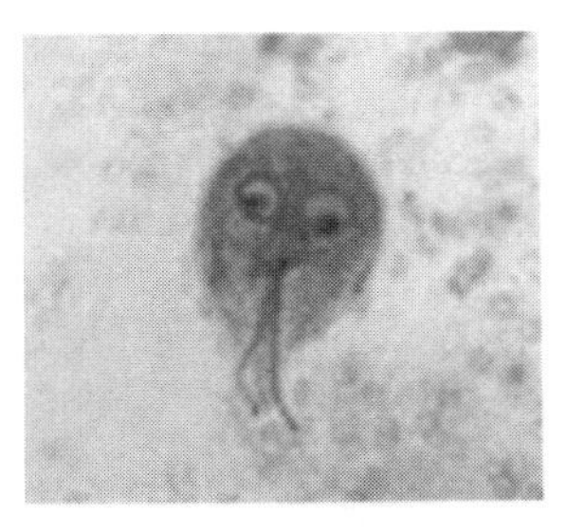
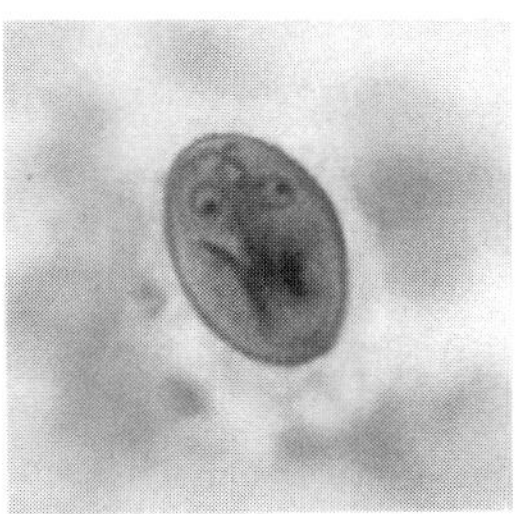
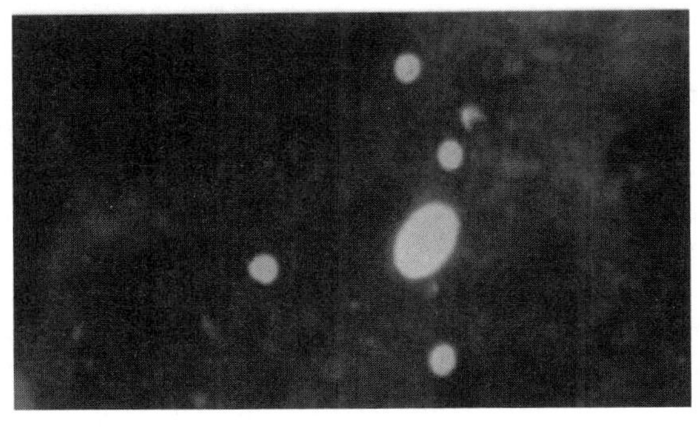

Figure 2. (Top left) *Giardia lamblia* trophozoite (note two nuclei, curved median bodies, and linear axonemes); (top right) *Giardia lamblia* cyst; (bottom) *Giardia lamblia* cyst (large) and *Cryptosporidium* sp. oocysts (small) (demonstrating fluorescence using the fecal fluorescent-antibody immunoassay; some will use a counterstain, while others will not).

Clinical aspects

There may be a sudden onset of explosive, watery, foul-smelling diarrhea, which may be accompanied by nausea, anorexia, malaise, low-grade fever, and chills. Other symptoms include epigastric pain, flatulence, and diarrhea, with increased amounts of fat and mucus in the stool but no blood. Weight loss is common. Occasionally, the gallbladder is also involved, causing gallbladder colic and jaundice. *G. lamblia* has also been identified in bronchoalveolar lavage fluid.

The acute phase is often followed by a chronic phase. Symptoms in these patients include recurrent, brief episodes of loose, foul-smelling stools; there may be increased distention and foul flatus. Between passing the mushy stools, the patient may have normal stools or may be constipated. Abdominal discomfort includes marked distention and belching with a rotten-egg taste. Chronic disease must be differentiated from amebiasis; diseases caused by other intestinal parasites, such as *Dientamoeba fragilis*, *Cryptosporidium* spp., *Cyclospora cayetanensis*, *Isospora belli*, and *Strongyloides stercoralis*; inflammatory bowel disease; and irritable colon. On the basis of symptoms such as upper intestinal discomfort, heartburn, and belching, giardiasis must also be differentiated from duodenal ulcer, hiatal hernia, and gallbladder and pancreatic disease.

Diagnosis

Routine procedures. Routine stool O&P examinations are usually recommended for the recovery and identification of intestinal protozoa (46, 64). However, because the organisms are attached so securely to the mucosa by means of the sucking disc, a series of even five or six stool samples may be examined without organism recovery. Also, organism numbers generally have no relationship to the patient's immune status. An Entero-Test capsule can be helpful in recovering the organisms, as can duodenal aspirates. Although cysts can often be identified on the wet stool preparation, many infections may be missed without the examination of a permanent stained smear (46, 64). If material from the string test (Entero-Test) or mucus from a duodenal aspirate is submitted, it should be examined as a wet preparation for motility; however, motility may be represented by nothing more than a slight flutter of the flagella because the organism will be caught up in the mucus. After diagnosis, the rest of the positive material can be preserved as a permanent stain.

Fluoroscopy may reveal hypermotility at the duodenal and jejunal levels, and X-rays may reveal mucosal defects. However, because giardiasis may not produce any symptoms at all, demonstration of the organism in symptomatic patients may not rule out other possibilities, such as peptic ulcer, celiac disease of some other etiology, strongyloidiasis, and possibly carcinoma.

Antigen detection. Fecal immunoassays for the detection of *Giardia* antigen in stool have dramatically improved the sensitivity of organism detection compared with that of routine O&P examination. Procedures include enzyme-linked immunosorbent assay (ELISA), fluorescent methods, and a newer cartridge format based on membrane flow methodology (46). Commercial reagent kits for the detection of *Giardia* can be used with formalin-based stool preservatives as well as with fresh or frozen specimens. Many of the cartridge format tests are now available to detect both *G. lamblia* and *Cryptosporidium* spp., provide an answer within 10 min, and are equal to or better than other immunoassay formats in their sensitivity and specificity. Many of these newer methods are being used to test patients suspected of having giardiasis or those who may be involved in an outbreak situation. The detection of antigen in stool or visual identification of organisms by using monoclonal antibody reagents indicates current infection. With the increase in *Giardia* infections and awareness of particular situations, such as nursery school settings, these detection assays are rapid and reliable immunodiagnostic procedures.

Antibody detection. Unfortunately, serodiagnostic procedures for giardiasis are not acceptable for wide clinical use, particularly since they may indicate either past or present infection. Since *Giardia* trophozoites rarely invade the tissues or stimulate the systemic immune response, serodiagnostic assays fail to show differences in serum antibody responses between symptomatic and asymptomatic patients and are not recommended.

Histology. Trophozoites can be seen in the duodenum and proximal jejunum of infected patients. Although there has been some suggestion of mucosal invasion by this parasite, mucosal invasion has generally been found in areas where necrosis or mechanical trauma was present. There can be a spectrum of change from normal villi to almost complete villus atrophy, with a greater density of inflammatory infiltrate in the lamina propria when villus atrophy is present. Patients with giardiasis tend to have reduced mucosal surface areas compared with uninfected patients, and the degree of abnormality correlates with the index of malabsorption. Enumeration of lamina propria plasma cells of different immunoglobulin classes reveals no clear patterns, suggesting that *Giardia* infection initially provokes IgM and IgE synthesis, with relative suppression of local IgA production.

Histologic changes in the mucosal architecture in immunodeficient patients with giardiasis also range from mild to severe villus atrophy. Nodular lymphoid hyperplasia has been reported, as well as mixed lesions. Again, the amount of villus damage seems to correlate with the degree of malabsorption. Giardiasis produces a more severe degree of villus damage in patients with hypogammaglobulinemia. In patients with AIDS, giardiasis does not appear to be an important pathogen, although the infection has certainly been found in this group and in homosexual men.

Therapy

If giardiasis is diagnosed, the patient should be treated. Current recommended treatments include the nitroheterocyclic drugs tinidazole, metronidazole, and furazolidone; the substituted acridine quinacrine; and the benzimidazole albendazole. Paromomycin is also used, as is nitazoxanide. Treatment failures have been reported with all the common antigiardial agents, and resistance to these drugs has been demonstrated in the laboratory. Clinical resistance has been reported and includes situations in which patients have failed both metronidazole and albendazole treatments (1, 46). Metronidazole is not recommended for pregnant women; although it is not absorbed and not highly effective, paromomycin may be used to treat giardiasis during pregnancy. Within the United States, metronidazole is the only member of the nitroimidazole class available to treat giardiasis.

Epidemiology and control

Although contaminated food or drink containing viable cysts may be the source of infection, intimate contact with an infected individual may also provide the infection mechanism. The *Giardia* organism tends to be found more frequently in children or in groups that live in close quarters. Often, there are outbreaks due to poor sanitation facilities or breakdowns, as evidenced by infections of travelers and campers. There is also an increase in the prevalence of giardiasis in the male homosexual population, probably because of anal and/or oral sexual practices.

Coccidia

Cryptosporidium spp.

Introduction. One of the more newly recognized parasitic organisms infecting humans, particularly immunocompromised patients, is *Cryptosporidium* spp., a cause of diarrhea. *Cryptosporidium* developmental stages do not occur deep within host cells but are confined to an intracellular, extracytoplasmic location. Each stage is within a parasitophorous vacuole of host cell origin. What was previously called *Cryptosporidium parvum* and was thought to be the primary *Cryptosporidium* species infecting humans is now classified as two separate species, *C. parvum* (infecting mammals, including humans) and *Cryptosporidium hominis* (infecting primarily humans) (94). Differentiation of these two species based on oocyst morphology is not possible.

These organisms are well recognized as human pathogens, particularly in those who are in some way immunosuppressed or immunodeficient. Cryptosporidiosis has been implicated as one of the more important diseases caused by opportunistic agents in patients with AIDS (17, 115, 119). Generally, a patient with a normal immune system will have a self-limited infection; however, patients who are immunocompromised may have a chronic infection with a wide range of symptoms (asymptomatic to severe) (152).

Based on questions regarding appropriate management and prognosis of inflammatory bowel disease patients infected with *Cryptosporidium* that arose from the 1993 Milwaukee outbreak, it is clear that cryptosporidiosis may present as an acute relapse of inflammatory bowel disease and responds to standard therapy. Antibiotics confer no benefit. It appears that immunosuppressive therapy does not predispose these patients to chronic or severe illness. Cryptosporidiosis may present with acute findings initially mimicking Crohn's disease.

Unfortunately, no totally effective therapy for cryptosporidiosis has been identified, despite testing of over 100 drugs. Detection of this parasite in immunocompromised hosts, especially those with AIDS, usually carries a poor prognosis. Also, reports of respiratory tract and biliary tree infections confirm that the developmental stages of this organism are not always confined to the gastrointestinal tract.

Organism description. The presence of a thin-walled, autoinfective oocyst in the life cycle may explain why a small inoculum can lead to an overwhelming infection in a susceptible host and why immunosuppressed patients may have persistent, life-threatening infections in the absence of documentation of repeated exposure to oocysts (Fig. 3). The stages found on the microvillous surface measure 1 μm, and the oocysts recovered in stool or sputum specimens measure 4 to 6 μm.

Each intracellular stage of *Cryptosporidium* resides within a parasitophorous vacuole within the microvillous region of the host cell. Oocysts undergo sporogony while they are in the host cells and, regardless of stool consistency, are immediately infective

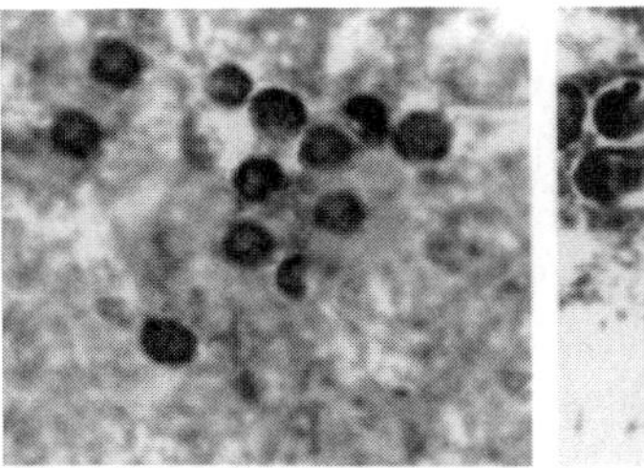
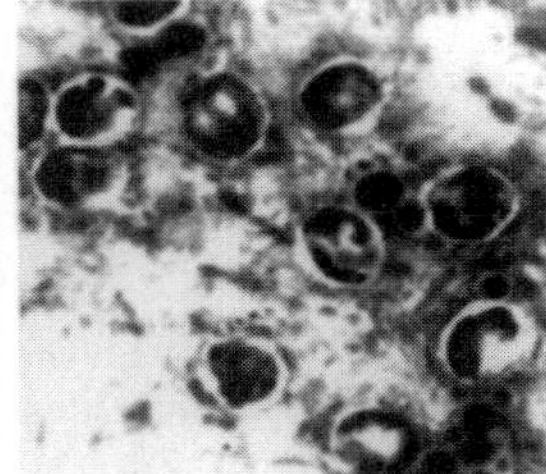
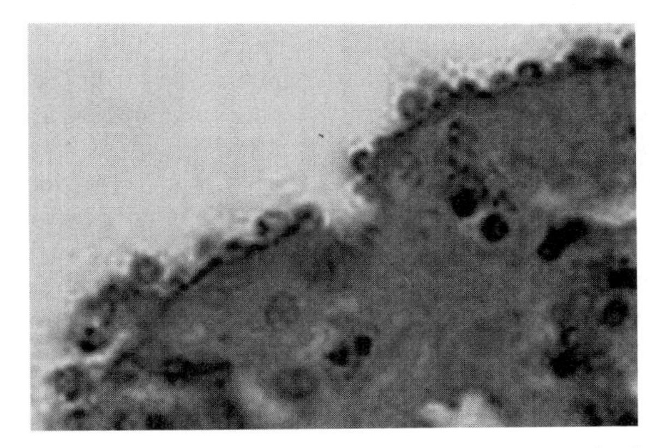

Figure 3. (Top left) *Cryptosporidium* sp. oocysts stained using modified acid-fast stain (note the spherical shape; oocysts measure 4 to 6 μm); (top right) *Cryptosporidium* sp. oocysts (note that sporozoites are visible in some of the oocysts, which are infective when passed, regardless of whether the sporozoites are visible or not); (bottom) histologic section of intestinal tissue showing organisms within parasitophorous vacuoles at the brush border. (Courtesy of Armed Forces Institute of Pathology.)

when passed in the stool. Approximately 20% of the oocysts of *Cryptosporidium* do not form the thick, two-layered, environmentally resistant oocyst wall. The four sporozoites within this autoinfective stage are surrounded by a single-unit membrane. After release of the oocyst from a host cell, this membrane ruptures and the invasive sporozoites penetrate the microvillous regions of other cells within the intestine and reinitiate the life cycle. As mentioned above, these thin-walled oocysts that can recycle are thought to be responsible for the development of severe, life-threatening disease in immunocompromised patients, even those who are no longer exposed to the environmentally resistant oocysts.

Clinical aspects. *Immunocompetent individuals.* Clinical symptoms include nausea, low-grade fever, abdominal cramps, anorexia, and 5 to 10 watery, frothy bowel movements per day, which may be followed by constipation. Other patients may have relatively few symptoms, particularly later in the course of the infection. In those with typical watery diarrhea, the stool specimen contains very little fecal material, mainly water and mucous flecks. Those who are immunocompetent will have a self-limited infection.

Occasionally, immunocompetent patients require fluid replacement and the diarrhea persists for more than 2 weeks. This is particularly true of infants, in whom excessive fluid loss may last for over 3 weeks. In general, when $CD4^+$ cells are present at levels of >200/μl, infections are acute and resolve in approximately 2 weeks; however, when the $CD4^+$ cell count drops below 200/μl, the infection may be chronic and may not resolve (99, 146).

Failure to thrive has also been attributed to chronic cryptosporidiosis in infants. Since diarrheal illness is a major cause of morbidity and mortality in young children living in developing countries, it is likely that cryptosporidiosis plays a major role in the overall health status of these children. Cryptosporidiosis has a lasting adverse effect on linear growth (height), especially if acquired during infancy and if children are stunted before they become infected. *Cryptosporidium* may be implicated in respiratory disease that often accompanies diarrheal illness in malnourished children. Those with respiratory infections tend to have coughing, wheezing, croup, hoarseness, and shortness of breath.

Immunocompromised individuals with AIDS. (i) Intestinal disease. Most severely immunocompromised patients cannot overcome the infection; the illness becomes progressively worse with time, and the sequelae may be a major factor leading to death. The length and severity of disease may also depend on the ability to reverse the immunosuppression. In these patients, *Cryptosporidium* infections are not always confined to the gastrointestinal tract; additional symptoms (respiratory problems, cholecystitis, hepatitis, and pancreatitis) have been associated with extraintestinal infections (121). Although the clinical features of sclerosing cholangitis secondary to opportunistic infections of the biliary tree in patients with AIDS are well known, the mechanisms by which pathogens such as *Cryptosporidium* actually cause disease remain unclear.

(ii) Biliary tract disease. Sclerosing cholangitis is well known as a complication of AIDS. Direct cytopathic effects are seen in infected monolayers (human biliary epithelial cell line), with widespread programmed cell death (apoptosis) beginning within hours after exposure to the organism. Specific cytopathic invasion of biliary epithelia by *Cryptosporidium* may be relevant to the pathogenesis and possible therapy of secondary sclerosing cholangitis seen in AIDS patients with biliary cryptosporidiosis. More severely immunocompromised individuals are more likely to exhibit biliary tract disease.

(iii) Pancreatitis. In AIDS patients with pancreatitis, cellular changes are generally not that severe, and these patients exhibit hyperplastic squamous metaplasia (20). These patients with pancreatitis usually present with abdominal pain resistant to analgesics, elevated serum amylase levels, and abnormalities on sonography and computed tomography. It is difficult to clarify the impact of cryptosporidiosis-related pancreatic disease, and this disease complication is apparently not linked to significant morbidity.

(iv) Respiratory tract disease. Because many patients with respiratory cryptosporidiosis also have other pathogens present, it is difficult to determine the significance of this complication in AIDS patients. Certainly, with respiratory symptoms, it is important to examine sputum as well as stool specimens to confirm the diagnosis.

Immunocompromised individuals with primary immunodeficiency diseases. Many different primary immunodeficiency diseases have been described, and in general, they can be categorized as follows: combined immunodeficiencies (impacting both T and B lymphocytes), antibody deficiencies, complement deficiencies, and defects in phagocytes (decreased number and function). In general, the number of these cases reported in relation to cryptosporidiosis is few;

however, the most serious immunodeficiency in terms of risk is severe combined immunodeficiency syndrome. Patients with this syndrome are at risk for disseminated disease, and often the prognosis is poor.

Immunocompromised individuals with malignant disease. Although cryptosporidiosis can be seen in malignant disease, infection with *Cryptosporidium* spp. does not appear to pose a special risk (51). Exceptions to this general statement seem to involve leukemia and other hematological malignancies. Generally, there is more interest in patients who are candidates for bone marrow transplantation, particularly in assessing possible risks related to cryptosporidiosis.

Organ transplant patients. Cryptosporidiosis has been reported for liver, kidney, and small bowel transplant recipients (12, 41, 47, 60, 78, 101, 145). However, in most cases, disease was not unusually severe and did not involve dissemination to extraintestinal sites.

Diagnosis. Although biopsy techniques have been used to recover and identify the organisms in tissue, it is possible that the specific area where the organisms reside may not be subjected to biopsy. The examination of stool material for oocysts by using a modified acid-fast technique or immunoassays allows one to screen a sample from the entire intestinal tract. *Cryptosporidium* oocysts in the stool range from 4 to 6 μm in diameter and can be very difficult to identify. One must also remember that even in patients with typical watery diarrhea, the number of organisms passed can be quite variable and may not correlate with the patient's immune status.

Diagnostic techniques include formalin sedimentation; acridine orange, auramine-rhodamine, iodine, and modified acid-fast staining; and immunoassay techniques. The most widely used techniques are modified acid-fast staining, Kinyoun's methods, and fecal immunoassays.

Because other organisms may also fluoresce when auramine or the auramine-rhodamine combination is used, presumptive positive specimens should be confirmed by using one of the modified acid-fast methods. Both *Cryptosporidium* and *Isospora* spp. fluoresce with auramine-rhodamine and are modified acid-fast stain positive. Oocysts are fully mature and infective at the time they are passed in the stool.

Respiratory cryptosporidiosis has also been reported for immunocompromised patients. Sputum specimens should be submitted in 10% formalin or as fresh specimens and can be examined by the same techniques as those used for stool samples.

Immunoassay procedures for the direct detection of *Cryptosporidium* antigen or oocysts in fecal specimens have proven to be much more sensitive than the modified acid-fast stains. Since stains do not always consistently stain all oocysts, the increased sensitivity and specificity of these immunoassays provide excellent screening methods. Enzyme immunoassay, solid-phase immunochromatographic assay in cartridge format, and fluorescent-antibody methods are currently available.

It is recommended that stool specimens be submitted in 5 or 10% formalin, sodium acetate-acetic acid-formalin, or the newer single-vial collection systems for processing. However, each laboratory must confirm that the fixatives selected are compatible with diagnostic procedures performed, including the O&P examination and fecal immunoassays (not yet approved for other specimens). Since fecal immunoassays are more sensitive than special stains, immunoassay kits are recommended, if available.

Therapy. Cryptosporidiosis is self-limiting in patients who have an intact immune system. In patients who are receiving immunosuppressive agents, one method of therapy is to discontinue such a regimen. Other approaches with specific therapeutic drugs have been tried, but to date the results are still somewhat controversial.

Highly active antiretroviral therapy leading to an increased CD4 count has had a dramatic impact on cryptosporidiosis in AIDS patients (24). Resolution of cryptosporidial diarrhea is apparently related to the enhanced CD4 count rather than to any change in the viral load or any therapeutic impact of the drugs. Thus, it appears that cellular immunity is critical in clearing *Cryptosporidium* infection in these patients.

Epidemiology and control. Studies have shown that calves and perhaps other animals may serve as potential sources of human infections. Kittens, rodents, and puppies are possible reservoir hosts, since they are easily infected with human *Cryptosporidium*. Person-to-person transmission is also likely and may occur through direct or indirect contact with stool material in the environment and the hospital setting.

The increase in the number of reported waterborne disease outbreaks associated with *Cryptosporidium* spp. can be attributed to improved techniques for oocyst recovery and identification (108). Unfortunately, *Cryptosporidium* oocysts are quite resistant to most commercial disinfectants, including iodine water purification tablets.

Cyclospora cayetanensis

Introduction. *Cyclospora cayetanensis* organisms are acid-fast-variable coccidia and have been found in the feces of immunocompetent travelers to developing countries, immunocompetent subjects with no travel history, and patients with AIDS (Fig. 4). Dual infections with *Cryptosporidium* spp. have also been seen in immunocompromised patients (56). The life cycle has been confirmed, but information on potential reservoir hosts has yet to be defined; however, it appears that in some areas, the human is the only host. Outbreaks linked to contaminated water and various types of fresh produce (raspberries, basil, baby lettuce leaves, and snow peas) have been reported.

Developmental stages of *C. cayetanensis* usually occur within epithelial cells of the jejunum and lower portion of the duodenum. *Cyclospora* infection reveals characteristics of a small bowel pathogen, including upper gastrointestinal symptoms, malabsorption of D-xylose, weight loss, and moderate to marked erythema of the distal duodenum. Histopathology in small bowel biopsy specimens reveals acute and chronic inflammation, partial villus atrophy, and crypt hyperplasia.

Organism description. It takes approximately 5 days or more for oocyst maturation, so the mature stage may not be seen in human fecal specimens. The oocyst contains two sporocysts, each containing two sporozoites, a pattern which places these organisms in the coccidian genus *Cyclospora*. Unsporulated oocysts are passed in the stool, and sporulation occurs within approximately 1 to 2 weeks. In patients who have *Cyclospora* oocysts in their stool specimens, two types of meronts and sexual stages have been found within the jejunal enterocytes.

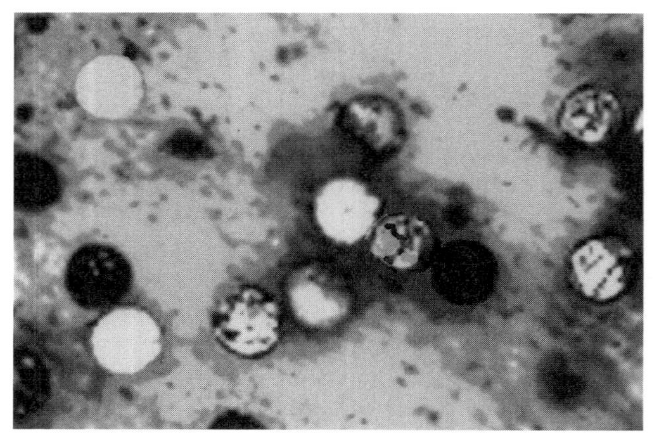

Figure 4. *Cyclospora cayetanensis* oocysts stained using modified acid-fast stain (note the spherical shape; oocysts measure 8 to 10 μm; some oocysts do not stain, and thus the organisms are said to be "modified acid-fast variable"). These oocysts are not infectious when passed, regardless of the stool consistency.

Clinical aspects. After one or more days of malaise and low-grade fever, there is a rapid onset of diarrhea of up to seven stools per day. Symptoms may also include fatigue, anorexia, vomiting, myalgia, and weight loss, with remission of self-limiting diarrhea in 3 to 4 days followed by relapses lasting from 4 to 7 weeks. In patients with AIDS, symptoms may persist for as long as 12 weeks; biliary disease has also been reported for this group. Diarrhea alternating with constipation has also been reported; this is not uncommon for a number of protozoal gastrointestinal infections. Clinical clues include unexplained prolonged diarrheal illness during the summer in any patient and in persons returning from tropical areas. The clinical presentation of patients infected with this organism is similar to that of patients infected with *Cryptosporidium* spp. and can also include pulmonary symptoms (62).

Diagnosis. In clean wet mounts, the organisms are seen as nonrefractile spheres and are acid-fast variable with the modified acid-fast stain; those that are unstained appear as glassy, wrinkled spheres. Modified acid-fast stains stain the oocysts light pink to deep red, and some oocysts contain granules or have a bubbly appearance. It is very important to be aware of these organisms when the modified acid-fast stain is used for *Cryptosporidium* spp. and other similar but larger structures (approximately twice the size of *Cryptosporidium* oocysts) are seen in the stained smear. All acid-fast oocysts should be measured, particularly if they appear to be somewhat larger than those of *Cryptosporidium* spp. The oocysts autofluoresce green (1+ to 3+; 450- to 490-nm dichroic mirror excitation filter) or blue (1+ to 3+; 365-nm dichroic mirror excitation filter) under UV epifluorescence. During concentration (formalin/ethyl acetate) of stool specimens, the centrifugation time and speed should be 10 min at $500 \times g$ (46, 64). Currently, no fecal immunoassays for *Cyclospora* are available.

Therapy. While some patients have been treated symptomatically with antidiarrheal preparations and have obtained some relief, the disease appears to be self-limiting within a few weeks. Trimethoprim-sulfamethoxazole (TMP-SMX) is currently the drug of choice; relief of symptoms has been seen in 1 to 3 days posttreatment. However, symptoms recur within

1 to 3 months posttreatment in over 40% of patients, particularly those who are immunocompromised. Although prophylaxis with TMP-SMX is not generally recommended, it might be worth considering for an HIV-positive person visiting an area of endemic infection, such as Nepal, during the summer, when the risk of infection tends to be the highest.

Epidemiology and control. Individuals of all ages, including immunocompetent and immunosuppressed individuals, can become infected. In Peru, infections with this organism have shown some seasonal variation, with peaks during April to June. This pattern is similar to that seen for *Cryptosporidium* infections in Peru. Preliminary data and extrapolation from what we currently know about cryptosporidiosis suggest that modes of transmission may be similar, particularly considering waterborne transmission.

Like *Cryptosporidium*, this parasite is not killed by routine chlorination; drinking water treated by halogenation may not be safe. Boiling is recommended. Hot drinks such as coffee or tea are recommended, rather than placing any reliance on iodine treatment of the water. Fresh fruits and vegetables should be washed thoroughly and/or peeled prior to being eaten.

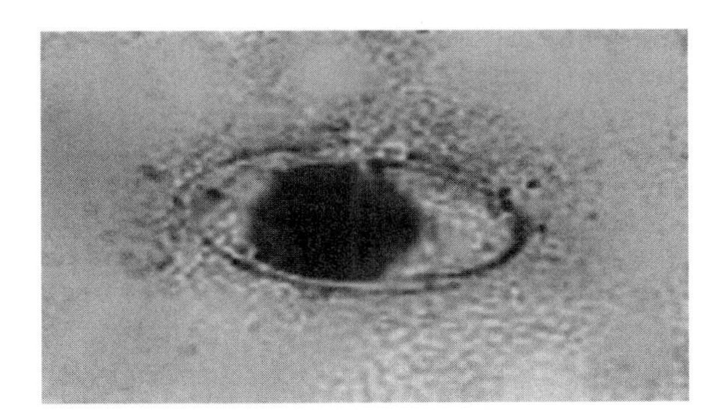

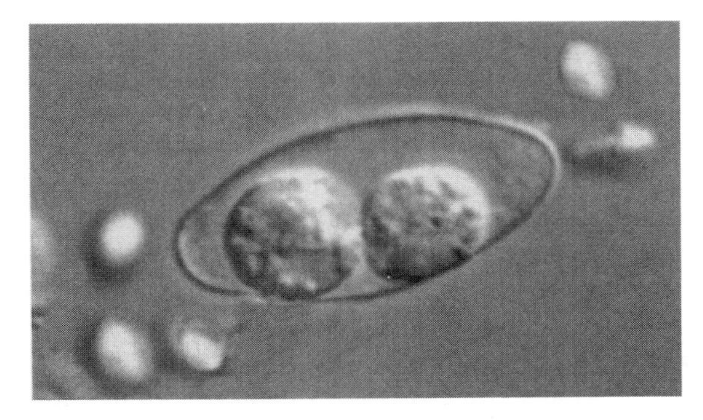

Figure 5. *Isospora* (*Cystoisospora*) *belli.* (Top) Immature oocyst (contains single sporocyst) stained using modified acid-fast stain; (bottom) more mature oocyst (contains two sporocysts) in a saline wet mount of stool concentration sediment.

Isospora (Cystoisospora) belli

Introduction. An organism similar to *Cryptosporidium* spp. that causes diarrhea in the immunocompromised host is *Isospora belli* (Fig. 5) (76). Transition to the new genus name (*Cystoisospora*) will probably occur during the next couple of years. Although *Isospora* infections have been found in many areas of the world, certain tropical areas in South America appear to contain some well-defined locations of endemicity. *Isospora* organisms can infect both adults and children. Intestinal tract involvement and symptoms are generally transient unless the patient is immunocompromised. In a group of AIDS patients in Brazil, 2% were found to be positive for *I. belli* infection. Prevalence data for this and other parasites in this group indicate that there does not appear to be any association between $CD4^+$ cell counts and any particular parasite. However, the data supported the value of using standard fecal examinations in this group of infected patients, even in the absence of diarrhea, since examinations can be performed easily, have a low cost, and can disclose treatable conditions (28).

Biopsies reveal a markedly abnormal mucosa, with short villi, hypertrophied crypts, and infiltration of the lamina propria with eosinophils, neutrophils, and round cells.

Extraintestinal infections in AIDS patients have been reported. Microscopic findings associated with *I. belli* infection were seen in the lymph nodes and walls of both the small and large intestines. *I. belli* has also been seen in the liver and spleen. Confirmation of merozoites within the lymphatic channels documents a means of dissemination to lymph nodes and other tissues.

Charcot-Leyden crystals derived from eosinophils have also been found in the stools of patients with *I. belli* infection. Diarrhea and other symptoms may continue in immunocompromised patients, even those on immunosuppressive therapy, when the regimen of therapy is discontinued.

Organism description. Schizogonic and sporogonic stages in the life cycle of *I. belli* have been found in human intestinal mucosal biopsy specimens. Development in the intestine usually occurs within the epithelial cells of the distal duodenum and proximal jejunum. Eventually, oocysts are passed in the stool; they are long and oval and measure 20 to 33 μm by 10 to 19 μm. Usually, the oocyst contains only one immature sporont, but two may be present. Continued development occurs outside the body, with the development of two mature sporocysts, each containing four sporozoites, which can be recovered from the fecal specimen.

Clinical aspects. Clinical symptoms include diarrhea, which may last for long periods (months to years), weight loss, abdominal colic, and fever; diarrhea is the main symptom. Bowel movements (usually 6 to 10 per day) are watery to soft, foamy, and offensive smelling, suggesting a malabsorption process. Eosinophilia is found in many patients, recurrences are quite common, and the disease is more severe in infants and young children.

Patients who are immunosuppressed, particularly those with AIDS, often present with profuse diarrhea associated with weakness, anorexia, and weight loss. It has been recommended that physicians consider *I. belli* in AIDS patients with diarrhea who have immigrated from or traveled to Latin America, are Hispanics born in the United States, are young adults, or have not received prophylaxis with TMP-SMX for *Pneumocystis jiroveci*. It has also been recommended that AIDS patients who travel to Latin America and other developing countries be advised of the waterborne and food-borne transmission of *I. belli* and may want to consider chemoprophylaxis.

Relapse tends to be common in both immunocompetent and immunosuppressed patients and is believed to be associated with extraintestinal stages (44). These unizoite tissue cysts in extraintestinal sites have been confirmed in the liver, spleen, and mesenteric lymphoid tissues in AIDS patients (44, 89).

Isospora belli infection is frequent in AIDS patients in tropical areas. It has also been reported for other immunodepressive diseases, such as lymphoblastic leukemia, adult T-cell leukemia, and Hodgkin's disease. Non-Hodgkin's lymphoma with chronic diarrhea due to *I. belli* has also been reported (118). *I. belli* can cause severe chronic diarrhea in patients with malignancies whose country of origin is in an area of endemicity. Isosporiasis should be suspected in HIV-infected patients from tropical countries with diarrhea, weight loss, eosinophilia, and low $CD4^+$ cell counts. Persistent nonbloody diarrhea, similar to that seen with microsporidiosis or cryptosporidiosis, is the major symptom seen in immunocompromised individuals.

Diagnosis. Wet mount examination of fresh stool material, either as a direct smear or as concentrated material, is recommended rather than a permanent stained smear. The oocysts are very pale and transparent and can easily be overlooked. Oocysts can also be difficult to see if the concentration sediment is from polyvinyl alcohol-preserved stool. The light level should be reduced, and additional contrast should be obtained with the microscope for optimal examination conditions. Because of the small numbers of organisms present, it is possible to have a positive biopsy specimen but not recover the oocysts in the stool. These oocysts are acid fast and can also be demonstrated by using auramine-rhodamine stains. Organisms tentatively identified by using auramine-rhodamine stains should be confirmed by wet smear examination or acid-fast stains, particularly if the stool contains other cells or excess artifact material (more normal stool consistency) (46).

Therapy. Organism eradication is effective using cotrimoxazole, TMP-SMX, pyrimethamine-sulfadiazine, primaquine phosphate-nitrofurantoin, and primaquine phosphate-chloroquine phosphate (1, 46). The drug of choice is TMP-SMX, which is classified as an investigational drug for treatment of this infection. In patients allergic to sulfonamides, pyrimethamine alone has cured infections. In immunosuppressed patients with recurrent or persistent infection, therapy must be continued indefinitely.

Epidemiology and control. *I. belli* is the only species of *Isospora* that infects humans, and no other reservoir hosts are known. Transmission is through ingestion of water or food contaminated with mature, sporulated oocysts. Sexual transmission by direct oral contact with the anus or perineum is also possible, although this mode of transmission is probably much less common. The oocysts are very resistant to environmental conditions and may remain viable for months if kept cool and moist. Oocysts mature within about 48 h following stool evacuation and are then infectious. Prevention involves improved personal hygiene measures and sanitary conditions to eliminate fecal-oral transmission from contaminated food, water, and possibly environmental surfaces.

Sarcocystis spp.

Introduction. *Sarcocystis* spp. require two hosts for completion of the life cycle. Two well-described species are *Sarcocystis bovihominis* (cattle) and *Sarcocystis suihominis* (pigs). When raw or poorly cooked meat from infected animals is ingested by humans, gamogony occurs within the intestinal cells, with the production of sporocysts in the stool. In this case, humans who have ingested meat containing the mature sarcocysts serve as the definitive hosts.

When humans serve as the accidental intermediate host, they accidentally ingest oocysts from other animal stool sources. The sarcocysts that develop in human muscle (schizogony) apparently do little, if any, harm (Fig. 6). There is essentially no inflammatory response to these stages in the muscle and no conclusive evidence of pathogenicity of the mature sarcocyst. A number of different morphologic types of skeletal and cardiac muscle sarcocysts have been recovered from humans; sarcocysts have also been found in muscles of the larynx, pharynx, and upper esophagus. Generally, histopathologic diagnoses are myositis with vasculitis and, sometimes, myonecrosis. No specific therapy is known for this type of infection.

Organism description. The sporocysts that are recovered in the stool are broadly oval, measuring 9 by 16 μm and containing four mature sporozoites and the residual body. Normally, two sporocysts are contained within the oocyst (similar to the case for *I. belli*); however, in *Sarcocystis* infections, the sporocysts are released from the oocyst and normally are seen singly. They tend to be larger than *Cryptosporidium* oocysts, which contain four sporozoites, and so there should be no confusion between the two (42).

Clinical aspects. When humans serve as the definitive host, there have been reports of fever, severe diarrhea, abdominal pain, and weight loss in immunocompromised hosts. Eosinophilic enteritis and ulcerative obstructive enterocolitis may be seen as occasional complications. In one group of Thai laborers, a 26.6% prevalence of *Sarcocystis* infection was seen. This high prevalence rate was indicative of the local habit of eating raw beef and pork; sarcocystosis could be a significant food-borne zoonotic infection in Thailand (153).

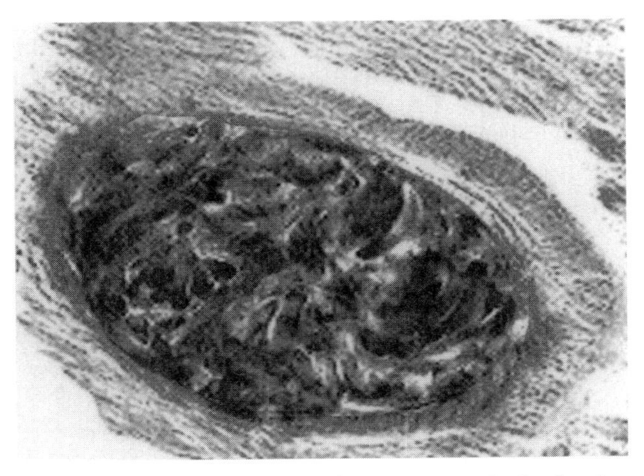

Figure 6. *Sarcocystis* sp. in muscle tissue (note the bradyzoites contained within the sarcocyst).

When humans serve as the accidental intermediate host, patients may present with lumps, pain in the limbs, or a discharging sinus. Histologic examination of biopsy specimens in all cases reveals characteristic cysts of *Sarcocystis* spp.

Diagnosis. Generally, histopathology with biopsy specimens is recommended. Oocysts would be recovered in the fecal concentration sediment from the routine O&P examination.

Therapy. Although a number of drugs have been tried, no specific therapy is known for the muscle stages of *Sarcocystis* spp. There is no treatment available for the intestinal symptoms with passage of sporocysts when the human serves as the definitive host.

Epidemiology and control. For infections in which humans serve as definitive hosts, prevention requires thorough cooking of beef and pork. For infections in which humans are intermediate hosts, preventive measures involve careful disposal of animal feces possibly containing infective sporocysts.

Microsporidia

Introduction

The microsporidia are obligate intracellular parasites that have been recognized in a variety of animals, particularly invertebrates; the organisms found in humans tend to be quite small, ranging from 1.5 to 2 μm (Fig. 7). With an increased understanding of AIDS within the immunosuppressed population, much greater attention has been focused on these organisms.

Phylogenetic studies have confirmed that the microsporidia evolved from the fungi, being most closely related to the zygomycetes (85). However, at this point, clinical and diagnostic issues and responsibilities may remain with parasitologists, and these organisms are maintained as a part of this presentation.

On the basis of serologic studies, it is very likely that immunocompetent humans have persistent or chronic infections with microsporidia. This is also supported by the fact that patients who are HIV seronegative with microsporidiosis show clinical resolution of their infections after a few weeks. As improved diagnostic methods, such as PCR, are developed and used, verification of human subclinical microsporidial infections should become possible.

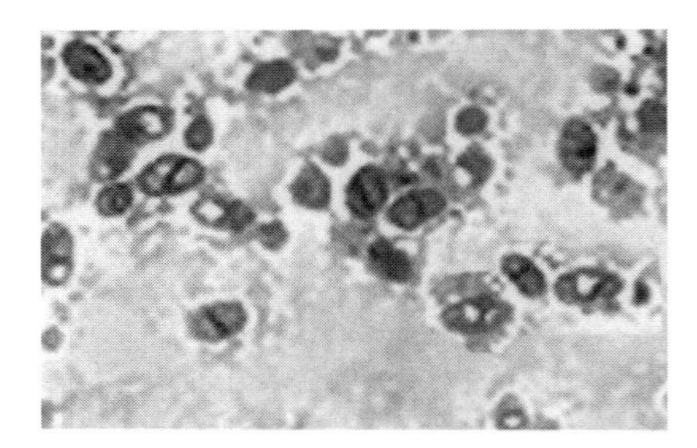

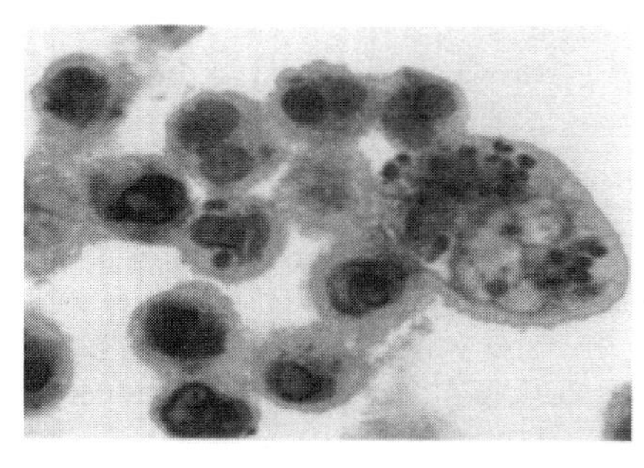

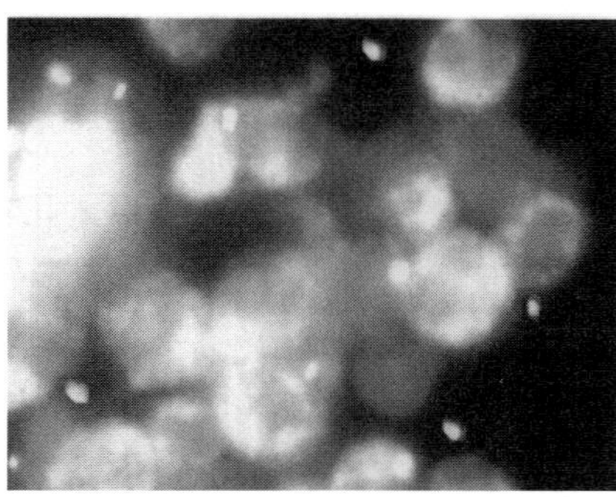

Figure 7. Microsporidia. (Top) Microsporidial spores seen in fecal specimen (concentration sediment) stained with Ryan blue modified trichrome stain (note the horizontal line through some of the spores, representing the presence of the polar tubule); (middle) spores stained with Gram stain reagents for the detection of intracellular spores; (bottom) spores within a urine sediment after being stained with an optical brightening agent (note that some of the spores are intracellular, while some are outside the cells).

In the immunocompromised host, microsporidial infection may lead to overwhelming disease and death (34, 36, 63, 123, 132). The first cases of human microsporidiosis were identified in children with impaired immune systems, and infections have been recognized widely and studied in individuals with AIDS, primarily those with CD4$^+$ T-lymphocyte counts of <100/μl. Microsporidial infections are increasingly being reported for patients following solid organ transplantation, where the main symptom has been diarrhea (23). The first case of pulmonary microsporidial infection in an allogeneic bone marrow transplant recipient in the United States, and only the second case in the world, was recently reported (102).

Organism description

Microsporidia are characterized by having spores containing a polar tubule, which is an extrusion mechanism for injecting the infective spore contents into host cells. To date, seven genera have been recognized in humans, including the more common genera, *Encephalitozoon* and *Enterocytozoon*, and the less common genera, *Brachiola*, *Pleistophora*, *Trachipleistophora*, *Vittaforma*, and *Microsporidium*, a catch-all genus for organisms that have not yet been classified (or may never be classified due to a lack of specimens).

Infection occurs with the introduction of infective sporoplasm through the polar tubule into the host cell. The life cycle includes repeated divisions by binary fission (merogony) or multiple fission (schizogony) and spore production (sporogony). Both merogony and sporogony can occur in the same cell at the same time. During sporogony, a thick spore wall is formed, thus providing environmental protection for this infectious stage of the parasite. An example of infection potential is illustrated by *Enterocytozoon bieneusi*, an intestinal pathogen. The spores are released into the intestinal lumen and are passed in the stool. These spores are environmentally resistant and can then be ingested by other hosts.

Clinical aspects

Enterocytozoon bieneusi. Chronic intractable diarrhea, fever, malaise, and weight loss are symptoms of *E. bieneusi* infection and are similar to those seen with cryptosporidiosis or isosporiasis. AIDS patients generally have four to eight watery, nonbloody stools each day, along with nausea and anorexia. There may be dehydration with mild hypokalemia and hypomagnesia, as well as D-xylose and fat malabsorption. Patients tend to be severely immunodeficient, with CD4 counts always below 200 and often below 100/μl (75). *E. bieneusi* infection has also been implicated in AIDS-related sclerosing cholangitis (30).

***Encephalitozoon* spp.** Both *Encephalitozoon cuniculi* and *Encephalitozoon hellem* have been isolated from human infections, with the first species being isolated from the central nervous system and the second being isolated from the eye (85).

A case of keratoconjunctivitis and chronic sinusitis due to infection with *E. cuniculi* was reported for an AIDS patient and confirmed by electron microscopy (85). Several eye infections with *E. hellem* have been reported for AIDS patients, including the first reported case, infecting not only sinuses and conjunctivae but also the nasal epithelium. In the first case of disseminated *E. hellem* infection in an AIDS patient, the autopsy revealed organisms in the eyes, urinary tract, and respiratory tract. The finding of numerous organisms within the lining epithelium of almost the entire length of the tracheobronchial tree suggested respiratory acquisition. Some of the eye infections have also suggested a topical route rather than dissemination.

Nodular cutaneous microsporidiosis has also been seen in a patient with AIDS; electron microscopy confirmed the genus as *Encephalitozoon*. The patient was treated successfully with long-term oral clindamycin (69).

Dual infections with *E. bieneusi* and *Encephalitozoon intestinalis* have been reported. *Encephalitozoon intestinalis* infects primarily small intestinal enterocytes, but infection does not remain confined to epithelial cells. *E. intestinalis* is also found in lamina propria macrophages, fibroblasts, and endothelial cells. Dissemination to the kidneys, lower airways, and biliary tract appears to occur via infected macrophages. Fortunately, these infections tend to respond to therapy with albendazole, unlike infections with *E. bieneusi*.

Other microsporidian species. *Brachiola connori* (formerly referred to as *Nosema connori*) was identified in human tissues from a 4-month-old infant with combined immunodeficiency disease with a disseminated, fatal infection. Parasites were found in the myocardium, diaphragm, arterial walls, kidney tubules, adrenal cortex, liver, and lungs. *Brachiola algerae* (formerly *Nosema algerae*) is apparently transmitted by the mosquito. Infections have been found in the eye and muscle, possibly from a disseminated case. Both immunocompromised (HIV negative, but receiving immunosuppressive therapy for rheumatoid arthritis) and immunocompetent patients have become infected. *Brachiola vesicularum* has also been associated with AIDS and myositis. Muscle tissue obtained at biopsy and examined by light and electron microscopy revealed organisms developing in direct contact with muscle cell cytoplasm and fibers.

Microsporidia in the genera *Pleistophora* and *Trachipleistophora* have rarely been identified in humans. However, when *Pleistophora* was found, atrophic and degenerating muscle fibers were full of spores, which were seen in clusters of about 12 organisms, with each cluster enclosed by an enveloping membrane, the pansporoblastic membrane. The disease was characterized by a 7-month history of progressive generalized muscle weakness and contractions, in addition to fever, generalized lymphadenopathy, and an 18-kg weight loss. The presence of HIV was never demonstrated in this patient. *Trachipleistophora hominis* causes severe myositis and sinusitis and was first described as infecting an AIDS patient. *Trachipleistophora antropophtera* is the most recently described microsporidian infecting humans and was found in brain tissues from two AIDS patients at autopsy; organisms were also found in the kidneys and heart.

Diagnosis

Although microsporidia can be identified in routine histologic preparations, the organisms do not tend to stain predictably. In formalin-fixed, paraffin-embedded, routine hematoxylin-and-eosin-stained sections, the spores may take on a refractile gold appearance. Since these changes in appearance may be due to the use of formalin, alternative fixatives are also being tried. Spores are occasionally seen very well by using the PAS stain, the methenamine silver stain, tissue Gram stains, or acid-fast stains. The spore has a small, PAS-positive posterior body, the spore coat will stain with silver, and the spores are acid-fast variable. Tissue examination by electron microscopic techniques is still considered the best approach; however, this option is not available to all laboratories and is not that sensitive due to the small amount of tissue sampled.

Touch preparations of fresh biopsy material that are air dried, methanol fixed, and Giemsa stained have been used; however, screening must be performed using a ×100 oil immersion objective. Another study, involving cytocentrifugation followed by Giemsa staining, found that 27% of 55 AIDS patients with chronic diarrhea were positive for *E. bieneusi*; all results were confirmed by electron microscopy.

Routine staining of stool, urine, or eye specimens can be performed using a modified trichrome stain in which the chromotrope 2R component added to the stain is 10 times the concentration normally used in the routine trichrome stain for stool. The stool preparations must be very thin; the staining time is 90 min, and the slide must be examined at a magnification of ×1,000 (or higher). Unfortunately, there tend to be many artifacts within stool material that mimic microsporidian spores, which are oval, stain pinkish with trichrome, and measure approximately 1.5 to 3 μm.

The newest approach for the identification of spores in clinical specimens uses antisera in an indirect fluorescent-antibody (IFA) procedure. Fluorescing microsporidial spores are distinguished by a darker cell wall and by internal visualization of the polar tubule as diagonal lines or cross lines within the cell. However, the reagents are not yet available commercially.

Another approach involves the use of chemofluorescent agents (optical brightening agents), such as calcofluor, Fungi-Fluor, and Uvitex 2B. These reagents are sensitive but nonspecific. Objects other than microsporidial spores will also fluoresce, and this is a problem in examining stool specimens. False-positive and false-negative results have been seen. When these reagents are used with other body fluids, particularly urine, the interpretation of results is much easier than when they are used with stool, which contains numerous artifacts (46). However, in spite of potential diagnostic problems, this method is recommended as a simple, sensitive screening method for the detection of microsporidial spores in stool specimens.

Although PCR methods have been used in the research setting, commercial products are not yet available (100, 138). For HIV-infected patients suspected of having microsporidiosis but with negative stool and biopsy results, PCR testing of intestinal biopsy specimens may be helpful in diagnosing the infection (148). Simultaneous detection of four human-pathogenic microsporidian species from human clinical samples has also been accomplished using an oligonucleotide microarray approach (150).

Serologic evidence strongly suggests that latent infections with microsporidia occur in many groups, ranging from patients with other infections, such as tuberculosis, typhoid fever, leprosy, malaria, schistosomiasis, toxoplasmosis, Chagas' disease, and toxocariasis, to healthy individuals and a group of homosexual men (59). Microsporidian infections in the immunocompetent patient may result in self-cure following mild symptoms over a short time frame, a situation similar to that seen with both cryptosporidiosis and isosporiasis.

Therapy

Metronidazole, itraconazole, octreotide, primaquine, lomotil, sulfasalazine, loperamide, and albendazole have been used in various patients. In some cases, the diarrhea subsided; however, biopsy specimens showed the continued presence of organisms, which were probably *Enterocytozoon bieneusi*. Evidence now indicates that a complete parasitological cure is possible with albendazole. However, these patients had disseminated infections with *Encephalitozoon* (*Septata*) *intestinalis*; albendazole appears to be very effective in treating this particular organism (1). Although static rather than parasiticidal effects are seen with *E. bieneusi* infections, treatment with albendazole results in diminished symptoms in as many as 50% of patients infected with this organism. Albendazole as a systemic agent is also recommended if the organisms have been confirmed in urine or nasal smears.

Fumagillin solutions applied topically have been used in corneal infections. However, the effects of this drug are static rather than parasiticidal, and relapses of infection occur when the treatment has been discontinued. Itraconazole can also be used to treat ocular, nasal, and paranasal sinus infections caused by *E. cuniculi* parasites when albendazole fails.

Epidemiology and control

Although sources of human infections are not yet totally defined, possibilities include human-to-human transmission and animal-to-human transmission. Primary infection can occur by inhalation or ingestion of spores from environmental sources or by zoonotic transmission. The presence of *Encephalitozoon intestinalis* has been confirmed for tertiary sewage effluent, surface water, and groundwater; *Enterocytozoon bieneusi* was confirmed in surface water; and *Vittaforma corneae* was confirmed in tertiary sewage effluent (32). The data certainly support the possibility that these parasites may be waterborne pathogens. Infective spores in human clinical specimens support the use of precautions in handling body fluids, and personal hygiene measures such as hand washing may be important in preventing primary infections in the health care setting.

Free-Living Amebae

Introduction

Free-living amebae are recognized as important parasitic pathogens, particularly in immunocompromised patients. Primary amebic meningoencephalitis (PAM), caused by *Naegleria fowleri*, and granulomatous amebic encephalitis (GAE), caused by *Acanthamoeba* spp., *Balamuthia mandrillaris* (including cases in patients with AIDS), and *Sappinia diploidia*, are now well recognized. When free-living, these organisms feed on bacteria and nutrients in moist soil and in fresh and marine waters.

Although both free-living and pathogenic protozoa harbor a variety of endosymbiotic bacteria, it is unclear what role these organisms play in terms of host survival, infectivity, and invasiveness. *Legionella pneumophila* can infect, multiply within, and kill both *Naegleria* and *Acanthamoeba* amebae.

These amebae may be natural hosts for *Legionella* as well as for other bacteria, such as *Listeria monocytogenes*, *Vibrio cholerae*, *Mycobacterium leprae*, and *Pseudomonas aeruginosa*. Because of their well-known resistance to chlorine, amebic cysts are considered to be vectors for these intracellular bacteria. This can have tremendous significance for any hospital where the water source is contaminated with free-living amebae. Evidence also suggests that the presence of amebae in domestic water supplies may provide growth conditions that enhance the pathogenicity of the organism for the human host. These amebae have also been found to be acceptable hosts for echoviruses and polioviruses. However, the epidemiology, immunology, protozoology, pathology, and clinical features of the infections produced by these free-living protozoa are quite different.

Naegleria fowleri

Introduction. Infections of the central nervous system (CNS) caused by free-living amebae have been recognized since the mid-1960s. *N. fowleri* causes PAM, a fulminant and rapidly fatal disease that affects mainly children and young adults (Fig. 8). The disease can mimic bacterial meningitis but is caused by this free-living ameba, an organism normally found in moist soil and freshwater habitats. Although PAM is usually associated with healthy, young individuals with a history of recent water-related sport activities, it is an important infection to consider in both immunocompetent and immunocompromised individuals; however, almost all infections have been seen in otherwise healthy individuals. Close to 200 cases of PAM have occurred worldwide, and approximately 90 of those cases have been reported from the United States.

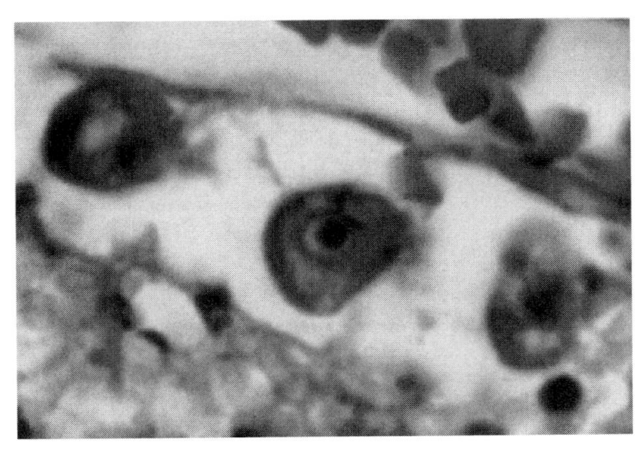

Figure 8. *Naegleria fowleri* trophozoites within brain tissue (note the large karyosome within the nucleus). (Courtesy of Armed Forces Institute of Pathology.)

Extensive tissue damage occurs along the path of amebic invasion; the nasopharyngeal mucosa shows ulceration, and the olfactory nerves are inflamed and necrotic. Hemorrhagic necrosis is concentrated in the region of the olfactory bulbs and the base of the brain. Organisms can be found in the meninges, perivascular spaces, and sanguinopurulent exudates.

Organism description. There are both trophozoite and cyst stages in the life cycle, and the stage depends primarily on environmental conditions. Trophozoites can be found in water or moist soil and can be maintained in tissue culture or other artificial media. The trophozoites can occur in two forms, ameboid and flagellate (46). The ameboid form (the only form recognized in humans) is elongated, with a broad anterior end and a tapered posterior end. The size ranges from 7 to 20 μm. The diameter of the rounded forms is usually 15 μm. There is a large central karyosome and no peripheral nuclear chromatin. The cytoplasm is somewhat granular and contains vacuoles. Ameboid-form organisms change to the flagellate form when they are transferred from culture or teased from tissue into water and maintained at a temperature of 27 to 37°C. The flagellate form is pear shaped, with two flagella at the broad end. Motility is typical, with either spinning or jerky movements. The flagellate-form organisms do not divide, but when the flagella are lost, the ameboid forms resume reproduction. Cysts from nature and from agar cultures look the same and have a single nucleus almost identical to that seen in the trophozoite. The cysts are generally round, measuring from 7 to 10 μm in diameter, and there is a thick double wall.

Clinical aspects. Amebic meningoencephalitis caused by *N. fowleri* is an acute, suppurative infection of the brain and meninges. With very rare exceptions, in humans the disease is rapidly fatal. The period between contact with the organism and onset of clinical symptoms, such as fever, headache, and rhinitis, may vary from 2 to 3 days to as long as 7 to 15 days.

Amebae enter the nasal cavity by inhalation or aspiration of water, dust, or aerosols. Through phagocytosis of the olfactory epithelial cells, the organisms penetrate the nasal mucosa and migrate via the olfactory nerves to the brain. Early symptoms include vague upper respiratory distress, headache, lethargy, and occasionally olfactory problems. Cysts of *N. fowleri* are generally not seen in brain tissue. The acute phase includes sore throat; stuffy, blocked, or discharging nose; and severe headache.

Progressive symptoms include pyrexia, vomiting, and stiffness of the neck. Mental confusion and coma usually occur approximately 3 to 5 days prior to death. The cause of death is usually cardiorespiratory arrest and pulmonary edema.

The first case of organ transplantation from a donor who had died of undiagnosed *N. fowleri* infection was reported in 1997 (125, 126). While no subsequent amebic infections occurred in the three organ recipients, this report emphasizes the need for adequate evaluation of the benefits and risks of transplanting tissues from persons whose illness might have been caused by an infectious agent. There was another instance in which kidneys were transplanted prior to determination that the donor was infected with *N. fowleri* (125, 126). The risk of transmission of *N. fowleri* by donor organs has not been clarified, and no practical test is available to ensure that donor organs are organism free. Also, no prophylactic drug regimen to treat transplant recipients has been established.

Diagnosis. PAM can mimic acute purulent bacterial meningitis and may be difficult to differentiate, particularly in the early stages. The cerebrospinal fluid (CSF) may have a predominantly polymorphonuclear leukocytosis, increased protein concentration, and decreased glucose concentration, like those seen with bacterial meningitis. If the CSF Gram stain is interpreted incorrectly (identification of bacteria as a false-positive result), the antibacterial therapy will have no effect on the amebae and the patient will usually die within several days.

Clinical and laboratory data usually cannot be used to differentiate pyogenic meningitis from PAM, so the diagnosis may have to be reached by a process of elimination. A high index of suspicion is often mandatory for early diagnosis. Most cases are associated with exposure to contaminated water. The rapidly fatal course of 4 to 6 days after the beginning of symptoms (with an incubation period of 1 day to 2 weeks) requires early diagnosis and immediate chemotherapy if the patient is to survive.

CSF analysis will show decreased glucose and increased protein concentrations. Leukocyte counts may range from several hundred to >20,000 cells/mm^3. Gram stains and bacterial cultures of CSF will be negative. A definite diagnosis could be made by demonstration of the amebae in the CSF or in biopsy specimens. Either CSF or sedimented CSF should be placed on a slide under a coverslip and observed for motile trophozoites; smears can also be stained with Wright's or Giemsa stain. CSF, exudate, or tissue fragments can be examined by light microscopy or phase-contrast microscopy. It is very easy to confuse leukocytes and amebae, particularly when using a counting chamber to examine the CSF, hence the recommendation to use a regular slide and coverslip. Motility may vary, so the main differential characteristic is the spherical nucleus with a large karyosome.

Most cases are diagnosed at autopsy; confirmation of these tissue findings should include culture and/or special staining by use of monoclonal reagents in IFA procedures. Organisms can also be cultured on nonnutrient agar plated with *Escherichia coli*. In cases of presumptive pyogenic meningitis in which no bacteria are identified in the CSF, the computed tomography appearance of basal arachnoiditis (obliteration of basal cisterns in the precontrast scan, with marked enhancement after the administration of intravenous contrast medium) should indicate the possibility of acute PAM.

The amebae can be identified in histology preparations by using indirect immunofluorescence and immunoperoxidase techniques. The organism in tissue sections with routine staining looks very much like an *Iodamoeba bütschlii* trophozoite, with a very large karyosome and no peripheral nuclear chromatin. In general, serologic tests have not been helpful in the diagnosis of this infection. The disease progresses so rapidly that the patient is unable to mount an immune response.

Acanthamoeba spp.

Introduction. Another type of meningoencephalitis, GAE, caused by freshwater amebae, may occur as a subacute or chronic disease with focal granulomatous lesions in the brain (18, 79). The *Acanthamoeba* group has also been found to cause keratitis and corneal ulceration. These published cases emphasize the need to consider acanthamoebic infection in the differential diagnosis of eye infections that fail to respond to bacterial, fungal, or viral therapy. These infections are often due to direct eye exposure to contaminated materials or solutions.

The pathogenesis of *Acanthamoeba* infection involves the ability of the organism to invade tissues. This capability depends on adherence to mucosal surfaces, migration through tissues, and the release of oxygen radicals and proteases that can destroy connective tissue. Studies have found that alkaline cysteine proteinases are more active in pathogenic strains, whereas serine proteases are found in both pathogenic and nonpathogenic strains (79). Data also suggest that obligate bacterial endosymbionts are able to enhance the amebic pathogenic potential in vitro, although the exact mechanisms have not yet been identified (11). *Acanthamoeba* spp. produce a variety of proteases, which may facilitate cytolysis of the corneal epithelium, invasion of the extracellular matrix, and dissolution of the corneal stromal matrix (126).

Organism description. Unlike *N. fowleri*, *Acanthamoeba* spp. do not have a flagellate stage in the life cycle, only the trophozoite and cyst stages (Fig. 9). Motile organisms have spinelike pseudopods; however, progressive movement is usually not very evident. There is a wide range in size, with the average diameter of the trophozoites being 30 μm. The nucleus has the typical large karyosome, like that seen in *N. fowleri*. This morphology can be seen on a wet preparation. The cysts are usually round, with a single nucleus, and also have the large karyosome seen in the trophozoite nucleus. The double wall is usually visible, with the slightly wrinkled outer cyst wall and what has been described as a polyhedral inner cyst wall. This cyst morphology can be seen in organisms cultured on agar plates.

Clinical aspects. The incubation period of GAE is unknown; several weeks or months are probably necessary to establish the disease. The clinical course is subacute or chronic and is usually associated with trauma or underlying disease, not swimming. Symptoms of GAE include confusion, dizziness, drowsiness, nausea, vomiting, headache, lethargy, stiff neck,

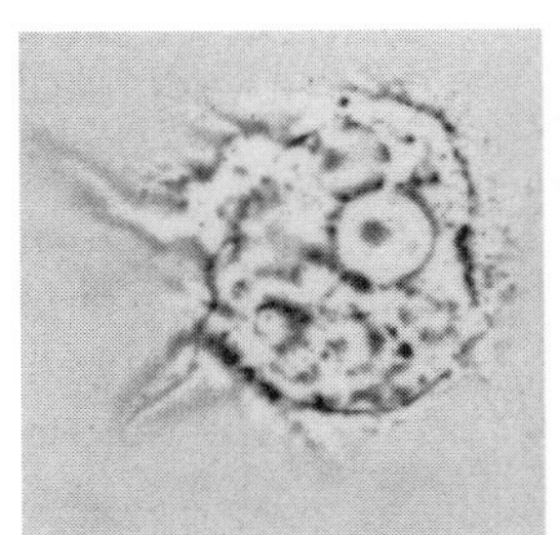

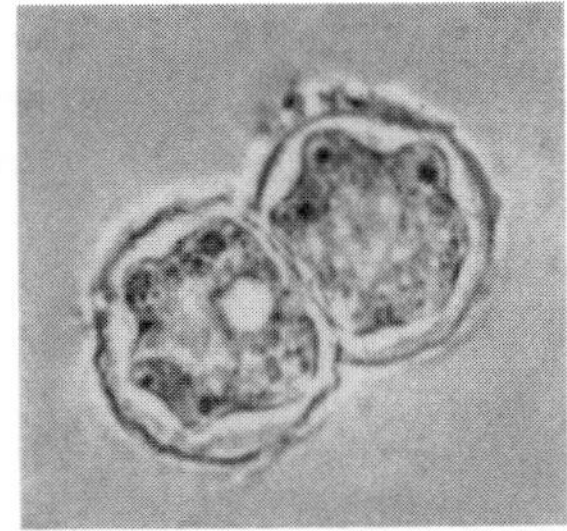

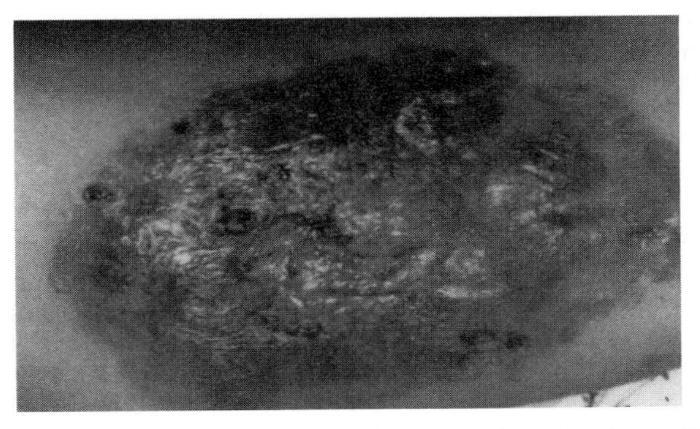

Figure 9. (Top left) *Acanthamoeba* sp. trophozoites (note the sharp, spiky pseudopodia); (top right) *Acanthamoeba* sp. cysts (note the hexagonal double wall); (bottom) skin lesion caused by *Acanthamoeba* sp. infection in an immunocompromised host.

seizures, and sometimes hemiparesis. Within the CNS, the cerebral hemispheres are the most likely tissue to be involved. There may be edema and hemorrhagic necrosis within the temporal, parietal, and occipital lobes. A chronic inflammatory exudate, comprised mainly of polymorphonuclear leukocytes and mononuclear cells, can be seen over the cortex.

Unlike the case for PAM caused by *N. fowleri*, both trophozoites and cysts are found throughout the CNS lesions. The route of CNS invasion is thought to be hematogenous, with the primary site being the skin or lungs. After several days of lethargy, acute onset of fever, headache, and pain in the neck may occur. Conditions associated with GAE include skin ulcers, liver disease, pneumonitis, diabetes mellitus, renal failure, rhinitis, pharyngitis, and tuberculosis (98, 137, 144). Predisposing factors include alcoholism, pregnancy, systemic lupus erythematosus, hematologic disorders, AIDS, chemotherapy, radiation therapy, and steroid treatment (73). This infection has become more widely recognized in AIDS patients, particularly those with low $CD4^+$ cell counts. Although *Acanthamoeba* infections stimulate a granulomatous response, the response in AIDS patients is minimal or absent, consistent with the poor immune response in these patients.

Keratitis, uveitis, and corneal ulceration have been associated with *Acanthamoeba* spp. Infections have been seen in both hard- and soft-lens wearers, and particular attention has been paid to soft-lens disinfection systems, including homemade saline solutions.

Decreased corneal sensation has contributed to the misdiagnosis of *Acanthamoeba* keratitis as herpes simplex keratitis; this mistake can also be attributed to the presence of irregular epithelial lesions, stromal infiltrative keratitis, and edema, which are commonly seen in herpes simplex keratitis. *Acanthamoeba* keratitis may be present as a secondary or opportunistic infection in patients with herpes simplex keratitis. Unfortunately, as a result, treatment can be delayed for weeks to months. Severe pain, nonhealing corneal ulcers, and the presence of ring infiltrates are also clinical signs that alert the ophthalmologist to the possibility of amebic infection.

It is important to consider *Acanthamoeba* infection in the differential diagnosis of uveitis in patients with AIDS (55). It is also important to remember that incisional keratotomy may predispose the cornea to delayed-onset infectious keratitis; *Acanthamoeba* should always be considered a possible cause of infection, and clinical specimens should be cultured in refractory cases.

Cutaneous infections are more common in patients with AIDS, regardless of the presence or absence of CNS involvement. The disease includes the presence of hard erythematous nodules or skin ulcers. Early lesions appear as firm papulonodules that drain purulent material; these lesions then develop into nonhealing indurated ulcers. Although disseminated skin lesions may be the first sign of *Acanthamoeba* infection, it is unclear whether these lesions represent a primary focus or may result from hematogenous spread from other body sites. The mortality rate for these individuals without CNS involvement is around 75%, but it increases to 100% if cutaneous infection is accompanied by CNS disease (106).

Diagnosis. Culture has not proven to be effective with GAE cases. However, for eye infections, the most effective approach uses nonnutrient agar plates with Page's saline and an overlay of *E. coli* on which the amebae feed. Specimens transported in ameba saline (5.0 ml) and filtered through 13-mm-diameter, 0.22-μm-pore-size cellulose acetate and nitrate filters (Millipore Corp., Bedford, MA) have also been acceptable for organism recovery. The filter is placed in the center of the nonnutrient agar plate seeded with *E. coli*. Tissue stains are also effective, and cysts can be stained with Gomori's silver methenamine, PAS stain, and calcofluor white (46).

Identification of acanthamoebae in ocular and other tissues can be difficult; in histologic preparations, the organisms look much like keratoplasts, as well as neutrophils and monocytes. Up to 70% of clinical *Acanthamoeba* keratitis cases are misdiagnosed as viral keratitis. Also, the mean time to diagnosis of keratitis due to *Acanthamoeba* infection can average 2.5 weeks longer for non-contact-lens wearers than for contact lens users. The availability of rapid, accurate, and relatively simple diagnostic tests would increase the timely application of appropriate chemotherapy. Genetic markers have also been developed and used to identify pathogenic *Acanthamoeba* strains associated with nonkeratitis infections.

Cutaneous lesion material can be inoculated onto various growth media, including nonnutrient agar plates with a bacterial overlay. Inoculation of mammalian cell culture monolayers is also possible. Histologic stains such as hematoxylin and eosin, PAS stain, and calcofluor white can be used. Also, DNA-based molecular methods, such as restriction fragment length polymorphism and randomly amplified DNA analysis, can be used.

Therapy. Trophozoites and cysts of *Acanthamoeba* are sensitive in vitro to ketoconazole, pentamidine, hydroxystilbamidine, paromomycin, flucytosine, polymyxin, sulfadiazine, TMP-SMX, azithromycin, and extracts of medicinal plants, especially with combinations of these drugs (1, 46, 79). Also, although steroids have been used in the past for cerebral edema and inflammation in CNS infections, this approach should not be used, since the steroids tend to exacerbate the *Acanthamoeba* infection.

Encephalitis has been treated with antimicrobial combinations, including sterol-targeting azoles (clotrimazole, miconazole, ketoconazole, fluconazole, and itraconazole), pentamidine isethionate, flucytosine, and sulfadiazine. The use of drug combinations helps to address resistance that may exist or occur during treatment (86).

If keratitis is correctly diagnosed early and the epithelium alone is involved, debridement may be sufficient for cure. However, if treatment is delayed and the organisms invade the cornea or tissues below the cornea, therapy may need to be continued for many months to a year or longer, and treatment failures tend to be fairly common. In vitro susceptibility testing may be very helpful in these cases (86).

Epidemiology and control. General preventive measures are similar to those for *N. fowleri*. *Naegleria* spp. are generally susceptible to swimming pool levels of chlorine, but *Acanthamoeba* spp. are more resistant (66). The recovery of *Acanthamoeba* spp. in nasal isolates and pharyngeal swabs may indicate human introduction of the organisms into swimming pools. A number of factors probably play a role in the increased incidence of infection, including the large number of HIV-infected individuals and more patients undergoing cancer chemotherapy or immunosuppressive therapy for organ transplantation.

Balamuthia mandrillaris

Introduction. *Balamuthia mandrillaris* was originally thought to be another harmless soil organism, unlikely or unable to infect mammals. However, in studies from the Centers for Disease Control and Prevention (CDC), antisera to the leptomyxid amebae have been shown to react in indirect immunofluorescence assays with tissues from cases of GAE in both animals and humans. There have been over 70 cases worldwide, some of which have been reported for AIDS patients (125, 126, 149).

No characteristic clinical symptoms, laboratory findings, or radiologic indicators are diagnostic for GAE. Neuroimaging findings show heterogeneous, hyperdense, nonenhancing, space-occupying lesions. Whether single or multiple, they involve mainly the cerebral cortex and subcortical white matter. These findings suggest a CNS neoplasm, tuberculoma, or septic infarcts.

Organism description. The disease is very similar to GAE caused by *Acanthamoeba* spp. and has an unknown incubation period (Fig. 10). Trophozoites and cysts of *B. mandrillaris* are found in many of the same CNS tissues as those of *Acanthamoeba*. Although differentiation of these two organisms in tissue by light microscopy is difficult, in some tissue sections *B. mandrillaris* appears to have more than one nucleolus in the nucleus. The trophozoites have extensive branching and a single nucleus (occasionally binucleate forms are seen) with a central karyosome, and they measure 15 to 60 μm. The cysts have a single nucleus (occasionally binucleate forms are seen), have the typical double wall, with the outer wall being thick and irregular, and measure 15 to 30 μm.

Clinical aspects. The clinical course is subacute or chronic and is usually not associated with swimming in freshwater. Patients complain of headaches, nausea, vomiting, fever, visual disturbances, dysphagias, seizures, and hemiparesis. There may also be a wide range in terms of the clinical course, from a few days to several months. Death can occur from a week to several months after the onset of stroke-like symptoms.

Diagnosis. Electron microscopy and histochemical methods are required for definitive identification of *B. mandrillaris*. Organism isolation from some human cases of GAE has not been successful, and these amebae do not grow well on *E. coli*-seeded nonnutrient agar plates. In the diagnostic laboratory, these organisms can be cultured using mammalian cell cultures; some success has been obtained with monkey kidney cells and with MRC, HEp-2, and diploid macrophage cell lines. Through the use of primary cultures of human brain microvascular endothelial cells, amebae have been isolated from the brain and CSF (66). Serum antibodies to *B. mandrillaris* have been found in both adults and children; however, testing is generally limited to the CDC, where PCR has also been used for the detection of mitochondrial 16S rRNA gene DNA from the amebae in clinical specimens (157).

B. mandrillaris, formerly regarded as having no pathogenic potential, will continue to be identified as the etiologic agent of fatal meningoencephalitis in humans as well as animals. This opportunistic infection may also continue to cause disease in individuals with AIDS.

Therapy. *B. mandrillaris* is apparently susceptible to pentamidine isethiocyanate, while other studies indicate that ketoconazole, propamidine isethionate, clotrimazole, and certain biguanides have amebicidal activity (46). For one immunocompetent patient who survived, treatment included combination antibiotics (pentamidine, sulfadiazine, fluconazole, and clarithromycin).

Epidemiology and control. Although information on the environment of *B. mandrillaris* is scarce, it is assumed that its habitat is similar to that of other free-living amebae. Attempts to isolate *B. mandrillaris* from soil samples related to previous

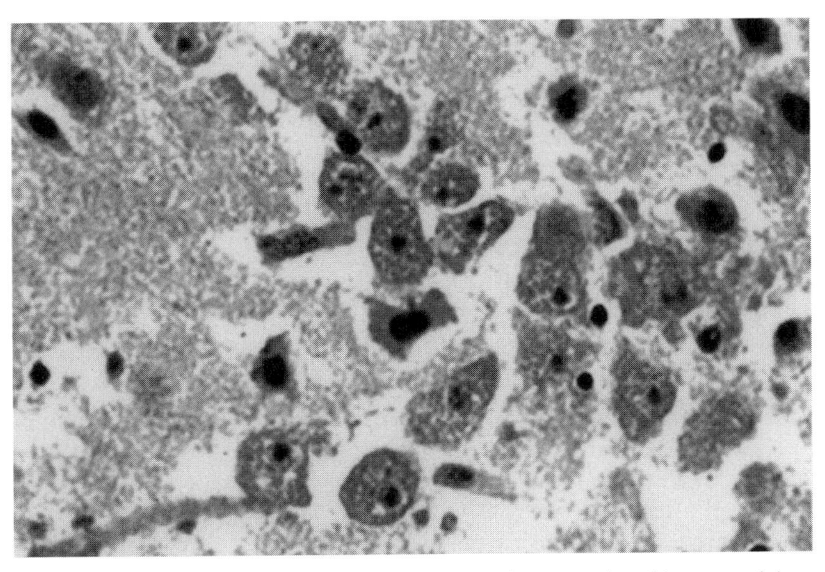

Figure 10. *Balamuthia mandrillaris* in brain tissue; both trophozoites and cysts are found in many of the same CNS tissues as those which harbor *Acanthamoeba* organisms. (Courtesy of Armed Forces Institute of Pathology.)

encephalitis cases have been unsuccessful. However, following the death of a 3-year-old northern California resident from amebic encephalitis caused by *Balamuthia*, environmental samples were collected around the child's home and outdoor play areas. An ameba consistent with *Balamuthia* was isolated from the soil found in a potted plant in the home (124). The isolation of *Balamuthia* amebae from soil affirms this organism's status as a free-living ameba; like *Acanthamoeba* and *Naegleria*, *Balamuthia* is an opportunistic pathogen and may also continue to cause disease in individuals with AIDS.

Toxoplasma gondii

Introduction

Toxoplasma gondii is found worldwide, and serologic data indicate that human infections are common, although most infections are benign or produce no symptoms. Serologic evidence from throughout the United States indicates that anywhere from 20 to 70% of the population is serologically positive for *T. gondii*. The percentage of serologically positive patients tends to be higher in the tropical areas of the world.

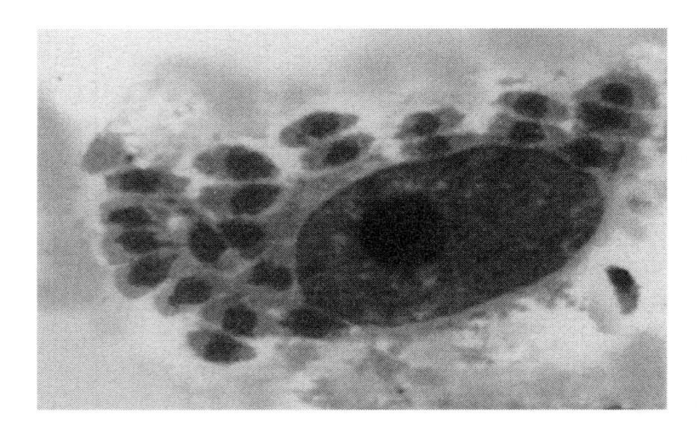

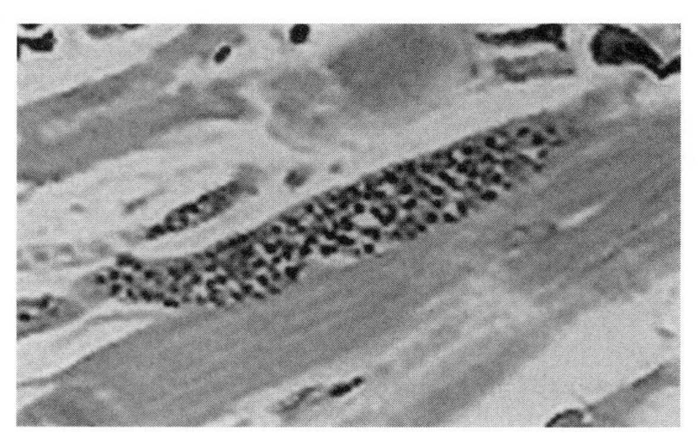

Figure 11. (Top) *Toxoplasma gondii* in tissue culture (note the somewhat crescent-shaped tachyzoites); (bottom) organisms in tissue.

Organism description

Humans usually become infected with *T. gondii* by ingesting raw or poorly cooked meats. Transmission also occurs from ingestion of the infected oocysts that are found in cat feces. In humans, *T. gondii* is found in two different stages. The actively proliferating intracellular forms are called trophozoites or tachyzoites and are crescent shaped (Fig. 11). Many different tissues may be parasitized by these organisms, particularly lung, heart, lymphoid organ, and CNS tissues. The resting forms or cyst stages are found in the tissues and contain the more slowly developing bradyzoites.

Pathogenesis

As tachyzoites proliferate, they invade adjacent cells from the original infected cell as it ruptures. Once the cysts are formed, the process becomes quiescent, with little or no multiplication and spread. In the immunocompromised or immunodeficient patient, a cyst rupture or primary exposure to the organisms often leads to lesions. The organisms can be disseminated via the lymphatics and the bloodstream to other tissues. Disintegration of cysts may give rise to clinical encephalitis in the presence of apparently adequate immunity.

Although an inflammatory process is observed during *Toxoplasma* encephalitis (TE) in HIV-positive patients, the major cause of CNS lesions is probably uncontrolled parasite multiplication rather than immunopathologic changes. Also, *T. gondii* is capable of enhancing HIV type 1 (HIV-1) replication within reservoir host cells, and simultaneously, HIV-1 itself undermines acquired immunity to the parasite, promoting reactivation of chronic toxoplasmosis.

Clinical aspects

Toxoplasmosis can be categorized into four groups, which include (i) disease acquired in the immunocompetent patient, (ii) acquired or reactivated disease in the immunosuppressed or immunodeficient patient, (iii) congenital disease, and (iv) ocular disease.

Toxoplasmosis varies from an asymptomatic, self-limiting infection to a fatal disease. In patients with congenital infections or in debilitated patients, underlying conditions may influence the final outcome of the infection (159, 160). In immunocompromised patients, the infection usually involves the nervous system, with diffuse encephalopathy, meningoencephalitis, or cerebral mass lesions. In a multiplex nested PCR analysis to genotype parasites in CSF, 8 of 10 HIV-positive patients had infections with type I strains, despite the fact that this lineage is normally uncommon in humans and animals (70, 77). Although rare, in some patients with AIDS, gastric toxoplasmosis may present as the primary manifestation (72, 88).

AIDS patients who become infected with *T. gondii* risk developing disease when their $CD4^+$ T-lymphocyte count falls below 100,000/ml. Fever and malaise usually precede the first neurologic symptoms; headache, confusion, seizures, or other focal signs strongly suggest the diagnosis of toxoplasmosis (45). In one study, disease was diagnosed 10 days to 74 months after the rise in *Toxoplasma* antibody titer. While the risk factors for development of toxoplasmosis remain incompletely defined, the importance of specific prophylaxis for patients with low CD4 cell counts and high *Toxoplasma* antibody titers is supported by these findings (9).

TE has been reported as a life-threatening opportunistic infection among patients with AIDS prior to the use of HAART. This condition is fatal if untreated. Most AIDS patients with TE will demonstrate IgG antibodies to *T. gondii*; however, $<5\%$ may not have antibody in their serum. Psychiatric manifestations of *T. gondii* are also seen in immunocompromised individuals with AIDS in whom latent infections have become reactivated. Altered mental status may occur in approximately 60% of patients, with symptoms including delusions, auditory hallucinations, and thought disorders (52).

In transplant recipients, disease severity depends on previous exposure to *T. gondii* by the donor and recipient, the type of organ transplanted, and the level of immunosuppression of the patient. Disease can occur from reactivation of a latent infection or from acute primary infection acquired directly from the transplanted organ. Stem cell transplant recipients are quite susceptible to severe toxoplasmosis, primarily due to reactivation of a latent infection (71, 83, 84, 161). If stem cell transplant patients have positive serology prior to transplantation, they are at risk for severe disseminated toxoplasmosis (37, 39, 87). All potential transplant recipients should be tested for *Toxoplasma*-specific IgG antibodies to determine their antibody status. An individual with acute acquired infection often produces detectable IgG and IgM antibodies, while those with reactivation may or may not have an increase in IgG antibodies and normally will not demonstrate an IgM response. Seronegative cardiac transplant recipients who receive an organ from a seropositive donor may develop toxoplasmic myocarditis; this disease presentation may also mimic organ rejection. Although it has been stated that prophylaxis and routine serologic evaluation of donors and recipients for *Toxoplasma* in noncardiac solid organ transplantation are not recommended, recent findings support a reconsideration of pretransplantation evaluation and prophylaxis strategies for solid organ transplant recipients (21).

Congenital infections result from the transfer of parasites from mother to fetus when the mother acquires a primary infection during pregnancy. In 90% of patients, no clinical symptoms are apparent during acute infection. If the mother acquires the infection during the first or second trimester of pregnancy, congenital infections may be particularly severe. At birth or soon thereafter, symptoms in these infants may include retinochoroiditis, cerebral calcification, and occasionally hydrocephalus or microcephaly. Symptoms of congenital CNS involvement may not appear until several years later.

Chorioretinitis in immunocompetent patients is generally due to an earlier congenital infection. Patients may be asymptomatic until the second or third decade, when cysts may rupture, with lesions then developing in the eye. The number of infected people who develop chorioretinitis later in life is unknown but may represent over two-thirds.

Chorea has been associated with cerebral toxoplasmosis and AIDS (109). Chorea is not unusual in AIDS, but the causes are variable, and careful neuroradiological and clinical evaluation is required to identify them. AIDS-related disease should be considered with young patients presenting with chorea without a family history of movement disorders.

Diagnosis

Toxoplasmosis can be diagnosed using various serologic procedures, histologic findings from the examination of biopsy specimens, and isolation of the organism, either in a tissue culture system or by animal inoculation. Individuals with positive serologic tests have been exposed to *Toxoplasma* organisms and may have the resting stages within their tissues. This is why histologic identification of the organism and recovery of the organism, either in a tissue culture system or by animal inoculation, may be misleading, since the organisms could be isolated yet might not be the etiologic agent of disease. Thus, serologic tests have been recommended as the diagnostic tests of choice (117, 128, 154).

In the case of reactivation, the value of serology is limited, but it can be used to diagnose a primary infection by showing seroconversion. In immunocompromised patients, the absence of specific antibodies does not exclude active disease. In some cases, IgG antibodies have a low avidity, which is thought to be indicative of recent infection. However, IgG can still be of low avidity after 10 months, which indicates that low-to-high-avidity switching in immunocompromised patients might be delayed (normally, 4 to 6 months), making interpretation of serology results even more difficult (156). When *T. gondii* IgG antibodies are present, archival serum samples can help to make a diagnosis of primary toxoplasmosis by demonstrating seroconversion, but antibodies can be absent early in the infection. For an immunodeficient or immunosuppressed patient, a presumptive diagnosis of toxoplasmosis can be made from an elevated serologic titer and the presence of the clinical syndrome, which would include neurologic symptoms. However, for certain patients with monoclonal gammopathies, titers to *T. gondii* may be extremely high but may not be the cause of the clinical condition (82). One of the most common diagnostic tests used for toxoplasmosis is the IFA procedure. Definitive diagnosis is usually accepted when there is a rising titer, an IgM IFA titer of at least 1:64, or an IgM ELISA titer of 1:256. An elevated serologic titer in conjunction with demonstration of the actual trophozoites or isolation of the organisms from spinal fluid is very significant.

Disseminated toxoplasmosis should be considered in the differential diagnosis of immunocompromised patients with culture-negative sepsis, particularly if it is combined with neurologic or respiratory symptoms or unexplained skin lesions. Examination of Wright's-stained peripheral blood smears or anti-toxoplasma immunoperoxidase studies of skin biopsy specimens may be diagnostic. Pulmonary toxoplasmosis in the absence of neurologic findings in HIV-positive patients emphasizes the importance of considering *Toxoplasma* as the cause of pulmonary disease in such patients with respiratory symptoms.

PCR has been used to detect *T. gondii* in deparaffinized ocular tissue sections, CSF, blood, and bronchoalveolar lavage fluid; however, the absence of *Toxoplasma* DNA does not exclude toxoplasmosis (71, 90). PCR can also be very helpful in evaluating certain infections; the sensitivity will be higher for confirming congenital toxoplasmosis from amniotic fluid but lower for patients with infection reactivation, such as HIV-positive patients (87). Despite the extensive cerebral lesions in some patients, PCR assays of brain biopsy and CSF samples have been negative. Sampling errors may be responsible. It has also been shown by PCR that *T. gondii* is only intermittently present in the CSF and that testing of repeated samples can elevate the assay's sensitivity (156).

Therapy

Drugs that have been found to be effective, either alone or in combination, include sulfonamides and pyrimethamine (for actively multiplying tachyzoites). Clindamycin and spiramycin have also been used under certain circumstances.

Although a combination of pyrimethamine and a sulfonamide is the standard regimen for treating TE in AIDS patients, this approach can be contraindicated for patients with a history of bone marrow suppression and severe allergic reactions to sulfonamides (7, 29, 120). The use of oral clindamycin and pyrimethamine is effective (1). Another combination regimen includes pyrimethamine-clarithromycin, which is comparable to the conventional therapeutic approach for acute TE in AIDS patients. Data also suggest that a combination of TMP-SMX (cotrimoxazole) may be as effective as pyrimethamine-sulfadiazine in AIDS patients with TE.

Epidemiology and control

Human infection can be acquired through ingestion or handling of infected meat or through ingestion of infective oocysts, which can remain viable within cool, moist soil for a year or longer. Certainly, hand washing is highly recommended when there has been potential exposure to the oocysts. It is recommended that meat be cooked until the internal temperature reaches 150°F (66°C). HIV-infected persons who are seronegative for *Toxoplasma* IgG should be counseled to protect themselves from primary infection by eating well-cooked meats and washing their hands after possible soil contact.

Cats kept as pets should be fed commercial or well-cooked food and should be kept indoors. Litter boxes should be cleaned daily, with feces disposed of in the toilet and the pans disinfected with boiling water (57). Pregnant women, unless they have serologic evidence of previous infection, should avoid contact with cats whose source of food is not controlled and should not empty litter boxes. Uterine or neonatal infections can give rise to a number of serious sequelae, including microcephaly and hydrocephalus.

BLOOD PROTOZOA

Leishmania spp.

Introduction

Visceral leishmaniasis (VL) continues to increase in frequency in patients who are HIV positive (81, 93). The majority of cases of leishmaniasis in AIDS patients have been seen in France, Spain, Italy, and Portugal (112). Like other opportunistic infections, subclinical VL can be found at any stage of HIV-1 infection, but symptomatic cases appear mainly when severe immunosuppression is present. HIV-*Leishmania* coinfection is being seen more and more frequently in the Mediterranean basin.

Once the organisms have been carried to the bone marrow, spleen, and liver, a granulomatous cell-mediated immune response occurs that can result in subclinical disease and self-cure or in the clinical syndrome of VL. The incubation period usually ranges from 2 to 6 months; however, it can take a few days or weeks to several years. In giving a history, many patients do not remember having a primary skin lesion. In areas where infection is endemic, the onset may be gradual, with vague symptoms and a general feeling of ill health. The onset in naïve patients (migrants, soldiers, and travelers to areas of endemic infection) may be acute, with high fever, chills, anorexia, malaise, weight loss, and frequently, diarrhea. This syndrome can easily be confused with typhoid fever, malaria, or other febrile illnesses caused by bacteria or viruses. Death may occur after a few weeks or several years. Generally, untreated VL will lead to death; secondary bacterial and viral infections are also common in these patients.

In VL, amastigotes can be found throughout the body and are not confined to the macrophages of the skin. Organisms inoculated into the bite site by the sand fly are engulfed by tissue macrophages, which then enter the bloodstream and are carried to reticuloendothelial centers, such as bone marrow, spleen, and liver (Fig. 12). Infection results in suppression of cell-mediated immunity, leading to uncontrolled reproduction of the parasite and dissemination.

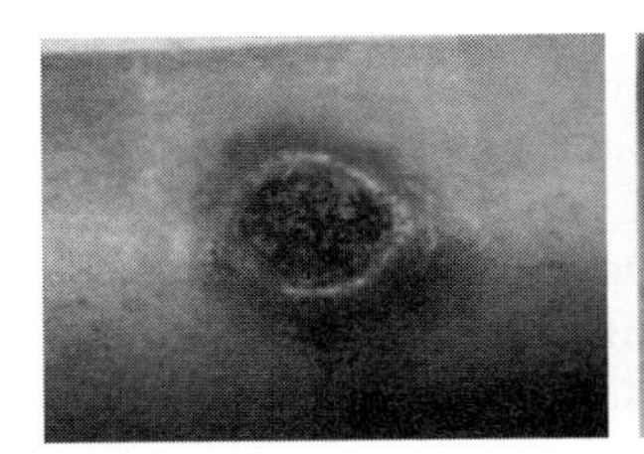
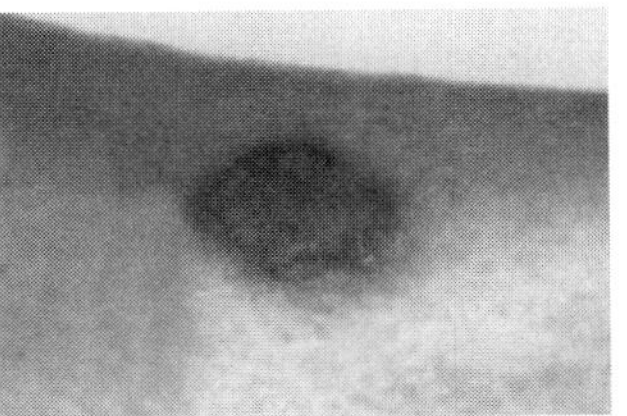
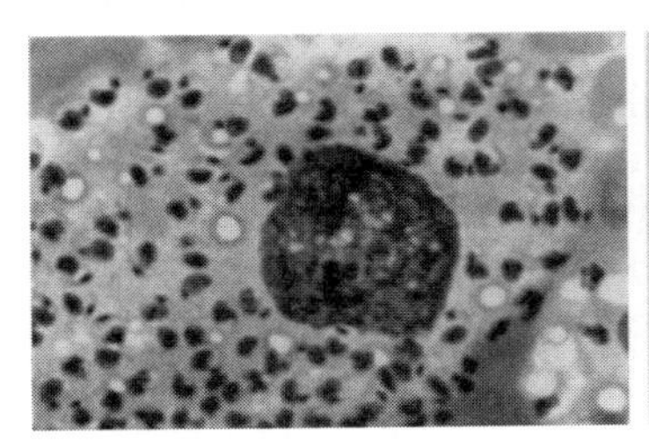
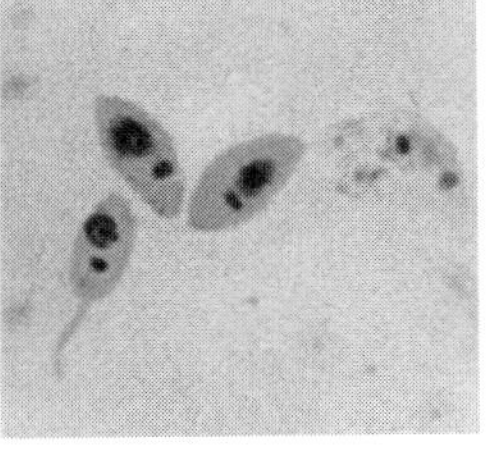

Figure 12. (Top left) cutaneous leishmaniasis (typical lesion); (top right) same lesion after therapy; (bottom left) *Leishmania donovani* in bone marrow (note the individual amastigotes, with each one containing a bar and nucleus); (bottom right) promastigotes from culture (specimen was stained using Giemsa stain).

Circulating immune complexes involving IgA, IgG, and IgM can be found in the serum. Delayed hypersensitivity responses to skin test antigens are suppressed or absent. With resolution of infection, the delayed hypersensitivity reaction returns, and patients recovering from infection are considered immune to reinfection.

Organism description

During a blood meal taken by the sand fly (genus *Phlebotomus*), amastigotes are ingested and transform into the promastigote stage. Promastigotes multiply by longitudinal fission in the insect gut. Stages found in the sand fly vary from rounded or stumpy forms to elongated, highly motile metacyclic promastigotes. The metacyclic promastigotes migrate to the hypostome of the sand fly, from which they are inoculated into humans when the sand fly takes its next blood meal. Various surface molecules on the promastigote, such as gp63 and lipophosphoglycans, help to bind the parasite to the host macrophage receptors, thus allowing it to be phagocytized. Depending on the species, the duration of the life cycle in the sand fly varies from 4 to 18 days.

Sand flies are pool feeders and possess cutting mouthparts that slice into the skin, allowing a small pool of blood to form. Thus, although they are small flies (2 to 3 mm), the bite can be painful. In most foci of endemicity, the vast majority of sand flies are uninfected; however, those that are infected are very efficient vectors. Infected flies probably remain so for life, although this is just a matter of weeks. Under optimal conditions, transformation of the amastigotes to promastigotes generally takes 5 to 7 days; by that time, the female phlebotomine fly is ready to take her next blood meal.

Clinical aspects

Over 850 coinfection cases have been recorded, with 7 to 17% of HIV-positive individuals with fever having amastigotes. This finding strongly suggests that individuals infected with *Leishmania* spp. without symptoms will express symptoms of leishmaniasis if they become immunosuppressed. Based on the patient's general health status and prior exposure to the infection, leishmaniasis may be seen as an early opportunistic infection or may be seen as a complication late in the course of AIDS. Clinical signs of VL usually occur in the later stages of AIDS. The majority of AIDS patients present with the classic picture of VL; however, asymptomatic cutaneous, mucocutaneous, diffuse cutaneous, and post-kala-azar dermal leishmaniases (PKDL; usually caused by *Leishmania infantum*) can be seen (19). Cutaneous lesions in VL are being reported increasingly frequently for patients with HIV infection, and their significance is still somewhat unclear (33, 35, 114). Lesions often do not present a uniform or specific appearance and have been seen as erythematous papules and hypopigmented macules on the dorsa of the hands, feet, and elbows; small subcutaneous nodules on the thighs; and erythematoviolaceous, scaly plaques on the face. Also, the digestive and respiratory tracts are often parasitized, as well as the pleura and peritoneum. Lesions can also be coinfected with other organisms, such as *Mycobacterium avium-intracellulare*.

In a bone marrow recipient, an opportunistic infection with an insect trypanosomatid was suspected when diagnostic tests did not confirm leishmaniasis as the coinfection. Further hybridization analyses against a panel of different monoxenous and heteroxenous trypanosomatids showed kinetoplast DNA cross-homology with *Leptomonas pulexsimulantis*, a trypanosomatid found in the dog flea (103).

Diagnosis

A clinical diagnosis of Old World VL is often made based on clinical findings and local epidemiologic factors. However, confirmation of the diagnosis requires demonstration of the amastigotes in tissues or clinical specimens or the promastigotes in culture. Detection of parasite genetic material or antigen detection will also suffice for confirmation. Leishmaniasis should be suspected in individuals who have resided in or traveled to areas where the disease is endemic. The diagnosis would be supported by findings of remittent fevers, hypergammaglobulinemia with anemia, circulating immune complexes, rheumatoid factors, weight loss, leukemia, and splenomegaly. The differential diagnosis of late-stage VL is limited to hematologic and lymphatic malignancies. The diagnosis of early disease is more difficult, and the differential diagnosis would include malaria, African trypanosomiasis, brucellosis, enteric fevers, bacterial endocarditis, generalized histoplasmosis, chronic myelocytic leukemia, Hodgkin's disease and other lymphomas, sarcoidosis, hepatic cirrhosis, and tuberculosis. Also, patients with multiple myeloma and Waldenström's macroglobulinemia have monoclonal hypergammaglobulinemia. Mixed cryoglobulinemia secondary to VL has also been documented.

Blood. Examination of buffy coat smears shows that there is a significant difference in recovery between samples collected during the day (46%) and those collected during the night (66%). Prior to an invasive procedure, blood specimens should be collected, keeping in mind the amastigote diurnal periodicity.

Tissue aspirates. Although splenic puncture yields the highest rate of organism detection (98%), it carries a high degree of risk for the patient. Deaths have occurred after this procedure, probably due to splenic laceration, and patients with coagulation disorders should not undergo splenic puncture. Other specimens include lymph node aspirates, liver biopsy specimens, sternal or iliac crest bone marrow, and buffy coat preparations of venous blood. Bone marrow aspirates are generally used in lieu of splenic puncture; however, smears from bone marrow usually contain fewer organisms, with a positivity rate of 80 to 85%. It is important to prepare multiple smears for examination. Although amastigotes can be seen in buffy coat smears in Indian and Kenyan patients with VL, it is unusual to make the diagnosis by this method.

Culture and animal inoculation. Culture and animal inoculation studies are also helpful. A method for rapidly isolating promastigotes from the culture fluid has been developed using the stain acridine orange. This type of vital staining combines the advantages of direct microscopic examination with more accurate morphological imaging of the stained parasites. Hamster inoculation is generally no longer available.

PCR and other antigen detection methods. PCR methods have excellent sensitivity and specificity for direct detection, for identification of causative species, and for assessment of treatment efficacy (46). A rapid immunochromatographic dipstick test using the recombinant K39 antigen has become available for the qualitative detection of total anti-*Leishmania* immunoglobulins. The test uses antigen-impregnated nitrocellulose paper strips. Results from one study were very promising, with 100% sensitivity and 98% specificity. The K39 strip test is ideal for rapid, reliable field diagnosis of VL. The test has high sensitivity and specificity, but it remains positive long after treatment (up to 3 years) (Kalazar Detect; InBios International, Inc., Seattle, WA). Another rapid test is now available (IT-Leish; DiaMed AG, Cressier sur Morat, Switzerland). It is important to remember that anti-K39 IgG can also persist in the serum for months after successful treatment for kala-azar; therefore, strip tests cannot be used to diagnose suspected relapse of VL (139). Blood may be acceptable for diagnosis of symptomatic cases; however, the sensitivity of the test may be higher with serum, which should preferably be used for strip testing.

Another strip test, the Dual-It L/M dipstick (DiaMed AG, Switzerland), which detects antibodies to both recombinant K39 antigen (for VL diagnosis) and specific plasmodial lactate dehydrogenase (for malaria diagnosis), is available.

FGT (for hypergammaglobulinemia). In patients with kala-azar, there is a characteristic hypergammaglobulinemia, including both IgG and IgM. This fact has been used as the basis for a diagnostic procedure called the Formol gel test (FGT), which is used in areas of endemic infection. The test can be used as a screen; most patients with active disease will be positive. A drop of formaldehyde is added to a test tube containing 1 ml of patient serum, which is observed for opacity and a stiff jelly consistency within 3 min to 24 h. In some studies, the jellification reaction was observed after 30 min, and the test was considered positive if a clot was seen at the bottom of the tube. However, for patients who do not tend to mount an immunoglobulin response, this test would perform with even less sensitivity and specificity.

The diagnostic performance of the FGT is often considered insufficient to recommend its use. Patients with a false-positive FGT result (hyperreactive malarial syndrome, a classical cause of polyclonal hypergammaglobulinemia, can be a diagnosis in non-VL patients) could wrongly be started on antileishmanial therapy, while 34% of true kala-azar patients, an unacceptably high proportion, would be missed by the FGT (26).

Serologic tests. A number of serologic diagnostic methods have been developed worldwide, although they are not widely available. In general, sensitivities are >90% but specificities are lower. False-positive results can be seen for patients with other infectious diseases; following treatment of VL patients, in whom antibodies persist for months; and for some asymptomatic VL patients. There are currently four tests in use, including the direct agglutination test (DAT), IFA test, counterimmunoelectrophoresis, and ELISA. DAT appears to be the best choice as a diagnostic tool, is very specific, and does not require expensive equipment or reagents. However, patients may still have positive sera 1.5 to 5 years after treatment. DAT has also been adapted to filter paper sampling, and the filter paper eluates compared well with their homologous sera and correlated strongly with respect to antibody titers. DAT based on axenic amastigote antigen provides 100% sensitivity and specificity, making it particularly useful for macular PKDL cases, which are often missed by the recombinant K39 strip test. Thus, DAT provides a simple, reliable, and inexpensive test for PKDL diagnosis with potential applicability under field conditions (46). However, newer studies using ELISA formats indicate excellent sensitivity and specificity and may be very helpful for diagnosing early infections.

Therapy

Patients should receive supportive care, especially if they are malnourished or have other infections, such as pneumonia, tuberculosis, or other chronic conditions. Since the mid-1930s, pentavalent antimony compounds have been the drugs of choice for the treatment of VL. However, with the first reports of primary treatment failures in the mid-1990s, additional drugs have been used and include lipid-associated amphotericin B for Mediterranean and Indian VL (1). VL in renal transplant recipients has been treated successfully with liposomal amphotericin B (AmBisome). The disadvantages of using sodium stibogluconate (Pentostan) and meglumine antimoniate (Glucantime) include the need for intravenous administration and hospitalization, cardiotoxicity, pancreatitis, treatment failures, and increasing resistance (46). The dosing regimen for treating VL is also associated with a high rate of serious side effects in HIV-infected patients. Therapy combining either paromomycin or allupurinol with pentavalent antimonials has also been used with some success.

There is no effective orally administered medication for any *Leishmania* infection. However, studies with miltefosine appear very promising. The use of orally administered miltefosine appears to be an effective treatment for Indian VL.

Epidemiology and control

CD4 T-lymphocyte counts in AIDS patients with VL are usually <50/mm^3 and are almost always <200/mm^3. Many physicians believe that VL does not cause severe symptoms in these patients but is just one of several opportunistic infections that may coinfect AIDS patients. However, VL can manifest atypical aspects in HIV-positive patients, depending on the degree of immunosuppression, and VL should be listed among AIDS-defining conditions (5). The most common symptoms are fever, splenomegaly, hepatomegaly, and pancytopenia.

Data from Spain provide some interesting outcome figures regarding *Leishmania* infections in persons with or without HIV infection. In spite of a good initial response to treatment for VL, 60.6% of patients had relapsed by the end of 1 year. Certainly in AIDS patients, VL is a recurrent disease that is highly prevalent and whose clinical course is modified by HIV. VL is a very prevalent disease among HIV-1-infected patients in southern Spain, with a large proportion of cases being subclinical. In HIV-infected patients, VL occurs in late stages of HIV disease and often has a relapsing course. Secondary prophylaxis reduces the risk of relapse.

For HIV-1-infected patients with active VL, the sensitivity of a peripheral blood smear is about 50% because of the high parasitemia in these patients. However, for patients with HIV-1 and subclinical VL, the sensitivity of a routine blood smear is <10%. Unfortunately, for these patients, serology and *Leishmania* skin tests have low sensitivities and are of little help. Cultures have proven to be effective for VL patients who are coinfected with HIV-1. Confirmation of Old World VL is discussed above.

Although the usual route of transmission is through the bite of infected sand flies, parasites can be transmitted in other ways, including blood transfusion, sexual contact, congenital transmission, occupational exposure, and shared needles. Apparently, *L. infantum* can circulate intermittently and at a low density in the blood of healthy seropositive individuals, who appear to be asymptomatic carriers; this certainly has implications for the safety of the blood supply in certain areas of the world, including southern France, where coinfection with HIV is a problem.

Trypanosoma cruzi

Introduction

American trypanosomiasis (Chagas' disease) is a zoonosis caused by *T. cruzi*, which was discovered in the intestine of a triatomid bug in Brazil in 1909 by Carlos Chagas. Approximately 100 million persons are at risk of infection; between 16 million and 18 million are actually infected. There are approximately 50,000 deaths per year due to Chagas' disease. In certain areas of endemic infection, approximately 10% of all adult deaths are due to Chagas' disease. The southern United States, particularly Texas, is also now identified as having a number of cases.

The disease is transmitted to humans through the bite wound caused by reduviid bugs (triatomids, kissing bugs, or conenose bugs). Humans are infected when parasites are released with the feces while the insect is taking a blood meal and the feces are rubbed or scratched into the bite wound or onto mucosal surfaces, such as the eyes or mouth, an action stimulated by the allergic reaction to the insect's saliva. The organisms can also be transmitted as congenital infections, by blood transfusion, or by organ transplantation.

In humans, *T. cruzi* can be found in two forms, namely, amastigotes and trypomastigotes (Fig. 13). The trypomastigote does not divide in the blood but carries the infection to all parts of the body. The amastigote form multiplies within virtually any cell, preferring cells of mesenchymal origin, such as reticuloendothelial, myocardial, adipose, and neuroglial cells. The parasites also occur in histiocytes of cutaneous tissue and in cells of the epidermis, as well as in the intestinal mucous membrane.

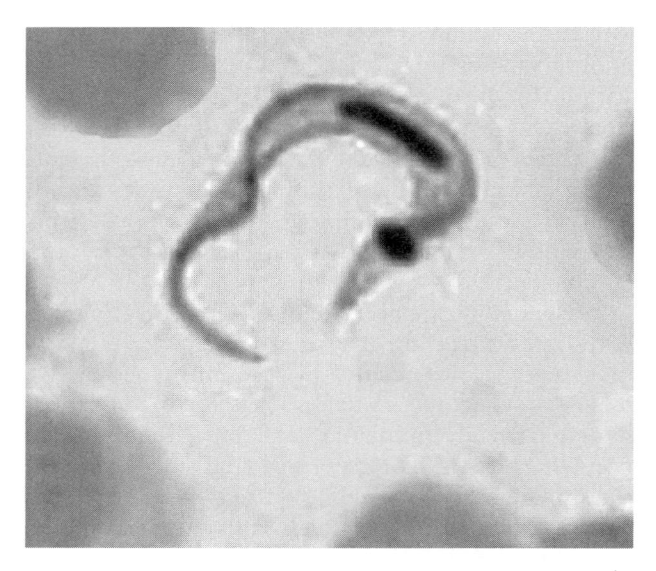

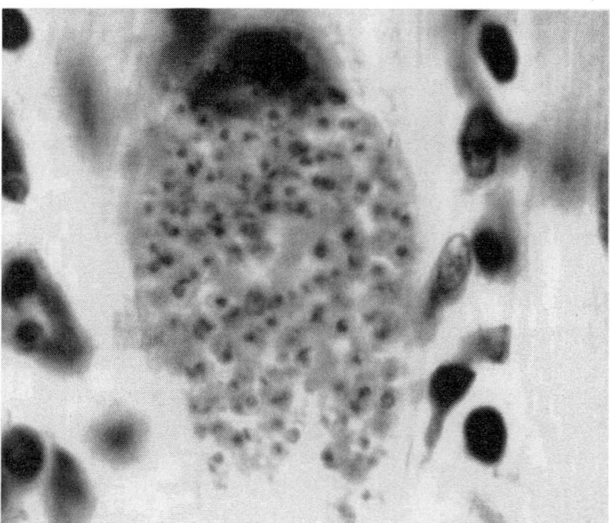

Figure 13. (Top) *Trypanosoma cruzi* trypomastigote (note the large kinetoplast at one end and the large central nucleus); (bottom) *T. cruzi* amastigotes in cardiac tissue.

Organism description

The trypomastigote is spindle shaped and approximately 20 μm long, and it characteristically assumes a C or U shape in stained blood films. Trypomastigotes occur in the blood in two forms, namely, a long slender form and a short stubby one. The nucleus is situated in the center of the body, with a large oval kinetoplast located at the posterior extremity. A flagellum arises from the blepharoplast and extends along the outer edge of an undulating membrane until it reaches the anterior end of the body, where it projects as a free flagellum.

Upon invasion of a cell, the trypomastigote loses its flagellum and undulating membrane and forms an amastigote, which is indistinguishable from those found in leishmanial infections. The amastigote continues to divide and eventually fills and destroys the infected cell. Both amastigote and trypomastigote forms are released from the cell.

Trypomastigotes are ingested by the reduviid bug as it obtains a blood meal. The trypomastigote transforms into an epimastigote that multiplies in the posterior portion of the midgut. After 8 to 10 days, metacyclic trypomastigotes develop from the epimastigotes and are passed in the feces, infecting humans if rubbed into the insect's puncture wound or onto exposed mucous membranes.

Clinical aspects

The clinical syndromes associated with Chagas' disease can be broken down into early, acute, indeterminate, and chronic stages. The acute stage is the result of the first encounter of the patient with the parasite, whereas the chronic phase is the result of late sequelae.

Early stage. An erythematous subcutaneous nodule (chagoma), seen most frequently on the face, may form. The chagoma is painful and may take 2 to 3 months to subside. In early stages of infection, amastigotes or trypomastigotes may be aspirated from the chagoma; if the route of inoculation is the ocular mucosa, edema of the eyelids and conjunctivitis may occur (Romaña's sign). Trypomastigotes will appear in the blood about 10 days after infection and will persist through the acute phase. These stages are rare or absent during the chronic phase.

Acute stage. Acute systemic signs occur around week 2 to 3 of infection and include high fevers, hepatosplenomegaly, myalgia, erythematous rash, acute myocarditis, lymphadenopathy, keratitis, and subcutaneous edema of the face, legs, and feet. There may be signs of CNS involvement, including meningoencephalitis, which has a very poor prognosis. Myocarditis is manifested by electrocardiographic changes, tachycardia, chest pain, and weakness. Amastigotes proliferate within the cardiac muscle cells and destroy the cells, leading to conduction defects and a loss of heart contractility. Death may occur due to myocardial insufficiency or cardiac arrest.

Indeterminate stage. Approximately 8 to 10 weeks after infection, the indeterminate stage begins, during which patients do not have any symptoms. Patients with low-virulence strains may remain infected for years with few to no problems.

Chronic stage. Approximately 30% of patients may develop chronic Chagas' disease, including cardiomyopathy, megacolon, and megaesophagus. Symptoms of the chronic stage are related to the damage sustained during the acute stage of the disease, the state of the host's immune system, and the inflammatory response. Other contributing factors include the host genetic background, environmental and social factors, the genetic composition of the parasite, mixed infections, and reinfection.

Immunocompromised patients. Chagas' disease has been reported for HIV-positive patients, who routinely have neurologic sequelae due to *T. cruzi*. Individuals previously infected with *T. cruzi* who later became positive for HIV are at risk of reactivation of their Chagas' disease. These patients develop a severe multifocal or diffuse meningoencephalitis with numerous tissue parasites. Meningoencephalitis is a rare event in non-AIDS patients but may be seen in immunocompetent children younger than 4 years. Because of the immunodeficiency associated with HIV infections, concomitant infection with *T. cruzi* may be difficult to recognize, particularly in individuals who have moved to areas where the disease is not endemic. *T. cruzi* infection may also have a protracted asymptomatic course in immunosuppressed HIV patients. Although a positive blood smear has been considered the key indicator of Chagas' disease reactivation in immunocompromised patients with chronic disease, this finding may occur late, rather than early, in the reactivation process.

Recrudescence of *T. cruzi* infections in immunosuppressed patients, particularly transplantation patients, is a serious problem. Transplant recipients can also become infected through receipt of infected organs (25, 147). In patients with end-stage Chagas' cardiomyopathy, heart transplantation is an option which has had variable success. Reactivation of the disease with the development of cutaneous lesions has been seen. Bone marrow transplant recipients are also at risk of Chagas' disease due to reactivation or transfusion. Prophylactic treatment of these patients has led to favorable outcomes.

Diagnosis

Blood and tissue. The definitive diagnosis depends on demonstration of trypomastigotes in the blood, amastigotes in tissues, or positive serologic reactions. Trypomastigotes may be detected easily in the blood of young children; however, in chronic disease, this stage is rare or absent except during febrile exacerbations. Trypomastigotes may be detected in blood by using thin and thick blood films or by buffy coat concentration techniques. The most sensitive methods for detection of trypomastigotes are concentration techniques such as buffy coat preparations.

In areas where kala-azar occurs, amastigote stages appear similar, and infections with *Leishmania donovani* and *T. cruzi* must be differentiated. This can be done by PCR, immunoblotting, culture (epimastigotes of *T. cruzi* versus promastigotes of *L. donovani*), serologic tests, animal inoculation, or xenodiagnosis.

Aspirates and biopsies. Aspirates from chagomas and enlarged lymph nodes can be examined for amastigotes and trypomastigotes. Histologic examination of biopsy specimens may also be done. Aspirates, blood, and tissues can also be cultured, which is valuable in detecting low-grade parasitemias. Some individuals with chronic Chagas' disease may have a depressed humoral immune response, being serologically negative.

PCR. Although not routinely available, PCR has been used to detect positive patients with as little as one trypomastigote in 20 ml of blood (26). PCR may be very useful for the diagnosis of patients with chronic Chagas' disease because of the lack of sensitivity and specificity of serologic tests and the lack of sensitivity with xenodiagnosis. PCR testing may be very useful for monitoring patients who have received therapy in order to validate cures.

Antigen detection. Although immunoassays have been used to detect antigens in sera and urine, this approach is not widely available. A prototype ELISA has excellent sensitivity and specificity for detection of antibodies to *T. cruzi* in donors. Moreover, among donations from Texas, observed seroprevalence was 0.03% (143).

Antibody detection. The diagnosis of chronic Chagas' disease is generally made by detecting antibodies to *Trypanosoma cruzi*. Most serological tests are based on lysates of whole parasites or semipurified antigen fractions from *T. cruzi* epimastigotes grown in culture. Inconclusive and false-positive results have been problems with the conventional assays, and there is no universally accepted gold standard for confirmation of positive test results. A new immunoblot assay showed a sensitivity of 100% ($n = 345$) and a specificity of 100% ($n = 500$). Additionally, all 269 potentially cross-reacting and *T. cruzi* antibody-negative specimens tested negative. This immunoblot assay has potential as a supplemental test for confirming the presence of antibodies to *T. cruzi* in blood specimens and for identifying false-positive results obtained with other assays (27).

Therapy

Although numerous drugs have been tried, few have proven to be effective for therapy of Chagas' disease. In acute and congenital Chagas' disease and infections caused by laboratory accidents, treatment should be administered as soon as possible, even though in some cases symptoms are self-limited. Drug therapy has little effect in reducing the progression of chronic Chagas' disease. Nifurtimox (Bayer 205 or Lampit) reduces the duration and severity of illness and decreases mortality due to acute and congenital Chagas' disease. Benznidazole (RO-7[hyphen]1051, Rochagan, or Radanil) is effective in reducing or suppressing parasites in the acute stages of disease but has limited capacity to produce a parasitic cure. Allopurinol, a purine analog, was found in limited clinical trials to be as effective as nifurtimox and benznidazole in treating Chagas' disease. All of these drugs have been associated with multiple side effects.

Symptoms associated with megaesophagus and megacolon may be treated with dietary measures or may require surgery. Patients with chronic chagasic heart disease may receive supportive therapy or may be managed, in some cases, with pacemakers.

Epidemiology and control

Human infections occur mainly in rural areas where poor sanitary and socioeconomic conditions and poor housing provide breeding places for reduviid bugs. These conditions allow maximum contact between the vector and humans. Although 12 species of reduviids occur within the United States, they have not adapted themselves to household habitation. Humans should avoid sleeping in thatch, mud, or adobe houses, and bed nets should be used by persons sleeping in these types of houses. Travelers staying in hotels, resorts, or other well-constructed housing facilities are not at high risk for contracting Chagas' disease.

INTESTINAL AND TISSUE HELMINTHS

Strongyloides stercoralis

Introduction

In the life cycle of the intestinal nematode *Strongyloides stercoralis*, the larvae migrate through the heart and lungs, pass up the trachea, are swallowed, and finally grow to maturity in the gastrointestinal tract. One significant feature of the *Strongyloides* life cycle is an internal autoinfection capability. Noninfective rhabditiform larvae, which normally pass out in the stool, may transform to infective filariform larvae while still within the gastrointestinal tract or perhaps on the perianal surface. These larvae can then penetrate the bowel wall or skin and reinitiate the life cycle. This cycle can maintain itself at a very low level over a period of >30 years in an individual without causing symptoms. The only unusual finding in the patient may be an unexplained eosinophilia (often at a low level). Patients who contracted the original infection many years earlier can become severely ill later, when they become immunosuppressed. In the compromised patient, the number of larvae and adult worms may increase rapidly, leading to hyperinfection syndrome and disseminated strongyloidiasis.

Patients with systemic *S. stercoralis* infections include those with various leukemias and lymphomas, chronic infections such as leprosy and tuberculosis, and miscellaneous conditions such as kidney disease, organ transplant, asthma, systemic lupus erythematosus, diabetes, hepatitis C, and various fungal infections (38, 58, 95, 96, 105, 107, 113, 122, 130, 133, 140, 142, 155, 158). Other conditions include malnutrition, alcoholism, chronic renal failure, and achlorhydria. In individuals with preserved immune function, direct development of *S. stercoralis* is favored, whereas in individuals with lesser immune function, indirect development is relatively more common. These results may explain the notable absence of disseminated strongyloidiasis in advanced HIV disease. Because disseminated infection requires the direct development of infective larvae in the gut, the observed favoring of indirect development in individuals immunosuppressed by advancing HIV disease is not consistent with the promotion of disseminated infection (129).

The prevalence of *S. stercoralis* is significantly higher in human T-cell leukemia virus type 1 (HTLV-1) carriers than in noncarriers in Japan and elsewhere (48, 49, 54). Since cellular immunity plays a major role in the host defense against strongyloidiasis, infection with HTLV-1 appears to change the immune system capability, leading to severe clinical manifestations and disseminated disease (4). Hypereosinophilia, which can be seen in patients with mild strongyloidiasis, is often not seen in these patients and may be inhibited by the HTLV-1 infection.

Organism description

The typical rhabditiform larvae of *S. stercoralis* are characterized by a short buccal capsule or mouth opening and the presence of a genital primordial packet of cells (Fig. 14). In contrast, the rhabditiform larvae of hookworm have a much longer buccal capsule and essentially no genital primordial cells present.

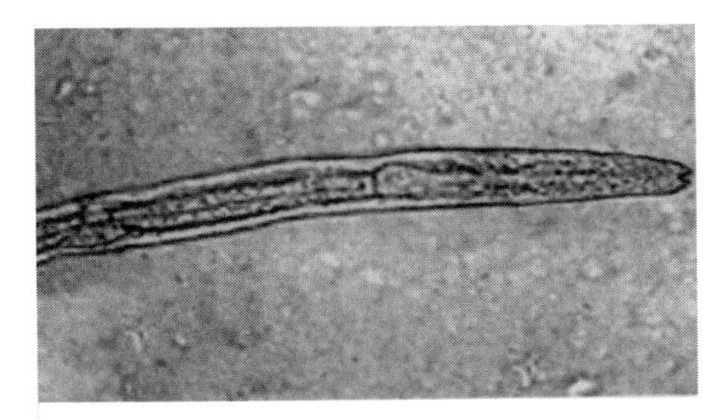

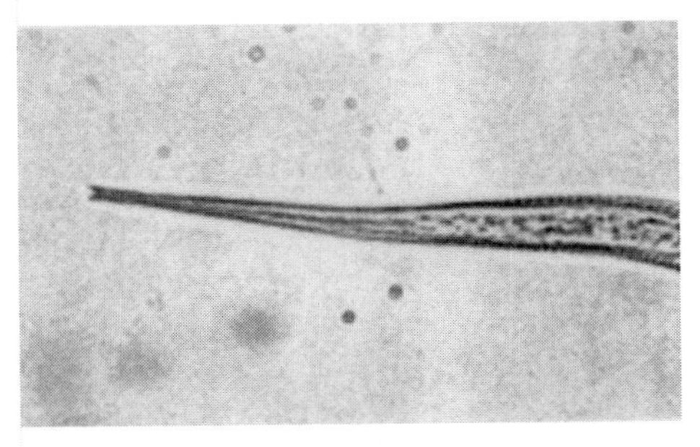

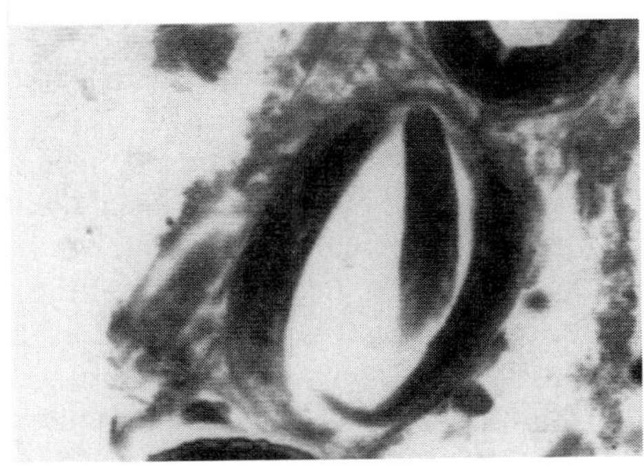

Figure 14. *Strongyloides stercoralis.* (Top) Rhabditiform larvae seen in fecal concentration sediment (note the short mouth opening/buccal capsule and the packet of genital primordial cells at the bottom left of the image); (middle) notched tail ("slit in tail") seen in filariform/infectious larvae; (bottom) larvae in tissue (disseminated infection at a higher magnification).

Clinical aspects

When migrating *Strongyloides* larvae begin to increase in numbers in the compromised host, the patient may present with abdominal complaints, such as colitis, and there may be repeated episodes of unexplained bacteremia or meningitis with enteric bacteria (6, 31, 43). This occurs when the larvae penetrate the bowel and reinitiate the cycle. In doing so, they carry members of the bowel flora with them. Pleural effusions are also seen in patients with disseminated disease (50, 80). The other consistent finding in these patients is an unexplained eosinophilia, which may range from 10 to 50%. In a chronic case, the infection can also present as generalized prurigo nodularis and lichen simplex chronicus (65).

Diagnosis

Like *G. lamblia*, *S. stercoralis* resides in the duodenum, and recovery of the larvae in the stool from a routine O&P examination is difficult for patients with low worm burdens. For this reason, additional techniques, such as the use of an Entero-Test string capsule and duodenal aspiration, may be used. Other techniques for concentrating and recovering *Strongyloides* larvae include the Harada-Mori and petri dish culture techniques, as well as the use of a Baermann apparatus (46).

Whenever one is working with material from a patient suspected of having strongyloidiasis, one should use extreme care because of possible filariform larvae in the specimen. Gloves must be worn to prevent skin penetration by these larval forms.

Treatment

The recommended drug of choice is ivermectin; however thiabendazole has been used very successfully for many years. In some cases, repeated therapy is necessary, and the cure rates vary. Patients with hyperinfection syndrome should be hospitalized during therapy for proper monitoring.

Side effects with thiabendazole occur in most patients and include nausea, foul-smelling urine, neuropsychiatric effects, malaise, and dizziness. Rectal administration of ivermectin has also been reported and may be helpful for patients who cannot absorb or tolerate oral therapy.

Mebendazole has also been used to treat strongyloidiasis; however, it is poorly absorbed. Albendazole has been used to treat patients with strongyloidiasis, as well as those with hyperinfection syndrome. Few side effects have been seen, and this drug is an excellent alternative to ivermectin. It is important to remember that HTLV-1 infection may be a possibility in patients with uncomplicated intestinal strongyloidiasis who do not respond to standard therapy.

Epidemiology and prevention

Contact with contaminated infective soil, feces, or surface water should be avoided. Communities where close living conditions and poor sanitation facilities exist, including both rural and urban areas in the developing world, often have high prevalence rates of strongyloidiasis. Also, closed communities, such as institutions for the mentally handicapped, may have high rates of infection and transmission. The geographic distribution is quite broad and includes both tropical and temperate climates. Remember that internal autoinfection allows for maintenance of the parasite within the host for years following the initial exposure, and thus a comprehensive patient history is mandatory. Since all infected individuals are at risk for hyperinfection and disseminated disease, the identification of people who may have contracted their infection many years earlier is of great clinical significance. Individuals found to have the infection should be treated. All patients who are going to receive immunosuppressive drugs should be screened for strongyloidiasis before therapy.

ARTHROPODS

Crusted (Norwegian) Scabies

Introduction

Scabies is a highly contagious infestation of the itch mite, *S. scabiei*. Symptoms are related to those of pruritic dermatitis, and transmission can occur through person-to-person direct skin contact or via fomites. Outbreaks in hospitals or institutions, particularly for the elderly or mentally retarded, have been documented worldwide. These outbreaks are particularly severe when associated with Norwegian or crusted scabies; such patients may be infected with thousands of mites on the skin. The number of documented reports, particularly for AIDS patients, suggests that crusted scabies may become more common in the future.

This infection has been linked with cellular immunodeficiencies, including AIDS. Crusted scabies should be considered in the differential diagnosis of a generalized cutaneous eruption in an HTLV-1-positive patient (10). Infection with HTLV-1 is an important cofactor related to Norwegian scabies in

Peru (13). The evaluation for HTLV-1 infection in all Norwegian scabies cases is highly recommended, especially when no other risk factors are apparent. These patients may be at increased risk of progressing to adult T-cell leukemia/lymphoma (10, 13). Definitive parasitic diagnosis can be difficult to obtain; thus, difficulties in management have led to renewed interest in both scabies and pediculosis. The diagnosis of scabies should always be considered for patients with advanced malignancies and associated pruritus.

Organism description

S. scabiei is microscopic and lives in cutaneous burrows, where the fertilized female deposits eggs (Fig. 15). The adult mites range from approximately 215 to 390 μm in length, depending on sex. The eggs are 170 μm long by 92 μm wide, and the fecal pellets are about 30 by 15 μm. The fecal pellets are yellow-brown.

Clinical aspects

Scabies is transmitted by close contact with infested individuals, including touching, shaking hands, sexual contact, and contact with those in day care centers for children and the elderly. The usual skin sites that are susceptible to infection are the interdigital spaces, backs of the hands, elbows, axillae, groin, breasts, umbilicus, penis, shoulder blades, small of the back, and buttocks. The outstanding clinical symptom is intense itching. Scratching commonly causes weeping, bleeding, and sometimes secondary infection. Because of the number of mites present, these patients are extremely contagious, from sloughing skin, direct contact, and environmental contamination. The symptoms mimic those of many other dermatologic conditions. Burrows may not be evident, and hyperkeratotic, crusted, scaling, fissured plaques are present over the scalp, face, and back, with associated gross nail thickening.

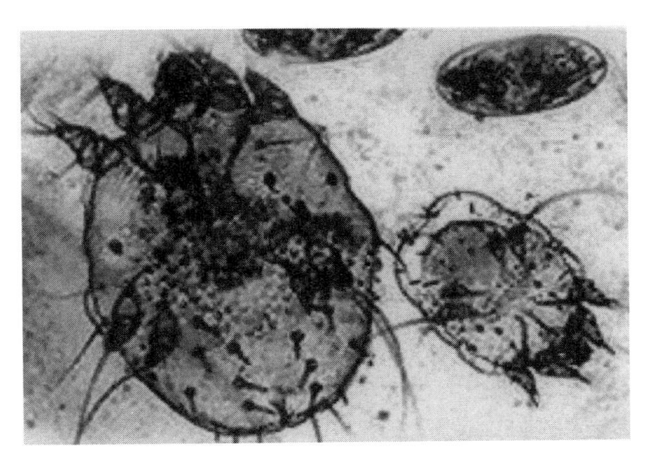

Figure 15. *Sarcoptes scabiei* "itch mite" (note the egg and immature and mature mites).

Diagnosis

The diagnosis of Norwegian scabies can easily be missed. Serpiginous tracks can be seen on the surface of Sabouraud's dextrose agar used for fungal culture of the skin scrapings from an infected individual. This unusual laboratory manifestation can alert the staff to the possible diagnosis of scabies. Although many microbiology laboratories are aware of these unusual findings, personnel can forget to consider scabies in such situations.

Skin-scraping technique. The diagnosis can be confirmed by demonstration of the mites, eggs, or scybala (fecal pellets). Because the mites are located under the surface of the skin, scrapings must be made from the infected area (46).

Plastic box or petri dish method. If mineral oil preparations of skin scrapings fail to demonstrate the mites, the encrusted skin scrapings, etc., can be placed in a small plastic box or small petri dish. The container should be left undisturbed at room temperature for 12 to 24 h. Away from the living host, the mites will drop to the bottom of the box or dish and can be seen with a magnifying glass or dissecting microscope (46).

Treatment

Ivermectin is increasingly being used to treat scabies, especially crusted (Norwegian) scabies. However, treatment failures, recrudescence, and reinfection can occur, even after multiple doses. As predicted, ivermectin resistance in scabies mites can develop after intensive ivermectin use.

REFERENCES

1. **Abramowicz, M. (ed.).** 2004. Drugs for parasitic infections. *Med. Lett.* **46:**1–12.
2. **Ackers, J. P., and D. Mirelman.** 2006. Progress in research on *Entamoeba histolytica. Curr. Opin. Microbiol.* **9:**367–373.
3. **Adams, V. J., M. B. Markus, J. F. Adams, E. Jordaan, B. Curtis, M. A. Dhansay, C. C. Obihara, and J. E. Fincham.** 2005. Paradoxical helminthiasis and giardiasis in Cape Town, South Africa: epidemiology and control. *Afr. Health Sci.* **5:** 276–280.
4. **Agape, P., M. C. Copin, M. Cavrois, G. Panelatti, Y. Plumelle, M. Ossondo-Landeau, D. Quist, N. Grossat, B. Gosselin,**

P. Fexaux, and E. Wattel. 1999. Implication of HTLV-I infection, strongyloidiasis, and p53 overexpression in the development, response to treatment, and evolution of non-Hodgkin's lymphomas in an endemic area (Martinique, French West Indies). *J. Acquir. Immune Defic. Syndr. Hum. Retrovirol.* **20**:394–402.

5. **Albrecht, H., I. Sobottka, C. Emminger, H. Jablonowski, G. Just, A. Stoehr, T. Kubin, B. Salzberger, T. Lutz, and J. van Lunzen.** 1996. Visceral leishmaniasis emerging as an important opportunistic infection in HIV-infected persons living in areas nonendemic for *Leishmania donovani*. *Arch. Pathol. Lab. Med.* **120**:189–198.
6. **Al Samman, M., S. Haque, and J. D. Long.** 1999. Strongyloidiasis colitis: a case report and review of the literature. *J. Clin. Gastroenterol.* **28**:77–80.
7. **Arendt, G., H. J. von Giesen, H. Hefter, E. Neuen-Jacob, H. Roick, and H. Jablonowski.** 1999. Long-term course and outcome in AIDS patients with cerebral toxoplasmosis. *Acta Neurol. Scand.* **100**:178–184.
8. **Barsoum, R. S.** 2004. Parasitic infections in organ transplantation. *Exp. Clin. Transplant.* **2**:258–267.
9. **Belanger, F., F. Derouin, L. Grangeot-Keros, L. Meyer, et al.** 1999. Incidence and risk factors of toxoplasmosis in a cohort of human immunodeficiency virus-infected patients: 1988–1995. *Clin. Infect. Dis.* **28**:575–581.
10. **Bergman, J. N., W. A. Dodd, M. J. Trotter, J. J. Oger, and J. P. Dutz.** 1999. Crusted scabies in association with human T-cell lymphotropic virus. *J. Cutan. Med. Surg.* **3**:148–152.
11. **Bhigjee, A. I., K. Naidoo, V. B. Patel, D. Govender, et al.** 1999. Intracranial mass lesions in HIV-positive patients—the KwaZulu/Natal experience. *S. Afr. Med. J.* **89**:1284–1288.
12. **Bjoro, K., E. Schrumpf, A. Bergan, T. Haaland, K. Skaug, and S. S. Froland.** 1998. Liver transplantation for endstage hepatitis C cirrhosis in a patient with primary hypogammaglobulinaemia. *Scand. J. Infect. Dis.* **30**:520–522.
13. **Blas, M., F. Bravo, W. Castillo, W. J. Castillo, R. Ballona, P. Navarro, J. Catacora, R. Cairampoma, and E. Gotuzzo.** 2005. Norwegian scabies in Peru: the impact of human T cell lymphotropic virus type I infection. *Am. J. Trop. Med. Hyg.* **72**:855–857.
14. **Bowley, D. M., J. Loveland, T. Omar, and G. J. Pitcher.** 2006. Human immunodeficiency virus infection and amebiasis. *Pediatr. Infect. Dis. J.* **25**:1192–1193.
15. **Brindicci, G., C. Picciarelli, L. Fumarola, S. Carbonara, F. Stano, E. Ciraci, M. Gramiccia, A. R. Sannella, M. Milella, D. De Vito, R. Monno, and L. Monno.** 2006. Amoebic hepatic abscesses in an HIV-positive patient. *AIDS Patient Care STDS* **20**:606–611.
16. **Bumb, R. A., and R. D. Mehta.** 2006. Amoebiasis cutis in HIV positive patient. *Indian J. Dermatol. Venereol. Leprol.* **72**:224–226.
17. **Burgner, D., N. Pikos, G. Eagles, A. McCarthy, and M. Stevens.** 1999. Epidemiology of *Cryptosporidium parvum* in symptomatic paediatric oncology patients. *J. Paediatr. Child. Health* **35**:300–302.
18. **Cabral, G. A., and F. Marciano-Cabral.** 2004. Cannabinoid-mediated exacerbation of brain infection by opportunistic amebae. *J. Neuroimmunol.* **147**:127–130.
19. **Calza, L., A. D'Antuono, G. Marinacci, R. Manfredi, V. Colangeli, B. Passarini, R. Orioli, O. Varoli, and F. Chiodo.** 2004. Disseminated cutaneous leishmaniasis after visceral disease in a patient with AIDS. *J. Am. Acad. Dermatol.* **50**:461–465.
20. **Calzetti, C., G. Magnani, D. Confalonieri, A. Capelli, S. Moneta, P. Scognamiglio, and F. Fiaccadori.** 1997. Pancreatitis caused by *Cryptosporidium parvum* in patients with severe immunodeficiency related to HIV infection. *Ann. Ital. Med. Int.* **12**:63–66.
21. **Campbell, A. L., C. L. Goldberg, M. S. Magid, G. Gondolesi, C. Rumbo, and B. C. Herold.** 2006. First case of toxoplasmosis following small bowel transplantation and systematic review of tissue-invasive toxoplasmosis following noncardiac solid organ transplantation. *Transplantation* **81**:408–417.
22. **Campos-Rodriguez, R., and A. Jarillo-Luna.** 2005. The pathogenicity of *Entamoeba histolytica* is related to the capacity of evading innate immunity. *Parasite Immunol.* **27**:1–8.
23. **Carlson, J. R., L. Li, C. L. Helton, R. J. Munn, K. Wasson, R. V. Perez, B. J. Gallay, and W. E. Finkbeiner.** 2004. Disseminated microsporidiosis in a pancreas/kidney transplant recipient. *Arch. Pathol. Lab. Med.* **128**:e41–e43.
24. **Carr, A., D. Marriott, A. Field, E. Vasek, and D. A. Cooper.** 1998. Treatment of HIV-1-associated microsporidiosis and cryptosporidiosis with combination antiretroviral therapy. *Lancet* **351**:256–261.
25. **Centers for Disease Control and Prevention.** 2006. Chagas' disease after organ transplantation—Los Angeles, California, 2006. *MMWR Morb. Mortal. Wkly. Rep.* **55**:798–800.
26. **Chappuis, F., Y. Mueller, A. Nguimfack, J. B. Rwakimari, S. Couffignal, M. Boelaert, P. Cavailler, L. Loutan, and P. Piola.** 2005. Diagnostic accuracy of two rK39 antigen-based dipsticks and the Formol gel test for rapid diagnosis of visceral leishmaniasis in northeastern Uganda. *J. Clin. Microbiol.* **43**: 5973–5977.
27. **Cheng, K. Y., C. D. Chang, V. A. Salbilla, L. V. Kirchhoff, D. A. Leiby, G. Schochetman, and D. O. Shah.** 2007. Immunoblot assay using recombinant antigens as a supplemental test to confirm the presence of antibodies to *Trypanosoma cruzi*. *Clin. Vaccine Immunol.* **14**:355–361.
28. **Cimerman, S., B. Cimerman, and D. S. Lewi.** 1999. Prevalence of intestinal parasitic infections in patients with acquired immunodeficiency syndrome in Brazil. *Int. J. Infect. Dis.* **3**:203–206.
29. **Colebunders, R., K. Depraetere, E. De Droogh, A. Kamper, B. Corthout, and E. Bottiau.** 1999. Obstructive nephropathy due to sulfa crystals in two HIV seropositive patients treated with sulfadiazine. *JBR-BTR* **82**:153–154.
30. **Conteas, C. N., O. G. Berlin, C. E. Speck, S. S. Pandhumas, M. J. Lariviere, and C. Fu.** 1998. Modification of the clinical course of intestinal microsporidiosis in acquired immunodeficiency syndrome patients by immune status and anti-human immunodeficiency virus therapy. *Am. J. Trop. Med. Hyg.* **58**: 555–558.
31. **Corsetti, M., G. Basilisco, R. Pometta, M. Allocca, and D. Conte.** 1999. Mistaken diagnosis of eosinophilic colitis. *Ital. J. Gastroenterol. Hepatol.* **31**:607–609.
32. **Cotte, L., M. Rabodonirina, F. Chapuis, F. Bailly, F. Bissuel, C. Raynal, P. Gelas, F. Persat, M. A. Piens, and C. Trepo.** 1999. Waterborne outbreak of intestinal microsporidiosis in persons with and without human immunodeficiency virus infection. *J. Infect. Dis.* **180**:2003–2008.
33. **Couppie, P., E. Clyti, M. Sobesky, F. Bissuel, P. Del Giudice, D. Sainte-Marie, J. P. Dedet, B. Carme, and R. Pradinaud.** 2004. Comparative study of cutaneous leishmaniasis in human immunodeficiency virus (HIV)-infected patients and non-HIV-infected patients in French Guiana. *Br. J. Dermatol.* **151**:1165–1171.
34. **Dascomb, K., R. Clark, J. Aberg, J. Pulvirenti, R. G. Hewitt, P. Kissinger, and E. S. Didier.** 1999. Natural history of intestinal microsporidiosis among patients infected with human immunodeficiency virus. *J. Clin. Microbiol.* **37**:3421–3422.

35. **Dedet, J. P., and F. Pratlong.** 2000. *Leishmania, Trypanosoma* and monoxenous trypanosomatids as emerging opportunistic agents. *J. Eukaryot. Microbiol.* **47**:37–39.
36. **Del Aguila, C., R. Navajas, D. Gurbindo, J. T. Ramos, M. J. Mellado, S. Fenoy, M. A. Munoz-Fermandex, M. Subirats, J. Ruiz, and N. J. Pieniazek.** 1997. Microsporidiosis in HIV-positive children in Madrid (Spain). *J. Eukaryot. Microbiol.* **44**:84S–85S.
37. **Dietrich, U., M. Maschke, A. Dorfler, M. Prumbaum, and M. Forsting.** 2000. MRI of intracranial toxoplasmosis after bone marrow transplantation. *Neuroradiology* **42**:14–18.
38. **Emad, A.** 1999. Exudative eosinophilic pleural effusion due to *Strongyloides stercoralis* in a diabetic man. *South. Med. J.* **92**:58–60.
39. **Ernst, T. M., L. Chang, M. D. Witt, H. A. Aronow, M. E. Cornford, I. Walot, and M. A. Goldberg.** 1998. Cerebral toxoplasmosis and lymphoma in AIDS: perfusion MR imaging experience in 13 patients. *Radiology* **208**:663–669.
40. **Espinosa-Cantellano, M., and A. Martinez-Paloma.** 2000. Pathogenesis of intestinal amebiasis: from molecules to disease. *Clin. Microbiol. Rev.* **13**:318–331.
41. **Faraci, M., B. Cappelli, G. Morreaie, E. Lanino, C. Moroni, R. Bandettini, M. P. Terranova, D. Di Martino, C. Coccia, and E. Castagnola.** 2007. Titazoxanice or CD3+/CD4+ lymphocytes for recovery from severe Cryptosporidium infection after allogeneic bone marrow transplant? *Pediatr. Transplant.* **11**:113–116.
42. **Fayer, R.** 2004. *Sarcocystis* spp. in human infections. *Clin. Microbiol Rev.* **17**:894–902.
43. **Foucan, L., I. Genevier, I. Lamaury, and M. Strobel.** 1997. Aseptic purulent meningitis in two patients co-infected by HTLV-1 and *Strongyloides stercoralis*. *Med. Trop.* **57**:262–264.
44. **Frenkel, J. K., M. B. Silva, J. Saldanha, M. L. de Silva, V. D. Correia Filho, C. H. Barata, E. Lages, L. E. Ramirez, and A. Prata.** 2003. *Isospora belli* infection: observation of unicellular cysts in mesenteric lymphoid tissues of a Brazilian patient with AIDS and animal inoculation. *J. Eukaryot. Microbiol.* **50**(Suppl.):682–684.
45. **Ganji, M., A. Tan, M. I. Maitar, C. M. Weldon-Linne, E. Weisenberg, and D. P. Rhone.** 2003. Gastric toxoplasmosis in a patient with acquired immunodeficiency syndrome. A case report and review of the literature. *Arch. Pathol. Lab. Med.* **127**:732–734.
46. **Garcia, L. S.** 2007. *Diagnostic Medical Parasitology,* 5th ed. ASM Press, Washington, DC.
47. **Gerber, D. A., M. Green, R. Jaffe, D. Greenberg, G. Mazariegos, and J. Reyes.** 2000. Cryptosporidial infections after solid organ trnsplantation in children. *Pediatr. Transplant.* **4**:50–55.
48. **Gotuzzo, E., C. Arango, A. de Queiroz-Campos, and R. E. Isturiz.** 2000. Human T-cell lymphotropic virus-I in Latin America. *Infect. Dis. Clin. N. Am.* **14**:211–239.
49. **Gotuzzo, E., A. Terashima, H. Alvarez, R. Tello, R. Infante, D. M. Watts, and D. O. Freedman.** 1999. *Strongyloides stercoralis* hyperinfection associated with human T cell lymphotropic virus type-1 infection in Peru. *Am. J. Trop. Med. Hyg.* **60**:146–149.
50. **Goyal, S. B.** 1998. Intestinal strongyloidiasis manifesting as eosinophilic pleural effusion. *South. Med. J.* **91**:768–769.
51. **Guarner, J., T. Matilde-Nava, R. Villasenor-Flores, and G. Sanchez-Mejorada.** 1997. Frequency of intestinal parasites in adult cancer patients in Mexico. *Arch. Med. Res.* **28**:219–222.
52. **Guk, S. M., M. Seo, Y. K. Park, M. D. Oh, K. W. Choe, J. L. Kim, M. H. Choi, S. T. Hong, and J. Y. Chai.** 2005. Parasitic infections in HIV-infected patients who visited Seoul National University Hospital during the period 1995–2003. *Korean J. Parasitol.* **43**:1–5.
53. **Hamzah, Z., S. Petmitr, M. Mungthin, S. Leelayoova, and P. Chavalitshewinkoon-Petmitr.** 2006. Differential detection of *Entamoeba histolytica, Entamoeba dispar,* and *Entamoeba moshkovskii* by a single-round PCR assay. *J. Clin. Microbiol.* **44**:3196–3200.
54. **Hayashi, J., Y. Kishihara, E. Yoshimura, N. Furusyo, K. Yamaji, Y. Kawakami, H. Murakami, and S. Sashiwagi.** 1997. Correlation between human T cell lymphotropic virus type-1 and *Strongyloides stercoralis* infections and serum immunoglobulin E responses in residents of Okinawa, Japan. *Am. J. Trop. Med. Hyg.* **56**:71–75.
55. **Heffler, K. F., T. J. Eckhardt, A. C. Reboli, and D. Stieritz.** 1996. Acanthamoeba endophthalmitis in acquired immunodeficiency syndrome. *Am. J. Ophthalmol.* **122**:584–586.
56. **Helmy, M. M., L. A. Rashed, and H. S. Abdel-Fattah.** 2006. Co-infection with *Cryptosporidium parvum* and *Cyclospora cayetanensis* in immunocompromised patients. *J. Egypt. Soc. Parasitol.* **36**:613–627.
57. **Hill, S. L., J. M. Cheney, G. F. Taton-Allen, J. S. Reif, C. Bruns, and M. R. Lappin.** 2000. Prevalence of enteric zoonotic organisms in cats. *J. Am. Vet. Med. Assoc.* **216**:687–692.
58. **Ho, P. L., W. K. Luk, A. C. Chan, and K. Y. Yuen.** 1997. Two cases of fatal strongyloidiasis in Hong Kong. *Pathology* **29**:324–326.
59. **Hollister, W. S., E. U. Canning, and A. Willcox.** 1991. Evidence for widespread occurrence of antibodies to *Encephalitozoon cuniculi* (*Microspora*) in man provided by ELISA and other serological tests. *Parasitology* **102**:33–43.
60. **Hong, D. K., C. J. Wong, and K. Gutierrez.** 2007. Severe cryptosporidiosis in a seven-year-old renal transplant recipient: case report and review of the literature. *Pediatr. Transplant.* **11**:94–100.
61. **Hung, C. C., H. Y. Deng, W. H. Hsiao, S. M. Hsieh, C. F. Hsiao, M. Y Chen, S. C. Chang, and K. E. Su.** 2005. Invasive amebiasis as an emerging parasitic disease in patients with human immunodeficiency virus type 1 infection in Taiwan. *Arch. Intern. Med.* **165**:409–415.
62. **Hussein, E. M., A. H. Abdul-Manaem, and S. L. el-Attary.** 2005. *Cyclospora cayetanensis* oocysts in sputum of a patient with active pulmonary tuberculosis, case report in Ismailia, Egypt. *J. Egypt. Soc. Parasitol.* **35**:787–793.
63. **Hutin, Y. J., M. N. Sombardier, O. Liguory, C. Sarfati, F. Derouin, J. Modai, and J. M. Molina.** 1998. Risk factors for intestinal microsporidiosis in patients with human immunodeficiency virus infection: a case-control study. *J. Infect. Dis.* **178**:904–907.
64. **Isenberg, H. D. (ed.).** 2004. *Clinical Microbiology Procedures Handbook,* 2nd ed. ASM Press, Washington, DC.
65. **Jacob, C. I., and S. F. Patten.** 1999. *Strongyloides stercoralis* infection presenting as generalized prurigo nodularis and lichen simplex chronicus. *J. Am. Acad. Dermatol.* **41**:357–361.
66. **Jayasekera, S., J. Sissons, J. Tucker, C. Rogers, D. Nolder, D. Warhurst, S. Alsam, J. W. White, E. M. Higgins, and N. A. Khan.** 2004. Post-mortem culture of *Balamuthia mandrillaris* from the brain and cerebrospinal fluid of a case of granulomatous amoebic meningoencephalitis, using human brain microvascular endothelial cells. *J. Med. Microbiol.* **53**:1007–1012.
67. **Joshi, M., A. S. Chowdhary, P. J. Dalal, and J. K. Maniar.** 2002. Parasitic diarrhoea in patients with AIDS. *Natl. Med. J. India* **15**:72–74.
68. **Kaminsky, R. G., R. J. Soto, A. Campa, and M. K. Baum.** 2004. Intestinal parasitic infections and eosinophilia in a

human immunedeficiency virus positive population in Honduras. *Mem. Inst. Oswaldo Cruz* **99**:773–778.

69. **Kester, K. E., G. W. Turiansky, and P. L. McEvoy.** 1998. Nodular cutaneous microsporidiosis in a patient with AIDS and successful treatment with long-term oral clindamycin therapy. *Ann. Intern. Med.* **128**:911–914.
70. **Khan, A., C. Su, M. German, G. A. Storch, D. B. Clifford, and L. D. Sibley.** 2005. Genotyping of *Toxoplasma gondii* strains from immunocompromised patients reveals high prevalence of type I strains. *J. Clin. Microbiol.* **43**:5881–5887.
71. **Khoury, H., D. Adkins, R. Brown, L. Goodnough, M. Gokden, T. Roberts, G. Storch, and J. DiPersio.** 1999. Successful treatment of cerebral toxoplasmosis in a marrow transplant recipient: contribution of a PCR test in diagnosis and early detection. *Bone Marrow Transplant.* **23**:409–411.
72. **Koeppel, M. C., R. Abitan, C. Angeli, J. Lafon, J. Pelletier, and J. Sayag.** 1998. Cutaneous and gastrointestinal mastocytosis associated with cerebral toxoplasmosis. *Br. J. Dermatol.* **139**:881–884.
73. **Koide, J., E. Okusawa, T. Ito, S. Mori, T. Takeuchi, S. Itoyama, and T. Abe.** 1998. Granulomatous amoebic encephalitis caused by *Acanthamoeba* in a patient with systemic lupus erythematosus. *Clin. Rheumatol.* **17**:329–332.
74. **Korelitz, B. I.** 1989. When should we look for amebae in patients with inflammatory bowel disease? *J. Clin. Gastroenterol.* **11**:373–375.
75. **Kotler, D. P., and J. M. Orenstein.** 1998. Clinical syndromes associated with microsporidiosis. *Adv. Parasitol.* **40**:321–349.
76. **Lewthwaite, P., G. V. Gill, C. A. Harat, and N. J. Beeching.** 2005. Gastrointestinal parasites in the immunocompromised. *Curr. Opin. Infect. Dis.* **18**:427–435.
77. **Mandell, L. A.** 1990. Infections in the compromised host. *J. Int. Med. Res.* **18**:177–190.
78. **Manz, M., and M. Steuerwald.** 2007. Cryptosporidiosis in a patient on PEG-interferon and ribavirin for recurrent hepatitis C after living donor liver transplantation. *Transpl. Infect. Dis.* **9**:60–61.
79. **Marciano-Cabral, F., and G. Cabral.** 2003. *Acanthamoeba* spp. as agents of disease in humans. *Clin. Microbiol. Rev.* **16**:273–307.
80. **Mariotta, S., G. Pallone, E. Li Bianchi, G. Gilardi, and A. Bisetti.** 1996. *Strongyloides stercoralis* hyperinfection in a case of idiopathic pulmonary fibrosis. *Panminerva Med.* **38**:45–47.
81. **Marlier, S., G. Menard, O. Gisserot, K. Kologo, and J. P. De Jaureguiberry.** 1999. Leishmaniasis and human immunodeficiency virus: an emerging coinfection? *Med. Trop.* **59**:193–200.
82. **Marra, C. M., M. R. Krone, L. A. Koutsky, and K. K. Holmes.** 1998. Diagnostic accuracy of HIV-associated central nervous system toxoplasmosis. *Int. J. STD AIDS* **9**:761–764.
83. **Maschke, M., U. Dietrich, M. Prumbaum, O. Kastrup, B. Turowski, U. W. Schaefer, and H. C. Diener.** 1999. Opportunistic CNS infection after bone marrow transplantation. *Bone Marrow Transplant.* **23**:1167–1176.
84. **Mathew, M. J., and M. J. Chandy.** 1999. Central nervous system toxoplasmosis in acquired immunodeficiency syndrome: an emerging disease in India. *Neurol. India* **47**:182–187.
85. **Mathis, A., R. Weber, and P. Deplazes.** 2005. Zoonotic potential of the microsporidia. *Clin. Microbiol. Rev.* **18**:423–445.
86. **McBride, J., P. R. Ingram, F. L. Henriquez, and C. W. Roberts.** 2005. Development of colorimetric microtiter plate assay for assessment of antimicrobials against *Acanthamoeba*. *J. Clin. Microbiol.* **43**:629–634.
87. **Menotti, J., G. Vilela, S. Romand, Y. J. Garin, L. Ades, E. Gluckman, F. Derouin, and P. Ribaud.** 2003. Comparison of PCR–enzyme-linked immunosorbent assay and real-time PCR assay for diagnosis of an unusual case of cerebral toxoplasmosis in a stem cell transplant recipient. *J. Clin. Microbiol.* **41**:5313–5316.
88. **Merzianu, M., S. M. Gorelick, V. Paje, D. P. Kottler, and C. Sian.** 2005. Gastric toxoplasmosis as the presentation of acquired immunodeficiency syndrome. *Arch. Pathol. Lab. Med.* **129**:e87–e90.
89. **Michiels, J. F., P. Hofman, E. Bernard, M. C. Saint Paul, C. Boissy, V. Mondain, Y. LeFichoux, and R. Loubiera.** 1994. Intestinal and extraintestinal *Isospora belli* infection in an AIDS patient. A second case report. *Pathol. Res. Pract.* **190**:1089–1094.
90. **Miller, R. F., M. A. Hall-Craggs, D. C. Costa, N. S. Brink, F. Scaravilli, S. B. Lucas, I. D. Wilkinson, P. J. Ell, B. F. Kendall, and M. J. Harrison.** 1998. Magnetic resonance imaging, thallium-201 SPET scanning, and laboratory analyses for discrimination of cerebral lymphoma and toxoplasmosis in AIDS. *Sex. Transm. Infect.* **74**:258–264.
91. **Mohandas, R. Sehgal, A. Sud, and N. Malla.** 2002. Prevalence of intestinal parasitic pathogens in HIV-seropositive individuals in northern India. *Jpn. J. Infect. Dis.* **55**:83–84.
92. **Moran, P., F. Ramos, M. Ramiro, O. Curiel, E. Gonzalez, A. Valadez, A. Gomez, G. Garcia, E. I. Melendro, and C. Ximenez.** 2005. Infection by human immunodeficiency virus-1 is not a risk factor for amebiasis. *Am. J. Trop. Med. Hyg.* **73**:296–300.
93. **Moreno-Camacho, A., R. Lopez-Velez, A. Munoz-Sanz, and P. Labarga-Echevarria.** 1998. Intestinal parasitic infections and leishmaniasis in patients with HIV infection. *Enferm. Infecc. Microbiol. Clin.* **16**:S52–S60.
94. **Morgan, U., R. Weber, L. Xiao, I. Sulaiman, R. C. Thompson, W. Ndiritu, A. Lal, A. Moore, and P. Deplazes.** 2000. Molecular characterization of *Cryptosporidium* isolates obtained from human immunodeficiency virus-infected individuals living in Switzerland, Kenya, and the United States. *J. Clin. Microbiol.* **38**:1180–1183.
95. **Mori, S., T. Konishi, K. Matsuoka, M. Deguchi, M. Ohta, O. Mizuno, T. Ueno, T. Okinaka, Y. Nishimura, N. Ito, and T. Nakano.** 1998. Strongyloidiasis associated with nephrotic syndrome. *Intern. Med.* **37**:606–610.
96. **Muller, A., G. Fatkenheuer, B. Salzberger, M. Schrappe, V. Diehl, and C. Franzen.** 1998. *Strongyloides stercoralis* infection in a patient with AIDS and non-Hodgkin lymphoma. *Dtsch. Med. Wochenschr.* **123**:381–385.
97. **Muller, N., and N. von Allmen.** 2005. Recent insights into the mucosal reactions associated with *Giardia lamblia* infections. *Int. J. Parasitol.* **35**:1339–1347.
98. **Murakawa, G. J., T. McCalmont, J. Altman, G. H. Telang, M. D. Hoffman, G. R. Kantor, and T. G. Berger.** 1995. Disseminated acanthamebiasis in patients with AIDS. A report of five cases and a review of the literature. *Arch. Dermatol.* **131**:1291–1296.
99. **Navin, T. R., R. Weber, D. J. Vugia, D. Rimland, J. M. Roberts, D. G. Addiss, G. S. Visvesvara, S. P. Walhquist, S. E. Hogan, L. E. Gallagher, D. D. Juranek, D. A. Schwartz, C. M. Wilcox, J. M. Stewart, S. E. Thompson III, and R. T. Bryan.** 1999. Declining CD4+ T-lymphocyte counts are associated with increased risk of enteric parasitosis and chronic diarrhea: results of a 3-year longitudinal study. *J. Acquir. Immune Defic. Syndr. Hum. Retrovirol.* **20**:154–159.
100. **Notermans, D. W., R. Peek, M. D. de Jong, E. M.Wentink-Bonnema, R. Boom, and T. van Gool.** 2005. Detection and identification of *Enterocytozoon bieneusi* and *Encephalito-*

zoon species in stool and urine specimens by PCR and differential hybridization. *J. Clin. Microbiol.* **43**:610–614.

101. **Ok, U. Z., M. Cirit, A. Uner, E. Ok, F. Akcicek, A. Basci, and M. A. Ozcel.** 1997. Cryptosporidiosis and blastocystosis in renal transplant recipients. *Nephron* **75**:171–174.
102. **Orenstein, J. M., P. Russo, E. S. Didier, C. Bowers, N. Bunin, and D. T. Teachey.** 2005. Fatal pulmonary microsporidiosis due to *Encephalitozoon cuniculi* following allogeneic bone marrow transplantation for acute myelogenous leukemia. *Ultrastruct. Pathol.* **29**:269–276.
103. **Pacheco, R. S., M. C. Marzochi, M. Q. Pires, C. M. Brito, M. de F. Madeira, and E. G. Barbosa-Santos.** 1998. Parasite genotypically related to a monoxenous trypanosomatid of dog's flea causing opportunistic infection in an HIV positive patient. *Mem. Inst. Oswaldo Cruz* **93**:531–537.
104. **Palau, L. A., and S. A. Kemmerly.** 1997. First report of invasive amebiasis in an organ transplant recipient. *Transplantation* **64**:936–937.
105. **Palau, L. A., and G. A. Pankey.** 1997. Strongyloides hyperinfection in a renal transplant recipient receiving cyclosporine: possible *Strongyloides stercoralis* transmission by kidney transplant. *Am. J. Trop. Med. Hyg.* **57**:413–415.
106. **Paltiel, M., E. Powell, J. Lynch, B. Baranowski, and C. Martins.** 2004. Disseminated cutaneous acanthamoebiasis: a case report and review of the literature. *Cutis* **73**:241–248.
107. **Parana, R., M. Portugal, L. Vitvitski, H. Cotrim, L. Lyra, and C. Trepo.** 2000. Severe strongyloidiasis during interferon plus ribavirin therapy for chronic HCV infection. *Eur. J. Gastroenterol. Hepatol.* **12**:245–246.
108. **Perz, J. F., F. K. Ennever, and S. M. Blancq.** 1998. *Cryptosporidium* in tap water: comparison of predicted risks with observed levels of disease. *Am. J. Epidemiol.* **147**:289–301.
109. **Piccolo, I., R. Causarano, R. Sterzi, M. Sberna. P. L. Oreste, C. Moioli, L. Caggese, and F. Girotti.** 1999. Chorea in patients with AIDS. *Acta Neurol. Scand.* **100**:332–336.
110. **Pillai, D. R., and K. C. Kain.** 2005. *Entamoeba histolytica*: identification of a distinct β_2-integrin-like molecule with a potential role in cellular adherence. *Exp. Parasitol.* **109**: 135–142.
111. **Pillai, D. R., P. S. K. Wan, W. C. W. Yau, J. I. Ravdin, and K. C. Kain.** 1999. The cysteine-rich region of the *Entamoeba histolytica* adherence lectin (170-kilodalton subunit) is sufficient for high-affinity Gal/GalNac-specific finding in vitro. *Infect. Immun.* **67**:3836–3841.
112. **Pineda, J. A., J. A. Gallardo, J. Macias, J. Delgado, C. Regordan, F. Morillas, F. Relimpio, J. Martin-Sanchez, A. Sanchez-Quijano, M. Leal, and E. Lissen.** 1998. Prevalence of and factors associated with visceral leishmaniasis in human immunodeficiency virus type 1-infected patients in southern Spain. *J. Clin. Microbiol.* **36**:2419–2422.
113. **Plumelle, Y., C. Gonin, A. Edouard, B. J. Bucher, L. Thomas, A. Brebion, and G. Panelatti.** 1997. Effect of *Strongyloides stercoralis* infection and eosinophilia on age at onset and prognosis of adult T-cell leukemia. *Am. J. Clin. Pathol.* **107**: 81–87.
114. **Posada-Vergara, M. P., J. A. Lindoso, J. E. Tolezano, V. L. Pereira-Chioccola, M. V. Silva, and H. Goto.** 2005. Tegumentary leishmaniasis as a manifestation of immune reconstitution inflammatory syndrome in 2 patients with AIDS. *J. Infect. Dis.* **192**:1819–1822.
115. **Pozio, E., G. Rezza, A. Boschini, P. Pezzotti, A. Tamburrini, P. Rossi, M. Di Fine, C. Smacchia, A. Schiasari, E. Gattei, R. Zucconi, and P. Ballarini.** 1997. Clinical cryptosporidiosis and human immunodeficiency virus (HIV)-induced immunosuppression: findings from a longitudinal study of HIV-positive and HIV-negative former injection drug users. *J. Infect. Dis.* **176**:969–975.
116. **Que, X., and S. L. Reed.** 2000. Cysteine proteinases and the pathogenesis of amebiasis. *Clin. Microbiol. Rev.* **13**:196–206.
117. **Raffi, F., J. Franck, H. Pelloux, F. Derouin, V. Reliquet, P. Ambroise-Thomas, J. P. Aboulker, C. Leport, and H. Dumon.** 1999. Specific anti-toxoplasmic IgG antibody immunoblot profiles in patients with AIDS-associated *Toxoplasma* encephalitis. *Diagn. Microbiol. Infect. Dis.* **34**:51–56.
118. **Resiere, D., J. M. Vantelon, P. Bouree, E. Chachaty, G. Nitenberg, and F. Blot.** 2003. *Isospora belli* infection in a patient with non-Hodgkin's lymphoma. *Clin. Microbiol. Infect.* **9**:1065–1067.
119. **Rezk, H., A. M. el-Shazly, M. Soliman, H. I. el-Nemr, I. M. Nagaty, and M. A. Fouad.** 2001. Coccidiosis among immuno-competent and -compromised adults. *J. Egypt. Soc. Parasitol.* **31**:823–834.
120. **Ribera, E., A. Fernandez-Sola, C. Juste, A. Rovira, F. J. Romero, L. Armadans-Gil, I. Ruiz, I. Ocana, and A. Pahissa.** 1999. Comparison of high and low doses of trimethoprim-sulfamethoxazole for primary prevention of toxoplasmic encephalitis in human immunodeficiency virus-infected patients. *Clin. Infect. Dis.* **29**:1461–1466.
121. **Rossi, P., F. Rivasi, M. Codeluppi, A. Catania, A. Tamburrini, E. Righi, and E. Pozio.** 1998. Gastric involvement in AIDS associated cryptosporidiosis. *Gut* **43**:476–477.
122. **Sandland, J. T., W. Kauffman, and P. M. Flynn.** 1997. *Strongyloides stercoralis* infection mimicking relapse in a child with small noncleaved cell lymphoma. *Am. J. Clin. Oncol.* **20**:215–216.
123. **Schmidt, W., T. Schneider, W. Heise, J. D. Schulzke, T. Weinke, R. Ignatius, R. L. Owen, M. Zeitz, E. O. Riecken, and R. Ullrich.** 1997. Mucosal abnormalities in microsporidiosis. *AIDS* **11**:1589–1594.
124. **Schuster, F. L., T. H. Dunnebacke, G. C. Booton, S. Yagi, C. K. Kohnmeier, C. Glaser, D. Vugia, A. Bakardjiev, P. Azimi, M. Maddux-Gonzalez, A. J. Martinez, and G. S. Visvesvara.** 2003. Environmental isolation of *Balamuthia mandrillaris* associated with a case of amebic encephalitis. *J. Clin. Microbiol.* **41**:3175–3180.
125. **Schuster, F. L., and G. S. Visvesvara.** 2004. Amebae and ciliated protozoa as causal agents of waterborne zoonotic disease. *Vet. Parasitol.* **126**:91–120.
126. **Schuster, F. L., and G. S. Visvesvara.** 2004. Free-living amoebae as opportunistic and non-opportunistic pathogens of humans and animals. *Int. J. Parasitol.* **34**:1001–1027.
127. **Selby, D. M., R. S. Chandra, T. A. Rakusan, B. Loechelt, B. M. Markle, and G. S. Visvesvara.** 1998. Amebic osteomyelitis in a child with acquired immunodeficiency syndrome: a case report. *Pediatr. Pathol. Lab. Med.* **18**:89–95.
128. **Sell, S.** 1987. *Immunology, Immunopathology and Immunity*, 4th ed. Elsevier Science Publishing, Inc., New York, NY.
129. **Setoyama, M., S. Fukumaru, T. Takasaki, H. Yoshida, and T. Kanzaki.** 1997. SLE with death from acute massive pulmonary hemorrhage caused by disseminated strongyloidiasis. *Scand. J. Rheumatol.* **26**:389–391.
130. **Shekhar, K. C., R. Krishnan, R. Pathmanathan, and C. S. Fook.** 1997. Gastric strongyloidiasis in a Malaysian patient. *Southeast Asian J. Trop. Med. Public Health* **28**:158–160.
131. **Silva, C. V., M. S. Ferreira, A. S. Borges, and J. M. Costa-Cruz.** 2005. Intestinal parasitic infections in HIV/AIDS patients: experience at a teaching hospital in central Brazil. *Scand. J. Infect. Dis.* **37**:211–215.

132. **Sobottka, I., D. A. Schwartz, J. Schottelius, G. S. Visvesvara, N. J. Pieniazek, C. Schmetz, N. P. Kock, R. Laufs, and H. Albrecht.** 1998. Prevalence and clinical significance of intestinal microsporidiosis in human immunodeficiency virus-infected patients with and without diarrhea in Germany: a prospective coprodiagnostic study. *Clin. Infect. Dis.* **26:**475–480.
133. **Sreenivas, D. V., A. Kimar, Y. R. Kumar, C. Bharavi, C. Sundaram, and K. Gayathri.** 1997. Intestinal strongyloidiasis—a rare opportunistic infection. *Indian J. Gastroenterol.* **16:**105–106.
134. **Stanley, S. L., Jr.** 2003. Amoebiasis. *Lancet* **361:**1025–1034.
135. **Stanley, S. L., Jr.** 1997. Progress towards development of a vaccine for amebiasis. *Clin. Microbiol. Rev.* **10:**637–649.
136. **Stauffer, W., and J. I. Ravdin.** 2003. *Entamoeba histolytica*: an update. *Curr. Opin. Infect. Dis.* **16:**479–485.
137. **Steinberg, J. P., R. L. Galindo, E. S. Kraus, and K. G. Ghanem.** 2002. Disseminated acanthamoebiasis in a renal transplant recipient with osteomyelitis and cutaneous lesions: case report and literature review. *Clin. Infect. Dis.* **35:**e43–e49.
138. **Subrungruang, I., M. Mungthin, P. Chavalitshewinkoon-Petmitr, R. Rangsin, T. Naaglor, and S. Leelayoova.** 2004. Evaluation of DNA extraction and PCR methods for detection of *Enterocytozoon bieneusi* in stool specimens. *J. Clin. Microbiol.* **42:**3490–3494.
139. **Sundar, S., R. Maurya, R. K. Singh, K. Bharti, J. Chakravarty, A. Paarekh, M. Rai, K. Kiman, and H. W. Murray.** 2006. Rapid, noninvasive diagnosis of visceral leishmaniasis in India: comparison of two immunochromatographic strip tests for detection of anti-K39 antibody. *J. Clin. Microbiol.* **44:**251–253.
140. **Suvajdzic, N., I. Kranjcic-Zee, V. Jovanovic, D. Popovic, and M. Colovic.** 1999. Fatal strongyloidosis following corticosteroid therapy in a patient with chronic idiopathic thrombocytopenia. *Haematologia* **29:**323–326.
141. **Tanyuksel, M., and W. A. Petri, Jr.** 2003. Laboratory diagnosis of amebiasis. *Clin. Microbiol. Rev.* **16:**713–729.
142. **Thomas, M. C., and S. A. Costello.** 1998. Disseminated strongyloidiasis arising from a single dose of dexamethasone before stereotactic radiosurgery. *Int. J. Clin. Pract.* **52:**520–521.
143. **Tobler, L. H., P. Contestable, L. Pitina, H. Groth, S. Shaffer, G. R. Blackburn, H. Warren, S. R. Lee, and M.P. Busch.** 2007. Evaluation of a new enzyme-linked immunosorbent assay for detection of Chagas antibody in U.S. blood donors. *Transfusion* **47:**90–96.
144. **Torno, M. S., Jr., R. Babapour, A. Gurevitch, and M. D. Witt.** 2000. Cutaneous acanthamoebiasis in AIDS. *J. Am. Acad. Dermatol.* **42:**351–354.
145. **Tran, M. Q., R. Y. Gohh, P. E. Morrissey, L. D. Dworkin, A. Gautam, A. P. Monaco, and A. F. Yango, Jr.** 2005. *Cryptosporidium* infection in renal transplant patients. *Clin. Nephrol.* **63:**305–309.
146. **Tumwine, J. K., A. Kekitiinwa, S. Bakeera-Kitaka, G. Ndeezi, R. Downing, X. Feng, D. E. Akiyoshi, and S. Tzipori.** 2005. Cryptosporidiosis and microsporidiosis in Ugandan children with persistent diarrhea with and without concurrent infection with the human immunodeficiency virus. *Am. J. Trop. Med. Hyg.* **73:**921–925.
147. **Valar, C., E. Keitel, R. L. Dal Pra, D. Gnatta, A. F. Santos, P. D. Bianco, T. C. Sukiennik, K. L. Pegas, A. E. Bittar, K. T. Oliveira, and V. D. Garcia.** 2007. Parasitic infection in renal transplant recipients. *Transplant. Proc.* **39:**460–462.
148. **Velasquez, J. N., S. Carnevale, W. M. Oelemann, C. Etchart, and J. M. Peralta.** 2005. Diagnosis of *Enterocytozoon bieneusi* by the polymerase chain reaction in archival fixed duodenal tissue. *Enferm. Infecc. Microbiol. Clin.* **23:**218–220.
149. **Walker, M. D., and J. R. Zunt.** 2005. Neuroparasitic infections: cestodes, trematodes, and protozoans. *Semin. Neurol.* **25:**262–277.
150 **Wang, Z., P. A. Orlandi, and D. A. Stenger.** 2005. Simultaneous detection of four human pathogenic microsporidian species from clinical samples by oligonucleotide microarray. *J. Clin. Microbiol.* **43:**4121–4128.
151. **Waywa, D., S. Kongkriengdaj, S. Chaidatch, S. Tiengrim, B. Kowadisaiburana, S. Chaikachonpat, S. Suwanagool, A. Chaiprasert, A. Curry, W. Bailey, Y. Suputtamongkol, and N. J. Beeching.** 2001. Protozoan enteric infection in AIDS related diarrhea in Thailand. *Southeast Asian J. Trop. Med. Public Health* **32**(Suppl. 2):151–155.
152. **Widmer, G., S. Tzipori, C. J. Fichtenbaum, and J. K. Griffiths.** 1998. Genotypic and phenotypic characterization of *Cryptosporidium parvum* isolates from people with AIDS. *J. Infect. Dis.* **178:**834–840.
153. **Wilairatana, P., P. Radomyos, B. Radomyos, R. Phraevanich, W. Plooksawasdi, P. Chanthavanich, C. Viravan, and S. Looareeuwan.** 1996. Intestinal sarcocystosis in Thai laborers. *Southeast Asian J. Trop. Med. Public Health* **27:**43–46.
154. **Wilson, M., and J. B. McAuley.** 2003. Clinical use and interpretation of serologic tests for *Toxoplasma gondii*. Document M36-A. NCCLS, Wayne, PA.
155. **Wong, T. Y., C. C. Szeto, F. F. Lai, C. K. Mak, and P. K. Li.** 1998. Nephrotic syndrome in strongyloidiasis: remission after eradication with antihelminthics agents. *Nephron* **79:**333–336.
156. **Wulf, M. W., R. van Crevel, R. Portier, C. G. Ter Meulen, W. J. Melchers, A. van der Ven, and J. M. Galama.** 2005. Toxoplasmosis after renal transplantation: implications of a missed diagnosis. *J. Clin. Microbiol.* **43:**3544–3547.
157. **Yagi, S., G. C. Booton, G. S. Visvesvara, and F. L. Schuster.** 2005. Detection of *Balamuthia* mitochondrial 16S rRNA gene DNA in clinical specimens by PCR. *J. Clin. Microbiol.* **43:**3192–3197.
158. **Yee, Y. K., C. S. Lam, C. Y. Yung, T. L. Que, T. H. Kwan, T. C. Au, and M. L. Szeto.** 1999. Strongyloidiasis as a possible cause of nephrotic syndrome. *Am. J. Kidney Dis.* **33:**e4.
159. **Yoshimura, K., T. Hara, H. Tsurumi, H. Goto, M. Tajika, Y. Fukutomi, N. Murakami, and H. Moriwaki.** 1999. Non-Hodgkin's lymphoma with Toxoplasma encephalitis. *Rinsho Ketsueki* **40:**563–567.
160. **Zamir, D., M. Amar, G. Groisman, and P. Weiner.** 1999. *Toxoplasma* infection in systemic lupus erythematosis mimicking lupus cerebritis. *Mayo Clin. Proc.* **74:**575–578.
161. **Zver, S., P. Cernelc, U. Mlakar, and J. Pretnar.** 1999. Cerebral toxoplasmosis—a late complication of allogeneic haematopoietic stem cell transplantation. *Bone Marrow Transplant.* **24:**1363–1365.

III. LABORATORY DIAGNOSIS: AN ORGAN SYSTEMS APPROACH

Diagnostic Microbiology of the Immunocompromised Host
Edited by Randall T. Hayden, Karen C. Carroll, Yi-Wei Tang, and Donna M. Wolk

Chapter 15

Lower Respiratory Tract Infections

KAREN C. CARROLL

LOWER RESPIRATORY TRACT INFECTIONS

Decades of advances in cancer treatments and transplantation immunology have expanded the population of severely immunocompromised patients. In addition, new therapies for the management of rheumatologic, autoimmune, and acquired immune diseases have reduced mortality among these patient groups. Pulmonary infections are the most common syndromes contributing to morbidity and mortality among immunosuppressed patients (2, 94, 102). Virtually any potential pathogen can result in significant illness, and pulmonary infiltrates may be caused by a variety of noninfectious syndromes as well. Management of pulmonary syndromes in these vulnerable populations is a challenge for both clinicians and microbiologists, as prompt diagnosis can prevent irreversible pulmonary complications and/or allow withdrawal of potentially toxic empiric therapies. Diagnostic approaches should consider the tempo of the pulmonary process, the extent of immunosuppression, and the radiographic patterns. In addition, the likelihood of a specific infection may be affected by recently administered prophylaxis or empiric therapies.

This chapter focuses on the infectious etiologies and more common noninfectious causes of lower respiratory tract syndromes among major immunosuppressed populations. The changing epidemiology of infections in the era of highly active antiretroviral therapy (HAART) in the case of human immunodeficiency virus (HIV)-positive patients and the impacts of both newer immunosuppressant therapies and anti-infective prophylaxis for other immunocompromised hosts are discussed. The chapter emphasizes diagnostic approaches and practice algorithms.

Pneumonia is defined as radiological evidence of a new or increasing pulmonary infiltrate(s) plus one or more of the following: fever, hypothermia, cough with or without sputum production, tachypnea, dyspnea, hemoptysis, wheezing, physical findings such as rales, and hypoxemia.

Community-Acquired Pneumonia

The pathogen distribution for community-acquired pneumonia in immunocompromised hosts is generally the same as that for healthy individuals. Eighty-five percent of cases are caused by *Streptococcus pneumoniae*, *Haemophilus influenzae*, or *Moraxella catarrhalis* (31). The remaining 15% are caused by *Legionella* spp., *Mycoplasma pneumoniae*, or *Chlamydophila pneumoniae*. *Legionella pneumophila* is the most important atypical agent in patients with impaired T-lymphocyte function and may cause either community- or hospital-acquired illness (31, 47). Both solid organ transplant (SOT) and hematopoietic stem cell transplant (HSCT) patients are particularly at risk for *Legionella* pneumonia (24), whereas the risk in HIV patients is not substantially higher than that seen in nonimmunocompromised patients (70). *L. pneumophila* and *Legionella micdadei* are the most common species, but other species cause infections in transplant patients, including *Legionella bozemanni*, *Legionella birminghamensis*, *Legionella dumoffii*, and *Legionella cincinnatiensis* (24). In addition, immunocompromised patients with legionellosis may have more severe presentations, characterized by rapidly expanding lesions and progression to pulmonary cavitation (24, 47). *Pseudomonas aeruginosa* pneumonia and pneumonia caused by other gram-negative bacilli are rare in healthy hosts. *P. aeruginosa* may cause a bacteremic infection in patients with neutropenia. *Klebsiella pneumoniae* pneumonia has been reported most frequently for patients with alcoholic liver

Karen C. Carroll • Department of Pathology, The Johns Hopkins University School of Medicine, and Division of Medical Microbiology, The Johns Hopkins Hospital, Baltimore, MD 21287.

disease. Nontyphoidal *Salmonella* pneumonia is rare and is most frequently associated with patients with HIV infection (16), although it has also been described for patients with underlying conditions such as malignancy, diabetes mellitus, corticosteroid therapy, and alcohol abuse (16). In one retrospective study of 51 HIV-positive patients with *Salmonella* bacteremia, 8 patients (15.7%) met the criteria for definite or probable *Salmonella* pneumonia (16).

Patients with defects in cellular immunity are at risk for infections caused by intracellular pathogens. These include *Mycobacterium* species, most notably *Mycobacterium tuberculosis*, *Mycobacterium avium-Mycobacterium intracellulare* complex (MAC), *Nocardia* sp., *Toxoplasma gondii*, *Pneumocystis jirovecii*, cytomegalovirus (CMV), *Cryptococcus neoformans*, and both opportunistic and pathogenic systemic molds.

The lung is usually the portal of entry for many fungal pathogens. Spores may remain dormant in the lung and present as clinical disease when the patient becomes immunosuppressed. Spores may also disseminate hematogenously. In a study comparing immunocompetent patients to immunocompromised patients with pulmonary cryptococcosis, immunocompromised patients had more extensive pulmonary abnormalities, often characterized by cavities, and higher serum cryptococcal antigen titers (21).

The discovery of several new respiratory viruses, namely, human metapneumovirus and human coronaviruses, has focused attention on respiratory viruses and their importance in this vulnerable population. Some reports indicate that some immunocompromised patients may have nasopharyngeal colonization with human metapneumovirus (32), while others describe severe pneumonia associated with transplant rejection, respiratory failure, and death (42, 59, 91). The type of infection (upper respiratory tract infection versus pneumonia) is influenced by the degree of immunosuppression and the timing of infection in relationship to transplantation. Rapid diagnosis with prompt administration of antiviral therapy, such as in cases of respiratory syncytial virus (RSV) and influenza virus infection, has been shown to reduce the severity of disease (59).

Nosocomial Pneumonia

Immunocompromised patients are particularly at risk for severe nosocomial infections related to a multiplicity of factors. These include alterations in natural host defenses due to immunosuppressant therapy and prescribed antibiotics in the setting of increased exposure to nosocomial pathogens from considerable time spent in health care settings (18). Intubation predisposes patients to ventilator-associated pneumonia, and catheters predispose them to hematogenous involvement of the lung for pathogens such as *Staphylococcus aureus*, aerobic actinomycetes, and resistant gram-negative bacilli. In one prospective study among patients in an oncology medical-surgical intensive care unit, pneumonia was the most common nosocomial infection, accounting for 28.9% of infections (112). Gram-negative infections, which are often multidrug resistant, are occurring with increased frequency and present therapeutic challenges. These include infections with pathogens such as extended-spectrum-β-lactamase-producing *Enterobacteriaceae*, *Acinetobacter*, *P. aeruginosa*, *Stenotrophomonas maltophilia*, and unusual nonfermenting gram-negative bacilli, such as *Chryseobacterium meningosepticum* (now called *Elizabethkingae meningoseptica*) (9, 95, 112). These bacterial pathogens figure prominently among all immunocompromised hosts. *Legionella* from contaminated water systems or construction has been problematic for many institutions, where the mode of transmission to patients is usually aspiration, but aerosolization or direct instillation into the lung may also occur (24). In one multivariate analysis, nosocomial acquisition and the development of pulmonary complications (e.g., cavitation [see above discussion]) were associated with increased mortality (107). The Centers for Disease Control and Prevention (CDC) recommends aggressive surveillance of water samples for centers that specialize in the care of immunocompromised patients (105). In the event of a single nosocomial laboratory-confirmed case, an epidemiologic and environmental investigation should occur (105). Several centers have discovered inapparent nosocomial transmission in their transplant units over an extended period after investigation of individual nosocomial cases (47).

Nosocomial transmission of tuberculosis, especially among AIDS patients, has been reported from numerous centers (17, 70, 89). In many instances, the transmitted strains were multidrug resistant and health care workers also became ill (89). Numerous factors have been cited for nosocomial spread, but delay in recognition due to atypical presentations (and hence delay in treatment) and inadequate infection control are the major factors for nosocomial transmission (70). Clusters of *Pneumocystis* infections among hospitalized patients suggest the possibility of nosocomial transmission, but this has been difficult to prove (70).

Nosocomial spread of RSV has been reported as a significant problem on pediatric units for several decades. More recently, outbreaks of this virus on adult and pediatric transplant units have been the subject of several reports (44, 59). Because of the significant morbidity and mortality associated with this virus among immunocompromised patients, especially HSCT recipients, strict adherence to published infection control measures (22, 105) and prompt treatment for those patients who become infected (22, 59) are mandatory strategies for hospitals with significant numbers of vulnerable populations. In addition, oftentimes patients infected with RSV and other community respiratory viruses will have superinfections caused by bacterial or fungal pathogens (22). The frequencies of opportunistic fungal infections caused by *Aspergillus* sp., *Fusarium* sp., and zygomycetes often increase in hospitals when construction is occurring.

HOST FACTORS AND SUBGROUPS

Pulmonary Infections Associated with Impaired Humoral Immunity

Patients with primary humoral immune disorders and acquired conditions, such as multiple myeloma and chronic lymphocytic leukemia, among others, are most at risk for pneumonia caused by encapsulated bacteria, such as *Streptococcus pneumoniae* and *Haemophilus influenzae*. Prophylaxis for these infections has prevented the consequences of bacteremia, such as disseminated intravascular coagulation. Other pathogens of importance to this group and patients with combined immunodeficiency syndromes include *P. jirovecii*, herpesviruses, and community-associated respiratory viruses (116).

Pulmonary Infections Associated with Neutropenia

Neutropenia may be a consequence of therapy for hematologic and other malignancies or may represent an inherited or acquired immunological disorder (see Chapter 1). In general, regardless of the cause, bacterial infections are most common in the early phases of neutropenia, whereas prolonged and severe neutropenia is associated most often with fungal disease and other opportunistic infections, such as nocardiosis. Bacterial causes of pneumonia are often polymicrobial, with gram-negative bacilli and *Staphylococcus aureus* predominating (83, 95). In the setting of persistent neutropenia (more than 7 days) in patients on broad-spectrum therapy, superinfections with resistant gram-negative bacilli, such as *Acinetobacter* sp., *Alcaligenes*, *Citrobacter* sp., *Enterobacter* sp., and *P. aeruginosa*, as well as other nonfermenters, such as *S. maltophilia* and *E. meningoseptica*, are a major concern (95). Many of these pathogens reach the lung via hematogenous spread. Fungal pathogens are also of concern for groups with prolonged neutropenia (83).

The frequency of aspergillosis is increasing among patients with hematologic malignancies. *Aspergillus fumigatus* is the most frequent pathogen involved, but other species, including *Aspergillus flavus*, *Aspergillus nidulans*, *Aspergillus niger*, *Aspergillus glaucus*, and *Aspergillus terreus*, are also important. The latter species is resistant to amphotericin B. Identification to the species level and differentiation of *Aspergillus* species from other hyaline molds are important for this patient population because of the variability in antifungal susceptibility. Infection is usually acquired by inhalation of spores, which increases in the setting of construction in proximity to the hospital (95). Zygomycetes and *Fusarium* sp. are filamentous fungi that may mimic aspergillosis both clinically and radiographically. In an extensive series by Kontoyiannis et al. (66), 71% of cancer patients who were infected with a zygomycete were neutropenic. *Trichosporon* spp. are less commonly encountered (83). In neutropenic patients with diffuse infiltrates, noninfectious causes are more likely, but in the appropriate clinical setting, pathogens such as *P. jirovecii*, *M. pneumoniae*, *Strongyloides stercoralis*, herpesviruses, and community respiratory viruses should be excluded (83).

Pulmonary Infections Associated with Defects in Cellular Immunity

Patients with impaired cellular immunity include ever-expanding numbers of individuals with lymphoproliferative disorders who are receiving chemotherapy with purine analogs, recipients of HSCT or SOT, and patients receiving corticosteroid therapy. Glucocorticoids are potent immunosuppressants and affect multiple cell lines, including T and B cells, macrophages, granulocytes, and monocytes (39). The infections that develop in these patients are different from those seen in neutropenic patients and patients with isolated humoral defects. In general, herpesviruses and respiratory viruses are significant causes of morbidity and mortality in this group of patients. Protozoan parasites such as *Toxoplasma gondii*, *Enterocytozoon bieneusi*, and the helminth *S. stercoralis* are also important. Other pathogens that play a more prominent role in patients with cell-mediated immune defects are *P. jirovecii*, mycobacteria, and opportunistic and systemic fungi. Detailed discussions of the specifics of infections as they pertain to major vulnerable groups are described in detail below.

HIV/AIDS

HIV causes significant impairment in lung host defenses. Several mechanisms have been elucidated and are reviewed by Beck (6). Several mechanisms of impairment are described briefly here. HIV decreases the numbers of cells that directly kill pathogens. In AIDS patients who are not on HAART, there is a decrease in $CD4^+$ T cells and an increase in $CD8^+$ T cells. The CD4/CD8 ratio may be lower in the lung than in the periphery. Production of T cells is also impaired. The increase in $CD8^+$ T cells may cause alveolitis (lymphoid interstitial pneumonia), whose intensity correlates with viral load. HIV also causes qualitative defects in metabolic and secretory functions of effector cells. For example, HIV infection may impair phagocytosis of organisms that commonly cause infections in normal hosts. Neutrophil defense appears to be impaired. HIV infection also interferes with the capacity of circulating lymphocytes and other cells to migrate to the lung to kill pathogens in alveolar spaces. Coinfections may also contribute to immunological impairment. For example, it has been shown that patients with *Pneumocystis* pneumonia have a higher rate of bacterial pneumonia and that patients with a prior history of CMV disease are at higher risk of developing non-Hodgkin's lymphoma than are those without such a history (104, 119).

HAART has changed the spectrum and epidemiology of pulmonary disease in the HIV-positive patient. Data from the CDC's HIV Outpatient Study, in which 7,300 patients have been monitored longitudinally since 1993, show significant declines in overall hospitalizations for pulmonary disease and decreases in pulmonary morbidity and mortality since 1994 (49). Studies have shown changes in causes of pulmonary infections, namely, a decline in infections due to *P. jirovecii*, *P. aeruginosa*, and opportunistic pathogens such as CMV, *Aspergillus*, and *Cryptococcus neoformans*, among others (37).

Few studies have measured the impact of HAART on bacterial pneumonia. In a study by Sullivan et al., the incidence of bacterial pneumonia declined after implementation of HAART (104). Factors significantly associated with bacterial pneumonia included injection drug use, lower CD4 cell counts, and prior *Pneumocystis* infection (104). Grubb et al. (49) reported an increase in community-acquired pneumonia and nosocomial pneumonia and a decline in *P. jirovecii* infection in the post-HAART era compared to the pre-HAART era. Some centers have reported a decline in Kaposi's sarcoma and non-Hodgkin's lymphoma (49), whereas others have documented a noticeable increase in these diseases in patients on HAART (37). The increases most likely represent proportional, not absolute, increases due to the decline in opportunistic infections and other respiratory diseases. The most important causes of bacterial pneumonia remain *S. pneumoniae*, *H. influenzae*, and *S. aureus*. The stage of immune disease is a major risk factor, although cigarette smoking has been shown to be a risk factor independent of CD4 cell counts (11).

Tuberculosis remains an important disease even in the HAART era, and HIV infection is still the largest risk factor for development of tuberculosis worldwide. Multidrug-resistant tuberculosis is also more common in HIV-positive patients (1).

An interesting consequence of HAART has been the development of immune restoration disease. This is believed to be related to an enhanced T-cell-mediated immunopathological response to infection. The patients most likely to develop this syndrome are those who have CD4 counts of <50 cells/μl and high viral loads and who have a good response to treatment (49). Infectious diseases that appear prominently during immune restoration are mycobacterial infections, including infections with *M. tuberculosis*, MAC, *Mycobacterium leprae*, and *Mycobacterium xenopi* (49).

Antiretroviral therapy itself should also be considered when other etiologies of pulmonary infiltrates have been excluded. For example, abacavir has been associated with a hypersensitivity reaction in 3.7% of patients who use the drug (54). Although gastrointestinal symptoms are the most prominent manifestations, patients may have fever, pharyngitis, cough, tachypnea, and pulmonary infiltrates (54).

In summary, although most of the published studies are observational, pulmonary infections in the HAART era have declined significantly, with most noticeable differences occurring in the rates of infection with opportunistic pathogens. Bacterial pneumonia, tuberculosis, and noninfectious problems such as malignancies and drug reactions continue to be significant causes of pulmonary syndromes among this patient population.

SOT patients

Pneumonia is the most frequent infectious complication after SOT (18) and is most frequent in the period immediately following transplantation (first 30 to 100 days) (18, 47, 99, 108). The etiology of pneumonia among SOT recipients varies at different times posttransplantation, but the spectra of pathogens are similar across transplantation groups. During the first month after transplantation, $>95\%$ of infections are due to bacteria that are nosocomial pathogens (18,

39, 47), and the infections are comparable to those in any postoperative patient. Reported gram-negative organisms include *Pseudomonas*, *Klebsiella*, *Escherichia*, *Legionella*, *Acinetobacter*, *Stenotrophomonas*, *Enterobacter*, *Serratia*, *Proteus*, and *Citrobacter* spp. Notable gram-positive organisms include *Staphylococcus*, *Corynebacterium*, *Enterococcus*, and *Streptococcus* spp. *Nocardia* and anaerobes are rare (39). Community-acquired pneumonia tends to occur any time after transplantation, although it most often appears 6 months after transplantation. *S. pneumoniae*, *H. influenzae*, and *Legionella* species are the most commonly identified pathogens (47). Nontuberculous mycobacteria (NTM) are uncommon causes of pulmonary infection, except in the lung transplant recipient (47, 77). Species commonly recovered from SOT recipients include MAC, *Mycobacterium kansasii*, *Mycobacterium fortuitum*, *Mycobacterium chelonae*, *Mycobacterium abscessus*, and *Mycobacterium haemophilum* (47, 77). *M. tuberculosis* is uncommon in developed countries (<2% of cases), but the risk is 70-fold greater than that for the general population (39). Infection with *M. tuberculosis* is very important in countries where tuberculosis is endemic (20, 47). Disseminated infection is more frequent in the transplant population than in healthy hosts. Finally, one must consider that in a significant percentage of cases an etiologic agent is never discovered (47).

In the period 1 to 6 months after SOT, infections caused by immunomodulating viruses, such as CMV, and those caused by opportunistic pathogens, such as *P. jirovecii*, *Listeria monocytogenes*, *Aspergillus* sp. (incidence, 18 to 22%), *Nocardia* sp., and *T. gondii*, are the most prevalent. However, routine prophylaxis with trimethoprim-sulfamethoxazole has significantly reduced the incidence of *P. jirovecii* and *T. gondii* infections in SOT recipients (18, 99). *Nocardia* species cause infection in 0.7 to 3% of transplant recipients. Concomitant infection with CMV, extreme immunosuppression, and hypogammaglobulinemia are important risk factors (47). Patients are at risk for dissemination from the lungs to other foci, particularly the brain. Reactivation of latent infection (with *M. tuberculosis*, *Histoplasma capsulatum*, or *Coccidioides immitis*) present before transplantation may occur during this period. As mentioned above, the incidence of tuberculosis is low in developed countries, but in areas where it is endemic tuberculosis causes significant morbidity and mortality.

During the late period (>6 months posttransplant), populations can be divided into several groups whose risk of infection varies, as follows. (i) For patients with good results from transplantation (about two-thirds of patients), the major risk is from community-associated bacterial pathogens (see above) and respiratory viruses (see Chapter 7 for detailed discussion). (ii) Ten to 15% of patients have chronic viral infections, such as hepatitis B or C virus infection. (iii) The remaining 5 to 10% of patients, who have poor allograft function, require chronic immunosuppression and are at greatest risk for opportunistic pathogens such as *Cryptococcus neoformans*, *P. jirovecii*, *L. monocytogenes*, and emerging filamentous fungi such as *Trichoderma*, *Pseudoallescheria boydii*, *Microascus*, *Penicillium*, zygomycetes, and especially *Rhizopus* and *Absidia*. Rare causes of infection in this group include *Strongyloides* and microsporidia (18, 39, 96, 100).

The incidence and timing of infections are affected by the type of organ transplanted, the degree of immunosuppression, the need for additional antirejection therapy, the occurrence of problems during surgery, exposure to a donor pathogen, and a variety of environmental and epidemiological factors (100). Infections with hepatitis C virus, CMV, Epstein-Barr virus (EBV), and HIV increase the likelihood of opportunistic infections. As mentioned above, the timing with which certain infections occur has historically been somewhat predictable. However, prophylactic regimens now make predictions unreliable. Figure 1 presents timelines from pre- and postimplementation of standard prophylactic regimens among SOT recipients. The following paragraphs briefly discuss issues peculiar to specific transplanted populations.

Lung transplantation. The main obstacles to long-term success with lung transplantation are chronic rejection in the form of obliterative bronchiolitis (OB) and infections (102). CMV is the main contributing factor to OB. Infectious complications are the most common cause of morbidity and mortality at all times following lung transplantation and are two times more common among lung transplant patients than among heart transplant recipients. Two-thirds of the infections involve the respiratory tract.

There are several predisposing factors unique to lung transplantation. The lung is the only transplant continuously exposed to the environment. Denervation of the allograft leads to abnormal ciliary clearance, and diminished cough reflex and ischemia lead to anastomotic narrowing and interruption of lymphatic drainage (102). Finally, the native lung may harbor occult infection that reactivates after immunosuppression (18, 39, 102). Like the case for other SOT recipients, the donor lung may also transmit infections (39, 102).

The incidence of bacterial pneumonia ranges from 35 to 66%. Early in the posttransplant period, the causative microorganisms originate from the

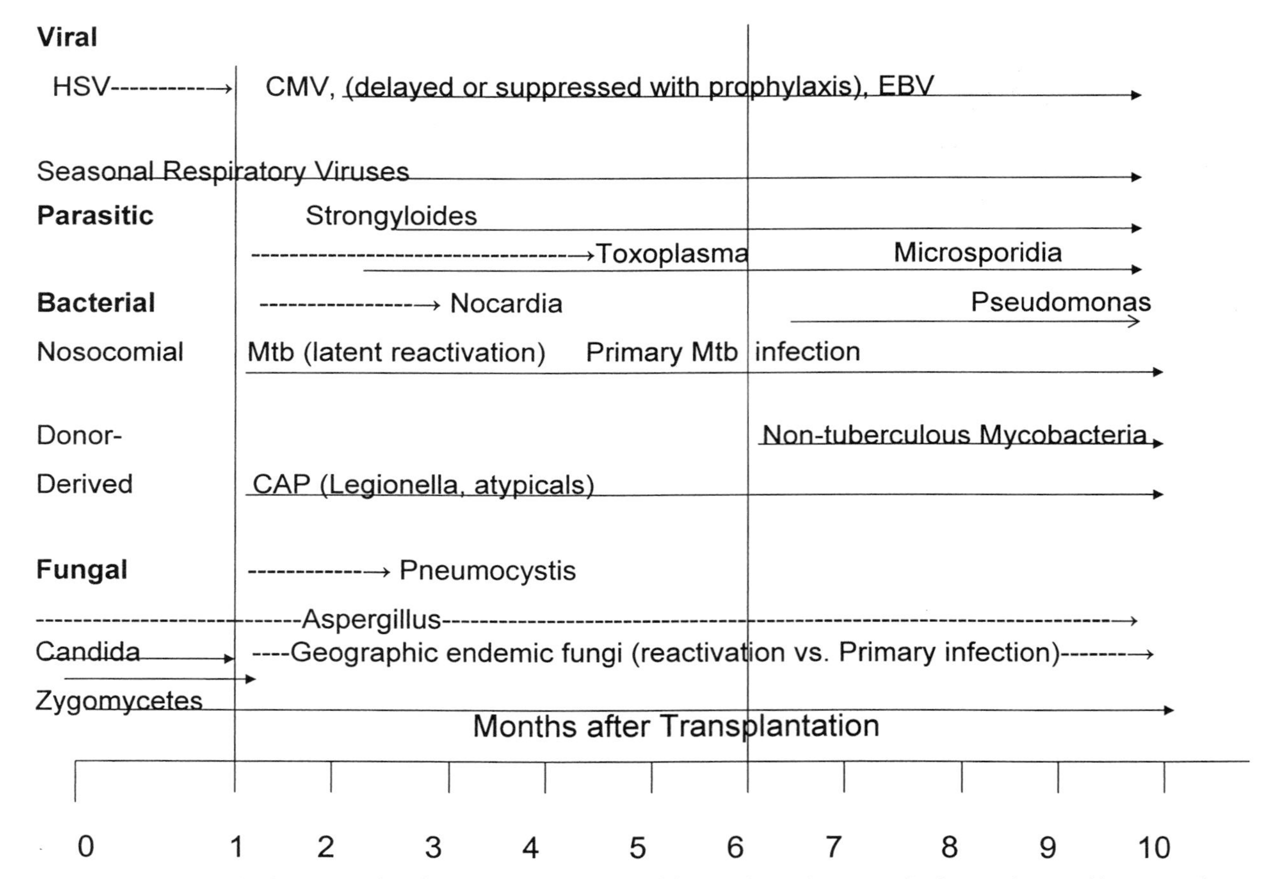

Figure 1. Proposed infection timeline for SOT recipients. Dotted lines indicate the onset of infection that would occur without prophylaxis; solid lines indicate the most common times to onset of infection for each pathogen. Time zero indicates the time of transplantation. CAP, community-acquired pneumonia; Mtb, *Mycobacterium tuberculosis*. (Modified with permission from reference 39.)

donor lung. However, the incidence of early episodes of bacterial pneumonia has decreased as a result of antimicrobial prophylaxis. Unfortunately, this has caused a shift of the occurrence of bacterial pneumonia to later in the posttransplant period. The majority (75%) of infections are caused by *Pseudomonas* species and *Enterobacteriaceae*; *S. aureus*, enterococci, and *H. influenzae* are also important. Patients with cystic fibrosis are often colonized with *Burkholderia cepacia*, *S. maltophilia*, and *Alcaligenes xylosoxidans*. *C. pneumoniae* infection has been associated with early graft rejection, bronchiolitis obliterans syndrome, and early mortality (48). *M. tuberculosis* may reactivate from remaining native lung (in cases of single organ transplantation) or may be transmitted occultly by the donor lung. In lung transplant recipients, infections with NTM may be common. In one large series, 8.8% of lung transplant recipients developed both pulmonary and extrapulmonary NTM infections (77). Many candidates for lung transplantation, such as patients with cystic fibrosis, are colonized with NTM pretransplantation, but it is unclear if this enhances the risk of posttransplantation infection (47).

Beyond the first month, viral pathogens are the most important group, representing the second most common cause of infections in lung transplant recipients (23 to 31% of all infections) (39). The frequency of CMV disease is higher than that in other SOT recipients; prophylaxis and monitoring are very important. Reactivation of herpes simplex virus (HSV) is usually prevented with antiviral prophylaxis, but severe pneumonia may occur in 10% of patients without prophylaxis (102). The incidence of EBV-related posttransplant lymphoproliferative disorders varies (2 to 33%) but is higher for EBV-negative recipients and patients transplanted for cystic fibrosis (93, 102). Infections caused by respiratory viruses range from asymptomatic illness to severe pneumonia; influenza A and B viruses, RSV, and adenovirus lead to significant pneumonitis (8). Parainfluenza viruses, adenoviruses, and RSV have been linked directly to OB in many patients (39, 59). Adenovirus infections are particularly severe in pediatric patients, with associated high mortality and morbidity (8).

Fungal infections are also common. *Aspergillus* colonization is common in this group (13). In a study by Mehrad et al. (79), 26 to 29% of patients developed airway colonization, and in the Cahill study (13), 46% of 151 transplant recipients had airway colonization. Since colonization can lead to invasive disease, empiric therapy with itraconazole at 200 mg twice a day for 6 months in patients with airway colonization is indicated and has been shown to successfully eradicate the organism (13, 79). Semi-invasive forms can occur at anastomotic sites and in the large airways, with the latter manifesting as tracheobronchitis (39). Isolated tracheobronchitis occurs in 4 to 5% of patients, is usually asymptomatic, and is often found on surveillance bronchoscopy. Ulcerations may be extensive (79). However, tracheobronchitis can be associated with fever, wheezing, cough, and hemoptysis. Invasive disease occurs in 5 to 8% of patients, usually within the first year, and presents in a manner similar to that for other immunocompromised hosts. Mortality is high, at approximately 60% (79).

The frequency of *P. jirovecii* infection varies; prevalence is high for patients not on prophylaxis. *T. gondii* infection occurs almost exclusively in heart-lung recipients.

Heart transplantation. The incidence of pneumonia has declined substantially over the past several decades, from 50 to 60% in the precyclosporine era to 14 to 21% in more recent series (25, 39, 73). In several series, pneumonia was the most common pulmonary complication and the most frequent infectious complication post-cardiac transplantation (25, 73). Pneumonias in the first 6 months are bacterial or fungal and nosocomial in origin, similar to the case for other SOT populations (see above discussion). The incidence of CMV and *P. jirovecii* infection has declined due to preemptive screening for viremia in the case of CMV and to prophylaxis for both pathogens. Prophylaxis for *P. jirovecii* with trimethoprim-sulfamethoxazole likely accounts for the decline in the incidence of nocardia pneumonia as well. Although *P. jirovecii* infection has declined in incidence, some centers have reported late-onset cases (>6 months posttransplantation) during periods of intensified immunosuppression, as in the case of treatment for acute rejection (72). During episodes of the latter, prophylaxis should be reinstituted. In the second 6 months posttransplantation, the etiology of pneumonia is similar to that for other patients with community-acquired pneumonia (73). The highest mortality rates are seen with *Aspergillus* and nosocomial pneumonias, at 50 to 75% and 33%, respectively (25, 73). Toxoplasmosis is highest among seronegative cardiac transplant recipients who receive an organ from a seropositive patient when prophylaxis has not been administered (81).

Liver transplantation. The lung is the second most common site of infection post-liver transplantation. Although the incidence of bacterial pneumonia has declined to <10% (108), bacteria remain the most common etiologic agents of pneumonitis following liver transplantation (92, 99, 108). *P. aeruginosa* was the first or second most common bacterial pathogen in three reported series (92, 99, 108). *Legionella* species were important causes of community and nosocomial pneumonia in the Pittsburgh series (99). Bacterial infections occur within the first 6 months posttransplantation but predominate in the first month, and the majority of these are nosocomial (99). Risk factors for bacterial pneumonia include older age, prolonged mechanical ventilation, and receipt of more intraoperative transfusions than normal (92, 108). Significant factors associated with increased risk for the development of pneumonia caused by any etiology include recurrent hepatitis C and severity of pretransplant disease (99). Opportunistic pathogens occur most frequently after the first month posttransplantation. CMV, *P. jirovecii*, and *Aspergillus* sp. are the most commonly reported opportunistic pathogens (108), although as noted above, CMV and *P. jirovecii* infections have declined significantly in recent series. Unfortunately, the contribution of fungi and their associated mortality remain significant, especially among those patients with risk factors (99).

Kidney transplantation. The incidence of pneumonia among renal transplant recipients is the lowest of all SOT recipients. Most occur in the first 12 months posttransplantation. The most common infections responsible for pneumonia during this time in one large study were bacterial and mixed bacterial infections (*S. aureus*, *S. pneumoniae*, and gram-negative bacilli predominated), fungal infections (*Cryptococcus* and *Aspergillus* spp.), and tuberculosis (20). Similar to what has been described for other SOT recipients, the incidence of *P. jirovecii* and CMV infection has declined, except in those patients with prolonged immunosuppression or in those patients noncompliant with prophylaxis (20). Mortality rates have declined in most series (20).

HSCT

Pulmonary infections are the most common infectious cause of death in HSCT recipients. As is true for SOT recipients, multiple factors other than immunosuppressant therapy predispose patients to infection. These factors include chemotherapy/ radiation-induced neutropenia, lung injury induced by conditioning regimens, and rejection in the form of graft-versus-host disease (GVHD). Autologous transplantation is associated with the least risk of infection. Higher rates of infection are seen with allogeneic transplantation, with matched unrelated donor transplants having the greatest risk of infections (2, 23, 47). Greater mismatch is associated with greater risk of GVHD, which results in impaired opsonization and reticuloendothelial function (47). In addition, immunosuppressants used to treat GVHD result in further defects in cellular immunity (23, 47, 121). Serious infection occurs in the initial 2-year period posttransplantation in 50% of uncomplicated transplant recipients with HLA-compatible siblings and in 80 to 90% of matched unrelated donors or histocompatible patients who develop GVHD (39). The posttransplant period is usually divided into three defined periods (Fig. 2). Phase 1 is the first 30 days after transplantation and includes the preengraftment period, phase 2 (early postengraftment) includes days 31 to 100, and phase 3 (late postengraftment) is more than 100 days posttransplantation.

In the preengraftment period, patients are most at risk for hospital-acquired bacterial infections, since this is the period of neutropenia and disruption of mucosal barriers (47, 121). The exact frequency of bacterial pneumonia is unknown, largely because patients are treated with broad-spectrum antimicrobial therapy at the first sign of neutropenic fever. When an etiologic agent is recovered, most often it is a resistant gram-negative bacterium, such as *P. aeruginosa* or *K. pneumoniae*. *Legionella* infections, usually in the setting of an outbreak related to contaminated potable water in hospitals (53), have also been reported, although they occur with less frequency than in SOT recipients (47). Infections with non-*L. pneumophila Legionella* species have been reported to have a better outcome than that of infections with *L. pneumophila* (53). Antibiotic prophy-

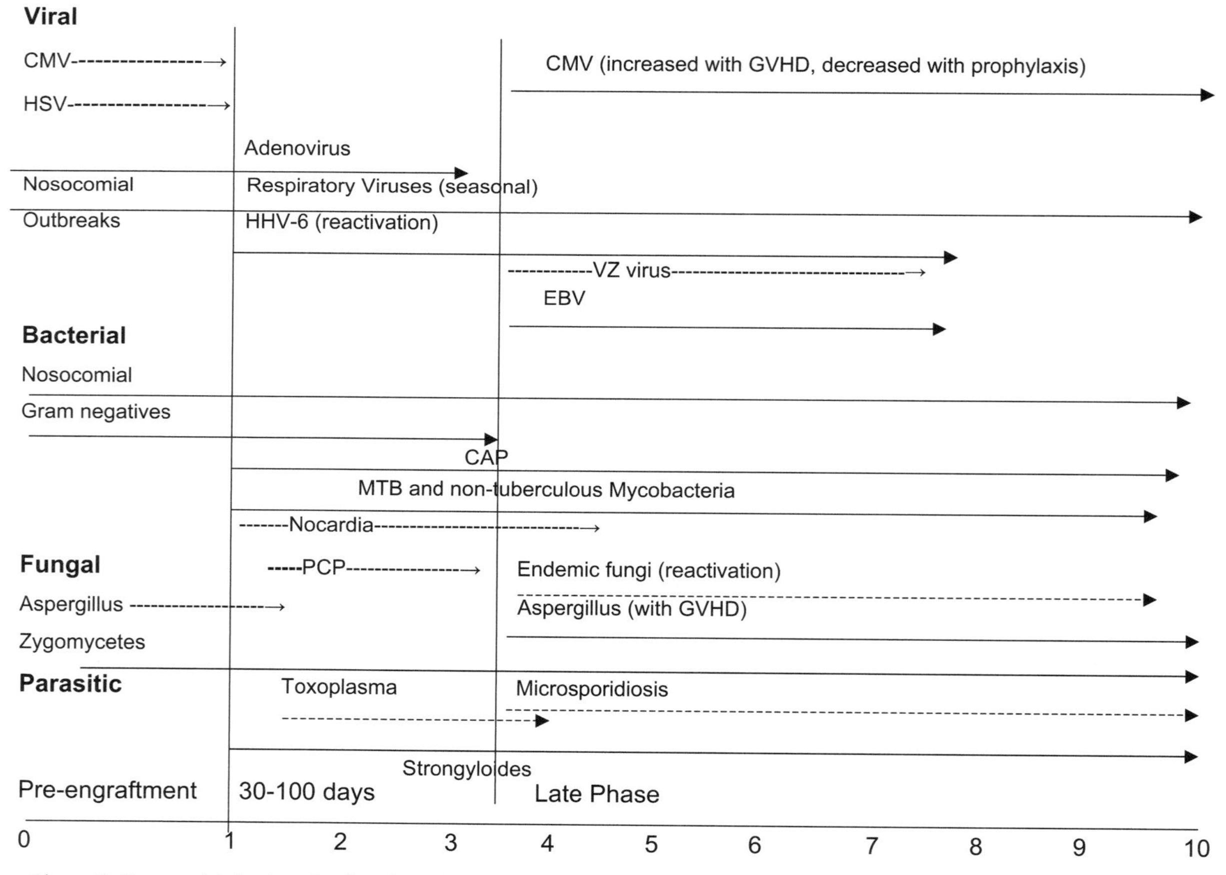

Figure 2. Proposed infection timeline for HSCT recipients, based on the use of common prophylaxis. Dotted lines indicate the onset of infection that would occur without prophylaxis; solid lines indicate the most common times to onset of infection for each pathogen. Time zero indicates the time of transplantation, and each number following indicates months posttransplantation. CAP, community-acquired pneumonia; MTB, *Mycobacterium tuberculosis*. (Modified with permission from reference 39.)

laxis has reduced the occurrence of opportunistic pathogens, such as *P. jirovecii*, herpesviruses, and opportunistic fungi. Despite this, *Aspergillus* remains an important pathogen and has a bimodal distribution (Fig. 2). The first peak occurs around 16 days posttransplantation, and the second peak is seen at the end of phase 2. Older male allogeneic patients appear to be most at risk, and infections seem to occur more frequently in the summer months and in institutions that have construction occurring in the vicinity of the hospital (121). Fever, dyspnea, dry cough, wheezing, pleuritic chest pain, and hemoptysis are clinical features consistent with invasive pulmonary aspergillosis (121). High-resolution computed tomography (HRCT) typically shows large (>10 mm) or small (<10 mm) nodules in almost all cases of pulmonary aspergillosis (43). HSV infections due to reactivation may be seen in patients who do not receive acyclovir prophylaxis. Most patients with HSV pneumonitis have obvious mucocutaneous infections in the form of either oropharyngeal or esophageal disease (121).

Noninfectious complications, such as pulmonary edema and diffuse alveolar hemorrhage, may also occur during this period and may be confused with infectious etiologies. Engraftment syndrome is a noninfectious pulmonary complication typically occurring at about 7 days posttransplantation, usually in patients with autologous or peripheral stem cell transplants (121).

Postengraftment bacterial infections are less common, unless the patient develops GVHD. Infections are more common in the early postengraftment period (phase 2 [30 to 100 days]) and less common in the late postengraftment period. Focal infiltrates on chest radiographs are likely to be bacterial, whereas interstitial infiltrates are likely to be CMV or other viral infections (Table 1).

Nocardia infections occur in the late postengraftment period and are seen almost exclusively in allogeneic bone marrow transplant (BMT) recipients. Frequently, *Nocardia* infections occur in patients coinfected with other opportunistic pathogens. Pulmonary infections with NTM are less commonly reported for HSCT recipients than for SOT recipients (47). Extrapulmonary infections, such as catheter-related infections, may occur more frequently. Tuberculosis is also uncommon. During phase 2, viral and fungal infections predominate. CMV and respiratory viruses, such as RSV, influenza virus, and parainfluenza virus, are of most concern. Patients with HSCT are at increased risk of CMV pneumonia in weeks 6 through 12 posttransplantation. Anti-CMV prophylaxis has changed the usual onset from the first 100 days to beyond the first 100 days, when prophylaxis is usually discontinued (Fig. 2). Patients with GVHD, allogeneic transplant recipients, and patients whose conditioning regimens included total body irradiation are at increased risk of CMV disease (121). Clinically, fever, dry cough, dyspnea, and hypoxemia are seen in the setting of interstitial infiltrates on chest radiographs. A variety of abnormalities may be seen on HRCT studies. Patchy ground-glass opacities are the most common features; however, patients with CMV pneumonitis may also have nodular infiltrates and air space consolidation (43).

Table 1. Radiographic appearances of pulmonary infiltrates in the immunocompromised host and likely etiologic agents[b]

Onset	Likely etiologic agent(s) for type of infiltrate[a]					
	Nodular	Perihilar	Multifocal	Diffuse	Consolidation	Cavitation
Acute	*Legionella* sp.	CHF, *P. jirovecii*	*Legionella* sp., *C. pneumoniae*, *M. pneumoniae*, CHF, pulmonary emboli	Respiratory viruses and herpesviruses, *P. jirovecii*, ARDS, alveolar hemorrhage	Bacterial pneumonia, *M. pneumoniae*, pulmonary infarct, CHF, *Aspergillus*	*P. aeruginosa*, *Klebsiella* sp., *S. aureus*
Subacute/ chronic	*Aspergillus*, other opportunistic fungi, dimorphic fungi, *Nocardia*, *Cryptococcus*, mycobacteria	RSV, CMV, *Cryptococcus*	*Aspergillus*, *Nocardia*, *M. tuberculosis* NTM	Invasive fungi, CMV, HHV-6, miliary tuberculosis, drug-induced pneumonitis, radiation, pneumonitis, leukoagglutinin reactions, lymphangitic spread of malignancies, BOOP	*Aspergillus*, *Nocardia*, *Cryptococcus*, carcinoma	*Nocardia*, *Aspergillus*, anaerobic lung abscess, *M. tuberculosis*, NTM

[a]ARDS, acute respiratory distress syndrome, BOOP, bronchiolitis obliterans organizing pneumonia; CHF, congestive heart failure.
[b]Modified with permission from reference 31.

Respiratory viruses in HSCT recipients may also be associated with severe morbidity and mortality. HSCT recipients are at greatest risk for severe infections. Patients infected preengraftment are most at risk for progression to pneumonia and subsequent death. Numerous transplant centers have reported the impact of RSV infections on morbidity and mortality in the HSCT population (59, 75). While the majority of infections appear to involve the upper respiratory tract, severe lower respiratory tract infections have been described (59, 75). In one large prospective study of respiratory virus infections among HSCT recipients in Europe, the majority of patients with severe RSV infections and all patients who died were allogeneic HSCT patients (75). Risk factors for the development of lower respiratory tract disease included lymphopenia, prior lung disease, and onset prior to HSCT or prior to engraftment (59, 75). Lymphopenia and allogeneic transplantation also appear to be significant risk factors for the development of influenza virus pneumonia, which also carries an attributable mortality ranging from 15.3% to 45% (59, 75). Upper respiratory tract symptoms, including cough, coryza, and rhinorrhea with associated fever, predominate in most patients who acquire parainfluenza virus infections. Patients with upper respiratory tract infections usually survive, but those who progress to pneumonia usually have a high mortality rate (59). Adenoviruses cause a variety of syndromes in HSCT recipients and other immunocompromised patients. These include respiratory, genitourinary, gastrointestinal, and central nervous system manifestations. Patients with pneumonia, with or without disseminated disease, have an extremely high mortality, of ~73 to 80% (59).

Several reports have emerged on the pathogenicity of human herpesvirus 6 (HHV-6) in this patient population. Fever, myelosuppression, delayed engraftment, rashes, interstitial pneumonitis, and encephalitis have been described (95, 122). There is some evidence demonstrating that severe interstitial pneumonia caused by HHV-6 occurs in the adult HSCT recipient, but data are less compelling for pediatric patients (122).

During phases 2 and 3, approximately 10 to 17% of allogeneic BMT patients develop idiopathic pneumonia syndrome (2, 23, 26, 62, 121). This is defined as widespread alveolar injury characterized by hypoxemia and bilateral pulmonary infiltrates in the absence of active lower respiratory tract infection (negative bronchoalveolar lavage [BAL] fluid and no response to antimicrobial therapy) (2, 26). The exact pathogenic mechanism is unclear, but this syndrome is seen most often in patients with risk factors such as high-dose total body irradiation and GVHD (121). Often, the clinical course of idiopathic pneumonia syndrome is complicated by superinfection with viruses and fungi (2). Mortality is high (70 to 90%) (2, 23, 62, 121).

In phase 3 (>100 days posttransplantation), the development of chronic GVHD often dictates the type of pulmonary complication and is an independent risk factor for the development of late pneumonia (23). CMV and fungal infections may occur due to termination of prophylaxis and prolonged immunosuppressive therapy (47, 121). Infection with other viruses, such as varicella-zoster virus, continued risk of infection from community-acquired respiratory virus pathogens, and EBV-related lymphoproliferative disease may emerge during phase 3. Most transplant centers report the elimination of *P. jirovecii* as a significant pathogen in this group due to prophylaxis. When it does occur, *P. jirovecii* infection is seen almost exclusively in patients not taking trimethoprim-sulfamethoxazole prophylaxis (23). Bacterial pneumonias with encapsulated bacteria (*S. pneumoniae* and *H. influenzae*) are common and, followed by gram-negative bacteria, are the most frequently recovered organisms in situations where a pathogen is recovered (23). OB is characterized by irreversible airflow obstruction on pulmonary function testing. Patients complain of dry cough, dyspnea, and wheezing. This syndrome is reported to occur most commonly in long-term survivors who have chronic GVHD (2, 121). The likely pathogenic mechanism is induction of bronchial epithelial injury by immunologic mechanisms (121). The frequency of superinfection with respiratory pathogens is high for patients with OB, as is mortality (121).

Other Vulnerable Populations

Patients with collagen vascular diseases

Collagen vascular diseases include rheumatoid arthritis, systemic lupus erythematosus, Sjogren's syndrome, polymyositis, dermatomyositis, systemic sclerosis, and granulomatous vasculitic conditions. These conditions are associated with immune dysregulation, as outlined in Chapter 1. These diseases can be associated with mild to moderate infections, but the therapies used to treat them often result in significant immunosuppression and subsequent severe infections, as mentioned above (51). Treatments include corticosteroids; methotrexate, which induces neutropenia; cyclophosphamide, which induces lymphopenia and suppresses T- and B-lymphocyte activity; and the newest agents, tumor necrosis factor alpha (TNF-α) inhibitors. The latter selectively inhibit

TNF, which subsequently down-regulates other cytokines in the proinflammatory immune pathways. TNF is the cytokine most essential for formation and maintenance of granulomas.

Patients on anti-TNF agents and lymphocyte-depleting monoclonal antibodies

The following two types of anti-TNF antibodies are licensed for clinical use: infliximab (Remicade; Centocor), a partially humanized monoclonal antibody, and adalimumab (Humira; Abbott Laboratories), a fully humanized monoclonal antibody. Etanercept (Enbrel) is a dimeric soluble TNF receptor. These agents are used to treat rheumatoid arthritis and other diseases in which TNF is believed to contribute to the pathophysiology of the disease. Most reports of infection are with infliximab treatment, which could be an artifact of usage and reporting or could reflect the fact that this is the most potent inhibitor of TNF-α.

A recent meta-analysis showed an increased risk for serious infection (odds ratio, 2.0; 95% confidence interval, 1.3 to 3.1) and malignancies (odds ratio, 3.3; 95% confidence interval, 1.2 to 9.1) in patients treated with infliximab and adalimumab (10). Tuberculosis is the most frequently reported serious infection, with a reversal of the normal pattern of disease, that is, more extrapulmonary disease (65%) than pulmonary disease (114). An increase in granulomatous infections due to NTM, *Nocardia* sp., *L. monocytogenes*, *H. capsulatum*, *C. neoformans*, *Aspergillus* species, *Candida* sp., and *C. immitis* has been noted as well (10, 114). The median time to onset of these infections was 40 days, suggesting reactivation of latent infection (114). In the same study, by Wallis et al. (114), granulomatous infections were 3.25-fold greater among patients who received infliximab than among those who received etanercept.

Currently, two monoclonal antibodies directed against antigen present on lymphocytes, namely, rituximab and alemtuzumab, are available for use in the management of hematologic malignancies. Alemtuzumab is a humanized monoclonal antibody that is directed against CD-52, a glycoprotein found on B and T lymphocytes. This antibody is used for the treatment of chronic lymphocytic leukemia and is also used frequently to prevent or treat acute allograft rejection in SOT recipients (78, 90). This agent causes profound lymphocyte depletion, and at least one large study has shown a significant increase in opportunistic infections among transplant patients treated for allograft rejection with alemtuzumab compared to patients to whom the drug was given for induction therapy (90). Many of the patients developed late infections (>200 days after transplantation), during a period when both routine monitoring for diseases such as CMV and antibiotic prophylaxis have been discontinued. This has implications for diagnostic and empirical treatment algorithms (90). Rituximab is a human monoclonal antibody directed against CD20 antigen present on B cells. It is frequently used to treat patients with a variety of B-cell malignancies (78). Similar to the experience with alemtuzumab, >50% of patients treated with this agent developed an opportunistic infection, usually with a viral or fungal pathogen (78).

Alcoholism

Chronic alcoholism predisposes patients to a number of serious impairments of normal respiratory host defense mechanisms, resulting in enhanced vulnerability to bacterial infections and *M. tuberculosis* (52). Both a decrease in saliva production and impairment in the normal acidic buffering capacity lead to gingival disease and consequent increased oral cavity colonization with anaerobes and gram-negative bacteria, such as *K. pneumoniae* (52). Altered mental status during periods of intoxication presents opportunities for aspiration of microbes into the lung. Once organisms reach the tracheobronchial tree, a combination of decreased ciliary activity and alcohol-induced impairments in innate and adaptive immunity prevents the host from resisting serious infections caused by organisms such as *S. pneumoniae*, *S. aureus*, *L. pneumophila*, *K. pneumoniae*, and *M. tuberculosis* (52). Complications such as bacteremia, abscess formation, and acute respiratory distress syndrome (ARDS) are a consequence of these bacterial infections in the alcoholic patient and result in extremely high mortality.

Patients with noninfectious pulmonary entities that may mimic pneumonia

A comprehensive discussion of the noninfectious entities that may mimic infectious causes of pulmonary infiltrates is beyond the scope of this chapter. However, these entities should be considered in the differential diagnosis of pulmonary syndromes among certain vulnerable populations. Noninfectious etiologies of pulmonary infiltrates account for 25 to 50% of abnormal radiographs (98). Some of these are mentioned in the sections above and are incorporated into the discussion on diagnosis below, but a brief summary of the more common conditions is again mentioned here. Cardiac and noncardiac pulmonary edema is often seen in the second or third week after BMT and has also been reported to be a problem in renal and liver transplant recipients and patients with hematologic malignancy (64, 98). Diffuse alveolar hemorrhage is characterized by the sudden onset of dyspnea, nonproductive cough, fever, and hypoxemia. Interstitial infiltrates on chest radiographs may be confused with opportunistic infections (64). Bronchoscopy is required to differentiate hemorrhage from infection. Likewise, radiation pneumonitis may also cause cough, fever, and abnormal chest radiographs. Pulmonary function tests showing a restrictive pattern and a reduced lung capacity, along with characteristic CT findings, are helpful in elucidating the diagnosis in patients at risk (64). Drug reactions can take the form of hypersensitivity pneumonitis, such as occurs with cytotoxic chemotherapy agents, or interstitial pneumonitis and pulmonary vasculitis, as reported for sirolimus, an agent frequently used in SOT patients (19, 98). Pulmonary embolic disease is often a problem in renal transplant patients, and veno-occlusive disease is a rare vascular complication following HSCT. In the latter circumstance, although it may be suspected on the basis of clinical features, open lung biopsy is often required to confirm the diagnosis (64). Histopathology demonstrates occlusive lesions of the small venules of the lung caused by fibrous proliferation of the intima (64). Primary malignancy and metastatic disease, such as lymphangitic spread of carcinoma, are important noninfectious causes of infiltrates in patients with AIDS (37).

DIAGNOSTIC APPROACHES AND INTERPRETATION

Patients with lower respiratory tract infections may present with a variety of clinical and radiographic manifestations, which may be altered by the underlying immune defect. Some of the clinical manifestations of the various syndromes have been elaborated upon in the sections above. In general, fever, dyspnea, chest pain, cough, and hypoxemia, even in the absence of other pulmonary symptoms, should prompt immediate diagnostic evaluation. Empiric therapy can often be guided initially by the type of radiographic appearance (Table 1), the pace of the disease and/or infiltrates, knowledge of preexisting exposures, and the status of the underlying immunodeficiency (such as CD4 count for patients with HIV disease) or immunocompromised state (elapsed time since transplantation). Finally, it is imperative to keep in mind that both infectious and noninfectious etiologies can produce fever, leukocytosis, and pulmonary infiltrates.

Radiography

Plain chest radiographs are often inadequate for the immunocompromised host but may be helpful in the setting of new infiltrates. In addition, although findings may be nonspecific, certain patterns may be suggestive of particular categories of infectious diseases and can be helpful in triaging both empiric therapy and further diagnostic management (Table 1). If the pattern is one of diffuse infiltrates or peribronchial infiltrates, the infectious causes include viral pneumonia and *P. jirovecii* (113). Noninfectious entities to be entertained include pulmonary hemorrhage, leukoagglutination reactions, ARDS, and congestive heart failure. Focal air space opacities, especially if they are acute, are likely to be caused by bacteria; if they are subacute or chronic, fungi, *Nocardia*, and mycobacteria are more likely. The differential diagnosis of multifocal air space opacities includes bacteria, *P. jirovecii*, and fungi such as *Cryptococcus* and *Aspergillus* (113). If the infiltrate has a nodular component, bacterial pathogens such as *S. aureus*, *Legionella*, and *Nocardia* should be sought, but in many cases, fungal or mycobacterial infections are responsible. Cavitary disease is most likely to be caused by fungi, *S. aureus*, *Klebsiella*, *P. aeruginosa*, *Nocardia*, and *M. tuberculosis*.

Plain chest radiographs are usually followed quickly by CT images. HRCT has replaced conventional CT imaging in many institutions. Oftentimes, patients who have normal-appearing plain chest radiographs will have abnormalities on HRCT, which is a highly sensitive technique. What follows is a brief description of HRCT findings for some of the more common opportunistic pathogens. Bacterial pneumonia usually causes asymmetric areas of consolidation, and in one report of HRCT in HSCT recipients, a significant number of patients (81%) also had nodular opacities (43). *P. jirovecii* causes ground-glass areas, and a negative HRCT result essentially rules out *P. jirovecii* infection (113).

Aspergillus spp. characteristically produce both large and small nodules that are surrounded by a halo of ground-glass attenuation (halo sign) or pleural wedge-shaped areas of consolidation which correspond to hemorrhagic infarcts (43, 113). The halo sign, while suggestive of aspergillosis, can also be seen in other conditions, such as zygomycosis, Kaposi's sarcoma, and HSV and CMV infection (113). Multifocal bilateral ground-glass opacities are the predominant abnormality seen in patients with CMV (43, 113). Various combinations of nodules, ground-glass attenuation, and consolidation have been described for pneumonia caused by respiratory viruses such as RSV (43). Although some radiographic patterns are quite suggestive of particular pathogens (such as large nodules and the halo sign with aspergillosis), most findings are nonspecific and require confirmation by histopathology and directed microbiological studies.

Blood Cultures

Any immunocompromised patient with fever should have blood cultures obtained. Adherence to current recommendations for specimen collection and appropriate volume are required to ensure adequate recovery and interpretation of results (27). Some recommend collection of 30 ml of blood per culture for this patient population (28). Blood cultures may be of limited value for patients on empiric antibiotics. However, for neutropenic patients and other populations at risk for bacterial pathogens, particularly *S. pneumoniae*, *S. aureus*, and *P. aeruginosa*, the blood cultures are often positive. In general, 15 to 30% of pneumonias in immunocompromised patients are associated with positive blood cultures (28). In SOT recipients, bacteremia is common in the early or late postoperative period, occurring in 13 to 29% of heart transplant recipients (25) and in up to 25% of patients with lung transplants. *S. aureus*, *P. aeruginosa*, and *Candida* spp. are the most common pathogens (102). Most yeasts, including *C. neoformans*, grow well in the nonradiometric, continuously monitored blood culture systems currently in use in clinical laboratories. When systemic or opportunistic fungi or mycobacterial infections are under consideration, the lysis centrifugation system, MycoF lytic medium (Becton Dickinson, Sparks, MD), or other media designed specifically to recover these organisms are recommended. Positive blood culture results may obviate the need for more invasive respiratory studies, particularly for the neutropenic patient and other immunocompromised hosts, when bacterial etiologies are highest in the differential diagnosis.

Urinary Antigen Studies

A variety of commercially available assays that detect antigen in the urine of patients infected with specific pathogens are available and, in many cases, are useful adjuncts to more aggressive diagnostic techniques. Urinary antigen assays for the detection of *L. pneumophila* serogroup 1 have reported sensitivities that range from 70% to 90% and, in general, have high specificities (99 to 100%) (55). The variability in sensitivity may be a factor of the population studied, the timing of specimen collection in relationship to onset of disease, and whether specimens have been concentrated or not (55). An immunochromatographic assay (Binax Now) provides results in 15 min. Sopena et al. (101) found that immunocompromised patients often excrete antigen for 60 days or longer and also take a longer time to defervesce. In geographic locations where the predominant serogroup is not type 1 and/or the species of *Legionella* is other than *L. pneumophila*, culture methods or nucleic acid amplification tests (NAATs), if available, are required for adequate diagnosis.

The *S. pneumoniae* urinary antigen test is another immunochromatographic assay that is FDA cleared for use on urine and cerebrospinal fluid (CSF) and has also been evaluated on positive blood cultures. To date, there have been no published studies that have evaluated this assay's performance exclusively in immunocompromised patients. A recent retrospective study was performed in a medical/surgical intensive care unit on patients admitted with severe community-acquired pneumonia (71). This study revealed a sensitivity and specificity of 72% and 90%, respectively. The performance of the immunochromatographic assay in this study was comparable to that in other published series (33, 84). In addition, it was found that prior use of antibiotics did not have an impact on test performance (71). In general, this assay is more sensitive for patients who have bacteremic pneumonia than for patients with positive sputum cultures. The specificity for use on specimens from adults is high, but false-positive results have been seen for children who have nasopharyngeal colonization in the absence of disease (35, 50).

Urine antigen tests also exist for the detection of histoplasmosis and blastomycosis (117). These tests are useful adjunctive diagnostic methods, particularly for patients with more extensive disease, and because urine samples are easy to obtain, often these tests may result in the initial diagnosis. However, neither test alone should be relied upon as definitive for the diagnosis. Ideally, cytology or histopathology of body fluids or tissues and culture methods should also be performed when the likelihood of infection with systemic mycoses is high. The comments made here about assay performance are specific to the enzyme-linked immunosorbent assay (ELISA) available through The Histoplasmosis Reference Laboratory (MiraVista Diagnostics, Indianapolis, Indiana), whose performance characteristics have been reviewed in the literature (117). In general, the sensitivity of the test is 90% for patients with disseminated histoplasmosis and is greater with urine than with serum (92% versus 82%). In addition, greater sensitivity has been observed for patients with AIDS than for patients with other types of immunosuppression (117). The sensitivity of the test for patients with acute pulmonary histoplasmosis is 75%. In addition to urine and serum, other specimen types that are amenable to testing include CSF and BAL specimens. Specificity is high, at 98%, although cross-reactivity with endemic mycoses, such as paracoccidioidomycosis, blastomycosis, African histoplasmosis, and *Penicillium marneffei* infection, has been described (117).

The immunoassay for *Blastomyces dermatitidis* has similar problems with cross-reactivity. In addition to the list of diseases mentioned above, intermittent cross-reactivity has also been seen with samples containing high concentrations of *C. neoformans* polysaccharide antigen and for a patient with aspergillosis (40). The overall sensitivity of the test in one study was 92.9% (40).

Tests on Other Nonrespiratory Sources

Cryptococcal antigen tests performed on serum and CSF have been quite useful in rapidly diagnosing untreated patients with this infection. The performances of available commercial assays have been reviewed (106). In that study, the sensitivities and specificities for the various assays performed on CSF were comparable; there were noticeable differences in sensitivity but not specificity among the assays in testing serum samples (106). In one study comparing immunocompetent to immunocompromised patients with pulmonary cryptococcosis, the serum cryptococcal antigen titers were higher, often reflective of more extensive radiographic findings, for the immunocompromised patients (20). At least one study has examined the utility of routine testing using BAL fluid, but this specimen source lacked sensitivity (71%) and the positive predictive value was 0.59 (17 false-positive samples) (69).

Serum galactomannan and 1,3-β-D-glucan tests are discussed extensively in chapter 10. Their performance in testing respiratory samples such as pleural fluid and BAL fluid is discussed below in the section on bronchoscopy.

Serological Tests

In general, as pointed out in other chapters on organism-specific diagnostic tests, serological tests are likely to be limited in diagnosing infections in immunocompromised patients. Tests for measurement of antibody are more useful for prescreening patients likely to be at risk for reactivation of infection or in the case of SOT and HSCT recipients, who are at risk for new infection from a seropositive donor. Tests that measure antibody are not timely because of the delay in mounting an immunological response or, in the case of some patients, the inability to do so. Falsely negative serological tests have been reported for immunocompromised patients with *Legionella* infections (55), histoplasmosis (103), and toxoplasmosis (81). A single serologic test should not be relied upon to make a diagnosis of any of these infections. In the case of toxoplasmosis, serology is most useful pretransplantation and in monitoring the HIV-positive patient by predicting who is likely to recrudesce in the case of positive serology and counseling patients who are seronegative. Alternative methods of diagnosis are required.

Respiratory Specimens

A variety of respiratory specimen types can be collected in an attempt to diagnose respiratory diseases. Regardless of whether these are obtained noninvasively or require bronchoscopy or biopsy, serious consideration must be given to adequate specimen collection, timely transport, and coordination between the clinical microbiology laboratory, cytology or surgical pathology laboratories, and the clinician taking care of the patient. Development of diagnostic algorithms with a consensus from pulmonologists, microbiologists, oncologists, and infectious disease specialists, among others, should be developed, particularly in centers with active transplantation programs.

Nasopharyngeal aspirates or washes are preferred to swabs for the initial diagnosis of respiratory virus infections, *Bordetella pertussis*, *Bordetella parapertussis*, and the "atypical pneumonia" agents *M. pneumoniae* and *C. pneumoniae*. These specimen types are not acceptable for other types of analysis. A comprehensive discussion of the various diagnostic modalities for respiratory virus pathogen detection is provided in Chapter 7. In general, rapid antigen ELISA and immunochromatographic membrane tests positive for influenza virus or RSV are useful for implementing treatment and infection control practices; however, if these tests are negative, then alternative testing using culture or nucleic acid amplification methods should be performed. It is important to note that the performances of these various assays have not been well studied for immunocompromised patients. Also available are immunofluorescence methods for a broader range of pathogen detection, such as adenovirus, parainfluenza viruses, RSV, influenza viruses, and metapneumovirus. In general, immunofluorescence microscopy is more sensitive than rapid antigen detection testing. Although listed as an acceptable source in many product package inserts for several rapid diagnostic tests, throat swabs are less desirable samples for respiratory virus diagnosis (30). In a comparative study of four specimen types for detection of influenza viruses by use of a rapid test and culture, Covalciuc et al. found that sputum and nasal aspirates gave the highest yields, whereas throat swabs were the least sensitive (30). Although both enzyme immunoassay and immunofluorescence methods performed on nasal washes are rapid and specific, in a study by Englund et al. the sensitivity of rapid antigen testing using nasal washes for RSV diagnosis was very low (15%) compared to the yield using endotracheal aspirates (71%) or BAL fluid (89%) (41).

Expectorated sputum, if available, may be useful for analysis in some immunocompromised patient populations, although in general, these specimens give low yields for this population. Attempts should be made to minimize oropharyngeal contamination by having the patient remove dentures and by rinsing the mouth with water prior to expectoration (97). Expectorated sputum should be transported to the laboratory as soon as possible after collection to minimize loss of fastidious pathogens, such as *S. pneumoniae*, and to prevent overgrowth of normal flora or nonfastidious pathogens. Specimens that cannot be sent and processed within 2 h of collection should be refrigerated (97). The standard criteria for specimen rejection (<10 epithelial cells/high-power field and >25 polymorphonuclear leukocytes/high-power field) are not valid for neutropenic patients. Early studies have shown decreased cellularity of pulmonary alveoli in neutropenic patients, including reductions in polymorphonuclear leukocytes compared to those in patients without neutropenia (29). Sputum analysis is most useful when organisms that are not part of the normal flora, such as *Legionella*, mycobacteria, or systemic fungi, are recovered.

Induced sputum

When patients appear incapable of spontaneous expectoration, specimens are often obtained by induction. This procedure should be performed by an experienced respiratory therapist. Patients must be able to cooperate with the procedure, which involves inhaling hypertonic saline (3%) through an ultrasonic nebulizer via a mouthpiece for up to 30 min. As with expectorated samples, induced sputa should likewise be collected in sterile containers and should be sent promptly to the laboratory. The utility of this procedure for diagnosing bacterial pneumonia in either immunocompetent or immunocompromised patients has not been well studied (4, 46). In general, the available studies (4, 46) show no appreciable increase in bacterial pathogen recovery compared to recovery by expectorated sputum, although stratification of patients by immunocompromised state was not performed. In several studies sputum induction appeared to have the best utility in the diagnosis of mycobacterial infections and *P. jirovecii* infection, although the results have been contradictory and some studies question the cost-effectiveness of this approach (74, 80, 109). One study examined the yield of repeated sputum induction in patients suspected of pulmonary tuberculosis who could not actively produce sputum (3). It was reported that the yields for acid-fast smear and culture were 64% and 70%, respectively, for one sample; 81% and 91%, respectively, for two samples; and 91% and 99%, respectively, for three induced samples (3). A prospective study performed among children in Africa in an area with high HIV prevalence demonstrated the feasibility of using induced sputum from children of less than 5 years of age (123). In that study, induced sputum was superior to gastric aspirates in the recovery of *M. tuberculosis* (123). Sputum induction was also useful in diagnosing *Pneumocystis* pneumonia (123).

In the pre-HAART era, performance of direct fluorescent-antibody (DFA) testing using monoclonal antibodies that detect both *Pneumocystis* cysts and trophozoites directly on induced sputa from susceptible HIV-positive patients significantly reduced the need to perform more invasive procedures (67). In this population, the organism burden was higher than in the non-AIDS patient (67, 109), and therefore this algorithm was more sensitive for the AIDS patient (67, 109). More recently, however, *Pneumocystis* pneumonia has significantly declined in HIV-positive patients and is diagnosed more often in non-HIV-infected immunocompromised patients in some centers (67). The quantities of *P. jirovecii* organisms in these patients are low, reducing the yields from induced specimens. Molecular techniques performed upon oral washes (45, 58, 67), induced sputum (15), or more invasively obtained specimens will likely soon replace DFA or less sensitive staining methods for diagnosing *Pneumocystis* pneumonia. Assays that use multicopy gene targets, such as mitochondrial rRNA or major surface glycoprotein genes, have the highest sensitivities (58, 67). There are currently no standardized FDA-cleared molecular assays, and the significance of a positive result for a patient without clinical disease needs clarification (67, 109).

In general, the best use of induced sputum may be for the patient incapable of expectoration for whom bronchoscopic techniques may not be readily available within 24 h of clinical suspicion of a progressive pulmonary process. Negative results should not be interpreted as definitive, and patients should be referred for bronchoscopy when induced sputum studies are not revealing.

Invasive Procedures

When rapid and noninvasive tests are not revealing within 24 to 48 h for the immunocompromised patient with a new pulmonary process, then quick progression to more definitive sampling of the lung is required. A suggested algorithm that includes empiric treatment and the use of noninvasive and invasive tests is depicted in Fig. 3. There are several diagnostic procedures that can be performed. These include fiber-optic bronchoscopy with BAL, transbronchial biopsy (TBB), transthoracic needle aspiration, video-assisted minithoracotomy, and open lung biopsy. Each of these is discussed below. In general, because they are more widely available, yield a greater volume of sample, and are somewhat safer than other methods, bronchoscopic techniques are usually performed first.

Fiber-optic bronchoscopy

The proper procedure for selecting the type of bronchoscope and protocols for performance of the procedure can be found in detail in standard pulmonary textbooks. It is very important, however, that standardization for sampling of the lung be agreed upon by microbiologists, pulmonologists, and clinicians caring for patients. This is extremely important for ensuring quality of the sample and optimal yield. In my unpublished experience, procedures can vary tremendously, even within the same institution, depending upon the training and experience of the pulmonologist. A variety of specimen types can be collected. A brief discussion of each is provided, with emphasis on BAL fluid.

Bronchial washings and bronchial brushings

Bronchial washings and bronchial brushings are usually obtained by installation of saline into a major airway through the bronchoscope channel and by aspiration of the secretions. These secretions are not representative of processes in the alveoli or small airways. The best use of bronchial washings is in the diagnosis of strict pathogenic organisms, such as *M. tuberculosis* and endemic fungi; these specimens are not appropriate for diagnosing bacterial pneumonia

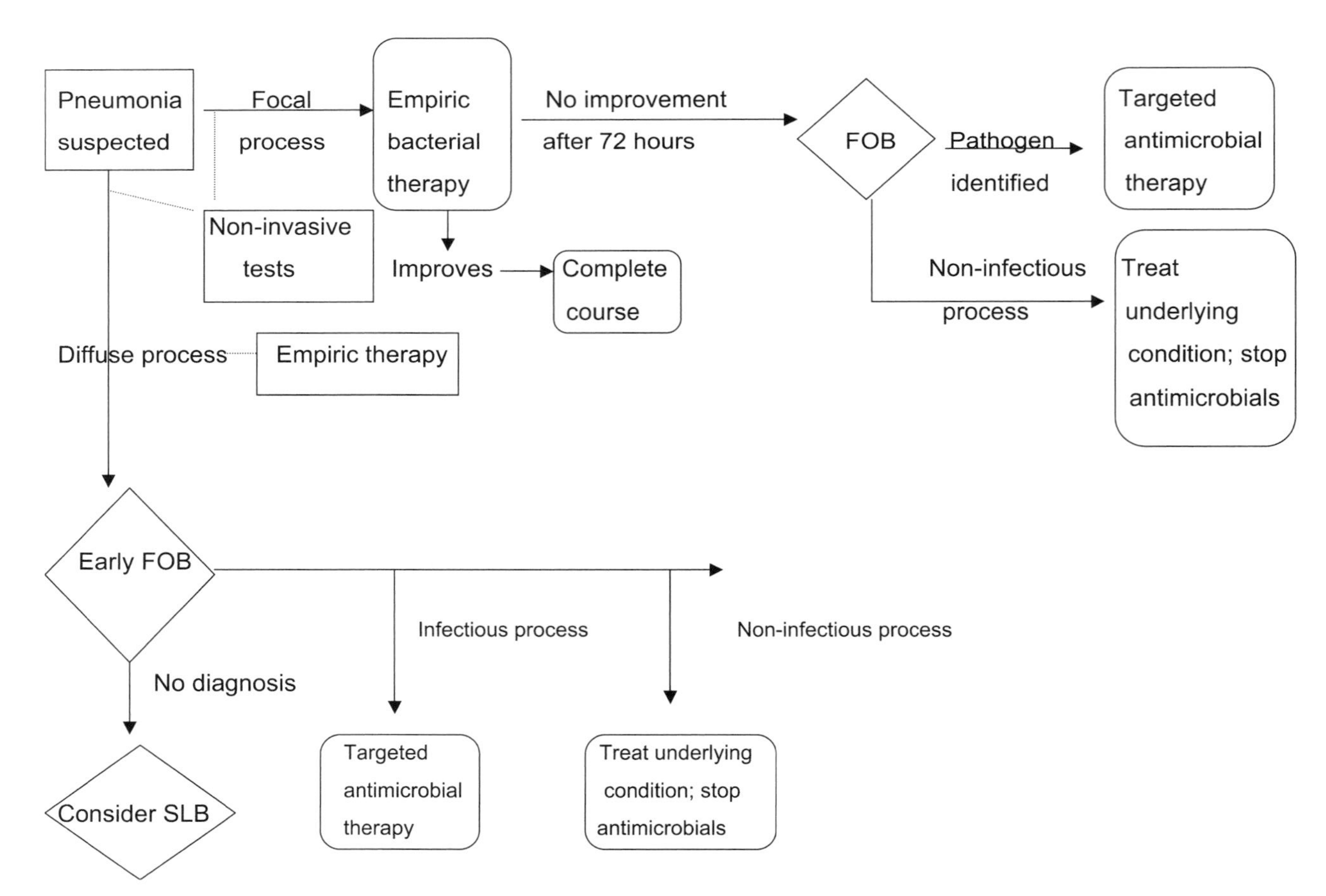

Figure 3 Diagnostic approach to pulmonary infiltrates in the immunocompromised host, stratified by type of process and response to empiric management. FOB, fiber-optic bronchoscopy. (Modified with permission from reference 18.)

(5). In general, this is not the preferred procedure for diagnosing pneumonia in immunocompromised patients, but on occasion, such samples may be the only samples available for unstable patients or when the BAL return volume is inadequate (5) (see below).

Bronchial brushings are usually performed to obtain cytologic samples for the diagnosis of malignancies, not for routine bacterial or other types of culture. Because cells are obtained, cytopathic changes caused by infectious agents, such as the type of inflammatory cell and viral inclusion bodies, may be seen as well.

Protected specimen brush samples are collected using two telescoping catheters, the outer one of which is occluded by a Carbowax plug which prevents secretions from entering the catheter during passage through the bronchoscope channel (5). After insertion of the bronchoscope, the device is passed through the bronchoscope channel, the inner catheter is advanced, and the Carbowax plug is expelled into the airway lumen, where it is absorbed. The brush is advanced past the tip of the inner catheter, where it has been protected from upper airway secretions (5). Usually, small-volume specimens are collected from the distal bronchioles. After the specimen is collected, the brush is retracted back into the inner catheter, which is retracted back into the outer catheter. After removal of the bronchoscope from the airway, the brush is carefully removed without contamination and placed into 1 ml of lactated Ringer's solution or some other diluent, after which it is submitted to the microbiology laboratory as soon as possible. This procedure was designed for specific diagnosis of bacterial pneumonia by Gram stain and quantitative culture. The volume of sample is such that its use in recovery of a broad range of pathogens is not possible.

BAL and TBB

In general, BAL is the preferred procedure for immunocompromised patients. In this procedure, the bronchoscope is carefully wedged into an airway lumen. In the patient with diffuse infiltrates, the scope is usually wedged in the right middle lobe or lingula. When the infiltrate is focal, it is important to wedge the scope in the pulmonary segment corresponding to the radiographic abnormality. Once the scope is wedged, a large volume of 0.9% saline (140 to 1,500 ml) is injected in 3 or 4 aliquots of 40 to 50 ml each (5). For the pediatric patient, this amount is usually 1 ml/kg of body weight. The amount returned is variable, but a returned volume of 40 to 60 ml is ideal. The first returned aliquot is usually contaminated with oropharyngeal and upper airway microbiota and should not be used for quantitative bacterial culture. It is usually discarded or used for detection of strict pulmonary pathogens, such as *Legionella*, fungi, or mycobacteria (5). When performed properly, this procedure is estimated to sample about 1 million alveoli.

In some situations, BAL fluid alone may be inadequate in the diagnosis of certain conditions where tissue is required for histopathology to clarify positive cultures from fluid or when noninfectious entities are being considered. In such cases, TBBs are often obtained at the same time that BAL is performed. TBB samples are obtained by passing a forceps through the working channel of the wedged bronchoscope under fluoroscopic guidance. Since the pieces of tissue obtained in this way may be quite small, several specimens are required.

TBB (or other procedures to acquire lung tissue) is the preferred method to diagnose acute rejection, CMV pneumonia, posttransplant lymphoproliferative disorder, *Candida* pneumonia, pneumonia caused by opportunistic fungi, and in some cases, disease caused by NTM (31, 60, 93, 102). Infiltrates caused by noninfectious etiologies, such as radiation and chemical pneumonitis and neoplasia, may also be diagnosed best by the inclusion of TBB (82).

Specimen handling and analysis of bronchoscopically obtained specimens

Bronchoscopic specimens should be transported to the laboratory in sterile leakproof containers as soon as possible after collection. Since multiple aliquots are usually obtained, the samples should be sent for cytologic, microbiologic, and chemical analyses. Microbiology laboratories need to work with transplant physicians, pulmonologists, infectious disease physicians, and oncologists to develop standardized algorithms that include not only the timing of procedures but also the workup of BAL and TBB specimens once they reach the clinical laboratory. Since a broad range of pathogens are possible in most immunocompromised hosts and since more than one process or infection may be present, multiple procedures should be performed routinely. Recommendations for the minimum approach, as well as some optional assays, are listed in Table 2. The emphasis should be on rapid processing using tests with short turnaround times, such as direct microscopy, antigen detection, and nucleic acid amplification methods, when available. For bacterial diagnosis, quantitative cultures obtained using either the calibrated loop method or a dilutional method should be performed (5). The diagnostic thresholds for significance of various quantities of bacteria are

Table 2. Recommended diagnostic studies for BAL, TBB, and SLB specimens

Type of pathogen	Procedure[a]
Bacteria	Cytospun Gram stain, quantitative bacterial culture, cytologic examination for intracellular pathogens, with cell count and differential
Mycoplasma pneumoniae[b]	NAAT, culture
Legionella pneumophila	Culture on BCYE medium, NAAT if available
Chlamydophila pneumoniae[b]	NAAT
Mycobacterium species	AFB smear, culture on selective media, NAAT for *M. tuberculosis*
Fungi	Calcofluor white stain on BAL, PAS and silver stains of any obtained tissue, culture on selective media, galactomannan assay on BAL, NAAT where available
Pneumocystis jirovecii	DFA, cytologic stains such as PAS stain and silver stain, NAAT if available
Respiratory viruses	Combination antigen detection and shell vial culture; NAAT for coronaviruses, metapneumovirus, and possibly others, if available
Herpesviruses	Combination antigen detection with shell vial culture, cytologic analysis and/or tissue analysis for interpretation of CMV and HSV results; NAAT is controversial
Noninfectious etiologies	Cytologic analysis (BAL) with cell count; examination for siderophages, gross hemorrhage, etc.; TBB and SLB tissue histopathology

[a]PAS, periodic acid-Schiff stain.
[b]Routine testing in BAL algorithms is probably not indicated, unless these pathogens have been detected in the community. NAAT may be done from throat swabs.

as follows. For protected specimen brush samples, quantities of bacteria that equal or exceed 10^3 CFU/ml should be identified and undergo susceptibility testing; for BAL samples, this number is 10^4 CFU/ml or greater. Direct stains that are useful include Gram stain, calcofluor white with KOH or other fungal stain, and acid-fast stains. Antigen detection methods for viruses, *Pneumocystis*, fungi (see below), and in selected cases, *Legionella* are suggested. Culture for viruses, mycobacteria, *Nocardia*, and fungi should be included routinely.

Additional tests that are useful to perform on BAL specimens

Galactomannan is a heat-stable heteropolysaccharide present in the cell walls of *Aspergillus* and *Penicillium* species. Galactomannan is released predominantly by *Aspergillus* hyphae during growth and, to a lesser extent, by conidia (65). Currently, two assays are available for the detection of galactomannan, including the Platelia ELISA (Bio-Rad, Hercules, CA), which is the most frequently used assay in the United States and Europe, and the Pastorex *Aspergillus* latex agglutination assay (Bio-Rad Laboratories, Hercules, CA), which is not available in the United States. These assays have been used primarily on serum samples, but literature has emerged on the utility of testing BAL fluid and fluids other than serum (7, 14, 65, 87; reviewed by Klont et al. in reference 65). The sensitivity of the Platelia ELISA is reported to be higher than that of the Pastorex assay in testing BAL fluid (85% to 100% versus 73% to 83%) (7, 14, 65), and both tests have a higher sensitivity in testing BAL than in testing serum (serum sensitivity, 47% and 41 to 45% for Platelia and Pastorex assay, respectively [65]). The Platelia assay was positive with BAL fluid for seven of nine (77.8%) patients with definite or probable aspergillosis and was the only indication of aspergillosis in three cases in a study by Hohenthal et al. (57). These tests may be of maximum utility when combined with suggestive findings on HRCT and/or in combination with PCR. Musher et al. (87) studied the utility of the Platelia assay and quantitative PCR (qPCR) performed on BAL fluid alone, in combination, and stratified by radiographic findings. For patients with nodular or focal infiltrates, the galactomannan assay had a sensitivity of 73 to 82%, using indexes for interpretation of 0.5 and 1.0. The sensitivity of qPCR was 73%. When both qPCR and galactomannan enzyme immunoassay with an index of 0.5 were used to diagnose invasive pulmonary aspergillosis in patients with nodular infiltrates, the overall sensitivity was 91%, and for culture-positive patients, this value was 100% (87). Note that almost two-thirds of patients who were not diagnosed by culture (44% were culture negative) had a positive galactomannan or qPCR assay result (87). It must be noted that the Platelia assay is not FDA cleared for testing of specimens other than serum.

NAATs are becoming available and will contribute to the diagnostic yield of bronchoscopic techniques. While there are few FDA-cleared assays at present, there is abundant literature on the performance of commercial analyte-specific reagents (ASR) and user-defined assays for specific pathogens. A comprehensive discussion is beyond the scope of this chapter, and details of pathogen-specific assays are covered in the early chapters of this text. However, some summary statements are included here. A review of this topic is also found in reference 86.

Several studies have evaluated the performance of PCR targeting the pneumolysin gene for detection of *S. pneumoniae* in respiratory samples (85, 86). Sensitivity on sputum samples ranges from 83 to 100%, but distinguishing colonization from true infection is a problem, and at present, PCR for this pathogen remains a research tool (86). In contrast, given the suboptimal performance of serological testing and the difficulty in cultivating *M. pneumoniae*, NAAT is the preferred diagnostic method for detection of this organism (55, 86). Throat swabs and/or nasopharyngeal specimens are the specimens of choice. Several studies have demonstrated that PCR is more sensitive than conventional methods for detecting *M. pneumoniae* (34, 55, 86, 115). Few data are available to assess the positive predictive value of NAATs for this infection. In a study of 92 children with respiratory infection and 74 control patients, 9 patients met diagnostic criteria for *M. pneumoniae* infection (34). None of the throat swabs from the 74 control patients were PCR positive (34), and follow-up throat swabs obtained from 4 of 7 PCR-positive patients were negative 4 to 12 weeks after the initial diagnosis (34). A separate study showed that 15% of patients who were positive by PCR continued to have *M. pneumoniae* DNA detected in their throats for 2 to 6 weeks following antibiotic therapy (115). While there are no FDA-cleared assays for this method, at least two commercial companies have ASRs available. Prodesse, Inc. (Waukesha, WI), has two assays, including *M. pneumoniae*—ProPneumo-1, for detection of *M. pneumoniae* and *C. pneumoniae* simultaneously, and a multiplex real-time assay called the Pneumoplex assay that detects *M. pneumoniae*, *L. pneumophila*, *L. micdadei*, and *B. pertussis* in one sample. The analytical performance of the Pneumoplex assay has been published (63). Cepheid (Sunnyvale, CA) likewise has an ASR assay available for *M. pneumoniae* detection. To date, there are no publications on the clinical performance of either the Prodesse or the Cepheid assay.

PCR detection of *Legionella* species by use of primers that amplify a variety of targets has been described. Various sample types, including urine and blood, have also been evaluated. The positivity rate for these nonrespiratory sources has been variable (55, 86). One of the difficulties with PCR for legionellae has been the presence of contaminating *Legionella* DNA in commercial extraction kits (110). At present, the development of standardized protocols that address the optimum specimen types, gene targets, and extraction methods is needed. Standard approaches have been recommended for NAAT for *C. pneumoniae* (36). PCR testing for *Pneumocystis* is discussed above. The ideal specimens are BAL and induced sputa. PCR is more sensitive than cytological methods and/or DFA and might be of greatest utility for patients with negative conventional studies and a high probability of clinical infection (86).

Development of molecular methods for detection of respiratory viruses has lead to the discovery of several new pathogens, namely, human metapneumoviruses and additional coronaviruses, which at this time appear to be detected best by NAATs. Several studies have demonstrated that PCR is more sensitive than a combination of culture, antigen detection, and serological testing (32, 42, 44, 59, 91, 111). Commercial assays for respiratory viruses are available. The ProDesse Hexaplex assay for detection of the typical respiratory pathogens has been available for several years as an ASR (56). The same company has available real-time assays for adenovirus (Adenoplex), human metapneumovirus (ProhMPV), influenza A and B viruses (ProFlu1), and severe acute respiratory disease coronavirus and other coronaviruses (Sars/Coronaplex). Recently the ProDesse ProFlu assay, which detects influenza A and B viruses and RSV, received FDA clearance. Cepheid also has a FluA/B assay and an RSV assay available as ASRs, which likewise have not been verified independently. Finally, the xTAG Respiratory Virus Panel (Luminex) detects six groups of viruses and has also received FDA clearance. None of these have been evaluated extensively in the literature.

Two FDA-cleared assays are available for detection of *Mycobacterium tuberculosis*, including the *Mycobacterium tuberculosis* direct test (Gen-Probe, Inc., San Diego, CA) and the Amplicor and Cobas *Mycobacterium tuberculosis* tests (Roche Molecular Systems, Pleasanton, CA). The performance characteristics of these tests have been well studied and are reviewed in reference 120 and in Chapter 12.

DIAGNOSTIC YIELD AND ASSAY INTERPRETATION

Studies reporting upon the diagnostic utility of BAL and TBB provide conflicting and often controversial findings. In general, the reported diagnostic yields of BAL alone range from 31% to 80% in several retrospective (20, 29, 57, 61, 72, 121) and prospective (60) studies across various immunocompromised host populations, but the majority report a diagnosis in roughly 60% of patients (20, 29, 121). Yield is higher when the patient has multifocal or diffuse pneumonia than when the patient has a focal process (29). In addition, bronchoscopy is more likely to establish an etiology when infection is the cause of the pulmonary process than when the cause is a noninfectious etiology (60). Noninfectious etiologies that may be diagnosed routinely by BAL include alveolar hemorrhage (presence of ≥20% hemosiderin-laden macrophages and/or progressive, diffusely bloody fluid), hypersensitivity pneumonitis, and pulmonary alveolar proteinosis (60, 64).

Fewer studies have examined the impact of BAL on patient management. In a retrospective study of 71 BMT patients who underwent bronchoscopy and BAL, the bronchoscopic findings resulted in changes in therapy for 65% of patients in whom an organism was identified and 22% of patients with completely negative results (38). A similar study performed on cardiac transplant patients found that changes in therapy occurred in 32% of patients (72).

In a study by Jain et al. (60), the performance of TBB with BAL increased the diagnostic yield significantly. Also in that study, TBB provided the sole source of diagnosis for 15 of 17 patients with noninfectious etiologies (60). None of the above-mentioned conditions, however, were included among the missed diagnoses. For lung transplant recipients, TBB helps to establish or exclude acute allograft rejection as a cause of pulmonary infiltrates (18). In contrast, other studies, primarily among HSCT patients, have not noted an improved diagnostic yield and do not recommend routine inclusion of TBB (88, 121). Yield may also be affected adversely by antimicrobial and antifungal treatments administered in the days prior to bronchoscopy. However, studies of neutropenic patients on broad-spectrum antibiotic or antifungal treatment have shown that BAL can yield important information, such as recovery of organisms resistant to the empiric therapy at the time of the procedure (88).

"False-positive" BAL specimens due to contaminating or colonizing organisms were seen most often with *Candida* species, coagulase-negative staphylococci, and CMV. As mentioned elsewhere, in these instances correlation with histopathology, cytology, or other positive tests (e.g. simultaneously positive CMV antigenemia test) is required. False-negative BAL specimens are seen most frequently for patients with fungal pneumonia (29).

Complications of Bronchoscopic Techniques

The risk of complications during fiber-optic bronchoscopy depends upon several important variables. These include the severity of the pneumonic process and the degree of hypoxemia, whether thrombocytopenia or a coagulopathy is present, and whether the patient is intubated. The main complications are hypoxemia, bleeding, pneumothorax, cardiac compromise, and rarely, transmission of infection from inadequately decontaminated bronchoscopes (60, 88). Patients with severe hypoxemia may require prophylactic intubation for successful and safe performance of the procedure. The addition of TBB does increase the complication rate.

SLB

When BAL with or without TBB is unrevealing and the patient is not responding to empiric treatment, surgical lung biopsy (SLB) is usually performed. In many cases, tissue can be obtained by video-assisted thoracic surgery, a minimally invasive procedure that uses a thoracoscope, camera, and trocars to obtain the tissues without the need for opening the chest cavity, or by minithoracotomy, thereby reducing procedure-related morbidity and mortality (76). Tissues can be processed and analyzed following the same algorithms as those for BAL samples (Table 2) and including special stains and interpretation by a surgical pathologist. Tissue specimens that are to be processed for fungal culture should not be ground, as this destroys the hyphal elements and may interfere with subsequent recovery in culture. Results from histopathology may frequently reveal an etiological agent or abnormal process before a pathogen is detected in the clinical microbiology laboratory.

The yield of SLB following a negative BAL sample has not been high for BMT patients (121). Studies performed among other patient populations have reported a change in therapy for a higher percentage of patients (~57% in a study by White et al.) (68, 118), but no studies have demonstrated improved outcomes as a result of SLB (28). The decision to perform SLB should be individualized and should include a realistic assessment of the likelihood of alterations in therapy and the risk involved in the procedure. If tissue is obtained, a comprehensive analysis as listed in Table 2 should be performed.

Other invasive techniques

For focal lesions that are accessible, CT-guided needle biopsies have acceptable yields. Transthoracic needle aspiration is rarely done because of the high incidence of pneumothorax and hemoptysis (28). At least one study compared the diagnostic yields from transthoracic needle aspiration, TBB, and SLB for a small number of granulocytopenic patients with pulmonary infiltrates. The diagnostic yields were 30%, 59%, and 94%, respectively (12). Regardless of the diagnostic method, recovered tissue should be processed for the pathogens listed in Table 2.

SUMMARY

Lower respiratory tract infections are the most common causes of morbidity and mortality among all groups of immunocompromised patients. In general, bacterial pathogens are the most frequent etiologic agents, but a vast array of opportunistic organisms may be responsible, depending upon the type, degree, and duration of immunosuppression. Radiographic images, particularly those visualized by HRCT scans, may be suggestive of a particular process, but there is enormous overlap. Therefore, diagnostic samples should be obtained as quickly as possible after new signs and symptoms appear. Fiber-optic bronchoscopy and BAL, with or without TBB, are the mainstays of diagnosis. For the lavage sample, a broad range of tests are recommended. Rapid antigen detection tests and NAATs are emerging and currently complement existing culture-based methods. Some pathogens require histopathology to determine colonization versus true infection. If carefully performed early in the course of infection, BAL can reveal an etiologic agent that results in a change in therapy in a substantial number of patients, leading to changes in empiric regimens in 30 to 40% of patients. If BAL is not revealing, SLB is usually the next step. The potential benefit provided by SLB must be weighed against the increased morbidity associated with this procedure.

REFERENCES

1. **Aaron, L., D. Saadoun, I. Calatroni, O Launay, N. Memain, V. Vincent, G. Marchal, B. Dupont, O Bouchaud, D. Valeyre, and O. Lortholary.** 2004. Tuberculosis in HIV-infected patients: a comprehensive review. *Clin. Microbiol. Infect.* **10:** 388–398.
2. **Afessa, B., and S. G. Peters.** 2006. Major complications following hematopoietic stem cell transplantation. *Semin. Respir. Crit. Care Med.* **27:**297–309.
3. **Al Zahrani, K., H. Al Jahdali, L. Poirier, P. Rene, and D. Menzies.** 2001. Yield of smear, culture, and amplification tests from repeated sputum induction for the diagnosis of pulmonary tuberculosis. *Int. J. Tuberc. Lung Dis.* **5:**855–860.
4. **Bandyopadhyay, T., D. A. Gerardi, and M. L. Metersky.** 1999. A comparison of induced and expectorated sputum for the microbiological diagnosis of community acquired pneumonia. *Respiration 67:*173–176.
5. **Baselski, V. S., and R. G. Wunderink.** 1994. Bronchoscopic diagnosis of pneumonia. *Clin. Microbiol. Rev.* **7:**533–558.
6. **Beck, J. M.** 2005. The immunocompromised host. HIV infection. *Proc. Am. Thorac. Soc.* **2:**423–427.
7. **Becker, M. J., E. J. Lugtenburg, J. J. Cornelissen, C. vanDer Schee, H. C. Hoogsteden, and S. de Marie.** 2003. Galactomannan detection in computerized tomography-based bronchoalveolar lavage fluid and serum in haematological patients at risk for invasive pulmonary aspergillosis. *Br. J. Haematol.* **121:**448–457.
8. **Billings, J. L., M. I. Hertz, and C. H. Wendt.** 2001. Community respiratory virus infections following lung transplantation. *Transpl. Infect. Dis.* **3:**138–148.
9. **Bloch, K. C., R. Nadarajah, and R. Jacobs.** 1997. *Chryseobacterium meningosepticum*: an emerging pathogen among immunocompromised adults. *Medicine* **76:**30–41.
10. **Bongartz, T., A. J. Sutton, M. J. Sweeting, I. Buchan, E. L. Matteson, and V. Montori.** 2006. Anti-TNF antibody therapy in rheumatoid arthritis and the risk of serious infections and malignancies. *JAMA* **295:**2275–2285.
11. **Boynton, R. J.** 2005. Infectious lung complications in patients with HIV/AIDS. *Curr. Opin. Pulm. Med.* **11:**203–207.
12. **Burt, M. E., M. W. Flye, B. L. Webber, and R. A. Wesley.** 1981. Prospective evaluation of aspiration needle, cutting needle, transbronchial biopsy, and open lung biopsy in patients with pulmonary infiltrates. *Ann. Thorac. Surg.* **32:**146–153.
13. **Cahill, B. C., J. R. Hibbs, K. Savik, B. A. Juni, B. M. Dosland, C. Edin-Stibbe, and M. I. Hertz.** 1997. Aspergillus airway colonization and invasive disease after lung transplantation. *Chest* **112:**1160–1164.
14. **Caillot, D., J. F. Couaillier, A. Bernard, O. Casasnovas, D. W. Denning, L. Mannone, J. Lopez, G. Couillault, F. Piard, O. Vagner, and H. Guy.** 2001. Increasing volume and changing characteristics of invasive pulmonary aspergillosis on sequential thoracic computed tomography scans in patients with neutropenia. *J. Clin. Oncol.* **19:**253–259.
15. **Caliendo, A. M., P. L. Hewitt, J. M. Allega, A. Keen, K. L. Ruoff, and M. J. Ferraro.** 1998. Performance of a PCR assay for detection of *Pneumocystis carinii* from respiratory specimens. *J. Clin. Microbiol.* **36:**979–982.
16. **Casado, J. L., E. Navas, B. Frutos, A. Moreno, P. Martin, J. M. Hermida, and A. Guerrero.** 1997. Salmonella lung involvement in patients with HIV infection. *Chest* **112:**1197–1201.
17. **Centers for Disease Control and Prevention.** 1990. Nosocomial transmission of multi-drug resistant tuberculosis to healthcare workers and HIV-infected patients in an urban hospital. *MMWR Morb. Mortal. Wkly. Rep.* **39:**718–722.
18. **Chakinala, M. M., and E. P. Trulock.** 2005. Pneumonia in the solid organ transplant patient. *Clin. Chest Med.* **26:**113–121.
19. **Champion, L., M. Stern, D. Israel-Biet, M. F. Mamzer-Bruneel, M. N. Peraldi, H. Kreis, R. Porcher and E. Morelon.** 2006. Brief communication: sirolimus-associated pneumonitis: 24 cases in renal transplant recipients. *Ann. Intern. Med.* **144:**505–509.
20. **Chang, G. C., C. L. Wu, S. H. Pan, T. Y. Yang, C. S. Chin, Y. C. Yang, and C. D. Chiang.** 2004. The diagnosis of pneumonia in renal transplant recipients using invasive and noninvasive procedures. *Chest* **125:**541–547.
21. **Chang, W. C., C. Tzao, H. H. Hsu, S. C. Lee, K. L. Huang, H. J. Tung, and C. Y. Chen.** 2006. Pulmonary cryptococcosis. Comparison of clinical and radiographic characteristics in immunocompetent and immunocompromised patients. *Chest* **129:**333–340.
22. **Chemaly, R. F., S. Ghosh, G. P. Bodey, N. Rohatgi, A. Safdar, M. J. Keating, R. E. Champlin, E. A. Aguilera, J. J. Tarrand, and I. I. Raad.** 2006. Respiratory viral infections in adults with hematologic malignancies and human stem cell transplantation recipients. *Medicine* **85:**278–287.
23. **Chen, C. S., M. Boeckh, K. Seidel, J. G. Clark, E. Kansu, D. K. Madtes, J. L. Wagner, W. R. P. Witherspoon, C. Anasetti, F. R. Appelbaum, W. I. Bensinger, H. J. Deeg, P. J. Martin, J. E. Sanders, R. Storb, J. Storek, J. Wade, M. Siadak, M. E. D. Flowers, and K. M. Sullivan.** 2003. Incidence, risk factors, and mortality from pneumonia developing late after hematopoietic stem cell transplantation. *Bone Marrow Transplant.* **32:**515–522.

24. **Chow, J. W., and V. L. Yu.** 1998. Legionella: a major opportunistic pathogen in transplant recipients. *Semin. Respir. Infect.* **13**:132–139.
25. **Cisneros, J. M., P. Munoz, J. Torre-Cisneros, M. Gurgui, M. J. Rodriguez-Hernandez, J. M. Aguado, A. Echaniz, and the Spanish Transplantation Study Group.** 1998. Pneumonia after heart transplantation: a multi-institutional study. *Clin. Infect. Dis.* **27**:324–331.
26. **Clark, J. G., J. A. Hansen, M. I. Hertz, R. Parkman, L. Jensen, and H. H. Peavy.** 1993. NHLBI workshop summary. Idiopathic pneumonia syndrome after bone marrow transplantation. *Am. Rev. Respir. Dis.* **147**:1601–1606.
27. **Clinical and Laboratory Standards Institute.** 2007. Principles and procedures of blood cultures; approved guideline. CLSI document M47-A. Clinical and Laboratory Standards Institute, Wayne, PA.
28. **Collin, B. A., and R. Ramphal.** 1998. Pneumonia in the compromised host including cancer patients and transplant patients. *Infect. Dis. Clin. N. Am.* **12**:782–805.
29. **Cordonnier, C., E. Escudier, F. Verra, L. Brochard, J. F. Bernaudin, and J. Fleury-Feith.** 1994. Bronchoalveolar lavage during neutropenic episodes: diagnostic yield and cellular pattern. *Eur. Respir. J.* **7**:114–120.
30. **Covalciuc, K. A., K. H. Webb, and C. A. Carlson.** 1999. Comparison of four clinical specimen types for detection of influenza A and B viruses by optical immunoassay (FLU OIA test) and cell culture methods. *J. Clin. Microbiol.* **37**:3971–3974.
31. **Cunha, B. A.** 2001. Pneumonias in the compromised host. *Infect. Dis. Clin. N. Am.* **15**:591–612.
32. **Debiaggi, M., F. Canducci, M. Sampaolo, M. C. Marinozzi, M. Parea, C. Terulla, A. A. Colombo, E. P. Alessandrino, L. Z. Bragotti, M. Arghittu, A. Goglip, R. Migliavacca, E. Romero, and M. Clementi.** 2006. Persistent symptomless human metapneumovirus infection in hematopoietic stem cell transplant recipients. *J. Infect. Dis.* **194**:474–478.
33. **Dominguez, J., N. Gali, S. Blanco, P. Pedroso, C. Prat, L. Matas, and V. Ausina.** 2001. Detection of *Streptococcus pneumoniae* antigen by a rapid immunochromatographic assay in urine. *Chest* **119**:243–249.
34. **Dorigo-Zetsma, J. W., S. A. J. Zaat, P. M. E. Wertheim-van Dillen, L. Spanjaard, J. Rijntjes, G. Van Waveren, J. S. Jensen, A. F. Angulo, and J. Dankert.** 1999. Comparison of PCR, culture, and serological tests for diagnosis of *Mycoplasma pneumoniae* respiratory tract infection in children. *J. Clin. Microbiol.* **37**:14–17.
35. **Dowell, S. F., R. L. Garman, G. Liu, O. S. Levine, and Y. H. Yang.** 2001. Evaluation of Binax NOW, an assay for detection of pneumococcal antigen in urine samples, performed among pediatric patients. *Clin. Infect. Dis.* **32**:824–825.
36. **Dowell, S. F., R. W. Peeling, J. Boman, G. M. Carlone, B. S. Fields, J. Guarner, M. R. Hammerschlag, L. A. Jackson, C. C. Kuo, M. Maass, T. O. Messmer, D. F. Talkington, M. L. Tondella, S. R. Zaki, and the C. pneumoniae Workshop Participants.** 2001. Standardizing Chlamydia pneumoniae assays: recommendations from the Centers for Disease Control and Prevention (USA) and the Laboratory Centre for Disease Control (Canada). *Clin. Infect. Dis.* **33**:492–503.
37. **Dufour, V., J. Cadranel, M. Wislez, A. Lavole, E. Bergot, A. Parrot, P. Rufat, and C. Mayaud.** 2004. Changes in the pattern of respiratory diseases necessitating hospitalization of HIV-infected patients since the advent of highly active antiretroviral therapy. *Lung* **182**:331–341.
38. **Dunagan, D. P., A. M. Baker, D. D. Hurd, and E. F Haponik.** 1997. Bronchoscopic evaluation of pulmonary infiltrates following bone marrow transplantation. *Chest* **111**:135–141.
39. **Duncan, M. D., and D. S. Wilkes.** 2005. Transplant-related immunosuppression. A review of immunosuppression and pulmonary infections. *Proc. Am. Thorac. Soc.* **2**:449–455.
40. **Durkin, M., J. Witt, A. LeMonte, B. Wheat, and P. Connolly.** 2004. Antigen assay with the potential to aid in diagnosis of blastomycosis. *J. Clin. Microbiol.* **42**:4873–4875.
41. **Englund, J. A., P. A. Piedra, A. Jewell, K. Patel, B. B. Baxter, and E. Whimbey.** 1996. Rapid diagnosis of respiratory syncytial virus infections in immunocompromised adults. *J. Clin. Microbiol.* **34**:1649–1653.
42. **Englund, J. A., M. Boeckh, J. Kuypers, G. Nichols, R. C. Hackman, R. A. Morrow, D. N. Fredricks, and L. Corey.** 2006. Brief communication: fatal human metapneumovirus infection in stem-cell transplant recipients. *Ann. Intern. Med.* **144**:344–349.
43. **Escuissato, D. L., E. L. Gasparetto, E. Marchiori, G. de Melo Rocha, C. Inoue, R. Pasquini, and N. L. Muller.** 2005. Pulmonary infections after bone marrow transplantation: high resolution CT findings in 111 patients. *Am. J. Roentgenol.* **185**:608–615.
44. **Falsey, A. R., and E. E. Walsh.** 2000. Respiratory syncytial virus infection in adults. *Clin. Microbiol. Rev.* **13**:371–384.
45. **Fischer, S., V. J. Gill, J. Kovacs, P. Miele, J. Keary, V. Silcott, S. Huang, L. Borio, F. Stock, G. Fahle, D. Brown, B. Hahn, E. Twonley, D. Lucey, and H. Masur.** 2001. The use of oral washes to diagnose *Pneumocystis carinii* pneumonia: a blinded prospective study using a polymerase chain reaction-based detection system. *J. Infect. Dis.* **184**:1485–1488.
46. **Fishman, J. A., R. S. Roth, E. Zanzot, E. J. Enos, and M. J. Ferraro.** 1994. Use of induced sputum specimens for microbiologic diagnosis of infections due to organisms other than *Pneumocystis carinii*. *J. Clin. Microbiol.* **32**:131–134.
47. **Gasink, L. B., and E. A. Blumberg.** 2005. Bacterial and mycobacterial pneumonia in transplant recipients. *Clin. Chest Med.* **26**:647–659.
48. **Glanville, A. R., M. Gencay, M. Tamm, P. Chhajed, M. Plit, P. Hopkins, C. Aboyoun, M. Roth, and M. Malouf.** 2005. *Chlamydia pneumoniae* infection after lung transplantation. *J. Heart Lung Transplant.* **24**:131–136.
49. **Grubb, J. R., A. C. Moorman, R. K. Baker, H. Masur, and the HOPS Investigators.** 2006. The changing spectrum of pulmonary disease in patients with HIV infection on antiretroviral therapy. *AIDS* **20**:1095–1107.
50. **Hamer, D. H., J. Egas, B. Estrella, W. B. MacLeod, J. K. Griffins, and F. Sempertegui.** 2002. Assessment of the Binax NOW *Streptococcus pneumoniae* urinary antigen test in children with nasopharyngeal pneumococcal carriage. *Clin. Infect. Dis.* **34**:1025–1028.
51. **Hamilton, C. D.** 2005. Immunosuppression related to collagen-vascular disease or its treatment. *Proc. Am. Thorac. Soc.* **2**:456–460.
52. **Happel, K. I., and S. Nelson.** 2005. Alcohol, immunosuppression and the lung. *Proc. Am. Thorac. Soc.* **2**:428–432.
53. **Harrington, R. D., A. E. Woolfrey, R. Bowden, M. G. McDowell, and R. C. Hackman.** 1996. Legionellosis in a bone marrow transplant center. *Bone Marrow Transplant.* **18**:361–368.
54. **Hewitt, R. G.** 2002. Abacavir hypersensitivity reaction. *Clin. Infect. Dis.* **34**:1137–1142.
55. **Hindiyeh, M., and K. C. Carroll.** 2000. Laboratory diagnosis of atypical pneumonia. *Semin. Respir. Infect.* **15**:101–113.
56. **Hindiyeh, M., D. R. Hillyard, and K. C. Carroll.** 2001. Evaluation of the Prodesse Hexaplex multiplex PCR assay for direct detection of seven respiratory viruses in clinical specimens. *Am. J. Clin. Pathol.* **116**:218–224.
57. **Hohenthal, U., M. Itala, J. Salonen, J. Sibila, K. Rantakokko-Jalava, O. Meurman, J. Nikoskelainen, R. Vainionpaa, and**

P. Kotilainen. 2005. Bronchoalveolar lavage in immunocompromised patients with haematological malignancy—value of new microbiological methods. *Eur. J. Haematol.* **74:**203–211.

58. **Huang, S. N., S. H. Fischer, E. O'Shaughnessy, V. J. Gill, H. Masur, and J. A. Kovacs.** 1999. Development of a PCR assay for diagnosis of *Pneumocystis carinii* pneumonia based on amplification of the multicopy major surface glycoprotein gene family. *Diagn. Microbiol. Infect. Dis.* **35:**27–32.
59. **Ison, M. G., and F. G. Hayden.** 2002. Viral infections in immunocompromised patients: what's new with respiratory viruses? *Curr. Opin. Infect. Dis.* **15:**355–367.
60. **Jain, P., S. Sandur, Y. Meli, A. C. Arroglia, J. K. Stoller, and A. C. Mehta.** 2004. Role of flexible bronchoscopy in immunocompromised patients with lung infiltrates. *Chest* **125:**712–722.
61. **Joos, L., P. N. Chhajed, J. Wallner, M. Battegay, J. Steiger, A. Gratwohl, and M. Tamm.** 2007. Pulmonary infections diagnosed by BAL: a 12-year experience in 1066 immunocompromised patients. *Respir. Med.* **101:**93–97.
62. **Kantrow, S. P., R. C. Hackman, M. Boeckh, D. Myerson, and S. W. Crawford.** 1997. Idiopathic pneumonia syndrome: changing spectrum of lung injury after marrow transplantation. *Transplantation* **63:**1079–1086.
63. **Khanna, M., J. Fan, K. Pehler-Harrington, C. Waters, P. Douglass, J. Stallock, S. Kehl, and K. J. Henrickson.** 2005. The pneumoplex assay, a multiplex PCR-enzyme hybridization assay that allows simultaneous detection of five organisms, *Mycoplasma pneumoniae*, *Chlamydia* (*Chlamydophila*) *pneumoniae*, *Legionella pneumophila*, *Legionella micdadei*, and *Bordetella pertussis*, and its real-time counterpart. *J. Clin. Microbiol.* **43:**565–571.
64. **Khurshid, I., and L. C. Anderson.** 2002. Non-infectious pulmonary complications after bone marrow transplantation. *Postgrad. Med. J.* **78:**257–262.
65. **Klont, R. R., M. A. S. H. Mennink-Kersten, and P. E. Verweij.** 2004. Utility of Aspergillus antigen detection in specimens other than serum specimens. *Clin. Infect. Dis.* **39:** 1467–1474.
66. **Kontoyiannis, D. P., V. C. Wessel, G. P. Bodey, and K. V. Rolston.** 2000. Zygomycosis in the 1990s in a tertiary-care cancer center. *Clin. Infect. Dis.* **30:**851–856.
67. **Kovacs, J. A., V. J. Gill, S. Meshnick, and H. Masur.** 2001. New insights into transmission, diagnosis, and drug treatment of *Pneumocystis carinii* pneumonia. *JAMA* **286:**2450–2460.
68. **Kramer, M. R., N. Berkman, B. Mintz, S. Godfrey, M. Saute, and G. Amir.** 1998. The role of open lung biopsy in the management and outcome of patients with diffuse lung disease. *Ann. Thorac. Surg.* **65:**198–202.
69. **Kravolic, S. M., and J. C. Rhodes.** 1998. Utility of routine testing of bronchoalveolar lavage fluid for cryptococcal antigen. *J. Clin. Microbiol.* **36:**3088–3089.
70. **Laing, R. B. S.** 1999. Nosocomial infections in patients with HIV disease. *J. Hosp. Infect.* **43:**179–185.
71. **Lasocki, S., A. Scanvic, F. Le Turdu, A. Restoux, H. Mentec, G. Bleichner, and J. P. Sollet.** 2006. Evaluation of the Binax NOW *Streptococcus pneumoniae* urinary antigen assay in intensive care patients hospitalized for pneumonia. *Intensive Care Med.* **32:**1766–1772.
72. **Lehto, J. T., V. J. Anttila, J. Lommi, M. S. Nieminen, A. Harjula, E. Taskinen, P. Tukiainen, and M. Halme.** 2004. Clinical usefulness of bronchoalveolar lavage in heart transplant recipients with suspected lower respiratory tract infection. *J. Heart Lung Transplant.* **23:**570–576.
73. **Lenner, R., M. L. Padilla, A. S. Teirstein, A. Gass, and G. J. Schilero.** 2001. Pulmonary complications in cardiac transplant recipients. *Chest* **120:**508–513.
74. **Li, L. M., L. Q. Bai, H. L. Yang, C. F. Xiao, R. Y. Tang, Y. F. Chen, S. M. Chen, S. S. Liu, S. N. Zhang, Y. H. Ou, and T. I. Niu.** 1999. Sputum induction to improve the diagnostic yield with suspected pulmonary tuberculosis. *Int. J. Tuberc. Lung Dis.* **3:**1137–1139.
75. **Ljungman, P., K. N. Ward, B. N. A. Crooks, A. Parker, R. Martino, P. J. Shaw, L. Brinch, M. Brune, R. De La Camara, A. Dekker, K. Pauksen, N. Russell, A. P. Schwarer, and C. Cordonnier.** 2001. Respiratory virus infections after stem cell transplantation: a prospective study from the Infectious Diseases Working Party of the European Group for Blood and Marrow Transplantation. *Bone Marrow Transplant.* **28:**479–484.
76. **Loddenkemper, R., and R. J. McKenna.** 2005. Pleuroscopy, thoracoscopy and other invasive procedures, p. 652. *In* R. J. Mason, J. F. Murray, V. C. Broaddus, and J. A. Nadel (ed.), *Murray and Nadel's Textbook of Respiratory Medicine*, 4th ed. Elsevier Saunders, Philadelphia, PA.
77. **Malouf, M. A., and A. R. Glanville.** 1999. The spectrum of mycobacterial infection after lung transplantation. *Am. J. Respir. Crit. Care Med.* **160:**1611–1616.
78. **Mavromatis, B., and B. D. Cheson.** 2003. Monoclonal antibody therapy of chronic lymphocytic leukemia. *J. Clin. Oncol.* **21:**1874–1881.
79. **Mehrad, B., G. Paccioco, F. J. Martinez, T. C. Ojo, M. D. Iannettoni, and J. P. Lynch III.** 2001. Spectrum of Aspergillus infection in lung transplant recipients. Case series and review of the literature. *Chest* **119:**169–175.
80. **Merrick, S. T., K. A. Sepkowitz, J. Walsh, L. Damson, P. McKinley, and J. L. Jacobs.** 1997. Comparison of induced versus expectorated sputum for diagnosis of pulmonary tuberculosis by acid fast smear. *Am. J. Infect. Control* **25:**463–466.
81. **Montoya, J. G., J. A. Kovacs, and J. S. Remington.** 2005. *Toxoplasma gondii*, p. 3177–3186. *In* G. L. Mandell, J. E. Bennett, and R. Dolin (ed.), *Principles and Practice of Infectious Diseases*, 6th ed., vol. 2. Elsevier Churchill Livingstone, Philadelphia, PA.
82. **Mulabecirovic, A., P. Gaulhofer, H. W. Auner, H. Popper, R. Krause, C. Hesse, and H. Sill.** 2004. Pulmonary infiltrates in patients with haematologic malignancies: transbronchial lung biopsy increases the diagnostic yield with respect to neoplastic infiltrates and toxic pneumonitis. *Ann. Hematol.* **83:**420–422.
83. **Mulinde, J., and M. Joshi.** 1998. The diagnostic and therapeutic approach to lower respiratory tract infections in the neutropenic patient. *J. Antimicrob. Chemother.* **41**(Suppl. D):51–55.
84. **Murdoch, D. R., R. T. R. Laing, G. D. Mills, N. C. Karalus, G. I. Town, S. Mirrett, and L. B. Reller.** 2001. Evaluation of a rapid immunochromatographic test for detection of *Streptococcus pneumoniae* antigen in urine samples from adults with community-acquired pneumonia. *J. Clin. Microbiol.* **39:**3495–3498.
85. **Murdoch, D. R., T. P. Anderson, K. A. Beynon, A. Chua, A. M. Fleming, R. T. R. Laing, G. I. Town, G. D. Mills, S. T. Chambers, and L. C. Jennings.** 2003. Evaluation of a PCR assay for the detection of *Streptococcus pneumoniae* in respiratory and non-respiratory samples from adults with community-acquired pneumonia. *J. Clin Microbiol.* **41:**63–66.
86. **Murdoch, D. R.** 2003. Nucleic acid amplification tests for the diagnosis of pneumonia. *Clin. Infect. Dis.* **36:**1162–1170.
87. **Musher, B., D. Fredricks, W. Leisenring, A. A. Balajee, C. Smith, and K. A. Marr.** 2004. *Aspergillus* galactomannan enzyme immunoassay and quantitative PCR for diagnosis of invasive aspergillosis with bronchoalveolar lavage fluid. *J. Clin. Microbiol.* **42:**5517–5522.

88. **Ninane, V.** 2001. Bronchoscopic invasive diagnostic techniques in the cancer patient. *Curr. Opin. Oncol.* **13**:236–241.
89. **Pearson, M. L., J. A. Jereb, and T. R. Frieden.** 1992. Nosocomial transmission of multidrug-resistant *Mycobacterium tuberculosis*. A risk to patients and healthcare workers. *Ann. Intern. Med.* **117**:191–196.
90. **Peleg, A. Y., S. Husain, E. J. Kwak, F. P. Silveira, M. Ndirangu, J. Tran, K. A. Shutt, R. Shapiro, N. Thai, K. Abu-Elmagd, K. R. McCurry, A. Marcos, and D. L. Paterson.** 2007. Opportunistic infections in 547 organ transplant recipients receiving alemtuzumab, a humanized monoclonal CD-52 antibody. *Clin. Infect. Dis.* **44**:204–212.
91. **Pene, F., A. Merlat, A. Vabret, F. Rozenberg, A. Buzyn, F. Dreyfus, A. Cariou, F. Freymouth, and P. Lebon.** 2003. Coronavirus 229E-related pneumonia in immunocompromised patients. *Clin. Infect. Dis.* **37**:929–932.
92. **Pirat, A., S. Ozgur, A. Torgay, S. Candan, P. Zeyneloglu, and G. Arslan.** 2003. Risk factors for postoperative respiratory complications in adult liver transplant recipients. *Transplant. Proc.* **36**:218–220.
93. **Preiksaitis, J. K., and S. Keay.** 2001. Diagnosis and management of post-transplant lymphoproliferative disorder in solid-organ transplant recipients. *Clin. Infect. Dis.* **33**(Suppl. 1):S38–S46.
94. **Rano, A., C. Agusti, O. Sibila, and A. Torres.** 2005. Pulmonary infections in non-HIV-immunocompromised patients. *Curr. Opin. Pulm. Med.* **11**:213–217.
95. **Rolston, K. V. I.** 2001. The spectrum of pulmonary infections in cancer patients. *Curr. Opin. Oncol.* **13**:218–223.
96. **Rubin, R. H., A. Schaffner, and R. Speich.** 2001. Introduction to the Immunocompromised Host Society consensus conference on epidemiology, prevention, diagnosis and management of infections in solid organ transplant patients. *Clin. Infect. Dis.* **33**(Suppl. 1):S1–S4.
97. **Sharp, S. E., A. Robinson, M. Saubolle, M. Santa Cruz, K. Carroll, and V. Baselski.** 2004. Cumitech 7B, Lower respiratory tract infections. Coordinating ed., S. E. Sharp. ASM Press, Washington, DC.
98. **Shorr, A. F., G. M. Susla, and N. P. O'Grady.** 2004. Pulmonary infiltrates in the non-HIV-infected immunocompromised patient. Etiologies, diagnostic strategies, and outcomes. *Chest* **125**:260–271.
99. **Singh, N., T. Gayowski, M. Wagener, I. R. Marino, and V. L. Yu.** 1996. Pulmonary infections in liver transplant recipients receiving tacrolimus: changing pattern of microbial etiologies. *Transplantation* **61**:396–401.
100. **Snydam, D. R.** 2001. Epidemiology of infections after solid-organ transplantation. *Clin. Infect. Dis.* **33**(Suppl. 1):S5–S8.
101. **Sopena, N., M. Sabria, M. L. Pedro-Botet, E. Reynaga, M. Garcia-Nunez, J. Domínguez, and L. Matas.** 2002. Factors related to persistence of Legionella urinary antigen excretion in patients with Legionnaires' disease. *Eur. J. Clin. Microbiol. Infect. Dis.* **21**:845–848.
102. **Speich, R., and W. van der Bij.** 2001. Epidemiology and management of infections after lung transplantation. *Clin. Infect. Dis.* **33**(Suppl. 1):S58–S65.
103. **Stevens, D. A.** 2002. Diagnosis of fungal infections: current status. *J. Antimicrob. Agents Chemother.* **49**(Suppl. S1):11–19.
104. **Sullivan, J. H., R. D. Moore, J. C. Keruly, and R. E. Chaisson.** 2000. Effect of antiretroviral therapy on the incidence of bacterial pneumonia in patients with advanced HIV infection. *Am. J. Respir. Crit. Care Med.* **162**:64–67.
105. **Tablan, O. C., L. J. Anderson, R. Besser, C. Bridges, and R. Hajjeh.** 2004. Guidelines for preventing health-care-associated pneumonia, 2003: recommendations of the CDC and the Healthcare Infection Control Practices Advisory Committee. *MMWR Recommend. Rep.* **53**(RR-3):1–36.
106. **Tanner, D. C., M. P. Weinstein, B. Fedorciw, K. L. Joho, J. J. Thorpe, and L. B. Reller.** Comparison of commercial kits for detection of cryptococcal antigen. *J. Clin. Microbiol.* **32**:1680–1684.
107. **Tkatch, L. S., S. Kusne, W. D. Irish, S. Krystofiak, and E. Wing.** 1998. Epidemiology of Legionella pneumonia and factors associated with Legionella-related mortality at a tertiary care center. *Clin. Infect. Dis.* **27**:1479–1486.
108. **Torres, A., S. Ewig, J. Insausti, J. M. Guergue, A. Xaubet, A. Mas, and J. M. Salmeron.** 2000. Etiology and microbial patterns of pulmonary infiltrates in patients with orthotopic liver transplantation. *Chest* **117**:494–502.
109. **Turner, D., Y. Schwarz, and I. Yust.** 2003. Induced sputum for diagnosing *Pneumocystis carinii* pneumonia in HIV patients: new data, new issues. *Eur. Respir. J.* **21**:204–208.
110. **Van der Zee, A., M. Peeters, C. de Jong, H. Verbakel, J. W. Crielaard, E. C. Claas, and K. E. Templeton.** 2002. Qiagen DNA extraction kits for sample preparation for *Legionella* PCR are not suitable for diagnostic purposes. *J. Clin. Microbiol.* **40**:1126.
111. **Van Elden, L. J. R., M. G. J. van Kraaij, M. Nijhuis, K. A. W. Hendriksen, A. W. Dekker, M. Rozenberg-Aska, and A. M. Van Loon.** 2002. Polymerase chain reaction is more sensitive than viral culture and antigen testing for the detection of respiratory viruses in adults with hematological cancer and pneumonia. *Clin. Infect. Dis.* **34**:177–183.
112. **Velasco, E., L. C. Thuler, C. A. Martins, L. M. Dias, and V. M. Goncalves.** 1997. Nosocomial infections in an oncology intensive care unit. *Am. J. Infect. Control* **25**:458–462.
113. **Waite, S., J. Jeudy, and C. S. White.** 2006. Acute lung infections in normal and immunocompromised hosts. *Radiol. Clin. N. Am.* **44**:295–315.
114. **Wallis, R. S., M. S. Broder, J. Y. Wong, M. E. Hanson, and D. O. Beenhouwer.** 2004. Granulomatous infectious diseases associated with tumor necrosis factor antagonists. *Clin. Infect. Dis.* **38**:1261–1265.
115. **Waring, A. L., T. A. Halse, C. K. Csiza, C. J. Carlyn, K. A. Musser, and R. J. Limberger.** 2001. Development of a genomics-based PCR assay for detection of *M. pneumoniae* in a large outbreak in New York State. *J. Clin. Microbiol.* **39**:1385–1390.
116. **Welsh, D. A., and C. M. Mason.** 2001. Host defense in respiratory infections. *Med. Clin. N. Am.* **85**:1329–1347.
117. **Wheat, L. J., T. Garringer, E. Brizendine, and P. Connolly.** 2002. Diagnosis of histoplasmosis by antigen detection based upon experience at the histoplasmosis reference laboratory. *Diagn. Microbiol. Infect. Dis.* **43**:29–37.
118. **White, D. A., P. W. Wong, and R. Downey.** 2000. The utility of open lung biopsy in patients with hematologic malignancies. *Am. J. Respir. Crit. Care Med.* **161**:723–729.
119. **Wolff, A. J., and A. E. O'Donnell.** 2001. Pulmonary manifestations of HIV infection in the era of highly active antiretroviral therapy. *Chest* **120**:1888–1893.
120. **Woods, G. L.** 2001. Molecular techniques in mycobacterial detection. *Arch. Pathol. Lab. Med.* **125**:122–126.
121. **Yen, K. T., A. S. Lee, M. J. Krowka, and C. D. Burger.** 2004. Pulmonary complications in bone marrow transplantation: a practical approach to diagnosis and treatment. *Clin. Chest Med.* **25**:189–201.
122. **Yoshikawa, T.** 2003. Human herpesvirus-6 and -7 infections in transplantation. *Pediatr. Transplant.* **7**:11–17.
123. **Zar, H. J., E. Tannebaum, D. Hanslo, and G. Hussey.** 2003. Sputum induction as a diagnostic tool for community-acquired pneumonia in infants and young children from a high HIV prevalence area. *Pediatr. Pulmonol.* **36**:58–62.

Diagnostic Microbiology of the Immunocompromised Host
Edited by Randall T. Hayden, Karen C. Carroll, Yi-Wei Tang, and Donna M. Wolk

Chapter 16

Genitourinary Tract Infections

BARBARA L. HALLER

Infections of the genitourinary tract can involve either the urinary system or the reproductive system and external genitalia. In immunocompromised hosts, infections of the urinary system, including the kidneys, bladder, and urethra, can contribute to morbidity and mortality. Other important infections of the genitourinary tract include prostatitis and epididymitis. While perinephric abscesses and intrarenal abscesses are uncommon, patients with diabetes mellitus or urinary tract calculi are at increased risk for development of these serious complications (90).

Infections of the reproductive system and external genitalia in immunocompromised hosts most commonly result from reactivation of latent viruses, leading to genital lesions and other significant sequelae, as seen with reactivation of herpes simplex virus (HSV) infection in transplant patients. Infections with human papillomavirus (HPV) occur frequently and are often severe in patients with primary and secondary immunodeficiencies or with human immunodeficiency virus (HIV) infection. HPV infection in these patients can present as severe anogenital warts, squamous intraepithelial lesions, or squamous cell carcinoma.

This chapter covers major genitourinary tract infections seen in different immunocompromised patient populations but does not address all possible infections of the reproductive system and external genitalia in these patients. Urinary tract infections (UTIs) in immunocompromised patients have been most well studied among transplant patients and HIV-infected patients.

AGENTS

UTIs are among the most common acute infectious diseases in humans (15). In immunocompetent patients, most UTIs are caused by gram-negative bacteria, with *Escherichia coli* being by far the most common uropathogen (15, 49). In immunocompromised patients, a broader range of organisms are associated with UTI. *Escherichia coli* and other *Enterobacteriaceae* are common urinary tract pathogens, but other organisms can also cause infections, including *Pseudomonas aeruginosa*, *Enterococcus* spp., and *Staphylococcus* spp., including *Staphylococcus saprophyticus* (10, 105). Other bacteria that can cause UTI in patients with severely compromised immune responses include *Ureaplasma urealyticum*, *Mycoplasma* spp., *Corynebacterium urealyticum*, and rarely, *Mycobacterium* spp. (10). *Candida* spp. are often considered contaminants if identified in urine cultures, but these organisms can be significant pathogens in the urinary tract in immunocompromised patients, including cancer patients treated with chemotherapy and transplant patients.

E. coli is the most common cause of acute bacterial prostatitis, with rare infections caused by *P. aeruginosa* and *Enterococcus* spp. (48, 90). Chronic bacterial prostatitis is also commonly caused by *E. coli*, but *Klebsiella* spp., *Enterobacter* spp., *Proteus mirabilis*, and *Enterococcus* spp. are also associated with this infection. Rare causes of prostatitis include *Candida* spp., *Blastomyces dermatitidis*, *Histoplasma capsulatum*, *Mycobacterium tuberculosis*, and nontuberculous mycobacteria. The prostate has also been identified as an important focus for subclinical *Cryptococcus neoformans* infection (90). Epididymitis in men older than age 35 is commonly caused by *Enterobacteriaceae* or *P. aeruginosa*. Epididymitis in younger men is most often sexually transmitted, with *Chlamydia trachomatis* and *Neisseria gonorrhoeae* being the most common pathogens (48). Perinephric abscesses usually result from infection with gram-negative enteric bacilli, or occasionally gram-positive cocci, if the infection is of a hematogenous origin (90). Renal abscesses can be polymicrobial, or rarely, fungi such as *Candida* spp. can be cultured from the abscess.

Barbara L. Haller • Department of Laboratory Medicine, University of California, San Francisco, San Francisco, CA 94143.

Adenoviruses have emerged as significant pathogens in hematopoietic stem cell transplant (HSCT) recipients, with upper and lower respiratory tract infection, gastrointestinal disease, hepatitis, or cystitis being seen most commonly in these patients (4, 40, 79). In renal transplant recipients, adenoviruses cause hemorrhagic cystitis, which can lead to graft failure and a high mortality rate (79). Hemorrhagic cystitis is most commonly associated with adenovirus type 11 and, less commonly, with subgroup B viruses (adenoviruses 34 and 35) (40). Adenoviruses can also be detected in urine or gastrointestinal tract specimens from AIDS patients, in whom infections can be asymptomatic or can result in colitis, parotitis, or encephalitis (4).

BK virus has been identified as the primary cause of both polyomavirus-associated nephropathy (PVAN) in renal transplant patients and hemorrhagic cystitis in HSCT recipients (25, 108).

Primary infection with HSV or reactivation of latent HSV infection in HIV-infected patients or patients with AIDS can lead to severe genital and perirectal ulcerations. In other immunocompromised patients, especially those with impaired cellular immunity, reactivation of genital herpes can lead to severe genital lesions and to life-threatening disseminated infection (16). HIV-infected men and women, transplant patients, patients with primary immunodeficiency, and patients with lymphoproliferative disorders are at increased risk of contracting anogenital and oral HPV infection (6, 65). In addition, HIV-infected patients and other immunocompromised patients are at increased risk for developing HPV-associated anogenital and oral cancers (13, 65, 67).

Strongyloides stercoralis is a parasite that can be detected in urine or stool of severely immunocompromised patients, including those treated with steroids. *Trichomonas* infection in HIV type 1 (HIV-1)-infected women in one South African study was associated with an increased prevalence of pelvic inflammatory disease (59). Another study demonstrated a trend toward more frequent detection of *Trichomonas* among HIV-seropositive women than among seronegative women (57).

PATHOPHYSIOLOGY

UTIs occur as a result of the interaction between virulence factors of uropathogenic bacteria and host defense systems. Since *E. coli* is the most common cause of UTI, many virulence factors have been identified in uropathogenic strains of *E. coli*, including O antigens, which are cell wall antigens; antiphagocytic capsular polysaccharide K antigens; and flagellar H antigens. Studies on bacteria isolated from immunocompetent hosts have demonstrated that distinct O, K, and H serotypes of *E. coli* are associated with cystitis and pyelonephritis and even correlate with the clinical severity of infection (44, 90). A more important virulence characteristic is the ability of bacteria to adhere to uroepithelium, which is critical for initiation of UTI (90, 100). This adherence is mediated by specific interactions between adhesins on the surfaces of uropathogenic bacteria and receptors on uroepithelial cell surfaces. The most important adhesion-receptor system involves the P fimbriae (also called P pili or type 2 pili) and glycosphingolipid cell receptors (44, 90, 100). These receptors are found on the uroepithelium and tissues of the urinary tract and kidney, which explains how uropathogenic bacteria with P fimbriae can ascend and invade the urinary tract. *E. coli* with P fimbriae resist phagocytosis and are the most important cause of pyelonephritis and urosepsis in immunocompetent hosts.

It is generally thought that expression of uropathogenic virulence factors by *E. coli* is required to initiate UTI in immunocompetent hosts, while expression of virulence factors is not needed to establish infection in immunocompromised hosts (43, 92). In support of this hypothesis, *E. coli* strains that lack P fimbriae are readily phagocytized in immunocompetent hosts, but these strains can be isolated from blood of severely neutropenic patients with pyelonephritis and patients with functional leukocyte defects (92, 100).

More recent studies suggest that uropathogenic virulence factors may play a significant role in UTIs in immunocompromised patients (77, 102). An important study evaluated virulence factors expressed by *E. coli* strains isolated from renal transplant patients presenting with lower or upper UTI symptoms more than 30 days posttransplant (77). Of the 39 *E. coli* isolates that could be serotyped, only 26% of isolates from patients with lower and upper UTIs belonged to the classic eight uropathogenic O serotypes (O1, O2, O4, O6, O7, O16, O18, and O75) seen in immunocompetent patients. The most common *E. coli* serotype (38%) in renal transplant patients with upper UTI symptoms was O25; the incidence of serotype O25 in nonimmunosuppressed patients with upper UTI is reportedly <2% (77). The significance of the increased incidence of serotype O25 *E. coli* in this group of renal transplant patients is not clear from this small study, but further evaluations of *E. coli* serotypes causing UTIs in other groups of immunocompromised patients and other transplant centers would be useful. Results from this study suggest that there is a difference in the *E. coli* serotypes causing UTIs in renal transplant patients, and serotype O25 was associated with increased virulence and acute allograft injury in patients with upper UTI. The expression of P fimbriae is reportedly less common on *E. coli* isolates from immunocompromised patients with upper UTI (100), and in a study of *E. coli* strains from renal transplant patients, fewer *E. coli* strains expressed P fimbriae than those among nonimmunocompromised patient isolates (77). However, 53% of *E. coli* strains associated with upper UTI did express P fimbriae, and 63% of patients with these strains had acute allograft injury. Further studies of virulence factors on *E. coli* strains causing UTIs in immunocompromised patients are needed to evaluate their role in incidence and severity of infection.

Other than the urethra, which is colonized with bacteria, urine and the urinary tract are normally sterile and relatively resistant to colonization and infection. Host defenses important in determining susceptibility to UTI include characteristics of the urine, normal anatomy and function of the urinary tract, and the immune response to infection (100). Urine characteristics such as high osmolality, high urea concentration, low pH, and urine flow during bladder emptying are important in prevention of UTI (90, 100). Factors that can lead to an increased risk for UTIs include obstruction to urine flow, vesicoureteral reflux, surgical alteration of the urinary tract, indwelling catheters, and incomplete bladder emptying (49).

Bacteria most commonly invade and spread in the urinary tract by an ascending route, where bacteria colonizing the urethra are introduced into the urinary tract. In both immunocompetent and immunocompromised patients, the presence of normal bacterial flora in the urethra, vaginal introitus, and periurethral region can decrease binding of uropathogens and decrease the risk of UTI (100). The use of indwelling urinary catheters, which can introduce bacteria into the sterile bladder and damage the bladder mucosa, greatly increases the risk for UTI (89, 101). Use of indwelling urinary catheters is common among immunocompromised patients during surgical or therapeutic interventions. A less common route of infection is through hematogenous spread of bacteria to the renal parenchyma to cause infection of the kidney, or pyelonephritis. Pyelonephritis can also result from movement of bacteria infecting the bladder into the renal pelvis and renal parenchyma. Patients with pyelonephritis can develop urosepsis, a severe bloodstream infection that can have high mortality in immunocompromised patients (90).

In immunocompetent individuals, UTIs most commonly occur in women of all ages due to the shorter female urethra and bacterial colonization of the vaginal introitus and periurethral area (15, 49, 89). UTIs in immunocompetent men are very rare, occurring most commonly in neonates and elderly men, where urinary tract obstruction can lead to infection. Among immunocompromised patients, women and men are both at risk for UTI. In the immune response to UTI, polymorphonuclear leukocytes have no role in preventing adherence or initiation of UTI, but they are important in limiting the extent of infection. Similarly, antibodies may limit kidney parenchyma scarring or provide some protection against reinfection. Cell-mediated immunity seems to have no role in host defenses against bacterial UTI (90, 100).

Bacterial prostatitis can develop as a result of bacterial spread through the hematogenous route, ascending from the urethra or through the rectal lymphatics (90). Other causes of prostatitis include urethral instrumentation and prostatic surgery. Patients who develop epididymitis usually have underlying urological pathology or have had recent manipulation of the genitourinary tract (48).

Host factors that may influence the etiology of genital tract infections include age, sexual activity, contraceptive use, genital tract instrumentation, vaginal flora, immune status, and prior infections (3). The organisms that cause genital lesions are well suited to multiply in squamous epithelial cells and on mucous membranes. These organisms are sexually transmitted, and a major risk factor is having multiple sexual partners. Impairment of host cell-mediated immunity, as seen in HIV-infected patients, can lead to reactivation of latent HSV infections, leading to genital lesions and disseminated disease (16). HPVs are DNA viruses that can infect epithelial cells in the basal cell layer of skin, larynx, and anogenital tissues, leading to lesions ranging from benign warts to cervical or anal squamous cell carcinoma (65, 66, 69, 96). HIV-infected patients, organ transplant recipients, and other immunosuppressed patients are at increased risk for HPV infection and progression to squamous intraepithelial lesions and squamous cell carcinoma due to an inability to control viral gene expression and to clear the virus (66). In HIV-infected patients, increasing immunosuppression, reflected by decreasing CD4 cell counts, correlates with abnormal cervical cytology in women, abnormal anal HPV-associated lesions in HIV-infected men who have sex with men, and higher rates of progression to high-grade squamous intraepithelial lesions (66).

SYNDROMES

UTIs present with different symptoms depending on the location of the infection. This description of upper and lower UTI symptoms is applicable to both immunocompetent and immunocompromised patients. The urinary tract is often divided into the upper portion, which includes the kidneys, renal pelvises, and ureters, and the lower portion of the tract, composed of the bladder and urethra. Infections of the upper tract most commonly result from ascending spread of organisms from the bladder through the ureters to the kidneys, resulting in acute pyelonephritis. Patients with acute pyelonephritis present with fever and pain or tenderness of the flank, often with other systemic symptoms, such as chills, headache, malaise, and nausea. It is important to collect blood cultures from patients with acute pyelonephritis, since 15 to 30% will be bacteremic and mortality can be as high as 8 to 30% (38). Less commonly, upper UTIs can result from hematogenous spread of bacteria to the kidneys in patients with bacteremia or sepsis, resulting in multiple renal abscesses and suppurative pyelonephritis.

Lower UTIs involve the urinary bladder and/or the urethra. Patients with cystitis, or infection of the bladder, usually present with dysuria (painful or burning urination), urgency or frequency of urination, and sometimes suprapubic pain or heaviness. However, similar symptoms can be seen in patients with urethritis due to infection with HSV or *Chlamydia trachomatis* or in those who have noninfectious urethral trauma or chemical irritation. The term "acute urethral syndrome" has been applied to another subset of patients who have similar symptoms but urine cultures with either no growth or only small numbers of uropathogenic bacteria. While some women with acute urethral syndrome have urethral irritation, not a true UTI, studies have shown that many dysuric women with low bacterial counts have a true bacterial UTI that requires therapy (49, 92). Children exhibiting symptoms of lower UTI with low bacterial counts on quantitative urine culture may also have a true UTI requiring therapy (22). While cystitis implies that bacterial infection is confined to the bladder, studies of bladder washouts and ureter catheterizations have demonstrated that 15 to 20% of patients with "cystitis" symptoms may have bacteria above the bladder (9, 42). Since patients presenting with "cystitis" symptoms can have true infection or urethral trauma or irritation, quantitative urine cultures, viral cultures, or molecular tests can be used to differentiate these possible etiologies.

Patients with acute bacterial prostatitis present with symptoms of lower UTI, with fever and lower abdominal or suprapubic discomfort. Results of urinalysis are abnormal and urine cultures are positive for these patients. Rectal examination will reveal an extremely tender prostate on palpation (48). Patients with chronic bacterial prostatitis typically experience recurrent UTIs with the same organism. Careful localization studies are required to confirm a diagnosis of chronic prostatitis. Patients with a perinephric abscess may present with symptoms suggestive of acute pyelonephritis or nonspecific urinary tract symptoms. Intrarenal abscesses usually result from hematogenous bacterial spread, often in the setting of severe pyelonephritis. A diagnosis of perinephric abscess or intrarenal abscess should be considered for any patient with fever and flank pain who responds slowly or does not respond to therapy for pyelonephritis (90).

Asymptomatic bacteriuria occurs when bacteria are isolated from the urine in significant quantities consistent with infection but patients have no local or systemic genitourinary signs or symptoms. Overall, the prevalence of asymptomatic bacteriuria in the general population is 3.5%, and prevalence increases with age (28). The incidence of asymptomatic bacteriuria among persons older than 70 is three times greater in women than in men. It is important that immunocompromised patients, such as renal transplant recipients, can have asymptomatic bacteriuria without pyuria that can lead to serious upper UTI symptoms. The clinical significance of asymptomatic bacteriuria and the need for treatment are controversial except for specific patient populations, such as children with vesicoureteral reflux, where kidney damage can result from silent infection; pregnant women, where there could be harm to the fetus with severe, symptomatic UTI; and immunocompromised patients, where antibiotic therapy may prevent more serious disease. In addition, patients who undergo an invasive procedure involving the urinary tract have an increased risk of urosepsis due to instrumentation if asymptomatic bacteriuria is present (28).

Hemorrhagic cystitis, characterized by hematuria, abdominal pain, dysuria, and urinary frequency, can be caused by adenoviruses in HSCT recipients and renal transplant patients (4, 12, 27). Adenovirus infections more frequently affect pediatric transplant recipients and can result from reactivation of latent viruses or from de novo infection (40, 79, 85). The onset of adenovirus infection typically occurs within the first 100 days after HSCT in children, while the onset of symptoms after 100 days has been documented for adult transplant patients (40). Graft failure or delayed engraftment can be seen with adenovirus infection in transplant recipients, and the mortality rate for symptomatic HSCT recipients may be as high as 26% if they are untreated (40). Clinical manifestations of adenovirus infection in HIV/AIDS patients are less severe in the highly active antiretroviral therapy (HAART) era (79).

Polyomavirus-associated nephropathy in renal transplant patients and hemorrhagic cystitis in HSCT patients result from BK virus reactivation during immunosuppression (20, 108).

Genital pustules and ulcers caused by primary infection with HSV or reactivation of latent virus are seen in both males and females and in both immunocompetent and immunocompromised patients. The most common microorganisms causing genital ulcers in HIV-infected patients include HSV and, in some geographical locations, *Treponema pallidum*, causing syphilis. Less common causes of genital ulcers in HIV-infected patients include *Haemophilus ducreyi* (chancroid), *Chlamydia trachomatis* serovars L1, L2, and L3 (lymphogranuloma venereum), and *Klebsiella granulomatis* (granuloma inguinale). The presence of genital warts, intraepithelial neoplastic lesions, or anogenital cancerous lesions suggests infection with HPV. Immunocompromised patients are at increased risk for HPV infection and progression of lesions to squamous cell carcinoma (66, 67).

HOST FACTORS AND PATIENT GROUPS

Genitourinary tract infections have been studied extensively for HIV/AIDS patients and recipients of solid organ transplants, especially renal transplant patients, where the risk of UTI is very high during the first 4 months posttransplant. Other immunocompromised patient groups are discussed briefly, since fewer documented studies of genitourinary tract infections in these patient groups are available. *Enterobacteriaceae* such as *E. coli* play a major role in UTIs in these patient groups, but other, less pathogenic bacteria or organisms, such as *Candida* spp., may also cause UTI, depending on the severity of immunosuppression and the part of the immune system that is impaired in a particular patient group.

HIV/AIDS

UTIs are an important health problem in HIV-infected persons, where the incidence is between 5% and 20% (37). Studies have shown that the incidence of UTIs is greater among men and women infected with HIV than among men and women who are seronegative for HIV (28, 84). This increased risk for bacteriuria correlates with the degree of immunosuppression, as reflected by the CD4 count. Most studies have demonstrated increased susceptibility to UTIs in HIV-infected patients with CD4 counts of <200 lymphocytes/mm^3 (28, 37, 84), although an earlier study of female sex workers in Africa failed to demonstrate a significant relationship between HIV status, CD4 count, and symptomatic UTI (63). In pre-HAART studies, the most common pathogen in patients with HIV was *Enterococcus* spp., in contrast to *E. coli* for seronegative patients (28, 47). More recent studies demonstrated a broad range of bacteria causing UTIs in HIV-infected patients, including the common uropathogens *Escherichia coli*, *Proteus* spp., *Enterobacter* spp., and *Klebsiella* spp.; nosocomial organisms, such as *Pseudomonas aeruginosa*, *Enterococcus*, and *Staphylococcus aureus*; and unusual microorganisms, including *Candida* spp., *Salmonella* spp., *Acinetobacter* spp., and cytomegalovirus (CMV) (2, 52, 84). HIV-infected patients develop complicated UTIs with an increase in severity of symptoms, increased risk of recurrent infection, and requirement for longer therapy (84). Several studies of hospitalization records of HIV-infected pregnant women in the United States demonstrated increased risk for puerperal sepsis, UTI, preterm labor, and premature delivery compared to that for uninfected pregnant women, even in the era of HAART (2, 47).

An association between HIV infection and sexually transmitted diseases is well documented. In a study of a large North American cohort of women infected with HIV, it was demonstrated that HIV-infected women were more likely to report a past history of syphilis, gonorrhea, and pelvic inflammatory disease (31). Both HIV status and CD4 counts were associated with physical evidence of genital ulcerations, warts, and vaginal candidiasis. There was also a high incidence and prevalence of chronic viral infections, including HSV and HPV infections, among the HIV-infected women (31). Overall, the incidence of sexually transmitted diseases, including urethritis caused by *Chlamydia trachomatis* or *N. gonorrhoeae*, HSV genital infection, and warts and squamous cell carcinoma caused by HPV infection, is higher for HIV-infected individuals than for the HIV-negative population (31). In addition, HIV-positive women who are infected with HPV have an increased risk of progression to high-grade cervical intraepithelial neoplasia and cervical cancer (8). For men, the incidence of bacterial prostatitis and epididymitis increases with HIV infection and progression to AIDS (52).

Renal Transplant and Other Solid Organ Transplant Recipients

The incidence of infection after solid organ transplantation varies with the type of organ transplanted, the amount of immunosuppression, the need for antirejection medications, and surgical complications (87). UTI is the most frequent bacterial infection seen in renal transplant recipients, with a reported incidence of between 26% and 58% (37). It is estimated that approximately 60% of bacteremias in renal transplant recipients originate in the urinary tract, which can lead to life-threatening sepsis in the early postoperative period. The first 3 months after renal transplantation represent the most critical time for UTIs, since pyelonephritis of the transplanted kidney can develop along with gram-negative bacteremia (37). Within the first month posttransplant, UTI usually results from surgical complications, exposure to nosocomial pathogens, and the presence of stents or a postoperative indwelling catheter (37). Removing the catheter within 48 hours after surgery decreased the incidence of UTI from 74% to 15% in one study (75). Other risk factors associated with development of post-renal transplant UTI include advanced age, female gender, reflux kidney disease prior to transplantation, and receipt of a cadaveric kidney (15). During the first month post-renal transplant, the most common viral infection is reactivated HSV infection in patients seropositive for HSV prior to transplantation or, more rarely, seronegative patients exposed to an HSV-seropositive donor kidney (87). Since HSV infection in these patients can result in severe mucocutaneous lesions and disseminated infection, patients undergoing solid organ transplantation or initiation of cancer chemotherapy are routinely given prophylactic acyclovir to prevent this complication (16). After the first month post-renal transplant, serious infections are more likely caused by opportunistic pathogens, such as *Pneumocystis jirovecii*, *Listeria monocytogenes*, or *Nocardia* spp. Candiduria has been documented for renal transplant patients, with *Candida glabrata* causing 53% of episodes and *Candida albicans* isolated from 35% of patients at one transplant center (81). While the presence of *Candida* spp. in urine may represent colonization and such infections in renal transplant patients can be asymptomatic, these patients are at risk for development of severe pyelonephritis and disseminated *Candida* infection (81). In the 2- to 6-month period after transplantation, adenovirus infection can result in hemorrhagic cystitis or disseminated disease in renal transplant patients (14). Polyomaviruses cause nephropathy in 2 to 8% of renal transplant patients, with BK virus being the most commonly identified virus, usually more than 6 months after transplantation (50, 87).

The threat of bacterial infections, especially UTIs, continues throughout the first year after renal transplantation. Prevention of UTIs in this patient population is critical, so trimethoprim-sulfamethoxazole (SXT) or ciprofloxacin may be recommended as prophylaxis to prevent UTI (88, 99). This strategy has significantly reduced the incidence of bacterial infections following renal transplantation, with the added bonus of effective prophylaxis against *Pneumocystis jirovecii* infection. If patients develop urosepsis in the late posttransplantation period, aggressive treatment and a search for obstruction, urolithiasis, or lower urinary tract dysfunction should be carried out (37).

For other transplanted organs, including the liver, heart, lung, and pancreas, the incidence of UTI during the first month posttransplant is similar to the UTI incidence seen in general surgery patients (87). Risk factors for UTI during this period include urinary catheterization, surgical complications causing obstruction, and poor bladder function (87). Also, during the first month after transplantation, infections caused by *Candida* spp. can be seen, especially in association with catheterization and immunosuppression (87, 90).

Primary Immunodeficiencies

In patients with deficiencies of late complement components (C6 to C9), *Neisseria* spp. are not lysed and killed. Patients can develop genitourinary infections with *Neisseria gonorrhoeae*, progressing to disseminated disease with skin lesions and arthritis (103). Patients with humoral specific primary immunodeficiency usually present with recurrent infections. In these patients, *Ureaplasma urealyticum* and *Mycoplasma* spp. can cause UTI (103). There are many cell-specific immunodeficiencies described, including phagocytic disorders such as chronic granulomatous disease. In this disease, severe infections of the skin, central nervous system, mouth, and gastrointestinal system are seen, and genitourinary tract infections caused by gram-negative organisms, *Staphylococcus* spp., and *Candida* spp. can develop (103). T-cell defects seen in severe combined immunodeficiency result in recurrent infections at a very early age, including infections with many bacterial types as well as viruses such as HSV, varicella-zoster virus (VZV), CMV, JC and BK polyomaviruses, and HPV (103).

Aplastic Anemia

Infections are the leading cause of death in patients with aplastic anemia, characterized by hypocellular bone marrow and pancytopenia (106). These patients can have infections in the bloodstream, respiratory tract, gastrointestinal tract, skin, soft tissue, and urinary tract. Most UTIs in these patients are caused by gram-negative bacteria, including *E. coli*, *Klebsiella pneumoniae*, and *Pseudomonas aeruginosa*, with symptoms of dysuria, frequency, urgency, and flank pain. Unusual organisms, including fungi such as *Candida albicans* and mycobacteria, can also cause UTI in these patients. Since patients with aplastic anemia have a longer duration of neutropenia, they can develop invasive fungal infections and fungal UTI (106).

Solid Tumors (Cancer)

Obstruction of the urinary tract, especially the ureters, by a solid tumor, such as a gynecological tumor or prostate cancer, can lead to an increased risk of UTI. Other factors that determine an increased risk for genitourinary infections in cancer patients include the complexity of surgical procedures, the nature and intensity of chemotherapy, and the use of radiation therapy (80). Major surgical procedures can lead to hospital-associated UTI caused by indwelling urinary catheters and to infections with *E. coli*, other *Enterobacteriaceae*, *Pseudomonas* spp., coagulase-negative staphylococci, enterococci, and yeasts (15). The major effects of cancer chemotherapy include myelosuppression and disruption of normal anatomic barriers. Damage to mucosal surfaces and neutropenia secondary to chemotherapy treatment can lead to an increased risk for UTI. With short-lived neutropenia secondary to chemotherapy, typical cystitis symptoms may be seen, but more intense chemotherapy for aggressive tumors can lead to acute pyelonephritis, with or without bacteremia, and to infection with gram-positive cocci or yeasts as well as gram-negative organisms (80). Radiation therapy can lead to tissue inflammation, myelosuppression, and impaired humoral and cell-mediated immunity. External pelvic radiation therapy can result in hemorrhagic cystitis or UTI, with 17% of patients undergoing pelvic radiation developing UTI during radiotherapy in one study (5). In a study of 36 patients undergoing external radiation therapy for gynecological malignancies, UTI was detected in 33% of patients (74), and in a study of patients with prostatic malignancy, 1 to 2% of patients developed UTI per week of radiation therapy (78). It is important that neutropenic patients may have a UTI but be asymptomatic and lack pyuria, so routine urine cultures or antibiotic prophylaxis during periods of neutropenia are recommended at many treatment centers (80). However, prolonged exposure to antibiotics increases the risk that resistant organisms will emerge, with resulting therapeutic complications.

Hematological Malignancies

As with solid tumors, patients with lymphoma or leukemia have an increased risk of UTI with obstruction of the urinary tract or with chemotherapy, which can lead to neutropenia and damage to mucosal surfaces (29). These patients can get symptoms of UTI with or without bacteremia or can be asymptomatic. In a survey of the most common sites of infections in febrile, neutropenic patients with hematologic malignancies, only 3% (17/1,049 episodes) had UTI (29). Organisms causing UTI in these patients include gram-negative bacilli, *Enterococcus* spp., coagulase-negative staphylococci, and *Candida* spp., as well as reactivation of HSV and VZV. Patients are usually treated prophylactically with antibiotics and should be monitored carefully for development of infectious complications (29).

HSCT

During the neutropenic period immediately after HSCT, patients can develop UTI caused by gram-negative bacilli, *Enterococcus* spp., coagulase-negative staphylococci, and *Candida* spp. (7). Patients present with typical symptoms, including dysuria and frequency and urgency of urination, often with development of fever and flank pain, indicating progression to pyelonephritis and bacteremia. During the second risk period, after engraftment, risk factors for severe infections include the presence of graft-versus-host disease and the use of steroids (7). Infections due to HSV, CMV, VZV, and adenovirus can be life-threatening, so prophylactic therapy during the months after HSCT are critical for patient survival. Adenoviruses have emerged as major pathogens in HSCT patients, causing symptomatic enteritis, severe hemorrhagic cystitis, pneumonia, and hepatitis (4, 40, 95). Approximately 10 to 20% of adenovirus-infected HSCT patients will have progression to disseminated disease, with very high mortality rates (40). BK virus is associated with hemorrhagic cystitis in HSCT recipients (25).

Diabetes Mellitus

While skin and soft tissue infections, such as cellulitis and diabetic foot ulcer, are common infections in diabetics, studies have shown that female patients with diabetes have an increased risk of UTI and bacteriuria (28, 62, 70, 71). Asymptomatic bacteriuria is the most common presentation of UTI in diabetic patients, with a prevalence of up to 29% in diabetic women, which is three times greater than that in nondiabetic women (71). In diabetic men, the reported overall prevalence of asymptomatic bacteriuria ranges from 1 to 11%, which is similar to the prevalence seen among nondiabetic men (71). Diabetic women with asymptomatic bacteriuria are at greater risk of developing symptomatic UTI and being hospitalized for urosepsis (71). However, no studies have shown that antibiotic therapy for asymptomatic bacteriuria is beneficial, so antibiotic treatment is not currently recommended for diabetic patients with asymptomatic bacteriuria (71, 90).

Diabetic patients can also present with cystitis or severe pyelonephritis. They are at increased risk for hospitalization for pyelonephritis and severe complications of UTI, such as renal abscesses, emphysematous cystitis, and emphysematous pyelonephritis (28, 60, 71). Emphysematous cystitis and emphysematous pyelonephritis are defined by the presence of gas within the bladder wall or renal parenchyma as a result of invasion by gas-producing enteric organisms or anaerobic organisms (19, 62, 71). Both emphysematous cystitis and emphysematous pyelonephritis infections can rapidly progress and become life-threatening in diabetic patients, with mortality rates reaching 70 to 90% (32, 58, 71).

Diabetic patients with UTI are commonly infected with organisms other than *E. coli*, such as *Klebsiella* spp., *Staphylococcus saprophyticus*, enterococci, group B streptococcus, and rarely, *Candida* spp. (28, 30, 71). Factors predisposing diabetic patients to UTI and serious complications include bladder dysfunction due to autonomic neuropathy with urine stagnation, increased use of catheterization, and increased glucose in urine, which can facilitate bacterial growth and impair phagocytosis (30, 71). In addition, specific defects in innate and adaptive immunity have been identified in vitro in diabetic patients, including decreased microbial killing by polymorphonuclear leukocytes and defects in T-lymphocyte function, especially in patients with poor glycemic control (71). Further studies are needed to define the impact of glycemic control on the prevalence and presentation of UTI in diabetic patients (62).

PROGNOSTIC FACTORS AND THERAPEUTIC CONSIDERATIONS

Renal transplant patients and HIV-infected patients are at greatest risk for UTI. These patients are more likely to develop severe pyelonephritis and to have organisms with increased antibiotic resistance. Prophylaxis with SXT or a fluoroquinolone may be appropriate for renal transplant patients. Other immunocompromised groups should have urine cultures and other studies, such as radiological imaging, performed if they present with UTI symptoms. If patients do not respond to antibiotics, infection with unusual organisms should be considered.

UTIs in immunocompromised patients should be considered complicated infections, since the state of immunosuppression predisposes these patients to UTI and can lead to severe complications, a broad spectrum of infecting microorganisms, and infections that are more difficult to treat (90, 105). Accurate diagnosis of the presenting syndrome, identification of the infecting organism(s), and accurate susceptibility testing are all critically important for immunocompromised patients. These factors allow the choice of appropriate antimicrobial agents for these patients. Making an accurate diagnosis requires a careful physical examination, with possible use of radiographic techniques for patients with symptoms of severe pyelonephritis or potential renal abscesses. To identify the infecting microorganism(s), laboratories can perform urine cultures to detect bacteria or yeast and viral culture or molecular techniques to detect HSV or adenovirus infections. Urine cytology or quantitative amplification assays can be used to identify BK virus and to quantitate viral loads to guide therapeutic interventions. Several antigen assays are available for detection of HPV infection. Testing organisms for antimicrobial resistance is also critical, since organisms isolated from patients with complicated UTI have a higher frequency of resistance (61).

OVERVIEW OF DIAGNOSTIC APPROACHES

Radiographic Methods

Radiographic imaging studies should be performed for immunocompromised patients with symptoms of severe pyelonephritis or if blockages caused by a tumor or renal calculi are suspected. Immunocompromised patients with UTI or pyelonephritis who fail to improve symptomatically after 72 hours of appropriate antimicrobial therapy should have imaging studies done to detect rare complications such as a perinephric or intrarenal abscess (90). A plain abdominal radiographic study can detect renal calculi or the presence of gas in renal parenchyma or perirenal spaces, which would suggest a diagnosis of emphysematous pyelonephritis (90). However, the most important imaging methods used for patients with UTI include renal ultrasonography (US), abdominal computerized tomography (CT) with contrast, and magnetic resonance imaging (MRI) (90). Ultrasound, which is portable and does not involve radiation or contrast, can detect increased renal size and increased echogenicity in patients with pyelonephritis, as well as masses which could represent perinephric or intrarenal abscesses. The most sensitive methods for detecting enlarged kidneys, focal defects, the presence of gas, or abscesses are CT with contrast and MRI. Both US and CT can be used to detect suppuration and to guide diagnostic needle aspiration of possible masses or abscesses in the urinary tract (90). Patients with acute bacterial prostatitis can develop prostatic abscesses. In these cases, transrectal US is the most specific diagnostic method; a CT or MRI scan of the pelvis can also be used to detect abscesses (90). CT with contrast is also highly specific and is the preferred method for detection of xanthogranulomatous pyelonephritis, which is a rare, chronic granulomatous process in the kidney induced by recurrent UTI (17).

Surgical Pathology Methods

Histopathological examination of abscesses or biopsy material from the urinary tract can identify the presence of inflammation. Bacteria appear as blue rods or cocci on routine hematoxylin and eosin staining. Specific stains can be used to identify pathogens, such as an acid-fast stain (FITE stain) to identify mycobacteria, Gomori methenamine silver stain to detect the presence of fungi, and immunofluorescent stains to detect viral antigens.

An important disease that results from reactivation of BK virus in renal transplant patients is PVAN. Clinically, patients with PVAN experience gradually decreasing renal function with BK viruria and viremia. Definitive diagnosis of PVAN requires an allograft biopsy, which will show enlargement of tubular epithelial cells, karyomegaly, and nuclear inclusion bodies (56). The diagnosis can be confirmed with immunohistochemical staining of paraffin sections with an antibody to the simian virus 40 large T antigen or by electron microscopy to detect the 40-nm BK virions (56). BK virus-infected tubular epithelial cells can be shed into the urine and can be identified using the Papanicolaou (Pap) stain on fixed urine (20, 36). These cells are called "decoy cells," since they can appear to be malignant cells with an enlarged nucleus occupied by a basophilic viral inclusion (21). Routine screening of urine specimens from renal transplant patients to detect decoy cells has been shown to be 100% sensitive, but only 71% specific, for diagnosis of PVAN (36). The low positive predictive value (29%) of decoy cell detection in urine has led to the development of molecular methods to detect BK viruria and viremia to screen for PVAN (36).

The Tzanck smear is a histopathological method used to presumptively identify HSV-infected cells from a genital lesion (1). For this smear, epithelial cells from the base of an ulcer are collected and smeared onto a slide, which is then stained with Giemsa stain. The slide is examined using light microscopy, and the presence of multinucleated giant cells with intranuclear inclusions indicates presumptive HSV-infected cells (1). Since both the sensitivity and specificity of the Tzanck smear method are poor, viral culture provides a more definitive diagnosis of HSV infection.

The Pap smear is a cytology method used to screen women for abnormal cervical lesions caused by infection with high-risk HPV. The cellular change consistent with HPV infection is the koilocyte, which represents a precancerous cell (1, 6). Koilocytes are squamous cells with enlarged, often binucleate, nuclei, hyperchromasia, and perinuclear clearing. Since individuals who are immunosuppressed as a result of organ transplantation or HIV infection are at increased risk for HPV-associated anogenital cancers (66, 86), frequent cervical and anal cancer screening with Pap smears and biopsies is recommended for these patients (13, 64, 96). Immunocompromised patients are also at increased risk for developing genital warts (condyloma acuminatum) caused by HPV infection (83). Usually, condyloma acuminatum can be diagnosed by appearance, but these lesions can be biopsied and examined for the presence of koilocytes to diagnose HPV infection.

Microbiological Method Overview

Immunocompromised patients are at risk for more severe UTI infections, are less likely to be treated with short-course therapy, and have a greater risk of treatment failures, with development of abscesses and other complications. For microbiological detection of UTIs in immunocompromised patients, rapid screening methods, such as the dipstick method for detection of leukocyte esterase (LE) or nitrate reduction in urine, have limited usefulness, since many of these patients lack pyuria, can be infected by organisms other than *E. coli*, and have low bacterial counts in urine (17, 107). Urine microscopy with either direct observation of bacteria in urine or use of a Gram stain may be a sensitive indicator of UTI if colony counts are $\geq 10^5$ CFU/ml but is insensitive with low colony counts (11, 107). The most important microbiological method for diagnosing UTI in immunocompromised patients is quantitative culture of bacteria from urine or biopsy specimens, with susceptibility testing to detect resistant organisms. Culture media, including chromogenic agars, should be selected to ensure detection of the broad range of microorganisms that can infect the genitourinary tract of immunocompromised patients, including bacteria and yeasts.

Viral culture using appropriate tissue culture cells can be set up to detect HSV in genital lesions or adenovirus in stool or blood specimens. Molecular methods such as PCR or quantitative real-time PCR are useful for detection of viruses causing genitourinary tract symptoms or for screening for increased levels of BK virus that can lead to allograft loss in renal transplant patients.

Sample Collection Issues

Methods for collection of urine specimens for diagnosis of UTI are similar for immunocompetent and immunocompromised patients. Collection of urine by performing suprapubic aspiration (SPA) is considered the "gold standard" method for urine collection (23). Under local anesthetic, a needle is inserted into the bladder above the symphysis pubis and urine is percutaneously aspirated. SPA collection avoids urethral commensal organisms, but this method is infrequently used because it is invasive and can be uncomfortable for the patient. SPA collection of urine may be useful with immunocompromised patients when a definitive identification and resistance profile of infecting microorganisms causing UTI symptoms are necessary. Collection of a midstream urine specimen, where the first part of the urine stream is voided and the next 10- to 20-ml portion is collected for testing, is common for both males and females. Collection of a "clean catch" midstream urine, where the patient cleanses the periurethral area, is no longer recommended by many experts (98). For patients unable to use the midstream collection method, collection of urine by insertion of a single catheter into the bladder (straight catheter method) can minimize urethral contamination and yield an acceptable specimen for culture. A disadvantage of this invasive technique, however, is that bacteria can be introduced into the bladder, precipitating a UTI or other complications. Many immunocompromised patients, especially those hospitalized for cancer or transplant surgery, have a chronic indwelling urethral catheter. These catheters typically become colonized with multiple uropathogens and commensal organisms, which can result in collection of urine containing many different organisms at high counts (23). Interpretation of urine cultures from patients with an indwelling urethral catheter is very challenging. Because of urethral contamination in most urine specimens, each laboratory will develop guidelines defining "significant bacteriuria" for the patient population that would require identification and susceptibility testing of uropathogens. To facilitate this culture workup and interpretation, it is very important that the collection method is indicated on the laboratory requisition slip. Urine specimens collected for culture should be plated within 2 hours after collection or refrigerated for no more than 24 hours to prevent bacterial overgrowth (98). For the outpatient setting, laboratories may provide transport systems with preservatives that can stabilize the number of CFU/ml to replace refrigeration and prevent bacterial growth (23).

For diagnosis of acute bacterial prostatitis, midstream urine collection should suffice. However, diagnosis of chronic bacterial prostatitis may require collection and quantitative culture of urethral urine, midstream urine, prostatic secretions expressed by massage, and urine voided after massage. Comparing the colony counts in these different specimens may allow localization of infection in the prostate and a diagnosis of chronic prostatitis (90).

Plasma or serum collected for adenovirus culture or molecular testing should be separated within several hours after collection to avoid hemolysis. Urine specimens should be kept at 2 to 8°C during transport. Specimens, virus isolates, and DNA extracts can be frozen at −70°C until they are tested (4, 79).

If genital lesions suggestive of HSV are present, swabs can be used to collect cells at the base of the lesions; cells are added to viral transport medium, to make slides for direct fluorescence detection of viral antigen or to inoculate viral cultures. Calcium alginate swabs or swabs with wooden shafts should not be used because these swabs decrease the infectivity of herpesviruses (41). Swabs for viral culture should be delivered to the laboratory immediately, or they can be refrigerated at 4°C for 24 hours in transport medium. Specimens for PCR testing of genital lesions can be collected with a Dacron swab, which is broken off into viral transport medium and stored at 4°C for up to 72 hours or frozen at −70°C for longer storage before PCR testing is done (41).

MICROBIOLOGICAL METHODS FOR IMMUNOCOMPROMISED PATIENTS

Rapid Screening Methods

Rapid screening methods for UTI have been used to provide clinicians with preliminary information about levels of bacteriuria and pyuria and to screen out urine specimens that would most likely be culture negative. The evaluation of a Gram stain of urine is a rapid, inexpensive method that may be helpful in guiding empiric therapy, especially for patients with symptoms of acute pyelonephritis or for immunocompromised patients, in whom a broad range of organisms may cause genitourinary tract infection. In uncentrifuged urine, the presence of ≥ 1 bacterial cell/oil immersion field ($\times 100$ objective) correlates with $>10^5$ CFU/ml by routine bacterial culture. The sensitivity for detection of UTI with uncentrifuged urine is at least 85% (15). The sensitivity of stained centrifuged urine ranges from 60 to 100%, with a specificity of 59 to 97% (15). However, bacterial concentrations of $<10^5$ CFU/ml are not reliably detected, using either uncentrifuged or centrifuged urine (15, 23). Therefore, it is recommended that the urine Gram stain be performed on specimens from patients with severe pyelonephritis or in cases where bacterial counts of $\geq 10^5$ CFU/ml are expected (11, 15, 107).

Urine microscopy for detection of pyuria reportedly has a sensitivity of 80 to 90% for predicting bacteriuria in immunocompetent patients, but the specificity is only 50 to 70% for UTI, since noninfectious conditions can also cause pyuria (23). Accurate measurement of pyuria requires use of a hemocytometer, which is labor-intensive, and the lack of pyuria in immunocompromised patients, such as neutropenic patients, limits the utility of urine microscopy.

A number of commercial rapid screening methods have been developed to detect enzymatic activities in urine, such as LE activity, indicating the presence of white blood cells (pyuria), and bacterial reduction of nitrate to nitrite (15, 23). These commercial dipstick tests are rapid and inexpensive. The sensitivity of the LE test is 75 to 96% for detection of $\geq 10^5$ CFU/ml, with a specificity of 94 to 98% (23). False-positive results have been observed with *Trichomonas* contamination, contamination with detergents or drugs, such as nitrofurantoin and gentamicin, or high levels of albumin in the urine (23). False-negative results for these two tests can occur with a small number of white blood cells, with frequent voiding that can dilute LE and nitrite in urine, or with the presence of *Enterococcus* spp. or *S. saprophyticus* in urine, since these organisms cannot reduce nitrate to nitrite (23, 98).

These rapid methods have limited utility for immunocompromised patients because pyuria may be absent and often small numbers of organisms are present in urine, leading to potential false-negative results. For immunocompromised patients, where all UTIs are complicated infections, urine cultures should always be set up regardless of results from screening tests (23). This is especially true for renal transplant patients, neutropenic patients, and specimens collected by invasive means, including SPA, bladder washouts, or surgery (23).

Cultures and Susceptibility Testing

Bacterial cultures

The major microbiological method for diagnosis of UTI, including cystitis and pyelonephritis, is the quantitative urine culture. Performance of urine cultures is relatively inexpensive, and results are usually reported within 24 to 48 hours after receipt in the laboratory. It is important that clinicians be aware of the limitations associated with each type of specimen collection method (potential for contamination) and the criteria used by the clinical microbiology laboratory to determine how many organisms will be identified and tested for drug susceptibilities for each specimen type. This is especially true for urine cultures from immunocompromised patients, since more diverse microorganisms can cause UTI in these patients, with a higher frequency of antibiotic resistance (61, 107).

A routine quantitative urine culture involves the inoculation of a 5% sheep blood agar plate with 0.01 or 0.001 ml of urine specimen, using a calibrated wire or plastic loop. An estimate of the CFU/ml in the original urine specimen can be determined from this plate. A MacConkey agar plate is usually also inoculated to detect gram-negative organisms. For urine specimens collected from patients where mixtures of commensal and pathogenic organisms may be present, such as catheterized patients, a colistin-nalidixic acid agar plate with 5% sheep blood may be included to inhibit growth of gram-negative organisms and to allow evaluation of gram-positive organisms in the urine specimen. These plates are incubated at 35°C for 24 to 48 hours. Interpretation of quantitative urine culture results can be challenging, given the urethral contamination in most urine specimens. Extensive guidelines that define "significant" bacteriuria, depending on the host characteristics, symptoms, and method of collection, have been developed to aid in interpretation of culture results (15, 23, 98). Typically, symptomatic patients with $\geq 10^5$ bacterial CFU/ml in a urine culture are considered to have a true UTI. Urine specimens with $<10^5$ CFU/ml bacteria may be considered the result of urethral contamination, but an important study demonstrated that approximately 40% of young women with true symptoms of cystitis can have bacterial concentrations in urine of $<10^5$ CFU/ml and as low as 10^2 CFU/ml (91). In addition, gram-positive organisms, fungi, and fastidious bacteria that may cause UTI in immunocompromised patients may not be present in numbers of $\geq 10^5$ CFU/ml. Other studies have shown that for persons with UTI symptoms, one quantitative culture result of 10^5 or more CFU/ml has a 95% probability of being a true UTI, whereas a result of $<10^5$ CFU/ml with UTI symptoms has a 33% chance of being a true bacterial UTI (23). For patients with symptoms of pyelonephritis, rapid identification of uropathogens and susceptibility testing are essential for appropriate antibiotic therapy. A urine Gram stain and collection of blood cultures are also important for guiding therapy in patients with pyelonephritis.

Guidelines for interpretation of quantitative urine cultures for immunocompromised patients are not well defined. Patients who are highly immunocompromised, such as HSCT patients, solid organ transplant recipients, or cancer patients on chemotherapy with prolonged neutropenia, commonly receive prophylactic antimicrobial agents to prevent infections, including UTI. If these patients did present with UTI symptoms, it is likely that low bacterial counts (10^2 or 10^3 CFU/ml) could represent cystitis or pyelonephritis that would require treatment (7, 80). One study reported that a colony count of $\geq 10^2$ CFU/ml was reliable for identifying symptomatic infection in chronically catheterized patients (93). Another study that compared culture of SPA specimens with intermittent straight catheter specimens demonstrated that a diagnostic criterion of $>10^2$ CFU/ml resulted in a sensitivity of 91% and a specificity of 97% for detection of bladder bacteriuria (23). These studies suggest that low colony counts in urine cultures from immunocompromised patients with urinary catheters should be considered clinically significant.

A recent advance in culture media for detection of UTI is the development of chromogenic media. These media contain enzyme substrates that release colored dyes upon hydrolysis by bacterial enzymes produced specifically by different uropathogenic bacteria. The agar offers presumptive identification of gram-positive and gram-negative organisms as well as yeasts. A major advantage of these new media is easy evaluation of mixed cultures with increased sensitivity, since uropathogens within a mixture of commensal flora can be identified more easily by color (34, 72, 82). Use of these agars for urine cultures can save time for technologists and offer sensitive detection of uropathogens as well as a presumptive identification of uropathogens. The HardyCHROM BluEcoli urine biplate is available through Hardy Diagnostics for detection of *E. coli* in urine cultures. BBL CHROMagar Orientation (Becton Dickinson) is another chromogenic medium designed to differentiate the major gram-negative and gram-positive organisms that cause UTI. In one study, the CHROMagar Orientation medium performed very well compared to routine culture in identification of *E. coli*, other *Enterobacteriaceae*, and *Enterococcus* spp., as well as gram-positive organisms such as *S. saprophyticus* (34). More organisms were recovered on the CHROMagar medium than on conventional media, and only 0.4% of *E. coli* strains failed to develop the expected pink color for *E. coli* (34). Another study demonstrated that bacterial colonies could be identified and transferred successfully directly from the chromogenic agar to a disk diffusion susceptibility test without the need for subculture (82). BBL CHROMagar Candida (Becton Dickinson) is a chromogenic agar medium that differentiates *Candida* spp. This chromogenic agar can be used to identify four important *Candida* spp. and is also useful for identifying *Candida* spp. in mixed cultures (72). Chromogenic agars offer distinct advantages over standard media, but these plates do cost more than conventional media used for UTI detection. Each laboratory will have to evaluate the cost-effectiveness of using these media for urine cultures.

Antimicrobial susceptibility testing

An important issue for UTI in immunocompromised patients is the isolation of uropathogens with increased antibiotic resistance (61, 107). The incidence of *Enterobacteriaceae* organisms producing extended-spectrum β-lactamases (ESBLs) has been increasing worldwide (76). A recent study documented the emergence of CTX-M-type ESBLs at a major health center in the United States, replacing the TEM- or SHV-type ESBLs that previously predominated in this center (53). Several different CTX-M-type ESBLs were found to be circulating, with some appearing in outpatients, where routine culture of uropathogens is not common. In this study, the CTX-M-type ESBL-producing organisms were found mostly among *E. coli* isolates from patients with UTI (53). It was suggested that microbiology laboratories should reconsider current practices and begin to screen *E. coli* and *Klebsiella* sp. isolates obtained from urine for ESBL production in order to track the spread of new ESBL enzymes (53). Another study documented an increase in colonization of high-risk patients with *Enterobacteriaceae* strains producing ESBLs, from 1.33% in 2000 to 3.21% in 2005 (76), by screening rectal swabs. The increased colonization rates were most marked in medical intensive care unit patients and the hematology/oncology and solid organ transplant units. Patients with increased colonization with ESBL-producing *Enterobacteriaceae* are at increased risk for infection with these organisms, and in this study, the number of bloodstream infections caused by ESBL-producing *Enterobacteriaceae* increased more than fourfold over 5 years (76).

Other resistance patterns seen in gram-negative bacteria include increasing resistance to quinolones, the spread of plasma-mediated AmpC β-lactamases, and the alarming recent appearance of *Klebsiella pneumoniae* carbapenemase-type β-lactamases in the New York City area (55). These carbapenemases can hydrolyze penicillins, cephalosporins, aztreonam, and carbapenems. Treatment with carbapenems has been the last defense against infections caused by highly resistant gram-negative bacteria.

Enterococcus spp. are important nosocomial pathogens that can cause bloodstream infections or UTIs in hospitalized patients (97). Vancomycin-resistant enterococci (VRE) have increasingly been reported for many patient populations, especially among high-risk patients. A recent study described an outbreak of VRE infections among patients with hematologic malignancy (109). Of 14 VRE infections in this patient population, 13 were bloodstream infections and 1 was a UTI. An analysis of risk factors for VRE infection identified the presence of neutropenia, neutropenia for ≥7 days, underlying diagnosis of acute myeloid leukemia, and receipt of vancomycin, metronidazole, or carbapenem antibiotics within 30 days prior to infection as major risk factors (109). Increasing colonization with VRE increases the risk for serious infections and mortality for hematology patients and other immunocompromised patient populations (109).

Fungal cultures

It is common to see funguria in hospitalized patients, and it can be difficult to determine if candiduria represents true infection, colonization, or contamination (46). An extensive study of funguria in hospitalized patients identified risk factors for funguria, including diabetes, urinary tract abnormalities, and malignancy, and 7.8% of patients who had neutropenia or were transplant recipients had funguria (45). Most patients with candiduria were asymptomatic. *Candida albicans* represented 52% of the yeast isolates, and *Candida glabrata* was found in 16% of patients. Non-*Candida* yeasts were very rare but included *Cryptococcus neoformans*, *Trichosporon beigelii*, systemic fungi, and *Saccharomyces cerevisiae*. *Candida* spp. usually grow on the 5% sheep blood agar plates used for routine bacterial urine culture, but it would probably be useful to order a urine fungal culture for severely immunocompromised patients, since the urine is plated on Sabouraud dextrose agar, with and without antibiotics, for recovery of yeasts or molds that may need the longer incubation time of 28 days for growth. CHROMagar Candida plates may be useful for identifying mixed cultures. Since many laboratories do not identify yeasts from urine specimens to the species level, a request for identification can usually be made for clinically significant isolates from high-risk patients (46). Identifying *Candida* spp. may be important for therapeutic decisions, since *C. glabrata* and *Candida krusei* can exhibit resistance to fluconazole, the first-line therapy for candiduria.

Viral Detection

HSV

Detection of HSV in genital lesions can be accomplished with a direct fluorescent-antibody assay on scrapings from the base of the pustular lesions, but the sensitivity of this method ranges from 10 to 87% compared to viral culture (41). Both HSV-1 and HSV-2 can be cultured from genital lesions, with cytopathic effect appearing within 5 days after cell inoculation (41).

HSV will grow on human diploid fibroblasts, such as MRC-5 cells or WI-38 cells, on mink lung cells, or on human epidermoid carcinoma cell lines, such as HEp-2 and A549 cells (41). Typically, HSV will grow in culture in 3 to 5 days. When cytopathic effect appears, the cells are removed from the tube and stained with a fluorescent-antibody stain to detect viral antigens. Shell vials can also be used for culture of HSV, CMV, and VZV. Pure monolayers of one of the cell lines may be used, or mixed-cell-line shell vials are also available. The H&V Mix Mixed FreshCell assay (Diagnostic Hybrids) consists of shell vials with a mixture of CV-1 cells and MRC-5 cells growing on a coverslip. For shell vial detection, a patient specimen is centrifuged onto the cell monolayer to increase infectivity and then incubated overnight. The shell vials are then stained with specific fluorescent antibodies to detect HSV, CMV, or VZV viral proteins (41). Shell vials offer rapid results, since results can be read after 16 to 48 hours of incubation. In one study, H&V Mix Mixed FreshCell shell vials demonstrated equivalent sensitivity to that of standard culture with CV-1 or MRC-5 cells (39). The sensitivities of shell vial cultures range from 71 to 97% compared to that of conventional viral culture (41).

A unique product designed for detection of HSV is called an enzyme-linked virus-inducible system (ELVIS; Diagnostic Hybrids). In this system, BHK cells have been engineered to have a reporter gene for β-galactosidase controlled by the promoter from the HSV-1 UL39 gene. Cell monolayers of engineered BHK cells are infected with the patient sample, incubated overnight, and examined using light microscopy. Cells infected with HSV-1 or HSV-2 express the reporter gene, resulting in a blue color change for infected cells. Other viruses do not transactivate this promoter (41). ELVIS allows HSV-positive cells to be tested for HSV-1 or HSV-2; it has been shown to be sensitive, specific, and rapid compared to tube culture

(68) and is comparable in performance to shell vial culture (18). An early study comparing ELVIS with shell vial culture demonstrated a sensitivity of 88%, specificity of >99%, positive predictive value of 99%, and negative predictive value of 99% (51). In 42% of the specimens, HSV was detected sooner by ELVIS than by shell vial culture (51).

Development of real-time PCR instruments has revolutionized how viruses are detected in the virology laboratory. Such methods offer rapid amplification and detection of viral pathogens within a few hours of specimen receipt. These are designed to be "closed systems," so the risk of contamination events is greatly decreased compared to that for PCR using more traditional, end-point detection methods. Analyte-specific reagents for HSV-1 and HSV-2 are available for real-time PCR instruments, including the LightCycler (Roche Diagnostics) and the SmartCycler (Cepheid). Applied Biosystems (API) also has a real-time instrument designed for high-throughput testing. User-developed assays may also be used to detect HSV in patient specimens. One study (73) compared the performance of the LightCycler instrument with that of the SmartCycler instrument for detection of HSV in 114 dermal or genital specimens. Both systems detected 35 HSV-positive specimens, with the SmartCycler typing 16 as HSV-1 and 19 as HSV-2 and the LightCycler typing 15 as HSV-1 and 19 as HSV-2, with one indeterminate result (73). The two assays performed equally well. Another study (26) demonstrated the LightCycler real-time HSV PCR assay to be more sensitive than MRC-5 shell vial cultures. Among 200 total specimens (160 from a genital source) tested by MRC-5 shell vials and the LightCycler assay, 88 HSV strains were detected, with 69 (78%) HSV strains detected by both methods and 19 specimens positive by LightCycler PCR only (26). An alternative PCR assay confirmed that these 19 specimens were truly positive. It was also reported that specimens could be extracted, amplified, and identified within 2 hours of receipt in the laboratory, for a rapid result (26).

HPV

Since HPV cannot be grown in culture, molecular assays have been developed to detect HPV-infected cells that can progress from precancerous lesions to squamous cell carcinoma. The Hybrid Capture II assay (Digene Corp.) uses liquid hybridization to screen for precancerous lesions or cervical cancer. The probe detects high-risk HPV types 16, 18, 31, 33, and others in cervical cytology specimens. The sensitivity of this assay is very high (93%), with a high negative predictive value, but the positive predictive value is reportedly low for detection of cervical intraepithelial neoplasia or invasive cancer (69). Another study (33) comparing the Hybrid Capture II assay with a newer Amplicor HPV test (Roche Molecular Systems) demonstrated respective accuracies of 94.8% and 91.9% for typing of high-risk HPV types in 271 patients with atypical cells of uncertain significance. Both assays detect 13 high-risk HPV types, and both demonstrated high negative predictive values, of 92.8% (Hybrid Capture II) and 87.7% (Amplicor). High accuracy and a high negative predictive value are both critical for the diagnosis of HPV cervical lesions, especially when results are used for management of women with atypical cells of uncertain significance, a precancerous lesion.

Adenoviruses

Immunofluorescent reagents are available to directly stain adenovirus antigens in material such as respiratory tract specimens or in viral cell cultures (79). However, the "gold standard" for detection of adenoviruses is viral culture. Nonenteric adenoviruses can be cultured on human epithelial cell lines, such as A549 cells, but it may take up to 21 days for cytopathic effect to appear (40, 79). Enteric adenoviruses are detected by nonculture methods because they grow poorly in tissue culture. The use of R-Mix shell vials (A549 cells and Mv1Lu cells) or R-Mix Too shell vials (A549 cells and MDCK cells) from Diagnostic Hybrids has shortened the time needed for growth (79). Adenovirus-specific immunofluorescent stains can be used in either traditional or shell vial cultures to detect growth of adenovirus.

Both conventional and real-time PCR assays have been designed for detection of adenoviruses. Most current adenovirus PCR assays are user-developed methods or labeled for "research use only." Chemicon International has developed an analyte-specific reagent called Adenovirus OligoDetect (79). Multiple primers and probes must be included in order to detect all serotypes. For immunocompromised patients, plasma, serum, urine, or stool can be tested for adenovirus. Quantitative real-time PCR assays have been developed to evaluate adenovirus viral loads in patients at greatest risk for hemorrhagic cystitis or severe adenovirus infection, including HSCT recipients and renal transplant recipients. Detection of viral replication, most commonly in plasma or serum, by quantitative real-time PCR while the patient is asymptomatic can identify patients who are at risk for developing severe, symptomatic adenovirus infection. While a level of viremia that would set in motion therapeutic intervention has not been agreed upon definitively, several studies have suggested potentially significant viral load levels. One study of patients who received T-cell-replete HSCT demonstrated a sensitivity for detection of adenoviremia in patients with proven adenovirus-associated disease of 87.5%, with a specificity of 86.8% (24). Based on the quantitative values for adenovirus in plasma, it was suggested that patients with a plasma DNA level of >1,000 copies/ml be treated in an attempt to prevent development of symptomatic adenovirus disease (24). Another study of pediatric recipients of HSCT or solid organ transplants recommended serial monitoring of adenovirus viral loads with quantitative PCR and treatment of these patients if there was a rise in the viral load of at least 10-fold, depending on the degree of immunosuppression and the risk of reduction of immunosuppression (85). The treatment to use for transplant patients thought to be at increased risk for severe adenovirus infections is somewhat controversial (12). The immunosuppressive regimen can be changed to decrease the level of immunosuppression. Cidofovir has been used safely and successfully to treat HSCT recipients who developed adenoviremia (110), but other studies have questioned the effectiveness of cidofovir (12, 95). Another antiviral agent, ribavirin, has had minimal success as therapy for adenovirus infections (12). Clearly, further studies of quantitative PCR for adenovirus DNA for transplant patients and evaluations of therapeutic interventions are needed to establish practice guidelines for these vulnerable patients. The role of PCR and quantitative adenovirus detection in the diagnosis and treatment of viral UTI has not been well defined.

BK virus

The majority of the population worldwide is infected at an early age with the polyomaviruses JC virus and BK virus (56). Reactivation of BK virus is associated with PVAN in renal transplant patients. The prevalence of PVAN in renal transplant recipients ranges from 2 to 10%, with graft loss in up to 60 to 80% of patients (35, 104). A definitive diagnosis of PVAN requires an allograft biopsy, and once the diagnosis is made, the only therapy is a reduction in the intensity of pharmacologic immunosuppression (35, 54, 104). Many transplant programs screen renal transplant patients for PVAN by urine cytology to detect the presence of decoy cells (104). However, studies have shown that cytology examination is less sensitive than nucleic acid amplification-based assays of urine or plasma (20). A recent study demonstrated that BK viruria was strongly associated with BK viremia, PVAN, and allograft loss (20). In a study of a cohort of renal transplant recipients with renal biopsy performed when decoy cells persisted for 2 months or serum creatinine increased, it was found that BK viremia of ≥10,000 copies/ml was significantly associated with PVAN (20). Another study evaluated the threshold of BK viruria and viremia associated with PVAN (104). A cohort of renal transplant recipients was screened for urine decoy cells, BK viruria, and BK viremia, and 114 patients also had an allograft biopsy. The prevalence of biopsy-proven PVAN was 4/114 patients, or 3.5%. Screening for decoy cells was shown to have a low sensitivity, of 25%, and a specificity of 84% for PVAN. However, PCR for BK viruria demonstrated a sensitivity, specificity, positive predictive value, and negative predictive value of 100%, 92%, 31%, and 100%, respectively, while PCR for viremia demonstrated values of 100%, 96%, 50%, and 100%, respectively (104). These results, comparing these screening methods with biopsy-proven PVAN, suggest that PCR testing for BK viruria or viremia is superior to cytopathology for predicting which renal transplant patients are at increased risk for PVAN and allograft loss. The significant levels of BK virus in urine and plasma were 2.5×10^7 copies/ml and 1.6×10^4 copies/ml, respectively (104). Current recommendations for screening for PVAN suggest that all renal transplant recipients be screened for BK virus every 3 months during the first 2 years and annually thereafter until the fifth year (20, 104). It appears that screening with molecular assays is superior to cytology screening in urine.

BK virus is also associated with hemorrhagic cystitis in HSCT recipients. It has been demonstrated that BK viremia is a marker for BK virus disease in HSCT recipients (25). However, BK viruria is common in asymptomatic immunocompromised patients due to reactivation of BK virus. This study demonstrated a higher peak of BK viremia for patients with hemorrhagic cystitis than that for controls, and in situ hybridization of bladder biopsies detected BK virus in the biopsies from two patients with severe cases of hemorrhagic cystitis with repeated high-level BK viremia (25). These data suggest that BK viremia could be used as a marker to detect BK virus disease in HSCT recipients (25).

Serology

Serology testing has no role in the diagnosis of UTI, but serological methods such as enzyme immunoassays (EIAs) are available to determine patient exposure to HSV, CMV, and VZV. It is most useful to perform this testing before the period of immunosuppression begins, if possible, since assays performed on severely immunocompromised patients can be insensitive, depending on the immunological defect. EIAs for immunoglobulin M (IgM) and IgG are commercially available for CMV and VZV. EIAs that detect IgG antibodies to specific glycoproteins of HSV can distinguish previous exposure to HSV-1 and HSV-2 (41). The EIA for detection of IgM antibodies to herpesviruses does not distinguish between HSV-1 and HSV-2, cannot distinguish new from established symptomatic lesions, and is not recommended (41). For patients with genital lesions and less severe states of immunocompromise, such as HIV patients or diabetic patients, an EIA that detects IgG against HSV-2-specific glycoprotein is available as an adjunct to viral culture of lesions. If the HSV, CMV, and VZV serostatus is known before transplantation or chemotherapy takes place, prophylactic therapy can be initiated to prevent devastating infections in these patients.

PROGNOSTIC FACTORS, THERAPY, AND MONITORING

In a discussion of prognostic factors, therapeutic considerations, and monitoring of UTIs in immunocompromised patients, it seems appropriate to consider cystitis and pyelonephritis as one entity. Immunocompromised patients are at higher risk for development of severe pyelonephritis, and it is likely that many immunocompromised patients exhibiting only cystitis symptoms also have infection of the upper urinary tract, as seen in a study of immunocompetent patients, where 30 to 50% of patients with cystitis symptoms had involvement of the upper urinary tract (9, 90). All immunocompromised patients with symptoms of cystitis or pyelonephritis should receive antibiotic therapy. An initial Gram stain of urine can be helpful in guiding empiric therapy. For gram-negative organisms, treatment with SXT or fluoroquinolones should be effective in most immunocompromised patients with cystitis or mild pyelonephritis, while ampicillin, amoxicillin, or cephalosporins should be effective against gram-positive organisms in these patients. Parenterally administered antibiotics are appropriate for cases of severe pyelonephritis, with or without bacteremia (90). Urine cultures with susceptibility testing of isolated uropathogens should be done to guide therapy for all immunocompromised patients with symptoms suggestive of cystitis or pyelonephritis. Detection of resistant organisms is critical, since infections with resistant organisms can lead to treatment failure and increased mortality for immunocompromised patients. However, it is important to remember that some antibiotics to which the organisms appear resistant in routine susceptibility testing may still be effective in vivo if they are concentrated in urine (90). In addition, the longer the duration of immunosuppression, the greater the risk that fungal UTI will develop. *Candida* infections can be identified by urine culture and treated with amphotericin B bladder irrigation, oral azole therapy, or systemic antifungal therapy, depending on the level of immunosuppression and severity of infection. Removal of indwelling urinary catheters will improve therapeutic outcomes, and follow-up urine cultures to ensure clearance of infection may be appropriate 1 to 2 weeks after completion of antibiotic therapy. Failure to improve symptomatically or to clear bacteriuria after 72 hours of appropriate therapy should lead to radiological studies, such as US, CT, or MRI, to look for urinary tract obstruction or abscess formation and possibly to change the antibiotic regimen. Prophylactic antibiotic therapy with SXT or fluoroquinolones during times of immunosuppression is very effective in preventing UTI and other infections. For genital lesions, culture or molecular methods for detection of HSV or adenovirus infection are very sensitive, while viral load testing for BK virus can guide adjustment of levels of immunosuppression to prevent development of hemorrhagic cystitis in transplant patients.

For primary immunodeficiencies, such as humoral deficiencies or T-cell deficiencies, and for aplastic anemia, urine culture should be performed to detect bacterial or fungal pathogens if patients are symptomatic for cystitis or pyelonephritis, and viral culture or molecular methods can be used to detect viral pathogens. Blood cultures should be done for febrile patients to identify patients with pyelonephritis and bacteremia. Broad antibiotic coverage should be instituted until culture and susceptibility testing are completed. If patients fail to improve on therapy, radiological studies and culture for unusual organisms, such as fungi or mycobacteria, should be carried out to identify obstruction, abscesses, or the presence of unusual organisms that could prevent effective clearance of infection. Prophylactic therapy should be considered for these patients with long-term immunosuppression and a very high risk for genitourinary tract infection.

Patients with cancer suffer damaging effects from radiation therapy and chemotherapy, which can lead to an increased risk of UTI. During treatment, patients should be treated aggressively if UTI symptoms are observed. Urinary catheters should be removed if possible, and appropriate antibiotic therapy should be initiated. If patients are febrile, blood cultures should be obtained, and if patients fail to improve symptomatically, radiological studies such as US, CT with contrast, or MRI should be carried out to identify blockages that may interfere with effective treatment.

HIV-infected patients are at increased risk for UTI, so urine Gram stain and urine culture are indicated if patients are symptomatic. UTI symptoms can be treated with either oral or parenteral antibiotics, depending on the severity of infection. Periodic evaluation of HIV-infected patients for HSV infection and other sexually transmitted infections is an important component of good clinical management (94).

Among transplant patients, renal transplant patients are at highest risk for UTI, especially early after transplantation. There is a high risk for loss of the renal allograft with UTI, so urine cultures and removal of urinary catheters as soon as possible are critical steps in the management of renal transplant patients. Prophylactic antibiotics after renal transplantation are recommended, since this treatment can effectively prevent UTI in these patients (14). Prophylaxis with either SXT or fluoroquinolones can be effective, but caution should be used with SXT, since up to 20% resistance has been seen in gram-negative organisms in some areas (90). For other solid organ transplant patients and HSCT patients, there is a high risk of UTI early after transplantation, so prophylaxis is also appropriate for these patients. All transplant patients should be monitored for UTI symptoms, and cultures should be performed if patients are symptomatic. Aggressive therapy and follow-up urine cultures to ensure resolution of infection are critical in the management of UTI in these vulnerable patients. Renal transplant patients should also be screened for asymptomatic bacteriuria by urine culture, and asymptomatic bacteriuria or candiduria in these patients should be treated to prevent scarring and loss of the renal allograft. In immunocompetent patients, diagnosis of asymptomatic bacteriuria requires two positive midstream urine samples, with isolation of the same organism in significant amounts (10, 90). There are no similar guidelines to define significant asymptomatic bacteriuria for immunocompromised patients, so current recommendations are to eliminate or suppress bacteriuria or candiduria in renal transplant patients. Renal transplant patients should be monitored by PCR for reactivation of BK virus, which can lead to hemorrhagic cystitis or nephropathy. Detection of adenovirus infections can be accomplished by culture or molecular assays.

It is currently not recommended that women with diabetes be screened periodically by urine culture for asymptomatic bacteriuria, and there is no evidence that antibiotic treatment in diabetic patients with asymptomatic bacteriuria will delay or decrease the frequency of symptomatic UTIs (71, 90). Diabetic patients who do develop UTI symptoms should be treated and have urine cultures with susceptibility testing to determine the infecting organisms. These patients should be monitored closely by bacterial culture for clearance of infection, since there is a high risk for the development of abscesses or emphysematous pyelonephritis. Radiological studies, such as CT with contrast or MRI, may be necessary to detect these complications. Since mortality in cases of emphysematous pyelonephritis can be as high as 70%, even with antibiotics and supportive care, immediate nephrectomy may be indicated for patients with this serious complication (32, 58, 71).

HSV genital lesions diagnosed by culture or molecular methods can be treated with oral acyclovir or valacyclovir in patients with low levels of immunosuppression, such as HIV-infected persons or diabetic patients. HSV infections in severely immunocompromised patients, however, require parenteral antiviral therapy and most likely represent disseminated infection. Prophylactic therapy with acyclovir is appropriate for patients during periods of severe immunosuppression, such as neutropenic patients and transplant patients.

REFERENCES

1. **Augenbraun, M. H.** 2005. Diseases of the reproductive organs and sexually transmitted diseases, p. 1338–1347. *In* G. L. Mandell, J. E. Bennett, and R. Dolin (ed.), *Principles and Practice of Infectious Diseases,* 6th ed. Elsevier, Philadelphia, PA.
2. **Bansil, P., D. J. Jamieson, S. F. Posner, and A. P. Kourtis.** 2007. Hospitalizations of pregnant HIV-infected women in the United States in the era of highly active antiretroviral therapy (HAART). *J. Womens Health* **16:**159–162.
3. **Baron, E. J., G. H. Cassell, L. B. Duffy, D. A. Eschenbach, J. R. Greenwood, S. M. Harvey, N. E. Madinger, E. M. Peterson, and K. B. Waites.** 1993. *Cumitech 17A, Laboratory Diagnosis of Female Genital Tract Infections.* Coordinating ed., E. J. Baron. American Society for Microbiology, Washington, DC.
4. **Baum, S.** 2005. Adenoviruses, p. 1835–1841. *In* G. L. Mandell, J. E. Bennett, and R. Dolin (ed.), *Principles and Practices of Infectious Diseases,* 6th ed. Elsevier, Philadelphia, PA.
5. **Bialas, I., E. M. Bessell, M. Sokal, and R. Slack.** 1989. A prospective study of urinary tract infection during radiotherapy. *Radiother. Oncol.* **16:**305–309.
6. **Bonnez, W., and R. C. Reichman.** 2005. Papillomaviruses, p. 1841–1856. *In* G. L. Mandell, J. E. Bennett, and R. Dolin (ed.), *Principles and Practice of Infectious Diseases,* 6th ed. Elsevier, Philadelphia, PA.
7. **Bowden, R. A.** 2000. Infections in blood and bone marrow transplant patients: allogeneic and autologous transplantation, p. 189–217. *In* M. P. Glauser and P. A. Pizzo (ed.), *Management of Infections in Immunocompromised Patients.* W. B. Saunders Company Ltd., London, United Kingdom.

8. **Branca, M., S. Costa, L. Mariani, F. Sesti, A. Agarossi, A. di Carlo, M. Galati, A. Benedetto, M. Ciotti, C. Giorgi, A. Criscuolo, M. Valieri, C. Favalli, P. Pava, D. Santini, E. Piccione, M. Alderisio, M. De Nuzzo, L. di Bonito, and K. Syrjanen.** 2004. Assessment of risk factors and human papillomavirus (HPV) related pathogenetic mechanisms of CIN in HIV-positive and HIV-negative women. Study design and baseline data of the HPV-Pathogen ISS study. *Eur. J. Gynaecol. Oncol.* **25:**689–698.
9. **Busch, R., and H. Huland.** 1984. Correlation of symptoms and results of direct bacterial localization in patients with urinary tract infections. *J. Urol* **132:**282–285.
10. **Calandra, T.** 2000. Practical guide to host defense mechanisms and the predominant infections encountered in immunocompromised patients, p. 3–16. *In* M. P. Glauser and P. A. Pizzo (ed.), *Management of Infections in Immunocompromised Patients.* W. B. Saunders Company Ltd., London, United Kingdom.
11. **Carroll, K. C., D. C. Hale, D. H. Von Boerum, G. C. Reich, L. T. Hamilton, and J. M. Matsen.** 1994. Laboratory evaluation of urinary tract infections in an ambulatory clinic. *Am. J. Clin. Pathol.* **101:**100–103.
12. **Chakrabarti, S.** 2007. Adenovirus infections after hematopoietic stem cell transplantation: still unraveling the story. *Clin. Infect. Dis.* **45:**966–968.
13. **Chin-Hong, P. V., and J. M. Palefsky.** 2005. Human papillomavirus anogenital disease in HIV-infected individuals. *Dermatol. Ther.* **18:**67–76.
14. **Chuang, P., C. R. Parikh, and A. Langone.** 2005. Urinary tract infections after renal transplantation: a retrospective review at two US transplant centers. *Clin. Transplant.* **19:**230–235.
15. **Clarridge, J. E., J. R. Johnson, and M.T. Pezzlo.** 1998. *Cumitech 2B, Laboratory Diagnosis of Urinary Tract Infections.* Coordinating ed., A. S. Weissfeld. American Society for Microbiology, Washington, DC.
16. **Corey, L.** 2005. Herpes simplex virus, p. 1762–1780. *In* G. L. Mandell, J. E. Bennett, and R. Dolin (ed.), *Principles and Practice of Infectious Diseases,* 6th ed. Elsevier, Philadelphia, PA.
17. **Craig, W. D., B. J. Wagner, and M. D. Travis.** 2008. Pyelonephritis: radiologic-pathologic review. *Radiographics* **28:**255–277.
18. **Crist, G. A., J. M. Langer, G. L. Woods, M. Procter, and D. R. Hillyard.** 2004. Evaluation of the ELVIS plate method for the detection and typing of herpes simplex virus in clinical specimens. *Diagn. Microbiol. Infect. Dis.* **49:**173–177.
19. **Cunha, B. A.** 1998. Infections in nonleukopenic compromised hosts (diabetes mellitus, SLE, steroids, and asplenia) in critical care. *Crit. Care Clin.* **14:**263–282.
20. **Drachenberg, C. B., H. H. Hirsch, J. C. Papadimitriou, R. Gosert, R. K. Walk, R. Munivenkatappa, J. Nogueira, C. B. Cangro, A. Haririan, S. Mendley, and E. Ramos.** 2007. Polyomavirus BK versus JC replication and nephropathy in renal transplant recipients: a prospective evaluation. *Transplantation* **83:**323–330.
21. **Drachenberg, C. B., J. C. Papadimitriou, and E. Ramos.** 2006. Histologic versus molecular diagnosis of BK polyomavirus-associated nephropathy: a shifting paradigm. *Clin. J. Am. Soc. Nephrol.* **1:**374–379.
22. **Dunne, W. M.** 1995. Laboratory diagnosis of urinary tract infection in children. *Clin. Microbiol. Newsl.* **17:**73–76.
23. **Eisenstadt, J., and J. A. Washington.** 1996. Diagnostic microbiology for bacteria and yeasts causing urinary tract infections, p. 29–66. *In* H. L. T. Mobley and J. W. Warren (ed.), *Urinary Tract Infections: Molecular Pathogenesis and Clinical Management.* ASM Press, Washington, DC.
24. **Erard, V., M.-L. Huang, J. Ferrenberg, L. Nguy, T. L. Stevens-Ayers, R. C. Hackman, L. Corey, and M. Boeckh.** 2007. Quantitative real-time polymerase chain reaction for detection of adenovirus after T cell-replete hematopoietic cell transplantation: viral load as a marker for invasive disease. *Clin. Infect. Dis.* **45:**958–965.
25. **Erard, V., H. W. Kim, L. Corey, A. Limaye, M.-L. Huang, D. Myerson, C. Davis, and M. Boeckh.** 2005. BK DNA viral load in plasma: evidence for an association with hemorrhagic cystitis in allogeneic hematopoietic cell transplant recipients. *Blood* **106:**1130–1132.
26. **Espy, M. J., J. R. Uhl, P. S. Mitchell, J. N. Thorvilson, K. A. Svien, A. D. Wold, and T. F. Smith.** 2000. Diagnosis of herpes simplex virus infections in the clinical laboratory by LightCycler PCR. *J. Clin. Microbiol.* **38:**795–799.
27. **Feuchtinger, T., P. Lang, and R. Handgretinger.** 2007. Adenovirus infection after allogeneic stem cell transplantation. *Leuk. Lymphoma* **48:**244–255.
28. **Foxman, B.** 2002. Epidemiology of urinary tract infections: incidence, morbidity, and economic costs. *Am. J. Med.* **113:**5S–13S.
29. **Glauser, M. P., and T. Calandra.** 2000. Infections in patients with hematological malignancies, p. 142–188. *In* M. P. Glauser and P. A. Pizzo (ed.), *Management of Infections in Immunocompromised Patients.* W. B. Saunders Company Ltd., London, United Kingdom.
30. **Goldstraw, M. A., M. G. Kirby, J. Bhardwa, and R. S. Kirby.** 2006. Diabetes and the urologist: a growing problem. *BJU Int.* **99:**513–517.
31. **Greenblatt, R. M., P. Bacchetti, S. Barkan, M. Augenbraun, S. D. Silver, R. Delapenha, P. Garcia, U. Mathur, P. Miotti, and D. Bruns.** 1999. Lower genital tract infections among HIV-infected and high-risk uninfected women: findings of the Women's Interagency HIV Study. *Sex. Transm. Dis.* **26:**143–151.
32. **Grupper, M., A. Kravtsov, and I. Potasman.** 2007. Emphysematous cystitis. *Medicine* **86:**47–53.
33. **Halfon, P., E. Trepo, G. Antoniotti, C. Bernot, P. Cart-Lamy, H. Khiri, D. Thibaud, J. Marron, A. Martineau, G. Penaranda, D. Benmoura, B. Blanc, et al.** 2007. Prospective evaluation of the hybrid capture 2 and Amplicor human papillomavirus (HPV) test for detection of 13 high-risk HPV genotypes in atypical squamous cells of uncertain significance. *J. Clin. Microbiol.* **45:**313–316.
34. **Hegstler, K. A., R. Hammann, and A.-M. Fahr.** 1997. Evaluation of BBL CHROMagar Orientation medium for detection and presumptive identification of urinary tract pathogens. *J. Clin. Microbiol.* **35:**2773–2777.
35. **Hirsch, H. H., C. B. Drachenberg, J. Steiger, and E. Ramos.** 2006. Polyomavirus-associated nephropathy in renal transplantation: critical issues of screening and management. *Adv. Exp. Med. Biol.* **577:**160–173.
36. **Hirsch, H. H., W. Knowles, M. Dickenmann, J. Passweg, T. Klimkait, M. J. Mihatsch, and J. Steiger.** 2002. Prospective study of polyomavirus type BK replication and nephropathy in renal-transplant recipients. *N. Engl. J. Med.* **347:**488–496.
37. **Hochreiter, W. W., and W. Bushman.** 1999. Urinary tract infection: a moving target. *World J. Urol.* **17:**364–371.
38. **Hsu, C.-Y., H. C. Fang, K. J. Chou, C. L. Chen, P. T. Lee, and H. M. Chung.** 2006. The clinical impact of bacteremia in complicated acute pyelonephritis. *Am. J. Med. Sci.* **332:**175–180.
39. **Huang, Y. T., S. Hite, V. Duane, and H. Yan.** 2002. CV-1 and MRC-5 mixed simultaneous detection of herpes simplex viruses and varicella zoster lesions. *J. Clin. Virol.* **24:**37–43.
40. **Ison, M. G.** 2006. Adenovirus infections in transplant recipients. *Clin. Infect. Dis.* **43:**331–339.

41. **Jerome, K. R., and R. A. Morrow.** 2007. Herpes simplex viruses and herpes B virus, p. 1523–1536. *In* P. R. Murray, E. J. Baron, J. H. Jorgenson, M. L. Landry, and M. A. Pfaller (ed.), *Manual of Clinical Microbiology,* 9th ed. ASM Press, Washington, DC.
42. **Johnson, J. R., and W. E. Stamm.** 1989. Urinary tract infections in women: diagnosis and treatment. *Ann. Intern. Med.* **111:**906–917.
43. **Johnson, J. R.** 1991. Virulence factors in *Escherichia coli* urinary tract infection. *Clin. Microbiol. Rev.* **4:**80–128.
44. **Johnson, J. R.** 2003. Microbial virulence determinants and the pathogenesis of urinary tract infection. *Infect. Dis. Clin. N. Am.* **17:**261–278.
45. **Kauffman, C. A., J. A. Vazquez, J. D. Sobel, H. A. Gallis, D. S. McKinsey, A. W. Karchmer, A. M. Sugar, P. K. Sharkey, G. J. Wise, R. Mangi, A. Mosher, J. Y. Lee, W. E. Dismukes, and the National Institute for Allergy and Infectious Diseases (NIAID) Mycoses Study Group.** 2000. Prospective multicenter surveillance study of funguria in hospitalized patients. *Clin. Infect. Dis.* **30:**14–18.
46. **Kauffman, C. A.** 2005. Candiduria. *Clin. Infect. Dis.* **41:** S371–S376.
47. **Kourtis, A. P., P. Bansil, M. McPheeters, S. F. Meikle, S. F. Posner, and D. J. Jamieson.** 2006. Hospitalizations of pregnant HIV-infected women in the USA prior to and during the era of HAART 1994–2003. *AIDS* **20:**1823–1831.
48. **Kreiger, J. N.** 2005. Prostatitis, epididymitis, and orchitis, p. 1381–1386. *In* G. L. Mandell, J. E. Bennett, and R. Dolin (ed.), *Principles and Practice of Infectious Diseases,* 6th ed. Elsevier, Philadelphia, PA.
49. **Kunin, C. M.** 1997. *Urinary Tract Infections: Detection, Prevention, and Management,* 5th ed. Williams & Wilkins, Baltimore, MD.
50. **Kwak, E. J., R. A. Vilchez, P. Randhawa, R. Shapiro, J. S. Butel, and S. Kusne.** 2002. Pathogenesis and management of polyomavirus infection in transplant patients. *Clin. Infect. Dis.* **35:**1081–1087.
51. **LaRocco, M. T.** 2000. Evaluation of an enzyme-linked viral inducible system for the rapid detection of herpes simplex virus. *Eur. J. Clin. Microbiol. Infect. Dis.* **19:**233–235.
52. **Lee, L. K., M. D. Dinneen, and S. Ahmad.** 2001. The urologist and the patient infected with human immunodeficiency virus or with acquired immunodeficiency syndrome. *BJU Int.* **88:**500–510.
53. **Lewis, J. S., II, M. Herrera, B. Wickes, J. E. Patterson, and J. H. Jorgensen.** 2007. First report of the emergence of CTX-M-type extended-spectrum β-lactamases (ESBLs) as the predominant ESBL isolated in a U.S. health care system. *Antimicrob. Agents Chemother.* **51:**4015–4021.
54. **Liptak, P., E. Kemeny, and B. Ivanyi.** 2006. Primer: histopathology of polyomavirus-associated nephropathy in renal allografts. *Nat. Clin. Pract. Nephrol.* **2:**631–636.
55. **Lomaestro, B. M., E. H. Tobin, W. Shang, and T. Gootz.** 2006. The spread of *Klebsiella pneumonia* carbapenemase-producing *K. pneumoniae* to upstate New York. *Clin. Infect. Dis.* **43:**e26–e28.
56. **Major, E. O., C. Ryschkewitsch, A. Valsamakis, and J. Hou.** 2007. Human polyomaviruses, p. 1612–1621. *In* P. R. Murray, E. J. Baron, J. H. Jorgenson, M. L. Landry, and M. A. Pfaller (ed.), *Manual of Clinical Microbiology,* 9th ed. ASM Press, Washington, DC.
57. **McClelland, R. S., L. Sangare, W.M. Hassan, L. Lavreys, K. Mandaliya, J. Kiarie, J. Ndinya-Achola, W. Jaoko, and J. M. Baeten.** 2007. Infection with *Trichomonas vaginalis* increases the risk of HIV-1 acquisition. *J. Infect. Dis.* **195:**698–702.
58. **Mokabberi, R., and K. Ravakhah.** 2007. Emphysematous urinary tract infections: diagnosis, treatment and survival. *Am. J. Med. Sci.* **333:**111–116.
59. **Moodley, P., D. Wilkinson, C. Connolly, J. Moodley, and A. W. Sturm.** 2002. Trichomonas vaginalis is associated with pelvic inflammatory disease in women infected with human immunodeficiency virus. *Clin. Infect. Dis.* **34:**519–522.
60. **Nicolle, L. E., D. Friesen, G. K. Harding, and L. L. Roos.** 1996. Hospitalization for acute pyelonephritis in Manitoba, Canada during the period from 1989–1992: impact of diabetes, pregnancy, and aboriginal origin. *Clin. Infect. Dis.* **22:**1051–1056.
61. **Nicolle, L. E.** 2002. Epidemiology of urinary tract infections. *Clin. Microbiol. Newsl.* **24:**135–140.
62. **Nicolle, L. E.** 2005. Urinary tract infection in diabetes. *Curr. Opin. Infect. Dis.* **18:**49–53.
63. **Ojoo, J., J. Paul, B. Batchelor, M. Amir, J. Kimari, C. Mwachari, J. Bwayo, F. Plummer, G. Gachihi, P. Waihaki, and C. Gilks.** 1996. Bacteriuria in a cohort of predominantly HIV-1 seropositive female commercial sex workers in Nairobi, Kenya. *J. Infect.* **33:**33–37.
64. **Ozsaran, A. A., T. Ates, Y. Dikmen, A. Zeytinoglu, C. Terek, Y. Erhan, T. Ozacar, and A. Bilgic.** 1999. Evaluation of the risk of cervical intraepithelial neoplasia and human papilloma virus infection in renal transplant patients receiving immunosuppressive therapy. *Eur. J. Gynaecol. Oncol.* **20:**127–130.
65. **Palefsky, J. M., M. L. Gillison, and H. D. Strickler.** 2006. HPV vaccines in immunocompromised women and men. *Vaccine* **24:**S140–S146.
66. **Palefsky, J. M., and E. A. Holly.** 2003. Immunosuppression and co-infection with HIV. *J. Natl. Cancer Inst. Monogr.* **31:** 41–46.
67. **Patel, H. S., A. R. J. Silver, and J. M. A. Northover.** 2007. Anal cancer in renal transplant patients. *Int. J. Colorectal Dis.* **22:**1–5.
68. **Patel, N., L. Kauffmann, G. Baniewicz, M. Forman, M. Evans, and D. Scholl.** 1999. Confirmation of low-titer, herpes simplex virus-positive specimen results by the enzyme-linked virus-inducible system (ELVIS) using PCR and repeat testing. *J. Clin. Microbiol.* **37:**3986–3989.
69. **Patterson, B. K.** 2007. Human papillomaviruses, p. 1601–1611. *In* P. R. Murray, E. J. Baron, J. H. Jorgenson, M. L. Landry, and M. A. Pfaller (ed.), *Manual of Clinical Microbiology,* 9th ed. ASM Press, Washington, DC.
70. **Patterson, J. E., and V. T. Andriole.** 1997. Bacterial urinary tract infections in diabetes. *Infect. Dis. Clin. N. Am.* **11:**735–750.
71. **Peleg, A. Y., T. Weerarathna, J. S. McCarthy, and T. M. E. Davis.** 2007. Common infections in diabetes: pathogenesis, management and relationship to glycaemic control. *Diabetes Metab. Res. Rev.* **23:**3–13.
72. **Perry, J. D., and A. M. Freydiere.** 2007. The application of chromogenic media in clinical microbiology. *J. Appl. Microbiol.* **103:**2046–2055.
73. **Podzorski, R. P.** 2006. Evaluation of the Cepheid herpes simplex virus typing real-time PCR assay using dermal or genital specimens. *Diagn. Microbiol. Infect. Dis.* **56:**173–177.
74. **Prasad, K. N., S. Pradhan, and N. R. Datta.** 1995. Urinary tract infection in patients of gynecological malignancies undergoing external pelvic radiotherapy. *Gynecol. Oncol.* **57:** 380–382.
75. **Rabkin, D. G., M. D. Stifelman, J. Birkhoff, K. A. Richardson, D. Cohen, R. Nowygrod, A. I. Benvenisty, and M. A. Hardy.** 1998. Early catheter removal decreases incidence of urinary tract infections in renal transplant recipients. *Transplant. Proc.* **30:**4314–4316.

76. **Reddy, P., M. Malczynski, A. Obias, S. Reiner, N. Jin, J. Juang, G. A. Noskin, and T. Zembower.** 2007. Screening for extended-spectrum β-lactamase-producing Enterobacteriaceae among high-risk patients and rates of subsequent bacteremia. *Clin. Infect. Dis.* **45**:846–852.
77. **Rice, J. C., T. Peng, Y.-F. Kuo, S. Pendyala, L. Simmons, J. Boughton, K. Ishihara, S. Nowicke, and B. J. Nowicki.** 2006. Renal allograft injury is associated with urinary tract infection caused by Escherichia coli bearing adherence factors. *Am. J. Transplant.* **6**:2375–2383.
78. **Roberts, F. J., J. Murphy, and C. Ludgate.** 1990. The value and significance of routine urine cultures in patients referred for radiation therapy of prostatic malignancy. *Clin. Oncol. (R. Coll. Radiol.)* **2**:18–21.
79. **Robinson, C., and M. Echavarria.** 2007. Adenoviruses, p. 1589–1600. *In* P. R. Murray, E. J. Baron, J. H. Jorgenson, M. L. Landry, and M. A. Pfaller (ed.), *Manual of Clinical Microbiology,* 9th ed. ASM Press, Washington, DC.
80. **Rolston, K. V. I.** 2000. Infections in patients with solid tumors, p. 118–140. *In* M. P. Glauser and P. A. Pizzo (ed.), *Management of Infections in Immunocompromised Patients.* W. B. Saunders Company Ltd., London, United Kingdom.
81. **Safdar, N., W. R. Slattery, V. Knasinski, T. E. Gangnon, L. Zhanhai, J. D. Pirsch, and D. Andes.** 2005. Predictors and outcomes of candiduria in renal transplant recipients. *Clin. Infect. Dis.* **40**:1413–1421.
82. **Samra, Z., M. Heifetz, J. Talmor, E. Bain, and J. Bahar.** 1998. Evaluation of use of a new chromogenic agar in detection of urinary tract pathogens. *J. Clin. Microbiol.* **36**:990–994.
83. **Schmook, T., I. Nindl, C. Ulrich, T. Meyer, W. Sterry, and E. Stockfleth.** 2003. Viral warts in organ transplant recipients: new aspect in therapy. *Br. J. Dermatol.* **149**(Suppl. 66):20–24.
84. **Schonwald, S., J. Begovac, and V. Skerk.** 1999. Urinary tract infections in HIV disease. *Int. J. Antimicrob. Agents* **11**:309–311.
85. **Seidemann, K., A. Heim, E. D. Pfister, H. Koditz, A. Beilken, A. Sander, M. Melter, K.-W. Sykora, M. Sasse, and A. Wessel.** 2004. Monitoring of adenovirus infection in pediatric transplant recipients by quantitative PCR: report of six cases and review of the literature. *Am. J. Transplant.* **4**:2102–2108.
86. **Sillman, F. H., S. Sentovich, and D. Shaffer.** 1997. Ano-genital neoplasia in renal transplant patients. *Ann. Transplant.* **2**:59–66.
87. **Snydman, D. R.** 2001. Epidemiology of infections after solid-organ transplantation. *Clin. Infect. Dis.* **33**(Suppl. 1):S5–S8.
88. **Soave, R.** 2001. Prophylaxis strategies for solid-organ transplantation. *Clin. Infect. Dis.* **33**(Suppl. 1):S26–S31.
89. **Sobel, J. D., and D. Kaye.** 1984. Host factors in the pathogenesis of urinary tract infections. *Am. J. Med.* **76**:122–130.
90. **Sobel, J. D., and D. Kaye.** 2005. Urinary tract infections, p. 875–905. *In* G. L. Mandell, J. E. Bennett, and R. Dolin (ed.), *Principles and Practice of Infectious Diseases,* 6th ed. Elsevier, Philadelphia, PA.
91. **Stamm, W. E., G. W. Counts, K. R. Running, S. Fihn, M. Turck, and K. K. Holmes.** 1982. Diagnosis of coliform infection in acutely dysuric women. *N. Engl. J. Med.* **307**:463–468.
92. **Stamm, W. E., T. M. Hooten, J. R. Johson, C. Johnson, A. Stapleton, P. L. Roberts, S. L. Moseley, and S. D. Fihn.** 1989. Urinary infections: from pathogenesis to treatment. *J. Infect. Dis.* **159**:400–406.
93. **Stark, R. P., and D. G. Maki.** 1984. Bacteriuria in the catheterized patient: what quantitative level of bacteriuria is relevant? *N. Engl. J. Med.* **311**:560–564.
94. **Strick, L. B., A. Wald, and C. Celum.** 2006. Management of herpes simplex virus type 2 infection in HIV type-1-infected persons. *Clin. Infect. Dis.* **43**:347–356.
95. **Symeonidis, N., A. Jakubowski, S. Pierre-Louis, D. Jaffe, E. Pamer, K. Sepkowitz, R. J. O'Reilly, and G. A. Papanicolaou.** 2007. Invasive adenoviral infections in T-cell-depleted allogeneic hematopoietic stem cell transplantation: high mortality in the era of cidofovir. *Transpl. Infect. Dis.* **9**:108–113.
96. **Tan, H. H., and C. L. Goh.** 2006. Viral infections affecting the skin in organ transplant recipients: epidemiology and current management strategies. *Am. J. Clin. Dermatol.* **7**:13–29.
97. **Teixeira, L. M., M. da Glória Siqueira Carvalho, and R. R. Facklam.** 2007. Enterococcus, p. 430–442. *In* P. R. Murray, E. J. Baron, J. H. Jorgenson, M. L. Landry, and M. A. Pfaller (ed.), *Manual of Clinical Microbiology,* 9th ed. ASM Press, Washington, DC.
98. **Thomson, R. B., Jr.** 2007. Specimen collection, transport, and processing: bacteriology, p. 291–333. *In* P. R. Murray, E. J. Baron, J. H. Jorgenson, M. L. Landry, and M. A. Pfaller (ed.), *Manual of Clinical Microbiology,* 9th ed. ASM Press, Washington, DC.
99. **Tolkoff-Rubin, N. E., A. B. Cosimi, P. S. Russell, and R. H. Rubin.** 1982. A controlled study of trimethoprim-sulfamethoxazole prophylaxis of urinary tract infection in renal transplant recipients. *Rev. Infect. Dis.* **4**:614–618.
100. **Tolkoff-Rubin, N. E., and R. H. Rubin.** 1997. Urinary tract infection in the immunocompromised host—lessons from kidney transplantation and the AIDS epidemic. *Infect. Dis. Clin. N. Am.* **11**:707–717.
101. **Trautner, B. W., and R. O. Darouiche.** 2004. Catheter-associated infections. *Arch. Intern. Med.* **164**:842–850.
102. **Tseng, C.-C., J.-J. Wu, H.-L. Liu, J.-M. Sung, and J.-J. Huang.** 2002. Roles of host and bacterial virulence factors in the development of upper urinary tract infection caused by Escherichia coli. *Am. J. Kidney Dis.* **39**:1–14.
103. **van der Meer, J. W. M., and T. W. Kuijpers.** 2000. Infections in patients with primary (congenital) immunodeficiencies, p. 47–88. *In* M. P. Glauser and P. A. Pizzo (ed.), *Management of Infections in Immunocompromised Patients.* W. B. Saunders Company Ltd., London, United Kingdom.
104. **Viscount, H. B., A. J. Eid, M. J. Espy, M. D. Griffin, K. M. Thomsen, W. S. Harmsen, R. R. Razonable, and T. F. Smith.** 2007. Polyomavirus polymerase chain reaction as a surrogate marker of polyomavirus-associated nephropathy. *Transplantation* **84**:340–345.
105. **Warren, J. W.** 1996. Clinical presentations and epidemiology of urinary tract infections, p. 3–27. *In* H. L. T. Mobley and J. W. Warren (ed.), *Urinary Tract Infections: Molecular Pathogenesis and Clinical Management.* ASM Press, Washington, DC.
106. **Weinberger, M.** 2000. Infections in patients with aplastic anemia, p. 89–115. *In* M. P. Glauser and P. A. Pizzo (ed.), *Management of Infections in Immunocompromised Patients.* W. B. Saunders Company Ltd., London, United Kingdom.
107. **Wilson, M. L., and L. Gaido.** 2004. Laboratory diagnosis of urinary tract infections in adult patients. *Clin. Infect. Dis.* **38**:1150–1158.
108. **Wong, A. S. Y., K. H. Chan, V. C. C. Cheng, Y. L. Kwong, and A. Y. H. Leung.** 2007. Relationship of pretransplantation polyoma BK virus serologic findings and BK viral reactivation after hematopoietic stem cell transplantation. *Clin. Infect. Dis.* **44**:830–837.

109. **Worth, L. J., K. A. Thursky, J. F. Seymour, and M. A. Slavin.** 2007. Vancomycin-resistant Enterococcus faecium infection in patients with hematologic malignancy: patients with acute myeloid leukemia are at high-risk. *Eur. J. Haematol.* 79:226–233.

110. **Yusuf, U., G. A. Hale, J. Carr, Z. Gu, E. Benaim, P. Woodard, K. A. Kasow, E. M. Horwitz, W. Leung, D. K. Srivastava, R. Handgretinger, and R. T. Hayden.** 2006. Cidofovir for the treatment of adenoviral infection in pediatric hematopoietic stem cell transplant patients. *Transplantation* **81:**1398–1404.

Diagnostic Microbiology of the Immunocompromised Host
Edited by Randall T. Hayden, Karen C. Carroll, Yi-Wei Tang, and Donna M. Wolk

Chapter 17

Gastrointestinal Infections

IRVING NACHAMKIN

Gastrointestinal infections in the immunocompromised host are caused by the common bacterial, viral, fungal, and parasitic agents that also cause infections in the immunocompetent host (Table 1). Of special consideration is that immunocompromised patients may be at increased risk for infection or experience increased disease severity caused by many common gastrointestinal pathogens. Patients with human immunodeficiency virus (HIV) infection, solid organ and bone marrow transplant recipients, patients with hematologic or other malignancies, diabetics, patients receiving immunosuppressive chemotherapy and corticosteroid therapy, patients with poor nutritional status, and those at the extremes of age are all at risk for conventional and opportunistic infections of the gastrointestinal tract. Additionally, certain microorganisms cause infection in the compromised host and rarely, if ever, are observed in the healthy, immunocompetent population (Fig. 1). Some infections may indicate an underlying immunodeficiency, such as *Candida* sp. esophagitis or microsporidial enteritis, which may be the first AIDS-defining illness experienced by patients with HIV infection.

Infections of the gastrointestinal tract can be categorized as either upper tract infections, affecting the oral cavity, esophagus, or stomach, or lower tract infections, affecting the small and large intestines. Oral mucositis resulting from chemotherapy or radiation therapy for a variety of malignancies can be a debilitating complication but does not have a particular infectious cause (18). Esophagitis may be characterized by dysphagia (difficulty in swallowing), odynophagia (painful swallowing), heartburn, and sometimes chest pain, which are common features of the common infectious causes of esophagitis in the compromised host, primarily with *Candida* spp. Gastritis may be either acute or chronic. Symptoms include epigastric pain, nausea, and vomiting. Infections such as salmonellosis or cytomegalovirus (CMV) infection may cause acute gastritis in the compromised host; *Helicobacter pylori* infection is the prototype infection for chronic gastritis (46). Necrotizing enterocolitis is a serious disease affecting low-birth-weight neonates, with high morbidity and mortality and without a known microbial etiology (60). Lower gastrointestinal infections may be associated with fever, abdominal pain, nausea, vomiting, and diarrhea. Microbial agents that cause diarrhea and enterocolitis are numerous and may cause direct damage to the intestinal mucosa or may be mediated through the production of toxins. Diarrhea of unknown etiology is common in the neutropenic host (neutropenic enterocolitis and typhilitis) and occurs in adult and pediatric populations, mainly associated with acute leukemia and use of cytotoxic drugs (2). Proctitis is mainly due to the manifestations of common sexually transmitted diseases, such as lymphogranuloma venereum, and may be associated with rectal pain, tenesmus, pruritis, and fever in invasive disease (1). Graft-versus-host disease is a common complication in patients undergoing hematopoietic stem cell transplants and also occurs in patients receiving solid organ transplants. Graft-versus-host disease may be associated with nausea, vomiting, diarrhea, and abdominal pain and can mimic a variety of infectious diseases (75).

Opportunistic infections in the compromised host usually occur during the first year following a decrease in host immunity. The risk after solid organ transplantation occurs during the first year, particularly in the first 6 months (26, 50). Opportunistic infections have dramatically decreased in patients with HIV receiving highly active antiretroviral therapy (ART) (42). Infections of the gastrointestinal tract in particular appear to be decreasing in HIV-positive patients, but infections, mainly *Candida* sp. esophagitis

Irving Nachamkin • Department of Pathology and Laboratory Medicine, University of Pennsylvania School of Medicine, Philadelphia, PA 19104-4823.

Table 1. Microbial agents causing more severe or complicated gastrointestinal infection in the immunocompromised host

Agent	Disease associations	Diagnostic approach[a]
Bacteria		
Salmonella spp.	Enterocolitis, bacteremia	Culture
Campylobacter spp.	Enterocolitis, bacteremia	Fecal Gram stain, culture
Shigella spp.	Enterocolitis	Culture
E. coli	Enterocolitis	Culture, fecal toxin assays
Aeromonas spp.	Diarrhea, serious extraintestinal disease	Culture
Plesiomonas shigelloides	Enterocolitis	Culture
Clostridium difficile	Diarrhea, pseudomembranous colitis	Fecal toxin assay
Listeria monocytogenes	Enterocolitis, bacteremia	Culture
Mycobacterium tuberculosis	Enterocolitis, esophagitis is less common	Histopathology, culture, NAAT
Mycobacterium spp.	Enterocolitis (MAC), esophagitis and gastritis are less common	Histopathology
Helicobacter spp.	Gastritis, no increased risk factors in HIV for *H. pylori*; other species may cause gastroenteritis	Histopathology, culture, serology, fecal antigen assay
Chlamydia trachomatis (LGV)	Proctitis	Culture, NAAT, serology
Fungi		
Candida spp.	Oropharyngeal, esophagitis	Gram or fungal stain, histopathology
Histoplasma capsulatum	Enterocolitis, esophagitis	Histopathology, culture, urinary antigen assay
Pneumocystis	Esophagitis is uncommon; extrapulmonary disease is rare	Histopathology, immunofluorescence
Cryptococcus neoformans	Upper and lower gastrointestinal disease, uncommon	Histopathology, culture
Aspergillus spp.	Localized infection is uncommon, usually in context with disseminated disease	Histopathology, culture, serum antigenemia
Penicillium marneffei	Diarrhea	Histopathology, culture
Zygomycetes (*Rhizopus* spp., *Mucor* spp., *Rhizomucor*, *Cunninghamella*, *Absidia*, *Apophysomyces*, *Saksenaea* spp., *Entomophthora* spp., *Conidiobolus* spp., *Basidiobolus* spp., *Cokeromyces* spp., *Syncephalastrum* spp.)	Upper and lower gastrointestinal tract	Histopathology, culture
Viruses		
CMV	Upper and lower gastrointestinal tract	Histopathology, culture, antigenemia, viral load
HIV	Esophagitis, uncommon	Serology, viral load
HSV	Esophagitis	Histopathology, culture, direct fluorescence assay
VZV	Esophagitis and enterocolitis, uncommon	Histopathology, culture, direct fluorescence assay
HHV-8	Gastric involvement complicating skin disease	Histopathology
HHV-6	Gastroduodenal disease, colitis	NAAT
Adenovirus	Hemorrhagic colitis	Immunoassay (enteric adenovirus types 40 and 41), culture, NAAT
EBV	Posttransplant lymphoproliferative disease, intestinal obstruction, bleeding oral hairy leukoplakia	Histopathology, viral load
Enterovirus	Prolonged diarrhea	Culture, NAAT
Norovirus	Severe diarrhea, dehydration	Electron microscopy, RT-PCR
Rotavirus	Severe diarrhea, dehydration	Immunoassay, RT-PCR
Human papillomavirus	Oropharyngeal lesions	Histopathology
Parasites		
Cryptosporidium	Diarrhea is common, gastritis is uncommon	Histopathology, immunoassay, O&P with special stains
Giardia lamblia	Chronic diarrhea with IgA deficiency	Immunoassay, O&P
Microsporidium	Enterocolitis	Histopathology, O&P with special stains
Entamoeba histolytica	Enterocolitis, extraintestinal disease	Histopathology, immunoassay, O&P
Leishmania	Upper and lower gastrointestinal disease	Histopathology, immunoassay
Strongyloides	Autoinfection and dissemination	O&P
Cyclospora cayetanensis	Severe diarrhea	Histopathology, O&P, special stains
Isospora belli	Severe diarrhea	Histopathology, O&P

[a]NAAT, nucleic acid amplification tests; RT-PCR, reverse transcriptase PCR; O&P, ova and parasite examination.

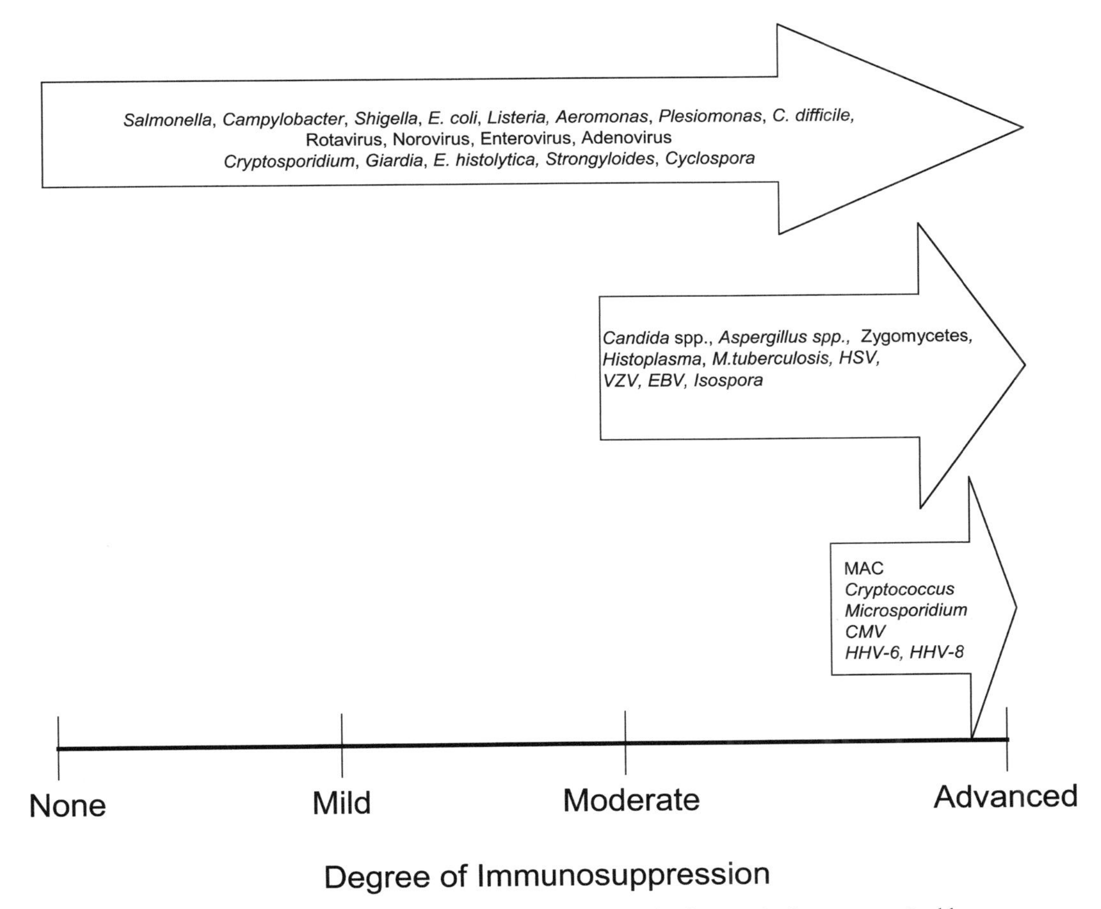

Figure 1. Microbial agents causing upper and lower gastrointestinal infections in the compromised host.

and CMV esophagitis or colitis, still occur (57). Noncompliance with medications is a major risk factor for development of opportunistic infections in this population (57).

Other risk factors for developing gastrointestinal infections in the compromised host include the use of cytokine antagonists and newer immunotherapies that deplete lymphocyte subpopulations. Among solid organ transplant patients at the University of Pittsburgh receiving alemtuzumab (anti-CD52), for example, 10% of patients developed opportunistic infections, including CMV gastrointestinal disease (enteritis and gastritis) and esophageal candidiasis (62). CMV ileitis and listeriosis have been reported following infliximab (anti-tumor necrosis factor alpha) therapy (44, 58).

AGENTS

Bacteria

The common microbial agents of food-borne diarrheal illness, i.e., *Salmonella* spp., *Campylobacter* spp., *Shigella* spp., *Escherichia coli*, and *Listeria monocytogenes*, all affect the compromised host, but with an increased risk of infection and severity. Salmonellosis may be self-limited; severe and persistent, with fever, bloody diarrhea, and weight loss; or septicemic (with or without gastrointestinal symptoms) (5). The incidence of salmonellosis in patients with HIV is 20 to 100 times greater than that in the general population without HIV (5). Recurrent bacteremia also occurs at a higher rate in patients with AIDS. Patients with campylobacteriosis may exhibit persistent diarrhea, bacteremia, and extraintestinal involvement (63). Patients with HIV infection, particularly men who have sex with men, are estimated to have a 39-fold increased risk for developing campylobacter infection over that of the general population (5). Development of antimicrobial resistance in *Campylobacter jejuni* during therapy appears to be more frequent in HIV-infected patients (5). *Campylobacter fetus* subsp. *fetus* causes serious infection in patients who are pregnant or in patients with other underlying disorders, such as cirrhosis, diabetes, hematologic malignancies, and HIV infection (76). Although gastroenteritis occurs with *C. fetus* subsp. *fetus*, the incidence is probably underestimated because the organism may not grow well in routine stool cultures used for *C. jejuni* (83). There is also an increased risk of shigellosis in men who have sex with men, and *Shigella* bacteremia is more common in HIV-positive patients (5). In the transplant population, listeriosis may occur with increased frequency, particularly during the first few months after transplantation (50). The most serious complications from *Listeria* infection are sepsis and infection of the central nervous system; however, patients with listeriosis may initially have abdominal cramps and diarrhea (72). *Aeromonas* spp. are a cause of diarrhea in the immunocompetent host but cause serious extraintestinal disease in the compromised host, particularly in patients with malignancies and hepatobiliary disease (11, 79). *Plesiomonas shigelloides* may cause more severe diarrhea in the compromised host, but this has not been well studied (70). *Clostridium difficile*-associated diarrhea and colitis occur in both competent and compromised hosts, and nosocomial infections in compromised hosts are common. Refractory *C. difficile* disease is associated with a lack of antibody response to *C. difficile* toxins and may respond to infusion of intravenous immunoglobulin (53), suggesting that patients with disorders affecting immunoglobulin production may be at risk for persistent disease.

The most common mycobacterial infections causing gastrointestinal disease in the compromised host are those caused by the *Mycobacterium avium-intracellulare* complex (MAC). The incidence of disseminated MAC infection in AIDS patients is 20 to 40% in the absence of ART or prophylaxis (5), with an incidence rate of 2 per 100 person-years for patients with $CD4^+$ T-cell counts of >100 to 200/μl who are receiving effective prophylaxis or responding to ART (5). In children with AIDS, disseminated MAC infection appears to be more common among children with hemophilia or transfusion-acquired HIV infection (~14%) than among those with perinatal HIV infection (5%) (55). Symptoms include fever, night sweats, weight loss, fatigue, chronic diarrhea, and abdominal pain. Gastric ulceration, enterocolitis, enteric fistulae, and intra-abdominal abscess and hemorrhage are common manifestations of disseminated MAC infection, with the duodenum being the most common site affected (74). Esophagitis and gastritis are uncommon manifestations of nontuberculous mycobacterial infections in this population. *Mycobacterium tuberculosis* may cause extrapulmonary disease affecting the gastrointestinal tract as a result of disseminated disease or primary intestinal involvement (35). Intestinal tuberculosis has increasingly been seen in immunocompromised patients in association with the HIV epidemic and in patients from outside the United States in areas where tuberculosis is endemic, such as the Indian subcontinent and Southeast Asia (39). The most common sites of infection are the ileocecal and jejunal-ileal sites and less commonly include the esophagus, stomach, and duodenum (33, 48). Patients may have signs suggestive of acute appendicitis or intestinal obstruction and can have rectal lesions presenting as perirectal abscesses, fistulae, or fissures (33).

H. pylori is the most common cause of chronic active gastritis in the competent host, and there is no evidence that patients with underlying immunodeficiencies are at risk for infection. Other species of *Helicobacter*, such as *Helicobacter cinaedi* and *Helicobacter fennelliae*, are associated with protocolitis and enteritis in the immunocompromised host and may be associated with bacteremia (22). A number of other *Helicobacter* species have been isolated infrequently from patients with gastroenteritis but may be difficult for laboratories to isolate (21). Lymphogranuloma venereum is associated with anorectal syndrome, with symptoms of anal pruritis, bloody mucopurulent rectal discharge, and localized tenderness (1). *Treponema pallidum* causes syphilis in both competent and immunocompromised patients and is a rare cause of gastrointestinal infection; gastric syphilis has been reported for patients with HIV infection, but whether there is an increased risk is unknown (29).

Fungi

Candida spp., most commonly *Candida albicans*, are the most common cause of esophagitis in the compromised host. Oral candidiasis is frequently associated with radiotherapy and chemotherapy for treatment of solid tumors and hematologic malignancies and is particularly common in HIV patients with low $CD4^+$ T-cell counts, occurring in about 10 to 15% of patients and associated with frequent recurrences (18, 19, 54). The incidence of *Candida* esophagitis has declined in the ART era, with increasing $CD4^+$ T-cell counts accounting for the dramatic improvement (54). Oral thrush is common among HIV-infected children. *Candida* esophagitis occurs in 12 to 16% of children of <13 years of age in the United States (55). A related species, *Candida dubliniensis*, has been reported with increasing frequency as a cause of oral candidiasis in HIV- and non-HIV-infected patients, in both primary and refractory infections (49, 73).

Gastrointestinal histoplasmosis, caused by *Histoplasma capsulatum*, occurs mostly in the competent host, but patients with HIV are at greater risk, and the disease is predominant in males (41). Gastrointestinal disease occurs in fewer than 10% of patients with AIDS and disseminated histoplasmosis (5). Esophageal histoplasmosis may occur following acute infection or may affect other parts of the gastrointestinal tract in disseminated disease (41). Esophageal ulcers with dysphagia or bleeding (41) are seen in patients with immunosuppression. The organism may affect all sites of the gastrointestinal tract but primarily affects the colon (4). There is high mortality associated with gastrointestinal histoplasmosis, and the disease is fatal in about one-fourth of cases (4).

Although patients with HIV are at increased risk for disseminated coccidioidomycosis, gastrointestinal involvement occurs uncommonly (87). Intestinal disease caused by *Penicillium marneffei*, a dimorphic fungus, is an uncommon manifestation of disseminated infection, which is endemic in Southeast Asia and has been reported for patients with HIV, renal transplantation, and corticosteroid treatment (43).

Disseminated cryptococcal infections affecting the esophagus, stomach, duodenum, and colon are rarely reported for the immunocompromised host (7). Localized cryptococcal infection of the esophagus and colon without evidence of disseminated disease has been reported for two patients with Job's syndrome (36, 38). Extrapulmonary *Pneumocystis jirovecii* infection affecting the gastrointestinal tract is rare among adults and children (55, 59).

Invasive aspergillosis of the gastrointestinal tract usually occurs in the context of disseminated disease in the immunocompromised host, affecting both the upper and lower gastrointestinal tract. In a postmortem study, approximately 30% of patients with disseminated invasive aspergillosis had evidence of gastrointestinal involvement (34). Localized infection of the gastrointestinal tract is uncommon (9). Gastrointestinal zygomycosis is relatively common in the immunocompromised host and has a very high mortality rate. Of 929 patients with zygomycosis, primarily low-birth-weight infants, patients with malnutrition, and patients receiving peritoneal dialysis, 7% had gastrointestinal infection (68). Among patients receiving solid organ transplants, gastrointestinal zygomycosis occurred in 11% of patients with zygomycotic infections (3).

Viruses

CMV is by far the most common viral agent causing gastrointestinal disease in the immunocompromised host and may cause esophagitis, gastritis, and enterocolitis. CMV enterocolitis is common in patients with advanced HIV disease, in solid organ, bone marrow, and hematopoietic stem cell transplant recipients, and in those with common variable immunodeficiency. Esophagitis may occur in <5 to 10% of patients with AIDS and CMV end organ disease, presenting with fever, odynophagia, nausea, and midepigastric or retrosternal discomfort (5). The most common site affected is the colon, with disease occasionally affecting more than one gastrointestinal site and sometimes becoming disseminated (77). The prevalence of gastrointestinal CMV disease appears to have been steady over the past decade (77). In a study of cancer patients over an 18-year period, the incidence of gastrointestinal disease caused by CMV was 20 per 100,000 cases, with a higher incidence among patients with hematologic cancers (102/100,000 cases) than that for patients with solid tumors (6/100,000 cases) (77). Patients receiving allogeneic stem cell transplants had a significantly higher incidence (608/100,000 cases) than did those receiving autologous transplants (58/100,000 cases). Of seropositive allograft recipients who developed late CMV disease, 26.2% had gastrointestinal disease (6). Colitis is the most common gastrointestinal manifestation among HIV-infected children with CMV end organ disease and also causes oral and esophageal ulcers, hepatic involvement, ascending cholangiopathy, and gastritis (55). Among patients receiving organ transplants, seronegative recipients receiving organs from seropositive donors are at the highest risk for severe acute disease, and thus CMV should be considered strongly in acute gastrointestinal disease in this setting (8).

Herpes simplex virus (HSV) can cause esophagitis in HIV-infected patients and solid organ transplant patients (61). It is usually seen in the oropharynx in patients receiving bone marrow transplants. There may be vesicles and punched-out ulcerations with an adherent pseudomembrane (46). Varicella-zoster virus (VZV) is a less common cause of gastrointestinal disease in the immunocompromised host and may cause esophagitis and enterocolitis. Human herpesvirus 8 (HHV-8), associated with Kaposi's sarcoma, may cause gastritis as a complication of primary skin disease. Epstein-Barr virus (EBV)-associated posttransplant lymphoproliferative disorder may have a gastrointestinal component, with a clinical presentation that may include gastrointestinal bleeding, obstruction, and perforation (45). The incidence of posttransplant lymphoproliferative disorder affecting the bowel in solid organ allograft recipients is estimated to be 11% overall but is 31% for children of less than 5 years of age (13). HHV-6 has been reported to cause colitis in patients following renal transplantation and is suggested as an etiologic agent of gastroduodenal disease and colitis following hematopoietic stem cell transplantation (17, 31).

Adenovirus infections usually cause self-limited gastrointestinal disease in the immunocompetent host. Adenovirus may cause diarrhea and hemorrhagic colitis in bone marrow transplant patients and solid organ transplant recipients, particularly small bowel transplant recipients, and may also lead to serious disseminated disease (37, 84). Norovirus may affect both immunocompetent and immunocompromised hosts but may cause significant disease in immunocompromised patients (52). The compromised host is susceptible to gastrointestinal infection with rotavirus, and infection may be associated with prolonged diarrhea and dehydration (64). Whether compromised patients are at greater risk for rotavirus infection is not clear.

Parasites

A number of parasitic agents cause disease in the immunocompromised host. Cryptosporidiosis is caused by *Cryptosporidium parvum* and *Cryptosporidium hominis*, which are morphologically indistinguishable (24). The incidence of cryptosporidiosis is <1 per 100 person-years in developed countries with good hygiene and available ART (5). The disease is not self-limited in the compromised patient and is seen mostly in HIV-infected patients, but other groups (Hodgkin's lymphoma, renal transplant, and other solid organ transplant patients) may also be at risk. The greatest risk is in HIV patients with $CD4^+$ T-cell counts of <100 cells/μl (5). Biliary disease is more common in HIV patients and is associated with low $CD4^+$ T-cell counts, i.e., <50 cells/μl (81). Watery and persistent diarrhea is the most common manifestation in children; fever and vomiting are also relatively common (55).

Microsporidiosis is unique to patients with profound immunosuppression, particularly patients with HIV infection. In the pre-ART era, 2 to 70% of HIV patients with diarrhea were infected with microsporidia, but the prevalence has dropped in the post-ART era. Infection is seen mostly when the $CD4^+$ T-cell count is <100/μl (5). Two species are particularly associated with gastrointestinal infection. *Enterocytozoon bieneusi* is associated with persistent diarrhea, fever, weight loss, malabsorption, and cholangitis. *Encephalitozoon intestinalis* is associated with diarrhea, disseminated infection, and superficial keratoconjunctivitis (24).

Giardia lamblia infections may have a chronic course in patients with immunodeficiencies, particularly patients with immunoglobulin A deficiency or other disorders associated with immunoglobulin deficiencies (10). *Entamoeba histolytica* causes enterocolitis and extraintestinal disease (liver abscess) and may be more common in the compromised patient, particularly in malnourished children (56). Acute necrotizing colitis is a rare complication with high mortality and is seen predominantly in developing countries (30). Toxic megacolon is also a rare complication and is associated with corticosteroid use (30). *Strongyloides* causes severe autoinfections in immunosuppressed patients—patients at risk include those with leukemia and lymphoma and solid organ transplant recipients (61). *Cyclospora cayetanensis* causes a food-borne illness causing persistent diarrhea in the compromised host but is usually self-limited in competent hosts. *Isospora belli* may cause severe diarrhea in AIDS patients, affecting primarily the small intestine, and is often associated with severe dehydration (5).

Leishmania spp. are an uncommon cause of gastrointestinal infection in the immunocompromised host. Disseminated visceral disease syndrome is the most common clinical presentation in AIDS patients (~70%) (5). Advanced immunosuppression in HIV-infected patients ($CD4^+$ T-cell counts of <100) is a risk factor for developing disease. With profound immunosuppression, there may be involvement of the upper and lower gastrointestinal tract, with stomach, duodenum, and colon sites affected (71). In a series of patients with HIV and *Leishmania* infections, 11 of 91 patients had a diagnosis of visceral leishmaniasis, based on the detection of organisms in the gastrointestinal tract (71).

DIAGNOSTIC APPROACHES

Because of the numerous microbial agents causing gastrointestinal infections in the immunocompromised patient, the diagnostic workup can be both prolonged and costly (Table 1). There are several approaches to differential diagnosis of immunocompromised patients with gastrointestinal infections. Patients with uncomplicated acute gastroenteritis may be classified as having either community-acquired, nosocomially acquired, or persistent infections, as suggested by the Infectious Diseases Society of America (28). Of course, the immunocompromised host is at risk not only for the common enteric pathogens seen in the competent host but also for a variety of infections not normally seen in the competent host, particularly infection with latent viruses, opportunistic mycobacterial and fungal infections, and parasitic infections (Fig. 1). In the algorithm of the Infectious Diseases Society of America, the differential diagnosis for the patient with an acute community-acquired diarrheal illness should first include the common community-acquired enteric bacterial pathogens. Agents such as *Salmonella* spp., *Campylobacter jejuni*, *Shigella* spp., *Aeromonas* spp., *Plesiomonas shigelloides*, and other selected pathogens should be considered depending upon the patient's history. Additional bacterial agents, such as *E. coli* O157 (or other *stx*-carrying *E. coli*) and *Vibrio* spp., should be considered based on history and the relative prevalence of organisms in the geographic region of the patient. Direct examination of stool samples by Gram stain for *Campylobacter* spp. may be a rapid, sensitive, and specific approach to diagnosing acute campylobacter infection but has no value for the diagnosis of other bacterial enteropathogens (20). Other rapid methods, such as detection of fecal white cells, have little to no value for the diagnosis of bacterial gastrointestinal pathogens (20). Stool culture is the primary method used for detecting bacterial enteropathogens, and for acute bacterial gastroenteritis, culture of up to two fecal samples has a high sensitivity (99%), although for single samples culture has a sensitivity of 96.9% (82). The detection of *L. monocytogenes* in stool cultures is problematic for clinical microbiology laboratories due to the lack of selective and differential medium for this organism, but at high concentrations, it may be detected on blood agar used as part of the routine setup for stool cultures. Commercial stool toxin tests for *stx*-carrying *E. coli* to detect hemorrhagic *E. coli* strains other than O157 strains are readily available (40). Nosocomially acquired infections with the above enteric pathogens, other than those with *C. difficile*, are extremely rare and should be considered in the context of suspected outbreaks. Patients with a history of antibiotic exposure who develop community-acquired or nosocomially acquired diarrhea may have *C. difficile* disease and should be tested for fecal toxin. Fecal toxin (TcdA and TcdB) assays for the diagnosis of *C. difficile* disease are widely available (80). The intestinal parasites generally cause persistent infections and should also be considered for patients with unremitting diarrheal illnesses.

Fungal infections of the gastrointestinal tract should be considered for patients with moderate to severe immunosuppression and may be detected through a combination of histopathologic analysis of tissue, culture of affected tissue, and immunoassays. There are limited data on the diagnosis of fungal infections in the gastrointestinal tract and on comparisons of diagnostic methods. Recovery of zygomycetes from tissue may be negative for histopathologically positive tissues, which may be due, in part, to poor preparation of tissues for culture, such as grinding of the tissue (65). A recent study showed that recovery of *Aspergillus* spp. and zygomycetes from respiratory tissue biopsies by culture was less sensitive than histopathology, with ~78% sensitivity for detecting *Aspergillus* spp. and 33% sensitivity for detecting zygomycetes, suggesting that similar results would be obtained with gastrointestinal tissues (66). In recent years, the use of antigen detection for diagnosis of disseminated histoplasmosis has become an important tool for diagnosis (41). There are few data on the performance of the histoplasma urinary antigen test specifically for gastrointestinal histoplasmosis, but the sensitivity of the urinary antigen test is approximately 90% for disseminated disease (41). An assay for *Aspergillus* antigen detection in serum recently became commercially available, and although the test is FDA cleared for routine use, the indications and optimal use of this test as an aid for detecting invasive aspergillosis are still undergoing investigation (32). Practice guidelines for the diagnosis and management of invasive aspergillosis are forthcoming from the Infectious Diseases Society of America.

The diagnosis of viral infections of the gastrointestinal tract may be accomplished by histopathologic analysis of tissue, conventional culture, and immunologic and molecular methods of detection (see the other chapters in this volume for specific viral agents). The approach to diagnosis of viral infection in a particular patient varies considerably. Histopathologic analysis of gastrointestinal tissues, particular those from the lower gastrointestinal tract, is the most common analysis performed for diagnosis of viral infection. Culture of viruses such as CMV from gastrointestinal tissues has a poor correlation with histopathologic findings and thus may be less specific for disease (12). Among HIV-infected patients with CMV gastrointestinal disease, characteristic inclusions of CMV were detected in 33 to 54% of patients by histopathologic analysis of duodenal/rectal biopsies, compared with 54 to 75% for culture and 100% using quantitative PCR (14). In patients with lesions in the oral cavity or esophagitis, HSV is the most common viral agent involved, and detection of HSV by direct immunofluorescence staining of smears along with culture of material from the lesions is an approach for rapid diagnosis. A comparison of cytologic and histopathologic examination with culture for detecting CMV and HSV esophageal ulceration in 15 HIV-positive patients showed that cytopathology was very insensitive (0%), whereas culture detected 26.7% of cases and histopathology (using immunohistochemical staining) was 100% sensitive (85). Tissue sampling may be an important issue for the diagnosis of viral esophagitis, as multiple biopsies are needed to increase the sensitivity for detection of CMV in HIV-infected patients (86).

In patients receiving solid organ transplants, the amount of time that has transpired posttransplantation is helpful in considering potential viral agents of disease. For patients receiving transplants within 1 month prior to developing illness, HSV is the most common herpesvirus to be considered. CMV disease usually occurs after 1 month or longer, with EBV and VZV disease occurring several months following transplantation (27). Herpesvirus infections (with CMV, EBV, HSV, and VZV) are related to the degree of immunosuppression in patients with HIV infection. The temporal appearance of these viruses in HIV-infected patients is related to low CD4$^+$ T-cell counts (27). For solid organ and hematopoietic stem cell transplant recipients, knowledge of the pretransplant serostatus for both the donor and recipient is helpful in estimating the risk of disease, particularly with CMV. Seronegative recipients who receive allografts from seropositive donors have a high risk for developing primary CMV disease. Seropositive recipients receiving seropositive or seronegative transplants are at risk for reactivation disease, and thus, testing for CMV infection in this population is usually initiated at the first signs of gastrointestinal infection. Viral load testing by molecular or antigenemia assays (described in more detail in Chapter 3) is highly sensitive for detecting early CMV disease and is used as an indicator for preemptive antiviral therapy. Interpretation of viral cultures may be problematic for some of the viral infections arising from reactivation. In patients suspected of having viral infections due to reactivation, such as with CMV, culturing of CMV from a biopsy or from cells from the colon or the esophagus may have low specificity for the diagnosis of CMV colitis or esophagitis because patients with low CD4$^+$ T-cell counts might be viremic and have positive cultures for CMV in the absence of clinical disease (5).

Testing for community-acquired viral infections, such as those with rotavirus and adenoviruses, can be performed using commercially available immunoassays for detection of viral antigen (51). Several commercial immunoassays for enteric adenovirus (types 40 and 41) have sensitivities of $>$90% compared with culture or electron microscopy, but the sensitivities of these assays may be reduced due to viral variants (67). More recent studies using reverse transcriptase PCR show that immunoassays for both rotavirus and adenovirus substantially underestimate the number of infections, and molecular testing would appear to be preferred in antigen-negative cases to rule out infection (47). Norovirus detection is accomplished by reverse transcriptase PCR assays, which are usually available in commercial reference laboratories or local or state public health laboratories. Several commercial immunoassays are available outside the United States, but they are inferior for detecting norovirus compared to current molecular methods (15, 16). Nosocomial transmission of rotavirus, adenovirus, astroviruses, and noroviruses may occur and should also be considered in the workup of nosocomially acquired enteric infections in the immunocompromised host (69, 78). Serologic tests for herpesviruses, other than for exposure status, are not useful for the diagnosis of viral infections in the compromised host. Similarly, detection of HHV-6 in blood samples may represent reactivation of the latent virus, without evidence of causation (31).

Patients with a persistent community-acquired diarrheal illness with negative bacterial stool studies should be evaluated for the presence of parasitic agents, particularly *Giardia* and *Cryptosporidium. E. histolytica* and other intestinal helminths, nematodes, and trematodes should be considered for patients with relevant exposures and/or a history of travel to areas where these parasites are endemic. Patients with abdominal pain and unexplained eosinophilia or travel to areas of endemicity should be tested for the presence of *Strongyloides*, since there is an increased risk for *Strongyloides* hyperinfection syndrome in the immunosuppressed patient. In patients with severe immunosuppression, such as patients with HIV and CD4$^+$ T-cell counts of $<$200, microsporidia should also be considered. Parasitic agents affecting the gastrointestinal tract are frequently diagnosed by examination of fecal samples by either conventional microscopic examination or immunoassay and may also be detected by histologic examination of tissue (24, 25). Examination of two or three fecal samples has a high sensitivity for detecting common intestinal parasites (82). Antigen detection-based immunoassays, particularly for detecting *E. histolytica*, *Giardia lamblia*, and *Cryptosporidium*, are widely used for diagnosis of these infections (23). Details about the performance characteristics of various parasitologic examinations can be found in Chapter 15.

SUMMARY

The approach to the diagnosis of gastrointestinal infections, particularly diarrheal illnesses, in the immunocompromised host is rather complex, and there is no simple diagnostic algorithm to follow. The compromised host is susceptible to all of the com-

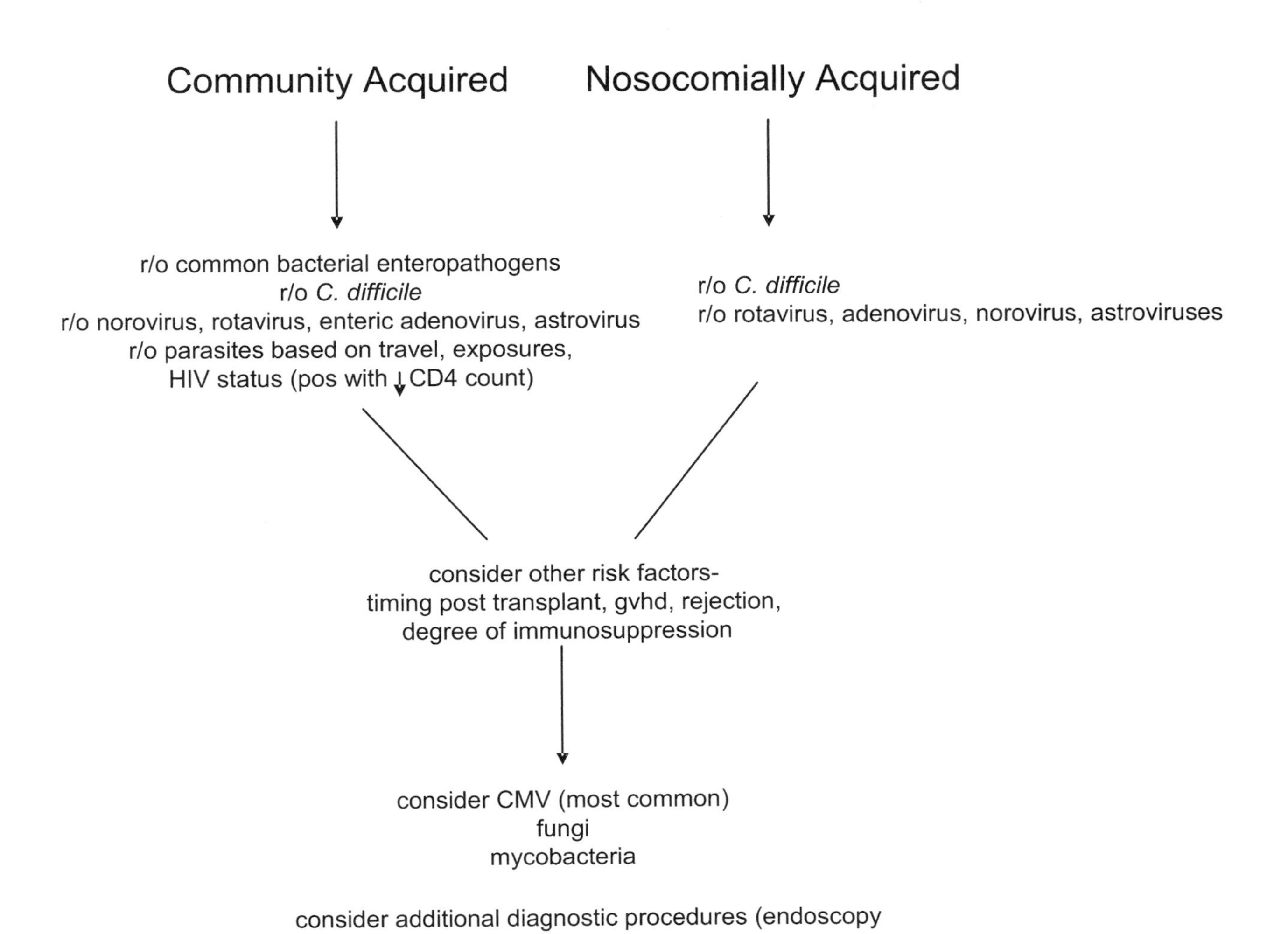

Figure 2. Algorithm for diagnostic approach to lower gastrointestinal infections in the compromised host.

mon bacterial, viral, and parasitic agents that cause disease in immunocompetent hosts, and thus, assessment of the patient for risk factors for specific infections will help to narrow the field of potential causative agents. An initial approach is shown in Fig. 2. In a patient presenting with acute lower gastrointestinal infection, consider whether it is a community- or health care-associated infection. It would be reasonable to perform bacterial stool cultures initially, as well as tests for *C. difficile* (if risk factors are present). Viral studies for norovirus and enteric adenoviruses may also be warranted as part of the initial workup. For patients with recent travel, within or outside the United States, with potential exposures putting the patient at risk for parasitic diseases, additional tests for relevant parasites should be ordered during the initial workup. Beyond the above studies, other agents, particularly viral, fungal, and mycobacterial agents, should be considered based on the patient's degree of immunosuppression (Fig. 1), time transpired since initial solid organ or hematopoietic stem cell transplantation, risk for graft-versus-host disease, and rejection. Consultation with the clinical microbiologist, infectious disease specialist, and gastroenterologist should be considered strongly for difficult-to-diagnose patients.

REFERENCES

1. **Ahdoot, A., D. P. Kotler, J. S. Suh, C. Kutler, and R. Flamholz.** 2006. Lymphogranuloma venereum in human immunodeficiency virus-infected individuals in New York City. *J. Clin. Gastroenterol.* **40:**385–390.
2. **Aksoy, D. Y., M. D. Tanriover, O. Uzun, P. Zarakolu, S. Ercis, S. Erguven, A. Oto, U. Kerimoglu, M. Hayran, and O. Abbasoglu.** 2007. Diarrhea in neutropenic patients: a prospective cohort study with emphasis on neutropenic enterocolitis. *Ann. Oncol.* **18:**183–189.
3. **Almyroudis, N. G., D. A. Sutton, P. Linden, M. G. Rinaldi, J. Fung, and S. Kusne.** 2006. Zygomycosis in solid organ transplant recipients in a tertiary transplant center and review of the literature. *Am. J. Transplant.* **6:**2365–2374.

4. **Assi, M., D. S. McKinsey, M. R. Driks, M. C. O'Connor, M. Bonacini, B. Graham, and F. Manian.** 2006. Gastrointestinal histoplasmosis in the acquired immunodeficiency syndrome: report of 18 cases and literature review. *Diagn. Microbiol. Infect. Dis.* **55:**195–201.
5. **Benson, C. A., J. E. Kaplan, H. Masur, A. Pau, and K. K. Holmes.** 2004. Treating opportunistic infections among HIV-infected adults and adolescents. *MMWR Morb. Mortal. Wkly. Rep.* **53**(RR-15):1–112.
6. **Boeckh, M., and K. A. Marr.** 2002. Infection in hematopoietic stem cell transplantation, p. 527–571. *In* R. H. Rubin and L. S. Young (ed.), *Clinical Approach to Infection in the Compromised Host.* Kluwer Academic/Plenum Publishers, New York, NY.
7. **Bonacini, M., J. Nussbaum, and C. Ahluwalia.** 1990. Gastrointestinal, hepatic, and pancreatic involvement with *Cryptococcus neoformans* in AIDS. *J. Clin. Gastroenterol.* **12:** 295–297.
8. **Buller, R. S., G. A. Storch, and M. Q. Arens.** 2003. Cumitech 38, Human cytomegalovirus. Coordinating ed., R. M. Jamison. ASM Press, Washington, DC.
9. **Chionh, F., K. E. Herbert, J. F. Seymour, H. M. Prince, M. Wolf, A. Zimet, C. Tam, and G. A. Kennedy.** 2005. Antemortem diagnosis of localized invasive esophageal aspergillosis in a patient with acute myeloid leukemia. *Leuk. Lymph.* **46:**603–605.
10. **Christenson, J. C., and H. R. Hill.** 2002. Infections complicating congenital immunodeficiency syndromes, p. 465–495. *In* R. H. Rubin and L. S. Young (ed.), *Clinical Approach to Infection in the Compromised Host.* Academic/Plenum Publishers, New York, NY.
11. **Clark, N. M., and C. E. Chenoweth.** 2003. *Aeromonas* infection of the hepatobiliary system: report of 15 cases and review of the literature. *Clin. Infect. Dis.* **37:**506–513.
12. **Clayton, F., E. B. Klein, and D. P. Kotler.** 1989. Correlation of in situ hybridization with histology and viral culture in patients with acquired immunodeficiency syndrome with cytomegalovirus colitis. *Arch. Pathol. Lab. Med.* **113:**1124–1126.
13. **Cosimi, A. B.** 2002. Surgical aspects of infection in the compromised host, p. 681–706. *In* R. H. Rubin and L. S. Young (ed.), *Clinical Approach to Infection in the Compromised Host.* Kluwer Academic/Plenum Publishers, New York, NY.
14. **Cotte, L., E. Drouet, F. Bissuel, G. A. Denoyel, and C. Trepo.** 1993. Diagnostic value of amplification of human cytomegalovirus DNA from gastrointestinal biopsies from human immunodeficiency virus-infected patients. *J. Clin. Microbiol.* **31:**2066–2069.
15. **de Bruin, E., E. Duizer, H. Vennema, and M. P. G. Koopmans.** 2006. Diagnosis of norovirus outbreaks by commercial ELISA or RT-PCR. *J. Virol. Methods* **137:**259–264.
16. **de Cal, I. W., A. Revilla, J. M. del Alamo, E. Roman, S. Moreno, and A. Sanchez-Fauquier.** 2006. Evaluation of two commercial enzyme immunoassays for the detection of norovirus in faecal samples from hospitalized children with sporadic acute gastroenteritis. *Clin. Microbiol. Infect.* **13:**341–343.
17. **Delbridge, M. S., M. S. Karim, B. M. Shrestha, and W. McKane.** 2006. Colitis in a renal transplant patient with human herpesvirus-6 infection. *Transpl. Infect. Dis.* **8:**226–228.
18. **Donnelly, J. P., L. A. Bellm, J. B. Epstein, S. T. Sonis, and R. P. Symonds.** 2003. Antimicrobial therapy to prevent or treat oral mucositis. *Lancet Infect. Dis.* **3:**405–412.
19. **Epstein, J. B., P. J. Hancock, and S. Nantel.** 2003. Oral candidiasis in hematopoietic cell transplantation patients: an outcome-based analysis. *Oral Surg. Oral Med. Oral Pathol. Oral Radiol. Endod.* **96:**154–163.
20. **Fitzgerald, C., and I. Nachamkin.** 2007. *Campylobacter* and *Arcobacter*, p. 933–946. *In* P. R. Murray, E. J. Baron, J. H. Jorgensen, M. L. Landry, and M. A. Pfaller (ed.), *Manual of Clinical Microbiology.* ASM Press, Washington, DC.
21. **Fox, J. G.** 2002. The non-*H. pylori* helicobacters: their expanding role in gastrointestinal and systemic diseases. *Gut* **50:**273–283.
22. **Fox, J. G., and F. Megraud.** 2007. *Helicobacter*, p. 947–962. *In* P. R. Murray, E. J. Baron, J. H. Jorgensen, M. L. Landry, and M. A. Pfaller (ed.), *Manual of Clinical Microbiology.* ASM Press, Washington, DC.
23. **Garcia, L. S.** 2007. Antibody and antigen detection in parasitic infections, p. 592–615. *In* L. S. Garcia (ed.), *Diagnostic Medical Parasitology.* ASM Press, Washington, DC.
24. **Garcia, L. S.** 2007. Parasitic infections in the compromised host, p. 506–548. *In* L. S. Garcia (ed.), *Diagnostic Medical Parasitology.* ASM Press, Washington, DC.
25. **Garcia, L. S.** 2007. Histologic identification of parasites, p. 616–669. *In* L. S. Garcia (ed.), *Diagnostic Medical Parasitology.* ASM Press, Washington, DC.
26. **Garrido, R. S. J., J. M. Aguado, C. Diaz-Pedroche, O. Len, M. Montejo, A. Moreno, M. Gurgui, J. Torre-Cisneros, F. Pareja, J. Segovia, M. Garcia, and C. Lumbreras.** 2006. A review of critical periods for opportunistic infection in the new transplantation era. *Transplantation* **82:**1457–1462.
27. **Griffiths, P. D.** 2002. The herpesviruses, p. 361–403. *In* R. H. Rubin and L. S. Young (ed.), *Clinical Approach to Infection in the Compromised Host.* Kluwer Academic/Plenum Publishers, New York, NY.
28. **Guerrant, R. L., T. Van Gilder, T. S. Steiner, N. M. Thielman, L. Slutsker, R. V. Tauxe, T. Hennessy, P. M. Griffin, H. L. DuPont, R. B. Sack, P. I. Tarr, M. Neill, I. Nachamkin, L. B. Reller, M. T. Osterholm, M. L. Bennish, and L. K. Pickering.** 2001. Practice guidelines for managing infectious diarrhea. *Clin. Infect. Dis.* **32:**331–351.
29. **Guerrero, A. F., T. M. Straight, J. Eastone, and K. Spooner.** 2005. Gastric syphilis in an HIV-infected patient. *AIDS Patient Care STDs* **5:**281–285.
30. **Haque, R., C. D. Huston, M. Hughes, E. Houpt, and W. A. Petri.** 2003. Amebiasis. *N. Engl. J. Med.* **348:**1565–1573.
31. **Hentrich, M., D. Oruzio, G. Jager, M. Schlemmer, M. Schleuning, X. Schiel, W. Hiddemann, and H. J. Kolb.** 2004. Impact of human herpesvirus-6 after haematopoietic stem cell transplantation. *Br. J. Haematol.* **128:**66–72.
32. **Hope, W. W., T. J. Walsh, and D. W. Denning.** 2005. Laboratory diagnosis of invasive aspergillosis. *Lancet Infect. Dis.* **5:**609–622.
33. **Hopewell, P. C., and R. M. Jasmer.** 2005. Overview of clinical tuberculosis, p. 15–31. *In* S. T. Cole, K. D. Eisenach, D. N. McMurray, and W. R. Jacobs, Jr. (ed.), *Tuberculosis and the Tubercle Bacillus.* ASM Press, Washington, DC.
34. **Hori, A., M. Kami, Y. Kishi, U. Machida, T. Matsumura, and T. Kashima.** 2002. Clinical significance of extra-pulmonary involvement of invasive aspergillosis: a retrospective autopsy-based study of 107 patients. *J. Hosp. Infect.* **50:**175–182.
35. **Horvath, K. D., and R. L. Whelan.** 1998. Intestinal tuberculosis: return of an old disease. *Am. J. Gastroenterol.* **93:**692–696.
36. **Hutto, J. O., C. S. Bryan, F. L. Greene, C. J. White, and J. I. Gallin.** 1988. Cryptococcus of the colon resembling Crohn's disease in a patient with the hyperimmunoglobulin E-recurrent infection (Job's) syndrome. *Gastroenterology* **94:**808–812.
37. **Ison, M. G.** 2006. Adenovirus infections in transplant recipients. *Clin. Infect. Dis.* **43:**331–339.
38. **Jacobs, D. H., A. M. Macher, R. Handler, J. E. Bennett, M. J. Collen, and J. I. Gallin.** 1984. Esophageal cryptococcosis in a

patient with the hyperimmunoglobulin E-recurrent infection (Job's) syndrome. *Gastroenterology* **87**:201–203.

39. **Jakubowski, A., R. K. Elwood, and D. A. Enarson.** 1988. Clinical features of abdominal tuberculosis. *J. Infect. Dis.* **158**:687–692.
40. **Johnson, K. E., C. M. Thorpe, and C. L. Sears.** 2006. The emerging clinical importance of non-O157 Shiga toxin-producing *Escherichia coli*. *Clin. Infect. Dis.* **43**:1587–1595.
41. **Kahi, C. J., L. J. Wheat, S. D. Allen, and G. A. Sarosi.** 2005. Gastrointestinal histoplasmosis. *Am. J. Gastroenterol.* **100**:220–231.
42. **Kaplan, J. E., D. Hanson, M. S. Dworkin, T. Frederick, J. Bertolli, M. Lindegren, S. Holmberg, and J. L. Jones.** 2000. Epidemiology of human immunodeficiency virus-associated opportunistic infections in the United States in the era of highly active antiretroviral therapy. *Clin. Infect. Dis.* **30**:S5–S14.
43. **Ko, C. I., C. C. Hung, M. Y. Chen, P. R. Hsueh, C. H. Hsiao, and J. M. Wong.** 1999. Endoscopic diagnosis of intestinal penicilliosis marneffei: report of three cases and review of the literature. *Gastrointest. Endosc.* **50**:111–114.
44. **Kohara, M. M., and R. N. Blum.** 2006. Cytomegalovirus ileitis and hemophagocytic syndrome associated with use of anti-tumor necrosis factor-alpha antibody. *Clin. Infect. Dis.* **42**:733–734.
45. **Kotton, C. N., and J. A. Fishman.** 2005. Viral infection in the renal transplant recipient. *J. Am. Soc. Nephrol.* **16**:1758–1774.
46. **Liu, C., and J. M. Crawford.** 2005. The gastrointestinal tract, p. 797–875. *In* V. Kumar, A. K. Abbas, and N. Fausto (ed.), *Pathologic Basis of Disease*. Elsevier Saunders, Philadelphia, PA.
47. **Logan, C., J. J. O'Leary, and N. O'Sullivan.** 2006. Real-time reverse transcription-PCR for detection of rotavirus and adenovirus as causative agents of acute viral gastroenteritis in children. *J. Clin. Microbiol.* **44**:3189–3195.
48. **Marshall, J. B.** 1993. Tuberculosis of the gastrointestinal tract and peritoneum. *Am. J. Gastroenterol.* **88**:989–999.
49. **Martinez, M., J. L. Lopez-Ribot, W. R. Kirkpatrick, B. J. Coco, S. P. Bachmann, and T. F. Patterson.** 2002. Replacement of *Candida albicans* with *C. dubliniensis* in human immunodeficiency virus-infected patients with oropharyngeal candidiasis treated with fluconazole. *J. Clin. Microbiol.* **40**:3135–3139.
50. **Marty, F. M., and R. H. Rubin.** 2006. The prevention of infection post-transplant: the role of prophylaxis, preemptive and empiric therapy. *Transplant. Int.* **19**:2–11.
51. **Matson, D. O., M. L. O'Ryan, X. Jiang, and D. K. Mitchell.** 2000. Rotavirus, enteric adenoviruses, caliciviruses, astroviruses, and other viruses causing gastroenteritis, p. 270–294. *In* S. Specter, R. L. Hodinka, and S. A. Young (ed.), *Clinical Virology Manual*. ASM Press, Washington, DC.
52. **Mattner, F., D. Sohr, A. Heim, P. Gastmeier, H. Vennema, and M. Koopmans.** 2006. Risk groups for clinical complications of norovirus infections: an outbreak investigation. *Clin. Microbiol. Infect.* **12**:69–74.
53. **McPherson, S., C. J. Rees, R. Ellis, S. Soo, and S. J. Panter.** 2006. Intravenous immunoglobulin for the treatment of severe, refractory, and recurrent *Clostridium difficile* diarrhea. *Dis. Colon Rectum* **49**:640–645.
54. **Mocroft, A., C. Oancea, J. van Lunzen, P. Vanhems, D. Banhegyi, A. Chiesi, E. Vinogradova, S. Maayan, A. N. Phillips, and J. Lundgren.** 2005. Decline in esophageal candidiasis and use of antimycotics in European patients with HIV. *Am. J. Gastroenterol.* **100**:1446–1454.
55. **Mofenson, L. M., J. Oleske, L. Serchuck, R. Van Dyke, and C. Wilfert.** 2004. Treating opportunistic infections among HIV-exposed and infected children. *MMWR Morb. Mortal. Wkly. Rep.* **53**(RR-14):1–63.
56. **Mondal, D., W. A. Petri, R. B. Sack, B. D. Kirkpatrick, and R. Haque.** 2006. Entamoeba histolytica-associated diarrheal illness is negatively associated with the growth of preschool children: evidence from a prospective study. *Trans. R. Soc. Trop. Med. Hyg.* **100**:1032–1038.
57. **Monkemuller, K. E., A. J. Lazenby, D. H. Lee, R. Loudon, and C. M. Wilcox.** 2005. Occurrence of gastrointestinal opportunistic disorders in AIDS despite the use of highly active antiretroviral therapy. *Dig. Dis. Sci.* **50**:230–234.
58. **Morelli, J., and F. A. Wilson.** 2000. Does administration of infliximab increase susceptibility to listeriosis? *Am. J. Gastroenterol.* **95**:841–842.
59. **Ng, V. L., D. M. Yajko, and W. K. Hadley.** 1997. Extrapulmonary pneumocystosis. *Clin. Microbiol. Rev.* **10**:401–418.
60. **Overturf, G. D.** 2006. Focal bacterial infections, p. 347–384. *In* J. S. Remington, J. O. Klein, C. B. Wilson, and C. J. Baker (ed.), *Infectious Diseases of the Fetus and Newborn Infant*. Elsevier Saunders, Philadelphia, PA.
61. **Patel, R., and C. V. Paya.** 1997. Infections in solid-organ transplant recipients. *Clin. Microbiol. Rev.* **10**:86–124.
62. **Peleg, A. Y., S. Husain, E. J. Kwak, F. P. Silveira, M. Ndirangu, J. Tran, K. A. Shutt, R. Shapiro, N. Thai, K. Abu-Elmagd, K. R. McCurry, A. Marcos, and D. L. Patterson.** 2007. Opportunistic infections in 547 organ transplant recipients receiving alemtuzumab, a humanized monoclonal CD-52 antibody. *Clin. Infect. Dis.* **44**:204–212.
63. **Perlman, D. M., N. M. Ampel, R. B. Schifman, D. L. Cohn, C. M. Patton, M. L. Aguirre, W.-L. L. Wang, and M. J. Blaser.** 1988. Persistent *Campylobacter jejuni* infections in patients infected with human immunodeficiency virus (HIV). *Ann. Intern. Med.* **108**:540–546.
64. **Rayani, A., U. Bode, E. Habas, G. Fleischhack, S. Engelhart, M. Exner, O. Schildgen, G. Bierbaum, A. Maria Eis-Hubinger, and A. Simon.** 2007. Rotavirus infections in paediatric oncology patients: a matched-pairs analysis. *Scand. J. Gastroenterol.* **42**:81–87.
65. **Ribes, J. A., C. L. Vanover-Sams, and D. J. Baker.** 2000. Zygomycetes in human disease. *Clin. Microbiol. Rev.* **13**:236–301.
66. **Rickerts, V., S. Mousset, E. Lambrecht, K. Tintelnot, R. Schwerdtfeger, E. Presterl, V. Jacobi, G. Just-Nubling, and R. Bialek.** 2007. Comparison of histopathological analysis, culture and polymerase chain reaction assays to detect invasive mold infections from biopsy specimens. *Clin. Infect. Dis.* **44**:1078–1083.
67. **Robinson, C., and M. Echavarria.** 2007. Adenoviruses, p. 1589–1600. *In* P. R. Murray, E. J. Baron, J. H. Jorgensen, M. L. Landry, and M. A. Pfaller (ed.), *Manual of Clinical Microbiology*. ASM Press, Washington, DC.
68. **Roden, M. M., T. E. Zaoutis, W. L. Buchanan, T. A. Knudsen, T. A. Sarkisova, R. L. Schaufele, M. Sein, T. Sein, C. C. Chiou, J. H. Chu, D. P. Kontoyiannis, and T. J. Walsh.** 2005. Epidemiology and outcome of zygomycosis: a review of 929 reported cases. *Clin. Infect. Dis.* **41**:634–653.
69. **Rodriguez-Baez, N., R. O'Brien, S. Qiu, and D. M. Bass.** 2002. Astrovirus, adenovirus, and rotavirus in hospitalized children: prevalence and association with gastroenteritis. *J. Pediatr. Gastroenterol. Nutr.* **35**:64–68.
70. **Rolston, K. V., and R. L. Hopfer.** 1984. Diarrhea due to *Plesiomonas shigelloides* in cancer patients. *J. Clin. Microbiol.* **20**:597–598.
71. **Rosenthal, E., P. Marty, P. del Giudice, C. Pradier, C. Ceppi, J. A. Gastaut, Y. Le Fichoux, and J. P. Cassuto.** 2000. HIV and *Leishmania* coinfection: a review of 91 cases with focus

on atypical locations of *Leishmania*. *Clin. Infect. Dis.* **31**:1093–1095.

72. **Rubin, R. H.** 2002. Infection in the organ transplant recipient, p. 573–679. *In* R. H. Rubin and L. S. Young (ed.), *Clinical Approach to Infection in the Compromised Host*. Academic/Plenum Publishers, New York, NY.
73. **Sullivan, D. J., G. P. Moran, and D. C. Coleman.** 2005. *Candida dubliniensis*: ten years on. *FEMS Microbiol. Lett.* **253**:9–17.
74. **Sun, H. Y., M. Y. Chen, M. S. Wu, S. M. Hsieh, C. T. Fang, C. C. Hung, and S. C. Chang.** 2005. Endoscopic appearance of GI mycobacteriosis caused by the *Mycobacterium avium* complex in a patient with AIDS: case report and review. *Gastrointest. Endosc.* **61**:775–779.
75. **Thompson, B., D. Salzman, J. Steinhauer, A. J. Lazenby, and C. M. Wilcox.** 2006. Prospective endoscopic evaluation for gastrointestinal graft-versus-host disease: determination of the best diagnostic approach. *Bone Marrow Transplant.* **38**: 371–376.
76. **Thompson, S. A., and M. J. Blaser.** 2000. Pathogenesis of *Campylobacter fetus* infections, p. 321–347. *In* I. Nachamkin and M. J. Blaser (ed.), *Campylobacter*, 2nd ed. ASM Press, Washington, DC.
77. **Torres, H. A., D. P. Kontoyiannis, G. P. Bodey, J. A. Adachi, M. A. Luna, J. J. Tarrand, G. M. Nogueras, I. I. Raad, and R. F. Chemaly.** 2005. Gastrointestinal cytomegalovirus disease in patients with cancer: a two decade experience in a tertiary care cancer center. *Eur. J. Cancer* **41**:2268–2279.
78. **Traore, O., G. Belliot, C. Mollat, H. Piloquet, C. Chamoux, H. Laveran, S. S. Monroe, and S. Billaudel.** 2000. RT-PCR identification and typing of astroviruses and Norwalk-like viruses in hospitalized patients with gastroenteritis: evidence of nosocomial infections. *J. Clin. Virol.* **17**:151–158.
79. **Tsai, M. S., C. Y. Kuo, M. C. Wang, H. C. Wu, C. C. Chien, and J. W. Liu.** 2006. Clinical features and risk factors for mortality in *Aeromonas* bacteremic adults with hematologic malignancies. *J. Microbiol. Immunol. Infect.* **39**:150–154.
80. **Turgeon, D. K., T. J. Novicki, J. Quick, L. Carlson, P. Miller, B. Ulness, A. Cent, R. Ashley, A. Larson, M. Coyle, A. P. Limaye, B. T. Bookson, and T. R. Fritsche.** 2003. Six rapid tests for direct detection of *Clostridium difficile* and its toxins in fecal samples compared with the fibroblast cytotoxicity assay. *J. Clin. Microbiol.* **41**:667–670.
81. **Vakil, N. B., S. M. Schwartz, B. P. Buggy, C. F. Brummitt, M. Kherellah, D. M. Letzer, I. H. Gilson, and P. G. Jones.** 1996. Biliary cryptosporidiosis in HIV-infected people after the waterborne outbreak of cryptosporidiosis in Milwaukee. *N. Engl. J. Med.* **334**:19–23.
82. **Valenstein, P., M. Pfaller, and M. Yungbluth.** 1996. The use and abuse of routine stool microbiology: a College of American Pathologists Q-probes study of 601 institutions. *Arch. Pathol. Lab. Med.* **120**:206–211.
83. **Vandenberg, O., A. Dediste, K. Houf, S. Ibekwem, H. Souayah, S. Cadranel, N. Douat, G. Zissis, J.-P. Butzler, and P. Vandamme.** 2004. *Arcobacter* species in humans. *Emerg. Infect. Dis.* **10**:1863–1867.
84. **Walls, T., A. G. Shankar, and D. Shingadia.** 2003. Adenorus: an increasingly important pathogen in paediatric bone marrow transplant patients. *Lancet Infect. Dis.* **3**:79–86.
85. **Wilcox, C. M., W. Rodgers, and A. Lazenby.** 2004. Prospective comparison of brush cytology, viral culture, and histology for the diagnosis of ulcerative esophagitis in AIDS. *Clin. Gastroenterol. Hepatol.* **2**:564–567.
86. **Wilcox, C. M., R. F. Straub, and D. A. Schwartz.** 1996. Prospective evaluation of biopsy number for the diagnosis of viral esophagitis in patients with HIV infection and esophageal ulcer. *Gastrointest. Endosc.* **44**:587–593.
87. **Woods, C. W., C. McRill, B. D. Plikaytis, N. E. Rosenstein, D. Mosley, D. Boyd, B. England, B. A. Perkins, N. M. Ampel, and R. A. Hajjeh.** 2000. Coccidioidomycosis in human immunodeficiency virus-infected persons in Arizona, 1994–1997: incidence, risk factors, and prevention. *J. Infect. Dis.* **181**:1428–1434.

Diagnostic Microbiology of the Immunocompromised Host
Edited by Randall T. Hayden, Karen C. Carroll, Yi-Wei Tang, and Donna M. Wolk

Chapter 18

Central Nervous System Infections

IGEN HONGO, KAREN C. BLOCH, AND YI-WEI TANG

DESCRIPTION OF THE AGENTS AND PATHOPHYSIOLOGY OF INFECTION

Infections of the central nervous system (CNS) are caused by a variety of infectious agents, including viruses, bacteria, fungi, protozoa, and prions. CNS infections are associated with significant morbidity and mortality, and identification of a causative agent(s) is of paramount importance for appropriate patient management. However, clinical manifestations of CNS infections caused by different pathogens overlap substantially, and identification of a specific pathogen requires laboratory or pathologic diagnosis. This chapter provides a framework for the assessment and laboratory evaluation of immunocompromised patients presenting with CNS infections. Detailed discussions of the individual agents are provided in the corresponding pathogen-specific chapters.

The CNS has several unique anatomic and immunologic characteristics that distinguish infections at this site from those involving other organ systems. The brain and spinal cord are contained within the bony confines of the calvarium and vertebral column. Because of the spatial limitations imposed by these structures, inflammation and edema rapidly progress to ischemia and, in severe cases, cerebral herniation.

The CNS is protected by a tight endothelial lining, the blood-brain barrier (BBB), which prevents blood-borne pathogens from entering the CNS. The BBB additionally serves to limit the entry of leukocytes, complement, and immunoglobulin into the CNS, and the absence of these mediators of innate immunity in the brain and cerebrospinal fluid (CSF) may explain the rapid progression and serious consequences of many CNS infections (140). Furthermore, because of this anatomic barrier, antibiotics used to treat CNS infections must have pharmacologic properties allowing them to penetrate the BBB, and higher doses are usually required for bactericidal activity.

In addition to infection with opportunistic pathogens, immunocompromised hosts are at risk for CNS infections with common microorganisms, although these may present atypically. This reality, combined with the potentially severe sequelae of CNS infection, necessitates both a high index of suspicion for infection and a low threshold to begin empiric treatment.

CLINICAL MANIFESTATIONS

Disease Spectrum

There are many ways to classify CNS infections. These include anatomic site of infection, type of pathogen (viral, bacterial, etc.), and disease duration (acute, subacute, chronic, or recurrent). Infections of the meninges, brain, and spinal cord result in meningitis, encephalitis, and myelitis, respectively. Infection may be limited to a single anatomic compartment or may involve multiple sites (e.g., meningoencephalitis or encephalomyelitis).

While this chapter is limited to CNS infections, noninfectious conditions may present in a similar fashion. Acute disseminated encephalomyelitis (ADEM) is a postinfectious or inflammatory demyelinating disease that can be confused with encephalitis. Reversible posterior leukoencephalopathy, which presents with headache, seizures, altered mental status, and cortical blindness, has been reported as a complication of both chemotherapy (132) and cyclosporine treatment (43). Both trimethoprim-sulfamethoxazole (54, 119), used for prophylaxis of *Pneumocystis jirovecii* pneumonia, and OKT3 (78), a monoclonal antibody used to reverse organ rejection, have been associated with aseptic meningitis. Immunocompromised patients are also at increased risk for CNS malignancies, including posttransplant lymphoproliferative disease and primary CNS lymphoma.

Igen Hongo • Department of Internal Medicine, Musashino Red Cross Hospital, Tokyo 180-8610, Japan.
Karen C. Bloch • Departments of Medicine and Preventive Medicine, Vanderbilt University School of Medicine, Nashville, TN 37232.
Yi-Wei Tang • Departments of Pathology and Medicine, Vanderbilt University School of Medicine, Nashville, TN 37232.

Meningitis

Acute meningitis is characterized by the onset of fever, mental status changes, headache, neck stiffness, and photophobia over a period of hours to days. While immunocompromised patients remain susceptible to the pathogens most frequently associated with meningitis in the immunocompetent population (enteroviruses, pneumococcus, meningococcus, etc.), additional infectious and noninfectious etiologies must be considered. Human immunodeficiency virus (HIV) itself can cause aseptic meningitis as part of the acute retroviral syndrome (119). Syphilis should be included in the differential diagnosis of culture-negative meningitis, particularly among HIV-infected patients (67). T-cell lymphopenia or impaired T-cell function increases the risk of *Listeria* meningitis (65). Gram-negative bacterial meningitis, including *Salmonella* meningitis, is unusual in immunocompetent hosts but may be seen in HIV-infected patients (71). Strongyloidiasis following organ transplantation can be complicated by acute meningitis caused by enteric organisms (85). Patients with CNS malignancies often require ventriculostomy for intraventricular administration of chemotherapy, and this subgroup is at particular risk for meningeal infection with both normal skin flora and nosocomially acquired bacteria (139).

Mycobacterial and fungal meningitides tend to cause subacute or chronic meningitis. Among HIV-infected individuals, tuberculous meningitis may represent either acute infection with extrapulmonary spread or reactivation disease (13, 51). In solid organ transplant (SOT) recipients, tuberculous meningitis is usually part of disseminated tuberculosis (124). Meningitis with *Cryptococcus neoformans* can be seen in either transplant (61) or HIV-infected patients. Among the latter group, infection usually occurs in HIV-infected patients with CD4 counts of $<100/mm^3$ (22). Depending on the geographic location, CNS infection with dimorphic fungi (*Histoplasma capsulatum*, *Blastomyces hominis*, and *Coccidioides immitis*) may be seen. In contrast, *Candida* is a relatively rare cause of meningitis in HIV patients (25). Enteroviruses can cause chronic meningitis in individuals with impaired humoral immunity (111).

Encephalitis and mass/space-occupying lesions of the brain

Encephalitis is characterized by inflammation of the brain parenchyma, and the clinical hallmark is alteration in mental status, ranging from lethargy to coma. Other frequent signs and symptoms include fever, motor or sensory deficits, and seizures. Acute encephalitis is most commonly caused by viruses in both the immunocompetent and immunocompromised host, but the array of organisms and spectrum of clinical presentation are broader for the latter group.

Herpes simplex virus type 1 (HSV-1) is the most common cause of acute encephalitis in the United States (141), but the incidence does not appear to be elevated in immunocompromised patients. A European study identified HSV-1 in only 4 of 918 (0.4%) CSF samples from HIV-infected patients with neurological symptoms (30). Determination of infection with other herpesviruses, such as Epstein-Barr virus (EBV) or cytomegalovirus (CMV), is complicated by the fact that these agents remain dormant in white blood cells following acute infection. Therefore, when one of these agents is identified, care must be taken to differentiate CNS infection from detection of latently infected leukocytes or macrophages. CMV encephalitis classically causes either periventricular demyelination or, less commonly, diffuse micronodular encephalitis (9). The significance of identifying EBV in the CSF is less clear. Reactivation of EBV may occur in the setting of CNS infection with an alternative agent (152). Finally, detection of EBV in the CSF of a patient with HIV may be a clue to the diagnosis of primary CNS lymphoma, with detection of virus often preceding diagnosis of malignancy by months to years (29).

Detection of other viruses is variably associated with infection. Human herpesvirus 6 (HHV-6) has been identified in brain tissues from bone marrow transplant recipients with encephalitis as well as from healthy individuals, but its etiological role in CNS infections remains poorly defined (36, 148). The combination of serological studies and CSF PCR for varicella-zoster virus (VZV) has been particularly helpful in identifying cases of VZV CNS infections without associated rash (sine herpete) (15). In HIV-infected patients, CSF PCR for VZV DNA may have utility in monitoring the therapeutic response and in predicting the outcome of VZV meningoencephalitis (28).

Identification of other viruses is pathognomonic of infection. JC virus, the etiologic agent of progressive multifocal leukoencephalopathy (PML), causes a subacute to chronic demyelinating disease. PML occurs almost exclusively in immunocompromised individuals, especially AIDS patients. Diagnosis of this infection has been facilitated greatly by development of a specific JC virus PCR (4, 150). West Nile virus has emerged in recent years as an important cause of encephalomyelitis in transplant patients, with infection occurring either due to implantation of tissue from an asymptomatically infected donor

(60) or due to natural infection following transplantation (70, 147).

The differential diagnosis of mass or space-occupying lesions in the immunocompromised host also differs from that for the immunocompetent counterpart. Cerebral toxoplasmosis is the most common CNS mass lesion in AIDS patients and may also be found among transplant patients, particularly following cardiac transplant. This disease is typically associated with reactivation of latent infection, although cases of cerebral toxoplasmosis have been reported for AIDS patients who are seronegative for *Toxoplasma gondii* (102). Granulomatous amoebic encephalitis in immunocompromised hosts is caused by *Acanthamoeba* species, free-living parasites that cause CNS infection characterized by multiple mass lesions throughout the brain (110). CNS microsporidiosis has been reported for two symptomatic AIDS patients (158). Mass lesions due to fungi are relatively common in both transplant recipients and HIV-infected hosts. Aspergillosis is by far the most frequent infection, but members of the class *Zygomycetes*, including the genera *Mucor*, *Rhizopus*, *Absidia*, and *Cunninghamella*, are also important causative agents (23, 123). Endemic fungi are less frequent causes of mass CNS lesions in this population. CNS cryptococcoma can masquerade as a brain abscess (101).

Bacterial causes of CNS mass lesions include brain abscesses (either polymicrobial or due to a single agent), *Nocardia* infection, and tuberculoma. While the risk of brain abscess is not appreciably higher in immunocompromised hosts, the last two etiologies are rarely found in individuals with intact immunity.

Myelitis

Clinical manifestations of myelitis include a variable combination of sensory loss, motor weakness, and bladder, bowel, and sexual dysfunction. Myelitis is often seen in combination with encephalitis or meningitis and can present as either an acute or chronic condition.

Acute myelitis is typically caused by viruses. CMV causes a distinctive syndrome of acute lumbosacral polyradiculopathy in AIDS patients, with progressive sensory and motor neuropathy (79). Characteristic CSF findings include neutrophilic pleocytosis, hypoglycorrhachia, and elevated CSF protein levels, but diagnosis is confirmed by PCR amplification of CMV from the CSF (82). Additional viral agents associated with myelitis in HIV infection include HSV-2, VZV, and HIV itself during acute infection (34).

Most cases of chronic viral myelopathy in HIV-infected patients are thought to be due to vacuolar myelopathy, a poorly understood entity characterized by lipid-laden macrophages causing microvacuolization of the spinal cord white matter (14). Human T-lymphocytic virus type 1 (HTLV-1)-associated myelopathy is a progressive myelopathy characterized by spasticity and lower extremity paraparesis, occurring in approximately 2% of individuals infected with HTLV-1. HIV coinfection may increase the incidence of HTLV-1-associated myelopathy, with symptomatic patients typically presenting with preserved CD4 T-cell counts (12, 93).

Infectious myelitis in transplant recipients can occur via transmission from an infected donor organ, via reactivation of latent infection, or via acute infection following transplantation. Subacute myelopathy due to HTLV-1 has been reported among multiple recipients of organs from an asymptomatically infected donor (47). Myelitis due to EBV, VZV, and CMV reactivation has rarely been reported for this population (10, 49, 127). There has been one report of HHV-7 identified in the CSF of an adult who presented with acute myelitis 11 months after bone marrow transplantation (149), but the significance of this finding remains uncertain. Acute flaccid paralysis due to West Nile virus has been reported for both SOT and stem cell transplant patients (70).

Immunocompromised Status and Syndromes

HIV/AIDS

HIV is a neurotropic virus, and CNS dysfunctions due to direct effects of HIV include aseptic meningitis in the acute retroviral syndrome, vacuolar myelopathy, and HIV-associated dementia. The risk of opportunistic infection among HIV-infected patients varies according to the level of immune impairment. The incidence and microbiology of CNS infections in individuals with CD4 T-cell counts of $>200/\mu l$ are similar to those seen for age-matched immunocompetent hosts. The risk of infection rises inversely with the decline of the CD4 count but may be modified by concurrent medications, particularly prophylaxis and highly active antiretroviral therapy. The most frequent HIV-associated CNS infections, i.e., PML, toxoplasma encephalitis, and cryptococcal meningitis, are uncommon among HIV patients with CD4 lymphocyte counts of $>100/mm^3$. Primary CNS lymphoma and CMV neurologic syndromes occur rarely in patients with CD4 counts of $>50/mm^3$ (76). Toxoplasma encephalitis is highly associated with reactivation of latent infection, but cases have been reported for AIDS patients who are seronegative for *Toxoplasma gondii* (4, 102). Trimethoprim-sulfamethoxazole prophylaxis, typically begun when the CD4 cell count declines below 200/mm^3, is protective against CNS infection due to *Toxoplasma*, as well as *Nocardia*, pneumococcus, and *Listeria*.

Transplantation

Infections in transplant recipients can occur through the following three mechanisms: (i) transmission from an infected donor organ, (ii) reactivation of latent infection, and (iii) acute infection following transplantation. Other important considerations in this population include type of organ transplanted, degree and length of immunosuppression, prophylactic medications, and time since transplantation.

The spectrum of infections acquired from infected donor tissues has expanded in the last decade. CMV disease acquired by transplantation of tissue from a seropositive donor to a seronegative recipient remains the primary concern, but CNS infections due to HTLV-1 (12), West Nile virus (60), rabies virus (128), and lymphocytic choriomeningitis virus (33) have recently been reported. These infections typically become apparent in the first month after transplantation.

Reactivation of dormant infection due to immunosuppression is associated with CNS infections due to CMV, HSV, VZV, *T. gondii*, and *Mycobacterium tuberculosis*. Because infection may have been acquired initially in the distant past, a thorough history focusing on geographic exposure to endemic fungi is important. The risk of reactivation or new acquisition of infection is modified by the use of prophylactic antimicrobial agents active against these various agents. Symptomatic infection due to reactivation of latent infection peaks during the first through sixth months following transplantation.

The specific organ transplanted is also an important determinant of CNS infection. The incidence of CNS infection is 0.5% for lung, 4% for heart, 6% for liver, and 2 to 8% for bone marrow transplant recipients (123). Among the latter group, autologous transplant recipients have the lowest risk of infection, as they have quicker resolution of neutropenia and require less immunosuppressive therapy, followed in increasing risk of infection by those receiving HLA-matched related-donor transplants, those receiving HLA-matched unrelated-donor transplants, and those receiving HLA-mismatched unrelated transplants.

The time elapsed since transplantation is also key in assessing the risk of various posttransplant CNS infections. In hematopoietic stem cell transplant recipients, the posttransplantation course can be divided into the following three time periods: the first month after transplant (preengraftment), the period between 30 days and 100 days posttransplant (postengraftment), and >100 days posttransplant (late phase) (3). CNS infections are relatively rare in the preengraftment stage. In the postengraftment phase, the immune system is impaired mainly by the extent of graft-versus-host disease and the degree of immunosuppressive therapy. Herpesviruses, especially CMV, and *Aspergillus* species are common pathogens in this period, but CNS involvement remains relatively infrequent. The incidence of CMV reactivation has been reduced markedly by laboratory surveillance and preemptive therapy. A French study reported cerebral toxoplasmosis as the most frequent cause of CNS infection in the hematopoietic stem cell transplant population, with a median time of onset of 46 days posttransplantation (33), but this is likely biased by the overall higher incidence of toxoplasmosis in that country. During the late posttransplant phase, cutaneous and visceral infections with CMV and VZV are common, but the CNS is typically spared. EBV-related posttransplant lymphoproliferative disorder is an important entity in this period, particularly if the allograft is T cell depleted (143).

Infections in SOT recipients follow a similar pattern, with the type of infection varying during the first month, the subsequent 1 to 6 months after transplantation, and more than 6 months after transplantation. CNS infections are relatively rare in the first month after SOT (96), although nosocomially acquired CNS aspergillosis is a concern in this time frame (84, 123). The degree of immunosuppression is usually highest in the second period. Fungal infections, including reactivation of latent endemic fungi, remain a concern in this phase (17). Reactivation of *Toxoplasma* is a concern during the second period after transplantation, particularly among cardiac transplant patients (41, 108). *Nocardia* infection can cause CNS diseases in SOT patients, with cases occurring most often between 1 and 6 months posttransplant (50, 94).

The overall risk of opportunistic CNS diseases declines at 6 months posttransplant, with some notable exceptions. Cryptococcal meningitis remains a risk throughout the late posttransplant period if immunosuppressant medications, including steroids, are continued (157). EBV-related posttransplant lymphoproliferative disorder peaks in the first year posttransplant (126). The peak incidence of PML, due to JC virus infection of the brain, occurs at 17 months post-SOT (122). *Listeria* meningitis, a food-borne infection, has been reported to occur both within the first 6 months after transplantation (129) and in the late posttransplant period (90).

Steroid use

Glucocorticoids have many effects on both innate and acquired immunity. The attributable risk of infection associated with steroids is difficult to quantify because it varies based on the patient's underlying medical problem and concurrent immunosuppressants. Despite this difficulty, several CNS infections are seen more commonly among patients taking corticosteroids than in immunocompetent individuals, including *Listeria* meningitis (97), coccidioidomycosis (113), and cryptococcal meningitis (95). Cerebral aspergillosis has been reported among patients on short courses of corticosteroids (83).

Other immunocompromising conditions

Diabetes mellitus, renal failure, liver failure, splenectomy, and malignancy all impair specific components of the immune system and have been associated with unusual CNS infections. Zygomycosis is an uncommon but often fatal infection caused by fungi of the class *Zygomycetes*. Rhinocerebral mucormycosis, a type of zygomycosis, is more common in diabetic patients, and ketoacidosis is the most important risk factor (46). Although group B streptococcus is an infrequent cause of meningitis in adults, in one large case series 19% of adults with group B streptococcus meningitis were diabetic (35). Other CNS infections seen more commonly in patients with alcohol dependence include *Campylobacter fetus* subsp. *fetus* meningitis (37) and *Listeria monocytogenes* meningitis or rhomboencephalitis (90). Malignancy is a recognized risk factor for toxoplasma encephalitis (59) and for cryptococcal meningitis (95). Patients with brain tumors or neurosurgical procedures, such as resection of CNS metastases, are at particular risk for meningitis or ventriculitis due to barrier disruption, neutrophil dysfunction, and corticosteroid use (104).

Prognostic Factors and Therapeutic Considerations

Prognostic factors and treatment regimens vary based on the specific pathogen involved and are discussed in detail in the relevant chapters. As a general rule, the risk for CNS infections and morbidity associated with infection correlate linearly with the degree of host immunosuppression. In addition to identification and treatment of the specific pathogen, therapy often revolves around bolstering the immune system. Consideration of exogenous granulocyte colony-stimulating factor is indicated among neutropenic patients with fungal infection of the CNS. Whenever possible, doses of immunosuppressive medications should be minimized, and glycemic control should be optimized in diabetic patients. In HIV patients with CNS infection, the antiretroviral regimen should be optimized to maximize the host immunologic response. Pretransplant screening of both donors and recipients for CMV and other viruses is important to identify latent infection and prevent new infection in naïve hosts.

DIAGNOSTIC APPROACHES AND INTERPRETATION

Nonmicrobiologic Diagnostic Approaches

Immunocompromised patients presenting with signs or symptoms of CNS disease should have emergent neuroimaging. While magnetic resonance imaging (MRI) is more sensitive than computed tomography (CT) scanning, it is also often more difficult to get on an emergent basis, and if there is a clinical concern for meningitis, radiologic evaluation should not be delayed. A noncontrast CT scan should always be performed on immunocompromised patients prior to lumbar puncture (LP), and empiric antibiotic therapy to cover the most likely bacterial pathogens should be given while awaiting the results (139).

Radiologic patterns may be helpful in generating a differential diagnosis for CNS infections in immunocompromised hosts. The presence of intracerebral mass lesions in an HIV-infected patient is suggestive of either toxoplasmosis or primary CNS lymphoma. Multiple lesions with surrounding vasogenic edema are more consistent with a diagnosis of toxoplasmosis but are not pathognomonic. Differentiation of these two entities requires assessing *Toxoplasma* serostatus, evaluating adherence to prophylactic antibiotic regimens with activity against *Toxoplasma*, and monitoring the radiologic response to empiric antiprotozoal therapy (4). In some cases, a diagnosis can be made only after biopsy of the involved area.

PML typically causes a focal CNS lesion without a significant mass effect or enhancement. It often occurs at the gray-white junction, with sparing of the cortical gray matter, and may have a scalloped appearance on MRI (11). Both HIV and CMV can cause diffuse parenchymal disease. Atrophy, white matter abnormalities, and periventricular hyperintensity on MRI are often seen with both conditions, and it is often difficult to distinguish between these two by neuroimaging (57). CNS manifestations of tuberculosis are diverse, but tuberculous meningitis is the most frequent form, with neuroimaging findings including basilar meningeal enhancement, hydrocephalus, and infarctions in the supratentorial brain parenchyma and/or brain stem (16). Radiologic findings in neurosyphilis include brain atrophy (37%), cerebral infarction (23%), and nonspecific white matter lesions (20%) (21). Almost one-third of patients with neurosyphilis have normal MRI scans.

Microbiologic Diagnostic Approaches

Technological revolutions in microbiology, immunology, and molecular biology have significantly expanded and improved the capabilities of diagnostic microbiology for CNS infections. The sensitivities of various microbiologic tests vary based on both the specific pathogen and the anatomic location of infection. As with other organ systems, CNS infections can be identified in multiple ways, including (i) cultivation of microorganisms by use of artificial media or living hosts, (ii) direct microscopic examination, (iii) measurement of microorganism-specific immune responses, and (iv) detection of microorganism-specific macromolecules, especially nucleic acids (73, 131, 159). These techniques are summarized in Table 1.

Sample collection and processing

Prompt and appropriate sample collection, transport, and processing are critically important in the diagnosis of CNS infections. The urgency is dictated by the lethality of acute bacterial and viral infections if left untreated, the morbidity if inadequately treated, and the curability if treated early with appropriate antimicrobial and antiviral agents. The yield of fungal and mycobacterial stains and cultures is improved with a larger volume of CSF (10 to 20 ml) (81, 138). Multiple specimens may be required to make a diagnosis. For instance, the sensitivity of CSF smear and culture for *M. tuberculosis* increases from 37% to 87% when four sequential examinations of CSF are performed (68).

Ideally, CSF for microbiologic studies should be placed in separate containers from those needed for other studies to expedite processing and to minimize contamination. CSF specimens should be delivered to the laboratory immediately after LP. Concentrations of CSF neutrophils degrade by 32% by 1 hour post-LP and by 50% by 2 hours post-LP (130). Several nucleic acid extraction kits, including the QIAamp MinElute virus kit, can effectively extract viral nucleic acids from CSF (121). If it is not rapidly processed, CSF should be incubated at 35°C or left at room temperature. The exception to this rule involves CSF for viral culture and molecular tests. These specimens may be refrigerated for less than 24 hours after collection or frozen at −80°C if a longer delay is anticipated.

If a diagnosis cannot be made through noninvasive testing, stereotactic brain biopsy should be considered. The yield of this procedure is highest when there is sampling of an area with a discrete radiologic abnormality. In one study of stereotactic biopsy among patients with HIV/AIDS, tissue evaluation was diagnostic in up to 98% of cases, and biopsy results changed treatment in as many as 88% of cases (44). Brain biopsy carries an inherent risk of bleeding, infection, or neurologic sequelae and may not be an option if a lesion is located in a critical anatomic area.

Direct microscopic examination

Although there is no single laboratory test of CSF that reliably distinguishes between bacterial and viral infections, the Gram stain, if positive, is

Table 1. Microbiological methods used for laboratory diagnosis of CNS infections

Test	Turnaround time	Result interpretation	Advantages	Disadvantages
Direct examination	1–3 h	Direct if correlated with symptoms	Rapid	Poor sensitivity and specificity; special skills are needed for interpretation
Culture	2–28 days	Definite	Phenotypic drug susceptibility testing	Time-consuming; poor sensitivity; limited to microorganisms that are culturable
Serology	4–6 h	Indirect	Automation	Results are generally retrospective; immunosuppressed hosts may be unable to mount a response; nonspecific cross-reactions
Molecular diagnostic test	1–2 days	Direct without knowing microbial viability	High sensitivity and specificity	Facility requirement; false-positive results due to carryover contamination and false-negative results due to inhibitors in specimen

virtually 100% diagnostic of bacterial meningitis. False-positive Gram stain results have, on occasion, been reported (89). Microscopic examination of a Gram-stained smear of CSF sediment obtained after cytocentrifugation provides a rapid and accurate identification of the causative microorganism for 60 to 90% of patients with bacterial meningitis and has a specificity of nearly 100%. The sensitivity depends on the bacterial pathogen: microorganisms have been observed by Gram stain in as many as 90% of infections with *Streptococcus pneumoniae* and in <50% of infections with *Listeria monocytogenes* (89). Rapid diagnosis may occasionally be made by visual detection of microorganisms by acid-fast bacillus (AFB) stain, modified AFB stain, dark-field examination, wet mount, and India ink stain. Amoebae are best observed by examining thoroughly mixed sediment as a wet preparation by phase-contrast microscopy. Latex agglutination or enzyme immunoassay testing for *Cryptococcus neoformans* capsular antigen is more sensitive than staining and extremely specific, such that antigen testing has largely replaced the India ink stain as a rapid diagnostic test for cryptococcal meningitis (27).

Rapid antigen testing

Rapid diagnosis may be made by specific antigen testing. Bacterial antigen testing with CSF has not been demonstrated to be clinically useful due to poor sensitivity and specificity (55, 98, 137). False-positive results may occur due to serologic cross-reactivity. One exception is that an immunochromatographic test for *S. pneumoniae* (Binax Now *Streptococcus pneumoniae* test; Binax Inc., Portland, ME) has been demonstrated to be rapid and highly sensitive for the diagnosis of meningitis (77, 115). In the early stages of infection, bacterial antigen may not leech into the surrounding CSF, rendering erroneously negative results. In contrast, detection of cryptococcal antigen in either CSF or sera of patients with HIV is both highly sensitive and specific (27), and a positive antigen test would be an indication to begin directed antifungal therapy. However, some strains of *C. neoformans* that infect patients with AIDS may not possess detectable capsules, so culture is also essential. Detection of *Aspergillus* galactomannan antigen in the CSF is suggestive of cerebral aspergillosis (145) and serves as a marker of the therapeutic response (75, 146). *Histoplasma* antigen testing of CSF has a sensitivity ranging from 38 to 67%, with a CSF-to-urine result ratio of >1 suggestive of intrathecal antigen production (154).

Culture

CSF culture is useful for detection of many bacterial, mycobacterial, and fungal causes of meningitis. The yield of CSF culture for bacterial meningitis is 70 to 85% for the untreated patient (19). Standard recommendations are to inoculate a 5% sheep blood agar plate, enriched chocolate agar, and enrichment broth for patients who have shunts or hardware. Viral cultures from CSF may be performed using several different cell lines. Cells are evaluated daily for cytopathic effect, and the findings can be confirmed by immunofluorescence staining. Culture possesses limited sensitivity in identification of viral pathogens in CSF (100). Culture sensitivity for enteroviruses from CSF is only 65 to 75% and requires a mean of 3.7 to 8.2 days for detection of cytopathic effect (52, 86, 111). Some serotypes of enteroviruses, especially coxsackievirus A strains, grow very poorly in tissue culture or are noncultivable (1, 31). HSV culture sensitivity from CSF is very poor, especially in recurrent cases of HSV-2 meningitis (133, 135).

A positive bacterial culture allows subsequent determination of the antimicrobial susceptibility of the causative agent. While the phenotypic techniques used for antimicrobial susceptibility testing for CNS infections remain the same, criteria for determining antibiotic resistance can differ from those used for other systems. Penicillin MIC breakpoints for pneumococcal CNS isolates, for example, are different from those used for non-CNS isolates.

Serology

Neurosyphilis may be diagnosed based on a positive serum rapid plasmin reagin test, enzyme immunoassay, immunoblot, or CSF VDRL test (66). While a positive serologic test for toxoplasma is not necessarily indicative of active infection, it is useful for guiding empiric therapy in a susceptible host with a compatible clinical picture. West Nile virus infection of the CNS is best diagnosed by identification of specific immunoglobulin M (IgM) antibody in the CSF (70). For other organisms, serologic testing has a limited role in the diagnosis of CNS infection. For example, enteroviruses lack a common antigen among various serotypes, making IgM levels inconsistent in sensitivity (56, 106). Local production of HSV IgG antibodies in CSF can be used for diagnosis; however, the presence of antibody is delayed until day 10 or 12 of infection and peaks at about day 20, delaying diagnosis beyond the time when treatment is beneficial (116).

To differentiate between peripheral and CNS infections, antibody titers in serum and CSF can be measured in parallel. With a peripheral infection, the antibody titer in the CSF should be <5% that in the peripheral blood, due to the protective effect of the BBB. Antibody titers in CSF of >5% those in serum are suggestive of intrathecal antibody synthesis, but peripheral blood contamination or BBB inflammation may skew this ratio and falsely suggest CNS infection. Examples of infections where the intrathecal antibody index may be useful for diagnosis include *Borrelia burgdorferi* infection and, if late in the course of infection, VZV infection (91).

Molecular assays

With the advent of molecular diagnostic clinical laboratory procedures, it is now possible to diagnose viral etiologies specifically, rapidly, and with high sensitivity and specificity. PCR has become the primary diagnostic modality for detection of several types of CNS infections, best exemplified by tests for herpesviruses and enteroviruses (125, 135). PCR analysis of HSV DNA in CSF represents the method of choice for the diagnosis and follow-up of treatment of herpes encephalitis in immunocompetent patients (103, 112). PCR amplification of viral nucleic acid in the CSF is sensitive and specific, and test turnaround time has been shortened significantly, to just a few hours, by incorporating either colorimetric enzyme immunoassay or real-time detection methods (39, 136). EBV DNA can be used as a tumor marker in cases of AIDS-related primary lymphoma of the CNS (5). The sensitivity and specificity of these assays for detecting EBV DNA in the CSF of persons who have primary CNS lymphomas are both very high (90% to 100%) (8, 72). PCR of CSF is more reliable than clinical features in the diagnosis of CMV-related neurological disease in patients with AIDS (48). One commercially available CMV PCR method for the diagnosis of neurologic CMV infections showed a sensitivity and specificity of 95 and 100%, respectively, versus an in-house PCR assay, indicating its utility in this clinical setting (153). PCR analysis allows recognition of both overt VZV disease of the CNS and subclinical reactivation of VZV infection in HIV-infected patients (24).

Enterovirus infections often do not require antiviral intervention, but the ability to rapidly differentiate enterovirus meningitis from bacterial illness has the potential for reductions of hospital stays and health care costs. Molecular assay allows rapid detection of enterovirus RNA in CSF, with excellent sensitivity and specificity (105). Assuming a mean time of 24 hours to diagnosis by PCR, between 1.2 and 2.8 days of hospitalization per patient could be saved over current averages by routine use of PCR (1, 106, 120). PCR systems amplify the highly conserved 5′-nontranslated region of the viral genome, detecting 60 of the 67 enterovirus serotypes, including the serotypes most commonly associated with meningitis (1, 111). In comparisons with viral culture, enterovirus PCR is more accurate, with a sensitivity of virtually 100% (107, 111). One study has shown that at least two-thirds of all CSF specimens from suspected aseptic meningitis patients who had negative enterovirus cell culture had positive results by reverse transcription-PCR (RT-PCR) (117).

Nucleic acid-based assays have been used successfully for detection of other pathogens in the CSF. Detection of JC virus DNA in CSF has been proven to be both sensitive and specific for the diagnosis of PML, a demyelinating disease of the CNS seen primarily in AIDS patients (40). RT-PCR has been used for the diagnosis of mumps virus CNS disease and may soon become the "gold standard" test for the diagnosis of this condition (99). While IgM serology is accepted as the test of choice for West Nile encephalitis, a real-time PCR method used to analyze specimens obtained during the 1999 New York outbreak detected the presence of West Nile virus sequences in CSF of all four individuals with fatal outcomes and in one of four who survived (20). PCR-based assays have been demonstrated to be useful in the diagnosis of meningitis, tuberculoma, brain abscess, myelopathy, and radiculopathy due to *M. tuberculosis*, although the sensitivity may be low (118). It has been reported that PCR-based assays yielded positive results from CSF for up to 40% of patients with an early stage of Lyme meningitis (92).

In certain situations, quantitative PCR can be used to monitor the response to treatment or progression of disease (7, 69). Quantitation of HHV-6 genomes in CSF may help to evaluate the viral replication level and the implication of disease (42). Molecular assays have been useful for detecting and confirming microorganism coinfections in the CNS (133, 152). PCR followed by direct sequencing has been used to detect mutations in the thymidine kinase gene of HSV and the phosphotransferase gene of CMV, which are related to acyclovir and ganciclovir resistance, respectively (26, 38).

Interpretation and Reporting of Results

Use of laboratory testing for prognosis, therapeutic decision-making, and monitoring varies based on the organism and syndrome and is covered in detail in the organism-specific chapters. Infectious agents causing meningitis, encephalitis, and myelitis among immunocompromised patients and their diagnostic methods are listed in Tables 2, 3, and 4.

Table 2. Major infectious causes of meningitis among immunocompromised hosts

Agent	Immune deficit or affected population	Diagnostic test
Viruses		
Enterovirus	Hypogammaglobulinemia	CSF PCR
HIV	HIV (acute infection)	RNA viral load
Bacteria		
S. pneumoniae	Hypogammaglobulinemia	CSF culture, antigen testing
L. monocytogenes	T-cell impairment	CSF culture
T. pallidum	HIV infection	CSF VDRL test
M. tuberculosis	T-cell impairment	*M. tuberculosis* CSF PCR, CSF mycobacterial culture
Fungi		
Cryptococcus neoformans	T-cell impairment	Cryptococcal antigen test (CSF), CSF fungal culture
H. capsulatum	T-cell impairment	CSF *Histoplasma* antigen test, CSF fungal culture
B. hominis	T-cell impairment	Urine *Blastomyces* antigen tests, CSF fungal culture
C. immitis	T-cell impairment	Serology, CSF fungal culture

Factors associated with a poor outcome among a cohort of nonimmunocompromised adults with community-acquired meningitis (4% with *Listeria*) included advanced age, low Glasgow coma scale value at admission, bacteremia, and low CSF white blood cell count (142). Indications for serial LP include infection with cephalosporin-resistant pneumococcus, shunt or ventriculostomy infection, and culture-negative meningitis to assess resolving inflammation with treatment.

Table 3. Infectious causes of encephalitis or mass lesions of the brain among immunocompromised hosts

Agent	Immune deficit or affected population	Diagnostic test
Viruses		
EBV[a]	HIV infection	EBV CSF PCR
HHV-6	Transplant recipients	HHV-6 CSF PCR
CMV	T-cell impairment	CMV CSF PCR
Rabies virus[b]	SOT recipients	Rabies CSF antibody test, rabies saliva PCR, DFA with nape-of-neck biopsy
Lymphocytic choriomeningitis virus[b]	SOT recipients	Serology
West Nile virus[b]	SOT or bone marrow transplant recipients	CSF IgM antibody test
JC virus	HIV infections transplant recipients	JC virus CSF PCR
Bacteria		
L. monocytogenes	T-cell impairment	CSF culture
T. pallidum	HIV infection	CSF VDRL test
M. tuberculosis	T-cell impairment	CSF PCR, mycobacterial culture, histopathology for tuberculoma
Nocardia	Neutropenia, transplant recipients	CSF culture, stain
Fungi		
Cryptococcus neoformans	T-cell impairment	Cryptococcal antigen test (CSF), CSF fungal culture
H. capsulatum	T-cell impairment	CSF *Histoplasma* antigen test, CSF fungal culture
B. hominis	T-cell impairment	Urine *Blastomyces* antigen test, CSF fungal culture
C. immitis	T-cell impairment	Serology, CSF fungal culture
Aspergillus spp.	Neutropenia, T-cell impairment	Culture, serum galactomannan assay, histopathology
Zygomycetes	Neutropenia, diabetes mellitus	Culture, histopathology
Protozoa		
Toxoplasma	HIV infections, transplant recipients	Histopathology, *Toxoplasma* antibodies (serum)
Acanthamoeba, *Naegleria fowleri*	HIV infections, transplant recipients	Cytology, histopathology, CSF wet mount

[a]Primary CNS lymphoma.
[b]Cases associated with transplant of solid organ from infected donor.

Table 4. Infectious causes of myelitis among immunocompromised patients

Agent	Immune deficit or affected population	Diagnostic test
Viruses		
West Nile virus[a]	Solid organ transplant recipients	CSF serology (IgM)
CMV	HIV infection, transplant recipients	CMV CSF PCR
HIV	HIV infection	Histopathology
EBV	Transplant recipients	EBV CSF PCR
VZV	HIV infection, transplant recipients	
HTLV-1	Transplant more than HIV infection	
Bacteria		
T. pallidum	HIV infection	RPR serum, CSF VDRL test
M. tuberculosis	HIV infection	Mycobacterial CSF PCR, mycobacterial culture, histopathology?

[a]Cases associated with transplant of solid organ from infected donor as well as acquired infection in previously immunocompromised host.

Meningitis must be excluded through CSF evaluation for any HIV-infected patient with positive blood cultures for *Cryptococcus* or elevated serum cryptococcal antigen (114). Elevated intracranial pressure is an important prognostic factor in this disease. Therefore, a follow-up LP 2 weeks after initiation of therapy to rule out increased pressure is recommended if the baseline opening pressure is <200 mm H_2O (114). If the baseline opening pressure is >250 mm H_2O, lumbar drainage to achieve a closing pressure of ≤200 mm H_2O should be performed (114). Daily LP with lumbar drainage should be performed until the CSF opening pressure is demonstrated to be within the normal range for several consecutive days. The degree of elevation of the cryptococcal antigen is also negatively associated with sterilization of the CSF with treatment (109). However, despite the inverse association between elevated cryptococcal antigen and outcome, serial monitoring of this measure to assess the response to antifungal agents is not beneficial for HIV patients with cryptococcal meningitis (6).

There are scant data on the use of laboratory testing to guide therapy or prognosis among patients with encephalitis. Much of the literature in this regard is pathogen specific, although 70% of cases remain undiagnosed despite extensive laboratory evaluation (45). Herpes simplex encephalitis (HSE) is the most frequent cause of endemic encephalitis in the United States. Although the incidence of HSE in immunocompromised patients does not differ from that in the immunocompetent host, infection among immunocompromised patients is often atypical, and the outcomes may be poorer (155). CSF PCR for HSV-1 is the gold standard for diagnosis of HSE, but false-negative results can occur early in the disease course. For patients with a high clinical suspicion of HSE based on temporal lobe focality, empiric therapy with acyclovir should be begun immediately and continued until a second CSF PCR 4 to 7 days later is negative (151, 155).

Immunosuppression and advanced age have been identified as poor prognostic features in West Nile virus encephalitis (18, 88). There are no laboratory markers that serve as prognostic indicators. Because West Nile virus IgM can remain positive for up to 500 days following CNS infection, there is no role for serial serology in assessing resolution of illness (110).

Toxoplasmic encephalitis typically occurs as reactivation in AIDS patients. Although it is recommended that HIV-infected individuals have antibodies to *Toxoplasma* assessed both at baseline and at the time of presentation with a CNS mass lesion, the sensitivity of serology is imperfect. Definitive diagnosis requires a pathologic review of tissue, but a radiology response after 2 weeks of therapy often serves as an indirect marker of infection (2). While not widely available, CSF PCR may have a role in diagnosis of *Toxoplasma* encephalitis among patients with AIDS in resource-limited settings, with a sensitivity as high as 67% (64).

Factors associated with prolonged survival among AIDS patients with PML include receipt of highly active antiretroviral therapy, a high CD4 cell count at time of diagnosis, an increase of CD4 cell count of >100 cells/μl, low HIV load, low JC virus levels in CSF, clearance of JC virus from CSF, and lack of neurologic progression 2 months after diagnosis (124). Therefore, serial sampling of CSF with quantitative JC virus PCR may play a role in assessment of prognosis with this disease.

Because syphilis may progress rapidly among HIV-infected individuals, the most recent treatment guidelines by the Centers for Disease Control and Prevention recommend CSF examination for HIV-infected patients with either late latent syphilis or syphilis of unknown duration (156). Patients with neurosyphilis should be monitored by serological evaluation 6, 12, 18, and 24 months after therapy. If, at any time, clinical symptoms develop, nontreponemal titers rise fourfold or more, or nontreponemal titers do not decline fourfold over 12 to 24 months, LP should be repeated and treatment administered accordingly (156).

ALGORITHMS FOR DIAGNOSIS, PROGNOSTICATION, AND PREDICTION OF RESPONSE TO THERAPY

The differential diagnosis of CNS infections in immunocompromised hosts is very broad, with crucial considerations including acuity of presentation, degree of immunocompromise, serostatus of the recipient and donor for organ transplants, use of prophylactic medications, and epidemiologic factors. The diagnostic algorithms presented in Fig. 1, 2, and 3 must be tempered by knowledge of the specific clinical presentation and epidemiology. Furthermore,

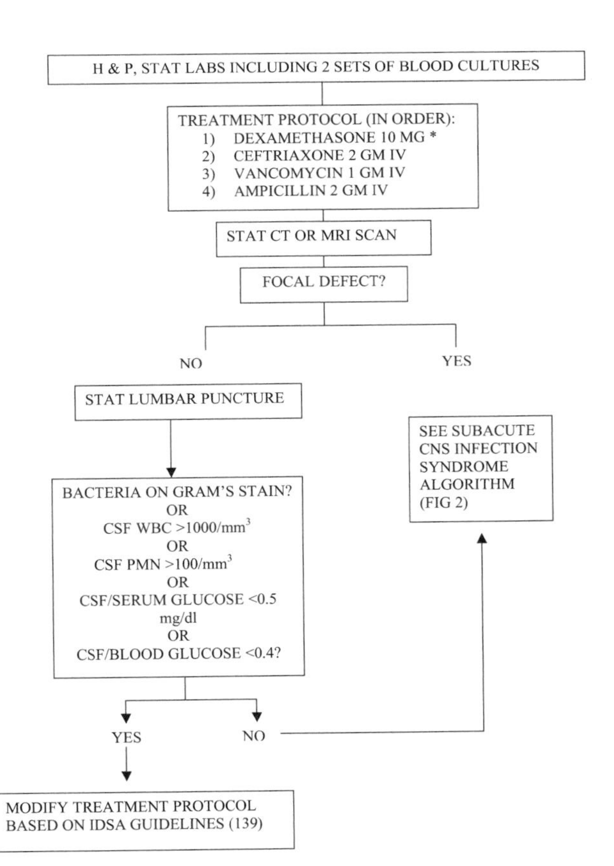

Figure 1. Management of immunocompromised adults with acute meningitis syndrome (fulminant course [<48 h], with fever, headache, impaired sensorium, or stiff neck). *, cultures for bacteria (aerobic/anaerobic), mycobacteria, and fungi; **, for patients on chronic corticosteroids, increase to stress dosing.

while immunocompromised patients are at risk for a unique spectrum of opportunistic infections, they also retain susceptibility to the organisms most frequently isolated in their immunocompetent counterparts.

A critical component in the evaluation of any patient with clinical suspicion of CNS infection is CSF analysis. Neuroimaging is recommended for all patients prior to LP, as this procedure is contraindicated for patients with radiologic evidence of mass effect (139). Thrombocytopenia poses a theoretical risk of LP-associated bleeding complications, and prophylactic platelet transfusion is recommended for adult patients with platelet counts of $<20 \times 10^9$/liter (144).

All immunocompromised patients should have CSF analyses of cell count, differential count, and glucose and protein levels, CSF Gram stain and AFB stain, and cultures for bacteria, fungi, and mycobacteria. Neutropenic patients with bacterial meningitis may have normal cell counts and chemistries, and empiric therapy should be extended to this population until culture results return negative (74). Cytology should be sought if there is concern for primary CNS lymphoma or other malignancy. A wet mount is useful to evaluate the presence of motile amoebae if there is clinical concern for granulomatous amoebic encephalitis. CSF serology or antigen testing can be useful for diagnosing neurosyphilis (VDRL), CNS fungal diseases (cryptococcal antigen), and some CNS viral diseases (IgM for West Nile virus). If there is concern for an endemic fungal infection, additional testing should be considered. CSF and urine antigen tests have sensitivities of up to 67% and 71%, respectively, for histoplasmosis (154). Detection of intrathecal antibody to *Coccidioides* by either complement fixation or immunodiffusion is diagnostic of this infection, but the sensitivities of these tests on CSF are lower than those with serum, and testing of both specimens should be performed in clinically suggestive cases (63).

Historically, identification of a microbiologic agent of CNS infection has been hindered by the low yield of CSF culture for viral and fastidious bacterial organisms, delays in production of organism-specific antibodies, and difficulties in obtaining brain biopsies. Viral cultures are of limited utility and, in general, rarely add useful information (100). Molecular methods performed on CSF are recognized as the new "gold standard" for some of the infections caused by microorganisms that are difficult to detect and identify, including HSV-1, HSV-2, VZV, CMV, EBV (in primary CNS lymphoma), HHV-6, enteroviruses, JC virus, and *Ehrlichia* and *Bartonella* species. Results of molecular diagnostic testing for

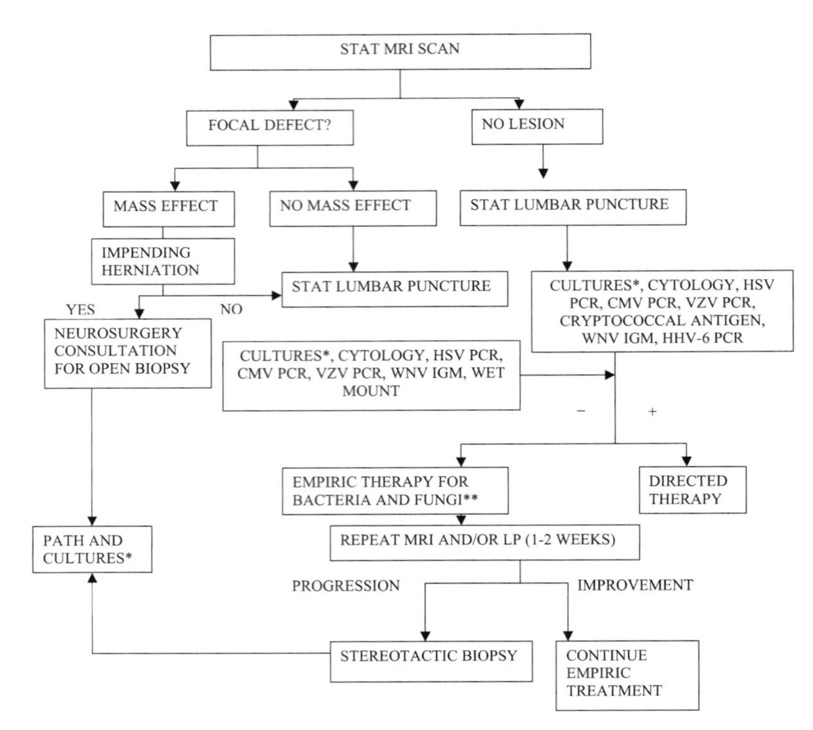

Figure 2. Management of patients with HIV/AIDS and subacute CNS infection syndrome (subacute illness [>3 to 7 days], with variable presence of fever, headache, progressive impairment of cognitive function, and/or focal defects). *, cultures for bacteria (aerobic/anaerobic), mycobacteria, and fungi; **, treatment decisions should factor in radiologic appearance, toxoplasma serology, and prophylaxis with trimethoprim-sulfamethoxazole. See the text for details.

CNS infections must be interpreted in the context of the individual patient presentation and clinical illness, and close cooperation between the laboratory and the clinician is required for optimal use of these new technologies. Since the volume of CSF may be limited, determination of which tests to order for an individual patient depends on epidemiologic (season and geographic location), demographic (age), immune status, and other clinical and radiologic data.

With implementation of new advanced assays, especially molecular assays, there is a tendency to request multiple tests in an effort to obtain a diagnostic result. The vast majority of these tests yield negative results, so cost-effective means of screening specimens to qualify them for PCR analyses are worth exploring, and a positive PCR should be interpreted in the context of the clinical presentation (32, 80, 87). For immunocompetent hosts, eliminating PCR tests for CSF samples with normal leukocyte and protein levels would save almost one-third of the cost associated with herpesvirus testing without having a significant impact on patient care (134). A logistic regression analysis of 2,233 consecutively tested CSF samples showed that fever, a virus-specific rash, and a CSF white cell count of 5/μl or more were independent predictors of a positive PCR result (62, 107). Whether this is true for immunocompromised hosts merits further investigation. An elevated CSF protein level was not a good predictor of RT-PCR positivity for enterovirus, and for patients aged >2 months, the absence of pleocytosis was highly predictive of a negative RT-PCR result (87). Clinicians are encouraged to use inexpensive screening tests, such as protein and leukocyte levels in CSF, prior to PCR testing and then to prioritize tests in order of the probability of a specific pathogen (53, 134). While it is important to obtain CSF cell count and chemistry tests because these may be supportive of an inflammatory process, until data are available regarding whether these inexpensive screening tests can be used to prioritize testing in the immunocompromised patient, the decision to perform nucleic acid-based tests should be based upon the probability of a specific pathogen.

The algorithms for management of immunocompromised adults with CNS infections do not take into account that additional testing of alternative specimens may be diagnostic. Serum antibody testing may provide important clues to reactivation of latent *Toxoplasma*, CMV, or *Coccidioides*. Additionally, paired-antibody testing is useful in identifying infection with rickettsias, ehrlichias, and arbo-

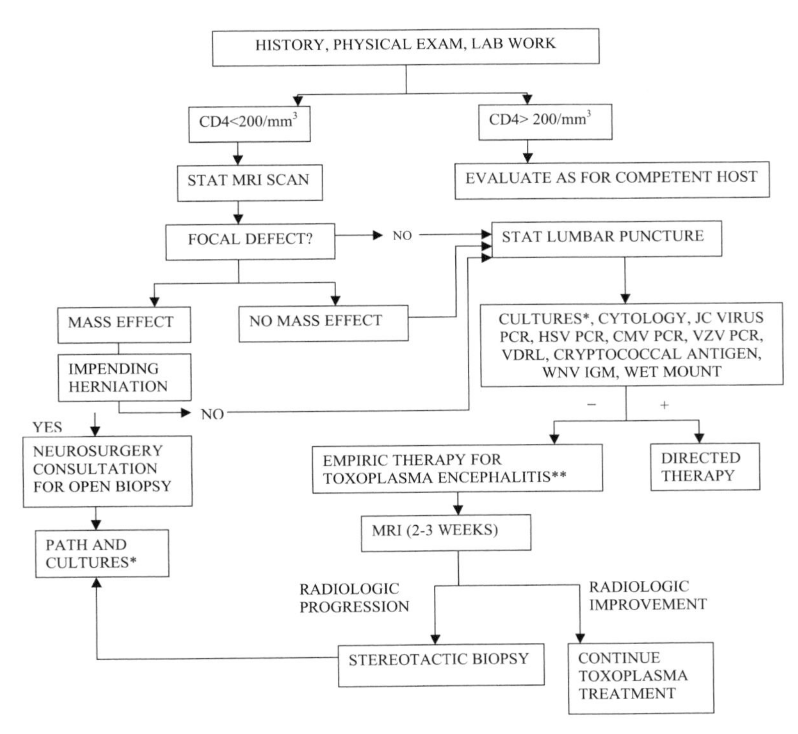

Figure 3. Management of transplant patients with subacute CNS infection syndrome (subacute illness [>3 to 7 days], with variable presence of fever, headache, progressive impairment of cognitive function, and/or focal defects). *, cultures for bacteria (aerobic/anaerobic), mycobacteria, and fungi; **, treatment decisions should factor in radiologic appearance, fungal antigen tests, and prophylactic antibiotic regimens. See the text for details.

viruses (58), although for the former two agents, treatment should not be delayed while awaiting confirmatory serology. Other potentially useful tests include serum antigen testing for *Cryptococcus*, urine antigen testing for *Histoplasma* or *Blastomyces*, *Aspergillus* galactomannan testing, nape-of-the-neck biopsy for detection of rabies virus direct fluorescent antigen, and purified protein derivative testing for exposure to *M. tuberculosis*. Additional sites of infection should be sought aggressively to further characterize any visualized abnormality that may yield the pathogen more quickly. For instance, for a patient with pulmonary symptoms, a CT scan of the chest followed by bronchoscopy or biopsy may more expeditiously yield an etiologic agent.

CNS infections remain important causes of morbidity and mortality in immunocompromised hosts and present a particular challenge for the clinician and microbiologist. While this chapter outlines many of the most common etiologic agents, with the emergence of new infections, the migration of infectious agents into new geographic niches, and the advent of an increasing array of medications with suppressive effects on the immune system, the field continues to evolve. Future goals for the clinical microbiologist include identifying diagnostic techniques to keep pace with this rapidly expanding field.

REFERENCES

1. **Ahmed, A., F. Brito, C. Goto, S. M. Hickey, K. D. Olsen, M. Trujillo, and G. H. McCracken, Jr.** 1997. Clinical utility of the polymerase chain reaction for diagnosis of enteroviral meningitis in infancy. *J. Pediatr.* **131:**393–397.
2. **Anonymous.** 1998. Evaluation and management of intracranial mass lesions in AIDS. Report of the Quality Standards Subcommittee of the American Academy of Neurology. *Neurology* **50:**21–26.
3. **Anonymous.** 2000. Guidelines for preventing opportunistic infections among hematopoietic stem cell transplant recipients. *MMWR Recommend. Rep.* **49:**1–125, CE1–CE7.
4. **Antinori, A., A. Ammassari, A. De Luca, A. Cingolani, R. Murri, G. Scoppettuolo, M. Fortini, T. Tartaglione, L. M. Larocca, G. Zannoni, P. Cattani, R. Grillo, R. Roselli, M. Iacoangeli, M. Scerrati, and L. Ortona.** 1997. Diagnosis of AIDS-related focal brain lesions: a decision-making analysis based on clinical and neuroradiologic characteristics combined with polymerase chain reaction assays in CSF. *Neurology* **48:**687–694.
5. **Antinori, A., G. De Rossi, A. Ammassari, A. Cingolani, R. Murri, D. Di Giuda, A. De Luca, F. Pierconti, T. Tartaglione, M. Scerrati, L. M. Larocca, and L. Ortona.** 1999. Value of combined approach with thallium-201 single-photon emission computed tomography and Epstein-Barr virus DNA polymerase chain reaction in CSF for the diagnosis

of AIDS-related primary CNS lymphoma. *J. Clin. Oncol.* **17:** 554–560.

6. **Antinori, S., A. Radice, L. Galimberti, C. Magni, M. Fasan, and C. Parravicini.** 2005. The role of cryptococcal antigen assay in diagnosis and monitoring of cryptococcal meningitis. *J. Clin. Microbiol.* **43:**5828–5829.
7. **Arribas, J. R., D. B. Clifford, C. J. Fichtenbaum, D. L. Commins, W. G. Powderly, and G. A. Storch.** 1995. Level of cytomegalovirus (CMV) DNA in cerebrospinal fluid of subjects with AIDS and CMV infection of the central nervous system. *J. Infect. Dis.* **172:**527–531.
8. **Arribas, J. R., D. B. Clifford, C. J. Fichtenbaum, R. L. Roberts, W. G. Powderly, and G. A. Storch.** 1995. Detection of Epstein-Barr virus DNA in cerebrospinal fluid for diagnosis of AIDS-related central nervous system lymphoma. *J. Clin. Microbiol.* **33:**1580–1583.
9. **Arribas, J. R., G. A. Storch, D. B. Clifford, and A. C. Tselis.** 1996. Cytomegalovirus encephalitis. *Ann. Intern. Med.* **125:**577–587.
10. **Au, W. Y., C. Hon, V. C. Cheng, and E. S. Ma.** 2005. Concomitant zoster myelitis and cerebral leukemia relapse after stem cell transplantation. *Ann. Hematol.* **84:**59–60.
11. **Bakshi, R.** 2004. Neuroimaging of HIV and AIDS related illnesses: a review. *Front. Biosci.* **9:**632–646.
12. **Beilke, M. A., S. Japa, C. Moeller-Hadi, and S. Martin-Schild.** 2005. Tropical spastic paraparesis/human T leukemia virus type 1-associated myelopathy in HIV type 1-coinfected patients. *Clin. Infect. Dis.* **41:**e57–e63.
13. **Berenguer, J., S. Moreno, F. Laguna, T. Vicente, M. Adrados, A. Ortega, J. Gonzalez-LaHoz, and E. Bouza.** 1992. Tuberculous meningitis in patients infected with the human immunodeficiency virus. *N. Engl. J. Med.* **326:**668–672.
14. **Berger, J. R., and A. Sabet.** 2002. Infectious myelopathies. *Semin. Neurol.* **22:**133–142.
15. **Bergstrom, T.** 1996. Polymerase chain reaction for diagnosis of varicella zoster virus central nervous system infections without skin manifestations. *Scand. J. Infect. Dis.* **100**(Suppl.)**:** 41–45.
16. **Bernaerts, A., F. M. Vanhoenacker, P. M. Parizel, J. W. Van Goethem, R. Van Altena, A. Laridon, J. De Roeck, V. Coeman, and A. M. De Schepper.** 2003. Tuberculosis of the central nervous system: overview of neuroradiological findings. *Eur. Radiol.* **13:**1876–1890.
17. **Blair, J. E., and J. L. Logan.** 2001. Coccidioidomycosis in solid organ transplantation. *Clin. Infect. Dis.* **33:**1536–1544.
18. **Bode, A. V., J. J. Sejvar, W. J. Pape, G. L. Campbell, and A. A. Marfin.** 2006. West Nile virus disease: a descriptive study of 228 patients hospitalized in a 4-county region of Colorado in 2003. *Clin. Infect. Dis.* **42:**1234–1240.
19. **Bohr, V., N. Rasmussen, B. Hansen, H. Kjersem, O. Jessen, N. Johnsen, and H. S. Kristensen.** 1983. 875 cases of bacterial meningitis: diagnostic procedures and the impact of preadmission antibiotic therapy. *J. Infect.* **7:**193–202.
20. **Briese, T., W. G. Glass, and W. I. Lipkin.** 2000. Detection of West Nile virus sequences in cerebrospinal fluid. *Lancet* **355:**1614–1615.
21. **Brightbill, T. C., I. H. Ihmeidan, M. J. Post, J. R. Berger, and D. A. Katz.** 1995. Neurosyphilis in HIV-positive and HIV-negative patients: neuroimaging findings. *Am. J. Neuroradiol.* **16:**703–711.
22. **Broom, J., M. Woods II, and A. Allworth.** 2006. Immune reconstitution inflammatory syndrome producing atypical presentations of cryptococcal meningitis: case report and a review of immune reconstitution-associated cryptococcal infections. *Scand. J. Infect. Dis.* **38:**219–221.
23. **Brown, J.** 2005. Zygomycosis: an emerging fungal infection. *Am. J. Health Syst. Pharm.* **62:**2593–2596.
24. **Burke, D. G., R. C. Kalayjian, V. R. Vann, S. A. Madreperla, H. E. Shick, and D. G. Leonard.** 1997. Polymerase chain reaction detection and clinical significance of varicella-zoster virus in cerebrospinal fluid from human immunodeficiency virus-infected patients. *J. Infect. Dis.* **176:**1080–1084.
25. **Casado, J. L., C. Quereda, J. Oliva, E. Navas, A. Moreno, V. Pintado, J. Cobo, and I. Corral.** 1997. Candidal meningitis in HIV-infected patients: analysis of 14 cases. *Clin. Infect. Dis.* **25:**673–676.
26. **Chou, S., S. Guentzel, K. R. Michels, R. C. Miner, and W. L. Drew.** 1995. Frequency of UL97 phosphotransferase mutations related to ganciclovir resistance in clinical cytomegalovirus isolates. *J. Infect. Dis.* **172:**239–242.
27. **Chuck, S. L., and M. A. Sande.** 1989. Infections with *Cryptococcus neoformans* in the acquired immunodeficiency syndrome. *N. Engl. J. Med.* **321:**794–799.
28. **Cinque, P., S. Bossolasco, L. Vago, C. Fornara, S. Lipari, S. Racca, A. Lazzarin, and A. Linde.** 1997. Varicella-zoster virus (VZV) DNA in cerebrospinal fluid of patients infected with human immunodeficiency virus: VZV disease of the central nervous system or subclinical reactivation of VZV infection? *Clin. Infect. Dis.* **25:**634–639.
29. **Cinque, P., M. Brytting, L. Vago, A. Castagna, C. Parravicini, N. Zanchetta, A. D'Arminio Monforte, B. Wahren, A. Lazzarin, and A. Linde.** 1993. Epstein-Barr virus DNA in cerebrospinal fluid from patients with AIDS-related primary lymphoma of the central nervous system. *Lancet* **342:**398–401.
30. **Cinque, P., L. Vago, R. Marenzi, B. Giudici, T. Weber, R. Corradini, D. Ceresa, A. Lazzarin, and A. Linde.** 1998. Herpes simplex virus infections of the central nervous system in human immunodeficiency virus-infected patients: clinical management by polymerase chain reaction assay of cerebrospinal fluid. *Clin. Infect. Dis.* **27:**303–309.
31. **Dagan, R.** 1996. Nonpolio enteroviruses and the febrile young infant: epidemiologic, clinical and diagnostic aspects. *Pediatr. Infect. Dis. J.* **15:**67–71.
32. **Davies, N. W., L. J. Brown, J. Gonde, D. Irish, R. O. Robinson, A. V. Swan, J. Banatvala, R. S. Howard, M. K. Sharief, and P. Muir.** 2005. Factors influencing PCR detection of viruses in cerebrospinal fluid of patients with suspected CNS infections. *J. Neurol. Neurosurg. Psych.* **76:**82–87.
33. **Denier, C., J. H. Bourhis, C. Lacroix, S. Koscielny, J. Bosq, R. Sigal, G. Said, and D. Adams.** 2006. Spectrum and prognosis of neurologic complications after hematopoietic transplantation. *Neurology* **67:**1990–1997.
34. **Denning, D. W., J. Anderson, P. Rudge, and H. Smith.** 1987. Acute myelopathy associated with primary infection with human immunodeficiency virus. *Br. Med. J. (Clin. Res. Ed.)* **294:**143–144.
35. **Domingo, P., N. Barquet, M. Alvarez, P. Coll, J. Nava, and J. Garau.** 1997. Group B streptococcal meningitis in adults: report of twelve cases and review. *Clin. Infect. Dis.* **25:**1180–1187.
36. **Drobyski, W. R., K. K. Knox, D. Majewski, and D. R. Carrigan.** 1994. Brief report: fatal encephalitis due to variant B human herpesvirus-6 infection in a bone marrow-transplant recipient. *N. Engl. J. Med.* **330:**1356–1360.
37. **Dronda, F., I. Garcia-Arata, E. Navas, and L. de Rafael.** 1998. Meningitis in adults due to Campylobacter fetus subspecies fetus. *Clin. Infect. Dis.* **27:**906–907.
38. **Englund, J. A., M. E. Zimmerman, E. M. Swierkosz, J. L. Goodman, D. R. Scholl, and H. H. Balfour, Jr.** 1990. Herpes simplex virus resistant to acyclovir. A study in a tertiary care center. *Ann. Intern. Med.* **112:**416–422.

39. **Espy, M. J., J. R. Uhl, P. S. Mitchell, J. N. Thorvilson, K. A. Svien, A. D. Wold, and T. F. Smith.** 2000. Diagnosis of herpes simplex virus infections in the clinical laboratory by LightCycler PCR. *J. Clin. Microbiol.* **38:**795–799.
40. **Fong, I. W., C. B. Britton, K. E. Luinstra, E. Toma, and J. B. Mahony.** 1995. Diagnostic value of detecting JC virus DNA in cerebrospinal fluid of patients with progressive multifocal leukoencephalopathy. *J. Clin. Microbiol.* **33:**484–486.
41. **Gallino, A., M. Maggiorini, W. Kiowski, X. Martin, W. Wunderli, J. Schneider, M. Turina, and F. Follath.** 1996. Toxoplasmosis in heart transplant recipients. *Eur. J. Clin. Microbiol. Infect. Dis.* **15:**389–393.
42. **Gautheret-Dejean, A., C. Manichanh, F. Thien-Ah-Koon, A. M. Fillet, N. Mangeney, M. Vidaud, N. Dhedin, J. P. Vernant, and H. Agut.** 2002. Development of a real-time polymerase chain reaction assay for the diagnosis of human herpesvirus-6 infection and application to bone marrow transplant patients. *J. Virol. Methods* **100:**27–35.
43. **Gijtenbeek, J. M., M. J. van den Bent, and C. J. Vecht.** 1999. Cyclosporine neurotoxicity: a review. *J. Neurol.* **246:**339–346.
44. **Gildenberg, P. L., J. C. Gathe, Jr., and J. H. Kim.** 2000. Stereotactic biopsy of cerebral lesions in AIDS. *Clin. Infect. Dis.* **30:**491–499.
45. **Glaser, C. A., S. Gilliam, D. Schnurr, B. Forghani, S. Honarmand, N. Khetsuriani, M. Fischer, C. K. Cossen, and L. J. Anderson.** 2003. In search of encephalitis etiologies: diagnostic challenges in the California Encephalitis Project, 1998–2000. *Clin. Infect. Dis.* **36:**731–742.
46. **Gonzalez, C. E., M. G. Rinaldi, and A. M. Sugar.** 2002. Zygomycosis. *Infect. Dis. Clin. N. Am.* **16:**895–914.
47. **Gonzalez-Perez, M. P., L. Munoz-Juarez, F. C. Cardenas, J. J. Zarranz Imirizaldu, J. C. Carranceja, and A. Garcia-Saiz.** 2003. Human T-cell leukemia virus type I infection in various recipients of transplants from the same donor. *Transplantation* **75:**1006–1011.
48. **Gozlan, J., M. el Amrani, M. Baudrimont, D. Costagliola, J. M. Salord, C. Duvivier, O. Picard, M. C. Meyohas, C. Jacomet, and V. Schneider-Fauveau.** 1995. A prospective evaluation of clinical criteria and polymerase chain reaction assay of cerebrospinal fluid for the diagnosis of cytomegalovirus-related neurological diseases during AIDS. *AIDS* **9:**253–260.
49. **Gruhn, B., A. Meerbach, R. Egerer, H. J. Mentzel, R. Hafer, F. Ringelmann, M. Sauer, J. Hermann, and F. Zintl.** 1999. Successful treatment of Epstein-Barr virus-induced transverse myelitis with ganciclovir and cytomegalovirus hyperimmune globulin following unrelated bone marrow transplantation. *Bone Marrow Transplant.* **24:**1355–1358.
50. **Gupta, S. K., K. S. Manjunath-Prasad, B. S. Sharma, V. K. Khosla, V. K. Kak, M. Minz, and V. K. Sakhuja.** 1997. Brain abscess in renal transplant recipients: report of three cases. *Surg. Neurol.* **48:**284–287.
51. **Hakim, J. G., I. T. Gangaidzo, R. S. Heyderman, J. Mielke, E. Mushangi, A. Taziwa, V. J. Robertson, P. Musvaire, and P. R. Mason.** 2000. Impact of HIV infection on meningitis in Harare, Zimbabwe: a prospective study of 406 predominantly adult patients. *AIDS* **14:**1401–1407.
52. **Hamilton, M. S., M. A. Jackson, and D. Abel.** 1999. Clinical utility of polymerase chain reaction testing for enteroviral meningitis. *Pediatr. Infect. Dis. J.* **18:**533–537.
53. **Hanson, K. E., B. D. Alexander, C. Woods, C. Petti, and L. B. Reller.** 2007. Validation of laboratory screening criteria for herpes simplex virus testing of cerebrospinal fluid. *J. Clin. Microbiol.* **45:**721–724.
54. **Harrison, M. S., S. J. Simonte, and C. A. Kauffman.** 1994. Trimethoprim-induced aseptic meningitis in a patient with AIDS: case report and review. *Clin. Infect. Dis.* **19:**431–434.
55. **Hayden, R. T., and L. D. Frenkel.** 2000. More laboratory testing: greater cost but not necessarily better. *Pediatr. Infect. Dis. J.* **19:**290–292.
56. **Hayward, R. A., M. F. Shapiro, and R. K. Oye.** 1987. Laboratory testing on cerebrospinal fluid. A reappraisal. *Lancet* **i:**1–4.
57. **Holland, N. R., C. Power, V. P. Mathews, J. D. Glass, M. Forman, and J. C. McArthur.** 1994. Cytomegalovirus encephalitis in acquired immunodeficiency syndrome (AIDS). *Neurology* **44:**507–514.
58. **Hongo, I., and K. C. Bloch.** 2006. Ehrlichia infection of the central nervous system. *Curr. Treat. Options Neurol.* **8:**179–184.
59. **Israelski, D. M., and J. S. Remington.** 1993. Toxoplasmosis in the non-AIDS immunocompromised host. *Curr. Clin. Top. Infect. Dis.* **13:**322–356.
60. **Iwamoto, M., D. B. Jernigan, A. Guasch, M. J. Trepka, C. G. Blackmore, W. C. Hellinger, S. M. Pham, S. Zaki, R. S. Lanciotti, S. E. Lance-Parker, C. A. DiazGranados, A. G. Winquist, C. A. Perlino, S. Wiersma, K. L. Hillyer, J. L. Goodman, A. A. Marfin, M. E. Chamberland, and L. R. Petersen.** 2003. Transmission of West Nile virus from an organ donor to four transplant recipients. *N. Engl. J. Med.* **348:**2196–2203.
61. **Jabbour, N., J. Reyes, S. Kusne, M. Martin, and J. Fung.** 1996. Cryptococcal meningitis after liver transplantation. *Transplantation* **61:**146–149.
62. **Jeffery, K. J., S. J. Read, T. E. Peto, R. T. Mayon-White, and C. R. Bangham.** 1997. Diagnosis of viral infections of the central nervous system: clinical interpretation of PCR results. *Lancet* **349:**313–317.
63. **Johnson, R. H., and H. E. Einstein.** 2006. Coccidioidal meningitis. *Clin. Infect. Dis.* **42:**103–107.
64. **Joseph, P., M. M. Calderon, R. H. Gilman, M. L. Quispe, J. Cok, E. Ticona, V. Chavez, J. A. Jimenez, M. C. Chang, M. J. Lopez, and C. A. Evans.** 2002. Optimization and evaluation of a PCR assay for detecting toxoplasmic encephalitis in patients with AIDS. *J. Clin. Microbiol.* **40:**4499–4503.
65. **Jurado, R. L., M. M. Farley, E. Pereira, R. C. Harvey, A. Schuchat, J. D. Wenger, and D. S. Stephens.** 1993. Increased risk of meningitis and bacteremia due to Listeria monocytogenes in patients with human immunodeficiency virus infection. *Clin. Infect. Dis.* **17:**224–227.
66. **Kaiser, R., and S. Rauer.** 1999. Serodiagnosis of neuroborreliosis: comparison of reliability of three confirmatory assays. *Infection* **27:**177–182.
67. **Katz, D. A., J. R. Berger, and R. C. Duncan.** 1993. Neurosyphilis. A comparative study of the effects of infection with human immunodeficiency virus. *Arch. Neurol.* **50:**243–249.
68. **Kennedy, D. H., and R. J. Fallon.** 1979. Tuberculous meningitis. *JAMA* **241:**264–268.
69. **Kimura, H., Y. Ito, M. Futamura, Y. Ando, Y. Yabuta, Y. Hoshino, Y. Nishiyama, and T. Morishima.** 2002. Quantitation of viral load in neonatal herpes simplex virus infection and comparison between type 1 and type 2. *J. Med. Virol.* **67:**349–353.
70. **Kleinschmidt-DeMasters, B. K., B. A. Marder, M. E. Levi, S. P. Laird, J. T. McNutt, E. J. Escott, G. T. Everson, and K. L. Tyler.** 2004. Naturally acquired West Nile virus encephalomyelitis in transplant recipients: clinical, laboratory, diagnostic, and neuropathological features. *Arch. Neurol.* **61:**1210–1220.
71. **Leonard, M. K., J. R. Murrow, R. Jurado, and R. Gaynes.** 2002. Salmonella meningitis in adults infected with HIV: case report and review of the literature. *Am. J. Med. Sci.* **323:**266–268.

72. **Limaye, A. P., M. L. Huang, E. E. Atienza, J. M. Ferrenberg, and L. Corey.** 1999. Detection of Epstein-Barr virus DNA in sera from transplant recipients with lymphoproliferative disorders. *J. Clin. Microbiol.* **37:**1113–1116.
73. **Lu, H. Z., K. C. Bloch, and Y. W. Tang.** 2002. Molecular techniques in the diagnosis of central nervous system infections. *Curr. Infect. Dis. Rep.* **4:**339–350.
74. **Lukes, S. A., J. B. Posner, S. Nielsen, and D. Armstrong.** 1984. Bacterial infections of the CNS in neutropenic patients. *Neurology* **34:**269–275.
75. **Machetti, M., M. Zotti, L. Veroni, N. Mordini, M. T. Van Lint, A. Bacigalupo, D. Paola, and C. Viscoli.** 2000. Antigen detection in the diagnosis and management of a patient with probable cerebral aspergillosis treated with voriconazole. *Transpl. Infect. Dis.* **2:**140–144.
76. **Mamidi, A., J. A. DeSimone, and R. J. Pomerantz.** 2002. Central nervous system infections in individuals with HIV-1 infection. *J. Neurovirol.* **8:**158–167.
77. **Marcos, M. A., E. Martinez, M. Almela, J. Mensa, and M. T. Jimenez de Anta.** 2001. New rapid antigen test for diagnosis of pneumococcal meningitis. *Lancet* **357:**1499–1500.
78. **Martin, M. A., R. M. Massanari, D. D. Nghiem, J. L. Smith, and R. J. Corry.** 1988. Nosocomial aseptic meningitis associated with administration of OKT3. *JAMA* **259:**2002–2005.
79. **McCutchan, J. A.** 1995. Clinical impact of cytomegalovirus infections of the nervous system in patients with AIDS. *Clin. Infect. Dis.* **21**(Suppl. 2)**:**S196–S201.
80. **McDermott, S. S., P. F. McDermott, J. Skare, M. Glantz, T. W. Smith, N. S. Litofsky, and L. D. Recht.** 2000. Positive CSF HSV PCR in patients with GBM: a note of caution. *Neurology* **54:**746–749.
81. **McGinnis, M. R.** 1983. Detection of fungi in cerebrospinal fluid. *Am. J. Med.* **75:**129–138.
82. **Miller, R. F., J. D. Fox, P. Thomas, J. C. Waite, Y. Sharvell, B. G. Gazzard, M. J. Harrison, and N. S. Brink.** 1996. Acute lumbosacral polyradiculopathy due to cytomegalovirus in advanced HIV disease: CSF findings in 17 patients. *J. Neurol. Neurosurg. Psych.* **61:**456–460.
83. **Monlun, E., F. de Blay, C. Berton, B. Gasser, A. Jaeger, and G. Pauli.** 1997. Invasive pulmonary aspergillosis with cerebromeningeal involvement after short-term intravenous corticosteroid therapy in a patient with asthma. *Respir. Med.* **91:**435–437.
84. **Morgan, J., K. A. Wannemuehler, K. A. Marr, S. Hadley, D. P. Kontoyiannis, T. J. Walsh, S. K. Fridkin, P. G. Pappas, and D. W. Warnock.** 2005. Incidence of invasive aspergillosis following hematopoietic stem cell and solid organ transplantation: interim results of a prospective multicenter surveillance program. *Med. Mycol.* **43**(Suppl. 1)**:**S49–S58.
85. **Morgan, J. S., W. Schaffner, and W. J. Stone.** 1986. Opportunistic strongyloidiasis in renal transplant recipients. *Transplantation* **42:**518–524.
86. **Muir, P., and A. M. van Loon.** 1997. Enterovirus infections of the central nervous system. *Intervirology* **40:**153–166.
87. **Mulford, W. S., R. S. Buller, M. Q. Arens, and G. A. Storch.** 2004. Correlation of cerebrospinal fluid (CSF) cell counts and elevated CSF protein levels with enterovirus reverse transcription-PCR results in pediatric and adult patients. *J. Clin. Microbiol.* **42:**4199–4203.
88. **Murray, K., S. Baraniuk, M. Resnick, R. Arafat, C. Kilborn, K. Cain, R. Shallenberger, T. L. York, D. Martinez, J. S. Hellums, D. Hellums, M. Malkoff, N. Elgawley, W. McNeely, S. A. Khuwaja, and R. B. Tesh.** 2006. Risk factors for encephalitis and death from West Nile virus infection. *Epidemiol. Infect.* **134:**1325–1332.
89. **Musher, D. M., and R. F. Schell.** 1973. False-positive Gram stains of cerebrospinal fluid. *Ann. Intern. Med.* **79:**603–604.
90. **Mylonakis, E., E. L. Hohmann, and S. B. Calderwood.** 1998. Central nervous system infection with Listeria monocytogenes. 33 years' experience at a general hospital and review of 776 episodes from the literature. *Medicine* (Baltimore) **77:**313–336.
91. **Nagel, M. A., B. Forghani, R. Mahalingam, M. C. Wellish, R. J. Cohrs, A. N. Russman, I. Katzan, R. Lin, C. J. Gardner, and D. H. Gilden.** 2007. The value of detecting anti-VZV IgG antibody in CSF to diagnose VZV vasculopathy. *Neurology* **68:**1069–1073.
92. **Nocton, J. J., B. J. Bloom, B. J. Rutledge, D. H. Persing, E. L. Logigian, C. H. Schmid, and A. C. Steere.** 1996. Detection of *Borrelia burgdorferi* DNA by polymerase chain reaction in cerebrospinal fluid in Lyme neuroborreliosis. *J. Infect. Dis.* **174:**623–627.
93. **Osame, M., K. Usuku, S. Izumo, N. Ijichi, H. Amitani, A. Igata, M. Matsumoto, and M. Tara.** 1986. HTLV-I associated myelopathy, a new clinical entity. *Lancet* **i:**1031–1032.
94. **Palomares, M., T. Martinez, J. Pastor, A. Osuna, J. A. Bravo, G. Alvarez, and C. Asensio.** 1999. Cerebral abscess caused by Nocardia asteroides in renal transplant recipient. *Nephrol. Dial. Transplant.* **14:**2950–2952.
95. **Pappas, P. G., J. R. Perfect, G. A. Cloud, R. A. Larsen, G. A. Pankey, D. J. Lancaster, H. Henderson, C. A. Kauffman, D. W. Haas, M. Saccente, R. J. Hamill, M. S. Holloway, R. M. Warren, and W. E. Dismukes.** 2001. Cryptococcosis in human immunodeficiency virus-negative patients in the era of effective azole therapy. *Clin. Infect. Dis.* **33:**690–699.
96. **Patchell, R. A.** 1994. Neurological complications of organ transplantation. *Ann. Neurol.* **36:**688–703.
97. **Paul, M. L., D. E. Dwyer, C. Chow, J. Robson, I. Chambers, G. Eagles, and V. Ackerman.** 1994. Listeriosis—a review of eighty-four cases. *Med. J. Aust.* **160:**489–493.
98. **Perkins, M. D., S. Mirrett, and L. B. Reller.** 1995. Rapid bacterial antigen detection is not clinically useful. *J. Clin. Microbiol.* **33:**1486–1491.
99. **Poggio, G. P., C. Rodriguez, D. Cisterna, M. C. Freire, and J. Cello.** 2000. Nested PCR for rapid detection of mumps virus in cerebrospinal fluid from patients with neurological diseases. *J. Clin. Microbiol.* **38:**274–278.
100. **Polage, C. R., and C. A. Petti.** 2006. Assessment of the utility of viral culture of cerebrospinal fluid. *Clin. Infect. Dis.* **43:**1578–1579.
101. **Popovich, M. J., R. H. Arthur, and E. Helmer.** 1990. CT of intracranial cryptococcosis. *Am. J. Roentgenol.* **154:**603–606.
102. **Potasman, I., L. Resnick, B. J. Luft, and J. S. Remington.** 1988. Intrathecal production of antibodies against Toxoplasma gondii in patients with toxoplasmic encephalitis and the acquired immunodeficiency syndrome (AIDS). *Ann. Intern. Med.* **108:**49–51.
103. **Powell, K. F., N. E. Anderson, R. W. Frith, and M. C. Croxson.** 1990. Non-invasive diagnosis of herpes simplex encephalitis. *Lancet* **335:**357–358.
104. **Pruitt, A. A.** 2004. Central nervous system infections in cancer patients. *Semin. Neurol.* **24:**435–452.
105. **Rabenau, H. F., A. Clarici, G. Mühlbauer, A. Berger, A. Vince, S. Muller, E. Daghofer, B. I. Santner, E. Marth, and H. H. Kessler.** 2002. Rapid detection of enterovirus infection by automated RNA extraction and real-time fluorescence PCR. *J. Clin. Virol.* **25:**155–164.
106. **Ramers, C., G. Billman, M. Hartin, S. Ho, and M. H. Sawyer.** 2000. Impact of a diagnostic cerebrospinal fluid

enterovirus polymerase chain reaction test on patient management. *JAMA* **283**:2680–2685.

107. **Read, S. J., and J. B. Kurtz.** 1999. Laboratory diagnosis of common viral infections of the central nervous system by using a single multiplex PCR screening assay. *J. Clin. Microbiol.* **37**:1352–1355.
108. **Renoult, E., E. Georges, M. F. Biava, C. Hulin, L. Frimat, D. Hestin, and M. Kessler.** 1997. Toxoplasmosis in kidney transplant recipients: report of six cases and review. *Clin. Infect. Dis.* **24**:625–634.
109. **Robinson, P. A., M. Bauer, M. A. Leal, S. G. Evans, P. D. Holtom, D. A. Diamond, J. M. Leedom, and R. A. Larsen.** 1999. Early mycological treatment failure in AIDS-associated cryptococcal meningitis. *Clin. Infect. Dis.* **28**:82–92.
110. **Rosenberg, A. S., and M. B. Morgan.** 2001. Disseminated acanthamoebiasis presenting as lobular panniculitis with necrotizing vasculitis in a patient with AIDS. *J. Cutan. Pathol.* **28**:307–313.
111. **Rotbart, H. A., M. H. Sawyer, S. Fast, C. Lewinski, N. Murphy, E. F. Keyser, J. Spadoro, S. Y. Kao, and M. Loeffelholz.** 1994. Diagnosis of enteroviral meningitis by using PCR with a colorimetric microwell detection assay. *J. Clin. Microbiol.* **32**:2590–2592.
112. **Rowley, A. H., R. J. Whitley, F. D. Lakeman, and S. M. Wolinsky.** 1990. Rapid detection of herpes-simplex-virus DNA in cerebrospinal fluid of patients with herpes simplex encephalitis. *Lancet* **335**:440–441.
113. **Rutala, P. J., and J. W. Smith.** 1978. Coccidioidomycosis in potentially compromised hosts: the effect of immunosuppressive therapy in dissemination. *Am. J. Med. Sci.* **275**:283–295.
114. **Saag, M. S., R. J. Graybill, R. A. Larsen, P. G. Pappas, J. R. Perfect, W. G. Powderly, J. D. Sobel, and W. E. Dismukes.** 2000. Practice guidelines for the management of cryptococcal disease. *Clin. Infect. Dis.* **30**:710–718.
115. **Saha, S. K., G. L. Darmstadt, N. Yamanaka, D. S. Billal, T. Nasreen, M. Islam, and D. H. Hamer.** 2005. Rapid diagnosis of pneumococcal meningitis: implications for treatment and measuring disease burden. *Pediatr. Infect. Dis. J.* **24**:1093–1098.
116. **Sauerbrei, A., U. Eichhorn, G. Hottenrott, and P. Wutzler.** 2000. Virological diagnosis of herpes simplex encephalitis. *J. Clin. Virol.* **17**:31–36.
117. **Sawyer, M. H., D. Holland, N. Aintablian, J. D. Connor, E. F. Keyser, and N. J. Waecker, Jr.** 1994. Diagnosis of enteroviral central nervous system infection by polymerase chain reaction during a large community outbreak. *Pediatr. Infect. Dis. J.* **13**:177–182.
118. **Scarpellini, P., S. Racca, P. Cinque, F. Delfanti, N. Gianotti, M. R. Terreni, L. Vago, and A. Lazzarin.** 1995. Nested polymerase chain reaction for diagnosis and monitoring treatment response in AIDS patients with tuberculous meningitis. *AIDS* **9**:895–900.
119. **Schacker, T., A. C. Collier, J. Hughes, T. Shea, and L. Corey.** 1996. Clinical and epidemiologic features of primary HIV infection. *Ann. Intern. Med.* **125**:257–264.
120. **Schlesinger, Y., M. H. Sawyer, and G. A. Storch.** 1994. Enteroviral meningitis in infancy: potential role for polymerase chain reaction in patient management. *Pediatrics* **94**:157–162.
121. **Sefers, S. E., J. Rickmyre, A. Blackman, H. Li, K. Edwards, and Y. W. Tang.** 2006. QIAamp MinElute virus kit effectively extracts viral nucleic acids from cerebrospinal fluids and nasopharyngeal swabs. *J. Clin. Virol.* **35**:141–146.
122. **Shitrit, D., N. Lev, A. Bar-Gil-Shitrit, and M. R. Kramer.** 2005. Progressive multifocal leukoencephalopathy in transplant recipients. *Transpl. Int.* **17**:658–665.
123. **Singh, N., and S. Husain.** 2000. Infections of the central nervous system in transplant recipients. *Transpl. Infect. Dis.* **2**:101–111.
124. **Singh, N., and D. L. Paterson.** 1998. Mycobacterium tuberculosis infection in solid-organ transplant recipients: impact and implications for management. *Clin. Infect. Dis.* **27**:1266–1277.
125. **Smalling, T. W., S. E. Sefers, H. Li, and Y. W. Tang.** 2002. Molecular approaches to detecting herpes simplex virus and enteroviruses in the central nervous system. *J. Clin. Microbiol.* **40**:2317–2322.
126. **Smith, J. M., K. Rudser, D. Gillen, B. Kestenbaum, S. Seliger, N. Weiss, R. A. McDonald, C. L. Davis, and C. Stehmen-Breen.** 2006. Risk of lymphoma after renal transplantation varies with time: an analysis of the United States Renal Data System. *Transplantation* **81**:175–180.
127. **Spitzer, P. G., D. Tarsy, and G. M. Eliopoulos.** 1987. Acute transverse myelitis during disseminated cytomegalovirus infection in a renal transplant recipient. *Transplantation* **44**:151–153.
128. **Srinivasan, A., E. C. Burton, M. J. Kuehnert, C. Rupprecht, W. L. Sutker, T. G. Ksiazek, C. D. Paddock, J. Guarner, W. J. Shieh, C. Goldsmith, C. A. Hanlon, J. Zoretic, B. Fischbach, M. Niezgoda, W. H. El-Feky, L. Orciari, E. Q. Sanchez, A. Likos, G. B. Klintmalm, D. Cardo, J. LeDuc, M. E. Chamberland, D. B. Jernigan, and S. R. Zaki.** 2005. Transmission of rabies virus from an organ donor to four transplant recipients. *N. Engl. J. Med.* **352**:1103–1111.
129. **Stamm, A. M., W. E. Dismukes, B. P. Simmons, C. G. Cobbs, A. Elliott, P. Budrich, and J. Harmon.** 1982. Listeriosis in renal transplant recipients: report of an outbreak and review of 102 cases. *Rev. Infect. Dis.* **4**:665–682.
130. **Steele, R. W., D. J. Marmer, M. D. O'Brien, S. T. Tyson, and C. R. Steele.** 1986. Leukocyte survival in cerebrospinal fluid. *J. Clin. Microbiol.* **23**:965–966.
131. **Storch, G. A.** 2000. Diagnostic virology. *Clin. Infect. Dis.* **31**:739–751.
132. **Tam, C. S., J. Galanos, J. F. Seymour, A. G. Pitman, R. J. Stark, and H. M. Prince.** 2004. Reversible posterior leukoencephalopathy syndrome complicating cytotoxic chemotherapy for hematologic malignancies. *Am. J. Hematol.* **77**:72–76.
133. **Tang, Y. W., M. J. Espy, D. H. Persing, and T. F. Smith.** 1997. Molecular evidence and clinical significance of herpesvirus coinfection in the central nervous system. *J. Clin. Microbiol.* **35**:2869–2872.
134. **Tang, Y. W., J. R. Hibbs, K. R. Tau, Q. Qian, H. A. Skarhus, T. F. Smith, and D. H. Persing.** 1999. Effective use of polymerase chain reaction for diagnosis of central nervous system infections. *Clin. Infect. Dis.* **29**:803–806.
135. **Tang, Y. W., P. S. Mitchell, M. J. Espy, T. F. Smith, and D. H. Persing.** 1999. Molecular diagnosis of herpes simplex virus infections in the central nervous system. *J. Clin. Microbiol.* **37**:2127–2136.
136. **Tang, Y. W., P. N. Rys, B. J. Rutledge, P. S. Mitchell, T. F. Smith, and D. H. Persing.** 1998. Comparative evaluation of colorimetric microtiter plate systems for detection of herpes simplex virus in cerebrospinal fluid. *J. Clin. Microbiol.* **36**:2714–2717.
137. **Tarafdar, K., S. Rao, R. A. Recco, and M. M. Zaman.** 2001. Lack of sensitivity of the latex agglutination test to detect bacterial antigen in the cerebrospinal fluid of patients with culture-negative meningitis. *Clin. Infect. Dis.* **33**:406–408.
138. **Thwaites, G., T. T. Chau, N. T. Mai, F. Drobniewski, K. McAdam, and J. Farrar.** 2000. Tuberculous meningitis. *J. Neurol. Neurosurg. Psych.* **68**:289–299.

139. **Tunkel, A. R., B. J. Hartman, S. L. Kaplan, B. A. Kaufman, K. L. Roos, W. M. Scheld, and R. J. Whitley.** 2004. Practice guidelines for the management of bacterial meningitis. *Clin. Infect. Dis.* **39:**1267–1284.
140. **Tunkel, A. R., and W. M. Scheld.** 1993. Pathogenesis and pathophysiology of bacterial meningitis. *Clin. Microbiol. Rev.* **6:**118–136.
141. **Tyler, K. L.** 2004. Update on herpes simplex encephalitis. *Rev. Neurol. Dis.* **1:**169–178.
142. **van de Beek, D., J. de Gans, L. Spanjaard, M. Weisfelt, J. B. Reitsma, and M. Vermeulen.** 2004. Clinical features and prognostic factors in adults with bacterial meningitis. *N. Engl. J. Med.* **351:**1849–1859.
143. **van Esser, J. W., B. van der Holt, E. Meijer, H. G. Niesters, R. Trenschel, S. F. Thijsen, A. M. van Loon, F. Frassoni, A. Bacigalupo, U. W. Schaefer, A. D. Osterhaus, J. W. Gratama, B. Lowenberg, L. F. Verdonck, and J. J. Cornelissen.** 2001. Epstein-Barr virus (EBV) reactivation is a frequent event after allogeneic stem cell transplantation (SCT) and quantitatively predicts EBV-lymphoproliferative disease following T-cell-depleted SCT. *Blood* **98:**972–978.
144. **Vavricka, S. R., R. B. Walter, S. Irani, J. Halter, and U. Schanz.** 2003. Safety of lumbar puncture for adults with acute leukemia and restrictive prophylactic platelet transfusion. *Ann. Hematol.* **82:**570–573.
145. **Verweij, P. E., K. Brinkman, H. P. Kremer, B. J. Kullberg, and J. F. Meis.** 1999. *Aspergillus* meningitis: diagnosis by non-culture-based microbiological methods and management. *J. Clin. Microbiol.* **37:**1186–1189.
146. **Viscoli, C., M. Machetti, P. Gazzola, A. De Maria, D. Paola, M. T. Van Lint, F. Gualandi, M. Truini, and A. Bacigalupo.** 2002. *Aspergillus* galactomannan antigen in the cerebrospinal fluid of bone marrow transplant recipients with probable cerebral aspergillosis. *J. Clin. Microbiol.* **40:**1496–1499.
147. **Wadei, H., G. J. Alangaden, D. H. Sillix, J. M. El-Amm, S. A. Gruber, M. S. West, D. K. Granger, J. Garnick, P. Chandrasekar, S. D. Migdal, and A. Haririan.** 2004. West Nile virus encephalitis: an emerging disease in renal transplant recipients. *Clin. Transplant.* **18:**753–758.
148. **Wang, F. Z., A. Linde, H. Hagglund, M. Testa, A. Locasciulli, and P. Ljungman.** 1999. Human herpesvirus 6 DNA in cerebrospinal fluid specimens from allogeneic bone marrow transplant patients: does it have clinical significance? *Clin. Infect. Dis.* **28:**562–568.
149. **Ward, K. N., R. P. White, S. Mackinnon, and M. Hanna.** 2002. Human herpesvirus-7 infection of the CNS with acute myelitis in an adult bone marrow recipient. *Bone Marrow Transplant.* **30:**983–985.
150. **Weber, T., R. W. Turner, S. Frye, W. Luke, H. A. Kretzschmar, W. Luer, and G. Hunsmann.** 1994. Progressive multifocal leukoencephalopathy diagnosed by amplification of JC virus-specific DNA from cerebrospinal fluid. *AIDS* **8:**49–57.
151. **Weil, A. A., C. A. Glaser, Z. Amad, and B. Forghani.** 2002. Patients with suspected herpes simplex encephalitis: rethinking an initial negative polymerase chain reaction result. *Clin. Infect. Dis.* **34:**1154–1157.
152. **Weinberg, A., K. C. Bloch, S. Li, Y. W. Tang, M. Palmer, and K. L. Tyler.** 2005. Dual infections of the central nervous system with Epstein-Barr virus. *J. Infect. Dis.* **191:**234–237.
153. **Weinberg, A., D. Spiers, G. Y. Cai, C. M. Long, R. Sun, and V. Tevere.** 1998. Evaluation of a commercial PCR kit for diagnosis of cytomegalovirus infection of the central nervous system. *J. Clin. Microbiol.* **36:**3382–3384.
154. **Wheat, L. J., C. E. Musial, and E. Jenny-Avital.** 2005. Diagnosis and management of central nervous system histoplasmosis. *Clin. Infect. Dis.* **40:**844–852.
155. **Whitley, R. J.** 2006. Herpes simplex encephalitis: adolescents and adults. *Antivir. Res.* **71:**141–148.
156. **Workowski, K. A., and S. M. Berman.** 2006. Sexually transmitted diseases treatment guidelines, 2006. *MMWR Recommend. Rep.* **55:**1–94.
157. **Wu, G., R. A. Vilchez, B. Eidelman, J. Fung, R. Kormos, and S. Kusne.** 2002. Cryptococcal meningitis: an analysis among 5,521 consecutive organ transplant recipients. *Transpl. Infect. Dis.* **4:**183–188.
158. **Yachnis, A. T., J. Berg, A. Martinez-Salazar, B. S. Bender, L. Diaz, A. M. Rojiani, T. A. Eskin, and J. M. Orenstein.** 1996. Disseminated microsporidiosis especially infecting the brain, heart, and kidneys. Report of a newly recognized pansporoblastic species in two symptomatic AIDS patients. *Am. J. Clin. Pathol.* **106:**535–543.
159. **Zunt, J. R., and C. M. Marra.** 1999. Cerebrospinal fluid testing for the diagnosis of central nervous system infection. *Neurol. Clin.* **17:**675–689.

Diagnostic Microbiology of the Immunocompromised Host
Edited by Randall T. Hayden, Karen C. Carroll, Yi-Wei Tang, and Donna M. Wolk

Chapter 19

Bloodstream Infections

Cathy A. Petti, Hafsa Hassan, and L. Barth Reller

The spectrum of microorganisms causing bacteremia and fungemia in immunocompromised hosts has changed over the last decade, owing in large part to widespread use of chemoprophylaxis, differences in immunosuppressive regimens, and significant increases in the use of long-term indwelling devices. Changes in clinical practice have caused a paradigm shift for bloodstream pathogens commonly associated with this patient population. Historically, patients with hematologic malignancies and febrile granulocytopenia constituted the majority of immunocompromised hosts, with gram-negative bacilli being the most frequently observed microorganisms (26, 59). Today, however, there are increasing numbers of other immunocompromised hosts, including those with hematopoietic stem cell and solid organ transplants. Many, if not most, of these patients have intravascular catheters, and gram-positive microorganisms, especially staphylococci, have become the most frequent causes of bloodstream infections. Additionally, less common microorganisms, such as nontuberculous mycobacteria and non-*Candida* yeasts, have emerged as important bloodstream pathogens. In this chapter, we discuss the approach to immunocompromised patients with suspected bacteremia and fungemia, with particular emphasis on culture-independent diagnostic methods. We briefly address the predilection of specific microorganisms to cause invasive bloodstream infections among certain immunocompromised populations and discuss individual risk factors associated with the development of bacteremia and fungemia.

PREDISPOSING FACTORS FOR BLOODSTREAM INFECTION

The absence of specific host immune responses (e.g., B-cell depletion from rituximab or complement deficiency in systemic lupus erythematosus) may predispose a patient to bloodstream infection, with the underlying reason for immunosuppression providing clues to the type of microorganism causing infection. For example, a complement-depleted patient with systemic lupus erythematosus who presents with sudden onset of septic shock is at increased risk for a bloodstream infection with *Neisseria meningitidis*. Similarly, unusual pathogens such as *Vibrio vulnificus* may cause bloodstream infections in patients with severe liver disease or liver transplantation. When anatomic barriers such as the skin or mucosa are compromised from graft-versus-host disease or cytotoxic chemotherapy, patients are more likely to become infected with microorganisms colonizing the skin or oropharyngeal and gastrointestinal tracts. The recovery of *Leptotrichia* sp. or *Capnocytophaga* sp. from blood culture may suggest disruption of the mucosal integrity of the oropharynx or gastrointestinal tract in a granulocytopenic patient with severe mucositis. One of the most important and common causes of disruptions in an anatomic barrier is the presence of indwelling intravascular devices, by which microorganisms gain access to the bloodstream via extraluminal migration along the catheter (e.g., nontunneled, noncuffed catheters) or by colonizing the endoluminal surface, bell, or hub of the port (e.g., tunneled or implantable catheters) (31, 56, 57, 64). Microorganisms

Cathy A. Petti • Departments of Medicine and Pathology, University of Utah School of Medicine, and Associated Regional and University Pathologists Laboratories, Salt Lake City, UT 84132. **Hafsa Hassan** • Department of Pathology, University of Utah School of Medicine, Salt Lake City, UT 84132. **L. Barth Reller** • Departments of Medicine and Pathology, Duke University School of Medicine, and Clinical Microbiology Laboratory, Duke University Medical Center, Durham, NC 27704.

such as *Staphylococcus aureus*, coagulase-negative staphylococci, and *Candida* sp. produce an exopolysaccharide-rich biofilm that enables them to adhere to and survive on the surface of the catheter (57). Hence, the recovery of these microorganisms from blood cultures often suggests an intravascular device as the potential source of infection.

In addition to underlying immunosuppression, immunocompromised patients have increased risk for infections from unusual or multidrug-resistant microorganisms owing to chemoprophylaxis regimens, extended hospitalizations, or prolonged stays in intensive care units. For example, patients receiving prophylaxis with fluoroquinolones are at risk for bloodstream infections with methicillin-resistant *S. aureus* (29) and multiply drug-resistant gram-negative bacilli (8). Additionally, candidemia is frequently observed among immunocompromised hosts with broad-spectrum or prolonged antibiotic therapy, parenteral nutrition, intravascular catheters, renal failure, and prolonged stays in an intensive care unit (4, 46, 69). Most patients with invasive aspergillosis have underlying hematological diseases or have undergone hematopoietic stem cell transplantation, and recognized risk factors are graft-versus-host disease, receipt of steroids, secondary neutropenia, age of >40 years, and a stem cell source (34). In contrast, invasive mold infections are unusual in patients with human immunodeficiency virus (24), who, depending on geography and CD4 count, more commonly present with mycobacteremia from *Mycobacterium avium-intracellulare* or *Mycobacterium tuberculosis* (2, 45). Bloodstream infections caused by nontuberculous mycobacteria are rapidly emerging among all hosts with altered immune responses and are often associated with an intravascular catheter (14, 16).

LABORATORY DIAGNOSIS

Culture-Dependent Methods

The culturing of blood from patients with suspected bloodstream infections has remained one of the most important diagnostic tests performed by the microbiology laboratory. Over the last several decades, improvements in blood culture media and continuously monitoring blood culture systems have greatly enhanced the diagnostic value of blood cultures for the detection of bacteremia and candidemia.

Types of blood culture systems

At initial evaluation of a patient with suspected bloodstream infection, blood should be cultured for both aerobic and anaerobic bacteria, as well as commonly encountered yeasts. If the clinician has a high suspicion of a dimorphic fungus or mycobacteria, then blood bottles specifically designed to recover these microorganisms should be obtained. Semiautomated systems as well as manual blood culture systems are still available but have largely been replaced by automated systems that enhance the detection of microorganisms and improve laboratory work flow (60). Examples of commercially available products are BacT/Alert charcoal-containing media (bioMérieux, Durham, NC) and Bactec resin-containing media (Becton Dickinson Diagnostic Instrument Systems, Sparks, MD). These systems have their unique relative strengths and limitations, but overall performances are comparable (38, 60). Since prior antibacterial therapy remains the most common cause of false-negative blood cultures (70), manufacturers have attempted to minimize this problem by adding activated charcoal or resins to broth media to neutralize the effect of antibacterial therapy (60, 61). Although continuously monitoring systems have emerged as the new gold standard, the Isolator tube system (Wampole Laboratories, Cranbury, NJ), a manual lysis-centrifugation system, remains the superior method for the detection of *Histoplasma capsulatum* and certain fastidious pathogens (60, 76).

Principles of blood culture collection

The importance of rigorously following guidelines for the proper collection of blood cultures, such as those published by the Clinical and Laboratory Standards Institute, cannot be overemphasized (10). At the initial evaluation for patients with signs and symptoms suggestive of bloodstream infection, an adequate volume of blood must be cultured to achieve overall test sensitivity. For nonfastidious bacterial pathogens in cases of no prior history of antibacterial therapy, one set of blood cultures (i.e., 20 ml of blood) has been shown to detect a bacterial or yeast bloodstream infection in approximately 65 to 91% of cases; use of two or three sets of blood cultures improves the overall yield of detecting the pathogens to 80 to 99% or 96 to >99% of cases, respectively (11, 68). Medical urgency often dictates prompt initiation of empiric antibacterial therapy in immunocompromised hosts, but two or three sets of blood cultures should be obtained within 5 minutes of each other before initiation of therapy. For patients with new skin lesions, skin biopsy with culture should also be performed (58). The optimal volume of blood for children can vary based on body weight, but similar to the case for adults, larger volumes of blood increase the likelihood of recovering microorganisms.

Aseptic technique in blood culture collection is imperative to prevent contamination from skin and other environmental flora. In the United States, the positivity rate for blood cultures is approximately 10%, but 30% to 40% of these positive results represent contaminating skin flora due to inadequate skin preparation or from colonized indwelling vascular catheters (68). Despite the American College of Physicians recommendations that blood cultures should be obtained through fresh venipunctures rather than indwelling intravascular devices (3), blood cultures from immunocompromised hosts are frequently collected from an indwelling central venous catheter. Several studies have shown that blood cultures drawn from indwelling catheters are less specific and have a lower positive predictive value and higher rates of contamination than those obtained from peripheral venipuncture (6, 15, 67). The exception is a newly inserted catheter, for which investigators have shown no increase in contamination rates for catheter-drawn blood cultures (21). The clinical significance of blood cultures drawn through an indwelling catheter, particularly when only one set of blood cultures is obtained, cannot be determined reliably, and true bacteremia cannot be distinguished from contaminating skin flora. For example, in a patient presenting with hypotensive shock, physicians may mistakenly infer causality from a single positive blood culture with enterococci that may result in misuse of antibiotics and failure to investigate alternative diagnoses.

Bacteremia

The sensitivity of blood cultures is excellent for the detection of nonfastidious bacterial pathogens and increases with the volume of blood collected. Only one microorganism per milliliter of blood is necessary for a blood culture to recover a microorganism (60), and culture remains more sensitive, on average, than nucleic acid amplification techniques. In fact, the most common cause of a falsely negative blood culture is administration of antibacterial therapy before blood collection. Clinicians are often concerned that their immunocompromised patients may be susceptible to invasive infections with more unusual, fastidious pathogens that may require unique incubation procedures. Microorganisms such as species of *Haemophilus*, *Actinobacillus*, *Cardiobacterium*, *Eikenella*, and *Kingella* can be recovered within 5 days of incubation in currently available automated blood culture systems; extended incubation is not necessary (52). Conversely, *Legionella* species grow poorly in routine blood culture media, require blind subculture from blood culture bottles, and need special supplementation for growth on solid media. With improved blood culture systems, the ability to detect *Bartonella* spp. has increased, but these organisms still require a prolonged incubation period, of up to 6 weeks, and subculture to freshly prepared medium (28). Serology, PCR, or both are preferred for detection of *Legionella* or *Bartonella*. When physicians suspect unusual bacterial pathogens, they should alert the clinical laboratory for optimal use of special media and incubation procedures to recover these microorganisms.

Candidemia and other medically important yeast infections

Candida spp. currently rank as the fourth most common cause of bloodstream infection in hospitalized patients (49, 77). Approximately 95% to 97% of all *Candida*-associated bloodstream infections are caused by the following five *Candida* species: *Candida albicans*, *C. glabrata*, *C. tropicalis*, *C. parapsilosis*, and *C. krusei* (49). Other yeasts, such as *Cryptococcus* sp., *Trichosporon* sp., *Geotrichum* sp., *Rhodotorula* sp., and *Hansenula* sp., have emerged as important bloodstream pathogens in the immunocompromised host (43, 54) and, with the exception of *Cryptococcus* sp., are often related to indwelling devices (20, 27). Standard blood culture bottles will recover medically important yeasts without the need for special fungal blood bottles (53). Reported sensitivities of blood culture for medically important yeasts are not clear, with experts suggesting that sensitivity falls from 80% in cases of line sepsis to 8 to 50% in cases of disseminated disease, where circulating organisms may not always be present.

Fungemia secondary to mold infections

Species of *Aspergillus* (*Aspergillus fumigatus*, *Aspergillus flavus*, *Aspergillus niger*, *Aspergillus terreus*), hyaline molds (e.g., *Fusarium* sp. and *Scedosporium* sp.), and a wide variety of dematiaceous fungi can cause invasive bloodstream infections (5, 17, 39, 66) and are associated with high mortality. The recovery of *Aspergillus* sp. from specific fungal blood culture bottles is infrequent despite disseminated disease (including endocarditis), owing to tissue tropism and size. Vessel invasion and skin infarcts preclude recovery from blood. Conversely, some molds and dematiaceous fungi, such as *Fusarium* sp. (5), *Paecilomyces* sp. (66), and *Wangiella* (*Exophiala*) *dermatitidis* (39), have been cultivated successfully from blood, especially from patients with indwelling lines. Overall, blood culture has limited sensitivity in recovering fungi compared with end organ or culture-independent methods. Skin or end organ biopsy for histopathological and microbiological investigation may be helpful in making early diagnosis of fungemia, particularly for disseminated *Fusarium* sp. or *Candida* sp. infections.

Mycobacteremia

The newer commercially available blood culture products, such as Bactec Myco/F Lytic medium and BacT/Alert MB medium, provide fast and sensitive methods to detect mycobacteria from blood cultures (13) and obviate the need for manual lysis-centrifugation systems. Compared with the use of continuously monitored blood culture systems, cultures from bone marrow do not significantly increase the likelihood of recovering mycobacteria from patients with disseminated *M. avium* complex and *M. tuberculosis* infections (25, 48). For the diagnosis of disseminated *M. tuberculosis* infections or nontuberculous mycobacterial infections (e.g., *Mycobacterium chelonae* or *Mycobacterium haemophilum*), other specimens (e.g., cultures of sputum or skin lesions) may improve the diagnostic yield (16, 72).

Rapid diagnosis from instrument-positive blood culture bottles

In addition to Gram staining and subculture, new technologies have emerged to reduce the time required for identification of microorganisms from blood culture, including PCR, fluorescence in situ hybridization (FISH), and rRNA probe matrices (23, 32, 75). Of all these techniques, FISH, an assay with peptide nucleic acid probes targeting specific 16S rRNA sequences of microorganisms, has the most promise. It is a rapid technique for the identification of multiple clinically relevant microorganisms, such as gram-positive cocci (*Staphylococcus aureus* and coagulase-negative staphylococci), gram-negative bacilli (*Escherichia coli*, *Klebsiella pneumoniae*, and *Enterobacter cloacae*), and yeasts (*C. albicans* and *C. tropicalis*). The sensitivity and specificity of FISH technology are reportedly between 95 and 100% (23, 50, 51). The approximate turnaround time of the FISH procedure is 2.5 to 3 hours, compared with >18 to 24 hours for identification of bacteria and yeast by conventional methods. Overall, FISH is a rapid, relatively inexpensive (about $20 per blood culture specimen), and well-validated method that has potential to enable more judicious use of antibacterial therapy owing to prompt identification of the microorganism (1). DNA target amplification and sequencing have been applied to instrument-positive blood cultures but are limited by expense and longer turnaround times than those with automated identification instruments (unpublished data).

Interpretation of positive blood culture

The two most useful parameters that aid in the interpretation of a positive blood culture result include the type of microorganism (e.g., nonenteric, gram-negative bacillus versus diphtheroid) detected (73) and the number of sets of blood cultures from which a putative pathogen is recovered (30, 73, 74). False-positive blood cultures can lead to diagnostic confusion and unnecessary antibiotic use for immunocompromised as well as immunocompetent patients. As previously described, the common contaminating microorganisms (e.g., coagulase-negative staphylococci or viridans group streptococci) can also be implicated in invasive disease. Conversely, positive blood cultures for yeasts must be assumed to represent disease, whereas negative results do not exclude it. Clinicians should interpret cautiously the clinical relevance of recovering molds or nontuberculous mycobacteria from blood cultures drawn through catheters because of possible environmental contamination with mold spores or asymptomatic colonization at the site of an indwelling line. Features in the clinical history (e.g., extensive immunosuppressive therapy, prolonged fever, and embolic events) may aid in clinical differentiation between infection and contamination. Collection of two sets of blood cultures from peripheral blood before initiation of antimicrobial therapy is the optimal approach for determining the clinical relevance of recovered microorganisms.

Culture-Independent Methods

Bacteria

Broad-range DNA target amplification and sequencing from blood. Broad-range gene PCR and sequencing from clinical material provide an alternative, culture-independent method for detecting pathogens. The promise of this technique has remained strong, especially for the bacterial domain, but its application has not been reliably reproducible owing in large part to issues with contamination and inadequate sensitivity (12). The gene target most commonly used to detect bacteria from a blood sample is the 16S rRNA gene, an ~1,500-base-pair gene that codes for a portion of the 30S ribosome. Although investigators have demonstrated the ability to detect bacteria directly from blood with 16S rRNA gene amplification and sequencing (63), they should not be performed routinely since culture remains a more sensitive diagnostic test. Also, broad-range PCR is associated with contamination and a positive result must be correlated with clinical history. When fastidious pathogens (*Bartonella* sp. or *Coxiella burnetii*) are suspected as the cause of bacteremia, serology and species-specific PCR from blood are the optimal diagnostic strategies. Finally, new rapid sepsis panels based on real-time PCR technology are in various stages of development for both in-house-developed and commercial assays. The advantage of this technology for immunocompromised hosts is its ability to rapidly detect and identify bacterial and fungal pathogens from whole blood samples by using broad-range primers. Its primary limitation is sensitivity, and until multicenter clinical trials are performed, it should be used as an adjunctive tool to routine blood culture.

Urinary antigen detection. Detection of *Legionella pneumophila* and *Streptococcus pneumoniae* in urine is a rapid and sensitive method for making an early diagnosis and enabling prompt initiation of targeted antibacterial therapy (19, 62); however, published performance characteristics of these assays for bacteremia are absent in the literature. The *Legionella* urinary antigen test detects only *Legionella pneumophila* type 1, which is responsible for the majority of invasive cases of *Legionella* sp. infection. False-positive tests may result with the pneumococcal antigen assay; its specificity is limited by cross-reactivity with members of the *Streptococcus mitis* group.

Fungi

Antigen-based methods. Detection of fungal cell wall and cytoplasmic antigens in serum can provide a rapid diagnosis of invasive fungal disease. The best examples are detection of polysaccharide antigens of *Cryptococcus neoformans* and *Histoplasma capsulatum* for the rapid diagnosis of cryptococcosis and disseminated histoplasmosis, respectively. These tests are extremely valuable for patients with advanced human immunodeficiency virus disease or recipients of solid organ transplants in whom disseminated disease is suspected. For the rapid diagnosis of other opportunistic fungi, commercial products are available to detect galactomannan for diagnosis of invasive *Aspergillus* infection and (1,3)-β-D-glucan (beta-glucan) for the diagnosis of invasive infections from *Candida* sp., *Aspergillus* sp., and other medically important fungi (*Fusarium* sp., *Trichosporon* sp., *Saccharomyces* sp., and *Acremonium* sp.) (37, 44, 46). Circulating galactomannan antigen can be detected in approximately two-thirds of patients with invasive aspergillosis at a mean of 8 days before diagnosis by conventional means (37). Test performance for galactomannan is variable. Sensitivity appears to be highest among hematopoietic stem cell recipients (33) and, although data are limited, appears significantly lower for solid organ transplant recipients and for patients receiving antifungal therapy (35). Additionally, test performance is limited by false-positive results that can occur in patients receiving piperacillin-tazobactam (71) and amoxicillin-clavulanate (36) therapy. Unlike galactomannan, beta-glucan is a cell wall component present in many medically important fungi, with the notable exceptions of *Cryptococcus* species and zygomycetes (44, 46). A single negative test result has been reported to have a high negative predictive value, whereas results can be falsely positive from poor handling of the specimen, hemodialysis with some cellulose membranes, exposure to certain types of gauze, and recent receipt of immunoglobulin or albumin products (44). When clinical suspicion of invasive aspergillosis or fungal infection is high, serial blood sampling for galactomannan or beta-glucan antigen testing should be performed to improve both test sensitivity and specificity. Positive tests should prompt consideration of broad-spectrum antifungal therapy and further diagnostic workup, including correlation with radiological and other microbiological investigations. Additionally, galactomannan and beta-glucan testing can be used as tests complementary to each other to improve the confidence in individual test results.

Molecular detection of fungi by PCR with blood may serve as an adjunctive tool for fungal diagnosis, but it has limited sensitivity and specificity. The target sequences vary widely; those used most commonly are ribosomal genes (18S rRNA gene) or internal transcribed spacer regions (22, 78). Sensitivity ranges from 78% to 100% for candidiasis and from 33% to 100% for patients with proven invasive aspergillosis (78). Similar to molecular diagnosis of bacteremia, further study is warranted in this area before recommending PCR use in routine clinical practice for the diagnosis of fungemia.

MANAGEMENT

When a clinically significant microorganism has been isolated from a patient with bacteremia, susceptibility testing should be performed according to the methods described by the Clinical and Laboratory Standards Institute (CLSI) (9). In the 24 to 48 hours when susceptibility results are not available, clinicians are encouraged to rely on their institution's antibiograms to help to guide their selection of antibacterial therapy. Although some organisms, such as viridans group streptococci, have predictable patterns and susceptibility testing may not be performed, clinicians should be aware that unusual resistance patterns can occur among immunocompromised hosts. For example, *Streptococcus mitis* and *Streptococcus salivarius*, two microorganisms considered to be very susceptible to penicillin, have been observed to have high penicillin MICs when recovered from patients with febrile neutropenia (7). Multidrug-resistant gram-negative bacillary infections are also becoming a serious problem in immunocompromised hosts (59). Depending on the institution's frequency of resistance owing to extended-spectrum beta-lactamases or *ampC*-mediated mechanisms, empirical therapy with a carbapenem may be warranted. For infections associated with intravascular devices, removal of the catheter is often necessary. Microorganisms can be embedded within a biofilm layer and thereby be resistant to the activity of antibiotics (55, 65), thwarting eradication of bloodstream infection without catheter removal (57). The presence of persistent bacteremia should raise suspicion of an endovascular source, such as endocarditis or a metastatic focus (e.g., vertebral osteomyelitis or renal abscess). Endocarditis in granulocytopenic patients is unusual, because inflammation-mediated processes and platelet deposition are critical for bacterial colonization of the endocardium. In other immunocompromised hosts, however, clinical suspicion of bacterial endocarditis should remain high. Further evaluation with transesophageal echocardiography may be required to assess the presence or absence of valvular vegetations. Depending on the clinical context, investigations for other metastatic foci may warrant radiological imaging, such as computed tomography or magnetic resonance imaging.

Susceptibility testing is appropriate for candidemia, particularly for cases of treatment failure or as supporting evidence that patients may be converted safely from intravenous to oral therapy. While awaiting testing results, susceptibility patterns can often be predicted based on the species of *Candida* isolated and the institution's specific antibiogram profiles. Fluconazole is appropriate as empiric therapy for institutions with low rates of non-*C. albicans* candidemia, except for patients with prior triazole exposure. Indications for voriconazole and echinocandins as empirical therapy are dictated by institution-specific trends for recovery of non-*C. albicans* yeast. Echinocandins generally are very effective for patients with candidemia, but notable exceptions include infections with *Candida guilliermondii* and *Trichosporon* sp., which reportedly have high echinocandin MICs.

CLSI has developed and published approved methods for broth dilution testing (CLSI approved standard M27-A2) (42) and disk diffusion testing (CLSI approved guideline M44-A) (40) of yeast. These methods are accurate and reproducible and are being used in clinical laboratories. Interpretative breakpoints are available only for *Candida* spp. treated with fluconazole, itraconazole, and flucytosine, and the clinical relevance of testing any other organism-drug combination remains uncertain. A "susceptible dose-dependent" category exists for species with elevated triazole MICs and denotes that higher doses may provide clinical efficacy. Susceptibility testing for filamentous fungi has been more challenging to standardize. CLSI has developed methods for MIC testing of filamentous fungi (CLSI document M38-A) (41), but more clinical data for correlating in vitro results with in vivo responses are necessary. Generally speaking, for filamentous molds, the term "resistance" testing is used informally as a means to predict therapeutic failures.

For fungemia, all central venous catheters should be removed when feasible. All patients should undergo at least one ophthalmologic examination to exclude endophthalmitis, and patients with fungal endocarditis require immediate cardiothoracic surgical evaluation. Breakthrough or persistence of candidemia (or other fungemia) during therapy suggests an infected intravascular device, profound immunosuppression, or microbiologic resistance. The duration of antifungal therapy is highly variable, depending on several factors, including the type of microorganism recovered (e.g., *Candida albicans* versus *Fusarium* sp.), degree of underlying immunosuppression, and presence of metastatic foci.

CONCLUSION

Over a century has passed since the introduction of blood cultures for the diagnosis of bloodstream infections. Significant advances have been made in improving blood culture media; to date, no molecular or antigen-based method has proven superior for the detection of bacteremia. Our ability to make a rapid and reliable diagnosis for fungemia, however, remains elusive. Laboratorians can continue to optimize the use of blood cultures for the diagnosis of bloodstream infections and to encourage their clinicians to obtain at least two sets of blood cultures by separate venipuncture before starting antimicrobial therapy. When blood cultures are obtained through indwelling lines or from a single venipuncture, positive blood culture results can be misleading and result in diagnostic confusion and injudicious use of antimicrobial therapy. Although blood cultures remain the best means of diagnosing bacteremia and candidemia, complementary testing with antigen tests, microbiologic investigations from other body sites, and histopathology can often aid in the diagnosis of disseminated disease.

REFERENCES

1. **Alexander, B. D., E. D. Ashley, L. B. Reller, and S. D. Reed.** 2006. Cost savings with implementation of PNA FISH testing for identification of Candida albicans in blood cultures. *Diagn. Microbiol. Infect. Dis.* **54:**277–282.
2. **Archibald, L. K., L. C. McDonald, S. Rheanpumikankit, S. Tansuphaswakikul, A. Chaovanich, B. Eampokalap, S. N. Banerjee, L. B. Reller, and W. R. Jarvis.** 1999. Fever and human immunodeficiency virus infection as sentinels for emerging mycobacterial and fungal bloodstream infections in hospitalized patients >15 years old, Bangkok. *J. Infect. Dis.* **180:**87–92.
3. **Aronson, M. D., and D. H. Bor.** 1987. Blood cultures. *Ann. Intern. Med.* **106:**246–253.
4. **Blumberg, H. M., W. R. Jarvis, J. M. Soucie, J. E. Edwards, J. E. Patterson, et al.** 2001. Risk factors for candidal bloodstream infections in surgical intensive care unit patients: the NEMIS prospective multicenter study. *Clin. Infect. Dis.* **33:**177–186.
5. **Boutati, E. I., E. Anaissie, and J. Fusarium.** 1997. A significant emerging pathogen in patients with hematologic malignancy: ten years' experience at a cancer center and implications for management. *Blood* **90:**999–1008.
6. **Bryant, J. K., and C. L. Strand.** 1987. Reliability of blood cultures collected from intravascular catheter versus venipuncture. *Am. J. Clin. Pathol.* **88:**113–116.
7. **Carratala, J., F. Alcaide, A. Fernandez-Sevilla, X. Corbella, J. Linares, and F. Gudiol.** 1995. Bacteremia due to viridans streptococci that are highly resistant to penicillin: increase among neutropenic patients with cancer. *Clin. Infect. Dis.* **20:**1169–1173.
8. **Carratala, J., F. Sevilla, F. Tubau, M. A. Dominguez, and F. Gudiol.** 1996. Emergence of fluoroquinolone-resistant *Escherichia coli* in fecal flora of cancer patients receiving norfloxacin prophylaxis. *Antimicrob. Agents Chemother.* **40:** 503–505.
9. **Clinical and Laboratory Standards Institute.** 2007. Performance standards for antimicrobial susceptibility testing; 17th informational supplement (M100-S17). Clinical and Laboratory Standards Institute, Wayne, PA.
10. **Clinical and Laboratory Standards Institute.** 2007. Principles and procedures for blood cultures; proposed guideline (M47-P). Clinical and Laboratory Standards Institute, Wayne, PA.
11. **Cockerill, F. R., J. W. Wilson, E. A. Vetter, K. M. Goodman, C. A. Torgerson, W. S. Harmsen, C. D. Schleck, D. M. Ilstrup, J. A. Washington, and W. R. Wilson.** 2004. Optimal testing parameters for blood cultures. *Clin. Infect. Dis.* **38:**1724–1730.
12. **Corless, C. E., M. Guiver, R. Borrow, et al.** 2000. Contamination and sensitivity issues with a real-time universal 16S rRNA PCR. *J. Clin. Microbiol.* **38:**1747–1752.
13. **Crump, J. A., D. C. Tanner, S. Mirrett, C. M. McKnight, and L. B. Reller.** 2003. Controlled comparison of BACTEC 13A, MYCO/F LYTIC, BacT/ALERT MB, and ISOLATOR 10 systems for detection of mycobacteremia. *J. Clin. Microbiol.* **41:** 1987–1990.
14. **De Groote, M. A., and G. Huitt.** 2006. Infections due to rapidly growing mycobacteria. *Clin. Infect. Dis.* **42:**1756–1763.
15. **DesJardin, J. A., M. E. Falagas, R. Ruthazer, J. Griffith, D. Wawrose, D. Schenkein, K. Miller, and D. R. Snydman.** 1999. Clinical utility of blood cultures drawn from indwelling central venous catheters in hospitalized patients with cancer. *Ann. Intern. Med.* **131:**641–647.
16. **Doucette, K., and J. A. Fishman.** 2004. Nontuberculous mycobacterial infection in hematopoietic stem cell and solid organ transplant recipients. *Clin. Infect. Dis.* **38:**1428–1439.
17. **Duthie, R., and D. W. Denning.** 1995. *Aspergillus* fungemia: report of two cases and review. *Clin. Infect. Dis.* **20:**598–605.
18. **Edmond, M. B., S. E. Wallace, D. K. McClish, M. A. Pfaller, R. N. Jones, and R. P. Wenzel.** 1999. Nosocomial bloodstream infections in United States hospitals: a three-year analysis. *Clin. Infect. Dis.* **29:**239–244.
19. **Guerrero, C., C. M. Toldos, G. Yague, C. Ramirez, T. Rodriguez, and M. Segovia.** 2004. Comparison of diagnostic sensitivities of three assays (Bartels enzyme immunoassay [EIA], Biotest EIA, and Binax NOW immunochromatographic test) for detection of *Legionella pneumophila* serogroup 1 antigen in urine. *J. Clin. Microbiol.* **42:**467–468.
20. **Hsueh, P. R., L. J. Teng, S. W. Ho, and D. T. Luh.** 2003. Catheter-related sepsis due to *Rhodotorula glutinis*. *J. Clin. Microbiol.* **41:**857–859.
21. **Isaacman, D. J., and R. B. Karasic.** 1990. Utility of collecting blood cultures through newly inserted intravenous catheters. *Pediatr. Infect. Dis. J.* **9:**815–818.
22. **Iwen, P. C., S. H. Hinrichs, and M. E. Rupp.** 2002. Utilization of the internal transcribed spacer regions as molecular targets to detect and identify human fungal pathogens. *Med. Mycol.* **40:**87–109.
23. **Kempf, V. A., K. Trebesius, and I. B. Autenrieth.** 2000. Fluorescent in situ hybridization allows rapid identification of microorganisms in blood cultures. *J. Clin. Microbiol.* **38:**830–838.
24. **Khoo, S. H., and D. W. Denning.** 1994. Invasive aspergillosis in patients with AIDS. *Clin. Infect. Dis.* **19:**S41–S48.
25. **Kilby, J. M., M. B. Marques, D. L. Jaye, P. B. Tabereaux, V. B. Reddy, and K. B. Waites.** 1998. The yield of bone marrow biopsy and culture compared with blood culture in the evaluation of HIV-infected patients for mycobacterial and fungal infections. *Am. J. Med.* **104:**123–128.
26. **Klastersky, J., S. H. Zinner, T. Calandra, H. Gaya, M. P. Glauser, F. Meunier, M. Rossi, S. C. Schimpff, M. Tattersall, and C. Viscoli.** 1988. Empiric antimicrobial therapy for febrile granulocytopenic cancer patients: lessons from four EORTC trials. *Eur. J. Cancer Clin. Oncol.* **24:** S35–S45.
27. **Kontoyiannis, D. P., H. A. Torres, M. Chagua, R. Hachem, J. J. Tarrand, G. P. Bodey, and I. I. Raad.** 2004. Trichosporonosis in a tertiary care cancer center: risk factors, changing spectrum and determinants of outcome. *Scand. J. Infect. Dis.* **36:**564–569.
28. **La Scola, B., and D. Raoult.** 1999. Culture of *Bartonella quintana* and *Bartonella henselae* from human samples: a 5-year experience (1993 to 1998). *J. Clin. Microbiol.* **37:** 1899–1905.
29. **LeBlanc, L., J. Pepin, K. Toulouse, M. F. Ouellette, M. A. Coulombe, M. P. Corriveau, and M. E. Alary.** 2006. Fluoroquinolones and risk of methicillin-resistant *Staphylococcus aureus*, Canada. *Emerg. Infect. Dis.* **12:**1398–1405.
30. **MacGregor, R. R., and H. N. Beaty.** 1972. Evaluation of positive blood cultures. Guidelines for early differentiation of contaminated from valid positive cultures. *Arch. Intern. Med.* **130:**84–87.

31. **Maki, D. G., S. M. Stolz, S. Wheeler, and L. A. Mermel.** 1997. Prevention of central venous catheter-related bloodstream infection by use of an antiseptic-impregnated catheter. A randomized, controlled trial. *Ann. Intern. Med.* **127:**257–266.
32. **Marlowe, E. M., J. J. Hogan, J. F. Hindler, I. Andruszkiewicz, P. Gordon, and D. A. Bruckner.** 2003. Application of an rRNA probe matrix for rapid identification of bacteria and fungi from routine blood cultures. *J. Clin. Microbiol.* **41:**5127–5133.
33. **Marr, K. A., S. A. Balajee, L. McLaughlin, M. Tabouret, C. Bentsen, and T. J. Walsh.** 2004. Detection of galactomannan antigenemia by enzyme immunoassay for the diagnosis of invasive aspergillosis: variables that affect performance. *J. Infect. Dis.* **190:**641–649.
34. **Marr, K. A., R. A. Carter, M. Boeckh, P. Martin, and L. Corey.** 2002. Invasive aspergillosis in allogeneic stem cell transplant recipients: changes in epidemiology and risk factors. *Blood* **100:**4358–4366.
35. **Marr, K. A., M. Laverdiere, A. Gugel, and W. Leisenring.** 2005. Antifungal therapy decreases sensitivity of the *Aspergillus* galactomannan enzyme immunoassay. *Clin. Infect. Dis.* **40:**1762–1769.
36. **Mattei, D., D. Rapezzi, N. Mordini, F. Cuda, C. L. Nigro, M. Musso, A. Arnelli, S. Cagnassi, and A. Gallamini.** 2004. False-positive *Aspergillus* galactomannan enzyme-linked immunosorbent assay results in vivo during amoxicillin-clavulanic acid treatment. *J. Clin. Microbiol.* **42:**5362–5363.
37. **Mennink-Kersten, M. A., J. P. Donnelly, and P. E. Verweij.** 2004. Detection of circulating galactomannan for the diagnosis and management of invasive aspergillosis. *Lancet Infect. Dis.* **4:**349–357.
38. **Mirrett, S., L. B. Reller, C. A. Petti, C. W. Woods, B. Vazirani, R. Sivadas, and M. P. Weinstein.** 2003. Controlled clinical comparison of BacT/ALERT standard aerobic medium with BACTEC standard aerobic medium for culturing blood. *J. Clin. Microbiol.* **41:**2391–2394.
39. **Nachman, S., O. Alpan, R. Malowitz, and E. D. Spitzer.** 1996. Catheter-associated fungemia due to *Wangiella (Exophiala) dermatitidis*. *J. Clin. Microbiol.* **34:**1011–1013.
40. **National Committee for Clinical Laboratory Standards.** 2004. Reference method for antifungal disk diffusion susceptibility testing of yeasts: approved standard M44-A. National Committee for Clinical Laboratory Standards, Wayne, PA.
41. **National Committee for Clinical Laboratory Standards.** 2002. Reference method for broth dilution antifungal susceptibility testing of filamentous fungi: approved standard M38-A. National Committee for Clinical Laboratory Standards, Wayne, PA.
42. **National Committee for Clinical Laboratory Standards.** 2002. Reference method for broth dilution antifungal susceptibility testing of yeasts: approved standard M27-A2, 2nd ed. National Committee for Clinical Laboratory Standards, Wayne, PA.
43. **Nucci, M., and K. A. Marr.** 2005. Emerging fungal diseases. *Clin. Infect. Dis.* **41:**521–526.
44. **Odabasi, Z., G. Mattiuzzi, E. Estey, H. Kantarijian, F. Saeki, R. J. Ridge, P. A. Ketchum, M. A. Finkelman, J. H. Rex, and L. Ostrosky-Zeichner.** 2004. Beta-D-glucan as a diagnostic adjunct for invasive fungal infections: validation, cutoff development, and performance in patients with acute myelogenous leukemia and myelodysplastic syndrome. *Clin. Infect. Dis.* **39:**199–205.
45. **Oplustil, C. P., O. H. M. Leite, M. S. Oliveira, S. I. Sinto, D. E. Uip, M. Boulos, and C. F. Mendes.** 2001. Detection of mycobacteria in the bloodstream of patients with acquired immunodeficiency syndrome in a university hospital in Brazil. *Braz. J. Infect. Dis.* **5:**252–259.
46. **Ostrosky-Zeichner, L., B. D. Alexander, D. H. Kett, J. Vazquez, P. G. Pappas, F. Saeki, P. A. Ketchum, J. Wingard, R. Schiff, H. Tamura, M. A. Finkelman, and J. H. Rex.** 2004. Multicenter clinical evaluation of the (1,3) beta-D-glucan assay as an aid to diagnosis of fungal infections in humans. *Clin. Infect. Dis.* **41:**654–659.
47. **Ostrosky-Zeichner, L., C. Sable, J. Sobel, B. D. Alexander, G. Donowitz, V. Kan, C. A. Kauffman, D. Kett, et al.** 2007. Multicenter retrospective development and validation of a clinical prediction rule for nosocomial invasive candidiasis in the intensive care setting. *Eur. J. Clin. Microbiol. Infect. Dis.* **26:**271–276.
48. **Pacios, E., L. Alcala, M. J. Ruiz-Serrano, D. G. de Viedma, M. Rodriguez-Creixems, M. Marin-Arriaza, J. Berenguer, and E. Bouza.** 2004. Evaluation of bone marrow and blood cultures for the recovery of mycobacteria in the diagnosis of disseminated mycobacterial infections. *Clin. Microbiol. Infect.* **10:**734–737.
49. **Pappas, P. G., J. H. Rex, J. Lee, R. J. Hamill, R. A. Larsen, W. Powderly, C. A. Kauffman, N. Hyslop, J. E. Mangino, S. Chapman, H. W. Horowitz, J. E. Edwards, W. E. Dismukes, et al.** 2003. A prospective observational study of candidemia: epidemiology, therapy, and influences on mortality in hospitalized adult and pediatric patients. *Clin. Infect. Dis.* **37:**634–643.
50. **Peters, R. P., P. H. M. Savelkoul, A. M. Simoons-Smit, S. A. Danner, C. M. Vandenbroucke-Grauls, and M. A. van Agtmael.** 2006. Faster identification of pathogens in positive blood cultures by fluorescence in situ hybridization in routine practice. *J. Clin. Microbiol.* **44:**119–123.
51. **Peters, R. P., M. A. van Agtmael, A. M. Simoons-Smit, S. A. Danner, C. M. Vandenbroucke-Grauls, and P. H. Savelkoul.** 2006. Rapid identification of pathogens in blood cultures with a modified fluorescence in situ hybridization assay. *J. Clin. Microbiol.* **44:**4186–4188.
52. **Petti, C. A., H. A. Bally, M. P. Weinstein, K. Joho, T. Wakefield, L. B. Reller, and K. C. Carroll.** 2006. Utility of extended blood culture incubation for isolation of *Haemophilus*, *Actinobacillus*, *Cardiobacterium*, *Eikenella*, and *Kingella* organisms: a retrospective multicenter evaluation. *J. Clin. Microbiol.* **44:**257–259.
53. **Petti, C. A., A. K. Zaidi, S. Mirrett, and L. B. Reller.** 1996. Comparison of Isolator 1.5 and BACTEC NR660 aerobic 6A blood culture systems for detection of fungemia in children. *J. Clin. Microbiol.* **34:**1877–1879.
54. **Pfaller, M. A., and D. J. Diekema.** 2004. Rare and emerging opportunistic fungal pathogens: concern for resistance beyond *Candida albicans* and *Aspergillus fumigatus*. *J. Clin. Microbiol.* **42:**4419–4431.
55. **Pfaller, M. A., S. A. Messer, and R. J. Hollis.** 1995. Variations in DNA subtype, antifungal susceptibility, and slime production among clinical isolates of *Candida parapsilosis*. *Diagn. Microbiol. Infect. Dis.* **21:**9–14.
56. **Raad, I., W. Costerton, U. Sabharwal, M. Sacilowski, E. Anaissie, and G. P. Bodey.** 1993. Ultrastructural analysis of indwelling vascular catheters: a quantitative relationship between luminal colonization and duration of placement. *J. Infect. Dis.* **168:**400–407.
57. **Raad, I. I., and H. H. Hanna.** 2002. Intravascular catheter-related infections: new horizons and recent advances. *Arch. Intern. Med.* **162:**871–878.
58. **Ramos-e-Silva, M., and A. L. Pereira.** 2005. Life-threatening eruptions due to infectious agents. *Clin. Dermatol.* **23:**148–156.

59. **Ramphal, R.** 2004. Changes in the etiology of bacteremia in febrile neutropenic patients and the susceptibilities of the currently isolated pathogens. *Clin. Infect. Dis.* **39:**S25–S31.
60. **Reimer, L. G., M. L. Wilson, and M. P. Weinstein.** 1997. Update on detection of bacteremia and fungemia. *Clin. Microbiol. Rev.* **10:**444–465.
61. **Rohner, P., B. Pepey, and R. Auckenthaler.** 1997. Advantage of combining resin with lytic BACTEC blood culture media. *J. Clin. Microbiol.* **35:**2634–2638.
62. **Roson, B., N. Fernandez-Sabe, J. Carratala, R. Verdaguer, J. Dorca, F. Manresa, and F. Gudiol.** 2004. Contribution of a urinary antigen assay (Binax NOW) to the early diagnosis of pneumococcal pneumonia. *Clin. Infect. Dis.* **38:**222–226.
63. **Rothman, R. E., M. D. Majmudar, G. D. Kelen, G. Madico, C. A. Gaydos, T. Walker, and T. C. Quinn.** 2002. Detection of bacteremia in emergency department patients at risk for infective endocarditis using universal 16S rRNA primers in a decontaminated polymerase chain reaction assay. *J. Infect. Dis.* **11:**1677–1681.
64. **Safdar, N., and D. G. Maki.** 2004. The pathogenesis of catheter-related bloodstream infection with noncuffed short-term central venous catheters. *Intensive Care Med.* **30:**62–67.
65. **Sheth, N. K., T. R. Franson, and P. G. Sohnle.** 1985. Influence of bacterial adherence to intravascular catheters on in-vitro antibiotic susceptibility. *Lancet* **ii:**1266–1268.
66. **Tan, T. Q., A. K. Ogeden, J. Tillman, G. J. Demmler, and M. G. Rinaldi.** 1992. *Paecilomyces lilacinus* catheter-related fungemia in an immunocompromised pediatric patient. *J. Clin. Microbiol.* **30:**2479–2483.
67. **Tonnesen, A., M. Peuler, and W. R. Lockwood.** 1976. Cultures of blood drawn by catheters versus venipuncture. *JAMA* **235:**1877.
68. **Towns, M. L., and L. B. Reller.** 2002. Current best practices and guidelines for isolation of bacteria and fungi in infective endocarditis. *Infect. Dis. Clin. N. Am.* **16:**363–376.
69. **Trick, W. E., S. K. Fridkin, J. R. Edwards, R. A. Hajjeh, R. P. Gaynes, et al.** 2002. Secular trend of hospital-acquired candidemia among intensive care unit patients in the United States during 1989–1999. *Clin. Infect. Dis.* **35:**627–630.
70. **Van Scoy, R. E.** 1982. Culture-negative endocarditis. *Mayo Clin. Proc.* **57:**149–154.
71. **Viscoli, C., M. Machetti, P. Cappellano, et al.** 2004. False-positive galactomannan Platelia Aspergillus test results for patients receiving piperacillin-tazobactam. *Clin. Infect. Dis.* **38:**913–916.
72. **von Gottberg, A., L. Sacks, S. Machala, and L. Blumberg.** 2001. Utility of blood cultures and incidence of mycobacteremia in patients with suspected tuberculosis in a South African infectious disease referral hospital. *Int. J. Tuberc. Lung Dis.* **5:**80–86.
73. **Weinstein, M. P., M. L. Towns, S. M. Quartey, S. Mirrett, L. G. Reimer, G. Parmigiani, and L. B. Reller.** 1997. The clinical significance of positive blood cultures in the 1990s: a prospective comprehensive evaluation of the microbiology, epidemiology, and outcome of bacteremia and fungemia in adults. *Clin. Infect. Dis.* **24:**584–602.
74. **Weinstein, M. P.** 2003. Blood culture contamination: persisting problems and partial progress. *J. Clin. Microbiol.* **41:** 2275–2278.
75. **Wellinghausen, N., B. Wirths, A. Essig, and L. Wassill.** 2004. Evaluation of the Hyplex BloodScreen multiplex PCR–enzyme-linked immunosorbent assay system for direct identification of gram-positive cocci and gram-negative bacilli from positive blood cultures. *J. Clin. Microbiol.* **42:**3147–3152.
76. **Wilson, M. L., T. E. Davis, S. Mirrett, J. Reynolds, D. Fuller, S. D. Allen, K. K. Flint, F. Koontz, and L. B. Reller.** 1993. Controlled comparison of the BACTEC high-blood-volume fungal medium, BACTEC Plus 26 aerobic blood culture bottle, and 10-milliter Isolator blood culture system for detection of fungemia and bacteremia. *J. Clin. Microbiol.* **31:**865–871.
77. **Wisplinghoff, H., T. Bischoff, S. M. Tallent, H. Seifert, R. P. Wenzel, and M. B. Edmond.** 2004. Nosocomial bloodstream infections in US hospitals: analysis of 24,179 cases from a prospective nationwide surveillance study. *Clin. Infect. Dis.* **39:**309–317.
78. **Yeo, S. F., and B. Wong.** 2002. Current status of nonculture methods for diagnosis of invasive fungal infections. *Clin. Microbiol. Rev.* **15:**465–484.

Diagnostic Microbiology of the Immunocompromised Host
Edited by Randall T. Hayden, Karen C. Carroll, Yi-Wei Tang, and Donna M. Wolk

Chapter 20

Soft Tissue and Prosthetic Device Infections

XIANG Y. HAN

Skin and soft tissue infections represent a potentially lethal risk for immunocompromised patients, yet the risks of such infections are often underappreciated. Key definitions related to soft tissue infections are worth noting, as follows. Soft tissues are the fibrous and adipose connective tissues that underline skin or surround solid organs. Soft tissue infections can occur in any host when a microorganism is introduced accidentally into these tissues. Prostheses are artificial devices intended to replace missing parts of the body; examples are prosthetic heart valves, breast implants, prosthetic joints, etc. Medical devices are instruments or devices that fulfill certain functions of or are used in medical interventions to the human body; examples are infusion pumps, Foley catheters, intravenous catheters, cardiac pacemakers, scalpels, etc. Prostheses and medical devices predispose the patient to infections due to their foreign body effects that disrupt anatomic integrity and favor microbial entry, colonization, or attachment to the device in all hosts, whether they are immunocompromised or not. However, infections in the immunocompromised host may warrant especially intense laboratory scrutiny, including further species identification for rare or unusual organisms or common commensal organisms, such as viridans group streptococci or coagulase-negative staphylococci (CoNS). Additional antimicrobial susceptibility testing may also be warranted. These data can be useful to guide prolonged antimicrobial therapy, increase our general knowledge about certain microorganisms and their pathogenic potential under conditions of predisposition, and discover novel microorganisms.

The interaction between the host and the microorganism determines whether infection takes place, its course, and its outcome. At the opposite ends of this delicate balance, rare but highly virulent microorganisms, such as *Yersinia pestis*, *Bacillus anthracis*, and *Francisella tularensis*, can strike any host, while common, minimally virulent microbes may infect only the most vulnerable hosts, such as those with immature or defective immune functions or those with prostheses or medical devices.

Several immunocompromised populations are vulnerable to soft tissue infections. First, patients with cancer have a number of immune and nonimmune defects that predispose them to infections. In addition, solid organ transplant recipients and allogeneic hematopoietic stem cell transplant (HSCT) recipients receive antirejection immunosuppressive therapies that make them vulnerable to many opportunistic infections. Moreover, graft-versus-host disease, associated with stem cell transplantation, may involve the skin, oral cavity, eyes, gastrointestinal tract, and other organs, further increasing infectious risk.

Unexpectedly, soft tissue infections and device-related infections are not particularly increased in individual patients with AIDS, even those who have a profound reduction in the number of $CD4^+$ lymphocytes and are prone to other unusual opportunistic infections. Similarly, neonates and the elderly are not at much higher risk for soft tissue and device-related infections. In light of these observations, this chapter focuses primarily on infections found in cancer patients and/or stem cell transplant recipients.

SOFT TISSUE INFECTIONS

Soft tissue infections generally refer to wound infection, cellulitis, and abscess formation. Surgery, wounds, body piercing, and accidental puncture injury are frequent risk factors. Among recipients of HSCT, graft-versus-host disease involving the skin may cause chronic wounds with frequent superinfections. Soft tissue infections are usually polymicrobial,

Xiang Y. Han • Department of Laboratory Medicine, Unit 84, The University of Texas M. D. Anderson Cancer Center, 1515 Holcombe Boulevard, Houston, TX 77030.

including aerobes and anaerobes, depending on the location of infection and mode of acquisition. Abdominal and pelvic wounds or abscesses commonly involve enteric organisms, such as members of the *Enterobacteriaceae*, enterococci, *Bacteroides*, *Clostridium*, *Peptostreptococcus*, and *Fusobacterium*. Head and neck wounds usually involve streptococci, *Haemophilus*, *Moraxella catarrhalis*, *Actinomyces*, and *Fusobacterium*. Skin wounds often harbor CoNS, *Staphylococcus aureus*, diphtheroids, and environmental organisms.

Cancer-associated infections can be caused by microorganisms originating either from body flora or from the outside environment. Opportunistic pathogens from either source are more commonly isolated than are obligate pathogens. Depending upon risk factors, the organisms involved vary accordingly, and underlying diseases frequently dictate the spectrum of infections, particularly relatively unique infections. For instance, patients with chemotherapy-associated mucositis frequently have infections with organisms from the oral or bowel flora, such as viridans group streptococci (10), *Enterobacteriaceae*, and enterococci.

Table 1 lists results of all wound cultures at our institution in 2002 from patients with cancer and/or HSCT. Nearly 80% of wound cultures contained at least one organism, and the skin flora was dominant by far, accounting for 37.3% of all isolates (1,266 of 3,396 isolates). Significant pathogens included *S. aureus*, enterococci, *Enterobacteriaceae*, anaerobes, streptococci, *Pseudomonas aeruginosa*, and yeasts (mostly *Candida*). Among all organisms, *S. aureus* was probably the most significant, considering its frequency of occurrence (14.8% of all isolates), pathogenicity, antibiotic resistance, and general hardiness.

The management of cellulitis, wound infections, and abscesses may require surgical incision, drainage, and debridement. Antibiotic treatment is frequently prolonged, particularly in the presence of neutropenia or other immune defects/weakness. Antimicrobial susceptibility tests are necessary to guide the treatment. Recognizing the possible etiologic agents by a direct Gram stain, such as in the cases of group A streptococcus or *Vibrio vulnificus* from necrotizing fasciitis and the gram-positive boxcar-shaped rod *Clostridium perfringens* from gangrene, can be life-saving. Such early characterization may allow prompt medical intervention, prior to culture-based diagnosis of the infectious agent.

Skin Infections Originating from Hematogenous Spread

Hematogenous spread and seeding of organisms from or to the soft tissue are favored in neutropenic patients during antineoplastic therapy, in HSCT recipients, and in other immunodeficient patients. We have reported cases in which hematogenous seeding of *Campylobacter curvus*, an anaerobic oral organism of low virulence, to sites of body piercing or to the liver in patients with cancer was likely (9). Similarly, formation of brain abscesses has been seen more frequently, in our experience (unpublished observation) and that of others, in patients with lymphoma and chronic lymphocytic leukemia. Presumably, the route of spread in such cases is also hematogenous; organisms such as *Nocardia* species and *Streptococcus intermedius* are common culprits.

Invasive fungal infections are also more common in immunocompromised patients. For instance, the plant fungal pathogen *Fusarium* rarely infects humans with intact immunity; however, immunocompromised individuals may develop *Fusarium* fungemia, which is associated with a high mortality rate. Seventy-two percent of all *Fusarium* infections involve the skin (20).

Lymph node dissection and removal may cause defective local drainage and defense, favoring cellulitis and spread of the causative organisms into the bloodstream. For instance, our unpublished preliminary data indicate that patients with breast cancer and head and neck cancer suffer more frequently from *Streptococcus pyogenes* cellulitis and bacteremia.

Wound Infections That Mimic a Mass or Tumor

Mycobacterial spindle cell pseudotumor is a described entity that mimics a soft tissue tumor. It occurs most commonly in patients with AIDS, either active or controlled. *Mycobacterium avium* is the most common pathogen in this setting, and the lesion may involve soft tissue, lymph nodes, and others. Immunohistochemical staining shows that the spindle cells in these lesions are derived from macrophages. An example is shown in Color Plate 10, depicting a Ziehl-Neelsen stain of a tissue biopsy from a female AIDS patient with a subcutaneous thigh mass; *M. avium* infection was diagnosed as a solitary lesion with innumerable acid-fast bacilli.

Bacillary angiomatosis, associated with *Bartonella henselae* or *Bartonella quintana*, can be viewed as a unique soft tissue infection in immunocompromised patients, particularly those with AIDS. This infection may be confused with other spindle cell soft tissue tumors, such as angioma, and diagnostic tissue biopsy is required. Diagnosis may be aided using PCR-based methods directed at the causative organisms, using either biopsy material or blood samples; such tests are now commercially available from several large reference laboratories.

Table 1. Results for 2,406 wound cultures at M. D. Anderson Cancer Center (2002) from patients with cancer or HSCT[a]

Dominant niche and organism	No. of isolates	% of all isolates
Skin, normal flora	1,266	37.3
Skin flora, unspecified mix	722	
CoNS	544	
Skin, *Staphylococcus aureus*	504	14.8
Methicillin sensitive	345	
Methicillin resistant	159	
Intestinal tract	443	13.0
Enterococci	400	11.8
Vancomycin-resistant enterococci	31	
Others	12	
Intestinal tract, *Enterobacteriaceae*	323	9.5
Enterobacter	81	
Escherichia coli	78	
Klebsiella	72	
Proteus	30	
Serratia marcescens	23	
Citrobacter	13	
Others	26	
Intestinal tract, anaerobes	116	3.4
Bacteroides	35	
Clostridium	27	
Peptostreptococcus	15	
Others	39	
Oral cavity and intestinal tract	403	11.9
Yeast, NOS	215	6.3
Streptococci, beta-hemolytic	102	3.0
Streptococci, alpha-hemolytic	57	
Candida	16	
Others	13	
Environment	267	7.9
Pseudomonas aeruginosa	163	4.8
Stenotrophomonas maltophilia	25	
Acinetobacter	16	
Others	63	
Unspecified	74	2.2
Total	3,396	100
No. of negative cultures	465	19.3
No. of positive cultures	1,941	80.7

[a]Duplicate cultures were removed.

Infections Diagnosed by FNA

Fine-needle aspiration (FNA) cytology has become a valuable diagnostic tool. A significant portion of lesions diagnosed by FNA (~25% in our experience) are of infectious etiology. This approach is particular useful for superficial lesions, such as skin or other soft tissue infections. Yet such samples can be quite challenging to the cytopathologist, the microbiologist, and the clinical microbiology laboratory. The tissue sample obtained using FNA is typically small and may be insufficient for a wide range of cultures; the second or third needle pass, intended for cultures, may not be representative of the lesion; the optimal approach for handling these FNA samples is not widely agreed upon; and cultures often fail to yield etiologic agents. Yields may be particularly poor for organisms such as *Aspergillus* species, dematiaceous molds, *Histoplasma capsulatum*, and fastidious bacteria. Thus, morphologic examination of the initial smear or on subsequent histopathology may be of primary importance. The cellular background and morphologic features of organisms may provide clues to how to utilize limited material for culture (for example, which media to inoculate). Appropriate empiric therapy may be suggested by features such as acid fastness, pseudohyphae, true hyphae, or the size or presence of budding yeast. In this regard, some experience in anatomic pathology may help microbiologists to recognize patterns of inflammatory reaction and may also provide clues as to the etiologic agent. Certainly, close interaction with cytopathologists may benefit all involved parties.

PROSTHETIC DEVICES

Prosthetic Heart Valves

Prosthetic valve endocarditis is an endovascular infection caused by bacteria or fungi on parts of a prosthetic or reconstructed native heart valve. Thanks to marked improvements in prostheses, such infections occur infrequently; the prevalence is reported to be 0.1 to 2.3% per patient year in a recent review by Piper et al. (23) and 1.5 to 6% in the work of Karchmer (14). The overall population-based incidence of endocarditis, for both native and prosthetic valves, is approximately 1.5 to 6.2 cases per 100,000 persons per year (14). Thus, prosthetic valves carry an infection risk of ~1,000 times that of native valves. The optimal diagnostic approach to suspected cases of endocarditis includes clinical manifestations, physical examination, blood culture, other laboratory investigations, and echocardiography. Years of experience show that transesophageal echocardiography is more sensitive for the detection of vegetations and yields more details on them than does transthoracic echocardiography.

Infective endocarditis produces constant shedding of bacteria or fungi into the bloodstream (27), and a positive blood culture has become one of the major criteria in establishing diagnosis (14). While there is little evidence to suggest that endocarditis specifically targets immunocompromised patients, the laboratory should be vigilant in assessing blood cultures for unusual microbes whenever immunocompromised host factors are involved. In this setting, close communication between the clinician and microbiologist is important in order to reach correct diagnosis. Culture-independent approaches, such as direct detection of microbial DNA in the blood or valve tissue through PCR, promise to become important in culture-negative cases, as reported by some large academic centers using this technology (11). This approach may be especially useful for the immunocompromised host, where low microbial loads or unusual microbes commonly present diagnostic challenges.

In addition, limitations of blood culture instruments must be considered. Based on 2 decades of experience, current automated blood culturing systems are robust enough to provide reliable culture results. Notwithstanding, some organisms, such as *Coxiella burnetii*, *Tropheryma whipplei*, *Bartonella* spp., and others, are difficult to detect by culture. Most pyogenic bacteria can be cultured readily; fastidious organisms, such as the HACEK (*Haemophilus*, *Actinobacillus*, *Cardiobacterium*, *Eikenella*, and *Kingella*) group and *Brucella*, may take 3 to 5 days to culture in an automated system.

Although infections are not necessarily more prevalent, limitations in patient immunity make knowledge of microbial causation important for selection of empiric therapy. Table 2 shows the microbial spectrum in 428 cases of prosthetic valve endocarditis from German and other European studies (23). CoNS, *S. aureus*, and streptococci were the most common organisms, followed by enterococci, HACEK organisms, fungi, and others. Culture-negative cases accounted for 7.5% of cases. CoNS, despite its relatively low virulence, causes most cases of prosthetic valve endocarditis due to its abundance as normal human bacterial flora, versatile metabolism, and biofilm-forming ability on prostheses. Compared to prosthetic valve endocarditis, native valve endocarditis is most frequently caused by organisms of higher pathogenicity, such as *S. aureus*, streptococci, and enterococci (14).

Breast Implants

Because reconstructive implants after breast cancer, mastectomy, and associated surgery or radiotherapy have a much higher infection rate than do cosmetic implants, assessments of infections are crit-

Table 2. Causative organisms in 428 cases of prosthetic valve endocarditis from German and other European studies[a]

Organism	No. of cases	% of cases
CoNS	121	28.3
Staphylococcus aureus	65	15.2
Streptococci	63	14.6
Enterococci	41	9.5
HACEK organisms	42	9.8
Fungi	22	5.0
Mixed organisms	6	1.4
Others	37	8.6
Culture negative	32	7.5
Total	428	100.1

[a]Reconstructed from reference 23.

ical. The infection complications for breast implants were reviewed recently by Pittet et al. (24). Salient features are summarized as follows. The overall incidence of infection is 2 to 2.5% for patients who undergo breast implantation. Preexisting tissue scarring, skin atrophy, postoperative ischemia, delayed wound healing, and axillary lymph node dissection are possible risks for an increased rate of infection. Most implant infections occur within days to weeks after surgery, although delayed infections (up to years) have been reported. The origins of infection include contaminated implants or saline, surgery itself or the surgical environment, the patient's underlying disease and skin or mammary ducts, and seeding from the bloodstream or far sites of infection. Skin flora (CoNS, diphtheroids, *Bacillus* spp., and *Propionibacterium acnes*) are the usual causative organisms. *S. aureus*, several types of gram-negative rods, and rapidly growing mycobacteria (RGM) have also been reported (12, 18, 24). In *S. aureus* cases, toxic shock syndrome can occur within one to a few days after surgery. Additionally, depending on the geographic area, implicated environmental organisms vary. For instance, in the southern coastal United States, RGM have been isolated relatively frequently from breast implant infections (31).

For microbiologic workup, infected wounds or exudates or removed implants should be stained and cultured for aerobes, anaerobes, and RGM. In our experience, RGM can be cultured within a few days from routine cultures of wounds or ground tissue (8). Thioglycolate broth, Trypticase soy broth, and sheep blood agar plates work well for RGM. The clinical management of breast implant infections aims to reduce the occurrence of infection by improving surgical technique, to institute perioperative antibiotic prophylaxis by risk assessment, to treat active infection, and to remove or replace the implants. Residual implant material may render antimicrobial treatment ineffective in clearing the infection.

Intravascular Catheters

Intravascular catheterization has become an important and routine part of modern medical intervention. Intravascular catheters provide a window to the bloodstream or to a target organ (such as the heart), where hemodynamics or organ functions may be measured and therapeutic drugs, nutrients, and radiological contrast materials are delivered. Unfortunately, rare adverse effects of intravascular catheters, most commonly in the form of catheter-related infections, may occur. The type of catheter, the location of insertion site, and the duration of placement all affect infection rates. In addition, hospital environment, such as hospital size, unit, and service type or underlying diseases, may also affect infection rates and the microbial spectrum of the infection. In a critical and systematic review of 200 prospective studies of intravascular catheter-related bloodstream infections (BSIs), Maki et al. (17) found that cuffed, tunneled, and surgically implanted long-term central venous catheters (CVCs) carry a very low infection risk on a daily basis (expressed as per device-day or 1,000 device-days), whereas cumulatively for the lifespan of these catheters (expressed as per device or 100 devices), they carry a relatively high infection rate. Proper procedures for the insertion and maintenance of catheters are critical to prevent catheter-related infections; such guidelines are published widely in a number of journals (21).

Colonization of catheters by microorganisms begins soon after insertion. Both microbial factors and host factors play roles in such colonization: microbes may creep along the insertion site into the catheter, whereas the host reacts to the foreign object and the increasing density of microbes through deposition of fibrin, thrombin, and other substances. Gradually, a mesh of microbes with host substances, i.e., a biofilm, forms, in which the microbes become less metabolically active and more resistant to antibiotics, and the colonized catheter becomes a source of symptomatic infection. A good example of an organism causing such infection is *P. aeruginosa*, a well-known biofilm-forming organism (3).

Clinical definitions of catheter-related infections are detailed in the 2002 publication *Guidelines for the Prevention of Intravascular Catheter-Related Infections* (21). Key clinical definitions are excerpted from this document as follows:

Localized catheter colonization: significant growth of an organism ($\geq$15 CFU) from the catheter tip, subcutaneous segment of the catheter, or the catheter hub

Exit site infection: erythema or induration within 2 cm of the catheter exit site, in the absence of a concomitant BSI and without concomitant purulence

Clinical exit site or tunnel infection: tenderness, erythema, or site induration of >2 cm from the catheter site along the subcutaneous tract of a tunneled (Hickmanor Broviac) catheter, in the absence of a concomitant BSI

Pocket infection: purulent fluid in the subcutaneous pocket of a totally implanted intravascular catheter that might or might not be associated with spontaneous rupture and drainage or necrosis of the overlapping skin, in the absence of a concomitant BSI

Infusate-related BSI: concordant growth of the same organism from the infusate and from blood cultures (preferably drawn percutaneously), with no other identifiable source of infection

Catheter-related BSI: bacteremia/fungemia in a patient with an intravascular catheter, with at least one positive blood culture obtained from a peripheral vein; clinical manifestations of infection (i.e., fever, chills, and/or hypotension) and no apparent source of the BSI except for the catheter; and one of the following—a positive semiquantitative (>15 CFU/catheter segment) or quantitative (>10^3 CFU/catheter segment) culture, whereby the same organisms (species and antibiogram) are isolated from the catheter segment and peripheral blood; simultaneous quantitative blood cultures with a >5:1 ratio for CVC versus peripheral blood; or differential period of CVC culture versus peripheral blood culture positive of >2 hours

Surveillance definitions for primary BSIs are different from clinical definitions and are defined as follows, within the context of the National Nosocomial Infections Surveillance System, now known as the National Healthcare Safety Network (NHSN). According to NHSN, catheter-associated BSIs are defined as follows:

- the catheter is a vascular access device that terminates at or close to the heart of one of the great vessels (an umbilical artery or vein catheter is considered a "central line")
- BSI is considered to be associated with a central line if the line was in during the 48-hour period before the development of the BSI; if the time frame between onset of infection and device use is >48 hours, there should be compelling evidence that the infection is related to the central line

In the 2002 report *Guidelines for the Prevention of Intravascular Catheter-Related Infections*, the CDC clearly states the following: "do not routinely culture catheter tips." Rather, culture of CVC tips should be reserved for occasions when a CVC-related BSI is suspected (1a, 21).

In light of the previous definitions, cultures of paired blood samples—one drawn from the catheter and another drawn from the peripheral vein—are essential for the diagnosis of catheter-related BSI (19). When quantitative cultures are performed, such as by the Isolator method (Wampole Laboratories, Princeton, NJ), colony counts from the catheter blood should be at least fivefold higher than those from peripheral blood. In automated culturing systems, the differential culturing time to positivity can be used, i.e., catheter blood should be positive at least 2 h sooner than the peripheral blood (2a, 19, 25). This method has also been found to be useful for pediatric oncology patients with catheters (2b).

Removed catheter tips can be cultured to track the rate of microbial colonization and to provide correlation data with blood culture results. However, the removed catheters are usually treated with antibiotics, which may reduce organism density and culture recovery. Culture of catheter tips is generally performed by the roll plate method (16) or sonication method (29). The roll plate method entails rolling the catheter tip on a blood agar plate three times; after incubation, the number of resultant colonies is counted. The result is expressed as CFU per tip. This method is simple and easy to perform and suits large or small medical centers or microbiology labs. A drawback is that the lumen side of the tip is not sampled, which can potentially miss important pathogens. The sonication method requires sonication of the catheter tip in a sterile aqueous bath to dislodge attached organisms into solution; a small fraction is then inoculated for culture. The result is expressed as CFU per ml of solution.

Table 3 summarizes catheter tip and catheter site culture results at M. D. Anderson Cancer Center. It is not surprising that catheter site cultures were frequently positive for skin flora, but recovery of *S. aureus*, enterococcus, and various gram-negative rods was also common. In both 2000 and 2005, 28% (303/1,203 sites) of CVC insertion sites were positive for at least one usually pathogenic organism, excluding the usual skin flora, such as CoNS, diphtheroids, α- and γ-hemolytic streptococci, *Bacillus* species, and molds. Catheter tips were cultured using the roll plate method in 2000 and the sonication method in 2005. Virtually all tips were cultured in 2000, whereas only about one-fifth of those were cultured in 2005. Overall, both methods picked up organisms typically regarded as cutaneous or environmental flora, likely finding their way onto catheters through the skin incision sites. The roll plate method picked up a larger proportion of skin flora than the sonication method did. The latter technique offers improved luminal sampling and may have a reduced yield of cutaneous flora due to dilutional effects of the sonicating solution. Comparing results from catheter sites versus tips, approximately 20 to 25% of insertion site cultures positive for gram-positive cocci or gram-negative rods also had positive corresponding catheter tip

Table 3. Cultures from CVC sites and catheter tips at M. D. Anderson Cancer Center

Category	CVC sites[a]		Catheter tips	
	2000	2005	2000	2005
Culture method	Wound	Wound	Roll plate	Sonication
No. of cultures	541	662	5,388	1,288
No. of cultures positive for any organism (% of cultures)	ND	ND	953 (17.7)	159 (12.3)
Polymicrobic (%)			138 (2.6)	26 (2.0)
Monomicrobic (%)			815 (15.1)	133 (10.3)
No. of cultures positive for one or more usual pathogens (%)[b]	136 (25.1)	167 (25.2)	276 (5.1)	88 (6.8)
No. of organisms (%)	159 (29.4)	176 (26.6)	1,124 (20.9)	194 (15.1)
Skin flora (%)	ND	ND	818 (15.2)	88 (6.8)
CoNS			669	67
Diphtheroids			87	16
Bacillus, α- and γ-hemolytic streptococci, molds			52	5
Normal flora, unspecified			10	
No. of gram-positive cocci (%)	67 (12.4)	92 (13.9)	140 (2.6)	51 (4.0)
MSSA[c]	20	54	70	19
MRSA[c]	38	27	14	13
Enterococci	8	7	34	17
Beta-hemolytic streptococci	1	3	2	2
Other cocci		1	20	
No. of gram-negative rods (%)	57 (10.5)	73 (11.0)	102 (1.9)	42 (3.3)
Pseudomonas aeruginosa	22	23	26	13
Enterobacter spp.	5	11	14	5
Klebsiella spp.	6	8	14	3
Escherichia coli	7	6	9	7
Stenotrophomonas maltophilia	4	7	6	5
Acinetobacter baumannii complex	2	4	10	1
Serratia marcescens	1	6	6	
Proteus mirabilis	2		3	2
Citrobacter spp.	1	2	3	1
Other rare gram-negative rods	6	6	13	5
No. of *Candida* spp.		1	13	4
No. of unspecified yeasts	35	10	51	9

[a] ND, not done. Nearly 60% of catheter site cultures contain skin flora.
[b] With or without skin flora (CoNS, diphtheroids, α- & γ-hemolytic streptococci, *Bacillus*, and molds).
[c] MSSA, methicillin-susceptible *S. aureus*; MRSA, methicillin-resistant *S. aureus*.

cultures, suggesting that successful colonization by these microbes from insertion site to catheter tip approaches 20 to 25%.

The microbial spectrum recovered from catheter tip cultures was similar to that for blood culture results in our institution (30), where skin and environmental flora, *S. aureus*, enterococcus, and enteric gram-negative rods comprised the vast majority of isolates. These data reflected the fact that in cancer patients, many BSIs are catheter related due to the frequent and prolonged use of such devices. They are different from the results of blood cultures seen in general hospitals, where skin and environmental flora are not as common (6).

A number of unusual organisms, such as *Moraxella osloensis*, *Roseomonas* spp., RGM, aerobic actinomycetes (*Streptomyces*, *Gordonia terrae*, and *Tsukamurella*), and a few newly described genera in the family *Xanthomonadaceae*, have also been reported to cause catheter-related BSIs (Table 4). *M. osloensis* is a mucosa-dwelling gram-negative bacillus, usually of low pathogenicity; in the presence of mucositis or sinusitis and a CVC, the organism may shower the bloodstream transiently, colonizing the catheter and leading to BSI (7). Other listed organisms are all of water or soil origin and likely find their way into catheters through insertion sites.

Other Implants or Devices

There are several other types of implants or devices frequently used in medical practice, such as Foley catheters, endotracheal tubes, pacemakers, orthopedic prostheses, peritoneal dialysis catheters,

Table 4. Some unusual bacterial species found in catheter-related bacteremia and other infections

Organism	Gram stain[a]	Usual niche	Risks	Reference
Moraxella osloensis	−, cocci to bacilli	Mucosa	Sinusitis, mucositis	7
RGM	+, acid-fast bacilli	Water and soil	Long-dwelling CVC	5, 8
Roseomonas	−, coccobacilli	Water	Long-dwelling CVC	2,4
Streptomyces	+, filamentous	Soil	Long-dwelling CVC	13
Gordonia terrae	+, branching bacilli	Soil	Long-dwelling CVC	22
Tsukamurella	+, branching bacilli	Soil	Long-dwelling CVC	28
Xanthomonadaceae	−, bacilli	Soil	Long-dwelling CVC	15

[a] +, gram positive; −, gram negative.

intrauterine contraceptive devices, intrathecal-peritoneal shunts, and others. Catheter-associated infections are commonly associated with environmental organisms, such as *Pseudomonas aeruginosa*, *Acinetobacter baumannii* complex, etc. Foley catheters, peritoneal dialysis catheters, and endotracheal tubes are associated with infections related to their anatomic positions more often than other types of devices or implants are; in principle, these concepts are similar to those discussed in the sections on intravascular catheters and breast implants and thus are covered here only in general terms.

Foley catheters are invariably colonized with enteric organisms upon insertion, and this colonization tends to occur at the anatomic point of insertion. Removing a colonized Foley catheter will usually relieve associated urinary tract infection. Thus, culturing Foley catheters or urine samples from them is inappropriate from the standpoint of patient management and resource utilization, and most laboratories do not accept them. Infections associated with peritoneal dialysis catheters are usually a reflection of the chronicity of underlying renal diseases, long durations of catheter placement, and chronicity of the infection itself. Methicillin-resistant *S. aureus*, with the potential to be predisposed to vancomycin resistance, remains a major challenge in this setting.

Endotracheal tubes are important sources of infection in all critical care patients. Microbial colonization of an endotracheal tube and the downward spread of microbes are the main causes of ventilator-associated pneumonias (VAPs) in critically ill patients. Environmental organisms and oropharyngeal flora are the usual VAP culprits. However, culture results from endotracheal aspirates have relatively poor predictive values for VAPs, and a clinical trial performed on immunocompetent intensive care unit patients showed that culture results from endotracheal aspirates (nonquantitative method) versus those from bronchoalveolar lavage fluid (quantitative method) did not result in any difference in clinical outcomes of VAPs or in targeted therapy (1); no comparison was made with an immunocompromised population.

CONCLUSION

Patients with prosthetic devices, cancer, or other immune defects or weakness are prone to various skin and soft tissue infections. In the immunocompromised population, the search for infectious agents must not end with the search for common organisms, since commensals and opportunistic pathogens can cause serious infection. Not surprisingly, the type of prosthesis, underlying disease, and organ or tissue involved with the infection have a major impact on the spectrum of microbes recovered. For example, biofilm formation is an important mechanism by which organisms attach to and colonize prosthetic devices. Opportunistic infections caused by organisms from within the body or from the environment comprise an increasing proportion of infections. Prevention of prosthetic device infection through improved surgical technique and postsurgical maintenance should be emphasized, rather than engaging in routine surveillance of such devices in the absence of symptoms.

In the immunocompromised population, antibiotic treatment is usually prolonged and empiric therapy must cover even unusual agents. Antimicrobial susceptibility tests may either confirm empiric treatment decisions or provide guidance on changes in therapy. The immunocompromised are especially prone to health care-associated infections, with *S. aureus* infection remaining a medical challenge due to its pathogenicity, versatile metabolism, abundance, and acquisition of antibiotic resistance.

Overall, the specialized environmental influences and circumstances surrounding prosthetic device and soft tissue infections in the immunocompromised host require an alert and well-informed laboratory staff so that potential pathogens are assessed appropriately and not disregarded as normal flora, as they might be in the culture of an immunocompetent patient. This awareness and accompanying responses must be accomplished while maintaining prudent use of valuable laboratory resources; therefore, laboratory communication with clinicians who treat immunocompromised patients as well as with infection prevention staff, who often monitor infections in these patients, is critical to the overall success of microbiological culture and other laboratory diagnostics in this population.

REFERENCES

1. **Canadian Critical Care Trials Group.** 2006. A randomized trial of diagnostic techniques for ventilator-associated pneumonia. *N. Engl. J. Med.* **355:**2619–2630.

1a. **Centers for Disease Control and Prevention.** 2002. Guidelines for the prevention of intravascular catheter-related infections. *MMWR Morb. Mortal. Wkly. Rep.* **51**(RR10)**:**1–26.

2. **De, I., K. V. Rolston, and X. Y. Han.** 2004. Clinical significance of Roseomonas species isolated from blood and catheter samples: analysis of 36 cases in patients with cancer. *Clin. Infect. Dis.* **38:**1579–1584.

2a. **Gaur, A. H., P. M. Flynn, D. J. Heine, M. A. Giannini, J. L. Shenep, and R. T. Hayden.** 2005. Diagnosis of catheter-related bloodstream infections among pediatric oncology patients lacking a peripheral culture, using differential time to detection. *Pediatr. Infect. Dis. J.* **24:**445–449.

2b. **Gaur, A. H., P. M. Flynn, M. A. Giannini, J. L. Shenep, and R. T. Hayden.** 2003. Difference in time to detection: a simple method to differentiate catheter-related from non-catheter-related bloodstream infection in immunocompromised pediatric patients. *Clin. Infect. Dis.* **37:**469–475.

3. **Greenberg, E. P.** 2003. Bacterial communication: tiny teamwork. *Nature* **424:**134.

4. **Han, X. Y., A. S. Pham, J. J. Tarrand, K. V. Rolston, L. O. Helsel, and N. P. Levett.** 2003. Bacteriological characterization of 36 strains of *Roseomonas* species and proposal of *Roseomonas mucosa* sp. nov. and *Roseomonas gilardii* subsp. *rosea* subsp. nov. *Am. J. Clin. Pathol.* **120:**256–264.

5. **Han, X. Y., A. S. Pham, J. J. Tarrand, P. K. Sood, and R. Luthra.** 2002. Rapid and accurate identification of mycobacteria by sequencing hypervariable regions of the 16S ribosomal RNA gene. *Am. J. Clin. Pathol.* **118:**796–801.

6. **Han, X. Y., and A. L. Truant.** 1999. The detection of positive blood cultures by the Accumed ESP-384 system: the clinical significance of three day testing. *Diagn. Microbiol. Infect. Dis.* 33:1–6.

7. **Han, X. Y., and J. J. Tarrand.** 2004. *Moraxella osloensis* blood and catheter infections: clinical and microbiological studies of 10 cases. *Am. J. Clin. Pathol.* **121:**581–588.

8. **Han, X. Y., I. De, and K. L. Jacobson.** 2007. Clinical and microbiologic studies of 115 cases of rapidly growing mycobacteria. *Am. J. Clin. Pathol.* **128:**612–621.

9. **Han, X. Y., J. J. Tarrand, and D. C. Rice.** 2005. Oral *Campylobacter* species involved in extraoral abscess: a report of three cases. *J. Clin. Microbiol.* **43:**2513–2515.

10. **Han, X. Y., M. Kamana, and K. V. Rolston.** 2006. Viridans streptococci isolated by cultures from blood of cancer patients: clinical and microbiologic analysis of 50 cases. *J. Clin. Microbiol.* **44:**160–165.

11. **Houpikian, P., and D. Raoult.** 2005. Blood culture-negative endocarditis in a reference center: etiologic diagnosis of 348 cases. *Medicine* (Baltimore) **84:**162–173.

12. **Johnson, L. B., M. J. Busuito, and R. Khatib.** 2000. Breast implant infection in a cat owner due to Pasteurella multocida. *J. Infect.* **41:**110–111.

13. **Kapadia, M., K. V. Rolston, and X. Y. Han.** 2007. Invasive *Streptomyces* infections: six cases and literature review. *Am. J. Clin. Pathol.* **127:**619–624.

14. **Karchmer, A. W.** 2001. Infective endocarditis, p. 809–816. *In* E. Braunwald, A. S. Fauci, D. L. Kasper, S. L. Hauser, D. L. Longo, and J. L. Jameson (ed.), *Harrison's Principles of Internal Medicine,* 15th ed. McGraw-Hill, Inc., New York, NY.

15. **LaSala, P. R., J. Segal, F. S. Han, J. J. Tarrand, and X. Y. Han.** 2007. First reported infections caused by three newly described genera in the family *Xanthomonadaceae*. *J. Clin. Microbiol.* **45:**641–644.

16. **Maki, D. G., C. E. Weise, and H. W. Sarafin.** 1977. A semiquantitative culture method for identifying intravenous-catheter-related infection. *N. Engl. J. Med.* **296:**1305–1309.

17. **Maki, D. G., D. M. Kluger, and C. J. Crnich.** 2006. The risk of bloodstream infection in adults with different intravascular devices: a systematic review of 200 published prospective studies. *Mayo Clin. Proc.* **81:**1159–1171.

18. **Memish, Z. A., M. Alazzawi, and R. Bannatyne.** 2001. Unusual complication of breast implants: *Brucella* infection. *Infection* **29:**291–292.

19. **Mermel, L. A., B. M. Farr, R. J. Sherertz, I. I. Raad, N. O'Grady, J. S. Harris, and D. E. Craven.** 2001. Guidelines for the management of intravascular catheter related infections. *Clin. Infect. Dis.* **32:**1249–1272.

20. **Nucci, M. and E. Anaissie.** 2002. Cutaneous infection by *Fusarium* species in healthy and immunocompromised hosts: implications for diagnosis and management. *Clin. Infect. Dis.* **35:**909–920.

21. **O'Grady, N. P., M. Alexander, E. P. Dellinger, J. L. Gerberding, S. O. Heard, D. G. Maki, H. Masur, R. D. McCormick, L. A. Mermel, M. L. Pearson, I. I. Raad, A. Randolph, and R. A. Weinstein.** 2002. Guidelines for the prevention of intravascular catheter-related infections. *MMWR Recommend. Rep.* **51:**1–31.

22. **Pham, A. S., I. Dé, K. V. Rolston, J. J. Tarrand, and X. Y. Han.** 2003. Catheter-related bacteremia caused by the nocardioform actinomycete *Gordonia terrae*. *Clin. Infect. Dis.* **36:**524–527.

23. **Piper, C., R. Korfer, and D. Horstkotte.** 2001. Prosthetic valve endocarditis. *Heart* **85:**590–593.

24. **Pittet, B., D. Montandon, and D. Pittet.** 2005. Infection in breast implants. *Lancet Infect. Dis.* **5:**94–106.

25. **Raad, I., H. A. Hanna, B. Alakech, I. Chatzinikolaou, M. M. Johnson, and J. Tarrand.** 2004. Differential time to positivity: a useful method for diagnosing catheter-related bloodstream infections. *Ann. Intern. Med.* **140:**18–25.

26. Reference deleted.

27. **Reimer, L. G., M. L. Wilson, and M. P. Weinstein.** 1997. Update on detection of bacteremia and fungemia. *Clin. Microbiol. Rev. 10:*444–465.

28. **Schwartz, M. A., S. R. Tabet, A. C. Collier, C. K. Wallis, L. C. Carlson, T. T. Nguyen, M. M. Kattar, and M. B. Coyle.** 2002. Central venous catheter-related bacteremia due to *Tsukamurella* species in the immunocompromised host: a case series and review of the literature. *Clin. Infect. Dis.* **35:**e72–e77.

29. **Sherertz, R. J., I. I. Raad, A. Belani, L. C. Koo, K. H. Rand, D. L. Pickett, S. A. Straub, and L. L. Fauerbach.** 1990. Three-year experience with sonicated vascular catheter cultures in a clinical microbiology laboratory. *J. Clin. Microbiol.* **28:**76–82.

30. **Tarrand, J. J., C. Guillot, M. Wenglar, J. Jackson, J. D. Lajeunesse, and K. V. Rolston.** 1991. Clinical comparison of the resin-containing Bactec 26 Plus and the Isolator 10 blood culturing systems. *J. Clin. Microbiol.* **29:**2245–2249.

31. **Wallace, R. J., L. C. Steele, A. Labidi, and V. A. Silcox.** 1989. Heterogeneity among isolates of rapidly growing mycobacteria responsible for infections following augmentation mammaplasty despite case clustering in Texas and other southern coastal States. *J. Infect. Dis.* **160:**281–288.

Diagnostic Microbiology of the Immunocompromised Host
Edited by Randall T. Hayden, Karen C. Carroll, Yi-Wei Tang, and Donna M. Wolk

Chapter 21

Hospital-Associated Infections

FANN WU, SUSAN WHITTIER, AND PHYLLIS DELLA-LATTA

The immunocompromised host is at increased risk of acquiring hospital-associated infections (HAIs), resulting in excessive morbidity and mortality, prolonged length of stay (LOS), and considerable cost to medical facilities. Patients with cancer, diabetes, or human immunodeficiency virus infection/AIDS, low-birth-weight infants, and patients having undergone solid organ or hematopoietic transplantation are particularly at risk for infection. The predisposing factors contributing to acquisition of HAI include the administration of potent immunosuppressive agents, extended antimicrobial therapy, and the use of indwelling medical devices, as well as complicated invasive procedures and impaired or immature immune systems.

Traditionally, infections are classified as either hospital acquired (nosocomial), when symptoms manifest after 72 hours of hospital admission, or community acquired, when infections occur before this time frame (18). However, due to the current expansion of the health care network to include nursing homes, rehabilitation centers, and other outpatient facilities, HAI is often considered a better descriptive term in these settings (17, 36, 53). The most life-threatening HAIs afflicting the immunocompromised host are sepsis and pneumonia. A summary of major pathogen distribution by infection site in immunocompromised patients is shown in Table 1. In addition, the alarming emergence of antimicrobial-resistant pathogens causing these HAIs limits therapeutic options and prompts aggressive infection control and surveillance measures. Early diagnosis of these infections results in rapid administration of the most appropriate therapy, which can significantly improve clinical outcomes for this vulnerable patient population.

This chapter presents an overview of the prominent emerging pathogens that cause sepsis and respiratory disease-related HAIs in immunocompromised patients. The clinical impact of evolving antimicrobial-resistant strains and their financial burden are discussed, along with the critical role of the diagnostic microbiology laboratory in providing tests that detect and identify these pathogens in a clinically relevant time frame.

HOSPITAL-ASSOCIATED BLOODSTREAM INFECTIONS

Hospital-associated bloodstream infections account for more than half (55.3%) of all bloodstream infections in the United States and are responsible for higher morbidity and mortality rates than those with community onset (52). A major source of bloodstream HAIs is intravascular devices. Microorganisms can gain access to the bloodstream via extraluminal or intraluminal catheters and form matrix-enclosed biofilms that adhere to catheter surfaces (13, 56). Examples of biofilms include extracellular poly-*N*-acetylglucosamine of *Staphylococcus aureus* and intercellular polysaccharide adhesion of *Staphylococcus epidermidis* (47). These biofilms are a major virulence factor that may contribute to the inability of antimicrobial agents to permeate cells, possibly affecting the concentration of drugs required to achieve bactericidal activity (63). In this regard, biofilms can provide a niche for recurrent bloodstream infections and emergence of bacterial drug resistance.

Microbial contamination of infusates, such as hematopoietic stem cell products, blood components (platelets and red blood cells [RBC]), and parenteral nutritional solutions, has been reported to cause bacteremias (26, 28). Recently, the Centers for Disease Control and Prevention (CDC) reported a multistate outbreak of *Pseudomonas fluorescens* bloodstream infections associated with the use of syringes

Fann Wu, Susan Whittier, and Phyllis Della-Latta • Clinical Microbiology Services, Columbia University Medical Center, New York-Presbyterian Hospital, 622 West 168th Street, CHC 3-325, New York, NY 10032.

Table 1. Major pathogen distribution by infection site in immunocompromised patients[a]

Site	Pathogen(s) or infection				
	Hematology/oncology		Transplantation	HIV infection	
	Children	Adults		Children	Adults
Blood	CoNS, *S. aureus*, *Klebsiella* species	CoNS, *S. aureus*, *Enterococcus* sp.	Bacteremia, fungemia	Encapsulated bacteria, *Mycobacterium avium* complex	*S. pneumoniae*
Respiratory tract	Gram-negative bacilli, *S. aureus*, fungi (predominantly *Candida* sp.)	Gram-negative bacilli, *S. aureus*, fungi (predominantly *Candida* sp.)	Bacterial, fungal, cytomegalovirus	*Pseudomonas* sp., *Pneumocystis jirovecii*, *S. pneumoniae*	*Pneumocystis jirovecii*, *Cryptococcus*, *Pseudomonas* sp.
Urinary tract	*E. coli*, *Enterobacter* sp.	*E. coli*, *Enterobacter* sp.	*E. coli*	*E. coli*	
Gastrointestinal tract	*Candida*, sp., herpes simplex virus, enteric bacteria	*Candida*, sp., herpes simplex virus, enteric bacteria	Herpes simplex virus, cytomegalovirus, *Candida* sp.	*Candida* sp., herpes simplex virus, cytomegalovirus, *Clostridium difficile*	*Candida* sp., herpes simplex virus, cytomegalovirus, *Clostridium difficile*
Cutaneous	*S. aureus*, *Pseudomonas* sp.	*S. aureus*, *Pseudomonas* sp.	*S. aureus*, *Pseudomonas* sp.	Varicella-zoster virus, herpes simplex virus	Varicella-zoster virus, cytomegalovirus

[a]Refer to reference 43.

preloaded with contaminated heparin for intravenous catheter flushes, used primarily in pediatric and adult oncology patients (8). Reports of bacterial contamination of blood products indicate a rate of up to 1.2% for platelet components, which is higher than that for RBC transfusions (<0.01%) (26). This is thought to be a result of the inability to store platelets at refrigerated temperatures, unlike the case for RBC, which are routinely refrigerated. Staphylococci are the predominant bacteria isolated from contaminated infusates. However, products contaminated with gram-negative bacilli, such as *Burkholderia cepacia* and *Serratia* spp., have been associated with more serious and even fatal outcomes in hematopoietic stem cell transplant patients (26).

Predominant Pathogens and Antimicrobial Resistance

During the past 2 decades, there have been changes in the causative pathogens of bloodstream HAIs in immunocompromised patients and in their antimicrobial susceptibility patterns (43). According to a 7-year study conducted from 1995 to 2002 by the Surveillance and Control of Pathogens of Epidemiological Importance (SCOPE) project, bloodborne pathogens causing HAIs have included gram-positive bacteria (65%), gram-negative bacteria (25%), and fungi (10%) (64). Data from neutropenic and nonneutropenic patients showed that 87% of bloodstream HAIs were monomicrobial, comprised of coagulase-negative staphylococci (CoNS) (30%), *S. aureus* (12%), enterococci (10%), and *Candida* species (9%) (Table 2). When neutropenic patients with bacteremia were compared to those in the immunocompetent population, two differences were observed. Enterococci were isolated more commonly from nonneutropenic patients (14%) than from those with neutropenia (6%), whereas viridans group streptococci were more common among patients with neutropenia (3%) than among those without neutropenia (1%) (65).

The isolation of CoNS from blood cultures of immunocompromised patients adds a level of clinical complexity to diagnosis of bloodstream infections compared to diagnosis in immunocompetent patients (3a, 21a). CoNS, along with other organisms, such as *Corynebacterium* sp. and *Propionibacterium acnes*, are often considered skin contaminants introduced into blood culture bottles during venipuncture of inadequately disinfected sites of immunocompetent patients. In the immunocompromised host, however, all bacteria should be considered potential pathogens. CoNS are increasingly recognized as a source of true bacteremia associated with central venous catheters, a common risk factor found in the immunocompromised patient (58). In addition, a certain species of CoNS, namely, *Staphylococcus lugdunensis*, is recognized as a truly virulent pathogen in all patients, regardless of immune status (40).

The rapid emergence of multidrug resistance (MDR) in pathogens causing HAIs limits the administration of optimal therapeutic agents essential to patients with compromised immune systems. Studies examining CoNS strains have demonstrated the emergence of resistance to the quinolones and tolerance and occasional resistance to vancomycin, the most widely used agent for treatment of staphylococcal septicemia. Alternative agents have been pro-

Table 2. Distribution of major pathogens in bloodstream infections in immunocompromised and immunocompetent patients[c]

Rank	Pathogen (% of bloodstream infections)		
	Total	Immunocompromised patients[a]	Immunocompetent patients[b]
1	CoNS (30.6)	CoNS (31.6)	CoNS (29.6)
2	*Staphylococcus aureus* (11.7)	*Staphylococcus aureus* (12.3)	*Enterococcus* species (13.9)
3	*Enterococcus* species (10.1)	*Streptococcus* species (9.1)	*Staphylococcus aureus* (11.1)
4	*Candida* species (8.5)	*Candida* species (8.6)	*Candida* species (8.4)
5	*Escherichia coli* (7.5)	*Escherichia coli* (7.3)	*Escherichia coli* (7.7)
6	*Streptococcus* species (6.9)	*Enterococcus* species (6.3)	*Klebsiella* species (6.8)
7	*Klebsiella* species (6.1)	*Klebsiella* species (5.4)	*Pseudomonas aeruginosa* (4.7)
8	*Pseudomonas aeruginosa* (4.2)	Anaerobes (4.8)	*Streptococcus* species (4.7)
9	Anaerobes (3.9)	*Pseudomonas aeruginosa* (3.6)	Anaerobes (2.9)
10	*Enterobacter* species (3.0)	*Enterobacter* species (3.1)	*Enterobacter* species (2.9)

[a]Activated neutrophil count, <1,000 neutrophils/μl.
[b]Activated neutrophil count, >1,000 neutrophils/μl.
[c]Data were adapted from reference 65.

posed for the prevention and treatment of device-related and glycopeptide-tolerant *S. epidermidis* infections, including minocycline, rifampin, and the oxazolidinones (44). While vancomycin resistance for *S. aureus* has remained low, anecdotal data suggest an increasing trend in MICs due to overuse of the antibiotic. Most automated systems perform poorly in the recognition of resistant, intermediate, or heteroresistant strains; therefore, alternative manual methods, such as Etest (AB Biodisk, Solna, Sweden) and other Clinical and Laboratory Standards Institute (CLSI)-recommended screening procedures, must be employed in order to improve vancomycin test accuracy (23).

Gram-positive bacteria

The proportion of bloodstream HAIs caused by gram-positive microorganisms has increased significantly, with some studies reporting rates as high as 70 to 81% (19, 49). In the SCOPE study (1995 to 2002) of bloodstream HAIs in patients with underlying malignancies, *Enterococcus faecium* isolates, including those which were vancomycin resistant, were isolated more frequently from patients without neutropenia. The role of neutropenia as a risk factor for vancomycin-resistant enterococci (VRE)-related bloodstream infections is unresolved (30, 32). Overall, there was a tendency for enterococci to be isolated earliest during the hospital stay (mean, day 10), as opposed to viridans group streptococci and *Candida* species (mean, days 17 and 18, respectively) (61). CoNS were the most commonly isolated organisms from both neutropenic and nonneutropenic hosts, and there were no differences in crude mortality rates (65) (Table 2).

Currently, the majority of HAIs are caused by methicillin-susceptible *S. aureus* (MSSA); however, the rise of methicillin-resistant *S. aureus* (MRSA) in both community and health care settings is alarming. One hospital's investigation of an MRSA outbreak demonstrated that 40% of its nosocomial bloodstream infections were caused by the same strain. Once the hospital instituted an active surveillance program to identify and isolate MRSA-positive patients, it was able to completely eradicate the outbreak strain from the hospital (57). Infections with community-associated MRSA (CA-MRSA) are increasing among patients who have never been hospitalized and have no other risk factors for MRSA acquisition (25, 67). CA-MRSA carries a different staphylococcal cassette chromosome *mec* gene from that of hospital-associated MRSA, and CA-MRSA isolates are resistant to fewer antibiotic classes (39). Table 3 summarizes major *S. aureus* toxins among CA-MRSA and health care-acquired MRSA strains.

MRSA strains present a major therapeutic challenge due to their resistance to β-lactam antibiotics

Table 3. *S. aureus* toxins[a]

Toxin	Origin	Gene
Toxic shock syndrome toxin (TSST-1)	Hospital and community	*tst*
Staphylococcal enterotoxin A	Hospital	*sea*
Enterotoxin gene cluster (*egc*) carrying five superantigenic enterotoxin genes	Hospital and community	*seg, sei, sem, sen, seo*
Panton-Valentine leukocidin	Community	*lukF/lukS-PV*

[a]All can be detected by PCR.

(the treatment of choice for MSSA) and the limited choice of remaining antimicrobial agents, e.g. vancomycin, linezolid, and daptomycin. Bacteremic patients with MRSA have longer LOS, higher rates of septic shock, and greater attributable costs. A recent report documented a threefold increase in the costs of treating bacteremia due to MRSA compared to those of treating bacteremia due to MSSA (4).

Gram-negative bacteria

Gram-negative bacteria causing bloodstream infections and other HAIs are especially critical in patients lacking intact immune mechanisms. *Klebsiella pneumoniae*, the second most common cause of gram-negative bacteremia, has been observed to be the most frequent underlying hospital-associated bloodstream pathogen (53%) in patients with neoplastic disease (60). Indwelling urinary catheters were present in 23% of these patients, and 17% were receiving immunosuppressive therapy. In the neonatal intensive care unit (NICU) setting, gram-negative bacteria are a considerable cause of morbidity and mortality. It was reported that 298 episodes of bloodstream infections occurred among 2,935 admissions (5.75 episodes per 1,000 patient-days), and 26% of these episodes were caused by gram-negative bacilli. The predominant gram-negative pathogen was *K. pneumoniae* (39%) (6, 29). Among neonates with very low birth weights, *K. pneumoniae* was responsible for approximately 4% of late-onset sepsis cases.

The daunting challenge of treating infections caused by gram-negative pathogens in the hospital environment is the emergence of MDR among strains with multiple resistance mechanisms that include drug-inactivating enzymes and porin mutations (41). Of particular concern in certain geographic areas is the rise in MDR within the *Enterobacteriaceae*, e.g., *K. pneumoniae*, and among intrinsically drug-resistant *Pseudomonas* spp. and *Acinetobacter* spp. The emergence of high-level transferable resistance to carbapenems, β-lactamase inhibitors, and expanded-spectrum cephalosporins has seriously limited therapeutic options. The extended-spectrum β-lactamases, e.g., those encoded by bla_{TEM}, bla_{SHV}, and $bla_{CTX\text{-}M}$ genes, as well as carbapenem resistance genes, such as bla_{IMP} and bla_{VIM}, convey resistance to all β-lactam drugs currently used (38). These MDR gram-negative pathogens oftentimes demonstrate in vitro susceptibility only to polymyxins and newer antimicrobials, such as tigecycline. The financial burden of outbreaks caused by extended-spectrum-β-lactamase-producing *K. pneumoniae* is considerable. It has been estimated that a total cost of $341,751 was attributable to a 4-month outbreak in a NICU of a major New York medical center, and the mean LOS for infected infants was 48.5 days longer than that for similarly stratified infants (55).

Inasmuch as MDR organisms cause HAIs associated with high mortality rates, particularly in immunosuppressed hosts, accurate microbiology tests are essential to the rapid administration of appropriate antimicrobial therapy. One report that examined early transplant-related mortality after cord blood transplantation from unrelated donors cited a rate close to 50%, mainly due to infectious complications. In this study, infections due to MDR *Acinetobacter* spp. demonstrated a higher mortality rate than those due to cytomegalovirus disease (50).

Outbreaks of gram-negative pathogens within hospitals have been linked to environmental sources. One study reported tobramycin-resistant *Pseudomonas aeruginosa* bacteremia in five bone marrow transplant patients that was linked to a contaminated sink (33). Another increase in the incidence of *P. aeruginosa* infection occurred in a NICU and was associated with staff colonization (16). Strict adherence to infection control guidelines and environmental decontamination are essential to break this chain of transmission.

Fungi

Candida is the most frequently isolated fungus from blood cultures, and candidemia has become a significant cause of morbidity and mortality in immunocompromised patients. Cases of candidemia among low-birth-weight infants are an increasing problem, and the incidence of infection has been reported to be as high as 25% in some NICUs (66). Data from the Pediatric Prevention Network of the CDC and the National Nosocomial Infections Surveillance system identified bloodstream infection as the most common HAI and *Candida* as the fourth most common pathogen causing bloodstream HAIs in the United States (42). In our institution over a 2-year period, 79 *Candida* isolates were recovered, including 43 from neonatal blood cultures and 36 recovered from the hands of NICU health care workers (15). The predominant isolates among neonates in this study were *Candida albicans*; however, *Candida parapsilosis* predominated among nurses in our NICUs.

In the adult population, neutropenia is the most important risk factor for invasive fungal infections. The groups that comprise the majority of cases include cancer patients who have received cytotoxic chemotherapy, bone marrow transplant recipients, and solid organ transplant recipients. These patients are commonly exposed to the usual risk factors for invasive candidiasis, including multiple broad-spectrum antibiotics, central venous catheters, and colonization with *Candida*. Neutropenic children and adults with persistent fever while receiving broad-spectrum antibiotics have a 20% risk of developing an invasive fungal infection (14).

Laboratory Diagnostic Modalities

Early and accurate detection of pathogens is the major goal of clinical microbiology laboratories. It is the cornerstone of rapid administration of optimal therapy and patient management.

Culture methods

Currently, we rely primarily on culture-dependent methodologies for the isolation, identification, and antimicrobial susceptibility testing of bloodstream pathogens. Both selective and nonselective media are utilized in order to optimize recovery of organisms. There are examples of specialized media that can be incorporated based on the initial Gram stain of the positive blood culture bottle. While the sensitivity and specificity of culture are excellent, the prolonged turnaround time and its impact on patient management are leading us to seek more rapid alternatives (Table 4). Although not currently FDA approved for sources other than nares, CHROMagar medium (BD Diagnostics, Sparks, MD) has several formulations which are selective and differential for *S. aureus*, MRSA, and the most commonly isolated yeast species. MRSASelect (BioRad, Hercules, CA) and Spectra MRSA (Remel, Lenexa, KS) are also selective and differential agar, which can be used to assess MRSA colonization status and to finalize reports in 24 hours.

Molecular assays

The advent of new molecular platforms for use in clinical microbiology laboratories to detect MRSA in real time is a major breakthrough. In cases of septicemia, rapid molecular detection of MRSA directly from blood cultures is vital to timely diagnosis, administration of appropriate therapy, and patient management and could significantly improve morbidity and mortality. The GeneOhm StaphSR assay (BD Diagnostics) offers rapid, definitive identification of *S. aureus* and MRSA directly from blood cultures. Home brew PCR assays for direct MRSA detection from blood cultures are labor-intensive, unstandardized, and challenged by the presence of nucleic acid amplification inhibitors. Molecular assays for the detection of MRSA from nasal swabs for surveillance purposes are available and include the BD GeneOhm MRSA (Becton Dickinson, Sparks, MD) and GeneXpert MRSA (Cepheid, Sunnyvale, CA) assays. The BD GeneOhm MRSA assay utilizes a SmartCycler real-time PCR instrument, which is technically easy to incorporate into the routine clinical microbiology laboratory. In our laboratory, we adapted this assay to determine its ability to detect MRSA directly from blood culture bottles newly positive for gram-positive cocci in clusters. Our results indicated that MRSA was isolated from 37 of 368 (10%) positive blood cultures tested throughout the study. The sensitivity, specificity, and positive and negative predictive values for BD GeneOhm MRSA assay compared to culture were 100%, 99.4%, 94.9%, and 100%, respectively (12a). Most importantly, MRSA detection using the modified BD GeneOhm assay was confirmed within 2 to 4 h, compared to 48 to 72 h by culture. A Cepheid test is also currently available as an analyte-specific reagent, which contains primers and a probe designed to detect a 98-bp region of the *Staphylococcus* protein A gene (*spaA*).

A new molecular platform, GeneXpert (Cepheid), holds the promise to detect both MSSA and MRSA directly from blood culture bottles, pending completion of its clinical trial. The Cepheid GeneXpert machine is a fully integrated and automated real-time PCR system, which incorporates sample preparation, nucleic acid extraction, amplification, and detection in one instrument. Using self-contained single-use cartridges, results are available within 30 to 70 minutes.

Table 4. Direct laboratory identification of *Candida* species

Major yeast species	Laboratory identification		First-line therapy
	CHROMagar (colony color)	PNA FISH (fluorescent color)	
C. albicans	Green	Green	Fluconazole
C. parapsilosis	White	Green	Fluconazole
C. tropicalis	Blue	Gold	Fluconazole
C. krusei	Pink	Red	Caspofungin
C. glabrata	Lavender	Red	Caspofungin

Table 5. Comparison of MRSA real-time PCR assays

Assay	Gene targets	Identification	Instrument	Controls	Sample preparation	Hands-on time
GeneOhm assay (BD)	*SCCmec* and *orfX*	MRSA and MSSA	SmartCycler	External (positive/negative) and internal	Manual	30
Xpert assay (Cepheid)	*spa, mecA, SCCmec* and *attB* gene	MRSA and MSSA	GeneXpert	Internal control	Automatic	10

An attribute comparison of the BD GeneOhm assay and the Cepheid Xpert assay is shown in Table 5.

PNA FISH

Rapid and accurate identification and differentiation of *Candida albicans* from non-*C. albicans* isolates in newly positive blood cultures that are smear positive for yeast are important in the early selection of optimal antifungal therapy. Conventional, multiday identification methods can lead to the initial administration of costly and inappropriate antifungal therapy, whereas early species identification enables optimal initial drug selection. The *C. albicans* peptide nucleic acid fluorescence in situ hybridization (PNA FISH) assay (AdvanDx, Woburn, MA; bioMerieux, Durham, NC) is a rapid assay (2.5 hours) that identifies *C. albicans* in smears made from positive blood cultures. The PNA FISH assay targets specific 26S rRNA sequences of *C. albicans* for species identification. A second-generation dual-color PNA FISH assay impacts clinically on the selection of appropriate antifungal therapeutic regimens by differentiating *C. albicans*, generally susceptible to fluconazole, from *Candida glabrata*, which can be resistant (51). The targets for the multicolor PNA FISH technology have been expanded in the newest assay, called Yeast Traffic Light (AdvanDx, Woburn, MA), to include not only *C. albicans* and *C. glabrata* but also *C. parapsilosis*, *Candida tropicalis*, and *Candida krusei*.

PNA FISH technology can also be applied for the identification of gram-positive microorganisms associated with bloodstream infections. The *Enterococcus* PNA FISH assay allows for rapid and accurate identification of enterococci directly from smear-positive blood cultures. The clinical importance of differentiating *E. faecium* from *Enterococcus faecalis* is that the former is often resistant to ampicillin and vancomycin, and appropriate targeted therapy could be initiated earlier. In addition, an *S. aureus* PNA FISH assay enables rapid identification, within hours, of *S. aureus* on smears made directly from newly positive blood culture bottles. The assay uses fluorescence-labeled PNA probes that target species-specific rRNA in *S. aureus*, causing green fluorescence.

Surveillance Activity

The CDC has historically promulgated that patients known to be colonized or infected with epidemiologically significant organisms be placed in contact isolation, with the obvious goal of reducing transmission to uncolonized inpatients. Rapid detection of targeted pathogens in at-risk patient populations is essential for this practice to be effective. However, despite surveillance efforts, whether they are targeted or hospitalwide, a system designed to monitor trends in organism prevalence and incidence is crucial to reducing hospital-associated transmission of pathogens. The immunocompromised populations within our medical centers need particular attention because a wide range of pathogens, including MDR bacteria, *Legionella* spp., and *Aspergillus* spp., have been implicated in outbreak scenarios (20). Our threshold for action and investigation must be lowered for patients most at risk. Careful review of unit-specific microbiology reports is critical. Real-time electronic surveillance is even more ideal, although probably beyond the scope of most hospital information systems.

The early detection of MRSA colonization in patients located in transplant and intensive care units is of paramount importance to avert HAI and transmission among patients. In order to curtail MRSA dissemination, nasal surveillance is being instituted by many hospitals, since the nares are considered the source of invasive MRSA infections. The advent of a rapid PCR assay, in lieu of routine culture, for detection of MRSA from the anterior nares has increased the sensitivity for detection and cut the turnaround time for results from days to hours.

Algorithms for routine surveillance of VRE and MDR gram-negative bacteria in immunocompromised patients are institution specific. There are few molecular assays for these pathogens, and laboratories must rely on culture or home brew molecular platforms. False-positive culture results can occur for patients who are colonized with enterococci other than *E. faecium* and *E. faecalis*, such as *Enterococcus gallinarum*. VRE infection is usually preceded by colonization of the gastrointestinal tract (48), and selective media, such as bile esculin azide agar containing 6 μg/ml of vancomycin or Enterococcosel agar containing vancomycin at 8 μg/ml (BD Diagnostics, Cockeysville, MD), have been used to differentiate *Enterococcus* species from other gastrointestinal flora. It is not well known how long this colonization can continue, but some patients can remain culture positive for up to 5 years (3). As opposed to the case for MRSA, active decolonization of VRE from the gastrointestinal tract is not routinely performed. Instead, strict infection control measures are taken in order to reduce the risk of transmission to other immunocompromised patients. Screening for drug-resistant gram-negative bacteria by use of rectal swabs is complex due to the lack of molecular assays and the lack of selective culture media to suppress growth of normal fecal flora.

HEALTH CARE-ASSOCIATED RESPIRATORY TRACT INFECTION

Respiratory tract infections are the second most common cause of HAIs. Immunocompromised hosts in health care settings may acquire respiratory infections with bacterial, viral, or fungal pathogens. The majority of such infections are ventilator-associated pneumonia (VAP), which is an important complication in patients with respiratory failure who undergo endotracheal intubation and mechanical ventilation. The illness is severe and has a high mortality rate and increased LOS. The incidence of pneumonia acquired in the ICU setting ranges from 10 to 65%. The spectrum of pathogens associated with VAP and their respective antimicrobial susceptibility patterns are well studied, as are the risk factors associated with their acquisition. One investigation reported the following three significant risk factors for the prediction of subsequent infections with potential MDR pathogens: duration of mechanical ventilation, prior antibiotic use, and prior use of broad-spectrum antimicrobial agents (59).

Predominant Pathogens and Antimicrobial Resistance

Immunocompromised patients are often intubated during their hospitalization, placing them at risk for pneumonia. In fact, patients who require ventilatory assistance account for 80 to 85% of health care-associated pneumonia cases in the United States (21). The majority of patients with VAP can be divided into two subgroups according to the time of onset of the disease. Early-onset VAP (occurring within the first 4 days of mechanical ventilation) is caused primarily by *Streptococcus pneumoniae*, *Haemophilus influenzae*, and *S. aureus* (34). These organisms probably originate from mouth and upper respiratory tract colonization acquired from the community. Late-onset VAP (occurring after 4 days of mechanical ventilation) is caused predominantly by gram-negative bacilli, such as *P. aeruginosa*, *Enterobacter* spp., *Acinetobacter baumannii*, and MRSA. The pathogen distinction between early and late onset varies with the severity of the patient's illness and with recent antimicrobial therapy.

Non-ventilator-related hospital-acquired pneumonia may occur in any health care facility, especially in immunocompromised patients. It has been reported since the 1970s that respiratory tract colonization with gram-negative bacteria contributes to the pathogenesis of hospital-associated pneumonia (22). Microorganisms that colonize the upper respiratory tract and the inner surfaces of endotracheal tubes are often aspirated into the lower respiratory tract, resulting in VAP. The endotracheal tube can breach first-line immune defense factors, provide airway access to inhaled particles, and eliminate the cough reflex.

A study comparing pneumonia due to MRSA versus that due to MSSA in ICU patients found that MRSA pneumonia resulted in higher rates of bacteremia and septic shock and had a significantly higher infection-associated mortality rate. Gram-negative bacteria, particularly *K. pneumoniae* and *A. baumannii*, account for approximately 30% of ICU pneumonias (45, 61). Most clinical laboratories are unable to perform molecular testing on these MDR strains, and therefore the precise mechanisms of resistance for hospital strains may remain unknown. However, an understanding of the epidemiology and transmission of these pathogens requires genetic characterization. Microbiology laboratories must implement the most current phenotypic detection methods as recommended by the most current CLSI guidelines in order to track local patterns of emerging resistance. It is imperative that the clinical laboratory have protocols and algorithms in place for these multi- or panresistant organisms in order to test newer antimicrobials, such as tigecycline, and drugs from the past, such as polymyxin B. Since these are not currently available on many automated susceptibility panels/cards, alternative test methods must be implemented and validated in accordance with the latest CLSI guidelines. For example, while polymyxin B and tigecycline do not need to be tested routinely, determination of specific MICs may be warranted in cases of carbapenem-resistant gram-negative organisms. These antimicrobial susceptibility testing decisions must be made in conjunction with the microbiology laboratory, infectious disease clinicians, and clinical pharmacists. Targeted strategies that have been employed with the aim of reduction of selection of MDR gram-negative pathogens include intense monitoring of trends in high-risk units, modifications of formularies, massive education of prescribers, and strict enforcement of infection control practices (46).

Both adult and pediatric immunocompromised patients are at an increased risk for acquiring viral infections. Common respiratory viruses, such as respiratory syncytial virus (RSV), influenza virus, parainfluenza virus, adenovirus, and human metapneumovirus, can lead to morbidity and mortality and are often hospital associated (12, 37). Acquisition can occur when infectious patients are not isolated appropriately or have not been tested for respiratory viruses or when an infected health care worker reports for duty instead of remaining at home. Transmission of RSV within hospital settings is common, but transmission has also been documented for influenza and parainfluenza viruses. Respiratory virus infections in pediatric populations with cancer or in very-low-birth-weight infants can lead to severe disease. Adenovirus is a significant cause of morbidity and mortality among transplant recipients, especially children who have received allogeneic human stem cell transplants. One report demonstrated that when more than one surveillance site was culture positive for adenovirus, patients almost always developed an infection due to adenovirus (31). In addition, patients with adenovirus in their blood have a lower survival rate than do those without viremia. Clinical adenovirus disease (pneumonia, enteritis, hemorrhagic cystitis, and encephalitis) can be predicted by isolation of adenovirus weeks before the clinical event. Currently, there are no licensed systemic drugs clearly effective for preventing or treating adenovirus disease. However, many compounds are under investigation as potential therapy in situations of life-threatening adenovirus infection, two of which are cidofovir and ribavirin (24). Studies indicate that cidofovir is somewhat effective in limiting adenovirus-related systemic complications and reducing viral loads in bone marrow transplant patients (5, 7).

In addition, *Legionella* infection among transplant recipients due to contamination of the hospital water supply, although rare, has caused outbreaks in a number of hospitals. While *Legionella pneumophila* serogroup 1 (LP1) may be the most common group associated with community-acquired legionellosis, hospital isolates may be quite varied. In addition to LP1, common colonizers of water systems include *Legionella micdadei*, *Legionella bozemanae*, and LP6 (27, 62). Oftentimes, clinical cases linked to these biofilm inhabitants are more severe than those caused by LP1.

Laboratory Diagnostic Modalities

The majority of clinical microbiology laboratories still rely on culture-dependent methodologies for the isolation, identification, and antimicrobial susceptibility testing of respiratory tract pathogens.

Culture methods

Pneumonia in the immunocompromised patient as well as in those who are ventilator dependent is life-threatening. In addition, the organisms which are associated with upper airway colonization are often resistant to most, if not all, antimicrobials. Thus, if infection is not accurately differentiated from colonization, the patient may be treated with broad-spectrum agents. The ideal specimen type for the diagnosis of pneumonia is a bronchoalveolar lavage sample, since it is the best representative of the lower respiratory tract. Unfortunately, this invasive procedure is not always practical or possible and less optimal specimen types, such as endotracheal aspirates, are collected. The value of the latter sample, which suffers from low specificity, decreases even more if the patient has received antimicrobial therapy within 72 hours prior to sample collection. On the other hand, negative tracheal aspirate cultures collected from untreated patients have a very good negative predictive value (94%) for VAP (1, 35).

The role of quantitative culture of lower respiratory secretions is under investigation. There is some evidence which indicates that organisms present at a concentration of $>10^4$ CFU/ml are more likely to represent true etiologic agents of pneumonia. Therapy can be targeted instead of relying on broad-spectrum antibiotics. Bacteria present below this threshold would most likely represent normal colonization of the upper respiratory tract, and therefore patients would not be treated. The reliability of quantitative cultures was recently reviewed, and unacceptable variations were found in sensitivity, ranging from 38% to 82%, as well as specificity, which ranged from 67% to 100% (11). In addition, this method is labor-intensive and requires specialized technical skills. False-negative results can be encountered if the patient has received therapy or as a result of sampling technique. Therefore, while the clinical utility of quantitative cultures is thought to be valuable by some, at present there appears to be insufficient data to recommend this as a routine practice.

Most bacterial pathogens can be recovered by utilizing a routine battery of agar media consisting of a sheep blood agar plate, chocolate agar plate, and MacConkey agar plate. If less common pathogens, such as *Legionella* sp. or *Nocardia*, are being considered, the laboratory must be notified so specialized medium (i.e., buffered charcoal yeast extract agar) can be included. However, many of the "atypical" etiologic agents associated with pneumonia in the immunocompromised patient are difficult to isolate and/or detect by routine methodologies. These include but are not limited to *Pneumocytis jirovecii*, Epstein-Barr virus, and *Chlamydia pneumoniae*. Implementation of molecular assays which are designed as multiplex platforms will allow rapid detection of the most common etiologic agents associated with pneumonia in the immunocompromised host.

Molecular assays

Currently, there are several molecular platforms under development that offer same-day detection of pathogens causing HAI. These include Prodesse's real-time PCR system, Roche Molecular's LightCycler real-time PCR technology, Cepheid's SmartCycler instrument, and bioMerieux's NASBA platform. While these test systems are very appealing from a diagnostic perspective, other considerations include space requirements, cost, and in many cases, specialized molecular training for microbiology technologists.

For the detection of viral pathogens, medical centers that care for immunocompromised patients must, at a minimum, offer rapid antigen detection. Currently, assays include detection of RSV, influenza A and B viruses, rotavirus, and adenovirus 40/41. In addition to the enzyme immunoassays available, some laboratories are able to offer direct fluorescent-antigen tests for the most common respiratory viruses. These assays are more labor-intensive than enzyme immunoassays and require a higher level of technical skill. The laboratory must be very aware of the performance issues and less-than-optimal sensitivities of these rapid tests. In the future, rapid molecular platforms will most likely replace the antigen detection tests we currently use. If the laboratory does not have a full-service virology section, it should consider performing rapid screens on-site and referring out negative samples for culture confirmation, particularly for immunocompromised patients. Currently, there are several molecular platforms, as previously mentioned, that offer the same-day detection of a panel of viruses, particularly respiratory viruses. However, there are a limited number of assays which target the detection of enteroviruses from cerebrospinal samples, e.g., the Xpert EV test for enteroviral meningitis (Cepheid, Sunnyvale, CA) and the NucliSENS assay (bioMerieux, Durham, NC).

Hospital-associated fungal infections in the immunocompromised host are primarily associated with *Aspergillus* spp. Invasive aspergillosis (IA) is difficult to diagnose and hard to treat in this patient population. Both clinical features and radiologic results are insensitive and nonspecific. The incidence has been determined to be 8 to 15% of patients undergoing allogeneic stem cell transplantation and 5 to 15% of patients undergoing solid organ transplantation (54). The mortality rate due to this life-threatening opportunistic mycosis is as high as 50 to 100%, and early diagnosis and prompt initiation of antifungal therapy remain the crucial factors that may aid in reduction of the mortality rate. Definitive diagnosis still relies on culture; however, isolation of *Aspergillus* spp. from blood cultures is extremely rare, and recovery from sputum may represent contamination or transient flora. Isolation of the fungus from multiple respiratory specimens, particularly those from the lower respiratory tract, is more representative of infection; however, growth may take as long as 4 weeks. As an alternative, detection of galactomannan antigen in serum and body fluids is available. The assay, called the Platelia *Aspergillus* galactomannan antigenemia test (Bio-Rad Laboratories, Redmond, WA), is a double-sandwich enzyme-linked immunosorbent assay that allows for early diagnosis of IA, if positive, and early exclusion of IA, if negative. It targets the polysaccharide galactomannan, an exoantigen component of the cell wall of aspergilli. The data demonstrated an overall sensitivity and specificity of 80.7% and 89.2%, respectively. An important limitation of the assay is the occurrence of false-positive results for patients who are being treated with piperacillin-tazobactam, amoxicillin, or amoxicillin-clavulanate because these antimicrobial agents contain high levels of galactomannan, which may persist for 5 days after stopping the antibiotic (2).

A final pathogen to consider in the immunocompromised patient who develops health care-associated pneumonia is *Legionella* spp. All hospital water systems are subject to colonization with a variety of *Legionella* spp., which thrive in the biofilms that develop within older pipes. Once *Legionella* is present, it is virtually impossible to eradicate. The best one can hope for is to keep the CFU to a low enough level that the potential for spread to patients is minimal. Even then, it is quite possible that immunocompromised patient populations may need additional safety guidelines put into place, such as bottled water and restricted showers. LP1 is responsible for 84% of community-acquired disease cases and is easily detected using a rapid immunochromatographic urine-based assay which detects LP1 soluble antigen in patients with current or past disease. However, if the patient is suspected to have hospital-associated legionellosis, lower respiratory tract specimens for culture are mandated because pneumonia is more often caused by LP4, LP6, *Legionella micdadei*, *Legionella bozemanae*, or other species; therefore, the urine antigen test would be ineffective. While there is no commercially available molecular assay for *Legionella*, many laboratories have developed and validated their own nucleic acid amplification tests (9). Molecular assays for detection of legionellosis from respiratory specimens are becoming available to clinical microbiology laboratories as multiplex systems.

This rapid technology is ideally suited for *Legionella* spp. because the organisms are fastidious; patients have often been treated, which lowers the bacterial burden to a level which will not be detected by culture; and serologic assays may reflect past exposure rather than current disease. Depending on the burden of *Legionella* spp. present within the plumbing system, it might be prudent for respiratory specimens collected from immunosuppressed patients to be cultured for this microorganism on a routine basis, using the specialized, selective, and nonselective buffered charcoal yeast extract media (Becton Dickinson, BBL, Sparks, MD).

Surveillance Activity

Efforts focused on the reduction/elimination of MDR pathogens from the hospital environment can reduce acquisition rates. This entails more than simply monitoring annual trends via the laboratory-generated antibiogram. While tracking infection rates on a monthly basis is a good strategy to flag aberrant patterns, a more proactive approach is necessary to achieve significant results. Active surveillance of patients, whether it be targeted by unit, by risk stratification, by incoming transfers, or by service, must be part of the hospital's overall plan for improving patient safety. Colonization/infection with microorganisms such as MRSA, VRE, and MDR gram-negative bacteria can be detected earlier, and patients can be placed in isolation and/or decolonized in a timely fashion. Efforts to reduce potential exposure to environmental aspergilli might include air sampling in areas of heavy construction, but routine testing in the absence of disease is not warranted. Also, routine testing of potable water is not necessary if there are no clinical cases of hospital-acquired legionellosis.

SUMMARY

Immunocompromised patients are at increased risk of acquiring HAI due to their inability to mount an effective antibody and/or cellular immune response. The upsurge of antimicrobial-resistant pathogens has complicated their treatment options and contributed to substantial increases in morbidity and mortality. Recent advances in real-time molecular diagnostic technologies for implementation in clinical microbiology laboratories offer the opportunity to detect targeted pathogens that colonize or infect at-risk populations. Limited laboratory resources and the rise in immunosuppressed hosts pose challenges to laboratorians, epidemiologists, and clinicians.

REFERENCES

1. **American Thoracic Society and Infectious Diseases Society of America.** 2005. Guidelines for the management of adults with hospital-acquired, ventilator-associated, and healthcare-associated pneumonia. *Am. J. Respir. Crit. Care Med.* **171:**388–416.
2. **Aubry, A., R. Porcher, J. Bottero, S. Touratier, T. Leblanc, B. Brethon, P. Rousselot, E. Raffpix, J. Menotti, F. Derouin, P. Ribaud and A. Sulahian.** 2006. Occurrence and kinetics of false-positive *Aspergillus* galactomannan test results following treatment with beta-lactam antibiotics in patients with hematological disorders. *J. Clin. Microbiol.* **44:**389–394.
3. **Baden, L. R., W. Thiemke, A. Skolnik, R. Chambers, J. Strymish, H. S. Gold, R. C. Moellering, Jr., and G. M. Eliopoulos.** 2001. Prolonged colonization with vancomycin-resistant *Enterococcus faecium* in long-term care patients and the significance of "clearance." *Clin. Infect. Dis.* **33:**1654–1660.

3a. **Bekeris, L. G., J. A. Tworek, M. K. Walsh, and P. N. Valenstein.** 2005. Trends in blood culture contamination: a College of American Pathologists Q-Tracks study of 356 institutions. *Arch. Pathol. Lab. Med.* **129:**1222–1225.

4. **Blot, S. I., K. H. Vandewoude, E. A. Hoste, and F. A. Colardyn.** 2002. Outcome and attributable mortality in critically ill patients with bacteremia involving methicillin-susceptible and methicillin-resistant *Staphylococcus aureus*. *Arch. Intern. Med.* **162:**2229–2235.
5. **Bordigoni, P., A. S. Carret, V. Venard, F. Witz, and A. Le Faou.** 2001. Treatment of adenovirus infections in patients undergoing allogeneic hematopoietic stem cell transplantation. *Clin. Infect. Dis.* **32:**1290–1297.
6. **Bradford, P. A., S. Bratu, C. Urban, M. Visalli, N. Mariano, D. Landman, J. J. Rahal, S. Brooks, S. Cebular, and J. Quale.** 2004. Emergence of carbapenem-resistant *Klebsiella* species possessing the class A carbapenem-hydrolyzing KPC-2 and inhibitor-resistant TEM-30 beta-lactamases in New York City. *Clin. Infect. Dis.* **39:**55–60.

7. **Carter B. A., S. J. Karpen, R. E. Quiros-Tejeira, I. F. Chang, B. S. Clark, G. J. Demmler, H. E. Heslop, J. D. Scott, P. Seu, and J. A. Goss.** 2002. Intravenous cidofovir therapy for disseminated adenovirus in a pediatric liver transplant recipient. *Transplantation* **74:**1050–1052.
8. **Centers for Disease Control and Prevention.** 2005. *Pseudomonas* bloodstream infections associated with a heparin/saline flush—Missouri, New York, Texas, and Michigan, 2004–2005. *MMWR Morb. Mortal Wkly. Rep.* **54:**269–272.
9. **Cloud, J. L., K. C. Carroll, P. Pixton, M. Erali, and D. R. Hillyard.** 2000. Detection of *Legionella* species in respiratory specimens using PCR with sequencing confirmation. *J. Clin. Microbiol.* **38:**1709–1712.
10. **Combes, A., C. Figliolini, J. L. Trouillet, N. Kassis, M. Wolff, C. Gibert, and J. Chastre.** 2002. Incidence and outcome of polymicrobial ventilator-associated pneumonia. *Chest* **121:** 1618–1623.
11. **Cook, D., and L. Mandell.** 2000. Endotracheal aspiration in the diagnosis of ventilator-associated pneumonia. *Chest* **117:**195–197.
12. **Couch, R. B., J. A. Englund, and E. Whimbey.** 1997. Respiratory viral infections in immunocompetent and immunocompromised persons. *Am. J. Med.* **102:**2–9.

12a. **Della-Latta, P., P. Pancholi, K. Pavletich and F. Wu.** 2005. Detection of MRSA directly from blood cultures using real-time PCR, abstr. P1391. ECCMID, Copenhagen, Denmark.

13. **Domingo, P., and A. Fontanet.** 2001. Management of complications associated with totally implantable ports in patients with AIDS. *AIDS Patient Care STDS* **15:**7–13.
14. **European Organization for Research and Treatment of Cancer and International Antimicrobial Therapy Cooperative Group.** 1989. Empiric antifungal therapy in febrile granulocytopenic patients. *Am. J. Med.* **86:**668–672.
15. **Feja, K. N., F. Wu, K. Roberts, M. Loughrey, M. Nesin, E. Larson, P. Della-Latta, J. Haas, J. Cimiotti, and L. Saiman.** 2005. Risk factors for candidemia in critically ill infants: a matched case-control study. *J. Pediatr.* **147:**156–161.
16. **Foca, M., K. Jakob, S. Whittier, P. Della-Latta, S. Factor, D. Rubenstein, and L. Saiman.** 2000. Endemic *Pseudomonas aeruginosa* infection in a neonatal intensive care unit. *N. Engl. J. Med.* **343:**695–700.
17. **Friedman, N. D., K. S. Kaye, J. E. Stout, S. A. McGarry, S. L. Trivette, J. P. Briggs, W. Lamm, C. Clark, J. MacFarquhar, A. L. Walton, L. B. Reller, and D. J. Sexton.** 2002. Health care-associated bloodstream infections in adults: a reason to change the accepted definition of community-acquired infections. *Ann. Intern. Med.* **137:**791–797.
18. **Garner, J. S., W. R. Jarvis, T. G. Emori, T. C. Horan, and J. M. Hughes.** 1988. CDC definitions for nosocomial infections. *Am. J. Infect. Control* **16:**128–140.
19. **Gonzalez-Barca, E., A. Fernandez-Sevilla, J. Carratala, A. Granena, and F. Gudiol.** 1996. Prospective study of 288 episodes of bacteremia in neutropenic cancer patients in a single institution. *Eur. J. Clin. Microbiol. Infect. Dis.* **15:**291–296.
20. **Gupta, A., P. Della-Latta, B. Todd, P. San Gabriel, J. Haas, F. Wu, D. Rubenstein, and L. Saiman.** 2004. Outbreak of extended-spectrum beta-lactamase-producing Klebsiella pneumoniae in a neonatal intensive care unit linked to artificial nails. *Infect. Control Hosp. Epidemiol.* **25:**210–215.
21. **Haley, R. W., D. H. Culver, J. W. White, W. M. Morgan, T. G. Emori, V. P. Munn, and T. M. Hooton.** 1985. The efficacy of infection surveillance and control programs in preventing nosocomial infections in U.S. hospitals. *Am. J. Epidemiol.* **121:**182–205.

21a. **Hall, K. K., and J. A. Lyman.** 2006. Updated review of blood culture contamination. *Clin. Microbiol. Rev.* **19:**788–802.

22. **Johanson, W. G., Jr., A. K. Pierce, J. P. Sanford, and G. D. Thomas.** 1972. Nosocomial respiratory infections with gram-negative bacilli: the significance of colonization of the respiratory tract. *Ann. Intern. Med.* **77:**701–706.
23. **Jones, R. N.** 2006. Microbiological features of vancomycin in the 21st century: minimum inhibitory concentration creep, bactericidal/static activity, and applied breakpoints to predict clinical outcomes or detect resistant strains. *Clin. Infect. Dis.* **42:**S13–S24.
24. **Kinchington, P. R., E. G. Romanowski, and Y. J. Gordon.** 2005. Prospects for adenovirus antivirals. *J. Antimicrob. Chemother.* **55:**424–429.
25. **King, M. D., B. J. Humphrey, Y. F. Wang, E. V. Kourbatova, S. M. Ray, and H. M. Blumberg.** 2006. Emergence of community-acquired methicillin-resistant *Staphylococcus aureus* USA 300 clone as the predominant cause of skin and soft-tissue infections. *Ann. Intern. Med.* **144:**309–317.
26. **Klein, M. A., D. Kadidlo, J. McCullough, D. H. McKenna, and L. J. Burns.** 2006. Microbial contamination of hematopoietic stem cell products: incidence and clinical sequelae. *Biol. Blood Marrow Transplant.* **12:**1142–1149.
27. **Knirsch, C. A., K. Jakob, D. Schoonmaker, J. A. Kiehlbauch, S. J. Wong, P. Della-Latta, S. Whittier, M. Layton, and B. Scully.** 2000. An outbreak of *Legionella micdadei* pneumonia in transplant patients: evaluation, molecular epidemiology, and control. *Am. J. Med.* **108:**290–295.
28. **Kuehnert, M. J., V. R. Roth, N. R. Haley, K. R. Gregory, K. V. Elder, G. B. Schreiber, M. J. Arduino, S. C. Holt, L. A. Carson, S. N. Banerjee, and W. R. Jarvis.** 2001. Transfusion-transmitted bacterial infection in the United States, 1998 through 2000. *Transfusion* **41:**1493–1499.
29. **Larson, E. L., J. P. Cimiotti, J. Haas, M. Nesin, A. Allen, P. Della-Latta, and L. Saiman.** 2005. Gram-negative bacilli associated with catheter-associated and non-catheter-associated bloodstream infections and hand carriage by healthcare workers in neonatal intensive care units. *Pediatr. Crit. Care Med.* **6:**457–461.
30. **Lautenbach, E., W. B. Bilker, and P. J. Brennan.** 1999. Enterococcal bacteremia: risk factors for vancomycin resistance and predictors of mortality. *Infect. Control Hosp. Epidemiol.* **20:**318–323.
31. **Lion, T., R. Baumgartinger, F. Watzinger, S. Matthes-Martin, M. Suda, S. Preuner, B. Futterknecht, A. Lawitschka, C. Peters, U. Potschger, and H. Gadner.** 2003. Molecular monitoring of adenovirus in peripheral blood after allogeneic bone marrow transplantation permits early diagnosis of disseminated disease. *Blood* **102:**1114–1120.
32. **Lucas, K. G., A. E. Brown, D. Armstrong, D. Chapman, and G. Heller.** 1996. The identification of febrile, neutropenic children with neoplastic disease at low risk for bacteremia and complications of sepsis. *Cancer* **77:**791–798.
33. **Lyytikainen, O., V. Golovanova, E. Kolho, P. Ruutu, A. Sivonen, L. Tiittanen, M. Hakanen, and J. Vuopio-Varkila.** 2001. Outbreak caused by tobramycin-resistant *Pseudomonas aeruginosa* in a bone marrow transplantation unit. *Scand. J. Infect. Dis.* **33:**445–449.
34. **Mayhall, C. G.** 1997. Nosocomial pneumonia. Diagnosis and prevention. *Infect. Dis. Clin. N. Am.* **11:**427–457.
35. **Mayhall, C. G.** 2001. Ventilator-associated pneumonia or not? Contemporary diagnosis. *Emerg. Infect. Dis.* **7:**200–204.
36. **McKibben, L., T. Horan, J. I. Tokars, G. Fowler, D. M. Cardo, M. L. Pearson, P. J. Brennan, et al.** 2005. Guidance on public reporting of healthcare-associated infections: recommendations of the Healthcare Infection Control Practices Advisory Committee. *Am. J. Infect. Control* **33:**217–226.

37. **Mendoza Sánchez, M. C., J. Ruiz-Contreras, J. L. Vivanco, F. Fernández-Carrión, M. Baro Fernández, J. T. Ramos, J. R. Otero, and D. Folgueira.** 2006. Respiratory virus infections in children with cancer or HIV infection. *J. Pediatr. Hematol. Oncol.* **28:**154–159.
38. **Moland, E. S., N. D. Hanson, J. A. Black, A. Hossain, W. Song, and K. S. Thomson.** 2006. Prevalence of newer beta-lactamases in gram-negative clinical isolates collected in the United States from 2001 to 2002. *J. Clin. Microbiol.* **44:**3318–3324.
39. **Naimi, T. S., K. H. LeDell, K. Como-Sabetti, S. M. Borchardt, D. J. Boxrud, J. Etienne, S. K. Johnson, F. Vandenesch, S. Fridkin, C. O'Boyle, R. N. Danila, and R. Lynfield.** 2003. Comparison of community- and health care-associated methicillin-resistant *Staphylococcus aureus* infection. *JAMA* **290:**2976–2984.
40. **Patel, R., K. E. Piper, M. S. Rouse, J. R. Uhl, F. R. Cockerill, and J. M. Steckelberg.** 2000. Frequency of isolation of *Staphylococcus lugdunensis* among staphylococcal isolates causing endocarditis: a 20-year experience. *J. Clin. Microbiol.* **38:**4262–4263.
41. **Paterson, D. L., L. Mulazimoglu, J. M. Casellas, W. C. Ko, H. Goossens, A. V. Gottberg, S. Mohapatra, G. M. Trenholme, K. P. Klugman, J. G. McCormack, and V. L. Yu.** 2000. Epidemiology of ciprofloxacin resistance and its relationship to extended-spectrum β-lactamase production in *Klebsiella pneumoniae* isolates causing bacteremia. *Clin. Infect. Dis.* **30:**473–478.
42. **Pfaller, M. A., R. N. Jones, S. A. Messer, M. B. Edmond, and R. P. Wenzel.** 1998. National surveillance of nosocomial blood stream infection due to *Candida albicans*: frequency of occurrence and antifungal susceptibility in the SCOPE Program. *Diagn. Microbiol. Infect. Dis.* 31:327–332.
43. **Pizzo, P. A.** 1999. Fever in immunocompromised patients. *N. Engl. J. Med.* **341:**893–900.
44. **Raad, I., A. Alrahwan, and K. Rolston.** 1998. *Staphylococcus epidermidis*: emerging resistance and need for alternative agents. *Clin. Infect. Dis.* **26:**1182–1187.
45. **Rahal, J. J.** 2000. Extended-spectrum beta-lactamases: how big is the problem? *Clin. Microbiol. Infect.* **6**(Suppl. 2)**:**2–6.
46. **Rahal, J. J., C. Urban, and S. Segal-Maurer.** 2002. Nosocomial antibiotic resistance in multiple gram-negative species: experience at one hospital with squeezing the resistance balloon at multiple sites. *Clin. Infect. Dis.* **34:**499–503.
47. **Rodriguez-Martinez, J. M., and A. Pascual.** 2006. Antimicrobial resistance in bacterial biofilms. *Rev. Med. Microbiol.* **17:**65–75.
48. **Rolston K. V., Y. Jiang, and M. Matar.** 2007. VRE fecal colonization/infection in cancer patients. *Bone Marrow Transplant.* **39:**567–568.
49. **Rubio, M., L. Palau, J. R. Vivas, E. del Potro, J. Diaz-Mediavilla, A. Alvarez, R. Martinez, and J. J. Picazo.** 1994. Predominance of gram-positive microorganisms as a cause of septicemia in patients with hematological malignancies. *Infect. Control Hosp. Epidemiol.* **15:**101–104.
50. **Saavedra, S., G. F. Sanz, I. Jarque, F. Moscardo, C. Jimenez, I. Lorenzo, G. Martin, J. Martinez, J. De La Rubia, R. Andreu, S. Molla, I. Llopis, M. J. Fernandez, M. Salavert, B. Acosta, M. Gobernado, and M. A. Sanz.** 2002. Early infections in adult patients undergoing unrelated donor cord blood transplantation. *Bone Marrow Transplant.* **30:**937–943.
51. **Safdar, A., V. Chaturvedi, B. S. Koll, D. H. Larone, D. S. Perlin, and D. Armstrong.** 2002. Prospective, multicenter surveillance study of *Candida glabrata*: fluconazole and itraconazole susceptibility profiles in bloodstream, invasive, and colonizing strains and differences between isolates from three urban teaching hospitals in New York City (*Candida* susceptibility trend study, 1998 to 1999). *Antimicrob. Agents Chemother.* **46:**3268–3272.
52. **Shorr, A. F., Y. P. Tabak, A. D. Killian, V. Gupta, L. Z. Liu, and M. H. Kollef.** 2006. Healthcare-associated bloodstream infection: a distinct entity? Insights from a large U.S. database. *Crit. Care Med.* **34:**2588–2595.
53. **Siegman-Igra, Y., B. Fourer, R. Orni-Wasserlauf, Y. Golan, A. Noy, D. Schwartz, and M. Giladi.** 2002. Reappraisal of community-acquired bacteremia: a proposal of a new classification for the spectrum of acquisition of bacteremia. *Clin. Infect. Dis.* **34:**1431–1439.
54. **Singh, N., and D. L. Paterson.** 2005. *Aspergillus* infections in transplant recipients. *Clin. Microbiol. Rev.* **18:**44–69.
55. **Stone, P. W., A. Gupta, M. Laughrey, P. Della-Latta, J. Cimiotti, E. Larson, D. Rubenstein, and L. Saiman.** 2003. Attributable costs and length of stay of an extended-spectrum beta-lactamase-producing *Klebsiella pneumoniae* outbreak in a neonatal intensive care unit. *Infect. Control Hosp. Epidemiol.* **24:**601–606.
56. **Tacconelli, E., M. Tumbarello, M. Pittiruti, F. Leone, M. B. Lucia, R. Cauda, and L. Ortona.** 1997. Central venous catheter-related sepsis in a cohort of 366 hospitalized patients. *Eur. J. Clin. Microbiol. Infect. Dis.* **16:**203–209.
57. **Thompson, R. L., I. Cabezudo, and R. P. Wenzel.** 1982. Epidemiology of nosocomial infections caused by methicillin-resistant *Staphylococcus aureus. Ann. Intern. Med.* **97:**309–317.
58. **Tokars, J. I.** 2004. Predictive value of blood cultures positive for coagulase negative staphylococci: implications for patient care and health care quality assurance. *Clin. Infect. Dis.* **39:**333–341.
59. **Trouillet, J. L., J. Chastre, A. Vuagnat, M. L. Joly-Guillou, D. Combaux, M. C. Dombret, and C. Gibert.** 1998. Ventilator-associated pneumonia caused by potentially drug-resistant bacteria. *Am. J. Respir. Crit. Care Med.* **157:**531–539.
60. **Tsay, R.-W., L. K. Siu, C.-P. Jung, and F.-Y. Chang.** 2002. Characteristics of bacteremia between community-acquired and nosocomial *Klebsiella pneumoniae. Infect. Arch. Intern. Med.* **162:**1021–1027.
61. **Urban, C., S. Segal-Maurer, and J. J. Rahal.** 2003. Considerations in control and treatment of nosocomial infections due to multidrug-resistant *Acinetobacter baumannii. Clin. Infect. Dis.* **36:**1268–1274.
62. **Visca, P., P. Goldoni, P. C. Luck, J. H. Helbig, L. Cattani, G. Giltri, S. Bramati, and M. Castellani Pastoris.** 1999. Multiple types of *Legionella pneumophila* serogroup 6 in a hospital heated-water system associated with sporadic infections. *J. Clin. Microbiol.* **37:**2189–2196.
63. **Vuong, C., and M. Otto.** 2002. *Staphylococcus epidermidis* infections. *Microbes Infect.* **4:**481–489.
64. **Wisplinghoff, H., T. Bischoff, S. M. Tallent, H. Seifert, R. P. Wenzel, and M. B. Edmond.** 2004. Nosocomial bloodstream infections in US hospitals: analysis of 24,179 cases from a prospective nationwide surveillance study. *Clin. Infect. Dis.* **39:**309–317.
65. **Wisplinghoff, H., H. Seifert, R. P. Wenzel, and M. B. Edmond.** 2003. Current trends in the epidemiology of nosocomial bloodstream infections in patients with hematological malignancies and solid neoplasmas in hospitals in the United States. *Clin. Infect. Dis.* **36:**1103–1110.
66. **Zafar, N., C. M. Wallace, P. Kieffer, P. Schroeder, M. Schootman, and A. Hamvas.** 2001. Improving survival of vulnerable infants increases neonatal intensive care unit nosocomial infection rate. *Arch. Pediatr. Adolesc. Med.* *155:*1098–1104.
67. **Zetola N., J. S. Francis, E. L. Nuermberger, and W. R. Bishai.** 2005. Community-acquired methicillin-resistant *Staphylococcus aureus*: an emerging threat. *Lancet Infect. Dis.* **5:**275–286.

IV. FUTURE TRENDS

Diagnostic Microbiology of the Immunocompromised Host
Edited by Randall T. Hayden, Karen C. Carroll, Yi-Wei Tang, and Donna M. Wolk

Chapter 22

Future Trends in Diagnosis of Infections in the Immunocompromised Population

Elizabeth M. Marlowe and Donna M. Wolk

Emerging technology justifiably carries the largest burden to improve the practice of health care and the fate of those who represent the most vulnerable of health care recipients, the immunocompromised. As each section of this book has illustrated, the focused and unique health care management of the immunocompromised patient has grown significantly over the last several decades. Improvements in testing paradigms continue to impact diagnostic options for detecting infections as well as treatment options for managing infections. As with all new knowledge, information derived from new technology also bestows new responsibilities on the clinical laboratory.

The diagnostic concepts are clear, especially for those most at risk and most expensive to the health care system; faster, more sensitive results can usually save lives. The diagnostic challenges are also clear; we must continue to balance our resources wisely and ensure clinical utility. Our ability to provide faster, more sensitive results continues to challenge our perceptions of current standards of care provided by the clinical laboratory; our goal is to critically validate the use of new technology and to use the technology with proven utility in a cost-beneficial manner. Decisions to include more costly technology must include discussion of the entire hospital budget, not just the laboratory budget. With new technology, benefits will include the capability to limit hospital-acquired infections, to guide appropriate therapy, and to minimize hospital stays and other related costs.

Since treatment of immunocompromised patients is extremely cost-intensive, assessment of the cost benefit of emerging technology for diagnosis and infection prevention must take into account the complexities and high cost of their routine care. For example, the U.S. Department of Health and Human Services guidelines for the use of antiretroviral agents for human immunodeficiency virus type 1 (HIV-1)-infected adults and adolescents now recommend the following testing: (i) genotypic drug resistance testing for all treatment-naïve patients; (ii) tropism testing prior to initiation of a CCR5 inhibitor and in the event of a CCR5 inhibitor failure; and (iii) genetic testing for HLA-B*5701, which is indicative of abacavir allergy, prior to initiation of abacavir therapy to limit the risk of a hypersensitivity reaction. HLA-B*5701-positive patients should not be prescribed abacavir, and an abacavir allergy should be recorded in their charts (34).

Today we are faced with powerful diagnostic options that substantially change the possibilities and add many scientific, ethical, and legal responsibilities to clinical practice. Laboratory medicine has a high likelihood of being targeted and affected by new technology, and it is within this framework that the technology is reviewed here. The future may hold the following concepts for clinical microbiology laboratories supporting cancer centers, transplant centers, children's hospitals, HIV clinics, and other foci of immunocompromised patients. As authors, we humbly submit our vision for the future of clinical microbiology and care of the most vulnerable of our patients.

DIAGNOSTICS WILL DRIVE FASTER AND MORE TARGETED INTERVENTIONS

Stat-PCR

While molecular diagnostics has revolutionized the way that medicine is practiced for 2 decades, the practice of molecular diagnostics remains tied to a

Elizabeth M. Marlowe • Southern California Permanente Medical Group, Regional Reference Laboratories, North Hollywood, CA 91605. **Donna M. Wolk** • Department of Pathology, University of Arizona School of Medicine, Tucson, AZ 85724.

highly skilled laboratory staff and batch processing. Emerging technology has the potential to change the reference lab-centric model of molecular diagnostics and to enable immediate (stat) testing, which allows rapid and targeted antimicrobial therapy and permits rapid decisions to be made regarding the isolation status of patients who are carriers of common hospital-associated pathogens. The future of molecular diagnostics is changing and is focused on producing rapid results within the facility. The concept of point-of-care molecular diagnostics is closer than ever thought possible.

One technology that has recently elevated molecular testing capability from "batch" to "stat" mode is the FDA-approved GeneXpert technology (Cepheid, Sunnyvale, CA). Using microfluidics, the GeneXpert instrument supports closed-system nucleic acid extraction and real-time PCR amplification and detection for a variety of pathogens, including group B streptococcus from vaginal or rectal swabs (57), enterovirus from cerebrospinal fluid (93), and methicillin-resistant *Staphylococcus aureus* (MRSA) from nares swabs.

These methods are rapid and simple to perform, with complexities buried behind the technology. Near point-of-care results are available in 1.5 to 2.5 h, as the single-unit cartridge automates and combines specimen processing, nucleic acid extraction, and target amplification and detection supported by software that automates data interpretation. The instrument is available in 4-bay and 16-bay configurations; a 72-bay random access configuration is planned. While the GeneXpert enterovirus method is rated a "highly complex" test, the GeneXpert group B streptococcus and GeneXpert MRSA methods are the first nucleic acid tests to be placed in the "moderate complexity" category (http://www.accessdata.fda.gov/scripts/cdrh/cfdocs/cfClia/Search.cfm). This designation allows testing by a broad variety of allied health personnel under the direction of medical technologists according to the Clinical Laboratory Improvement Act (http://www.cms.hhs.gov/CLIA/).

The implications of this technology are enormous for clinics and for urgent care and emergency medicine facilities that serve immunocompromised populations, who are at the highest risk for infections. With relatively high technology costs, focused use includes infection prevention in high-risk populations and detection of other severe infections, such as meningitis. As cost-benefit models of the rapid technology emerge, the clinical utility is likely to expand. Applications could include diagnostic-guided treatment for any infectious disease. As the range of test targets expands, patients most at risk for demise from infectious disease will reap the advantages of rapid diagnosis and subsequent targeted therapy that could most improve their chances of survival. Future applications of such technology in a high-risk clinical setting will have clear impact on the clinician's ability to prescribe targeted antibiotics more quickly and to avoid overuse of broad-spectrum antibiotics. In essence, stat-PCR may become a reality as molecular laboratories integrate their practices into the 24/7/365 venue of the clinical laboratory.

While there are currently no other such FDA-approved devices on the market, there is expected pressure from competitors. For example, HandyLab, Inc. (Ann Arbor, MI), is developing a proprietary microfluidic, nucleic acid- and protein-based cartridge. Iqumm, Inc. (Allston, MA), is developing the Lab-in-a-Tube (Liat), a portable molecular amplification platform which uses a flexible tube as a sample vessel, with all reagents prepackaged in the tube segments. Like the GeneXpert system, the HandyLab cartridge and Liat platforms are fully automated, and thus the need for specially trained personnel and separate pre- and post-PCR rooms is eliminated. Such features are well suited for low-volume/sporadically tested stat-PCR needs.

High-volume assays, including viral load methods, such as those for cytomegalovirus and Epstein-Barr virus, could be performed with the high-volume versions of such technology; rapid results of viral load assays could rapidly impact decisions about immunosuppressive therapy. As these stat molecular assays increase in number, the capacity to rapidly screen donor organs and recipients for cytomegalovirus and other viruses should expand. The assays may also prove to be useful for monitoring the development of posttransplantation disease, as well as the success of antiviral therapy. While these devices show much promise, to date, only a qualitative molecular assay is commercially available. The capability of these devices to accurately and consistently quantify a viral load remains to be seen.

Stat Biomarkers

Biomarkers of the immune response, such as procalcitonin (PCT), are evolving in clinical utility for the diagnosis of severe bacterial infections such as sepsis, for prediction of sepsis complications, and for therapeutic follow-up. High plasma levels of PCT are observed in situations characterized by the onset of organ dysfunction and other symptoms of "severe sepsis" or "septic shock" (8, 25, 60, 155, 166, 169).

The use of biomarker assays, such as the Brahms PCT LIA assay (Brahms Diagnostica, LLC), for determination of PCT in human serum and plasma, supports early diagnosis and the intensive therapy that can reduce morbidity and mortality while saving unnecessary spending for critically ill patients. This test is FDA cleared and is followed by the Brahms PCT-Q, a rapid assay available in research-use-only (RUO) format. In addition, the Vidas Brahms PCT, a CE-marked PCT measurement test for the Vidas system (bioMérieux), is available. Clinical trials are ongoing for use of rapid testing of PCT via automated instruments (Brahms Diagnostica, LLC).

Clinical use of these tests requires a short turnaround time and places the microbiology laboratory in a position to either provide 24/7 support for these assays or share medical guidance with a stat laboratory where testing is performed. Since there are diagnostic limitations with all biomarkers (for example, in cases where only a small quantity of PCT is induced and simultaneous induction occurs due to trauma), clinical consultation for interpretation of these results is critical. Therapy may be initiated or altered based on infectious disease biomarkers, and clinical microbiology laboratories will be challenged to expand test menus to accommodate new biomarker technology.

THERAPY WILL DRIVE DIAGNOSTICS

Genetic predisposition to infection is a proven reality for a select group of viral and bacterial infections (5, 17, 24, 197). In some instances, treatment response to antibiotic therapy may be altered by genetic predispositions (117). System-based disease may also be affected. In one study (64), pulmonary infections were shown to be more prevalent in allogeneic stem cell transplant recipients of a particular genotype.

Genetic screening of patients is likely to evolve from its beginnings in the arenas of toxicology and cancer therapy to the purview of molecular microbiology, as new evidence-based, population-based studies emerge to impact clinical practice. Implementation of genetic screening is likely to move more slowly than genetic testing of microbial pathogens. Difficulties in the statistical analysis and risk factor assessment of population-based studies will limit the utility of data; it may be several years until this information can find utility in the clinical microbiology laboratory. Research related to genetic predisposition of infection has merit and will provide the basis for future clinical utility studies.

Another aspect of genetic prediction may focus on the discipline of antibiotic testing, where testing of genetic predisposition to antibiotic toxicity may be prudent in certain patient populations. Likewise, genetic testing for responders and nonresponders will continue to evolve as our understanding of effective antibiotic prescription expands (138, 171, 204).

Because of the complexity and costs of new drug development, drug-diagnostic test codevelopment is already impacting cardiology and coagulation therapeutics. The FDA has been very supportive of this strategy (http://www.fda.gov/ohrms/dockets/dailys/04/aug04/080304/04n-0279-c00001-vol1.pdf), which will impact microbiology laboratories in the future. Toxic antibiotics, such as isoniazid, may be the first to receive attention (138, 171, 204).

SINGLE-ORGANISM DETECTION WILL EVOLVE TO MULTIPLE-ORGANISM DETECTION

In many clinical laboratories, real-time PCR is a mainstay of core molecular technologies. One of the limitations to both end-point and real-time PCR technologies is the number of targets that can be amplified and detected in a single well. Innovative technologies, designed to perform multiplex PCR and to detect multiple targets simultaneously, could have clinical utility if limitations in analytical sensitivity, which are inherent to multiplex PCR, can be overcome. Improvements in extraction methods to ensure nucleic acid purity will be critical in order to optimize any of the emerging multiplex PCR technologies; however, use of multiplex assays is inevitable and will be critical to the rapid diagnosis and treatment of immunocompromised hosts in a variety of circumstances, as detailed below.

Microbial and fungal infections are among the most common causes of high morbidity and mortality in critically ill patients in the United States; mortality rates can exceed 80%. Routine cultures for bacteria, fungi, and sometimes viruses may require subculture for agent identification. The entire process can take days to weeks to produce a final result. Since routine methods are relatively slow to produce useful results and are ineffective for detection of some pathogens, there is a significant need to improve multiplex assay speed and sensitivity in order to provide a more accurate diagnostic representation of infections. Within this framework, molecular methods have merit, as do chromogenic growth media that can definitively or presumptively identify common bacteria or fungi (28, 44, 84, 95, 122, 167, 170).

Molecular multiplex methods are discussed first, as they are rapidly emerging for use in the clinical laboratory. Although likely to be cost-prohibitive for use on immunocompetent patients, the development of multiple-pathogen identification systems would include gram-positive and gram-negative bacterial and fungal infections and would enable more rapid and targeted antimicrobial interventions for disease in immunocompromised or critically ill patients. In the critically ill, if rapid detection of infection is followed by appropriate antimicrobial therapy, morbidity and mortality can be reduced. For instance, with bloodstream infections (BSIs), rapid determination of Gram stain status as well as bacterial or fungal classification could be life-saving, since infection can lead to sepsis and often to death. In such cases, rapid intervention with appropriate antimicrobial therapy is essential (111, 112, 125, 157).

Another example of the usefulness of multiplex assays is the rapid identification of viral infections. While antigen detection and molecular single-target assays have been useful for rapid detection of infections, methods that can identify multiple respiratory or bloodstream pathogens would have utility for assessing the infection status of immunocompromised hosts. Likewise, system-based "panels" aimed at central nervous system infections or gastrointestinal infections would have merit for both immunocompromised and immunocompetent patients.

Moreover, molecular detection is likely to advance our understanding of infections with difficult-to-cultivate pathogens and of mixed infections. An enhanced understanding of the infections caused by difficult-to-cultivate bacterial pathogens, such as *Mycoplasma* sp. and *Chlamydia* sp., is also likely to unfold. The understanding of mixed viral infections is sure to evolve, as those infections have historically been overlooked in routine viral cultures because of the inherent limitations in the detection of multiple viral etiologies once a cell line was infected by the predominant or more robust viral etiology.

The diagnostic future will include identification of mixed bacterial, viral, and fungal infections as technology provides the ability to focus on system-based approaches rather than agent-based approaches to laboratory testing.

Several molecular platforms are slated to play a role in advancing the diagnostic capabilities of clinical laboratories. Only time, cost-benefit studies, and practical evidence-based reviews will reveal the ultimate utility of these methods for diagnosis of infections in the immunocompromised population. Some examples of the key emerging technologies are described below.

Real-Time Multiplex PCR with Fluorescent Probe Detection

Real-time detection instruments typically have two to four detector channels that allow for the discrimination of dye-labeled probes to detect amplified targets and controls, but several real-time PCR manufacturers have announced potential improvements that could allow more targets to be identified.

For example, the Applied Biosystems 7500 instrument (Applied Biosystems, Foster City, CA) is a five-channel real-time PCR instrument capable of detecting four targets and an internal control for laboratory-designed assays. In addition, Roche Molecular Diagnostics (Pleasanton, CA) has introduced the LightCycler 2.0 system, which is a six-channel instrument to be used with the LightCycler Septi*Fast* test, potentially allowing for the direct detection (without prior blood culture incubation) and identification of the top 25 pathogens in BSIs. Detection and identification are achieved by multiwell amplification and melting curve analysis, with results available in roughly 5 hours. The system has been evaluated and is commercially available and CE marked in Europe. U.S. clinical trials are expected and may demonstrate cost-effectiveness in certain circumstances, such as immunocompromised patient populations.

Users of multiplex technology are faced with multiple challenges. Instrument users should be advised that not all real-time detection systems can detect all dye-labeled probes and that, in multiplex assays, bleedthrough of dye into the next dye channel is common. Dye bleedthrough can make it difficult to distinguish positive results from background and will erroneously make a sample in the bleedthrough channel appear positive (96, 214). For single-target as well as multiple-target assays, use of the appropriate fluorescent filter combinations is critical. Furthermore, a multiplex assay cannot necessarily be performed on all real-time PCR instruments, and it is extremely critical to perform rigorous method verification when real-time PCR assays are performed by off-label procedures.

In addition, there are technical challenges that are inherent to multiplex PCR; amplification of multiple targets has been known to lead to overamplification of one target to the detriment of another present at a lower concentration. These challenges can be mitigated by (i) improvements in nucleic acid purity that will allow better amplification of low-concentration targets; (ii) improvements in primer design or PCR process, which can improve target specificity and minimize cross-amplification of unwanted targets; and (iii) improvements in amplicon detection technology. It seems that nucleic acid extraction chemistry, microarray technology, and bioinformatics hold solutions to the challenges presented by multiplex PCR. Therefore, these approaches are detailed in the following sections.

Multiplex PCR with Microarrays

To overcome some of the limitations of multiplex PCR amplification, microarray-based systems offer solutions. Microarrays may be used (i) when one needs more information than a single-target assay can provide, (ii) to screen for a panel of infectious agents, or (iii) when more information is needed about a particular infectious agent. Such technology has had the greatest impact on academic research and the pharmaceutical industry (183), but the technology is evolving rapidly to include clinical microbiology.

Multiplex PCR with liquid microarray detection of amplicons

Liquid microarrays are rapidly emerging as a common detection platform for multiplex PCR. Liquid bead-based multiplexing microarray detection systems have potential for the rapid and accurate identification and detection of pathogens as well as multiple mutations associated with viruses, drug-resistant bacteria, mycobacteria, and fungi. There is currently one major instrument platform for liquid microarray technology; the Luminex 200 flow cytometer system supports several commercial RUO applications, such as Qiagen's RespPlex panels (Qiagen, Inc., Germantown, MD), based on novel QIAplex technology, a target-enriched multiplex PCR technology (20, 21, 68, 101), and Eragen's MultiCode-PLx system (Madison, WI) (96, 104, 113, 129). The Luminex Diagnostics (formerly Tm Biosciences) xTAG respiratory virus panel is FDA cleared (20, 27, 54, 89, 96, 99, 106, 114, 121, 129, 146).

The Qiagen technology holds particular promise, as the template enrichment step increases specificity by using very small amounts of gene-specific primers, therefore reducing the chance of primer-dimer formation and background amplification. After template enrichment, PCR increases sensitivity by using a pair of high-concentration superprimers that have a high affinity for *Taq* polymerase, thus maximizing efficiency for target amplification. In its current format, the technology is among the most rapid of the multiplex platforms, at 5 hours.

Multiplex PCR with solid microarray detection of amplicons

Both technical issues and costs present some obstacles to the routine use of traditional solid microarrays. There are three basic categories of solid array technology, including (i) synthesized arrays, (ii) printed arrays, and (iii) electronic arrays. Each is discussed briefly.

Synthesized arrays. Synthesized arrays are made by in situ synthesis. Only short oligonucleotides can be synthesized effectively, and only a few chemistries are available. They can be prepared at a very high density and, in that format, typically require dedicated equipment from the manufacturer in order to process the array. Synthesized arrays are manufactured and sold premade. Both Affymetrix (Santa Clara, CA) and Nimblegen Systems, Inc. (Madison, WI), offer custom synthesis on a solid chip.

Affymetrix was the first company to develop a commercially available chip for the detection of HIV drug resistance mutations and *Mycobacterium* species identification as well as for rifampin resistance testing. Evaluations of the first generation of HIV chips demonstrated comparable results to those of conventional sequencing and line probe assay (LiPA) techniques (211). *Mycobacterium* studies also showed good correlation with conventional methods (195). These chips, once clinically tested, have the potential to speed diagnosis, determine correct mycobacterial therapy options, and enable rapid and targeted therapy to HIV patients and others to whom *Mycobacterium tuberculosis* may be life-threatening.

Printed arrays. Printed arrays utilize presynthesized or purified nucleic acids that are printed in discrete areas. They tend to be printed at a lower density than synthesized arrays and can be purchased premade or printed in one's own laboratory. Typically, arrays are printed on glass slides and probes are bound with coupling chemistry. Printed arrays can use longer oligonucleotides, enabling more protocol variety and flexibility. Companies that supply printed array technology include General Electric (Piscataway, NJ), Agilent (Santa Clara, CA), and Corning (Corning, NY).

Autogenomics (Carlsbad, CA) offers a BioFilm-Chip microarray, a printed array in a porous aqueous environment; automated processing is performed on their Infiniti analyzer. Autogenomics sells multiplex PCR primer sets to be used with premade microarray cartridges. The Infiniti MTBDR assay is available for detection of *Mycobacterium tuberculosis* and susceptibility testing with first-line drugs, such as rifampin, isoniazid, and pyrazinamide, via *rpoB*, *katG*, and *pncA* gene mutations, respectively. Future applications include an Infiniti system assay respiratory viral panel, with 25 individual targets within eight viral groups.

Multiplex Analysis via PCR-MS

High-performance mass spectrometry (MS) has recently been adapted and developed for use in conjunction with multiplex PCR for rapid identification and strain typing of emerging pathogens (41, 164, 200) and a variety of community-acquired infectious agents (4, 20-22), including *S. aureus* genetic targets (214a). No other current system identifies such a broad range of pathogens and strain-type microbes within the same testing method; therefore, the utility of this method could range from a rapid replacement for viral cultures to a rapid intervention in a suspected hospital-acquired pathogen outbreak in a transplant unit.

In the PCR-MS approach, broad-range primers, rather than primers targeted to single targets or species, are used to amplify PCR products from groupings of microbes. Electrospray ionization–time-of-flight MS is used to determine the nucleic base composition (the numbers of adenosines [A], cytidines [C], guanosines [G], and thymidines [T]) of the PCR amplicon, and this base composition signature is used to identify the pathogens present. A dedicated mass spectrometer is required because contaminants interfere with negative-ion-mode ionization of the PCR amplicons. Unlike nucleic acid probe assays or microarrays, MS does not require prior knowledge of products analyzed but simply measures the masses of the nucleic acids present in the sample. The analog signal of mass is converted to a digital signal of base composition.

This novel technology could impact the way that clinical microbiology laboratories approach the diagnosis of life-threatening diseases in vulnerable hosts, as this technology has the potential to rapidly identify multiple targets of relatively unknown origins, much like we currently use the broad growth capabilities of a blood agar plate or a tube of primary monkey kidney cells. The PCR-MS platform is significantly different from previous technologies in its ability to detect virtually all microbes from a family, and even microbes that have mutated significantly. This is accomplished through the use of several novel strategies. First, the employed primer target sites are highly conserved in all members of a microbe family. Second, T5000 PCR conditions (Ibis Technologies, Carlsbad, CA, and Abbott Molecular, Des Plains, IL) are, by design, permissive and thus tolerant to mismatches. Third, inosine and other "wildcard" nucleotides are used in primers to facilitate mispaired PCR. Fourth, due to MS analysis, the sequence of the amplicon need not be known to enable detection.

Another advantage is that the mass spectrometer is an extremely sensitive analytical tool; small quantities of nucleic acid amplicons within a complex mixture can be detected post-PCR, with a throughput of about one sample per minute. The Ibis T5000 system is comprised of two independent decks that are HEPA enclosed and fully automated, with a high-throughput capacity. This system includes a rigorous desalting and purification process, which is essential for optimal amplification of multiple genetic signature targets and optimal detection of the amplicon in the mass spectrometer (5).

Multiple-Pathogen Detection via Chromogenic Agar

Development and creative utilization of chromogenic agar are useful endeavors whenever rapid identification of isolates provides an option for a therapeutic advantage. Improvements to ensure identification of multiple organisms will provide the most utility for clinical laboratories. Examples are detailed below.

Several species of *Candida* are relatively resistant to the commonly used antifungal drugs, and therefore rapid identification of the species involved can speed effective treatment. Conventional identification methods are not timely, and several chromogenic agars are available and are reviewed more extensively in Chapter 12 (28, 44, 84, 122, 167, 170).

For identification of patients colonized with vancomycin-resistant enterococci (VRE), rapid identification methods will be central to active surveillance efforts. A prototype chromogenic agar medium (VRE-BMX; bioMérieux, Marcy l'Etoile, France) is designed to recover either vancomycin-resistant *Enterococcus faecium* or *Enterococcus faecalis* from clinical specimens, based on distinct colony colors. VRE-BMX compares with bile esculin azide agar supplemented with vancomycin ($n = 147$ stool samples). After 24 hours of incubation, VRE-BMX provided improved recovery of VRE from stool specimens, with the added advantage of being able to differentiate between vancomycin-resistant *E. faecalis* and *E. faecium*. The sensitivities and positive predictive values were as follows: for bile esculin azide agar supplemented with vancomycin, 75.7% and 74.6%, respectively; and for VRE-BMX, 95.5% and 91.3%, respectively (95).

For MRSA, two agars are gaining prominence. MRSA*Select* (12, 26, 28, 76, 95, 182, 201), from Bio-Rad (Marnes la Coquette, France), and CHROMagar MRSA (35, 46, 69, 78, 126, 139, 143), from Becton Dickinson (Cockeysville, MD), appear to have similar MRSA detection capacities; however, MRSA*Select* can report results up to 30 hours faster than CHROMagar MRSA. There are rare reports of false-positive cultures on chromogenic agar, but the specificity is very high (46, 124, 203).

SEQUENCE-BASED IDENTIFICATION METHODS WILL CONTINUE TO EVOLVE

DNA Sequencing

DNA sequencing has had an impact on the clinical laboratory. In larger laboratories, the 16S rRNA sequences of bacteria and 18S rRNA sequences of fungi are utilized to identify unusual isolates. Viral mutations are most commonly detected by direct sequencing of the specific viral reading frames, which encode the proteins that are targeted by currently available antiviral drugs. Examples include resistance to polymerase and protease inhibitors for HIV, acyclovir and penciclovir resistance in herpes simplex virus, acyclovir resistance in varicella-zoster virus, ganciclovir resistance in cytomegalovirus, famciclovir and lamivudune resistance in hepatitis B virus, and amantidine resistance in influenza A virus (172). One of the drawbacks to sequencing is that it can take 2 days or more before results are available.

Pyrosequencing (Biotage, Uppsala, Sweden) is a rapid method for sequencing based on the detection of pyrophosphate released during DNA synthesis. This method's advantages include speed and ease of use in comparison to traditional sequencing technology; disadvantages include the short lengths of sequences that can currently be analyzed. Pyrosequencing provides short sequence information, roughly 30 to 50 bases; it is useful for short-read DNA and mutation/SNP analysis. It is ideally suited for applied genomic research, including molecular applications for disease diagnosis, clinical prognosis, and pharmacogenomic testing. After PCR amplification and amplicon cleanup, run times approach 1 hour for 96 samples, with approximately 30 to 45 minutes for sequence analysis applications. Clinical applications include bacterial (including *Mycobacterium*), fungal, and viral identification as well as resistance detection (6, 13, 30, 55, 59, 83, 86, 88, 91, 123, 128, 147, 194, 196, 216-218).

Galor et al. (55) applied pyrosequencing to aid in the diagnosis of *Nocardia* keratitis after only a few days of bacterial growth, in contrast to conventional methods, which provided identification within 14 days of isolation. Because *Nocardia* species have various antibiotic susceptibilities, the rapid pyrosequencing identification of *Nocardia abscessus*, *Nocardia arthritidis*, and *Nocardia asiatica* allowed appropriate treatment to be initiated, and the infection rapidly cleared, with minimal scarring. Susceptibility results obtained later supported antibiotic treatment decisions. As more protocols become available and DNA databases expand, the application of this real-time short-read sequencing technology could have a much larger impact on the diagnosis of the immunocompromised.

FISH

When sequencing is not available or practical, oligonucleotide probes can serve to provide a rapid, direct answer. Peptide nucleic acid fluorescence in situ hybridization (PNA FISH) involves the use of fluorescent PNA probes that hybridize to rRNA. These probes are commercially available from AdvanDx (Woburn, MA) for the direct identification of *S. aureus*, *Candida albicans*, and *E. faecalis*/other enterococci from positive blood cultures. Probe kit utilization is driven by the Gram stain result; thus, identification of gram-positive cocci in clusters would utilize the *S. aureus* probe, identification of yeast would utilize the *C. albicans* probe, and identification of gram-positive cocci in pairs and chains would utilize the dual probe for *E. faecalis*/other enterococci. PNA FISH performance data have demonstrated the following sensitivities and specificities: for *S. aureus*, 98.8% and 99.6%; for *C. albicans*, 99.3% and 100%; for *E. faecalis*, 96.3% and 98.3%; and for other enterococci, 93.1% and 99.3% (6, 62, 132, 133, 135, 156, 178, 181, 209). The compatibility of these probes with various blood culture media (including FAN and resin bottles) has also demonstrated good sensitivity and specificity (7, 77, 132, 133, 135, 209).

The application and benefits of PNA FISH probes include the following: (i) for *S. aureus*, the ability to curtail unnecessary antibiotic use due to contaminated blood cultures; (ii) support of antifungal selection for candidemia due to *C. albicans* or non-*C. albicans* species; and (iii) support of antibiotic selection for bacteremia due to *E. faecalis* or other enterococci. Forrest et al. (51) demonstrated that the rapid differentiation of *S. aureus* from coagulase-negative staphylococcus (a common skin contaminant) in blood cultures by use of PNA FISH significantly reduced the median hospital stay, by 2 days ($P < 0.05$), and resulted in less vancomycin usage, with an estimated cost savings per patient of roughly $4,000.

The ability of PNA FISH to support appropriate antifungal selection has been evaluated (7, 50, 51). *Candida* spp. are the fourth most common cause of nosocomial BSIs, with *C. albicans* accounting for 55% of candidemias (7). Infections are especially troublesome in the immunocompromised host population. Per Infectious Diseases Society of America guidelines, fluconazole can be used as the initial therapy for *C. albicans* without prior azole use, but broad-spectrum agents should be considered for non-*C. albicans* species due to possible fluconazole resistance (140). PNA FISH has the ability to provide direct identification from positive blood cultures in 2.5 h, in contrast to 1 to 5 days by conventional culture methods. In two studies, the cost savings realized with the *C. albicans* PNA FISH method was $1,729 to $1,837 per patient; the majority of the cost savings were realized in antifungal expenditures (7, 50, 51).

The *E. faecalis* PNA FISH probe can also be utilized to reduce the time to correct antibiotic therapy (193). *E. faecalis* isolates are often vancomycin susceptible, with other enterococci (i.e., *E. faecium*) making up the majority of VRE. Using an algorithm with the *E. faecalis* PNA FISH probes, Toombs et al. (193) demonstrated that appropriate treatment for VRE and non-VRE can be directed for BSI. The PNA FISH results showed a significant reduction in the number of days to change to appropriate therapy, from 2.5 to 1.4 days ($P < 0.05$). In turn, there was a trend toward less mortality (36% versus 14%) ($P < 0.05$). What is clearly demonstrated in these studies is that by applying PNA FISH to BSI, diagnostics can direct therapy and can drive down hospital costs and save lives. In addition, AdvanDx now offers an *S. aureus*–coagulase-negative staphylococcus dual probe and a *C. albicans-Candida glabrata* dual probe.

Bioinformatics: Understanding the Basis of Sequence-Based Technology

Bioinformatics merges biology, computer science, and information technology. It is the cornerstone of molecular diagnostic assay design and data analysis. From the early stages of primer and probe design to the data management of sequence information, each aspect of the molecular process is a building block that affects the others. Historically, bioinformatic challenges lay within the limitations in our ability to create, store, and maintain databases of nucleotide and amino acid sequences.

Databases such as GenBank (http://www.ncbi.nlm.nih.gov/Genbank/index.html) have solved initial database limitations and allow microbiologists to view and query thousands of known microbial sequences and view taxonomic databases such as the NCBI taxonomy database (http://www.ncbi.nlm.nih.gov/Taxonomy/taxonomyhome.html/) (15, 208). The Ribosomal Differentiation of Microorganisms (RIDOM) project (http://www.ridom.de/) was created to overcome the drawbacks of phenotypic identification systems and publicly accessible sequence databases, performing quality control on both strands of the 5′ end of the 16S rRNA gene (14, 16, 72-75, 118-120, 163, 185). The Ribosomal Database Project release 9.56 (http://rdp.cme.msu.edu/) consists of 451,545 aligned and annotated 16S rRNA sequences, along with seven online analysis tools (94, 108-110).

The creation and development of such databases involve a series of complex interfaces, which allow molecular microbiologists to access, submit, or revise microbial genomic data. Moreover, the accompanying software allows these data to be combined to provide a comprehensive snapshot of microbial biodiversity. Scientists can then select genetic targets of interest for primer and probe design (183). A description of bioinformatic tools is useful for the understanding of molecular assays currently in use, as they are always designed based on the current genomic knowledge of microbial diversity at the time of design (http://www.ncbi.nlm.nih.gov/About/primer/bioinformatics.html).

New sequence data are posted to genetic databases on a daily basis, and most postings do not affect our ability to use molecular assays for pathogen detection; however, the potential exists for nearly all molecular assays to completely miss the identification of some organisms. This understanding is critical in assessing infections in immunocompromised patients, where antibiotic pressure could force mutations to occur or where immune vulnerability could permit infections with less virulent, and as yet unknown, strains of an organism. Molecular methods have limitations as genomes change or geographical differences in microbes are discovered, and laboratory staff must be cognizant of the potential limitations. For assays that detect rapidly mutating agents, such as HIV-1, international databases can be surveyed; some commercial systems, such as Roche Molecular Diagnostics, integrate changes in genomes as they arise, and this concept may become more and more critical as time goes by.

Since changes in DNA or RNA sequences may alter the accuracy, sensitivity, specificity, and precision of molecular methods, the application of computational approaches to facilitate the understanding of changing microbial genomes will become increasingly imperative. Drifts and shifts in genetic sequences may require a more global perspective in the assay design phase. The emerging technology of database mining will evolve to track changes in pathogen genomes for assays in which either genetic or antibiotic pressures may alter target sequences even among the most conserved genomic regions. It is also clear that bioinformatics will pave the way to the next wave of technology advances, which will impact the prediction, diagnosis, and treatment of human infection; however, databases will need to be regulated, updated, and maintained to allow for accurate assessments in real time.

Primer/probe design is a bioinformatic challenge to the design of multitarget molecular technologies, as it can dramatically affect the sensitivity, specificity, and clinical predictive value of the assay. There are a number of challenges which make the design process difficult, including natural genomic variations, the potential for natural or engineered threats, the variable and very large total genome size of the competing background (host DNA), and biochemical factors that affect hybridization properties.

Many primer design algorithms are used (e.g., Beacon Designer 3.0 and Oligo 6) (19, 78a, 100, 102, 150, 160, 175, 191, 192). To assess genomic signatures, some programs show promise (13, 45, 79) but are limited in the ability to refine primers when known strains evolve, new strains are discovered, or unique genetic regions erode due to microbial evolution. "Rigorous" computing algorithms, as opposed to the traditional heuristic algorithms, represent a novel way to improve primer/probe design for multiplex assays (48, 49, 151, 152). Rigorous computing algorithms must be considered in order to mitigate limitations of other primer/probe design schemes and to allow quick design of unique primers and probes for multiplex methods, which can easily be updated with new and changing pathogen genomes. Molecular methods are becoming commonplace; therefore, upkeep of primer and probe sets will be essential for the future of molecular microbiology.

MOLECULAR CHARACTERIZATION WILL GAIN FURTHER IMPORTANCE

Genotyping and Molecular Epidemiology

Microbial outbreak investigation is a key aspect of successful infection control and prevention efforts in healthcare units that house immunocompromised patients. In these units, implementation of rapid and focused outbreak investigation and targeted interventions are necessary in order to quickly intervene and to stop further transmission of the offending pathogen. Molecular genotyping is the preferred tool to characterize bacterial and fungal strains so that the transmission source(s) can be identified.

Molecular genotyping methods are essential for complete assessment of point sources, transmission modalities, and the success of intervention measures (38, 179) and are becoming the new standard of care in high-risk patient care areas. Examples of the successful use of microbial genotyping for high-risk patient care units include typing of *Aspergillus* sp. isolates from bone marrow transplant units (159, 187) and of MRSA (18) and VRE isolates from dialysis patients (90, 115, 142, 198).

Pulsed-field gel electrophoresis (PGFE) is the historic reference method for microbial genotyping; however, many different methods can be used, and each method has its own particular advantages and limitations (61, 131, 189). While PGFE is considered to be reliable, discriminatory, and reproducible, it can be tedious, expensive, time-consuming, and difficult for most hospitals to use as a routine form of support for infection control practitioners involved in outbreak investigations (174, 184, 206). Due to limitations in testing conditions, there is inherent gel-to-gel variability with PGFE; it is difficult to compare interinstitutional electrophoresis results. It is also difficult and sometimes cost-prohibitive to acquire near-real-time longitudinal results over the course of an outbreak.

For most health care facilities, access to rapid microbial genotyping is lacking. For many immunocompromised patient care units, the gap between identification of a hospital-acquired outbreak and subsequent life-saving interventions must be shortened drastically; rapid assessment of the outbreak's molecular epidemiology is crucial for clinical microbiologists and infection control practitioners to make informed, cost-effective decisions.

Although other microbial genotyping methods have been used to characterize the epidemiology of outbreaks, most remain unstandardized or impractical for many hospital laboratories (116, 153). Multilocus sequence typing (70) and automated ribotyping (144) are sometimes used. No matter what choice of method, it is important to compare results only within the method of choice, as all methods will vary in their strengths and limitations (61, 188).

Emerging technology for hospital laboratories includes platforms for repetitive element sequence-based PCR (rep-PCR), a bacterial and fungal genotyping method that has been used successfully as a rapid molecular tool for outbreak investigation and identification of MRSA (31, 32, 199, 202) and other hospital-acquired pathogens (23, 148, 149, 161, 212). The rep-PCR method utilizes the presence of repetitive genetic elements interspersed throughout the microbial genome at specific positions among different strains of microorganisms. PCR primers designed to amplify targeted genetic regions between these repetitive elements result in variably sized PCR products (amplicons) (215), each of which will produce a unique gel banding pattern (fingerprint). Electrophoresis of these differently sized products, derived from different organisms, allows strain-to-strain comparison and differentiation of outbreak strains. Advancements and improvements to the Diversilab system (bioMérieux, Durham, NC [formerly Bacterial Bar Codes]), a commercial rep-PCR fragment analysis system using capillary electrophoresis, enable standardized and reproducible performance of rep-PCR. This system provides a rapid and cost-effective option for outbreak investigation in hospitals and communities. rep-PCR compares favorably with PFGE in published comparisons (212, 215). While techniques are not always completely identical due to differences inherent to the methods and the respective genetic targets, they are equally as discriminatory. Healy et al. described 95% concordance with PFGE when rep-PCR was challenged with VRE isolates from around the world; Woods et al. found a concordance of 87% (212, 215).

Molecular Testing of Drug Resistance

Rapid and accurate determination of microbial drug susceptibility is critical to various aspects of patient therapy and is essential to facilitate successful antimicrobial therapy of any person with immune function impairment. Rapid testing for genetic resistance markers is an emerging clinical practice that can not only identify the potential for drug resistance but also help distinguish ambiguous breakpoints associated with susceptibility testing. Rapid detection of well-characterized resistance genes can be used to monitor their epidemiological spread or to identify prudent choices for antimicrobial therapy. Despite the scarcity of evidence-based studies to assess the utility of these markers, the application of molecular diagnostic methods to detect drug resistance is evolving as routine practice for some laboratories. Testing facilitates educated choices for therapy, which can be initiated early in diagnosis to impact patient outcomes. These test methods are particularly useful for slow-growing organisms, such as *Mycobacterium* sp. (82), but the use of molecular methods to detect resistance can be applied to bacteria, viruses, and fungi. For example, a QIAplex amplification system was recently used for simultaneous detection of 24 anti-tuberculosis drug resistance-related mutations (58). PCR mass spectrometry (Ibis and Abbott) can also be useful for identification of drug resistance in *Mycobacterium* spp.

Among bacteria, antimicrobial resistance genes include resistance genes for β-lactams, aminocyclitols, aminoglycosides, chloramphenicol, fluoroquinolones, glycopeptides, isoniazids, macrolides, mupirocin, rifampin, sulfonamides, tetracylines, and trimethoprim (47, 154, 190). For a list of PCR primers used to target such resistance markers, see the work of Tenover and Rasheed (190).

Perhaps the most well-documented applications of testing for bacterial drug resistance genes are those applied to MRSA and VRE. For MRSA, the *mecA* gene mediates oxacillin resistance in most MRSA strains, while the *vanA* and *vanB* genes mediate acquired vancomycin resistance in most VRE. Commercially available tests for MRSA include latex agglutination tests for PBP2a (the product of *mecA*) and PCR for the detection of *mecA* in *S. aureus*. PCR has been used to detect both MRSA and VRE (4, 29, 52, 63, 67, 81, 85, 103, 105, 137, 165, 176, 186). Together, MRSA and VRE are the two most important drug-resistant pathogens in U.S. hospitals, and their rapid detection remains critical as antimicrobial resistance continues to increase in the United States and worldwide (36). Success for active surveillance efforts in high-risk populations has been described, and the use of rapid PCR technology could play an important role in identifying carriers upon hospital admission and could aid in the prevention and control efforts for MRSA and VRE (53, 80, 136, 141, 162, 176, 180, 205). Routine use of these tests is evolving as a cost-effective approach with proven infection prevention studies (92).

The detection of viral mutations associated with drug resistance has been well documented. Examples include resistance to polymerase and protease inhibitors for HIV, acyclovir and penciclovir resistance in herpes simplex virus, acyclovir resistance in varicella-zoster virus, ganciclovir resistance in cytomegalovirus, famciclovir and lamivudune resistance in hepatitis B virus, and amantidine resistance in influenza A virus (172). Viral mutations are most commonly detected by direct sequencing of the specific viral reading frames, which encode the proteins that are targeted by currently available antiviral drugs.

Genotypic resistance testing to manage HIV-1-infected patients is widely used by physicians. There are currently two commercially available FDA-cleared sequencing assays that include reagent kits and software. These are the Truegene HIV-1 genotyping kit and OpenGene DNA sequencing system (Bayer Corp., Tarrytown, NY) and the ViroSeq HIV-1 genotyping system (Abbott, Abbott Park, IL). Genotype testing requires a skilled laboratory staff which is proficient in sequencing, alignments, editing, mutation detection, and interpretation of sequences.

Commercially available LiPAs, using a reverse hybridization method, allow laboratories which are proficient in PCR and have limited sequencing capabilities to detect mutations without the need to perform sequencing (33, 168). Comparison studies of these HIV-1 genotyping methods have proven them to be reliable and accurate. However, the LiPA is designed to identify known primary mutations associated with high-level drug resistance, while direct sequencing can detect new mutations (22, 43, 65, 71).

Rapid detection of antifungal resistance is useful, primarily due to the increase in fungal infections among immunocompromised patients. Current antifungal assays rely on fungal susceptibility testing, which is dependent on growth. The practical application of antifungal molecular testing has yet to be seen, as there is still much to learn about the genetic markers which mediate resistance. The genetic information needed to examine fungal resistance at the molecular level is much more complex than that for viruses and could involve the evaluation of fungal gene expression. For a review of molecular mechanisms of antifungal resistance, see reference 42.

The full potential of molecular diagnostics for drug resistance testing in microbiology has not been reached, and its application is still in its infancy. As the molecular mechanisms of antimicrobial resistance are described, newer technologies will enhance the utility of marker testing. Microarray technology has the promise to impact the rapid and accurate detection of multiple mutations associated with resistant bacteria, mycobacteria, viruses, and fungi. As with all molecular diagnostics, laboratories that perform molecular resistance testing need to ensure quality control of specimens.

Currently, there is still a need to cultivate organisms for further testing of other antimicrobials or typing for epidemiological studies. Thus, it is important to retain specimens or cultures until the laboratory can be sure that a result is negative and the specimen can be discarded (36). Until the full potential of drug resistance markers is understood, rapid molecular antimicrobial testing must still be combined with traditional microbial cultivation.

PROTEOMICS WILL EVENTUALLY IMPACT MICROBIOLOGY LABORATORIES

Protein Biomarker Detection

The future may hold changes that bring us back to the detection of functional proteins, coming full circle to historical methods that relied on detection of cell wall proteins and enzymes. Improvements in the speed by which proteins can accurately be characterized will impact our practices; instead of fluorescent-antibody detection of a pathogen's proteins, we will be able to identify and characterize pathogen proteins within minutes, perhaps even faster than PCR. Functional analysis of key proteins will be crucial to the diagnosis of infections and determination of drug susceptibility of pathogens in the immunocompromised or critically ill.

Proteins are the result of gene expression, and they often undergo posttranslational modifications, yielding tens of thousands of different types of proteins in a given specimen, such as blood or tissue. The presence and levels of key proteins tell the unmitigated story of microbial and human cellular activity; genetic detection is simply a precursor to the detailed story that the protein profile can reveal. Although there is presently a wide gap between biomarker discovery and its implementation in the clinical microbiology laboratory, the future will most assuredly hold an integration of molecular and protein-based detection methods for vulnerable hosts.

Though in its infancy, proteomic technology has the potential to play a key role in the future of clinical microbiology diagnostics as techniques become more rapid and affordable and the list of applicable biomarkers expands. MS and two-dimensional gel electrophoresis have historically been the two primary techniques for detection and identification of microbial proteins (39); however, advances in protein separation technology, microfluidics, and MS with integrated optics (219) should enable identification of many proteins simultaneously to assess the physiological state of living organisms as well as human biomarkers of infection. The power of proteomics could be placed into the clinical microbiology laboratory to assess the severity of potentially life-threatening infections and immune response to infection.

A mass spectrometer can take proteins and further separate them by producing charged particles (ions) (173). The mass spectrometer differentially moves ionized molecules, separated by their mass-to-charge (*m/z*) ratio, through a vacuum by means of an electromagnetic field. For the sake of discussion, if one assumes that each component of the mixture has a different molecular weight, then the mass spectrum contains unique "peaks" for each compound that is present. For more information about the different types of MS, refer to the work of Douglas (39).

A few reports have begun to surface in the clinical microbiology literature, describing how proteomic methods may impact laboratories in the future. In one report, matrix-assisted laser desorption ionization–time-of-flight MS was used to rapidly identify fungal proteins that evoked a specific human immune response, which may prove to be linked to active infection and outcome (145). In another study, matrix-assisted laser desorption ionization–time-of-flight MS, gel electrophoresis, and tandem MS were used to identify intra-amniotic proteins, which could lead to the discovery of novel human biomarkers for human intra-amniotic infection (66). Ultimately, these tools will help to elucidate the interaction of proteins with protein precursors, DNA, and mRNA and add to the understanding of pathogenesis and disease. Out of this understanding, novel biomarkers for early detection of disease or disease outcomes are expected to be found.

One emerging technology, the Luminex xMap system, can identify multiple immune proteins, such as serotype-specific antibodies, in a single-well or tube multiplex format. It has been used to identify multiple immune proteins (87) and bacterial DNA (40), but routine applications in the clinical laboratory will require further translational efforts.

In addition, biomarkers of the immune response, such as PCT, are useful for the diagnosis of severe bacterial infections such as sepsis, for prediction of sepsis complications, and for therapeutic follow-up. High plasma levels of PCT are observed in situations characterized by the onset of organ dysfunction and other symptoms of "severe sepsis" or "septic shock" (8, 25, 60, 155, 166, 169). Further details about one of these methods, the Brahms PCT LIA assay (Brahms Diagnostica, LLC) for determination of PCT in human serum, may be found in a previous section of this chapter.

SUMMARY

Today and in the future, molecular diagnostics will continue to impact and drive screening, diagnosis, and treatment decisions for immunocompromised and vulnerable patients. Patient test results that used to take days can now be completed in merely hours or minutes. We no longer perform microbial testing just because we need to in order to make a diagnosis; now, we explore issues such as infection prevention, guided therapy decisions, overall health care utility, and economics for our decisions to perform microbial testing. Immunocompromised patients are a unique population, for which extreme measures may be required to make relevant diagnoses, prevent infection, or prevent overuse of health care resources. For this population, it is not difficult to foresee a future in which clinicians not only rely on patients' vital signs and traditional laboratory diagnostics but also rely on patients' DNA profiles to determine predisposition to infection or to predict adverse effects of therapy. It is not difficult to foresee a future in which the costs of rapid diagnosis are small in comparison to the costs of waiting for results, both in terms of humanity and in terms of dollars. Near-point-of-care molecular testing is now reality, and it is a matter of time before new technology takes results out of the realm of days to weeks and into the realm of same-day testing of microbes and antibiotic susceptibility. What was once seen in science fiction TV shows is becoming our reality. We are faced with unique challenges as the future of clinical microbiology unfolds, first within the context of the immunocompromised host and then within the context of the general population.

REFERENCES

1. Reference deleted.
2. Reference deleted.
3. Reference deleted.
4. **Aarestrup, F. M., P. Ahrens, M. Madsen, L. V. Pallesen, R. L. Poulsen, and H. Westh.** 1996. Glycopeptide susceptibility among Danish *Enterococcus faecium* and *Enterococcus faecalis* isolates of animal and human origin and PCR identification of genes within the VanA cluster. *Antimicrob. Agents Chemother.* **40:**1938–1940.
5. **Abbott, W., E. Gane, I. Winship, S. Munn, and C. Tukuitonga.** 2007. Polymorphism in intron 1 of the interferon-gamma gene influences both serum immunoglobulin E levels and the risk for chronic hepatitis B virus infection in Polynesians. *Immunogenetics* **59:**187–195.
6. **Adelson, M. E., M. Feola, J. Trama, R. C. Tilton, and E. Mordechai.** 2005. Simultaneous detection of herpes simplex virus types 1 and 2 by real-time PCR and pyrosequencing. *J. Clin. Virol.* **33:**25–34.
7. **Alexander, B. D., E. D. Ashley, L. B. Reller, and S. D. Reed.** 2006. Cost savings with implementation of PNA FISH testing for identification of Candida albicans in blood cultures. *Diagn. Microbiol. Infect. Dis.* **54:**277–282.
8. **al-Nawas, B., and P. M. Shah.** 1996. Procalcitonin in patients with and without immunosuppression and sepsis. *Infection* **24:**434–436.
9. Reference deleted.
10. Reference deleted.
11. Reference deleted.
12. **Athanasopoulos, A., P. Devogel, C. Beken, C. Pille, I. Bernier, and P. Gavage.** 2007. Comparison of three selective chromogenic media for methicillin-resistant Staphylococcus aureus detection. *Pathol. Biol.* (Paris) **55:**366–369.
13. **Beck, R. C., D. J. Kohn, M. J. Tuohy, R. A. Prayson, B. Yen-Lieberman, and G. W. Procop.** 2004. Detection of polyoma virus in brain tissue of patients with progressive multifocal leukoencephalopathy by real-time PCR and pyrosequencing. *Diagn. Mol. Pathol.* **13:**15–21.
14. **Becker, K., D. Harmsen, A. Mellmann, C. Meier, P. Schumann, G. Peters, and C. von Eiff.** 2004. Development and evaluation of a quality-controlled ribosomal sequence database for 16S ribosomal DNA-based identification of *Staphylococcus* species. *J. Clin. Microbiol.* **42:**4988–4995.
15. **Benson, D., D. J. Lipman, and J. Ostell.** 1993. GenBank. *Nucleic Acids Res.* **21:**2963–2965.

16. **Blackwood, K. S., C. Y. Turenne, D. Harmsen, and A. M. Kabani.** 2004. Reassessment of sequence-based targets for identification of *Bacillus* species. *J. Clin. Microbiol.* **42:**1626–1630.
17. **Bochud, P. Y., A. S. Magaret, D. M. Koelle, A. Aderem, and A. Wald.** 2007. Polymorphisms in TLR2 are associated with increased viral shedding and lesional rate in patients with genital herpes simplex virus type 2 infection. *J. Infect. Dis.* **196:**505–509.
18. **Bogut, A., M. Koziol-Montewka, I. Baranowicz, L. Jozwiak, A. Ksiazek, Z. Al-Doori, D. Morrison, D. Kaczor, and J. Paluch-Oles.** 2007. Characterisation of Staphylococcus aureus nasal and skin carriage among patients undergoing haemodialysis treatment. *New Microbiol.* **30:**149–154.
19. **Brouqui, P., and D. Raoult.** 2001. Endocarditis due to rare and fastidious bacteria. *Clin. Microbiol. Rev.* **14:**177–207.
20. **Brunstein, J., and E. Thomas.** 2006. Direct screening of clinical specimens for multiple respiratory pathogens using the Genaco respiratory panels 1 and 2. *Diagn. Mol. Pathol.* **15:**169–173.
21. **Brunstein, J. D., C. L. Cline, S. McKinney, and E. Thomas.** 2008. Evidence from multiplex molecular assays for complex multipathogen interactions in acute respiratory infections. *J. Clin. Microbiol.* **46:**97–102.
22. **Caliendo, A. M., and D. Yen-Lieberman.** 2004. Viral genotyping, p. 489–499. *In* D. H. Persing, F. C. Tenover, J. Versalovic, Y. Tang, E. R. Unger, D. A. Relman, and T. J. White (ed.), *Molecular Microbiology: Diagnostic Principles and Practice.* ASM Press, Washington, DC.
23. **Carretto, E., D. Barbarini, C. Farina, A. Grosini, P. Nicoletti, and E. Manso.** 2008. Use of the DiversiLab(R) semiautomated repetitive-sequence-based polymerase chain reaction for epidemiologic analysis on Acinetobacter baumannii isolates in different Italian hospitals. *Diagn. Microbiol. Infect. Dis.* **60:**1–7.
24. **Chapman, S. J., C. C. Khor, F. O. Vannberg, A. Frodsham, A. Walley, N. A. Maskell, C. W. Davies, S. Segal, C. E. Moore, S. H. Gillespie, P. Denny, N. P. Day, D. W. Crook, R. J. Davies, and A. V. Hill.** 2007. IkappaB genetic polymorphisms and invasive pneumococcal disease. *Am. J. Respir. Crit. Care Med.* **176:**181–187.
25. **Chawes, B. L., C. Rechnitzer, K. Schmiegelow, and M. Tvede.** 2007. Procalcitonin for early diagnosis of bacteraemia in children with cancer. *Ugeskr. Laeger* **169:**138–142.
26. **Cherkaoui, A., G. Renzi, P. Francois, and J. Schrenzel.** 2007. Comparison of four chromogenic media for culture-based screening of methicillin-resistant Staphylococcus aureus. *J. Med. Microbiol.* **56:**500–503.
27. **Chun, J. Y., K. J. Kim, I. T. Hwang, Y. J. Kim, D. H. Lee, I. K. Lee, and J. K. Kim.** 2007. Dual priming oligonucleotide system for the multiplex detection of respiratory viruses and SNP genotyping of CYP2C19 gene. *Nucleic Acids Res.* **35:**e40.
28. **Ciok-Pater, E., P. Zalas, and E. Gospodarek.** 2006. Evaluation of Agar Candida ID2 (bioMerieux) a chromogenic medium for yeasts differentiation. *Med. Dosw. Mikrobiol.* **58:**247–251.
29. **Clark, N. C., R. C. Cooksey, B. C. Hill, J. M. Swenson, and F. C. Tenover.** 1993. Characterization of glycopeptide-resistant enterococci from U.S. hospitals. *Antimicrob. Agents Chemother.* **37:**2311–2317.
30. **Clarke, S. C.** 2005. Pyrosequencing: nucleotide sequencing technology with bacterial genotyping applications. *Expert Rev. Mol. Diagn.* **5:**947–953.
31. **Del Vecchio, V. G., J. M. Petroziello, M. J. Gress, F. K. McCleskey, G. P. Melcher, H. K. Crouch, and J. R. Lupski.** 1995. Molecular genotyping of methicillin-resistant *Staphylococcus aureus* via fluorophore-enhanced repetitive-sequence PCR. *J. Clin. Microbiol.* **33:**2141–2144.
32. **Deplano, A., A. Schuermans, J. Van Eldere, W. Witte, H. Meugnier, J. Etienne, H. Grundmann, D. Jonas, G. T. Noordhoek, J. Dijkstra, A. van Belkum, W. van Leeuwen, P. T. Tassios, N. J. Legakis, Z. A. van der, A. Bergmans, D. S. Blanc, F. C. Tenover, B. C. Cookson, G. O'Neil, M. J. Struelens, et al.** 2000. Multicenter evaluation of epidemiological typing of methicillin-resistant *Staphylococcus aureus* strains by repetitive-element PCR analysis. *J. Clin. Microbiol.* **38:**3527–3533.
33. **Descamps, D., V. Calvez, G. Collin, A. Cecille, C. Apetrei, F. Damond, C. Katlama, S. Matheron, J. M. Huraux, and F. Brun-Vezinet.** 1998. Line probe assay for detection of human immunodeficiency virus type 1 mutations conferring resistance to nucleoside inhibitors of reverse transcriptase: comparison with sequence analysis. *J. Clin. Microbiol.* **36:**2143–2145.
34. **DHHS Panel on Antiretroviral Guidelines for Adults and Adolescents—a Working Group of the Office of AIDS Research Advisory Council (OARAC).** 2007. Guidelines for the use of antiretroviral agents in HIV-1-infected adults and adolescents. Office of AIDS Research Advisory Council, Bethesda, MD.
35. **Diederen, B., I. van Duijn, A. van Belkum, P. Willemse, P. van Keulen, and J. Kluytmans.** 2005. Performance of CHROMagar MRSA medium for detection of methicillin-resistant *Staphylococcus aureus. J. Clin. Microbiol.* **43:**1925–1927.
36. **Diekema, D. J., K. J. Dodgson, B. Sigurdardottir, and M. A. Pfaller.** 2004. Rapid detection of antimicrobial-resistant organism carriage: an unmet clinical need. *J. Clin. Microbiol.* **42:**2879–2883.
37. Reference deleted.
38. **Diekema, D. J., and M. A. Pfaller.** 2003. Infection control epidemiology and clinical microbiology, p. 129–138. *In* P. M. Murray, E. J. Baron, J. H. Jorgenson, M. A. Pfaller, and R. H. Yolken (ed.), *Manual of Clinical Microbiology.* ASM Press, Washinton, DC.
39. **Douglas, J. F.** 2004. Elucidations of diagnostic and vaccine candidates by mass spectroscopy and proteomics, p. 307–319. *In* D. H. Persing, F. C. Tenover, J. Versalovic, Y. Tang, E. R. Unger, D. A. Relman, and T. J. White (ed.), *Molecular Microbiology: Diagnostic Principles and Practice.* ASM Press, Washington, DC.
40. **Dunbar, S. A., C. A. Vander Zee, K. G. Oliver, K. L. Karem, and J. W. Jacobson.** 2003. Quantitative, multiplexed detection of bacterial pathogens: DNA and protein applications of the Luminex LabMAP system. *J. Microbiol. Methods* **53:**245–252.
41. **Ecker, D. J., R. Sampath, L. B. Blyn, M. W. Eshoo, C. Ivy, J. A. Ecker, B. Libby, V. Samant, K. A. Sannes-Lowery, R. E. Melton, K. Russell, N. Freed, C. Barrozo, J. Wu, K. Rudnick, A. Desai, E. Moradi, D. J. Knize, D. W. Robbins, J. C. Hannis, P. M. Harrell, C. Massire, T. A. Hall, Y. Jiang, R. Ranken, J. J. Drader, N. White, J. A. McNeil, S. T. Crooke, and S. A. Hofstadler.** 2005. Rapid identification and strain-typing of respiratory pathogens for epidemic surveillance. *Proc. Natl. Acad. Sci. USA* **102:**8012–8017.
42. **Edlind, T. D.** 2004. Molecular detection of antifungal resistance, p. 569–575. *In* D. H. Persing, F. C. Tenover, J. Versalovic, Y. Tang, E. R. Unger, D. A. Relman, and T. J. White (ed.), *Molecular Microbiology: Diagnostic Principles and Practice.* ASM Press, Washington, DC.
43. **Erali, M., S. Page, L. G. Reimer, and D. R. Hillyard.** 2001. Human immunodeficiency virus type 1 drug resistance testing: a comparison of three sequence-based methods. *J. Clin. Microbiol.* **39:**2157–2165.

44. **Eraso, E., M. D. Moragues, M. Villar-Vidal, I. H. Sahand, N. Gonzalez-Gomez, J. Ponton, and G. Quindos.** 2006. Evaluation of the new chromogenic medium Candida ID 2 for isolation and identification of *Candida albicans* and other medically important *Candida* species. *J. Clin. Microbiol.* **44:**3340–3345.
45. **Fitch, J. P., et al.** 2002. Rapid development of nucleic acid diagnostics, p. 1708–1720. *In* Proc. IEEE. IEEE, Los Alamitos, CA.
46. **Flayhart, D., J. F. Hindler, D. A. Bruckner, G. Hall, R. K. Shrestha, S. A. Vogel, S. S. Richter, W. Howard, R. Walther, and K. C. Carroll.** 2005. Multicenter evaluation of BBL CHROMagar MRSA medium for direct detection of methicillin-resistant *Staphylococcus aureus* from surveillance cultures of the anterior nares. *J. Clin. Microbiol.* **43:**5536–5540.
47. **Fluit, A. C., M. R. Visser, and F. J. Schmitz.** 2001. Molecular detection of antimicrobial resistance. *Clin. Microbiol. Rev.* **14:**836–871.
48. **Fofanov, Y., Y. Luo, C. Katili, J. Wang, Y. Belosludtsev, T. Powdrill, C. Belapurkar, V. Fofanov, T. B. Li, S. Chumakov, and B. M. Pettitt.** 2004. How independent are the appearances of n-mers in different genomes? *Bioinformatics* **20:**2421–2428.
49. **Fofanov, Y., and B. M. Pettitt.** 2002. Reconstruction of the genetic regulatory dynamics of the rat spinal cord development: local invariants approach. *J. Biomed. Inform.* **35:**343–351.
50. **Forrest, G. N., K. Mankes, M. A. Jabra-Rizk, E. Weekes, J. K. Johnson, D. P. Lincalis, and R. A. Venezia.** 2006. Peptide nucleic acid fluorescence in situ hybridization-based identification of *Candida albicans* and its impact on mortality and antifungal therapy costs. *J. Clin. Microbiol.* **44:**3381–3383.
51. **Forrest, G. N., S. Mehta, E. Weekes, D. P. Lincalis, J. K. Johnson, and R. A. Venezia.** 2006. Impact of rapid in situ hybridization testing on coagulase-negative staphylococci positive blood cultures. *J. Antimicrob. Chemother.* **58:**154–158.
52. **Francois, P., D. Pittet, M. Bento, B. Pepey, P. Vaudaux, D. Lew, and J. Schrenzel.** 2003. Rapid detection of methicillin-resistant *Staphylococcus aureus* directly from sterile or nonsterile clinical samples by a new molecular assay. *J. Clin. Microbiol.* **41:**254–260.
53. **Francois, P., G. Renzi, D. Pittet, M. Bento, D. Lew, S. Harbarth, P. Vaudaux, and J. Schrenzel.** 2004. A novel multiplex real-time PCR assay for rapid typing of major staphylococcal cassette chromosome *mec* elements. *J. Clin. Microbiol.* **42:**3309–3312.
54. **Freymuth, F., A. Vabret, D. Cuvillon-Nimal, S. Simon, J. Dina, L. Legrand, S. Gouarin, J. Petitjean, P. Eckart, and J. Brouard.** 2006. Comparison of multiplex PCR assays and conventional techniques for the diagnostic of respiratory virus infections in children admitted to hospital with an acute respiratory illness. *J. Med. Virol.* **78:**1498–1504.
55. **Galor, A., G. S. Hall, G. W. Procop, M. Tuohy, M. E. Millstein, and B. H. Jeng.** 2007. Rapid species determination of nocardia keratitis using pyrosequencing technology. *Am. J. Ophthalmol.* **143:**182–183.
56. Reference deleted.
57. **Gavino, M., and E. Wang.** 2007. A comparison of a new rapid real-time polymerase chain reaction system to traditional culture in determining group B streptococcus colonization. *Am. J. Obstet. Gynecol.* **197:**388.e1–388.e4.
58. **Gegia, M., N. Mdivani, R. E. Mendes, H. Li, M. Akhalaia, J. Han, G. Khechinashvili, and Y. W. Tang.** 2008. Prevalence of and molecular basis for tuberculosis drug resistance in the Republic of Georgia: validation of a QIAplex system for the detection of drug resistance-related mutations. *Antimicrob. Agents Chemother.* **52:**725–729.
59. **Gharizadeh, B., E. Norberg, J. Loffler, S. Jalal, J. Tollemar, H. Einsele, L. Klingspor, and P. Nyren.** 2004. Identification of medically important fungi by the pyrosequencing technology. *Mycoses* **47:**29–33.
60. **Giamarellos-Bourboulis, E. J., A. Mega, P. Grecka, N. Scarpa, G. Koratzanis, G. Thomopoulos, and H. Giamarellou.** 2002. Procalcitonin: a marker to clearly differentiate systemic inflammatory response syndrome and sepsis in the critically ill patient? *Intensive Care Med.* **28:**1351–1356.
61. **Goering, R. V.** 2000. Molecular strain typing for the clinical laboratory. *Clin. Microbiol. Newsl.* **22:**169–173.
62. **Gonzalez, V., E. Padilla, M. Gimenez, C. Vilaplana, A. Perez, G. Fernandez, M. D. Quesada, M. A. Pallares, and V. Ausina.** 2004. Rapid diagnosis of Staphylococcus aureus bacteremia using S. aureus PNA FISH. *Eur. J. Clin. Microbiol. Infect. Dis.* **23:**396–398.
63. **Gordts, B., H. Van Landuyt, M. Ieven, P. Vandamme, and H. Goossens.** 1995. Vancomycin-resistant enterococci colonizing the intestinal tracts of hospitalized patients. *J. Clin. Microbiol.* **33:**2842–2846.
64. **Granell, M., A. Urbano-Ispizua, J. I. Arostegui, F. Fernandez-Aviles, C. Martinez, M. Rovira, J. Rius, S. Plaza, A. Gaya, A. Navarro, C. Talarn, E. Carreras, M. Monzo, E. Montserrat, and J. Yague.** 2006. Effect of NOD2/CARD15 variants in T-cell depleted allogeneic stem cell transplantation. *Haematologica* **91:**1372–1376.
65. **Grant, R. M., D. R. Kuritzkes, V. A. Johnson, J. W. Mellors, J. L. Sullivan, R. Swanstrom, R. T. D'Aquila, M. Van Gorder, M. Holodniy, J. R. Lloyd, Jr., C. Reid, G. F. Morgan, and D. L. Winslow.** 2003. Accuracy of the TRUGENE HIV-1 genotyping kit. *J. Clin. Microbiol.* **41:**1586–1593.
66. **Gravett, M. G., M. J. Novy, R. G. Rosenfeld, A. P. Reddy, T. Jacob, M. Turner, A. McCormack, J. A. Lapidus, J. Hitti, D. A. Eschenbach, C. T. Roberts, Jr., and S. R. Nagalla.** 2004. Diagnosis of intra-amniotic infection by proteomic profiling and identification of novel biomarkers. *JAMA* **292:**462–469.
67. **Grisold, A. J., E. Leitner, G. Muhlbauer, E. Marth, and H. H. Kessler.** 2002. Detection of methicillin-resistant *Staphylococcus aureus* and simultaneous confirmation by automated nucleic acid extraction and real-time PCR. *J. Clin. Microbiol.* **40:**2392–2397.
68. **Han, J., D. C. Swan, S. J. Smith, S. H. Lum, S. E. Sefers, E. R. Unger, and Y. W. Tang.** 2006. Simultaneous amplification and identification of 25 human papillomavirus types with Templex technology. *J. Clin. Microbiol.* **44:**4157–4162.
69. **Han, Z., E. Lautenbach, N. Fishman, and I. Nachamkin.** 2007. Evaluation of mannitol salt agar, CHROMagar Staph aureus and CHROMagar MRSA for detection of methicillin-resistant Staphylococcus aureus from nasal swab specimens. *J. Med. Microbiol.* **56:**43–46.
70. **Hanage, W. P., E. J. Feh, A. B. Breuggmann, and B. G. Spratt.** 2004. Multilocus sequence typing: strain characterization, population biology, and patterns of evolutionary descent, p. 235–243. *In* D. H. Persing, F. C. Tenover, J. Versalovic, Y. Tang, E. R. Unger, D. A. Relman, and T. J. White (ed.), *Molecular Microbiology: Diagnostic Principles and Practice*. ASM Press, Washington, DC.
71. **Hanna, G. J., and R. T. D'Aquila.** 2001. Clinical use of genotypic and phenotypic drug resistance testing to monitor antiretroviral chemotherapy. *Clin. Infect. Dis.* **32:**774–782.
72. **Harmsen, D., H. Claus, W. Witte, J. Rothganger, H. Claus, D. Turnwald, and U. Vogel.** 2003. Typing of methicillin-resistant *Staphylococcus aureus* in a university hospital setting by using novel software for *spa* repeat determination and database management. *J. Clin. Microbiol.* **41:**5442–5448.

73. **Harmsen, D., S. Dostal, A. Roth, S. Niemann, J. Rothganger, M. Sammeth, J. Albert, M. Frosch, and E. Richter.** 2003. RIDOM: comprehensive and public sequence database for identification of Mycobacterium species. *BMC Infect. Dis.* **3**:26.
74. **Harmsen, D., J. Rothganger, M. Frosch, and J. Albert.** 2002. RIDOM: Ribosomal Differentiation of Medical Micro-Organisms Database. *Nucleic Acids Res.* **30**:416–417.
75. **Harmsen, D., C. Singer, J. Rothganger, T. Tonjum, G. S. de Hoog, H. Shah, J. Albert, and M. Frosch.** 2001. Diagnostics of *Neisseriaceae* and *Moraxellaceae* by ribosomal DNA sequencing: ribosomal differentiation of medical microorganisms. *J. Clin. Microbiol.* **39**:936–942.
76. **Harriau, P., F. Ruffel, and J. B. Lardy.** 2006. Use of BioRad plating agar MRSASelect for the daily detection of methicillin resistant staphylococci isolated from samples taken from blood culture bottles. *Pathol. Biol.* (Paris) **54**:506–509.
77. **Hartmann, H., H. Stender, A. Schafer, I. B. Autenrieth, and V. A. Kempf.** 2005. Rapid identification of *Staphylococcus aureus* in blood cultures by a combination of fluorescence in situ hybridization using peptide nucleic acid probes and flow cytometry. *J. Clin. Microbiol.* **43**:4855–4857.
78. **Hedin, G., and H. Fang.** 2005. Evaluation of two new chromogenic media, CHROMagar MRSA and S. aureus ID, for identifying *Staphylococcus aureus* and screening methicillin-resistant *S. aureus*. *J. Clin. Microbiol.* **43**:4242–4244.
78a. **Heller, A.** 1 February 2007, accession date. On the front lines of biodefense. http://www.eurekalert.org/features/doe/2004-04/dlnl-otf041204.php.
79. **Hohl, M., S. Kurtz, and E. Ohlebusch.** 2002. Efficient multiple genome alignment. *Bioinformatics* **18**(Suppl. 1):S312–S320.
80. **Huletsky, A., R. Giroux, V. Rossbach, M. Gagnon, M. Vaillancourt, M. Bernier, F. Gagnon, K. Truchon, M. Bastien, F. J. Picard, A. van Belkum, M. Ouellette, P. H. Roy, and M. G. Bergeron.** 2004. New real-time PCR assay for rapid detection of methicillin-resistant *Staphylococcus aureus* directly from specimens containing a mixture of staphylococci. *J. Clin. Microbiol.* **42**:1875–1884.
81. **Hussain, Z., L. Stoakes, V. Massey, D. Diagre, V. Fitzgerald, S. El Sayed, and R. Lannigan.** 2000. Correlation of oxacillin MIC with *mecA* gene carriage in coagulase-negative staphylococci. *J. Clin. Microbiol.* **38**:752–754.
82. **Inderleid, C. B., and G. E. Pfyffer.** 2003. Susceptibility test methods: mycobacteria, p. 1149–1177. *In* P. M. Murray, E. J. Baron, J. H. Jorgenson, M. A. Pfaller, and R. H. Yolken (ed.), *Manual of Clinical Microbiology*. ASM Press, Washington, DC.
83. **Isola, D., M. Pardini, F. Varaine, S. Niemann, S. Rusch-Gerdes, L. Fattorini, G. Orefici, F. Meacci, C. Trappetti, O. M. Rinaldo, and G. Orru.** 2005. A pyrosequencing assay for rapid recognition of SNPs in Mycobacterium tuberculosis embB306 region. *J. Microbiol. Methods* **62**:113–120.
84. **Iyampillai, T., J. S. Michael, E. Mathai, and M. S. Mathews.** 2004. Use of CHROMagar medium in the differentiation of Candida species: is it cost-effective in developing countries? *Ann. Trop. Med. Parasitol.* **98**:279–282.
85. **Jonas, D., M. Speck, F. D. Daschner, and H. Grundmann.** 2002. Rapid PCR-based identification of methicillin-resistant *Staphylococcus aureus* from screening swabs. *J. Clin. Microbiol.* **40**:1821–1823.
86. **Jonasson, J., M. Olofsson, and H. J. Monstein.** 2002. Classification, identification and subtyping of bacteria based on pyrosequencing and signature matching of 16S rDNA fragments. *APMIS* **110**:263–272.
87. **Jones, L. P., H. Q. Zheng, R. A. Karron, T. C. Peret, C. Tsou, and L. J. Anderson.** 2002. Multiplex assay for detection of strain-specific antibodies against the two variable regions of the G protein of respiratory syncytial virus. *Clin. Diagn. Lab. Immunol.* **9**:633–638.
88. **Jureen, P., L. Engstrand, S. Eriksson, A. Alderborn, M. Krabbe, and S. E. Hoffner.** 2006. Rapid detection of rifampin resistance in *Mycobacterium tuberculosis* by pyrosequencing technology. *J. Clin. Microbiol.* **44**:1925–1929.
89. **Kaye, M., S. Skidmore, H. Osman, M. Weinbren, and R. Warren.** 2006. Surveillance of respiratory virus infections in adult hospital admissions using rapid methods. *Epidemiol. Infect.* **134**:792–798.
90. **Khudaier, B. Y., R. Tewari, S. Shafiani, M. Sharma, R. Emmanuel, M. Sharma, and N. Taneja.** 2007. Epidemiology and molecular characterization of vancomycin resistant enterococci isolates in India. *Scand. J. Infect. Dis.* **39**:662–670.
91. **Kobayashi, N., T. W. Bauer, M. J. Tuohy, I. H. Lieberman, V. Krebs, D. Togawa, T. Fujishiro, and G. W. Procop.** 2006. The comparison of pyrosequencing molecular Gram stain, culture, and conventional Gram stain for diagnosing orthopaedic infections. *J. Orthop. Res.* **24**:1641–1649.
92. **Kola, A., I. F. Chaberny, F. Mattner, U. Reischl, R. P. Vonberg, K. Weist, C. Wendt, W. Witte, S. Ziesing, S. Suerbaum, and P. Gastmeier.** 2006. Control of methicillin-resistant S. aureus by active surveillance. Results of a workshop held by the Deutsche Gesellschaft fur Hygiene und Mikrobiologie. *Anaesthesist* **55**:778–783.
93. **Kost, C. B., B. Rogers, M. S. Oberste, C. Robinson, B. L. Eaves, K. Leos, S. Danielson, M. Satya, F. Weir, and F. S. Nolte.** 2007. Multicenter beta trial of the GeneXpert enterovirus assay. *J. Clin. Microbiol.* **45**:1081–1086.
94. **Larsen, N., G. J. Olsen, B. L. Maidak, M. J. McCaughey, R. Overbeek, T. J. Macke, T. L. Marsh, and C. R. Woese.** 1993. The ribosomal database project. *Nucleic Acids Res.* **21**:3021–3023.
95. **Ledeboer, N. A., K. Das, M. Eveland, C. Roger-Dalbert, S. Mailler, S. Chatellier, and W. M. Dunne.** 2007. Evaluation of a novel chromogenic agar medium for isolation and differentiation of vancomycin-resistant *Enterococcus faecium* and *Enterococcus faecalis* isolates. *J. Clin. Microbiol.* **45**:1556–1560.
96. **Lee, W. M., K. Grindle, T. Pappas, D. J. Marshall, M. J. Moser, E. L. Beaty, P. A. Shult, J. R. Prudent, and J. E. Gern.** 2007. High-throughput, sensitive, and accurate multiplex PCR-microsphere flow cytometry system for large-scale comprehensive detection of respiratory viruses. *J. Clin. Microbiol.* **45**:2626–2634.
97. **Lee, W. M., K. Grindle, T. Pappas, D. J. Marshall, M. J. Moser, E. L. Beaty, P. A. Shult, J. R. Prudent, and J. E. Gern.** 2007. High-throughput, sensitive, and accurate multiplex PCR-microsphere flow cytometry system for large-scale comprehensive detection of respiratory viruses. *J. Clin. Microbiol.* **45**:2626–2634.
98. Reference deleted.
99. **Lee, W. M., C. Kiesner, T. Pappas, I. Lee, K. Grindle, T. Jartti, B. Jakiela, R. F. Lemanske, P. A. Shult, and J. E. Gern.** 2007. A diverse group of previously unrecognized human rhinoviruses are common causes of respiratory illnesses in infants. *PLoS ONE* **2**:e966.
100. **Lenhof, H. P., B. Morgenstern, and K. Reinert.** 1999. An exact solution for the segment-to-segment multiple sequence alignment problem. *Bioinformatics* **15**:203–210.
101. **Li, H., M. A. McCormac, R. W. Estes, S. E. Sefers, R. K. Dare, J. D. Chappell, D. D. Erdman, P. F. Wright, and Y. W. Tang.** 2007. Simultaneous detection and high-throughput identification of a panel of RNA viruses causing respiratory tract infections. *J. Clin. Microbiol.* **45**:2105–2109.

102. **Li, P., K. C. Kupfer, C. J. Davies, D. Burbee, G. A. Evans, and H. R. Garner.** 1997. PRIMO: a primer design program that applies base quality statistics for automated large-scale DNA sequencing. *Genomics* **40:**476–485.
103. **Louie, L., J. Goodfellow, P. Mathieu, A. Glatt, M. Louie, and A. E. Simor.** 2002. Rapid detection of methicillin-resistant staphylococci from blood culture bottles by using a multiplex PCR assay. *J. Clin. Microbiol.* **40:**2786–2790.
104. **Louie, L., A. E. Simor, S. Chong, K. Luinstra, A. Petrich, J. Mahony, M. Smieja, G. Johnson, F. Gharabaghi, R. Tellier, B. M. Willey, S. Poutanen, T. Mazzulli, G. Broukhanski, F. Jamieson, M. Louie, and S. Richardson.** 2006. Detection of severe acute respiratory syndrome coronavirus in stool specimens by commercially available real-time reverse transcriptase PCR assays. *J. Clin. Microbiol.* **44:**4193–4196.
105. **Maes, N., J. Magdalena, S. Rottiers, G. Y. De, and M. J. Struelens.** 2002. Evaluation of a triplex PCR assay to discriminate *Staphylococcus aureus* from coagulase-negative staphylococci and determine methicillin resistance from blood cultures. *J. Clin. Microbiol.* **40:**1514–1517.
106. **Mahony, J., S. Chong, F. Merante, S. Yaghoubian, T. Sinha, C. Lisle, and R. Janeczko.** 2007. Development of a respiratory virus panel test for detection of twenty human respiratory viruses by use of multiplex PCR and a fluid microbead-based assay. *J. Clin. Microbiol.* **45:**2965–2970.
107. Reference deleted.
108. **Maidak, B. L., J. R. Cole, T. G. Lilburn, C. T. Parker, Jr., P. R. Saxman, R. J. Farris, G. M. Garrity, G. J. Olsen, T. M. Schmidt, and J. M. Tiedje.** 2001. The RDP-II (Ribosomal Database Project). *Nucleic Acids Res.* **29:**173–174.
109. **Maidak, B. L., J. R. Cole, T. G. Lilburn, C. T. Parker, Jr., P. R. Saxman, J. M. Stredwick, G. M. Garrity, B. Li, G. J. Olsen, S. Pramanik, T. M. Schmidt, and J. M. Tiedje.** 2000. The RDP (Ribosomal Database Project) continues. *Nucleic Acids Res.* **28:**173–174.
110. **Maidak, B. L., N. Larsen, M. J. McCaughey, R. Overbeek, G. J. Olsen, K. Fogel, J. Blandy, and C. R. Woese.** 1994. The Ribosomal Database Project. *Nucleic Acids Res.* **22:** 3485–3487.
111. **Marra, A. R., K. Bar, G. M. Bearman, R. P. Wenzel, and M. B. Edmond.** 2006. Systemic inflammatory response syndrome in adult patients with nosocomial bloodstream infection due to Pseudomonas aeruginosa. *J. Infect.* **53:**30–35.
112. **Marra, A. R., M. B. Edmond, B. A. Forbes, R. P. Wenzel, and G. M. Bearman.** 2006. Time to blood culture positivity as a predictor of clinical outcome of *Staphylococcus aureus* bloodstream infection. *J. Clin. Microbiol.* **44:**1342–1346.
113. **Marshall, D. J., E. Reisdorf, G. Harms, E. Beaty, M. J. Moser, W. M. Lee, J. E. Gern, F. S. Nolte, P. Shult, and J. R. Prudent.** 2007. Evaluation of a multiplexed PCR assay for detection of respiratory viral pathogens in a public health laboratory setting. *J. Clin. Microbiol.* **45:**3875–3882.
114. **Martins, T. B.** 2002. Development of internal controls for the Luminex instrument as part of a multiplex seven-analyte viral respiratory antibody profile. *Clin. Diagn. Lab. Immunol.* **9:**41–45.
115. **Mascini, E. M., A. C. Gigengack-Baars, R. J. Hene, T. E. Kamp-Hopmans, A. J. Weersink, and M. J. Bonten.** 2000. Epidemiologic increase of various genotypes of vancomycin-resistant Enterococcus faecium in a university hospital. *Ned. Tijdschr. Geneeskd.* **144:**2572–2576.
116. **Mathema, B., and B. N. Kreiswirth.** 2004. Genotyping bacteria by using variable-number tandem repeats, p. 223–234. *In* D. H. Persing, F. C. Tenover, J. Versalovic, Y. Tang, E. R. Unger, D. A. Relman, and T. J. White (ed.), *Molecular Microbiology: Diagnostic Principles and Practice.* ASM Press, Washington, DC.
117. **McLeod, H. L.** 2005. Pharmacogenetic analysis of clinically relevant genetic polymorphisms. *Clin. Infect. Dis.* **41**(Suppl. 7):S449–S452.
118. **Mellmann, A., J. L. Cloud, S. Andrees, K. Blackwood, K. C. Carroll, A. Kabani, A. Roth, and D. Harmsen.** 2003. Evaluation of RIDOM, MicroSeq, and GenBank services in the molecular identification of Nocardia species. *Int. J. Med. Microbiol.* **293:**359–370.
119. **Mellmann, A., A. W. Friedrich, F. Kipp, F. Hinder, U. Keckevoet, and D. Harmsen.** 2005. Evidence-based infection control methods using spa genotyping for MRSA spread in hospitals. *Dtsch. Med. Wochenschr.* **130:**1364–1368.
120. **Mellmann, A., A. W. Friedrich, N. Rosenkotter, J. Rothganger, H. Karch, R. Reintjes, and D. Harmsen.** 2006. Automated DNA sequence-based early warning system for the detection of methicillin-resistant Staphylococcus aureus outbreaks. *PLoS Med.* **3:**e33.
121. **Molenkamp, R., A. van der Ham, J. Schinkel, and M. Beld.** 2007. Simultaneous detection of five different DNA targets by real-time Taqman PCR using the Roche LightCycler480: application in viral molecular diagnostics. *J. Virol. Methods* **141:**205–211.
122. **Murray, C. K., M. L. Beckius, J. A. Green, and D. R. Hospenthal.** 2005. Use of chromogenic medium in the isolation of yeasts from clinical specimens. *J. Med. Microbiol.* **54:** 981–985.
123. **Naas, T., L. Poirel, and P. Nordmann.** 2006. Pyrosequencing for rapid identification of carbapenem-hydrolysing OXA-type beta-lactamases in Acinetobacter baumannii. *Clin. Microbiol. Infect.* **12:**1236–1240.
124. **Nahimana, I., P. Francioli, and D. S. Blanc.** 2006. Evaluation of three chromogenic media (MRSA-ID, MRSA-Select and CHROMagar MRSA) and ORSAB for surveillance cultures of methicillin-resistant Staphylococcus aureus. *Clin. Microbiol. Infect.* **12:**1168–1174.
125. **Nguyen, H. B., E. P. Rivers, B. P. Knoblich, G. Jacobsen, A. Muzzin, J. A. Ressler, and M. C. Tomlanovich.** 2004. Early lactate clearance is associated with improved outcome in severe sepsis and septic shock. *Crit. Care Med.* **32:**1637–1642.
126. **Nguyen Van, J. C., M. D. Kitzis, A. Ly, A. Chalfine, J. Carlet, A. A. Ben, and F. Goldstein.** 2006. Detection of nasal colonization methicillin-resistant Staphylococcus aureus: a prospective study comparing real-time genic amplification assay vs selective chromogenic media. *Pathol. Biol.* (Paris) **54:**285–292.
127. Reference deleted.
128. **Nilsson, I., I. Shabo, J. Svanvik, and H. J. Monstein.** 2005. Multiple displacement amplification of isolated DNA from human gallstones: molecular identification of Helicobacter DNA by means of 16S rDNA-based pyrosequencing analysis. *Helicobacter* **10:**592–600.
129. **Nolte, F. S., D. J. Marshall, C. Rasberry, S. Schievelbein, G. G. Banks, G. A. Storch, M. Q. Arens, R. S. Buller, and J. R. Prudent.** 2007. MultiCode-PLx system for multiplexed detection of seventeen respiratory viruses. *J. Clin. Microbiol.* **45:**2779–2786.
130. Reference deleted.
131. **Olive, D. M., and P. Bean.** 1999. Principles and applications of methods for DNA-based typing of microbial organisms. *J. Clin. Microbiol.* **37:**1661–1669.
132. **Oliveira, K., S. M. Brecher, A. Durbin, D. S. Shapiro, D. R. Schwartz, P. C. De Girolami, J. Dakos, G. W. Procop, D. Wilson, C. S. Hanna, G. Haase, H. Peltroche-Llacsahuanga,**

K. C. Chapin, M. C. Musgnug, M. H. Levi, C. Shoemaker, and H. Stender. 2003. Direct identification of *Staphylococcus aureus* from positive blood culture bottles. *J. Clin. Microbiol.* 41:889–891.

133. Oliveira, K., G. Haase, C. Kurtzman, J. J. Hyldig-Nielsen, and H. Stender. 2001. Differentiation of *Candida albicans* and *Candida dubliniensis* by fluorescent in situ hybridization with peptide nucleic acid probes. *J. Clin. Microbiol.* 39: 4138–4141.
134. Reference deleted.
135. Oliveira, K., G. W. Procop, D. Wilson, J. Coull, and H. Stender. 2002. Rapid identification of *Staphylococcus aureus* directly from blood cultures by fluorescence in situ hybridization with peptide nucleic acid probes. *J. Clin. Microbiol.* 40:247–251.
136. Ornskov, D., B. Kolmos, H. P. Bendix, N. J. Nederby, I. Brandslund, and P. Schouenborg. 2008. Screening for methicillin-resistant Staphylococcus aureus in clinical swabs using a high-throughput real-time PCR-based method. *Clin. Microbiol. Infect.* 14:22–28.
137. Padiglione, A. A., E. A. Grabsch, D. Olden, M. Hellard, M. I. Sinclair, C. K. Fairley, and M. L. Grayson. 2000. Fecal colonization with vancomycin-resistant enterococci in Australia. *Emerg. Infect. Dis.* 6:534–536.
138. Palenzuela, L., N. M. Hahn, R. P. Nelson, Jr., J. N. Arno, C. Schobert, R. Bethel, L. A. Ostrowski, M. R. Sharma, P. P. Datta, R. K. Agrawal, J. E. Schwartz, and M. Hirano. 2005. Does linezolid cause lactic acidosis by inhibiting mitochondrial protein synthesis? *Clin. Infect. Dis.* 40:e113–e116.
139. Pape, J., J. Wadlin, and I. Nachamkin. 2006. Use of BBL CHROMagar MRSA medium for identification of methicillin-resistant *Staphylococcus aureus* directly from blood cultures. *J. Clin. Microbiol.* 44:2575–2576.
140. Pappas, P. G., J. H. Rex, J. D. Sobel, S. G. Filler, W. E. Dismukes, T. J. Walsh, and J. E. Edwards. 2004. Guidelines for treatment of candidiasis. *Clin. Infect. Dis.* 38:161–189.
141. Paule, S. M., D. M. Hacek, B. Kufner, K. Truchon, R. B. Thomson, Jr., K. L. Kaul, A. Robicsek, and L. R. Peterson. 2007. Performance of the BD GeneOhm methicillin-resistant *Staphylococcus aureus* test before and during high-volume clinical use. *J. Clin. Microbiol.* 45:2993–2998.
142. Pearman, J. W. 2006. 2004 Lowbury Lecture: the Western Australian experience with vancomycin-resistant enterococci—from disaster to ongoing control. *J. Hosp. Infect.* 63:14–26.
143. Perry, J. D., A. Davies, L. A. Butterworth, A. L. Hopley, A. Nicholson, and F. K. Gould. 2004. Development and evaluation of a chromogenic agar medium for methicillin-resistant *Staphylococcus aureus*. *J. Clin. Microbiol.* 42: 4519–4523.
144. Pfaller, M. A., and R. J. Hollis. 2004. Automated ribotyping, p. 245–258. *In* D. H. Persing, F. C. Tenover, J. Versalovic, Y. Tang, E. R. Unger, D. A. Relman, and T. J. White (ed.), *Molecular Microbiology: Diagnostic Principles and Practice*. ASM Press, Washington, DC.
145. Pitarch, A., J. Abian, M. Carrascal, M. Sanchez, C. Nombela, and C. Gil. 2004. Proteomics-based identification of novel *Candida albicans* antigens for diagnosis of systemic candidiasis in patients with underlying hematological malignancies. *Proteomics* 4:3084–3106.
146. Plouzeau, C., M. Paccalin, A. Beby-Defaux, G. Giraudeau, C. Godet, and G. Agius. 2007. Diagnosis and epidemiological surveillance of influenza and respiratory syncytial virus infections: interest of multiplex PCR. *Med. Mal. Infect.* 37:728–733.
147. Poirel, L., T. Naas, and P. Nordmann. 2006. Pyrosequencing as a rapid tool for identification of GES-type extended-spectrum beta-lactamases. *J. Clin. Microbiol.* 44:3008–3011.
148. Pounder, J. I., D. Hansen, and G. L. Woods. 2006. Identification of *Histoplasma capsulatum*, *Blastomyces dermatitidis*, and *Coccidioides* species by repetitive-sequence-based PCR. *J. Clin. Microbiol.* 44:2977–2982.
149. Pounder, J. I., S. Williams, D. Hansen, M. Healy, K. Reece, and G. L. Woods. 2005. Repetitive-sequence-PCR-based DNA fingerprinting using the Diversilab system for identification of commonly encountered dermatophytes. *J. Clin. Microbiol.* 43:2141–2147.
150. Proutski, V., and E. C. Holmes. 1996. Primer Master: a new program for the design and analysis of PCR primers. *Comput. Appl. Biosci.* 12:253–255.
151. Putonti, C., S. Chumakov, R. Mitra, G. E. Fox, R. C. Willson, and Y. Fofanov. 2006. Human-blind probes and primers for dengue virus identification. *FEBS J.* 273:398–408.
152. Putonti, C., Y. Luo, C. Katili, S. Chumakov, G. E. Fox, D. Graur, and Y. Fofanov. 2006. A computational tool for the genomic identification of regions of unusual compositional properties and its utilization in the detection of horizontally transferred sequences. *Mol. Biol. Evol.* 23:1863–1868.
153. Rademaker, J. L., and P. Savelkoul. 2004. PCR amplification-based microbial typing, p. 197–221. *In* D. H. Persing, F. C. Tenover, J. Versalovic, Y. Tang, E. R. Unger, D. A. Relman, and T. J. White (ed.), *Molecular Microbiology: Diagnostic Principles and Practice*. ASM Press, Washington, DC.
154. Rasheed, J. K., and F. C. Tenover. 2003. Detection and characterization of antimicrobial resistance genes in bacteria, p. 1196–1213. *In* P. R. Murray, E. J. Murray, E. J. Baron, M. A. Pfaller, F. C. Tenover, and R. H. Yolken (ed.), *Manual Of Clinical Microbiology*. ASM Press, Washington, DC.
155. Reingardene, D. 2004. Procalcitonin as a marker of the systemic inflammatory response to infection. *Medicina* 40:696–701.
156. Rigby, S., G. W. Procop, G. Haase, D. Wilson, G. Hall, C. Kurtzman, K. Oliveira, S. Von Oy, J. J. Hyldig-Nielsen, J. Coull, and H. Stender. 2002. Fluorescence in situ hybridization with peptide nucleic acid probes for rapid identification of *Candida albicans* directly from blood culture bottles. *J. Clin. Microbiol.* 40:2182–2186.
157. Rivers, E., B. Nguyen, S. Havstad, J. Ressler, A. Muzzin, B. Knoblich, E. Peterson, and M. Tomlanovich. 2001. Early goal-directed therapy in the treatment of severe sepsis and septic shock. *N. Engl. J. Med.* 345:1368–1377.
158. Reference deleted.
159. Rodriguez, E., F. Symoens, P. Mondon, M. Mallie, M. A. Piens, B. Lebeau, A. M. Tortorano, F. Chaib, A. Carlotti, J. Villard, M. A. Viviani, F. Chapuis, N. Nolard, R. Grillot, J. M. Bastide, et al. 1999. Combination of three typing methods for the molecular epidemiology of Aspergillus fumigatus infections. *J. Med. Microbiol.* 48:181–194.
160. Rose, T. M., E. R. Schultz, J. G. Henikoff, S. Pietrokovski, C. M. McCallum, and S. Henikoff. 1998. Consensus-degenerate hybrid oligonucleotide primers for amplification of distantly related sequences. *Nucleic Acids Res.* 26:1628–1635.
161. Ross, T. L., W. G. Merz, M. Farkosh, and K. C. Carroll. 2005. Comparison of an automated repetitive sequence-based PCR microbial typing system to pulsed-field gel electrophoresis for analysis of outbreaks of methicillin-resistant *Staphylococcus aureus*. *J. Clin. Microbiol.* 43:5642–5647.

162. **Rossney, A. S., C. M. Herra, M. M. Fitzgibbon, P. M. Morgan, M. J. Lawrence, and B. O'Connell.** 2007. Evaluation of the IDI-MRSA assay on the SmartCycler real-time PCR platform for rapid detection of MRSA from screening specimens. *Eur. J. Clin. Microbiol. Infect. Dis.* **26:**459–466.
163. **Rothganger, J., M. Weniger, T. Weniger, A. Mellmann, and D. Harmsen.** 2006. Ridom TraceEdit: a DNA trace editor and viewer. *Bioinformatics* **22:**493–494.
164. **Sampath, R., S. A. Hofstadler, L. B. Blyn, M. W. Eshoo, T. A. Hall, C. Massire, H. M. Levene, J. C. Hannis, P. M. Harrell, B. Neuman, M. J. Buchmeier, Y. Jiang, R. Ranken, J. J. Drader, V. Samant, R. H. Griffey, J. A. McNeil, S. T. Crooke, and D. J. Ecker.** 2005. Rapid identification of emerging pathogens: coronavirus. *Emerg. Infect. Dis.* **11:**373–379.
165. **Satake, S., N. Clark, D. Rimland, F. S. Nolte, and F. C. Tenover.** 1997. Detection of vancomycin-resistant enterococci in fecal samples by PCR. *J. Clin. Microbiol.* **35:**2325–2330.
166. **Sauer, M., K. Tiede, R. Volland, D. Fuchs, and F. Zintl.** 2000. Procalcitonin in comparison to C-reactive protein as markers of the course of sepsis in severely immunocompromised children after bone marrow transplantation. *Klin. Padiatr.* **212:**10–15.
167. **Saunte, D. M., L. Klingspor, S. Jalal, J. Arnau, and M. C. Arendrup.** 2005. Four cases of Candida albicans infections with isolates developing pink colonies on CHROMagar Candida plates. *Mycoses* **48:**378–381.
168. **Schmit, J. C., L. Ruiz, L. Stuyver, K. Van Laethem, I. Vanderlinden, T. Puig, R. Rossau, J. Desmyter, E. De Clercq, B. Clotet, and A. M. Vandamme.** 1998. Comparison of the LiPA HIV-1 RT test, selective PCR and direct solid phase sequencing for the detection of HIV-1 drug resistance mutations. *J. Virol. Methods* **73:**77–82.
169. **Schneider, H. G., and Q. T. Lam.** 2007. Procalcitonin for the clinical laboratory: a review. *Pathology* **39:**383–390.
170. **Sendid, B., N. Francois, A. Standaert, E. Dehecq, F. Zerimech, D. Camus, and D. Poulain.** 2007. Prospective evaluation of the new chromogenic medium CandiSelect 4 for differentiation and presumptive identification of the major pathogenic Candida species. *J. Med. Microbiol.* **56:**495–499.
171. **Sesti, F., G. W. Abbott, J. Wei, K. T. Murray, S. Saksena, P. J. Schwartz, S. G. Priori, D. M. Roden, A. L. George, Jr., and S. A. Goldstein.** 2000. A common polymorphism associated with antibiotic-induced cardiac arrhythmia. *Proc. Natl. Acad. Sci. USA* **97:**10613–10618.
172. **Shafer, R. W., and S. Chou.** 2003. Mechanisms of resistance to antiviral agents, p. 1625–1635. *In* P. R. Murray, E. J. Murray, E. J. Baron, M. A. Pfaller, F. C. Tenover, and R. H. Yolken (ed.), *Manual of Clinical Microbiology*. ASM Press, Washington, DC.
173. **Shevchenko, A., M. Wilm, O. Vorm, and M. Mann.** 1996. Mass spectrometric sequencing of proteins silver-stained polyacrylamide gels. *Anal. Chem.* **68:**850–858.
174. **Shopsin, B., and B. N. Kreiswirth.** 2001. Molecular epidemiology of methicillin-resistant Staphylococcus aureus. *Emerg. Infect. Dis.* **7:**323–326.
175. **Slezak, T., T. Kuczmarski, L. Ott, C. Torres, D. Medeiros, J. Smith, B. Truitt, N. Mulakken, M. Lam, E. Vitalis, A. Zemla, C. E. Zhou, and S. Gardner.** 2003. Comparative genomics tools applied to bioterrorism defence. *Brief. Bioinform.* **4:**133–149.
176. **Sloan, L. M., J. R. Uhl, E. A. Vetter, C. D. Schleck, W. S. Harmsen, J. Manahan, R. L. Thompson, J. E. Rosenblatt, and F. R. Cockerill III.** 2004. Comparison of the Roche LightCycler vanA/vanB detection assay and culture for detection of vancomycin-resistant enterococci from perianal swabs. *J. Clin. Microbiol.* **42:**2636–2643.
177. Reference deleted.
178. **Sogaard, M., H. Stender, and H. C. Schonheyder.** 2005. Direct identification of major blood culture pathogens, including *Pseudomonas aeruginosa* and *Escherichia coli*, by a panel of fluorescence in situ hybridization assays using peptide nucleic acid probes. *J. Clin. Microbiol.* **43:**1947–1949.
179. **Soll, D. R., S. R. Lockhart, and C. Pujol.** 2003. Laboratory procedures for the epidemiological analysis of microorganisms, p. 139–161. *In* P. M. Murray, E. J. Baron, J. H. Jorgenson, M. A. Pfaller, and R. H. Yolken (ed.), *Manual of Clinical Microbiology*. ASM Press, Washington, DC.
180. **Stamper, P. D., M. Cai, C. Lema, K. Eskey, and K. C. Carroll.** 2007. Comparison of the BD GeneOhm VanR assay to culture for identification of vancomycin-resistant enterococci in rectal and stool specimens. *J. Clin. Microbiol.* **45:**3360–3365.
181. **Stender, H.** 2003. PNA FISH: an intelligent stain for rapid diagnosis of infectious diseases. *Expert Rev. Mol. Diagn.* **3:**649–655.
182. **Stoakes, L., R. Reyes, J. Daniel, G. Lennox, M. A. John, R. Lannigan, and Z. Hussain.** 2006. Prospective comparison of a new chromogenic medium, MRSASelect, to CHROMagar MRSA and mannitol-salt medium supplemented with oxacillin or cefoxitin for detection of methicillin-resistant *Staphylococcus aureus*. *J. Clin. Microbiol.* **44:**637–639.
183. **Stover, A., E. Jeffery, J. Xu, and D. H. Persing.** 2004. Hybridization array technology, p. 619–639. *In* D. H. Persing, F. C. Tenover, J. Versalovic, Y. Tang, E. R. Unger, D. A. Relman, and T. J. White (ed.), *Molecular Microbiology: Diagnostic Principles and Practice*. ASM Press, Washington, DC.
184. **Stranden, A., R. Frei, and A. F. Widmer.** 2003. Molecular typing of methicillin-resistant *Staphylococcus aureus*: can PCR replace pulsed-field gel electrophoresis? *J. Clin. Microbiol.* **41:**3181–3186.
185. **Strommenger, B., C. Kettlitz, T. Weniger, D. Harmsen, A. W. Friedrich, and W. Witte.** 2006. Assignment of *Staphylococcus* isolates to groups by *spa* typing, SmaI macrorestriction analysis, and multilocus sequence typing. *J. Clin. Microbiol.* **44:**2533–2540.
186. **Strommenger, B., C. Kettlitz, G. Werner, and W. Witte.** 2003. Multiplex PCR assay for simultaneous detection of nine clinically relevant antibiotic resistance genes in *Staphylococcus aureus*. *J. Clin. Microbiol.* **41:**4089–4094.
187. **Symoens, F., S. Bertout, M. A. Piens, J. Burnod, F. Renaud, N. Nolard, F. Chapuis, and R. Grillot.** 2001. A longitudinal study of lung transplant recipients infected with Aspergillus: genetic polymorphism of A fumigatus. *J. Heart Lung Transplant.* **20:**970–978.
188. **Tenover, F. C., R. D. Arbeit, R. V. Goering, et al.** 1997. How to select and interpret molecular strain typing methods for epidemiological studies of bacterial infections: a review for healthcare epidemiologists. *Infect. Control Hosp. Epidemiol.* **18:**426–439.
189. **Tenover, F. C., R. D. Arbeit, R. V. Goering, P. A. Mickelsen, B. E. Murray, D. H. Persing, and B. Swaminathan.** 1995. Interpreting chromosomal DNA restriction patterns produced by pulsed-field gel electrophoresis: criteria for bacterial strain typing. *J. Clin. Microbiol.* **33:**2233–2239.
190. **Tenover, F. C., and J. K. Rasheed.** 2004. Detection of antimicrobial resistance genes and mutations associated with antimicrobial resistance in microorganisms, p. 391–406. *In* D. H. Persing, F. C. Tenover, J. Versalovic, Y. Tang, E. R. Unger, D. A. Relman, and T. J. White (ed.), *Molecular*

Microbiology: Diagnostic Principles and Practice. ASM Press, Washington, DC.

191. **Thompson, J. D., D. G. Higgins, and T. J. Gibson.** 1994. CLUSTAL W: improving the sensitivity of progressive multiple sequence alignment through sequence weighting, position-specific gap penalties and weight matrix choice. *Nucleic Acids Res.* **22:**4673–4680.
192. **Thompson, J. D., D. G. Higgins, and T. J. Gibson.** 1994. Improved sensitivity of profile searches through the use of sequence weights and gap excision. *Comput. Appl. Biosci.* **10:**19–29.
193. **Toombs, L., E. Weeks, G. N. Forrest, D. Licalis, J. K. Johnson, and R. A. Venezia.** 2007. Impact of peptide nucleic acid (PNA) fluorescence in situ hybridization (FISH) for enterococcal blood stream infections, abstr. 131. *Abstr. IDSA 2006 Annu. Meet.* Infectious Disease Society of America, Arlington, VA.
194. **Trama, J. P., E. Mordechai, and M. E. Adelson.** 2005. Detection of *Aspergillus fumigatus* and a mutation that confers reduced susceptibility to itraconazole and posaconazole by real-time PCR and pyrosequencing. *J. Clin. Microbiol.* **43:**906–908.
195. **Troesch, A., H. Nguyen, C. G. Miyada, S. Desvarenne, T. R. Gingeras, P. M. Kaplan, P. Cros, and C. Mabilat.** 1999. *Mycobacterium* species identification and rifampin resistance testing with high-density DNA probe arrays. *J. Clin. Microbiol.* **37:**49–55.
196. **Tuohy, M. J., G. S. Hall, M. Sholtis, and G. W. Procop.** 2005. Pyrosequencing as a tool for the identification of common isolates of Mycobacterium sp. *Diagn. Microbiol. Infect. Dis.* **51:**245–250.
197. **Ueta, M., C. Sotozono, T. Inatomi, K. Kojima, K. Tashiro, J. Hamuro, and S. Kinoshita.** 2007. Toll-like receptor 3 gene polymorphisms in Japanese patients with Stevens-Johnson syndrome. *Br. J. Ophthalmol.* **91:**962–965.
198. **van der Steen, L. F., M. J. Bonten, E. van Kregten, J. J. Harssema-Poot, R. Willems, and C. A. Gaillard.** 2000. Vancomycin-resistant Enterococcus faecium outbreak in a nephrology ward. *Ned. Tijdschr. Geneeskd.* **144:**2568–2572.
199. **van der Zee, A., H. Verbakel, J. C. van Zon, I. Frenay, A. van Belkum, M. Peeters, A. Buiting, and A. Bergmans.** 1999. Molecular genotyping of *Staphylococcus aureus* strains: comparison of repetitive element sequence-based PCR with various typing methods and isolation of a novel epidemicity marker. *J. Clin. Microbiol.* **37:**342–349.
200. **Van Ert, M. N., S. A. Hofstadler, Y. Jiang, J. D. Busch, D. M. Wagner, J. J. Drader, D. J. Ecker, J. C. Hannis, L. Y. Huynh, J. M. Schupp, T. S. Simonson, and P. Keim.** 2004. Mass spectrometry provides accurate characterization of two genetic marker types in Bacillus anthracis. *Biotechniques* **37:**642–648.
201. **van Loo, I. H., S. van Dijk, I. Verbakel-Schelle, and A. G. Buiting.** 2007. Evaluation of a chromogenic agar (MRSASelect) for the detection of methicillin-resistant Staphylococcus aureus with clinical samples in The Netherlands. *J. Med. Microbiol.* **56:**491–494.
202. **Versalovic, J., T. Koeuth, and J. R. Lupski.** 1991. Distribution of repetitive DNA sequences in eubacteria and application to fingerprinting of bacterial genomes. *Nucleic Acids Res.* **19:**6823–6831.
203. **Vinh, D. C., K. A. Nichol, F. Rand, and J. A. Karlowsky.** 2006. Not so pretty in pink: *Staphylococcus cohnii* masquerading as methicillin-resistant *Staphylococcus aureus* on chromogenic media. *J. Clin. Microbiol.* **44:**4623–4624.
204. **Vuilleumier, N., M. F. Rossier, A. Chiappe, F. Degoumois, P. Dayer, B. Mermillod, L. Nicod, J. Desmeules, and D. Hochstrasser.** 2006. CYP2E1 genotype and isoniazid-induced hepatotoxicity in patients treated for latent tuberculosis. *Eur. J. Clin. Pharmacol.* **62:**423–429.
205. **Warren, D. K., R. S. Liao, L. R. Merz, M. Eveland, and W. M. Dunne, Jr.** 2004. Detection of methicillin-resistant *Staphylococcus aureus* directly from nasal swab specimens by a real-time PCR assay. *J. Clin. Microbiol.* **42:**5578–5581.
206. **Weller, T. M.** 2000. Methicillin-resistant Staphylococcus aureus typing methods: which should be the international standard? *J. Hosp. Infect.* **44:**160–172.
207. **Westin, L., C. Miller, D. Vollmer, D. Canter, R. Radtkey, M. Nerenberg, and J. P. O'Connell.** 2001. Antimicrobial resistance and bacterial identification utilizing a microelectronic chip array. *J. Clin. Microbiol.* **39:**1097–1104.
208. **Wheeler, D. L., T. Barrett, D. A. Benson, S. H. Bryant, K. Canese, V. Chetvernin, D. M. Church, M. Dicuccio, R. Edgar, S. Federhen, M. Feolo, L. Y. Geer, W. Helmberg, Y. Kapustin, O. Khovayko, D. Landsman, D. J. Lipman, T. L. Madden, D. R. Maglott, V. Miller, J. Ostell, K. D. Pruitt, G. D. Schuler, M. Shumway, E. Sequeira, S. T. Sherry, K. Sirotkin, A. Souvorov, G. Starchenko, R. L. Tatusov, T. A. Tatusova, L. Wagner, and E. Yaschenko.** 2007. Database resources of the National Center for Biotechnology Information. *Nucleic Acids Res.* **35:**D5–D12.
209. **Wilson, D. A., M. J. Joyce, L. S. Hall, L. B. Reller, G. D. Roberts, G. S. Hall, B. D. Alexander, and G. W. Procop.** 2005. Multicenter evaluation of a *Candida albicans* peptide nucleic acid fluorescent in situ hybridization probe for characterization of yeast isolates from blood cultures. *J. Clin. Microbiol.* **43:**2909–2912.
210. Reference deleted.
211. **Wilson, J. W., P. Bean, T. Robins, F. Graziano, and D. H. Persing.** 2000. Comparative evaluation of three human immunodeficiency virus genotyping systems: the HIV-GenotypR method, the HIV PRT GeneChip assay, and the HIV-1 RT line probe assay. *J. Clin. Microbiol.* **38:**3022–3028.
212. **Wise, M. G., M. Healy, K. Reece, R. Smith, D. Walton, W. Dutch, A. Renwick, J. Huong, S. Young, J. Tarrand, and D. P. Kontoyiannis.** 2007. Species identification and strain differentiation of clinical Candida isolates using the DiversiLab system of automated repetitive sequence-based PCR. *J. Med. Microbiol.* **56:**778–787.
213. Reference deleted.
214. **Wittwer, C. T., M. G. Herrmann, C. N. Gundry, and K. S. Elenitoba-Johnson.** 2001. Real-time multiplex PCR assays. *Methods* **25:**430–442.

214a. **Wolk, D. M.** 2006. PCR mass-spectrometry characterization of community-acquired oxacillin-resistant *Staphylococcus aureus* isolates, abstr. K-1187. *Abstr. 46th Intersci. Conf. Antimicrob. Agents Chemother.* American Society for Microbiology, Washington, DC.

215. **Woods, C. R., J. Versalovic, T. Koeuth, and J. R. Lupski.** 1993. Whole-cell repetitive element sequence-based polymerase chain reaction allows rapid assessment of clonal relationships of bacterial isolates. *J. Clin. Microbiol.* **31:**1927–1931.
216. **Yang, Z. J., M. Z. Tu, J. Liu, X. L. Wang, and H. Z. Jin.** 2006. Comparison of amplicon-sequencing, pyrosequencing and real-time PCR for detection of YMDD mutants in patients with chronic hepatitis B. *World J. Gastroenterol.* **12:**7192–7196.
217. **Zhao, J. R., Y. J. Bai, Y. Wang, Q. H. Zhang, M. Luo, and X. J. Yan.** 2005. Rapid detection of rifampin resistance of

Mycobacterium tuberculosis using high-throughput pyrosequencing technique. *Zhonghua Jie He He Hu Xi Za Zhi* 28:297–300.

218. **Zhao, J. R., Y. J. Bai, Q. H. Zhang, Y. Wang, M. Luo, and X. J. Yan.** 2005. Pyrosequencing-based approach for rapid detection of rifampin-resistant Mycobacterium tuberculosis. *Diagn. Microbiol. Infect. Dis.* **51:**135–137.

219. **Zheng, S., E. Ross, M. A. Legg, and M. J. Wirth.** 2006. High-speed electroseparations inside of silica colloidal crystals. *J. Am. Chem. Soc.* **128:**9016–9017.

INDEX